T0227599

second edition

Handbook of
OPTICAL
METROLOGY

Principles and
Applications

Edited by
Toru Yoshizawa

NPO 3D ASSOCIATES, YOKOHAMA, JAPAN

CRC Press
Taylor & Francis Group
Boca Raton London New York

CRC Press is an imprint of the
Taylor & Francis Group, an **informa** business

CRC Press
Taylor & Francis Group
6000 Broken Sound Parkway NW, Suite 300
Boca Raton, FL 33487-2742

First issued in hardback 2017

© 2015 by Taylor & Francis Group, LLC
CRC Press is an imprint of Taylor & Francis Group, an Informa business

No claim to original U.S. Government works

ISBN-13: 978-1-4665-7359-8 (hbk)
ISBN-13: 978-1-138-89363-4 (pbk)

Library of Congress Cataloging-in-Publication Data

Handbook of optical metrology : principles and applications / editor, Toru Yoshizawa. -- Second edition.
     pages cm
Includes bibliographical references and index.
ISBN 978-1-4665-7359-8 (alk. paper)
  1. Optical measurements--Handbooks, manuals, etc. 2. Metrology--Handbooks, manuals, etc. I. Yoshizawa, Toru, 1939-

QC367.H36 2015
681'.25--dc23

2014042945

Visit the Taylor & Francis Web site at
http://www.taylorandfrancis.com

and the CRC Press Web site at
http://www.crcpress.com

Contents

Preface to the Second Edition...vii
Preface to the First Edition ix
Contributors xi

SECTION I   Fundamentals of Optical Elements and Devices

Chapter 1   Light Sources...3

Natalia Dushkina

Chapter 2   Lenses, Prisms, and Mirrors 87

Peter R. Hall

Chapter 3   Optoelectronic Sensors...107

Motohiro Suyama

Chapter 4   Optical Devices and Optomechanical Elements 139

Akihiko Chaki and Kenji Magara

SECTION II   Fundamentals of Principles and Techniques for Metrology

Chapter 5   Propagation of Light...161

Natalia Dushkina

Chapter 6   Interferometry 197

David A. Page

Chapter 7   Holography 225

Giancarlo Pedrini

Chapter 8   Speckle Methods and Applications...249

Nandigana Krishna Mohan

Chapter 9   Moiré Metrology 273

Lianhua Jin and Toru Yoshizawa

Chapter 10 Optical Heterodyne Measurement Method 299

Masao Hirano

Chapter 11 Diffraction 343

Toru Yoshizawa

Chapter 12 Light Scattering 355

Lev T. Perelman

Chapter 13 Polarization 373

Michael Shribak

Chapter 14 Near-Field Optics 389

Wenhao Huang, Xi Li, and Guoyong Zhang

SECTION III *Practical Applications*

Chapter 15 Length and Size 405

René Schödel

Chapter 16 Displacement 433

Akiko Hirai, Mariko Kajima, and Souichi Telada

Chapter 17 Straightness and Alignment 451

Ruedi Thalmann

Chapter 18 Flatness 461

Toshiyuki Takatsuji and Youichi Bitou

Chapter 19 Surface Profilometry 475

Toru Yoshizawa and Toshitaka Wakayama

Chapter 20 Three-Dimensional Shape Measurement 517

Frank Chen, Gordon M. Brown, and Mumin Song

Chapter 21 Fringe Analysis 541

Jun-ichi Kato

Chapter 22 Photogrammetry..555
                Nobuo Kochi

Chapter 23 Optical Methods in Solid Mechanics...583
                Anand Asundi

Chapter 24 Optical Methods in Flow Measurement...605
                Sang Joon Lee

Chapter 25 Polarimetry..627
                Baoliang (Bob) Wang

Chapter 26 Birefringence Measurement..677
                Yukitoshi Otani

Chapter 27 Ellipsometry..705
                Hiroyuki Fujiwara

Chapter 28 Optical Thin Film and Coatings 725
                Cheng-Chung Lee and Shigetaro Ogura

Chapter 29 Film Surface and Thickness Profilometry 743
                Katsuichi Kitagawa

Chapter 30 Optical Coherence Tomography for Industrial Applications 769
                Tatsuo Shiina

Chapter 31 Interference Microscopy for Surface Structure Analysis.........................791
                Peter J. de Groot

Chapter 32 Noncontact Dimensional and Profile Metrology by Video Measurement 829
                Hiroo Tsumuraya and Shuichi Sakai

Chapter 33 Optical Metrology in Manufacturing Technology 853
                Rainer Tutsch

Chapter 34 On-Machine Measurements 867
                Takashi Nomura and Kazuhide Kamiya

Index..889

Preface to the Second Edition

I am very happy for this chance to revise the handbook first published in 2009. Up to now, numerous books and handbooks with such titles as *Optics and Optical Engineering* have been published. However, it was still not easy for people to apply those optical principles to their research or for industrial purposes. In this handbook, I, as the editor, have attempted to bridge the gap that exists between academic research principles and the methods used for practical applications.

To my great delight, we received a number of favorable responses to the previous edition, especially from researchers and engineers in industrial fields, in addition to people in academic areas. Therefore, my intention here is to update the ten chapters whose authors needed to revise on the basis of recent development and progress in their specific fields. In one instance, devices such as CCD (charge-coupled device)/CMOS (complementary metallic-oxide-semiconductor) cameras in imaging technology have made remarkable progress since 2009, which enabled downsizing in shape and high definition in resolution. Consequently, combining optical principles and new device technologies has expanded the areas of application. This created a kind of synergistic effect that new principles are born inspired by new elements and devices. At the same time, based on requests from readers in related areas, four new chapters have been added. Naturally, we would have liked to include even more topics, but as always it is difficult to cover everything due to various practical restrictions.

The very rapid progress in fundamental research, applied techniques, and device technology has, I found, made it difficult to achieve ideal contents. In a short period, many novel ideas may be reported and instruments and equipment for practical use commercialized before it is clear whether or not they are successful in the market.

I will be delighted if this update makes the handbook more useful for people who are seeking solutions to their problems using optical principles and methods described here. I shall gratefully receive any comments, suggestions, and advice from readers.

Toru Yoshizawa

Preface to the First Edition

Over the years, many optical principles that have the potential for metrological applications have been found, and many related techniques and devices have been invented. On the basis of these principles and techniques, numerous interesting applications have been reported in academic communities. In addition, a number of optical equipment and systems have already been commercialized.

So we may ask, "What are the features of optical metrology?" First of all, that it is a "noncontact approach" in metrological applications should be cited. This feature causes neither deformation nor damage to the specimen due to the measuring force. Also, flexibility in setting up measuring systems is also brought by this noncontact arrangement. Moreover, "full-field measurement" is possible. We are not limited to "point-wise measurement." This feature, essentially inherent to optical methods such as photography and video, is currently summarized as "imaging." Full-field data with abundant information can be instantly captured by imaging devices at a single moment and then analyzed by computational processing in a short time.

Furthermore, compatibility of light with electricity is another important factor for incorporating optical techniques to practical applications. "Light" is easily converted into an electrical quantity (electric current, voltage, etc.) using various detectors and sensing devices, and this means computational analysis of captured data is easily attainable due to recent remarkable advances in electronic devices and computer technology. Optical metrology can thus be seen to offer a variety of advantages.

Because of these superior and attractive features of light and/or optical methods, many research papers and technical reports on a wide range of applications have appeared over the years. However, this does not mean that compiling these results in a book on optical metrology is easy. With due consideration to recent trends in optical metrology, the title *Handbook of Optical Metrology: Principles and Applications* was chosen. The editor's fundamental concept for this handbook was inspired by the following phrase found in the book *Zen of Vegetable Roots* (written by Zicheng Hong) published around 1600: "An academic learning is just like a daily meal necessary for our life. However, at the same time, it should be practically useful when needed."

When the editorial work for this book started, some people suggested solicitously that a few similar books might have been published already. However, as far as I checked, there are no comparable books that have been compiled with the same intention as ours. Admittedly, some books titled *Optical Metrology* do exist, but most of them are focused on a specified area or restricted topic. In this book, we ensure that a wide range of applications using optical methods are included, in addition to the fundamental principles and techniques.

This book is basically written for beginners, but, in addition to introductory description, each chapter contains further suggestions, carefully prepared by the contributor to meet the needs of people with advanced knowledge.

I am pleased and relieved that I could accomplish the goal we set through the collaboration of many people.

I thank all contributors and other people who supported the publication of this book, especially Dr. Michael Shribak with whom the fundamental concept was discussed at a restaurant in Boston, Dr. Natalia Dushkina who made a tremendous effort to cover two fundamental chapters, and

Dr. Peter Hall who warmly encouraged the editor through his English sense of humor. Kind suggestions by Dr. Toshiyuki Takatsuji, editorial secretary, were helpful for editing chapters relating to dimensional metrology. Dr. Toshitaka Wakayama, research associate, and Megumi Asano, private secretary, are also warmly appreciated for their editorial assistance.

I hope this handbook will be useful for researchers and engineers in industrial fields, as well as for academicians and students.

Toru Yoshizawa

Contributors

Anand Asundi
School of Mechanical and Aerospace
   Engineering
Nanyang Technological University Singapore
Singapore

Youichi Bitou
National Metrology Institute of Japan
National Institute of Advanced Industrial
   Science and Technology
Tsukuba, Japan

Gordon M. Brown
Optical Systems Engineering
Naples, Florida

Akihiko Chaki
Chuo Precision Industrial Co., Ltd.
Ageo, Japan

Frank Chen
Ford Motor Company
Dearborn, Michigan

Peter J. de Groot
Zygo Corporation
Middlefield, Connecticut

Natalia Dushkina
Department of Physics
Millersville University
Millersville, Pennsylvania

Hiroyuki Fujiwara
Department of Electrical and Electronic
   Engineering
Gifu University
Gifu, Japan

Peter R. Hall
Mindsmeet Electro-Optics Ltd.
Cambridgeshire, United Kingdom

and

BJR Systems Ltd.
Hertfordshire, United Kingdom

Akiko Hirai
National Metrology Institute of Japan
National Institute of Advanced Industrial
   Science and Technology
Tsukuba, Japan

Masao Hirano
Photon Probe Inc.
Hino, Japan

Wenhao Huang
Department of Precision Machinery and
   Instrumentation
University of Science and Technology of China
Hefei, Anhui, People's Republic of China

Lianhua Jin
Department of Information and Mechatronics
University of Yamanashi
Kofu, Japan

Mariko Kajima
National Metrology Institute of Japan
National Institute of Advanced Industrial
   Science and Technology
Tsukuba, Japan

Kazuhide Kamiya
Department of Intelligent Systems Design
   Engineering
Toyama Prefectural University
Imizu, Japan

Jun-ichi Kato
Nanophotonics Laboratory
RIKEN, Advanced Science Institute
Wako, Japan

Katsuichi Kitagawa
Independent Consultant
Moriyama, Japan

Nobuo Kochi
Imaging and Measuring Laboratory
Corporate R&D Center
Topcon Corporation
Tokyo, Japan

Cheng-Chung Lee
Department of Optics and Photonics
National Central University
Chung-Li, Taiwan, People's Republic of China

Sang Joon Lee
Department of Mechanical Engineering
Center for Biofluid and Biomimic Research
Pohang University of Science and Technology
Pohang, South Korea

Xi Li
Department of Precision Machinery and
  Instrumentation
University of Science and Technology of China
Hefei, Anhui, People's Republic of China

Kenji Magara
Chuo Precision Industrial Co., Ltd.
Ageo, Japan

Nandigana Krishna Mohan
Applied Optics Laboratory
Department of Physics
Indian Institute of Technology Madras
Chennai, Tamil Nadu, India

Takashi Nomura
Department of Intelligent Systems Design
  Engineering
Toyama Prefectural University
Imizu, Japan

Shigetaro Ogura
Graduate School
Kobe Design University
Kobe, Japan

Yukitoshi Otani
Center for Optical Research and Education
Utsunomiya University
Utsunomiya, Japan

David A. Page
Stevenage MBDA U.K. Ltd.
Hertfordshire, United Kingdom

Giancarlo Pedrini
Institute of Applied Optics
University of Stuttgart
Stuttgart, Germany

Lev T. Perelman
Biomedical Imaging and Spectroscopy
  Laboratory
Beth Israel Deaconess Medical Center
and
Harvard University
Boston, Massachusetts

Shuichi Sakai
Industrial Instruments Development
  Department
NIKON Corporation Instruments Company
Yokohama, Japan

René Schödel
Interferometry on Material Measures
Physikalisch-Technische Bundesanstalt
Braunschweig, Germany

Tatsuo Shiina
Department of Information and Image
  Sciences
Chiba University
Chiba, Japan

Michael Shribak
Marine Biological Laboratory
Woods Hole, Massachusetts

Mumin Song
Ford Research Laboratory
Ford Motor Company
Dearborn, Michigan

Motohiro Suyama
Electron Tube Division
Hamamatsu Photonics K.K.
Iwata, Japan

Toshiyuki Takatsuji
National Metrology Institute of Japan
National Institute of Advanced Industrial
  Science and Technology
Tsukuba, Japan

Souichi Telada
National Metrology Institute of Japan
National Institute of Advanced Industrial
  Science and Technology
Tsukuba, Japan

Ruedi Thalmann
Federal Office of Metrology METAS
Wabern, Switzerland

Hiroo Tsumuraya
Industrial Instruments Development
   Department
NIKON Corporation Instruments Company
Yokohama, Japan

Rainer Tutsch
Institute for Production Measurement Technology
Technical University of Braunschweig
Braunschweig, Germany

Toshitaka Wakayama
Department of Biomedical Engineering
Saitama Medical University
Hidaka, Japan

Baoliang (Bob) Wang
Hinds Instruments Inc.
Hillsboro, Oregon

Toru Yoshizawa
3D Associates
Yokohama, Japan

Guoyong Zhang
Department of Precision Machinery and
   Instrumentation
University of Science and Technology of China
Hefei, Anhui, People's Republic of China

Section I

Fundamentals of Optical Elements and Devices

1 Light Sources

Natalia Dushkina

CONTENTS

1.1 Radiometry and Photometry...4
    1.1.1 Differences between Radiometry and Photometry...4
    1.1.2 Human Eye 5
    1.1.3 Radiometric and Photometric Quantities and Units 6
        1.1.3.1 Radiometric Quantities and Units...7
        1.1.3.2 Lambert's Law and Lambertian Sources..10
        1.1.3.3 Photometric Quantities and Units...11
        1.1.3.4 Conversion between Radiometric and Photometric Units 13
        1.1.3.5 Choosing a Light Source...16
    1.1.4 Photometric Measurement Techniques...21
    1.1.5 Colorimetry...23
1.2 Light Emission...26
    1.2.1 Blackbody Radiation..26
    1.2.2 Luminescence..28
        1.2.2.1 Photoluminescence 28
        1.2.2.2 Fluorescence...29
        1.2.2.3 Phosphorescence 29
        1.2.2.4 Multiphoton Photoluminescence 29
        1.2.2.5 Electroluminescence...32
        1.2.2.6 Cathodoluminescence..33
        1.2.2.7 Sonoluminescence 33
        1.2.2.8 Chemiluminescence...33
        1.2.2.9 Bioluminescence..34
        1.2.2.10 Triboluminescence 35
        1.2.2.11 Scintillation..35
1.3 Conventional Light Sources..36
    1.3.1 Incandescent Lamps and Tungsten Halogen Lamps 36
    1.3.2 Gas Discharge Lamps...37
        1.3.2.1 Low-Pressure Discharge Lamps 38
        1.3.2.2 High-Pressure Discharge Lamps 39
1.4 Light-Emitting Diodes..41
    1.4.1 LED Basics 41
    1.4.2 LED Material Systems..46
    1.4.3 White LEDs...48
    1.4.4 Surface-Emitting LEDs and Edge-Emitting LEDs 51
    1.4.5 Organic LEDs...53
    1.4.6 LED Metrology...54
        1.4.6.1 Measuring the Luminous Intensity of LEDs 54
        1.4.6.2 Measuring the Luminous Flux of LEDs...55

             1.4.6.3 Mapping the Spatial Radiation Pattern of LEDs 56
             1.4.6.4 LED Measuring Instrumentation..56
      1.4.7 LED Applications 57
             1.4.7.1 LEDs Illumination 58
             1.4.7.2 LED-Based Optical Communications 59
             1.4.7.3 Applications of the LED Photovoltaic Effect 60
             1.4.7.4 High-Speed Pulsed LED Light Generator for Stroboscopic
                       Applications...64
             1.4.7.5 LED Applications in Medicine and Dentistry 64
             1.4.7.6 LED Applications in Horticulture, Agriculture, Forestry, and Fishery.......65
1.5   Lasers...66
      1.5.1 Stimulated Emission and Light Amplification 67
      1.5.2 Essential Elements of a Laser...69
             1.5.2.1 Gain Medium..69
             1.5.2.2 Pump 70
             1.5.2.3 Laser Cavity 71
             1.5.2.4 Cooling System 72
      1.5.3 Operating Characteristics of a Laser 72
             1.5.3.1 Laser Modes..72
             1.5.3.2 Gaussian Laser Beams..73
             1.5.3.3 Continuous Wave and Pulsed Operation 75
      1.5.4 Characteristics of Laser Light 75
             1.5.4.1 Monochromaticity...75
             1.5.4.2 Coherence 76
             1.5.4.3 Divergence 78
             1.5.4.4 Focusing a Laser Beam...79
             1.5.4.5 Laser Irradiance 79
      1.5.5 Laser Systems 79
             1.5.5.1 Gas Active Medium 79
             1.5.5.2 Liquid Active Medium..80
             1.5.5.3 Solid-State Active Medium..80
             1.5.5.4 Semiconductor Active Medium 80
References...81

This chapter offers a review of radiometry, photometry, and light sources in terms of the basic notions and principles of measurement of electromagnetic radiation that are most relevant for optical metrology. It explains the difference between radiometry and photometry, and their quantities and units. More information on these topics can be found in specialized literature (e.g., see Refs. [1–8]). This chapter also includes a brief survey of conventional light sources. Special attention is paid to modern light sources such as lasers, light emitting diodes (LEDs), and super luminescent diodes (SLDs), which are widely used in the contemporary techniques of optical metrology. There is an enormous amount of literature on this topic, but the reader may want to start with the detailed reviews of LED that are given in some introductory books (e.g., Refs. [9–12] and references therein).

1.1 RADIOMETRY AND PHOTOMETRY

1.1.1 DIFFERENCES BETWEEN RADIOMETRY AND PHOTOMETRY

Radiometry is the science of measurement of electromagnetic radiation within the entire range of optical frequencies between 3×10^{11} and 3×10^{16} Hz, which corresponds to wavelengths between

0.01 and 1000 µm. The optical range includes the ultraviolet (UV), the visible, and the infrared (IR) regions of the electromagnetic spectrum. Radiometry explores the electromagnetic radiation as it is emitted and detected, and measures quantities in terms of absolute power as measured with a photodetector. It considers the related quantities such as energy and power, as independent of wavelength. The relevant quantities are designated as radiant and their units are derivatives of joules.

Photometry deals with the part of the electromagnetic spectrum that is detectable by the human eye, and therefore, it is restricted to the frequency range from about 3.61×10^{14} to roughly 8.33×10^{14} Hz (which is the wavelength range from about 830 to 360 nm). These limits of the visible part of the spectrum are specified by the International Commission on Illumination (CIE; Commission *Internationale de l'Éclairage*). The emission of light results from rearrangement of the outer electrons in atoms and molecules, which explains the very narrow band of frequencies in the spectrum of electromagnetic radiation corresponding to light. Photometry is the science of measurement of light, in terms of its perceived brightness to the human eye; thus, it measures the visual response. Therefore, photometry considers the wavelength dependence of the quantities associated with radiant energy and new units are introduced for the radiometric quantities. The photometric quantities are designated as luminous. Typical photometric units are lumen, lux, and candela.

When the eye is used as a comparison detector, we speak about visual photometry. Physical photometry uses either optical radiation detectors that mimic the spectral response of the human eye or radiometric measurements of a radiant quantity (i.e., radiant power), which is later weighted at each wavelength by the luminosity function that models the eye brightness sensitivity. Photometry is a basic method in astronomy, as well as in lighting industry, where the properties of human vision determine the quantity and quality requirements for lighting.

Colorimetry is the science and measurements of color. The fundamentals of colorimetry are well presented, for example, in Refs. [13,14].

1.1.2 HUMAN EYE

The human eye is not equally sensitive to all the wavelengths of light. The eye is most sensitive to green–yellow light, the wavelength range where the sun has its peak energy density emission, and the eye sensitivity curve falls off at higher and lower wavelengths. The eye response to light and color depends also on light conditions and is determined by the anatomical construction of the human eye, described in detail in *Encyclopedia Britannica*, 1994. The retina includes rod and cone light receptors. Cone cells are responsible for the color perception of the eye and define the light-adapted vision, that is, the photopic vision. The cones exhibit high resolution in the central part of the retina, the foveal region (fovea centralis), which is the region of greatest visual acuity. There are three types of cone cells, which are sensitive to red, green, and blue light. The second type of cells, the rods, are more sensitive to light than cone cells. In addition, they are sensitive over the entire visible range and play an important role in night vision. They define the scotopic vision, which is the human vision at low luminance. They have lower resolution ability than the foveal cones. Rods are located outside the foveal region, and therefore, are responsible for the peripheral vision. The response of the rods at high-ambient-light levels is saturated and the vision is determined entirely by the cone cells (see also Refs. [5,15]). Photometry is based on the eye's photopic response, and therefore, photometric measurements will not accurately indicate the perceived brightness of sources in dim lighting conditions.

The sensitivity of eye to light of different wavelength is given in Figure 1.1. The light-adapted relative spectral response of the eye is called the spectral luminous efficiency function for photopic vision, $V(\lambda)$. The dark-adapted relative spectral response of the eye is called the spectral luminous efficiency function for scotopic vision, $V'(\lambda)$. Both are empirical curves. The curve of photopic vision, $V(\lambda)$, was first adopted by CIE in 1924. It has a peak of unity at 555 nm in air (in the yellow–green region), reaches 50% at near 510 and 610 nm and decreases to levels below 10^{-5} at about 380 and 780 nm. The curve for scotopic vision, $V'(\lambda)$, was adopted by the CIE in 1951. The maximum

FIGURE 1.1 Spectral sensitivity of the human eye. The solid line is the curve $V(\lambda)$ for photopic vision from 1924 CIE, and the dashed line is the curve for scotopic vision $V'(\lambda)$ from 1951 CIE. (Žukauskas, A., Shur, M.S., and Gaska, R.: *Introduction to Solid-State Lighting.* 2002. Copyright Wiley-VCH Verlag GmbH & Co. KGaA. Reproduced with permission.)

of the scotopic spectrum is shifted to smaller wavelengths with respect to the photopic spectrum. The $V'(\lambda)$ curve has a peak of unity at 507 nm, reaches 50% at about 455 and 550 nm, and decreases to levels below 10^{-3} at about 380 and 645 nm. Photopic or light-adapted cone vision is active for luminances greater than 3 cd/m². Scotopic or dark-adapted rod vision is active for luminances lower than 0.01 cd/m². The vision in the range between these values is called mesopic and is a result of varying amounts of contribution of both rods and cones.

As most of the human activities are at high-ambient illumination, the spectral response and color resolution of photopic vision have been extensively studied. Recent measurements of the eye response to light have made incremental improvement of the initial CIE curves, which underestimated the response at wavelengths shorter than 460 nm [16–20]. The curves $V(\lambda)$ and $V'(\lambda)$ can easily be fit with a Gaussian function by using a nonlinear regression technique, as the best fits are obtained with the following equations [21]:

$$V(\lambda) \cong 1.019 e^{-285.4(\lambda-0.559)^2} \quad \text{for photopic vision}$$

$$V'(\lambda) \cong 0.992 e^{-321.9(\lambda-0.503)^2} \quad \text{for photopic vision}$$

However, the fit of the scotopic curve is not quite good as the photopic curve. The Gaussian fit with the functions above is only an approximation that is acceptable for smooth curves but is not appropriate for narrow wavelength sources, like LEDs.

1.1.3 RADIOMETRIC AND PHOTOMETRIC QUANTITIES AND UNITS

Radiometric quantities characterize the energy content of radiation, whereas photometric quantities characterize the response of the human eye to light and color. Therefore, different names and units are used for photometric quantities. Every quantity in one system has an analogous quantity in the other system. Many radiometric and photometric terms and units have been used in the optics literature; however, we will consider here only the SI units.

A summary of the most important photometric quantities and units with their radiometric counterparts are given in Table 1.1. Radiometric quantities appear in the literature either without subscripts or with the subscript e (electromagnetic). Photometric quantities usually are labeled with a subscript λ or v (visible) to indicate that only the visible spectrum is being considered. In our text,

TABLE 1.1

Radiometric and Photometric Quantities and Units

	Radiometric Quantities	Units	Photometric Quantities	Units
Energy	Radiant energy Q	J (joule)	Luminous energy Q_λ	lm·s (or talbot)
Energy density	Radiant energy density $w = \dfrac{dQ}{dV}$	$\dfrac{J}{m^3}$	Luminous density $w_\lambda = \dfrac{dQ_\lambda}{dV}$	$\dfrac{lm·s}{m^3}$
Power	Radiant flux or radiant power $\Phi = \dfrac{dQ}{dt}$	W or $\dfrac{J}{s}$	Luminous flux $\Phi_\lambda = \dfrac{dQ}{dt}\lambda$	lm = cd sr
Power per area	Radiant exitance $M = \dfrac{d\Phi}{da}$ (for emitter) or irradiance $E = \dfrac{d\Phi}{da}$ (for receiver)	$\dfrac{W}{m^2}$	Luminous exitance $M_\lambda = \dfrac{d\Phi_\lambda}{da}$ (light emitter) or illuminance (light receiver) $E_\lambda = \dfrac{d\Phi_\lambda}{da}$	Lux or $\dfrac{lm}{m^2}$
Power per area per solid angle	Radiance $L = \dfrac{d\Phi}{d\Omega\, da\cos\alpha} = \dfrac{dI}{da\cos\alpha}$	$\dfrac{W}{sr·m^2}$	Luminance $L_\lambda = \dfrac{dI_\lambda}{da\cos\alpha}$	$\dfrac{cd}{m^2} = \dfrac{lm}{m^2·sr}$ (or nit)
Intensity	Radiant intensity $I = \dfrac{d\Phi}{d\Omega}$	$\dfrac{W}{sr}$	Luminous intensity $I_\lambda = \dfrac{d\Phi}{d\Omega}$	Candela or $\dfrac{lm}{sr}$

we use the subscript λ for the symbols of photometric quantities and no subscript for the radiometric quantities. Subsequently, we define in more detail the terms in Table 1.1 and their units.

1.1.3.1 Radiometric Quantities and Units

Most of the theoretical models build on the supposition of a point light source that emits in all directions, we have to define first the term "solid angle," which is measured in steradians (sr). According to NIST SP811, the steradian is defined as follows: "One steradian (sr) is the solid angle that, having its vertex in the center of a sphere, cuts off an area on the surface of the sphere equal to that of a square with sides of length equal to the radius of the sphere." The solid angle is thus the ratio of the spherical area to the square of the radius. The spherical area is a projection of the object of interest onto a unit sphere, and the solid angle is the surface area of that projection. If we divide the surface area of a sphere by the square of its radius, we find that there are 4π sr of solid angle in a sphere. One hemisphere has 2π sr. The solid angle, as illustrated in Figure 1.2, is defined by the area the cone

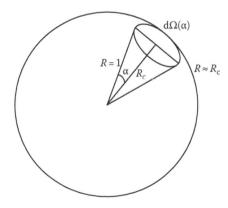

FIGURE 1.2 Definition of solid angle. For a small solid angle, the spherical elemental section is approximated by a flat section and $R \approx R_c$.

cuts out from a sphere of radius $R = 1$ (Figure 1.2). For small solid angles, the spherical section can be approximated with a flat section, and the solid angle is

$$d\Omega(\alpha) = \frac{\pi(R\sin\alpha)^2}{R^2} = \pi\sin^2\alpha \qquad (1.1)$$

Radiant energy, Q, measured in joules (J) and radiant energy density, w (J/m^3), are basic terms that need no further explanation. Radiant flux or radiant power, Φ, is the time rate of flow of radiant energy measured in watts (W = J/s). The radiant energy can be either emitted, scattered, or reflected from a surface (the case of an emitter), or incident onto a surface (the case of a receiver). Thus, different symbols and names may be used for the same quantity for the cases of emitter and receiver.

Radiant intensity, I, is defined as the radiant flux emitted by a point source per unit solid angle, Ω, in a given direction:

$$I = \frac{d\Phi}{d\Omega} \qquad (1.2)$$

It is expressed in watts per steradian (W/sr). For example, the intensity from a sphere emitting radian flux Φ (W) uniformly in all directions is $\Phi/4\pi$ (W/sr).

In general, the direction of propagation of radiant energy is given by the Poynting vector. The radiant flux density (W/m^2) emitted from a surface with a surface area da is called radiant exitance, M,

$$M = \frac{d\Phi}{da}\left[\frac{W}{m^2}\right] \qquad (1.3a)$$

The radiant flux density of radiation incident onto a surface with a given area da' is called irradiance, E. Thus, the irradiance is defined as the radiant flux incident onto a unit area perpendicular to the Poynting vector

$$E = \frac{d\Phi}{da'}\left[\frac{W}{m^2}\right] \qquad (1.3b)$$

For an emitter with a given area and a constant radiant flux, the irradiance is constant. For example, the irradiance of the Sun is $E = 1.37 \times 10^3$ W/m^2. This numerical value is called the solar constant. It describes the solar radiation at normal incidence on an area above the atmosphere. In space, the solar radiation is practically constant, while on Earth it varies with the time of day and year as well as with the latitude and weather. The maximum value on the Earth is between 0.8 and 1.0 kW/m^2.

Radiant efficiency describes the ability of the source to convert the consumed electric power (IV) into radiant flux Φ:

$$\text{Radiant efficiency} = \Phi/(IV) \qquad (1.4)$$

Here I and V are the electric current through and potential difference across the light source, respectively. The radiant efficiency is dimensionless and may range from 0 to 1.

Figure 1.3 represents an emitter with a flat circular area for which the diameter is small compared to the distance R to the receiver. If the area of the emitter is da and a cone of light emerges

FIGURE 1.3 Opening of the cone toward the receiver determined by the solid angle $d\Omega(\alpha) = da'/R^2$ and the opening of the cone toward the emitter determined by the solid angle $d\Omega(\alpha') = da/R^2$.

from each point of this area, the opening of the cone is determined by the area of the receiver da' (solid lines). The solid angle of the cone, $d\Omega(\alpha)$, is given as

$$d\Omega(\alpha) = \frac{da'}{R^2} \tag{1.5}$$

where
  α is the half angle of the cross section
  da' is the area of the receiver

Each point of this area receives a cone of light, and the opening of this cone is determined by the area of the emitter da (the cone with dashed lines in Figure 1.3). The solid angle of the cone, $d\Omega(\alpha')$, is given as

$$d\Omega(\alpha') = \frac{da}{R^2} \tag{1.6}$$

where
  α' is the half angle of the cross section
  da is the emitter's area

The solid angles $d\Omega(\alpha)$ and $d\Omega(\alpha')$ illustrate the cone under which the emitter sees the receiver and the receiver sees the emitter.

The radiant flux arriving at the receiver is proportional to the solid angle and the area of the emitter

$$\Phi = L \, da \, d\Omega(\alpha) = L \, da \frac{da'}{R^2} \, [\text{W}] \tag{1.7}$$

The constant of proportionality L is called radiance and is equal to the radiant flux arriving at the receiver per unit solid angle and per unit projected source area:

$$L = \frac{\Phi}{da \, d\Omega(\alpha)} \left[\frac{\text{W}}{\text{m}^2 \cdot \text{sr}} \right] \tag{1.8}$$

Radiance and irradiance are often confusingly called intensity.

If both the emitter and receiver areas are small compared to their distance from one another, and in addition, if they are perpendicular to the connecting line, Equations 1.1 and 1.7 can be combined to express the radiant flux as

$$\Phi = L(\pi \sin^2 \alpha) da \tag{1.9}$$

The radiant flux from the emitter arriving at the receiver may also be expressed as

$$\Phi = L\,da\,d\Omega(\alpha) = L\,da\frac{da'}{R^2} = L\,da'\frac{da'}{R^2} = L\,da'\,d\Omega(\alpha') = \Phi' \tag{1.10}$$

The left side represents the power proportional to the area da of the emitter multiplied by the solid angle $d\Omega(\alpha)$, while the right side gives the power proportional to the area da' of the receiver multiplied by the solid angle $d\Omega(\alpha')$. Both sides should be equal if there are no energy losses in the medium between the emitter and the receiver.

The power arriving at the receiver with area da' can be expressed by using Equations 1.3 and 1.10 as

$$\Phi = E\,da' = L\,da'd\Omega(\alpha') \tag{1.11}$$

Equation 1.11 gives the relation between irradiance and radiance:

$$E = L\,d\Omega(\alpha') \tag{1.12}$$

The radiance L (W/m²·sr) multiplied by the solid angle results in the irradiance E. The radiance L of Sun can be calculated using Equation 1.12 if we consider Sun as a small, flat emitter and if we bear in mind that Sun is seen from Earth under an angle $\alpha = 0.25° = 0.004$ rad. For a small angle, $\sin\alpha \cong \alpha$ and we can write

$$d\Omega(\alpha) = \pi\sin^2(0.25°) = \pi(0.004\ \text{rad})^2 = 6\times10^{-5}\ \text{sr}$$

and

$$L = \frac{E}{d\Omega} = \frac{1.35\times10^3\ \text{W}/\text{m}^2}{6\times10^{-5}\ \text{sr}} = 2.25\times10^7\ \frac{\text{W}}{\text{m}^2\cdot\text{sr}}$$

Thus, the radiance L of Sun is constant and has the value of 2.25×10^7 W/m²·sr.

1.1.3.2 Lambert's Law and Lambertian Sources

When we consider an emitter or a receiver, we often refer to them as isotropic and Lambertian surfaces. These two terms are often confused and used interchangeably because both terms involve the implication of the same energy in all directions.

Isotropic surface means a spherical source that radiates the same energy in all directions, that is, the intensity (W/sr) is the same in all directions. A distant star can be considered as an isotropic point source, although an isotropic point source is an idealization and would mean that its energy density would have to be infinite. A small, uniform sphere, such as a globular tungsten lamp with a milky white diffuse envelope, is another good example for an isotropic surface.

Lambertian surface refers to a flat radiating surface, which can be an active surface or a passive, reflective surface. According to Lambert's law, the intensity for a Lambertian surface falls off as the cosine of the observation angle with respect to the surface normal and the radiance (W/m²·sr) is independent of direction. An example could be a surface painted with a good matte white paint. If it is uniformly illuminated, like from Sun, it appears equally bright from whatever direction you view it. Note that the flat radiating surface can be an elemental area of a curved surface.

Lambert's law defines the distribution of the emitted power when the area of the emitter and the receiver are tilted with respect to each other, as shown in Figure 1.4. Let us assume that the emitter

FIGURE 1.4 Illustration of Lambert's law application. The emitter is tilted in respect with the receiver.

and the receiver have small areas da and da', respectively. The area da is tilted through the angle θ with respect to the line connecting the areas da and da'. In 1769, Johann Lambert found experimentally that when one of both, the emitter or the detector, is tilted through an angle θ with respect to the other, the emitted power depends on θ via $\cos\theta$ factor. Therefore, this law is known as cosine law of Lambert. The mathematical representation of Lambert's law is

$$\Phi(\theta) = L\frac{da'}{R^2}da \qquad \cos\theta = L\frac{da'}{R^2}da_\perp \qquad (1.13)$$

where $da_\perp = da\cos\theta$ is the projection of the area da on the plane perpendicular to the connecting line between emitter and receiver. According to Lambert's law, the area da, when looked at the angle θ, has the same radiance L as the smaller area $da\perp$.

If we consider Sun as a sphere consisting of small areas da emitting power $\Phi(\theta)$ into a solid angle $d\Omega(\alpha) = da'/R^2$, we can use Equation 1.13 to express the radiance of Sun as $L = (\Phi(\theta))/(d\Omega\ da)_\perp$. If Lambert's law is fulfilled for all equal sections da of the spherical emitting surface, each small section $da_\perp$ would emit the same power $\Phi(\theta)$. That is why we see Sun as a uniformly emitting disk.

Sources that obey Lambert's law are called Lambertian sources. The best example for a Lambertian source is a blackbody emitter. In contrary, a laser with its directional beam is an example of extremely poor Lambertian source.

Until now, we assumed small areas for both the emitter and the detector. Let us consider now the case when either the emitter or the detector has a large area. If we consider a small emitter area and a large receiver area, in order to find the total power arriving at the detector, we have to integrate over the total area of the receiver. This leads to $\Phi = L(\pi\sin^2\alpha)da$, which is the same as Equation 1.9 obtained for the case of small emitter and receiver areas, both perpendicular to the connecting line. We get similar results when we consider large emitter area and small receiver area, $\Phi = L(\pi\sin^2\alpha')$ da'. The area involved here is that of the receiver, the angle opens toward the emitter, and L is the radiance of the emitter. Thus, for both cases, Lambert's law did not have to be considered. Detailed calculations and more specific cases can be found in Ref. [22].

If we consider a Lambertian sphere illuminated by a distant point source, the maximum radiance will be observed at the surface where the local normal coincides with the incoming beam. The radiance will decrease with a cosine dependence to zero at viewing angles close to 90°. If the intensity (integrated radiance over area) is unity when viewing from the source, then the intensity when viewing from the side is $1/\pi$. This is a consequence of the fact that the ratio of the radiant exitance, defined by Equation 1.3a, to the radiance, via Equation 1.12, of a Lambertian surface is a factor of π and not 2π. As the integration of the radiance is over a hemisphere, one may intuitively expect that this factor would be 2π, because there are 2π sr in a hemisphere. Though, the presence of $\cos\pi$ in the definition of radiance results in π for the ratio, because the integration of the Equation 1.13 includes the average value of $\cos\alpha$, which is 1/2.

1.1.3.3 Photometric Quantities and Units

Photometry weights the measured power at each wavelength with a factor that represents the eye sensitivity at that wavelength. The luminous flux, Φ_λ (also called luminous power or luminosity), is the light power of a source as perceived by the human eye. The SI unit of luminous flux is the lumen, which is defined as "a monochromatic light source emitting an optical power of 1/683 W at 555 nm has a luminous flux of 1 lumen (lm)." For example, high performance visible-spectrum LEDs have a luminous flux of about 1.0–5.0 lm at an injection current of 50–100 mA.

The luminous flux is the photometric counterpart of radiant flux (or radiant power). The comparison of the watt (radiant flux) and the lumen (luminous flux) illustrates the distinction between radiometric and photometric units.

We illustrate this with an example. The power of light bulbs is indicated in watts on the bulb itself. An indication of 60 W on the bulb can give information how much electric energy will be consumed for a given time. But the electric power is not a measure of the amount of light delivered by this bulb. In a radiometric sense, an incandescent light bulb has an efficiency of about 80%, as 20% of the energy is lost. Thus, a 60 W light bulb emits a total radiant flux of about 45 W. Incandescent bulbs are very inefficient lighting source, because most of the emitted radiant energy is in the IR. There are compact fluorescent bulbs that provide the light of a 60 W bulb while consuming only 15 W of electric power. In the United States (US), already for several decades, the light bulb packaging also indicates the light output in lumens. For example, according to the manufacturer indication on the package of a 60 W incandescent bulb, it provides about 900 lm, as does the package of the 15 W compact fluorescent bulb.

The luminous intensity represents the light intensity of an optical source as perceived by the human eye and is measured in units of candela (cd). The candela is one of the seven basic units of the SI system. A monochromatic light source emitting an optical power of 1/683 W at 555 nm into the solid angle of 1 sr has a luminous intensity of 1 cd. The definition of candela was adopted by the 16th General Conference on Weights and Measures (CGPM) in 1979: "The candela is the luminous intensity, in a given direction, of a source that emits monochromatic radiation of frequency 540×10^{12} Hz and that has a radiant intensity in that direction of 1/683 W/sr."

The choice of the wavelength is determined by the highest sensitivity of the human eye, which is at wavelength 555 nm, corresponding to frequency 5.4×10^{14} Hz (green light). The number 1/683 was chosen to make the candela about equal to the standard candle, the unit of which it superseded. Combining these definitions, we see that 1/683 W of 555 nm green light provides 1 lm. And respectively, 1 candela equals 1 lumen per steradian (cd = lm/sr). Thus, an isotropical light source with luminous intensity of 1 cd has a luminous flux of 4π lm = 12.57 lm. The concept of intensity is applicable for small light sources that can be represented as a point source. Usually, intensity is measured for a small solid angle ($\sim 10^{-3}$ sr) at different angular position and the luminous flux of a point source is calculated by integrating numerically the luminous intensity over the entire sphere [23]. Luminous intensity, while often associated with an isotropic point source, is a valid specification for characterizing highly directional light sources such as spotlights and LEDs. If a source is not isotropic, the relationship between candelas and lumens is empirical. A fundamental method used to determine the total flux (lm) is to measure the luminous intensity (cd) in many directions using a goniophotometer, and then numerically integrate over the entire sphere. In the case of an extended light source, instead of intensity, the quantity luminance L_λ is used to describe the luminous flux emitted from an element of the surface da at an angle a per unit solid angle. Luminance is measured in candela per square meter (cd/m^2). Photopic vision dominates at a luminance above 10 cd/m^2, whereas luminance below 10^{-2} cd/m^2 triggers the scotopic vision.

The illuminance E_λ in photometry corresponds to the radiometric irradiance and is given by

$$E_\lambda = \frac{d\Phi_\lambda}{da} = \frac{I_\lambda \cos\alpha'}{r^2} \qquad (1.14)$$

where
    da is the small element of the surface
    α' is the angle of incidence
    r is the distance from a point source to the illuminated surface

The illuminance is the luminous flux incident per unit area and is measured in lux (lux = lm/m^2). Lux is an SI unit. The higher the illuminance, the higher the resolution of the eye, as well as its ability to distinguish small difference in contrasts and color hues. The illuminance of full moon is 1 lux, and that of direct sunlight is 100,000 lux. The annual average of total irradiance of Sun just outside Earth's atmosphere is the solar constant. It has a value of 1350 W/m^2, but the photometric illuminance of Sun is 1.2×10^5 lm/m^2. Dividing this value of illuminance by 683 gives radiometric irradiance of 0.17×10^3 W/m^2. This number is about eight times smaller than 1.3×10^3 W/m^2, which shows that the wavelength range for the photometric consideration is smaller than the range in radiometry.

The relationship between the illuminance at a distance r from a point light source, measured in lux, and the luminous intensity, measured in candela, is $E_\lambda r^2 = I_\lambda$. The relationship between the illuminance at the same distance r from the source, measured in lux, and the luminous flux, measured in lumen, is

$$E_\lambda 4\pi r^2 = \Phi_\lambda$$

The illumination provided by a given source depends on the shape of the source and the receiver, and on their relative position. A summary of the formulae for the luminous flux and illuminance for few particular cases are presented in Table 1.2. The detailed derivation of the formulae are given in Ref. [22]. If a point source illuminates an area, and the area is tilted by the angle θ with respect to the normal of the source area, the projected area must be used. The mathematical solution of the problem of a line source, such as fluorescent lamp, and a point receiver is similar to the derivation of the Biot–Savart law for magnetisms. The luminous flux and the illuminance, at a point receiver provided by a line source, obey $1/R$ law of attenuation of the emitted light toward a point receiver.

1.1.3.4 Conversion between Radiometric and Photometric Units

The relation between watts and lumens is not just a simple scaling factor, because the relationship between lumens and watts depends on the wavelength. One watt of IR radiation (which is where most of the radiation from an incandescent bulb falls) is worth zero lumens. The lumen is defined only for 555 nm, that is, the definition of lumen tells us that 1 W of green 555 nm light is worth 683 lm.

The conversion between radiometric and photometric units for the other wavelengths within the visible spectrum is provided by the eye sensitivity function, $V(\lambda)$, shown in Figure 1.1. The eye has its peak sensitivity in the green at 555 nm, where the eye sensitivity function has a value of unity, that is, $V(555 \text{ nm}) = 1$. For wavelengths smaller than 390 nm and larger than 720 nm, the eye sensitivity function falls off below 10^{-3}.

The conversion from watts to lumens at any other wavelength involves the product of the radiant power and the $V(\lambda)$ value at the wavelength of interest. The spectral density of the radiant flux, $\Phi(\lambda) = d\Phi/d\lambda$, is the light power emitted per unit wavelength and measured in watts per nanometer, and is also called spectral power distribution. The luminous flux, Φ_λ, is related to the spectral density of the radiant flux $\Phi(\lambda)$ through the luminous efficiency function $V(\lambda)$ measured in lumen:

$$\Phi_\lambda = 683 \left[\frac{\text{lm}}{\text{W}} \right] \int_\lambda V(\lambda)\Phi(\lambda)\, d\lambda \tag{1.15}$$

This equation suggests that radiant flux of 1 W at a wavelength of 555 nm produces a luminous flux of 683 lm. This equation could also be used for other quantity pairs. For instance, luminous intensity (cd) and spectral radiant intensity (W/sr·nm), illuminance (lux) and spectral irradiance (W/m^2·nm), or luminance (cd/m^2) and spectral radiance (W/m^2·sr·nm).

TABLE 1.2

Illumination by Sources with Different Geometrical Shapes

Type of Source	Luminous Flux	Illuminance
Point source and area receiver 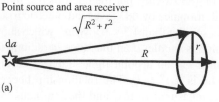 (a)	$\Phi_\lambda = I_\lambda \dfrac{\pi r^2}{r^2 + R} \cong I_\lambda \dfrac{\pi r^2}{R^2}$	$E_\lambda = \dfrac{\Phi_\lambda}{\pi r^2} = L_\lambda \dfrac{da}{r^2 + R^2}$
Area source and point receiver 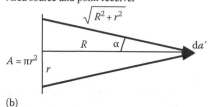 (b)	$\Phi_\lambda = L_\lambda \pi \sin^2 \alpha\, da'$	$E_\lambda = \dfrac{\Phi_\lambda}{da'} = L_\lambda \pi \sin^2 \alpha$
Line source (fluorescent lamp) and point receiver 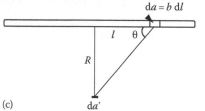 (c)	$\Phi_\lambda = L_\lambda da'\left(b\dfrac{2}{R} \right)$, where b is the width of the line source, and $(2b/R)$ gives the solid angle in steradians.	$E_\lambda = L_\lambda \left(b\dfrac{2}{R} \right)$
Oblique illumination: point source and point receiver (d)	$\Phi_\lambda = L_\lambda da\left(\dfrac{da'}{R^2} \right)\cos\theta$	$E_\lambda = L_\lambda \dfrac{da\cos\theta}{R^2}$
Oblique illumination: point source and area receiver (e)	$\Phi_\lambda = L_\lambda da\, \pi \sin^2 \alpha \cos\theta$	$E_\lambda = L_\lambda \pi \sin^2 \alpha \cos\theta$

Equation 1.15 represents a weighting, wavelength by wavelength, of the radiant spectral term by the visual response at that wavelength. The wavelength limits can be set to restrict the integration only to those wavelengths where the product of the spectral term $\Phi(\lambda)$ and $V(\lambda)$ is nonzero. Practically, this means that the integration is over the entire visible spectrum, because out of the visible range, $V(\lambda) \cong 0$ and the integral tends to zero. As the $V(\lambda)$ function is defined by a table of empirical values, the integration is usually done numerically. The optical power emitted by a light source is then given by

$$\Phi = \int_\lambda \Phi(\lambda)\, d\lambda \tag{1.16}$$

Let us compare the luminous flux of two laser pointers both with radiometric power of 5 mW working at 670 and 635 nm, respectively. At 670 nm, $V(\lambda) = 0.032$ and the luminous flux is 0.11 lm (683 lm/W $\times$ 0.032 $\times$ 0.005 W = 0.11 lm). At 635 nm, $V(\lambda) = 0.217$ and the laser pointer has luminous flux is 0.74 lm (683 lm/W $\times$ 0.217 $\times$ 0.005 W = 0.74 lm). The shorter wavelength (635 nm) laser pointer will create a spot that is almost seven times as bright as the longer wavelength (670 nm) laser (assuming the same beam diameter). Using eye sensitivity function, $V(\lambda)$, one can show also that 700 nm red light is only about 4% as efficient as 555 nm green light. Thus, 1 W of 700 nm red light is worth only 27 lm.

The International Commission on Weights and Measures (CGPM) has approved the use of the CIE curve for photopic (light conditions) vision $V(\lambda)$ and the curve for scotopic (dark-adapted) vision $V'(\lambda)$ for determination of the value of photometric quantities of luminous sources. Therefore, for studies at lower light levels, the curve for scotopic (dark-adapted) vision $V'(\lambda)$ should be used. This $V'(\lambda)$ curve has its own constant of 1700 lm/W, the maximum spectral luminous efficiency for scotopic vision at the peak wavelength of 507 nm. This value was chosen such that the absolute value of the scotopic curve at 555 nm coincides with the photopic curve, at the value 683 lm/W.

Converting from lumens to watts is far more difficult. As we want to find out $\Phi(\lambda)$ that was already weighted and placed inside an integral, we must know the spectral function Φ_λ of the radiation over the entire spectral range where the source emits not just the visible range.

The luminous efficacy of optical radiation is the conversion efficiency from optical power to luminous flux. It is measured in units of lumen per watt of optical power. The luminous efficacy is a measure of the ability of the radiation to produce a visual response and is defined as

$$\text{Luminous efficacy} = \frac{\Phi_\lambda}{\Phi} = \frac{683 \left[\dfrac{\text{lm}}{\text{W}}\right] \displaystyle\int_\lambda V(\lambda)\Phi(\lambda)\,d(\lambda)}{\displaystyle\int_\lambda \Phi(\lambda)\,d\lambda} \left[\frac{\text{lm}}{\text{W}}\right] \tag{1.17}$$

For monochromatic light ($\Delta\lambda \rightarrow 0$), the luminous efficacy is equal to the eye sensitivity function $V(\lambda)$ multiplied by 683 lm/W. Thus, the highest possible efficacy is 683 lm/W. However, for multicolor light sources and especially for white light sources, the calculation of the luminous efficacy requires integration over all wavelengths.

The luminous efficiency of a light source is the ratio of luminous flux of the light source to the electrical input power (IV) of the device and shows how efficient the source is in converting the consumed electric power to light power perceived by the eye:

$$\text{Luminous efficiency} = \Phi_\lambda / (IV) \tag{1.18}$$

The luminous efficiency is measured in units of lumen per watt, and the maximum possible ratio is 683. Combining Equations 1.4, 1.17, and 1.18 gives that the luminous efficiency is a product of luminous efficacy and electrical-to-optical power conversion efficiency (radiant efficiency):

$$\text{Luminous efficiency} = \text{luminous efficacy} \times \text{radiant efficiency} \tag{1.19}$$

For an ideal light source with a perfect electrical-to-optical power conversion, the luminous efficiency is equal to the luminous efficacy. The luminous efficiency of common light sources varies from about 15 to 25 lm/W for incandescent sources such as tungsten filament light bulbs and quartz halogen light bulbs, from 50 to 80 lm/W for fluorescent light tubes, and from 50 to about 140 lm/W for high-intensity discharge sources such as mercury-vapor light bulbs, metal-halide light bulbs, and high-pressure sodium (HPS) vapor light bulbs [12].

As an example, let us calculate the illuminance and the luminous intensity of a standard 60 W incandescent light bulb with luminous flux of 900 lm. The luminous efficiency, that is, the number of lumen emitted per watt of electrical input power, of the light bulb is 15 lm/W. The illuminance on a desk located 1.5 m below the bulb would be $E_\lambda = 31.8$ lux and would be enough for simple visual tasks (orientation and simple visual tasks require 30–100 lux as recommended by the Illuminating Engineering Society of North America) [24,25]. The luminous intensity of the light bulb is $I_\lambda = 71.6$ lm/sr = 71.6 cd.

In addition to the luminous efficiency, another highly relevant characteristic for an LED is the luminous intensity efficiency. It is defined as the luminous flux per steradian per unit input electrical power and is measured in candela per watt. In most applications of LEDs, the direction of interest is normal to the chip surface. Thus, the luminance of an LED is given by the luminous intensity emitted along the normal to the chip surface divided by the chip area. The efficiency can be increased significantly by using advanced light-output-coupling structures [26], but it is on the expense of low luminance because only a small part of the chip is injected with current. LEDs emit in all directions, as the intensity depends on the viewing angle and the distance from the LED. Therefore, in order to find the total optical power emitted by an LED, the spectral intensity $I(\lambda)$ (W/m$^2 \cdot$nm) should be integrated over the entire surface area of a sphere A:

$$\Phi = \iint_{A\ \lambda} I(\lambda)\ d\lambda dA \tag{1.20}$$

1.1.3.5 Choosing a Light Source

Several important factors determine the best type of a light source for a given application: spectral distribution, output (needed irradiance), F-number, source size and shape, and the use of a grating monochromator.

Spectral distribution: The light source should have intense output in the spectral region of interest. Arc lamps are primarily UV to visible light sources, which make them useful for UV spectroscopy and UV photochemistry applications. Mercury lamps have very strong peaks in the UV. Mercury (xenon) lamps have strong mercury lines in the UV with xenon lines in the IR, and enhancement of the low-level continuum in the visible range. Xenon flashlamps are pulsed sources of UV near-IR radiation, which are flexible, but not complicated. Figure 1.5 compares the average irradiance from a xenon DC arc lamp and a large bulb xenon flashlamp with pulse duration 9 μs. Quartz tungsten halogen lamps are a good choice for long-wave visible to near-IR applications. In addition, quartz tungsten halogen lamps deliver smooth spectra without spikes. Figure 1.6 shows the spectral irradiance of mercury, xenon, and halogen lamps.

Source power: The needed irradiance determines the choice of the source output. Besides lasers, short arc lamps are the brightest sources of DC radiation. A high power lamp is best for applications that require irradiation of a large area and when the used optics imposes no limitation. For a small target or when a monochromator with narrow slit is used, a small arc lamp may produce more irradiance because of its high radiance within a very small effective arc area. When the illuminated area is small, both the total flux and the arc dimensions are of importance. Narrow spectrograph slits require the use of collection and imaging system, which involves calculation of F-numbers.

F-number (*power/nm*): Lamps are used barely without lamp house containing a collimator lens. Figure 1.7 shows a lens of clear aperture D, collecting light from a source and collimating it. The source is placed at the focal length f of the lens. As most sources radiate at all angles, it is obvious that increasing D or decreasing f allows the lens to capture more light. The F-number concept

FIGURE 1.5   Comparison of average irradiance at 0.5 m from the 6251 75 W xenon DC arc lamp and the 6427 large bulb flashlamp running at 60 Hz (60 W average and pulse duration 9 μs). (Courtesy of Newport Corporation, Richmond, VA.)

FIGURE 1.6   Spectral irradiance of Hg, Xe, and halogen lamp. (Courtesy of Newport Corporation, Richmond, VA.)

FIGURE 1.7 Lens collecting and collimating light from a source. (Courtesy of Newport Corporation, Richmond, VA.)

puts these two parameters together in order to enable a quick comparison of optical elements. The F-number is defined as

$$F/\# = \frac{1}{2n\sin\theta} \tag{1.21}$$

where

n is the refractive index of the medium in which the source is located (normally air, therefore $n = 1$)

θ is the angle as shown in Figure 1.7

However, the approximation is valid only for the paraxial rays (small angles θ). Therefore,

$$F/\# = f/D \tag{1.22}$$

is widely used as F-number, even at large values of θ. The $F/\#$ is useful to convert from the irradiance data in Figure 1.6 to beam power. For example, the irradiance value for the 150 W xenon lamp at 540 nm is 2 µW/cm²/nm (see Figure 1.6). If $F/1$ condenser is used, the conversion factor is 0.82. The conversion factor is written in the manual of the lamp housing and might vary between different lamp suppliers. Multiplying the irradiance value with the conversion factor gives the result in milliwatt per nanometer, which for the given example is (2 µW/cm² · nm) (0.82) = 1.64 mW/nm. In a typical lamp housing, a reflector is mounted behind the lamp, which increases the lamp output by 60%. Hence, the output is (1.64 mW/nm) (1.6) = 2.6 mW/nm.

Source size and shape: The size and shape of the source, as well as the available optics, also determine the illumination of the sample. If the application requires a monochromatic illumination, a monochromator is normally placed between the lamp and the target. The object size of an arc is considerably smaller than that of a coil. Hence, with an arc lamp, more light can be guided through a monochromator in comparison with a quartz tungsten halogen lamp of the same power.

Grating monochromators: A grating monochromator permits the selection of a narrow line (quasi-monochromatic light) from a broader spectrum of incident optical radiation. Used with a white light source, a monochromator provides selectable monochromatic light over the range of several hundred nanometers for specific illumination purposes. Two processes occur within the monochromator. First, it is the geometrical optical process wherein the entrance slit is imaged on the exit slit. The second process is the dispersion process on the rotating grating. Table 1.3 shows the selecting criteria for a proper grating based on the most important operating parameters of a monochromator. The resolution is the minimum detectable difference between two spectral lines

TABLE 1.3
Selection Criteria of Gratings

Groove density (or groove frequency): The number of grooves contained on a grating surface (lines/mm)	Groove density affects the mechanical scanning range and the dispersion properties of a system. It is an important factor in determining the resolution capabilities of a monochromator. Higher groove densities result in greater dispersion and higher resolution capabilities. Select a grating that delivers the required dispersion when using a charge-coupled device (CCD) or array detector, or the required resolution (with appropriate slit width) when using a monochromator.
Mechanical scanning range: The wavelength region in which an instrument can operate	Refers to the mechanical rotation capability (not the operating or optimum range) of a grating drive system with a specific grating installed. Select a grating groove density that allows operation over your required wavelength region.
Blaze wavelength: The angle in which the grooves are formed with respect to the grating normal, often called blaze angle	Diffraction grating efficiency plays an important role in monochromator or spectrograph throughput. Efficiency at a particular wavelength is largely a function of the blaze wavelength if the grating is ruled, or a modulation if the grating is holographic. Select a blaze wavelength that encompasses the total wavelength region of your applications, and, if possible, favors the short wavelength side of the spectral region to be covered.
Quantum wavelength range: The wavelength region of highest efficiency for a particular grating	Normally determined by the blaze wavelength. Select a grating with maximum efficiency over the required wavelength region for your application.

(peaks) provided by a given monochromator. While resolution can be theoretically approximated by multiplying slit width (mm) × dispersion (nm/mm), the actual resolution rarely equals this due to aberrations inherent in monochromators. The dispersion of a monochromator indicates how well the light spectrum is spread over the focal plane of the exit plane. It is expressed in the amount of wavelengths (nm) over a single millimeter of the focal plane. The aperture (or F-number) is the ratio of the focal length of the monochromator to the grating diameter, equivalent to Equation 1.22. The focal length is the distance between the slit and the focusing component of the monochromator. The bandpass (nm) is the actual resolution of a monochromator as a function of the slit width. At zero order, the grating acts as a mirror, and all incident wavelengths are reflected. For this reason, it is often helpful to use zero-order reflection when setting up an optical system. The grating linearly polarizes the light passing through the monochromator, and the degree of polarization depends on the wavelength.

Typical system configurations of light sources and monochromators for transmittance, reflectance, and emission measurements using different detectors are shown in Figure 1.8. As the emission intensity of a light source and the efficiency of a grating versus wavelength are not constant (see Figure 1.5), it is necessary to make corrections to the measured spectra. For this purpose, the response of the illuminating system, that is, monochromator and lamp, is measured with a calibrated detector. The corrected response of the measured spectrum is given by

$$R_{\text{corr}} = (I_{\text{m}}/I_{\text{D}}) \times R_{\text{D}} \tag{1.23}$$

where
I_{m} is the measured response
I_{D} is the reference measurement with the detector
R_{D} is the response function of the detector

FIGURE 1.8 Typical system configurations of light sources and monochromators. (Courtesy of Newport Corporation, Richmond, VA.)

1.1.4   PHOTOMETRIC MEASUREMENT TECHNIQUES

Photometric measurements employ photodetectors (devices that produce an electric signal when exposed to light), and the choice of the photometric method depends on the wavelength regime under study.

Photometric measurements are frequently used within the lighting industry. Simple applications include switching luminaires (light fixtures) on and off based on ambient light conditions and light meters used to measure the amount of light incident on a surface. Luminaires are tested with gonio-photometers or rotating mirror photometers. In both the cases, a photocell is kept stationary at a sufficient distance such that the luminaire can be considered as a point source. The rotating mirror photometers use a motorized system of mirrors to reflect the light from the luminaire in all directions to the photocell, while the goniophotometers use a rotating table to change the orientation of the luminaire with respect to the photocell. In either case, luminous intensity is recorded and later used in lighting design.

Three-dimensional photometric measurements use spherical photometers to measure the directional luminous flux produced by light sources. They consist of a large-diameter globe with the light source mounted at its center. A photocell measures the output of the source in all directions as it rotates about the source in three axes.

Photometry finds broad applications also in astronomy, where photometric measurements are used to quantify the observed brightness of an astronomical object. The brightness of a star is evaluated by measuring the irradiance E (the energy per unit area per unit time) received from the star or when the measurements are done with the same detecting system (telescope and detector)— the radiant flux (energy per unit time). The ancient studies of stellar brightness (by Hipparchus, and later around AD 150 by Ptolemy) divided the stars visible to the naked eye into six classes of brightness called magnitudes. The first-magnitude stars were the brightest and the sixth-magnitude stars the faintest. The later quantitative measurements showed that each jump of one magnitude corresponded to a fixed flux ratio, not a flux difference. Thus, the magnitude scale is essentially a logarithmic scale, which can be explained with the logarithmic response of the eye to brightness. It was found that a difference of 5 magnitudes (mag) corresponds to a factor of 100 in energy flux. If the radiant fluxes of two stars are Φ_1 and Φ_2, and m_1 and m_2 are the corresponding magnitudes, the mathematical expression of this statement is

$$\frac{\Phi_1}{\Phi_2} = 100^{(m_2 - m_1)/5} = 10^{(m_2 - m_1)/2.5} \qquad (1.24)$$

This equation is used to calculate the brightness ratio for a given magnitude difference. It shows that each time $m_2 - m_1$ increases by 5, the ratio Φ_1/Φ_2 decreases by a factor of 100; thus, increasing the irradiance decreases the magnitude. The magnitude difference for a given brightness ratio is given by

$$m_2 - m_1 = 2.5 \log\left(\frac{\Phi_1}{\Phi_2}\right) \qquad (1.25)$$

The magnitude range for stars visible to the naked eye is 1–6 mag. This corresponds to a flux ratio $\Phi_1/\Phi_2 = 10^{(6-1)/2.5} = 10^2$. The largest ground-based telescopes allow to extend the magnitude range from 6 to 26 mag, which corresponds to a flux ratio $\Phi_1/\Phi_2 = 10^{(26-6)/2.5} = 10^8$. This way, using telescopes allows creating a continuous scale that extends the original 6 mag scale to fainter or brighter objects. In this new scale, brighter objects than 1 mag can have 0 mag, or even negative magnitudes.

Photometric techniques in astronomy are based on measuring the radiant flux of the stellar object through various standard filters. The radiation from a star is gathered in a telescope, passing it through specialized optical filters (passbands), and the light energy is then captured and recorded with a photosensitive element. The set of well-defined filters with known sensitivity to a given incident radiation is called a photometric system. The sensitivity depends on the implemented optical system, detectors, and filters. The first-known standardized photometric system for classifying stars according to their colors is the Johnson–Morgan UBV (ultraviolet, blue, and visual) photometric system, which employs filters with mean wavelengths of response functions at 364, 442, and 540 nm, respectively [27]. The Johnson–Morgan technique used photographic films to detect the light response. Those films were sensitive to colors on the blue end of the spectrum, which determined the choice of the UBV filters.

Nowadays, there are more than 200 photometric systems in use, which could be grouped according to the widths of their passbands in three main groups. The broadband systems employ wide passbands, wider than 30 nm, as the most frequently used system is the Johnson–Morgan UBV system. The intermediate band and the narrow band systems have passbands widths of 10–30 and less than 10 nm, respectively. Most optical astronomers use the so-called Johnson–Cousins photometric system, which consists of ultraviolet, blue, visual, red, and infrared (UBVRI) filters and photomultiplier tubes [28,29].

For each photometric system, a set of primary standard stars is used. The light output of the photometric standard stars is measured carefully with various passbands of a photometric system. The light flux received from other celestial objects is then compared to a photometric standard star to determine the exact brightness or stellar magnitude of the object of interest. It is assumed that the direct observations have been corrected for extinction. The most often used current set of photometric standard stars for UBVRI photometry is that published by Landolt [30,31].

The apparent brightness of a star depends on the intrinsic luminosity of the star and its distance from us. Two identical stars at different distances will have different apparent brightnesses. This requires correcting the apparent brightness for the distance to the star. Starlight is also affected by interstellar extinction. Therefore, in all cases in which the distance to an astronomical object is a concern, the observer should correct for the effects of interstellar absorption. In order to compare two stars, we have to know their total luminosities (power output). If the total radiant power of a star is Φ, and if no radiation is absorbed along the way, all energy per second leaving the surface of the star will cross a sphere at a distance d in the same time. The radiant exitance of the star $M = \Phi/(4\pi r^2)$ falls off inversely as the square of the distance. Unfortunately, distances to astronomical objects are generally hard to determine. Distances to nearby stars could be determined by a simple trigonometric method called trigonometric parallax. It is based on triangulation from two different observing points [32]. With current ground-based equipment, parallax can be measured within a few thousandths of an arc second. Parallax measurements are useful for the few thousand nearest stars. Therefore, they are the starting point for a very complex system for determining distances to astronomical objects.

Photometry was intensively used in the 1960s to determine the red shift in the spectrum of a distant astronomical object such as a galaxy or quasar. The red shift is a measure of the rate at which any galaxy is receding from us and hence, through Hubble's law, of the distance to it [32,33]. The technique relies on the assumption that the radiation spectrum emitted by the object has strong features that can be detected by the relatively crude filters. This technique was largely replaced in the 1970s by more reliable spectroscopic determinations of red shifts [34]. Photometry is also used to measure the light variations of objects such as variable stars, minor planets, active galactic nuclei, and supernovae, or to detect transiting extrasolar planets. These measurements can be used, for example, to determine the rotation period of a minor planet or a star, the orbital period and the radii of the members of an eclipsing binary star system, or the total energy output of a supernova.

The traditional photometric methods employed photoelectric photometers, devices that measured the light intensity of a single object by focusing its light on to a photosensitive cell. The modern photometric systems use CCD cameras, which can simultaneously image multiple objects and allow to measure accurately the signal spread over many pixels due to broadening caused by the optics in the telescope and the star twinkling. The methods used to extract data from the raw CCD image depend on the observed objects. Aperture photometry is a method used to determine the radiant flux of a distant star that can be considered as a point source (an object with an angular diameter that is much smaller than the resolution of the telescope). It consists in adding up the pixel counts within a circle centered on the object and subtracting off the pixel count of the background light. When the target is a globular cluster, the individual fluxes of the overlapping sources are obtained via de-convolution techniques such as point spread function fitting. After determining the flux of an object in counts, one must calibrate the measurement. For this purpose, differential, relative, or absolute photometry can be used, depending on the type of photometric measurements, as all techniques require corrections for the atmospheric extinction when the object is observed from the ground.

Differential photometry measures the changes in the brightness of an object over time and requires corrections for temporal variations in the sensitivity of the instrument. For this purpose, a number of comparison stars, which are assumed to be constant, are simultaneously observed together with the target. In general, differential photometry provides the highest precision of photometric measurements.

Relative photometry compares the apparent brightnesses of multiple objects relative to each other. This technique requires corrections for spatial changes in the sensitivity of the instrument. If the objects being compared are too far apart to be observed simultaneously, additional corrections for their temporal variations have to be done.

Absolute photometry measures the apparent brightness of an object with a standard photometric system. Thus, in addition to the other factors mentioned above, one has to account also for the differences between the effective passband through which the object is observed and the passband used to define the standard photometric system. To provide correction for this effect, the target should be observed through multiple filters, simultaneously with a number of photometric standard stars. Due to those complications, absolute photometry is the most difficult to do with high precision.

More information about astronomical photometry could be found in specialized books and literature [35–42].

1.1.5 COLORIMETRY

Colorimetry deals with measurement of color. The fundamentals of colorimetry are presented in detail in Refs. [13,14]. Colorimetry deals with basic concepts such as tristimulus values, chromaticity coordinates, color temperature, and color rendering. Colorimetry describes colors in numbers (tristimulus values) and provides a physical color match using a variety of measurement instruments. Colorimetry utilizes the standard color science calculations and codes provided by the 1931 CIE. The description of colors by numbers is based on the fact that all colors can be represented as a combination of the three primary colors (stimuli): red, green, and blue. The tristimulus values X, Y, and Z give the amounts of each stimuli in a given color represented by the spectral power distribution $\Phi(\lambda)$. The numerical value of a given color is obtained by integrating the spectrum with the standard color-matching functions $\bar{x}(\lambda)$, $\bar{y}(\lambda)$, and $\bar{z}(\lambda)$ based on a 1931 CIE standard observer:

$$X = \int \bar{x}(\lambda)\Phi(\lambda)\, d\lambda \tag{1.26a}$$

$$Y = \int \bar{y}(\lambda)\Phi(\lambda)\, d\lambda \tag{1.26b}$$

$$Z = \int \bar{z}(\lambda)\Phi(\lambda)\, d\lambda \tag{1.26c}$$

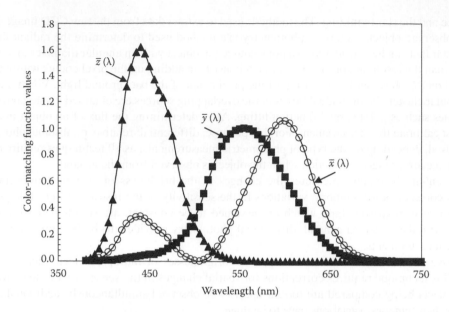

FIGURE 1.9 1978 CIE color matching functions: $\bar{x}(\lambda)$, O; $\bar{y}(\lambda)$, ■; and $\bar{z}(\lambda)$, ▲. (Adapted from Schubert, E.F., *Light-Emitting Diodes*, Cambridge University Press, Cambridge, U.K., 2003.)

The O $\bar{x}(\lambda)$, ■ $\bar{y}(\lambda)$, and ▲ $\bar{z}(\lambda)$ color-matching functions are shown on Figure 1.9. The graph plots the tabulated values of the 1978 CIE color-matching functions, also called the Judd–Vos-modified 1978 CIE two-degree color matching functions for point sources [14,16–20]. The ■ color-matching function, $\bar{y}(\lambda)$, is identical to the eye sensitivity function $V(\lambda)$ for photopic vision.

The chromaticity coordinates (x, y) are functions of the tristimulus values X, Y, and Z and describe quantitatively the color of a radiating light source with a spectrum $\Phi(\lambda)$:

$$x = \frac{X}{X + Y + Z} \tag{1.27a}$$

$$y = \frac{Y}{X + Y + Z} \tag{1.27b}$$

$$z = \frac{Z}{X + Y + Z} \equiv 1 - x - y \tag{1.27c}$$

As the coordinate z can be expressed by the other two, every color could be represented by only two independent chromaticity coordinates (x, y) in a two-dimensional graph called chromaticity diagram. The 1931 CIE chromaticity diagram [5,43,44] provides simple means of color mixing. A set of n primary sources with chromaticity coordinates (x_i, y_i) and radiant fluxes Φ_i will produce a color with chromaticity coordinates (x_c, y_c), which are given by

$$x_c = \frac{\sum_{i=1}^{n} x_i \Phi_i}{\sum_{i=1}^{n} \Phi_i} \quad \text{and} \quad y_c = \frac{\sum_{i=1}^{n} y_i \Phi_i}{\sum_{i=1}^{n} \Phi_i} \tag{1.28}$$

The radiant fluxes Φ_i of the primary sources used in these equations should be normalized so that the sum $X_i + Y_i + Z_i$ equals unity for all i.

Colorimetry is a basic method for describing light sources in lighting industry. Depending on the nature of the emission processes, different light sources emit different spectra. As a way of evaluating the ability of a light source to reproduce the colors of various objects being lit by the source, the CIE introduced the color rendering index (CRI; 1974 CIE and updated in 1995), which gives a quantitative description of the lighting quality of a light source [45,46]. After reflecting from an illuminating object, the original radiation spectrum is altered by the reflection properties of the object, that is, the object's reflectivity spectrum. This results in a shift of the chromaticity coordinates, the so-called colorimetric shift. As reflectivity spectra will produce different colorimetric shifts for different spectral composition of the source, colors may appear completely different from, for instance, colors produced by sunlight. To estimate the quality of lighting, eight test samples are illuminated successively by the source to be tested and by a reference source. The reflected spectra from the samples are determined and the chromatic coordinates are calculated for both the tested and the reference sources. Then, the colorimetric shifts are evaluated and graded with respect to chromatic adaptation of the human eye using the 1960 CIE uniform chromaticity scale diagram and the special CRI R_i for each sample is calculated. The general CRI R_a is obtained by averaging of the values of the eight special CRI R_i:

$$R_a = \frac{1}{8} \sum_{i=1}^{8} R_i \qquad (1.29)$$

This method of grading of colorimetric shifts obtained in test samples is called CIE test-color method, and is applicable for light sources that have chromaticity close to the reference illuminant. The CRI is a measure of how well balanced the different color components of white light are. A standard reference illuminant has a general CRI $R_a = 100$. The two primary standard illuminants are the CIE Standard Illuminant A (representative of tungsten-filament lighting with color temperature of 2856 K) and the CIE Standard Illuminant D65 (representative of average daylight with color temperature of 6500 K). Color rendering is a very important property of cold illuminants such as LEDs and discharge lamps, whose emission spectrum contains certain wavelengths. More information about the test-color samples, standard colorimetric illuminants and observers, and CRIs can be found in related references [3,5,24,45] and CIE reports [47–50], as well as in the literature on LEDs [9–11].

The measurements of color light sources (lamps, LEDs, and displays) involve spectroradiometers and tristimulus colorimeters. In order to characterize the color, lamps are usually measured for spectral irradiance, whereas displays are measured for spectral radiance. When a spectroradiometer is designed to measure spectral irradiance (W/m^2·nm), it is equipped with a diffuser or a small integrating sphere as input optics, while a spectroradiometer, which is designed to measure spectral radiance (W/sr·m^2·nm), employs imaging optics. A spectroradiometer could be one of a mechanical scanning type, which are more accurate but slow, or of diode-array type, which are fast, but less accurate devices. Spectroradiometers are calibrated against spectral irradiance and radiance standards [49], which suggest that their measurement uncertainty is determined first by that of the reference source. Other factors influencing the measurement uncertainty are wavelength error, detector non-linearity, bandwidth, stray light of monochromator, measurement noise, etc. [44]. Tristimulus colorimeters are used to calibrate display devices and printers by generating color profiles used in the workflow. Accurate color profiles are important to ensure that screen displays match the final printed products. Tristimulus colorimeters are low-cost and high-speed devices, but their measurement uncertainty tend to be higher than spectroradiometers.

A colorimeter can refer to one of the several related devices. In chemistry, a colorimeter generally refers to a device that measures the absorbance of a sample at particular wavelengths of light. Typically, it is used to determine the concentration of a known solute in a given solution by the application of the Beer–Lambert law, which states that the concentration of a solute is proportional to the absorbance. Colorimetry finds wide area of applications in chemistry and in industries such as color printing, textile manufacturing, paint manufacturing, and in the food industry.

Colorimetry is also used in astronomy to determine the color index of an astronomical object. The color index is a simple numerical expression of the color of an object, which in the case of a star gives its temperature. To measure the color index, the object is observed successively through two different sets of standard filters such as U and B, or B and V, centered at wavelengths λ_1 and λ_2, respectively. The difference in magnitudes found with these filters is called U–B or B–V color index. For example, the B–V color index corresponding to the wavelengths λ_B and λ_V is defined by

$$m_B - m_V = 2.5 \log \left[\frac{I(\lambda_V)}{I(\lambda_B)} \right] + \text{constant} \qquad (1.30)$$

where $I(\lambda_V)$ and $I(\lambda_B)$ are the intensities averaged over the filter ranges. If we suppose that an astronomical object is radiating like a blackbody, we need to know only the ratio of brightnesses at any two wavelengths to determine its temperature. The constant in Equation 1.30 is adjusted so that $m_B - m_V$ is zero for a particular temperature star, designated as A0. The ratio of blue to visible intensity increases with the temperature of the object. This means that the $m_B - m_V$ color index decreases. The smaller the color index, the more blue (or hotter) the object is. Conversely, the larger the color index, the more red (or cooler) the object is. This is a consequence of the logarithmic magnitude scale, in which brighter objects have smaller magnitudes than dimmer objects. For example, the yellowish Sun has a B − V index of 0.656 ± 0.005 [51], while the B − V color index of the bluish Rigel is −0.03 (its B magnitude is 0.09 and its V magnitude is 0.12, B − V = −0.03) [52]. The choice of filters depends on the object's color temperature: B − V are appropriate for mid-range objects, U − V for hotter objects, and R − I for cool objects. Bessel specified a set of filter transmissions for a flat response detector, thus quantifying the calculation of the color indices [53].

1.2 LIGHT EMISSION

Light can be emitted by an atomic or molecular system when the system, after being excited to a higher energy level, decays back to a lower energy state through radiative transitions. The excitation of the system can be achieved by different means: thermal agitation, absorption of photons, bombarding with high-speed subatomic particles, and sound and chemical reactions. Additional excitation mechanisms such as radioactive decay and particle–antiparticle annihilation result in emission of very-high-energy photons (γ-rays). The following description of emission processes and light sources is not intended to be comprehensive. More details about the different excitation mechanisms, the resultant emission processes, and light sources could be found in specialized books [54–63].

1.2.1 BLACKBODY RADIATION

The most common light sources are based on thermal emission, also called incandescence (from Latin "to grow hot"). Incandescence is the radiation process, which results from the de-excitation of atoms or molecules after they have been thermally excited. A body at a given temperature emits a broad continuum of wavelengths, for example, sunlight, incandescent light bulbs, and glowing solid particles in flames.

A blackbody is a theoretical model that closely approximates many real objects in thermodynamic equilibrium. A blackbody is an ideal absorber (absorbs all the radiation impinging on it, irrespective of wavelength or angle of incidence), as well as a perfect emitter. A blackbody at constant temperature radiates energy at the same rate as it absorbs energy; thus, the blackbody is in thermodynamic equilibrium. The balance of energy does not mean that the spectrum of emitted radiation should match the spectrum of the absorbed radiation. The wavelength distribution of the emitted radiation is determined by the temperature of the blackbody and was first described accurately by Max Planck in 1900. The empirical formula for the spectral exitance M_λ of a blackbody at temperature T(K), derived by Planck is

$$M_\lambda = \frac{2\pi hc^2}{\lambda^5}\left[\frac{1}{\exp(hc/\lambda k_B T)-1}\right] \tag{1.31}$$

The spectral exitance M_λ is the power per unit area per unit wavelength interval emitted by a source. The constant h is called Planck's constant and has the numerical value $h = 6.62606876 \times 10^{-34}$ J·s, $k_B = 1.3806503 \times 10^{-23}$ J/K $= 8.62 \times 10^{-5}$ eV/K is Boltzmann's constant, and $c = 2.99792458 \times 10^8$ m/s is the speed of light in vacuum. In order to provide a theoretical background for this relation, Planck postulated that a blackbody can only radiate and absorb energy in integer multiples of $h\nu$. That is, the energy could only be emitted in small bundles, or quanta, each quanta having energy $h\nu$ and frequency ν. The explanation of the quantization in the process of radiation and absorption was proposed by Albert Einstein in 1905.

Figure 1.10 shows blackbody spectra for six temperatures. At any wavelength, a hotter blackbody gives off more energy than a cooler blackbody of the same size. Also, as the temperature increases, the peak of the spectrum shifts to shorter wavelengths. The relation between the wavelength at which the peak $\lambda_{\max}$ occurs and the temperature T is given by Wien's displacement law:

$$\lambda_{\max} T = \frac{hc}{5k_B} = 2.898 \times 10^3 \ \mu\mathrm{m}\cdot\mathrm{K} \tag{1.32}$$

This relation can be derived from Equation 1.31 by differentiating M_λ with respect to λ and setting it equal to zero. The curve illustrating Wien's law is shown on Figure 1.10 with a dashed line.

Integrating the spectral exitance from Equation 1.31 over all wavelengths gives the total radiant exitance, M, which is also the area under the blackbody radiation curve at temperature T:

$$M = \int_0^\infty M_\lambda d\lambda = \sigma T^4 \tag{1.33}$$

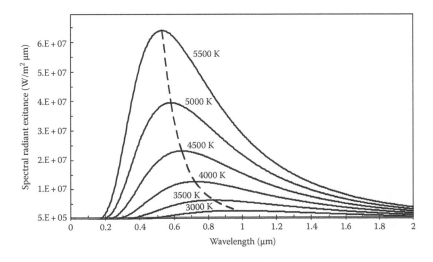

FIGURE 1.10 Blackbody radiation at six temperatures (K). Solid curves show the spectral distribution of the radiation. The dashed curve connecting the peaks of the six solid curves illustrates Wien's displacement law.

where $\sigma = 5.6703 \times 10^{-8}$ W/m$^2 \cdot$K^4 is the Stefan–Boltzmann constant. According to Equation 1.33, known as Stefan–Boltzmann law, the radiant exitance (the total energy per unit time per unit surface area) given off by a blackbody is proportional to the fourth power of the temperature.

Sun and stars, as well as blackened surfaces, behave approximately as a blackbody. Blackbody sources (blackened cavities with a pinhole) are commercially available at operating temperatures from $-196°$C (liquid nitrogen) to 3000°C. Sun emits thermal radiation covering all wavelengths of the electromagnetic spectrum. The surface temperature of Sun is around 6000 K and the emitted radiation peaks in the visible region. Other stellar objects, like the great clouds of gas in the Milky Way, also emit thermal radiation but as they are much cooler, their radiation is in the IR and radio regions of the electromagnetic spectrum. In astronomy, the radiant flux, Φ, or the total power given off by a star is often referred as luminosity. The luminosity of a spherical star with a radius R and total surface area $4\pi R^2$ is

$$\Phi = \int M \, da = (4\pi R^2)\sigma T^4 \tag{1.34}$$

The luminosity of Sun, calculated for surface temperature about 6000 K and radius of 7×10^5 km, is about 4×10^{26} W. This quantity is called a solar luminosity and serves as a convenient unit for expressing the luminosity of other stars.

The color of light sources is measured and expressed with the chromaticity coordinates. But when the hue of white light source has to be described, it is more practical to use the correlated color temperature (CCT) in Kelvin instead of the chromaticity coordinates. The term "color temperature" originates from the correlation between temperature and emission spectrum of Planckian radiator. The radiation from a blackbody is perceived as a white light for a broad range of temperatures from 2,500 to 10,000 K but with different hues: for example, a blackbody at 3,000 K radiates white light with reddish hue, while at 6,000 K the light adopts a bluish hue. The CCT is used to describe the color of white light sources that do not behave as blackbody radiators. The CCT of a white source is the temperature of a Planckian radiator whose perceived color is closest to the color of the white light source at the same brightness and under specified viewing conditions [44,50]. For example, 2800 K is associated with the warm color of incandescent lamps, while the color temperature of quartz halogen incandescent lamps is in the range from 2800 to 3200 K [45].

1.2.2 LUMINESCENCE

Luminescence is a nonthermal radiation process that occurs at low temperatures and is thus a form of cold body radiation. Nonthermal radiators are known as luminescent radiators. Luminescence differs from incandescence in this that it results in narrow emission spectra at shorter wavelengths, which correspond to transitions from deeper energy levels. Luminescence can be manifested in different forms, which are determined by the mechanism or the source of excitation: photoluminescence (PL), electroluminescence, cathodoluminescence, chemiluminescence, sonoluminescence, bioluminescence, triboluminescence, etc.

1.2.2.1 Photoluminescence

PL results from optical excitation via absorption of photons. Many substances undergo excitation and start glowing under UV light illumination. PL occurs when a system is excited to a higher energy level by absorbing a photon and then spontaneously decays to a lower energy level, emitting a photon. When the luminescence is produced in a material by bombarding with high-speed subatomic particles such as β-particles and results in x- or γ-ray radiation, it is called radioluminescence. An example of a common radioluminescent material is the tritium-excited luminous paints used on watch dials and gun sights, which replaced the previously used mixture of radium and copper-doped zinc sulfide paints.

1.2.2.2 Fluorescence

Fluorescence is a luminescence phenomenon in which atom de-excitation occurs almost spontaneously ($<10^{-7}$ s) and in which emission ceases when the exciting source is removed. In fluorescent materials, the radiative transitions are one-step process that takes place between energy states with the same spin (with equal multiplicity):

$$E^* \rightarrow E + h\nu \tag{1.35}$$

Here the transition from an excited state E^* to the ground or lower energy state E consists of emission of a photon of frequency ν and energy $h\nu$ (h is the Planck's constant). The couple of states E^* and E could be singlet–singlet or triplet–triplet. The amplification of fluorescence by stimulated emission is the principle of laser action. The quantum yield of a fluorescent substance is defined by the number of reemitted photons to the total number of absorbed photons:

$$\phi = \frac{\text{number of reemitted photons}}{\text{number of absorbed photons}} \tag{1.36}$$

1.2.2.3 Phosphorescence

Phosphorescence is a luminescence phenomenon from spin-forbidden transitions (e.g., triplet–singlet), and the emission of light persists after the exciting source is removed. Phosphorescence is a two-step process that could be described as

$$S_0 + h\nu \rightarrow S_1 \rightarrow T_1 \rightarrow S_0 + h\nu' \tag{1.37}$$

where
   S and T denote a singlet and a triplet state, respectively
   $h\nu'$ is the emitted photon
   $h\nu$ is the absorbed photon

Subscripts "0" and "1" indicate the ground state and the first excited state, which is given here only for simplicity because transitions can also occur to higher energy levels, respectively.

The excited electrons become trapped in the triplet state (a metastable state) for prolonged period of time, which causes substantial delay between excitation and emission. Although most phosphorescent compounds de-excite rather quickly (on the order of milliseconds), there are certain crystals as well as some large organic molecules, which remain in a metastable state sometimes as long as several hours. These substances effectively store light energy, and if the phosphorescent quantum yield is high, they release significant amounts of light over long time, creating an afterglow effect. The metastable states could be populated also via thermal excitation; therefore, phosphorescence is temperature-dependent process. Common pigments used in phosphorescent materials include zinc sulfide and strontium oxide aluminate. Since strontium oxide aluminate has a luminance approximately 10 times greater than zinc sulfide, it is used widely in safety-related signage (exit signs, pathway marking, and other).

1.2.2.4 Multiphoton Photoluminescence

Multiphoton PL occurs when a system is excited by absorption of two or more photons of the same or different energies, followed by emission of one photon of which energy can be greater than the energy of one of the exciting photons. Two-photon PL is the most often used process. In this process, two identical photons of energy $h\nu_1$ are absorbed by the system, which spontaneously de-excites emitting a photon with energy larger than the energy of the absorbed photons ($h\nu_2 = 2h\nu_1$). If the decay includes a non-radiative relaxation to an intermediate energy level, the energy of the fluorescence photon is $h\nu_2 < 2h\nu_1$. Two-photon absorption is a nonlinear optical process that depends on the impinging intensity and occurs preferentially where the intensity of light in the sample is the greatest. When two-photon PL is used to study biological samples, one should have in mind that the longer wavelength of the excitation photons allows for greater depth of penetration into the sample. Therefore, in order to avoid damage of biological samples, the average intensity of the excitation beam should be sufficiently low. At the same time, the peak intensity of the excitation beam should be high enough to ensure two-photon absorption. Such conditions can be provided by a mode-locked laser working in femtosecond pulse regime; the ultrashort optical pulses provide very short time of interaction of light with the biological tissue, as well as high-peak power but low-average power of the excitation beam.

A popular imaging technique known as two-photon laser scanning fluorescence microscopy employs two-photon fluorescence to obtain higher resolution than ordinary microscopy [57,58]. The method uses a fluorescent probe (fluorophore) linked to specific location in the sample. The excitation beam is focused on the fluorescent probe providing conditions for two-photon absorption. The probe absorbs a pair of photons incident on it and emits a single photon of larger energy, which is detected. The two-photon absorption rate at the position of the probe, along with the emission rate, is proportional to the square of the impinging intensity [58–62]. This results in a stronger signal from the area in focus and much weaker fluorescence background from the out-of-focus area. In addition, the image is sharper and the resolution is higher than the ordinary microscopy based on single-photon absorption.

Another application of multiphoton absorption is the three-dimensional micro-lithography. The method employs specially designed transparent polymers with strong polymerization nonlinearity. A micro-object can be created in the volume of the polymer matrix by focusing high-power optical pulses at the location of interest. The intensity of the focused beam is high enough to initiate multiphoton polymerization only in the vicinity of the focal point. A computer-generated three-dimensional microstructure can be transferred into the polymer matrix by moving the focal point of the lens to follow the desired shape.

Figure 1.11 illustrates upconversion fluorescence process used to convert IR photons to visible photons. The system, usually phosphors doped with rare-earth ions, is excited by two-photon absorption of an auxiliary photon of energy $h\nu_1$ and an IR photon of low energy $h\nu_2$. A rare-earth ion such as Er^{3+} can trap the electron excited by the first photon till the arrival of the second photon, which brings the electron to the upper energy level via sequential absorption. If there is no nonradiative relaxation involved, the energy of the resultant luminescence photon is equal to the sum of energies of both absorbed photons, $h\nu_3 = h\nu_1 + h\nu_2$. One practical application of upconversion fluorescence is for viewing an IR laser beam using an IR sensor card. The card can be reflective or transmissive and consists of upconverting phosphor powder laminated between a pair of stiff transparent plastic sheets. Viewing the spatial distribution of an IR beam is important when aligning IR lasers or optical systems using such lasers.

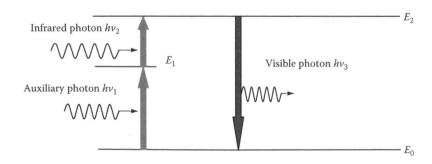

FIGURE 1.11   Upconversion fluorescence process.

PL occurs in all forms of matter, and it is a very useful tool for investigating the optical properties such as temperature dependence of absorption coefficient and energy position of the bandgap
of semiconductor materials. In fact, PL measurements have been used to study absorption and
emission of CdS and GaAs [63,64]. The model used to explain the experimental data underlies the
detailed balance principle and is expressed by the modified Roosbroeck–Shockley equation considering self-absorption [64]:

$$I(T, h\nu) \propto (h\nu)^2 \{1 - \exp[-\alpha(T, h\nu)d]\} \exp(-h\nu/kT_c) \tag{1.38}$$

where
  I is the emitted light intensity
  $h\nu$ is the photon energy
  d is the active sample thickness, which corresponds to the diffusion length of the minority carriers
  k is the Boltzmann constant
  α is the absorption coefficient
  T is the lattice temperature
  T_c is the actual carrier temperature (the carrier temperature might exceed considerably the lattice
    temperature $T_c \gg T$)

Figure 1.12 shows PL spectra (solid lines) of bulk p-type ($\approx 10^{18}$ cm^{-3} GaAs). The PL spectra were
measured in the temperature range 5–300 K [64], using an argon laser as optical excitation and
the data were recorded with a Fourier transform infrared Bomen spectrometer. The broken lines
represent the fits to the PL spectra using Equation 1.38, while the absorption coefficient has been
modeled with density of states and Urbach's rule. Figure 1.13 shows the variation of the bandgap
according to the PL spectra shown in Figure 1.12. The solid line is calculated with a vibronic
model, pointing to the fact that the band-gap energy in semiconductors depends much stronger on
the electron–lattice interaction rather than on volume changes. The squares correspond to the PL
maxima in Figure 1.12 and the solid dots to the values calculated from Equation 1.38 in conjunction
with the modeled absorption coefficient [64].

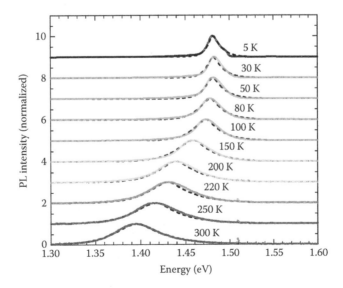

FIGURE 1.12 PL spectra (solid line) and fits (broken line) according to Equation 1.38. For clarity, the spectra
are normalized and shifted. (From Ullrich, B. et al., *Semicond. Sci. Technol.*, 22, 1174, 2007. With permission.)

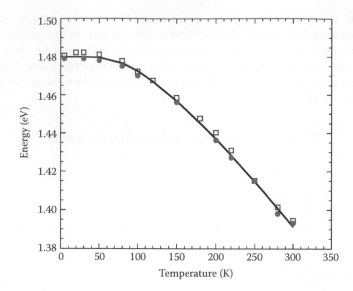

FIGURE 1.13 Bandgap energy of *p*-type GaAs versus temperature. The solid line represents the fit with the vibronic model used in Ref. [56], whereas the squares correspond to the PL maxima in Figure 1.12, and the dots are the bandgap values found from Equation 1.38. (From Ullrich, B. et al., *Semicond. Sci. Technol.*, 22, 1174, 2007. With permission.)

Because of the strong luminescence that direct band-gap semiconductors such as CdS, ZnS, GaAs, or GaN exhibit, they are most easily examined by luminescence techniques. But indirect semiconductors such as silicon can be examined as well. Although they emit light at much weaker level, the luminescence of dislocated silicon is different from intrinsic silicon and can be used to localize defects in integrated circuits.

1.2.2.5 Electroluminescence

Electroluminescence is the light emission when electric field is applied to a material. Electroluminescence in semiconductors results from radiative recombination of electrons from the conduction band with holes from the valence band. Prior to recombination, electrons and holes are separated either by doping of the material to form a *p–n* junction (as in electroluminescence devices such as LEDs) or through excitation by impact of high-energy electrons accelerated by a strong electric field. When the emission is triggered by an electric current injected into a forward-biased *p–n* junction, as in the case of LEDs, the process is called injection electroluminescence [65,66]. Examples of electroluminescent materials include III–V semiconductors, such as InP, GaAs, and GaN, powder zinc sulfide doped with copper or silver, thin-film (TF) zinc sulfide doped with manganese, organic semiconductors, natural blue diamond (diamond with boron as a dopant), etc. In crystals such as diamonds, the source of luminescence can be lattice defects.

Electroluminescence devices have lower power consumption than neon signs and find broad applications in advertising boards and safety signs. Electroluminescent panels based on powder phosphor provide a gentle, even illumination and are frequently used as backlights to liquid crystal displays (LCDs). The low-electric-power consumption makes them convenient for battery-operated devices such as pagers, wristwatches, and computer-controlled thermostats. Recently, blue, red, and green emitting TF electroluminescent materials have been developed that offer the potential for long life and full color electroluminescent displays. Electroluminescent lamps can be made in any color. However, the most commonly used are of greenish color because they provide the greatest luminous efficiency (the greatest apparent light output for the least electrical power input).

1.2.2.6 Cathodoluminescence

Cathodoluminescence is the emission of light after excitation by impact of high-energy electrons. The impingement of energetic electrons generated by a cathode-ray tube, like in the case of a TV screen or an image intensifier (used in night vision devices), onto a luminescent material (phosphor or a semiconductor) results in the promotion of electrons from the valence band into the conduction band, leaving behind a hole. The recombination of an electron and a hole can be radiative process with emission of a photon or non-radiative process with creation of a phonon. The energy (color) of the emitted photon and the probability that a photon will be emitted, not a phonon, depend on the material and its purity.

Cathodoluminescence finds applications in geology, mineralogy, materials science, and semiconductor engineering. It can be implemented in an optical or electron microscope with the proper accessories to examine the optical properties of nonmetallic materials such as semiconductors, minerals, ceramic, glass, etc. The depth of penetration into a sample can be modified by changing the electron energy. As different components in a material produce emission at different wavelengths, cathodoluminescence is used to study the material structure in order to get information on its composition, growth, and quality. For this purpose, a scanning electron microscope with specialized optical detectors or an optical cathodoluminescence microscope is most often used [67–70]. The choice of the microscope technique depends on the application. An electron microscope with a cathodoluminescence detector has much higher magnification and versatility but is more complicated and more expensive compared to an optical cathodoluminescence microscope, which is easy to use and shows actual visible color features immediately through the eyepiece. Other advantages of the scanning electron microscope are high resolution (10–20 nm), the ability to detect the entire spectrum at each point (hyperspectral imaging) if a CCD camera is used, and the ability to perform nanosecond to picosecond time-resolved spectroscopy if the electron beam can be chopped into nanosecond or picosecond pulses. These advantages are important when examining low-dimensional semiconductor structures, such as quantum wells or quantum dots.

1.2.2.7 Sonoluminescence

Sonoluminescence is the emission of short bursts of light from a gaseous cavity (bubbles) induced in a liquid by high-intensity sound [71,72]. The light emission occurs when the bubbles quickly collapse and reach minimum size. Sonoluminescence can be observed from clouds of bubbles and, under certain circumstances, from a single bubble [73–76]. The bubble may already exist in the liquid or may be created through cavitation, the process of formation, growth, and collapse of bubbles under the influence of high-intensity sound or ultrasound. A single bubble can be trapped at a pressure antinode of the acoustic standing wave formed in the liquid. It will periodically expand and collapse, emitting a flash of light each time it collapses. Single-bubble flashes can have very stable periods and positions. The resonant frequencies depend on the shape and size of the container. The light pulses from the bubbles are extremely short (between 35 and a few hundred picoseconds) [76]. Adding a small amount of noble gas (helium, argon, or xenon) to the gas in the bubble increases the intensity of the emitted light. The collapsed bubbles are about 1 μm in diameter depending on the ambient fluid (e.g., water) and on the gas content of the bubble (e.g., atmospheric air). When the bubble collapses, the pressure increases significantly and the temperature in the interior of the bubble can reach around 10,000 K. This causes ionization of the noble gas present in the bubble [74]. Such high temperatures provide conditions for thermonuclear fusion. This possibility makes sonoluminescence especially interesting subject for investigation [77].

1.2.2.8 Chemiluminescence

Chemiluminescence is the emission of light as a result of excitation via a chemical reaction. Under certain circumstances, the energy released during a chemical reaction might be sufficient to excite the reaction product, which fluoresces as it decays to a lower energy level. For example, one typical chemiluminescence reaction is the luminol reaction:

$$\text{Luminol} + H_2O_2 \rightarrow \text{3-APA*} \rightarrow \text{3-APA} + \text{light} \qquad (1.39)$$

Luminol in an alkaline solution with hydrogen peroxide (H_2O_2) in the presence of a suitable catalyst (iron or copper, or an auxiliary oxidant) produces chemiluminescence. The energy released in the reaction is sufficient to populate the excited state of the reaction product 3-aminophthalate (3-APA*), which de-excites to 3-APA emitting a photon. In the ideal case, the number of the emitted photons should correspond to the number of molecules of the reactant. However, the actual quantum efficiency of nonenzymatic reactions, as in this case, seldom exceeds 1%. The luminal reaction is an example of liquid-phase reaction [78,79].

One of the oldest known chemoluminescent reactions is that of elemental white phosphorus oxidizing in moist air, producing a green glow (gas-phase reaction). Another type of chemical reactions produces chemiluminescence from a fluorescent dye or fluorophor, which is incorporated in the chemical mixture of reactants. The reaction can be described as Reactant 1 + Reactant 2 + dye → Products + dye*, where dye* is the excited state of the dye. The most common example of this type of reaction is the glow of lightsticks. They emit steady glow for up to 12 h and are used for emergency indications, as well as for illumination in underwater and military environment. The chemical reaction by which the lightsticks work is

$$\text{Cyalume} + H_2O_2 + \text{dye} \rightarrow \text{Phenol} \rightarrow 2CO_2 + \text{dye*} \qquad (1.40)$$

The chemicals are kept in separated compartments (for storage purposes). Once the reactants are permitted to mix, the energy of the reaction activates the fluorescent dye, which decays to a lower energy level emitting light. This is the most efficient nonenzymatic chemiluminescence reaction known. A typical organic system useful for chemiluminescence exhibits a quantum efficiency of about 23% compared with about 3% for the best-known aqueous systems [78–80]. The color, intensity, and duration of the emitted light are determined by the fluorescent dye. Some of the most common dyes are 9,10-diphenylanthracene emits blue light, tetracene emits yellow–green light, rhodamine 6 G gives bright orange light with moderate duration of emission, while rhodamine B emits red light.

Chemiluminescence is an important tool for gas analysis, for instance, for determining the content of ozone, nitric oxides, sulfur compounds, or small amounts of impurities or poisons in air. In environmental air-quality testing, the concentration of nitric oxide can be detected down to 1 ppb by using chemiluminescence techniques. Other applications of chemiluminescence include analysis of inorganic species in liquid phase, analysis of organic species, DNA sequencing using pyrosequencing, etc. Enhanced chemiluminescence is a common technique in biology. It allows detection of minute quantities of a biomolecule and femtomole quantities of proteins, which are well below the detection limit for most assay systems.

1.2.2.9 Bioluminescence

Bioluminescence is the chemiluminescence produced by living organisms such as bacteria, fireflies, glowworms, and jellyfish. Some studies suggest that 90% of deep-sea marine organisms produce bioluminescence in one form or another. Although, certain fish emit red and IR lights, most marine light-emission is in blue region of the spectrum where seawater is transparent. Non-marine or land bioluminescence is less distributed, as the two best-known examples are fireflies and New Zealand glowworms. Bioluminescence results from enzyme-catalyzed chemiluminescence reactions that can reach the highest known quantum efficiency for a chemiluminescence reaction (88% for the firefly), whereas synthetic reactions can provide only a 23% efficiency at best [80,81]. Bioluminescent organisms are intensively studied for better understanding of behavioral mechanisms (means of communication), as well as because of potential applications of engineered bioluminescence such as new methods for detecting bacterial contamination of meats and other foods, bio-identifiers, etc.

1.2.2.10   Triboluminescence

Triboluminescence is the light generated via the breaking of asymmetrical bonds in a crystal when that material is scratched, crushed, or rubbed (from the Greek tribein "to rub"). For example, many minerals, such as quartz, glow when scratched; sucrose emits a blue light when crushed; and a diamond may begin to glow while being sawn during the cutting process or a facet is being ground. Diamonds may fluoresce blue or red. The mechanism of triboluminescence is not fully understood yet, but the research in this area suggests that charge is separated upon fracture of asymmetrical materials and when the charges recombine, the electric discharge ionizes the surrounding air, causing a flash of light [82].

1.2.2.11   Scintillation

Scintillation is the emission of short light pulses from certain chemical compounds after being excited by charged particles or by photons of high energy. Scintillation is an inherent molecular property of conjugated and aromatic organic molecules but can occur also in inorganic materials. Ionizing particles passing through organic scintillators excite the molecules to deep energy levels, which upon relaxation emit UV light. The emitted spectra are characteristic for the electronic structure of the molecules. By adding a sensitizer (wavelength-shifting molecules), the UV photons are converted into visible light. Examples of organic scintillators include crystals such as anthracene ($C_{14}H_{10}$) and naphthalene ($C_{10}H_8$); organic liquids (usually have brand names such as PPO, POPOP, PBD, etc.); and plastics—a solid solvent, doped with aromatic compounds (TP, PPO, or PBD) or with wavelength shifters, such as polysterene and polyvinyltuolene. Examples of inorganic scintillators are crystals such as NaI(Tl), CsI(Tl), and BaF_2, and gaseous and liquid scintillators on the base of Xe, Kr, Ar, He, or N. Inorganic materials find application in high-precision calorimetry. In chemistry, calorimetry is used to determine the heat released or absorbed in a chemical reaction and specific heat capacities of liquids. In particle physics, a calorimeter is an experimental apparatus that measures energy of charged particles.

Scintillation counters measure ionizing radiation and are used in high-energy physics experiments for timing, for fast event selection, or for measuring the energy of charged particles by total absorption in sampling calorimeters. Each charged particle passing through the scintillator produces a flash, as the intensity of the light flash depends on the energy of the charged particles. The scintillation counter apparatus consists basically of a scintillator and a photomultiplier with the counter circuits. A scintillation counter can be used also as a spectrometer to determine the energy of x-ray or γ-photons causing scintillation by measuring the intensity of the flash (the number of lower energy photons produced by the high-energy photon). Cesium iodide (CsI) in crystalline form is used as the scintillator for the detection of protons and α-particles, while sodium iodide (NaI) containing a small amount of thallium is used as scintillator for the detection of γ-particles.

Light yield (or light gain) is an important parameter of scintillation counters. It is a measure of the scintillation efficiency, and shows how much of the energy of a charged particle or a high-energy photon in the scintillation counter is converted into visible light. For historical reasons, anthracene is used as a standard for light gain. The light yield of anthracene crystals is of the order of 5% for blue light or about two photons per 100 eV for high-energy particles. For comparison, the inorganic crystal NaI has a light yield of about 230% of anthracene, while typical plastic scintillators exhibit light yield of 50%–60% [83].

1.3 CONVENTIONAL LIGHT SOURCES

Conventional light sources employ either incandescence or electrical discharge in gases. Both phenomena come along with large energy losses, due to the inherent involvement of high temperature and large Stokes shifts (the emitted photons are with lower energy than the absorbed photons), which result in relatively low luminous efficiencies.

1.3.1 INCANDESCENT LAMPS AND TUNGSTEN HALOGEN LAMPS

Incandescent lamps utilize electric current to heat a material (tungsten filament) and produce light by incandescence. The spectrum of incandescent lamps depends on the operating temperature of the tungsten filament. The peak of the tungsten spectrum is shifted to smaller wavelengths providing higher luminous efficacy than a blackbody radiator. This blue shift, though, does not affect the color rendering, which remains almost the same as a Planckian radiator. The main portion of the radiation is in the IR part of the spectrum. The radiation losses in addition to heat losses (in the lead wires from the wire surface into the surrounding gas) reduce the luminous efficiency of this type of light sources. The overall efficiency of a 120 V/100 W lamp is only about 7%, which means that only about 7% of the consumed energy is converted into useful visible radiation and the remainder as IR radiation. Light output (lm) depends both on the filament temperature and the electrical input power (wattage). A 120 V incandescent lamp working at CCT of 2790 K and consuming 60 W electric power provides a luminous flux of 865 lm and luminous efficiency about 14.4 lm/W [5]. An incandescent lamp has a CRI R_a = 100 and exhibits color rendering properties identical to the standard reference illuminant. The efficiency of the incandescent lamps depends on the voltage: the higher the voltage, the less efficient the lamp is. The incandescent lamps have very short lifetime, typically less than 1000 h. During operation, tungsten gradually evaporates from the filament and deposits as a dark film on the inner bulb surface. This can decrease the light output by as much as 18% during the life of the lamp. The efficiency and the lifetime can be increased by reducing the tungsten evaporation rate by the addition of halogen vapor (iodine or bromine) to the gas filling of the lamp. The halogen bounds with the evaporated tungsten and forms tungsten halide, which diffuses back to the filament, reducing the evaporation effect and providing higher operational temperatures (as high as 3450 K). A typical tungsten halogen lamp has at least twice longer lifetime of an incandescent lamp at the same efficiency. Figure 1.14 compares the spectral exitance of an Oriel quartz tungsten halogen lamp with a blackbody radiator. A 120 V/50 W tungsten halogen lamp at operation temperature of 2750 K, provides a luminous flux of 590 lm and has luminous efficiency of about 11.8 lm/W and CRI R_a = 100 [5]. Incandescent and tungsten halogen lamps are of interest mainly for residential lighting and motor vehicle headlights. The tungsten filament lamp is also a popular light source for optical instrumentation designed to use continuous radiation in the visible and into the IR regions.

More recently, modified FEL 1000 W quartz halogen lamps were tested for suitability as photometric transfer standards, and they were shown to be stable within 0.2%–0.6% over 60 h of operation. Of special interest for calibration is the quartz tungsten halogen lamp standard of Gamma Scientific, Inc., Model 5000-16C 1000 W FEL lamp standard. It provides absolute calibration of spectral radiance from 250 nm to 2.5 μm. This lamp standard has been selected by the National Institute of Standards and Technology (NIST) to replace the 1000 W, DXW standard. It is calibrated to the 1973 NIST spectral irradiance scale, while the DXW standard is calibrated to the 1965 NIST scale [84]. The lamp calibration is directly traceable to NIST by incorporating two NIST spectral irradiance scales. This results in estimated accuracies of 2.4% at 250 nm, 1.8% at 450 nm, and 1.5% over the range from 555 to 1600 nm. The 1965 NIST scale of spectral irradiance is used over the spectral range from 1700 to 2500 nm, as NIST uncertainty over this wavelength region is approximately 3%.

Quartz tungsten halogen lamps emit down to 240 nm. At the operating high wattages, the emitted UV radiation is hazardous. Therefore, the user of these sources has to wear UV protective glasses and gloves.

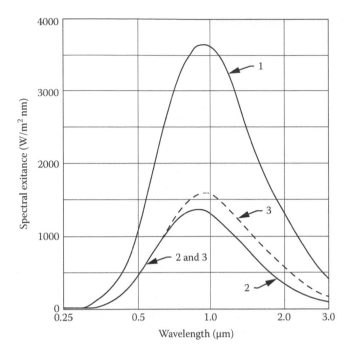

FIGURE 1.14 Spectral exitance of (1) blackbody radiator, (2) a tungsten surface, and (3) a gray body with emissivity of 0.425, all at 3100 K. (Courtesy of Newport Corporation, Richmond, VA.)

1.3.2 GAS DISCHARGE LAMPS

Gas discharge lamps utilize an electrical discharge through an ionized gas to generate light. Typically, such lamps are filled with a noble gas (argon, neon, krypton, and xenon) or a mixture of these gases, as most lamps contain also vapors of mercury, sodium, and metal halides. Sodium and mercury are efficient low-pressure discharge emitters. When a large enough electric field is applied to the gas in the lamp, the gas partially ionizes producing conductive plasma, which contains free electrons, ions, and neutral particles. The character of the gas discharge depends on the frequency of the electric current. The gas discharge could be stabilized by limiting the electric current via a ballast. The inert gas determines the electrical characteristics of the gas discharge, mainly by controlling the mean path traveled by electrons between collisions. The source of the electrons may be a heated cathode (thermionic emission), a strong field applied to the cathode (field emission), or the impact of positive ions on the cathode (secondary emission). The electrons, accelerated by the applied electric field, collide inelastically and excite the gas and metal vapor atoms, which may relax by emitting visible or UV light. To increase the luminous efficiency, the inner surface of the glass tube of some lamps is covered with a fluorescent coating of varying blends of metallic and rare-earth phosphor salts, which converts the UV radiation to visible light.

Gas discharge lamps working at high pressure and high current produce broad spectral continuum with pronounced spectral line characteristics of the vapor. At lower pressure and current, the spectral output consists of sharper spectral lines with minimal background continuum. As each gas has a characteristic emission spectrum of certain wavelengths, different gas discharge lamps may reproduce differently the colors of various surfaces being lit by the source. An important characteristic of cold illuminants is the CRI, which is used to evaluate the color reproduction ability of a light source. Gas discharge lamps have indexes below 100, which means that the colors appear different from the colors seen under sunlight or other standard illuminant. In general, gas discharge lamps offer longer life and higher light efficiency than incandescent lamps, but exhibit poorer color rendering and are more complicated to manufacture.

1.3.2.1 Low-Pressure Discharge Lamps

1.3.2.1.1 Fluorescent Lamps

Fluorescent lamps are low-pressure gas discharge lamps, commonly used in lighting industry. A fluorescent lamp consists of a glass or quartz tube with two electrodes at each end, filled with an inert gas (most often argon or neon, and sometimes xenon or krypton) containing low-pressure mercury vapor as an active ingredient. A typical mixture contains about 1% or even as little as 0.1% metal vapor, and 99%–99.9% inert gas. The pressure of the metal vapor is around 0.5–0.8 Pa. Mercury emits in the UV at wavelengths of 253.7 nm (about 65% of the emitted light) and 185 nm (about 10%–20%). White light is produced by stimulated fluorescence in a phosphor coating on the inner surface of the tube. The UV photons emitted by the mercury vapors are absorbed by the phosphors, which fluoresce giving off lower frequency photons that combine to produce white light.

The spectral output and color-rendering properties of the fluorescent lamps depend on the emission spectra of the used phosphors. For example, daylight fluorescent lamps use a mixture of zinc beryllium silicate and magnesium tungstate. The phosphors should exhibit high quantum efficiency (more than 85%) in addition to other requirements for structural and chemical stability. There are three main groups of fluorescent lamps based on the employed phosphors: halophosphate phosphors, triphosphors, and multiband phosphors. The oldest fluorescent lamps employ calcium halophosphate phosphor, $Ca_5(PO_4)_3(Cl, F):(Sb^{3+}, Mn^{2+})$. The spectrum contains mainly blue and yellow light (480 and 580 nm), corresponding to the emission lines of the antimony ions and the manganese ions, as the green and red content is very limited [5,25]. To the eye, this mixture appears white, but the spectrum is incomplete, which explains the poor color rendering ($R_a = 50$–76). The triphosphors, developed in the 1970s, comprise three components that provide the three primary colors, the combination of which gives white light. A variety of matrices doped with rare-earth europium and terbium ions are used: for example, $Y_2O_3:Eu^{3+}$ emits red; $CeMgAl_{11}O_{19}:Tb^{3+}$, $GdMgB_5O_{10}:Ce^{3+},Tb^{3+}$, and $LaPO_4:Ce^{3+},Tb^{3+}$ emit green; and $BaMg_2Al_{16}O_{27}:Eu^{2+}$ and $Sr_{5-x-y}Ba_xCa_y(PO_4)_3Cl:Eu^{2+}$ emit blue color. A triphosphor fluorescent lamp yields higher quantum efficiency, longer lifetime, and better color rendering ($R_a = 80$–85) than halophosphates, but still distorts colors because of the three narrow emission lines [25]. The multiband phosphors comprise up to five phosphors—triphosphors (for blue and green colors), halophosphates, and Mn^{2+}-activated pentaborate, which emits a broad red band peaking at about 620 nm. The multiband phosphors have the highest CRIs ($R_a > 90$) [5]. The poorer color rendering of real-life fluorescent lamps to incandescent lamps ($R_a = 100$) is compensated by the higher luminous efficiency. Low-power fluorescent lamps (4–5 W) have luminous efficiency in the range from 35 to 50 lm/W (for comparison, an incandescent lamp of 10 W has luminous efficiency of 8 lm/W), while high-power linear lamps (70–125 W) reach luminous efficiency from 70 to 100 lm/W.

The overall efficiency of conversion of electric energy consumed in a fluorescent lamp to visible light at optimal conditions is about 28%. Due to losses as heat in the discharge, the walls, and the electrodes, only about 63% of the consumed electric energy is converted to UV radiation, and due to non-radiative transitions and large Stokes shift, only about 40% of the UV radiation reaching the phosphor is converted to visible light. Although low, the efficiency of a fluorescent lamp is about four times higher than a 100 W incandescent bulb of an equivalent brightness. However, the light output of a fluorescent lamp depends strongly on the ambient temperature. Fluorescent lamps operate best around room temperature as the optimal working range is between 10°C and 35°C, where the normalized luminous flux reaches about 90% [5,25]. At temperatures below freezing, it falls off to about 25% (at −20°C) and standard fluorescent lamps may not start. Therefore, special fluorescent lamps might be needed for outdoor use in cold weather. The lifetime of fluorescent lamps depends on the construction and ranges from 5,000 to 24,000 h, which is about 10–20 times as long as an equivalent incandescent lamp. A 32 W fluorescent triphosphor lamp at an operation temperature of 4100 K provides a luminous flux of 2850 lm, luminous efficiency of 84 lm/W, and CRI R_a = 78 [5]. Fluorescent lamps are the most common lamps in office lighting, and the general lighting resource in industrial and commercial environment.

1.3.2.1.2   Compact Fluorescent Lamps

Compact fluorescent lamps (CFLs) are well known as effective energy-saving light source. They are of the same size as an incandescent bulb and are made to fit the regular incandescent sockets. They utilize electronic ballasts, which do not produce light flicker. The higher initial cost is compensated by lower energy consumption over its life. Therefore, nowadays, CFLs become very popular in residential lighting. However, there is one inconvenience: the common CFL cannot be connected to a standard dimmer switch used for incandescent lamps. A 15 W CFL at an operation temperature of 2700 K provides a luminous flux of 900 lm, luminous efficiency of 51 lm/W, and CRI R_a = 82 [5].

1.3.2.1.3   Low-Pressure Sodium and Mercury Lamps

The low-pressure sodium (LPS) arc lamps are filled with neon at a pressure of 400–2000 Pa (3–15 torr) as the sodium vapor pressure is kept at 0.7–1 Pa ($5–8 \times 10^{-3}$ torr) at operating temperatures. These lamps need a long warm-up time. Upon switching on, the discharge starts in the neon and the sodium vapor pressure reaches optimal value after 10–15 min. To facilitate starting, small amounts of argon, xenon, or helium are added to the neon. The LPS lamps emit almost monochromatic light. The spectrum consists of a doublet (D-line) of two very close yellow lines at wavelengths 589.0 and 589.6 nm. Because of the high monochromaticity of the D-line, LPS lamps are commonly used for calibration and research. A 90 W LPS lamp at an operation temperature of 1,800 K provides a luminous flux of 12,750 lm, luminous efficiency of 123 lm/W, and CRI R_a = −44 [5].

The low-pressure mercury discharge lamps when used with filters can also provide strong monochromatic radiation at wavelengths of 404.7 and 435.8 nm (violet), 546.1 nm (green), and 577.0 and 579.1 nm (yellow). The low-pressure mercury and LPS lamps are used in applications that require high spectral purity.

The luminous efficacy of the D-line of the LPS lamps is 530 lm/W. The high monochromaticity of the D-line suggests extremely high theoretical efficiency of the LPS lamp. Although, 60%–80% of the input power is lost in heat and IR radiation, the LPS lamps have the highest luminous efficiencies (100–200 lm/W) among the present practical lamps. The light output of LPS lamps is in the range of 1,800–33,000 lm for wattage of 18–180 W. However, the LPS lamps exhibit very poor color rendering (R_a = −44). Therefore, from the point of view of lighting applications, LPS lamps are appropriate only for street illumination and security lighting. Lifetime is typically 14,000–18,000 h.

1.3.2.2   High-Pressure Discharge Lamps

High-pressure discharge lamps, also called high-intensity discharge (HID) lamps produce light by means of an electric arc between tungsten electrodes housed inside a transparent fused quartz or fused alumina tube. The tube is filled with inert gas and metal salts. The operating pressure is around 1 atm and higher. At such high pressure, the temperature of the plasma reaches 4000–6000 K, the interaction between the heavy particles (atoms and ions) in the plasma is enhanced, resulting in high rate of elastic collisions and collision broadening of the emission spectral lines. The HID lamps exhibit continuous spectral output with spectral lines characteristic for the metal vapors, and therefore, considerably improved color rendering in comparison to the low-pressure discharge lamps. This group includes high-pressure mercury-vapor (HMPV) lamps (the oldest type of high-pressure lamps), metal-halide lamps, HPS discharge lamps, xenon and mercury–xenon arc lamps, etc.

1.3.2.2.1   High-Pressure Mercury-Vapor Lamps

High-pressure mercury-vapor lamps work at operating pressures from 0.2 to 1 MPa (2–10 atm). The spectrum emitted from the mercury vapor at such high pressure consists of a broad continuous background overlaid with spectral lines of longer wavelengths (404.7, 435.8, 546.1, 577.0, and 579.1 nm). Because of deficiency of red color, a clear mercury lamp has a very poor CRI ($R_a = 16$). A phosphorus coating on the inner wall of the lamp, typically europium-activated yttrium vanadate, improves the color rendering (up to 50), as well as the luminous efficiency. In HPMV lamps, more than half of the consumed power is lost in heat and UV photon down-conversion. A 250 W high-pressure mercury lamp at an operation temperature of 3,900 K provides a luminous flux of 11,200 lm, luminous efficiency of 34 lm/W, and CRI $R_a = 50$ [5]. Lifetime may reach 24,000 h, and failure is usually due to loss of emission from the electrodes. Because of their lower luminous efficiency, HPMV lamps have been replaced in most applications by metal-halide lamps and HPS lamps.

1.3.2.2.2 Metal-Halide Lamps

Metal-halide lamps are high-pressure mercury lamps with added metal halides to improve the luminous efficiency and color rendering. As the excitation energy of the additive metals (about 4 eV) is lower than that of mercury (7.8 eV), the metal halides can produce a substantial amount of light and thus alter the emitted spectrum. The emission spectrum can be modified by varying the halide composition: for example, line spectra can be obtained by introducing sodium, scandium, thallium, indium, cesium, and rare-earth iodides, whereas tin or tin–sodium halides (iodides with addition of bromides and chlorides) produce more continuous spectra. The luminous efficiency is 70–110 lm/W, depending on the wattage (from 20 to 18,000 W) and additive metals used. CRIs are typically above 60 and may reach 95 for some rare-earth compounds. A 400 W fluorescent triphosphor lamp at an operation temperature of 4100 K provides a luminous flux of 2850 lm, luminous efficiency of 84 lm/W, and CRI $R_a = 78$ [5]. The lifetime is 2,000–30,000 h. HID lamps are typically used when high light levels over large areas are required. Since metal-halide lamps produce well-balanced white light with high brightness, they are widely used in indoor lighting of high buildings, parking lots, shops, roadways, and sport terrains.

1.3.2.2.3 High-Pressure Sodium Discharge Lamps

HPS discharge lamps contain sodium at pressure about 7000 Pa. The higher sodium pressure improves the color rendering but decreases the luminous efficiency and lifetime. The luminous efficiency of HPS lamps ranges from 60 to 130 lm/W for a typical output of 50–1000 W, respectively. Color rendering ($R_a = 20$–25), although higher than the LPS lamps, is still too poor. Typical lifetime is about 24,000 h. HPS lamps produce a broad spectrum, which contains a dip at the wavelengths of the D-line (self-reversed spectral lines). Because of the broader spectrum peaking in the range of the highest sensitivity of the human eye, these lamps are used mainly for street lighting, as well as for artificial photo-assimilation for growing plants.

1.3.2.2.4 Xenon and Mercury–Xenon Arc Lamps

Xenon arc lamps provide smooth continuum from near UV through the visible and into the near IR with pronounced xenon lines between 750 and 1000 nm. As xenon lamps produce bright white light that closely mimics natural daylight, they are appropriate for applications involving solar simulation, in projection systems, as well as absorbance and fluorescence research that require high-intensity broadband sources. The mercury–xenon arc lamps have essentially the mercury spectrum with xenon's contribution to the continuum in the visible and some strong xenon lines in the near IR. Xenon and mercury–xenon arc lamps operate at very high pressure and temperature, and therefore, have to be handled with care to avoid contamination of the bulb and thermal stress. They emit strong UV radiation and require proper safety measures: protective goggles and gloves, and the lamp must be operated in a fully enclosed housing.

1.3.2.2.5 Flash Tubes

Flash tubes are pulsed high-output broadband UV–near-IR sources for which radiation results from a rapid discharge of stored electrical energy in a tube containing typically a xenon gas.

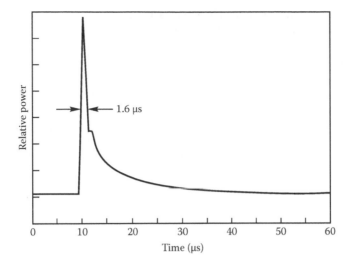

FIGURE 1.15 Pulse shape of 6426 Oriel xenon flashlamp. (Courtesy of Newport Corporation, Richmond, VA.)

For example, Oriel xenon flashlamps emit short microsecond pulses (Figure 1.15) with relatively high UV output, can operate from low frequencies up to 100 Hz, and provide maximum pulse energy of 5 J at frequencies up to 12 Hz. The short pulses find applications in lifetime and kinetic studies. Xenon flash lamps and arc lamps are often used for optical pumping of solid-state lasers (i.e., Nd:YAG).

More detailed information about the construction, principle of operation, transport mechanism, advantages, and disadvantages of the lighting sources considered above can be found in specialized books [5,24,25,85,86].

1.4 LIGHT-EMITTING DIODES

LEDs are small, rugged, and bright light sources with a huge potential to become the dominant light source in the future. Nowadays, LEDs are the most efficient sources of colored light in the visible range of the spectrum. White LEDs already surpassed incandescent lamps in performance and undergo continuous improvement of their efficiency. Today, almost a quarter of the electric energy used is spend for lighting, and perhaps half of this energy could be saved by the employment of efficient and cold solid-state lighting sources [5]. It seems that "Optopia" is on its way and solid-state light sources are at the forefront of the ongoing lighting revolution.

At present, commercially available LEDs cover the spectrum from near-UV through visible to near-IR regions. Apart from the lighting industry, LEDs find numerous applications in automotive and aviation industry, in large area displays and communication systems, in medicine and agriculture, in amusement media industry, and other in everyday life consumer products.

A considerable amount of books and publications is dedicated to the fundamentals, technology, and physics of LEDs, their electrical and optical properties, and advanced device structures (e.g., see Refs. [9–12,23,26], and the references thereafter). We give here only a brief summary of the LED characteristics relevant to optical metrology and some applications.

1.4.1 LED BASICS

LEDs are semiconductor devices, which generate light on the base of electroluminescence due to carrier injection into a p–n junction. Basically speaking, an LED works as a common semiconductor diode: the application of a forwardly directed bias drives a current across the p–n junction.

The excess electron–hole pairs are forced by the external electric field to enter the depletion region at the junction interface where recombination takes place. The recombination process can be either a spontaneous radiative process or a non-radiative process with energy release to the lattice of the material in the form of vibrations (called phonons). This dual process of creation of excess carriers and subsequent radiative recombination of the injected carriers is called injection electroluminescence. LEDs emit fairly monochromatic but incoherent light owing to the statistical nature of spontaneous emission based on electron–hole recombinations.

The energy of the emitted photon and the wavelength of the LED light, depends on the band-gap energy of the semiconductor materials forming the p–n junction. The energy of the emitted photon is approximately determined by the following expression:

$$h\nu \approx E_g \tag{1.41}$$

where

h is the Planck's constant
ν is the frequency of the emitted light
E_g is the gap energy, that is, the energy difference between the conduction band and the valence band of the semiconductor used

The average kinetic energy of electrons and holes according to the Boltzmann distribution is the thermal energy kT. When $kT \ll E_g$, the emitted photon energy is approximately equal to E_g, as shown by Equation 1.41, whereas wavelength of the emitted light is

$$\lambda = \frac{hc}{E_g} \tag{1.42}$$

where c is the speed of light in vacuum. For example, E_g of GaAs at room temperature is 1.42 eV and the corresponding wavelength is 870 nm. Thus, the emission of a GaAs LED is in the near-IR region. It is known from the literature [12,87,88] that the emission intensity of an LED is determined by the values of E_g and kT. In fact, the intensity $I(E)$ as a function of photon energy E is given by the following simple expression:

$$I(E) \propto \sqrt{E - E_g} \exp\left(-\frac{E}{kT}\right) \tag{1.43}$$

The maximum intensity of the theoretical emission spectrum of an LED given by Equation 1.43 occurs at energy

$$E = E_g + \frac{1}{2}kT \tag{1.44}$$

and the full width at half maximum (FWHM) of the spectral emission corresponds to

$$\Delta E = 1.8kT \tag{1.45}$$

For detailed derivation of these expressions, see Section 5.2 in Ref. [12].

In general, the spectral power distribution of an LED tends to be Gaussian with a specific peak wavelength and a FWHM of a couple of tens of nanometers. At room temperature, $T = 293$ K,

kT = 25.3 meV, and Equation 1.45 gives the theoretical FWHM of an only thermally broadened emission band of an LED, ΔE = 46 meV. The FWHM expressed in wavelength, $\Delta\lambda$, is defined by

$$\Delta\lambda = \lambda \frac{\Delta E}{E_g} \tag{1.46}$$

For example, the theoretical spectral linewidth at room temperature of a GaAs LED emitting at λ = 870 nm is $\Delta\lambda$ = 28 nm. The output spectrum depends not only on the semiconductor band-gap energy but also on the dopant concentration levels of the p–n junction. Random fluctuations of the chemical composition of the active material additionally broaden the spectral line (alloy broadening). Therefore, realistically, the emission broadening ΔE of LEDs is between 2.5 and $3kT$. A typical output spectrum of a red GaAsP LED (655 nm) has a linewidth of 24 nm, which corresponds to an energy spectrum of about $2.7kT$ of the emitted photons at room temperature [88]. Nevertheless, despite the broadening, the typical emission spectrum of an LED is fairly narrow and appears to be monochromatic to the human eye.

The peak wavelength of an LED is the wavelength at the maximum intensity of the LED emission spectrum and is generally given in LED data sheets. However, the peak wavelength has little significance for practical purposes since two LEDs may have the same peak wavelength but different color perception. Therefore, the dominant wavelength, which is a measure of the color sensation produced by the LED in the human eye, should be also specified. Figure 1.16 illustrates how the dominant wavelength of an LED can be determined from the CIE color diagram: a straight line is taken through the color coordinates of a reference illuminant, generally the equal energy point E and the LED measured color coordinates F. The intersection point S of the straight line with the boundary of the color diagram gives the value of the dominant wavelength. Purity is another important colorimetric characteristic of LEDs, which is defined by the ratio of the distance E–F from the equal energy point E to the LED color coordinate F and the distance E–S from the equal energy

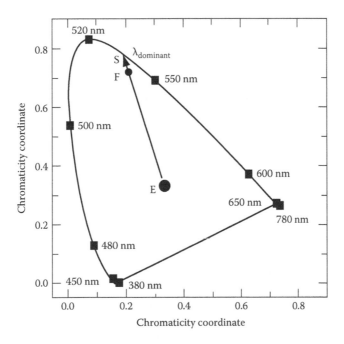

FIGURE 1.16 Dominant wavelength of an LED determined from the 1931 CIE color diagram for 2° observer. (From Schubert, E.F., *Light-Emitting Diodes*, Cambridge University Press, Cambridge, U.K., 2003. With permission.)

point E to the intersection point S in the color diagram. Most LEDs are narrow-band radiators with a purity of nearly 100%, that is, the color is perceived as monochromatic light.

The output light intensity is determined by the current through the p–n junction. As the LED current increases, so does the injection minority carrier concentration and the rate of recombination, thus the emitted light intensity. For small currents, the LED light power depends linearly on the injection current. At high current levels, however, at strong injection of minority carriers, the recombination time depends on the injected carrier concentration and hence on the current itself, which leads to a nonlinear recombination rate with current. The turn-on voltage from which the LED current increases very steeply with voltage is typically about 1.5 V. The turn-on voltage depends on the semiconductor and generally increases with the energy bandgap E_g. For example, for a blue LED it is about 3.5–4.5 V, for a yellow LED it is about 2 V, and for a GaAs IR LED it is around 1 V.

The LED emission output is a function of the forward current and the compliance voltage. Therefore, LEDs are normally operated in a regime of a constant current. Following the LED switch-on, the temperature in the p–n junction gradually rises due to electrical power consumed by the LED chip and then stabilizes when a stable forward voltage is attained. The stabilization process can last several seconds and, in the case of white LEDs, might be influenced by the properties of the phosphor. Different types of LEDs have different temperature stabilization times.

As the heat from the p–n junction must be dissipated into the surroundings, the emission intensity of LEDs also depends on the ambient temperature. An increase in the temperature causes a decrease in the intensity of the emitted light due to non-radiative deep-level recombinations, surface recombinations, or carrier losses at interfaces of hetero-paired materials. The III–V nitride diodes have less sensitive temperature dependence than AlGaInP LEDs. As the ambient temperature increases, the required forward voltage for all the three diodes (blue, green, and red) decreases due to the decrease of the band-gap energy. Thus, if the ambient temperature rises, the entire spectral power distribution is shifted in the direction of the longer wavelengths. The shift in peak wavelength is typically about 0.1–0.3 nm/K. Therefore, current and temperature stabilizations are very important for attaining constant spectral properties.

In addition to radiometric and photometric characteristics, the description of the optical properties of an LED also includes the quantum performance of the LED determined by its internal and external quantum efficiencies and extraction efficiency.

The internal quantum efficiency of an LED is defined by the number of photons generated from a given active region (inside the semiconductor) divided by the number of injected electrons. The internal quantum efficiency can also be expressed with measurable quantities such as the optical power Φ_{int} (radiant flux) emitted from the active region and the injected current I:

$$\eta_{int} = \frac{\Phi_{int}/(h\nu)}{I/e} \tag{1.47}$$

where
$h\nu$ is the energy of the emitted photon
$\Phi_{int}/(h\nu)$ gives the number of photons emitted from a given active region per second
$e = 1.6022 \times 10^{-19}$ C is the elementary charge
I/e is the number of electrons injected into the LED per second

An ideal active region would have a quantum efficiency of unity, because every injected electron would recombine with a hole through a radiative transition producing a photon. In reality, the internal quantum efficiency of an LED is determined by the competition between radiative and non-radiative recombination processes. High material quality, low defect density, and low trap concentration are prerequisites for large internal efficiency. Double heterostructures, doping of active region, doping of the confinement regions, lattice matching in double heterostructures, and displacement of the *p–n* junction into the cladding layer, all these are new LED designs that allow to increase the internal efficiency [89].

The extraction efficiency is determined by the escape probability of the photons generated in the active region, thus, by the number of the photons that leave the LED as emitted in the space per number of photons generated by the active region:

$$\eta_{\text{extraction}} = \frac{\Phi_{\text{air}}/(h\nu)}{\Phi_{\text{int}}/(h\nu)} \tag{1.48}$$

where Φ_{air} is the optical power emitted by the LED into free space. In an ideal LED, all photons emitted from the active region would escape in the space, and the extraction efficiency would be unity. In a real LED, there are losses of light due to reabsorption of the emitted light within the LED structure itself or light absorbed by the metallic contact surface or the effect of total internal reflection (TIR) from the parallel sides of the junction, which traps the emitted light into the junction region. Those are inherent losses due to the principal structure of an LED and are very difficult to reduce or avoid without major and costly changes in the device fabrication processes or LED geometry.

The refractive index of most semiconductors is quite high (>2.5) and the critical angle of TIR at the interface semiconductor/air is less than 20°. Only the light enclosed in the cone determined by the critical angle (escape cone) can leave the semiconductor. Even at normal incidence, the surface reflectivity is too high. For example, GaAs has an index of refraction 3.4, and the reflection losses for vertical incidence on the GaAs/air interface are about 30%. Thus, only a few percent of the light generated in the semiconductor is able to escape from a planar LED. There is a simple expression for the relation between the extraction efficiency and the refractive indices of the semiconductor n_{s} and air n_{air} [12]:

$$\eta_{\text{extraction}} = \left(\frac{1}{2} \frac{n_{\text{air}}}{n_{\text{s}}} \right)^2 \tag{1.49}$$

The problem is less significant in semiconductors with small refractive index and polymers, which have refractive indices on the order of 1.5. For comparison, the refractive indices of GaAs, GaN, and light-emitting polymers are 3.4, 2.5, and 1.5, respectively, and the extraction efficiencies are 2.2%, 4.2%, and 12.7%, respectively. If the GaAs planar LED is encapsulated in a transparent polymer, the extraction efficiency would be about a factor of 2.3 higher. Thus, the light extraction efficiency can be enhanced two to three times by dome-shaped epoxy encapsulants with a low refractive index, typically between 1.4 and 1.8. Due to the dome-shape, TIR losses do not occur at the epoxy–air interface. Besides improving the external efficiency of an LED, the encapsulant can also be used as a directing spherical lens for the emitted light.

The photon escape problem is essential especially for high-efficiency LEDs. To achieve an efficient photon drag out of LEDs is one of the main technological challenges. The extraction efficiency optimization is based on modifications and improvement of the device geometry. The most common approaches that have allowed a significant increase of the extraction efficiency are encapsulation in a transparent polymer; shaping of LED dies (using nonplanar surfaces, dome-shape, LED chip shaped as parabolic reflector, and truncated-inverted-pyramid LED); thick-window chip geometry can increase the quantum efficiency to about 10%–12% if the top layer has thickness of 50–70 μm, instead of few micrometers; current-spreading layer (also known as window layer); transparent contacts; double heterostructures, which reduce reabsorption of light by the active region of the LED; antireflection optical coatings; distributed Bragg reflectors; and TF LED with microprisms, flip-chip (FC) packaging, etc. [5,12,90]. Many commercial LEDs, especially GaN/InGaN, use also sapphire substrate transparent to the emitted wavelength and backed by a reflective layer increasing the LED efficiency.

The external quantum efficiency is the ratio of the total number of photons emitted from an LED into free space (useful light) per number of injected electron–hole pair:

$$\eta_{external} = \frac{\Phi/(h\nu)}{I/e} = \eta_{int}\eta_{extraction} \tag{1.50}$$

The relation incorporates the internal efficiency of the radiative recombination process and the efficiency of photon extraction from the device. For indirect bandgap semiconductors, $\eta_{external}$ is generally below 1%, while, on the other hand, for direct bandgap semiconductors with appropriate device structure, $\eta_{external}$ can be as high as 20%. The radiant efficiency (also called wall-plug efficiency) of the LED is given by $\eta = \Phi/(IV)$, where the product IV represent the electrical power delivered to the LED.

Planar LEDs of high refractive-index semiconductors behave like a Lambertian light source; thus, the luminous intensity depends on the viewing angle, θ, according to a cosine law and have a constant isotropic luminance, which is independent of direction of observation. For Lambertian far-field pattern, the light intensity in air at a distance r from the LED is given by [12]

$$I_{air} = \frac{\Phi}{4\pi r^2} \frac{n_{air}^2}{n_s^2} \cos\theta \tag{1.51}$$

where
   Φ is the total optical power of the LED
   $4\pi r^2$ is the surface area of a sphere with radius r

Equation 1.51 suggests that maximum intensity is observed normal to the semiconductor surface, $\theta = 0°$, and the intensity decreases to half of the maximum value at $\theta = 60°$. The Lambertian pattern is shown schematically in Figure 1.17 together with two other far-field radiation patterns. Nowadays, LEDs can be fabricated in a wide range of designs that allow achieving a particular spatial radiation pattern. Hemispherically shaped LEDs produce an isotropic emission pattern, while a strongly directed radiation pattern can be obtained with parabolically shaped LED surface. Although, curved, polished LED surfaces are feasible, such LEDs are difficult to fabricate and are expensive. In addition, lenses, mirrors, or diffusers can be built into the package to achieve specific spatial radiation characteristics.

Integrating the intensity, Equation 1.51, over the entire hemisphere gives the total power of the LED emitted into air:

$$\Phi_{air} = \frac{\Phi}{4} \frac{n_{air}^2}{n_s^2} = \eta_{external}\Phi \tag{1.52}$$

Equation 1.52 does not account for losses from Fresnel reflection at the interface semiconductor or air.

1.4.2 LED MATERIAL SYSTEMS

Commercially available and commonly used LED material systems for the visible range of the spectrum are as follows: GaAsP/GaAs and AlInGaP/GaP emitting red–yellow light; GaP:N emitting yellow–green; SiC, GaN, and InGaN emitting green–blue; GaN emitting violet; and AlGaN

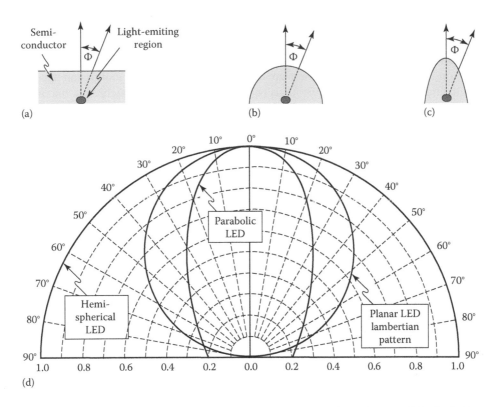

FIGURE 1.17 LEDs with (a) planar, (b) hemispherical, and (c) parabolic surfaces, and (d) far-field patterns of the different types of LEDs. At an angle of $\theta = 60°$, the Lambertian emission pattern decreases to 50% of its maximum value occurring at $\theta = 0°$. The three emission patterns are normalized to unity intensity at $\theta = 0°$. (After Schubert, E.F., *Light-Emitting Diodes*, Cambridge University Press, Cambridge, U.K., 2003.)

emitting in the UV range. For high efficiency LEDs, direct bandgap III–V semiconductors are used. There are various direct bandgap semiconductors that can be readily n- or p-doped to make commercial p–n junction LEDs, which emit radiation in the red and IR spectral ranges. A further important class of commercial semiconductor materials that cover the visible spectrum is the III–V ternary alloy based on alloying GaAs and GaP denoted as $GaAs_{1-y}P_y$). The popularity and success of that alloy is historically based on the fact that the maturing GaAs technology paved the way for the production of these alloy LEDs. We note here that GaP has an indirect fundamental transition and is therefore a fairly poor light emitter. As a consequence, at certain GaP content, the $GaAs_{1-y}P_y$ alloy possesses an indirect transition as well. For $y < 0.45$, however, $GaAs_{1-y}P_y$ is a direct bandgap semiconductor and the rate of recombination is directly proportional to the product of electron and hole concentration. The GaP content y determines color, brightness, and the internal quantum efficiency. The Emitted wavelengths cover visible red light from about 630 nm for $y = 0.45$ ($GaAs_{0.55}P_{0.45}$) to the near-IR at 870 nm for $y = 0$ (GaAs).

$GaAs_{1-y}P_y$ alloys with $y > 0.45$, which include GaP, are indirect bandgap semiconductors. The electron–hole pair recombination process occurs through recombination centers and involves excitation of phonons rather than photon emission. The radiative transitions can be enhanced by incorporating isoelectronic impurities such as nitrogen into the crystal lattice, which serve as special recombination centers. Nitrogen is from the same Group V in the periodic table as P with similar outer electronic structure. Therefore, N atoms easily replace P atoms and introduce electronic traps. As the traps are energy levels typically localized near the conduction band edge, the energy of the emitted photon is only slightly less than E_g. The radiative recombination depends on the nitrogen doping and is not as efficient as direct recombination [91,92]. The reason for the lower internal quantum efficiency for LEDs based on GaAsP and GaP:N is the mismatch to the GaAs substrate [93] and indirect impurity-assisted radiative transitions, respectively. Nitrogen-doped indirect bandgap $GaAs_{1-y}P_y$ alloys are widely used in inexpensive green, yellow, and orange LEDs. These LEDs are suitable only for low-brightness applications only, but the green emission appears to be fairly bright and useful for many applications because the wavelength matches the maximum sensitivity of the human eye. The main application of commercial low-brightness green GaP:N LEDs is indicator lamps [94].

High-brightness green and red LEDs are based on GaInN and AlGaAs, respectively, and are suitable for traffic signals and car brake lights because their emission is clearly visible under bright ambient conditions. GaAs and $Al_yGa_{1-y}As$ alloys for Al content $y < 0.45$ are direct-band semiconductors, and indirect for $y > 0.45$ [95]. The most-efficient AlGaAs red LEDs are double-heterostructure transparent-substrate devices [96,97]. The reliability and lifetime of AlGaAs devices are lower than that of AlGaInP because high-Al-content AlGaAs layers are subject of oxidation and corrosion. Being developed in the late 1980s and early 1990s, AlGaInP material system has reached technological maturity and it is today the material of choice for high-brightness LEDs emitting in the red, orange, and yellow wavelength ranges [11,98–100].

Traditionally, red, yellow, and, to a certain extent, green emissions were covered fairly well from the beginning of the LED history. The realization of blue LEDs, however, has been very cumbersome. Back in the 1980s, blue LEDs based on the indirect semiconductor SiC appeared on the market. However, due to low efficiency and not less important due to the high expense of these devices, SiC LEDs never achieved a breakthrough in the field of light emission. Besides SiC, direct semiconductors such as ZnSe and GaN with a bandgap at 2.67 and 3.50 eV, respectively, have been in the focus of researchers and industry. After almost 20 years of research and industrial development, the decision came in the favor of GaN, although it seems during the early 1990, that ZnSe-based device structures fabricated by SONY will finally provide blue light LEDs. The activities of Shuji Nakamura at Nichia Chemical Industries Ltd., Tokushima, Japan changed—it is fair to say overnight—everything. At the end of 1993, after gaining control over n- and p-type doping of GaN with a specifically adapted high-temperature (1000°C) two-gas-flow metal–organic chemical vapor deposition setup, Nakamura finally paved the way for commercially available blue LEDs using GaN as core material. The research on the ZnSe-based devices was abandoned because the GaN-based products were brighter by 1 order of magnitude and exhibited a much longer lifetime already at that early stage of the development. In the following, the GaInN hetero-pairing was developed to such a stage that blue and green light emitting devices became commercially available in the second half of the 1990s [101–103]. Nowadays, GaInN is the primary material for high-brightness blue and green LEDs.

In summary, the typical emission ranges, applications, and practical justifications of various LED material combinations are as follows: yellow (590 nm) and orange (605 nm) AlGaInP, and green (525 nm) GaInN LEDs are excellent choices for high luminous efficiency devices; Amber AlGaInP LEDs have higher luminous efficiency and lower manufacturing cost compared with green GaInN LEDs, and are preferred in applications that require high brightness and low power consumption such as solar cell-powered highway signals; GaAsP and GaP:N LEDs have the advantage of low power and low cost but posses much lower luminous efficiency, and are therefore not suitable for high-brightness applications. In general, one has to bear in mind that not only the luminous efficiency but also the total power emitted by an LED is of considerable importance for many applications. A detailed review of the optical and electrical characteristics of high-brightness LEDs is given in Refs. [5,12].

1.4.3 WHITE LEDs

Most white-light LEDs use the principle of phosphorescence, that is, the short wavelength light emitted by the LED itself pumps a wavelength converter, which reemits light at a longer wavelength.

As a result, the output spectrum of the LED consists of at least two different wavelengths. Two parameters, the luminous efficiency and the CRI, are essential characteristics for white LEDs. For example, for signage applications, the—eye catching—luminous efficiency is of primary importance and the color rendering is less important, while for illumination applications, both luminous efficiency and CRI are of equal importance. White-light sources using two monochromatic complimentary colors produce the highest possible luminous efficiency (theoretically, more than 400 lm/W) [104,105]. However, the CRI of dichromatic devices is lower than that of broadband emitters.

Most commercially available white LEDs are modified blue GaInN/GaN LEDs coated with a yellowish phosphor made of cerium-doped yttrium aluminum garnet (Ce^{3+}:YAG) powder suspended in epoxy resin, which encapsulates the semiconductor die. The blue light from the LED chip ($\lambda = 450$–470 nm) is efficiently converted by the phosphor to a broad spectrum centered at about 580 nm (yellow). As yellow color is the complement of blue color, the combination of both produces white light. The resulting pale yellow shade is often called lunar white. This approach was developed by Nichia and has been used for the production of white LEDs since 1996 [101]. The present Nichia white LED NSSW108T (chromaticity coordinates 0.310/0.320), based on a blue LED with special phosphor, has a luminous intensity of 2.3 cd at forward voltage 3.5 V and driving current 20 mA and is intended to be used for ordinary electronic equipment (such as office equipment, communication equipment, measurement instruments, and household appliances).

The materials for effective wavelength conversion include phosphors, semiconductors, and dyes. Phosphors are stable and reliable and, therefore, most commonly used. Depending on the type, phosphors can exhibit quantum efficiencies close to 100%. The quantum efficiency of Ce^{3+}:YAG is reported to be 75% [106]. A common phosphor consists of an inorganic host material doped with an optically active element. A well-known host material is the YAG, while the optically active dopant is a rare-earth element, typically cerium, but terbium and gadolinium are also used. The spectrum of white LED consists of the broad phosphorescence band and clearly shows the blue emission line originating from the LED chip. In order to optimize the luminous efficiency and the color-rendering characteristics of the LED, the contribution of each band can be tuned via the concentration of the phosphor in the epoxy resin and the thickness of epoxy encapsulant. The spectral output can also be tailored by substituting the cerium with other rare-earth elements, and can even be further adjusted by substituting some or all of the aluminum in the YAG with gallium.

Another phosphor-based white LED group employs tricolor phosphor blend as an optically active element and an AlGaInN UV LED emitting at 380–400 nm LED as a pump source. The tricolor phosphor blend consists of high-efficiency europium-based red and blue emitting phosphors green emitting copper and aluminum-doped zinc sulfide (ZnS:Cu, Al). Figure 1.18 shows the spectra of two white LEDs based on different pump sources: tricolor phosphor white LED (InGaAlN) and warm white LED (InGaN). The UV-pumped phosphor-based white LEDs exhibit high CRIs, typically between 60 and 80 [107]. The visible emission is solely due to phosphor, while the exact emission line of the pumping LED is not of fundamental importance. The UV-pumped white LEDs yields light with better spectral characteristics and color rendering than the blue LEDs with YAG:Ce phosphor but are less efficient. The lower luminous efficiency is due to the large Stokes shift, more energy is converted to heat in the UV-into-white-light conversion process. Because of the higher radiative output of the UV LEDs than of the blue LEDs, both approaches yield comparable brightness. However, the UV light causes photodegradation to the epoxy resin and many other materials used in LED packaging, causing manufacturing challenges and shorter lifetimes.

The third group of white LEDs, also called photon-recycling semiconductor LED (PRS-LED), is based on semiconductor converters, which are characterized by narrow emission lines, much narrower than many phosphors and dyes. The PRS-LED consists of a blue GaInN/GaN LED (470 nm) as the primary source and an electrically passive AlGaInP/GaAs double heterostructure LED as the secondary active region. The blue light from the GaInN LED is absorbed by the AlGaInP LED and reemitted or recycled as lower energy red photons [108]. The spectral output of the PRS-LED consists of two narrow bands corresponding to the blue emission at 470 nm from the GaInN/GaN LED

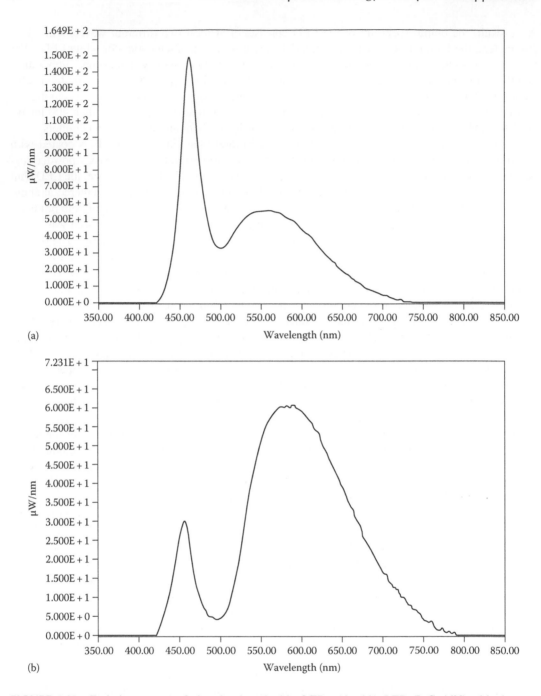

(a)

(b)

FIGURE 1.18 Emission spectra of phosphor-based white LEDs: (a) white LED (InGaAlN) with chromaticity coordinates $x = 0.29$ and $y = 0.30$ and (b) warm white LED (InGaN) with chromaticity coordinates $x = 0.446$ and $y = 0.417$. (Courtesy of Super Bright LEDs, Inc., St. Louis, MO.)

and the red emission at 630 nm from the AlGaInP LED. Therefore, the PRS-LED is also called dichromatic LED. In order to obtain white light, the intensity of the two light sources must have a certain ratio that is calculated using the chromaticity coordinates of the Illuminant C standard [12]. In order to improve the color rendering properties of a PRS-LED, a second PRS can be added to the structure; thus, adding a third emission band and creating a trichromatic LED.

The theoretical luminous efficiency, assuming unit quantum efficiency for the devices and the absence of resistive power losses, of different types of white LEDs ranges as follows: 300–340 lm/W for dichromatic PRS-LED, 240–290 lm/W for trichromatic LED, and 200–280 lm/W for phosphor-based LED [12]. As mentioned above, on the expense of the CRI, the dichromatic PRS-LEDs have the highest luminous efficiency as compared to spectrally broader emitters.

White LEDs can also be fabricated using organic dye molecules as a wavelength converter. The dyes can be incorporated in the epoxy encapsulant [109] or in optically transparent polymers. Although dyes are highly efficient converter (with quantum efficiencies close to 100%), they are less frequently used because their lifetime is considerably shorter than the lifetime of semiconductor or phosphor wavelength converters. Being organic molecules, dyes bleach out and become optically inactive after about 10^4 to 10^6 optical transitions [12]. Another disadvantage of dyes is the relatively small Stokes shift between the absorption and the emission bands. For example, the Stokes shift for the dye coumarin 6 is just 50 nm, which is smaller than the Stokes shift of about 100 nm or more required for dichromatic white LEDs.

A research in progress involves coating a blue LED with quantum dots that glow yellowish white in response to the blue light from the LED chip.

1.4.4 SURFACE-EMITTING LEDs AND EDGE-EMITTING LEDs

There are two basic types of LED emitting configurations: surface-emitting LEDs and edge-emitting LEDs. Schematic illustrations of both structures and the corresponding emission patterns are given in Figure 1.19. The light output of surface-emitting LEDs exits the device through a surface that is parallel to the plane of the active region, while an edge-emitting LED emits light from the edge of the active region. A quick comparison between both LED configurations shows that surface emitters have relatively simpler structure and are less expensive but have much larger emitting area (circular area with diameters of typically 20–50 μm). Therefore, the total LED optical output power is as high as or higher than the edge-emitting LEDs. However, the larger emitting area in addition to the Lambertian radiation emission pattern (light is emitted in all directions) and the low-to-moderate operating speeds impose limitations on the use of surface-emitting LEDs in fiber-optic communication systems, which require fast response time and high coupling efficiency to the optical fiber.

High-brightness LEDs, or power LEDs, are surface-emitting LEDs with optimized heat removal, allowing much higher power levels. High-brightness LEDs operate at high forward current. The higher driving current results in slight red shift of the peak wavelength caused by heating the chip. Power LEDs produce significant amounts of heat, which can reduce lumen output or cause device failure. Therefore, the LED design has to implement careful thermal control and effective

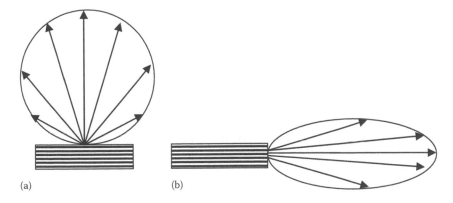

(a) (b)

FIGURE 1.19 (a) Surface-emitting LED and (b) edge-emitting LED, and the corresponding far-field radiation patterns.

heat removal. The power LEDs packaging incorporates a heat sink slug, which transfers heat from the semiconductor chip to the external heat sink with high efficiency. Liquid cooling allows the most powerful LEDs to run at their limits, while safeguarding operating life and maximizing instantaneous output. In power LEDs, fully integrated smart control manages the current to avoid overdrive.

Resonant-cavity LEDs (RCLEDs) are essentially highly optimized surface-emitting LEDs with a small area of the active region. RCLEDs have more complicated construction involving an active region of multiquantum-well structure and two distributed Bragg reflectors, which form an optical cavity. GaInP/AlGaInP RCLEDs emit at 650 nm and are appropriate for optical communications using plastic optical fibers that can have core diameters as large as 1 mm. In comparison with conventional LEDs, RCLEDs exhibit high brightness, narrow spectral width, higher spectral purity, low beam divergence, and therefore, higher coupling efficiency and substantially higher fiber-coupled intensity [110,111]. High-speed transmission of 250 Mbit/s over plastic optical fibers has been obtained with RCLEDs emitting at 650 nm [112].

Edge-emitting LEDs offer high-output power levels and high-speed performance but are more complex and expensive, and typically are more sensitive to temperature fluctuations than surface-emitting LEDs. The multilayered structure of an edge-emitted LED acts as a waveguide for the light emitted in the active layer. Current is injected along a narrow stripe of active region producing light that is guided to the edge, where it exits the device in a narrow angle. As the light emanates from the edge of the active area, the emitting spot is very small ($<20 \ \mu m^2$), the LED output power is high, and a strongly directed emission pattern is obtained. Typically, the active area has a width less than 10 μm and thickness not more than 2 μm. The light emitted at $\lambda = 800$ nm from such an area would have a divergence angle about 23°, which is poor in comparison with a strong directional laser beam, but much higher than surface-emitting LEDs. Edge-emitting LEDs deliver optical power at most several microwatts, compared to milliwatts for laser diodes (power collected by a collection numerical aperture of 0.1). However, the small emission area and relatively narrow emission angles result in brightness levels 1–2 orders of magnitude higher than comparable surface-emitting LEDs. As an example, a surface-emitting LED with an area diameter of 100 μm delivering 5 mW of total optical power would put less than 0.05 μW into an optical fiber with numerical aperture of 0.1. The high-brightness, small emitting spot and narrow emission angle allow effective light coupling to silica multimode fibers with typical core diameters of 50–100 μm. The light intensity emitted by the LED is directly proportional to the length of the waveguide (the stripe of the active region). However, the electrical current required to drive the LED also increases with the stripe length, and when the current reaches sufficiently high level, stimulated emission takes place.

Superluminescent diodes (SLDs) are edge-emitting LEDs, which operate at such high current levels that stimulated emission occurs. The current densities in SLD are similar to that of a laser diode ($\sim kA/cm^2$). The emission in a SLD begins with a spontaneous emission of a photon because of radiative electron–hole recombination. Sufficiently, strong current injection creates conditions for stimulated emission. Then, the spontaneously emitted photon stimulates the recombination of electron–hole pair and the emission of another photon, which has the same energy, propagation direction, and phase as the first photon. Thus, both photons are coherent. In contrast to LEDs, the light emitted from an SLD is coherent, but the degree of coherence is not so high compared to laser diodes and lasers. SLDs are a cross between conventional LEDs and semiconductor laser diodes. At low current levels, SLDs operate like LEDs, but their output power increases super linearly at high currents. The optical output power and the bandwidth of SLDs are intermediate between that of an LED and a laser diode. The narrower emission spectrum of the SLDs results from the increased coherence caused by the stimulated emission. The FWHM is typically about 7% of the central wavelength.

SLDs comprise a multiquantum-well structure and the active region is a narrow stripe, which lies below the surface of the semiconductor substrate. In an SLD, the rear facet is polished and highly reflective. The front facet is coated with antireflection layer, which reduces optical feedback and allows light emission only through the front facet. The semiconductor materials are selected such that the cladding layers have greater bandgap energies than the bandgap energy of the active layer, which achieves carrier confinement, and smaller refractive indices than the refractive index of the active layer, which provides light confinement. If a photon, generated in the active region, strikes the surface between the active layer and the cladding layer at an angle smaller than the critical angle of TIR, it will leave the structure and will be lost. The photons that undergo TIR are confined and guided in the active layer like in a waveguide. SLDs are similar in geometry to semiconductor lasers but lack the optical feedback required by laser diodes.

The development of SLDs was in response to the demand for high-brightness LEDs for higher bandwidth systems operating at longer wavelengths, and that allow for high-efficiency coupling to optical fibers for longer distance communications. SLDs are suitable as communication devices used with single-mode fibers, as well as high-intensity light sources for the analysis of optical components [113]. SLDs are popular for fiber-optic gyroscope applications, in interferometric instrumentation such as optical coherence tomography (OCT), and in certain fiber-optic sensors. SLDs are preferred to lasers in these applications, because the long coherence time of laser light can cause troublesome randomly occurring interference effects.

1.4.5 ORGANIC LEDs

Organic LEDs (OLEDs) employ an organic compound as an emitting layer, which can be small organic molecules in a crystalline phase or conjugated polymer chains. To function as a semiconductor, the organic emitting material must have conjugated π-bonds. Polymer LEDs (PLEDs) can be fabricated in the form of thin and flexible light-emitting plastic sheets and are also known as flexible LEDs.

The OLED structure features an organic heterostructure sandwiched between two inorganic electrodes (typically, calcium and indium tin oxides). The heterostructure represents two thin ($\approx$100 nm) organic semiconductor films, a p-type transport layer made of triphenyl diamine derivative and an n-type transport layer of aluminum tris(8-hydroxyquinoline) [58]. A glass substrate may carry several heterostructures that emit different wavelengths to provide a multicolor OLED. A white OLED uses phosphorescent dopants to create green and red lights and a luminescent dopant to create blue light, and also to enhance the quantum efficiency of the device. The three colors emerge simultaneously through the transparent anode and glass substrate. White OLEDs exhibit nearly unity quantum efficiency and good color rendering properties. Higher brightness requires a higher operating current and, thus, a trade-off in reliability and lifetime. An improved OLED structure uses a microcavity tandem in order to boost the optical output and reduce the operating current of OLEDs [114]. The result was an enhancement of the total emission by a factor of 1.2 and of the brightness by a factor of 1.6. This seems significant, especially when considering the simplicity of the design change compared with methods such as incorporation of microlenses, microcavities, and photonic crystals.

PLEDs have similar structure, and the manufacturing process uses a simple printing technology, by which pixels of red, green, and blue materials are applied on a flexible substrate of polyphenylene vinylene. Compared to OLEDs, PLEDs are easier to fabricate and have greater efficiencies, but offer limited range of colors. In comparison with inorganic LEDs, LEDs are lighter, flexible, rollable, and generate diffuse light over large areas, but have substantially lower luminous efficacy. The combination of small organic molecules with polymers in large ball-like molecules with a heavy-metal ion core is called a phosphorescent dendrimers [58]. The connoisseurs distinguish between the technological processes, device structures and characteristics, and use the acronym OLEDs only when they refer to small-molecules OLEDs, while in the mass media and daily life the term OLEDs is used also for PLEDs.

OLEDs have amazing potential for multicolor displays, because they provide practically all colors of the visible spectrum, high resolution, and high brightness achieved at low drive voltages/current densities, in addition to a large viewing angle, high response speeds, and full video capability.

For example, the Kodak display AM550L on a 2.2 in. screen features 165° viewing angle that is up to 107% larger than the LCDs on most cameras. The operating lifetime of OLEDs exceeds 10,000 h.

OLEDs are of interest at low-cost replacements for LCDs because they are easier to fabricate (fewer manufacturing steps and, more importantly, fewer materials used) and do not require a back-light to function, potentially making them more energy efficient. The backlighting is crucial to improve brightness in LCDs and requires extra power, which, for instance, in a laptop translates into heavy batteries. In fact, OLEDs can serve as the source for backlighting in LCDs.

OLEDs have been used to produce small-area displays for portable electronic devices such as cell phones, digital cameras, MP3 players, and computer monitors. The OLED display technology is promising for thin TVs, flexible displays, transparent monitors (in aviation and car navigation systems), and white-bulb replacement, as well as decoration applications (wall decorations, luminous clothing, accessories, etc.). Larger OLED displays have been demonstrated, but are not practical yet, because they impose production challenges in addition to still too short lifetime (<1000 h). Philips' TF PolyLED technology is promising for the production of full color less than 1 mm thick information displays.

In a most recent announcement [115], OSRAM reported record values of efficiency and lifetime and a simultaneous improvement of these two crucial OLED characteristics while maintaining the brightness of a white OLED. OSRAM reported that under laboratory conditions, warm white OLED achieved efficiency of 46 lm/W (CIE color coordinates x/y of 0.46/0.42) and a 5000 h lifetime, at a brightness of 1000 cd/m^2. The large-scale prototype lights up an area of nearly 90 cm^2. With this improvement, flat OLED light sources are approaching the values of conventional lighting solutions and open new opportunities for application.

1.4.6 LED METROLOGY

The breakthroughs in LED technologies and the following rapid expansion of the LED market demanded accurate photometric techniques and photometric standards for LED measurement. LED metrology is an important tool for product quality control, as well as a prerequisite for reliable and sophisticated LED applications. CIE is currently the only internationally recognized institution providing recommendations for LED measurements. The basics of LED metrology are outlined in the CIE Publication 127 Measurements of LEDs [116]. Practical advices and extensive description of LED measurements and error analysis are presented also in Ref. [117]. Four important conditions must be met when performing light measurements on LEDs with accuracies better than 10%: CIE-compatible optical probe for measuring the relevant photometric parameter, calibration equipment traceable to a national calibration laboratory, high-performance spectroradiometer (with high dynamic measuring range and precision), and proper handling. The characteristics of optical measuring instruments, photometers, and spectroradiometers are presented in Chapter 3. We limit the considerations here only to measuring optical characteristics, and omit measurements of electrical parameters.

1.4.6.1 Measuring the Luminous Intensity of LEDs

The most frequently measured parameter of an LED is luminous intensity. The definition of intensity and the underlying concept for measuring radiant and luminous intensity assumes a point light source. Although an LED has a small emitting surface area, it cannot be considered as a point source, because the LED area appears relatively large compared to the short distance between the detector and the LED that is typically used for the intensity measurements. Thus, the inverse square law that holds for point source does no longer hold for an LED, and cannot be used for calculating radiant intensity from irradiance. Therefore, the CIE has introduced the concept of averaged LED intensity, which relates to a measurement of illuminance at a fixed distance [116]. The CIE specifies the conditions for measuring luminous intensity in different laboratories irrespective of the design of the LED. The LED should be positioned in such a way that its mechanical axis is directly in line with the center point of a round detector with an active area of 1 cm², and the surface of the detector is perpendicular to this axis. The distance between the LED and the detector surface should be measured always from the front tip of the LED. The CIE recommends two geometries for measuring luminous intensity. Condition A sets the distance between LED tip and detector equal to 316 mm and a solid angle of 0.001 sr, while Condition B uses distance between LED tip and detector of exactly 100 mm and a solid angle of 0.01 sr. Condition B is suitable also for weak LEDs, and therefore, is used more often than Condition A. Both geometries require the use of special intensity probes. For example, the LED-430 measuring adapter developed by Instrument Systems LED-430 is used with Condition B (100 mm), whereas the LED-440 probe conforms to Condition A for bright LEDs with a very narrow emission angle.

1.4.6.2 Measuring the Luminous Flux of LEDs

Two principal methods are used for measuring the luminous flux of LEDs: the integrating sphere, which integrates the total luminous flux, and the goniophotometer, which measures the radiation beam of the LED at different θ and φ angles with subsequent calculation of total luminous flux.

The first method, shown on Figure 1.20, employs a hollow sphere with a diameter of 80 or 150 mm with a port for the LED and a baffled port for the detector positioned at 90° with respect to the LED port. The interior of the sphere is coated with a very stable material that ensures diffuse reflection of the LED light. After multiple reflections, the light is captured by the detector, and the measured irradiance E is proportional to the launched total radiant flux Φ. This applies only to the ideal case when the interior of the sphere has a Lambertian characteristic with constant reflectance over the entire interior of the sphere and constant spectral properties, the detector has perfect cosine correction, and there are no absorbing surfaces in the sphere [118]. However, in reality, the diffuse reflector is not perfect; also, the spectral characteristics of the coating and the size of the ports are sources of error. The main factor determining the luminous-flux measurement accuracy is the wide range of radiation characteristics of LEDs. An accuracy of about 5% can be obtained for LEDs with diffuse emission, but deviations of more than 10% are possible for LEDs with narrow emission angle. The larger sphere is used when it is important to keep measurement errors to a

FIGURE 1.20 Cross-section of an integrating sphere.

minimum, because the ratio of the sphere area to the size of the ports and the LED is more favorable [117]. However, the larger area results in a loss of intensity.

The second method, the goniophotometer, uses a cosine-corrected detector that moves on an imaginary sphere of radius r enclosing the LED. The detector measures the irradiance E as the partial radiant flux $d\Phi$ incident on a detector area dA as a function of θ and φ. The angles θ and φ vary from $0°$ to $360°$. The total radiant power Φ is obtained by integrating the irradiance over the entire sphere surface. Alternatively, instead of moving the detector, which requires mechanical adjustments, the LED can be rotated about its tip. The CIE recommended distance LED—detector is 30 cm, the area of the detector should be 1 cm^2 in the case of diffuse LEDs, and should be reduced for measurements of narrow-angled LEDs. The goniophotometer provides greater accuracy than the integrating sphere, which includes numerous geometric and spectral sources of error, in particular the wide range of radiation characteristics of LEDs.

1.4.6.3 Mapping the Spatial Radiation Pattern of LEDs

Different packages and types of LEDs exhibit different far-field radiation patterns. The spatial distribution of the emitted radiation is an important LED characteristic for many applications, in particular, for full-color (red, green, and blue) LED displays in which color balance can change when the display is observed off axis. Careful analysis of the radiation pattern is important also for white LED applications: the color coordinates of a white LED often show a significant blue shift due to the angle dependence of the light path through the phosphor. The method used to map the LED radiation pattern involves a goniometer. The LED is pivoted about its tip and the intensity is recorded at different angles providing at first the profile of the radiated beam in one plane. After that, the LED is rotated about its mechanical axis in order to obtain the two-dimensional radiation pattern.

1.4.6.4 LED Measuring Instrumentation

LED metrology is based on two measuring procedures: the spectral resolution method based on a spectroradiometer and the integration method based on a photometer. In the first method, a spectroradiometer measures the total spectral power distribution of the LED and the photometric value of interest is determined from the measured spectrum using standard CIE tables and a special software. In the second method, a photometer, which employs a broadband detector in conjunction with a $V(\lambda)$ filter, is used. Photometers are calibrated for measuring a specific photometric quantity, that is, the output current of the detector is directly proportional to the photometrically measured value. For example, a photometer for luminous intensity is calibrated in candela per photocurrent. The $V(\lambda)$ filters are optimized for the spectral radiation distribution of a standard illuminant A light source (a Planckian radiator with 2850 K color temperature), which is maximum in the IR region. LEDs, however, have a completely different spectral power distribution, with respect to emission line shape, FWHM, and specific peak wavelength. Because of the inadequate correction of the $V(\lambda)$ filter in the visible and the blue part of the spectrum, industrial photometers are not recommended for testing blue, deep red, and white LEDs. Therefore, spectroradiometers are more suitable for LED metrology. The absolute calibration of the spectroradiometer should be done with a standard LED traceable to a national calibration laboratory. The spectral resolution of a spectrometer (the bandpass) should be approximately 1/5 of the FWHM of the LED.

Only an LED can be used as a reference for absolute calibration of the intensity probe for LED measurements. An accurate calibration of the measuring instrumentation designed for a given type of light sources can be obtained only with a reference standard source from the same group with similar characteristics. In general, an absolute calibration of a detector for irradiance employs a broadband light source, such as a standard halogen lamp, traceable to a national calibration laboratory, and calculates the radiant intensity from irradiance using the inverse square law. However, LEDs are not point light sources under CIE standard measuring conditions and the inverse square law does not hold. In addition, their spectral distribution and radiation characteristics differ considerably from those of a halogen lamp. Therefore, such lamps are not suitable for absolute calibration of LED measuring instrumentation. For this purpose, standard temperature-stabilized LEDs with Lambertian radiation characteristics are used [117]. The value for luminous intensity or radiant intensity of these standards has to be determined by a national or international calibration authority.

Apart from the spectrometer, other external factors that influence strongly the measuring accuracy of LEDs are the ambient temperature, forward voltage stabilization, and careful handling. The temperature stabilization time of an LED depends on the type of the LED and the ambient temperature. To assure high accuracy, the measurements should be performed after a steady state of the LED has been attained, which can be identified by the forward voltage of the LED. The stability and accuracy of the current source should be taken into account, because a deviation of 2% in the current, for example, causes more than 1% change in the value for luminous intensity of a red LED. Careful handling and precise mechanical setup are essential for high accuracy and reproducibility of the LED measurements. A deviation of 2 mm in the distance LED detector leads to an error in the luminous intensity measurement of approximately ±4%. The quality of the test socket holding the LED is very important especially for clear, narrow-angled LEDs, where reproducible alignment of the mechanical axis of the LED is crucial for reproducible measurement of luminous intensity [117].

The LED characterization in the production flow imposes additional requirements to detectors and measurement time. Array spectroradiometers are now preferable for production control of LEDs because of significant increase in sensitivity and quick response time due to employment of high-quality back-illuminated CCD sensors. The compact array spectrometer CAS140B from Instrument Systems GmbH is an efficient measuring tool that permits luminous intensity (cd), luminous flux (lm), color characteristics (color coordinates, color temperature, and dominant wavelength), and the spatial radiation pattern of LEDs to be determined using just one instrument. It features measuring times of a few tens of milliseconds and high level of precision necessary for evaluating the optical parameters of LEDs. It can be used for development and quality assurance in areas where LEDs need to be integrated in sophisticated applications (e.g., in the automotive and avionics sector and in displays and the lighting industry).

Usually, the test time of an LED under production conditions is in the order of a few milliseconds. This period is shorter than the temperature stabilization time for most LED types and is not sufficient to guarantee a measurement of the LED in a steady state. Although, the values measured under production conditions differ from those obtained under constant-current conditions, there is generally a reproducible correlation between them, which is used for proper correction of the values measured in production testing.

1.4.7 LED Applications

During the last decades, the LED concept underwent a tremendous development. The LED market has grown over the past 5 years at an average rate of 50% per year and forecasts to reach $90 million by 2009 [119]. Because of the increasing brightness and decreasing cost, LEDs find countless everyday applications—from traffic signal lights to cellular phone and digital camera backlighting, in automotive interior lighting and instrument panels, giant outdoor screens, aviation cockpit displays, LED signboards, etc. In 2005, the cell-phone applications held the highest share (52%) of the high-brightness LED market, followed by LCD backlighting (11.5%), signage (10.5%), and automotive applications (9.5%). Besides the rapid progress and recent advance of white LEDs, the LED illumination market is still low at 5.5%, because the consumer is not familiar yet with the advantages and possibilities of solid state lighting. LEDs in traffic lights held 4.5%, and 7% of the LED market included other niche applications [120]. The five manufacturers who currently dominate the market are Lumileds Lighting LLC, Osram Opto Semiconductors, Nichia Corp., Cree Inc., and Toyoda Gosei Co. Ltd.

The applications of LEDs can be grouped roughly by the emission spectral range. LEDs emitting in the IR range ($\lambda > 800$ nm) find applications in communication systems, remote controls, and optocouplers. White LEDs and colored LEDs in the visible range are of main importance for general illumination, indicators, traffic signal lights, and signage. UV LEDs ($\lambda < 400$ nm) are used as pump source for white LEDs, as well as in biotechnology and dentistry. Superluminescent LEDs (SLEDs) were developed originally for the proposes of large luminescent displays and optical communication, but quickly found numerous novel applications as an efficient light source in the fields of medicine, microbiology, engineering, traffic technology, horticulture, agriculture, forestry, fishery, etc. SLEDs are the latest trend in general and automotive lightings.

For a comprehensive introduction to lighting technology and applications of solid-state sources, see Žukauskas et al. [5]. Here, we briefly cover few more high-brightness LED illumination applications that were not included in the previously mentioned review.

1.4.7.1 LEDs Illumination

The main potential of white LEDs lies in general illumination. Penetration of white LEDs into the general lighting market could translate (globally) into cost savings of $10 [11] or a reduction of power generation capacity of 120 GW [90].

The strong competition in the field has fueled the creation of novel device architectures with improved photon-extraction efficiencies, which in turn have increased the LED's brightness and output power. In September 2007, Cree, Inc. (Durham, NC) demonstrated light output of more than 1000 lm from a single R&D LED—an amount equivalent to the output of a standard household light bulb. A cool-white LED at 4 A delivered 1050 lm with efficacy of 72 lm/W. The LED operated at substantially higher efficacy levels than those of today's conventional light bulbs.

In response to this challenge, in October 2007, Philips Lumileds launched the industry's first 1 A LED called LUXEON K2 with thin-film flip chip (TFFC) LED [121]. This cool-white LED is designed, binned, and tested for standard operation at 1000 mA and capable of being driven at 1500 mA. LUXEON K2 with TFFC operates only at 66% of its maximum power rating and delivers unprecedented performance for a single 1 mm² chip: light output of 200 lm and efficacy over 60 lm/W (5 W) at 1,000 mA, and after 50,000 h of operation at 1,000 mA, it retains 70% of the original light output. The TFFC technology combines the advantages of InGaN/GaN FCdesign by Philips Lumileds, with a TF structure to create a higher-performance TFFC LED [122]. At present, LUXEON K2 with TFFC is the most robust and powerful LED available on the market, offering the lowest cost of light with the widest operating range. Another LED product of Philips Lumileds, LUXEON Rebel LED (typical light output of 145 lm of cool-white light at 700 mA, CCT 4100 K, and Lambertian radiation pattern), is the smallest surface mountable power LED available today. With thickness of the package of only 2.6 mm, the ultracompact LUXEON Rebel is ideal for both space-constrained and conventional solid-state lighting applications.

Nowadays, white LEDs are used as a substitution for small incandescent lamps and of strobe light in digital cameras, in flashlights and lanterns, reading lamps, emergency lighting, marker lights (steps and exit ways), scanners, etc. Definitely, LEDs will widely displace incandescent light bulbs because of the comparably low power consumption and versatility of technological adaptation. The luminous flux per LED package has increased by about four orders of magnitude over a period of 30 years [5,90], and the performance of commercial white LEDs marked a tremendous progress in the last few years. However, in times of growing environmental and resource saving challenges, it is still far from complete satisfaction of the constantly increasing requirements of the general illumination market.

Other applications of white LEDs are for backlighting in LCD, cellular phones, switches, keys, illuminated advertising, etc. White LEDs, like high-brightness colored LEDs, can be used also for signage and displays, but only in low ambient-illumination applications (night-light).

Power LEDs enable the design of high-intensity lighting systems for industrial machine-vision applications such as line lights for inspection of printed circuit boards, wafers, glass products, and high-speed security paper. The illumination systems in machine vision typically use high-intensity discharge bulbs, which provide several thousand lumens of light, but also generate a lot of heat. Therefore, the HID lamps in these systems are always combined with a fiber optic to distance the target from the bulb. When the application requires monochromatic lighting, filters have to be used, which reduce the intensity output. In addition, HID bulbs are expensive, last only few thousand hours, and suffer intensity and color changes during their lifetimes. Nowadays, many of the HID-lamp illumination systems in machine vision are replaced with power LED-based lighting systems that are able to provide intensities exceeding those of HID systems. LEDs-based systems are cheaper to buy, maintain, and run. LEDs are capable of 100,000 h of operation and their intensity remains constant during their lifetime. In comparison with other light sources, which require as much as 100 V for operation, LEDs have nominal voltage typically of 1.5 V for a nominal current of 100 mA. Thanks to many advantages LED lighting offers, it is becoming the standard in metrology applications.

1.4.7.2 LED-Based Optical Communications

LEDs can be used for either free-space communication or short- and medium-distance (<10 km) optical fiber communications. The basics of free-space communication and optical fiber communications are outlined in Refs. [123,124]. Free-space communication is usually limited to direct line of sight applications and includes the remote control of household appliances, data communication via IR port between a computer and peripheral devices, etc. The light should be invisible to prevent distraction; hence, GaAs/GaAs (~870 nm) or GaInAs/GaAs (~950 nm) LEDs emitting in the near IR are appropriate for this application. The total light power and the far-field emission pattern are the important LED parameters for free-space communication applications. The LED output determines the transmission distance, usually less than 100 m, although distances of several kilometers are also possible at certain conditions. The Lambertian emission pattern provides wider span of the signal and more convenience for the user, because it reduces the requirement of aiming the emitter toward the receiver.

Optical-fiber communications employ two types of optical fibers: single-mode and multimode fibers. Only fibers with small core diameter (in the order of a few micrometers) and small numerical apertures can operate as a single-mode fiber. For example, a silica-glass fiber with refractive index $n = 1.447$ and numerical aperture NA = 0.205 can operate as a single-mode fiber at wavelength $\lambda = 1.3$ μm, if the core diameter of the fiber is less than 4.86 μm. The small-core diameter and numerical aperture impose stringent requirements to the light beam diameter, divergence, and brightness. High coupling efficiency of the light emanating from the LED to the fiber can be attained if the light emitting spot is smaller than the core diameter of the optical fiber. Although SLEDs are occasionally used with single-mode fibers, LEDs do not provide sufficiently high LED-fiber coupling efficiency to compete with lasers. However, LEDs can meet the requirements for multimode (graded-index or step-index) fiber applications. Silica multimode fibers have core diameters of 50–100 μm, while the plastic fibers diameter could be as large as 1 mm. Thus, the LEDs that have typically circular emission regions with diameters of 20–50 μm are suitable for devices with multimode fibers. Surface-emitting RCLEDs based on AlGaInP/GaAs material system and emitting in the range 600–650 nm are useful for plastic optical fiber communications. Edge-emitting superluminescent InGaAsP/InP LEDs emitting at 1300 nm are used with graded-index silica fibers for high-speed data transmission.

The LED exitance (power emitted per unit area) is useful figure of merit for optical-fiber communication applications. It determines the transmission distance in the fiber. LED-based optical-fiber communication systems are suited for low and medium data-transmission rates (<1 Gbit/s) over distances of a few kilometers. The limitation imposed on the transmission rate is related to the LED response time.

Response time is a very important characteristic, especially for LEDs used in optical communication applications. A light source should have short enough response time in order to meet the bandwidth limits of the system. The response time is determined by the source's rise (switch on) or fall (switch off) time of the signal. The rise time is the time required for the signal to go from 10% to 90% of the peak power. The turn-off time of most LEDs is longer than the turn-on time.

Typical values for LED turn-off times are 0.7 ns for the electrical signal and 2.5 ns for the optical signal. SLEDs used in optical communications have a very small active area, much smaller than the die itself (thus, small diode capacitance), and the response time is determined by the spontaneous recombination lifetime (the fall time). For a resonant-cavity LED at room temperature, the response time ranges from about 3 to 1.1 ns for voltage swing $V_{on} - V_{off} = 0.4-1.4$ V [12,125]. The response time of about 1 ns in highly excited semiconductors limits the maximum transmission rate attainable with LEDs below 1 Gbit/s. However, transmission rates of several hundred megabits per second are satisfactory for most local-area communication applications. Lasers and laser diodes are used for higher bit rates and longer transmission distances.

1.4.7.3 Applications of the LED Photovoltaic Effect

1.4.7.3.1 Photovoltaic Effect

In addition to high efficiency and brightness of several candelas, superbright LEDs (SBLEDs) exhibit a remarkably large photovoltaic effect, as large as a conventional Si photodiode, which suggests that a SBLED has an excellent ability to function as a photodiode [126]. Figure 1.21 shows the photoemission and photoreception spectra of red LED at room temperature. The photovoltaic effect was demonstrated with a pair of two identical red (660 nm) SBLEDs (Stanley H-3000, brightness of 3 cd, and rated current of 20 mA). As shown in Figure 1.22, one of the LEDs was the emitter, the other played the role of a receiver. Figure 1.23 compares the photocurrent I_p of a SBLED H3000 with the photocurrent of a typical Si photodiode (TPS708): the LED photocurrent is approximately 1/3 of the photocurrent for the Si photodiode. However, this does not mean that the SBLED as photodetector is inferior to the Si photodiode. Since the LED photovoltage is 1.5 V for a current through the first LED $I_1 = 20$ mA and about three times of that of the Si photodiode (0.5 V), the LED has almost an equal ability with respect to the conversion of optical radiation to electrical energy, at least at this wavelength (660 nm). In addition, the light-receiving ability is approximately proportional to the brightness (cd) of LED and GaAlAs-system red LED exhibits the largest light-receiving ability.

1.4.7.3.2 LED/LED Two-Way Communication Method

On the basis of the LED photovoltaic effect, a new two-way LED/LED system for free-space communication was developed [127]. Each LED in the two-way optical transmission system plays a twofold role: of a light source and of a photo-receiver. The system, consisting of a pair of SBLEDs

FIGURE 1.21 Photoemission and photoreception spectra of red LED (Toshiba TLRA190P) at room temperature. (From Okamoto, K., *Technical Digest, The International PVSEC-5*, Kyoto, Japan, 1990. With permission.)

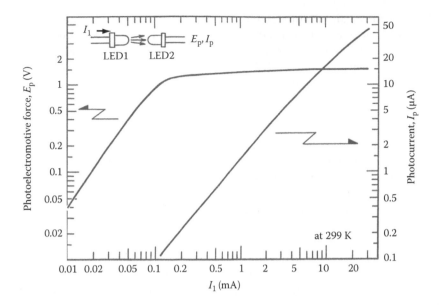

FIGURE 1.22 Photovoltaic effect between two LEDs. (From Okamoto, K., *Technical Digest, The International PVSEC-5*, Kyoto, Japan, 1990. With permission.)

FIGURE 1.23 Comparison of photocurrent I_p between SBLED H3000 and a typical Si $p-n$ photodiode TPS708. (From Okamoto, K., *Technical Digest, The International PVSEC-5*, Kyoto, Japan, 1990. With permission.)

(660 nm GaAlAs) and a pair of bird-watching telescopes, was demonstrated for free-space optical transmission of an analogue audio signal (radio signal) at a distance of 5 km. Instead of telescopes, a pair of half-mirror reflex cameras were used later in a versatile reciprocal system for long-distance optical transmission of analogue signals [128]. The employment of the reflex cameras eases the collimation of the light beam and alignment of the optical axis, and as a result, multiple communications are possible.

1.4.7.3.3 LED Solar Cell

Using both, the high-efficient luminescence and the large photovoltaic effect of the SBLEDs, Okamoto developed a new type of solar cell, named LED CELL, which was the world-first reversible solar cell [126]. A later version, the 7 W LED CELL III [129], consisted of 3060 pieces of red SBLEDs (Toshiba TLRA190P, GaAlAs system, light intensity of 7 cd, wavelength of 660 nm, and diameter of 10 mm) and provided an open circuit voltage $V_o = 1.6$ V and a short circuit current $I_s = 2$ mA for sunshine of 120 klux. This electric power could drive as many as several hundreds of wrist watches. The cell is a stand-alone one, equipped with batteries, solar sensor, sun-tracking mechanism, and control circuit. The output voltage of the cell is adjustable between 1.5 and 12 V, and the maximum electric power is about 5 W. The LED cell not only generates electricity but also emits dazzling red-color light in a reversible manner. Figure 1.24 shows the I–V curve of LED CELL III. The LED solar cell exhibits a good rectangular I–V curve with fill factor of about 0.85, which is better than a Si-crystal solar cell of about the same electric power. The conversion efficiency of the single LED (7 cd, GaAlAs, and 10 mm) is 4.1%, whereas the efficiency of LED CELL III is 1.5% [128].

1.4.7.3.4 Photo-Coupler

Using the photovoltaic effect of SBLED, a new type of photo-coupler or photo-relay can be realized. Because of the large induced photovoltage, the effect is especially favorable for the voltage-drive type photo-coupler or photo-relay. Figure 1.25 shows a two-way (reversible) photo-coupler composed of SBLEDs only [126]. In this photo-coupler, each side has three SBLEDs connected in series. The photo-coupler produces a photovoltage as high as $1.5 \times 3 = 4.5$ V for a small primary input current of 1 mA or less, therefore the output can drive a CMOS-IC directly. The photo-coupler has a particularly convenient feature: it works in any direction and there is no need to pay attention to its direction upon mounting onto a socket or attaching it onto a circuit plate.

FIGURE 1.24 Comparison of a Si solar cell and LED CELL III: (a) I–V curve of a crystal Si solar cell for sunlight of 120,000 lux and (b) I–V curve of the 3060 pieces LED CELL III. (From Okamoto, K. and Tsutsui, H., *Technical Digest, The International PVSEC-7*, Nagoya, Japan, 1993. With permission.)

FIGURE 1.25 Two-way (reversible) photo-coupler composed only of LEDs capable of direct drive of CMOS-IC. (From Okamoto, K., *Technical Digest, The International PVSEC-5*, Kyoto, Japan, 1990. With permission.)

1.4.7.3.5 LED Translator

The SBLED emits light with enough intensity even for a low-level input current of the order of several hundreds microamperes. On the other hand, it generates a large photovoltage. Therefore, if two SBLEDs are connected in parallel (with the same polarity) and one of them is illuminated by light with enough intensity, then the other SBLED will light up. This was implemented in a signal translator with two-way mode as shown in Figure 1.26 [126]. Similarly, if three SBLEDs are connected in parallel like in Figure 1.27, optical signals or data can be transmitted freely among the three LED terminals.

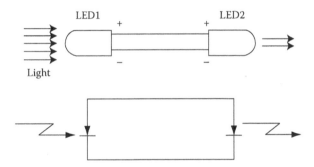

FIGURE 1.26 Light signal translator using two SBLEDs. (From Okamoto, K., *Technical Digest, The International PVSEC-5*, Kyoto, Japan, 1990. With permission.)

FIGURE 1.27 Data bus using three SBLEDs. (From Okamoto, K., *Technical Digest, The International PVSEC-5*, Kyoto, Japan, 1990. With permission.)

1.4.7.4 High-Speed Pulsed LED Light Generator for Stroboscopic Applications

Due to their large dc power capacity, SBLEDs offer large and stable intensity output when operated in ultrashort pulsed mode. On this basis, a high-speed pulsed light generator system was developed for optical analysis and imaging of fast processes in fluid dynamics [130]. The system was demonstrated with a red Toshiba SBLED (TLRA190P) in pulsed operation at charging voltage of 200 V and driving current 30 A, pulse duration of 30 ns and pulse repetition rate 100 Hz. The pulse repetition rate is up to 1 kHz and the pulse width can be adjusted in the range 20–100 ns. Under pulsed operation, this diode emits at about 660 nm with FWHM of about 25 nm. The maximum light power generated by the LED is 1.5 W. Due to the pronounce coherence, this type of LEDs are especially suited for stroboscopic methods. An integrated lens reduces the aperture of the light beam to approximately 10°. If the object to be illuminated is easily accessible, the LED can be installed without any additional optics. The pulsed light generator system was used in stroboscopic applications such as spray formation and evaporation of a jet of diesel fuel injected under diesel engine conditions into a chamber (typical velocities of such a flow field are in the range of 100–200 m/s).

1.4.7.5 LED Applications in Medicine and Dentistry

The most common applications of LEDs in medicine and dentistry are in custom-designed modules for replacement of mercury lamps. These modules employ UV and deep-UV LEDs grown on sapphire substrates and emitting in the range of 247–385 nm. For example, Nichia's NCSU034A UV LED with peak wavelength at 385 nm offers optical power of 310 mW at typical forward voltage 3.7 V and driving current of 500 mA and has an emission angle of 240°. The LED systems are used for surgical sterilization and early detection of teeth- or skin-related problems, in biomedical and laboratory equipment, for air and water sterilizations. Nowadays, high-brightness blue LED illumination is widely used for photoinduced polymerization of dental composites [131].

The U.S. firm Lantis Laser has developed a new noninvasive imaging technique, OCT, which employs an InP-based superluminescent UV LED. The OCT instrument is used by dentists to detect early stages of teeth disease before they show up on an x-ray; hence, less-damaging treatment of diseased teeth. This is possible, because the spatial resolution of the three-dimensional images produced by OCT is 10 times better than an x-ray. One reason for this is that x-ray detection of tooth decay depends on a variation in material density, a sign that damage has already been done, whereas OCT relies on more subtle variations in optical characteristics.

An original research has shown that irradiation with blue and green SBLEDs may suppress the division of some kind of leukemia and liver cancer cells, for instance, the chronic myelogenous cell, K562, and the human acute myelogenous leukemia cell, KG-1 [132,133]. The commonly used method of photodynamic therapy (PDT) aims to destroy cancerous cells through photochemical reactions induced by a laser light in photosensitive agents, such as metal-free porphyrins, which have an affinity toward cancerous cells. There are two typical medical porphyrins: porfimer sodium (Photofrin) is used with an excimer laser and talaporfin sodium (Laserphyrin) is used with a laser diode with wavelengths 630 and 660 nm, respectively. The Laserphyrin is newer and milder than Photofrin with respect to skin inflammation caused by ambient light (natural light) and remaining porphyrin in the skin. In vitro experiments of photodynamic purging of leukemia cells employed a high-power InGaN LED by Nichia (peak wavelength of 525 nm and power density of 5 W/m^2). The green LED irradiation suppressed drastically the proliferation of K562 leukemia cells in the presence of a small quantity of Photofrin, which is traditionally used in PDT of cancers [132].

Treatment of neonatal jaundice is another original medical application of high-brightness LEDs developed by Okamoto et al. [134]. Neonatal jaundice is caused by the surplus of bilirubin in the bloodstream, which exists in the blood serum. Bilirubin is most sensitive to blue light with wavelength 420–450 nm. Under blue (420–450 nm) and green (500–510) lights, the original bilirubin is transformed from oleaginous bilirubin to water-soluble bilirubin, which is easier to excrete by the liver and kidneys. The conventional method of phototherapy for hyperbilirubinemia utilizes bluish-white or bluish-green fluorescent lamps; few fluorescent lamps are placed 40–50 cm above the newborn laid in an incubator. The LED phototherapy apparatus of Okamoto uses Nichia blue (450 nm) and bluish-green (510 nm) InGaN LEDs. Seidman et al. [135] performed clinical investigation of the LED therapeutic effect on 69 newborns, which showed that LED phototherapy is as efficient as conventional phototherapy, but the LED source has the advantages of being smaller, lighter, and safer (no glass parts, no UV radiation), in addition to low DC voltage supply, long lifetime, and durability. Another advantage of the LED source is the easy control of the LED light output by the driving current to correspond to the necessary treatment.

1.4.7.6   LED Applications in Horticulture, Agriculture, Forestry, and Fishery

1.4.7.6.1   Plant Growth under LED Illumination

Chlorophyll, which plays a dominant role of photosynthesis in green plants, absorbs selectively red (at 650–670 nm) and blue (at 430–470 nm) lights [136]. The red light contributes to the photosynthesis, while the blue light is crucial for the plant's morphologically healthy growth. As seen in Figure 1.28, these two photo-absorption peaks coincide perfectly with the peak emission wavelengths of 660 and 450 nm for the GaAlAs superbright red LED and the InGaN blue LED, respectively. Because of the small mass and volume, solid construction, light wavelength efficiency, low heat dissipation, and long lifetime, LEDs have attracted immediately the attention as possible light source for cultivation of plants and vegetables in a tightly controlled environment such as space-based plant culture system [137,138]. The world-first successful sound plant growth (lettuce, *Lactuca sativa*) under artificial illumination by blue (450 nm) and red (660 nm) SBLEDs was realized by Okamoto and Yanagi in 1994 [139]. The plant growth system consisted of 3060 red and 6120 blue LEDs: red (Toshiba TLRA190P, 7 cd, diameter of 10 mm, wavelength range of 620–700 nm, and peak wavelength at 660 nm) and blue LEDs (Nichia NLPB520, 2 cd, diameter of 5 mm, wavelength range of 400–550 nm, and peak wavelength at 450 nm). This principle was implemented in various LED plant-growth systems, in which the LED light intensity can be easily controlled by changing the dc driving current. The light quality can be controlled by the utilization of LEDs of different colors. The LED illumination plant cultivation method was demonstrated also for in vitro growth of different plantlets [140,141], pepper plant [142], wheat plant [143], and other plants [5]. Later, the authors demonstrated simultaneous plant growth and growth sensing using an illumination panel of white and red LEDs [144]. In growth mode, both white and red LEDs are lighting. In sensing mode, only white LEDs are lighting and red LEDs detect the green light from the leaves

FIGURE 1.28   Photoemission spectra of red and blue LEDs and light absorption spectrum of chlorophyll. (From Okamoto, K. et al., *Acta Hortic.*, 440, 111, 1996. With permission.)

as photodiodes. In this application, red InGaAlP-system red LEDs emitting at 644 nm were used together with GaN-system white LEDs. Because these white LEDs have also a principle photoemission peak at 450 nm in the blue region, they successfully replaced the blue LED. The spectrum of the white LED contains a green component that crosses the absorption curve of the red LED, which allows for the red LED to receive the green light component as a photodiode. The plant growth and sensing method enables unmanned fully automatic plant cultivation. It does not require a television camera or other image sensor for monitoring the plant growth. The LED plant growth method was implemented also in horticulture and forestry, because it enables mass production of seedlings of superior quality through tissue culture method or micro-propagation [145].

1.4.7.6.2   Suppression of Mold's Growth

During this research, Okamoto found also that blue LED light drastically suppresses the growth and propagation of blue mold, and on this basis, he developed a prototype of an anti-mold storage chamber, which is used for preserving food [145]. It consists of 2240 pieces of high-brightness blue (450 nm) LEDs. Blue mold spoils many foods and is normally treated with chemical agents. Inhibition of bluish mold propagation by 450 nm blue LED has important applications in the field of antibacteria and anti-mold technology, microbiology, food warehouses, antibacterial and antifungal facilities in hospitals, crops transportation facilities, etc.

1.4.7.6.3   Blue LED Fishing Lamp

Most marine species including squids show regular phototaxis—fish is attracted by light and proceed toward the light source. There is a logarithmic relationship between the catch and the light intensity, and therefore, squid fishing is done during the night using very powerful fishing lamps. A typical fishing lamp consists of 60 metal-halide bulbs, each with electric power consumption of 3 kW. Hence, the total electric power consumption of the fishing-lamp system exceeds 180 kW per ship. In the peak season (June and July), more than 10,000 squid fishing boats of Japan and South Korea work around the Sea of Japan and the Korean Strait. More than 90% of the squid-fishing boats are small vessels, under 10 tons, and less than 20 m in length, and such a ship consumes as much as 500–1000 L petroleum per night. The total energy consumption attains 1 GW, which can be compared to the electric capacity of an atomic power station, and results in emission of huge amounts of gases of significant environmental concern such as CO_2, SO_2, and NO_2. The light from the fishing lamps is intolerably strong and dazzling, and in addition, contains intensive UV component that is harmful to fishermen and marine life. The development and the subsequent wide implementation in the squid-fishing of a new type, the LED fishing lamp [146,147], have drastically reduced many of the problems discussed above. The LED fishing lamp has many advantages over the conventional metal-halide fishing lamp: 30 times less energy consumption, high efficiency, long life, compact, and environment friendly. Another unquestionable advantage is the opportunity for underwater implementation of the LED fishing lamp [148].

1.5   LASERS

The laser is a device that generates a highly collimated, high-intensity beam of monochromatic and coherent radiation. What distinguish lasers from the conventional light sources are the unique properties of laser light: coherence, monochromaticity, directionality, polarization, and high intensity. The most common lasers today can deliver continuous wave (cw) or pulsed power output with wavelengths that vary from 193 nm (deep UV) to 10.6 μm (far IR), with cw power outputs that vary from 0.1 mW to 20 kW, or pulses with duration as short as a few femtoseconds and pulse energies as high as 10^4 J, and whose overall efficiencies (laser energy out divided by pump energy in) vary from less than 0.1% to 20% [15]. Lasers are used in applications that require light with very specific properties, which cannot be provided by other light sources. Many high-precision techniques used in optical metrology, and described later in this book, employ phenomena and effects that originate from the unique properties of laser light. Therefore, we briefly review here some laser characteristics and parameters and some laser systems that are important for optical metrology. For a specific method, a laser system, or an application, the reader should refer to specialized books and the references thereafter describing in details the laser fundamentals, the different laser systems, and their area applications [149–153].

1.5.1 Stimulated Emission and Light Amplification

Light interact with matter via three processes, which provide thermodynamic equilibrium between light and matter: stimulated absorption (or simply absorption), spontaneous emission, and stimulated emission. Einstein showed that Planck's empirical formula that describes blackbody radiation at thermodynamic equilibrium could be derived from the quantum theory of radiation only if all these three processes are considered.

Since the process of interaction of light with matter must conserve energy, only specific energy levels that are separated by the photon energy $h\nu$ are involved in the absorption and emission of light. Absorption is the process by which light transfers energy to matter. For the absorption process, the principle of conservation of energy implies that an atom can absorb an incident photon of energy $h\nu$ and undergo a transition from a lower energy state E_1 to a higher energy state E_2, only if $E_2 - E_1 = h\nu$. If the material is left undisturbed, the atom will eventually de-excite via a radiative or non-radiative transition to a lower energy state. If a radiative transition takes place, it will result in spontaneous emission of a photon, which has random phase and propagation direction. Stimulated emission occurs when an incident-stimulating photon of energy equal to the energy difference between the upper and the lower energy levels ($h\nu = E_2 - E_1$) interacts with the excited atom before spontaneous emission occurs. As a result, the atom de-excites via stimulated emission of a photon, which is an exact copy of the stimulating photon: it has the same frequency, phase, propagation direction, and polarization. In other words, both photons are spatially and temporally coherent.

The probability for absorption or stimulated emission process to occur depends on the photon energy density $u(\nu)$ in the frequency range driving the transition (measured in J s/m^2). Spontaneous emission does not require a photon to occur; therefore, the corresponding probability is independent of the photon energy density $u(\nu)$. The rates of occurrence of these processes are given by the Einstein A_{21}, B_{12}, and B_{21} coefficients as follows: the probability of spontaneous emission of a photon is A_{21}, the probability of absorption of a photon is the product $B_{12}u(\nu)$, and the probability of stimulated emission of a photon is $B_{21}u(\nu)$.

The number of atoms per unit volume that is in a given energy state defines the population density of that state. Let us denote the population densities of the lower E_1 and higher E_2 energy states with N_1 and N_2, respectively. The rate of change of the population of a given energy state or the transition rate depends on both the transition probability and the population of that state. Thus, the decay of the population N_2 of the higher energy state E_2 due to both the spontaneous and the stimulated emissions can be written as

$$\frac{dN_2}{dt} = -\left[A_{21} + B_{21}u(\nu)\right]N_2 \tag{1.53}$$

The transition rate for absorption is

$$\frac{dN_1}{dt} = \left[B_{12}u(\nu)\right]N_1 \tag{1.54}$$

At thermodynamic equilibrium, the rate of emission must be equal to the rate of absorption $(dN_2/dt) = -(dN_1/dt)$, thus

$$\frac{N_2}{N_1} = \frac{B_{12}u(\nu)}{\left[A_{21} + B_{21}u(\nu)\right]} \tag{1.55}$$

The population N_i of a higher energy state E_i is given by the Maxwell–Boltzmann distribution

$$N_i = N_0 e^{-E_i/k_B T} \tag{1.56}$$

where
  N_0 is a constant for a given temperature
  T is the equilibrium temperature
  k_B is Boltzmann's constant

The higher the energy state, the fewer atoms will be in that state. By substituting N_1 and N_2 in Equation 1.55 with Equation 1.56 for $i = 1, 2$ and using $E_2 - E_1 = h\nu$ yield

$$u(\nu) = \frac{A_{21}}{B_{12}e^{h\nu/k_B T} - B_{21}} \tag{1.57}$$

Einstein was able to derive a relation between the coefficients A_{21}, B_{12}', and B_{21} from the condition of thermodynamic equilibrium between a radiation field and an assembly of atoms. The quantum theory of blackbody radiation gives the following expression for the spectral energy density in an electromagnetic field in thermodynamic equilibrium with its surroundings at temperature T [15,154]:

$$u(\nu) = \frac{8\pi h\nu^3}{c^3} \frac{1}{e^{h\nu/k_B T} - 1} \tag{1.58}$$

Equation 1.57 can be reduced to Equation 1.58 only if the following relations between the Einstein coefficients hold:

$$B_{12} = B_{21} \tag{1.59}$$

and

$$\frac{A_{21}}{B_{21}} = \frac{8\pi h\nu^3}{c^3} \tag{1.60}$$

This implies that as the energy density $u(\nu)$ becomes large, the rate of spontaneous emission becomes negligible, and the probability of stimulated emission is equal to the probability of stimulated absorption. The equality of the Einstein coefficients in Equation 1.59 leads to equality of the ratios of the rates for stimulated emission R_{stim} to that of absorption R_{abs} and the population density of the upper and lower energy levels, N_2 and N_1, respectively:

$$\frac{R_{stim}}{R_{abs}} = \frac{N_2}{N_1} \tag{1.61}$$

At thermodynamic equilibrium, the population N_1 is much larger than N_2. When $E_2 - E_1 = 1$ eV, N_2/N_1 is typically of the order of 10^{-17}. This means that practically all the atoms are in the ground state, and the overall rate of absorption is always higher than the rate of stimulated emission. Therefore, absorption is always the dominant process for a collection of atoms in thermodynamic equilibrium; thus, light amplification cannot occur at equilibrium condition.

Amplification of radiation can take place only if population inversion is created and maintained in the atomic system, that is, the population of a higher energy level is kept larger than the population of a lower energy level. The probability for a photon with energy equal to the difference of the inverted levels to initiate stimulation emission is higher than the probability to be absorbed. As the stimulating and stimulated photons travel in the material, more stimulated emissions than absorptions will take place, and the number of photons will increase exponentially with the distance along the beam in the material. So will the irradiance of the beam, and as the beam has very high degree of directionality, this will be valid also for the beam intensity $I(z)$:

$$I(z) = I_0 e^{Gz} \tag{1.62}$$

where
I_0 is the initial intensity of the radiation at $z = 0$
z is the distance along the beam direction
G (1/m) is the gain of the medium

The gain gives the number of photons created by stimulated emission per unit length of the traveled distance.

1.5.2 ESSENTIAL ELEMENTS OF A LASER

The first condition for amplification of light is the presence of a gain medium (also called active medium) capable of sustaining a preferential population inversion that leads to stimulated emission.

In order for an assembly of atoms to amplify an incident radiation, the atoms have to be driven out of thermal equilibrium and population inversion has to be created. This is possible only if energy from outside (pump energy) is supplied to the atomic system that selectively populates an excited state. A nonequilibrium condition cannot be reached and maintained simply by increasing the temperature of the system. Thus, the second condition for light amplification is the continuous supply of pump energy that creates and maintains a preferential population inversion leading to stimulated emission. Most laser materials have very low gain, typically between 0.01 and 0.0001 cm^{-1}. In order to produce a large amplification, the light has to pass a long length of the laser medium. This can be achieved by placing the gain medium between two mirrors that bounce the light back and forth through the active medium. The gain medium and both mirrors form the resonant laser cavity.

The essential elements of a laser are the gain medium, the pump, and the laser cavity or resonator. Lasers with medium- and high-power outputs require also a cooling system.

1.5.2.1 Gain Medium

The gain medium can be an assembly of atoms, ions, or molecules in a gas- (or plasma), liquid-, or solid-state phase. The gain medium of some solid-state lasers consists of a host and active atoms. For example, in the Nd:YAG laser, the YAG crystal is the host and the trivalent neodymium ions are the active element; in the ruby laser, the Al_2O_3 crystal is the host, and chromium ions Cr^{3+} (about 0.05%) are the active element. In order to produce laser emission, the gain medium must have proper set of energy states that can be effectively pumped to create population inversion and can serve as laser levels. The upper laser level should be a long-lived metastable state. A metastable state has typical lifetime of the order of 10^{-3} s, whereas most excited levels in an atom might

FIGURE 1.29 Four-level energy diagram.

decay in times of the order of 10^{-8} s. Figure 1.29 shows a four-level energy diagram. The excitation energy pumps electrons from the ground level to the excited state E_3, which decays very rapidly to the metastable state E_2 creating a population inversion. The laser emission can start with a spontaneously emitted photon via the transition $E_2 \rightarrow E_1$, followed by an avalanche-stimulated emission; because of the long lifetime of the metastable state, the probability for stimulated emission is much higher than the probability for spontaneous emission. In general, laser emission can take place between the metastable state E_2 and the ground state E_0 (as in the case of the Ruby laser). Since the ground level has the largest population at thermodynamic equilibrium, a substantial pumping energy is required to create population inversion. In the four-level laser, however, the lower laser level is not the ground state. As soon as some electrons are pumped to the upper laser level E_2, population inversion is achieved. This requires less pumping energy than does a three-level laser. In the four-level system, the lower laser level E_1 is an ordinary level that depopulates rapidly by a relaxation process returning the atoms in the ground state. Thus, the population inversion is continuously maintained.

1.5.2.2 Pump

The pump is a source of external energy that provides the required population inversion in the gain medium. The most commonly used pump mechanisms are optical pumping and electrical discharge (for gas lasers). Lasers with solid-state and liquid (dyes) gain media use optical pumping (by a flashlamp or another laser, most often diode lasers). Semiconductor lasers use electric current and optical pumping. Xenon and krypton lamps are most often used, because of their high efficiency of conversion of electrical input to light output, $\approx 60\%$ and 50%, respectively. Krypton lamps serve for optical pumping of Nd:YAG lasers. All gas lasers use electric discharge, as all excimer lasers use short-pulse electric discharge. In monoatomic or monomolecular gas lasers, the free electrons generated in the discharge process collide with and excite the atoms, ions, or molecules directly. In a two-component gas mixture, the gases are chosen to have coinciding excited states and the excitation results from inelastic atom–atom (or molecule–molecule) collisions. This is the case of a He–Ne laser. Helium has an excited energy state that is 20.61 eV above the ground state, whereas neon has an excited state that is 20.66 eV above its ground state and just 0.05 eV above the first excited state of the helium. The free electrons generated in the gas discharge collide with and transfer energy to the helium atoms, which excite the neon atoms by collision and resonant transfer of energy. The kinetic energy of the helium atoms provides the extra 0.05 eV of energy needed to excite the neon atoms to the 20.66 eV metastable state.

1.5.2.3 Laser Cavity

The laser resonator converts the gain medium into an oscillator and hence into a light generator. The purpose of the laser cavity is threefold: (1) to confine the emission in the active medium, thus increasing the photon density and the probability for stimulated emission; (2) to enhance the amplification; and (3) to shape the beam. The laser cavity represents a Fabry–Perot resonator of two carefully aligned plane or curved mirrors [155,156]. A highly reflecting mirror (reflectivity as close to 100% as possible) serves as a back mirror, and a partially reflecting mirror (about 99%) is the output coupler, which allows a certain fraction of the laser light to escape the cavity and to form the external laser beam. The resonator provides the optical feedback by directing the photons back and forth through the active medium. The multiple passes ensure the amplification of the radiation in the direction of the beam, which accounts for the amazing degree of collimation of the laser beam. As the photon density in the cavity grows, the rate of stimulated emission increases causing a decrease of the population inversion in the gain medium. In addition, a fraction of the radiation exits through the output coupler at every round-trip and is lost for the cavity. The intensity of the light after a round-trip in the cavity can be expressed from Equation 1.62 as

$$I(2L) = R_1 R_2 I_0 e^{G2L} \tag{1.63}$$

where
L is the length of the laser medium within an optical cavity
R_1 and R_2 are the reflectances of the two mirrors

To maintain the amplification of the stimulated emission, it is required that

$$I(2L) > I_0 \tag{1.64}$$

which leads to the condition for lasing:

$$G \geq -\frac{1}{2L} \ln(R_1 R_2) \tag{1.65}$$

Equation 1.65 is used in designing the laser cavity. The reflectance of the output mirror can be calculated for a given length of the active medium. Let us assume that the length of a laser tube is 150 mm and the gain factor of the laser material is 0.0005 cm^{-1}. If $R_1 = 100\%$, the reflectance of the output coupler R_2 as calculated using Equation 1.65 should be equal or higher than 98.5% in order to maintain the lasing.

The amount of stimulated emission grows on each pass until it reaches an equilibrium level. When the gain equals the losses per round-trip of the radiation through the cavity (condition of gain saturation), the laser settles into steady-state continuous wave operation.

1.5.2.4 Cooling System

The overall efficiency of a laser system is the ratio of the optical output power of the laser to the total power required to pump the laser. Typical efficiencies range from fractions of percent to about 25%, as for the most commonly used lasers they are below 2%. Carbon dioxide (CO_2) lasers have efficiencies between 5% and 15%, dye lasers ~1% and 20%, whereas the most efficient are the semiconductor lasers ~1% and 50%. The losses are mainly thermal losses, and therefore, for high-power lasers the cooling is essential. Argon ion gas laser is one of the most commonly used lasers in real-time holography. The overall efficiency of a 10 W argon laser is 0.05%, and the total power used is 2×10^4 W. Of this power, 99.95% is wasted as heat energy that if not removed, will damage the components of the system. Argon and krypton lasers use water cooling or forced air depending on the length of the tube and output power.

The cooling method depends mainly on the laser output power. The low-power HeNe, HeCd, and erbium:fiber lasers are air cooled and do not need external cooling. Gas lasers that produce medium- to high-level output powers, in particular, CO_2 laser can reach up to 20 kW, are cooled by air, water, or forced air. The solid-state lasers use air or water cooling. The Nd:glass laser uses only water cooling, while semiconductor lasers are designed to work with air cooling and heat sink.

1.5.3 OPERATING CHARACTERISTICS OF A LASER

1.5.3.1 Laser Modes

The stability of the Fabry–Perot cavity is determined by the radius of curvature and the separation between the mirrors. For a stable cavity, the radius of curvature should be several times the distance between the mirrors. Fabry–Perot cavity will support standing wave modes of wavelengths λ_m and frequencies ν_m that satisfy the condition

$$\nu_m = m \frac{c}{2nL} \tag{1.66}$$

where
  L is the separation of the mirrors
  c is the speed of light in vacuum
  n is the refractive index of the active medium
  m is a large positive integer

The longitudinal cavity modes ν_m of oscillations (called also axial modes) correspond to standing waves set up along the cavity or z-axis. Consecutive modes are separated by a constant difference, which is the free spectral range of the Fabry–Perot etalon [155,156]:

$$\nu_{m+1} - \nu_m = \Delta\nu = \frac{c}{2nL} \tag{1.67}$$

In this equation, $\Delta\nu$ also corresponds to the inverse of the round-trip time. Using $d\lambda = (\lambda^2/c)d\nu$, the corresponding distance between two adjacent modes $\Delta\lambda$ is expressed by

$$\Delta\lambda_{\text{mode}} = \frac{\lambda^2}{2nL} \tag{1.68}$$

The gain of the laser medium is determined by the bandwidth of the spontaneous atomic transition and is also a function of the wavelength. As shown in Figure 1.30, the resonant modes of the

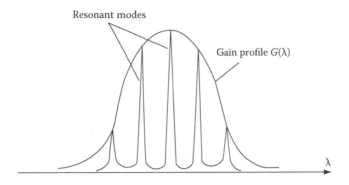

FIGURE 1.30 Several resonant modes can fit within the gain profile.

cavity are much narrower than the gain profile. The actual wavelength content of the laser beam is the product of the envelope of longitudinal oscillation modes and the gain profile. Therefore, the cavity can select and amplify only certain modes, or even only one if desired. Hence, the resonator acts as a frequency filter, which explains the extreme quasi-monochromaticity of the laser light. For example, a typical He–Ne laser, $\lambda = 632.8$ nm and $n = 1$, has a cavity length of 30 cm, and the mode separation $\Delta\lambda_{mode}$ is

$$\Delta\lambda_{mode} = \frac{\lambda^2}{2nL} = \frac{(632.8 \times 10^{-9} \text{ m})^2}{2 \times 1 \times 0.3 \text{ m}} = 6.7 \times 10^{-13} \text{ m}$$

Considering that the half-width of the gain profile of He and Ne gases is about 2×10^{-12} m, three longitudinal modes can be excited.

As indicated by the example above, several wavelengths satisfy the oscillation condition. The coexistence of multiple modes reduces the monochromaticity of the laser light and should be avoided in many applications. One possible way to generate a single mode in the cavity would be to decrease the length of the cavity (see Equation 1.68) in order to make the mode separation exceed the transition band. However, this limits the length of the amplifying medium and thus the output power of the laser. The most common technique is to insert a Fabry–Perot etalon, a pair of reflective surfaces, between the laser mirrors to form a second resonant cavity. Of the multiple longitudinal modes of the laser cavity, only the one that coincides with the etalon mode will continue to oscillate; all other modes will be canceled by destructive interference when they pass through the etalon.

The cavity can sustain also transverse modes, which determine the transverse irradiance pattern of the beam corresponding to the distribution of the electromagnetic field within the laser cavity. While the longitudinal modes are governed by the axial dimensions of the resonant cavity, the transverse modes are determined by the cross-sectional dimensions of the laser cavity. The last depends on the construction of the resonator cavity and the mirror surfaces. The transverse modes are denoted as TEM_{mn} (from transverse electric and magnetic) modes, where n and m are integers, which give the number of nodes in the x- and y-direction across the emerging beam. The larger the values of m and n, the more bright spots are contained in the laser beam.

1.5.3.2 Gaussian Laser Beams

The lowest order transverse mode or the fundamental mode TEM_{00} represents a Gaussian beam and has special characteristics, because of which lasers operating in a single TEM_{00} mode are widely used in high-precision measurement systems: a Gaussian beam is completely spatially coherent, provides the smallest beam divergence, and can be focused down to the smallest-sized spot, limited only by diffraction and lens aberrations. Most laser cavities are designed to produce only the TEM_{00} mode. This can be achieved by placing a small circular aperture that is slightly larger than the spot size of the TEM_{00} mode in the middle of the laser cavity. Alternatively, by making the bore diameter of the laser tube just slightly larger than the spot size of the TEM_{00} mode. However, lasers that are designed to maximize the output power operate in higher order modes.

The distribution of the irradiance across a Gaussian beam is described by a Gaussian function, a bell-shaped curve that is symmetrical around the central axis:

$$I = I_0 e^{-2r^2/w^2} \tag{1.69}$$

where
r is the radial distance measured from the central axis
I_0 is the irradiance at the center of the profile

At $r = w$, Equation 1.69 becomes $I = I_0 e^{-2} = 0.14 I_0$, and the beam irradiance is only 14% of the maximum value I_0.

In order for the laser cavity to be stable or marginally stable, one or both mirrors have to be curved. The curved mirrors tend to focus the beam. As a result, the cross section of the beam changes along the optical axis of the cavity and reaches a minimum value somewhere between the mirrors. The smallest beam cross section is called beam waist. The radius of the beam waist w_0 is usually given in the technical data sheet of the commercial lasers. The radius and exact location of the beam waist are determined by the design of the laser cavity and depend on the radii of curvature of the two mirrors and the distance between the mirrors. For example, the beam waist in a confocal resonator is halfway between the mirrors. (A confocal resonator is a marginally stable resonator with radius of curvature of both mirrors equal to the spacing between the mirrors.) The size of the beam waist determines the external divergence of the laser beam, which is a continuation of the beam divergence out from this waist. The half-width of the beam at any point along the z-axis can be derived from a rigorous analysis of the electromagnetic field in the cavity, when the beginning of the coordinate system $z = 0$ is set at the beam waist and is given by [157]

$$w(z) = w_0 \left[1 + \left(\frac{\lambda z}{\pi w_0^2} \right)^2 \right]^{1/2} \tag{1.70}$$

Besides the beam waist, the Rayleigh range z_R is another important characteristic of a Gaussian beam. It is defined as

$$z_R = \frac{\pi w_0^2}{\lambda} \tag{1.71}$$

At $z = z_R$, Equation 1.71 yields $w(z) = \sqrt{2} w_0$. Hence, the Rayleigh range is the distance at which the cross-sectional area of the Gaussian beam doubles. In addition, the Rayleigh range is the distance at which the curvature of the beam wavefront is minimum. As the beam propagates along z, the curvature of the beam wavefront varies with z according to

$$R(z) = z + \frac{z_R^2}{z} \tag{1.72}$$

At the beam waist, $z = 0$ and $R(0) \rightarrow \infty$, and the wavefronts are plane. Within the Rayleigh range $|z| < z_R$, the beam remains collimated. At $z = z_R$, the curvature of the beam wavefront has the smallest value $R(z_R) = 2z_R$. The far-field beam divergence is determined by the full-angular width via the relation

$$\Theta = \frac{2\lambda}{\pi w_0} = 0.637 \frac{\lambda}{w_0} \tag{1.73}$$

Equation 1.73 shows that the far-field divergence of a Gaussian beam is inversely proportional to the beam waist w_0; the smaller the beam waist w_0 is, the smaller the Rayleigh range (see Equation 1.71), and the faster the beam diverges.

If the cavity consists of plane mirrors, the beam will be aperture limited via diffraction. The divergence of an aperture limited circular laser beam of diameter D is

$$\Theta = 2.44\lambda/D \tag{1.74}$$

1.5.3.3 Continuous Wave and Pulsed Operation

A laser can operate in cw or in pulsed mode. A laser in cw mode delivers a beam of constant irradiance, whereas in pulsed mode, a laser can deliver pulses with durations as small as a few femtoseconds. The laser output can be pulsed by different methods, as the most commonly used are the Q-switching and mode locking [149]. He–Ne and He–Cd lasers deliver only cw output, while the nitrogen, Nd:glass, and excimer gas lasers deliver only pulsed output. Argon and krypton lasers can operate in cw or in pulsed modes achieved by mode locking of the cw output. The following lasers can also be operated in both cw and pulsed modes: CO_2 (long pulse), dye (ultrashort pulse), Nd:YAG, Ti:sapphire (ultrashort pulse), alexandrite and erbium:fiber, and all semiconductor lasers. The pulsed output allows controlling the laser energy delivery in material processing applications. Pulsed lasers are used also in time-resolved spectroscopy, holography of moving objects, in time-of-flight distance measurements, and other applications.

1.5.4 Characteristics of Laser Light

1.5.4.1 Monochromaticity

In general, the degree of monochromaticity of a light source is defined by the linewidth Δv, which is the FWHM of the spectral emission of the source. The spectral bandwidth $\Delta\lambda$, which is found by interference theory [157], is given by

$$\Delta\lambda = \frac{\lambda^2}{2\pi L}\left(\frac{1-R}{\sqrt{R}}\right) \tag{1.75}$$

where R is the reflectance of the output mirror.

The laser output results primarily from stimulated emission of photons with identical frequencies. However, some spontaneously emitted photons with random phase can also contribute to the laser output; thus, causing the finite linewidth of the laser emission. Besides this inherent effect, other external factors such as mechanical vibrations, which change the cavity length, or variations of the index of refraction of the gain medium additionally broaden the linewidth of a laser. Nevertheless, the lasers provide the highest degree of monochromaticity, almost reaching the ideal limit. For instance, the spectral bandwidth of the ruby laser is only $\Delta\lambda = 0.53$ nm ($\lambda = 694.3$ nm). The reason for the high degree of laser light monochromaticity is the additional frequency-filtering effect of the Fabry–Perot laser cavity.

1.5.4.2 Coherence

Coherence is the most prominent characteristic of laser light. The degree of coherence is determined by the degree of correlation between the phases of the electromagnetic field at different locations (spatial coherence) and different times (temporal coherence). Temporal coherence is a measure of the degree of monochromaticity of the light, whereas spatial coherence is a measure of the uniformity of phase across the optical wavefront. An ideal monochromatic point source, illustrated in Figure 1.31, produces perfectly coherent light. As the point source emits an ideal sinusoidal wave, it is easy to predict the phase of the radiation at a given point at a given time t_2, if we know what was the phase at the same point at an earlier time t_1. This is perfect temporal coherence. The longitudinal spatial coherence requires correlation between the phases in a given moment at two points P_1 and P_2 located along a radius line and is similar to temporal coherence. Therefore, the temporal coherence is also called longitudinal spatial coherence. Now, let us consider two points P_1 and P_3 along the same wavefront. Perfect transverse spatial coherence means that if we know the phase of the wave at point P_1 at time t_1, we can predict the phase of the field at this same moment at point P_3 along the same wavefront.

A conventional light source, such as a tungsten lamp, produces incoherent light. In order to obtain partially coherent light from a tungsten lamp, a filter and a pinhole have to be used, but this reduces drastically the light intensity. In the laser active medium, neighboring point sources produce light of the same frequency and with correlated phase. Thus, the high degree of temporal and spatial coherences of laser light is an inherent property originating from the process of generation of laser light via stimulated emission. A single-mode laser beam exhibits the highest degree of temporal and spatial coherences, as the transverse spatial coherence extends across the full width of the beam.

1.5.4.2.1 Temporal Coherence

The temporal coherence is determined by the natural linewidth Δv of the emitted radiation, and thus is a manifestation of spectral purity. The radiative electron transitions that result in spontaneous emission of photons have finite durations on the order of 10^{-8} to 10^{-9} s. This is the time extent during which a single-photon wavetrain will be nicely sinusoidal. Because the emitted wavetrains are finite, there will be a spread in the frequencies, which determine the natural linewidth Δv of the emitted radiation. The coherence time t_c is the same order of magnitude as the reciprocal of the natural linewidth Δv.

$$t_c \approx \frac{1}{\Delta v} \tag{1.76}$$

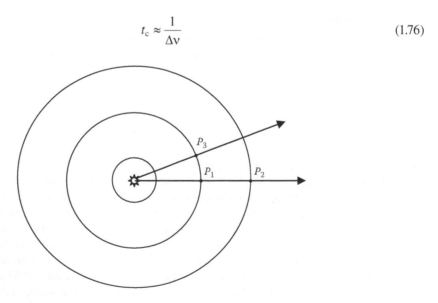

FIGURE 1.31 Monochromatic point source emits perfectly coherent light.

The coherence time t_c is determined by the average time interval over which the phase of the wave at a given point in the space can be reliably predicted. The coherence length l_c is the spatial extend over which the phase of the wave remains unchanged, or it is simply the distance traveled by the light within the coherence time:

$$l_c = ct_c \tag{1.77}$$

where c is the speed of light in vacuum. The coherence length depends on the nominal wavelength and the spectral bandwidth of the laser beam, as given by

$$l_c = \frac{\lambda^2}{2\Delta\lambda} \tag{1.78}$$

The coherence length is a very useful measure of temporal coherence, because it tells us how far apart two points along the light beam can be and still remain coherent with each other. In interferometry and holography, the coherence length is the maximum optical path difference that can be tolerated between the two interfering beams.

The broad continuous spectrum of white light has a bandwidth about 0.3×10^{15} Hz, the coherence time is roughly 3×10^{-15} s, and the coherence length is only few wavelengths long. The narrow emission line from a low-pressure mercury lamp ($\lambda = 546$ nm) is an example of quasi-monochromatic light; it has a linewidth of about 10^9 Hz, which corresponds to coherence time of about 1 ns, and coherence length of approximately 0.3 m. Typically, frequency-stabilized lasers with long laser cavities have coherence lengths that can be several meters. For a single-mode He–Ne laser, the coherence time is of the order of microsecond and the coherence length is hundreds of meters.

A high degree of coherence is essential for interferometry and holography, which are the foundations of a large group of optical metrology techniques for noninvasive and nondestructive testing. The coherence length of a monochromatic light source can be measured easily using a Michelson interferometer [155,156]. One of the mirrors is mounted on a translation stage, so that the optical path difference ΔL between the two beams in the arms of the interferometer can be varied. Whenever the optical path difference ΔL is equal to an integer number of wavelengths, the two beams interfere constructively resulting in a bright central spot. (Circular fringes will be observed only if the optical surfaces of the mirrors constructing the interferometer are perfectly flat; imperfect surfaces result in distorted fringe pattern.) The visibility V of the fringe pattern is determined by the maximum and minimum intensities, I_{max} and I_{min}, of the bright and dark fringes, respectively:

$$V = \frac{I_{max} - I_{min}}{I_{max} + I_{min}} \tag{1.79}$$

As ΔL increases, the visibility V of the fringe pattern begins to decrease. The coherence length l_c is defined as the value of the optical path difference ΔL at which the visibility of the fringes is one half. This way, the fringe visibility can be used as a measure of coherence: $V = 1$ corresponds to complete coherence, while $V = 0$ indicates lack of any coherence or complete incoherence. If $0 < V < 1$, there is partial coherence.

1.5.4.2.2 Transverse Spatial Coherence

The Van Cittert–Zernike theorem [158] defines the area of spatial (or lateral) coherence within the central maximum of the Fraunhofer diffraction of a circular aperture. If an extended light source is considered as an aperture of the same size and shape, the central maximum of the diffraction pattern produced by this aperture will correspond to the region of coherence of the extended light source. According to the Van Cittert–Zernike theorem, a thermal source such as a star with an angular diameter θ produces a region of spatial coherence with width l_s given by

$$l_s = \frac{1.22\lambda}{\theta} \tag{1.80}$$

where θ is measured in radians.

The classical double-slit experiment of Young can be used for demonstration of the transverse spatial coherence. The two slits act as two point sources of quasi-monochromatic light with wavelength λ separated by a distance d. The light waves from the two slits interfere constructively and destructively to produce bright and dark fringes observed on a screen. The interference pattern on the screen will consist of stationary fringes only when the light exiting the slits is correlated in phase. At the location of zero path difference, the constructive interference results in a bright fringe. The area of constructive interference, that is, the central bright fringe, defines the region of spatial coherence. The spacing of the interference fringes depends on the double-slit separation d, whereas their visibility depends on the spatial coherence width l_s. As l_s approaches the limit specified by Equation 1.80, the visibility decreases to zero and the fringes disappear when Equation 1.80 is satisfied.

When Young's experiment uses white light, the central fringe of the resultant interference pattern is white, and the fringes on both sides of it are rainbows of colors. All frequencies from the white spectrum have zero optical path difference in the area of the central fringe; they interfere constructively there and form the white fringe. This shows that even white light and light from thermal sources that emit a broad spectrum exhibit some degree of spatial coherence. For example, the spatial coherence width for Sun is $l_s = 0.0768$ mm, as calculated with Equation 1.80 when assuming a circular cross section with angular diameter $\theta \approx 0.5°$ and average wavelength of 550 nm. Sun is an extended thermal source, and it illuminates a properly oriented observing screen uniformly. However, if two tiny apertures are placed at a distance less than $l_s = 0.0768$ mm, they will produce interference fringes. To see interference fringes from an extended source with more widely spaced apertures, the light from the source has to pass through a single small aperture initially in order to create an area of coherence that is large enough to contain the two-slit aperture.

A manifestation of high degree of coherence is the laser speckle effect that can be easily observed when a laser beam is diffusely reflected from a rough surface such as painted wall. The coherent laser light is scattered from the surface granules, which act as small point sources of coherent wavelets that combine and produce a stationary interference pattern in the surrounding space and are projected on the retina of the eye. Speckle is not observed if the same surface is illuminated with natural light. This is because the coherence length of natural white is much less than the optical path difference between two wavelets scattered from different surface bumps and arriving at a given point in the space. The speckle effect will be presented in more details in Chapter 4.

1.5.4.3 Divergence

No other light source can provide a beam with such a degree of collimation like the laser. The high directionality of the laser beam originates from the geometrical design of the laser cavity and of the fact that the stimulated emission process produces identical photons propagating in the same direction. A single-mode laser beam exhibits smaller divergence than a multimode beam. The divergence angle of a single-mode (TEM_{00}) laser output is given by Equation 1.73. The smaller the beam waist w_0, the larger the divergence angle. For a He–Ne laser (632.8 nm) with an internal beam waist of radius about 0.25 mm, the divergence angle of the beam is 8×10^{-4} rad. Typical values for the beam divergence of the He–Ne lasers are 0.5–3 mrad. Argon and krypton ion gas lasers and various dye lasers exhibit the lowest divergence of 0.4–1.5 and 0.3–2 mrad, respectively, while the semiconductor lasers have an oval beam with the largest divergence 200–600 mrad. The beam divergence of molecular gas lasers is 1–7 mrad, the excimer gas lasers have divergence of 2–6 mrad, and for solid-state lasers it varies up to 25 mrad.

1.5.4.4 Focusing a Laser Beam

Conventional light sources have finite dimensions and emit strongly divergent light. Therefore, the image of a lightbulb, for instance, formed by a lens (or a mirror) has a finite size, and only the light that is intercepted by the lens contributes to the image irradiance, as most of the light output of the bulb is lost. A laser beam, in contrast, has a small diameter and is highly collimated. Thus, a lens can focus all the laser power in a spot on the order of the laser wavelength, as the spot's size is limited only by lens aberrations and diffraction. This way, a very high energy density can be achieved in a very small spot, and can be used for laser-material processing (drilling, cutting, material removal, melting, welding, etc.) in many industrial applications, as well as in surgery or other medical applications.

1.5.4.5 Laser Irradiance

Due to both small divergence and small diameter of the laser beam, the irradiance (power per unit area) of a laser is far greater than the conventional light sources. Let us compare the irradiance of a He–Ne laser with an output power of 1 mW and a lightbulb with output power of 100 W. Because of the small divergence of the laser beam, the beam radius is still about 2 mm at a distance 1 m from the laser, and the irradiance would be

$$\text{Irradiance} = \frac{\text{output power}}{\text{illuminated area}} = \frac{0.001\,[\text{W}]}{\pi(0.002\,[\text{m}])^2} = 79.6\ \text{W/m}^2$$

The lightbulb, though, emits in all directions, so the illuminated area is a sphere with an area $4\pi r^2$. The irradiance 1 m from the 100 W lightbulb would be

$$\text{Irradiance} = \frac{\text{output power}}{\text{illuminated area}} = \frac{100\,[\text{W}]}{4\pi(1\,[\text{m}])^2} = 7.96\ \text{W/m}^2$$

Hence, the irradiance 1 m from the lightbulb is a factor of 10 times smaller at 10^5 times larger output power compared with the He–Ne laser.

1.5.5 Laser Systems

In general, lasers are classified by the type of the active media. The most commonly used lasers today are listed below their main emission wavelengths, output power, and mode of operation given in parentheses [15].

1.5.5.1 Gas Active Medium

- Atomic gas lasers include the He–Ne laser ($\lambda = 632.8$ nm, etc.; 0.1–50 mW; and cw) and the He–Cd laser ($\lambda = 325$, 441.6 nm, etc.; 5–150 mW; and cw).
- Ion gas lasers include the argon laser (main lines at $\lambda = 488$ and 514.5 nm, and others from 350 to 530 nm; 2 mW to 20 W; and cw or mode-locked) and the krypton laser ($\lambda = 647.1$ nm and others from 350 to 800 nm; 5 mW to 6 W; cw or mode-locked).
- Molecular gas lasers include the carbon dioxide laser ($\lambda = 10.6\ \mu$m, 3 W–20 kW, and cw or long pulse) and the nitrogen laser ($\lambda = 337.1$ nm).

- Excimer gas lasers include the argon fluoride (λ = 193 nm and up to 50 W average), the krypton fluoride (λ = 248 nm and up to 100 W average), the xenon chloride (λ = 308 nm, etc.; and up to 150 W average), and the xenon fluoride (λ = 351 nm, etc.; and up to 30 W average), all in pulsed output.

In general, gas lasers have better coherence characteristics and are cheaper than other types of lasers. The He–Ne laser is the most economical choice for measurement and control applications based on low-power cw laser output. It operates on a single line at 632.8 nm and the output can be strongly polarized. In addition, it does not require external cooling system; it is very simple to operate and has a long life. However, commercial He–Ne lasers oscillate in two to five longitudinal modes, depending on the power and have limited coherence length. Argon and krypton ion lasers deliver high-power output with extended coherence length. A single line can be selected from the multiline output of the Ar^+ or Kr^+ lasers, if the end mirror of the laser cavity is replaced by a prism, while a single longitudinal mode can be isolated by incorporating a Fabry–Perot etalon in the cavity.

1.5.5.2   Liquid Active Medium

The group of liquid lasers consists of various dye lasers (tunable λ = 300–1000 nm, 20 mW–1 W average, and cw or ultrashort pulsed). With a given dye, the operating wavelength can be tuned within a range of 50–80 nm by a wavelength selector, such as a diffraction grating or birefringent filter, incorporated in the cavity. Changing the wavelength over a wide range is obtained by switching dyes.

1.5.5.3   Solid-State Active Medium

The group of solid-state lasers include the Nd:YAG laser (λ = 1.064 μm, up to 10 kW average, and cw or pulsed), the Nd:glass laser (λ = 1.06 μm), Ti-sapphire laser (tunable λ = 660–1000 nm, ~2 W average power, and cw or ultrashort pulsed), and the erbium:fiber laser (λ = 1.55 μm, 1–100 W, and cw or pulsed). A Nd:YAG laser is pumped optically with flashlamp, arc lamp, or diode laser. The advantage of using a diode laser to pump a Nd:YAG with a frequency doubling crystal is that such a system is more compact and lighter. In addition, it delivers a cw output of green light (λ = 532 nm) with a coherence length of few meters and output power up to 100 mW. The ruby laser (λ = 694.3 nm) was the first laser. It is pumped by a flashlamp with pulse durations ranging from a fraction of a millisecond to a few milliseconds. It delivers output with pulse energy up to 100 J/pulse. It can operate in single mode or multimode and can also be Q-switched. The ruby laser is still the most widely used pulsed laser for optical holography, because of the large output energy and the match of the laser wavelength to the peak sensitivity of commercial photographic materials for holography [159].

1.5.5.4   Semiconductor Active Medium

Although semiconductor lasers employ solid-state media, they are usually presented in a separate group because of their specific construction, operation, and properties. Typical representatives of this group of lasers are GaAs and GaAlAs (both emitting at λ = 780–900 nm and 1 mW to several watts), and InGaAsP lasers (λ = 1100–1600 nm and 1 mW to ~1 W). They operate on the principle of edge-emitting LEDs and surface-emitting LEDs, with the main difference that the semiconductor lasers have polished ends or use the cleaved surfaces at the ends of the laser crystal, providing optical feedback, thus a laser cavity. The semiconductor lasers operate with relatively low power input and are very efficient. They are very small, diffraction limited point source, and therefore do not require a spatial filter. They can be made to operate in a single longitudinal mode and deliver cw or pulsed output. The emission wavelength range depends on the alloy composition. The main disadvantage of this group of lasers is the large beam divergence in comparison with the other types of lasers.

The choice of a laser depends on the specific application. Most of the optical metrology techniques employ lasers with light output in the visible range such as He–Ne, He–Cd (441.6 μm), Ar^{3+}, Kr^{3+}, tunable dye, Ti:sapphire, and alexandrite (both in the deep red range) lasers. Techniques that are based on interferometry and holography can employ pulsed lasers in order to eliminate problems connected with the stability of the recording system. Optical sensing and communication applications involving fibers use mainly the Nd:YAG laser. The most powerful CO_2 laser is widely used in industrial applications such as material processing (laser cutting, welding, drilling, etc.), and almost all of the lasers presented above find applications in different medical procedures. The list of laser applications involving interaction of laser light with matter and laser processing of information is very long and diverse and is out of the scope of this text. The reader can find abundant information in books devoted on laser applications [149,159–162].

REFERENCES

1. Boyd, R.W., *Radiometry and Detection of Radiation*, Wiley, New York, 1983.
2. DeCusatis, C., *Handbook of Applied Photometry*, American Institute of Physics (AIP) Press, Springer Verlag, New York, 1997.
3. Rea, M. (Ed.), *Lighting Handbook: Reference and Application*, 8th edn., Illuminating Engineering Society of North America (IESNA), New York, 1993.
4. International Commission on Illumination (CIE), *The Basis of Physical Photometry*, Publication No. 18.2, Technical Report, Paris, France, 1983.
5. Žukauskas, A., Shur, M.S., and Gaska, R., *Introduction to Solid-State Lighting*, John Wiley & Sons, Inc., New York, 2002, Chapter 2.
6. *American National Standard Nomenclature and Definitions for Illuminating Engineering*, ANSI Standard ANSI/IESNA RP-16 96, 1996.
7. *The International System of Units (SI)*, 8th edn., International Bureau of Weights & Measures (BIPM), STEDI Media, Paris, France, 2006.
8. Cromer, C.L. et al., The NIST detector-based luminous intensity scale, *Journal of Research of the National Institute of Standards and Technology*, **101**, 109, 1996.
9. Bergh, A.A. and Dean, P.J., *Light-Emitting Diodes*, Clarendon Press, Oxford, U.K., 1976.
10. Gillessen, K. and Schaiper, W., *Light-Emitting Diodes: An Introduction*, Prentice Hall, Englewood Cliffs, NJ, 1987.
11. Stringfellow, G.B. and Craford, M.G. (Eds.), *High Brightness Light Emitting Diodes*, Semiconductors and Semimetals, Vol. 48, Series Eds. R.K. Willardson and E.R. Weber, Academic Press, San Diego, CA, 1997.
12. Schubert, E.F., *Light-Emitting Diodes*, Cambridge University Press, Cambridge, U.K., 2003.
13. MacAdam, D.L. (Ed.), *Colorimetry—Fundamentals*, SPIE Optical Engineering Press, Bellingham, WA, 1993.
14. Wyszecki, G. and Stiles, W.S., *Color Science: Concepts and Methods, Quantitative Data and Formulae*, Wiley, New York, 2000.
15. Pedrotti, F.L., Pedrotti, L.S., and Pedrotti L.M., *Introduction to Optics*, 3rd edn., Pearson Prentice Hall, Upper Saddle River, NJ, 2007, p. 421.
16. Judd, D.B., Report of U.S. Secretariat Committee on colorimetry and artificial daylight, *Proceedings of the 12th Session of the CIE*, Bureau Central de la CIE, Stockholm, Paris, Vol. 1, 1951, p. 11.
17. Vos, J.J., Colorimetric and photometric properties of a 2-deg fundamental observer, *Color Research and Application*, **3**, 125–128, 1978.
18. Brindley, G.S., The color of light of very long wavelength, *Journal of Physiology*, **130**, 35–44, 1955.
19. CIE, *Commission Internationale de l'Eclairage Proceedings, 1931*, Cambridge University Press, Cambridge, U.K., 1932.
20. Stiles, W.S. and Burch, J.M., Interim report to the Commission Internationale de l'Éclairage Zurich, 1955, on the National Physical Laboratory's investigation of colour-matching, *Optica Acta*, **2**, 168–181, 1955.
21. Palmer, J.M., Radiometry and photometry: Units and conversions, in *Handbook of Optics*, 2nd edn., Part III, Ed. M. Bass, McGraw-Hill, New York, 2001.
22. Möller, K.D., *Optics*, University Science Books, Mill Valley, CA, 1988, p. 384.

23. Hodapp, M.V., Application of high-brightness light-emitting diodes, *High Brightness Light Emitting Diodes*, Eds. G.B. Stringfellow and M.G. Craford, Semiconductors and Semimetals, Vol. 48, Series Eds. R.K. Willardson and E.R. Weber, Academic Press, New York, 1997.

24. Rea, M.S. (Ed.), *The IESNA Lighting Handbook*, IESNA, New York, 2000.

25. Coaton, J.R. and Marsden, A.M. (Eds.), *Lamps and Lighting*, Arnold, London, U.K., 1997, p. 546.

26. Schmid, W. et al., High-efficiency red and infrared light-emitting diodes using radial outcoupling taper, *IEEE Journal of Selected Topics in Quantum Electronics*, **8**, 256, 2002.

27. Johnson, H.L. and Morgan, W.W., Fundamental stellar photometry for standards of spectral type on the revised system of the Yerkes spectral atlas, *The Astrophysical Journal*, **117**, 313–352, 1953.

28. Cousins, A.W.J., Revised Zero Points and UBV photometry of stars in the Harvard E and F regions, *Monthly Notices of the Royal Astronomical Society*, **166**, 711, 1974.

29. Cousins, A.W.J., Standard stars for VRI photometry with S25 response photocathodes, *Monthly Notes of the Astronomical Society of Southern Africa*, **33**, 149, 1974.

30. Landolt, A.U., UBVRI photometric standard stars in the magnitude range $11.5 < V < 16.0$ around the celestial equator, *The Astrophysical Journal*, **104**, 340, 1992.

31. Landolt, A.U., Broadband UBVRI photometry of the Baldwin-Stone southern hemisphere spectrophotometric standards, *The Astrophysical Journal*, **104**, 372, 1992.

32. Kutner, M.L., *Astronomy: A Physical Perspective*, Harper & Row Publishers, New York, 1987, Chapter 18.

33. Baum, W.A., in *Problems of Extra-Galactic Research*, Ed. G.C. McVittie, IAU Symposium No. 15, Mac Millan Press, New York, 1962, p. 390.

34. Bolzonella, M., Miralles, J.-M., and Pelló, R., Photometric redshifts based on standard SED fitting procedures, *Astronomy and Astrophysics*, **363**, 476–492, 2000.

35. Eccles, M.J., Sim, M.E., and Tritton, K.P., *Low Light Level Detectors in Astronomy*, Cambridge University Press, Cambridge, U.K., 1983.

36. Hall, D.S., Genet, R.M., and Thurston, B.L., *Automatic Photoelectric Telescopes*, The Fairborn Press, Mesa, AZ, 1986.

37. Henden, A.A. and Kaitchuk, R.H., *Astronomical Photometry*, Van Nostrand Reinhold Company, Inc., New York, 1982.

38. Wolpert, R.C. and Genet, R.M., *Advances in Photoelectric Photometry*, Vol. 1, Fairborn Observatory, Patagonia, AZ, 1983.

39. Golay, M., *Introduction to Astronomical Photometry*, Reidel Publishing Company, Boston, MA, 1974.

40. Low, F. and Rieke, G.H., The instrumentation and techniques of infrared photometry, *Methods of Experimental Physics*, **12**, 415, 1974.

41. Robinson, L.J., The frigid world of IRAS—I, *Sky and Telescope*, **69**(1), 4, 1984.

42. Schorn, R.A., The frigid world of IRAS—II, *Sky and Telescope*, **69**(2), 119, 1984.

43. CIE Publication No. 15.2, Colour and Vision, *Colorimetry*, 2nd edn., 1986.

44. Ohno, Y., CIE fundamentals for color measurements, *Proceedings of the IS&T NIP16 Conference*, Vancouver, CA, October 16–20, 2000.

45. CIE Publication No. 13.3, *Method of Measuring and Specifying Colour Rendering Properties of Light Sources*, 1995.

46. Ohno, Y., Photometry and standards, *OSA/AIP Handbook of Applied Photometry*, Optical Society of America, Washington DC, 1997, Chapter 3, pp. 55–99.

47. ISO/CIE 10526-1991, CIE standard colorimetric illuminants, 1991.

48. ISO/CIE 10527-1991, CIE standard colorimetric observers, 1991.

49. Walker, J. et al., Spectral irradiance calibrations, *NIST Special Publication, 250-20*, National Institute of Standards and Technology, Gaithersburg, MD, 1987.

50. CIE Publication No. 17.4/IEC Publication 50(845), *International Lighting Vocabulary*, 1987.

51. Gray, D.F., The inferred color index of the Sun, *Publications of the Astronomical Society of the Pacific*, **104**(681), 1035–1038, 1992.

52. The Simbad Astronomical Database Rigel page, http://simbad.u-strasbg.fr/simbad/sim-basic?Ident=Rigel&submit=SIMBAD+search.

53. Bessell, M.S., *Publications of the Astronomical Society of the Pacific*, **102**, 1181, 1990.

54. Williams, E.W. and Hall, R., *Luminescence and the Light Emitting Diode*, Pergamon Press, New York, 1978, p. 241.

55. Weber, M.J. (Ed.), *Selected Papers on Phosphors, Light Emitting Diodes, and Scintillators: Applications of Photoluminescence, Cathodoluminescence, Electroluminescence, and Radioluminescence*, Milestone Series, Vol. 151, SPIE Optical Engineering Press, Bellingham, WA, 1998.

56. Lakowicz, J.R., *Principles of Fluorescence Spectroscopy*, 3rd edn., Springer, New York, 2006.
57. Masters, B.R., *Selected Papers on Multiphoton Excitation Microscopy*, Milestone Series, Vol. 175, SPIE Optical Engineering Press, Bellingham, WA, 2003.
58. Saleh, B.E.A. and Teich, M.C., *Fundamentals of Photonics*, 2nd edn., John Wiley & Sons, Inc., Hoboken, NJ, 2007.
59. Fox, M., *Quantum Optics: An Introduction*, Oxford University Press, New York, 2006.
60. Klauder, J.R. and Sudarshan, E.C.G., *Fundamentals of Quantum Optics*, Benjamin, Inc., New York, 1968; Dover Publications, Inc., Mineola, New York, reissued 2006.
61. Louisell, W.H., *Quantum Statistical Properties of Radiation*, John Wiley & Sons, Inc., Hoboken, NJ, 1973; reprinted 1990.
62. Saleh, B., *Photoelectron Statistics*, Springer-Verlag, New York, 1978.
63. Ullrich, B. et al., The influence of self-absorption on the photoluminescence of thin film CdS demonstrated by two-photon absorption, *Optics Express*, **9**(3), 116–120, 2001.
64. Ullrich, B., Munshi, S.R., and Brown, G.J., Photoluminescence analysis of p-doped GaAs using the Roosbroeck-Shockley relation, *Semiconductor Science and Technology*, **22**, 1174–1177, 2007.
65. Mueller, G. (Ed.), *Electroluminescence I*, Semiconductors and Semimetals, Vol. 64, Series Eds. R.K. Willardson and E.R. Weber, Academic Press, New York, 2000.
66. Mueller, G. (Ed.), *Electroluminescence II*, Semiconductors and Semimetals, Vol. 65, Series Eds. R.K. Willardson and E.R. Weber, Academic Press, New York, 2000.
67. Yacobi, B.G. and Holt, D.B., *Cathodoluminescence Microscopy of Inorganic Solids*, Plenum, New York, 1990.
68. Norman, C.E., Reaching the spatial resolution limits of SEM-based CL and EBIC, *Microscopy and Analysis*, **16**(2), 9–12, Cambridge, U.K., 2002.
69. Galloway, S.A., Miller, P., Thomas, P., and Harmon, R., Advances in cathodoluminescence characterisation of compound semiconductors with spectrum imaging, *Physica Status Solidi* (C), **0**(3), 1028–1032, 2003.
70. Parish, C.M. and Russell, P.E., Scanning cathodoluminescence microscopy, in *Advances in Imaging and Electron Physics*, Vol. 147, Ed. P.W. Hawkes, Academic Press, San Diego, CA, 2007, p. 1.
71. Frenzel, H. and Schultes, H., Luminescenz in ultraschallbeschickten Wasser, *Zeitschrift für Physikalische Chemie*, **B27**, 421, 1934.
72. Young, F.R., *Sonoluminescence*, CRC Press, Boca Raton, FL, 2005.
73. Gaitan, D.F. et al., Sonoluminescence and bubble dynamics for a single, stable, cavitation bubble, *Journal of the Acoustical Society of America*, **91**, 3166, 1992.
74. Putterman, S.J., Sonoluminescence: Sound into light, *Scientific American*, **272**(2), 46–51, 1995.
75. Matula, T.J. and Crum, L.A., Evidence for gas exchange in single-bubble sonoluminescence, *Physical Review Lett*ers, **80**, 865–868, 1998.
76. Brenner, M., Hilgenfeldt, S., and Lohse, D., Single bubble sonoluminescence, *Reviews of Modern Physics*, **74**, 425, 2002.
77. Taleyarkhan, R.P. et al., Evidence for nuclear emissions during acoustic cavitation, *Science*, **295**, 1868, 2002.
78. Rauhut, M.M., Chemiluminescence, in *Kirk-Othmer Concise Encyclopedia of Chemical Technology*, 3rd edn., Ed. M. Grayson, John Wiley & Sons, New York, 1985, p. 247.
79. Barnett, N.W. et al., New light from an old reagent: Chemiluminescence from the reaction of potassium permanganate with sodium borohydride, *Australian Journal of Chemical Education*, **65**, 29–31, 2005.
80. Bollyky, L.J. and Rauhut, M.M., American Cyanamid, U.S. Patent No. **3,597,362**, Generation of light by the reaction of esters of oxalic-type acids, 1971.
81. Tsuji, A. et al. (Eds.), Bioluminescence and chemiluminescence: Progress and perspectives, *Proceedings of the 13th International Symposium*, World Scientific Publishing Co., Singapore, 2005.
82. Sweeting, L.M. et al., Crystal structure and triboluminescence II. 9-Anthracencecarboxylic acid and its esters, *Chemistry of Materials (ACS)*, **9**, 1103–1115, 1997.
83. Wheelon, A.D., *Electromagnetic Scintillation*, Cambridge University Press, Cambridge, U.K., 2001.
84. Tsai, K.B., Developments for a new spectral irradiance scale at the national institute of standards and technology, *Journal of Research of the National Institute of Standards and Technology*, **102**, 551–558, 1997.
85. Waymouth, J.F., *Electric Discharge Lamps*, The MIT Press, Cambridge, MA, 1971.
86. Elenbaas, W., *The High Pressure Mercury Vapor Discharge*, North Holland Publishing Co., Amsterdam, the Netherlands, 1951.
87. Pankove, J.I., *Optical Processes in Semiconductors*, Dover Publications, Inc., New York, 1971.

88. Kasap, S.O., *Principles of Electronic Materials and Devices*, 2nd edn., McGraw Hill, New York, 2002.
89. Schubert, E.F. and Hunt, N.E.J., 15,000 hours stable operation of resonant-cavity light-emitting diodes, *Applied Physics A*, **66**, 319, 1998.
90. Grundmann, M., *The Physics of Semiconductors*, Springer-Verlag, Berlin, Germany, 2006.
91. Craford M.G. et al., Radiative recombination mechanism in GaAsP diodes with and without nitrogen doping, *Journal of Applied Physics*, **43**, 4075, 1972.
92. Groves, W.O., Herzog, A.H., and Craford, M.G., GaAsP electroluminescent device doped with isoelectronic impurities, US Patent Re. 29,845, 1978.
93. Holonyak, N., Jr. et al., The "direct-indirect" transition in Ga(AsP) *p–n* junctions, *Applied Physics Letters*, **3**, 47, 1963.
94. Logan, R.A., White H.G., and Wiegmann W., Efficient green electroluminescent junctions in GaP, *Solid State Electronics*, **14**, 55, 1971.
95. Steranka, F.M., AlGaAs red light-emitting diodes, in *High Brightness Light Emitting Diodes*, Eds. G.B. Stringfellow and M.G. Craford, Semiconductors and Semimetals, Vol. 48, Academic Press, San Diego, CA, 1997.
96. Ishiguro, H. et al., High-efficient GaAlAs light-emitting diodes of 660 nm with double heterostructure on a GaAlAs substrate, *Applied Physics Letters*, **43**, 1034, 1983.
97. Ishimatsu, S. and Okuno, Y., High-efficiency GaAlAs LED, *Optoelectronics—Devices and Technologies*, **4**, 21, 1989.
98. Chen, C.H. et al., OMVPE growth of AlGaInP for high-efficiency visible light-emitting diodes, in *High Brightness Light Emitting Diodes*, Eds. G.B. Stringfellow and M.G. Craford, *Semiconductors and Semimetals*, Vol. 48, Academic Press, San Diego, CA, 1997.
99. Kish, F.A. and Fletcher, R.M., AlGaInP light-emitting diodes, in *High Brightness Light Emitting Diodes*, Eds. G.B. Stringfellow and M.G. Craford, Semiconductors and Semimetals, Vol. 48, Academic Press, San Diego, CA, 1997.
100. Krames, M.R. et al., High-brightness AlGaInN light emitting diodes, *Proceedings of the SPIE*, **3938**, 2, 2000.
101. Nakamura, S. and Fasol, G., *The Blue Laser Diode: GaN Based Light Emitters and Lasers*, Springer, Berlin, Germany, 1997.
102. Nakamura, S. and Chichibu, S.F. (Eds.), *Introduction to Nitride Semiconductor Blue Lasers and Light Emitting Diodes*, Taylor & Francis, London, U.K., 2000.
103. Strite, S. and Morkoc, H., GaN, AlN, and InN: A review, *The Journal of Vacuum Science and Technology*, **B10**, 1237, 1992.
104. MacAdam, D.L., Maximum attainable luminous efficiency of various chromaticities, *Journal of the Optical Society of America*, **40**, 120, 1950.
105. Ivey, H.F., Color and efficiency of luminescent light sources, *Journal of the Optical Society of America*, **53**, 1185, 1963.
106. Schlotter, P. et al., Facbrication and characterization of GaN/InGaN/AlGaN double heterostructure LEDs and their application in luminescence conversion LEDs, *Materials Science and Engineering*, **B59**, 390, 1999.
107. Kaufmann, U. et al., Ultraviolet pumped tricolor phosphor blend white emitting LEDs, *Physica Status Solidi* (a), **188**, 143, 2001.
108. Guo, X., Graff, J.W., and Schubert, E.F., Photon-recycling semiconductor light-emitting diode, *IEDM Technical Digest*, **IEDM-99**, 600, 1999.
109. Schlotter, P., Schmidt R., and Schneider, J., Luminescence conversion of blue light emitting diodes, *Applied Physics*, **A64**, 417, 1997.
110. Schubert, E.F. et al., Resonant-cavity light-emitting diode, *Applied Physics Letters*, **60**, 921, 1992.
111. Schubert, E.F. et al., Highly efficient light-emitting diodes with microcavities, *Science*, **265**, 943, 1994.
112. Streubel, K. and Stevens, R., 250 Mbit/s plastic fiber transmission using 660 nm resonant cavity light emitting diode, *Electronics Letters*, **34**, 1862, 1998.
113. Liu, Y., Passive components tested by superluminescent diodes, *WDM Solutions*, February, 41, 2000.
114. Wu, C.-C., *Applied Physics Letters*, February 20, 081114, 2006, and March 13, 111106, 2006.
115. Osram Opto Semiconductors, *Laser Focus World*, Santa Clara, CA, March 2008, p. 102.
116. Commission Internationale de l'Éclairage, Measurements of LEDs, CIE Publication 127, 1997.
117. *Handbook of LED Metrology*, Instrument Systems GmbH, Version 1.1, München, Germany. http://www.instrumentsystems.com/fileadmin/editors/downloads/products/LED_Handbook_e.pdf.
118. Ohno, Y., Fundamentals in photometry and radiometry II—Photometers and integrating spheres, *CIE LED Workshop and Symposium*, Vienna, Austria, 1997.

119. Market Reports, High-brightness LEDs: The new trend in illumination, *Photonics Spectra*, January, 92, 2006.
120. Market Reports, The bright side of the LED market, *Photonics Spectra*, January, 90, 2006.
121. Philips LumiLeds LUXEON K2 with TFFC Technical Datasheet DS60, October 2008.
122. Shchekin, O. and Sun, D., Evolutionary new chip design targets lighting systems, *Compound Semiconductor*, **13**, 2, 2007.
123. Hecht, J., *Understanding Fiber Optics*, Prentice Hall, Upper Saddle River, NJ, 2001.
124. Keiser, G., *Optical Fiber Communications*, 3rd edn., McGraw-Hill, New York, 1999.
125. Schubert, E.F. et al., Temperature and modulation characteristics of resonant-cavity light-emitting diodes, *IEEE Journal of Lightwave Technology*, **14**, 1721, 1996.
126. Okamoto, K., Application of super-bright light emitting diodes to new type solar cell, *Technical Digest, The International PVSEC-5*, Kyoto, Japan, 1990.
127. Okamoto, K., A new system of two-way optical communication using only light-emitting diodes, The Institute of Electronics, Information and Communication Engineers, Japan, *Third Symposium on Optical Communication Systems, OCS*, **89–1S-11S**, 21–26, 1989.
128. Oyama, T. and Okamoto, K., Two-way visible light LED/LED optical communication system using a pair of half-mirror reflex cameras, *Extended Abstarcts, Optics and Photonics 2007*, Osaka, Japan, 2007, pp. 408–409.
129. Okamoto, K. and Tsutsui, H., Light-emitting solar cell with sun-tracking mechanism, *Technical Digest, The International PVSEC-7*, Nagoya, Japan, 1993.
130. Miyazaki, E. and Okamoto, K., High speed light pulse generator using a superbright LED, *Proceedings of the 33rd SICE (Society of Instrument and Control Engineers) Annual Conference*, Tokyo, Japan, July 26–28, 1994, pp. 801–806.
131. Mills, R.W., Blue light-emitting diodes—Another method of light curing, *British Dental Journal*, **178**(5), 169, 1995.
132. Kamano, H. et al., Photodynamic purging of Leukemia cells by high-brightness Light Emitting Diode and Gallium-Metal Porphyrin, *IEEE Proceedings of the CLEO/Pacific Rim*, Piscataway, NJ, 1999, pp. 1006–1007.
133. Okamoto, K., Kamano, H., and Sakata, I., Development of a LED apparatus for photodynamic purging of Leukemia cells in hematopoietic stem-cell transplantation, *Transplantation Proceedings*, Elsevier, **32**, 2444–2446, 2000.
134. Okamoto, K., Kameda R., and Obinata, K., Development of novel phototherapy system for neonatal jaundice using superbright blue and bluish green light emitting diodes, *Japanese Journal of Medical Electronics and Biological Engineering*, **35**(Suppl.), 144, 1997.
135. Seidman, D.S. et al., A new blue light-emitting phototherapy device: A prospective randomized controlled study, *Journal of Pediatrics*, **136**(6), 771–774, 2000.
136. McCree, K.J., The action spectra, absorbance and quantum yield of photosynthesis in crop plants, *Agricultural Meteorology*, **9**, 191–196, 1992.
137. Bula, R.J. et al., Light-emitting diodes as a radiation source for plants, *HortScience*, **26**, 203–205, 1991.
138. Barta, D.J. et al., Evaluation of light-emitting diode characteristics for a space-based plant irradiation source, *Advances in Space Research*, **12**, 141–149, 1992.
139. Okamoto, K. and Yanagi, T., Development of light source for plant growth using blue and red superbright LEDs (in Japanese), *1994 Shikoku-Section Joint Convention Record of the Institute of Electrical and Related Engineers*, 1994, p. 109.
140. Okamoto, K., Yanagi, T., and Takita, S., Development of plant growth apparatus using blue and red LED as artificial light source, *Proceedings of the International Symposium of Plant Production in Closed Ecosystems, Acta Horticulturae, International Society for Horticultural Science*, **440**, 111–116, 1996.
141. Tanaka, M. et al., In vitro growth of Cymbidium plantlets cultured under superbright red and blue light-emitting diodes (LEDs), *Journal of Horticultural Science and Biotechnology*, **73**(1), 39–44, 1998.
142. Brown, C.S., Schuerger, A.C., and Sager, J.C., Growth and photomorphogenesis of pepper plants under red light-emitting diodes with supplemental blue or far-red lighting, *Journal of the American Society of Horticultural Science*, **120**(5), 808–813, 1995.
143. Tripathy, B.C. and Brown, C.S., Root-shoot interaction in the greening of wheat seedlings grown under red light, *Plant Physiology*, **107**(2), 407–411, 1995.
144. Okamoto, K., Baba, T., and Yanagi, T., Plant growth and sensing using high-brightness white and red light-emitting diodes, *CLEO 2000 Conference Edition*, IEEE/OSA, San Francisco, CA, May 7–12, 2000, pp. 450–451.

145. Okamoto, K., Novel applications of high-brightness blue/green light-emitting diodes in the fields of horticulture, agriculture, forestry, fishery, medical science, microbiology, and traffic technology, *Record of the 16th Electronic Materials Symposium*, Osaka, Japan, July 9–11, 1997, pp. 269–272.
146. Okamoto, K. et al., Development of fishing lamp using bluish-color light emitting diode, *National Convention Record of The Institute of Electrical Engineers of Japan*, **1**, 373, 2001.
147. Hashimoto, Y. et al., Development of squid-fishing lamp using blue LED and its experiment on the ocean, *Extended Abstracts Optics Japan 2003, Optical Society of Japan*, Shizuoka, Japan, December 8–9, 2003.
148. Fujita, J. et al., Development of LED fishing lamp with low energy consumption, *Extended Abstracts Optics Japan 2005, Optical Society of Japan*, Tokyo, Japan, November 23–25, 2005.
149. Silfvast, W.T., *Laser Fundamentals*, 2nd edn., Cambridge University Press, Cambridge, U.K., 2004.
150. Siegman, A.E., *Lasers*, University Science, Mill Valley, CA, 1986.
151. Milonni, P.W. and Eberly, J.H., *Lasers*, John Wiley & Sons, Inc., New York, 1988.
152. Willett, C.S., *Introduction to Gas Lasers*, Pergamon Press, New York, 1974.
153. Splinter, R. and Hooper B.A., *An Introduction to Biomedical Optics*, Taylor & Francis Group, LLC, New York, 2007.
154. Scully, M.O. and Zubairy, M.S., *Quantum Optics*, Cambridge University Press, Cambridge, U.K., 1997.
155. Hecht, E., *Optics*, 4th edn., Pearson Education, Inc., Addison Wesley, San Francisco, CA, 2002.
156. Bennett, C.A., *Principles of Physical Optics*, John Wiley & Sons, Inc., Hoboken, NJ, 2008.
157. Klein, M.V. and Furtak, T.E., *Optics*, 2nd edn., John Wiley & Sons, Inc., Hoboken, NJ, 1986, p. 475.
158. Born, M. and Wolf, E., *Principles of Optics*, 6th edn., Pergamon Press, New York, 1980.
159. Hariharan, P., *Optical Holography, Principles, Techniques, and Applications*, 2nd edn., Cambridge University Press, Cambridge, U.K., 1996.
160. Waynant, R.W., *Lasers in Medicine*, CRC Press, Boca Raton, FL, 2001.
161. Weber, M.J., *Handbook of Lasers*, CRC Press, Boca Raton, FL, 2000.
162. Ready, J.F. (Editor in Chief), *LIA Handbook of Laser Materials Processing*, 1st edn., Laser Institute of America, Magnolia Publishing, Inc., Orlando, FL, 2001.

Websites

http://www.nichia.com/product/index.html.
http://www.superbrightleds.com/specs/w18015_specs.htm.
http://www.cree.com/products/led_docs.asp.
http://www.lantislaser.com/oct_dentistry.asp.
http://www.lumileds.com/products.

2 Lenses, Prisms, and Mirrors

Peter R. Hall

CONTENTS

2.1 Introduction 87
2.2 Wave-Front and Wave-Front-Aberration 87
2.3 Primary Aberrations and Design Start Points 88
2.4 Optical Design 88
2.5 Optical Design Tools 89
2.6 Optical Design Theory 89
2.7 Primary Aberrations 91
2.8 Thin Lens Seidel Aberration Formulae 92
2.9 Lenses 94
2.10 Optical Materials 95
2.11 Stock Tooling 97
2.12 Production Tolerances 97
2.13 Aspheric Surfaces 98
2.14 Testing and Single-Wavelength Systems 99
2.15 Prisms 99
2.16 Image Orientation 100
2.17 Porro Prism 101
2.18 Tunnel Diagram 102
2.19 Pechan and Abbe Prisms 103
2.20 Relay Lenses 104
2.21 Telecentric Systems 104
2.22 Conclusions 105
References 105

2.1 INTRODUCTION

This chapter attempts to give the reader an understanding of the underlying theory that enables the design of lenses, with a practical introduction to how the optical industry operates. It provides design techniques for analyzing and designing prism systems. The ultimate cost, shape, and performance of an instrument have as much to do with the design of its optics as with the concept underlying the technique. It is therefore important to understand not just the purpose of the various optical components but also the limitations of a simplistic design approach.

2.2 WAVE-FRONT AND WAVE-FRONT-ABERRATION

Optical engineers use the word aberration to refer to both the defects of an optical system and the defects in the image of transmitted light beam that they cause.

When light is emitted from a point, that is, an infinitely small area of a source, it propagates as sinusoidal waves emitted from that point on the surface. Until they encounter optical components the waves are spherical, centered on the point source.

We may choose the peak of any one of the waves and follow it through the system until it forms an image. This single wave peak, chosen at random and frozen in time, is what we think of as a wave-front, and consists of points that are all at a constant optical path distance from the start point. Each point on the wave-front propagates in the direction of the normal to the wave-front at that point. Therefore, to be heading toward a single perfectly formed image point, all places on the wave-front must have surface normals that are pointing toward the same single point, and for this to be true the wave-front must be perfectly spherical. Any departure from perfect spherical form will deviate light from its path to the perfect image.

Departure from perfect sphericity of the wave-front is known as the wave-front aberration or optical path difference (OPD) and is a fundamental measure of the system's defects. It is expressed in wavelengths of the light that is used.

To interpret the wave-front aberration, it may be expanded as a power series in aperture and field. Such a power series can take many forms, such as that which yields the conventional Seidel aberrations (the lowest order terms of one form of the series) or the more mathematical representations in Zernike or Buchdahl notations. The notation used is not particularly important (although Buchdahl is a career rather than a design tool). In general, designers use only the lowest order terms of the series, and understanding their role is an essential step in the creation of any new design.

2.3 PRIMARY ABERRATIONS AND DESIGN START POINTS

The builders of early optical instruments observed that the aberrations were not random but had nameable forms that could be identified as having specific characteristics. This resulted from the symmetries that exist about the axis of an optical system. Many forms of aberration cannot exist in a rotationally symmetric lens system because they lack the correct symmetry properties relative to the axis. At any monochromatic wavelength, symmetry reduces the number of lowest-term (primary) aberrations to just five types. Another two primary aberrations describe the effects of wavelength on the system's focal plane position and magnification. If the performance is bad enough, other more complex errors become large enough to be visible. However, if a system is to be of high performance then the five primary aberrations must be well corrected. If they are, then it is likely that the higher-order terms will also be reasonably well controlled. Correction beyond this level of performance is largely the domain of the automated optical design program. However it must be understood that it is the designer's responsibility to provide a reasonably well-corrected start point for the automatic optimization program. The reader may ask, "why is it important that the starting point be good, surely achieving a good result is what the optimization program is for?" The optimization program is a navigation system. The design starts off at one point and the design program seeks a route to the most perfect solution possible. It does this by continually improving the system. It can never make it worse.

The difficulty is then analogous to setting off from the top of a mountain in search of the sea. The steepest route down the mountain will not necessarily lead to the sea. If the summit was a volcano, it is possible to head the wrong way initially and simply arrive at the bottom of the crater. The program can only guide us downhill, it can never take us uphill again even if the destination is only a mountain pass away. If we are lucky, the program may lead us eventually to the sea. On the other hand, we may end up in a lake with no outlet, a local minimum from which there is no route to either the sea or to the best design, we may even be trapped in just a slight hollow.

2.4 OPTICAL DESIGN

Optical design started in the nineteenth century and many generations of designers have created thousands of lenses, and many have been published or patented. Thus in books [1,2] and patents, and in databases (compiled largely from patents in the public domain), the designer may find good starting points. The design may still need extra components or a change in emphasis from angular coverage to wider aperture, but at least the sea can be seen from the starting point.

The problem occurs when a truly new design is needed. The first person to design a submarine periscope and the people tasked with correcting the Hubble telescope all needed some reliable design tools, some mountaineering skills to get them off the mountain in one piece.

2.5   OPTICAL DESIGN TOOLS

The tools that are used came into existence nearly a century ago. From the beginning, lens design has been a computation-intensive process. Lens designs were among the first civilian products to benefit from the use of digital computers. By the early 1950s, datasets were already flown from Britain to the United States to be analyzed on the Manhattan Project's computers at Los Alamos. Today, much more powerful computers can be found on every engineer's desk and their use in optical design has passed from analysis, almost without thought, to optimization. Immensely powerful computer programs such as ZEMAX (Zemax Development Corporation) and CODE-V (Optical Research Associates) are able to optimize designs in minutes that would have taken months or years in earlier times. However, their very speed risks the classic problem of rubbish-in–rubbish-out. Although tremendous progress has been made in the development of high-speed ray tracing and efficient optimization algorithms, very little has been achieved in creating software to synthesize starting points. This is still the responsibility of the designer.

Before digital computers, optical designers had to decide exactly what they needed to know, before they could pass the problem to a room of people who performed laborious calculations with pencil and paper, taking days or weeks. The designer had spare time to think about the design, to understand its underlying defects, and correct them in as few design iterations as possible. This is still a desirable design strategy.

If an automatic design program is permitted to choose expensive glass types, create hemispherical surfaces, or produce a design with extremely tight tolerances, it is still a bad idea. Fifty years on, it is still better to study drawings of the system and first-order analyses, in an attempt to understand where in the system problems are arising.

The drawings that the programs can create are particularly useful to the designer. If surfaces are becoming hemispherical or the rays are meeting surfaces at grazing incidence, then the design is almost certainly in trouble (or the user is an experienced designer working on a very difficult problem, and does not need advice!)

2.6   OPTICAL DESIGN THEORY

All optical design depends on the ability to trace rays through a system, and for that we require Snell's law:

$$n_1 \sin i_1 = n_2 \sin i_2. \tag{2.1}$$

A ray incident at an incidence angle i_1 (to the surface normal) in a medium of refractive index n_1, refracts into a medium of refractive index n_2 at an angle i_2 to the surface normal, see Figure 2.1 and Equation 2.1.

Equation 2.1, when generalized into three dimensions, is the basis of all ray tracing and as might be expected, an expression in terms of the sines of angles in three dimensions leads to very little insight as to what is happening in the design. For that we need a simplification. We need the paraxial approximation.

If the ray is considered to be so close to the optical axis that its angles, and incidence heights are so small that the angles and their sines are proportional to their tangents, then we are in

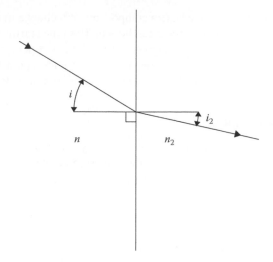

FIGURE 2.1 Refraction at a surface, Snell's law.

what is known as the paraxial region and many formulae can be simplified to much more interpretable paraxial forms. Snell's law, for example, can be represented as in the following equation:

$$n_1 i_1 = n_2 i_2. \tag{2.2}$$

Paraxial approximations are potentially very useful. To use them, the angles need not actually be small, provided we remember that if they are not small then this is truly an approximation and the results cannot be relied upon for total accuracy. However, formulae based on this approximation will generally give a good prediction of where the focal planes will be, of image dimensions, and of approximately how large the lens apertures will need to be.

The paraxial approximation does give an accurate computation of the primary aberrations. Even better, for each of the seven primary aberrations, the total for the whole lens is the sum of individual surface contributions. The paraxial approximation permits these individual surface contributions to be calculated, enabling the designer to see where in the system problems are arising.

If in addition we assume that not only is the paraxial approximation true but also that the lenses have zero center thickness, then this is known as the thin lens approximation. The height of the paraxial ray is the same at both surfaces of a lens element and further simplifications result. Systems can be sketched as a series of simple line objects as shown in Figure 2.2, where we see two lenses of power k (the power k is the reciprocal of the focal length, and they are subscripted 1 and 2). The lenses are separated by a distance d_1. A ray is incident at a height h_1 and at an angle u_1 at the first

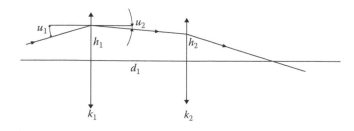

FIGURE 2.2 Refraction of a ray at a thin lens.

lens where it is refracted and leaves lens 1 at an angle u_2. The path of the ray through lens 1 arriving at lens 2 at a height h_2 can be calculated using the following simple formulae:

$$h_1 k_1 = u_1 - u_2 \tag{2.3}$$

$$h_2 = h_1 + d_1 u_2. \tag{2.4}$$

The notation is Cartesian. A good account of this theory and that of paraxial aberrations, paraxial theory, and the Seidel aberrations may be found in Welford [3].

The thin lens layout is an ideal point in the synthesis of the design for the designes to confirm that they have provided sufficient space for fold mirrors and prisms, that any windows or filters will not be uneconomically large, and a variety of other fundamental considerations.

2.7 PRIMARY ABERRATIONS

In principle, the five monochromatic primary aberration totals for the whole lens describing its performance over the whole field and for all apertures. The mix of aberrations at a particular point of the field of view, and for a given F/number of the lens will result from the power series variables in aperture and field.

The designer only needs to minimize the five aberration totals, and then all variants of aperture and field of view should be corrected. Just five numbers describe the whole monochromatic lens performance. It is this remarkable simplification that permits designers to understand even complex systems.

In reality, higher-order aberrations may well become important at large field angles and apertures, but if the designer has the five primary aberrations under control, they can fairly and safely leave the optimization program to attempt to correct the higher orders.

The five aberrations are spherical aberration, coma, astigmatism, field curvature, and distortion. Although they are all of fourth order (in aperture and field), the extent to which the fourth order consists of aperture terms or field terms changes as we go through the list.

Spherical aberration is independent of field, and is constant over the whole field. By contrast, distortion is independent of aperture, so it does not matter whether the lens is used at $F/1$ or $F/32$, the distortion will not change. Coma is less dependent on aperture but is linearly dependent on field. Astigmatism and field curvature are dependent on the square of aperture (being essentially defocus in effect) and on the square of field angle.

For an image of a very small star-like source, spherical aberration is the only aberration that can occur on axis, it appears as a symmetrical circular blur of light. Depending on the sign of the aberration total, it will be a small disk on one side of focus and a small bright ring on the other.

Coma manifests as a small comet tail. The nominal image point is at the point of the comet tail. The tail gets wider as it grows away from the nominal image point. These wider regions are light that came from the outer zones of the lens aperture, so that if the lens is closed down from say $F/1$ to $F/4$, the coma seen will greatly reduce. The point of the tail remains fixed but the extent of the tail decreases.

As has been mentioned, the primary aberrations are the only series terms that have the correct symmetry properties, so that if the small comet tail points upward in the upper half of the scene then it points downward in the lower half, maintaining symmetry. Any term that would not have done this would have been excluded from the power series. Thus the coma term is linear in field so that it changes symmetrically across the optical axis.

Astigmatism and field curvature are often grouped as a pair because their origins and consequences intertwine.

Consider a test chart consisting of radial lines and concentric circles. If astigmatism is present and there is no field curvature, then as we move out across the field we find increasingly that we can either focus the radial lines or we can focus concentric circles, but we cannot get both to focus at the same time. One astigmatic focus will be nearer to the lens, and the other slightly further away relative to the nominal focal plane. Which way around these are dependent on the sign of the astigmatism total.

Now consider the same chart for a lens that has zero astigmatism but a nonzero field curvature total. Now as we progress across the field, we find that we can focus both circles and radial lines at the same time, but as we move outward across the field, the place of common focus moves away from the nominal focal plane. A focal surface of good imagery exists, but it is either concave or convex. That is, there is a field curvature. Whether the surface is convex or concave depends on the sign of the field curvature aberration total for the lens. Now we shall see why these two terms are often combined.

Consider a lens in which the astigmatism and field curvature totals are both nonzero. In this case, we find two curved focal surfaces of different curvatures. On one curved focal surface, the radial lines are in focus, and on the other the concentric circles are in focus. It is useful to note that if we cannot correct astigmatism or field curvature fully, we may introduce the other in just sufficient quantity to cause the surfaces to be equally concave and convex. That is, the two surfaces depart from the flat nominal focal plane, symmetrically, one in front of the plane and the other behind it. This compromise generally gives the best result when either astigmatism or field curvature cannot be fully corrected. Field curvature is also commonly known as Petzval curvature or Petzval sum.

Distortion is a variation of magnification with field angle. If we double the field angle, then the distance of the image from the optical axis, across the focal surface, should also approximately double. The effect of distortion is to cause the image to be further (or less) from the center of the image plane than it should be. The effect of distortion increases as the field angle increases. It may be unnoticeable for small field angles but become very pronounced at the edge of the field. The effect is not necessarily small and may be very intrusive. Zoom lenses for news gathering often have large amounts of distortion and can create a characteristic seasickness effect as the camera is swept across the scene, the motion of the scene increases as it reaches the edge of the field of view. It is such a large effect that a television zoom lens was once marketed as ×13 because inward distortion at narrow angle combined with outward distortion at wide angle to give ×13 change of field angle, even though the change in focal length was only ×10.

The above description of the primary monochromatic aberrations is intended to give the reader a feel for what is meant by the terms used by designers. There are many textbooks, such as those by Welford [3] and Born and Wolf [4] that the reader can refer to for a more mathematical account. Unfortunately the subject is one where the mathematics tends to obscure rather than illumine. This short chapter cannot cover the subject adequately. There are very few good texts on practical optical design. Laikin [1] is excellent as is Kingslake [2] although Kingslake's notation is archaic. The conventional routes into the subject are either the small number of master degree courses (Imperial, London and University of Arizona typically) or to be taught on the job by an experienced designer.

A small number of engineers find it an interesting challenge to teach themselves and I admire them, but am grateful to Charles Wynne and Gordon Cook who taught me!

2.8   THIN LENS SEIDEL ABERRATION FORMULAE

Formulae exist, derived by Wynne [3,5], from which the primary aberrations of a thin lens system can be calculated, based on glass types and basic constructional data. They were derived by Wynne [5] nearly a century after Seidel derived the surface-by-surface formulae. Optical design is not a subject that changes quickly.

In these equations, the radii of curvature of each lens become a single variable known as the shape or bend of the lens. The resulting sets of biquadratic equations in the bend describe the available solutions. Solving large sets of simultaneous biquadratic equations does not lead to clarity of understanding. In fact, analytical solution of the equations is impractical for systems more complex than a doublet. A triplet can be solved, but requires an iteration to choose the glass type for the middle lens. It lies beyond the scope of this chapter but the author has found that this form of theory is much more useful when studying aspheric lens systems. Because the equation sets are simpler (those for the aspheric terms are linear rather than quadratic), generic equations that describe whole families of useful designs can be derived.

The equations for individual lens elements do describe the way in which the aberration totals for that element will vary with shape, depending on whether they are before or after the stop of the system. Novice designers can program these and will find them helpful. Later, they will not need to actually use the formulae explicitly, because knowing what they would predict in general terms is sufficient.

The fundamental design variable in these formulae is the shape or bend of the lens. If a lens has surface radii of curvature r_1 and r_2 then (where $c = 1/r$) the shape or bend B is given by

$$B = \frac{c_1 + c_2}{c_1 - c_2}.\tag{2.5}$$

Figure 2.3 shows the cross section of a lens altering with bend (it also illustrates why designers refer to the process as bending a lens element).

Frequently, the aberration totals for an optical element are quadratic in bend. That is, they are represented by a parabolic curve. This means that there are different bends at which each aberration has an extremum (a maximum or minimum as the case may be) and these extrema cannot be exceeded. Very often, zero does not lie within the scope of the parabola. So that, for example, the spherical aberration of a single element lens cannot generally be set to zero by choosing its bend. To achieve zero, either one surface must become aspheric or a positive lens must be paired with a negative lens, and their shapes chosen so that the negative contribution of one cancels the positive contribution of the other, and so the path to system complexity begins.

Although laying out a complex system (consisting of separated groups of simple lenses) as a series of thin lenses (each representing a complete group, see, e.g., Figures 2.15 and 2.16 later in the chapter) can be helped by paraxial and thin lens formulae, it is still too great an approximation to assist in laying out a complex lens such as a microscope objective or camera lens.

This degree of simplification neither will predict the resolution of the system nor will it cope accurately with ray paths at large angles to the axis. However, it will always be true that if the paraxial ray paths in a thin lens layout predict excessively large angles and apertures, then the design problems that will result will be nontrivial and other layouts should be considered before proceeding.

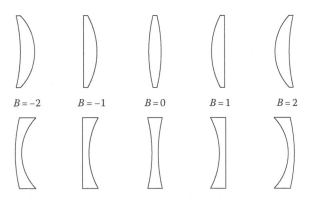

FIGURE 2.3   Shape or bend B of a lens element.

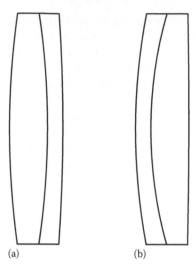

(a)   (b)

FIGURE 2.4   (a) Fraunhofer and (b) Steinheil doublets.

The calculation of the Seidel aberrations, the lowest terms of the polynomial that describes the wave-front as it exits the lenses, is provided within all well-known optical design programs. It shows very reliably the surfaces creating largest problems, helping greatly with the process of altering the structure in search of a better starting point.

We have said that we cannot rely on the program to correct a poorly chosen starting point. A simple example may be sufficient to demonstrate the problem. The dispersion is a measure of the rate of variation of refractive index with wavelength, the Abbe number, see below.

Figure 2.4 shows two lenses of identical focal length and aperture. In one lens, the high-index high-dispersion element (the negative component) leads, and in the other it comes second. These are both achromatic doublets and in both the two most important aberrations (spherical aberration and coma) have been corrected. Whether the negative lens first (the Steinheil doublet) or the negative lens second (the Fraunhofer doublet) is the better solution depends on many factors, sometimes one is better, sometimes the other.

The important thing to appreciate is that once one lens or the other has become the starting point, the optimization program is unable to reach the other solution. To do so it would need to reverse the glass types and there is no way to do so without passing through a situation where either the refractive indices or dispersions are equal. Optimization programs are constrained to choose paths along which designs get better, rather than ones that get worse and then get better.

So, if in this simplest of cases we cannot get from even a reasonably well-chosen starting point to a better one, then it is equally likely that a more complex system will have little chance of overcoming a fundamentally badly chosen starting point, but we shall have much less chance of guessing why.

It is the designer who must make manual drastic changes to the basic design structures. If glass types need reversing, or lens elements need to be added, then that must be done manually as a result of insight on the part of the designer. Hammer and Global optimization routines that use random number strategies to attempt to locate better solutions exist, but may be better left until a designer has no idea what to do next.

2.9 LENSES

A lens may be as simple as a magnifying glass, or as complex as a zoom lens containing 20 or more individual lenses. The individual lens components are more properly referred to as lens elements, and may be thick or thin, and formed of plastic, glass, or crystal. Their surfaces may be plano (flat), spherical, or nonspherical (aspheric). In GRIN materials, the refractive index varies radially. Finally, the lens may not actually be present but may have its effect resulting from a hologram.

In an introductory chapter such as this, we cannot cover all of these possibilities. By the time the more esoteric options are needed, the reader will inevitably be in contact with specialist suppliers and will have access to all the information and advice that they need.

2.10   OPTICAL MATERIALS

The materials that can be used for lenses and prisms are determined by the wavelength to be used. Lenses and prisms for the infrared and ultraviolet are generally made from crystals, the choice is limited, and data for them are tabulated in the design programs. For the visible waveband, they utilize the optical glasses that have been developed specifically for the purpose. It has been known for centuries that adding lead oxide to the recipe for a glass, produces a much denser glass that has a higher refractive index and a greater dispersion. The dispersion of a glass is the rate at which its refractive index changes with wavelength, and is important because it determines the element's contribution to chromatic aberration. Similarly, when used in a prism it is the dispersion that causes the well-known prismatic effect, the rainbow spectrum that was first explained by Isaac Newton.

There is an advantage in using a slightly higher refractive index in total-internal-reflection (TIR) prisms, because the higher index increases the range of angles at which TIR can occur. Increasing the index of the glass is generally a limited option because the higher index glasses are more expensive and the pieces required for prisms are quite large.

Higher index glasses are generally less expensive if the designer can accept that the dispersion increases. Glass suppliers provide charts showing their selection of glass types, plotted against refractive index n and dispersion (the Abbe number V that decreases as the dispersion increases) as shown in Figure 2.5. The glasses shown are those of the Japanese company Hoya. The chart illustrates the effort required to create the glasses at the top left of the chart. The natural line for glasses to follow is that of the right-hand boundary. This boundary line results from gradually adding more and more lead oxide to a basically window-glass recipe. To extend types to the upper left of the chart requires exotic additives such as Lanthanum, resulting in the expensive glass types that optical designers prize.

To create such a plot, it is first necessary to choose three wavelengths, the first refractive index n_1 is in the middle of the waveband of interest and two more are representative of the waveband at

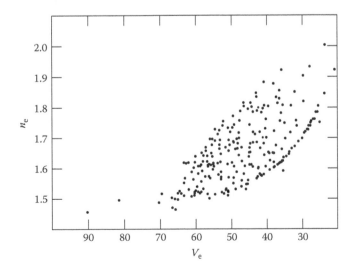

FIGURE 2.5 n–V glass chart.

shorter and longer wavelengths. Glass manufacturers conventionally use the spectral d or e lines for the first wavelength. However, if designing for the near infrared or the ultraviolet you will need to recalculate (and perhaps replot) the n and V values because changes of waveband can change the relationships between glasses quite dramatically.

The Abbe number V is defined by numbering the wavelengths referred to as 1, 2, and 3, then,

$$V = \frac{n_1 - 1}{n_2 - n_3}. \tag{2.6}$$

In lenses too, a higher refractive index is often useful. In most cases, the contribution that a lens element (or surface) makes to the total aberration reduces as the index (and cost) increases. The automatic optimization program tends to increase the index if it can. To avoid increasing the chromatic aberration, the program may also alter the glass type to have both a high index and a low dispersion, increasing the material cost.

Most design programs permit the designer to set boundaries to the types of glass that it may choose, and it is possible to manually block the use of specific glasses. Glass suppliers can provide a list of guide prices, normalized relative to the classic crown glass 517642. This list should be at your right hand. It can be very difficult to revert to cheaper glass types later.

While discussing glass types, it is worth noting that it is wise for the lens designer to have good relations with the glass supplier's local agent. The availability of glass types varies with time. Glass is made in batches depending on demand, and it is quite a commonplace for even a common glass type to be in worldwide shortage. Glass types may also be discontinued, without notice and for no apparently good reason. However, that glass type may be in sufficient stock worldwide that its termination will not have any effect for a long time. The supplier's agent is your best source of information on all such matters.

Moldings are often cheaper than sawn block, because sawn block involves wastage and additional shaping processes at First-World labor rates. Molding allows the glass supplier to minimize waste, and may be carried out cheaply by a Third-World company.

Finally, beware of equivalents. Many glass types are intended to be equivalents such that one supplier's glass can be used in place of another's. This equivalence is often only partially true. Two glasses may have the same refractive index n_e and the same V value. However, a detailed check may reveal that the partial dispersion, thermal expansion, density, chemical durability, or spectral-transmission curve may be different. This is even true for a specific supplier if they have changed the recipe of the glass, for example, while attempting to reduce the lead content.

There is a great advantage in being able to visit your usual lens supplier to discuss glass types, coating options, and details of the design (such as the edge thickness and the ratio of diameter to lens thickness). This is not necessarily the cheapest way to procure lenses but it is certainly a way to avoid expensive misunderstandings. The cheapest way is to cooperate with Far Eastern suppliers, such as those in the hi-tech zones in China. Careful choice of contact can result in very good prices for lenses in bulk. However, these suppliers frequently require a large minimum order, almost always 20 pieces and often a 100 piece minimum order.

This is suitable for production quantities, but creates the problem of where to buy the small quantities needed for the prototype. Bearing in mind that a prototype is supposed to represent the final product, to build the prototype with components that are not made by the production supplier is risky for two reasons. The first is that new problems at the production stage can turn that into a 100 item prototype batch. The second is more complex and will be covered shortly—the subject of stock tooling.

Experience suggests that if the designer is confident that the design will proceed to production then it may be better to accept the minimum order quantity for the prototype rather than build the prototype with one supplier and then start full-scale production with an as yet untested new supplier.

However, if the system has high performance and contains extreme or expensive components, it may be better to forego low prices in favor of a local supplier with easier communication and a smaller minimum batch.

2.11   STOCK TOOLING

Traditionally, optical polishers use small lens-like glass tools to test components while still in place on the polishing machine [6]. These tools are polished plano on one side and on the other is a very precisely created optical surface. By placing it on the work piece, Newton's rings reveal the difference between the work surface and the test-tool surface. The tools are polished in convex–concave pairs until they are near-perfect spheres (to $\lambda/20$ or so). Any departure from perfection when used on a work piece is then seen to be the current error in the work surface and the polisher uses this information to create a better surface (or figure), a process known as figuring.

In creating these highly precise tools, a great deal of skill is required to arrive at perfection at the same moment as the radius of curvature is the exact value requested by the designer. They are expensive to make and become part of the company's assets. Optical designers are encouraged to use existing stock tools wherever possible.

Optical companies publish their lists of tools, frequently as datasets within ZEMAX, CODE-V, etc. Designers adjust their final design to suit their chosen supplier's stock-tool list.

This then is the second reason why using different suppliers at the prototype and production stages creates problems. The designer has to fit a new set of stock tools, and the new design may be sufficiently different to require requalifying or the alteration of metal components.

The severity of the problem is also dependent on the sophistication with which the two suppliers are able to calibrate the radii of curvature. Interferometric measurement enables the radii to be known to almost micron precision, certainly better accuracy than the designer requires. Smaller polishing shops may still depend on an older technique, spherometry. In spherometry, a micrometer is used to measure the camber (height) of the surface over a known diameter.

A simple but ill-conditioned equation then allows the radius to be determined. The resulting radii are not known to be much better than 0.01% in the best circumstances and may well be in error by 1%.

In modern practice, both in the West and the Far East, surface testing can be done by interferometry. The radius is dialed into the instrument with an accuracy much better than spherometry. The figure is then tested relative to the designer's data radius. In general, however, it is still more convenient and cheaper for glass tools to be used, and the designer is still encouraged to use stock radii. One or two surfaces with difficult stock tool problems may be left for interferometric testing.

Unusually in engineering practice, the designer will generally specify a radius of curvature with no tolerance assigned to it. By doing so, they are specifying not just a radius but a specific tool set.

2.12   PRODUCTION TOLERANCES

The designer will of course need to assign tolerances for all other dimensions. We might expect the errors in manufacture to have a normal distribution, that is, a bell-shaped error curve centered on the nominal value. That would misunderstand the nature of the role of tolerances and dimensional errors in the traditional lens-making process.

Optical craftsmen work to accuracies that are much better than those on the drawings. The larger tolerances on the drawing are not there because the process is imprecise, they are there for a completely different reason—to reduce scrap.

For example, a thickness may be specified as 10.1–9.9 mm. What then is the most likely value if a batch is measured? A production engineer would say 10.0. In fact, the glass polisher is working to

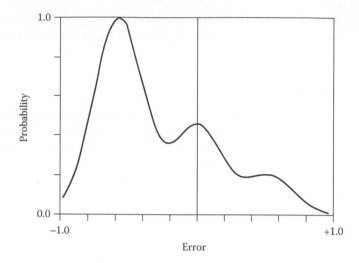

FIGURE 2.6   Distribution of yield within optical tolerance limits.

personal tolerances of tens of micrometers. They know that glass is prone to scratching and that if scratched at the moment that the batch of components is finished, the scrap item will need to be put aside, to become part of a rework batch, all of which will be slightly thinner than those that came out of the first round. There may even be a second rework batch that are even thinner.

The polisher therefore aims for the first attempt to get a thickness just under the top limit, the second attempt to be around the mean dimension, and the third attempt to be just inside the lower limit.

The distribution is thus, not a bell-shaped normal distribution centered on the mean but a stepped distribution with the greatest number just under the top limit, see Figure 2.6. A designer who trusts the supplier to supply mainly first-attempt items may decide to bias the tolerances so that the upper limit is nearer to the actual value required. The lower limit is then lower than the value that can really be accepted, but is unlikely to occur. The designer accepts that there will be occasional unacceptable systems that need rectification, but the component cost will be kept down overall, and better systems will result because each component is on average better than if unbiased tolerances were used. This strategy works well when companies are polishing for their own products, but requires unusual rapport between workshop and designer. It does not work well if the purchaser or seller has adopted an aggressive strategy.

2.13   ASPHERIC SURFACES

Rubbing two surfaces together randomly produces spherical surfaces by a process that has not changed significantly in seven centuries. To create a nonspherical (aspheric) surface requires much more dexterity in polishing or expensive machine tools. The use of aspheric surfaces can significantly reduce the number of components in a lens system [7–9], and their use should be seriously considered when the optical material is expensive, or expensive mold tools need to be made. Molded aspheric surfaces in glass and plastic have been used in millions in a DVD player. They are also used in glass optics to control higher-order aberration and distortion in expensive photographic and television objectives, provided they are manufactured in sufficient quantity for economies of scale to reduce the tooling costs to an acceptable level. They are common in cheap plastic optics because they reduce the number of mold tools required. Aspheric-surface costs do not matter so much for molds because even spherical surfaces may need aspheric mold surfaces, to allow for differential shrinkage.

2.14 TESTING AND SINGLE-WAVELENGTH SYSTEMS

Lens systems tend to be specified in terms of resolution. The equipment required to make this measurement is often built around commonly available lasers such as the HeNe 632.8 nm. A problem then arises when designing a system that only has a very good performance at the single wavelength of use, if the wavelength of use is a long way away from 632.8 nm. This can make it difficult or impossible to test the system adequately with any of the equipment that the company already possesses. The method of testing should always be in the designer's mind at an early stage. If it will be very difficult, expensive, or impossible to test the lens system then this is a serious consideration.

2.15 PRISMS

The simplest prisms disperse light into a spectrum or act as beam splitters. These are well known and will not be dealt further here. More importantly, prisms can displace or rotate the direction of light propagation. This may simply be for design convenience, or it may be to invert, revert, or rotate the orientation of an image.

An advantage of a prism (relative to a mirror) is that it can often reduce the volume occupied by the system. This is shown in Figure 2.7. Here, the requirement is to alter the direction of propagation by 45°. A mirror is seen to be large compared with an equivalent Schmidt prism. Not only are the prism surfaces smaller because reflections within the prism are at less oblique angles but also the light beam is folded, reducing the space required for the system.

Clean prisms have almost zero loss at their TIR faces, while rear surface mirrors suffer absorption losses caused by the metal–glass interface. Front-surface mirror-coatings have higher reflectivity but the cost is greater and the surfaces are much less durable.

Prisms are inevitably expensive, and this is partly because many surfaces are used at oblique angles. Surfaces that are aligned normal to the optical axis are cheaper to polish. The reason is that a small change in focus can correct the effects of the usual surface tolerance of 1.5 wavelengths of curvature error. The optical polisher, using glass contact tools then only needs to get to within 1.5 wavelengths of the correct form to be able to detect 0.25 wavelength of elliptical error, the part of the tolerance that affects resolution.

However, this is not true of an oblique surface. For example, the mirror shown in Figure 2.7 is at 67.5° so the circular beam has an elliptical footprint on the surface, the major axis being 2.6 times the minor axis of the ellipse. Thus to a first order, a spherical form error of 1.5λ across the major axis would be decreased by a factor of 6.8 (i.e., 2.6^2) to 0.22λ across the minor axis. As a result, after reflection, one section of the beam has 6.8 times less wave-front error than the other section. The resulting astigmatism cannot be corrected by refocusing. A much tighter surface-figure tolerance should be specified for an oblique surface, with a consequent increase in cost.

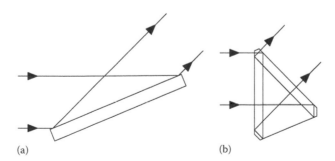

(a)   (b)

FIGURE 2.7   Comparison of (a) mirror and (b) Schmidt prism.

Introducing a prism into the light path affects the design of the rest of the system. When light is focused through a prism, such as a prism in the back-focal space of a lens, the prism will introduce; chromatic aberration, spherical aberration, coma, and astigmatism, and these need to be allowed for in the design of the lens itself.

2.16   IMAGE ORIENTATION

Despite their cost, prisms have a vital role to play in correctly orientating the beams, in ways that can be difficult to reproduce using only lenses. We therefore need to understand how prisms operate, and also how to interpret what they are doing in our system.

If the deviation of light in a right-angle prism is in a horizontal plane then the image, just as with a mirror, remains upright but is reverted side to side. This is intuitively correct because the right-angle prism is just a mirror with glass in front of it.

But, how can we determine the image orientation in a more complex prism system? The right-angle prism is a good place to start. Many texts show prisms with a letter F entering and exiting a prism, thereby defining the beam orientation on exit for that prism type.

It is important to note that the direction in which the F is viewed is important. Both pictures in Figure 2.8 are correct. However, they appear to contradict. The problem is that Figure 2.8a is oriented so that the left-hand (entry) F is shown from behind while the right-hand (exit) F is shown in front. In Figure 2.8b, a more suitable orientation of the prism shows both the Fs from the direction in which they are propagating and the side-to-side reversion that would be expected is now seen. Extending this technique to a drawing of a more complex prism can be difficult.

An alternative simple way to determine the orientation of an image uses no more than a long narrow strip of paper that has a line drawn down one edge. Lay the paper strip on a drawing of the prism, and at each prism surface fold the paper along the reflection surface of the prism. The new direction of the strip at each fold is the direction of the beam after reflection. Eventually, at the point where the strip exits the prism, the position of the line along the edge of the paper defines whether or not the image has inverted. The technique is shown in use for the right-angle prism in Figure 2.9.

If the side-to-side reversion has been determined by the first folding, then any top–bottom inversion of the image can be interpreted from a second folding as shown in Figure 2.9b. From these, it is clear that the image has not inverted vertically but has only reverted horizontally. This technique requires practice and a little dexterity but is very useful for interpreting practical systems.

If the beam is actually focusing toward an image plane, with perhaps intermediate images within the train of prisms, then it may be necessary to create a scale print of the straightened optical system (i.e., with reflections removed but suitable blocks of glass to represent the presence of the prism in

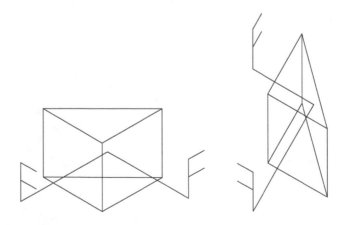

FIGURE 2.8   Image orientation in a right-angle prism.

FIGURE 2.9 Folded paper-strip technique.

the system). The printout can be folded, as above, leading to a clear understanding of how the prisms interact with the optical system.

Ultimately, a fully three-dimensional nonsequential ray-trace can be set up, which will create enormous amounts of information about the passage of light through the prisms. The reader is cautioned, however, that enormous amounts of information are best created after the designer understands their system.

2.17 PORRO PRISM

Almost all pairs of binoculars contain four Porro prisms. They create the familiar cranked shape. This common prism illustrates key design techniques.

A simple telescope (i.e., just an objective lens and an eyepiece) creates an image that is inverted vertically and reverted horizontally, the image has been rotated by 180°. The telescope becomes even longer if a relay stage is added to correct the orientation of the final image. During the Napoleonic wars, ship's telescopes were known as "bring-'em nears" and the image was in fact inverted.

Using Porro prisms to rotate the image has the additional benefit of folding the system to make it more compact. The Porro prism, shown in Figure 2.10a is simply a right-angle prism in which the light enters normal to the hypotenuse surface, is totally internally reflected at the two smaller faces, and exits via the hypotenuse surface. The Porro prism is shown in detail in Figure 2.10b. The paper-strip diagram in Figure 2.10a shows that in passing through a Porro prism whose long axis is vertical, the image has inverted. An image that required both inversion and reversion would now only require reversion. Conveniently, the light is now traveling away from the eyepiece. Therefore, a second Porro prism is placed with its long axis horizontal (Figure 2.10c), the image is reverted and when seen by the eyepiece, the scene has been rotated right way up and right way round, and the light is also traveling toward the eyepiece. The Porro-prism pair has folded the telescope into the familiar binocular shape, a much more convenient shape for use.

Unlike the right-angle prism, that conventionally has square apertures, the Porro prism normally has hemi-cylindrically shaped ends, as seen in Figure 2.10b. This feature has a number of useful functions. Firstly, the removal of the glass reduces weight and volume. Secondly, the rounded ends are much more robust in an instrument that is frequently dropped. Thirdly, it is also easier to

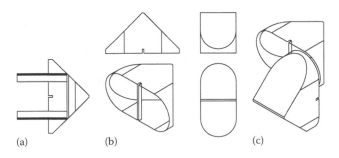

FIGURE 2.10 Porro prism.

machine the housing to have rounded sockets to suit the ends of a Porro prism than it is to make a socket for a square prism. The difference may be small, but over hundreds of thousands of items the small savings add up, and it can be molded.

There is a chamfer along the line where the two small faces intersect. This protects the corner from chipping and provides a stronger edge across which to place a sprung metal strip to hold the prism in place. Commonly, the prisms are initially held just by the metal strips. Then they are adjusted until both arms of the binocular are looking in exactly the same direction. Finally, the prisms are secured in place with spots of glue.

Thus, the shape of the prism has been determined by a strong, convenient, and cheap method of assembly and adjustment.

2.18 TUNNEL DIAGRAM

The Porro prism has a slot, cut across the center of the hypotenuse face, apparently dividing the hypotenuse into an entry face and an exit face. In fact, it is doing rather more than that, but to understand and analyze its function, it is helpful to review another useful technique available to the designer—the tunnel diagram.

Figure 2.11a shows a general ray propagating through a Porro prism. However, if we were to look in through the entry face we would not see fold after fold, instead, we would see an apparently unfolded space, looking toward an exit face beyond which is the outside world (Figure 2.11b).

The designer initially determines the total glass path within the prism, and enters it in the optical design program as a simple block of glass, that fully represents the prism, as it relates to resolution, aberrations, etc.

However, to understand the relationship between the light beams and the physical features of the prism we need to do the opposite of the paper-strip technique. That is, instead of folding paper to represent light's folded path through the prism, we may instead unfold the prism, so that light appears to pass through each reflection surface into another piece of glass. Just as with the paper strip, the reflection surface acts as a fold line and the shape of the next glass space within the prism (or at least a mirror image of it) is created, as what is known as a tunnel diagram (Figure 2.11b).

Let us use Figure 2.11b to consider the complex shape that has been developed. On viewing through the entry face, a glass–air interface is seen to one side, parallel to the line of sight, and on the other side two triangular glass spaces, each of which has a number of alternative and potentially incorrect exits from the prism, are seen. Various features also exist, created by the chamfers that are cut into the prism. This tunnel really does exist; it is not just an engineering fiction. If a Porro prism is available, the reader can immediately see that this really is what the prism looks like from the viewpoint of the light that is passing through it.

The tunnel diagram is useful for checking the relationship between the margins of the beam and the chamfers. The chamfers not only need to be far enough out from the beam to allow it to pass

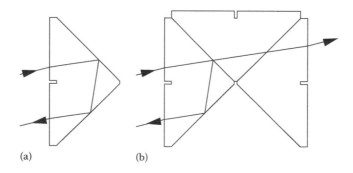

(a)   (b)

FIGURE 2.11 Porro prism tunnel diagram.

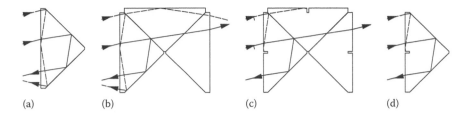

(a)　　　　　　(b)　　　　　　　　　(c)　　　　　　　　(d)

FIGURE 2.12 Ghost path within the Porro prism.

but also need to be far enough in to cut off access to the triangular side spaces through which light could pass creating ghost images. In particular, this diagram explains the slot across the middle of the hypotenuse face.

Figure 2.12a shows a Porro prism that has no slot, and we see that if a ray enters too close to the outer edge of the aperture, it can reflect at grazing incidence on the inside of the hypotenuse face, only after that does it reach the other small surface to reflect and exit through the hypotenuse exit face, but traveling at an angle to its correct path and creating a false or ghost image. These paths could be constructed on the drawing of a single prism, with many reflections to be constructed to gain a full understanding of which rays can follow this false path.

However, if we use the tunnel diagram (Figure 2.12b) it is easy to determine the rays that can reach the hypotenuse surface to follow the false path. There are no complex reflections to construct because light follows a simple path within the tunnel. It immediately becomes clear that the slot is a baffle to trap the false path at the point where it reaches the hypotenuse surface (Figure 2.12c and d). The slot can be drawn on the tunnel diagram to trap as much of the false path as necessary, while ensuring that the slot does not obstruct the straight-through correct path.

In principle, the tunnel diagram can be reduced in length so that its length is equal to the equivalent path in air. This allows the designer to ignore refraction at the entry and exit faces. This sounds convenient. However, any serious optical design requires that the aberrations due to the presence of the glass be taken into account in the design of the associated lenses. The reduced tunnel is also much more complicated to draw. The author has always found the glass tunnel to be much more useful than the reduced tunnel. The folded-strip technique and the tunnel diagram between them solve most of the problems involved in obtaining a working understanding of a prism system. The concepts can be extended to the folding of complete optical systems using large-scale prints from the optical design program.

2.19　PECHAN AND ABBE PRISMS

Modern binoculars use a variety of straight-through prisms to create small compact systems. An early version of this type of prism is the Pechan. It is shown in Figure 2.13, together with its paper-strip diagram. It has a diagonal airspace just thick enough to support TIR. It can be seen that a very large amount of a system's length can be folded into this compact prism, leading to very small systems, which is why many World War I officers carried these as vest-pocket binoculars.

(a)　　　　　　　　　　　　　　　(b)

FIGURE 2.13 Pechan prism.

FIGURE 2.14   Abbe prism.

Figure 2.14 shows the Abbe prism. It serves a similar function, and shows how two prisms having the same function can create very different profiles for a system.

2.20   RELAY LENSES

Simply creating an image with the correct focal length optics is sometimes not enough to create the ideal system. This is particularly true of lenses that relay image planes. It is also true of those that measure size. In both cases, the designer can significantly improves the system by adding a little sophistication to the layout.

Figure 2.15a shows a series of relay lenses. Clearly, as the light is relayed, less of it is caught by the second lens. Simply relaying the image plane was not sufficient. In Figure 2.15b we see a system in which an extra lens at the intermediate image plane relays the image of the lens aperture. That is, we now have two interleaved relay systems, one relaying the image and the other relaying the lens aperture. It requires no further explanation to appreciate the advantage of adding the second set of lenses. A lens placed at an image plane to relay the aperture is known as a field lens. Incidentally, the surfaces of a field lens should not be quite in the image plane, because any scratches or dust will be in focus and will be added to the scene. The lens should therefore be close to, but not actually at, the image plane.

2.21   TELECENTRIC SYSTEMS

Optical catalogs contain telecentric lenses for measuring systems. Telecentricity is important, both of itself and also as an example of how a measuring system can be improved by insight into how it operates. Telecentric lenses are almost always much larger than a simple camera lens of the same focal length. Often, they are also of much larger F/number. A camera lens may be F/1.4 but F/4 is quite a good aperture for a telecentric lens. They are more expensive, so a designer may well opt for the cheap lens with the bigger signal. Such a simplistic choice is shown as a thin lens layout in Figure 2.16a.

(a)

(b)

FIGURE 2.15   Relay systems.

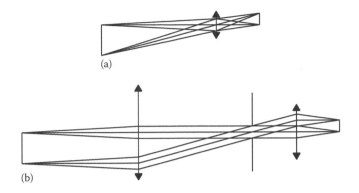

FIGURE 2.16 Telecentric lenses.

Here we see a simple lens (perhaps a doublet or a cheap camera lens). It is imaging an object with a magnification of −0.5. The designer hopes that they have a perfect system for measuring the object by determining its size on a camera chip. Measuring a reference object and correcting for any distortion, the test object may be measured to a few tens of microns—if it is in focus!

From Figure 2.16a we see that if the object were slightly too near the camera it would appear slightly larger, and if the camera chip were slightly too near the lens, the object would appear slightly too small. Of course, in both cases the image would be blurred by defocus, but from a measurement point of view it is the error in size that is more important. Figure 2.16b shows a much more complex system in which the light from the object to the lens and from the lens to the camera chip both travel substantially parallel to the axis of the instrument. As a result, an out-of-focus image will be blurred but it will not have changed in size, calibration will be correct, and only the resolution would have been degraded.

Photogrammetry lenses for aerial survey normally have the film held by vacuum against a glass plate, to define the film's position accurately. If the vacuum is imperfect the film may be slightly rumpled. That is why all good photogrammetry lenses should have telecentric illumination at the film plane, so that images remain in the correct place. Telecentricity then is the condition in which light leaves an object or approaches the image with its chief (central) ray normal to the object or focal plane.

Incidentally, Figures 2.15 and 2.16 are good examples of the value of using thin lens layouts to think through the underlying problems of a system.

2.22 CONCLUSIONS

In addition to introducing the problems involved in designing an optical system, the opportunity has been taken to provide a practical toolbox of techniques, and things to consider when setting out to design an optical system. Other concepts such as the modulation transfer function and theoretical resolution limits, coherence and partial-coherence, etc., have been omitted. An introductory chapter could not have treated them with any degree of rigor, and it is hoped that the material presented instead is of more practical assistance.

REFERENCES

1. Laikin, M., *Lens Design*, Dekker, New York, 1990.
2. Kingslake, R., *Lens Design Fundamentals*, Academic Press, New York, 1978.
3. Welford, W.T., *Aberrations of Optical Systems*, Adam Hilger Ltd., London, U.K., 1986.
4. Born, M. and Wolf, E., *Principles of Optics*, 6th edn., Pergamon, London, U.K., 1993.
5. Wynne, C.G., Thin lens aberration theory, *Optica Acta*, 8(3), 255–265, 1961.
6. Horne, D.F., *Optical Production*, 2nd edn., Institute of Physics, London, U.K., 1982.

7. Hall, P.R., Use of aspheric surfaces in infra-red optical design, *Optical Engineering*, 26, 1102–1111, 1987.
8. Altman, R.M. and Lytle, J.D., Optical design techniques for polymer optics, *International Lens Design Conference 1980*, *Proc. SPIE*, 237, 380–385, 1980.
9. Kuttner, P., Optical systems with aspheric surfaces for the correction of spherical aberration as main error in the infra-red and visible spectral regions, *Aspheric Optics, Design, Manufacture and Testing Conference*, London, U.K., 1981, *Proc. SPIE*, 235, 7–17, 1981.

3 Optoelectronic Sensors

Motohiro Suyama

CONTENTS

3.1 Introduction 107
3.2 Operating Principle of Optoelectronic Sensors 108
    3.2.1 Conversion of Light to Electrons 108
    3.2.2 Spectral Response 111
3.3 Point Sensors 113
    3.3.1 Semiconductor Sensors for High-Light-Level Applications 113
    3.3.2 Point Sensors for Low-Light-Level Applications 114
    3.3.3 Readout Electronics 120
3.4 Multi-Pixel and Position-Sensitive Sensors 121
    3.4.1 Multi-Pixel Sensors 121
    3.4.2 Position-Sensitive Sensors 122
3.5 Image Sensors and Cameras 123
    3.5.1 Image Sensors 124
    3.5.2 Image Sensors and Cameras for Low-Light-Level Applications 126
    3.5.3 High-Speed Cameras 130
3.6 X-Ray or Gamma-Ray Sensors 131
    3.6.1 Physics of Detection 131
    3.6.2 Practical Sensors 132
3.7 Selection Guide for Optoelectronic Sensors 135
References 136

3.1 INTRODUCTION

Optoelectronic sensors detect electromagnetic radiations such as x-rays, ultraviolet (UV), visible, and infrared (IR) light, ranging from short to long wavelengths. These radiations are converted to electrons by the sensors and output as an electric current. The output current is generally amplified by an amplifier, converted to a digital number by an analog-to-digital converter (ADC), and recorded by a computer, for example. The optoelectronic sensors, the most essential part of this detection process, are discussed in this chapter.

This chapter is written to help readers select the most appropriate sensor for their application. To make a proper selection, readers must know the operating principles of optoelectronic sensors, which are discussed in Section 3.2, including the mechanisms for converting radiation to electrons and the spectral responses of sensors from UV to IR. Sections 3.3 through 3.5 discuss point sensors, multi-pixel/position-sensitive sensors, and image sensors, respectively, for the detection of UV–IR light. In each section, sensors with high gain of the order of 10^6 for low light level as well as those for high light level are described. Since x-rays behave differently from other lights, x-ray sensors are described in an independent section (Section 3.6). By the

FIGURE 3.1 Critical factors in selecting an optoelectronic sensor discussed in this chapter for an application.

end of this chapter, readers should understand the need to take several factors into account such as spectral response, gain, and spatial and time resolutions for the selection of proper sensors, as shown in Figure 3.1. A selection guide to sensors is described in Section 3.7 instead of a summary of this chapter.

3.2 OPERATING PRINCIPLE OF OPTOELECTRONIC SENSORS

Optoelectronic sensors convert light information to electrical signal. Thus, the operation is based on a mechanism for converting light to an electric current. The mechanism and efficiency of the conversion are described in this section.

3.2.1 Conversion of Light to Electrons

Since the smallest units of light and electricity are expressed by a photon and an electron, respectively, we can describe the detection process in those terms. Photons are converted to electrons by optoelectronic sensors and measured as an electric current. More precisely, when a photon is absorbed in a semiconductor material, for example, an electron is excited from a valence band to a conduction band and extracted to a readout circuit as an output signal.

Three processes are used to extract electrons from materials: photovoltaic, photoconductive, and photoelectric effects. The photovoltaic effect occurs in a semiconductor sensor having p- and n-type regions facing each other to produce a p–n junction, as shown in Figure 3.2a. By absorbing photons, electrons are excited to a conduction band while holes are left behind in a valence band, thus generating electron–hole pairs. The electrons diffuse or drift in the material toward the p–n junction and eventually reach the n-type region due to a potential difference built between the n- and p-type regions, as shown in Figure 3.2b [1]. The holes, on the other hand, reach the p-type region because of their positive charge. As a result, an electric current is provided to the outer circuit. This type of optoelectronic sensor is called a photodiode (PD).

Because of its operating principle, a PD can detect photons having energy higher than the bandgap of the material, where the bandgap is the potential difference between the conduction

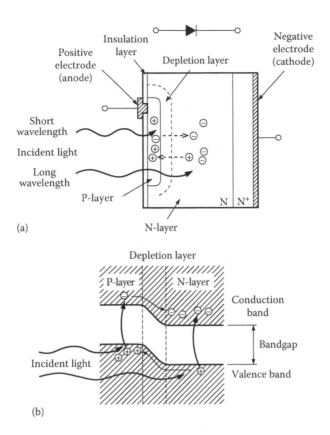

FIGURE 3.2 (a) Structure of a photodiode (PD) using photovoltaic effect. (b) In response to incident photons, electron–hole pairs are generated and separated by potential difference at the p–n junction to output a current to a readout circuit.

and valence bands. The energy of a photon (E [J or eV]) can be described as a function of its wavelength (λ [m]) by the following formula:

$$E = \frac{h*c}{\lambda} \ (J)$$

$$= \frac{h*c}{\lambda*e} \ (eV) \tag{3.1}$$

where
  h is Planck's constant (6.6×10^{-34} Js)
  c is the speed of light (3.0×10^{8} m/s)
  e is the charge of an electron (1.6×10^{-19} C)

In the case of silicon (Si), the most widely used material for PDs, photons of less than 1100 nm can be detected because Si's bandgap is 1.1 eV. InGaAs or InSb is used for longer wavelengths in the IR because of its smaller bandgap. Spectral responses of sensors using these materials are discussed in Section 3.2.2.

The photoconductive effect also generates electron–hole pairs in a semiconductor material in response to incident photons. However, in this case, generated charges decrease the resistivity of the material. By applying a bias voltage to both ends of an effective area, the input light power can be

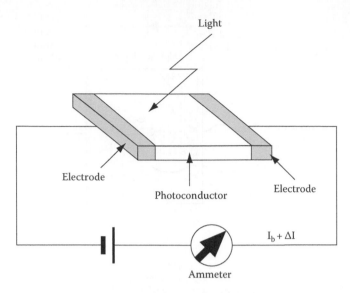

FIGURE 3.3 Schematic drawing of a sensor using photoconductive effect, which varies its resistivity in response to input light. Thus, input light power is measured by an electric current to the output circuit. (From Dakin, J.P. and Brown, R.G., *Handbook of Optoelectronics*, 1, Copyright 2006. Reproduced by permission of Taylor & Francis Group, LLC, a division of Informa plc, Boca Raton, FL.)

measured as a variation of a current to a readout circuit, as shown in Figure 3.3 [2]. An optoelectronic sensor using the photoconductive effect is called a photoconductor, a photoconductive cell, or a photoresistor. Photoconductors made of materials with a small bandgap are mostly used to detect IR light.

The photoelectric effect is an emission of electrons from a material to vacuum in response to input photons when the energy of the photon is larger than the work function of the material. The work function is a potential difference between Fermi and vacuum levels. Although the photoelectric effect occurs in pure metals such as gold (Au) or silver (Ag), the conversion efficiency from a photon to an electron is very low and limited to the UV region. Composite materials generally called photocathodes, which more readily convert photons to electrons, are used in sensors. An alkali–antimonide photocathode made of antimony (Sb) and alkali materials such as sodium (Na), potassium (K), and cesium (Cs) is fabricated on an input surface of a vacuum tube, as shown in Figure 3.4, to detect photons from UV to near-IR light. Photocathodes are used in photomultiplier tubes (PMTs) as discussed in Section 3.3.2.

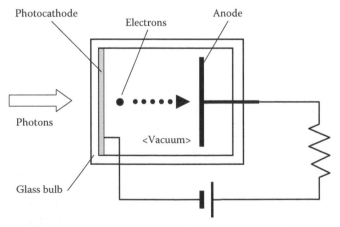

FIGURE 3.4 A phototube has a photocathode, where photoelectric effect occurs. Electrons emitted from the photocathode are received by an anode and output to a readout circuit.

3.2.2 SPECTRAL RESPONSE

As mentioned in the previous section, the conversion of photons to electrons is a key process in opto-electronic sensors. The conversion efficiency, which depends upon wavelength, is called a spectral response and expressed as quantum efficiency (QE). The spectral response is also described as radiant sensitivity, a ratio of the output current and the corresponding input light power. The QE (%) at a specific wavelength (λ [nm]) is related to the radiant sensitivity (S [A/W]) as shown in the following formula:

$$QE = \frac{S*1240*100}{\lambda} \ (\%) \tag{3.2}$$

QEs of photovoltaic and photoelectric sensors for UV, visible, and near-IR regions are summarized in Figure 3.5 as a function of wavelength [1,3]. As shown in this figure, a Si-PD shows a QE of approximately 70% from UV to near-IR, whereas photocathodes are less sensitive than that because the emission probability of electrons to vacuum is small. Recently, the emission probability has been drastically improved, and a QE of 45% around 400 nm is achieved by an "ultra bialkali (UBA)" photocathode. GaAsP photocathodes made from semiconductor crystals reach a QE of 50% in the visible region, as shown in Figure 3.6 [4]. GaAs and InP/InGaAs photocathodes, as shown in the same figure, have high sensitivity in the near-IR region [5,6].

In the case of optoelectronic sensors for IR, D* (D-star) is generally used as a figure of merit to compare the sensitivity of different types of sensors such as PDs and photoconductors. D* is the sensitivity indicating the signal-to-noise ratio (S/N) for AC signal per unit light power, as shown in the following formula:

$$D* = \frac{(S/N)*\sqrt{\Delta f}}{P*\sqrt{A}} \ (cm \ Hz^{0.5}/W) \tag{3.3}$$

where
$\quad$ Δf (Hz) is the bandwidth of the output circuit
$\quad$ P (W/cm^2) is the input light power density
$\quad$ A (cm^2) is the effective area

D* values of several photoconductors and PDs are summarized in Figure 3.7 [7]. Spectral response of mercury cadmium telluride (MCT) depends on its composition, as shown here. The D* of a

FIGURE 3.5 Spectral responses of a Si-PD (photovoltaic effect) and several photocathodes (photoelectric effect) from UV to near-IR region.

FIGURE 3.6 Spectral responses of new photocathodes, such as "ultra bialkali (UBA)," GaAsP, GaAs, and InP/InGaAs photocathodes.

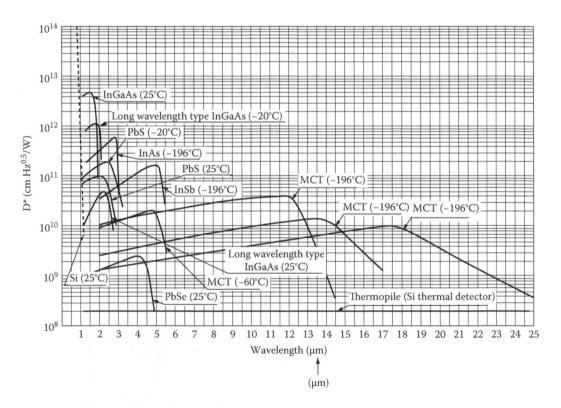

FIGURE 3.7 D* of PDs (photovoltaic effect) and photoconductors (photoconductive effect) with various compound materials are shown. D* is used to compare sensitivities of sensors for IR region.

thermopile, which converts the energy of photons to heat and the variation of temperature is measured by a current, is shown as well.

3.3   POINT SENSORS

A variety of optoelectronic sensors are available for a variety of applications. Point sensors are used for applications that do not require any position resolution. In this section, point sensors for low to high light levels and some electronics for these sensors are described.

3.3.1   SEMICONDUCTOR SENSORS FOR HIGH-LIGHT-LEVEL APPLICATIONS

For the detection of high light level, the most widely used optoelectronic sensor is a PD made of Si, which detects light from 200 to 1100 nm, as shown in Figure 3.5. Based on mature process technology, Si-PDs show a high QE of 70% in the visible region. Si-PDs with effective areas of up to 7.8 cm^2 are available [8]. However, a smaller effective area is better in terms of dark current and response time. The response time (fall time), T_f (s), of a PD relates to the bandwidth, B (Hz), as expressed by the following equations:

$$T_f = \frac{0.35}{B} \text{ (s)} \tag{3.4}$$

$$B = \frac{1}{2*\pi*C*R} \text{ (Hz)} \tag{3.5}$$

where
  R (Ω) is an impedance of a readout circuit
  C (F) is the capacitance of a PD determined by the following equation:

$$C = \frac{\varepsilon_0 * \varepsilon_r * A}{d} \text{ (F)} \tag{3.6}$$

where
  ε_0 is the electric constant (8.9×10^{-12} F/m)
  ε_r is a relative dielectric constant (11.9 for Si)
  A (m^2) is the effective area
  d (m) is the thickness of the depletion layer

The smaller the effective area (A), the faster the time response (T_f). Another way to ensure fast time response is to increase the thickness of the depletion layer, and PIN-PDs are an example of this approach. PIN-PDs have a thick intrinsic (low impurity level) layer that is depleted readily to reduce the capacitance. The bandwidths of PIN-PDs are 1.0 GHz and 60 MHz for effective areas of 0.13 mm^2 [9] and 13 mm^2 [1], respectively.

Because of their stability, reliability, robustness, and price, varieties of Si-PDs are commercialized for various applications such as analytical instruments, medical, illuminometer, and academic research, as shown in Figure 3.8. Si-PDs are also used in cars for automatic head lights, automatic air conditioner, and automatic wipers.

For the detection of IR light whose wavelength is longer than 1000 nm, PDs and photoconductors made of various compound materials are used. Spectral response (D*) is determined by the material, as shown in Figure 3.7. InGaAs shows high sensitivity at wavelengths shorter than 2 μm, InSb is the best around 5 μm, and MCT shows the best D* at wavelengths longer than 5 μm. It is

FIGURE 3.8   Si-PDs commercialized for various applications.

recommended to choose the material showing the highest D* at the wavelength of interest. In general, PDs show better D* and more stable operation than photoconductors. On the other hand, PDs require cooling to reduce dark current, making them more expensive than photoconductors. One of the major applications of IR sensors is in optical communication at 1.3 or 1.55 μm [10], where a loss of optical fiber is minimal. High-speed PDs made of InGaAs are used for this application. IR sensors are also used to measure temperature because thermal objects emit IR radiation [11]. Gas detection is another application in the IR region [11] because each gas has keen absorption in specific wavelengths depending upon its composition or a chemical bond.

3.3.2   Point Sensors for Low-Light-Level Applications

It is very difficult for a PD to detect a single photon because one photon generates only one electron–hole pair in a sensor and the charge of an electron (1.6×10^{-19} C) is very small. An amplifier lowers the detection limit; however, it is still in the order of 10^8 photons/s as determined by the noise of the amplifier, assuming a noise of 1 pA/Hz$^{0.5}$ at a bandwidth of 10 kHz. To lower the detection limit further, sensors should have an internal gain.

An avalanche photodiode (APD) is a sensor with internal gain [12]. As shown in Figure 3.9, the APD has a region of high electric field around the p–n junction where an avalanche multiplication occurs. In this region, charges are accelerated rapidly by a high electric field to produce another electron–hole pair upon interaction with other atoms. This phenomenon is called impact ionization, and successive occurrences lead to an avalanche multiplication. The impact ionization depends heavily on the electric field [13]; thus, the gain of an APD is determined by a bias voltage.

Since the ionization coefficient of an electron is much larger than a hole's [13], Si-APDs are designed so that electrons, not holes, enter the avalanche region. APDs are used at the gain of less than 100 for stable operation, and temperature dependence of the gain should be considered during operation. Many Si-APDs are used as sensors for light detection and ranging, for example, where pulsed light is irradiated, and reflected light from an object is measured by an APD to determine the distance to the object. Large-area APDs, such as 5 × 5 mm, are used in high-energy physics experiments [14].

When a bias voltage to an APD is increased, leakage current starts to rise rapidly. This voltage is called a breakdown voltage, which is specified at the leakage current of 1 μA. APDs are generally used in a linear mode at voltages lower than the breakdown voltage, where an output current is linearly proportional to incident light power. However, APDs are sometimes used in Geiger mode [15], where a higher voltage than the breakdown voltage is applied, for photon counting. In this mode, an electron triggers an avalanche breakdown, which produces high output current to be detected. Since the avalanche breakdown causes damage to the APD, a bias circuit is designed

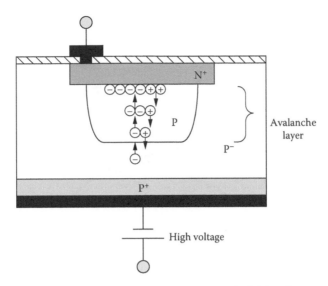

FIGURE 3.9 Schematic drawing of an APD having a high-electric-field region around the p–n junction, where avalanche multiplication occurs for internal gain.

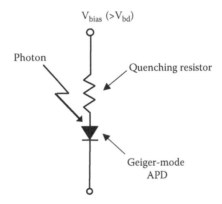

FIGURE 3.10 An equivalent circuit of Geiger-mode APD to detect single photons. *Step 1*: The bias voltage is set at larger than the breakdown voltage (V_{bd}). *Step 2*: When a photon arrives, avalanche breakdown occurs, and large pulse is output to the circuit. *Step 3*: A bias voltage is reduced at once by the quenching resistor to stop avalanche breakdown and returns to the original point waiting for the arrival of the next photon.

to reduce the voltage to quench the breakdown by a resistor. An equivalent circuit explaining this detection process of Geiger-mode APD is schematically shown in Figure 3.10. Since dark current generated spontaneously at room temperature triggers the avalanche breakdown, cooling is necessary, and the effective diameter of the APD should be small, as small as 0.175 mm, to lower the dark count rate [16]. Owing to its photon-counting capability and high QE of 65% around 700 nm, Geiger-mode APDs are used for photon-counting applications in fluorescent measurement. According to the operating principle, only one pulse is output even when many photons arrive at the same time.

A new type of Geiger-mode APD capable of counting simultaneous photons, shown in Figure 3.11, has been developed [17]. This new sensor, called a "multi-pixel photon counter (MPPC)" [18] or a "silicon photomultiplier (Si-PM)" [19], has a large number of pixels constructed on a monolithic Si chip. Each pixel has its own quenching resistor. When two photons arrive at different pixels simultaneously, an output pulse twice as large as that for one photon is obtained. Thus, the number of

FIGURE 3.11 Operating principle of an MPPC, which consists of effective pixels that are connected to quenching resistors. When two photons arrive at different pixels simultaneously, an output pulse that is twice as large as that of a single photon appears on the output circuit; thus, the MPPC has both photon-counting capability and linearity.

incident photons can be measured, as shown in Figure 3.11. The linearity is limited by the number of pixels on a chip because the pulse height of output pulses starts saturating when more than one photon tends to hit one pixel at the same time. MPPCs with 1×1 mm and 3×3 mm effective area are available in the market with a pixel pitch from 25 to 100 μm. A 3×3 mm MPPC with pixel pitch of 25 μm has 14,400 pixels [18]. Because of its photon-counting capability and fast response time, MPPCs are suitable for high-energy physics experiments [20] and positron emission tomography (PET) [21]. Because MPPCs are also insensitive to high magnetic fields, they are especially suitable for magnetic resonance imaging PET.

In terms of photon counting, PMTs [3] are still the most commonly used optoelectronic sensors in various applications. In addition to a photocathode (shown in Figure 3.4), a PMT has dynodes in a vacuum tube to multiply electrons, as shown in Figure 3.12 [22].

In response to incident photons, electrons are emitted from the photocathode. The emitted electrons are then multiplied at each dynode by secondary electron multiplication, where a primary electron generates additional 5–10 electrons when striking a dynode. With 8 or 10 stages of dynodes, the total gain reaches 10^6–10^7, which is optimal to detect single photons. A typical gain characteristic of a PMT is shown in Figure 3.13 [22] as a function of a bias voltage. Contrary to Geiger-mode APDs, PMTs detect continuous light as well as pulsed light. Since the multiplication occurs in vacuum, temporal information of the input photon is not degraded. Typical rise and fall times of a PMT are 1 and 10 ns, respectively.

In addition to high gain and fast time response, PMTs with a large effective diameter (up to 500 mm) are available at a reasonable price. A low dark count of 0.01–0.1 count/s/mm² or 1.6×10^{-21} to 1.6×10^{-22} A/mm² for bialkali photocathodes is another feature of PMTs. In general, QEs of

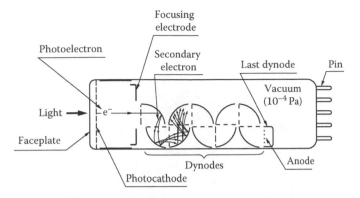

FIGURE 3.12 Sectional drawing of a PMT, which consists of a photocathode and dynodes in a vacuum tube. In response to incident photons, electrons are emitted from the photocathode and multiplied by the dynodes. Since each dynode multiplied electrons by a factor of 5–10, the gain reaches 10^6–10^7 in total, high enough to detect single photons.

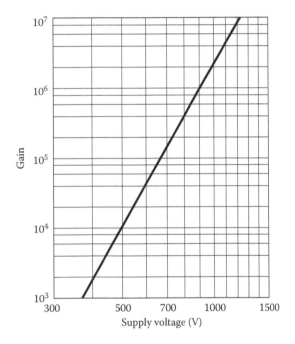

FIGURE 3.13 Typical gain characteristics of a PMT as a function of bias voltage.

photocathodes are lower than those of sensors using the photovoltaic effect as shown in Figure 3.5. However, a drastic improvement in QE is achieved in a "UBA" and a semiconductor crystal photocathode, reaching QEs of approximately 50%, as discussed in Section 3.2.2.

As shown in Figure 3.14 [22], PMTs with various photocathode materials, shapes (round, square, hexagonal), and dynode structures are commercialized for various low-light-level applications.

In general, PMTs are bulkier than semiconductor sensors; however, a "flat panel PMT" is available in the market [23]. Because of "metal-channel dynodes" made of thin metal plates, the outer dimension of this PMT is 52 × 52 × 15 mm, thinner than any other PMTs. The flat panel PMT is discussed in Section 3.4.1 in detail. PMTs are typically used in analytical instruments like spectrometers [24], in biology and medical fields to detect fluorescent light [25,26], and in medical imaging

FIGURE 3.14 PMTs are designed with various photocathode materials, shapes, and dynode structures for various applications.

such as PET [27] and gamma cameras [28]. Now PMTs can also be used for portable instruments, such as micro-total analysis systems (μTAS) and environmental monitors, due to a significant reduction in detector size. A fusion of micro-electro-mechanical systems and vacuum technology, a miniaturized PMT with outer dimensions of $13 \times 10 \times 2$ mm, and a 1×3 mm bialkali photocathode has been commercialized, as shown in Figure 3.15 [29]. Even with this small body, a gain of 10^6 and time response in the nanoseconds range are obtained.

A hybrid photodetector (HPD) is another optoelectronic sensor that detects single photons. Instead of dynodes, the HPD has a Si avalanche diode (AD) to multiply electrons, as shown in Figure 3.16 [30]. Electrons emitted from the photocathode are accelerated up to 8 keV and are focused onto the AD. Then, an electron deposits its kinetic energy and generates approximately 1000 electron–hole pairs in Si; this is called electron-bombarded gain. The electrons generated in the AD drift toward the p–n junction for avalanche multiplication; thus, the total gain of the HPD reaches 10^5. Due to the large electron-bombarded gain, the noise figure of electron multiplication is almost ideal, close to one; thus, the S/N is quite high. This is why the HPD can distinguish up to five electrons by its output pulse height, as shown in Figure 3.17. An HPD incorporating a small AD shows faster response time than conventional PMTs [30,31]; the rise and fall times are 400 ps,

FIGURE 3.15 A miniaturized PMT on a fingertip appeals high performance, not compromised for its small size.

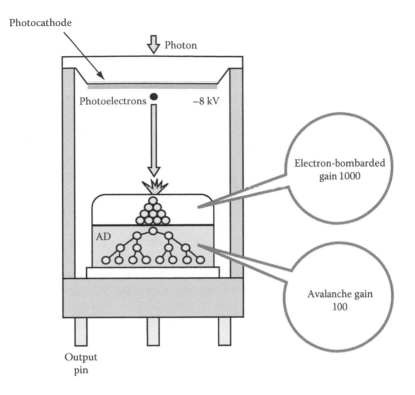

FIGURE 3.16 An operating principle of an HPD is shown, where electrons from the photocathode are accelerated and focused on the AD to make an electron-bombarded gain of 1000. The electrons are further multiplied by avalanche multiplication for a total gain of 10^5, high enough to detect single photons.

FIGURE 3.17 Pulse height spectrum of an HPD for irradiation of several photons to the photocathode on average. Several peaks corresponding to one, two, and three electrons from the photocathode are clearly observed.

and the timing resolution for a single photon is less than 90 ps. In addition to good timing resolution, the HPD shows a small probability of afterpulsing, a false signal triggered by the real signal and appearing a certain period after the real signal. Smaller afterpulsing than PMTs and Geiger-mode APDs makes HPDs attractive for use in fluorescence correlation spectroscopy and time-correlated single-photon counting (TCSPC) [32,33].

3.3.3 READOUT ELECTRONICS

Outputs of optoelectronic sensors should be recorded for analysis. Recording methods are discussed here for recording a waveform, counting photons, measuring a pulse-height spectrum, and timing of the arrival of photons.

To record a waveform, a current-to-voltage amplifier or a transimpedance amplifier is used for front-end electronics to feed voltage signal to an ADC. The ADC digitizes the signal, and a computer records it, as shown in Figure 3.18a. A digital oscilloscope works this way.

To count photons, a high-speed AC-coupled amplifier is used as front-end electronics to distinguish arriving photons individually. The output signal is compared to a threshold level by a comparator, and a number of output pulses are measured by a counter, as shown in Figure 3.18b.

The intensity of pulsed light is measured with a charge-sensitive amplifier, which accumulates input charge for a certain period and converts it to voltage. The output is further amplified and shaped by a shaping amplifier, and recorded by a multichannel analyzer, as shown in Figure 3.18c. The result obtained by this method is called a pulse height spectrum, as shown in Figure 3.17.

It is very useful to time the arrival of photons because the timing resolution of optoelectronic sensors is roughly 1/10th of the rise and fall times of the sensors. TCSPC uses this method to achieve time resolution in the order of 10 ps and a wide dynamic range of five orders [33]. To measure the timing of each photon, the output pulse of a high-speed amplifier is fed to a constant fraction discriminator (CFD), which outputs timing signal accurately even though pulse height fluctuates. The output signal of the CFD and trigger output of the light source are used for the start and stop signals, respectively, of a time-to-amplitude converter (TAC), where the TAC's output is proportional to the time difference between start and stop signals. The output signal from the TAC is analyzed and recorded by a multichannel analyzer, as shown in Figure 3.18d.

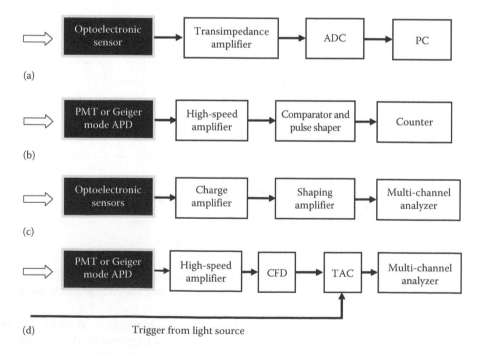

FIGURE 3.18 Readout electronics recording methods of signal from optoelectronic sensors for (a) recording waveform, (b) counting photons, (c) measuring pulse height spectrum, and (d) timing of the arrival of photons.

3.4 MULTI-PIXEL AND POSITION-SENSITIVE SENSORS

In the case of measuring multiple wavelengths or spectrum of light, multi-pixel sensors are required to measure them all at once; otherwise, it is necessary to scan the wavelength over a slit to be detected by a point sensor. On the other hand, for trigonometry, position-sensitive sensors are useful in measuring the position of a spot of light. In this section, multi-pixel sensors and position-sensitive sensors are discussed.

3.4.1 Multi-Pixel Sensors

Semiconductor sensors with plural pixels are easy to design and fabricate by a semiconductor process. A variety of multi-pixel Si sensors linearly arranged from 2 to 46 pixels are available in the market. An example shown in Figure 3.19 has a 46-pixel array of 0.9 × 4.4 mm [34]. A 2D array of 2 × 2 pixels (2.5 × 2.5 mm/pixel) is also available [35]. For near-IR light, InGaAs array sensors up to 16 pixels (0.45 × 1 mm) [36] can be used. These are some examples of multi-pixel sensors for high light level, which are mostly used for spectrometers. Any modifications on number, size, and shape of pixels are technically possible.

One of the merits of multi-pixel sensors is high-speed readout because the signal from each pixel is read out in parallel. However, the parallel readout is sometimes a demerit because the complexity of readout electronics is proportional to the number of pixels. For more than 100 pixels, specifically designed complementary metal-oxide semiconductor (CMOS) readout electronics are used to read out every pixel sequentially. In this case, readout speed of whole pixels is limited. The multi-pixel sensors with CMOS readout electronics are categorized as image sensors and discussed in Section 3.5.1.

For low light level, multi-pixel PMTs with special dynodes called "metal-channel dynodes" [3] are used. Metal-channel dynodes, bars of thin metal shaped somewhat like the letter "S," are arranged in rows and stacked on top of each other (with a pitch as small as 0.5–1 mm) to create channels for multiplied electrons, as shown in Figure 3.20. Electrons from the photocathode are multiplied in the dynodes while preserving their position information and are output from plural anodes located behind the dynodes; thus, a multi-pixel sensor is realized. Flat panel PMT with metal-channel dynodes is shown in Figure 3.20 [23]. This PMT has 16 × 16 (256 in total) anodes with an effective area of 49 × 49 mm. The PMT is very thin, just 15 mm thick, owing to the

FIGURE 3.19 Dimensional outline of 46-linear array of Si-PD [34]. Size of each pixel is 0.9 × 4.4 (mm).

FIGURE 3.20   Photograph of a "flat panel PMT" and the structure of "metal-channel dynodes."

metal-channel dynodes. Because of their high speed and high gain, multi-pixel PMTs are used in spectrometers for confocal microscopes [37] and high-energy physics experiments [38].

3.4.2   POSITION-SENSITIVE SENSORS

A trigonometry is used to measure distance for auto-focusing of a camera, for example. As shown in Figure 3.21 [11], this system needs to measure the position of a spot light on a sensor; thus, a

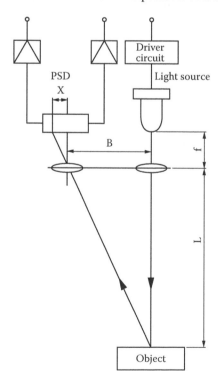

FIGURE 3.21   Schematic drawing of a trigonometry to measure distance, where an irradiated position is measured by a PSD.

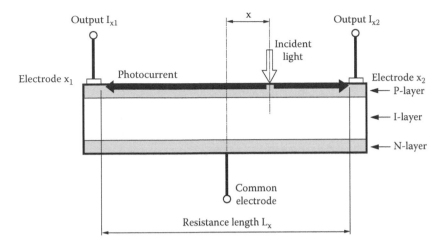

FIGURE 3.22 Sectional drawing of a one-dimensional PSD, having resistive layer (p-layer) and two terminals at both ends. An irradiated position is determined by the ratio of output currents from the terminals.

position-sensitive detector (PSD), shown in Figure 3.22 [39], with a resistive layer (p-layer) and two output terminals at both ends, is used. An irradiated position (x, from the center) is determined by the following equation:

$$x = \frac{L_x}{2*(I_{x2}-I_{x1})/(I_{x2}+I_{x1})} \tag{3.7}$$

where
  L_x is the length of the effective area
  I_{x1} and I_{x2} are output currents from both terminals

Because of its operating principle, a small spot of less than 200 μm is required to irradiate PSDs. One-dimensional-type PSDs are available with effective lengths up to 37 mm [39]. A 2D type works in the same manner and up to 14 × 14 mm are available in the market [39]. One of the merits of PSDs is a good position resolution of 10 μm at a response time in the order of 1 MHz. These are achieved because signals from just two terminals are read out for one dimension and four terminals for two dimensions. It is more difficult to determine the position of a spot light using multi-pixel sensors or image sensors at these resolutions in position and time. It is worth noting that the response time and the position resolution depend upon the size, resistivity, and structure of the PSD as well as power and wavelength of the input light.

For low light level, multi-pixel PMTs are used as position-sensitive sensors by connecting each anode with resistors [40], as shown in Figure 3.23. The output current splits to four terminals through the resistor chain according to Equation 3.7. Thus, the irradiated position is determined. Putting segmented scintillator on a faceplate, a PMT with 16 × 16 pixels and a resistor chain are used for the state-of-the-art PET scanner [41].

3.5 IMAGE SENSORS AND CAMERAS

To capture fine images, a large number of pixels—more than 250,000 pixels in general—are required. Then, some electronics are necessary to read out the signal of each pixel sequentially because parallel readout is impossible for so many pixels. Optoelectronic sensors having sequential-readout electronics are categorized as image sensors in this chapter and discussed in this section.

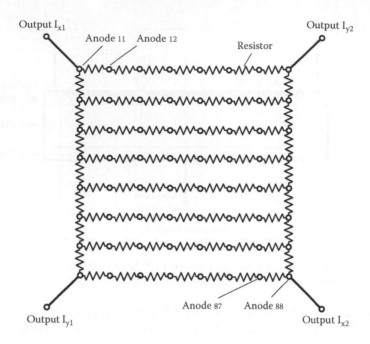

FIGURE 3.23   Resistor chain of 8 × 8 pixel PMT for position-sensitive operation.

3.5.1   IMAGE SENSORS

Historically, there have been plenty of image sensors starting from image pickup tubes. However, two types of image sensors are mainly used today. One is a charge-coupled device (CCD), and the other is a CMOS image sensor.

A CCD has many light-sensitive and charge-transfer pixels monolithically, where charges are transferred from one pixel to the next one, like a bucket relay. Figure 3.24a [42] schematically shows the structure of an interline transfer CCD. Plenty of PDs work as light-sensitive pixels, converting incident photons to electrons and accumulating them. After an exposure time of 33 ms, for example, for TV-rate operation, charges in each PD are transferred to the charge transfer pixels (vertical shift register) nearby in a short time for readout. In a readout procedure, the charges are transferred one line down toward the horizontal shift register followed by repetitive transfers in horizontal shift registers to the readout circuit. Repeating this process achieves readout of all pixels. While reading out, signals for the next frame are accumulated in the PDs. Interline CCDs are used in scientific applications due to their performance and reasonable cost; however, sensitivity is limited because about half of the area is used for the charge transfer and thus insensitive to photons. Micro-lenses fabricated on PDs help to collect light [42], but the sensitivity is not as high as back-illuminated CCDs as discussed in Section 3.5.2.

The structure of another type of CCD, a frame transfer CCD, is shown in Figure 3.24b [42]. The effective area consists of a transparent shift register; thus, photons are converted to electrons in Si below the register. Once charges are accumulated, they are transferred vertically to the storage area in a very short time, from which charges are read out sequentially as described in interline CCDs. The frame transfer structure is used for a back-illuminated CCD and an electron-multiplying CCD as discussed in Section 3.5.2.

CMOS image sensors, on the other hand, consist primarily of plural PDs as light-sensitive pixels and MOS transistors that act as switches, turning on/off the connection between each PD and the readout circuit, as shown in Figure 3.25 [43]. A control circuit (vertical and horizontal shift registers) is fabricated monolithically on a chip to switch MOS transistors properly for sequential readout. Recently, an active pixel sensor having not only a switch but also an amplifier for each pixel

FIGURE 3.24 (a) A pixel layout of an interline-transfer CCD, where plural PDs accumulate charges, while CCDs (vertical and horizontal shift registers) transfer the charges. (b) Pixel layout of a frame-transfer CCD, where vertical transfer from effective area to storage area is carried out in a very short period, followed by a pixel-to-pixel readout from storage area.

FIGURE 3.25 Schematic drawing of CMOS image sensor, where MOS transistors act as switches to connect PDs sequentially to the output electronics for readout.

is commonly used both to reduce noise and to increase frame rate [44]. The advantages of CMOS image sensors compared to CCDs are as follows:

1. The fabrication process is simple.
2. All electronics necessary for operation are fabricated monolithically, making operation simple.
3. Power consumption is low.

These are the reasons why CMOS image sensors are widely used in digital cameras and cell phones. In scientific applications, a CCD was a main player for a long time because of its high sensitivity and low noise. However, a scientific CMOS described in the following section is taking over the position. For practical use, an image sensor is assembled as a camera and generally connected to a computer for recording and analyzing images.

3.5.2 Image Sensors and Cameras for Low-Light-Level Applications

CCDs can be used for low-light-level imaging with some modifications. Since the detection limit at low light levels is determined by an S/N, increasing the signal is essential and simple to achieve. Signals can be increased by improving QE or extending the exposure time. Back-illuminated CCDs help to increase signal due to improved QE. In a front-illuminated CCD, as shown in Figure 3.26 [45], photons enter the Si substrate through shift registers (polysilicon electrodes), which absorb or reflect part of the incident photons. As a result, the QE is 40% at the peak wavelength and zero below 400 nm, as shown in Figure 3.27 [46]. On the other hand, the back-illuminated CCD shows almost ideal spectral response, shown in the same figure, because there is nothing but effective Si layer on the back. Manufacturing the back-illuminated CCD requires a complicated process to thin the substrate down to approximately 10 μm. Thinning is necessary for high spatial resolution; otherwise, electrons generated at the back surface split to the neighboring pixels while they diffuse to the front surface for transfer.

The other method to increase signal is to extend the exposure time. However, the dark current of a CCD is accumulated as well and generates noise. Cooling is useful to decrease the dark current because cooling of every 20°C decreases the dark current by one order of magnitude. Therefore, the lower the temperature, the better in terms of the dark current, as shown in Figure 3.28 [42]. To achieve a minimum temperature of −90°C, a CCD and a Peltier cooler are

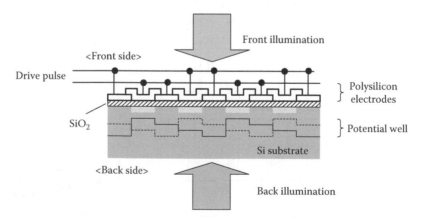

FIGURE 3.26 Sectional drawing of a frame-transfer CCD for front and back illumination, where the front surface is the processed surface with layers of polysilicon electrodes (shift registers), which absorb and reflect part of incident light. (From Dakin, J.P. and Brown, R.G., *Handbook of Optoelectronics*, 1, Copyright 2006. Reproduced by permission of Taylor & Francis Group, LLC, a division of Informa plc, Boca Raton, FL.)

FIGURE 3.27 Spectral responses of front- and back-illuminated CCDs. The back-illuminated CCD shows high QE of 90% in visible region.

FIGURE 3.28 A dark current of a CCD as a function of temperature, where the dark current is decreased to 1/10th with cooling of every 20°.

packed in a vacuum chamber, as shown in Figure 3.29 [42]. Various cooled CCDs are supplied as cameras in the market.

A cooled CCD camera with long exposure time cannot be used to observe dim and fast-moving objects such as specimen under a microscope, for example. The solutions to this problem are either to get an internal gain in a sensor or to reduce noise further. In the case of Si, the impact ionization discussed in Section 3.3.2 can be used to obtain a gain. An electron-multiplying CCD (EM-CCD) uses this effect at the multiplier register by applying enough high voltage to the transfer gates, as shown in Figure 3.30 [42]. Although the gain of each transfer is small (such as 1.05), the total gain after 500 transfers reaches 1000. Cooling should be carried out the same way as a cooled CCD because dark current is also multiplied at the registers.

FIGURE 3.29 A vacuum chamber containing a CCD and a Peltier cooler to cool the CCD down to −90°C.

FIGURE 3.30 Structure of an EM-CCD to obtain gain at multiplier resistors by impact ionization.

The other approach to get a better image at short exposure times is to reduce noise by using a CMOS sensor whose every column has its own readout circuit, as shown in Figure 3.31 [43]. Because there are thousands of readout circuits, the bandwidth of each circuit could be reduced to make the readout noise less than 2 electrons (root mean square), which is one-fifth or less of a CCD's readout noise. Furthermore, the S/N is readily preserved because the signal is digitized by an on-chip ADC. The additions of readout circuits and ADC are made possible by state-of-the-art CMOS technology that shrinks the feature size and allows the fabrication of circuits on silicon. This type of CMOS image sensor is called scientific CMOS (sCMOS), and its development has expanded CMOS applications to include scientific fields [47,48].

Another method to obtain a high gain is to combine a conventional CCD with an image intensifier. As shown in Figure 3.32 [49], the image intensifier consists of a photocathode,

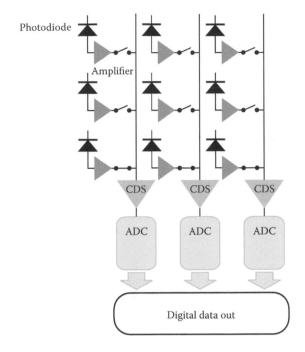

FIGURE 3.31 Structure of sCMOS, where a correlated double sampling (CDS) reducing electrical noise and ADC are fabricated; thus, digital signal is output from the sensor.

FIGURE 3.32 Structure and operating principle of an image intensifier is shown, where electrons emitted from the photocathode are multiplied by a factor of 1000 at the MCP; thus, input light image is intensified.

a microchannel plate (MCP), and a phosphor screen. An MCP is a thin glass plate with millions of small through-holes whose diameter is typically 6 μm and the thickness is 300 μm, as schematically shown in Figure 3.33 [50]. By applying bias voltage of typically 1 kV between the input and output surfaces of the MCP, incident electrons are multiplied by successive secondary electron multiplication while preserving their position information within a hole.

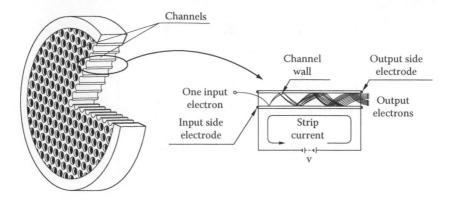

FIGURE 3.33 Schematic drawing of an MCP, which is a thin glass plate with a multitude of through holes. By applying voltage between the input and output surfaces, electrons are multiplied within the holes by a factor of 1000.

A proximity-focused structure preserves the spatial information of electrons entering the MCP, where electrons are multiplied by a factor of 1000. The multiplied electrons are accelerated and strike the phosphor screen to emit light. Thus, the input image is intensified. The output image is connected to a CCD through a fiber-optic plate (FOP) [51], a glass plate made of a bundle of optical fibers, which transmits the image while preserving spatial information. The coupling of images by an FOP shows better efficiency by one order of magnitude than that by a lens. This type of imaging system is called an intensified CCD (I-CCD). Using two layers of MCPs, the gain reaches 10^5. This gain value makes I-CCDs suitable for single-photon counting. Another feature is a very short exposure time (in nanoseconds) controlled by a gating operation, where a short forward bias voltage corresponding to the exposure time is applied between the photocathode and the MCP. In response to incident light image, an electron image is emitted from the photocathode.

3.5.3 HIGH-SPEED CAMERAS

Short exposure time is available with gate-operated I-CCDs. However, a CCD camera's frame rate is still limited, typically 30 frame per second (fps). For higher frame rates, a specially designed CMOS image sensor similar to sCMOS with multiple output terminals is used. A C-MOS sensor with 512 output terminals and ADCs sharing light-sensitive pixels is reported [52] for this purpose. Then, parallel readout is carried out at 2 mega-pixels/s for each output terminal; thus, approximately 3500 fps is available with 512 × 512 pixels. In the market, 12,500 fps with 1024 × 1024 pixels is available [53]. Since a short exposure time means a very small number of photons per frame, an image intensifier is coupled to a C-MOS sensor to intensify light in many applications such as fluorescence microscopy [54].

A multi-port CCD [55] is another solution for high-speed imaging. However, this sensor is mostly used for time-delayed integration (TDI). In TDI operation, the CCD performs vertical transfer of charges repeatedly while matching the speed of images. Thus, charges are integrated, and their spatial information is preserved while moving on the CCD and finally read out, as shown in Figure 3.34 [55]. In this way, the TDI-CCD acts as a line sensor with extremely high sensitivity. Since the vertical transfer rate is limited by the pixel readout rate, a multi-port is useful to speed up the transfer. A TDI-CCD with 16 terminals for 2048 horizontal pixels is available, as shown in Figure 3.35 [55]. Since this CCD has output terminals in both upper and lower ends, bidirectional operation transferring charges both upward and downward is available.

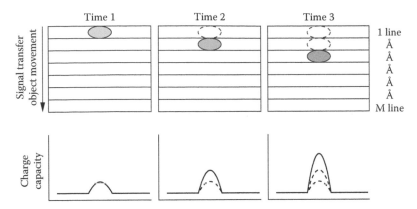

FIGURE 3.34 An operating principle of a TDI-CCD is shown, where the speed of vertical transfer of charges matches that of objects; thus, charges are integrated while transferring.

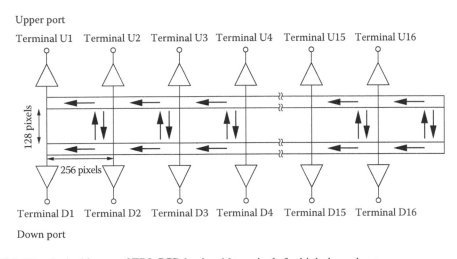

FIGURE 3.35 A pixel layout of TDI-CCD having 16 terminals for high throughput.

3.6 X-RAY OR GAMMA-RAY SENSORS

An x-ray or a gamma ray is also an electromagnetic radiation with a much shorter wavelength than UV photons, shorter than 10 nm. X-rays behave differently from the other photons: they transmit through materials easily or generate a lot of electron–hole pairs when absorbed. In this section, the physics for detecting x-rays and practical sensors are described.

3.6.1 PHYSICS OF DETECTION

To detect x-rays, they must first be absorbed by a material. The transmission characteristic of x-rays should be taken into account because transmission through a material depends heavily on an x-ray's energy. As an example, the absorption characteristics of x-rays by Si are shown in Figure 3.36 as a parameter of thickness. This indicates that Si with a thickness of 0.5 mm absorbs x-rays of lower than 20 keV efficiently, meaning Si is sensitive only for low-energy x-rays. Once an x-ray is absorbed by Si, an electron having almost the same energy as that of the incident x-ray is generated, deposits its energy in Si, and generates a number of electron–hole pairs depending upon the energy

FIGURE 3.36   Absorption characteristics of x-rays by Si are shown as a parameter of the thickness.

of the x-ray (E_x [eV]). The average number (N_{eh}) of generated electron–hole pairs is calculated according to the following equation:

$$N_{eh} = \frac{E_x}{E_{eh}} \qquad (3.8)$$

where E_{eh} (eV) is the average energy needed to generate one electron–hole pair in a material. For Si, E_{eh} is 3.6 eV. The generated charges can be read out in the same way as a PD or a CCD. It is worth noting that x-rays are commonly expressed not by wavelength but by energy. Use Equation 3.1 to convert between energy and wavelength.

To absorb high-energy x-rays, specially designed lithium (Li)-drifted Si or germanium (Ge) detectors with a thick substrate are used [56]. They show excellent energy resolution for incident x-rays. However, they are not easy to use because cooling at liquid nitrogen temperature (–196°C) is required. Cadmium tellurium (CdTe) sensor is reported to be operated at room temperature [57] for small gamma cameras [58].

Another—and the most common solution—for the detection of high-energy x-rays is a combination of a scintillator and an optoelectronic sensor. Scintillators are made of materials with high atomic number and density, thus suitable to absorb high-energy x-rays. In a scintillator, an x-ray is converted to an electron with high kinetic energy, which excites many other electrons when depositing its energy. When relaxed, these electrons emit photons in the visible region to be detected by an optoelectronic sensor. Table 3.1 [59] lists several scintillators available in the market and includes their composition, density, decay time, emission wavelength, and major applications.

3.6.2   PRACTICAL SENSORS

A practical point sensor for x-rays consists of a scintillator and a PMT, as shown in Figure 3.37 [3].

A single x-ray photon is detected with this system because of the high sensitivity of a PMT. Since the number of emitted photons is proportional to the energy of the x-ray, the energy can be analyzed by the pulse height obtained by the readout electronics shown in Figure 3.18c. This type of sensor is widely used for a scintillation counter. A PET scanner [27] or a gamma camera [28] also uses a combination of a scintillator and a PMT. On the other hand, PDs are widely used to receive light from a scintillator, for example, for x-ray CT, when single-photon sensitivity is not required.

For imaging, scintillation light can be observed by a CCD camera through a lens [60], as shown in Figure 3.38, and the scintillator is deposited, for example, on an amorphous carbon plate [61].

TABLE 3.1

Characteristics and Typical Applications of Scintillators

	Density (g/cm³)	Relative Emission Intensity[a]	Decay Time (ns)	Emission Peak Wavelength (nm)	Applications
NaI (Tl)	3.67	100	230	410	Survey meter, area monitor, gamma camera, SPECT
BGO	7.13	15	300	480	PET, x-ray CT
CsI (Tl)	4.51	45–50	1000	530	Survey meter, area monitor, x-ray CT
Pure CsI	4.51	<10	10	310	High-energy physics experiment
BaF₂	4.88	20	0.9/630	220/325	FOF, PET
GSO:Ce	6.71	20	30	310/430	Area monitor, oil well logging
Plastic	1.03	25	2	400	Area monitor, neutron detector
LSO	7.35	70	40	420	PET
PWO	8.28	0.7	15	470	High-energy physics experiment
YAP	5.55	40	30	380	Survey meter, gamma camera

Source: Dakin, J.P. and Brown, R.G., *Handbook of Optoelectronics*, 1, Copyright 2006. Reproduced by permission of Taylor & Francis Group, LLC, a division of Informa plc, Boca Raton, FL.

[a] Normalized to NaI(Tl) as 1000.

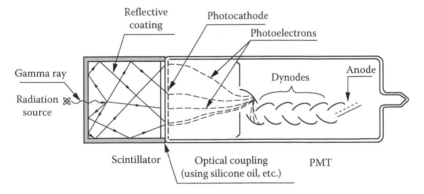

FIGURE 3.37 Configuration for scintillation counting, where a scintillator converts gamma rays to light to be detected by a PMT.

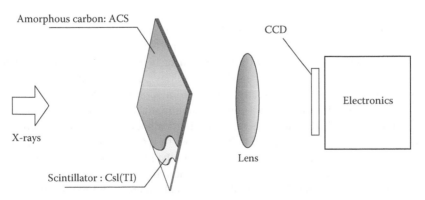

FIGURE 3.38 An imaging system for radiography consists of a scintillator on an amorphous carbon plate and CCD camera coupled by a lens.

FIGURE 3.39   A high-sensitive sensor for radiography, where a scintillator is deposited on an FOP coupled to a C-MOS image sensor.

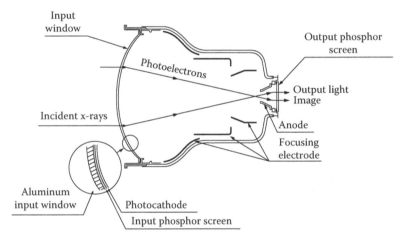

FIGURE 3.40   Configuration of an x-ray image intensifier, where x-rays are converted to light by a scintillator deposited on a photocathode. Electrons from the photocathode are demagnified and focused on a phosphor screen to be detected by a CCD camera.

This system is called CCD-digital radiography (CCD-DR). To improve coupling efficiency of light by the lens, a large C-MOS image sensor has been developed to receive the light from the scintillator through an FOP, as shown in Figure 3.39 [62]. This sensor is mechanically thin and distortion free without a lens, and this sensor is used for medical radiology. An x-ray image intensifier (x-ray II), another sensor, uses a photocathode directly coupled to a scintillator deposited on a faceplate, as shown in Figure 3.40 [63]. In response to x-rays, the scintillator emits photons, which are converted to electrons by the photocathode. Electrons from the photocathode are accelerated, reduced by a factor of 5, and focused on a phosphor screen to be read out by a CCD camera. The x-ray II shows the best sensitivity because of its efficient optical coupling to the photocathode and a gain of the conversion at the phosphor screen. Demerits are its large size (including a readout camera) and the distortion caused by an electron lens. X-ray IIs are used for medical imaging and nondestructive inspection. Since optics for x-rays are still not available for practical use, x-ray image sensors are relatively large, such as 125 × 125 mm for C-MOS image sensors [62] and 73 × 55 mm for x-ray IIs [63].

3.7 SELECTION GUIDE FOR OPTOELECTRONIC SENSORS

As an alternative to a summary of this chapter, this section describes how to select the most appropriate sensor for each application:

1. Specify the wavelength of light to be detected. Refer to Section 3.2.2 for spectral responses for UV to IR and Section 3.6 for x-rays. Select the best one in terms of sensitivity.
2. Specify position resolution. Select a point sensor, a multi-pixel/position-sensitive sensor, or an image sensor, described in Sections 3.3 through 3.5, respectively.
3. Check the power of incident light. Sensors for high light and low light levels are described in Sections 3.3 through 3.5 as well.
4. Specify the response time. Point and multi-pixel sensors have fast time response as mentioned in Sections 3.3 and 3.4, whereas image sensors are slow in general. High-speed cameras are available as described in Section 3.5.3.

Figure 3.41 categorizes every sensor by wavelength and number of pixels. In each bold box, sensors in the upper left have high sensitivity and fast response, while those in lower right have low sensitivity and slow response. For optical measurement, selection of the most appropriate sensor for your application is the first step, and this chapter helps in the selection.

	X-ray		Ultraviolet, visible, near-infrared (<1 μm)		Infrared (>1 μm)	
Point sensor	Scintillator + PMT (3.6.2)	Scintillator + PIN-PD (3.6.2)	PMT, HPD, MPPC, Small APD (3.3.2)	Small Si-PD (3.3.1)	InP/InGaAs PMT (3.2.2), InGaAs-APD[a]	Compound PD (3.3.1)
		Scintillator + PD (3.6.2)		Large PD (3.3.1)		Compound photoconductor (3.3.1)
Position-sensitive multi-pixel sensor			Multi-pixel PMT, PMT with resistor chain (3.4)	Multi-pixel PD, PSD (3.4)		Compound PD array (3.4)
Image sensor			sCMOS, ICCD, TDI-CCD, II + high-frame-rate C-MOS (3.5.2, 3.5.3)	High-frame-rate C-MOS (3.5.3)		
	X-ray image intensifier, scintillator + CMOS (3.6.2)	Scintillator + CCD (3.6.2)	EM-CCD, cooled CCD (3.5.2)	CCD, C-MOS (3.5.1)	InP/ InGaAs-ICCD[a]	Infrared camera[a]

High sensitivity **Fast** response	Low sensitivity Fast response
High sensitivity Slow response	**Low** sensitivity **Slow** response

[a] Not described in this chapter.

FIGURE 3.41 Photon sensors categorized by wavelength and number of pixels.

REFERENCES

1. Si Photodiodes, Catalogue of Hamamatsu Photonics K.K., 2011.
2. Dakin, J. P. and Brown, R. G. W., *Handbook of Optoelectronics*, Taylor & Francis, Boca Raton, FL, Vol. 1, 418, 2006.
3. Hakamata, T. et al., *Photomultiplier Tubes: Basics and Applications*, 3rd edn., Hamamatsu Photonics K.K., Hamamatsu City, Japan, 2006.
4. Suyama, M., PoS (PD07), http://pos.sissa.it/archive/conferences/051/018/PD07_018.pdf (accessed January, 2007).
5. Niigaki, M. et al., Field-assisted photoemission from InP/InGaAsP photocathode with p/n junction, *Appl. Phys. Lett.*, 71(17), 2439, 1997.
6. NIR Photomultiplier Tubes, Catalogue of Hamamatsu Photonics K.K., 2005.
7. Infrared Detectors, Catalogue of Hamamatsu Photonics K.K., 2013.
8. Si PIN photodiode, S3204/S3584 Series, Catalogue of Hamamatsu Photonics K.K., 2012.
9. Si PIN photodiode, S5751, S5752, S5973 Series, Catalogue of Hamamatsu Photonics K.K., 2012.
10. Optical Communication Device, Catalogue of Hamamatsu Photonics K.K., 2008.
11. Yamamoto, K. et al., *Handbook of Optoelectronic Devices*, Hamamatsu Photonics K.K., Hamamatsu City, Japan, 2004 (in Japanese) https://www.hamamatsu.com/resources/pdf/ssd/e06_handbook_compound_semiconductor.pdf (accessed November 26, 2014).
12. McIntyre, R. J., Recent developments in silicon avalanche photodiodes, *Measurement*, 3(4), 146, 1985.
13. Sze, S. M., *Physics of Semiconductor Devices*, 2nd edn., Wiley-Interscience, New York, Vol. 47, 1981.
14. Organtini, G., Avalanche photodiodes for the CMS electromagnetic calorimeter, *IEEE Trans. Nucl. Sci.*, 46(3), 243, 1999.
15. Cova, S., Ripamonti, G., and Lacaita, A., Avalanche semiconductor detector for single optical photons with a time resolution of 60 ps, *Nucl. Instrum. Methods*, A253, 482, 1987.
16. SPCM-AQR, Catalogue of Perkin Elmer.
17. Golovin, V. and Saveliev, V., Novel type of avalanche photodetector with Geiger mode operation, *Nucl. Instrum. Methods*, A518, 560, 2004.
18. MPPC, Catalogue of Hamamatsu Photonics K.K., 2012.
19. MiniSM, Silicon Photomultiplier Modules, Catalogue of SensL, 2013.
20. Gomi, S. et al., Development of multi-pixel photon counters, *Nucl. Instrum. Methods*, 581(1), 57, 2007.
21. Otte, N. et al., The SiPM—A new photon detector for PET, *Nucl. Phys. B—Proc. Suppl.*, 150, 417, 2006.
22. Photomultiplier Tubes and Related Products, Catalogue of Hamamatsu Photonics K.K., 2010.
23. Flat Panel Type Multi-Anode Photomultiplier Tube Assembly H9500, H9500-03, Catalogue of Hamamatsu Photonics K.K., 2007.
24. Agilent Technologies, http://www.chem.agilent.com/Library/technicaloverviews/Public/si-1368.pdf (accessed November 26, 2014).
25. Multi Photon Laser Scanning Microscope, Catalogue of Olympus Corporation.
26. *Introduction to Flow Cytometry: A Learning Guide of BD Biosciences*, Manual Part Number 11-11032-01, 2000.
27. Muehllehner, G. and Karp, J. S., Positron emission tomography, *Phys. Methods Biol.* 51, 117, 2006.
28. GE Healthcare, http://www3.gehealthcare.com/en/Products/Categories/Nuclear_Medicine/General_Purpose_Cameras (accessed November 26, 2014).
29. Watase, K., μPMT is key to high-performance portable devices, *Laser Focus World*, 5, 60, 2013.
30. High Speed Compact HPD Module R10467U-40, Catalogue of Hamamatsu Photonics K.K., 2012.
31. Fukasawa, A. et al., High speed HPD for photon counting, *IEEE Trans. Nucl. Sci.*, 55(2), 758, 2008.
32. Michalet, X. et al., Hybrid photodetector for single-molecule spectroscopy and microscopy, *Proc. SPIE*, 6862, 12, 2008.
33. Becker, W., *Advanced Time-Correlated Single Photon Counting Techniques*, Springer, New York, 2005.
34. Si Photodiode Array S4111/4114 Series, Catalogue of Hamamatsu Photonics K.K., 2011.
35. Si PIN Photodiode S5980, S5981, S5870, Catalogue of Hamamatsu Photonics K.K., 2010.
36. InGaAs PIN Photodiode Array G7150-16, Catalogue of Hamamatsu Photonics K.K., 2007.
37. Nikon Enhanced Spectral Detection Unit, http://www.microscopyu.com/articles/confocal/spectralimaging.html (accessed November 26, 2014).
38. Tagg, N. et al., Performance of Hamamatsu 64-anode photomultipliers for use with wavelength-shifting optical fibres, *Nucl. Instrum. Methods*, A539, 668, 2005.
39. PSD, Catalogue of Hamamatsu Photonics K.K., 2010.

40. Siegel, S. et al., Simple charge division readouts for imaging scintillator arrays using a multi-channel PMT, *IEEE Trans. Nucl. Sci.*, 43, 1634, 1996.
41. Inadama, N. et al., Performance evaluation for 120 four-layer DOI block detectors of the jPET-D4, *Radiol. Phys. Technol.*, 1, 75, 2008.
42. Image EM, Technical Note of Hamamatsu Photonics K.K., 2009.
43. ORCA-Flash 2.8, Technical Note of Hamamatsu Photonics, 2010.
44. Fossum, E. R., Active pixel sensors: Are CCD's dinosaurs? *Proc. SPIE*, 1900, 2, 1993.
45. Dakin, J. P. and Brown, R. G. W., *Handbook of Optoelectronics*, Taylor & Francis, Boca Raton, FL, Vol. 1, 451, 2006.
46. Image Sensors, Catalogue of Hamamatsu Photonics K.K., 2011.
47. Fullerton, S., ORCA-Flash 4.0, White Paper of Hamamatsu Photonics, 2012.
48. The Online Home of Scientific CMOS Technology, http://www.scmos.com/ (accessed November 26, 2014).
49. Image Intensifiers, Catalogue of Hamamatsu Photonics K.K., 2009.
50. MCP and MCP Assembly, Selection guide of Hamamatsu Photonics K.K., 2013.
51. Fiber Optic Plates, Catalogue of Hamamatsu Photonics K.K., 2005.
52. Furuta, M., A high-speed, high-sensitivity digital C-MOS image sensor with a global shutter and 12-bit column-parallel cyclic A/D converters, *IEEE J. Solid-State Circ.*, 42(4), 766, 2007.
53. Fastcam SAX-2, Catalogue of Photron Limited, 2013.
54. Focuscope SV200-i, Catalogue of Photoron Limited, 2007.
55. TDI-CCD Image Sensors, Catalogue of Hamamatsu Photonics K.K., 2012.
56. Ortec, http://www.ortec-online.com/Solutions/RadiationDetectors/semiconductor-photon-detectors.aspx (accessed November 26, 2014).
57. Tomita, Y. et al., X-ray color scanner with multiple energy discrimination capability, *Proc. SPIE*, 5922, 40, 2005.
58. Tsuchimochi, M. et al., A prototype small CdTe gamma camera for radioguided surgery and other imaging applications, *Eur. J. Nucl. Med. Mol. Imaging*, 30(12), 1605, 2003.
59. Dakin, J. P. and Brown, R. G. W., *Handbook of Optoelectronics*, Taylor & Francis, Boca Raton, FL, Vol. 1, 422, 2006.
60. Seibert, J. A. et al., Evaluation of an optically coupled CCD digital radiography system, *Proc. SPIE*, 5745, 458, 2005.
61. X-ray Scintillator, Catalogue of Hamamatsu Photonics K.K., 2009.
62. Flat Panel Sensors, Selection Guide of Hamamatsu Photonics K.K., 2013.
63. X-Ray I.I. Digital Camera Unit Series, Catalogue of Hamamatsu Photonics K.K., 2012.

4 Optical Devices and Optomechanical Elements*

Akihiko Chaki and Kenji Magara

CONTENTS

4.1 Introduction 140
4.2 Infrared Optical Element 140
4.3 Ultraviolet Optical Element 141
4.4 X-Ray Optical Element 142
4.5 Filter 143
    4.5.1 Photographic Filter 143
    4.5.2 Spatial Filter, Spatial Frequency Filter 143
4.6 Pinhole 144
4.7 Light Control Device 145
4.8 Optical Modulator 145
4.9 Spatial Light Modulator 145
4.10 Optical Fiber 146
4.11 Optical Polariscope and Optical Scanner 147
4.12 Optical Integrated Circuit and Optoelectronic Integrated Circuit 147
4.13 Attenuator 148
4.14 Shutter 148
    4.14.1 Camera Shutter 148
    4.14.2 Shutter Array 148
4.15 Mechanism Element 149
4.16 Mount 149
4.17 Optical Bench 150
4.18 Tabletop 151
4.19 Vibration Isolation System 152
4.20 Micromotion Element 153
4.21 Positioning Element 154
4.22 Calibration Specimen 156
4.23 Optical Flat 156
4.24 Optical Parallel 157

---

* Some parts of this chapter are reproduced from the chapter "Optomechanical Elements and Devices" in the Japanese book *Practical Optical Keywords Book*, 1999. With permission from Asakura Publishing Co., Ltd.

4.1 INTRODUCTION

Various devices and elements are utilized for the handling of optical elements and the operation of light tuned to a target experiment. With infrared light and ultraviolet light outside the visible light range, some visible light other than the wavelength to be used transmits through it or a special specification for capability to withstand high output is used. Optical elements such as lenses, prisms, and mirrors are not used in naked style, but are handled after being placed in a frame for holding and fixing by suitable means. Articles corresponding to this frame are mounts and holders. An optical bench and a tabletop are used for disposing and fixing the optical element placed in the holder according to the purpose of the experiment, and a vibration isolation stand is used to perform experiments under stable conditions, without being affected by vibrations and other disturbances.

The optical element fixed to the holder constitutes an optical system to meet the purpose of the experiment. Adjustments to the optical system are carried out using a micromotion element and a positioning element so that conditions such as distance, position, inclination, and angle are satisfied while light from a light source is confirmed.

In the optical system, light-related operations, such as strength adjustment of light intensity, taking out of specific wavelength light, and separation of specific wavelength, are performed causing light to pass through the optical system at the required time, blocking the light at the required time, or spatial filtering such as noise component cutting. An attenuator, a filter, and a pinhole are used for these operations. Furthermore, an optical modulator such as an acoustic optical modulator (AOM) and a light deflector become necessary for adjustments of light intensity modulation and deflection angle, and a scanning equipment, such as a galvanometer or a scanner, is used for two-dimensional light operations. For infrared light that is not visible light, adjustments are performed in many cases using visible light in the optical system, and special optical elements for infrared light, such as a lens that has high transmissivity in the infrared region and allows transmission of even visible light, are used. Visible light for adjustments should be used also for ultraviolet light, and in this case, quartz is used for materials of the optical element.

4.2 INFRARED OPTICAL ELEMENT

Most materials such as glass and crystal normally used in the visible light region are transparent also in the near-infrared region and are used for lenses, prisms, and window materials. Silica glass can be used up to a wavelength of around 4 μm and even Pyrex can be used up to a wavelength of around 3 μm. In the infrared region of a wavelength longer than that, use of materials such as halide single crystals, oxide single crystals, glass, chalcogenide glass, and semiconductors is necessary. In optical communication, silica-glass fiber, in which the OH-group that is responsible for absorption is reduced as much as possible, is used. In infrared spectrometry where the wavelength region used is broader, a reflecting optical system is used frequently, whereas with a temperature measurement device and an imaging device such as a night-vision device where a wavelength of 3–5 and 8–14 μm, referred to as the window of atmosphere is used primarily, refractive lens employing Ge and Si of the semiconductor is used and achromatizing lens and zoom lens are manufactured to be used.

In infrared spectrometry, optical elements such as prisms, window materials, and cells are necessary, while appropriate materials are selected taking the applicable wavelength limit, workability, deliquescence, and stability into consideration. Halide single crystals are transparent to wavelength ranging from the ultraviolet region to the infrared region. Although MgF_2 and CaF_2 are stable, their transparency region is up to around 12 μm. The transparency region of NaCl, KBr, and CsI is long—up to 20, 30, and 70 μm, respectively—but these materials need caution in handling due to their deliquescence. In addition, NaCl and KBr exhibit cleavage. The mixed crystal KRS-5 of TlBr and TlI is transparent up to a wavelength of 50 μm and exhibits a little deliquescence. Oxide crystals

FIGURE 4.1 Infrared optical elements and materials in a wavelength range.

(e.g., sapphire) are thermally and chemically stable, and their transparency region is from ultraviolet light to a wavelength of around 7 μm. The refraction factor is from about 1.5 to 2.0. Chalcogenide glass such as $Ge_{10}As_{20}Te_{70}$ is transparent in wavelength range of 1–20 μm, has excellent chemical stability, and is used as the infrared transmission fiber. The refraction factor is as high as 2.5–3.0. In addition, an interference filter using a dielectric body multilayer film and a diffracting grating using a metallic mesh are used (Figure 4.1).

4.3 ULTRAVIOLET OPTICAL ELEMENT

With ordinary optical glasses, with wavelength less than about 350 nm, transmissivity is reduced and cannot be used for long. Therefore in the ultraviolet region, as materials for lens and prisms, silica glass (quartz), Al_2O_3 (sapphire), CaF_2 (fluorite), halide such as MgF_2, and LiF are used. The shortest wavelength of these materials in the transparent region is about 160, 150, 125, 115, and 105 nm, respectively. Due care should be given, if a strong ultraviolet light is irradiated to silica glass or LiF because defects such as color center are generated, thereby reducing the transmissivity (solarization). Performances of silica glass have improved; LiF exhibits a slight deliquescence.

For optical elements used in the ultraviolet region, reducing lens for lithography is mentioned. For window materials of the resonator of KrF excimer laser (wavelength 248 nm), ArF excimer laser (wavelength 193 nm), and F_2 excimer laser (wavelength 157 nm) as the light source, halide crystal is used, while silica glass is used for reducing lens due to the homogeneity of the refraction factor and distortion-related problems. Since chromatic aberration correction is not possible with a single glassy material, several plans are proposed including an excimer laser of a plateau with a narrower band made by etalon, prism, and diffracting grid and combination with a mirror optical system. Among ultraviolet regions, in the vacuum ultraviolet region with a wavelength less than 200 nm, circumstances are significantly different. Since an absorption band, referred to as the Schmann region (wavelength 120–190 nm) due to oxygen molecule, is present, the optical system is placed in vacuum where handling of the optical system becomes troublesome. In some cases, replacement by He gas and the like is possible, though this depends on the intended application. In addition, direct incident reflection of a metal rarely takes place with wavelength less than 40 nm, where some new ideas should be used (see Section 4.4).

For a deflection element, Wollaston prism has been produced using MgF_2 that has birefringence or synthetic quartz as the base material. Although filters are produced by fixing metallic thin films on the metallic gauze, filters are easily broken and therefore, due care should be given. A diffracting grid is used for the spectroscope of the vacuum ultraviolet application, while an oblique incidence-type grid (such as a runner-type diffracting grid) using total reflection is used in the wavelength region less than 50 nm.

4.4   X-RAY OPTICAL ELEMENT

X-ray has a short wavelength and the refraction factor is nearly 1 for every substance; therefore, optical elements that are used under ordinary visible light cannot be used. Normally, spectral elements using Bragg reflection by a single crystal are used. Suppose that the grid interval of the crystal is represented by d, the oblique incidence angle to x-ray crystal (angle from crystal plane) by θ, wavelength by λ, and when $n\lambda = 2d \sin\theta$ (n is integer) is satisfied, a diffraction light can obtain a reinforcing reflection. Further, an x-ray filter is available, which is made of a thin film for selecting character x-ray from x-rays generated in x-ray tubes. Of elements that have been attracting attention currently, an optical element that is able to reflect x-ray or to form an image in the wavelength region referred to as soft x-ray (wavelength several 1/10 ~ several tens nm) is mentioned. Since the refraction factor of substances in the soft x-ray region is slightly smaller than 1, contrary to visible light, total reflection is caused when soft x-ray is incident to substances. However, since there is absorption, 100% reflectance is not obtained as observed with visible light. Figure 4.2a shows a representative Wolter-type optical system. This system consists of a hyperboloid of revolution in which one focal point is shared and an ellipsoid of revolution (or paraboloid of revolution). Since incidence to a mirror surface should be made with an extremely small oblique incidence angle, an imaging

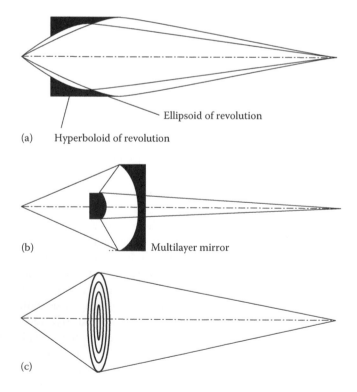

(a)   Hyperboloid of revolution

Ellipsoid of revolution

(b)   Multilayer mirror

(c)

FIGURE 4.2   Soft x-ray imaging element: (a) Wolter type, (b) Schwarzchild type, and (c) Fresnel zone plate.

system with a large numerical aperture is not produced, but chromatic aberration is less and this optical system can be used in a broader wavelength region. Furthermore, the light from an x-ray light source of continuous wavelength can be used effectively.

There is a multilayer film optical system in which substances each having different optical constants are laminated in several tens to several hundred layers to increase reflectance. Use of a multilayer film enables construction of a direct incident optical system, and an optical system with excellent resolving power and broader field of vision can be realized. As for the reflection characteristic of a multilayer mirror, although reflectance as high as several tens of percent is obtained with a wavelength of more than 13 nm, reflectance remains less than 20% with a short wavelength less than that. At present, practical application of an x-ray exposure device has been promoted and a higher efficiency light focusing method has been investigated. Figure 4.2b shows a representative Schwarzschild type. However, with an element of this type, the wavelength of soft x-ray is short, film thickness becomes thin, and interfacial roughness of the multilayer film is not negligible compared to wavelength, and therefore, realization of a multilayer film optical system with a short wavelength is difficult. Deflection elements using a multilayer film and a beam splitter have also been developed.

A Fresnel zone plate that is a diffracting grid in concentric shape is also used as the imaging element (Figure 4.2c). Because this plate has a chromatic aberration, monochromaticity is required for the x-ray to be incident. The resolving power of the zone plate is nearly equal to the width of the outermost shell zone and does not depend on the wavelength. This plate has such features that the resolving power of several tens of nanometers is obtained by an electron beam drawing device and lithography technology and utilization down to a relatively shorter wavelength of several nanometers is possible.

4.5 FILTER

A filter is a component that is provided in an optical path and extracts, enhances, or weakens specific components of light or image as needed and causes the components to pass through.

4.5.1 Photographic Filter

The photographic filter is designed so that its transmissivity changes primarily according to the wavelength of the light. As typical optical filters, ultraviolet protection filter, infrared protection filter, color conversion filter, chromatic compensation filter, short-wavelength light sharp cut filter, and beam attenuator filter are mentioned. The primary object of these filters is to remove light undesirable for obtaining an image or to allow analysis by only a required wavelength.

A deflection filter that selectively passes straight deflection in a particular direction is also used for removal of light reflected by a surface, such as an aqueous surface or a glass surface. Normally, single-lens reflex cameras and video cameras are equipped with a filter mounting section of threaded type at the front frame of the photographing lens.

4.5.2 Spatial Filter, Spatial Frequency Filter

A spatial or spatial frequency filter is an optical component that is provided on the image surface and has a predetermined transmissivity distribution. Filters in which a 100% transmission part and an absorption part are provided alternately and one in which transmissivity changes in sinusoidal wave fashion are available, and they are used for extraction of features of an image or detection of movements. A filter in which a spatial filter is provided in optical Fourier transform plane is referred to as spatial frequency filter, which is used for noise elimination from image, correction, feature extraction, and accentuation and is effective for image processing and analysis (Figure 4.3).

FIGURE 4.3 Functions of spatial filter: (a) without spatial filter, (b) spatial filter under adjustment, and (c) with spatial filter completely adjusted.

4.6 PINHOLE

A pinhole is an optical component produced such that a small hole for passing light is drilled to an optically opaque substrate. The following pinholes are mentioned according to the application:

1. A component used in the optical system using a laser for elimination of unnecessary optical noises (interference stripes or speckle patterns caused in the luminous flux). This is marketed as "pinhole centering device" being disposed at the position where laser light is focused by microscope lens in combination with a micromotion device in three-dimensional directions. In this case, this device is referred to as a spatial filter.
2. Used in combination with a dark box to obtain images. This is used for "camera obscura" (an old painting device) in which such a phenomenon is used that, from a hole drilled to the wall of a dark room, the outside landscape is reflected to the opposite wall. A pinhole camera in which photographic sensitive material is disposed in the dark box has come into practical use. As for the size of the pinhole, there is an optimum diameter with regard to the blurring of image due to diffraction and geometric optics balance.
3. A small hole provided to the dark box constituting the camera to allow passing of light. This results in light trailing of the camera or photographic fog of the film. This sort of trouble occurs frequently with folding-type camera with bellows.

4.7 LIGHT CONTROL DEVICE

An element inserted in the optical path of an optical device, which controls the state of the light arbitrarily, may be referred to as a light control device. In particular, this is meant in many cases by the element that is able to change a controlled variable by adding a certain operation from the outside. For operations from the outside, electric field, magnetic field, sound (ultrasonic sound), and the like are used. In addition, in some cases, the controlled variable is regulated by exchanging a plurality of elements. The light control device includes, in addition to various optical components, a diffracting grid, an optical crystal, and a liquid crystal. Also, an optical shutter for controlling laser oscillation in a Q-switched laser may also be considered as a type of light control device. As an optical shutter for Q-switch, spin mirror and the like were used in conventional technology, whereas in recent years, acousto-optic elements are used. Characteristics of the light to be controlled are light volume, direction of light traveling, frequency, phase, in-plane light volume distribution, spectroscopic energy distribution, and deflection. If attention is given to target the modulation component such as the light volume and deflection angle, it is handled as another type of element even if it is the same element. As an example of such elements, an acousto-optic element is taken into consideration. In the acousto-optic element, an ultrasonic planar wave goes straight into the medium and allows a laser light to be incident to a diffracting grid generated three dimensionally corresponding to a crude wave of the sound. If attention is given to this, the straight light volume of laser is reduced, and the element concerned may be used as the attenuator. Moreover, if the fact that the light volume of primary diffraction light changes according to the intensity of the ultrasonic wave in the medium is utilized, the element is used as the optical modulator. In addition, because the deflection angle of the diffraction light changes according to the changes in frequency of ultrasonic wave (pitch of the diffracting grid), it is used as the light deflector. Also, if the fact that the frequency of the diffraction light changes as much as the frequency of ultrasonic wave is applied, it may be used as the light control device for frequency shift or phase creation.

4.8 OPTICAL MODULATOR

The optical modulator means an optical element that controls light volume, phase, and polarization state by a signal given from the outside. With liquid crystal or optical crystal, the polarization state or phase of the light to be injected changes with the electric field applied (voltage applied across electrodes). It is also possible to change the light volume by disposing an appropriate polarization filter at the incident side and the injection side. With an acousto-optic element, the distribution of the refraction factor generated by traveling of an ultrasonic wave forms a diffracting grid in the medium. Phase modulation or frequency shift is caused to the light injected from here, and light intensity modulation can be performed by changing the frequency of the ultrasonic wave.

4.9 SPATIAL LIGHT MODULATOR

An optical element in which optical modulators are disposed in two-dimensional fashion is referred to as a spatial light modulator (SLM) and is used for liquid crystal televisions and projectors. Historically, the concept originated in two-dimensional arrangement of a Kerr cell in the 1910s. Realization of television was attempted through parallel arrangement of signal lines from selenium photoelectric conversion elements being arranged in two-dimensional fashion on an image face of the imaging optical system. Today, liquid crystal and DMD (digital mirror device) type are used as the SLM, while speeding up of light response time and higher performances are sought with recent techniques and development in optical information and optical communication fields.

4.10 OPTICAL FIBER

An optical fiber is an optical component that uses the fact that light, being incident to one end of a transparent optical component, is output from the other end at high efficiency after total reflection or refraction are repeated inside. This is used for transmission and image transmission in information communication and energy fields. Optical fibers are classified as shown in Table 4.1, according to constituent materials, distribution of refraction factor, and propagation mode.

Figure 4.4 shows different conditions of light propagation depending on types of optical fibers. With a step-index-type optical fiber, light is propagated by repeating total reflection at the interface of high-refraction factor core part and low-refraction factor clad part. With graded-index-type

TABLE 4.1

Classification of Optical Fibers

Classification by constituent materials	Silica-based glass optical fiber (SiO_2 + Ge, B, F)
	Multicomponent-based glass optical fiber ($SiO_2 + B_2O_3 + Na_2O + \cdots$)
	Plastic optical fiber (PMMA and others: for short-distance application)
	Glass fluoride optical fiber (Ultralow loss at $\lambda = 3\ \mu m$)
Classification by propagation mode	Single-mode optical fiber (Type SM)
	Multimode optical fiber (Type MM)
Classification by distribution of refraction factor	Step-index optical fiber (Type SI)
	Graded-index optical fiber (Type GI)

FIGURE 4.4 Propagation of light through optical fiber: (a) optical fiber of step-index type (multimode), (b) optical fiber of graded-index type (multimode), and (c) optical fiber of single-mode type (single mode).

optical fiber, light is propagated with smooth meandering according to the distribution of the refraction factor. Although an arbitrary number of optical paths can be set from geometric optics viewpoint, only one with the same phase as the first light is present in one cycle of reflection or meandering in the optical fiber, due to the relationship with the interference with light, and is transmitted. This is referred to as optical fiber mode. With a multimode optical fiber, since the propagation rate is different in every mode, light being incident to the input end becomes a time-wise expanded pulse at the output end due to the time difference for every mode. In the meantime, with single-mode-type optical fiber, transmission of a large amount of signals is possible because there is no modal dispersion. Further, optical fibers with a hole inside, referred to as holey fibers, are available to be used for large-capacity and long-distance transmission. Typical dimensions of the optical fiber are core diameter 50 µm/clad and outside diameter 125 µm for multimode and core diameter 10 µm/clad and outside diameter 125 µm for single mode. When the optical fiber is used for optical path of the interferometer, single-mode optical fiber is used because light with single phase is handled.

With a step-index-type optical fiber, when core diameter is made large and the difference of the refraction factor between core and clad is made large, light being incident from wider angle in a wider area can be taken in efficiently. Plastic optical fiber used for short-distance information transmission is designed as mentioned. A step-index-type optical fiber is also used in optical fibers for illumination light. When step-index-type optical fibers are arranged regularly in two-dimensional fashion, an image that is incident to one end can be conveyed exactly to other end. A bundle of flexible optical fibers is used in endoscopes for medical and industrial use. Furthermore, an optical fiber faceplate in which a bundle of optical fibers is fixed in plate shape is used for image transmission.

4.11 OPTICAL POLARISCOPE AND OPTICAL SCANNER

An optical component that changes the direction of light traveling is referred to as an optical polariscope or optical scanner. The principle of light deflection is classified as follows:

1. *Light reflection*: Law of light reflection is used. Because the incident angle and reflection angle are identical, when the reflection plane is turned by angle θ, the direction of reflection light changes as much as 2θ. A galvano-mirror and polygon mirror are mentioned as the optical polariscopes using this principle and are used for laser printers and scanners.
2. *Refraction of light*: When light is incident to a prism, the direction of traveling of injection light changes according to the apex angle and the refraction factor of the prism. Pentaprism provided in the finder of the single-lens reflex camera and Porro prism for correction of images of the binocular are examples.
3. *Diffraction of light*: When the grid pitch of the diffracting grid is changed, the direction of travel of diffraction light is changed. With an ultrasonic wave optical polariscope using acousto-optic effects, the grid pitch is changed by changing the frequency of the ultrasonic wave, thereby performing light deflection.

4.12 OPTICAL INTEGRATED CIRCUIT AND OPTOELECTRONIC INTEGRATED CIRCUIT

This is a component with such a construction that optical circuit elements performing transmission and treatment of optical signal are integrated on a single substrate such as a dielectric crystal, in analogous fashion of ordinary IC (integrated circuit), in which electric elements are integrated on a single substrate such as silicon. With this component, a signal is generated, transmitted, and treated in the original form of light.

For optical circuit elements used for light IC, laser light source, light detector, modulator (intensity or phase), polariscope, optical mode converter, and the like are mentioned in addition to waveguide, lens, diffracting grid, branching filter, mode separator, and modal synthesizer.

Compared to conventional optical instruments using individual optical element such as lenses, prisms, shutters, and the like, features of this component are small-sized, lightweight, stable, and have quick response. This component is distinguished from OEIC (optoelectronic integrated circuit), which processes and transmits a light signal after converting it to an electrical signal. OEIC is an integrated circuit on which optical semiconductor elements and electric semiconductor elements are integrated on the same semiconductor substrate and are associated with each other. In the OEIC, the electronic circuit portion excels at signal logic operation and high-speed access memory functions and light circuit portion excels at signal transmission, high-speed processing, and image input/ output functions. A new function is realized by combining these features.

These components are largely classified into those in which luminous element such as semiconductor laser and field-effect transistor for driving it are integrated and those in which a light receiving element such as photodiode and field-effect transistor for amplification and signal processing are integrated (charge-coupled device (CCD), etc.). The primary application is transmission and receiving of optical communication. Compound semiconductors such as gallium–arsenic and indium–phosphorous and mixed crystals have been attracting attention as a material.

4.13 ATTENUATOR

An attenuator is a device provided in an optical path to be used for adjustments of light volume (attenuation device). This device is capable of maintaining the quality of light. Various types are available, which include a type where an optical wedge with changing transmissivity depending on positions is moved mechanically for adjustments, a type in which light volume is regulated using light modulation element by liquid crystal or crystal, and a type in which regulation is performed using light volume loss in the optical waveguide or optical fiber.

4.14 SHUTTER

A shutter is an optical device disposed in an optical path to shift transmission and isolation of light. In general, a shutter that has a mechanical construction is referred to as a mechanical shutter; one that has an electrical construction as represented by Kerr cell and liquid crystal is referred to as an electrical shutter; one that controls the exposure time of imaging element as represented by CCD and complementary metal-oxide semiconductor directly and electrically to realize shutter function is referred to as an electronic shutter. Ways of discrimination between electric shutter, electronic shutter, and electrically controlled mechanical shutter are diverse depending on the purpose intended, and they are frequently mixed up.

4.14.1 CAMERA SHUTTER

A camera shutter is used to impart appropriate exposure to photosensitive materials or imaging element. A camera shutter disposed in the vicinity of lens is referred to as lens shutter; one that is disposed in the vicinity of image formation is referred to as focal plane shutter.

4.14.2 SHUTTER ARRAY

An array in which a liquid crystal shutter, a Lanthanum-modified lead zirconate Titanate (PLZT) shutter, or the like is disposed in one-dimensional or two-dimensional fashion is referred to as a shutter array. When image formation is made on a member with photosensitivity through imaging optics, a printer can be realized. Further, a combination of back light and two-dimensional liquid crystal shutter enables the creation of liquid crystal display.

FIGURE 4.5   Stages with translation and rotation. (Courtesy of Chuo Precision Industrial, Tokyo, Japan.)

4.15   MECHANISM ELEMENT

Mechanism element is normally referred to as a stage (Figure 4.5) and is used for movement and positioning in horizontal $(X \cdot Y)$, vertical (Z), rotation (θ), and tilted directions. For stage guiding method, dovetail mechanism, hard steel wire and steel ball, V-groove rail and steel ball, V-groove rail and cross roller, ball bush, ball way, rotation bearing, and the like are mentioned, and selection is made depending on load, movement amount, movement direction, and accuracy.

1. *Dovetail mechanism*: Used for guiding of sliding contact instead of rolling contact and suited for guiding of a stage for mere positioning, which does not need good accuracy.
2. *Hard steel wire and steel ball*: Suited for guiding of light-load stage, which does not need high-accuracy positioning.
3. *V-groove rail and steel ball*: Suited primarily for guiding of high-accuracy light-load precision positioning stage.
4. *V-groove rail and cross roller*: Equipped with high load capacity for up/down and transverse loads and suited for guiding of high-accuracy heavy-load precision positioning stage.
5. *Ball bush*: Equipped with precision shaft and ball slide bearing as the rail track stand and therefore, a relatively longer stroke is made possible. Suited for simplified type and light-load stage.
6. *Ball way*: Equipped with rail track stand and circulating type bearing and therefore, suited for guiding of long-stroke, heavy-load, and relatively high-precision positioning stage.

For movement mechanism, rack and pinion, worm gear, micrometer head, PZT element, micromotor (DC motor), stepping motor, and linear motor are used.

4.16   MOUNT

Used for holding laser, shutter, optical accessories such as aperture, and optical elements such as lenses, mirrors, and prisms. Many are designed to be disposed on an optical tabletop using a magnetic stand by providing a shaft with diameter of 12 or 20 mm at the lower part (Figure 4.6). In many

FIGURE 4.6   Mirror mount. (Courtesy of Chuo Precision Industrial, Tokyo, Japan.)

cases, the mount is equipped with a mechanism for adjusting the angle of the component to be held in horizontal and vertical directions. The direction of light traveling can be manipulated or adjusted by adjusting the component angle. For fixing of the optical element, the following methods are used: (1) direct adhesion to holder, (2) element pressed against holder using presser ring, (3) fixing by V-groove and presser, and (4) element supported by three supporting bars at three points. As for angle adjustment mechanism, lever type, plate spring, and gimbal type are available. Fine screw thread and micrometer head are used for angle adjustments. Electrically driven micrometer head is used when remote operation is required.

4.17   OPTICAL BENCH

An optical bench is a support bed used in combination with a carrier that can be slid and fixed on the rail fixed to the tabletop at an arbitrary position in the optical axis direction for optical performance experiments. Experiments can be performed while optical elements such as lens and filter are disposed on the straight line.

As for profiles of the optical bench, triangle type, X-type, and Y-type are available. Triangle type has been used conventionally and has an optical element straddled across the optical bench through the carrier, and therefore, positional repeatability after mounting and dismounting, and stability are good. If a scale is mounted to the side face of the rail and reading of the scale is taken from the carrier, this reading can be used for position adjustment and measurements.

X-type has the same sectional profile as that of the prototype meter, has high rigidity, and is capable of corresponding to a longer size article (Figure 4.7). For materials, aluminum alloy is mainly used, contributing to the lightweight feature. The carrier can be fixed to the side face and the upper face, allowing two-dimensional or three-dimensional combination.

Triangle type and X-type optical benches are fixed using a dedicated stand or mount. Y-type has such a profile that X-type rail is halved, which allows reduction in height of the optical axis and direct mounting to the tabletop.

FIGURE 4.7   X-type rail. (Courtesy of Chuo Precision Industrial, Tokyo, Japan.)

FIGURE 4.8   X-type rail carrier. (Courtesy of Chuo Precision Industrial, Tokyo, Japan.)

The carrier for disposing the optical element on the rail is fixed by pressing the fixing screw provided to the carrier side face against the rail side face. A holder for optical elements and screw holes for fixing the micromotion mechanism for optical axis adjustment are provided in a carrier (Figure 4.8).

As an optical system to be disposed on the straight line, the optical bench is used for Fourier transform optical system, focal length measurement, and photoelasticity.

4.18   TABLETOP

A base upper face of a tabletop is a precision planar surface, and this is used as the reference surface for various experiments (Figure 4.9). This is used for many optical experiments in addition to precision measurements. The features required are the flatness of the tabletop surface should be

FIGURE 4.9   Optical tabletop. (Courtesy of Chuo Precision Industrial, Tokyo, Japan.)

satisfactory for stable arrangement and fixing of the optical element; rigidity should be high enough to prevent positional displacement of the optical element being fixed; tabletop itself does not act as the vibration source due to vibration exerted to the tabletop from the outside; and attenuation of the tabletop is satisfactory against vibration exerted. If mark-off lines in grid pattern are marked on the tabletop surface, they are conveniently used in positioning of the optical element.

For the tabletop, stone materials (granite, etc.), cast iron, aluminum alloy casting, and steel materials are used. Although advantages of stone material are good surface flatness, good electrical insulation properties, and high thermal capacity, this material is nonmagnetic and therefore, fixing of the optical element using a magnetic base is not possible. Another disadvantage is that processing of screw holes is difficult. Cast iron is manufactured by casting. Although deformation due to cooling after manufacturing is a problem, no distortion is caused if sufficient annealing is provided at production. However, dust proof processing is required. Aluminum alloy casting is light in weight and drilling of screw holes is easy, while flatness is slightly poor since the surface remains as machine finished. As for steel materials, the steel plate is pasted to the upper and lower faces of the honeycomb structure. A stainless steel plate (SUS430) that has magnetism and is hardly rusted is used for the upper faceplate to allow use of magnetic base, while steel honeycomb core or aluminum honeycomb core is used for the core of honeycomb structure. In case magnetism is not desirable, a stainless steel plate (SUS403) that is nonmagnetic is used. With honeycomb structure, hexagonal cores are closely bonded to form a honeycomb profile, and high rigidity is maintained though light in weight. However, flatness is slightly deteriorated.

Tabletop is not used solely, but optical elements are fixed onto it to be used for various optical experiments, and in many cases, tabletop is mounted on the vibration isolation stand to reduce influences of vibrations from the floor face.

4.19 VIBRATION ISOLATION SYSTEM

A vibration isolation system is used to perform optical experiments in a stable manner avoiding vibrations from the floor face or is used to attenuate vibrations conveyed to the optical element or tabletop. Though the vibration isolation system is used to prevent vibrations to the tabletop or optical element fixed to the table plate surface as much as possible, it is not able to reduce vibrations to zero level. The principle of vibration isolation is explained by the vibration system based on springs in such a way that for vibration frequency f of the vibration system exerted from the outside such as floor vibration, the smaller the vibration frequency f_n of the vibration isolation system including optical element and tabletop, and the greater the f/f_n, the better vibration isolation effects are obtained. If vibration frequency f of the vibration system is equal to vibration frequency f_n of the vibration isolation system, vibration increases due to resonance, and stable experiments are not possible. Vibrations caused from the floor face are in more than a vibration region of 2–4 Hz, and almost all vibration energy is concentrated in 5–15 Hz. For this reason, to develop vibration isolation effects, the frequency f_n of the vibration isolation system including the vibration isolation stand (Figure 4.10) should be in a range of less than 1–3 Hz, which is lower than the frequency region of the vibration caused from the floor face.

Many of the vibration isolation systems employ a pneumatic spring for vibration isolation. The pneumatic spring has an airflow restriction mechanism by a small throttle or an orifice in a sealed tank provided to the leg part of the vibration isolation stand. When a displacement is caused in the vertical direction due to vibrations from the floor face, the orifice in the sealed tank and air in the tank partitioned by the orifice serve as a spring, thereby absorbing and attenuating the vibrations. The mass of the tabletop and optical element is supported with a metallic receiver by pneumatic pressure in the tank. Accordingly, the pneumatic spring needs a pneumatic source. In addition, an automatic pressure regulating mechanism for correcting tilting of the tabletop due to mass of the optical element mounted and a regulator for pressure of the air supplied are added.

In order to suppress the resonance phenomenon or to shorten vibration attenuation time, a vibration sensor and position variation sensor are mounted on the vibration isolation stand or floor, and

FIGURE 4.10 Optical table consisting of vibration isolator and bread board with optical arrangement for holography. (Courtesy of Chuo Precision Industrial, Tokyo, Japan.)

vibrations and position variations are detected and fed back to the vibration isolation mechanism, thereby obtaining significant vibration isolation effects. This component is used for observation of interference and minute structure, device for processing, and optical system.

4.20 MICROMOTION ELEMENT

A fine-pitch screw or micrometer head are used for a micromotion element (Figure 4.11). The principle of the micrometer is such that displacement of a length is enlarged by the rotational angle of the screw and diameter of the thimble, and the construction used is such that the male screw cut to the spindle is engaged with the female screw fixed to the frame (Figure 4.12).

FIGURE 4.11 Micrometer head.

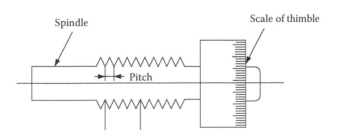

FIGURE 4.12 Screw inside micrometer.

Suppose that the spindle turns angle α and measurement end of the spindle then moves X (mm), a relationship represented by $X = P\alpha/2\pi$ (P, pitch of screw [mm]; α, rotational angle of screw [rad]) is established between X and α. If the pitch is made small, the amount of movement of one turning of the spindle becomes small, and the same movement can be expressed by a large angle, which results in increased accuracy. Even if the rotational angle is the same, if the radius of a graduated cylinder (thimble) is made large, the width of the graduation is widened, and it improves the accuracy. The pitch of screw of an ordinary micrometer is 0.5 mm, and less than that is divided into 50 equal parts in circumferential direction while the minimum scale is set to be 0.01 mm. Pitch of screw of a fine micrometer is 0.1 mm and the minimum scale is set to be 0.002 mm. In addition, as micrometers with special mechanism, a micrometer with vernier that ensures better sensitivity and accuracy than those of ordinary ones, differential micrometer, micrometer with dial gauge, micrometer with numerical scale, micron micrometer, and digital micrometer are mentioned.

4.21 POSITIONING ELEMENT

For positioning elements, piezoelectric element (PZT element), micromotor (DC coreless motor), and stepping motor are mentioned. Piezoelectric element is an element that generates a voltage when a stress is applied and generates a strain when a voltage is applied (piezoelectric effect), and denotes a certain crystal material. Several substances found through nature also exhibit such phenomena, while the material of the piezoelectric element used at present is multicrystal ceramics such as lead zirconium titanate (PZT) (Figure 4.13). With conventional piezoelectric element, a linear displacement of several hundred micrometers is obtained if a voltage is supplied. The actual amount of displacement depends on the materials themselves constituting an actuator, number of layers of materials, and voltage supplied to each layer. Piezoelectric element is capable of responding to a voltage with time constant of a microsecond unit and its displacement resolution is restricted by noises of the electrical supply source. In addition, since a piezoelectric element involves hysteresis effects (Figure 4.14) at actuation, it is not possible to perform linear motion thoroughly. However, in recent years, a piezoelectric element that has a strain gauge at its body has appeared. High linearity driving is now possible if feedback control of the position is carried out using such a sensor. When the stage is driven using a piezoelectric element, the simplest

FIGURE 4.13 Construction of piezoelectric element for positioning.

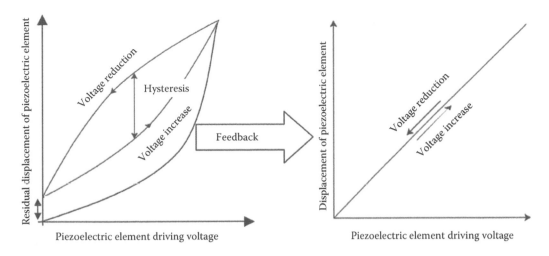

FIGURE 4.14 Hysteresis effects in driving piezoelectric element.

structure is micromotion/coarse motion composite mechanism in which a micrometer head is used for coarse motions and piezoelectric element is used for micromotions. As representative of the stage using a plurality of piezoelectric elements, inchworm motion mechanism is mentioned. A micromotor is a low-output precision motor that uses a permanent magnet for the stator, and many are of coreless construction. For example, a DC coreless motor that is one type of micromotor has the following features:

1. Housing structure consists of a magnetic shield.
2. Low-noise and longer service life are possible by adoption of precious metal brush.
3. Electric noises are low (some models incorporate a noise killer).

When the micromotor is used as the positioning element, two methods are available. One is to push out the spindle through the gear in a similar manner as the micromotor head for positioning, and the other is to perform direct positioning by the gear. In addition, positioning accuracy can be improved with the use of an encoder.

Stepping motors are used widely as the positioning element thanks to their features as enumerated below.

1. Because turns are synchronized in a stepwise fashion with the input pulse, high moving resolution is obtained easily when built into a machine.
2. High-precision positioning is possible with open loop.
3. Response at starting and stopping is excellent.
4. Angular error at stopping is not accumulated.
5. High torque is obtained at low-speed operation.
6. Construction of motor is simple allowing ease of maintenance.
7. Suited for mass production because motors can be procured at inexpensive price.

The stepping motor has a big retaining power even at stopping and is able to maintain stop position without depending on a mechanical brake. When a stage is driven using the stepping motor, the feed screw is turned through coupling. Moving resolution of the stage at this point is determined by pitch of the feed screw and rotational resolution of the motor. In recent years, however, resolution is made small using electrical microstepping.

4.22   CALIBRATION SPECIMEN

Calibration is used primarily for sensitivity calibration and performance tests as the reference of measurements of measurement equipment or those of test equipment. Normally, small-shaped or small-type standard* and reference instruments† are referred to as calibration specimen. Further, an optical flat is used frequently as the flatness and straightness measurement calibration specimen, and a block gauge or a step gauge is used as step measurement calibration specimen. Typical calibration specimen includes the following:

1. Calibration specimen for sensing pin-type surface roughness measurement machine: Calibration specimen for calibration of indication of surface roughness measurement machine or longitudinal magnification. Shape, scattering of surface roughness, and indication of surface roughness (Ra or Rz) are defined.
2. Standard for calibration of roundness measurement machine: A cylindrical instrument with one or more than one small planar surface or curved surface on the outer circumference to give dynamic and constant displacement to a probe.
3. Calibration specimen for rotational accuracy inspection of roundness measurement machine: A standard with spherical, semispherical, or cylindrical shape.
4. Calibration specimen for calibration of sensing pin-type shape measurement machine: A calibration specimen with both shapes of length (X direction) and step (Z direction) reference.
5. Calibration specimen for ultrasonic flaw detection: A specimen for which material, shape, and dimensions are defined and also verified for ultrasonic wave, used for performance test of a flaw detector or sensitivity adjustment.

4.23   OPTICAL FLAT

An optical flat is an optical element in disk shape with a flat measurement surface, made of optical glass or silica glass, and is used for flatness measurement of an optically polished surface or precision finished surface, such as a block gauge, using light wave interference. Measurement surface of the optical flat is placed on the measurement object surface to generate interference fringes. These fringes are referred to as Newton ring, and flatness is obtained from the number of these fringes considering that the interval of fringes is 1/2 of wavelength λ. The poorer the surface of the measurement object, the more fringes are generated, and if the surface is distorted, fringes are also distorted. With a desirable surface, interference fringes are in unicolor, and the flatness F is obtained from the ratio of the amount of bending a with regard to the center distance of interference fringes b, while one end of the optical flat is used as the contact and the other end is floated to generate parallel interference fringes, as shown in Figure 4.15.

$$F = \frac{\lambda}{2} \times \frac{b}{a}$$

Meanwhile, ratio of b/a is used as it is. For example, if this is 1/2, it is referred to as 1/2 fringe from the number of fringes or as flatness of $\lambda/4$ while wavelength λ is referenced.

If the fringe is concave to the contact side, measurement target surface is judged to be a convex surface, and if in convex shape, judged to be a concave surface. For observation of interference fringes, use of a short-wavelength spectral lamp provides good contrast; white lamp illumination is

---

* *Standard*: One that is used as the reference of the measurement and represents a magnitude of a volume expressed by a certain unit.
† *Reference instrument*: One that is used as the reference of weighing in official verification or inspection at the manufacturer.

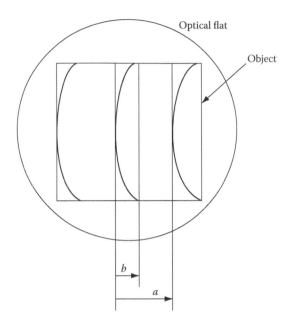

FIGURE 4.15 Flatness measurement.

normally used and in this case, stripes are colored and judgment is made by their red color. If the wavelength λ is 0.64 μm, measurement is possible with 0.32 μm unit. Since reflectance of metals is higher than that of glasses, if increasing reflection coating is provided to a flat measurement surface, the contrast of fringes is enhanced and it becomes easy to see. It is preferable that, at measurement, the measurement object surface and optical flat are cleaned and the temperature is maintained constant for a while before measurement is commenced.

According to JIS (Japanese Industrial Standards) B7430 (1977) "Optical flat," single surface measurement plane and both surface measurement plane are specified excluding those used as components for the interferometer and other optical measurement instrument, and the outside diameter is specified to be 45, 60, 80, 100, and 130 mm. Performances are classified into three categories. For example, allowable value of Class 1 flatness is 0.05 μm. (Class 0 is specified as reference class and performances are half of it.) If this is expressed based on wavelength λ 0.64 μm, λ/12 is obtained.

4.24 OPTICAL PARALLEL

An optical parallel is an optical element in disk shape with a parallel and flat measurement surface, made of optical glass or silica glass and is used for measurements of flatness and parallelism of the measurement surface of the micrometer using light wave interference and of periodical error of its spindle. Screw pitch of the micrometer is 0.5 mm; one set of four sheets, each with thicknesses different by 0.12, 0.25, and 0.37 mm with regard to the reference, is used; thickness is provided to allow measurements every 1/4 turning. For flatness of the measurement surface of an anvil or a spindle, the optical parallel (Figure 4.16) is placed on it to generate interference fringes, and the number of fringes is counted. For parallelism, the optical parallel is sandwiched by the anvil and spindle, the measurement surface of the anvil is observed from the spindle side and the measurement surface of the spindle is observed from the anvil side, and the number of fringes is counted. For observation of the interference fringes, white lamp illumination is normally used and judgment is made by red color fringes, as mentioned in the optical flat section. Since the interval of fringes is 1/2 of wavelength λ, then wavelength λ = 0.64 μm, and flatness and parallelism can be obtained every 0.32 μm unit.

FIGURE 4.16 Optical parallel. (Courtesy of Nikon, Tochigi, Japan.)

According to JIS B7431 (1977) "Optical parallel," outside diameter is ϕ 30 mm, thickness is 12 and 24 mm, performances are classified into two categories, and, for example, allowable value of Class 1 is flatness 0.10 µm and parallelism 0.2 µm. (Class 0 is specified as reference class and precision of flatness and parallelism is half of those, respectively.)

Section II

Fundamentals of Principles and Techniques for Metrology

5 Propagation of Light

Natalia Dushkina

CONTENTS

5.1 Properties of Light... 162
    5.1.1 Maxwell's Equations.. 162
    5.1.2 Harmonic Waves and the Wave Equation.. 164
    5.1.3 Nature of Light 165
        5.1.3.1 Light Is a Three-Dimensional Plane Electromagnetic Wave.................... 166
        5.1.3.2 Light Is a Transverse Wave 166
        5.1.3.3 Electric and Magnetic Vectors Are Perpendicular to Each Other............ 167
        5.1.3.4 Light Polarization 167
        5.1.3.5 Energy Flow and Poynting Vector... 168
        5.1.3.6 Irradiance.. 169
        5.1.3.7 Photons, Radiation Pressure, and Momentum... 170
        5.1.3.8 Electromagnetic Spectrum.. 171
5.2 Propagation of Light in Isotropic Media 172
    5.2.1 Propagation of Light in Dielectrics.. 172
    5.2.2 Reflection and Refraction 172
        5.2.2.1 Boundary Conditions.. 172
        5.2.2.2 Laws of Reflection and Refraction.. 173
        5.2.2.3 Fresnel Formulae 174
        5.2.2.4 Normal Incidence.. 179
        5.2.2.5 Degree of Polarization 179
        5.2.2.6 Brewster Angle 179
        5.2.2.7 Phase Relations 180
    5.2.3 Total Internal Reflection of Light 181
        5.2.3.1 Reflected Wave.. 181
        5.2.3.2 Evanescent Wave... 184
    5.2.4 Dispersion 185
        5.2.4.1 Dispersion in Dielectrics... 185
        5.2.4.2 Normal Dispersion.. 187
        5.2.4.3 Anomalous Dispersion.. 188
        5.2.4.4 Group Velocity of a Wave Packet 189
        5.2.4.5 Color of Objects 190
    5.2.5 Propagation of Light in Metamaterials... 190
References.. 193

This chapter presents a brief description of the fundamental properties of light as described by Maxwell's electromagnetic (EM) theory and the laws governing the propagation of light in isotropic materials. The goal is to summarize some of the most often used concepts and expressions such as Maxwell's equations, wave equation, energy flow, the Poynting vector, irradiance, reflection, refraction, and dispersion. We confine our discussion to isotropic and homogeneous materials, and mainly to nonconducting materials with linear properties. Comprehensive description of propagation of light in anisotropic media, as well as reflection and propagation of light in conducting media is given in Refs. [1,2]. This chapter concentrates on the wave character of light and does not consider the quantum nature of light. Coherence and polarization of light are not discussed in details here, because they are extensively described in Chapters 1 and 5, respectively. The detailed derivation of the final formulae and equations presented here can be found in Refs. [1–5]. For introduction to the EM theory of light one can refer, to Refs. [4–10], while more detailed description of the concepts of modern optics are given in Refs. [11–15]. The fundamentals of electrodynamics, which is the basis of Maxwell's EM theory and some of the mathematical methods used are presented in Refs. [16–19]. The books of Shen [20] and Boyd [21] are possible introductory text to the principles of nonlinear optics and some introductory articles on metrological methods that employ nonlinear optics [20–24]. For the specifics of interaction of light with holographic media, as well as principles, techniques, and applications of optical holography, one could refer to the text of Hariharan [25]. For an introduction to biomedical optics and propagation of light in live tissues, one can start with the book of Splinter and Hooper [26]. References [27–49] are only a short list of the literature devoted to the principles of nano-optics, evanescent fields, near-field optical methods, and other optical methods for characterization of thin films or nano-structures. At the end of this chapter, the reader finds a brief introduction to the newly emerging and quickly developing field of artificially constructed metamaterials (MMs), which are of considerable interest because of their extraordinary EM properties [50–74]. There is a large amount of publications in all these areas and the ones referred to here are only a small fraction.

5.1 PROPERTIES OF LIGHT

5.1.1 Maxwell's Equations

Light is an EM wave. Its properties are described by a set of four equations, known as Maxwell's equations [1–3]. Maxwell's EM theory explains all optical phenomena except certain quantum phenomena, for instance, absorption and emission of light. Maxwell's equations, given in a general form below, summarize the four basic laws of classical electrodynamics [16–18]:

$$\text{div } \mathbf{D} = \rho \tag{5.1a}$$

$$\text{div } \mathbf{B} = 0 \tag{5.1b}$$

$$\text{curl } \mathbf{E} = -\frac{\partial \mathbf{B}}{\partial t} \tag{5.1c}$$

$$\text{curl } \mathbf{H} = \mathbf{J} + \frac{\partial \mathbf{D}}{\partial t} \tag{5.1d}$$

where
  $\mathbf{E}$ and $\mathbf{H}$ are the electric and magnetic field vectors, respectively
  ρ is the free charge density
  $\mathbf{J}$ is the free current density

Vector fields $\mathbf{D}$ and $\mathbf{B}$ are the electric displacement and magnetic induction, respectively which describe the behavior of material media in the presence of external electric and magnetic fields. Equation 5.1a represents Gauss's law for electric field (free space): the outward electric flux integrated over a closed surface is proportional to the net electric charge enclosed by the surface. Equation 5.1b is Gauss's law for magnetic field (free space): the outward magnetic flux integrated over a closed surface is zero. It implies the experimental fact that no isolated magnetic poles (magnetic monopoles) have ever been observed. Equation 5.1c is the mathematical representation of Faraday's induction law for free space: a time-varying magnetic field induces an electric field. Equation 5.1d is Ampere's law for free space, which states that a time-varying electric field induces a magnetic field. Equations 5.1a through d are valid for time-varying fields such as the electric and magnetic fields constituting an EM wave and relate the electric and magnetic fields with the sources (currents and charges) that produce them in matter. Maxwell's equations do not account for fields that are constant in time and space and any such additional field should be added to the $\mathbf{E}$ and $\mathbf{H}$ fields. Since the electric field $\mathbf{E}$ and the magnetic induction $\mathbf{B}$ determine forces acting on a charge (or current), the actual physical observables are forces. The force exerted on a charge q moving with velocity $\mathbf{v}$ in an EM field $(\mathbf{E}, \mathbf{B})$ is given by the Lorentz force law:

$$\mathbf{F} = q(\mathbf{E} + \mathbf{v} \times \mathbf{B}) \tag{5.2}$$

The EM properties of the medium are described by the macroscopic polarization $\mathbf{P}$ and magnetization $\mathbf{M}$, which are linked to the vectors $\mathbf{E}$, $\mathbf{D}$, $\mathbf{H}$, and $\mathbf{B}$ according to

$$\mathbf{D}(\mathbf{r}, t) = \varepsilon_0 \mathbf{E}(\mathbf{r}, t) + \mathbf{P}(\mathbf{r}, t) \tag{5.3a}$$

$$\mathbf{H}(\mathbf{r}, t) = \mu_0^{-1} \mathbf{B}(\mathbf{r}, t) - \mathbf{M}(\mathbf{r}, t) \tag{5.3b}$$

These general equations are always valid because they do not impose conditions on the medium. The constants ε_0 and μ_0 are the electric permittivity and the permeability of free space, respectively, with values

$$\varepsilon_0 = 8.8542 \times 10^{-12} \frac{\mathrm{C}^2}{\mathrm{N} \cdot \mathrm{m}^2} \tag{5.4a}$$

and

$$\mu_0 = 4\pi \times 10^{-7} \frac{\mathrm{N} \cdot \mathrm{s}^2}{\mathrm{C}^2} = 1.2566 \times 10^{-6} \frac{\mathrm{N} \cdot \mathrm{s}^2}{\mathrm{C}^2} \tag{5.4b}$$

The behavior of matter under the influence of electric and magnetic fields is described by the material equations, which for a nondispersive, isotropic, and continuous medium have the form

$$\mathbf{D} = \varepsilon \mathbf{E} \tag{5.5a}$$

$$\mathbf{B} = \mu \mathbf{H} \tag{5.5b}$$

$$\mathbf{J}_c = \sigma \mathbf{E} \tag{5.5c}$$

$$\mathbf{P} = \varepsilon_0 \chi_e \mathbf{E} \tag{5.5d}$$

$$\mathbf{M} = \chi_m \mathbf{H} \tag{5.5e}$$

where
   $\mathbf{J}_c$ denotes the induced conduction current density
   χ_e and χ_m are the electric and magnetic susceptibility, respectively
   Material parameters ε, μ, and σ are the permittivity, magnetic permeability, and specific conductivity, respectively

The permittivity ε and permeability μ describe the response of a homogeneous material to the electric and magnetic components, respectively, of a radiation of a given frequency. The permittivity and the magnetic permeability of the substance are expressed as $\varepsilon = \varepsilon_r \varepsilon_0$ and $\mu = \mu_r \mu_0$ respectively, where $\varepsilon_r = \varepsilon/\varepsilon_0$ and $\mu_r = \mu/\mu_0$ are the values relative to vacuum. The relative permittivity of the material ε_r is also referred to as the dielectric constant. For nonmagnetic materials $\mu_r = 1$. In this chapter, we consider only nonmagnetic materials, thus $\mu = \mu_0$.

An isotropic and homogeneous medium is one, whose properties at each point are independent of direction. Hence, the material parameters ε, μ, and σ of such a medium are constants. For anisotropic materials, their values in Equations 5.5 are replaced by tensors. Equations 5.5 describe a linear medium. In the case of nonlinear material, the right-hand sides contain terms of higher power. If the material parameters are functions of frequency, it said that the material exhibits temporal dispersion. In linear, isotropic, homogenous, and source-free media, the EM field is entirely defined by two scalar fields.

In a nonmagnetic, nonconducting material ($\mathbf{J} = 0$, $\sigma = 0$) with no free charges ($\rho = 0$), the four Maxwell's equations can be expressed only in terms of the $\mathbf{E}$ and $\mathbf{B}$ fields. In Cartesian coordinates, they become [1,2]

$$\nabla \cdot \mathbf{E} = 0 \tag{5.6a}$$

$$\nabla \cdot \mathbf{B} = 0 \tag{5.6b}$$

$$\nabla \times \mathbf{E} = -\frac{\partial \mathbf{B}}{\partial t} \tag{5.6c}$$

$$\nabla \times \mathbf{B} = \mu_0 \varepsilon \frac{\partial \mathbf{E}}{\partial t} \tag{5.6d}$$

where the divergence and curl of a vector, for instance of the vector $\mathbf{E}$, are

$$\operatorname{div} \mathbf{E} = \nabla \cdot \mathbf{E} = \frac{\partial \mathbf{E}_x}{\partial x} + \frac{\partial \mathbf{E}_y}{\partial y} + \frac{\partial \mathbf{E}_z}{\partial z} \tag{5.7a}$$

and

$$\operatorname{curl} \mathbf{E} = \nabla \times \mathbf{E} = \left(\frac{\partial \mathbf{E}_z}{\partial y} - \frac{\partial \mathbf{E}_y}{\partial z} \right) \hat{\mathbf{x}} + \left(\frac{\partial \mathbf{E}_x}{\partial z} - \frac{\partial \mathbf{E}_z}{\partial x} \right) \hat{\mathbf{y}} + \left(\frac{\partial \mathbf{E}_y}{\partial x} - \frac{\partial \mathbf{E}_x}{\partial y} \right) \hat{\mathbf{z}} \tag{5.7b}$$

where $\hat{\mathbf{x}}$, $\hat{\mathbf{y}}$, and $\hat{\mathbf{z}}$ are unit vectors in the x, y, and z directions.

5.1.2 Harmonic Waves and the Wave Equation

Harmonic waves are produced by sources that oscillate in a periodic fashion and can be described by sine or cosine functions. A general harmonic wave in three-dimensions can be represented in a complex form as

$$\psi(x,y,z,t) = A(x,y,z)e^{-i(\omega t - k \cdot r + \varphi_0)}, \tag{5.8}$$

where $A(x, y, z)$ is the amplitude of a wave with wavelength λ, frequency f, and a period $T = 1/f$. The wavelength and frequency determine the propagation constant $k = 2\pi/\lambda$ and the angular frequency $\omega = 2\pi f$ of the wave. Since k is related to λ in the same way as the angular frequency

$\omega = 2\pi/T$ is related to the wave temporal period T, the wavelength λ can be considered as the spatial period of the wave, and the propagation constant k as the spatial frequency of the wave. In this context, the reciprocal of the wavelength $1/\lambda$ is called wave number. The argument of the exponent $\varphi = \omega t - \mathbf{k} \cdot \mathbf{r} + \varphi_0$ is the phase of the wave with φ_0 being the initial phase at $t = 0$ and r = 0. The phase φ depends on the spatial coordinates and time. Equation 5.8 describes a plane wave propagating in arbitrary direction determined by the propagation vector $\mathbf{k}$, often also called wave vector. If the components of the propagation vector are (k_x, k_y, k_z), and (x, y, z) are the components of a point in space where the displacement ψ is considered, then the dot product is given by $\mathbf{k} \cdot \mathbf{r} = k_x x + k_y y + k_z z$. The propagation vector $\mathbf{k}$ is normal to planes of constant phase defined by $\varphi = $ const. and called wave fronts. If the phase φ is equal to integer multiples of 2π on these planes, a three-dimensional plane wave can be visualized as a train of wave fronts separated by one wavelength λ and moving with the wave speed v in direction given by $\mathbf{k}$.

If the wave propagation is along the z direction, $\mathbf{k} \cdot \mathbf{r}$ becomes kz. During a time interval dt, the plane of constant phase moves a distance dz. The magnitude of the phase velocity $\mathbf{v}$ of the wave moving in z direction is

$$v = \frac{dz}{dt} = \frac{\omega}{k} = \lambda f \tag{5.9}$$

The behavior of the phase difference $\Delta\varphi$ at a given time between two points on the wave that are separated by a certain distance is important for evaluating the degree of coherence of the wave. If a monochromatic wave travels along the z direction with a wave vector $\mathbf{k}$, then the phase difference between two points along z separated by distance Δz is $\Delta\varphi = \mathbf{k}\Delta z$, ωt is the same at each point. If the phase difference $\Delta\varphi = \mathbf{k}\Delta z$ is equal to 0 or integer multiples of 2π, then these two points are in phase. A wave that maintains a constant phase difference in space and time exhibits high degree of coherence.

A wave described by Equation 5.8 satisfies the partial differential equation:

$$\nabla^2 \Psi = \frac{1}{v^2} \frac{\partial^2 \psi}{\partial t^2}, \tag{5.10}$$

which is the general form of the wave equation. Here v is the wave propagation velocity. The wave equation can be derived by employing the chain rule of partial differentiation to find the first and second spatial and temporal derivatives and equating the results for the two second derivatives. Plane waves have rectangular symmetry and are solutions of the wave equation (Equation 5.10), where the Laplacian operator is given in Cartesian coordinates by

$$\nabla^2 = \frac{\partial^2}{\partial x^2} + \frac{\partial^2}{\partial y^2} + \frac{\partial^2}{\partial z^2} \tag{5.11}$$

5.1.3 NATURE OF LIGHT

The four Maxwell's equations for free space can be combined and manipulated to give as a final result the following two equations for the electric and magnetic vectors [1,4]:

$$\nabla^2 \mathbf{E} = \varepsilon_0 \mu_0 \frac{\partial^2 \mathbf{E}}{\partial t^2} \tag{5.12a}$$

$$\nabla^2 \mathbf{B} = \varepsilon_0 \mu_0 \frac{\partial^2 \mathbf{B}}{\partial t^2} \tag{5.12b}$$

These equations have the same form as the wave equation (Equation 5.10). In addition, the Laplacian operator acts on each component of **E** and **B**; hence, each component of the EM field obeys the scalar wave equation (Equation 5.10). This fact suggests the existence of harmonic EM waves that propagate in free space with velocity

$$c = \frac{1}{\sqrt{\varepsilon_0 \mu_0}} = 2.998 \times 10^8 \text{ m/s} \tag{5.13}$$

The speed of light in vacuum is a fundamental constant with the exact value $c = 299{,}792{,}458$ m/s, which was defined by the *Conference Générale des Poids et Mesures* in 1983.

5.1.3.1 Light Is a Three-Dimensional Plane Electromagnetic Wave

This property follows from the fact that each component of **E** and **B** satisfies the differential wave equation (Equation 5.10). For EM waves (including light), ψ can be either of the varying electric or magnetic fields that constitute the wave. Thus, the variations of the electric and magnetic fields at point **r** on a plane perpendicular to the propagation vector **k** can be described with Equation 5.8 as

$$\mathbf{E}(\mathbf{r},t) = \mathbf{E}_0(\mathbf{r})e^{-i(\omega t - \mathbf{k} \cdot \mathbf{r} + \varphi_0)} \tag{5.14a}$$

and

$$\mathbf{B}(\mathbf{r},t) = \mathbf{B}_0(\mathbf{r})e^{-i(\omega t - \mathbf{k} \cdot \mathbf{r} + \varphi_0)}, \tag{5.14b}$$

where $\mathbf{E}_0(\mathbf{r})$ and $\mathbf{B}_0(\mathbf{r})$ are the complex amplitudes of the electric and magnetic fields, respectively.

Equations 5.14a and b describe harmonic waves of the same frequency f (same ω) and propagate in arbitrary direction defined by the same propagation vector **k**; hence, have the same wavelength and speed in a given dielectric material

$$v = \frac{1}{\sqrt{\varepsilon \mu_0}} \tag{5.15}$$

5.1.3.2 Light Is a Transverse Wave

EM waves are transverse waves, because both vectors **E** and **B** lie in a plane that is perpendicular to the direction of propagation, determined by the direction of **k**. This follows from Maxwell's equations (Equations 5.6a and b) when applied to the filed vectors **E** and **B** given by Equations 5.14a and b:

$$\frac{\partial E_x}{\partial x} + \frac{\partial E_y}{\partial y} + \frac{\partial E_z}{\partial z} = i(k_x E_x + k_y E_y + k_z E_z) = i\mathbf{k} \cdot \mathbf{E} = 0$$

and

$$\frac{\partial B_x}{\partial x} + \frac{\partial B_y}{\partial y} + \frac{\partial B_z}{\partial z} = i(k_x B_x + k_y B_y + k_z B_z) = i\mathbf{k} \cdot \mathbf{B} = 0$$

Since $\mathbf{k} \cdot \mathbf{E} = 0$ and $\mathbf{k} \cdot \mathbf{B} = 0$, both **E** and **B** must be perpendicular to **k**, which indicates that EM waves are transverse waves.

5.1.3.3 Electric and Magnetic Vectors Are Perpendicular to Each Other

This follows from the third Maxwell equation (Equation 5.6c). Assuming a wave traveling in positive z direction, with an electric vector $\mathbf{E}$ along the x-axis, we get from Equation 5.6c that $\hat{\mathbf{y}}(\partial E_x / \partial z) - (\partial B_y / \partial t) = i\omega B_y$; thus, the only nonzero component of the magnetic field is along the y-axis. We can generalize this result using the cross product

$$\mathbf{k} \times \mathbf{E} = \omega \mathbf{B} \tag{5.16}$$

From Maxwell's equations it follows that the harmonic variations of the electric and magnetic fields are always perpendicular to one another and to the direction of propagation given by $\mathbf{k}$. In addition, in an isotropic nonconducting medium, the field magnitudes are related by

$$E = \frac{\omega}{k} B = vB \tag{5.17}$$

or in free space, this equation becomes

$$E = cB \tag{5.18}$$

In vacuum and in air, the $\mathbf{E}$ and $\mathbf{B}$ fields oscillate in phase at any point in space and at any time. In a dissipative medium, however, a phase shift between the fields takes place. In good conductors, the magnetic field is much larger than the electric field and exhibits a phase delay of approximately 45°. In non-dissipative media, like glass for visible light, the $\mathbf{E}$ and $\mathbf{B}$ fields behave similar as in vacuum and are in phase.

5.1.3.4 Light Polarization

Since the two fields are interdependent, once one of them is specified, the direction and magnitude of the other can be obtained from the relations between them, that is, Equations 5.16 through 5.18. It is customary to specify the electric field and the direction of the electric field determines the polarization of the wave.

Let us consider again a wave traveling in positive z direction, with $\varphi_0 = 0$, and electric vector $\mathbf{E}$ along the x-axis, that is, the wave is polarized in the x direction. Such a wave can be represented by

$$\mathbf{E} = E_0 \sin(kz - \omega t) \tag{5.19}$$

Using Equation 5.18, the magnetic field associated with this wave in free space is given by

$$\mathbf{B} = \frac{1}{c} E_0 \sin(kz - \omega t) \tag{5.20}$$

According to Lorentz force law, Equation 5.2, the direction of the force exerted on a moving charge by the EM filed is defined by the direction of the electric vector $\mathbf{E}$, thus, by the polarization of the wave. Since $\mathbf{E}$ and $\mathbf{B}$ lie in one plane which is perpendicular to $\mathbf{k}$, we say that light is a plane polarized wave with polarization perpendicular to the direction of wave propagation. Equations 5.2 and 5.20 show that the electric force acting on the charge in general must be much larger than the magnetic force, which underlines the importance of the wave polarization in many optical phenomena and applications. The wave is said to be linearly polarized if the direction of the electric vector $\mathbf{E}$ remains constant in a given direction. The electric field vector, Equation 5.19, can be represented as the vector sum of two perpendicular waves $\mathbf{E}(z, t) = \mathbf{E}_x(z, t) + \mathbf{E}_y(z, t)$. For example, the light would be linearly polarized along the x direction, if the electric field oscillates always along the x direction. For a linear polarization, the amplitude and the direction of the electric field oscillation remain constant over time. If one component of the electric field, for instance this along the y direction, is always out of phase by $\pi/2$ with the x-component, but the amplitudes of both components are fixed, the polarization is said to be circular polarization. Both linear and circular polarizations are particular cases of the general case of elliptical polarization. The light is elliptically polarized when the electric field vector $\mathbf{E}(z, t)$ rotates and changes its magnitude over time. Natural light and light from sources other than lasers is randomly polarized, also called unpolarized light.

Since Chapter 5 is devoted entirely to polarization of light, we will neither discuss here the nature of the polarization of EM waves, nor will we describe here the ways of production of the different types of wave polarization.

5.1.3.5 Energy Flow and Poynting Vector

EM waves carry energy as they travel. An expression for the wave energy can be derived by using the analogy with a capacitor storing energy as an electric field and an inductor (a solenoid) storing energy as a magnetic field. By this analogy, the energy densities u_E and u_B (J/m^3), associated with the static electric field of an ideal capacitor and the static magnetic field of an ideal solenoid are

$$u_E = \frac{1}{2}\varepsilon_0 E^2 \quad \text{and} \quad u_B = \frac{1}{2}\frac{B^2}{\mu_0} \text{(in free space)} \tag{5.21}$$

for an EM wave, $u_E = u_B$. This result can be easily obtained from Equation 5.21 by replacing B with E/c from Equation 5.18:

$$u_B = \frac{1}{2}\frac{B^2}{\mu_0} = \frac{1}{2}\frac{1}{\mu_0}\left(\frac{E}{c}\right)^2 = \frac{1}{2}\frac{\varepsilon_0\mu_0}{\mu_0}E^2 = \frac{1}{2}\varepsilon_0 E^2 = u_E \tag{5.22}$$

The total energy density of the wave is the sum of the energy densities of its constituent electric and magnetic fields

$$u = u_E + u_B = 2u_E = 2u_B \tag{5.23}$$

which is equivalent to

$$u = \varepsilon_0 E^2 = \frac{B^2}{\mu_0} = \varepsilon_0 c E B \theta \tag{5.24}$$

These two equations define the energy density of the EM field, that is, the energy per unit volume of the space where the EM field exists. A traveling wave carries energy and the rate at which energy is transported by the EM field defines the power of the EM wave.

The total energy transferred by the wave in a time Δt through an area A perpendicular to the direction of wave propagation is $u\Delta V$. Here, ΔV is the volume determined by the cross section A and the length $c\Delta t$, that is, the distance traveled by the wave in a time Δt. Thus, the power of the wave can be expressed by

$$\text{Power} = \frac{\text{Energy}}{\Delta t} = \frac{u\Delta V}{\Delta t} = \frac{u(c\Delta t A)}{\Delta t} = ucA \tag{5.25}$$

The power per unit area (W/m²), or the energy flow per unit area per unit time, defines the energy flux S:

$$S = \frac{ucA}{A} = uc \qquad (5.26)$$

By using Equation 5.24, the energy flux can also be expressed as

$$S = (\varepsilon_0 E^2)c = \frac{1}{\mu_0} EB = \varepsilon_0 c^2 EB \qquad (5.27)$$

As the energy flow is in the direction of wave propagation, the energy flux S can be considered as a vector with magnitude given by Equation 5.27 and direction—the direction of wave propagation. Using the fact that $\mathbf{E} \times \mathbf{B}$ points in the direction of energy flow, the vector $\mathbf{S}$ can be expressed by the vectors of the electric and magnetic fields as

$$\mathbf{S} = \varepsilon_0 c^2 \mathbf{E} \times \mathbf{B} = \frac{1}{\mu_0} \mathbf{E} \times \mathbf{B} \qquad (5.28)$$

The vector $\mathbf{S}$ is called the Poynting vector.

5.1.3.6 Irradiance

In Chapter 1, we defined the irradiance as the radiant flux incident onto a unit area perpendicular to the direction of propagation of radiant energy. Irradiance is a measurable quantity and involves a measurement over some time interval, which is much larger than the period of variation of the EM field of optical frequencies (in the order of 10^{14} Hz for the visible spectrum). Since the electric and magnetic fields, therefore the magnitude of the Poynting vector, are rapidly varying functions of time, the measurement of the energy flux S inevitably involves time averaging. Hence, the irradiance is the time average of the magnitude of the Poynting vector:

$$I = \langle |\mathbf{S}| \rangle = \varepsilon_0 c^2 \left\langle E_0 B_0 \sin^2(\mathbf{k} \cdot \mathbf{r} - \omega t) \right\rangle = \frac{1}{2} \varepsilon_0 c^2 E_0 B_0 \ [\text{W/m}^2] \qquad (5.29)$$

An alternative representation of irradiance is based on the alternative forms of energy density for free space, Equation 5.24:

$$I = \frac{1}{2} \varepsilon_0 c E_0^2 = \frac{1}{2} \frac{c}{\mu_0} B_0^2 \ (\text{free space}) \qquad (5.30)$$

which within linear homogeneous isotropic material becomes

$$I = \frac{1}{2} \varepsilon v E_0^2 \qquad (5.31)$$

Usually, irradiance is denoted by E, as it was in Chapter 1. However, to avoid confusion of electric field with irradiance, we use here the symbol I for irradiance. Equation 5.29 has to employ the real representation of harmonic waves by sine and cosine functions, because the definition of Poynting vector, Equation 5.28, involves the product of two harmonic fields ($\mathbf{E} \times \mathbf{B}$) and does not hold if they are given in complex form.

In summary, light is a three-dimensional plane polarized transverse EM wave, which carries energy and exerts a force on charges in the wave path; the magnitudes of the electric and magnetic fields are related by Equation 5.17 or 5.18, and the vectors **E** and **B** are oriented so that (**E** × **B**) points in the direction of energy flow.

5.1.3.7 Photons, Radiation Pressure, and Momentum

Maxwell's EM theory cannot explain the processes of absorption and emission of light. These are quantum phenomena, which involve electronic transitions to and from quantized energy states, which lead to quantization of the absorbed and emitted radiations also. This reveals the quantum nature of light. In some optical phenomena, light behaves as a harmonic EM wave, while in others, it has to be considered a stream of energy quanta called photons. The phenomena absorption and emission of light are described in more details in Chapter 1, where it is shown that the energy of the absorbed or emitted photon is determined by the difference between the energies E_i and E_j of the energy levels involved in the electron transition. The photon energy cannot be subdivided and is given by

$$E_{\mathrm{ph}} = hf = h\frac{c}{\lambda} \tag{5.32}$$

where

h is Planck's constant ($h = 6.626 \times 10^{-34}\,\mathrm{J \cdot s} = 4.136 \times 10^{-15}\,\mathrm{eV \cdot s}$)
f is the frequency
λ is the wavelength of the radiation

Specifically in semiconductor optics, the energy is expressed in electron volt rather than in joules. Hence, it is convenient to apply the following relation to find the energy into electron volt corresponding to a given wavelength in nanometer (10^{-9} m),

$$E_{\mathrm{ph}}\,[\mathrm{eV}] = \frac{1240}{\lambda\,[\mathrm{nm}]} \tag{5.33}$$

Although photons have zero mass, they have nonzero momentum, which is determined by the photon energy

$$p = \frac{E}{c} = \frac{h}{\lambda} = \hbar k \tag{5.34}$$

where $\hbar = h/2\pi$. The vector form of Equation 5.34, $\mathbf{p} = \hbar\mathbf{k}$, with $\mathbf{k}$ being the propagation vector reflects the fact that photon momentum is in the direction of energy flow.

When EM radiation impinges on an object's surface, it interacts with the matter and exerts a force on the constituent charges. Since the force exerted on unit area defines pressure, the interaction of radiation with the surface results in radiation pressure [13]. Maxwell showed that the radiation pressure is determined by the energy density of the EM wave; thus, it can be expressed in terms of the magnitude of the Poynting vector. The average radiation pressure, which is the time average of the magnitude of the Poynting vector divided by the speed of light in vacuum, can be expressed with the irradiance of the impinging beam by

$$\langle P(t)\rangle = \frac{\langle S(t)\rangle}{c} = \frac{I}{c}\left[\frac{\mathrm{N}}{\mathrm{m}^2}\right] \tag{5.35}$$

On the other hand, force is the time rate of change of momentum, $F = \Delta p/\Delta t$; hence, radiation pressure is also related to the change of EM momentum of the wave reflected from or absorbed by the surface $\langle P \rangle = \Delta p/A\Delta t$, where Δp is the amount of momentum transferred to the surface area A during the time interval Δt. The momentum transferred to the object per photon Δp is twice the photon momentum for total reflection from the surface and equal to the photon momentum for complete absorption. Since a beam of light with irradiance I has a photon flux per unit area equal to I/hv, the pressure exerted on the surface when the light is perfectly reflected from it can be expressed by

$$\langle P(t) \rangle = \frac{I}{hv}\left(\frac{2h}{\lambda}\right) = \frac{2I}{c} \tag{5.36}$$

Similarly, the pressure for complete absorption is given by $\langle P(t) \rangle = I/c$, which is the same as Equation 5.35.

5.1.3.8 Electromagnetic Spectrum

The EM waves cover an extremely broad spectrum of wavelengths. We can detect only a very small segment of this spectrum directly through our sense of sight from approximately 750–400 nm.

EM waves are produced by oscillating charge distributions; thus, the frequency of the EM wave is determined by the frequency of the source of the wave. Equation 5.9 links the wave frequency and wavelength with the speed of propagation of the wave through a given medium (in vacuum, $v = c$). On the basis of the energy distribution among the EM waves of different frequencies, the spectrum is divided into six conceptual categories, as described below:

1. Radio waves are low-frequencies (up to 10^{10} Hz), low-energy (up to 10^{-3} eV), long-wavelength (from kilometers for long radio waves to millimeters for microwaves) EM waves. This range includes AM and FM radio waves, broadcast television, and microwaves. Sometimes, microwaves, which extend from about 10^9 up to about 3×10^{11} Hz, and corresponding wavelengths of about 1 mm to 30 cm, are presented as a separate category. The signals are detected directly with antennas.

2. Infrared (IR) radiation includes frequencies roughly from 3×10^{11} to about 4×10^{14} Hz corresponding to wavelengths from 1 mm to about 750 nm. Photon energies range from 10^{-3} to about 1 eV. IR radiation is produced via thermal agitation of the molecules and any object at a temperature above the absolute zero will emit IR radiation. For example, a blackbody at room temperature emits radiation with wavelength maximum at around 10 μm.

3. Visible range is the part of the EM spectrum that is detectable by the human eye. The term "light" refers to the visible EM radiation. The International Commission on Illumination has specified its limits to the frequency range from about 3.61×10^{14} to roughly 8.33×10^{14} Hz (which is the wavelength range from about 830 to 360 nm). Many texts, though, give narrower boundaries for the visible range, roughly from about 400 (violet) to 750 nm (red), corresponding to the visual response of the human eye. The emission of light results from rearrangement of the outer electrons in atoms and molecules. Photon energies are about 1–3 eV, which is sufficiently large to allow single-photon detection.

4. Ultraviolet (UV) range covers frequencies approximately from 8×10^{14} to about 3.4×10^{16} Hz, with wavelengths from 380 down to 10 nm. The photon energies range from roughly 3.2 to 100 eV. These energy levels are large enough to ionize atoms in the upper atmosphere, to initiate photochemical reactions or to cause possible damage to living tissues. UV radiation results from electron transitions from highly excited states.

5. X-rays are EM waves with wavelengths from 10 to about 10^{-4} nm (which is less than the atomic size) and frequencies from roughly 2.4×10^{16} to about 5×10^{19} Hz, and photon energies from about 100 eV up to 0.2 MeV. X-rays are produced from inner-shell atomic transitions when a metal target is bombarded with high-energy electrons.

6. γ-Rays originate from nuclear transitions and have the shortest wavelengths (from 0.1 to 10^{-6} nm) and the highest energies (10^4 to about 10^{19} eV). Because of these high energies, detection of single photons is possible. In addition, exposure of living tissues to x-rays and γ-rays can be extremely damaging.

5.2 PROPAGATION OF LIGHT IN ISOTROPIC MEDIA

5.2.1 PROPAGATION OF LIGHT IN DIELECTRICS

We consider here only a homogeneous isotropic dielectric media, the permittivity of which is independent of the coordinates as well as of time. For such a medium, Maxwell equations and all results derived for EM waves in vacuum are valid when ε_0 is replaced by ε. The wave velocity in a dielectric is

$$v = \frac{1}{\sqrt{\varepsilon_r \varepsilon_0 \mu_0}} = \frac{c}{\sqrt{\varepsilon_r}} = \frac{c}{n} \tag{5.37}$$

where

$$n = \sqrt{\varepsilon_r} = \frac{c}{v} \tag{5.38}$$

is the index of refraction and $\varepsilon_r = \varepsilon/\varepsilon_0$ is the relative permittivity. For magnetic materials, $n = \sqrt{\varepsilon_r, \mu_r}$, the wavelength in the dielectric is $\lambda = \lambda_0/n$ and the wave number is $k = 2\pi/\lambda$. The energy density is given by

$$u = \varepsilon E^2 \tag{5.39}$$

while the energy flux for a dielectric is

$$S = uv = v\varepsilon E^2 \tag{5.40}$$

The irradiance of light within a homogeneous linear isotropic material is given by Equation 5.31.

5.2.2 REFLECTION AND REFRACTION

5.2.2.1 Boundary Conditions

The behavior of light at the interface between two different media is determined by the boundary conditions for the field vectors of the wave. In nonmagnetic materials with no free charges and conduction currents, these conditions require continuity of the tangential components of the electric and magnetic fields across the boundary [1]:

$$E_{2\tau} = E_{1\tau} \quad \text{and} \quad H_{2\tau} = H_{1\tau} \tag{5.41a}$$

followed automatically by the conditions for the normal components of **D** and **B**

$$D_{2n} = D_{1n} \quad \text{and} \quad B_{2n} = B_{1n} \tag{5.41b}$$

Here n and τ denote the normal and tangential component of the vectors, respectively. When the boundary conditions are imposed on the electric vectors of the incident, reflected, and refracted plane waves, the boundary conditions require that the strength of the electric field in the first medium, which consists of the field of the incident and reflected waves, is equal to the strength of the field in the second medium:

$$\left[\mathbf{E}_{0i}e^{-i(\omega_i t - \mathbf{k}_i \cdot \mathbf{r})} + \mathbf{E}_{0r}e^{-i(\omega_r t - \mathbf{k}_r \cdot \mathbf{r})} \right]_\tau = \left[\mathbf{E}_{0t}e^{-i(\omega_t t - \mathbf{k}_t \cdot \mathbf{r})} \right]_\tau \tag{5.42}$$

Here the subscripts i, r, and t denote the incident, reflected, and refracted (transmitted) waves. Equation 5.42 can be satisfied for arbitrary and independent variations of time and radius vector only if

$$\omega_i t = \omega_r t = \omega_t t \tag{5.43a}$$

$$\mathbf{k}_i \cdot \mathbf{r} = \mathbf{k}_r \cdot \mathbf{r} = \mathbf{k}_t \cdot \mathbf{r} \tag{5.43b}$$

Thus, $\omega_i = \omega_r = \omega_t$, that is, the frequency of an EM wave, does not change as a result of reflection or refraction.

5.2.2.2 Laws of Reflection and Refraction

In geometrical optics, the direction of propagation of energy, which is in the direction of the wave propagation vector $\mathbf{k}$, is depicted as a ray. The rays are always perpendicular to the wave fronts. Let us consider a ray impinging on the interface between two media with refractive indices n_1 and n_2, respectively. The origin of the radius vector $\mathbf{r}$ on the interface can be chosen in such a way that $\mathbf{r}$ is perpendicular to the vector $\mathbf{k}_i$, that is, $\mathbf{k}_i \cdot \mathbf{r} = 0$. Consequently, from Equation 5.43b, $\mathbf{k}_r \cdot \mathbf{r} = \mathbf{k}_t \cdot \mathbf{r} = 0$, which means that the vectors of the incident, reflected, and refracted rays, $\mathbf{k}_i$, $\mathbf{k}_r$, and $\mathbf{k}_t$, all lie in the same plane. The plane specified by $\mathbf{k}_i$ and the normal to the boundary $\mathbf{n}$ is called plane of incidence. In Figure 5.1, the plane of incidence is the xz plane, the origin O of the rectangular Cartesian coordinate system is chosen at the point of incidence of the ray, and the direction of the z-axis is toward the second medium in which the refracted ray propagates. The vectors $\mathbf{k}_i$, $\mathbf{k}_r$, and $\mathbf{k}_t$ are applied at point O. The angles that the incident, reflected, and refracted rays make with Oz are θ_i, θ_r, and θ_t,

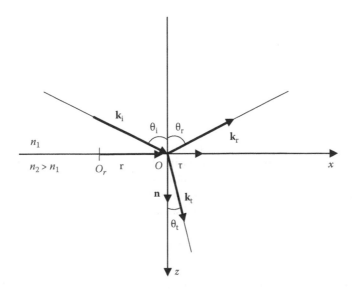

FIGURE 5.1 Incident, reflected, and refracted rays at a boundary between two dielectric media for external reflection ($n_2 > n_1$).

respectively. The unit vector $\mathbf{n}$ is directed into the second medium along the normal to the interface. The tangential unit vector τ lies on the interface along the x-axis.

The laws of reflection and refraction follow from the boundary conditions. The law of reflection states that the wave vector $\mathbf{k}_r$ of the reflected wave lies in the plane of incidence and the angle of reflection is equal to the angle of incidence, or $\theta_r = -\theta_i$, when the angles are measured from the normal to the surface.

It follows from the boundary conditions that the ratio $(\sin \theta_i / \sin \theta_t)$ is a constant and equal to the ratio of the wave velocities in both media as follows:

$$\frac{\sin \theta_i}{\sin \theta_t} = \frac{v_1}{v_2} = \frac{n_2}{n_1} = n_{12} \tag{5.44}$$

This equation can be rewritten as $n_1 \sin \theta_i = n_2 \sin \theta_t$ and together with the statement that the refracted wave vector $\mathbf{k}_t$ is in the plane of incidence, constitute the Snell's law of refraction. When a wave impinges on the interface between two media from the side of the optically rare medium, $n_2 > n_1$, the reflection is called external reflection. In this case, for every angle of incidence θ_i there is a real angle of refraction θ_t, which, by Equation 5.44 and $\sin \theta_t < \sin \theta_i$, is smaller than the incident angle $\theta_t < \theta_i$.

The refractive index is always greater than unity and depends not only on the substance but also on the wavelength of the light and temperature. For gases, it depends on the pressure also. Under normal conditions, the refractive index of gases differs from unity only by 10^{-3} or 10^{-4}. For example, for air $n = 1.0003$, but if a very high accuracy is not required one can use for air $n = 1$ and the speed of light in vacuum. Equation 5.44 is valid when the magnetic properties of the media on different sides of the interface are identical $\mu_1 = \mu_2$. If this condition is not fulfilled and $\mu_1 \neq \mu_2$, then Equation 5.44 becomes

$$\frac{Z_1}{Z_2} = \frac{\mu_1 \sin \theta_i}{\mu_2 \sin \theta_t} \tag{5.45}$$

where

$$Z = \sqrt{\frac{\mu}{\varepsilon}} \tag{5.46}$$

is the wave impedance of the medium. For vacuum, $Z_0 = \sqrt{\mu_0/\varepsilon_0} = 377\Omega$.

5.2.2.3　Fresnel Formulae

A plane wave linearly polarized in an arbitrary direction can be represented as a sum of two plane waves, one of which is polarized in the plane of incidence (denoted by subscript ∥) and the other polarized in a plane perpendicular to the plane of incidence (denoted by subscript ⊥). The sum of the energy flux densities of these two waves should be equal to the energy flux density of the initial wave. Then, the incident, reflected, and refracted waves would be given by

$$E_i = E_{i\parallel} + E_{i\perp}, \quad E_r = E_{r\parallel} + E_{r\perp}, \quad \text{and} \quad E_t = E_{t\parallel} + E_{t\perp} \tag{5.47}$$

According to the boundary conditions (Equations 5.41), the tangential components of the electric and magnetic fields should be continuous across the boundary. Hence, Equation 5.41a can be written for each polarization separately.

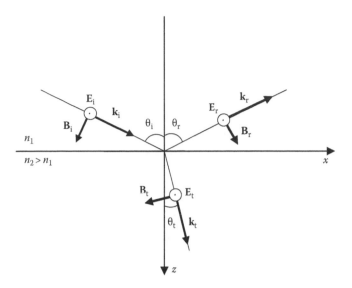

FIGURE 5.2   The incident wave is polarized perpendicularly to the plane of incidence.

Electric vector is perpendicular to the plane of incidence: This case is shown in Figure 5.2. It is seen that the tangential components of all electric field vectors are *y*-components, while for all magnetic field vectors they are *x*-components. The boundary condition for the electric vectors can be written as

$$E_{0i} + E_{0r} = E_{0t} \tag{5.48a}$$

and for the magnetic vectors

$$-B_{0i} \cos \theta_i + B_{0r} \cos \theta_i = -B_{0t} \cos \theta_t \tag{5.48b}$$

Dividing both sides of Equation 5.48a by E_{0i} and using Equation 5.18 to substitute B in Equation 5.48b with $(n/c)E$, and then rearranging the terms, gives the Fresnel's amplitude ratios for the reflected and transmitted waves:

$$\left(\frac{E_{0r}}{E_{0i}} \right)_\perp = \frac{n_1 \cos \theta_i - n_2 \cos \theta_t}{n_1 \cos \theta_i + n_2 \cos \theta_t} \tag{5.49a}$$

$$\left(\frac{E_{0t}}{E_{0i}} \right)_\perp = \frac{2 n_1 \cos \theta_i}{n_1 \cos \theta_i + n_2 \cos \theta_t} \tag{5.49b}$$

When the law of refraction (Equation 5.44) is used, Equation 5.49 can also be written in the following alternative form, which does not contain the refractive indices:

$$\left(\frac{E_{0r}}{E_{0i}} \right)_\perp = -\frac{\sin(\theta_i - \theta_t)}{\sin(\theta_i + \theta_t)} \tag{5.50}$$

$$\left(\frac{E_{0t}}{E_{0i}} \right)_\perp = \frac{2 \sin \theta_t \cos \theta_i}{\sin(\theta_i + \theta_t)} \tag{5.51}$$

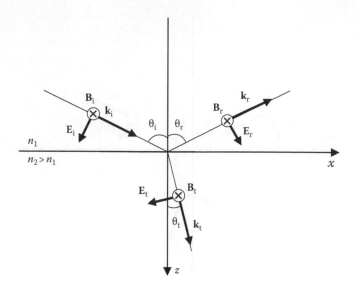

FIGURE 5.3 The incident wave is polarized parallel to the plane of incidence.

Electric vector is parallel to the plane of incidence: This case is shown in Figure 5.3. This time the tangential components of all magnetic field vectors are y-components, while the tangential components of all electric field vectors are x-components. The boundary condition for the electric vectors in this case is

$$E_{0i} \cos \theta_i + E_{0r} \cos \theta_i = E_{0t} \cos \theta_t \tag{5.52a}$$

and for the magnetic vectors

$$B_{0i} - B_{0r} = B_{0t} \quad \text{which is also} \quad n_1 E_{0i} - n_1 E_{0r} = n_2 E_{0t} \tag{5.52b}$$

Again, dividing both sides of Equation 5.52a and b by E_{0i} and rearranging, gives the Fresnel's formulae for the amplitudes of the parallel polarization:

$$\left(\frac{E_{0r}}{E_{0i}} \right)_\| = \frac{n_2 \cos \theta_i - n_1 \cos \theta_t}{n_2 \cos \theta_i + n_1 \cos \theta_t} \tag{5.53}$$

$$\left(\frac{E_{0t}}{E_{0i}} \right)_\| = \frac{2 n_1 \cos \theta_i}{n_2 \cos \theta_i + n_1 \cos \theta_t} \tag{5.54}$$

And the alternative forms are

$$\left(\frac{E_{0r}}{E_{0i}} \right)_\| = \frac{\tan(\theta_i - \theta_t)}{\tan(\theta_i + \theta_t)} \tag{5.55}$$

$$\left(\frac{E_{0t}}{E_{0i}} \right)_\| = \frac{2 \sin \theta_t \cos \theta_i}{\sin(\theta_i + \theta_t) \cos(\theta_i - \theta_t)} \tag{5.56}$$

If both media are transparent (no absorption), according to the principle of conservation of energy, the energies in the refracted and reflected waves should add up to the energy of the incident wave.

The redistribution of the energy of the incident field between the two secondary fields is given by the reflection and transmission coefficients. The amount of energy in the primary wave, which is incident on a unit area of the interface per second, is given by

$$J_i = I_i \cos\theta_i = \frac{\varepsilon_0 c}{2} n_1 |E_{0i}|^2 \cos\theta_i, \tag{5.57a}$$

And similarly, the energies of the reflected and refracted waves leaving a unit area of the interface per second are given by

$$J_r = I_r \cos\theta_i = \frac{\varepsilon_0 c}{2} n_1 |E_{0r}|^2 \cos\theta_i \tag{5.57b}$$

$$J_t = I_t \cos\theta_t = \frac{\varepsilon_0 c}{2} n_2 |E_{0t}|^2 \cos\theta_t \tag{5.57c}$$

where I_i, I_r, and I_t are the irradiances of the incident, reflected, and refracted waves, as defined by Equation 5.31. The ratios

$$\mathscr{R} = \frac{J_r}{J_i} = \frac{|E_{0r}|^2}{|E_{0i}|^2} \tag{5.58}$$

and

$$\mathscr{T} = \frac{J_t}{J_i} = \frac{n_2 \cos\theta_t}{n_1 \cos\theta_i} \frac{|E_{0r}|^2}{|E_{0i}|^2} \tag{5.59}$$

are called reflectivity and transmissivity (or coefficients of reflection and transmission), respectively. It can be easily verified that for nonabsorbing media the energy flux density of the incident wave is equal to the sum of the energy flux densities of the reflected and transmitted waves. Hence, the law of conservation of energy is satisfied

$$\mathscr{R} + \mathscr{T} = 1 \tag{5.60}$$

The reflectivity and transmissivity depend on the polarization of the incident wave. If the vector $\mathbf{E}_i$ of the incident wave makes an angle α_i with the plane of incidence, then $E_{i\parallel} = E_i \cos\alpha_i$ and $E_{i\perp} = E_i \sin\alpha_i$ and Equations 5.58 and 5.59 can be rewritten using Equations 5.47 as

$$\mathscr{R} = \frac{J_r}{J_i} = \frac{J_{r\parallel} + J_{r\perp}}{J_i} = \mathscr{R}_\parallel \cos^2\alpha_i + \mathscr{R}_\perp \sin^2\alpha_i \tag{5.61}$$

and

$$\mathscr{T} = \frac{J_t}{J_i} = \mathscr{T}_\parallel \cos^2\alpha_i + \mathscr{T}_\perp \sin^2\alpha_i, \tag{5.62}$$

where

$$\mathcal{R}_{\parallel} = \frac{J_{r\parallel}}{J_{i\parallel}} = \frac{\left|E_{0r\parallel}\right|^2}{\left|E_{0i\parallel}\right|^2} = \frac{\tan^2(\theta_i - \theta_t)}{\tan^2(\theta_i + \theta_t)} \tag{5.63a}$$

$$\mathcal{R}_{\perp} = \frac{J_{r\perp}}{J_{i\perp}} = \frac{\left|E_{0r\perp}\right|^2}{\left|E_{0i\perp}\right|^2} = \frac{\sin^2(\theta_i - \theta_t)}{\sin^2(\theta_i + \theta_t)} \tag{5.63b}$$

$$\mathcal{T}_{\parallel} = \frac{J_{t\parallel}}{J_{i\parallel}} = \frac{n_2}{n_1}\frac{\cos\theta_t}{\cos\theta_i}\frac{\left|E_{0t\parallel}\right|^2}{\left|E_{0i\parallel}\right|^2} = \frac{\sin 2\theta_i \sin 2\theta_t}{\sin^2(\theta_i + \theta_t)\cos^2(\theta_i - \theta_t)} \tag{5.64a}$$

$$\mathcal{T}_{\perp} = \frac{J_{t\perp}}{J_{i\perp}} = \frac{n_2}{n_1}\frac{\cos\theta_t}{\cos\theta_i}\frac{\left|E_{0t\perp}\right|^2}{\left|E_{0i\perp}\right|^2} = \frac{\sin 2\theta_i \sin 2\theta_t}{\sin^2(\theta_i + \theta_t)} \tag{5.64b}$$

Again we may verify that the reflection and transmission coefficients satisfy the law of energy conservation: $R_{\parallel} + T_{\parallel} = 1$ and $R_{\perp} + T_{\perp} = 1$. The dependence of the reflectivity and transmissivity on the angle of incidence θ_i, given by Equations 5.63 and 5.64 for external reflection, is shown in Figure 5.4.

When the magnetic properties of the media on the different sides of the interface are not identical, $\mu_1 \neq \mu_2$, then n_1 and n_2 in Equations 5.63 and 5.64 should be replaced with Z_2 in Z_1 respectively as defined by Equation 5.46. As a result, the reflectivity and transmissivity become

$$\mathcal{R}_{\parallel} = \left(\frac{Z_1\cos\theta_i - Z_2\cos\theta_t}{Z_1\cos\theta_i + Z_2\cos\theta_t}\right)^2$$

$$\mathcal{R}_{\perp} = \left(\frac{Z_2\cos\theta_i - Z_1\cos\theta_t}{Z_2\cos\theta_i + Z_1\cos\theta_t}\right)^2 \tag{5.65}$$

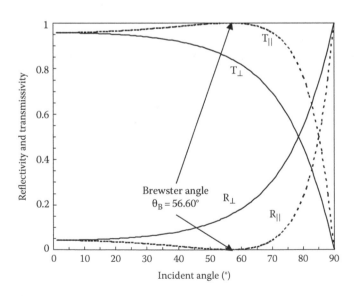

FIGURE 5.4 Fresnel's reflection and transmission from glass in air, $n_1 < n_2$: $n_1 = 1.0003$ (glass BK 7) and $n_2 = 1.5167$ (for the sodium D line $\lambda = 589.3$ nm). Brewster angle $\theta_B = 56.60°$.

$$\mathcal{T}_{\parallel} = \frac{4Z_1 Z_2 \cos\theta_i \cos\theta_t}{(Z_1 \cos\theta_i + Z_2 \cos\theta_t)^2}$$

$$\mathcal{T}_{\perp} = \frac{4Z_1 Z_2 \cos\theta_i \cos\theta_t}{(Z_2 \cos\theta_i + Z_1 \cos\theta_t)^2} \tag{5.66}$$

5.2.2.4 Normal Incidence

In the case of normal incidence, $\theta_i = 0$, consequently $\theta_t = 0$, and the expressions for the reflectivity and transmissivity are identical for both polarization components:

$$\mathcal{R} = \left(\frac{n_2 - n_1}{n_2 + n_1}\right)^2 \tag{5.67}$$

$$\mathcal{T} = \frac{4n_1 n_2}{(n_2 + n_1)^2} \tag{5.68}$$

It is obvious that the smaller the difference in the optical densities of the two media, the less energy is carried away by the reflected wave.

For media with magnetic properties, the coefficients (Equations 5.67 and 5.68) for normal incidence become

$$\mathcal{R} = \left(\frac{\mu_1 n_2 - \mu_2 n_1}{\mu_1 n_2 + \mu_2 n_1}\right)^2 \tag{5.69}$$

$$\mathcal{T} = \left(\frac{2\mu_2 n_1}{\mu_1 n_2 + \mu_2 n_1}\right)^2 \tag{5.70}$$

5.2.2.5 Degree of Polarization

The variation in the polarization of light upon reflection or refraction is described by the degree of polarization:

$$P = \frac{I_\perp - I_\parallel}{I_\perp + I_\parallel} \tag{5.71}$$

where $I_\perp$ and $I_\parallel$ are the irradiances of the perpendicular and parallel polarization components constituting the reflected or refracted beam, respectively. For unpolarized light, $P = 0$. For completely polarized light, $P = 1$, if the vector $\mathbf{E}$ is perpendicular to the plane of incidence and $P = -1$, if the vector $\mathbf{E}$ lies in the plane of incidence.

5.2.2.6 Brewster Angle

There is a special angle of incidence for which the sum of the incident angle and the refracted angle is 90°. When $\theta_i + \theta_t = \pi/2$, the denominator in Equation 5.55 turns to infinity, since tan $(\theta_i + \theta_t) \to \infty$, and consequently $R_\parallel = 0$, while Equation 5.56 gives $T_\parallel = 1$. This angle of incidence $\theta_i = \theta_B$ is called Brewster angle. Since the angle of reflection is equal to the angle of incidence, we can write $\theta_r + \theta_t = \pi/2$; from where it follows that when light strikes a surface at the angle of Brewster, the reflected and transmitted rays are also perpendicular to each other. Using this relation and the law of refraction gives

$$\tan \theta_B = \frac{n_2}{n_1} \tag{5.72}$$

The zero value of the curve for the parallel component in Figure 5.4 corresponds to the Brewster angle for glass $\theta_B = 56.60°$. If light is incident at Brewster angle θ_B, the reflected wave contains only the electric field component perpendicular to the plane of incidence, since $R_\parallel = 0$. Thus, the reflected light is completely polarized in the plane of incidence. Figure 5.4 shows that if the electric vector of the incident wave lies in the plane of incidence, all the energy of the wave incident at Brewster angle will be transferred into the transmitted wave, since there is no reflected wave with parallel polarization. Reflection at the Brewster angle is one of the ways of obtaining linearly polarized light.

Brewster angle can be experimentally determined by measuring the intensity of the parallel component of the reflected light by varying the angle of incidence continuously around the Brewster angle. Measurements of zero type (monitoring the intensity) are more precise and convenient than precise monitoring of the angle of incidence. At $\theta_i = \theta_B$, the intensity of the reflected wave with parallel polarizations should turn to zero according to Fresnel's formulae for reflection. However, in reality, the intensity of the parallel component of the reflected wave does not become zero. The reason for this is that Fresnel's formulae are derived when reflection and refraction take place at a mathematical boundary between two media. Actually, reflection and refraction take place over a thin surface layer with thickness about the interatomic distances. The properties of the transition layer differ from the properties of the bulk material, and therefore, the phenomena of reflection and refraction do not result in an abrupt and instantaneous change of the parameters of the primary wave, as described by the Fresnel's formulae. This is also a consequence of the fact that Maxwell's equations are formulated under the assumption that the medium is continuous and cannot describe accurately the process of reflection and refraction in the transition layer. As a result, for an incident wave of only perpendicular polarization, the wave reflected at Brewster angle is found to be elliptically polarized, instead of only linear polarization (perpendicular polarization) as Fresnel's formulae predict. The deviations from Fresnel's formulae due to the transition layer for angles other than the Brewster angle are hard to observe.

5.2.2.7 Phase Relations

The alternative forms (Equations 5.50, 5.51, 5.55, and 5.56) are more convenient when analyzing the phase of the reflected and refracted waves in respect with the incident wave.

Refraction: Since $E_{t\parallel}$ and $E_{t\perp}$ have the same signs as $E_{i\parallel}$ and $E_{i\perp}$, there is no change in the phase of the refracted waves for both polarizations. This means that under all conditions the refracted wave has the same phase as the incident wave.

Reflection: Depending on the conditions of reflection, the reflected wave may exhibit a change in phase:

1. External reflection (a wave is incident at an interface from the optically rarer medium, $n_1 < n_2$ and $\theta_t < \theta_i$)
   a. *Perpendicular polarization*: At any angle of incidence, the phase of the perpendicular component of the electric field in the reflected wave changes by π, that is, $E_{r\perp}$ and $E_{i\perp}$ have opposite phases at the interface. This follows from Equation 5.50, which shows that at $\theta_t < \theta_i$ the signs of $E_{r\perp}$ and $E_{i\perp}$ are different and the phases therefore differ by π. This means that the direction of the $E_{r\perp}$ is opposite to that shown in Figure 5.2.
   b. *Parallel polarization*: For electric vector oscillating in the plane of incidence, the phase change depends upon the angle of incidence. The ratio $(E_{0r}/E_{0i})_\parallel$ in Equation 5.55 is

positive for $\theta_i < \theta_B$, and negative for $\theta_i > \theta_B$ ($\theta_i + \theta_t > \pi/2$). This means that at Brewster angle the phase of the parallel component $E_{r\parallel}$ abruptly changes to π. For incident angles smaller than the Brewster angle, $\theta_i < \theta_B$, the phase of the reflected wave (parallel polarization) is opposite to the phase of the incident wave, and does not change for $\theta_i > \theta_B$; thus, the reflected and incident waves have the same phases when $\theta_i > \theta_B$ (i.e., at $\theta_i > \theta_B$ the phase of the magnetic field changes by π).

2. Internal reflection (a wave is incident at the interface from an optically denser medium, $n_1 > n_2$). The phase change is just the opposite of external reflection:

   a. *Perpendicular polarization*: At any angle of incidence, the phase of the perpendicular component of the electric field of the wave remains the same (i.e., in this case, the phase of the magnetic field changes by π).

   b. *Parallel polarization*: The phase of the parallel component remains unchanged for angles of incidence smaller than Brewster angle, $\theta_i < \theta_B$, and changes by π for angles of incidence exceeding θ_B (thus, the phase of the magnetic filed vector in the reflected wave does not change).

5.2.3 TOTAL INTERNAL REFLECTION OF LIGHT

When light impinges on the interface from the optically denser medium, $n_1 > n_2$, there is an angle of incidence, θ_i for which according to Equation 5.44 the refracted ray emerges tangent to the surface, $\theta_t = 90°$. This angle is called critical angle θ_c. The critical angle is determined from Snell's law by substituting $0_t - 90°$:

$$\sin\theta_c = \frac{n_2}{n_1} = n_{12} \quad (n_2 < n_1) \tag{5.73}$$

In this case, the angles of refraction θ_t have real values only for those incident angles θ_i, for which $\sin\theta_i < n_{12}$. When the angle of incidence is larger than the critical angle, total internal reflection takes place. For a glass/air interface with $n_{glass} = 1.52$, $\theta_c = 41.1°$, the fact that θ_c is less than 45° makes it possible to use the hypotenuse of a triangular prism with angles 45°, 45°, and 90° as a totally reflecting surface.

5.2.3.1 Reflected Wave

For incidence angles smaller than the critical angle $\theta_i < \theta_c$, the Fresnel's formulae (Equations 5.50, 5.55, and 5.63) are used, and the phase relations discussed for internal reflection above are valid. The case of total internal reflection, when the angle of incidence exceeds the critical angle, requires a special consideration.

Amplitude ratios: The law of refraction $\sin\theta_t = \sin\theta_i/n$ does not give a real value for θ_t, when $\theta_i > \theta_c$. Therefore, $\cos\theta_t$ in the Fresnel's formulae is a complex quantity, represented as

$$\cos\theta_t = \pm i\sqrt{\sin^2\theta_t - 1} = \pm i\sqrt{\frac{\sin^2\theta_i}{n^2} - 1} \tag{5.74}$$

where for clarity and simplicity n_{12}. is replaced with n. The modified Fresnel's amplitude ratios for internal reflection when $\theta_1 > \theta_c$ have the form:

$$\left(\frac{E_{0r}}{E_{0i}}\right)_{\parallel} = \frac{n^2\cos\theta_i - i\sqrt{\sin^2\theta_i - n^2}}{n^2\cos\theta_i + i\sqrt{\sin^2\theta_i - n^2}}$$

$$\left(\frac{E_{0r}}{E_{0i}}\right)_{\perp} = \frac{\cos\theta_i - i\sqrt{\sin^2\theta_i - n^2}}{\cos\theta_i + i\sqrt{\sin^2\theta_i - n^2}} \tag{5.75}$$

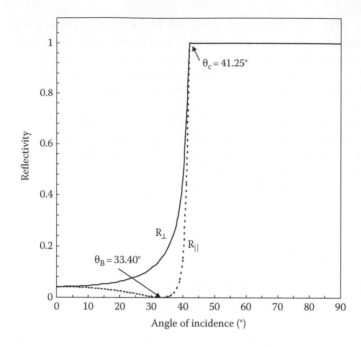

FIGURE 5.5 Internal reflection from glass/air interface: $n_1 = 1.5167$ (for the sodium D line $\lambda = 589.3$ nm, glass BK 7) and $n_2 = 1.0003$.

The reflectivity, which gives the ratio of the intensities of the reflected and incident waves, in this case would be the product of the amplitude ratio (Equation 5.75) and its complex conjugate, which leads to $R_{\parallel} = R_{\perp} = 1$. Hence, the irradiance of the reflected light is equal to the irradiance of the incident light, independent of polarization; therefore, it is said that the wave is totally reflected in the first medium. The reflectivity for internal incidence is shown in Figure 5.5.

Phase relations: When the incident angle exceeds the critical angle $\theta_i > \theta_c$, we can say that the amplitudes of the reflected and incident waves are equal,

$$\left|\frac{E_{0r}}{E_{0i}}\right|_{\parallel} = 1 \quad \text{or} \quad \left|E_{0r\parallel}\right| = \left|E_{0i\parallel}\right|$$

$$\left|\frac{E_{0r}}{E_{0i}}\right|_{\perp} = 1 \quad \text{or} \quad \left|E_{0r\perp}\right| = \left|E_{0i\perp}\right| \tag{5.76}$$

and set

$$\frac{E_{r\parallel}}{E_{i\parallel}} = e^{i\delta_{\parallel}} \quad \text{and} \quad \frac{E_{r\perp}}{E_{i\perp}} = e^{i\delta_{\perp}} \tag{5.77}$$

The formulae for the phase relation between the reflected and incident waves would be

$$\tan\frac{\delta_{\parallel}}{2} = -\frac{\sqrt{\sin^2\theta_i - n^2}}{n^2\cos\theta_i}$$

$$\tan\frac{\delta_{\perp}}{2} = -\frac{\sqrt{\sin^2\theta_i - n^2}}{\cos\theta_i} \tag{5.78}$$

The relative phase difference between both polarization states $\delta = \delta_\perp - \delta_\parallel$ is then given by

$$\tan\frac{\delta}{2} = \frac{\cos\theta_i\sqrt{\sin^2\theta_i - n^2}}{\sin^2\theta_i} \tag{5.79}$$

Maximum value of the relative phase difference can be found by differentiating Equation 5.79 with respect to the incident angle θ_i and setting it zero, $\dfrac{d}{d\theta_i}\left(\tan\dfrac{\delta}{2}\right) = 0$. This leads to a relation between θ_i and the relative refractive index, that has to be satisfied in order to get the maximum value of the relative phase difference:

$$\sin^2\theta_i = \frac{2n^2}{1+n^2} \tag{5.80}$$

And the maximum value of the relative phase difference is

$$\tan\frac{\delta_{max}}{2} = \frac{1-n^2}{2n} \tag{5.81}$$

The case $\delta = \delta_\perp - \delta_\parallel = 90°$ is of particular interest for practical applications. It allows producing circularly polarized light from linearly polarized light that undergoes total internal reflection. For this purpose, the incident light should be linearly polarized at an angle of 45° with respect to the incidence plane. This guarantees equal amplitudes of the electric field for both polarization components of the incident wave, $|E_{0i\parallel}| = |E_{0i\perp}|$, and by Equation 5.76 equal amplitudes of both polarizations in the reflected wave, $|E_{r\parallel}| = |E_{r\perp}|$. In order to obtain relative phase difference of $\pi/2$ ($\delta = \delta_\perp - \delta_\parallel = 90°$) with a single reflection from the boundary, the relative refractive index should be less than 0.414 (from $\tan(\pi/4) = 1<(1-n^2)/2n$, i.e., $n_{12} < \sqrt{2}-1 = 0.414$). If one of the media is air, then the other should have a refractive index of at least 2.42. Although there are some transparent materials with such a large refractive index, it is cheaper and more practical to use two total reflections on glass. Fresnel used a glass rhomb with an angle $\alpha = 54°37'$, which provided an angle of incidence θ_i also equal to $54°37'$ and two total reflections to demonstrate this method.

For materials with magnetic properties, the amplitude and phase relations for total internal reflection become

$$\begin{aligned}\left(\frac{E_r}{E_i}\right)_\parallel &= \frac{Z_1\cos\theta_i - iZ_2(s/k_2)}{Z_1\cos\theta_i + iZ_2(s/k_2)} \\[2mm] \left(\frac{E_r}{E_i}\right)_\perp &= \frac{Z_2\cos\theta_i - iZ_1(s/k_2)}{Z_2\cos\theta_i + iZ_1(s/k_2)}\end{aligned} \tag{5.82}$$

$$\begin{aligned}\tan\frac{\delta_\parallel}{2} &= -\frac{Z_2(s/k_2)}{Z_1\cos\theta_i} \\[2mm] \tan\frac{\delta_\perp}{2} &= -\frac{Z_1(s/k_2)}{Z_2\cos\theta_i}.\end{aligned} \tag{5.83}$$

where s is explained in the next section.

5.2.3.2 Evanescent Wave

The modified Fresnel's ratios suggest that the energy of the incident wave is totally reflected back to the first medium. However, the EM field of the incident wave penetrates into the second medium and generates a wave that propagates along the interface of the two media. The energy carried by this wave flows to and fro, and although there is a component of the Poynting vector in the direction normal to the boundary, its time average vanishes. This implies that there is no lasting flow of energy into the second medium.

Let us consider the x-axis along the interface between the two media, and z-axis along the normal to the interface. Let the wave that penetrates into the optically less dense medium propagate along the positive direction of the x-axis. Such a wave can be described by [28]:

$$E_t = E_{0t} \exp(-sz) \exp\left[-i(\omega t - k_1 x \sin \theta_i)\right] \tag{5.84}$$

where

$$s = k_2 \left[\left(\frac{n_1}{n_2}\right)^2 \sin^2 \theta_i - 1\right]^{1/2} = \frac{n_2 \omega}{c} \left[\left(\frac{\sin \theta_i}{\sin \theta_c}\right)^2 - 1\right]^{1/2} \tag{5.85}$$

The amplitude $E_{0t} \exp(-sz)$ of the wave decreases exponentially upon moving from the interface into the second medium. Therefore, this wave is called evanescent wave. The distance at which the amplitude of the wave decreases by a factor of e is called penetration depth, d_p:

$$d_p = s^{-1} = \frac{\lambda_2}{2\pi} \left[\left(\frac{n_1}{n_2}\right)^2 \sin^2 \theta_i - 1\right]^{-1/2} \tag{5.86}$$

The effective depth of penetration is on the order of a wavelength ($d_p \sim \lambda_2/2\pi$). The evanescent wave has very special properties. It is not a plane wave because its amplitude is not constant in a plane perpendicular to the direction of its propagation. This wave is not transversal wave either since the component of the electric vector in the direction of propagation does not vanish [1]. The phase velocity of the penetrating disturbance

$$v_{2x} = \frac{\omega}{k_1 \sin \theta_i} = \frac{\sin \theta_c}{\sin \theta_i} v_2 \tag{5.87}$$

is also not constant. The phase velocity v_{2x} is smaller than the velocity of the wave in the second medium v_2. The radiation penetrating into the second medium can be demonstrated by using frustrated total reflection [29] produced by two prisms separated by a distance of about a quarter of a wavelength, as shown in Figure 5.6. This way, the energy of the evanescent wave can couple into the second prism before damping to zero. The width of the gap determines the amount of the transmitted energy, which can be measured for quantitative description of the effect. In addition, the ray returns to the first medium at a point, which is displaced parallel to the interface by a distance d relative to the point of incidence, as shown in Figure 5.7. This displacement is called Goos–Hänchen shift and is less than a wavelength [30]. The frustrated total reflection allows also measurement of the Goos–Hänchen shift. Quantitative comparison of the theory of frustrated total reflection with experiment for the optical wavelengths shows that the values d/λ range from 0.25 to 1 approximately,

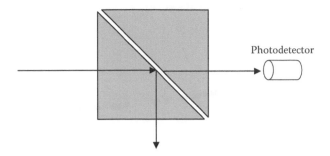

FIGURE 5.6   Frustrated total reflection.

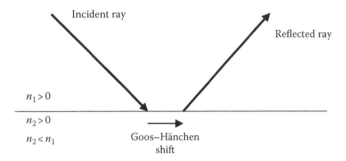

FIGURE 5.7   Positive Goos–Hänchen shift on total reflection. (After Lakhtakia, A., *Int. J. Electron. Commun.*, 58, 229, 2004.)

which corresponds to values from 80% to 0.008% for the transmission of the evanescent wave into the second prism [29]. The significance of the Goos–Hänchen shift becomes evident with the emergence of near-field microscopy and lithography. Another way to demonstrate the penetrating radiation is by observing directly the luminescence from a thin fluorescent layer deposited on the second prism in the same configuration.

In summary, in the case of total reflection, the energy flux of the incident wave is totally reflected and the components of the electric field parallel and perpendicular to the plane of incidence undergo different phase changes. The wave penetrates the second medium to a depth of about a wavelength, and propagates in the second medium along the interface with a phase velocity which is smaller than the velocity of the wave in the second medium. The totally reflected ray returns to the first medium at a point, which is displaced relative to the point of incidence.

5.2.4   Dispersion

5.2.4.1   Dispersion in Dielectrics

The dielectric constant is ε_r; hence, the index of refraction and velocity of EM waves in a dielectric depend on frequency. This phenomenon is called dispersion. Since the various frequencies that constitute the wave propagate with different speeds, the effect of dispersion can be easily demonstrated with white light refracted by a prism resulting in a spread of colors. Figure 5.8 shows the variation of refractive index with wavelength within the visible range for different optical materials. The value of n typically decreases with increasing wavelength and thus increases with increasing frequency. Light of longer wavelength usually has greater speed in a material than light of shorter wavelength.

Dispersion is a consequence of the frequency dependence of atomic polarization of the material, which is determined by the dipole moments of the constituent atoms or molecules. A dipole is a pair

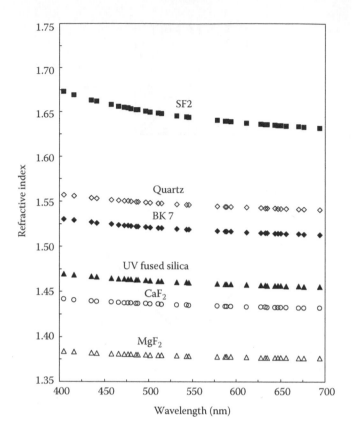

FIGURE 5.8 Variation of refractive index with wavelength. (Courtesy of Newport Corporation, Richmond, VA.)

of a negative $-Q$ and a positive $+Q$ charges with equal magnitudes and separated by a distance s. The dipole moment is a vector with magnitude Qs and direction pointing from $-Q$ to $+Q$:

$$\mathbf{p} = Qs \tag{5.88}$$

An external electric field can induce dipole moments in dielectrics with symmetrical constituent molecules, or can align the permanently existing and randomly oriented dipole moments in dielectrics composed by asymmetrical molecules. An example for asymmetrical molecule with permanent dipole moment is the water molecule. The macroscopic polarization of a dielectric material is defined by the dipole moment per unit volume [1–3]:

$$\boldsymbol{P} = Np = N\frac{e^2 E}{m(\omega_0^2 - \omega^2 - i\gamma\omega)} \tag{5.89}$$

where
$p = e^2 E/[m(\omega_0^2 - \omega^2 - i\gamma\omega)]$ is the dipole moment of an atom induced by the electric field of a harmonic wave with frequency ω
e and m are the charge and the mass of an electron, respectively
N (m^{-3}) is the number of dipole moments per unit volume, which is the number of electrons per unit volume with natural frequency ω_0
γ describes the damping of oscillations

The macroscopic polarization $\mathbf{P}$ is a complex quantity with frequency-dependent and time-varying behavior. For linear dielectrics, it is proportional to the electric field $\mathbf{E}$ via the material Equation 5.5d, $\mathbf{P} = \varepsilon_0 \chi_1 \mathbf{E}$. Combining this expression with Equation 5.89 gives the linear complex dielectric susceptibility as

$$\chi_1 = \frac{e^2 N}{\varepsilon_0 m (\omega_0^2 - \omega^2 - i\gamma\omega)} \tag{5.90}$$

Taking into account Equations 5.3a and 5.5a, the frequency-dependent permittivity can be written as

$$\varepsilon_\omega = \varepsilon_0 + \frac{e^2}{\varepsilon_0 m} \cdot \frac{N}{\omega_0^2 - \omega^2 - i\gamma\omega} \tag{5.91}$$

Consequently, the refractive index would be

$$\hat{n}_\omega^2 = \varepsilon_{r\omega} = 1 + \frac{e^2}{\varepsilon_0 m} \cdot \frac{N}{\omega_0^2 - \omega^2 - i\gamma\omega} \tag{5.92}$$

where $\varepsilon_{r\omega} = \varepsilon_\omega / \varepsilon_0$ is the frequency-dependent relative permittivity. Equation 5.92 shows that the index of refraction and hence the velocity of EM waves in a given material depend upon frequency. The refractive index is a complex quantity:

$$\hat{n}_\omega = n_\omega + i\xi_\omega \tag{5.93}$$

where n_ω and ξ_ω are the real and imaginary part of $\hat{n}_\omega$ as both depend upon frequency $\omega = 2\pi f$. If in addition to the electrons with natural oscillation frequency ω_0, there are also electrons with other natural oscillation frequency ω_{0i}, they also contribute to the dispersion and have to be taken into account. If N_i is the number density of the electrons with natural oscillation frequency ω_{0i}, the refractive index can be written as the sum:

$$n_\omega^2 = 1 + \frac{e^2}{\varepsilon_0 m} \sum_i \frac{N_i}{\omega_{0i}^2 - \omega^2} \tag{5.94}$$

The expressions about the refractive index above represent the case of dispersion caused by oscillating electrons, which is manifested in the visible range. However, dispersion can be caused also by the oscillations of ions. Because of the large mass of ions, their natural frequencies are much lower than for electrons and lie in the far IR region of the spectrum. Therefore, the oscillations of ions do not contribute to the dispersion in the visible range. However, they influence the static permittivity, which may differ from the permittivity in the visible range. For example, the refractive index of water for optical frequencies is $n_\omega = \sqrt{\varepsilon_r} = 1.33$, while the static value is $\sqrt{\varepsilon_r} \cong 9$. This anomaly is explained by the contribution of ions to the dispersion [2].

5.2.4.2 Normal Dispersion

For rarefied gases, the refractive index is close to unity and Equation 5.94 can be simplified by using the approximation $n_\omega^2 - 1 = (n_\omega - 1)(n_\omega + 1) \cong 2(n_\omega - 1)$:

$$n_\omega = 1 + \frac{e^2}{2\varepsilon_0 m} \sum_i \frac{N_i}{\omega_{0i}^2 - \omega^2} \tag{5.95}$$

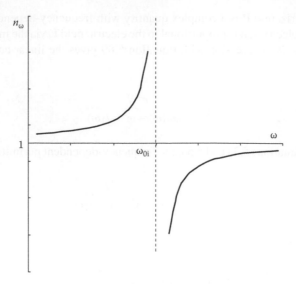

FIGURE 5.9 Normal dispersion.

When the refractive index n_ω increases with frequency ω, as shown in Figure 5.9, the dispersion is called normal. Normal dispersion is caused by the oscillations of electrons and is observed in the entire visible region.

For low frequencies $\omega \ll \omega_{0i}$, Equation 5.95 assumes the form:

$$n = 1 + \frac{e^2}{2\varepsilon_0 m} \sum_i \frac{N_i}{\omega_{0i}^2} \tag{5.96}$$

which gives the static value of the refractive index. As it was discussed above, the static value may differ significantly from the refractive index for optical frequencies.

For very high frequencies such as x-rays, $\omega \gg \omega_{0i}$, then Equation 5.95 becomes

$$n_\omega = 1 - \frac{e^2}{2\varepsilon_0 m \omega^2} \sum_i N_i \tag{5.97}$$

This expression shows that the refractive index tends to, but remains, less than unity. This means that for a radiation of very high frequencies, a dielectric behaves as an optically less dense material than vacuum. The reason for this effect is that at very high frequencies, the nature of the electron bonds in the atom does not play any role, and as per Equation 5.97, the refractive index depends only on the total number of oscillating electrons per unit volume.

5.2.4.3 Anomalous Dispersion

The dispersion Equation 5.94 was derived from Equation 5.92 when $\gamma = 0$, that is, the damping of oscillations was not taken into consideration. For this case, the refractive index $n\omega$ turns to infinity at the resonant frequency $\omega = \omega_{0i}$. If damping of oscillations is not ignored, $\gamma \neq 0$, the dispersion curve becomes continuous in the neighborhood of the natural oscillation frequencies ω_{0i} and does not turn to infinity at $\omega = \omega_{0i}$. For rare gases, when the refractive index differs slightly from unity, Equation 5.92 can be written as

$$\hat{n}_\omega = \sqrt{\varepsilon_{r\omega}} = n_\omega + i\xi_\omega \cong 1 + \frac{e^2}{2\varepsilon_0 m} \frac{N}{\omega_{0i}^2 - \omega^2 - i\gamma\omega} \tag{5.98}$$

Writing separately the real and imaginary parts

$$n_\omega = 1 + \frac{e^2}{2\varepsilon_0 m} \cdot \frac{N(\omega_0^2 - \omega^2)}{(\omega_{0i}^2 - \omega^2) + \gamma^2 \omega^2} \tag{5.99}$$

and

$$\xi_\omega = \frac{e^2}{2\varepsilon_0 m} \cdot \frac{N\gamma\omega}{(\omega_{0i}^2 - \omega^2) + \gamma^2 \omega^2} \tag{5.100}$$

The imaginary part $\xi\omega$ of the refractive index describes the absorption in the material. Equation 5.99 shows that near the resonance frequency $\omega = \omega_{0i}$, the refractive index decreases with increasing frequency. This phenomenon is called anomalous dispersion. Normal dispersion is observed throughout the region where a material is transparent, while anomalous dispersion takes place in the absorption regions.

The electric vector of a plane wave propagating in the direction of the z-axis can be presented as

$$\mathbf{E}(z,t) = \mathbf{E}_0 e^{-i(\omega t - kz)} = \mathbf{E}_0 e^{-i\xi_\omega z/c} e^{-i(\omega t - \omega n_\omega z/c)} \tag{5.101}$$

where the wave vector

$$k = \frac{\omega}{v} = \omega \frac{\sqrt{\varepsilon_{r\omega}}}{c} = \omega \frac{n_\omega}{c} + i\omega \frac{\xi_\omega}{c} \tag{5.102}$$

was expressed with the complex form of the refractive index from Equation 5.93, $\hat{n}_\omega = \sqrt{\varepsilon_{r\omega}} = n_\omega + i\xi_\omega$. Equation 5.101 shows that due to absorption ($\gamma \neq 0$) in the material, the amplitude of the wave decreases exponentially with the traveled distance in the medium, that is, the imaginary part ξ_ω of the refractive index describes the damping of a plane wave in a dielectric. At frequencies near the resonance frequencies $\omega = \omega_{0i}$, the wave energy is consumed, on exciting forced oscillations in electrons. In turn, the oscillating electrons reemit EM waves of the same frequencies but in random directions. This results in scattering of the EM waves upon passing through a dielectric. The scattering is insignificant when γ is small.

5.2.4.4 Group Velocity of a Wave Packet

A wave packet or a pulse, is a superposition of waves whose frequencies and wave numbers are confined to a quite narrow interval. A wave packet of waves with the same polarization can be represented in the form:

$$E(z,t) = \frac{1}{2\pi} \int_{-\infty}^{+\infty} F(k) e^{-i(\omega t - kz)} dk \tag{5.103}$$

The amplitude $F(k)$ describes the distribution of the waves over wave number and hence over frequencies. Thus, $F(k)$ is the Fourier transform of $E(z,0)$ at $t = 0$:

$$F(k) = \int_{-\infty}^{+\infty} E(z,0) e^{-ikz} dz \tag{5.104}$$

The amplitude of the wave packet is nonzero only in the narrow interval of values of k near k_0 and has a sharp peak near k_0. Usually, when we speak of the velocity of a signal, we mean the group velocity at a frequency corresponding to the maximum amplitude in the signal. The shape of the pulse does not change when the pulse moves with a group velocity:

$$v_g = \frac{d\omega_0}{dk_0} = \frac{d[2\pi c/(n\lambda)]}{d(2\pi/\lambda)} = c\frac{(nd\lambda + \lambda dn)}{d\lambda/\lambda^2} = \frac{c}{n}\left(1 + \frac{\lambda}{n}\frac{dn}{d\lambda}\right) \tag{5.105}$$

It is obvious that the group velocity of the pulse depends on the degree of monochromaticity of the wave packet.

5.2.4.5 Color of Objects

The coloration of bodies results from the selective absorption of light at the resonance frequencies ω_{0i}. When the natural oscillation frequencies ω_{0i} lie in the UV region, the materials appear almost colorless and transparent, because there is no absorption in the visible part of the spectrum. For example, glass is transparent for wavelengths in the visible range, but strongly absorbs UV. The absorption may occur in the bulk of a body or in its surface layer. When light is reflected from a surface, the most strongly reflected wavelengths are those which are absorbed most strongly upon passing through the bulk of the material. The color due to selective reflection and the color due to selective absorption are complementary (added together they produce white light). For example, a chunk of gold has a reddish-yellow color due to selective reflection. However, a very thin foil of gold exhibits a deep blue color when viewed through, which is due to selective absorption.

5.2.5 Propagation of Light in Metamaterials

MMs are artificially constructed materials or composites that exhibit negative index of refraction [50–53]. MMs are also called negative-index materials, double negative media, backward media, left-handed materials, or negative phase-velocity (NPV) materials, the last being the least ambiguous of all suggested names. First reported in 2000 [50], MMs are the most recent example of a qualitatively new type of material. In comparison with their constituent materials, MMs have distinct and possibly superior properties. Their exotic properties and potential for unusual applications in optics, photonics, electromagnetism, material science, and engineering have garnered significant scientific interest and provide the rationale for ongoing investigation of this rapidly evolving field. The MMs comprise an array of subwavelength-discrete and subwavelength-independent elements designed to respond preferentially to the electric or magnetic component of an external EM wave. Thus, they have a remarkable potential to control EM radiation.

Already in 1968, on the base of theoretical considerations, Veselago predicted many unusual properties of substances with simultaneously negative real permittivity and negative real permeability, including inverse refraction, negative radiation pressure (change from radiation pressure to radiation tension), and inverse Doppler effect [54]. But it was the first observation of the negative refraction by Smith et al. [50] in 2000 that made the breakthrough and initiated intensive research in this area. The phenomenon was first demonstrated with a composite medium, based on a periodic array of interspaced conducting nonmagnetic split ring resonators and continuous wires that exhibit simultaneously negative values of effective permittivity and permeability in the microwave frequency region. The current implementations of negative refractive index media have occurred at microwave to near-optical frequencies (from about 10 GHz to about 100 THz) and in a narrow frequency band [55–57]. The extension of MMs to the visible range and the implementation of structures with tunable and reconfigurable resonant properties would be significant for at least two reasons: the opportunity to study and begin to understand the intrinsic properties of negative media and possible future applicability [58–64]. MMs, although fabricated with a microstructure, are effectively homogenous in the frequency range wherein negative refraction is observed. Negative refraction by periodically inhomogeneous substances such as photonic crystals has also been demonstrated theoretically as well as experimentally. Some composite media with simple structure, consisting of insulating magnetic spherical particles embedded in a background matrix, can also exhibit negative permeability and permittivity over significant bandwidths [64]. Due to some practical difficulties in achieving negative refraction in isotropic dielectric–magnetic materials, attention has been directed toward more complex materials such as anisotropic materials [65] as well as isotropic chiral materials [66–69].

The phase velocity of light propagating in materials with real relative permittivity and real relative permeability both being simultaneously negative is negative. The phase velocity of an EM wave is described as negative, if it is directed opposite to the power flow given by the time-averaged Poynting vector. Negative phase velocity results in negative refraction, which is shown in Figure 5.10. This phenomenon does not contradict Snell's law. A comprehensive theoretical description of propagation of light in isotropic NPV materials is presented in a series of works of Lakhtakia: for example, propagation of plane wave in a linear, homogeneous, isotropic, and dielectric–magnetic medium [60,67,69–72]; propagation of a plane wave in anisotropic chiral materials [66,73]; and diffraction of a plane wave from periodically corrugated boundary of vacuum and a linear, homogeneous, uniaxial, and dielectric–magnetic NPV medium [67,74]. The study of diffraction from a NPV medium is especially important, because all experimental realizations of NPV MMs so far are based on periodical structures of composite materials, with size of the unit cell smaller than the wavelength. On another hand, gratings made of NPV materials offer quite different properties from the traditional gratings.

The two material parameters that govern the response of homogeneous media to external EM field are the permittivity ε and the magnetic permeability μ. In dielectrics, the role of permeability is ignored. For NPV MMs though, the permeability is of the same importance as the permittivity. In general, both ε and μ are frequency-dependent complex quantities, given by

$$\varepsilon(\omega) = \varepsilon'(\omega) + i\varepsilon''(\omega) \tag{5.106a}$$

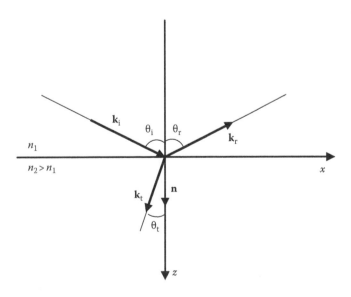

FIGURE 5.10 Negative refraction at a boundary between two isotropic media ($n_2 > n_1$).

and

$$\mu(\omega) = \mu'(\omega) + i\mu''(\omega) \tag{5.106b}$$

Thus, there are in total four parameters that completely describe the response of an isotropic material to EM radiation at a given frequency.

The theoretical description of negative refraction is out of the scope of this chapter. Moreover, there is no universal model that would fit the description of all particular cases of microstructures that have been observed to exhibit negative refraction. However, let us briefly discuss here as an example, the particular case of total internal reflection when the optically rarer medium has negative real permittivity and negative real permeability, as presented by Lakhtakia [72]. Consider a planar interface of two homogeneous media with relative permittivity ε_1 and ε_2, and relative permeability μ_1 and μ_2, respectively, defined by Equations 106a and b at a given angular frequency ω. The first medium is non-dissipative ($\varepsilon_1'' = 0$ and $\mu_1'' = 0$), while dissipation in the second medium is very small ($\varepsilon_2'' \ll \varepsilon_2'$ and $\mu_2'' \ll \mu_2'$). A plane wave is incident on the interface through the first medium and reflected back. The amplitude reflection coefficients in this case are given by

$$r_\parallel = \frac{1 - (\alpha_1/\alpha_2)(\varepsilon_2/\varepsilon_1)}{1 + (\alpha_1/\alpha_2)(\varepsilon_2/\varepsilon_1)} \tag{5.107a}$$

for electric vector oscillating in the plane of incidence, and

$$r_\perp = -\frac{1 - (\alpha_1/\alpha_2)(\mu_2/\mu_1)}{1 + (\alpha_1/\alpha_2)(\mu_2/\mu_1)} \tag{5.107b}$$

for perpendicular polarization. In these equations,

$$\alpha_1 = k_0\sqrt{\varepsilon_1\mu_1 - \varepsilon_1\mu_1 \sin^2\theta_i} \quad \text{and} \quad \alpha_2 = k_0\sqrt{\varepsilon_2\mu_2 - \varepsilon_1\mu_1 \sin^2\theta_i} \tag{5.108}$$

where
k_0 is the free-space wave number
θ_i is the angle of incidence

The second medium must be a passive medium, thus the imaginary part of α_2 must be positive or equal to zero.

For the phase velocity to oppose the direction of the time-averaged Poynting vector, the inequality

$$\left(|\varepsilon_2| - \varepsilon_2'\right)\left(|\mu_2| - \mu_2'\right) > \varepsilon_2''\mu_2'' \tag{5.109}$$

has to be satisfied. Thus arises the conclusion that both ε_2' and μ_2' do not have to be simultaneously negative for the phase velocity to be negative [72].

Total internal reflection occurs at angles of incidence larger than the critical angle. In dielectrics, where only permittivity plays a role, the Goos–Hänchen shift is positive, as shown in Figure 5.7. For NPV materials though, because of equal involvement of permeability and permittivity, the reversal of the signs of both results in a reversal of the direction of the Goos–Hänchen shift. Figure 5.11 shows a negative Goos–Hänchen shift. Furthermore, if both the real permittivity and the real permeability of the optically rarer medium are negative, the negative Goos–Hänchen shifts occur for both perpendicular and parallel polarized beams [72].

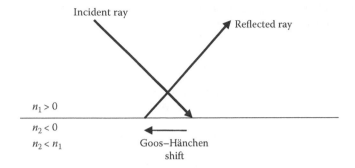

FIGURE 5.11 Negative Goos–Hänchen shift on total reflection. (After Lakhtakia, A., *Int. J. Electron. Commun.*, 58, 229, 2004.)

The area of artificial MMs is a subject of huge scientific interest with exponentially growing number of publications. The brief introduction here is aimed to provide some initial information for those who might be interested in expanding their knowledge in this direction.

REFERENCES

1. Born, M. and Wolf, E., *Principles of Optics*, 6th edn., Pergamon Press, New York, 1980.
2. Matveev, A.N., *Optics*, Mir Publishers, Moscow, Russia, 1988.
3. Smith, F.G., King, T.A., and Wilkins, D., *Optics and Photonics*, 2nd edn., John Wiley & Sons, Inc., Hoboken, NJ, 2007.
4. Hecht, E., *Optics*, 4th edn., Pearson Education, Inc., Addison Wesley, San Francisco, CA, 2002.
5. Bennett, C.A., *Principles of Physical Optics*, John Wiley & Sons, Inc., Hoboken, NJ, 2008.
6. Pedrotti, F.L., Pedrotti, L.S., and Pedrotti L.M., *Introduction to Optics*, 3rd edn., Pearson Prentice Hall, Upper Saddle River, NJ, 2007, p. 421.
7. Möller, K.D., *Optics*, University Science Books, Mill Valley, CA, 1988, p. 384.
8. Klein, M.V. and Furtak, T.E., *Optics*, 2nd edn., John Wiley & Sons, Inc., Hoboken, NJ, 1986, p. 475.
9. Strong, J., *Concepts of Classical Optics*, W.H. Freeman and Co., San Francisco, CA, 1958; or Dover Publications, New York, 2004.
10. Halliday, D., Resnick, R.P., and Walker, J., *Fundamentals of Physics*, 7th edn., John Wiley & Sons, Hoboken, NJ, 2004.
11. Fowles, G.R., *Introduction to Modern Optics*, 2nd edn., Dover Publications, New York, 1989.
12. Guenther, R.D., *Modern Optics*, John Wiley & Sons, Hoboken, NJ, 1990.
13. Saleh, B.E.A. and Teich, M.C., *Fundamentals of Photonics*, 2nd edn., John Wiley & Sons, Inc., Hoboken, NJ, 2007.
14. Meyer-Arendt, J.R., *Introduction to Classical and Modern Optics*, 4th edn., Prentice-Hall, Inc., Englewood Cliffs, NJ, 1995.
15. Scully, M.O. and Zubairy, M.S., *Quantum Optics*, Cambridge University Press, Cambridge, U.K., 1997.
16. Griffithz, D.J., *Introduction to Electrodynamics*, 3rd edn., Prentice-Hall, Inc., Englewood Cliffs, NJ, 1999.
17. Jackson, J.D., *Classical Electrodynamics*, 3rd edn., John Wiley & Sons, Inc., New York, 1998.
18. Grant, I.S. and Phillips, W.R., *Electromagnetism*, 2nd edn., John Wiley & Sons, Inc., New York, 1990.
19. Boas, M.L., *Mathematical Methods in the Physical Sciences*, 3rd edn., John Wiley & Sons, Inc., Hoboken, NJ, 2006.
20. Shen, Y.R., *The Principles of Nonlinear Optics*, John Wiley & Sons, Inc., New York, 1984.
21. Boyd, R.W., *Nonlinear Optics*, 2nd edn., Academic Press, London, U.K., 2003.
22. Sheik-Bahae, M. et al., Sensitive measurement of optical nonlinearities using a single beam, *IEEE Journal of Quantum Electronics* **26**, 760, 1990.
23. Balu, M. et al., White-light continuum Z-scan technique for nonlinear materials characterization, *Optics Express* **12**, 3820, 2004.
24. Dushkina, N.M. et al., Influence of the crystal direction on the optical properties of thin CdS films formed by laser ablation, *Photodetectors: Materials and Devices IV*, San Jose, CA; *Proceedings of SPIE* **3629**, 424, 1999.

25. Hariharan, P., *Optical Holography, Principles, Techniques, and Applications*, 2nd edn., Cambridge University Press, Cambridge, U.K., 1996.

26. Splinter, R. and Hooper B.A., *An Introduction to Biomedical Optics*, Taylor & Francis Group, LLC, New York, 2007.

27. Novotny, L. and Hecht, B., *Principles of Nano-Optics*, Cambridge University Press, Cambridge, U.K., 2006.

28. de Fornel, F., *Evanescent Waves*, Springer, Berlin, Germany, 2001.

29. Zhu, S. et al., Frustrated total internal reflection: A demonstration and review, *American Journal of Physics* **54**, 601–607, 1986.

30. Lotsch, H.K.V., Beam displacement at total reflection. The Goos-Hänchen effect I, *Optik* **32**, 116–137, 1970.

31. Otto, A., Excitation of nonradiative surface plasma waves in silver by the method of frustrated total reflection, *Zeitschrift für Physik* **216**, 398, 1968.

32. Raether, H., Surface plasma oscillations and their applications, in: *Physics of Thin Films, Advances in Research and Development*, G. Hass, M.H. Francombe, and R.W. Hofman (Eds.), Vol. 9, Academic Press, New York, 1977, pp. 145–261.

33. Ehler, T.T. and Noe, L.J., Surface plasmon studies of thin silver/gold bimetallic films, *Langmuir* **11**(10), 4177–4179, 1995.

34. Homola, J., Yee, S.S., and Gauglitz, G., Surface plasmon resonance sensors: Review, *Sensors and Actuators B* **54**, 3, 1999.

35. Mulvaney, P., Surface plasmon spectroscopy of nanosized metal particles, *Langmuir* **12**, 788, 1996.

36. Klar, T. et al., Surface plasmon resonances in single metallic nanoparticles, *Physical Review Letters* **80**, 4249, 1998.

37. Markel, V.A. et al., Near-field optical spectroscopy of individual surface-plasmon modes in colloid clusters, *Physical Review B* **59**, 10903, 1999.

38. Kotsev, S.N. et al., Refractive index of transparent nanoparticle films measured by surface plasmon microscopy, *Colloid and Polymer Science* **281**, 343, 2003.

39. Dushkina, N. and Sainov, S., Diffraction efficiency of binary metal gratings working in total internal reflection, *Journal of Modern Optics* **39**, 173, 1992.

40. Dushkina, N.M. et al., Reflection properties of thin CdS films formed by laser ablation, *Thin Solid Films* **360**, 222, 2000.

41. Schuck, P., Use of surface plasmon resonance to probe the equilibrium and dynamic aspects of interactions between biological macromolecules, *Annual Review of Biophysics and Biomolecular Structure* **26**, 541–566, 1997.

42. Dushkin, C.D. et al., Effect of growth conditions on the structure of two-dimensional latex crystals: Experiment, *Colloid and Polymer Science* **227**, 914–930, 1999.

43. Tourillon, G. et al., Total internal reflection sum-frequency generation spectroscopy and dense gold nanoparticles monolayer: A route for probing adsorbed molecules, *Nanotechnology* **18**, 415301, 2007.

44. He, L. et al., Colloidal Au-enhanced surface plasmon resonance for ultrasensitive detection of DNA hybridization, *Journal of the American Chemical Society* **122**, 9071–9077, 2000.

45. Krenn, J.R. et al., Direct observation of localized surface plasmon coupling, *Physical Review B* **60**, 5029–5033, 1999.

46. Liebsch, A., Surface plasmon dispersion of Ag, *Physical Review Letters* **71**, 145–148, 1993.

47. Prieve, D.C. and Frej, N.A., Total internal reflection microscopy: A quantitative tool for the measurement of colloidal forces, *Langmuir* **6**, 396–403, 1990.

48. Perner, M. et al., Optically induced damping of the surface plasmon resonance in gold colloids, *Physical Review Letters* **78**, 2192–2195, 1993.

49. Tsuei, K.-D., Plummer, E.W., and Feibelman, P.J., Surface plasmon dispersion in simple metals, *Physical Review Letters* **63**, 2256–2259, 1989.

50. Smith, D.R. et al., Composite medium with simultaneously negative permeability and permittivity, *Physical Review Letters* **84**, 4184–4187, 2000.

51. Smith, D.R., Pendry, J.B., and Wiltshire, M.C.K., Metamaterials and negative refractive index, *Science* **305**, 788, 2006.

52. Lakhtakia, A., McCall, M.W., and Weiglhofer, W.S., Negative phase-velocity mediums, in: *Introduction to Complex Mediums for Optics and Electromagnetics*, Weiglhofer, W.S. and Lakhtakia, A. (Eds.), SPIE Press, Bellingham, WA, 2003, pp. 347–363.

53. Ramakrishna, S.A., Physics of negative refractive index materials, *Reports on Progress in Physics* **68**, 449–521, 2005.

54. Veselago, V.G., The electrodynamics of substances with simultaneously negative values of ε and μ, *Soviet Physics Uspekhi* **10**, 509–514, 1968.

55. Shelby, R.A., Smith, D.R., and Schultz, S., Experimental verification of a negative index of refraction, *Science* **292**, 77–79, 2001.

56. Grbic, A. and Eleftheriades, G.V., Experimental verification of backward-wave radiation from a negative index metamaterial, *Journal of Applied Physics* **92**, 5930–5935, 2002.

57. Houck, A.A., Brock, J.B., and Chuang, I.L., Experimental observations of a left-handed material that obeys Snell's law, *Physical Review Letters* **90**, 137401, 2003.

58. Padilla, W.J., Basov, D.N., and Smith, D.R., Negative refractive index metamaterials, *Materials Today* **9**, 28, 2006.

59. Shalaev, V.M. and Boardman, A. Focus issue on metamaterials, *Journal of Optical Society of America B* **23**, 386–387, 2006.

60. McCall, M.W., Lakhtakia, A., and Weiglhofer, W.S., The negative index of refraction demystified, *European Journal of Physics* **23**, 353, 2002.

61. Lakhtakia, A. and McCall, M.W., Focus on negative refraction, *New Journal of Physics* **7**, 191–193, 2005.

62. Enrich, C. et al., Magnetic metamaterials at telecommunication and visible frequencies, *Physical Review Letters* **95**, 203901, 2005.

63. Pendry, J.B., Negative refraction, *Contemporary Physics* **45**, 191–202, 2004.

64. Holloway, C.L. et al., A double negative (DNG) composite medium composed of magnetodielectric spherical particles embedded in a matrix, *IEEE Transactions on Antennas and Propagation* **51**, 2596, 2003.

65. Hu, L. and Lin, Z., Imaging properties of uniaxially anisotropic negative refractive index materials, *Physics Letters A* **313**, 316–24, 2003.

66. Mackay, T.G. and Lakhtakia, A., Plane waves with negative phase velocity in Faraday chiral mediums, *Physical Review E* **69**, 026602, 2004.

67. Lakhtakia, A., Reversed circular dichroism of isotropic chiral mediums with negative real permeability and permittivity, *Microwave and Optical Technology Letters* **33**, 96–97, 2002.

68. Pendry, J.B., A chiral route to negative refraction, *Science* **306**, 1353–1355, 2004.

69. Mackay, T.G., Plane waves with negative phase velocity in isotropic mediums, *Microwave and Optical Technology Letters* **45**, 120–121, 2005; a corrected version is available at: http://www.arxiv.org/abs/physics/0412131.

70. Lakhtakia, A. and Mackay, T.G., Fresnel coefficients for a permittivity-permeability phase space encompassing vacuum, anti-vacuum, and nihility, *Microwave and Optical Technology Letters* **48**, 265–270, 2006.

71. Lakhtakia, A., On planewave remittances and Goos-Hänchen shifts on planar slabs with negative real permittivity and permeability, *Electromagnetics* **23**, 71–75, 2003.

72. Lakhtakia, A., Positive and negative Goos-Hänchen shifts and negative phase-velocity mediums (alias left-handed materials), *International Journal of Electronics and Communications* (AEÜ) **58**, 229–231, 2004.

73. Mackay, T.G. and Lakhtakia, A., Negative phase velocity in a material with simultaneous mirror-conjugated and racemic chirality characteristics, *New Journal of Physics* **7**(165), 1–23, 2005.

74. Depine, R.A. and Lakhtakia, A., Diffraction by a grating made of a uniaxial dielectric-magnetic medium exhibiting negative refraction, *New Journal of Physics* **7**, 1–23, 2005.

6 Interferometry

David A. Page

CONTENTS

6.1 Introduction 197
6.2 Interfering Waves 198
6.3 Early Examples of Interference 204
    6.3.1 Young's Slits 204
    6.3.2 Fresnel's Biprism 204
    6.3.3 Lloyd's Experiment 205
6.4 Optical Sources 205
6.5 Simple Interferometer 206
6.6 Fizeau Interferometer 207
6.7 Twyman Green Interferometer 209
    6.7.1 Mach–Zender Interferometer 210
    6.7.2 Shearing Interferometer 210
6.8 Infrared Interferometry 211
6.9 Test Arrangements 214
    6.9.1 Testing of Flats 214
    6.9.2 Measurement of Materials 214
    6.9.3 Testing of Spherical Surfaces 214
    6.9.4 Testing of Lenses 215
    6.9.5 Testing of Domes 215
    6.9.6 Testing Telescopes 216
6.10 Fringe Analysis 218
6.11 Aberration Types 220
    6.11.1 Spherical Aberration 220
    6.11.2 Coma 220
    6.11.3 Astigmatism 220

6.1 INTRODUCTION

Interferometry employs the ability of two beams of electromagnetic radiation to interfere with one another provided certain criteria of coherence are achieved. The technique can be used to sense, very accurately, disturbances in one of the beams provided the other beam's properties are accurately known. There are many variants of interferometers but this chapter discusses the application of interferometers to the testing of optical components and systems, and is restricted to the commonest types of interferometers commercially available.

Interferometry is arguably one of the most powerful diagnostic test tools in the optical toolbox. Interferometric techniques are used for a myriad of purposes ranging from the control of CNC machines of nanometer accuracy to the search for dark matter through gravitational lensing on a cosmological scale. Between these extremes is the use of interferometry in the optical workshop during lens manufacture and system assembly and test, and this subject is the main thrust of this chapter. Interferometers used in optical manufacture range from the very simple, for example, test plates, to the more complex systems used for accurate measurement of components and systems. In common with many other forms of test equipment, interferometric analysis has taken advantage of the developments in computing both in terms of speed and graphical presentation.

The performance of an interferometer depends on the quality of the components used, be they the projecting or collecting optics or the qualities of the radiation source used. The coherence properties of this latter factor are key to the performance of the interferometer both in terms of accuracy and convenience of use.

6.2 INTERFERING WAVES

Interferometers are instruments that can measure directly the distortions of a wavefront resulting from aberrations in optical systems, badly manufactured optical components, inhomogeneities in materials, etc. This is achieved by the measurement of the complex amplitude distribution in an electromagnetic wave. The measurement of the complex amplitude is performed by mixing the distorted wavefront with a nominally perfect wavefront with which it is mutually coherent. Points on the wavefront which are in phase with the perfect wavefront will exhibit a greater concentration of energy than points which are out of phase. The former will thus be bright and the latter dark as shown in Figure 6.1.

The waveforms represent the complex amplitude of the electromagnetic wavefront. The interfering wavefronts are shown as having the same amplitude, so that when they are out of phase by π, there is perfect cancellation. When in phase the complex amplitude of the resultant waveform is double the complex amplitude of the individual waveforms. Note that the time averaged complex amplitude in both cases will still be zero because the electromagnetic wave's electric and magnetic vectors can take positive and negative values. Fortunately detectors and the human eye do not respond to complex amplitude but to intensity, which as the square of the complex amplitude has the effect of rectifying the waveform. Such detectors then see a time averaged intensity of the rectified waveform; hence the in phase interference will produce a bright area and the out of phase interference a dark area.

Figure 6.2 shows a similar situation except that one of the beams has only half the amplitude of the other. It can be seen that perfect cancellation does not occur so any fringe seen will not have optimum contrast. Additionally the maximum amplitude is reduced. This serves to highlight the

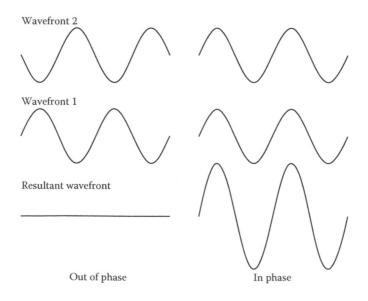

Wavefront 2

Wavefront 1

Resultant wavefront

Out of phase   In phase

FIGURE 6.1 Summation of complex amplitude.

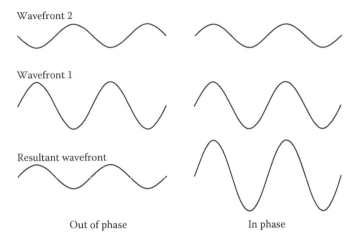

Wavefront 2

Wavefront 1

Resultant wavefront

Out of phase

In phase

FIGURE 6.2 Summation with non-equal amplitude.

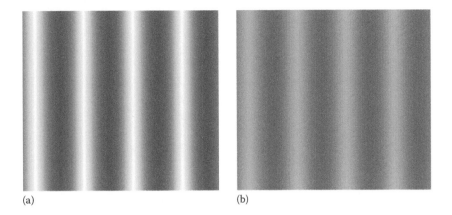

(a)

(b)

FIGURE 6.3 (a) High- and (b) low-contrast fringes.

need for approximately equal intensity in the interfering wavefronts. A pictorial representation of high- and low-contrast fringes is shown in Figure 6.3.

Figure 6.4 shows the variation in contrast for different ratios of wavefront intensity. The contrast is defined as

$$C_I = \frac{I_{\max} - I_{\min}}{I_{\max} + I_{\min}}$$

where
$I_{\max}$ is the maximum intensity of the fringe pattern
$I_{\min}$ is the minimum intensity

The contrast is expressed in terms of intensity since that is the condition we can see and detect. $I_{\max}$ is defined by the sum of the amplitudes of the two interfering beams when they are in phase converted to an intensity and time averaged. $I_{\min}$ is the condition when they are out of phase.

The simplest interferometer is simply a plate of glass. The reflectance of the glass will be around 4% from the front and rear surfaces. The setup of such a device is shown in Figure 6.5.

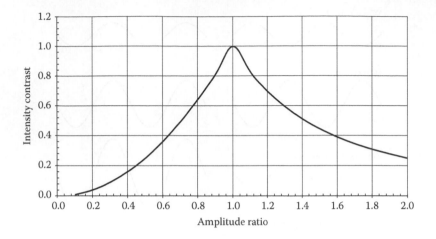

FIGURE 6.4 Contrast for non-equal amplitudes.

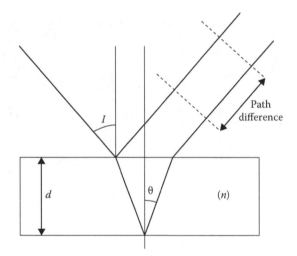

FIGURE 6.5 Two beam interference.

The glass is of thickness d and refractive index n. A plane parallel wavefront is incident on the plate at an angle I. Refraction makes the internal angle different, in this case labeled θ. The path difference between the two wavefronts is given by

$$Pd = 2nd \cos \theta$$

or in terms of phase units

$$Ph = 2nd \cos \theta \frac{2\pi}{\lambda}$$

where λ is the wavelength of the electromagnetic radiation.

It can be seen that the phase difference between the reflections from the front and rear faces of the glass depends on two parameters that are not single valued. The angle within the plate will depend on the angular subtense of the source. This is normally a pinhole located at the focus of a collimating lens or mirror illuminated by a thermal source. If a screen is placed in the reflected beams then they will produce interference. The phase of the interference will depend on the position within the pinhole from where the particular pair of wavefronts originate. There will effectively be a continuum of interference patterns all of different phase. Since these wavefronts are originating at different parts of a random source, they will not be coherent. This means that all the individual interference patterns will add together in intensity rather than complex amplitude. The result is that the overall contrast is reduced.

If the wavefront from the center of the pinhole is incident normally on the glass plate, then at the center of the beam the path difference between the wavefronts reflected from the front and rear surfaces is

$$P_0 = 2nd$$

If we now consider a wavefront that originates from the edge of the pinhole, then this will arrive at an incidence angle I. As in Figure 6.5, the angle within the glass plate will be θ. The path difference in this case will be

$$P_\theta = 2nd \cos \theta$$

The change in path difference is given by,

$$\Delta P = 2nd(1 - \cos \theta)$$

If we assume that θ is small then the expression can be expanded to

$$\Delta P = 2nd \left[1 - \left(1 - \frac{\theta^2}{2} \right) \right]$$

$$\Delta P = nd\theta^2$$

If we now consider the wavefronts to have originated at a pinhole of diameter a in the focal plane of a collimator of focal length f, then the maximum angle within the glass plate will be

$$\theta = \frac{a}{nf}$$

giving a change in path difference of

$$\Delta P = \frac{da^2}{nf^2}$$

As a rule of thumb the maximum shift between the fringe patterns must not exceed a quarter of a wavelength (after Lord Rayleigh). Substituting this for ΔP and rearranging gives the maximum value for the angular subtense of the pinhole as

$$\text{Subtense} = \frac{a}{f} = \sqrt{\frac{n\lambda}{4d}}$$

Of course, the corollary is true; if you have effectively zero path difference then useful fringes can be achieved with an extended source. This is the case when optical components are contact tested with test plates. The path length is effectively zero so an extended non-collimated source can be used; normally, a sodium lamp.

Similarly, it is impossible to have a thermal source with zero waveband; there will always be a spread of wavelengths. This has exactly the same effect. Each wavelength produces interference at a different phase, which again reduces the contrast in the fringes. This effect may be reduced by reducing the path difference, so some interferometers which employ relatively broadband sources have a means of introducing phase delay or advancement in one of the beams to ensure that the path difference is as close to zero as possible for all wavelengths. In this way, it is not necessary to define a limiting value as in the case of the pinhole, since the effect may be tuned out, and when it comes to spectral line widths of thermal sources, there is very little that can be done to change it.

Consider the same arrangement as before with a waveband $\Delta\lambda$ centered on wavelength λ and normal incidence. At the center wavelength, the path difference, in terms of number of wavelengths, will be

$$N = \frac{2nd}{\lambda}$$

At the extreme of the waveband, the number of wavelengths of path difference will be

$$N' = \frac{2nd}{\lambda + \Delta\lambda/2}$$

If, as before, we make the difference between the two values equal to a quarter of a wavelength, and assuming $\Delta\lambda$ is much smaller than λ then after rearranging,

$$\Delta\lambda = \frac{\lambda^3}{4nd}$$

This then describes the maximum bandwidth allowed to achieve interference over a given physical path difference d of index n.

Any departures from flatness of either of the surfaces will lead to a change in the shape of the wavefront reflected from that surface, which will then be evident in the interference fringes. However, the quality of the input wavefront will also have a bearing on the fringe shape. If the input beam wavefront is curved, that is, it is either converging or diverging, then because of the path difference this will show up on the interference fringe.

If the radius of curvature of the input beam is R at the first surface, and the beam is incident normally on the glass plate, then the reflected wavefront will also have a radius R. The beam reflected from the bottom surface will have traversed an additional optical path $2d/n$ and thus its radius will be $R + 2d/n$. The curvature of the fringe pattern will be dependent on the difference in the sag of the two wavefronts. If we wished to determine the radius of curvature of one of the reflecting surfaces, these fringes due to the curvature of the input wavefront would be misleading. If it is assumed that we have an aperture of radius h, which is small compared to R, then the sag of the first wavefront is

$$S = \frac{h^2}{2R}$$

and that of the second surface is

$$S' = \frac{h^2}{2\left(R + \dfrac{2d}{n}\right)}$$

If we then specify a minimum value that this change should be in order to satisfy the measurement accuracy required,

$$N\lambda = S - S'$$

And assuming that d is much smaller than R, then a minimum value of R can be calculated as

$$R \approx h\sqrt{\frac{d}{nN\lambda}}$$

If we take a typical case, most visible interferometers have an aperture of 0.1 m, the propagation is in air and the wavelength is 632.8 nm. To produce a wavefront error of 0.05λ with a propagation distance of 1 m, the radius of curvature of the input wavefront should be 281 m. To make measurements of this order of accuracy, then the error on the wavefront must be significantly less than this, an order of magnitude improvement is generally assumed.

As we shall discuss later, with the advent of software analysis this is not the problem it once was, since software, particularly phase-shifting software, can be calibrated to remove any artifacts due to the quality of the input wavefront.

Another possible area of error is with the collecting optics. Typically this will be an achromatic lens. If the test requires that the operator input a large number of tilt fringes, then the two interfering beams can take different paths through the collecting optics. If the optics are poor, then the aberrations will be imprinted onto the two wavefronts. However, with different paths taken, it is possible that the two beams will suffer differently. An example is if the lens suffers from spherical aberration. A wavefront aberration will be introduced into both the test and reference beams dependent on the position in the pupil of the collection lens.

For the test beam, considering only the direction of tilt

$$W = ay^4$$

For the reference beam

$$W' = a(y + \delta y)^4$$

This latter equation may be expanded to

$$W' \approx a\left(y^4 + 4y^3\delta y + \frac{12}{2!}y^2\delta y^2 + \cdots \right)$$

We assume that powers of δy greater than 1 can safely be ignored. If we now look at the additional apparent aberration introduced into the test beam,

$$W' - W \approx 4ay^3\delta y$$

this has the appearance of coma. The aberration does not exist in the item under test, but is introduced by the failure of the collecting optics to properly handle the pupil shear caused by introducing tilted fringes.

6.3 EARLY EXAMPLES OF INTERFERENCE

Several very simple experiments were performed to demonstrate the principle of interference of electromagnetic radiation. Although seemingly simple now, at that time they were groundbreaking experiments. The aims of the experiments were to demonstrate that interference could take place and hence demonstrate that light behaves as wave rather than a particle. Modern quantum mechanics tells us that the picture is much more complex than the simplistic view taken at the time and that the photon can paradoxically demonstrate wave and particle characteristics.

6.3.1 YOUNG'S SLITS

Perhaps the simplest arrangement was devised by Young. This comprised a light source, two slits and a screen (Figure 6.6). The slits were narrow enough such that the light would diffract into a cone. In the shaded region, where the two cones overlap, straight-line fringes are seen.

6.3.2 FRESNEL'S BIPRISM

Other experiments are variations on the above theme. Fresnel achieved interference using an optical component, a biprism, as shown in Figure 6.7. In this case, the manipulation of the beam, to provide two overlapping beams, is performed using refraction rather than diffraction. Energetically this is a much more efficient system.

FIGURE 6.6   Young's slits.

FIGURE 6.7   Fresnel biprism.

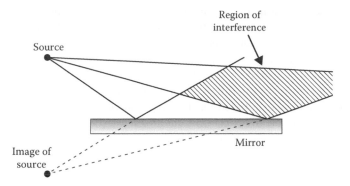

FIGURE 6.8 Lloyd's mirror.

6.3.3 Lloyd's Experiment

Another version uses reflection to split the beam into two. This is Lloyds experiment and is shown in Figure 6.8.

6.4 OPTICAL SOURCES

Traditionally, the sources used in interferometers have been arc lamps, which display a strong narrowband emission line such as sodium or mercury lamps. As described above, the performance of the interferometer will depend on the narrowness of the emission line and the size of the source. The source size is normally controlled by a pinhole so some optimization is available by reducing the pinhole diameter but at the cost of reducing the amount of illumination available for viewing and recording.

In recent years, lasers have been exclusively used in commercially available interferometers. In particular, the helium–neon (HeNe) laser operating at 632.8 nm is extensively used for testing of glass optical components and systems. The laser has a much narrower spectral bandwidth than thermal sources and thus has a much longer coherence length; rather than the millimeters of path difference permitted, useful interference fringes can be obtained over path differences of many meters. Lasers do, however, have their own problems. In particular, because of the very long coherence length, the unwanted returns from scattering and reflecting surfaces other than those intended can also give unwanted fringes in the final image. These tend to be of much lower contrast than the intended interference fringes and can normally be accounted for.

A greater problem is the stability of the laser cavity itself, coupled with the fact that many lasers will emit several longitudinal modes at the same time. Because of the nature of the cavity design, wavefronts separated by one cavity length will be in phase, because the ends of the cavity represent nodes in the wavefront and this will be true of each longitudinal mode. Outside the cavity, any single mode will interfere with copies of itself produced by reflections representing long path differences. When multiple modes are present, then each mode will produce an interferogram, which has a different phase to the others. However, because of the nature of propagation within a cavity, all these interferogram will be in matched phase at intervals equivalent to multiples of the cavity length. The result is that good contrast fringes may be achieved by tuning the path difference to a multiple of the cavity length. It is safest to use a single-mode stabilized laser; however, this can come at significantly increased cost, and depending on the wavelength chosen can be difficult to procure commercially.

The best laser that the author has used for interferometry has been at 10.6 μm using a CO_2 laser. These are generally single-mode devices and can produce high contrast fringes over many tens of meters. On the other hand, at 3.39 μm a HeNe laser is used. At about 1 m long, this laser will produce either two or three modes within the gain bandwidth of the cavity, the number often varying with temperature. When operating in three modes, the central mode is dominant and good contrast fringes can be achieved over many meters. When operating in two modes, they have roughly equal power. Both modes will produce interference fringes, but at intervals along the axis, the two sets of interference fringes are exactly out of phase and cancel out each other. Midway between these positions, the two interference patterns have the same relative phase and thus add together. These lasers often need several minutes to achieve a steady temperature so that there is no thermal mode hopping. The cavity length is then tuned for the environment to produce a dominant longitudinal mode. The lasers are then stable over long time durations.

In recent years, advances in the quality of laser diodes, have made them suitable for long path difference interferometry, though it is often necessary to spatially filter the laser, usually by transmitting it through a single-mode optical fiber. Laser diodes have been used at 850 nm and 1.55 μm, very successfully with path differences of several meters.

6.5 SIMPLE INTERFEROMETER

Figure 6.9 demonstrates the method that is employed to achieve the interference fringes. An incoming beam of monochromatic radiation is split into two at a beam splitter. In the example shown, part of the beam is retroreflected (we will call it "R") and part is transmitted "T." The reflected component R propagates directly toward the detection system. The transmitted beam T carries on toward the optical system under test, in this case a simple reflector, and is retroreflected back through the optical system toward the beam splitter. In this case the surface under test also performs the retroreflection.

FIGURE 6.9 Simple interferometry.

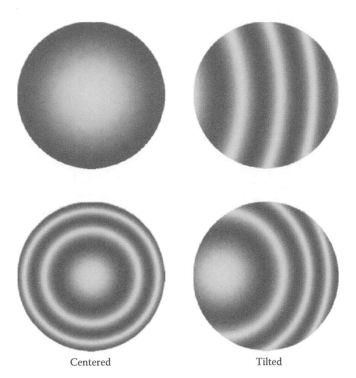

Centered   Tilted

FIGURE 6.10 Interference fringes of a curved wavefront.

Part of the beam passes through the beam splitter where it is free to interfere with the original component R. The probe or test beam effectively passes through the component under test twice, thus the system is said to be a double-pass interferometer in which the apparent aberrations are doubled. The separation between adjacent bright areas in the interference pattern thus represents λ/2 of aberration. In this way, a useful doubling of sensitivity can be achieved.

The reflected beam from the beam-splitting surface may be tilted by tilting the beam splitter in order to introduce tilt fringes that can help to clarify the aberration shape for small aberrations. Figure 6.10 shows an example of how this can be of use.

The centered (Figure 6.10a) interferogram cannot be reliably assessed because it cannot be determined what proportion of a fringe the variation in intensity represents. By introducing tilt fringes as in tilted (Figure 6.10a), then a measurement is possible.

When there are a large number of fringes as in Figure 6.10b, then information may be obtained from both centered and tilted fringe patterns. In severe cases it may not be possible to remove the central closed fringe from the pattern without introducing so many fringes that visibility becomes a problem.

6.6 FIZEAU INTERFEROMETER

Figure 6.11 shows the optical arrangement of a Fizeau interferometer. This is one of the simplest arrangements. In the above form, however, it does have certain limitations, in that for thermal sources the spectral lines are not sufficiently narrow to provide a very high level of coherence, thus the beam-splitting surface must be close to the surface under test. This makes it difficult to test complete optical systems. Additionally, in the infrared (IR), thermal sources are not available with sufficient intensity to stimulate a sensor such as a pyroelectric vidicon, which is relatively insensitive. More sensitive sensors are expensive and require cooling. The arrangement shown in the figure is only slightly more sophisticated than using optical test gauges in a polishing shop.

This problem is solved by substituting a laser for the thermal source (Figure 6.12). We now have a long coherence length that will enable much more complex systems to be tested, and we have

FIGURE 6.11 Fizeau interferometer.

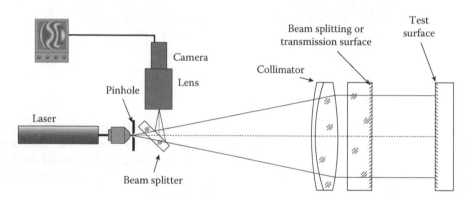

FIGURE 6.12 Laser Fizeau interferometer.

an abundance of power to stimulate the sensor. For systems employing CO_2 lasers, the problem is usually one of how to remove much of the laser power before it impinges on the sensor. Great improvements in the sensors, particularly focal plane arrays, have meant that much less laser power is required to produce good quality images. At most, a few tens of milliwatts are needed in the IR. Visible systems typically have 5 mW HeNe lasers.

The use of lasers does have a disadvantage in that the long coherence length enables interference to take place between any of the many beams that are reflected around an optical system due to imperfections in coatings and the like. For this reason it is necessary to be very careful regarding antireflection coatings on all optical components after the pinhole. The pinhole itself acts as a spatial filter for all optics that precedes it and thus removes all coherent noise from focusing optics such as microscope objectives used in visible or near-IR systems.

The Fizeau arrangement has one major advantage over other optical arrangements—that all of the optical system of the interferometer is used identically by both the test beam and the reference beam, except for the single surface that performs the beam splitting. This means that aberrations in the interferometer optical system only have a low order effect on the shape of the fringe pattern ultimately observed. The beam-splitting surface, however, must be of very high quality to avoid having a major effect on fringe shape.

In order to measure non-flat surfaces, the beam from the interferometer must be modified using a transmission sphere as shown in Figure 6.13. The beam from the interferometer is collimated as usual, but a lens is affixed to the front of the interferometer. This focuses down with a *f*/number suitable for

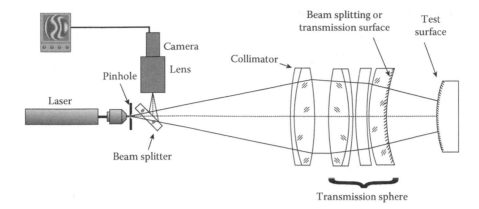

FIGURE 6.13 Testing of spherical surfaces.

testing the component for test. The final surface of the transmission sphere is concentric with the focus of the transmission sphere so that the rays of light are incident at normal incidence. This surface is uncoated so that a reflection is obtained back into the interferometer—this represents the reference beam. The reflectivity of the surface under test must be similar in order to obtain high contrast fringes. Tilt is normally introduced by tilting the whole transmission sphere. Interferometers normally have a range of transmission spheres with different f/numbers, which enable the optimization of test arrangements.

6.7 TWYMAN GREEN INTERFEROMETER

Another very common type of interferometer used in optical laboratories and workshops is the Twyman Green interferometer (Figure 6.14). When using a thermal source such as a mercury lamp, this arrangement has significant advantage as the lengths of the two arms can be equalized by mounting the reference mirror on rails and moving it. Until the advent of laser sources in interferometers, this system was very widely used for the assessment of complete optical systems in preference to the Fizeau interferometer, which with a thermal source has very limited application. Using laser sources, this system can still be used effectively; however, the number of components that must be made to high accuracy increase. Both sides of the beam splitter, together with the reference mirror, now need to be very flat. The optics used for collimating the laser beam and for projecting the combined beams to the camera need not be accurate to the same extent.

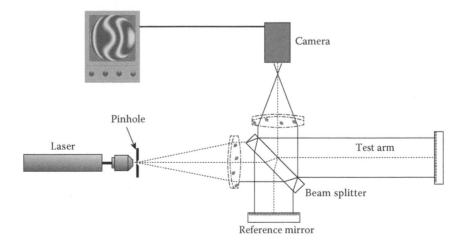

FIGURE 6.14 Twyman Green interferometer.

6.7.1 MACH–ZENDER INTERFEROMETER

The Mach–Zender interferometer is shown in Figure 6.15. This is different from the interferometers already shown in that the test and reference arms take completely different paths.

The Mach–Zender is a single-pass interferometer; the two beams are recombined after the test beam has passed through the optical system under test. The advantage of this over the double-pass interferometer is that apertures that may cause diffraction are only passed once, thus it is much easier to focus on the diffracting aperture to remove the effects of diffraction. This type of system is invaluable in situations where diffraction is important, such as the testing of microlenses.

6.7.2 SHEARING INTERFEROMETER

Figure 6.16 shows a shearing interferometer consisting of a single plate of glass. This duplicates the optical beam complete with any defects it has picked up after passing through an optical system. The two beams are then interfered with one another. This is equivalent to differentiating the wavefront in the direction of the shear.

FIGURE 6.15   Mach–Zender interferometer.

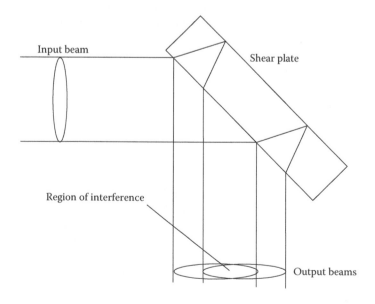

FIGURE 6.16   Shear plate interferometer.

Considering an incoming beam with a focus defect, then the wave aberration is described as

$$W = ay^2$$

The other pupil is displaced by an amount δy so that its aberration is from a slightly different part of the wavefront.

$$W' = a(y + \delta y)^2$$

The resulting fringe pattern represents contours of wavefront given by

$$W' - W = a(2y\delta y + \delta y^2)$$

or

$$W' - W \approx 2ay\delta y$$

This is linear with y, so a defect of focus in the input beam will produce straight-line fringes. The spacing of the fringes will depend on the amount of defocus and the amount of shear between the two pupils.

This type of interferometer is very useful for focusing optical systems, in particular collimating systems for other interferometers. Because the fringes are straight, it is very important that there are no other sources of straight-line fringes in the shear plate, that is, the two faces must be parallel to one another.

6.8   INFRARED INTERFEROMETRY

Interferometry is used as a routine test for optical components and systems at wavelengths ranging from the visible to above 10 µm. In the IR, the main wavebands used are from 3 to 5 µm and from 8 to 12 µm. Typical laser sources used are HeNe (3.39 µm) and HeXe (3.49 µm) in the shorter waveband. CO_2 lasers (10.6 µm or tuneable) are used in the longer waveband. IR interferometry presents some problems that require that special care is taken in the setting up of the measurement, and that the results are analyzed in as rigorous a manner as possible.

Infrared optical systems are being produced with smaller optical apertures than previously for reasons of weight, cost, and configurability. This trend has been aided by improvements in the detector technology. The result is that the IR lens for a camera system may have an aperture comparable in size to a visible camera system. However, the wavelength in the IR system may be as much as 20 times greater than in the visible band. This makes the effects of diffraction much more dominant in IR interferometry, requiring that optical pupils are well focused onto the interferometer's camera sensor surface. Because of the double-pass nature of most commercially available interferometers the aperture is "seen" twice by the interferometer—if the two images of the optical aperture are not conjugate with one another, then some diffraction is inevitable. The effects of this must be minimized to enable accurate analysis.

Figure 6.17 shows two ways of analyzing a lens with an interferometer. One test uses a spherical reference mirror while the other uses a flat reference mirror but requires an additional high-quality focusing lens (or reference sphere) in order to make the measurement. From the point of view of diffraction, the measurement using the reference sphere is preferred since the return mirror can be very close to the pupil of the lens under test, and therefore the pupil and its image via the mirror can both be very close to the focus on the interferometer's camera faceplate. This is not possible using the other arrangement, diffraction will be present especially if the spherical mirror radius is small so that it is physically far from the lens under test. Many tests in the IR use high grade ball bearings as a return mirror—it is important that any analysis of the interferograms takes into account the limitations of the method.

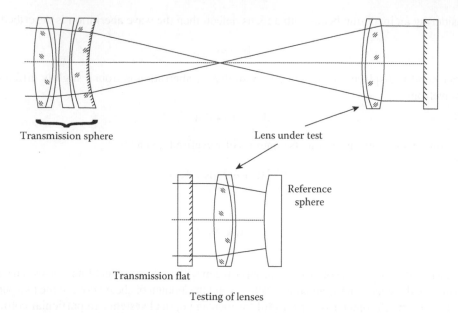

Transmission sphere   Lens under test

Reference
sphere

Transmission flat

Testing of lenses

FIGURE 6.17   Reflection configurations.

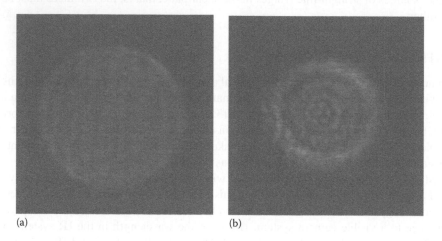

(a)   (b)

FIGURE 6.18   Pupil imagery. (a) Focused and (b) unfocused pupil.

Figure 6.18 shows the effect of diffraction on the pupil imagery in the interferometer. In the unfocused case, diffraction rings can clearly be seen inside the pupil. This has the effect of breaking up the interference pattern such that the fringes are discontinuous or vary in width due to contrast changes. Any automatic fringe analysis program must be used with caution when analyzing such fringes. In general, the peak-to-valley wavefront aberration reported will be worse than the actual for the lens under test.

While Figure 6.18 shows the effect of diffraction within the optical pupil, the other common case is when the diffracted radiation appears outside the pupil. This has the effect of blurring the pupil edge and so its size is uncertain and because the diffracted radiation has a different curvature to the main beam, any interference fringes will curve within this zone. The effect is very similar to that seen with spherical aberration especially that of higher order and if not corrected can lead to a large departure in measured versus actual performance. This situation is shown schematically in Figure 6.19.

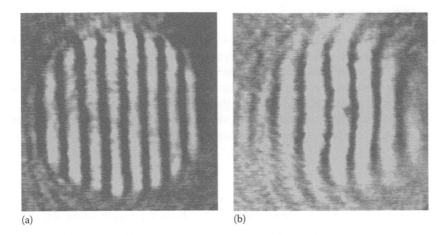

(a)   (b)

FIGURE 6.19 Pupil imagery effects on fringes. (a) Focused and (b) unfocused pupil.

In this case, if information is known about the size of the pupil within the interferogram, it is possible to mask out much of the effects of diffraction simply by ignoring those parts outside the pupil.

There will be cases where it is impossible to achieve good pupil imagery without going to great expense. In such circumstances, double-pass interferometry can still provide a confidence test on system performance, but ultimate proof may require a single-pass test, either by interferometry or by noncoherent broadband modulation transfer function (MTF) testing or similar.

Infrared interferometry relies on sensing using an IR camera, either staring array cadmium mercury telluride (CMT) or a pyroelectric vidicon tube. The cameras exhibit random noise on the video signal. This is most obvious in the 3–5 μm waveband, where available, affordable lasers tend to be of low power (typically 5–10 mW). High signal gain is required to provide sufficient contrast for recording fringe patterns or analyzing fringe patterns using either static or phase-shifting software. It is also of relevance when laser power is marginal is the profile of the laser beam. As much of the beam as possible must be used to maximize contrast but this does mean that, due to the profile, the contrast varies across the interferogram. This is particularly noticeable when using HeNe lasers at 3.39 μm where perhaps only 5–10 mW of power is available. This can have the effect of making fringes appear both fainter and thinner toward the edge of the pupil. When using static fringe analysis packages, care needs to be taken to check that the digitized fringes accurately represent the real shape of the fringe pattern.

The problems raised by laser power profiles are greater when using phase-shifting systems. Phase-shifting systems involve a threshold contrast below which phase will not be measured. This normally shows as a blank space in the resulting phase map. This can be remedied by reducing the threshold. However, using pyroelectric vidicons with marginal laser power represents a very noisy environment. Reducing the threshold too far means that the system can pick up on random noise in the picture and allocate phase values that are significantly larger than any other area of the pupil. This has the effect of a gross distortion of the results. A fine balance needs to be trodden between these two situations such that phase information is available over the whole pupil without random noise causing a problem. Generally with 3.39 μm systems, the threshold needs to be set at around 2.5%–3.5%.

At 10.6 μm, these problems do not occur. The main problem in the upper IR waveband is finding ways of dumping excess energy. This is normally performed with CaF_2 filters. At 10.6 μm 4 mm thickness of CaF_2 will absorb 75% of the radiation. With tuneable systems a wire grid polarizer is necessary since the CaF_2 does not absorb so strongly at the shorter wavelengths. Because so much power is available, the central part of the beam can be used ensuring a much flatter profile, and sufficient power can be allowed through in order to enable the camera to be set to low gain, thereby reducing noise to an insignificant level.

6.9 TEST ARRANGEMENTS

We have already seen how a single flat optical surface can be measured using an interferometer. Other surfaces, components, and systems often require some specialized equipment in addition to the interferometer. These will include transmission spheres, which are high-quality lenses used to condition the beam from the interferometer so that it can be focused in the focal plane of a lens or system of lenses; beam expanders, which enable the size of the test beam to be enlarged to accommodate larger optical apertures; and attenuators, which allow an operator to match the amplitudes of test beams and reference beams thereby maximizing fringe contrast.

6.9.1 Testing of Flats

A simple arrangement for testing flats is shown in Figure 6.20. The flat for measurement simply replaces the return mirror. The reference beam arises from the transmission flat that is shown mounted in a tilt stage, which permits horizontal and vertical tilt fringes to be introduced.

6.9.2 Measurement of Materials

An arrangement for measuring materials is shown in Figure 6.21. The material is simply placed within the test arm. Various properties of the material can produce distortions of the fringe pattern, these being departures from flatness of the two surfaces and inhomogeneity of the refractive index of the material. Before the advent of highly accurate phase-shifting interferometers, in order to assess the homogeneity of a material, the two sides of the material blank had to be polished very flat such that they would not contribute to fringe distortion. Modern systems can measure the homogeneity of materials by combining a number of measurements which record the total fringe distortion of the blank combined with individual measurements of the two surfaces.

6.9.3 Testing of Spherical Surfaces

Concave surfaces are measured by using a transmission sphere focused down at the center of curvature of the concave surface (Figure 6.22).

Convex surfaces are shown in Figure 6.23. It is immediately obvious that there is a significant problem with the measurement of convex surfaces. In order to measure the surface, the output optical aperture of the transmission sphere must be greater than the diameter of the spherical surface

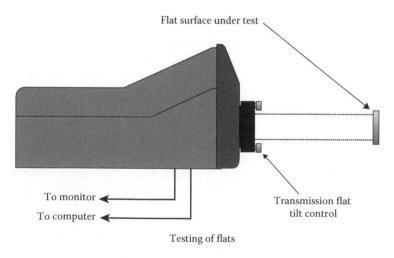

Testing of flats

FIGURE 6.20 Testing of flats.

FIGURE 6.21   Testing of material blanks.

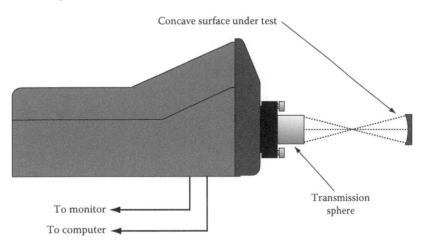

FIGURE 6.22   Testing of concave surfaces.

under test. It is often necessary to expand the beam from the interferometer so that a larger diameter transmission sphere can be used. Convex surface do not suffer from this problem since the transmission spheres focuses through the center of curvature so that as long as the *f*/numbers of the transmission spheres and surface under test are matched, the whole surface can be covered.

6.9.4   TESTING OF LENSES

Figure 6.24 shows the arrangement for testing a lens. The transmission sphere focuses at the focus of the lens under test. A flat return mirror is then used to reflect the radiation back into the interferometer.

6.9.5   TESTING OF DOMES

A common component in surveillance systems and aerial photography is a protective dome behind which the camera is free to move so that it can be pointed. The test arrangement for such a component is shown in Figure 6.25. A high-quality convex return mirror is used to return the radiation to the interferometer.

FIGURE 6.23 Testing of convex surfaces.

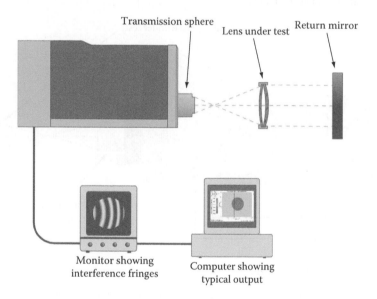

FIGURE 6.24 Testing of lenses.

6.9.6 TESTING TELESCOPES

The testing of a telescope is shown in Figure 6.26. The telescope being afocal can be measured from either direction with a collimated beam. The configuration shown has obvious advantages in that the table area required to perform the measurement is small and the relative cost of the mounting and optical components required when compared with large aperture systems is also reduced. However, there are other advantages with regard to pupil focusing at high magnifications. The effective depth of focus for the pupil imagery can be large in this arrangement thereby reducing the effects of diffraction. For low magnification system or systems where the return mirror is placed at the eyepiece end, the positioning of the return mirror with respect to the position of the entrance pupil becomes much more important.

Telescopes often have zoom or switched magnification arrangements. It is often found that at the high magnifications the optical pupil of the telescope is near the front lens. At the smallest

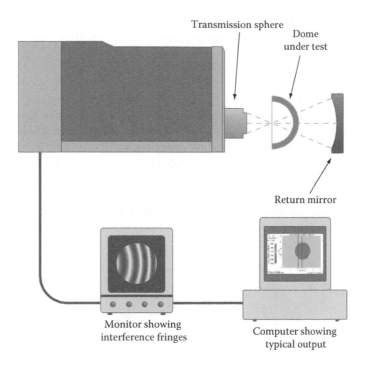

FIGURE 6.25 Testing of domes.

FIGURE 6.26 Testing of telescopes.

magnification, the pupil is often buried within the telescope. This makes it impossible to focus the optical pupil efficiently for all magnifications and some diffraction effects are likely to occur.

6.10 FRINGE ANALYSIS

Once an interferogram has been obtained it is necessary to analyze it. This can take several forms, the simplest being with a straight edge on a photograph, the most complex being computer-aided fringe analysis. It is not the intent here to describe in detail the fringe analysis software but simply to describe their main points and their availability. Assessing wavefront error from a photograph can be performed as shown in Figure 6.27. This is a relatively crude analysis but is adequate enough for a many applications, though, of course, information from only one fringe is calculated.

Figure 6.28 demonstrates the operation of a simple static fringe digitization software package. Such packages are readily available and are quite adequate for most tasks. However, there are some limitations. The number of digitized points is quite low being typically of the order of 200–300 points.

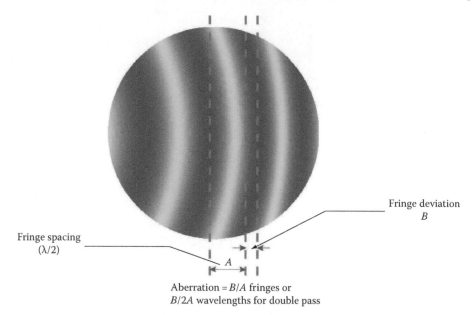

FIGURE 6.27 Simple fringe analysis.

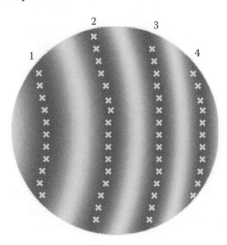

FIGURE 6.28 Digitization of fringe centers.

Additionally there are areas of the optical pupil where there are no digitized points. The software will interpolate for points within the pupil but must extrapolate at the edge of the pupil. The extrapolation can be performed in a number of ways.

- A Zernike polynomial fit is made to the obtained data points, which then provide the wavefront shape. The polynomial coefficients can be used for the extrapolation, but this has the danger that the high-order polynomial could roll off rapidly giving a pessimistic result.
- A linear extrapolation can be performed; this however could ignore any roll off and could provide an optimistic result.

In both the cases mentioned above, it would be unwise to give too much credence to a peak-to-valley wavefront value since it could be out by a large amount. Such figures can still be quoted as results but a hard copy of the interferogram should always be provided. A much more reliable parameter is the root mean square (RMS) wavefront deviation. Normally, information is available over the majority of the optical pupil, therefore effect of roll off due to the polynomial fitting will not have such a large effect. This is fortuitous in that many of the parameters that may be required, particularly for complete optical systems, are based on the RMS wavefront aberration, in particular the Strehl ratio and the MTF. Once the shape of the wavefront is known, then it possible to calculate all geometrical properties of an optical system, MTF, point spread function, encircled energy function, etc.

An important limitation of the above process is that although the fringe shape is known, the sense of that shape is not. That is, we do not know whether a fringe perturbation is a wavefront retardation or advance. This problem is solved, together with the fringe extrapolation problem by using phase-scanning software. This method captures the fringe data over the whole of the optical pupil and by changing phase several times and analyzing the resulting data can determine the initial phase of any pupil point. This provides much greater accuracy particularly where peak-to-valley wavefront deviation is important. A further advantage is that the measurement of material refractive index inhomogeneity can be reliably measured as all other parameters (surface aberrations, etc.) can be measured to high accuracy.

The following curves (Figures 6.29 and 6.30) show that the phase-shifting technique is by far the more accurate. Each curve represents 100 measurements. The two software packages were provided by the same supplier and used the same hardware, for example, computer, interferometer, frame grabber, etc., although the actual component measured was different.

The spread of results with the phase-shifting system is much smaller than with the static system. The static system varies considerably with time as the environment changes. The phase-shifting

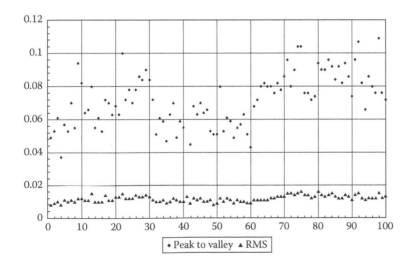

FIGURE 6.29 Variation of measurement with static fringe analysis software.

FIGURE 6.30 Variation of measurement with phase-shifting software.

system was not so badly affected mainly because of the ability to make several measurements and display an average. This has the tendency to smooth out time-dependent variation such as air turbulence, which may be frozen into a static fringe acquisition. It may be expected that the phase-shifting system may suffer more from vibration effects since the measurement process takes longer—the static system simply grabs one frame and thereby freezes the fringe. However, if the fringes are "fluffed out" the software can cope with small vibrations, if the fringes are not fluffed out then more serious effects can occur as they move about.

6.11 ABERRATION TYPES

6.11.1 SPHERICAL ABERRATION

Figure 6.31 shows the fringe patterns obtained for spherical aberration. The pattern on the left is closest to the paraxial focus of the lens under test. The best focus is represented by the central fringe pattern which is also shown three dimensional. This is the position at which the peak-to-valley wavefront error is minimized.

The three positions can be seen again but with tilt fringes added in Figure 6.32.

6.11.2 COMA

The wavefront of coma is shown in Figure 6.33. The best focus is the symmetrical pattern shown in the center. Figure 6.34 shows the coma with tilt fringes, again the symmetry of the best focus position is evident.

6.11.3 ASTIGMATISM

The familiar saddle shape of astigmatism is shown in Figure 6.35. Again the best focus position is shown at the center. At this point, referenced from the center of the interferogram, the aberration is the same in the horizontal and vertical directions but in opposite directions. Figure 6.36 shows the same situation but with tilt fringes.

FIGURE 6.31   Spherical aberration.

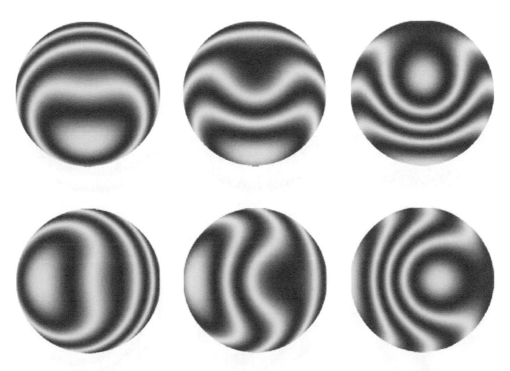

FIGURE 6.32   Spherical aberration with tilt fringes.

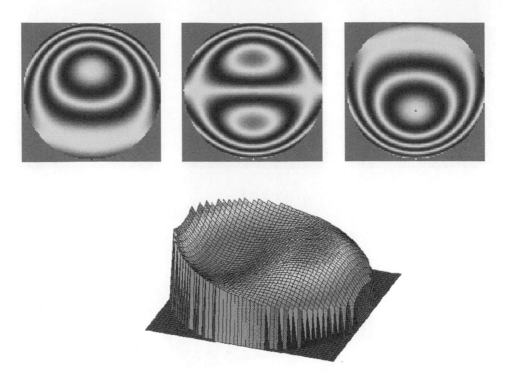

FIGURE 6.33   Coma through focus.

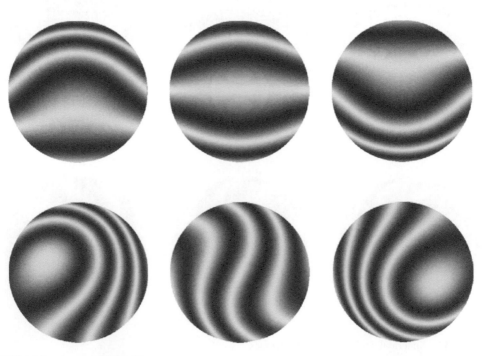

FIGURE 6.34   Coma with tilt fringes.

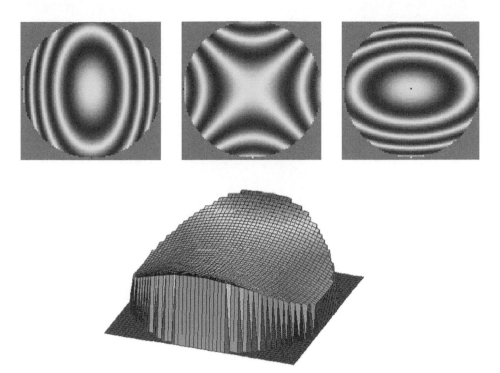

FIGURE 6.35   Astigmatism through focus.

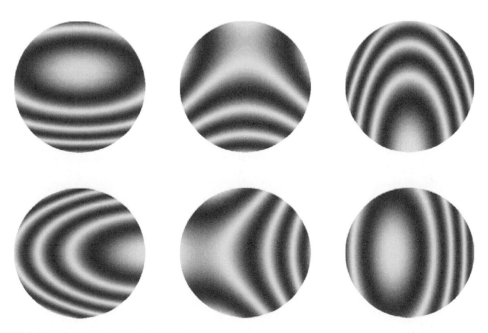

FIGURE 6.36   Astigmatism with tilt fringes.

7 Holography

Giancarlo Pedrini

CONTENTS

7.1 Introduction 225
7.2 Basic Arrangement for Recording and Reconstructing Holograms....................226
7.3 Fourier Analysis of the Hologram 228
7.4 Digital Holography 230
    7.4.1 Historical Remarks and Technological Developments............................230
    7.4.2 Limitations due to the Recording of Holograms on an Electro-Optical Device 231
    7.4.3 Mathematical Tools for the Digital Reconstruction of a Hologram 232
    7.4.4 Some Arrangements for Digital Holography...234
7.5 Applications of Digital Holography..236
    7.5.1 Digital Holographic Microscopy 237
    7.5.2 Digital Holographic Sectioning by Using a Short Coherence Light Source 237
    7.5.3 Digital Holographic Interferometry...241
        7.5.3.1 Measurement of Object Deformations....................................241
        7.5.3.2 Pulsed Digital Holographic Interferometry for the Measurement of Vibration and Defect Recognition 242
        7.5.3.3 Digital Holographic Interferometry for the Investigation of Microsystems 244
References..245

7.1 INTRODUCTION

Holography is an interferometric technique that has proved to be convenient to store and reconstruct the complete amplitude and phase information of a wavefront. The popular interest in this technique is due to its unique property to obtain 3D images of an object, but there are other less known but very useful applications that are concerned with microscopy, deformation measurement, aberration correction, beam shaping, optical testing, and data storage, just to mention a few. Since the technique was invented by Gabor [1], as an attempt to increase the resolution in electron microscopy, an extensive research has been done to overcome the original limitations and to develop many of its applications. Gabor demonstrated that if a suitable coherent reference wave is present simultaneously with the light scattered from an object, the information about both the amplitude and the phase of the object wave could be recorded, although the recording media registers only the light intensity. He called this "holography," from the Greek word holos (the whole), because it contained the whole information. The first concept of holography developed in 1948 was as in-line arrangement using a mercury arc lamp as the light source; it allowed to record only holograms of transparent objects and contained an extraneous twin image. For the next 10 years, holographic techniques and applications did not develop further until the invention of the laser in 1960, whose coherent light was ideal for making holograms. By using a laser source, in 1962, Leith and Upatnieks developed the off-axis reference beam method, which is most often used today [2]. It permitted the making of holograms of solid objects by reflected light.

There are several books describing the principles of holography and their applications; in particular, classical references [3–9] written in the years from 1969 to 1974 are interesting because together with a clear description of the technique, they predict some developments. In Refs. [6,8], possible commercial applications in the 3D television and information storage are reported. Forty years later, we are still quite far from the holographic television since the electro-optical modulators, for example, liquid crystal displays do not have the necessary spatial resolution, but investigation to circumvent this limitation is in progress. The holographic storage predicted by Lehmann [6] is still not commercially available, but new materials fulfilling the requirements for this application are under development.

In 1966, Brown and Lohmann had described a method for generating holograms by computer and reconstructing images optically [10]; the emerging digital technology had made it possible to create holograms by numerical simulations and print the calculated microstructure on a plate that when illuminated forms a wavefront of a 3D object that does not possess physical reality. This technique is now called computer-generated holography and has useful applications, for example, for correcting aberrations or producing a shaped beam.

The first holograms were recorded on photographic emulsions; these materials have an excellent spatial resolution (typically 2000 lp/mm) and are still used in particular for large-size recording; other high-resolution recording materials are photorefractive crystals and thermoplastic films. Already, in 1967, Goodman [11] has shown that holograms may be recorded on optoelectronic devices, and the physical reconstruction may be replaced by a numerical reconstruction using a computer. This technique, which is now called digital holography, had an impressive development in the last years due in particular to the availability of CCD or CMOS sensors with an increased number of pixels and computer resources. Different arrangements based on digital holography are actually used for applications like microscopic investigations of biological samples, deformation measurement, vibration analysis, and particle visualization.

In this chapter, we will at first review some general concepts about holography; later, the recording of holograms on an electronic device and the digital reconstruction by using a computer are discussed. Some arrangements and their applications for the investigation of microscopic samples, shape contouring, and measurement of deformations will be presented.

7.2 BASIC ARRANGEMENT FOR RECORDING AND RECONSTRUCTING HOLOGRAMS

Basically in holography, two coherent wavefronts, a reference and an object wave, are recorded as an interference pattern (Figure 7.1). We represent the light wavefronts by using scalar functions (scalar wave theory) and consider only purely monochromatic waves; this approximation is sufficient for describing the principle of holography. Let $|U_o|$, $|U_R|$, ϕ_o, ϕ_R be the amplitudes and phases of the object and the reference waves, respectively. In Figure 7.1, we use a plane wave to represent the reference, but in the general case, a spherical or any other nonuniform wave may be used for

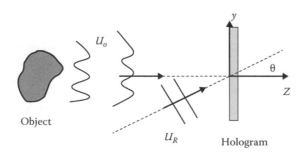

FIGURE 7.1 Hologram recording.

recording the hologram. The complex amplitude of reference and object waves may be written in the exponential form as follows:

$$U_R(x,y,z,t) = |U_R(x,y,z)| \exp\{-i\phi_R(x,y,z)\} \exp\{i2\pi vt\}, \tag{7.1a}$$

$$U_o(x,y,z,t) = |U_o(x,y,z)| \exp\{-i\phi_o(x,y,z)\} \exp\{i2\pi vt\}. \tag{7.1b}$$

In these equations, $\exp\{i2\pi vt\}$ is a time-varying function, where v is the frequency, which for visible light has a value in the range 3.8×10^{14} to 7.7×10^{14} Hz. The time dependence can be omitted in the description of the holographic process, and the interference between U_o and U_R takes the form

$$U_{oR}(x,y,z) = U_o(x,y,z) + U_R(x,y,z). \tag{7.2}$$

The intensity recorded by a detector is given by

$$I = (U_R + U_o)\left(U_R^* + U_o^*\right) = U_R U_R^* + U_o U_o^* + U_R^* U_o + U_R U_o^*, \tag{7.3}$$

where "*" denotes a complex conjugation. Here and what follows to simplify the notation, we have omitted the dependence on (x, y, z). $|U_o|^2 = U_o U_o^*$ and $|U_R|^2 = U_R U_R^*$ are the intensities of the two wavefronts and do not contain any information about the phases; on the other side, we may write the two last terms in the equation as

$$U_R^* U_o + U_R U_o^* = |U_R||U_o|\left[\exp\{-i\phi_R\}\exp\{i\phi_0\} + \exp\{i\phi_R\}\exp\{-i\phi_0\}\right]$$

$$= 2|U_R||U_o|\cos(\phi_R - \phi_0). \tag{7.4}$$

Their sum depends on the relative phase, thus the recorded intensity contains information about amplitude and phase of reference and object beams, which is encoded in the interference pattern in the form of a fringe modulation. We may assume that this intensity is recorded on a photosensitive material, for example, a film that was the first used holographic support. After development, we obtain a transparency (we call it hologram) with transmittance proportional to the recorded light intensity. By illuminating the hologram by using the reference wave U_R, we obtain the transmitted wave field

$$U_R I = U_R |U_R|^2 + U_R |U_o|^2 + U_R U_R^* U_o + U_R U_R U_o^*. \tag{7.5}$$

The first term $U_R|U_R|^2$ represents the reference wave attenuated by the factor $|U_R|^2$; in the second term $U_R|U_o|^2$, the reference is modulated by the intensity of the object beam; the third term is nothing else as the object wavefront modulated by the intensity of the reference beam (in principle, we may choose a constant-intensity reference). Since the reconstruction of the third term has exactly the same properties as the original wave, an observer perceives it as coming from a virtual image of the object located exactly at the original object location (see Figure 7.2). The last term contains the complex conjugate of the object beam U_o^*; the best way for reconstructing this field is to illuminate the hologram with a conjugate of the reference wave U_R^*, as shown in Figure 7.3. In the in-line arrangement proposed by Gabor [1], the four reconstructions were superimposed; in the off-axis configuration illustrated in Figures 7.1 through 7.3, the angle θ between reference and object wave allows an easy spatial separation of the waves.

FIGURE 7.2   Reconstruction of the virtual image of the subject.

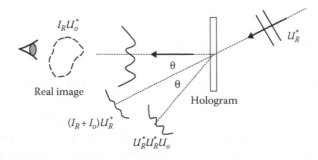

FIGURE 7.3   Reconstruction of the real image of the subject.

The holographic process is not limited to visible light; thus, in principle, hologram may be recorded by using UV and X-rays for which frequency is higher (10^{15}–10^{18} Hz) compared to the visible radiation.

7.3   FOURIER ANALYSIS OF THE HOLOGRAM

The Fourier transform (FT) is one of the most important mathematical tools and is frequently used to describe several optical phenomena like wave propagation, filtering, image formation, and the analysis of holograms. We will see later how it can be used for the digital reconstruction of holographic wave fields. The formulas relating a 2D function $f(x, y)$ and its FT $\tilde{f}(f_x, f_y)$ are the following:

$$\tilde{f}\left(f_x, f_y\right) = \int\limits_{-\infty}^{\infty}\int\limits_{-\infty}^{\infty} f\left(x, y\right)\exp\left\{2\pi i\left(xf_x + yf_y\right)\right\}dxdy, \tag{7.6a}$$

$$f\left(x, y\right) = \int\limits_{-\infty}^{\infty}\int\limits_{-\infty}^{\infty} \tilde{f}\left(f_x, f_y\right)\exp\left\{-2\pi i\left(xf_x + yf_y\right)\right\}df_xdf_y. \tag{7.6b}$$

Sometimes, we will use the notations $\tilde{f}(f_x, f_y) = FT\{f(x, y)\}$ and $f(x, y) = FT^{-1}\{\tilde{f}(f_x, f_y)\}$, where FT and FT^{-1} are the FT and its inverse, respectively. The convolution (denoted by the symbol $\otimes$) between two functions $f(x, y)$ and $g(x, y)$

$$h(x', y') = f \otimes g = \int\limits_{-\infty}^{\infty}\int\limits_{-\infty}^{\infty} f(x, y)g\left(x' - x, y' - y\right)dxdy \tag{7.7}$$

is another important mathematical tool that we need for our analysis.

The Fourier transform of the holographically reconstructed wavefront is equal to the sum of the FT of four individual terms:

$$FT\{U_R I\} = FT\{U_R |U_R|^2\} + FT\{U_R |U_o|^2\} + FT\{U_R U_R^* U_o\} + FT\{U_R U_R U_o^*\}. \tag{7.8}$$

We consider a hologram recorded by using a plane reference wave having constant amplitude $|U_R|$ propagating in the direction $(0, \sin\theta, \cos\theta)$

$$U_R(x,y,z) = |U_R(x,y,z)| \exp\{2\pi i [y\sin\theta + z\cos\theta]/\lambda\}, \tag{7.9}$$

where λ is the laser wavelength; in writing the waves, we shall normally omit the propagation constant for the z direction. The spectrum of the hologram distribution (absolute value of the FT of the reconstructed wavefront) is schematically shown in Figure 7.4a. Since the reference is a plane wave,

FIGURE 7.4 Spectrums of holograms. Off-axis (a) and in-line (b). Asymmetrical spectrums of an off-axis hologram (c).

the FT of $U_R|U_R|^2$ is a delta function $\delta(f_x, f_y - f_r)$, centered at $f_x = 0, f_y = f_r$, where f_x, f_y are the spatial frequencies and $f_r = \sin\theta/\lambda$; in the frequency domain, this term will be concentrated on a single point. By using the convolution theorem, which states that a multiplication in the spatial domain equals convolution in the spatial-frequency domain, we obtain for the second term the relation

$$FT\left\{U_R(x,y)U_o^*(x,y)U_o(x,y)\right\} = \tilde{U}_R(f_x,f_y) \otimes \tilde{U}_o^*(f_x,f_y) \otimes \tilde{U}_o(f_x,f_y). \tag{7.10}$$

If the object has frequency components not higher than B, then the convolution term expressed by Equation 7.10 has an extension equal to $2B$ and like the first term is centered at $f_x = 0, f_y = f_r$.

The third and the fourth terms in Equation 7.8 are given by

$$FT\left\{U_R(x,y)U_R^*(x,y)U_o(x,y)\right\} = \tilde{U}_R(f_x,f_y) \otimes \tilde{U}_R^*(f_x,f_y) \otimes \tilde{U}_o(f_x,f_y) = \tilde{U}_o(f_x,f_y), \tag{7.11a}$$

$$FT\left\{U_R(x,y)U_R(x,y)U_o^*(x,y)\right\} = \tilde{U}_R(f_x,f_y) \otimes \tilde{U}_R(f_x,f_y) \otimes \tilde{U}_o^*(f_x,f_y)$$

$$= \delta(f_x, f_y - 2f_r) \otimes \tilde{U}_o^*(f_x,f_y). \tag{7.11b}$$

These relations were obtained by considering that $\tilde{U}_R(f_x,f_y) = \delta(f_x, f_y - f_r)$ and $\tilde{U}_R^*(f_x,f_y) = \delta(f_x, f_y + f_r)$, respectively. In order to obtain a spatial separation of the object reconstructions terms from the self-interference term, it is necessary that $f_r \geq 3B$.

The maximal angle between the reference and the object wave θ_{max} determines the maximal spatial frequency f_{max} in the hologram:

$$f_{max} = \frac{2}{\lambda}\sin\left[\frac{\theta_{max}}{2}\right]. \tag{7.12}$$

The resolution of the recording device should be able to record this spatial frequency in order to obtain a spatial separation of the reconstructed waves. Figure 7.4b illustrates the spectrum of an in-line Gabor hologram recorded by using a reference plane wave propagating in the same direction as the object field; thus, in Equation 7.9, the angle θ is equal to 0 and $\tilde{U}_R(f_x,f_y) = \tilde{U}_R^*(f_x,f_y) = \delta(f_x,f_y)$. The four terms are overlapped; we will see later that by using digital holography together with a phase-shifting technique, it is possible to separate these terms. Until now, we considered a plane reference wave because this makes the mathematical treatment easier; if we use another reference (e.g., a spherical wave), we will obtain asymmetrical spectra like that shown in Figure 7.4c. The spectra in Figure 7.4 describe the separation of the terms far away from the hologram (Fraunhofer diffraction or equivalently far field); at a shorter distance, the light distributions will be different, but if the reconstructions are separated at the infinity, it is usually possible to obtain their separation at a shorter distance.

7.4 DIGITAL HOLOGRAPHY

7.4.1 Historical Remarks and Technological Developments

In classical holography, the interferences between reference and object waves are stored on high-resolution photographic emulsions, occasionally on photo-thermo plastic or photorefractive material. For the reconstruction, a coherent wave, usually a laser, is used. The process is time consuming, particularly with respect to the development of the photographic film.

In 1967, Goodman and Lawrence [11] reported the first case of detection of a hologram by means of an electro-optical device; they used a vidicon to sample the hologram on a 256 × 256 array, and the digital reconstruction was done by using a PDP-6 computer. For this kind of reconstruction, fast Fourier transform (FFT) algorithms have been used. In a conference paper [12], Goodman explained that this experiment was motivated by the desire to form images of actively illuminated satellites free from the disturbing effects produced by the earth atmosphere. Interesting is to notice that a 256 × 256 FFT transform took about 5 min in 1967 by using one of the most powerful computers available at that time; now (2014) a typical personal computer performs this computation in only few milliseconds.

The procedures for digitally reconstructing holograms were further developed in the seventies years in particular by Yaroslavskii and others [13,14]. They digitized one part of a conventional hologram recorded on a photographic plate and fed those data to a computer; the digital reconstruction was obtained by a numerical implementation of the diffraction integral.

During the last 30 years, we had an impressive development of the charge-coupled devices (CCDs) that were invented in 1969 but became commercially available only since 1980. Complementary metal–oxide–semiconductor sensors (CMOS) are another type of electro-optical sensors having almost the same sensitivity and signal-to-noise ratio as the CCDs. Both CCD and CMOS can have several thousands of pixels in each direction (e.g., 4000 × 4000), a typical pixel size of 5 × 5 μm^2 and a dynamic range of 8–10 bits, which may be increased to 12–14 bits by using a cooling system. Those devices have become an attractive alternative to photographic film, photo–thermoplastic, and photorefractive crystals as recording media in holography. The time-consuming wet processing of photographic film is avoided, and the physical reconstruction of holograms can be replaced by a numerical reconstruction using a computer, normally an ordinary PC. Haddad et al. [15] were probably the first to report the recording of a hologram on a CCD. They used a lensless arrangement for the recording of a wavefront transmitted by a microscopic biological sample and were able to obtain reconstructions with a resolution of 1 μm. During the past 20 years, several arrangements have been developed for the investigation of transmission and reflection of both technical and biological samples. Digital holographic microscopy [15–40] allows high-resolution 3D representations of microstructures and provides quantitative phase contrast images. Recording of digital holograms of macroscopic objects is reported in Refs. [41–46]; most of these published works combine the holographic technique with interferometry in order to measure deformations of mechanical components and biological tissues. Excellent books describing the methods of digital holography and digital holographic interferometry are available [47–50].

7.4.2 LIMITATIONS DUE TO THE RECORDING OF HOLOGRAMS ON AN ELECTRO-OPTICAL DEVICE

One disadvantage, however, of using electro-optical sensors for hologram recording is the relatively low spatial resolution, compared to that of photographic material or photorefractive crystals. The recording of holograms on low-resolution media has been discussed in many books and publications [4–9,47–50]. The spatial resolution of commercially available CCD and CMOS is typically 200 lines/mm, which is much poorer than photographic films (5000 lines/mm); this limits the spatial frequency that can be recorded using these devices. To record a hologram of the entire object, the resolution must be sufficient to resolve the fringes formed by the reference wave and the wave coming from any object point. Consider that the intensity described by Equation 7.3 is recorded on a 2D array of sensors ($M \times N$ cells) of dimension $\Delta x \times \Delta y$. For the most common case of equidistant cells, we have $\Delta x = \Delta y = \Delta$, and the discrete recorded intensity can be written in the form $I(m\Delta, n\Delta)$, where n and m are integer numbers. From the sampling theorem (see, e.g., Ref. [50]), it is necessary to have at least two sampled points for each fringe; therefore, the maximum spatial frequency that can be recorded is

$$f_{max} = \frac{1}{2\Delta}. \tag{7.13}$$

By inserting this in Equation 7.12, we obtain the condition for θ_{max}: $\sin(\theta_{max}/2) \leq \lambda/(4\Delta)$ and for small angles $\theta_{max} \leq \lambda/(2\Delta)$.

7.4.3 MATHEMATICAL TOOLS FOR THE DIGITAL RECONSTRUCTION OF A HOLOGRAM

From the diffraction theory, we know that if the amplitude and phase of a wavefront are known at a plane (x_1, y_1), it is possible to calculate the propagation of such wavefront in the plane (x_2, y_2) (see Figure 7.5). For calculating the diffracted wavefront, there are different approximations. If we consider the Rayleigh–Sommerfeld diffraction integral (see, e.g., Ref. [51]), the wave fields at the plane (x_2, y_2) located at a distance d from the plane (x_1, y_1) are given by

$$U_2(x_2, y_2) = \frac{i}{\lambda} \iint U_1(x_1, y_1) \frac{e^{ikr_{12}}}{r_{12}} \cos \alpha \, dx_1 dy_1, \tag{7.14}$$

where

$r_{12} = \sqrt{d^2 + (x_2 - x_1)^2 + (y_2 - y_1)^2}$

$k = 2\pi/\lambda$

$\cos \alpha$ is the obliquity factor

In the following, the obliquity factor will be neglected. The diffraction integral can be written as a convolution

$$U_2(x_2, y_2) = \frac{i}{\lambda} \iint U_1(x_1, y_1) g(x_2 - x_1, y_2 - y_1) dx_1 dy_1 = U_1 \otimes g, \tag{7.15}$$

where

$$g(x_2 - x_1, y_2 - y_1) = \frac{i \exp\left\{ik\sqrt{d^2 + (x_2 - x_1)^2 + (y_2 - y_1)^2}\right\}}{\lambda \sqrt{d^2 + (x_2 - x_1)^2 + (y_2 - y_1)^2}} \tag{7.16}$$

is the impulse response of the free-space propagation. According to the Fourier theory, the convolution integral between two functions may be calculated by multiplying their individual FTs followed by an inverse FT, and thus, Equation 7.15 may be rewritten in the form

$$U_2(x_2, y_2) = FT^{-1}\{FT\{U_1(x_1, y_1)\}FT\{g(x_1, y_1)\}\}. \tag{7.17}$$

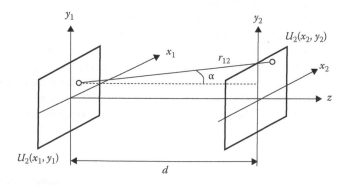

FIGURE 7.5   Diffraction geometry.

If the distance d is large compared with $(x_1 - x_2)$ and $(y_1 - y_2)$, the binominal expansion of the square root up to the third term yields

$$r_{12} \approx d \left[1 + \frac{(x_1 - x_2)^2}{2d} + \frac{(y_1 - y_2)^2}{2d} \right]. \tag{7.18}$$

By inserting Equation 7.18 in Equation 7.15, we obtain the Fresnel approximation of the diffraction integral:

$$U_2(x_2, y_2) = \frac{e^{ikd} e^{ik\frac{(x_2^2 + y_2^2)}{2d}}}{i\lambda d} \iint U_1(x_1, y_1) e^{ik\frac{(x_1^2 + y_1^2)}{2d}} e^{-ik\frac{(x_1 x_2 + y_1 y_2)}{d}} dx_1 dy_1. \tag{7.19}$$

This can be seen as a FT of $U(x_1, y_1)\exp\{ik[x_1^2 + y_1^2]/(2d)\}$ multiplied by a quadratic factor.

In digital holography, the intensity is recorded on a discrete array of sensors $I(m\Delta x, n\Delta y)$, and for the digital calculation of the propagation, it is necessary to use a discrete form of Equations 7.14 through 7.19. If we consider a digitized wave field (e.g., formed by $M \times N = 1000 \times 1000$ points), it would take a very long time to compute the convolution between the two functions U_1 and $\exp\{ikr_{12}\}/r_{12}$ described by Equation 7.14; the direct use of such equation is thus not recommended. A more suitable way for the wavefront calculation is to use the 2D discrete FTs. The discrete forms of Equation 7.6 are

$$\tilde{f}(r,s) = \sum_{m=0}^{M-1} \sum_{n=0}^{N-1} f(m,n) \exp\left\{ 2\pi i \left[\frac{mr}{M} + \frac{ns}{N} \right] \right\}, \tag{7.20a}$$

$$f(m,n) = \sum_{r=0}^{M-1} \sum_{s=0}^{N-1} \tilde{f}(r,s) \exp\left\{ -2\pi i \left[\frac{mr}{M} + \frac{ns}{N} \right] \right\}, \tag{7.20b}$$

where r, s, m, n are integer numbers. Fortunately, the 2D discrete FT can be calculated in a short time by using the FFT algorithm (some milliseconds for $M \times N = 1000 \times 1000$) allowing the fast computation of the wavefront propagation. A rigorous mathematical description of the reconstruction of discrete wavefronts by using FFT transforms may be found in Refs. [14, 47–50]; here, we just give the discrete form of Equation 7.19 representing the Fresnel propagation from a plane (x_1, y_1), where the input is discretized on $M \times N$ array of pixels having size $(\Delta x_1, \Delta y_1)$, to an $M \times N$ array of pixels having size $(\Delta x_2, \Delta y_2)$ in the plane (x_2, y_2):

$$U_2(r\Delta x_2, s\Delta y_2) = e^{ikd} e^{\frac{i\pi}{d\lambda}\{\Delta x_2^2 r^2 + \Delta y_2^2 s^2\}} \sum_{m=0}^{M-1} \sum_{n=0}^{N-1} U_1(m\Delta x_1, n\Delta y_1) e^{\frac{i\pi}{d\lambda}\{m^2 \Delta x_1^2 + n^2 \Delta y_1^2\}} e^{-i2\pi\{\frac{mr}{M} + \frac{ns}{N}\}} \tag{7.21}$$

The pixel size in the two planes is given by

$$\Delta x_2 = \frac{d\lambda}{M\Delta x_1}, \quad \Delta y_2 = \frac{d\lambda}{N\Delta y_1}. \tag{7.22}$$

For this special relation, the last exponential term in Equation 7.21 has exactly the form of the last exponent in the discrete FT 7.20, and the Fresnel propagation can be calculated by using a fast algorithm (FFT). The numerical calculation of the propagation according to the discrete form of Equation 7.17 (sometimes this is simply called convolution algorithm [49,50]) employs three FFTs and is thus more time consuming compared with the simple discrete Fresnel transform described by Equation 7.21. By using the convolution algorithm, the pixel spacing is always the same, while in the Fresnel algorithm, it proportionally depends on the product of the wavelength and the reconstruction distance. Two-step Fresnel reconstruction algorithms have been developed (double Fresnel transform algorithm [52]) allowing to adjust the pixel spacing in the reconstructed wavefront.

After this mathematical review of the formula of wave propagation, we may now consider some recording geometries and the corresponding reconstruction method.

7.4.4 SOME ARRANGEMENTS FOR DIGITAL HOLOGRAPHY

Classical and digital holograms may be classified according to the geometry used for the recording. We consider at first a lensless off-axis setup (Figure 7.6a). The only difference between the lensless arrangement used for the recording of a classical and a digital hologram is that for the digital recording, the angle between object and reference should be small in order to allow the low-resolution digital sensor to sample the interference pattern. The recorded intensity $I(m\Delta x, n\Delta y)$ is multiplied with a numerical representation of the reference wave followed by numerical calculation of the resulting diffraction field. Usually, plane or spherical uniform waves are used for recording digital holograms, since these waves can be easily simulated during the reconstruction process. Figure 7.6b through d shows three reconstructions of small objects (5 cm); the digital focusing allows, without any mechanical movements, to focalize any plane along the z axis. For recording the hologram, a spherical reference originating from a point source located close to the object has been used. The reconstructions display the characteristic image pair described by $U_R U_R^* U_o^*$ and $U_R U_R U_o^*$ flanking a central bright spot representing the spherical wave $U_R |U_R|^2$ focusing at the object plane. The convolution term that according to Figure 7.4a has twice the extension of the object term is spread over the whole field. For recording this hologram, the condition $f_r \geq 3B$ was not satisfied, but the reference wave was much stronger compared with the object wave; this decreases the magnitude of the convolution term compared with the reconstructed image as shown schematically in Figure 7.6e. For the hologram recording, the distance between the CCD and the object was 100 cm.

If the object is large and the sensor has limited resolution, it is necessary to locate the object far away from the recording device in order to satisfy the sampling theorem. We illustrate this by a simple example where an object having dimension $L_o = 1$ m is holographically recorded on a CCD having pixel size $\Delta = 5$ μm by using light with a wavelength of 0.5 μm. The maximal allowable angle between reference and object ($\theta_{max} \approx L_o/a$) should satisfy the condition $\theta_{max} < \lambda/(2\Delta)$; thus, a should be larger than $2L_o\Delta/\lambda = 20$ m. In practice, the lensless arrangement is not suitable for recording digital holograms of large objects by using CCDs or CMOS. This problem can be solved by inserting a lens and an aperture having dimension L_A between the object and the sensor (Figure 7.7a). L_A is usually few millimeters, and the sampling condition is satisfied when the distance b between aperture and sensor is some centimeters.

For recording digital holograms of microscopical samples, the arrangement shown in Figure 7.7b is usually used; in this last case, a microscope objective images the object on the sensor or at any other plane.

For both the arrangements shown in Figure 7.7a and b, the reference is a spherical wave originating from a point located close to the aperture; by performing a FT of the recorded intensity, we obtain two reconstructions (primary and secondary) together with the central convolution term. When these terms are well separated (as in the case of Figure 7.7c), we can select one reconstruction where the information about the amplitude and phase of the wavefront at the plane of the aperture is included (see Figure 7.7d). After filtering, we propagate the wavefront at any other plane to obtain digital focused images.

FIGURE 7.6 Lensless arrangement for digital holography (a). Digital reconstructions of small objects at three different planes (b, c, d). Spectra of the hologram (e).

A disadvantage of an off-axis configuration is that the tilted reference, necessary for introducing the carrier fringes and separating the reconstructions, increases the spatial frequencies contained in the hologram, and this prohibits the effective use of all the pixels of the array sensor. This problem can be reduced by using a phase-shifting technique, in which the phase and amplitude information of the object beam are obtained through an in-line setup (see Figure 7.8), with no inclination between object and reference beams [53]. A mirror mounted on a piezoelectric transducer is used to shift stepwise the phase of the reference beam. Several interferograms with mutual phase shift are recorded and used for calculating the phase of the object wave. The disadvantage of the method is that at least three interferograms need to be recorded; this makes it not well suited for investigations of dynamical events, restricting its application to slow-varying phenomena. The in-line arrangement

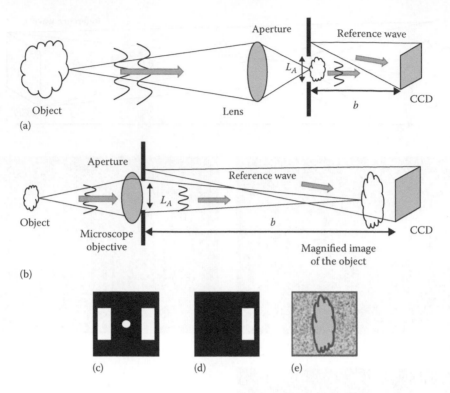

FIGURE 7.7 Arrangement for recording hologram of large (a) and microscopical (b) objects where a lens and an aperture are inserted between object and sensor. Reconstruction of amplitude and phase in the plane of the aperture by FFT of the recorded intensity (c), filtering (d) and image reconstruction by digital propagation of the wavefront (e).

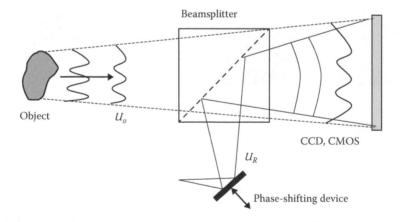

FIGURE 7.8 Arrangement for the recording of a lensless in-line digital hologram.

may be modified by inserting a lens in the object wavefront; this can be particularly useful for microscopic investigations.

7.5 APPLICATIONS OF DIGITAL HOLOGRAPHY

Digital holography has been proposed for several applications; in this section, we report few results showing that it can be used for the investigations of microscopical samples, for shape contouring and for the measurement of displacements.

7.5.1 DIGITAL HOLOGRAPHIC MICROSCOPY

Digital holography has been broadly applied to microscopy; some groups use a lensless setup where the sensor records the interference between the light scattered by the object and the reference; others prefer to introduce a lens (e.g., a microscope objective) in order to image the sample to a plane behind the sensor (see Figure 7.7b).

Being a scan-less and also a label-free technique, digital holographic microscopy can be used to extract the 3D information of the organism using a single image (digital hologram). The quantitative phase and amplitude information of the object wavefront can be digitally obtained along the depth of the object from the recorded hologram, which makes it possible to digitally focus on different layers of the specimen and reconstruct a 3D profile of the optical thickness of the organism using the quantitative phase value. Figure 7.9 shows that digital holography can be used for bright- and dark-field reconstructions of amplitude and phases of a sea urchin larva. Because of having no background in dark-field imaging, the noise coming from the coherent light in the background is suppressed, leading to a significant contrast enhancement [38].

Like in conventional microscopy, the spatial resolution of digital holographic microscopy is given by

$$\delta x = \frac{\kappa_1 \lambda}{NA_{imag} + NA_{illum}}, \tag{7.23}$$

where

NA_{imag} and NA_{illum} are the numerical aperture of the illumination and imaging systems, respectively
λ is the wavelength
κ_1 is the factor determined by experimental parameters, such as coherent noise level, signal-to-noise ratio of the detector [54]

The spatial resolution can thus be improved by using a short wavelength or synthesizing a larger NA, which can be obtained by off-axis illumination or projecting a periodic pattern onto a specimen.

An arrangement that uses a short wavelength is reported in Ref. [40], and thus according to Equation 7.23, the resolution is increased. The light source was an excimer laser emitting radiation at 193 nm; all the optical elements (beamsplitter, lenses, waveplates, and polarizers) were made of fused silica. These materials have a good transmission in the ultraviolet light. A custom-design objective with large numerical aperture $NA_{imag} = 0.75$ and 1 mm focal length was used. To further improve the resolution, oblique illumination allowing to increase the numerical aperture NA_{illum} is implemented in the setup. A digital camera is commercially available for this wavelength with high UV sensitivity, which can perform direct imaging with no filter or fluorescent. Figure 7.10a shows a scanning electron microscope image of the nano-structured template designed to examine the resolution of the holographic system. The template is made of 35 nm of gold layer coated by ion-beam sputtering on a fused silica substrate. It includes square and line structures, ranging from 100 to 500 nm in width. The logo of the Institut für Technische Optik "ito" is also included in the template in which its elements have a width of ~300 nm. The image taken by an optical microscope (Zeiss Axiovert 200, with × 100 objective, NA = 0.7) is shown in Figure 7.10b. The reconstructed amplitude and phase obtained by digital holography are shown in Figure 7.10c and d, respectively. In Figure 7.10c, the "ito" logo is clear, and even the line structures with the width of 250 nm are well resolved.

7.5.2 DIGITAL HOLOGRAPHIC SECTIONING BY USING A SHORT COHERENCE LIGHT SOURCE

Several imaging techniques, such as coherence radar [55] or digital light-in-flight holography [56], take advantage of low temporal coherence to examine 3D objects. Furthermore, optical coherence tomography (OCT) [57] provides high-resolution cross-sectional images of the internal microstructure of materials and biological samples. As has been reported in Refs. [58,59], the use of short coherence digital holography allows optical sectioning and has the ability to discriminate light

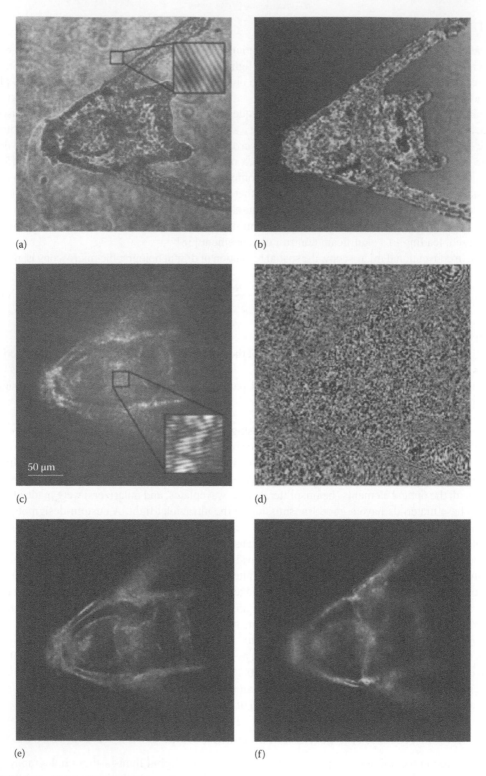

(a)

(b)

(c)

(d)

(e)

(f)

FIGURE 7.9 Bright-field hologram of a sea urchin larva illuminated by a short coherence source (a) and reconstructed phase (b). Dark-field hologram (c), phase (d) and intensity reconstructions at two focus planes (e and f). The holograms are taken by a bright/dark-field objective (20×, NA = 0.45). (From Faridian, A. et al., *J. Biomed. Opt.*, 18(8), 086009, 2013.)

(a)   (b)   (c)   (d)

FIGURE 7.10 Scanning electron microscope image of the nano-structured template (a); image taken by a conventional optical microscope with NA = 0.75, × 100 objective (b); the reconstructed amplitude (c) and phase (d) of the object. (From Faridian, A. et al., *Opt. Express*, 18(13), 14159, 2010.)

coming from different planes of a sample. In Ref. [60], the light source was a short coherence laser diode, and a lensless in-line arrangement like that shown in Figure 7.8 has been used. The phase of the wavefront was recovered by using a phase-shifting method. Due to the short coherence of the laser, the interference is just observed when the optical path lengths of reference and object beams are matched within the coherence length of the laser; the numerical reconstruction of the hologram theoretically corresponds to a defined layer, and reconstruction gets rid of any spurious light coming from other parts of the sample. The selection of the plane of interest can be simply performed by mechanical shifting of a mirror in the experimental setup that changes the path difference between object and reference. Some advantages of this method over other imaging techniques allowing optical sectioning, like confocal microscopy or OCT, are the simplicity of the optical arrangement and the possibility to record at once the whole information about the plane of interest, without any need of lateral scanning. In Figure 7.11 are shown the images of different layers of a fly with an interval of 0.1 mm between each image.

FIGURE 7.11 Reconstructed images of a fly. (From Martínez-León, L. et al., *Appl. Opt.*, 44, 3977, 2005.)

In Ref. [61], it was demonstrated how short temporal coherence digital holography with a femtosecond frequency comb laser source may be used for multilevel optical sectioning. In contrast to conventional approaches, each hologram reconstructs not only the section of the object where reference and object beam matches, but as well other planes with equidistant intervals. Figure 7.12 shows an experimental result, where the object used for the investigation was a metallic cone with half angle of 12°, 80 mm coned length and 36 mm base diameter (see Figure 7.12a). Figure 7.12b through d shows three numerical reconstructions obtained from a single hologram with digital

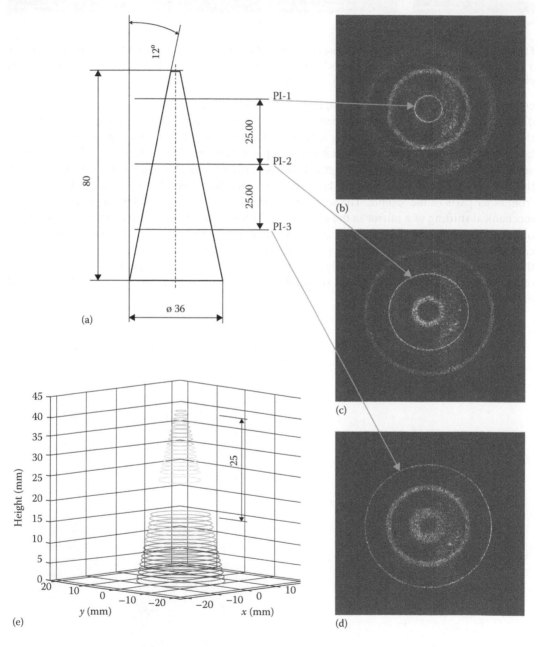

FIGURE 7.12 Schematic of the rough metallic cone used for the investigation (a). Reconstruction of holograms at three different planes separated by 25 mm (b through d). Numerical reconstruction of a part of the cone shape (e). (From Körner, K. et al., *Opt. Express*, 20(7), 7237, 2012.)

focusing in three different planes, each separated by 25 mm. In Figure 7.12b, we see a ring with small radius at the center that corresponds to the intersection of the cone with a plane (we will call this PL-1); in the same figure, there are two other rings with larger radius describing intersections of the object with two planes (PL-2 and PL-3) axially spaced by 25 and 50 mm with respect to PL-1. Three rings appear because we use a frequency comb laser having a periodical coherence function displaying peaks spatially separated by 50 mm. A conventional short coherence or a femtosecond laser provides single sectioning only. In Figure 7.12b, the intersection lines of the cone with PL-2 and PL-3 are blurred since the reduced depth of focus does not allow to reconstruct sharp profiles located at different planes. Figure 7.12c and d shows that digital focusing of the hologram allows to obtain sharp profile lines at PL-3 and PL-2, respectively. The three reconstructed profiles obtained from a single hologram do not allow to retrieve the complete 3D shape of the object. More sections would be necessary for this purpose, which could be obtained by using an fc-laser having larger frequency comb spacing and thus a reduced axial plane separation. The scanning method was used to obtain more section profiles, where the path length of the reference beam was changed by shifting the reference mirror, and a sequence of holograms was recorded. Such holograms allow the reconstruction of different sections of the object as shown in Figure 7.12e.

7.5.3 Digital Holographic Interferometry

One of the most significant applications of classical holography is holographic interferometry, which was first developed in 1965 by Stetson and Powell [62,63]. This technique makes it possible to measure displacement, deformation, and vibration of rough surface with an accuracy of a fraction of a micrometer. It has been extensively used for metrological application since it allows the interferometric comparison of wavefronts of different states of the object that are not present simultaneously. In classical holographic interferometry, two or more holograms of the object under investigation are recorded usually by a photographic plate. The subsequent interference of these wave fields when they are physically reconstructed produces a fringe pattern, from where the information of the deformations or change of refraction index of the object between the exposures can be obtained. There exists a large literature describing the applications of classical holographic interferometry; particularly recommended are the books from C. Vest W. Schumann, Jones, and Wykes [64–66]. Electronic speckle interferometry (ESPI) [67,68] allowed to avoid the processing of films, which represents the major disadvantage of classical holographic interferometry. In the first ESPI systems, a vidicon device recorded the interference between light reflected by a diffusing object and a reference wave, but starting from 1980, CCDs and CMOS devices were used leading to a dramatic improvement of the results. ESPI and digital holographic interferometry are very similar techniques; both can use in-line [69] or off-axis [70,71] arrangement. The only difference is that for ESPI, the object is imaged on the recording device, and thus, a reconstruction by digital calculation of the wavefront propagation is not necessary. When the interferograms are digitally recorded (ESPI or digital holography), the phase of the wavefront can be determined by processing the intensity pattern, and the comparison between wavefronts recorded at different times is easy and fast. Digital holography has been extensively used for the measurement of deformations of large and microscopic objects [72–75], vibrations analysis [76–81], defect recognition [82], contouring [83], particle size and position measurements [84], and for the investigation of the change of refraction index of liquids and gases [85]. Here, we will show some results where the technique has been used to measure deformations.

7.5.3.1 Measurement of Object Deformations

We consider two digital holograms of the object recorded on two separate frames of a CCD or CMOS detector: the first one with the object in its initial state and the second one after the object has undergone a deformation that shifts the optical phase compared to the one in the first recording.

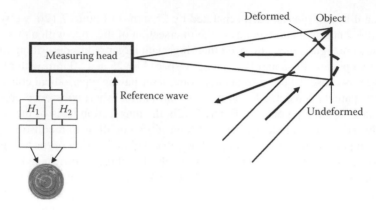

FIGURE 7.13 Digital holographic interferometry.

Denoting the two digitally reconstructed wave fields with U_{O1} and U_{O2}, respectively, their phases ϕ_{o1} and ϕ_{o2} can be directly calculated:

$$\phi_{oj}\left(m\Delta x, n\Delta y\right) = \arctan\left[\frac{\text{Im}\left[U_{oj}\left(m\Delta x, n\Delta y\right)\right]}{\text{Re}\left[U_{oj}\left(m\Delta x, n\Delta y\right)\right]}\right], \tag{7.24}$$

where $j = 1, 2$. ϕ_{o1} and ϕ_{o2} are wrapped phases $(0 < \phi_{oj} < 2\pi)$, and their difference

$$\Delta\phi_w = \phi_{o2} - \phi_{o1} \tag{7.25}$$

needs to be processed in order to eliminate the 2π incertitude and obtain $\Delta\phi$. The relation between the deformation $\vec{d}$ of an arbitrary point of the surface and the phase change is

$$\Delta\phi = \frac{2\pi}{\lambda}\vec{d}\cdot\vec{s}. \tag{7.26}$$

The vector $\vec{s} = \vec{k}_o - \vec{k}_i$ (where $\vec{k}_o$ and $\vec{k}_i$ are the unit vectors of observation and illumination, respectively) is referred to as the sensitivity vector, which is given by the geometry of the recording arrangement. It lies along the bisector of the angle between the illumination and viewing directions. Equation 7.26 is the basis of all quantitative deformation measurements of solid objects by holographic interferometry. Figure 7.13 shows the recording geometry and the procedure for obtaining the phase difference from two digitally recorded holograms H_1 and H_2.

A single measurement yields one vector component of the deformation $\vec{d}$ (projection to the sensitivity vector). In general, for determining the full 3D deformation vector $\vec{d}$, we need to perform at least three measurements with different sensitivity vectors $(\vec{s}_1, \vec{s}_2, \vec{s}_3)$. For each sensitivity vector, the phase difference $(\Delta\phi_1, \Delta\phi_2, \Delta\phi_3)$ between the two deformation states is evaluated, and the 3D deformation field is determined by decomposing the sensitivity vectors into their scalar components and solving a linear system of equations [79].

7.5.3.2 Pulsed Digital Holographic Interferometry for the Measurement of Vibration and Defect Recognition

Figure 7.14 illustrates the measurement of a vibrating plate having a diameter of 15 cm. The plate was excited at the resonance frequency of 4800 Hz by using a shaker. A pulsed ruby laser, which emits two high-energy short pulses (pulse width 20 ns) separated by few microseconds, has been

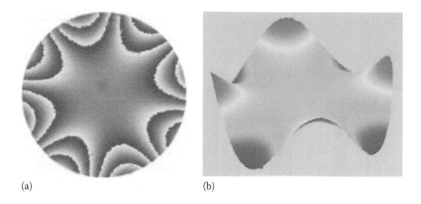

(a)   (b)

FIGURE 7.14   Plate (15 cm diameter) vibrating at 4800 Hz. Phase map (a) and pseudo-3D representation (b).

used for the recording of the two holograms on a CCD; in this specific case, the pulse separation was 10 μs. The measuring head consists of an off-axis arrangement with lens as shown in Figure 7.7, where the object was directly imaged on the CCD. The two digital holograms were processed, and their phase difference is shown in Figure 7.14a. By using an unwrapping procedure, it was possible to obtain the vibration shape shown in Figure 7.14b. The sensitivity vector was parallel to the observation direction, and the out-of-plane vibration of the plate was measured.

Figure 7.15 shows another example where pulsed digital holography interferometry has been used to investigate a metal plate submitted to a shock. The laser pulses were fired shortly after the shock produced by a metal ball fell on the plate. After calculation and subtraction of the phases of the two reconstructed digital holograms, we obtain a phase map with concentric fringes around the point of the impact. The shock wave propagating at the surface of the plate is disturbed by a small defect that can be clearly seen in the phase map. This shows the principle for defect detection in mechanical components.

The acquisition speeds of the CCD and CMOS semiconductor detectors have increased continuously (cameras recording 5000 image/s with a resolution of 500 × 500 pixels are commercially available), and methods for measuring dynamic events have been developed in which a sequence of digital holograms are recorded on a high-speed sensor. By unwrapping the phase in the temporal domain, it is possible to retrieve the displacement of an object as a function of time. The use of high-speed cameras for the investigation of dynamic events is reported in Refs. [86–90].

(a)   (b)

FIGURE 7.15   Object submitted to a shock excitation (a). Phase map obtained by double pulse exposure digital holography (b). (From Tiziani, H.J. et al., Interferometry, in: *Landolt-Börnstein New Series*, Laser Physics and Applications, Springer-Verlag, Berlin, Germany, 2006, pp. 221–284.)

7.5.3.3 Digital Holographic Interferometry for the Investigation of Microsystems

The trend toward miniaturization of micro-electro-mechanical systems (MEMS) and micro-opto-electro-mechanical systems continues to lead to more and more compact devices [91]. Applications cover a wide range of fields, from optical telecommunications to medicine; furthermore, MEMS-based accelerometers and pressure sensors are used in the aeronautical and automotive industry. Part of the functionality and reliability of MEMS devices is based on the displacement and deformation of micromechanical parts under mechanical, thermal, magnetic, or electrostatic loads. The measurement of the shape and deformation of such systems is thus of great importance for confirming analytical and finite element models, accessing material and device properties, detecting potential defects, and determining performance [92–95]. These data can then be used in further device optimization. Since typical structures exhibit dimensions in the order of some micrometers, it is necessary to measure their deformation with accuracies down to the lower nanometer range. Digital holography allows measurements of nanometric deformations and is thus well suited for the characterization of the micro devices.

A system is described in Ref. [96] to precisely determine the deformation of MEMS; the setup uses a stabilized laser and high-quality optical components with antireflection coating and was fixed on a vibration-isolated table. The CCD camera used has low noise and high dynamics (12 bits). The test object was the MEMS shown in Figure 7.16, designed to have a very accurate out-of-plane displacement (parallel to the observation direction) produced by applying an electrical voltage to the structure. Notice that the process of loading by applying a voltage to the MEMS and the acquisitions of the digital holograms are synchronized and done dynamically. Figure 7.16b through d shows 3D representations of the measured out-of-plane deformations when 15, 30, and 50 V are applied to the MEMS shown in Figure 7.16a. In order to compare the results, we consider the deformation at the center of the moving element as a function of the applied voltage. In Figure 7.17a and b, it is not

FIGURE 7.16 Photo (a) and out-of-plane deformations of a MEMS obtained by applying 15 V (b), 30 V (c), and 50 V (d). (From Pedrini, G. et al., *Opt. Eng.*, 50, 101504-1, 2011.)

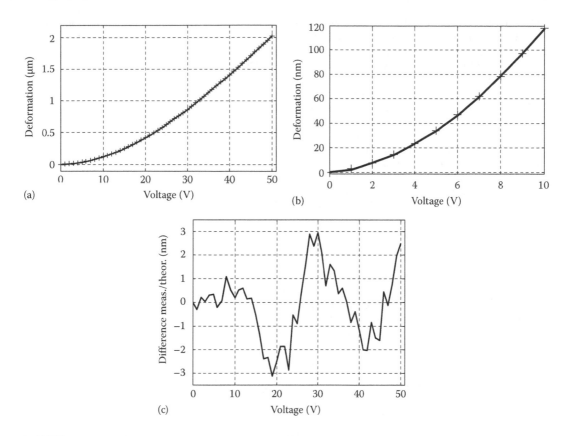

FIGURE 7.17 (a) Measured (+) and expected (solid line) deformation as a function of the applied voltage; (b) detail of (a); (c) difference between measured and expected results. (From Pedrini, G. et al., *Opt. Eng.*, 50, 101504-1, 2011.)

possible to see any difference between measured (+) and expected (solid line) deformations. Their difference is reported in Figure 7.17c; the standard deviation is 1.4 nm, thus very small compared with the total deformation of 2 μm.

REFERENCES

1. D. Gabor, A new microscopic principles, *Nature*, **161**, 777–778 (1948).
2. E. N. Leith, J. J. Upatnieks, Wavefront reconstruction with diffused illumination and three-dimensional objects, *Opt. Soc. Am.*, **54**, 1295 (1964).
3. G. W. Stroke, *An Introduction to Coherent Optics and Holography*, 2nd edn., Academic Press, New York (1969).
4. H. M. Smith, *Principles of Holography*, Wiley-Interscience, New York (1969).
5. H. J. Caulfield, S. Lu, *The Applications of Holography*, Wiley-Interscience, New York (1970).
6. M. Lehmann, *Holography, Technique and Practice*, The Focal Press, London, U.K. (1970).
7. R. J. Collier, C. B. Burckhard, L. H. Lin, *Optical Holography*, Academic Press, New York (1971).
8. J. N. Butters, *Holography and Its Technology*, Peter Peregrinus Ltd., Published on behalf of the Institution of Electrical Engineering, London, U.K. (1971).
9. W. T. Cathey, *Optical Information Processing and Holography*, John Wiley & Sons, New York (1974).
10. B. R. Brown, A. W. Lohmann, Complex spatial filtering with binary masks, *Appl. Opt.*, **5**, 967–969 (1966).
11. J. W. Goodman, R. W. Lawrence, Digital image formation from electronically detected holograms, *Appl. Phys. Lett.*, **11**, 77–79 (1967).

12. J. W. Goodman, Digital image formation from holograms: Early motivation modern capabilities, in: *Adaptive Optics: Analysis and Methods/Computational Optical Sensing and Imaging/Information Photonics/Signal Recovery and Synthesis Topical Meetings on CD-ROM*, OSA Technical Digest (CD), paper JMA1, Optical Society of America, Washington, DC (2007).

13. M. A. Kronrod, N. S. Merzlyakov, L. P. Yaroslavskii, Reconstruction of a hologram with a computer, *Sov. Phys. Tech. Phys.*, **17**, 333–334 (1972).

14. L. P. Yaroslavskii, N. S. Merzlyakov, *Methods of Digital Holography*, Consultants Bureau, New York (1980).

15. W. S. Haddad, D. Cullen, J. C. Solem, J. W. Longworth, A. McPherson, K. Boyer, C. K. Rhodes, Fourier-transform holographic microscope, *Appl. Opt.*, **31**, 4973–4978 (1992).

16. Y. Takaki, H. Ohzu, Fast numerical reconstruction technique for high-resolution hybrid holographic microscopy, *Appl. Opt.*, **38**, 2204–2211 (1999).

17. G. Pedrini, P. Froning, H. J. Tiziani, F. Mendoza-Santoyo, Shape measurement of microscopic structures using digital holograms, *Opt. Commun.*, **164**, 257–268 (1999).

18. G. Pedrini, S. Schedin, H. J. Tiziani, Aberration compensation in digital holographic reconstruction of microscopic objects, *J. Mod. Opt.*, **48**, 1035–1041 (2001).

19. L. Xu, X. Peng, J. Miao, A. K. Asundi, Studies of digital microscopic holography with applications to microstructure testing, *Appl. Opt.*, **40**, 5046–5051 (2001).

20. I. Yamaguchi, J. Kato, S. Ohta, J. Mizuno, Image formation in phase-shifting digital holography and applications to microscopy, *Appl. Opt.*, **40**, 6177–6185 (2001).

21. G. Pedrini, H. J. Tiziani, Short-coherence digital microscopy by use of a lensless holographic imaging system, *Appl. Opt.*, **41**, 4489–4496 (2002).

22. W. Xu, M. H. Jericho, I. A. Meinertzhagen, H. J. Kreuzer, Digital in-line holography of microspheres, *Appl. Opt.*, **41**, 5367–5375 (2002).

23. M. Gustafsson, M. Sebesta, B. Bengtsson, S. G. Pettersson, P. Egelberg, T. Lenart, High-resolution digital transmission microscopy: A Fourier holography approach, *Opt. Lasers Eng.*, **41**, 553–563 (2004).

24. D. Carl, B. Kemper, G. Wernicke, G. Von Bally, Parameter optimized digital holographic microscope for high resolution living cell analysis, *Appl. Opt.*, **43**, 6536–6544 (2004).

25. P. Marquet et al., Digital holographic microscopy: A noninvasive contrast imaging technique allowing quantitative visualization of living cells with subwavelength axial accuracy, *Opt. Lett.*, 30(5), 468–470 (2005).

26. B. Kemper, G. von Bally, Digital holographic microscopy for live cell applications and technical inspection, *Appl. Opt.*, **47**(4), A52–A61 (2008).

27. A. Anand, V. K. Chhaniwal, B. Javidi, Imaging embryonic stem cell dynamics using quantitative 3-D digital holographic microscopy, *IEEE Photon. J.*, **3**(3), 546–554 (2011).

28. B. Kemper et al., Label-free quantitative cell division monitoring of endothelial cells by digital holographic microscopy, *J. Biomed. Opt.*, **15**(3), 036009 (2010).

29. S. Witte et al., Short-coherence off-axis holographic phase microscopy of live cell dynamics, *Biomed. Opt. Express*, **3**(9), 2184–2189 (2012).

30. F. Dubois et al., Partial spatial coherence effects in digital holographic microscopy with a laser source, *Appl. Opt.*, **43**(5), 1131–1139 (2004).

31. Y. Park et al., Speckle-field digital holographic microscopy, *Opt. Express*, **17**(15), 12285–12292 (2009).

32. F. Dubois, P. Grosfils, Dark-field digital holographic microscopy to investigate objects that are nanosized or smaller than the optical resolution, *Opt. Lett.*, **33**(22), 2605–2607 (2008).

33. F. Verpillat et al., Dark-field digital holographic microscopy for 3D-tracking of gold nanoparticles, *Opt. Express*, **19**(27), 26044–26055 (2011).

34. M. K. Kim, Principles and techniques of digital holographic microscopy, *SPIE Rev.*, **1**(1), 018005 (2010).

35. P. Marquet, B. Rappaz, P. J. Magistretti, E. Cuche, Y. Emery, T. Colomb, C. Depeursinge, Digital holographic microscopy: A noninvasive contrast imaging technique allowing quantitative visualization of living cells with subwavelength axial accuracy, *Opt. Lett.*, **30**, 468–470 (2005).

36. J. Garcia-Sucerquia, W. Xu, S. K. Jericho, P. Klages, M. H. Jericho, H. J. Kreuzer, Digital in-line holographic microscopy, *Appl. Opt.*, **45**, 836–850 (2006).

37. T. Colomb, F. Montfort, J. Kühn, N. Aspert, E. Cuche, A. Marian, F. Charrière, S. Bourquin, P. Marquet, C. Depeursinge, Numerical parametric lens for shifting, magnification, and complete aberration compensation in digital holographic microscopy, *J. Opt. Soc. Am. A*, **23**, 3177–3190 (2006).

38. A. Faridian, G. Pedrini, W. Osten, High-contrast multilayer imaging of biological organisms through dark-field digital refocusing, *J. Biomed. Opt.*, **18**(8), 086009 (2013).

39. G. Pedrini, F. Zhang, W. Osten, Digital holographic microscopy in the deep (193 nm) ultraviolet, *Appl. Opt.*, **46**, 7829–7835 (2007).

40. A. Faridian et al., Nanoscale imaging using deep ultraviolet digital holographic microscopy, *Opt. Express*, **18**(13), 14159–14164 (2010).

41. U. Schnars, Direct phase determination in hologram interferometry with use of digitally recorded holograms, *J. Opt. Soc. Am. A*, **11**, 2011–2015 (1994).

42. G. Pedrini, Y. L. Zou, H. J. Tiziani, Digital double pulse-holographic interferometry for vibration analysis, *J. Mod. Opt.*, **42**, 367–374 (1995).

43. U. Schnars, Th. Kreis, W. P. Jueptner, Digital recording and numerical reconstruction of holograms: reduction of the spatial frequency spectrum, *Opt. Eng.*, **35**(4), 977–982 (1996).

44. G. Pedrini, Ph. Froening, H. Fessler, H. J. Tiziani, In-line digital holographic interferometry, *Appl. Opt.*, **37**, 6262–6269 (1998).

45. G. Pedrini, S. Schedin, H. Tiziani, Lensless digital-holographic interferometry for the measurement of large objects, *Opt. Comm.*, **171**, 29–36 (1999).

46. G. Pedrini, H. J. Tiziani, Digital holographic interferometry, in: *Digital Speckle Pattern Interferometry and Related Techniques*, P. K. Rastogi, ed., Wiley, Chichester (2001), pp. 337–362.

47. L. Yaroslavsky, *Digital Holography and Digital Image Processing: Principles, Methods, Algorithms*, Kluwer Academic Publishers, Boston (2004).

48. U. Schnars, W. Jueptner, *Digital Holography: Digital Hologram Recording, Numerical Reconstruction, and Related Techniques*, Springer, Berlin, Germany (2005).

49. Th. Kreis, *Handbook of Holographic Interferometry (Optical and Digital Methods)*, Wiley-VCH, Weinheim, Germany (2005).

50. Th. Kreis, *Holographic Interferometry, Principles and Methods*, Akademie Verlag, Berlin, Germany (1996).

51. W. Goodman, *Introduction to Fourier Optics*, McGraw-Hill, New York (1996).

52. F. Zhang, I. Yamaguchi, L. P. Yaroslavsky, Algorithm for reconstruction of digital holograms with adjustable magnification, *Opt. Lett.*, **29**, 1668–1670 (2004).

53. I. Yamaguchi, T. Zhang, Phase-shifting digital holography, *Opt. Lett.*, **22**, 1268–1270 (1997).

54. W. Osten, N. Reingand, *Optical Imaging and Metrology*, Wiley-VCH, Berlin (2010).

55. T. Dresel, G. Häusler, H. Venzke, Three-dimensional sensing of rough surfaces by coherence radar, *Appl. Opt.*, **31**, 919–925 (1992).

56. B. Nilsson, T. E. Carlsson, Direct three-dimensional shape measurement by digital-in-flight holography, *Appl. Opt.*, **37**, 7954–7959 (1998).

57. J. Schmitt, Optical coherence tomography (OCT): A review, *IEEE J. Sel. Top. Quantum Electron.*, **5**, 1205–1215 (1999).

58. G. Indebetouw, P. Klysubun, Optical sectioning with low coherence spatio-temporal holography, *Opt. Commun.*, **172**, 25–29 (1999).

59. E. Cuche, P. Poscio, C. Depeursinge, Optical tomography by means of a numerical low-coherence holographic technique, *J. Opt.*, **28**, 260–264 (1997).

60. L. Martínez-León, G. Pedrini, W. Osten, Applications of short-coherence digital holography in microscopy, *Appl. Opt.*, **44**, 3977–3984 (2005).

61. K. Körner, G. Pedrini, I. Alexeenko, T. Steinmetz, R. Holzwarth, W. Osten, Short temporal coherence digital holography with a femtosecond frequency comb laser for multi-level optical sectioning, *Opt. Express*, **20**(7), 7237–7242 (2012).

62. R. L. Powell, K. A. Stetson, Interferometric vibration analysis by wavefront reconstructions, *J. Opt. Soc. Am.*, **55**, 1593–1598 (1965).

63. K. A. Stetson, R. L. Powell, Interferometric hologram evaluation and real-time vibration analysis of diffuse objects, *J. Opt. Soc. Am.*, **55**, 1694–1695 (1965).

64. C. M. Vest, *Holographic Interferometry*, John Wiley & Sons, New York (1979).

65. W. Schumann, M. Dubas, *Holographic Interferometry*, Springer Verlag, Heidelberg, Germany (1979).

66. R. Jones, C. Wykes, *Holographic and Speckle Interferometry*, 2nd edn., Cambridge University, Cambridge, U.K. (1989).

67. J. N. Butters, J. A. Leendertz. Holographic and video techniques applied to engineering measurement, *J. Meas. Control*, **4**, 349–354 (1971).

68. A. Macovski, D. Ramsey, L. F. Schaefer, Time lapse interferometry and contouring using television systems, *Appl. Opt.*, **10**, 2722–2727 (1971).

69. K. Creath, Phase-shifting speckle interferometry, *Appl. Opt.*, **24**, 3053 (1985).

70. G. Pedrini, B. Pfister, H. J. Tiziani, Double pulse electronic speckle interferometry, *J. Mod. Opt.*, **40**, 89–96 (1993).
71. G. Pedrini, H. J. Tiziani, Double-pulse electronic speckle interferometry for vibration analysis, *Appl. Opt.*, **33**, 7857 (1994).
72. U. Schnars, W. P. O. Juptner, Digital recording and reconstruction of holograms in hologram interferometry and shearography, *Appl. Opt.*, **33**, 4373 (1994).
73. C. Wagner, S. Seebacher, W. Osten, W. Jüpner, Digital recording and numerical reconstruction of lensless Fourier holograms in optical metrology, *Appl. Opt.*, 38, 4812–4820 (1999).
74. S. Sebacher, W. Osten, W. Jueptner, Measuring shape and deformation of small objects using digital holography, *Proc. SPIE*, **3479**, 104–115 (1998).
75. S. Seebacher, W. Osten, T. Baumbach, W. Jüptner, The determination of material parameters of microcomponents using digital holography, *Opt. Laser Eng.*, **36**, 103–126 (2001).
76. G. Pedrini, H. Tiziani, Y. Zou, Digital double pulse-TV-holography, *Opt. Laser Eng.*, 26, 199–219 (1997).
77. G. Pedrini, Y. Zou, H. J. Tiziani, Simultaneous quantitative evaluation of in-plane and out-of-plane deformations using multi directional spatial carrier, *Appl. Opt.*, **36**, 786–792 (1997).
78. G. Pedrini, H. J. Tiziani, Quantitative evaluation of two dimensional dynamic deformations using digital holography, *Opt. Laser Tech.*, **29**, 249–256 (1997).
79. S. Schedin, G. Pedrini, H. J. Tiziani, F. M. Santoyo, Simultaneous three-dimensional dynamic deformation measurements with pulsed digital holography, *Appl. Opt.*, **38**, 7056–7062 (1999).
80. C. Perez-Lopez, F. Mendoza Santoyo, G. Pedrini, S. Schedin, H. J. Tiziani, Pulsed digital holographic interferometry for dynamic measurement of rotating objects with an optical derotator, *Appl. Opt.*, **40**, 5106–5110 (2001).
81. H. J. Tiziani, N. Kerwien, G. Pedrini, Interferometry, in: *Landolt-Börnstein New Series*, Laser Physics and Applications, Springer-Verlag, Berlin, Germany (2006), pp. 221–284.
82. S. Schedin, G. Pedrini, H. J. Tiziani, A. K. Aggarwal, M. E. Gusev, Highly sensitive pulsed digital holography for built-in defect analysis with a laser excitation, *Appl. Opt.*, **40,** 100–103 (2001).
83. G. Pedrini, Ph. Fröning, H. J. Tiziani, M. E. Gusev, Pulsed digital holography for high speed contouring employing the two-wavelength method, *Appl. Opt.*, **38**, 3460–3467 (1999).
84. M. Adams, Th. Kreis, W. Jueptner, Particle size and position measurements with digital holography, in: *Optical Inspection and Micromeasurements*, Ch. Gorecki, ed., *Proceedings of SPIE*, The International Society for Optical Engineering, Vol. 3098 (1997), pp. 234–240.
85. P. Gren, S. Schedin, X. Li, Tomographic reconstruction of transient acoustic fields recorded by pulsed TV holography, *Appl. Opt.*, **37**, 834–840 (1998).
86. G. Pedrini, I. Alexeenko, W. Osten, H. Tiziani, Temporal phase unwrapping of digital hologram sequences, *Appl. Opt.*, **42**, 5846–5854 (2003).
87. G. Pedrini, W. Osten, M. E. Gusev, High-speed digital holographic interferometry for vibration measurement, *Appl. Opt.*, **45**, 3456–3462 (2006).
88. G. Pedrini, W. Osten, Time resolved digital holographic interferometry for investigations of dynamical events in mechanical components and biological tissues, *Strain*, **43**(3), 240–249 (2007).
89. Y. Fu, G. Pedrini, W. Osten, Vibration measurement by temporal Fourier analyses of a digital hologram sequence, *Appl. Opt.*, **46**, 5719–5727 (2007).
90. Y. Fu, R. M. Groves, G. Pedrini, W. Osten, Kinematic and deformation parameter measurement by spatiotemporal analysis of an interferogram sequence, *Appl. Opt.*, **46**, 8645–8655 (2007).
91. J. Korvink, O. Paul, eds., *MEMS: A Practical Guide to Design, Analysis and Applications*, William Andrew Publishing, Norwich, NY (2006).
92. W. Osten, ed., *Optical Inspection of Microsystems*, CRC Press, Boca Raton, FL (2006).
93. W. Osten, ed., Optical microsystems metrology—Part I, *Opt. Lasers Eng.*, **36**, 75 (2001).
94. W. Osten, ed., Optical microsystems metrology—Part II, *Opt. Lasers Eng.*, **36**, 401 (2001).
95. G. Coppola, P. Ferraro, M. Iodice, S. De Nicola, A. Finizio, S. Grilli, A digital holographic microscope for complete characterization of microelectromechanical systems, *Meas. Sci. Technol.*, **15**, 529–539 (2004).
96. G. Pedrini, J. Gaspar, M. Schmidt, I. Alekseenko, O. Paul, W. Osten, Measurement of nano/micro out-of-plane and in-plane displacements of micromechanical components by using digital holography and speckle interferometry, *Opt. Eng.*, **50**, 101504-1–101504-10 (2011).

8 Speckle Methods and Applications

Nandigana Krishna Mohan

CONTENTS

8.1 Introduction to Speckle Phenomenon 249
8.2 Speckle Photography 250
8.3 Speckle Interferometry 251
    8.3.1 Out-of-Plane Displacement Measurement 252
    8.3.2 In-Plane Displacement Measurement 253
8.4 Speckle Shear Interferometry 255
    8.4.1 Slope Measurement 255
    8.4.2 Curvature Measurement 256
8.5 Digital Speckle and Speckle Shear Interferometry 257
    8.5.1 Fringe Analysis 257
    8.5.2 Optical Systems in Digital Speckle and Speckle Shear Interferometry 262
    8.5.3 Application to Microsystems Analysis 264
8.6 Conclusions 266
References 266

8.1 INTRODUCTION TO SPECKLE PHENOMENON

The speckle phenomenon came into prominence after the advent of laser. When an optically rough surface is illuminated with a coherent beam, a high-contrast granular structure, known as speckle pattern, is formed in space. Speckle pattern is a three-dimensional (3D) interference pattern formed by the interference of secondary, dephased wavelets scattered by an optical rough surface or transmitted through a scattered medium that imposes random phases of the wave. The speckle pattern formed in the space is known as the objective speckle pattern. It can also be observed at the image plane of a lens and it is then referred as subjective speckle pattern. It is assumed that the scattering regions are statistically independent and uniformly distributed between $-\pi$ and π. The pioneering contributions of Leendertz [1] and Archbold et al. [2] in 1970 demonstrating that the speckles in the pattern undergo both positional and intensity changes when the object is deformed, triggered the origin of speckle work. The randomly coded pattern that carries the information about the object deformation provided to develop a wide range of methods, which can be classified into three broad categories: speckle photography, speckle interferometry, and speckle shear interferometry. Speckle photography includes all those techniques where positional changes of the speckle are monitored, whereas speckle interferometry includes methods that are based on the measurement of phase changes and hence intensity changes. If instead of phase change, we measure its gradient, the technique falls into the category of speckle shear interferometry. All these methods can be performed using digital/electronic detection using a charge-coupled device (CCD) and image processing system. The attraction to speckle methods relies their applicability for optical metrology: (1) static and dynamic deformations and their gradient measurements, (2) shape of the 3D objects, (3) surface roughness, and (4) nondestructive evaluation (NDE) of engineering and biomedical specimens.

The range of the objects that can be evaluated using the speckle methods is of the order of few hundred microns such as micro-electro-mechanical systems (MEMS) to few meters such as space craft engineering structures. The prominent advantages of the speckle methods with the real-time operation systems (photorefractive crystals, CCD, and image processing systems) are (1) noncontact measurement, (2) real-time whole field operation, (3) data storage and retrieval for analysis, (4) quantitative evaluation using phase shifting, (5) variable measuring sensitivity, and also (6) possibility to reach remote areas.

8.2 SPECKLE PHOTOGRAPHY

Speckle photography is based on recording of objective or subjective speckle patterns, before and after the object is subjected to load. This is one of the simplest method in which a laser-lit diffuse surface is recorded at two different states of stress [1,2]. Speckle photography permits one to measure in-plane displacements and derivatives of deformation. The speckle photography is exploited for metrology by using either a photographic plate or a CCD detector as the recording medium.

An imaging geometry arrangement for recording speckle pattern on a photographic plate is shown in Figure 8.1. An expanded laser beam illuminates the object that scatters light, and two exposures are made on a single photographic plate placed at the image plane, and the object is given an in-plane displacement between the exposures. The recorded and processed photographic plate consists of speckled image but mutually translated due to displacement. These two traditionally registered speckle patterns are identical, except that one is displaced with respect to the other by an amount that depends on the magnitude of the displacement of the object and imaging geometry. The recorded speckle patterns after development can be analyzed by means of two classical filtering methods [3,4]. The first method, a point-by-point approach, consists of interrogating the speckle pattern by means of a narrow laser beam to generate a diffraction halo in the far field at each data point. The halo is modulated by a system of Young's fringes. The fringe spacing and orientation are governed by the displacement vector. The fringe orientation is always perpendicular to the direction of displacement, while the fringe spacing is inversely proportional to the magnitude of displacement. The Young's fringes are observed when the two identical speckles are separated by a distance equal to or more than the average speckle size. The second method based on spatial filtering or Fourier filtering of the speckle pattern provides a direct whole field visualization of the distribution of interest. This approach also offers a degree of latitude in sensitivity selection during the reconstruction process. In short, point-wise filtering yields absolute value of displacement, while the Fourier filtering gives fringes representing incremental change. The advantages of speckle photography such as its relatively lower sensitivity to external vibrations and a reasonably low initial cost of developing a speckle photography-based system advocate strongly in favor of a widespread

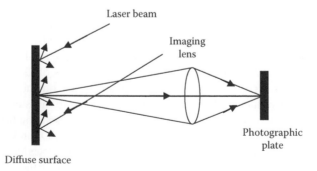

FIGURE 8.1 Schematic for recording a subjective speckle pattern on a photographic plate using an imaging lens.

use of the method in nonlaboratory environment. On the other hand, some important drawbacks that have prevented the widespread use of speckle photography reside in its inherent requirement of a two-step process: exposing two speckle patterns pertaining to initial and displaced states of the object, respectively, on a photographic plate, and adopting the filtering methods to extract the information. The disadvantage related with the photographic plate has been addressed by the use of a CCD detector for recording the speckle pattern and quantitative analysis by means of digital processing (digital speckle photography [DSP]).

DSP, also called as speckle correlation (SC), electronic speckle photography (ESP), or digital image correlation (DIC) is the digital/electronic version of speckle photography for the evaluation of displacement fields [5–11]. A basic DSP system has relatively low complexity in the hardware as the photographic plate is replaced with a CCD camera. The registration of the speckle pattern is made by video/digital recording of the object and the evaluation process is performed on a PC. The analysis is much faster as no film processing or reconstruction (filtering) is needed. The method is based on a cross-correlation analysis of two speckle patterns, one stored before and another after the object is deformed. The method does not follow the movement of each speckle, but rather the movement of a number of speckles acting together as a subimage.

Photorefractive crystals with their unique properties of higher resolution, low intensity operation, and the capability of real-time response have been used to record and analyze the SC fringes [12–16]. The basic concept here is that the two beams interfere inside the photorefractive material and produce a dynamic grating. At the observation plane, one can have two speckle fields: (1) the first one constitutes the initial state of the object, that is, due to the diffraction of the pump beam from the grating formed at the crystal plane, and (2) the second scattered beam that represents the directly transmitted object beam that constitutes the deformed state of the object. The overlap of these two identical but displaced speckles give rise to the SC fringe pattern in real time.

8.3 SPECKLE INTERFEROMETRY

Speckle interferometry is a powerful measurement method involving coherent addition (interference) of the scattered wave from the object surface with a reference wave that may be a specular or scattered field [1]. Unlike speckle photography, this technique has a sensitivity comparable to that of holographic interferometry. The phase changes within one speckle due to the object displacement coded by the reference wave. This displacement is extracted by correlating two speckle patterns one taken before and another after object displacement. Figure 8.2 shows a schematic

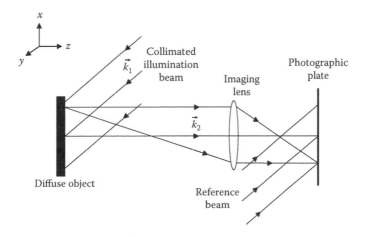

FIGURE 8.2 Schematic of a speckle interferometer.

of a speckle interferometer. Here the speckled image of the object is superposed with the reference beam at the image plane of the imaging lens and they add coherently to form a resultant speckle pattern.

The irradiance of a speckle pattern at any point in the image plane, before object deformation can be written as

$$I_B = I_O + I_R + 2\sqrt{I_O I_R} \cos \phi \tag{8.1}$$

where

I_O and I_R are the irradiances of the object and reference waves, respectively, at the image plane
ϕ is the random phase difference between them

The irradiance at the same point after deforming the object can be written as

$$I_A = I_O + I_R + 2\sqrt{I_O I_R} \cos(\phi + \Delta\phi) \tag{8.2}$$

where $\Delta\phi$ is the additional phase change introduced due to the object deformation. The total intensity recorded by the double exposure is given by

$$I = I_A + I_B = 2(I_O + I_R) + 4\sqrt{I_O I_R} \cos\left(\phi + \frac{\phi}{2}\right)\cos\left(\frac{\Delta\phi}{2}\right) \tag{8.3}$$

The speckle pattern would be unchanged if $\Delta\phi = 2n\pi$ and would be totally de-correlated when $\Delta\phi = (2n + 1)\pi$. The correlated areas appear as bright fringes and uncorrelated areas appear as dark fringes on filtering. The phase change because of object deformation can be expressed as

$$\Delta\phi = (\vec{k}_2 - \vec{k}_1) \cdot \vec{L} \tag{8.4}$$

where $\vec{k}_2$ and $\vec{k}_1$ are the propagation vectors in the directions of illumination and observation, respectively, as shown in Figure 8.2. $\vec{L}(u, v, w)$ is the displacement vector, u, v, and w being the displacement components along x, y, and z directions, respectively. Since the fringes are loci of constant phase difference, the deformation vector components can be measured using appropriate optical configurations.

If one of the surfaces is moved along the line of sight, the speckle pattern undergoes an irradiance change. However, if the surface is displaced through a distance of $n\lambda/2$, where n is an integer, λ is the wavelength of the light used, the resultant speckle pattern undergoes a phase change of $2n\pi$ and returns to the original texture. The resultant speckle pattern in the original and the displaced states of the surfaces are correlated. For very large displacements, the correlation property fails to exist, as the whole speckle structure changes.

8.3.1 OUT-OF-PLANE DISPLACEMENT MEASUREMENT

If the displacement varies across the surface of the object, the resultant speckle pattern will remain correlated in the areas where the displacement along the line of sight is an integral multiple of $\lambda/2$ (which corresponds to a net change of λ or a phase change of 2π) and they will be uncorrelated in other areas. It is possible to detect these positions of correlation and the interferometer can be used for measuring the change in phase (or the surface movement). A straightforward way to doing this is to adopt the photographic mask method. The resultant speckle pattern is recorded on a photographic plate, and the processed plate is replaced exactly in its original position. Since the processed plate is a negative of the recorded speckle pattern, the dark areas of the plate correspond to bright speckles and vice versa. The overall light transmitted by the plate will thus be very small. If the surface is deformed such that the displacement varies across the surface, the phase change also varies across the surface. The mask will transmits light in those parts of the image where the speckle pattern is not correlated with it, but does not transmit light in correlated area. This gives rise to an apparent fringe pattern covering the image. The fringes are termed as correlation fringes, which are obtained by the addition of two or more speckle patterns. However, we often use subtraction correlation fringes in electronic/digital speckle pattern interferometry (ESPI/DSPI). The phase change can be related to out-of-plane displacement component as Ref. [1]

$$\Delta\phi = \left(\frac{4\pi}{\lambda}\right)w \tag{8.5}$$

A comparative speckle interferometric configuration in which two objects (master and test objects) are compared with each other for nondestructive testing is also reported in literature [17,18].

8.3.2 IN-PLANE DISPLACEMENT MEASUREMENT

Leendertz [1] proposed a novel method for in-plane measurements that consists of symmetrically illuminating the diffuse object under test by two collimated beams at an angle θ on either side of the surface normal as shown in Figure 8.3. If this surface is imaged on to a photographic plate, the image will consist of two independent, coherent speckle patterns due to each beam.

If we assume the illuminating beams to lie symmetrically with respect to normal in the x–z plane, following Equation 8.4 one can write the phase difference as [1]

$$\Delta\phi = \left(\frac{2\pi}{\lambda}\right)2u\sin\theta \tag{8.6}$$

where u is the x-component of the displacement. The arrangement is intrinsically insensitive to the out-of-plane displacement component; the y component (v) is not sensed because of the illumination beams lying in the x–z plane. Physically, the out-of-plane displacement of the object introduces identical path changes in both of the observation beams, and the v component of in-plane

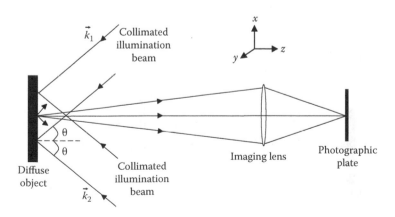

FIGURE 8.3 Leendertz's arrangement for in-plane displacement measurement.

displacement does not introduce any path change in either of the observation beams. Bright fringes are formed when

$$2u \sin \theta = m\lambda \qquad (8.7)$$

where m is an integer.

The fringe spacing corresponds to an incremental displacement of $\lambda/2 \sin \theta$. The sensitivity of the setup can be varied by changing the angle between the illuminating beams. The orthogonal component of in-plane displacement (v) can be obtained by rotating either the object or the illuminating beams by 90° about the normal to the object. Further, two pairs of symmetrical beams orthogonal to each other can be used to obtain the u and v components simultaneously. An optical configuration for measuring the in-plane displacement component with twofold sensitivity is also proposed [19,20]. In this arrangement, the object is illuminated symmetrically with two beams, and the observation is made along the same directions of illumination, and combined them at the image plane of the imaging system to yield twofold measuring sensitivity.

A method for measuring in-plane displacement has been described by Duffy [21,22]. It consists of imaging the object surface through a lens fitted with a double aperture located symmetrically with respect to the optical axis as shown in Figure 8.4. The lens focused on the surface sees an arbitrary point on the object surface along two different directions. As a result of the coherent superposition of the beams emerging from the two apertures, each speckle is modulated by a grid structure running perpendicular to the line joining the apertures. The pitch of the fringe system embedded in each speckle depends on the separation of the two apertures. During filtering, the grid structure diffracts light in well-defined directions. Because there is considerably more light available in these halos, there is an enormous gain in the signal-to-noise ratio, resulting in almost unit contrast fringes [23].

The in-plane displacement (u) can be expressed in terms of optical geometry as

$$\Delta\phi = \left(\frac{2\pi}{\lambda}\right) 2u \sin \alpha \qquad (8.8)$$

Equation 8.8 is similar to Equation 8.6 except that θ is replaced by α. The angle α is governed by the aperture of the lens, and hence the sensitivity of Duffy's arrangement is inherently poor. Khetan and Chang [23] extended this technique by using four apertures, two along x-axis and two along the y-axis. A modified arrangement for measuring in-plane displacement equal to Leendertz's dual beam illumination method [1] is also reported [24]. Novelty of the multiaperture

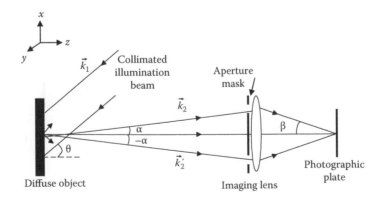

FIGURE 8.4 Duffy's arrangement for in-plane displacement measurement.

methods is in multiplexing using theta modulation or frequency modulation [25]. All these interferometers have been extensively used with electronic/digital speckle interferometry for various
applications of optical metrology.

8.4   SPECKLE SHEAR INTERFEROMETRY

Speckle shear interferometry or shearography is also a full-field, noncontact optical technique,
usually used for the measurement of changes in the displacement gradient caused by surface and
subsurface defects [26–28]. It is an optical differentiation technique where the speckle pattern is
optically mixed with an identical but displaced, or sheared, speckle pattern using a shearing device,
and viewed through a lens, forming a speckle shear interferogram at the image plane. It is analogous to the conventional shear interferometry where two sheared wavefronts are made to interfere
in order to measure the rate of change of phase across the wavefront. A double exposure record is
made with the object displaced between the exposures. The correlation fringes formed on filtering
provide displacement gradient in the shear direction. This is a more useful data for experimental
stress analysis as this quantity is proportional to the strain.

8.4.1   SLOPE MEASUREMENT

Speckle shear interferometry is useful for measurement of first-order derivative of the out-of-plane
displacement or the slope. Figure 8.5 shows the optical configuration of the speckle shear interferometer for slope measurement. It is a Michelson interferometer, where a collimated laser beam
is used for object illumination. A lateral shear is introduced between the images by tilting the
mirror M_2. Assuming a shear Δx in the x-direction, any point in the image plane receives contributions from the neighboring object points (x, y) and $(x + \Delta x, y)$. The phase change due to object
deformation can be written as

$$\Delta\phi = \Delta\phi_2 - \Delta\phi_1 \tag{8.9}$$

where $\Delta\phi_1$ and $\Delta\phi_2$ are the phase changes occurring due to object deformation at the two sheared
object points.

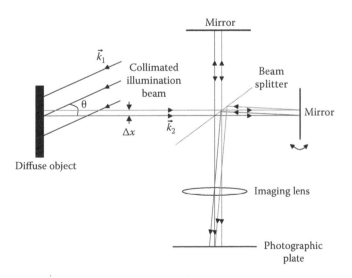

FIGURE 8.5   Speckle shear interferometer for slope measurement.

The phase changes $\Delta\phi_1$ and $\Delta\phi_2$ are defined by

$$\Delta\phi_1 = \left(\vec{k}_2 - \vec{k}_1\right) \cdot \vec{L} \tag{8.10}$$

$$\Delta\phi_2 = \left(\vec{k}_2 - \vec{k}_1\right) \cdot \vec{L}' \tag{8.11}$$

where
  $\vec{L} = \vec{L}(u,v,w)$ and $\vec{L}' = \vec{L}(u+\Delta u, v+\Delta v, w+\Delta w)$ are the deformation vectors at the points (x, y)
    and $(x + \Delta x, y)$, respectively
  $\vec{k}_1$ and $\vec{k}_2$ are the wave vectors in the illumination and observation directions, respectively

If the illumination angle is θ, then the phase change $\Delta\phi$ can be written as

$$\Delta\phi = \frac{2\pi}{\lambda}\left[\frac{\partial u}{\partial x}\sin\theta + \frac{\partial w}{\partial x}(1+\cos\theta)\right]\Delta x \tag{8.12}$$

For normal illumination, $\theta = 0$, the phase can be related to first-order derivative of out-of-plane displacement component (w) as

$$\Delta\phi = \frac{4\pi}{\lambda}\frac{\partial w}{\partial x}\Delta x \tag{8.13}$$

Thus the fringes formed are contours of $\partial w/\partial x$. Similarly by introducing a shear along y-direction, $\partial w/\partial y$ fringes can be obtained.

Multiaperture speckle shear interferometry is also used for slope measurement [28]. The main advantages of image plane shear using multiaperture arrangements are (1) large depth of focus/field, (2) the interferograms of unit contrast, and (3) its ability to store and retrieve multiple information pertaining to partial derivatives of an object deformation from a single specklegram. Further, the setups are inexpensive, simple, and require little vibration isolation. The limitations of these interferometers are that they are intrinsically sensitive to both the in-plane displacement component and its derivatives. The fringe pattern thus contains information of (1) in-plane displacement, (2) its derivative, and (3) the derivative of the out-of-plane displacement component (slope) [28–30]. Apart from lateral shear, radial, rotational, inversion, and folding shear have also been introduced in speckle shear interferometry [28].

8.4.2 Curvature Measurement

For plate bending problems and flexural analysis, the second-order derivatives of the out-of-plane displacements are required for calculating the stress components. Hence it is necessary to perform differentiation of the slope data to obtain curvature information. Multiple aperture configurations [31,32] can yield curvature information by proper filtering of the halos. The configuration is shown in Figure 8.6, where a plate with three holes drilled in a line is used as a mask in front of the imaging lens. Two identical small angled wedges are placed in front of the two outer apertures, while a suitable flat glass plate is placed in front of the central aperture to compensate for the optical path. Each point on the image plane gets contribution from three neighboring points on the object. Thus, three different processes of coherent speckle addition take place simultaneously in the interferometer. As a result, the speckle pattern is modulated by three systems of grid structures running perpendicular to the lines joining the apertures. Two exposures are made of the object illuminated with laser light, once before and once after loading. When subjected to Fourier filtering, the double exposed speckle shearogram results in five diffraction halos in the Fourier plane. By filtering the first-order halos one can get the curvature fringes as Moiré pattern [31]. The information along any other direction can be obtained by orienting the wedge device along that particular direction.

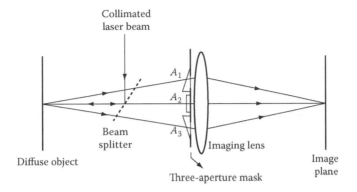

FIGURE 8.6   Three-aperture speckle shear interferometer for curvature measurement.

The holographic optical elements (HOEs) are also extensively used in speckle interferometry and speckle shear interferometry configurations for the measurement of displacement and its derivatives [33–35].

8.5   DIGITAL SPECKLE AND SPECKLE SHEAR INTERFEROMETRY

The average speckle size is typically in the range 5–100 μm. A standard television (TV) camera can therefore be used to record the intensity pattern instead of a photographic plate. The video processing of the signal may be used to generate SC fringes equivalent to those obtained by photographic processing. The suggestion of replacing the photographic plate with a detector and an image processing unit for real-time visualization of correlation speckle fringes was first reported by Leendertz and Butters [36]. This pioneering contribution is today referred in literature as ESPI, DSPI, phase-shifting speckle interferometry (PSSI), or TV holography (TVH) [37–40]. Starting from these basic ideas, contributions on various refinements, extensions, and novel developments have been proposed. They have continued to pour in over the years from researchers across the globe, and have contributed to the gradual evolution of the concept into a routinely used reliable and versatile technique for quantitative measuring displacement components of a deformation vector of diffusely scattering object under static or dynamic load. The ability to measure 3D surface profiles by generating contours of constant depth of an object is yet another major achievement of DSPI. The last three decades have seen a significant increase in the breadth and scope of applications of DSPI in areas such as experimental mechanics, vibration analysis, nondestructive testing, and evaluation. Digital shearography (DS) or digital speckle shear pattern interferometry (DSSPI) is an outgrowth of speckle shear interferometry in concept, and involves optical and electronic principle akin to those used in DSPI [38]. It has been widely used for measuring the derivatives of surface displacements of rough objects undergoing deformation. The method has found a prominent use for nondestructive testing of structures. Digital/electronic speckle and speckle shear configurations have been combined with phase-shifting interferometry (PSI) [41] for a wide range of quantitative data analysis of engineering structures.

8.5.1   FRINGE ANALYSIS

Figure 8.7 represents widely used in-line DSPI arrangement for the measurement of out-of-plane displacement. The beam from a laser source is split into an object beam and a reference beam. The scattered object beam is combined with a reference beam using a beam splitter (BS) at the CCD plane. The complex amplitudes of the object, A_O and the reference, A_R beams at any point (x, y) on the CCD array can be expressed as

$$A_O(x, y) = a_O(x, y) \exp[i\phi_O(x, y)] \qquad (8.14)$$

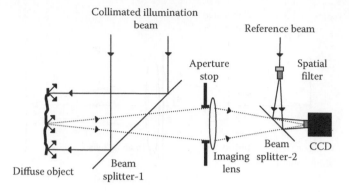

FIGURE 8.7 DSPI arrangement sensitive to out-of-plane displacement.

and

$$A_R(x,y) = a_R(x,y)\exp[i\phi_R(x,y)] \tag{8.15}$$

where
    $a_O(x, y)$ is the amplitude
    $\phi_O(x, y)$ is the phase of the light scattered from the object in its initial state
    $a_R(x, y)$ is the amplitude
    $\phi_R(x, y)$ is the phase of the reference beam

The irradiance at any point (x, y) on the CCD array can be expressed as

$$\begin{aligned}
I_B(x,y) &= |A_O(x,y) + A_R(x,y)|^2 \\
&= |A_O(x,y)|^2 + |A_R(x,y)|^2 + 2|A_O(x,y)||A_R(x,y)|\cos[\phi_B(x,y)] \\
&= I_O(x,y) + I_R(x,y) + 2\sqrt{I_O(x,y)I_R(x,y)}\,\cos[\phi_B(x,y)]
\end{aligned} \tag{8.16}$$

where
    $\phi_B(x, y) = \phi_O(x, y) - \phi_R(x, y)$ represents the random speckle phase
    $I_O(x, y)$ and $I_R(x, y)$ are the irradiances of the object and reference beams, respectively

Assuming the output camera signal to be proportional to the irradiance, the video signal is then

$$v_B \propto I_B(x,y) \tag{8.17}$$

The analog video signal v_B from the CCD is sent to an analog-to-digital converter, which samples the video signal at the TV rate and records it as a digital frame in the memory of a computer for subsequent processing.

If the object is deformed, the relative phase of the two fields will change, thus causing a variation in the irradiance of the speckle pattern. The deformation gradients applied on the object surface are assumed to be small. Following the application of the load, the complex amplitude of the object wavefront can be expressed as

$$A'_O(x,y) = a_O(x,y)\exp\left\{i\left[\phi_O(x,y) + \Delta\phi(x,y)\right]\right\} \tag{8.18}$$

where $\Delta\phi(x, y)$ is the phase difference introduced in the original object beam due to object deformation.

The irradiance distribution after object deformation can be written as

$$I_A(x,y) = I_O(x,y) + I_R(x,y) + 2\sqrt{I_O(x,y)I_R(x,y)}\cos[\phi_A(x,y)] \tag{8.19}$$

where $\phi_A(x, y) = \phi_B(x, y) + \Delta\phi(x, y) - \phi_R(x, y)$.

A common method to produce SC fringes is based on direct electronic subtraction of the intensities in the initial and deformed states of the object. The video signal ν_A is digitized and then subtracted from the digital frame corresponding to ν_B.

The subtracted video signal ν is then given by

$$\nu = (\nu_A - \nu_B) \propto (I_A - I_B) = \Delta I = 4\sqrt{I_O I_R}\sin\left(\phi_B + \frac{\Delta\phi}{2}\right)\sin\left(\frac{\Delta\phi}{2}\right) \tag{8.20}$$

This signal has both positive and negative values. The TV monitor, however, displays the negative going signals as black areas. To avoid this loss of signal, it is rectified before being displayed on the monitor. The brightness on the monitor is then proportional to $|\Delta I|$ and can be expressed as

$$B = C|\Delta I| = C\left|4\sqrt{I_O I_R}\sin\left(\phi_B + \frac{\Delta\phi}{2}\right)\sin\left(\frac{\Delta\phi}{2}\right)\right| \tag{8.21}$$

where C is a proportionality constant. The term $(\phi_B + \Delta\phi/2)$ represents the speckle noise which varies randomly between 0 and 1. Equation 8.21 describes the modulation of the high-frequency noise by a low-frequency interference pattern related to the phase-difference term $\Delta\phi$.

The brightness will be minimized when

$$\Delta\phi = 2n\pi, \quad n = 0,1,2,\ldots \tag{8.22}$$

This condition corresponds to dark fringe and denotes all of those regions where the speckle patterns, both before and after deformation, are correlated. On the other hand, the brightness will be maximized when

$$\Delta\phi = (2n+1)\pi, \quad n = 0,1,2,\ldots \tag{8.23}$$

This condition corresponds to bright fringe and denotes all of those regions where the speckle patterns, both before and after deformation, are uncorrelated. As a result, SC fringes representing contours of constant displacement are seen to appear on the monitor in the form of bright and dark bands modulating the object image.

For in-plane displacement and slope measurements, the configurations shown in Figures 8.3 and 8.5 are widely adopted by replacing the photographic plate with a CCD. Figure 8.8a and b shows the real-time out-of-plane and slope correlation fringes from a rigidly clamped circular diaphragm loaded at the center.

The procedure described above provides the qualitative analysis of the fringes. It is, in general, impossible to obtain a unique phase distribution from a single interferogram. For example, substitution of $-\phi_A$ for ϕ_A in Equation 8.19 results in the same intensity map, indicating that positive displacement cannot be distinguished from negative displacements without further information. The near-universal technique of solving this problem is to add to the phase function a known phase ramp, or carrier, which is linear in either time or position. In the former case, the resulting intensity distributions are sampled at discrete time intervals and the technique is known as temporal phase

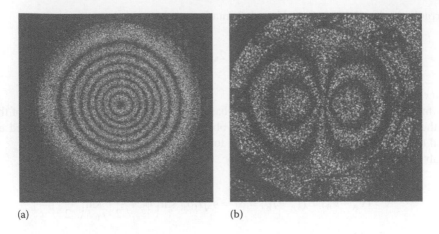

(a)   (b)

FIGURE 8.8 SC fringe patterns obtained for a rigidly clamped circular diaphragm loaded at the center from (a) DSPI arrangement and (b) DS arrangement.

shifting (TPS) [41]; in the latter case, a single intensity distribution is sampled at discrete points in the image, in which case it is referred to as spatial phase shifting (SPS) [42]. TPS technique involves introduction of an additional controlled phase change, ϕ_0, between the object and reference beams. Normally a linear ramp $\phi_0 = \alpha t$, is used, where α represents the phase shift introduced between two successive frames, and t is the time variant [43,44]. Equation 8.16 then becomes

$$I_B(x,y) = I_O(x,y) + I_R(x,y) + 2\sqrt{I_O(x,y)I_R(x,y)} \cos\left[\phi_B(x,y) + \alpha t\right] \qquad (8.24)$$

Equation 8.24 can also be written as

$$I_B(x,y) = I_b(x,y) + \gamma(x,y)\cos[\phi_B(x,y) + \alpha t] \qquad (8.25)$$

where
  $I_b = I_O + I_R$ is the bias intensity
  $\gamma = 2\sqrt{I_O I_R}$ is the visibility of the interference pattern

The temporal phase-shifting technique involves analyzing data from each pixel independently of all of the other pixels in the image, using different phase-shifting algorithms.

Most widely used algorithm for phase calculation is the four-step algorithm [38,45]. A $\pi/2$ phase shift is introduced into the reference beam between each of the sequentially recorded interferograms. So ϕ_0 now takes on four discrete values as $\phi_0 = 0, \pi/2, \pi, 3\pi/2$. Substituting each of these four values into Equation 8.25 results in four equations describing the four measured intensity patterns as

$$I_1 = I_b + \gamma \cos\phi_B \qquad (8.26)$$

$$I_2 = I_b + \gamma \sin\phi_B \qquad (8.27)$$

$$I_3 = I_b + \gamma \cos\phi_B \qquad (8.28)$$

$$I_4 = I_b + \gamma \sin\phi_B \qquad (8.29)$$

The phase ϕ thus can be calculated as

$$\phi_B = \arctan\left(\frac{I_2 - I_4}{-I_1 + I_3}\right) \tag{8.30}$$

A phase-shifting algorithm that is less sensitive to reference phase-shift calibration errors is provided by Hariharan et al. [46]. The phase ϕ_B can be obtained from

$$\phi_B = \arctan\left[\frac{2(I_2 - I_4)}{-I_1 + 2I_3 - I_5}\right] \tag{8.31}$$

The subtraction of the phase obtained before (ϕ_B) and after loading (ϕ_A) the object yields the phase map $\Delta\phi$ related to object deformation [37].

The most common method used to introduce the phase shift in a PSI is to translate one of the mirrors in the interferometer with a piezoelectric translator. Piezoelectric translator consists of lead–zirconium–titanate (PZT) or other ceramic materials, which expand or contract with an externally applied voltage. Shifts from submicrometer to several micrometers can be achieved using PZT. However, the miscalibration, nonlinearity, and hysteresis of PZTs are the error sources, which influence the accuracy of the phase measurements [47,48]. Polarization components such as rotating half-wave plate (HWP), quarter-wave plate (QWP), and polarizers are also used as phase shifters [49]. The other schemes available for introducing phase shifts are liquid crystal phase shifters, source wavelength modulation [37,50], air-filled quartz cell [51], etc.

Spatial phase-shifting technique provides a useful alternative to TPS when dynamic events are being investigated, or when vibration is a problem, because all the data needed to calculate a phase map are acquired at the same instant [42]. Since the phase shift then has to take place in space instead of time, this approach is quite generally called SPS. In principle, it gets possible to track the object phase at the frame rate of the camera, with the additional benefit that any frame of the series can be assigned the new reference image.

A simpler technique involves recording of just two images, once before and once after the object deformation, in which a spatial carrier is introduced within the speckle. The time variation in Equation 8.25 is then replaced by the following spatial variation:

$$I_B(x,y) = I_b(x,y) + \gamma(x,y)\cos[\phi_B(x,y) + \alpha x] \tag{8.32}$$

In the case of out-of-plane displacement, the simplest way of generating a carrier fringe within a speckle with a linear variation is to use off-axis reference wave with a small angular offset between the object and reference beams, unlike the conventional DSPI system [52,53]. The spatial carrier is apparent as grid-like structure, and the random variations in speckle to speckle manifest themselves as lateral shifts of the carrier. It is common to set the carrier fringe so that there is a 90° or 120° phase shift between the adjacent pixels of the CCD detector and to use three or more adjacent pixels to evaluate the phase of the interference. The alternative approach to generate the spatial carrier fringe is by adopting a double aperture mask in front of the imaging system [54,55] as shown in Figure 8.4.

The "phase-of-differences" method [37] represents a useful way of generating visible phase-shifted fringe patterns. There is an alternative method known as "difference-of-phases," where the phase ϕ evaluated from the phase shifted "before" load images is subtracted from the phase $\phi + \Delta\phi$ evaluated from the phase shifted "after" load images to yield the phase $\Delta\phi$ [37,38]. As the calculated phase values lie in the principle range $(-\pi, \pi)$, $\Delta\phi$ lies in the range $(-2\pi, 2\pi)$. This would normally be wrapped back on to the range [56]. The phase-change distribution $\Delta\phi$ due to object deformation obtained from Equation 8.30 or 8.31 yields only a sawtooth function showing the phase modulo 2π. These 2π phase jumps can be eliminated by using the phase unwrapping method [56,57] so that a continuous function of the relative phase change $\Delta\phi$, i.e., the unwrapped phase map, is generated. The phase unwrapping method can be performed only for a noise reduced sawtooth image. As a result of electronic and optical noise mainly due to speckle decorrelation effects, it is difficult to distinguish between real phase jumps, and jumps due to noise because they fulfill the same mathematical requirements. Consequently, there must be a relationship between phase edge preservation and noise removal in the design of the filter. Many filtering techniques have been proposed to condition the phase map before the application of the unwrapping procedure [56,57]. The known filter operations in the field of digital image processing can be broadly classified into two categories: those used for linear operations, i.e., low-pass filtering or average filtering, and for nonlinear operations, for example, median filtering. Since both the speckle noise and the 2π discontinuities in phase fringe patterns are characterized by high spatial frequencies, applying a filter not only reduces the noise but also smears out the discontinuities. This problem is solved by calculating the sine and cosine of the wrapped phase map $\Delta\phi$, which leads to the continuous fringe patterns sin $\Delta\phi$ and cos $\Delta\phi$ and then applying filtering, known as sine–cosine filtering scheme [58]. Another interesting type of filter is the phase filter [38] that is a modified version of average filter. There are many other sophisticated filters like scale-space filter [59], adaptive median filter [60], special averaging filter [61], and histogram-data-orientated filter [62], in making noise reduced phase map before unwrapping. All the filters have relative merits and demerits, and usefulness of a particular filter depends upon the application.

Unwrapping is the procedure which removes these 2π phase jumps and the result is converted into desired continuous phase function, thus phase unwrapping is an important final procedure for DSPI fringe analysis process. The process is carried out by adding or subtracting 2π each time the phase map presents a discontinuity. Even though these methods are well suited for unwrapping noise-free data, the process becomes more difficult when the absolute phase difference between the adjacent pixels with no discontinuities is greater than π. These spurious jumps may be caused by noise, discontinuous phase jumps, and regional under-sampling in the fringe pattern. If any of such local phase inconsistencies are present, an error appears which is propagated along the unwrapping path. To avoid this problem, sophisticated algorithms have been proposed [57]. These methods include unwrapping by the famous branch-cut technique [63] and with discrete cosine transform (DCT) [64]. The former method detects error inducing locations in the phase map and connects them to each other by a branch-cut, which must not be crossed by a subsequent unwrapping with a simple path dependent method. The DCT approaches the problem of unwrapping by solving the Poisson equation relating wrapped and unwrapped phases by 2D DCT. This way, any path dependency is evaded and error propagation does not appear "localized" as for scanning methods. The advantage of this technique is that noise in wrapped data has less influence, and that the whole unwrapping is performed in a single step (no step function is created), though the time consumed is rather high compared to path dependent methods. The multigrid algorithm is an iterative algorithm that adopts a recursive mode of operation and forms the basis for most multigrid algorithms [57,65]. Multigrid methods are a class of techniques for rapidly solving partial differential equations (PDEs) on large grids. These methods are based on the idea of applying Gauss–Seidel relaxation schemes on coarser, smaller grids. After application of any phase unwrapping algorithm, "true" phase jumps in unwrapped phase data higher than 2π cannot be identified correctly, as these are interpreted as changes of order. Figure 8.9 shows the speckle fringe analysis on a centrally loaded, rigidly clamped circular diaphragm using the TPS "difference-of-phases" method [37] and DCT unwrapping algorithm [57].

8.5.2 Optical Systems in Digital Speckle and Speckle Shear Interferometry

Several optical systems in digital speckle and speckle shear interferometry have also been developed for the measurement of displacement and its derivates [37,38]. For measurement of the out-of-plane

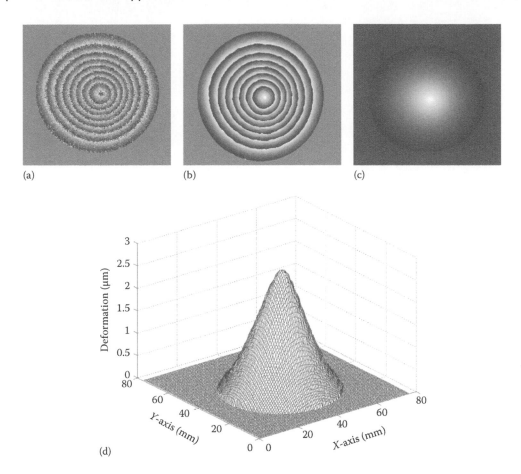

FIGURE 8.9 Speckle fringe analysis using temporal phase shifting "difference-of-phases" method: (a) raw phase map, (b) filtered phase map, (c) corresponding unwrapped 2D, and (d) 3D plots of the out-of-plane displacement.

displacement component of a deformation vector, the optical systems are based on either (1) in-line reference wave [66,67] or (2) off-axis reference wave [68,69]. A desensitized system for large out-of-plane displacement is also reported [70–72]. Further, portable systems have been reported with the possibility for the system to be implemented in remote conditions [73–77].

For in-plane displacement measurement, Leendertz's dual beam illumination and normal observation configuration are widely adopted (Figure 8.3). The other configurations emerged in recent years are (1) oblique illumination–observation [78], (2) normal illumination–dual direction observation [79], and (3) dual beam symmetric illumination–observation [79–84]. The dual beam symmetric illumination–observation systems had an additional advantage of measuring the in-plane displacement component with twofold sensitivity [82–84]. The in-plane sensitive systems are also used to determine the shape or surface profile of the 3D objects [37]. Some novel systems for evaluating displacement components from a single optical head are also described [85–88].

Digital speckle shear interferometry or DS has been widely used for the measurement of the spatial derivatives of object deformation. The DS system has found a prominent application in aerospace industry for nondestructive testing of spacecraft structures [89–91]. A Michelson-type of shear interferometer as shown in Figure 8.5 is extensively employed to introduce lateral shear in an optical configuration [92–95]. NDE analysis on a polymethyl methacrylate (PMMA) panel clamped at the bottom and subjected to thermal stressing using DS is shown in Figure 8.10. A universal shear interferometer has been proposed to implement lateral, radial, rotational, and inversion

FIGURE 8.10 NDE of PMMA panel subjected to thermal stressing with DS: (a) fringe pattern, (b) filtered phase map, (c) 2D plot, and (d) 3D plots of the slope.

shear evaluations [96]. Dual-beam symmetric illumination in a Michelson shear arrangement for obtaining the partial derivatives of in-plane displacements (strain) is also reported [97–99]. The other prominent shearing devices reported are split lens shear element [100], birefringent crystals [101,102], and HOE such as holo-shear lens (HSL) [103] and holo-gratings (HG) [104]. Waldner and Brem [105] developed a portable, compact shearography system. A multiwavelength DS is also reported for extracting the in-plane and out-of-plane strains from a single system [106,107]. A large area composite helicopter structures for non destructive testing (NDT) inspection is also examined by Pezzoni and Krupka [108], to show the potential application of the shearography for large size structures. Of the many salient developments, the two that one could probably cite here concern the measurement of out-of-plane displacement and their derivatives [109–115], and the measurement of second-order partial derivatives [116,117] of out-of-plane displacement (curvature and twist). The shearography is also exploited for slope contouring of 3D objects either by varying the refractive index, wavelength, or rotation of the 3D surface between the frames [118–120].

8.5.3 Application to Microsystems Analysis

Microsystems such as MEMS and micro-opto-electromechanical systems (MOEMS) combining micro-size electrical and mechanical mechanism are finding increasing applications in various fields. In recent years, the technology of fabricating MEMS has been adopted by a number of industries to produce sensors, actuators, visual display components, AFM probe tips, semiconductor devices, etc. Aided by photolithography and other novel methods, large number of MEMS devices are fabricated simultaneously. So, high-quality standards are a must for the manufacturers in MEMS technology. The materials behavior in combination with new structural design cannot be predicted by theoretical simulations. A possible reason for wrong predictions made by finite element method (FEM) and other model analysis with respect to the loading conditions or in practical applications of microdevices is, for instance, the lack of reliable materials data and boundary conditions in microscale. The MEMS devices require measurement and inspection techniques that are fast noncontact and robust. The reasons for such a technology is obvious, properties determined on much larger specimens cannot be scaled down from the bulk material without any experimental evaluation. Further on, in microstructures, the materials behavior is noticeably affected by the production technology. Therefore, simple and robust optical measuring methods to analyze the shape and deformation of the microcomponents under static and dynamic conditions are highly desired [121]. The optical measuring system for the inspection and characterization of M(O)EMS should have the following conditions. First, it should not alter the integrity and the mechanical behavior of the device. Since the microcomponents have an overall size up to a millimeter, a high spatial resolution measuring system is required. The measuring system should be able to perform both static and dynamic measurements. As the deflections, motions, or vibration amplitudes of MEMS are typically in the nanometer to a few microns range, high sensitivity optical methods are highly desired. Finally, the measurement must be quantitative and reliable with a low sensitivity to environmental conditions. Microscopic systems for microcomponents analysis are mostly dependent on the optical system such as long working distance microscope with the combination of different magnification objective lens. Lokberg et al. [122] proposed feasibility of microscopic system on microstructures. Various microscopic imaging systems have been developed for microsystems characterization [123–128]. A schematic diagram of a microscopic digital speckle interferometric system is shown in Figure 8.11. In this arrangement, the collimated laser beam illuminates the object and a PZT-driven reference mirror via a cube BS. The scattered object wave from the microspecimen and the smooth reference wave are recombined coherently onto the CCD plane via the same cube BS and the microscopic imaging system. As an example, Figure 8.12 shows the application of the microscopic digital speckle interferometric system for deflection characterization on a MEMS pressure sensor. An endoscopic DSPI system is also reported in literature to study intracavity inspection of biological specimen such as in vitro and in vivo (porcine stomach, mucous membrane) [129–131].

FIGURE 8.11 Schematic of a microscopic DSPI system.

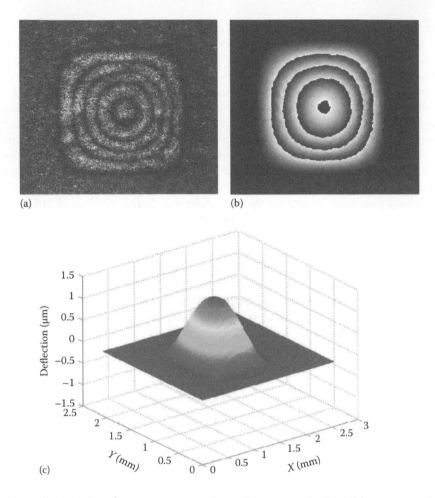

FIGURE 8.12 MEMS (15 mm²) pressure sensor subjected to pressure load: (a) fringe pattern, (b) wrapped phase map, and (c) 3D plot.

8.6 CONCLUSIONS

The chapter covers the speckle methods and their applications to optical metrology. Speckle methods can be broadly classified into three categories: speckle photography, speckle interferometry, and speckle shear interferometry (shearography). The digital speckle methods with high-speed cameras and modern imaging processing systems will find a widespread use in the field of microsystem engineering, aerospace engineering, biological and medical sciences.

REFERENCES

1. Leendertz, J.A., Interferometric displacement measurement on scattering surfaces utilizing speckle effect, *J. Phys. E: Sci. Instrum.*, 3, 214–218, 1970.
2. Archbold, E., Burch, M., and Ennos, A.E., Recording of in-plane surface displacement by double-exposure speckle photography, *Opt. Acta*, 17, 883–889, 1992.
3. Archbold, E. and Ennos, A.E., Displacement measurement from double exposure laser photographs, *Opt. Acta*, 19, 253–271, 1992.
4. Khetan, R.P. and Chang, F.P., Strain analysis by one-beam laser speckle interferometry. 1. Single aperture method, *Appl. Opt.*, 15, 2205–2215, 1976.
5. Sutton, M.A., Wolters, W.J., Peters, W.H., Ranson, W.F., and McNeill, S.R., Determination of displacements using an improved digital correlation method, *Image Vision Comput.*, 1, 133–139, 1983.

6. Chen, D.J. and Chiang, F.P., Optimal sampling and range of measurement in displacement-only laser-speckle correlation, *Exp. Mech.*, 32, 145–153, 1992.

7. Noh, S. and Yamaguchi, I., Two-dimensional measurement of strain distribution by speckle correlation, *Jpn. J. Appl. Phys.*, 31, L1299–L1301, 1992.

8. Sjödahl, M. and Benckert, L.R., Electronic speckle photography analysis of an algorithm giving the displacement with sub pixel accuracy, *Appl. Opt.*, 32, 2278–2284, 1993.

9. Sjödahl, M. and Benckert, L.R., Systematic and random errors in electronic speckle photography, *Appl. Opt.*, 33, 7461–7471, 1994.

10. Sjödahl, M., Electronic speckle photography: Increased accuracy by non integral pixel shifting, *Appl. Opt.*, 33, 6667–6673, 1994.

11. Andersson, A., Krishna Mohan, N., Sjödahl, M., and Molin, N.-E., TV shearography: Quantitative measurement of shear magnitudes using digital speckle photography, *Appl. Opt.*, 39, 2565–2568, 2000.

12. Tiziani, H.J., Leonhardt, K., and Klenk, J., Real-time displacement and tilt analysis by a speckle technique using $Bi_{12}SiO_{20}$ crystal, *Opt. Commun.*, 34, 327–331, 1980.

13. Liu, L., Helmers, H., and Hinsch, K.D., Speckle metrology with novelty filtering using photorefractive two-beam coupling in $BaTiO_3$, *Opt. Commun.*, 100, 301–305, 1993.

14. Sreedhar, P.R., Krishna Mohan, N., and Sirohi, R.S., Real-time speckle photography with two-wave mixing in photorefractive $BaTiO_3$ crystal, *Opt. Eng.*, 33, 1989–1995, 1994.

15. Krishna Mohan, N. and Rastogi, P.K., Phase shift calibration in speckle photography—A first step to practical real time analysis of speckle correlation fringes, *J. Mod. Opt.*, 50, 1183–1188, 2003.

16. Krishna Mohan, N. and Rastogi, P.K., Phase shifting whole field speckle photography technique for the measurement of object deformations in real-time, *Opt. Lett.*, 27, 565–567, 2002.

17. Joenathan, C., Ganesan, A.R., and Sirohi, R.S., Fringe compensation in speckle interferometry—Application to non destructive testing, *Appl. Opt.*, 25, 3781–3784, 1986.

18. Rastogi, P.K. and Jacquot, P., Measurement of difference deformation using speckle interferometry, *Opt. Lett.*, 12, 596–598, 1987.

19. Sirohi, R.S. and Krishna Mohan, N., In-plane displacement measurement configuration with twofold sensitivity, *Appl. Opt.*, 32, 6387–6390, 1993.

20. Krishna Mohan, N., Measurement of in-plane displacement with twofold sensitivity using phase reversal technique, *Opt. Eng.*, 38, 1964–1966, 1999.

21. Duffy, D.E., Moiré gauging of in-plane displacement using double aperture imaging, *Appl. Opt.*, 11, 1778–1781, 1972.

22. Duffy, D.E., Measurement of surface displacement normal to the line of sight, *Exp. Mech.*, 14, 378–384, 1974.

23. Khetan, R.P. and Chang, F.P., Strain analysis by one-beam laser speckle interferometry. 2. Multiaperture aperture method, *Appl. Opt.*, 18, 2175–2186, 1979.

24. Sirohi, R.S., Krishna Mohan, N., and Santhanakrishnan, T., Optical configurations for measurement in speckle interferometry, *Opt. Lett.*, 21, 1958–1959, 1996.

25. Joenathan, C., Mohanty, R.K., and Sirohi, R.S., Multiplexing in speckle shear interferometry, *Opt. Acta*, 31, 681–692, 1984.

26. Leendertz, J.A. and Butters, J.N., An image-shearing speckle pattern interferometer for measuring bending moments, *J. Phys. E: Sci. Instrum.*, 6, 1107–1110, 1973.

27. Hung, Y.Y. and Taylor, C.E., Speckle-shearing interferometric camera—A tool for measurement of derivatives of surface displacement, *Proc. SPIE*, 41, 169–175, 1973.

28. Sirohi, R.S., Speckle shearing interferometry—A review, *J. Opt.*, 33, 95–113, 1984.

29. Krishna Mohan, N. and Sirohi, R.S., Fringe formation in multiaperture speckle shear interferometry, *Appl. Opt.*, 35, 1617–1622, 1996.

30. Krishna Mohan, N., Double and multiple-exposure methods in multi-aperture speckle shear interferometry for slope and curvature measurement, *Proceedings of Laser Metrology for Precision Measurement and Inspection in Industry*, Florianópolis, Brazil, Module 4, Holography and Speckle Techniques, pp. 4.87–4.98, 1999.

31. Sharma, D.K., Sirohi, R.S., and Kothiyal, M.P., Simultaneous measurement of slope and curvature with a three-aperture speckle shearing interferometer, *Appl. Opt.*, 23, 1542–1546, 1984.

32. Sharma, D.K., Krishna Mohan, N., and Sirohi, R.S., A holographic speckle shearing technique for the measurement of out-of-plane displacement, slope and curvature, *Opt. Commun.*, 57, 230–235, 1986.

33. Shaker, C. and Rao, G.V., Speckle metrology using hololenses, *Appl. Opt.*, 23, 4592–4595, 1984.

34. Joenathan, C., Mohanty, R.K., and Sirohi, R.S., Hololenses in speckle and speckle shear interferometry, *Appl. Opt.*, 24, 1294–1298, 1985.

35. Joenathan, C. and Sirohi, R.S., Holographic gratings in speckle shear interferometry, *Appl. Opt.*, 24, 2750–2751, 1985.
36. Leendertz, J.N. and Butters, J.N., Holographic and video techniques applied to engineering measurements, *J. Meas. Control.*, 4, 349–354, 1971.
37. Rastogi, P.K., Ed., *Digital Speckle Pattern Interferometry and Related Techniques*, John Wiley, Chichester, U.K., 2001.
38. Steinchen, W. and Yang, L., *Digital Shearography: Theory and Application of Digital Speckle Pattern Shearing Interferometry*, PM100, SPIE, Optical Engineering Press, Washington, DC, 2003.
39. Doval, A.F., A systematic approach to TV holography, *Meas. Sci. Technol.*, 11, R1–R36, 2000.
40. Krishna Mohan, N., Speckle techniques for optical metrology, in *Progress in Laser and Electro-Optics Research*, Arkin, W.T., Ed., Nova Science Publishers Inc., New York, 2006, Chapter 6, pp. 173–222.
41. Creath, K., Phase-measurement interferometry techniques, in *Progress in Optics*, Wolf, E., Ed., North-Holland, Amsterdam, the Netherlands, 1988, Vol. XXV, pp. 349–393.
42. Kujawinska, M., Spatial phase measurement methods, in *Interferogram Analysis-Digital Fringe Pattern Measurement Techniques*, Robinson, D.W. and Reid, G.T., Eds., Institute of Physics, Bristol, U.K., 1993, Chapter 5, pp. 141–193.
43. Creath, K., Phase-shifting speckle interferometry, *Appl. Opt.*, 24, 3053–3058, 1985.
44. Robinson, D.W. and Williams, D.C., Digital phase stepping speckle interferometry, *Opt. Commun.*, 57, 26–30, 1986.
45. Wyant, J.C., Interferometric optical metrology: Basic principles and new systems, *Laser Focus*, 5, 65–71, 1982.
46. Hariharan, P., Oreb, B.F., and Eiju, T., Digital phase shifting interferometry: A simple error compensating phase calculation algorithm, *Appl. Opt.*, 26, 2504–2505, 1987.
47. Schwider, J., Burow, R., Elsner, K.E., Grzanna, J., Spolaczyk, R., and Merkel, K., Digital wavefront measuring interferometry: Some systematic error sources, *Appl. Opt.*, 24, 3421–3432, 1983.
48. Ali, C. and Wyant, J.C., Effect of piezoelectric transducer non linearity on phase shift interferometry, *Appl. Opt.*, 26, 1112–1116, 1987.
49. Bhaduri, B., Krishna Mohan, N., and Kothiyal, M.P., Cycle path digital speckle shear pattern interferometer: Use of polarizing phase shifting method, *Opt. Eng.*, 45, 105604-1-6, 2006.
50. Tatam, R.P., Optoelectronic developments in speckle interferometry, *Proc. SPIE*, 2860, 194–212, 1996.
51. Krishna Mohan, N., A real time phase reversal read-out system with BaTiO$_3$ crystal as a recording medium for speckle fringe analysis, *Proc. SPIE*, 5912, 63–69, 2005.
52. Burke, J., Helmers, H., Kunze, C., and Wilkens, V., Speckle intensity and phase gradients: Influence on fringe quality in spatial phase shifting ESPI-systems, *Opt. Commun.*, 152, 144–152, 1998.
53. Burke, J., Application and optimization of the spatial phase shifting technique in digital speckle interferometry, PhD dissertation, Carl von Ossietzky University, Oldenburg, Germany, 2000.
54. Sirohi, R.S., Burke, J., Helmers, H., and Hinsch, K.D., Spatial phase shifting for pure in-plane displacement and displacement-derivative measurements in electronic speckle pattern interferometry (ESPI), *Appl. Opt.*, 36, 5787–5791, 1997.
55. Basanta Bhaduri, Krishna Mohan, N., Kothiyal, M.P., and Sirohi, R.S., Use of spatial phase shifting technique in digital speckle pattern interferometry (DSPI) and digital shearography (DS), *Opt. Exp.*, 14(24), 11598–11607, 2006.
56. Huntley, J.M., Automated analysis of speckle interferograms, in *Digital Speckle Pattern Interferometry and Related Techniques*, Rastogi, P.K., Ed., John Wiley, Chichester, U.K., 2001, Chapter 2, 59–140.
57. Ghiglia, D.C. and Pritt, M.D., *Two-Dimensional Phase Unwrapping*, John Wiley, New York, 1998.
58. Waldner, S., Removing the image-doubling in shearography by reconstruction of the displacement fields, *Opt. Commun.*, 127, 117–126, 1996.
59. Kaufmann, G.H., Davila, A., and Kerr, D., Speckle noise reduction in TV holography, *Proc. SPIE*, 2730, 96–100, 1995.
60. Capanni, A., Pezzati, L., Bertani, D., Cetica, M., and Francini, F., Phase-shifting speckle interferometry: A noise reduction filter for phase unwrapping, *Opt. Eng.*, 36, 2466–2472, 1997.
61. Aebischer, H.A. and Waldner, S., A simple and effective method for filtering speckle-interferometric phase fringe patterns, *Opt. Commun.*, 162, 205–210, 1999.
62. Huang, M.J. and Sheu, W.H., Histogram-data-orientated filter for inconsistency removal of interferometric phase maps, *Opt. Eng.*, 44, 045602-1–045602-11, 2005.
63. Goldstein, R., Zebker, M.H., and Werner, C.L., Satellite radar interferometry-two dimensional phase unwrapping, *Radio Sci.*, 23, 713–720, 1987.

64. Ghiglia, D.C. and Romero, L.A., Robust two-dimensional weighted and unweighted phase unwrapping that uses fast transforms and iterative methods, *J. Opt. Soc. Am. A*, 1, 107–117, 1994.

65. Botello, S., Marroquin, J.L., and Rivera, M., Multigrid algorithms for processing fringe pattern images, *Appl. Opt.*, 37, 7587–7595, 1998.

66. Joenathan, C. and Khorana, R., A simple and modified ESPI system, *Optik*, 88, 169–171, 1991.

67. Joenathan, C. and Torroba, R., Modified electronic speckle pattern interferometer employing an off-axis reference beam, *Appl. Opt.*, 30, 1169–1171, 1991.

68. Santhanakrishnan, T., Krishna Mohan, N., and Sirohi, R.S., New simple ESPI configuration for deformation studies on large structures based on diffuse reference beam, *Proc. SPIE*, 2861, 253–263, 1996.

69. Flemming, T., Hertwig, M., and Usinger, R., Speckle interferometry for highly localized displacement fields, *Meas. Sci. Technol.*, 4, 820–825, 1993.

70. Krishna Mohan, N., Saldner, H.O., and Molin, N.-E., Combined phase stepping and sinusoidal phase modulation applied to desensitized TV holography, *J. Meas. Sci. Technol.*, 7, 712–715, 1996.

71. Krishna Mohan, N., Saldner, H.O., and Molin, N.-E., Recent applications of TV holography and shearography, *Proc. SPIE*, 2861, 248–256, 1996.

72. Saldner, H.O., Molin, N.-E., and Krishna Mohan, N., Desensitized TV holography applied to static and vibrating objects for large deformation measurement, *Proc. SPIE*, 2860, 342–349, 1996.

73. Hertwig, M., Flemming, T., Floureux, T., and Aebischen, H.A., Speckle interferometric damage investigation of fiber-reinforced composites, *Opt. Lasers Eng.*, 24, 485–504, 1996.

74. Holder, L., Okamoto, T., and Asakura, T.A., Digital speckle correlation interferometer using an image fibre, *Meas. Sci. Technol.*, 4, 746–753, 1993.

75. Joenathan, C. and Orcutt, C., Fiber optic electronic speckle pattern interferometry operating with a 1550 laser diode, *Optik*, 100, 57–60, 1993.

76. Paoletti, D., Schirripa Spagnolo, G., Facchini, M., and Zanetta, P., Artwork diagnostics with fiber-optic digital speckle pattern interferometry, *Appl. Opt.*, 32, 6236–6241, 1993.

77. Gülker, G., Hinsch, K., Hölscher, C., Kramer, A., and Neunaber, H., ESPI system for in situ deformation monitoring on buildings, *Opt. Eng.*, 29, 816–820, 1990.

78. Joenathan, C., Sohmer, A., and Burkle, L., Increased sensitivity to in-plane displacements in electronic speckle pattern interferometry, *Appl. Opt.*, 34, 2880–2885, 1995.

79. Krishna Mohan, N., Andersson, A., Sjödahl, M., and Molin, N.-E., Optical configurations in TV holography for deformation and shape measurement, *Lasers Eng.*, 10, 147–159, 2000.

80. Krishna Mohan, N., Andersson, A., Sjödahl, M., and Molin, N.-E., Fringe formation in a dual beam illumination-observation TV holography: An analysis, *Proc. SPIE*, 4101A, 204–214, 2000.

81. Krishna Mohan, N., Svanbvo, A., Sjödahl, M., and Molin, N.-E., Dual beam symmetric illumination–observation TV holography system for measurements, *Opt. Eng.*, 40, 2780–2787, 2001.

82. Solmer, A. and Joenathan, C., Twofold increase in sensitivity with a dual beam illumination arrangement for electronic speckle pattern interferometry, *Opt. Eng.*, 35, 1943–1948, 1996.

83. Basanta Bhaduri, Kothiyal, M.P., and Krishna Mohan, N., Measurement of in-plane displacement and slope using DSPI system with twofold sensitivity, *Proceedings of Photonics 2006*, Hyderabad, India, 2006, pp. 1–4.

84. Basanta Bhaduri, Kothiyal, M.P., and Krishna Mohan, N., Digital speckle pattern interferometry (DSPI) with increased sensitivity: Use of spatial phase shifting, *Opt. Commun.*, 272, 9–14, 2007.

85. Shellabear, M.C. and Tyrer, J.R., Application of ESPI to three dimensional vibration measurement, *Opt. Laser Eng.*, 15, 43–56, 1991.

86. Bhat, G.K., An electro-optic holographic technique for the measurement of the components of the strain tensor on three-dimensional object surfaces, *Opt. Lasers Eng.*, 26, 43–58, 1997.

87. Bowe, B., Martin, S., Toal, T., Langhoff, A., and Wheelan, M., Dual in-plane electronic speckle pattern interferometry system with electro-optical switching and phase shifting, *Appl. Opt.*, 38, 666–673, 1999.

88. Krishna Mohan, N., Andersson, A., Sjödahl, M., and Molin, N.-E., Optical configurations for TV holography measurement of in-plane and out-of-plane deformations, *Appl. Opt.*, 39, 573–577, 2000.

89. Hung, Y.Y., Electronic shearography versus ESPI for non destructive testing, *Proc. SPIE*, 1554B, 692–700, 1991.

90. Steinchen, W., Yang, L.S., Kupfer, G., and Mäckel, P., Non-destructive testing of aerospace composite materials using digital shearography, *Proc. Inst. Mech. Eng.*, 212, 21–30, 1998.

91. Hung, Y.Y., Shearography a novel and practical approach to non-destructive testing, *J. Nondestructive Eval.*, 8, 55–68, 1989.

92. Krishna Mohan, N., Saldner, H.O., and Molin, N.-E., Electronic shearography applied to static and vibrating objects, *Opt. Commun.*, 108, 197–202, 1994.

93. Steinchen, W., Yang, L.S., Kupfer, G., Mäckel, P., and Vössing, F., Strain analysis by means of digital shearography: Potential, limitations and demonstrations, *J. Strain Anal.*, 3, 171–198, 1998.

94. Yang, L. and Ettemeyer, A., Strain measurement by 3D-electronic speckle pattern interferometry: Potential, limitations and applications, *Opt. Eng.*, 42, 1257–1266, 2003.

95. Hung, Y.Y., Shang, H.M., and Yang, L., Unified approach for holography and shearography in surface deformation measurement and nondestructive testing, *Opt. Eng.*, 42, 1197–1207, 2003.

96. Ganesan, A.R., Sharma, D.K., and Kothiyal, M.P., Universal digital speckle shearing interferometer, *Appl. Opt.*, 27, 4731–4734, 1988.

97. Rastogi, P.K., Measurement of in-plane strains using electronic speckle and electronic speckle-shearing pattern interferometry, *J. Mod. Opt.*, 43, 403–407, 1996.

98. Patroski, K. and Olszak, A., Digital in-plane electronic speckle pattern shearing interferometry, *Opt. Eng.*, 36, 2010–2015, 1997.

99. Aebischer, H.A. and Waldner, S., Strain distributions made visible with image-shearing speckle pattern interferometry, *Opt. Lasers Eng.*, 26, 407–420, 1997.

100. Joenathan, C. and Torroba, R., Simple electronic speckle-shearing-pattern interferometer, *Opt. Lett.*, 15, 1159–1161, 1990.

101. Hung, Y.Y. and Wang, J.Q., Dual-beam phase shift shearography for measurement of in-plane strains, *Opt. Lasers Eng.*, 24, 403–413, 1996.

102. Nakadate, S., Phase shifting speckle shearing polarization interferometer using a birefringent wedge, *Opt. Lasers Eng.*, 6, 31–350, 1997.

103. Krishna Mohan, N., Masalkar, P.J., and Sirohi, R.S., Electronic speckle pattern interferometry with holo-optical element, *Proc. SPIE*, 1821, 234–242, 1992.

104. Joenathan, C. and Bürkle, L., Electronic speckle pattern shearing interferometer using holographic gratings, *Opt. Eng.*, 36, 2473–2477, 1997.

105. Waldner, S. and Brem, S., Compact shearography system for the measurement of 3D deformation, *Proc. SPIE*, 3745, 141–148, 1999.

106. Kastle, R., Hack, E., and Sennhauser, U., Multiwavelength shearography for quantitative measurements of two-dimensional strain distribution, *Appl. Opt.*, 38, 96–100, 1999.

107. Groves, R., James, S.W., and Tatam, R.P., Polarisation-multiplexed and phase-stepped fibre optic shearography using laser wavelength modulation, *Proc. SPIE*, 3744, 149–157, 1999.

108. Pezzoni, R. and Krupka, R., Laser shearography for non-destructive testing of large area composite helicopter structures, *Insight*, 43, 244–248, 2001.

109. Krishna Mohan, N., Saldner, H.O., and Molin, N.-E., Electronic speckle pattern interferometry for simultaneous measurement of out-of-plane displacement and slope, *Opt. Lett.*, 18, 1861–1863, 1993.

110. Fomitchov, P.A. and Krishnaswamy, S., A compact dual-purpose camera for shearography and electronic speckle-pattern interferometry, *Meas. Sci. Technol.*, 8, 581–583, 1997.

111. Basanta Bhaduri, Krishna Mohan, N., and Kothiyal, M.P., A dual-function ESPI system for the measurement of out-of-plane displacement and slope, *Opt. Lasers Eng.*, 44, 637–644, 2006.

112. Basanta Bhaduri, Krishna Mohan, N., and Kothiyal, M.P., A TV holo-shearography system for NDE, *Laser Eng.*, 16, 93–104, 2006.

113. Basanta Bhaduri, Krishna Mohan, N., and Kothiyal, M.P., (5, N) phase shift algorithm for speckle and speckle shear fringe analysis in NDT, *Holography Speckle*, 3, 18–21, 2006.

114. Basanta Bhaduri, Krishna Mohan, N., and Kothiyal, M.P., (1,N) spatial phase shifting technique in DSPI and DS for NDE, *Opt. Eng.*, 46, 051009-1–051009-7, 2007.

115. Basanta Bhaduri, Krishna Mohan, N., and Kothiyal, M.P., Simultaneous measurement of out-of-plane displacement and slope using multi-aperture DSPI system and fast Fourier transform, *Appl. Opt.*, 46, 23, 5680–5686, 2007.

116. Rastogi, P.K., Measurement of curvature and twist of a deformed object by electronic speckle shearing pattern interferometry, *Opt. Lett.*, 21, 905–907, 1996.

117. Basanta Bhaduri, Kothiyal, M.P., and Krishna Mohan, N., Curvature measurement using three-aperture digital shearography and fast Fourier transform method, *Opt. Lasers Eng.*, 45, 1001–1004, 2007.

118. Saldner, H.O., Molin, N.-E., and Krishna Mohan, N., Simultaneous measurement of out-of-plane displacement and slope by electronic holo-shearography, *Proceedings of the SEM Conference on Experimental Mechanics*, Lisbon, Portugal, 1994, pp. 337–341.

119. Huang, J.R., Ford, H.D., and Tatam, R.P., Slope measurement by two-wavelength electronic shearography, *Opt. Lasers Eng.*, 27, 321–333, 1997.

120. Rastogi, P.K., An electronic pattern speckle shearing interferometer for the measurement of surface slope variations of three-dimensional objects, *Opt. Lasers Eng.*, 26, 93–100, 1997.

121. Osten, W., Ed., *Optical Inspection of Microsystems*, CRC Press, Boca Raton, FL, 2007.

122. Lokberg, O.J., Seeberg, B.E., and Vestli, K., Microscopic video speckle interferometry, *Opt. Lasers Eng.*, 26, 313–330, 1997.

123. Kujawinska, M. and Gorecki, C., New challenges and approaches to interferometric MEMS and MOEMS testing, *Proc. SPIE*, 4900, 809–823, 2002.

124. Aswendt, P., Micromotion analysis by deep-UV speckle interferometry, *Proc. SPIE*, 5145, 17–22, 2003.

125. Patorski, K., Sienicki, Z., Pawlowski, M., Styk, A.R., and Jozwicka, A., Studies of the properties of the temporal phase-shifting method applied to silicone microelement vibration investigations using the time-average method, *Proc. SPIE*, 5458, 208–219, 2004.

126. Yang, L., and Colbourne, P., Digital laser microinterferometer and its applications, *Opt. Eng.*, 42, 1417–1426, 2003.

127. Paul Kumar, U., Bhaduri, B., Somasundaram, U., Krishna Mohan, N., and Kothiyal, M.P., Development of a microscopic TV holographic system for characterization of MEMS, *Proceedings of Photonics 2006*, Hyderabad, India, 2006, MMM3, pp. 1–4.

128. Paul Kumar, U., Basanta Bhaduri, Krishna Mohan, N., and Kothiyal, M.P., 3-D surface profile analysis of rough surfaces using two-wavelength microscopic TV holography, *International Conference on Sensors and Related Networks (SEENNET07)*, VIT Vellore, India, 2007, pp. 119–123.

129. Kemper, B., Dirksen, D., Avenhaus, W., Merker, A., and VonBally, G., Endoscopic double-pulse electronic-speckle-pattern interferometer for technical and medical intracavity inspection, *Appl. Opt.*, 39, 3899–3905, 2000.

130. Kandulla, J., Kemper, B., Knoche, S., and VonBally, G., Two-wavelength method for endoscopic shape measurement by spatial phase-shifting speckle-interferometry, *Appl. Opt.*, 43, 5429–5437, 2004.

131. Dyrseth, A.A., Measurement of plant movement in young and mature plants using electronic speckle pattern interferometry, *Appl. Opt.*, 35, 3695–3701, 1996.

9 Moiré Metrology

Lianhua Jin and Toru Yoshizawa

CONTENTS

9.1 Introduction 273
9.2 In-Plane Moiré Method and Measurement of Strain...274
    9.2.1 Pattern Formation of In-Plane Moiré 274
    9.2.2 Application to Strain Measurement..276
        9.2.2.1 Linear Strain Measurement 276
        9.2.2.2 Shear Strain Measurement...277
9.3 Out-of-Plane Moiré Method and Profilometry 278
    9.3.1 Shadow Moiré and Contour..279
    9.3.2 Intensity of Shadow Moiré Pattern 280
    9.3.3 Application to 3D Profile Measurement 282
        9.3.3.1 Phase-Shifting Method 282
        9.3.3.2 Frequency Sweeping Method..284
    9.3.4 Projection Moiré 287
9.4 Reflection Moiré 288
9.5 Diffraction Grating, Interference Moiré, and Application 288
9.6 Recent Topics..290
    9.6.1 Development in Shape Measurement 291
    9.6.2 High-Speed Measurement 295
9.7 Conclusion 297
References...297

9.1 INTRODUCTION

The French word *moiré* originates from a type of textile, traditionally of silk, with a grained or watered appearance. Now moiré is generally used for a fringe that is created by superposition of two (or more) patterns such as dot arrays and grid lines (see Figure 9.1).

Moiré phenomenon can be observed in our everyday surroundings, like the folded netting. Some moiré patterns need to be got rid of as an undesired artifact, for example, the patterns produced during scanning a halftone picture. Some moiré patterns, on the contrary, act as very useful phenomena for different application fields. For instance, in the textile industry, designers intentionally generate beautiful moiré patterns with silk fabrics; in health-care field, the moiré is applied to diagnostic test of scoliosis, which is more common in teenage females.

In this chapter, applications of the moiré phenomenon to the optical metrology are described. In moiré metrology, the moiré fringe is resulted from the superposition of two periodic gratings structured with 1D lines. One of these gratings is called reference grating, and the other one is object grating, which is to be distorted by a structure whose deformation or shape is represented by the resulting moiré fringes. The moiré fringe created by these two superposed gratings in the same plane is termed in-plane moiré and by two gratings in different planes is out-of-plane moiré.

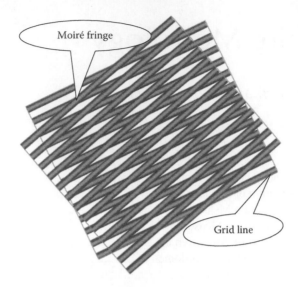

FIGURE 9.1   Superposition of two grid lines and moiré fringe.

In the following sections, we will introduce principle of pattern formation of in-plane and out-of-plane moirés and basic applications to strain analysis and profilometry.

9.2   IN-PLANE MOIRÉ METHOD AND MEASUREMENT OF STRAIN

9.2.1   PATTERN FORMATION OF IN-PLANE MOIRÉ

In-plane moiré pattern is obtained by superposing the reference and the object gratings by direct contact or by imaging one onto the other via a lens. The reference grating has a constant period and fixed spatial orientation, and the object grating is either printed or attached to the object surface. Before deformation of the object, the period and orientation of the object grating are identical to that of the reference grating. Here, let us consider that the object grating is rotated by an angle θ with respect to the reference grating, as shown in Figure 9.1. Seen from a far distance, one no longer can resolve the grating elements, and only see dark and pale bands—moiré pattern. The pale bands correspond to the lines of nodes, namely, elements passing through the intersections of two gratings.

How are the orientation and interval of these moiré patterns in Figure 9.1? Let p and p' be the periods of the reference and object gratings and φ and d the orientation and interval of moiré pattern, respectively. From geometry, in Figure 9.2,

$$OA = \frac{p}{\cos\left(\varphi - \dfrac{\pi}{2}\right)} = \frac{p}{\sin\varphi}, \tag{9.1}$$

$$OA = \frac{p'}{\cos\left(\varphi - \dfrac{\pi}{2} - \theta\right)} = \frac{p'}{\sin(\varphi - \theta)}. \tag{9.2}$$

Therefore,

$$p'\sin\varphi = p\sin(\varphi - \theta). \tag{9.3}$$

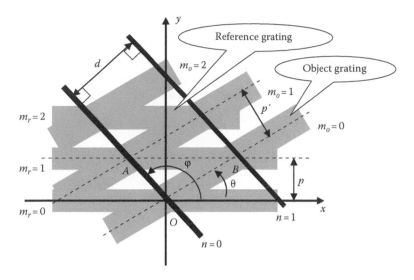

FIGURE 9.2 Order n, orientation φ, and interval d of in-plane moiré pattern; $n = m_r - m_o$ (m_r, m_o are number of reference and object grating elements, respectively).

Rearranging by using geometric functions, we obtain

$$\varphi = \tan^{-1} \frac{p \sin \theta}{p \cos \theta - p'}. \tag{9.4}$$

From Figure 9.2, we have

$$OB = \frac{p}{\sin \theta}, \tag{9.5}$$

and

$$d = OB \cos\left(\varphi - \frac{\pi}{2} - \theta\right) = OB \sin(\varphi - \theta). \tag{9.6}$$

Substituting Equations 9.3 and 9.5 into Equation 9.6 leads to

$$d = \frac{p' \sin \varphi}{\sin \theta}. \tag{9.7}$$

Rearranging Equation 9.4 by using geometric relationship and then substituting into Equation 9.7 yields

$$d = \frac{pp'}{\sqrt{p^2 \sin^2 \theta + (p \cos \theta - p')^2}}. \tag{9.8}$$

From Equations 9.4 and 9.8, it is obvious that once p, p', and θ are given, the orientation φ and interval d of the moiré fringe are uniquely decided. Conversely, from the measured values of φ and d, and a given period p of the reference grating, we can obtain the period p' and orientation θ of the object grating whose structure may be deformed by strains.

9.2.2 Application to Strain Measurement

Strain is the geometric expression of deformation caused by the action of stress on an object. Strain is calculated by assuming a change in length or in angle between the beginning and final states. If strain is equal over all parts of the object, it is referred to as homogeneous strain; otherwise, it is inhomogeneous strain. Here, we apply in-plane moiré to measurement of homogeneous linear strain and shear strain.[1–4]

9.2.2.1 Linear Strain Measurement

The linear strain ε, according to Eulerian description, is given by

$$\varepsilon = \frac{\delta\ell}{\ell_f} = \frac{\ell_f - \ell_o}{\ell_f},$$ (9.9)

where ℓ_o and ℓ_f are the original and final lengths of the object, respectively. (According to Lagrangian description, the denominator is the original length ℓ_o instead of ℓ_f; for the application of moiré methods, here we use Eulerian description. When the deformation is very small, the difference between the two descriptions is negligible.) The extension $\delta\ell$ is positive if the object has gained length by tension and negative if it has reduced length by compression. Since ℓ_f is always positive, the sign of linear strain is always the same as the sign of the extension.

In measuring the uniaxial linear strain, the object grating will be printed onto the object surface with its elements perpendicular to the direction of the strain, and the reference grating is superposed on it with the same orientation, as shown in Figure 9.3. Figure 9.3c shows the resultant moiré patterns across the object deformed by the tension. In this case, the period of the object grating will be changed from p to p', and the orientation of it unchanged, namely, the rotation angle θ in Figure 9.2 is zero.

Substituting θ = 0 into Equation 9.4, the interval d of the moiré fringe becomes

$$d = \left| \frac{pp'}{p - p'} \right|.$$ (9.10)

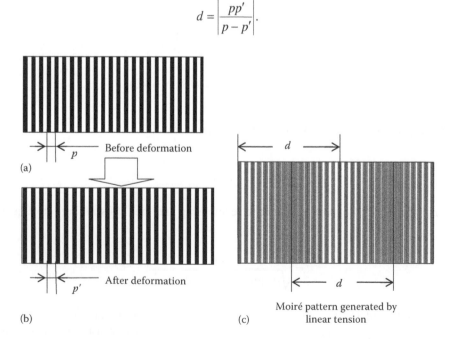

FIGURE 9.3 The reference gating (a) before and (b) after deformation of the object by linear tension and (c) the resultant moiré pattern.

Thus, from Equations 9.9 and 9.10, we obtain the linear strain

$$|\varepsilon| = \left| \frac{\ell_f - \ell_o}{\ell_f} \right| = \left| \frac{p - p'}{p'} \right| = \frac{p}{d}. \tag{9.11}$$

Since both p and d are positive magnitudes, the sign of absolute value is introduced to the strain. This means that the appearance of moiré fringe will not explain whether it is a result of tensile or compression strain. To determine the sign of moiré fringe as well as the resulting strain, such techniques as mismatch method and fringe shifting method by translating the reference grating are available.

9.2.2.2 Shear Strain Measurement

The shear strain γ is defined as the angular change between any two lines in an object before and after deformation, assuming that the line lengths are approaching zero. In applying the in-plane moiré method to measure the shear strain, the object grating should be so oriented that its principal direction is parallel to the direction of shear as shown in Figure 9.4a. Figure 9.4b shows typical moiré patterns across the object deformed by the shear. In this deformation, the period p' of the object grating is assumed same as p before deformation, and the orientation was changed from zero to θ (the quantity of θ is very small). Thus, from Equation 9.8, the interval of the moiré pattern is

$$d = \frac{p}{\theta}. \tag{9.12}$$

Comparing Figure 9.4b with Figure 9.2, one can easily see that the shear strain γ is the resulting rotation angle θ of the object grating; hence, it can be expressed by

$$\gamma = \frac{p}{d}. \tag{9.13}$$

FIGURE 9.4   The reference gating (a) before and (b) after deformation of the object by shear and (c) the resultant moiré pattern.

FIGURE 9.5   Strain analysis from 2D moiré fringe.

Although linear and shear strains are discussed independently, in a general 2D deformation, the object grating undergoes rotation as well as change of period with magnitudes varying from place to place, as depicted in Figure 9.5. From this 2D moiré pattern, we obtain linear and shear strains given by

$$\varepsilon_x = p\frac{\partial n_x}{\partial x},\qquad\qquad(9.14a)$$

$$\varepsilon_y = p\frac{\partial n_y}{\partial y},\qquad\qquad(9.14b)$$

$$\gamma = p\left(\frac{\partial n_x}{\partial y} + \frac{\partial n_y}{\partial x}\right),\qquad\qquad(9.14c)$$

where n_x, n_y are the fringe order at point P along x, y directions, respectively

Gratings used for strain analysis with in-plane moiré method are 20–40 lines/mm and formed by holographic interference technique, e-beam writing, x-ray lithography, phase mast, etc. Object grating is transferred to the object (metal) by lithography using photosensitive coating or photoresist or dichromate gelatin on the object.

9.3   OUT-OF-PLANE MOIRÉ METHOD AND PROFILOMETRY

As mentioned in the previous section, in-plane moiré pattern is generated by superposing the reference and object gratings in the same plane. In applying out-of-plane moiré method to the contour mapping, the object grating formed across the object is distorted in accordance with object profile. This out-of-plane moiré method is also termed moiré topography. The moiré topography is categorized mainly into shadow moiré and projection moiré methods. Shadow moiré method is known as the first example of applying the moiré phenomenon to 3D measurement. (In the early 1970s,

H. Takasaki and D. Meadows et al. applied shadow moiré as a technique to the observation of contours on object surface successfully.[5,6]

Started with the principle of moiré pattern formation of shadow moiré, following sections introduce its application to whole-field measurement and introduce projection moiré.

9.3.1 SHADOW MOIRÉ AND CONTOUR

Shadow moiré pattern is formed between the reference grating and its shadow (the object grating) on the object. The arrangement for shadow moiré is shown in Figure 9.6. The point light source and the detector (the aperture of detector lens is assumed a point) are at a distance l from the reference grating surface and their interseparation is s. Period of the reference grating is p ($p \ll l$ and $p \ll s$). Without loss of generality, we may assume that a point O on the object surface is in contact with the grating. The grating lying over the object surface is illuminated by the point source, and its shadow is projected onto the object. The moiré pattern observed from the detector is the result of superposition between the grating elements contained in OB of the reference grating and the elements contained in OP of the objected grating, which is the shadow of elements contained in OA of the reference grating. Assuming that OA and OB have i and j grating elements, respectively, from the geometry

$$AB = OB - OA = jp - ip = np, \quad n = 0, 1, 2, 3, \ldots \tag{9.15}$$

$$AB = h_n (\tan \alpha + \tan \beta), \tag{9.16}$$

where
  n is the order of the moiré pattern
  h_n is the depth of nth-order moiré pattern as measured from the reference grating ($n = 6$ in Figure 9.6)

Hence,

$$h_n = \frac{np}{\tan \alpha + \tan \beta}. \tag{9.17}$$

FIGURE 9.6 Optical arrangement of shadow moiré.

From Figure 9.6, we also have

$$\tan\alpha = \frac{P_x}{l+h_n} \quad \text{and} \quad \tan\beta = \frac{s-x_P}{l+h_n},$$

(9.18)

where x_P is x component of OP

Substituting these equations in Equation 9.17 and rearranging lead to

$$h_n = \frac{npl}{s-np}.$$

(9.19)

From Equation 9.19, it is seen that the moiré fringes are contours of equal depth measured from the reference grating.

Figure 9.6b shows moiré contour lines on a convex object. Besides these contours, intensity distribution of these moiré fringes also causes interesting result. In practice, knowing the order n of a contour line, we can approximately guess the location of points on the object, and knowing further more about the intensity of that fringe, we can exactly plot measurement point.

9.3.2 Intensity of Shadow Moiré Pattern

Before mathematical development of intensity of moiré fringes, let us review about the square wave grating shadow moiré deals with. In Fourier mathematics, it is known that all types of periodic functions (including square wave function) can be described as a sum of simple sinusoidal functions. In shadow moiré, the amplitude transmittance of the square wave grating is considered as that of a sinusoidal grating.

$$T(x,y) = \frac{1}{2} + \frac{1}{2}\cos\left(\frac{2\pi}{p}x\right).$$

(9.20)

Then, the resulting intensity at point P is proportional to the product $T_A(x_A, y)\cdot T_B(x_B, y)$.

$$I(x,y) = \left[\frac{1}{2} + \frac{1}{2}\cos\left(\frac{2\pi}{p}x_A\right)\right]\left[\frac{1}{2} + \frac{1}{2}\cos\left(\frac{2\pi}{p}x_B\right)\right].$$

(9.21)

From Figure 9.6, it is seen that $x_A = lx_P/(l+h_P)$, $x_B = (sh_P+lx_P)/(l+h_P)$. Substituting these equations in Equation 9.21 and rearranging it, we have following normalized intensity:

$$I(x,y) = 1 + \cos\frac{2\pi}{p}\left(\frac{lx_P}{l+h_n}\right) + \cos\frac{2\pi}{p}\left(\frac{sh_n+lx_P}{l+h_n}\right) + \frac{1}{2}\cos\frac{2\pi}{p}\left(\frac{2lx_P+sh_n}{l+h_n}\right) + \frac{1}{2}\cos\frac{2\pi}{p}\left(\frac{sh_n}{l+h_n}\right).$$

(9.22)

The last term in Equation 9.22 is solely dependent on height and is the contour term. The other three cosine terms representing the reference grating, although height-dependent, are also dependent on x (location) and, hence, do not represent contours. The patterns corresponding to these three terms can obscure the contours as shown in Figure 9.7a. Intensity distribution along cross-sectional line clearly shows the disturbance of the reference grating itself. To remove these unwanted patterns, H. Takasaki proposed, during exposure, to translate the grating in azimuth. The resultant intensity is proportional to

$$I(x,y) = K(x,y)\left[1 + \frac{1}{2}\cos\frac{2\pi}{p}\left(\frac{sh_n}{l+h_n}\right)\right] = a(x,y) + b(x,y)\cos\left[\frac{2\pi}{p}\left(\frac{sh_n}{l+h_n}\right)\right],$$

$$= a(x,y) + b(x,y)\cos\varphi(x,y) \qquad (9.23)$$

where
   $a\,(=K)$ is the intensity bias
   $b\,(=K/2)$ is the amplitude
   φ is the phase related to the temporal phase shift of this cosine variation

The resultant moiré fringes of this equation are shown in Figure 9.7b. Compared with Figure 9.7a, it is obviously seen in Figure 9.7b that the unwanted noise patterns are "mixed out" due to the averaging effect of the reference grating.

When we take pictures, during the exposure, if the object moves, we will get unclear picture. Here, this effect helps us, on the contrary, have clear contour image. Say again, in-plane moiré

(a)   Before translating the grating   (b)   After translating the grating

Intensity along cross-sectional line A–A'   Intensity along cross-sectional line B–B'

FIGURE 9.7   Shadow moiré pattern (a) before and (b) after translating the reference grating.

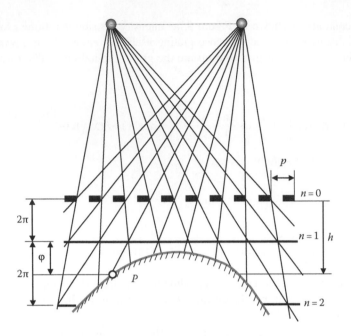

FIGURE 9.8 Relation between phase and the shadow moiré pattern.

method, translating the reference grating is one of important techniques to shift the moiré fringe and then determine the sign and order of the fringe.

Figure 9.8 shows the relations between the phase φ and the moiré contour. The depth $h(x, y)$ of any point $P(x, y)$ on the object surface, then, can be got from the following equation:

$$h(x, y) = h_n + (h_{n+1} - h_n)\left[\frac{\varphi(x, y)}{2\pi}\right]. \tag{9.24}$$

9.3.3 Application to 3D Profile Measurement

To map 3D profiles of objects with shadow moiré topography, phase distribution encoded in intensity distribution can be retrieved from the moiré pattern by using a phase-shifting method.[7–18]

9.3.3.1 Phase-Shifting Method

In electromagnetic wave interference, the phase-shifting method is well done to get the phase information from modulated intensity. The four-step algorithm is particularly famous for phase calculation in image processing. Here, we apply this algorithm to mechanical interference—moiré. The four-step algorithm to obtain the phase φ in intensity Equation 9.23 is given by

$$\varphi = \tan^{-1}\frac{I_3 - I_1}{I_0 - I_2}, \tag{9.25}$$

where I_k is the intensity of the moiré fringes across the object and is described as follows:

$$I_k = a + b\cos\left(\varphi + \frac{\pi}{2}k\right), \quad k = 0, 1, 2, 3. \tag{9.26}$$

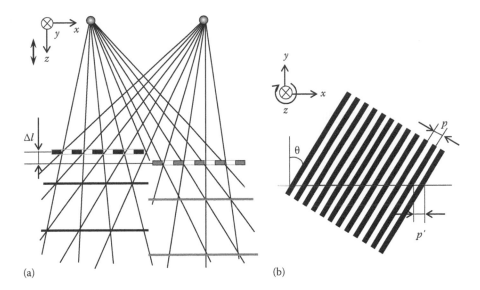

FIGURE 9.9 (a) Up-and-down movement of the grating and the resultant moiré pattern shift and (b) rotation of the grating.

In shadow moiré, a vertical movement of the grating results in a change of moiré pattern as well as a shift in the phase (see Figure 9.9a). The distance Δh between adjacent moiré fringes (i.e., $\Delta n = 1$) can be deduced from Equation 9.19:

$$\Delta h_{n,n-1} = h_n - h_{n-1} = \frac{dpl}{(s-np)[s-(n-1)p]}.$$ (9.27)

Hence, when the quantity of a vertical movement of the grating is Δl, the shifted phase can be expressed as follows:

$$2\pi \frac{\Delta l}{\Delta h_{n,n-1}} = \frac{2\pi \Delta l(d-np)\left[d-(n-1)p\right]}{dpl}.$$ (9.28)

It is seen that this quantity is not constant but decreases with the order n of moiré contours. Therefore, it is impossible to attain a constant phase change merely by vertical movement of the grating. To keep the phase shift in every order as constant as possible, it is available to rotate the reference grating in addition to move it vertically. Figure 9.9b shows the reference grating rotated, and this rotation results in the variation of the measurement period by

$$p' = \frac{p}{\cos\theta},$$ (9.29)

where θ is the rotation angle of the reference grating.

The combination of the vertical movement and rotation of the reference grating results in two equations: first, the distance h_n' between the moiré contour of nth order and the reference grating plane, which is transformed from h_n, is given as follows:

$$h_n' = \frac{n(l+\Delta l)[p/\cos\theta]}{s-n[p/\cos\theta]}.$$ (9.30)

Second, the depth h at the point $P(x, y)$ can be expressed with h_n' and Δl under the condition that h_n' exists between h_n and h_{n+1}:

$$h = h_n + (h_{n+1} - h_n)\varphi / 2\pi$$
$$= h_n' + \Delta l. \tag{9.31}$$

From Equation 9.31, to shift the phase φ by $\pi/2$, π, $3\pi/2$, the needed quantities of vertical movement Δl and rotation angle θ of the reference grating are

$$\Delta l = \frac{\varphi p l}{2\pi(s - p)}, \tag{9.32}$$

$$\theta = \cos^{-1}\left(\frac{l}{l + \Delta l}\right). \tag{9.33}$$

Figure 9.10 shows example of four images with phase-shifted moiré patterns and the analyzed results expressed with wire frame.

The limitation of this phase evaluation method is that it cannot be applied to those objects with discontinuous height steps and/or spatially isolated surfaces, because the discontinuities and/or the surface isolations hinder the unique assignment of fringe orders and the unique phase unwrapping.

9.3.3.2 Frequency Sweeping Method

In laser interferometric area, the wavelength shift method is used to measure 3D shapes of objects. Similar concept can be applied to moiré pattern analysis. With this concept, named frequency sweeping, the distance of object from the reference grating plane can be measured by evaluating temporal carrier frequency instead of the phase. Different from the wavelength shift method, this technique changes the grating period p by rotating the grating in different intervals and produces the spatiotemporal moiré patterns.

Let's think again about Equation 9.23,

$$I(x, y) = a(x, y) + b(x, y)\cos\left[\frac{2\pi}{p}\left(\frac{sh(x, y)}{l + h(x, y)}\right)\right]. \tag{9.34}$$

Here, $2\pi/p$ is defined as the virtual wave number, which is analogous to the wave number $k = 2\pi/\lambda$ (λ: wavelength).

As stated in earlier section, when the reference grating is rotated, the measurement grating period of p is changed, namely, the virtual wave number $g = 2\pi/p$ is changed with time t. By controlling the amount of the rotation angle θ, we can get the following quasi-linear relationship between different virtual wave number and time.

$$g(t) = g_0 + C \cdot t, \tag{9.35}$$

where
C is a constant showing the variation of the virtual wave number
g_0 is the initial virtual wave number $2\pi/p$

Then, Equation 9.34 can be rewritten in the following time-varying form:

$$I(x, y; t) = a(x, y; t) + b(x, y; t)\cos\left[g(t)H(x, y)\right], \tag{9.36}$$

$$H(x, y) = \left[\frac{dh(x, y)}{l + h(x, y)}\right]. \tag{9.37}$$

FIGURE 9.10 (a–d) Four images with modulated intensity distribution (a through d) and analyzed result. (With permission of Japan Society for Precision Engineering [35].)

Substituting Equation 9.26 into Equation 9.24 results in the following equation:

$$I(x, y; t) = a(x, y; t) + b(x, y; t) \cos\left[CH(x, y)t + g_0 H(x, y)\right].$$

Here, let us define the temporal carrier frequency $f(x, y)$ as

$$f(x, y) = \frac{C \cdot H(x, y)}{2\pi}, \tag{9.38}$$

and the second term, initial phase, as

$$\varphi_0 = g_0 \cdot H(x, y), \tag{9.39}$$

then,

$$I(x, y; t) = a(x, y; t) + b(x, y; t) \cos[2\pi f(x, y)t + \varphi_0(x, y)]. \tag{9.40}$$

As the virtual wave number varies with time t, the intensities at different points vary as shown in Figure 9.11. In this sinusoidal variation, it is obvious that the amounts of the modulated phase φ_0 and the temporal carrier frequency $f(x, y)$ depend on the distance $h(x, y)$ of the object. This means, for the latter, the further the distance, the higher the frequency. Therefore, the height distribution of an object can be obtained from the temporal carrier frequency:

$$h(x, y) = \frac{l}{dC/2\pi f(x, y) - 1}. \tag{9.41}$$

The frequency f in Equation 9.41 is available by applying fast Fourier transform (FFT) method. Since this frequency sweeping method does not involve phase φ, it is not necessary to carry out the phase-unwrapping process.

Figure 9.12 shows a measurement result, by using the method given earlier, of two objects (ring and rectangle in shape) separated from each other. The period of the grating used for shadow moiré is about 1.0–2.0 mm. The problem of shadow moiré method is that a big grating is necessary for the

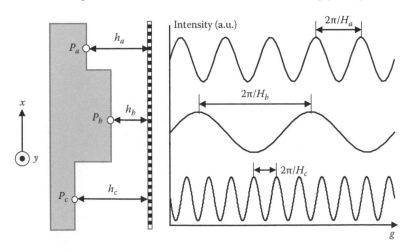

FIGURE 9.11 Relation between the distance h and spatial frequency g.

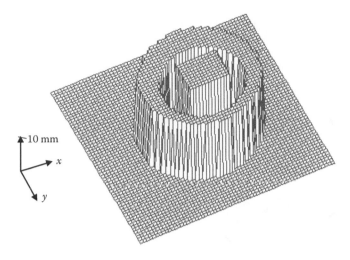

FIGURE 9.12 Measurement result by means of frequency sweeping method. (From Jin, L. et al., *Opt. Eng.*, 40(7), 1383, 2001. With permission.)

measurement and the size is dependent on that of objects. The projection moiré method can solve this problem very flexibly by using two gratings.

9.3.4 PROJECTION MOIRÉ

Projection moiré uses two exactly same gratings: one for projection and the other for reference as shown in Figure 9.13. The projection grating is projected across the object, and the detector captures an image through the reference grating. From Figure 9.13, it is obvious that the principle of

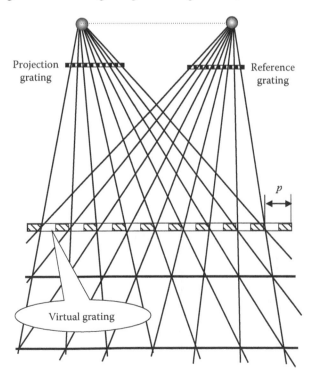

FIGURE 9.13 Principle of projection moiré.

moiré pattern formation is completely same as that of shadow moiré. The concepts of phase-shift and frequency sweeping methods are also valuable for projection moiré. For these methods, it just needs to move one of two gratings. The gratings can be easily made of liquid crystal panel plate, or without using any physical gratings, one just projects grating patterns, which is designed with the aid of an application software, through a liquid crystal projector, and then superposes the same grating patterns, inside of the analysis system, on the image with deformed grating patterns.[19] All processes such as shifting the grating and modulating the period can be easily carried on with computer programming.

9.4 REFLECTION MOIRÉ

In preceding sections, the moiré technique is performed for strain or profile measurement of diffuse objects. The reflection moiré method can be applied on the objects with mirror-like surfaces.[20–23] According to how the reflection moiré fringe is obtained, there are several types of reflection moiré method. Figure 9.14 shows one of examples of reflection moiré. The mirror-line surface of the object makes the virtual mirror image of the grating. Then, through the lens, the image of the mirror image of the grating is observed. The moiré fringe can be obtained by exposing double times, before and after deformation of the object, or by placing a reference grating in the image plane of lens. The resulting reflection moiré pattern can be applied to the analysis of slope deformation.

9.5 DIFFRACTION GRATING, INTERFERENCE MOIRÉ, AND APPLICATION

When a holographic grating is illuminated by a monochromatic parallel light beam under an angle of incidence θ_0, between the grating and the object, there occur light beams of zeroth and diffracted higher orders, as shown in Figure 9.15. Namely, the grating acts as diffraction grating. Each beam then interferes with all others. In practice, due to the technique of holographic grating generation, only the zeroth and ±1st orders interfere. The interference pattern formed by the zeroth and first diffraction order has the same period as that of the diffraction grating p. It extends in the direction θ:

$$\theta = \frac{\theta_0 + \theta_1}{2},$$
(9.42)

where θ_1 is the propagation direction of the first diffraction order beam.

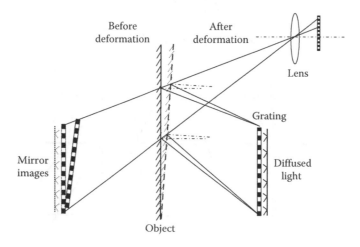

FIGURE 9.14   Principle of reflection moiré.

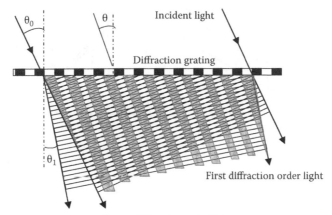

FIGURE 9.15   Interference pattern by diffracted light waves.

According to diffraction theory, the angle θ_1 is given by

$$\theta = \sin^{-1}\left(\frac{\lambda}{p} + \sin\theta_0\right),\tag{9.43}$$

where λ is the wavelength of incident light.

As depicted in Figure 9.16, the interference pattern is reflected at the surface of the object. The reflected interference pattern will be deformed by the flatness of the object. Through a detector, one can observe moiré fringes superposed by the reflected interference pattern and the diffraction grating (reference grating). From Figure 9.16, for an optical arrangement with parallel illumination and parallel detecting, the distance Δh between two contours can be expressed as

$$\Delta h = \frac{p}{2\tan\theta}.\tag{9.44}$$

The sensitivity of this method, see from Equation 9.44, will be changed by the incidence angle. This method is also applicable for coherent and incoherent monochromatic light sources, and due to its high sensitivity, moiré method is very attractive for flatness measurement field of such objects as, computer

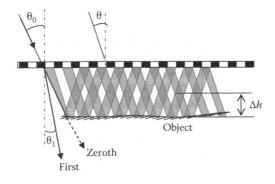

FIGURE 9.16   Generation of moiré pattern.

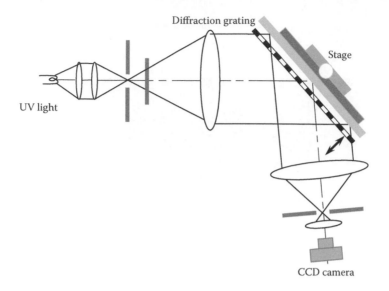

FIGURE 9.17   Flatness measurement system using moiré method.

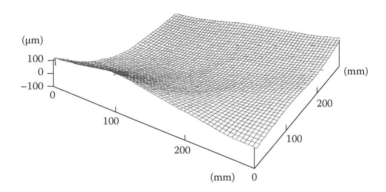

FIGURE 9.18   Measurement result of flatness of a soda glass substrate. (From Fujiwara, H. et al., *Proc. SPIE*, 2862, 172, 1996. With permission.)

disks, wafers, and smoothly polished glass substrates with highly processed. Figure 9.17 shows a system applying the earlier method to measure flatness of a soda glass substrate. The moiré pattern can be shifted by moving the grating perpendicularly to the objects. Here, the phase-shift method introduced in the previous section can be applied for pattern analysis. In this system, to cancel the influence of the light reflected from the back side of the substrate, an ultraviolet (UV) light ($\lambda = 313$ nm) is used as the light source. Part of the UV light will be reflected on the surface of the soda glass, and the left part transmitted into the glass will be absorbed by it. Figure 9.18 shows measurement results by using this system. The grating employed was 10 lines/mm, and the incident angle 60°.

9.6   RECENT TOPICS

As described in the previous sections, moiré phenomenon has been a good topic of research in various fields. Still now we can enjoy moiré produced by overlapping line gratings or random patterns, for example, on the website "http://switzernet.com/people/emin-gabrielyan/070306-paper-moire-basics/paper-eps.pdf (The basics of line moire patterns and optical speedup by Emin Gabrielyan)."

At the same time, such an interesting application of moiré as "moiré-based optical tweezers" has been reported. Microparticles and bacteria were trapped and rotated by this idea.[24]

Regarding moiré applications, we had an opportunity of writing one chapter "Moiré Measurement" in the book.[26] We would like to suggest readers who are interested in optical methods for dimensional measurements to read this handbook.

9.6.1 DEVELOPMENT IN SHAPE MEASUREMENT

In electromagnetic wave interference, the phase-shifting method is well done to get the phase information from modulated intensity. The four-step algorithm is particularly famous for phase calculation in image processing. As mentioned in Section 9.3.3.1, mere vertical movement of the grating is not able to invoke a constant phase. To solve this problem, in the early 1990s, Yoshizawa and Tomisawa[12] originally proposed to move the grating vertically and at the same time to translate the light source vertically.

In the arrangement in Figure 9.19, the interval Δh (0.65 mm) is too large to measure the surface profile of the coin (0.25 mm in maximum height). Therefore, phase-shifting technique (that was not popular until 1974) is required to be applied to realize higher sensitivity. They moved the grating vertically by Δl and the light source horizontally by Δd, then four images with mutual phase difference of $\pi/2$ are acquired as shown in Figure 9.20. Let us note that one or less than one contour is observed in this case. However, computer processing of these four images (a, b, c, d) using four-step phase-shifting algorithm can produce a representation of the result shown in Figure 9.20e. This method was applied to check damage or wear of a score cap (Figure 9.21a), which is used for die stamping on the top of cans. For the calibration of measurement (Figure 9.21b), cross-sectional profiles obtained by this method (Figure 9.21c) are shown with the result obtained by a mechanical

FIGURE 9.19 Arrangement for the surface measurement of a coin.

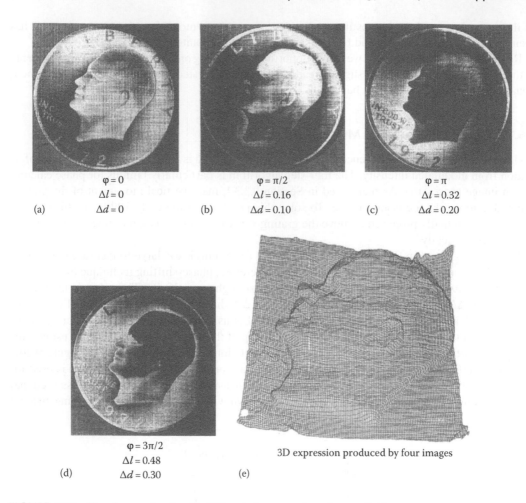

	$\varphi = 0$		$\varphi = \pi/2$		$\varphi = \pi$
	$\Delta l = 0$		$\Delta l = 0.16$		$\Delta l = 0.32$
(a)	$\Delta d = 0$	(b)	$\Delta d = 0.10$	(c)	$\Delta d = 0.20$

	$\varphi = 3\pi/2$		3D expression produced by four images
	$\Delta l = 0.48$		
(d)	$\Delta d = 0.30$	(e)	

FIGURE 9.20   Four images (a–d) with different phases produce the result (e).

contact profilometer. The two results agree well under the condition that these measurements have to be performed differently.

This method was the first trial to apply phase-shifting technique to shadow moiré. However, it was not so easy to use this interesting idea for practical purposes mainly because the devices such as the light source and the grating had to be moved mechanically and manually at that time.

In the case of projection moiré, phase-shifting technique is easily applicable by moving one of separated two gratings, and to get better quality of moiré images, one pair of gratings should be translated at the same speed during the exposure. A primitive proposal was made by Yoshizawa and Yonemura in 1976.[26] However, it was not until a few years ago that elements and devices such as liquid crystal gratings, moving stages, and small CCD or CMOS cameras became easily available in addition to PCs. One trial is shown in Figure 9.22, where two linear gratings (G_1 and G_2) are mounted on the motor-controlled translation stages (M_1 and M_2). Relative difference of the phase of gratings that move at constant speed produces heterodyne moiré signal.[27] Various ideas were integrated in this arrangement, and one measured example is shown in Figure 9.23. One cross section (Figure 9.24) of the coin proves that accuracy of this method is very high in comparison with the result by a mechanical stylus. Height error was said to be as small as 4.3 μm.

Similar and very refined system was reported by J.J. Dirckx et al. by incorporating such digital devices as the liquid crystal grating and CCD camera. In their arrangement shown in Figure 9.25,[28,29]

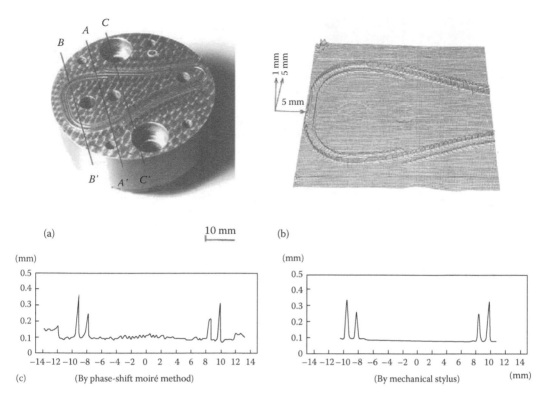

(a)

|——| 10 mm

(b)

(c)   (By phase-shift moiré method)

(By mechanical stylus)   (mm)

FIGURE 9.21 Checking of score cap: (a) score cap, (b) measurement result, and (c) cross section AA' by moiré and contact methods.

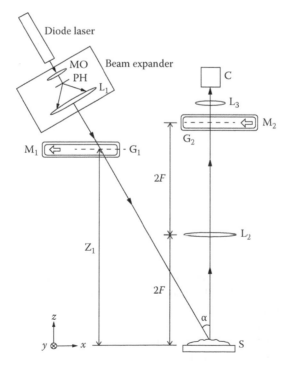

FIGURE 9.22 Optical configuration. S, sample, G_1 and G_2, linear gratings, M_1 and M_2, motorized translation stage [27].

FIGURE 9.23   (a) Sample coin and measured results, (b) moiré fringes, (c) phase map, and (d) reconstructed coin image [27].

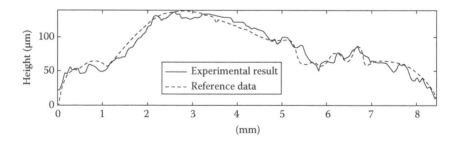

FIGURE 9.24   Cross section *AA'*. Experimental result and reference data by a stylus method [27].

the green grating (reference grating) is projected on to the sample through the projector lens, and the deformed grating image is formed on the blue-channel liquid crystal light modulating matrix placed outside of the projector with two orthogonal polarizers (second grating). Produced moiré pattern is captured by the CCD camera. In addition to easy phase shift realized by moving one grating, these paired gratings are translated simultaneously to remove grating noise and averaging the "ruling error" of the gratings. This sophisticated system solved various problems inherent to the classical moiré measurement method.

FIGURE 9.25 Setup for digital moiré topography using LCD projector [28,29].

9.6.2 HIGH-SPEED MEASUREMENT

One of the requests to optical methods for industrial purposes is high-speed or, if possible, real-time measurement. In moiré applications, such trial shown in Figure 9.19 may be high in accuracy, but too tedious in operation for practical usage. Not a few techniques have been proposed to solve this problem. Recently, one idea was born that is called as "sampling moiré method."[31–33] The fundamental principle of this method was reported as early as in 1997 by Arai et al.,[30] which aimed to analyze phase distribution of moiré images produced from one-shot image of the deformed grating pattern. At first, this idea was thought just one of "one-shot methods" classified into spatial phase-shifting methods. However, in a subsequent experimental study, the phase distribution of phase-shifted moiré was proved to be analyzed with higher accuracy and higher speed than other methods like FFT method in some cases. This method was applied to analyze 2D grating image. If this principle is used for analyzing 1D grating, it is actually same with the spatial phase analysis based on sampling technique. This sampling method was applied to a dynamically moving object and succeeded in shape measurement as well as strain measurement.

In the sampling moiré method, a specimen grating pattern pasted or drawn on an object is recorded by a digital camera. A specified camera system for this purpose was also developed.[31] Though the digitized image shows grating, a moiré fringe pattern appears by thinning out the pixels, that is, by sampling the image with a constant pixel pitch. Figure 9.26 illustrates the appearance of moiré fringe patterns by the sampling moiré method. Figure 9.26a shows the deformed grating pattern attached to the specimen. The pitch of the grating in the figure is 1.125 times larger than that of the sampling points. The recorded image is shown in Figure 9.26b, in which no moiré fringe pattern can be discerned. Figure 9.26c shows the moiré fringe patterns when the recorded image is sampled at every Nth pixel (in this case, $N = 4$). The four images in Figure 9.26c are obtained by using the first, second, third, and fourth pixels of Figure 9.26b as the sampling start point, respectively. This process corresponds to the phase shifting of the fringe pattern. The sampled images shown in Figure 9.26c are interpolated using neighboring data. Then, the four phase-shifted moiré images shown in Figure 9.26d are obtained from one single picture in Figure 9.26b. Phase distribution analyzed from Figure 9.26d is shown in Figure 9.26e. This principle is applicable to the

FIGURE 9.26   Principle of sampling moiré method [31,34].

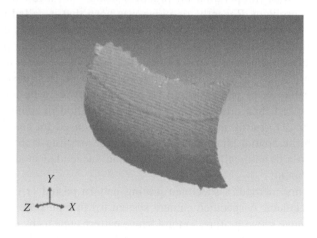

FIGURE 9.27   3D shape measurement of a rotating tire [33].

measurement of strain as well as surface shape. Examples are shown in Figures 9.27 and 9.28.[32,33] This technique is applicable to capture dynamic objects like a rotating tire and vibrating plates. In the case of mechanical movement of stages or devices, the sampling speed is limited by their moving speed. But in this case, much higher speed is attainable because no mechanical device is used to shift the pattern, that is, to sampling the moiré fringes with continuously changing phase. The next request from industrial fields is the realization of portable and compact equipment, and more and more improvement will be added. From the viewpoint of factory automation, the theses[34] are useful and interesting in that it covers such high-speed optical techniques as pattern projection methods (structured light method) in addition to moiré method.

FIGURE 9.28 Measurement result of the strain distribution: (a) in *x*-direction, (b) in *y*-direction, and (c) shear strain [33].

9.7 CONCLUSION

Moiré method and applications are described with an emphasis on dimensional measurement and, at the same time, how moiré (originally classical physical phenomenon) has been staying alive for a long time because of its varied characteristics. More different topics can be found in our other review work.[25]

REFERENCES

1. W. F. Riley and A. J. Durelli, Application of Moiré methods to the determination of transient stress and strain distributions, *J. Appl. Mech.*, 29(1), 23, 1962.
2. F. P. Chiang, On Moiré method applied to the determination of two-dimensional dynamic strain distributions, *J. Appl. Mech., Trans. ASME*, 39(Series E, 3), 829–830, 1972.
3. F. P. Chiang, V. J. Parks, and A. J. Durelli, Moiré fringe interpolation and multiplication by fringe shifting, *Exp. Mech.*, 8(12), 554–560, 1968.
4. F. P. Chiang, Moiré methods of strain analysis, *Exp. Mech.*, 19(8), 290–308, 1979.
5. H. Takasaki, Moiré topography, *Appl. Opt.*, 9(6), 1467–1472, 1970.
6. D. M. Meadows, W. O. Johnson, and J. B. Allen, Generation of surface contours by moiré patterns, *Appl. Opt.*, 9(4), 942–947, 1970.
7. J. B. Allen and D. M. Meadows, Removal of unwanted patterns from moiré contour maps by grid translation techniques, *Appl. Opt.*, 10(1), 210–213, 1971.

8. L. Pirodda, Shadow and projection moire techniques for absolute or relative mapping of surface shapes, *Opt. Eng.*, 21(4), 640–649, 1982.
9. H. E. Cline, W. E. Lorensen, and A. S. Holik, Automatic moire contouring, *Appl. Opt.*, 23(10), 1454–1459, 1984.
10. V. Srinivasan, H. C. Liu, and M. Halioua, Automated phase-measuring profilometry of 3-D diffuse objects, *Appl. Opt.*, 23(18), 3105–3108, 1984.
11. J. J. J. Dirckx, W. F. Decraemer, and G. Dielis, Phase shift method based on object translation for full field automatic 3-D surface reconstruction from moire topograms, *Appl. Opt.*, 27(6), 1164–1169, 1988.
12. T. Yoshizawa and T. Tomisawa, Shadow moire topography by means of the phase-shift method, *Opt. Eng.*, 32(7), 1668–1674, 1993.
13. G. Mauvoisin, F. Brémand, and A. Lagarde, Three-dimensional shape reconstruction by phase-shifting shadow moiré, *Appl. Opt.*, 33(11), 2163–2169, 1994.
14. Y. Arai, S. Yokozeki, and T. Yamada, Fringe-scanning method using a general function for shadow moiré, *Appl. Opt.*, 34(22), 4877–4882, 1995.
15. X. Xie, J. T. Atkinson, M. J. Lalor, and D. R. Burton, Three-map absolute moiré contouring, *Appl. Opt.*, 35, 6990–6995, 1996.
16. X. Xie, M. J. Lalor, D. R. Burton, and M. M. Shaw, Four-map absolute distance contouring, *Opt. Eng.*, 36(9), 2517–2520, 1997.
17. L. Jin, Y. Kodera, Y. Otani, and T. Yoshizawa, Shadow moiré profilometry using the phase-shifting method, *Opt. Eng.*, 39(8), 2119–2213, 2000.
18. L. Jin, Y. Otani, and T. Yoshizawa, Shadow moiré profilometry using the frequency sweeping, *Opt. Eng.*, 40(7), 1383–1386, 2001.
19. A. Asundi, Projection moire using PSALM, *Proc. SPIE*, 1554B, 254–265, 1991.
20. F. K. Lichtenberg, The moire method: A new experimental method of the determination of moments in small slab models, *Proc. SESA*, 12, 83–98, 1955.
21. F. P. Chiang and J. Treiber, A note on Ligtenberg's reflective moiré method, *Exp. Mech.*, 10(12), 537–538, 1970.
22. W. Jaerisch and G. Makosch, Optical contour mapping of surfaces, *Appl. Opt.*, 12(7), 1552–1557, 1973.
23. H. Fujiwara, Y. Otani, and T. Yoshizawa, Flatness measurement by reflection moiré technique, *Proc. SPIE*, 2862, 172–176, 1996.
24. P. Zhang, D. Hernandez, D. Cannan, Y. Hu, S. Fardad, S. Huang, J. C. Chen, D. N. Christodoulides, and Z. Chen, Trapping and rotating microparticles and bacteria with moire-based optical propelling beams, *Biomed. Opt. Express*, 3(8), 1891–1897, 2012.
25. K. Harding, ed., *Handbook of Optical Dimensional Metrology*, CRC Press (Taylor & Francis), Boca Raton, FL, 2013.
26. T. Yoshizawa and M. Yonemura, Moiré topography with lateral movement of a grating, *J. Jpn. Soc. Prec. Eng.*, 43(5), 556–561, 1976 (in Japanese).
27. W. Chang, F. Hsu, K. Chen, J. Chen, and K. Hsu, Heterodyne moiré surface profilometry, *Opt. Express*, 22(3), 2845–2852, 2014.
28. J. Buytaert and J. Dirckx, Fringe generation and phase shifting with LCDs in projection moiré topography, *Proc. SPIE*, 7522, 75220H-1–75220H-8, 2010.
29. J. Dirckx, J. Buytaert, and S. Van der Jeught, Implementation of phase-shifting moiré profilometry on a low cost commercial data projector, *Opt. Lasers Eng.*, 48(2), 244–250, 2010.
30. Y. Arai, S. Yokozeki, K. Shiraki, and T. Yamada, High precision two-dimensional spatial fringe analysis method, *J. Mod. Opt.*, 44(4), 739–751, 1997.
31. M. Fujigaki, Y. Morimoto, and Q. Gao, Shape and displacement measurement by phase-shifting scanning moiré method using digital micro-mirror device, *Proc. SPIE*, 4537, 362–365, 2002.
32. M. Fujigaki, Y. Sasatani, A. Masaya, H. Kondo, M. Nakabo, T. Hara, Y. Morimoto, D. Asai, T. Miyagi, and N. Kurokawa, Development of sampling moiré camera for real-time phase analysis, *Appl. Mech. Mater.*, 83, 48–53, 2011.
33. M. Fujigaki, K. Shimo, A. Masaya, and Y. Morimoto, Dynamic shape and strain measurements of rotating tire using a sampling moiré method, *Opt. Eng.*, 50(10), 101506-1–101506-8, 2011.
34. D. Asai, Speeding up 3D measurement with phase shifting grating projection in factory automation, Doctor thesis, Wakayama University, Wakayama, Japan, March 2014 (in Japanese).
35. Y. Kodera, L. Jin, Y. Otani, and T. Yoshizawa, Shadow moire profilometry using phase shifting method, *J. Precision Eng.*, 65(10), 1455, 1999 (in Japanese).

10 Optical Heterodyne Measurement Method

Masao Hirano

CONTENTS

10.1 Introduction 300
10.2 Heterodyne Method 300
    10.2.1 High-Precision Measurement: Principle and Defects 300
    10.2.2 Construction of the Cancelable Circuit: An Introduction of Two Signals
        by Two Beams 302
    10.2.3 Construction of the Cancelable Circuit: Generation and Selection
        of Two Lights 306
    10.2.4 Construction of the Cancelable Circuit: Symmetrical Layout in the Circuit 307
    10.2.5 Relation between the Doppler Method and the Heterodyne Method 308
10.3 Accuracy and Noise Reduction...309
    10.3.1 Deterioration of Accuracy Caused by the Environment 309
    10.3.2 Deterioration of Accuracy Caused by Optical Parts............................310
        10.3.2.1 AOM and Its RF Driver 310
        10.3.2.2 Frequency-Stabilized HeNe Laser..311
        10.3.2.3 Interference Circuit and Parts...312
        10.3.2.4 Sample Surface and Sample Holder 312
        10.3.2.5 Detection Method and System...313
10.4 Attention for Static and Dynamic Measurement 314
    10.4.1 Static Measurement..314
    10.4.2 Dynamic Measurement 314
10.5 Applications...314
    10.5.1 Displacement, Vibration Frequency, and Amplitude............................315
    10.5.2 Positioning with Stage 315
    10.5.3 Thick Measurement 315
    10.5.4 Profilometry 315
    10.5.5 Refractometry 316
    10.5.6 Photothermal Interferometry 317
    10.5.7 Thermal Expansion Coefficient 317
    10.5.8 Young's Modulus of Thin Film...318
    10.5.9 Birefringence Index and Polarimeter...318
    10.5.10 Fiber Sensor 320
    10.5.11 Dynamical Surface Measurement...320
10.6 Optical Doppler Measurement...320
    10.6.1 Introduction..320
    10.6.2 Principle and Defect...320

    10.6.3 Comparison between Doppler Method and Optical Heterodyne Method 321
            10.6.3.1 Accuracy 321
            10.6.3.2 Measurement of Velocity and Displacement 322
10.7 Measurement of Transient Phenomena 322
    10.7.1 Application of the Optical Heterodyne Method 322
    10.7.2 Attention to the Design of Transient Measuring System 323
    10.7.3 Transient Measurement Method 325
            10.7.3.1 Signal Processing 325
            10.7.3.2 Photoelectric Conversion Circuit 326
            10.7.3.3 Period Detection 327
            10.7.3.4 Phase Detection and Displacement 328
            10.7.3.5 Analysis of Error Factor 329
    10.7.4 Examples of Transient Measurement 331
            10.7.4.1 Tuning Fork and Cantilever Vibration 331
            10.7.4.2 Concentration Fluctuation 335
References 337

10.1   INTRODUCTION

The optical heterodyne method is remarkable as a noncontact method that measures physical values with high precision. To achieve high precision, there are many restraints and conditions. In this chapter, first, the principle of the method is described, and the defects of the method are also shown. Second, the restraints and conditions are explained. These require the introduction of a cancelable optical circuit as a necessary condition. Third, several desirable conditions for optical parts and mechanical parts used in the circuit are given in detail. Finally, some applications are listed.

Utilizing the merits of the heterodyne method, heterodyne technology has also been used in similar fields. This chapter is confined mainly to a description of the basic view of the heterodyne method. Some heterodyne detection techniques and their applications are omitted: for example, OTDR [1], laser radar [2], spectrophotometry [3–5], optical CT [6–8], nerve bundle measurement [9], gravitational wave detection [10], interferometer in astronomy [11,12], heterodyne speckle interferometer [13,14], optical bistability measurement [15], fiber gyro [16,17], interferometer in atomic physics [18,19], and measurements for many physical variables.

10.2   HETERODYNE METHOD

10.2.1   HIGH-PRECISION MEASUREMENT: PRINCIPLE AND DEFECTS

The optical heterodyne method consists of an interferometer with two laser beams whose wavelengths are shifted by modulators from a laser wavelength.

These beams are represented with the wavelengths of λ_A and λ_B, as follows:

$$A \exp\left(i2\pi \frac{ct}{\lambda_A} + \phi_A \right) \quad \text{and} \quad B \exp\left(i2\pi \frac{ct}{\lambda_B} + \phi_B \right),$$

where
    A and B are optical strengths
    c and t are the light velocity and time, respectively
    ϕ_A and ϕ_B are initial phases of two beams

The interference signal of the two beams is represented with the times elapsed from a laser to a detector, t_A and t_B, as follows:

$$\left| A\exp\left(i2\pi\frac{c(t-t_A)}{\lambda_A}+\phi_A \right)+B\exp\left(i2\pi\frac{c(t-t_B)}{\lambda_B}+\phi_B \right) \right|^2$$

$$= \text{const} + 2AB\sin\left(2\pi ct\left(\frac{1}{\lambda_A}-\frac{1}{\lambda_B}\right)-2\pi c\left(\frac{t_A}{\lambda_A}-\frac{t_B}{\lambda_B}\right)+\phi_A-\phi_B \right). \tag{10.1}$$

The frequency of the interference signal (= beat frequency f) is represented as the difference of two wavelengths, $2\pi c(1/\lambda_A - 1/\lambda_B)$. The phase of the beat signal is described with the sum of two parts: (1) the difference of the initial phases and (2) the difference of optical path lengths in the circuit. This optical path length is given by the relation of *optical path length = Σ (the geometric distance of the path × the refractive index of the path)*. So, when the optical path lengths are $L_A (= ct_A)$ and $L_B (= ct_B)$ as shown in Figure 10.1, the phase of the beat signal changes by

$$\varphi = 2\pi\left(\frac{L_A}{\lambda_A}-\frac{L_B}{\lambda_B} \right). \tag{10.2}$$

When sample targets move further by ΔL_A and ΔL_B, the phase of the beat signal is

$$\varphi+\Delta\varphi = 2\pi\left(\frac{L_A+\Delta L_A}{\lambda_A}-\frac{L_B+\Delta L_B}{\lambda_B} \right), \quad \text{that is,} \quad \Delta\varphi = 2\pi\left(\frac{\Delta L_A}{\lambda_A}-\frac{\Delta L_B}{\lambda_B} \right). \tag{10.3}$$

Equation 10.2 shows that, under an ordinary condition of $L_A, L_B \gg \lambda$, the value of φ is very large, so it cannot be known; therefore, the value of L cannot be confirmed. However, Equation 10.3 shows that, under a condition $\Delta L_A, \Delta L_B < \lambda$, the value of $\Delta\varphi$ can give ΔL (either ΔL_A or ΔL_B) as follows:

$$\Delta L = \frac{\Delta\phi}{2\pi}\lambda. \tag{10.4}$$

This proportional relation between ΔL and λ means (1) the ruler (measuring scale) in the method is the wavelength and (2) the displacement ΔL can be measured in wavelength units. This is the

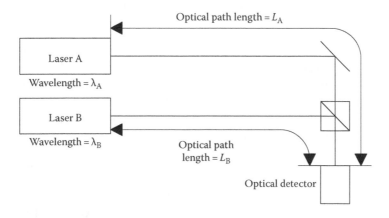

FIGURE 10.1 Wavelength and optical path length for one detector.

basic relation in the heterodyne method that achieves a high-precision measurement. In addition, Equations 10.1, 10.3, and 10.4 show that two beams must satisfy the following conditions:

1. The difference of the two beams' frequencies is a small value, which can be detected and handled with an electric circuit. For example, $\lambda_1 = 633.00000$ and $\lambda_2 = 633.00001$ produce a beat signal of $f = 7.5$ MHz. And beams need to be a single mode for both a longitudinal mode and a transverse mode.
2. The wavelength (i.e., frequency f) has to have high stabilization, which is represented with $\Delta f / f \leq 10^{-9}$, where Δf is a frequency fluctuation (e.g., $0.00001/633 = 1.6 \times 10^{-8}$).

These conditions limit the selection of a useful laser. Semiconductor lasers have many longitudinal modes and multi-transverse mode, and frequency stabilization is difficult to achieve with Ar and Kr lasers. HeNe laser has been attracted because its frequency fluctuation in a free-running operation is about 10^{-6}. No lasers can satisfy the condition in ordinary operations, but by developing a frequency-stabilized HeNe laser, the conditions are almost satisfied. Therefore, the frequency-stabilized HeNe laser is mainly used in the heterodyne measurement, which will be described in a later section.

It should be noted that Equation 10.4 also shows some of the defects in the heterodyne method.

Defect (1): When ΔL is over the wavelength within a sampling time, a correct displacement cannot be known because of its uncertain factor.
Defect (2): The displacement for the direction perpendicular to the beam direction cannot be measured.

Defect (1) can be improved by shortening the sampling time. However, Defect (2) is unescapable. Even if scattering light is gathered, no displacement information can be obtained because the scattering angle distribution cannot be measured. By using one more measurement system, Defect (2) will be improved.

10.2.2 Construction of the Cancelable Circuit: An Introduction of Two Signals by Two Beams

To achieve high-precision measurement with the heterodyne method, a new cancelable optical circuit has been proposed to keep the method's advantages. In this and the following sections, some essential points are shown.

The interference of two beams makes one ordinarily signal. For only one beat signal, one displacement is calculated with Equation 10.4. Two optical paths (L_1 and L_2) will be fluctuated independently by different causes. Therefore, to obtain a high-precision displacement (ΔL_1 or ΔL_2), the fluctuation has to be suppressed much less than the displacement. For example, the optical length of $L_1 = 20$ cm demands that the expansion ratio on the path is under 5×10^{-8} for $\Delta L_1 = 10$ nm resolution. This is very difficult to achieve. To make this strict condition easier, the system introduces two interference signals. As shown in Figure 10.2, both laser beams λ_A and λ_B are divided into two parts, and four optical paths, LA_1, LA_2, LB_1, and LB_2, make two pairs, and two detectors get two beat signals of pairs as written by

$$\text{Detector 1 SIG1} = C_1 \sin\left((\omega_A - \omega_B)t + \phi_A - \phi_B + \omega_A \times \frac{LA_1}{c} - \omega_B \times \frac{LB_1}{c} \right),$$

$$\text{Detector 2 SIG2} = C_2 \sin\left((\omega_A - \omega_B)t + \phi_A - \phi_B + \omega_A \times \frac{LA_2}{c} - \omega_B \times \frac{LB_2}{c} \right),$$

(10.5)

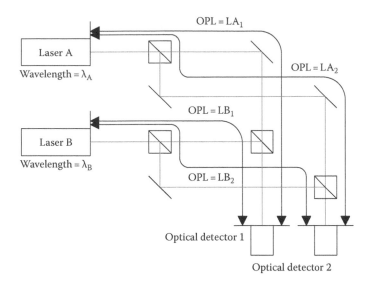

FIGURE 10.2 Wavelength and optical path length for two detectors.

where

  C_1 and C_2 are the strength of two beat signals
  ϕ_A and ϕ_B are the initial phase of two beams
  ω_A and ω_B are $2\pi c/\lambda_A$ and $2\pi c/\lambda_B$, respectively

The difference of two signal phases is given as follows:

$$\varphi + \Delta\varphi = \omega_A \frac{LA_1 + \Delta LA_1 - LA_2 - \Delta LA_2}{c} - \omega_B \frac{LB_1 + \Delta LB_1 - LB_2 - \Delta LB_2}{c}$$

$$\therefore \Delta\varphi = 2\pi \left(\frac{\Delta LA_1 - \Delta LA_2}{\lambda_A} - \frac{\Delta LB_1 - \Delta LB_2}{\lambda_B} \right). \tag{10.6}$$

Equation 10.6 is very similar to Equation 10.4. However, in Equation 10.4, ΔL is the displacement of an optical path, while in Equation 10.6, $\Delta LA_1 - \Delta LA_2$ is the difference of the displacements of two optical paths. The difference between Equations 10.4 and 10.6 is essential in a cancelable circuit. Equation 10.6 represents that the phase fluctuation can be reduced even if both the optical paths fluctuate. In particular, if two optical paths fluctuate for the same reason, the fluctuations do not affect $\Delta\varphi$ at all (i.e., the influences are canceled). In addition, Equation 10.6 also shows that several circuit layouts are possible as listed in Table 10.1.

By discussing these five cases in Table 10.1, the merits of two beat signals will be understood.

For case (a), the following conditions are required:

  1. The optical path lengths, LA_1, LB_1, and LB_2, never vary.
  2. The wavelength never varies.
  3. The distortion of the wave front never appears.

As these conditions are very hard to satisfy, there is no merit.

For the case of (b), the difference, $\Delta LA_1 - \Delta LA_2$, is obtained. This case fails one of the conditions, "LA_2 never vary." When LA_1 and LA_2 include a common path (having nearly the same length), a fluctuation on the common path between LA_1 and LA_2 affects equally and simultaneously.

TABLE 10.1

Resolutions for Circuit Layouts

Case	Variable	Constant	$\Delta\varphi$	Requested Result
(a)	LA_1	LA_2, LB_1, LB_2	$2\pi\Delta LA_1/\lambda_A$	ΔLA_1
(b)	LA_1, LA_2	LB_1, LB_2	$2\pi(\Delta LA_1 - \Delta LA_2)/\lambda_A$	$\Delta LA_1 - \Delta LA_2$
(c)	LA_1, LB_1	LA_2, LB_2	$2\pi(\Delta LA_1/\lambda_A - \Delta LB_1/\lambda_B)$	$\Delta LA_1 - \Delta LB_1$
			$\cong 2\pi(\Delta LA_1 - \Delta LB_1)/\lambda_A$	
(d)	LA_1, LB_2	LA_2, LB_1	$2\pi(\Delta LA_1/\lambda_A + \Delta LB_1/\lambda_B)$	$\Delta LA_1 + \Delta LB_1$
			$\cong 2\pi(\Delta LA_1 + \Delta LB_1)/\lambda_A$	
(e)	LA_1, LA_2, LB_1, LB_2	None	$2\pi\{(\Delta LA_1 - \Delta LA_2)/\lambda_A + (\Delta LB_1 - \Delta LB_2)/\lambda_B\}$	$\Delta LA_1 - \Delta LA_2 +$
			$\cong 2\pi(\Delta LA_1 - \Delta LA_2 + \Delta LB_1 - \Delta LB_2)/\lambda_A$	$\Delta LB_1 - \Delta LB_2$

Notes: The $\Delta\varphi$ values in cases (c) through (e) are approximate. However, the difference between λ_A and λ_B is extremely small, nearly $\lambda_A/\lambda_B = 1.00000002$ or less, so the measurement error is far more below a detectable error. The sign of results is determined by the displacement direction.

Therefore, the result can exclude or cancel the influence of the fluctuation. This effect is also obtained in a one-beam signal system.

For cases (c) through (e), the system of the two beat signals is more effective. Example circuits are illustrated in Figure 10.3 (a sample is set upon a basement) and Figure 10.4. A signal phase of the beam reflected on the sample fluctuates by movements of both the basement and the sample. A signal phase of the beam reflected on the basement fluctuates by only the movement of the basement. Therefore, the operation of $\Delta LA_1 - \Delta LB_1$ makes the movement of the basement disappear (cancel) and gives the displacement of the sample only. In the same discussion for Figure 10.4, beams reflected at the front and at the rear of sample cancel a vibration of the sample. Therefore, an operation of $\Delta LA_1 + \Delta LB_2$ makes the movement of the basement disappear (cancel), and the expansion/contraction of the sample only is measured. For this reason, they are called a cancelable optical circuit in this chapter.

The accuracy of this measurement depends on the degree to which these conditions are satisfied. The relationship between accuracy and these conditions will be described in Section 10.3.

FIGURE 10.3 Cancelable circuit: simultaneous measurement of sample and basement displacement. A fluctuation of basement can be out of consideration.

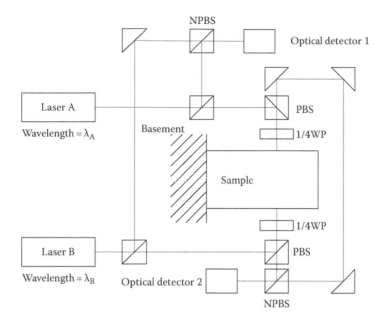

FIGURE 10.4 Cancelable circuit: thickness variation measurement of sample only.

Why does a cancelable optical circuit achieve better measurement? The answer is very simple. Even if several external noises or error factors occur in φ_1 and φ_2, the influences can be reduced. Two examples are discussed.

Case 1: a frequency fluctuation occurs from ω_A to $\omega_A + \delta\omega_A$.
The phases are modified in Equation 10.6 as follows:

$$\varphi_1 + \delta\varphi_1 = (\omega_A + \delta\omega_A)\frac{LA_1}{c} + \omega_B\frac{LB_1}{c} + \phi_A - \phi_B,$$

$$\varphi_2 + \delta\varphi_2 = (\omega_A + \delta\omega_A)\frac{LA_2}{c} + \omega_B\frac{LB_2}{c} + \phi_A - \phi_B.$$

$$(10.7)$$

The difference of the phases (= output signal) is given by

$$\varphi = (\varphi_1 + \delta\varphi_1) - (\varphi_2 + \delta\varphi_2) = \varphi_0 + \delta\omega_A\frac{LA_1 - LA_2}{c}. \tag{10.8}$$

In a one-beat signal system, the phase fluctuation is expressed by the relation, $\delta\varphi = \delta\omega \times LA/c$. Even if $\delta\omega$ is very small, a large LA does not result in a small $\delta\varphi$. In a two-beat signal system, the phase fluctuation is expressed with Equation 10.8. Smaller $LA_1 - LA_2$ results in lower fluctuation and better resolution. Usually, $\delta\omega_B$ will occur at one time.

Case 2: an environment fluctuation occurs from LA_1 and LA_2 to $LA_1 + \delta LA_1$ and $LA_2 + \delta LA_2$, respectively. The output signal is rewritten as

$$\varphi = \varphi_0 + \omega_A\frac{(\delta LA_1 - \delta LA_2)}{c}. \tag{10.9}$$

Equation 10.9 indicates that the condition $\delta LA_1 = \delta LA_2$ is desired to diminish the fluctuation. To obtain the same fluctuation for two optical paths, all of the paths should be close to place them in the same environmental conditions: temperature, pressure, and humidity.

The cancelable optical circuit and the introduction of two signals are a powerful method for obtaining good measurement results even when many kinds of fluctuations occur.

10.2.3 CONSTRUCTION OF THE CANCELABLE CIRCUIT: GENERATION AND SELECTION OF TWO LIGHTS

In the heterodyne method, ω_2 is very close to ω_2. This condition is required to obtain a beat signal whose frequency can be processed with an electric circuit. A modulator is used to change the optical frequency. An acousto-optic modulator (AOM) [20–22] and an electro-optic modulator (EOM) [23] are usually used. AOM and EOM are based on the acousto-optic effect and the electro-optic effect, respectively. Both modulators shift the optical frequency using an input RF signal with a frequency in the range of 30 MHz–100 GHz. As a usage of the modulator in the cancelable optical circuit, there are two schemes:

(S1) Only one beam is modulated.
(S2) Both the beams are modulated by using two modulators.

Obviously, an insertion of the modulator on the path causes several fluctuations and results in imprecision. The selection of the modulator is one of the most significant factors in treating the cancelable optical circuit. So, the characteristics of the modulator have to be ascertained.

AOM is commonly used because of its highly stable performance. To understand why the frequency fluctuation occurs, the principle of AOM will be described briefly as follows and in Figure 10.5. Ferroelectric thin plates ($LiNbO_3$, etc.) are mounted on an acousto-optic medium ($PbMoO_3$, TeO_2, etc.) with a metallic cement adherently. An electrode is pasted on the plate by using the vacuum evaporation. When RF power (frequency f_1: 30–400 MHz) is added to the electrode, the plate distorts with a stationary wave form based on the frequency. The distortion propagates into the acousto-optic medium with an ultrasound (nearly) plane wave. In the medium, this compressional wave makes a diffraction grating. An incident optic beam (frequency f_0) is diffracted by the grating. The diffracted beam's frequency is able to build up theoretically many types of $f_0 + nf_1$ ($n = \ldots, -3, -2, -1, 0, 1, 2, 3, \ldots$) (Bragg diffraction or Raman–Nath diffraction).

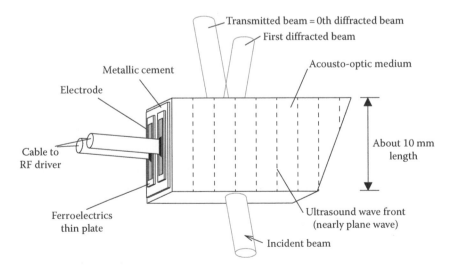

FIGURE 10.5   Structure of AOM. Ultrasound wave generates and propagates into medium.

Usually, the diffraction type $n = 1$ or $n = -1$ is used. In this diffraction process, the frequency fluctuation arises from the following points:

1. The ferroelectric plate is small: about 2 mm × 10 mm. Large-sized plates are difficult to produce because of the plate's characteristics. This means that only an approximate (not ideal) plane wave only propagates.
2. The metallic cement and the plate do not always have an ideal uniform thickness. Ideal distortion is not always produced.
3. The RF power spectrum is broad. A high-frequency power source easily catches external noises from other instruments.
4. A thermal distribution occurs by adding RF power (about 1 W) to the medium (the temperature of the medium rises by 5°–15°).

The frequency fluctuation of AOM will help selecting (S_1) or (S_2). These four points show that the frequency fluctuation cannot be taken away and cannot be discussed ideally. Therefore, (S_1) is not a desirable condition. To get merits in (S_2), the following selection points are requested: (1) use of the same acousto-optic media, (2) use of the same production of a company, (3) use of one electric RF source that branches out two outputs, and (4) use of a cooler for the medium.

EOM is utilized with the Pockels effect or Kerr effect caused in the piezoelectric crystal (KDP, $LiNbO_3$, $LiTaO_3$, etc.). The merits are that the modulation frequency extends to 100 GHz, and the distortion of the wave front is smaller than AOM. The demerits are that the modulation is not always stable in a long time measurement, and a modulation rate (= diffraction efficiency) is smaller than AOM. In high-precision measurement, the low beat frequency described in the previous section is desirable. Therefore, EOM is not adopted in this chapter.

10.2.4 CONSTRUCTION OF THE CANCELABLE CIRCUIT: SYMMETRICAL LAYOUT IN THE CIRCUIT

In an optical circuit, several fluctuations also occur as follows: (1) variation of measuring length, (2) disorder of wave front, and (3) fluctuation of beam direction. To fully resolve these issues, the optical circuit has to use a good layout for two beams. In the layout, attention must be paid to the selection of mechanical parts and optical parts because of the following issues. For point (1), these are the following causes:

1. HeNe laser beam varies its direction slightly and slowly.
2. Refractive indexes of optical parts and surrounding air have temperature dependence.
3. The effect of the surface roughness of optical parts depends on the beam diameter.
4. Mechanical parts expand and contract thermally.

For (2), these are the following causes:

1. The distortion (including optical rotation) of the optical isolator (which is not always used, but its use is desirable to suppress the frequency fluctuation of the laser) and optical parts.
2. The medium of the optical parts (usually quartz, fused silica, or BK7 glass) shows rarely irregularity of the refractive index, which has polarization dependence.
3. A mismatch of the polarization direction between the beam and the optical parts (especially, wave plate and PBS). An oblique incidence and multi-reflection usually appear.

For (3), these are the following causes:

1. A fluctuation of HeNe laser beam's direction, but this has been minimized by the manufacturers.
2. Mechanical parts relax its configuration from the setup position very slowly, so that the reflection points and directions in the layout are shifted slowly.

By understanding that all parts have many defects and do not have ideal characteristics, a suitable means should be conceived to avoid the issues given earlier. In conclusion, the cancelable circuit requires the following points: (1) all parts are used as a pair, (2) a symmetrical arrangement and layout are taken in the circuit, (3) all parts are compact and layout is simple, and (4) optical path length is as short as possible. These are based on the fact that the same parts affect the measurement with the same factor, and the relation $\Delta LA_1 = \Delta LA_2$ in Equation 10.6 is the most desirable. In addition, the cancelable circuit is based on an accurate knowledge of optical path length difference (OPD).

10.2.5 RELATION BETWEEN THE DOPPLER METHOD AND THE HETERODYNE METHOD

Reflection light from a moving target shifts its frequency (Doppler effect). In the Doppler method, the displacement can be calculated by integration with the shift value and the measuring time.

Considering the effect, Equation 10.5 is modified and described as a function of time as follows:

$$SIG_1(t) = C_1 \sin((\omega_{A1}(t) - \omega_{B1}(t))t + \phi_A - \phi_B),$$

$$SIG_2(t) = C_2 \sin((\omega_{A2}(t) - \omega_{B2}(t))t + \phi_A - \phi_B).$$

The phase shift for time variation from t to $t + \Delta t$ is given within the first-order approximation by

$$\varphi_1(t + \Delta t) - \varphi_1(t) - (\varphi_2(t + \Delta t) - \varphi_2(t))$$

$$= (\omega_{A1}(t + \Delta t) - \omega_{B1}(t + \Delta t))(t + \Delta t) - (\omega_{A1}(t) - \omega_{B1}(t))t$$

$$- (\omega_{A2}(t + \Delta t) - \omega_{B2}(t + \Delta t))(t + \Delta t) - (\omega_{A2}(t) - \omega_{B2}(t))t$$

$$= [\omega_{A1}(t) - \omega_{B1}(t) - \omega_{A2}(t) + \omega_{B2}(t)]\Delta t$$

$$+ t\left[\frac{d\omega_{A1}(t)}{dt} - \frac{d\omega_{B1}(t)}{dt} - \frac{d\omega_{A2}(t)}{dt} + \frac{d\omega_{B2}(t)}{dt}\right]\Delta t$$

As a simple example, it is assumed that only LA_1 moves. The total phase shift $\Phi(t_1, t_2)$ from t_1 to t_2 is given by

$$\Phi(t_1, t_2) = \int_{t1}^{t2} dt\left[\omega_{A1}(t) + t\frac{d\omega_{A1}(t)}{dt} - \omega_{A20}\right] = \omega_{A1}(t_2)t_2 - \omega_{A1}(t_1)t_1 - \omega_{A20}(t_2 - t_1),$$

where ω_{A20} is constant because the target A_2 does not move. When the average speed of the target A_1 is V, the relation, $\omega_{A1}(t_2) = c/(c + V) \times \omega_{A1}(t_1)$, is satisfied. And by substituting the initial condition, $\omega_{A1}(t_1) = \omega_{A20}$, $\Phi(t_1, t_2)$ is given as

$$\Phi(t_1, t_2) = 2\pi\frac{V(t_2 - t_1)}{\lambda_1} = 2\pi\frac{\Delta L}{\lambda_1}. \tag{10.10}$$

Equation 10.10 coincides with the result based on the Doppler effect.

When considering these relations, it is understood that the heterodyne measurement bases on the difference between the phases at the start time and at the stop time, and it calculates directly the displacement. In the Doppler method, a displacement is indirectly calculated through an integral of the target velocity.

10.3 ACCURACY AND NOISE REDUCTION

It was described earlier that the cancelable circuit is essential in the heterodyne method. In this chapter, several significant factors that affect measuring accuracy are shown, and some countermeasures against these factors are discussed.

It seems that an errorless signal processing and a noiseless phase variation will give the highest accuracy and the highest resolution for measurement. How high can a resolution be achieved in a real circuit? How high can a laser frequency's stabilization be achieved with commercial instruments? How low can noise be suppressed in a real electric circuit? To obtain better results, the level or grade to which a real system can respond to the demands of these questions must be considered. The answers will reveal the limits of the accuracy, the resolution, and the selection of the optical circuit and the electric circuit.

In this section, several leading factors are discussed. For other factors omitted in this section, other reports have discussed with the miniaturization of the circuit [24], the parts selection [25], the symmetry layout [26], the introduction of the fiber [27], the nonlinearity [28–30], the path length [31,32], the structure [33], the three AOM circuit, and the signal processing [34–36].

10.3.1 DETERIORATION OF ACCURACY CAUSED BY THE ENVIRONMENT

The refractive index of air depends on temperature, pressure, vapor pressure, constituents, convection of air, etc. When beams propagate in air, the optical path length is changed by the refractive index, that is, by these factors. The refractive index of air has been proposed in some papers [37–41], for example, Edlén's refractive index formula shown by

$$n - 1 = (n-1)_{st} \frac{0.00138823 \times P}{1 + 0.003671 \times T}, \tag{10.11}$$

where
    n is the refractive index of air
    n_{st} is the refractive index of air on the standard condition
    P is the pressure in Torr
    T is the temperature in °C

The value of n_{st} (including 0.03% CO_2) for HeNe laser (wavelength $\simeq$ 633 nm) is 1.0002765. (The dependence on a vapor pressure is discussed in other papers.) Equation 10.11 indicates that variations of temperature and/or pressure change an optical path length and result in virtual displacement as if a target moves. When temperature changes by just 1°C, the refractive index changes only 0.00000038. When a geometric path length is 3 cm, the virtual displacement is 11.5 nm. If high-precision measurement is desired, this serious problem cannot be neglected. The effort of measuring the phase with a high resolution disappears ineffectively. To avoid this problem as effectively as possible, the optical system needs to satisfy the following points:

1. Optical path length is short to the best of the system's ability.
2. Strict control of temperature and humidity in the laboratory. When the weather is bad or unstable, measurement should be postponed until the atomic pressure is stable.
3. Conditions of the cancelable optical circuit, $LA_1 = LA_2$ and $LB_1 = LB_2$ (i.e., OPD = 0), should be satisfied as the first priority. Even if the relation $LA_1 - LA_2 = 1$ mm is set, it is not sufficient. This is because the change in atomic pressure from 1013 hPa (standard pressure) to 1033 or 980 hPa yields the virtual displacement of 5.4 or −9.1 nm, respectively.
4. When higher accuracy is desired, the optical circuit has to be set in a vacuum chamber.

10.3.2 Deterioration of Accuracy Caused by Optical Parts

In optical circuits, many optical parts are used, but, unexpectedly, some of them bring about desirable results to the circuits. When a designer handles them, he must understand their characteristics satisfactorily. Related to these characteristics, one of the reports shows that nonlinearity in a measurement system occurs also from nonideal polarization effects of the optics [42].

10.3.2.1 AOM and Its RF Driver

The mechanism and the structure of AOM were described briefly in the previous section. The medium of AOM is crystal or glass (the amount supplied is not sufficient) as shown in Table 10.2. $PbMoO_4$ and TeO_2 are the main mediums used recently, and a GaInSe crystal in the infrared region [43] and a liquid crystal [44] have been proposed.

As the crystal has a crystallographic axis, it does not always have an isotropy for the polarization. This anisotropy results in some limitations for an incident light's polarization and the other conditions. The angle of diffraction χ satisfies the following relation approximately:

$$\chi \cong N\lambda \frac{nf_d}{V}, \tag{10.12}$$

where N, λ, n, f_d, and V are the order of the diffraction, the wavelength, the refractive index of the medium, the frequency of the RF driver, and the acoustic velocity of the medium, respectively. In two AOM systems that have the RF frequencies f_{d1} and f_{d2} ($\neq f_{d1}$), the designer must recognize that the difference between the angles may yield an unexpected OPD.

Is the sentence of "the interference of two beams produces the beat frequency $|f_1 - f_2|$" always right? The sentence assumes that $(f_0)_1 = (f_0)_2$, which means that the laser frequency before the

TABLE 10.2
Characteristics of AOM Materials

Material	Refractive Index at 633 nm	Figure of Merit $(10^{-15} \text{ s}^3/\text{kg})$	Acoustic Velocity (km/s)	Thermal Expansion Coefficient $(10^{-6}/°C)$	Density (10^3 kg/m^3)	Used Wavelength (nm)
TeO_2	Ne = 2.412	25.6(∥)	L; 4.20	27	5.99	360–1550
	No = 2.260	34.5(⊥)	S; 0.617			
$PbMoO_4$	Ne = 2.262	36.3(∥)	3.63	58	6.95	442–1550
	No = 2.386	36.1(⊥)				
$LiNbO_3$	2.203	7.0	6.57	2.0	4.64	420–1550
$LiTaO_3$	2.180	1.37	6.19	—	7.45	400–1550
Fused silica	1.45702	1.57	L; 5.96	0.55	2.203	325–650
			S; 3.76			
AOT-40	1.942	19.5(∥)22.3(⊥)	3.16	−1.54	4.92	440–780
AOT-5	2.090	23.9(∥)21.0(⊥)	3.47	10.3	5.87	633–850
N-SF6	1.79885	—	3.509	0.93	3.36	440–633

Notes: Interaction strength coefficient η in AOM is given by

$$\eta = \frac{\pi^2}{\cos^2 \theta} M_e P_a \frac{L}{H} \frac{1}{\lambda^2},$$

where θ, M_e, P_a, L, H, and λ are Bragg angle, figure of merit, applied acoustic power, transducer length, transducer height, and wavelength, respectively.

Refractive index: No = the index for ordinary ray, Ne = the index for extraordinary ray.

Acoustic velocity: L = longitudinal mode, S = slow shear mode.

$LiNbO_3$ and $LiTaO_3$ are hardly used. AOT-40 and AOT-5 were made but are not produced now.

diffraction of the first beam is equal to the laser frequency before the diffraction of the second beam. The frequency of the frequency-stabilized laser (even if $\Delta f/f = 10^{-9}$) fluctuates by $\Delta f = 4.74 \times 10^{14} \times 10^{-9} = 474 \times 10^{3}$ [Hz]: this is not small compared to the RF frequency (= about 80 MHz). When making device arrangements that satisfy this sentence, the following points should be watched and should be observed.

1. How long are the sampling time and the measuring gate time (= detecting time gated for one signal)? In the case of the gate time interval of 10 ms or more, the fluctuations of HeNe laser and RF driver on the beat signal can be eliminated [45] by an averaged calculation.
2. Are the beam position (including incident angle), polarization direction, and beam radius that are written in the AOM manufacturers' manual kept? These points affect the AOM operation.
3. Is stable cooling designed in the surrounding area of AOMs? Because RF power is about 1 W and the heat generated in AOM changes the refractive index of the medium and distorts the wave front. The change results in the fluctuation of the beam as known from Equation 10.12.
4. Use one RF driver that makes two RF frequencies simultaneously. Only one oscillator should be used on the driver. For the beat signal, the frequency fluctuation is canceled.
5. Connect AOMs to RF driver as short as possible with coaxial cables. The cable leaks a high-powered RF frequency noise. The leaks affect the detectors and equipment directly, and a virtual beat signal sometimes happens in spite of no optical signal.
6. Do not use the beat signal generated by the two RF frequencies as a reference signal. The signals have to be generated through the interference of the optical beams.

10.3.2.2 Frequency-Stabilized HeNe Laser

In the previous sections, it was proven that the laser frequency has to be stabilized. The available products are a solid laser and a frequency-stabilized HeNe laser [46] (including I_2 absorbed type [47]). The latter (not including I_2 absorbed type) has better cost performance than the former. Although attempts have been made to stabilize semiconductor lasers, satisfactory results have not been obtained yet.

A free-running type of HeNe laser has a stability of $\Delta f/f \approx 10^{-6}$, which is the best stabilization value under a free-running operation. However, the value is not enough for high-precision measurement. The commercial frequency-stabilized HeNe laser has about $\Delta f/f \approx 2 \times 10^{-9}$ [48]. On a manufacturer side, the value is upper limit under a normal control.

Why does the limitation arise? Laser operation occurs within a frequency band induced by a gain curve that is calculated with the energy levels of He and Ne gases. Output beam of HeNe laser has only a few longitudinal modes; the frequencies of these modes are restricted by a stationary wave condition for the laser cavity length. The frequencies of the modes fluctuate with the variations of the cavity length and of the refractive index of He and Ne gases. Under the assumption that the refractive index does not vary and 30 cm cavity length and 632.9913 nm wavelength, the wave numbers of the stationary wave, 47393, 47394, and 47395, are valid. These values correspond to the cavity length 29.999356, 29.999989, and 30.000622 cm, respectively. If the cavity length coincides with one of them, the wavelength is recalculated with the cavity length and the stationary wave condition. A 30.000000 cm cavity and the number of stationary wave 47394 require a wavelength of 632.9915 nm. A shrinkage of only 11 nm cavity length yields a variation of 0.0002 nm in wavelength, which corresponds to $\Delta f/f \approx 3.16 \times 10^{-7}$. Laser makers achieved a value of about 2×10^{-9} using their original techniques.

Is this stability of 2×10^{-9} sufficient for high-precision measurement? The answer is no. Therefore, as the second best solution, a countermeasure must be found under the assumption that this unsatisfactory laser is used. A frequency fluctuation is usually random. In spite of the random characteristics, for short gate time t_G (= length of the detector's opening time in which one signal is outputted), the fluctuation will not always be small by integrating signals because it depends on the detector's characteristics. So, the sampling time t_S (= integration time of the output signal = time interval of output data) controls the result. A short t_S increases the time resolution but worsens the S/N ratio of the data.

Multiple longitudinal modes are usually simultaneously outputted. Only the largest one is used, and the others must be cut. If the transverse mode is multiple, only TEM00 mode should be utilized because an unexpected interference shifts the beat frequency.

A laser is very weak against a back reflection light. An optical isolator [49] should be inserted perpendicular to the path after the laser. The isolation of 30 dB or more is desirable.

Attention should be paid to the following points:

1. Maintain the high-frequency stability and do not add any thermal influence. To protect the laser, a cover made of an insulator should be used.
2. Check the longitudinal mode. When it is multimode, only main mode is used with a suppression ratio of 20 dB or more.
3. Maintain an extinction ratio of 25 dB or more. The optical isolator should be used to suppress the frequency fluctuation.

10.3.2.3 Interference Circuit and Parts

In optical circuit, the intensity and the polarization of the beam are controlled mainly with several optical parts, mirror, PBS (polarized beam splitter), NPBS (nonpolarized beam splitter), wave plates, corner reflector, optical fiber, polarizer, and lens. These parts change the characteristics of the beam. When designing the circuit, attention should be paid to the following three points:

1. Suppress the direction fluctuation of the optic axis.
2. Match the polarization with an extinction ratio of at least 25 dB.
3. Notice temperature dependences of the refractive index of the parts.

The direction fluctuation occurs mainly from the laser beam's fluctuation and is sometimes based on the flatness of the parts and a relaxation of the mechanical parts. To suppress the laser beam's fluctuation, the use of a fiber collimator [27,50] may be a good idea because the output hardly fluctuates.

These optical parts are made mainly from crystal, quartz, fused silica, and glass as the medium [51]. They are usually polished surfaces with a flatness of $\lambda/20$–$\lambda/200$. The coating film of the parts is mainly a metallic monolayer or a multilayer of a dielectric substance that is manufactured using vacuum evaporation or similar method. The transmission factor of the antireflection coating is about 99% at best, and the maximum reflectivity is about 99%. All parts have a distortion that differs slightly from the ideal form, for example, a perpendicularity of $90° \pm 1''$ for the best quality. The refractive index and other characteristics are listed in Table 10.3. Further, quartz has the optical rotatory power. That is to say, no optical parts are ideal. Therefore, for the optical parts, the qualities of the parts should be enhanced carefully as much as possible.

Mechanical parts also have fluctuations in the optical beam direction. The fluctuations are usually larger than that for the optical parts. The main reason for the fluctuations is a heat flow. The heat is produced in active parts: for example, AOM and laser, and is added to every part in the circuit. If the heat does not flow, the influence is too small because a thermal balance rises and disorders stop. Therefore, for the mechanical parts, the suppression of the fluid is the most important scheme for high-precision measurement.

10.3.2.4 Sample Surface and Sample Holder

Only reflected light has information of a target sample. Therefore, the detection method and detection device decide the accuracy and the S/N ratios of the measuring signals. Attention should be

TABLE 10.3

Characteristics of Optical Glass

Materials	Refractive Index n at $\lambda = 633$ nm	dn/dT $(10^{-6}/K)$ $\lambda = 633$ nm	Photoelasticity Coefficient $(10^{-12}/Pa)$	Thermal Conductivity $(w/m/K)$	Thermal Expansion Coefficient $(10^{-6}/K)$	Density (kg/m^3)
Quartz	Ne = 1.5517	−5	—	10.66(c‖)	7.97(c‖)	2.65
	No = 1.5427	−6		6.18(c ⊥)	13.17(c ⊥)	
Fused silica	1.45702	12.8	—	1.3	0.55	2.203
Sapphire	Ne = 1.760	—	—	25.1(c‖)	7.7(c‖)	3.98
	No = 1.768			23.0(c ⊥)	7.0(c ⊥)	
Pyrex	1.473	—	—	7740	3.25	2.23
Zerodur™	1.5399	—	—	1.46	0.0 ± 0.02	2.53
	($\lambda = 643.8$)					
N-BK7	1.51509	0.7	2.82	1.21	7.2	2.52
N-PK52A	1.49571	−7.1	0.74	0.837	15.5	3.70
N-SF6	1.79885	−1.5	2.64	0.96	9.3	3.37
N-SF4	1.74968	−0.2	—	0.95	8.7	3.23
N-LAF3	1.71389	−0.3	2.71	—	8.0	4.2

Notes: Assignment of N-BK7, N-PK52A, N-SF6, N-SF4, and N-LAF3 are based on Schott Glass, Inc. The terms of (c‖) and (c ⊥) show that the optical axis is parallel and perpendicular to c axis of the crystal, respectively. In optical isolator, Faraday media are mainly YIG (Yttrium Iron Garnet) or TGG (Terbium Gallium Garnet). Reflection rate at $\lambda = 633$ nm of the metal material coated with the vacuum vapor deposition: Al: 90.8%, Au: 95.0%, Ag: 98.3%, Cu: 96.0%.

paid to the flatness and/or the roughness of the surface because they can easily change the optical path length and result in scattered light. To obtain high S/N ratios, much scattered light must be condensed. To get good condensation rate, attention should be also paid to a kind condensation lens, NA value of the lens, the incident beam radius, and the period of the roughness. Particularly when the roughness is bad, the surface should be improved by repolishing or depositing metal.

When the sample is gas or liquid, the material and characteristics of a holder or a cell should be selected carefully to maintain the beam quality. Especially, a thin glass plate brings about multiple reflections from both the front and the rear of the plate.

10.3.2.5 Detection Method and System

The displacement resolution is determined by a phase resolution of the beat signal, but laser noises and electric noises in signal processing (including detector) worsen the quality. The former noises are discussed in the earlier sections. For the latter noises, there are two dominant factors: (1) a photocurrent fluctuation induced by a laser amplitude modulation and (2) characteristics of the electric circuit including memory, processor, amplifier, and filter. To reduce them, (1) the detector should have an extremely wide temporal bandwidth and a highly linear response and (2) the circuit should have a low shot noise and a high S/N ratio with an electric shield. Photodiode (PD) and avalanche photodiode (APD) are usual used as the detector. A larger frequency band needs a smaller detector diameter. To decrease the diameter, a severe condensation of the beam and a small direction fluctuation of light are required. The frequency band of the detector, Fb [MHz], satisfies the following relation with the beat frequency fb [MHz], and the displacement resolution Dr [nm], roughly (HeNe laser use).

$$Fb > fb \frac{633}{Dr}. \tag{10.13}$$

This condition gives a limit for *Dr*. Therefore, the selection of the detector has to take many factors given earlier into consideration. In addition, some detectors depend on polarization. In using them, it is desirable for a wave plate of $\lambda/4$ to be inserted in front of the condensed lens.

10.4 ATTENTION FOR STATIC AND DYNAMIC MEASUREMENT

10.4.1 STATIC MEASUREMENT

In this case, the sample hardly moves (almost stationary). All variations drift slowly with a long span. Control of these drifts is absolutely essential. The cancelable optical circuit with zero OPD is required of necessity. Attention must be paid to the following points:

1. Control of the laboratory's temperature, pressure, humidity, and sound
2. Prohibition of all movement (including air convection) near the measurement system
3. Temperature control of the basement of the optical circuit
4. Noise reduction in the AOM's RF driver and in the power source

10.4.2 DYNAMIC MEASUREMENT

In this case, sample vibrates and/or translates with the function of time. However, the defect of the heterodyne method (= defect 1) limits for the sampling time t_S, for which the following relations are required:

$$\lambda > F(t+t_S) - F(t) \quad \text{for all } t \text{ in a straight measurement case.}$$

$$\frac{\lambda}{2} > F(t+t_S) - F(t) \quad \text{for all } t \text{ in a reflection measurement case.} \tag{10.14}$$

These cases are classified by the relations between the detector and the sample. The latter case has double the resolution of the former case. Considering "which case is effective" is important for designing the optical circuit and the other factors.

In addition, the relation between the gate time t_G and the sampling time t_S is a key point. Considering three restrictions, (1) Equation 10.14, (2) $t_S \geq t_G$, and (3) t_G conditions described in Section 10.2.2, both the time resolution and the displacement resolution are varied complementally. Especially, for short t_S, the following points are required:

1. High-frequency stability for both the HeNe laser and the RF driver
2. Small diameter for the detector
3. High-frequency band in the electric signal processing

10.5 APPLICATIONS

Basically, the physical variable measured in the heterodyne method is a displacement. By considering why the displacement occurs, the displacement result can be applied to the detection of other physical variables.

The displacement does not always originate in the movement of the sample: to be accurate, it originates in a change of the optical path length. This change also occurs because of the variation of the refractive index of the path. So, when a gas, a liquid, a crystal, etc., are set on the path, their characteristics can be investigated such as gaseous pressure, kinds, sort, temperature, disorder degree, anisotropy effect, birefringence index, and optical rotatory power. By extending these ideas, in many fields of thermodynamics, hydromechanics, electrodynamics, magnetohydrodynamics, plasma physics, and statistical mechanics, the heterodyne method will be utilized as one of superior metrology. This method is effective even for dynamic measurement within an extremely short time. For example, with a femto-second pulse laser, a two-photon absorption process is also reported [52].

The focusing of the laser beam with a microscope [53–56] can be used to measure the displacement in an extremely confined region of 1–30 μm. By accumulating an AOM and an optical path on a crystal, several miniaturized circuits are demonstrated [57–61].

Some companies commercially produce many instruments and systems that base on the heterodyne method: for example, a displacement meter, a vibrometer, a polarimeter, a profilometer, a phasemeter, many typed optical sensors (including fiber sensor), a laser radar, a micro-scanning system, and other interferometers.

10.5.1 Displacement, Vibration Frequency, and Amplitude

Displacement measurement is achieved with ultrahigh resolution [62]. Several reports point out a possibility of sub-nm or pm order resolution under laboratory conditions [63,64].

Vibration amplitude and frequency are directly measured and calculated through an integration of the displacement. A combination of heterodyne method and Doppler effect method is effective for a synchronous measurement of variations because the Doppler effect measures the frequency directly. A piezoelectric transducer vibrates within 10 kHz to 10 GHz. For high-frequency vibration, the Doppler method is effective under the heterodyne detection. For low-frequency vibration, the heterodyne method gets superior results.

By utilizing this basic displacement measurement, the temperature of unburned gases is measured by its refractive index variation as a function of a density, a concentration, a type, and a Gladstone–Dale constant of gas [65,66].

With superluminescent diode, 3D measurement is proposed in the full-field heterodyne white-light interferometry and the full-field step-and-scan white-light interferometry.

10.5.2 Positioning with Stage

Mechanical stages are utilized to measure the shape and size of a 3D structure sample. By translating and rotating the sample by the stages, a displacement is apparently generated. Therefore, the resolution is determined by qualities of the stages, combination, backlash, lost motion, and electrical feedback control of stages, which seriously affect the reproducibility and linearity. In using a microscope or condensing lenses mounted on the stages, the alignment error between optical axes of the lenses and those of the probe lights has to be compensated for even if the stages move [67]. Also, when the focal length and the focal depth of the lens do not match the surface shape, the signal quality and the resolution get worse. In particular, in the field of the optical lithography and the x-ray lithography, an extremely high-grade stage is required [68,69].

10.5.3 Thick Measurement

Thin film is usually measured by the ellipsometry. By using the heterodyne method as shown in Figure 10.6, the thickness of the film is measured. The incident angle, thickness, and refractive index of the film decide the OPDs of the two beams that are reflected at the top surface and the bottom surface of the film. By analyzing the relation of the OPD vs. the incident angle, the thickness is obtained. A 300 μm thick amorphous silicon is investigated with a beat frequency of 5 kHz [70].

10.5.4 Profilometry

Flatness, roughness, and structure on a sample surface (namely, quantitative 2D surface profile) are usually measured by two means: (1) electromechanical beam scanning and (2) camera detection of

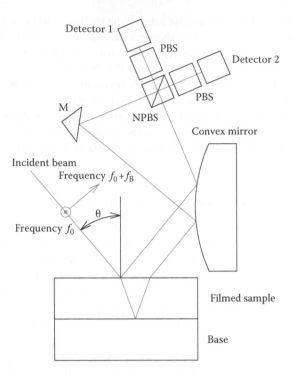

FIGURE 10.6 Measurement of thickness of filmed sample. Incident lights have different wavelengths for two polarizations. The f_0 light reflects at the sample surface, and the $f_0 + f_B$ light reflects at the base surface. By interfering two beams with two detectors, the sample thickness is analyzed with varying angle θ as a parameter.

the expanding beam. The resolution in a direction perpendicular to the beam (transverse resolution) is about 1 μm to 1 mm, which is much larger than the resolution in a direction parallel to the beam (longitudinal resolution). The limitation of the transverse resolution is based on the beam size and the wave front quality.

For means (1), the laser beam diameter is small: about 0.3–2 mm. (When the microscope is used, the beam waist is about 1–30 μm.) By shifting the beam or moving the target with stages, a displacement distribution in a broad area of the target can be measured. The transverse resolution is limited by the beam diameter and the stage shift control. The performance of the stages and the sensitivity in the shift operation severely affect profile measurement. Several instruments and systems that use with this means are sold commercially.

For means (2), a laser beam is expanded, and a CCD camera is used. The resolution and the measurable region are limited by the size of the CCD's pixels, the number of pixels, and the focus lens system. A tunable diode laser is also used [71]. A normal photo-detector is used instead of CCD [72].

Other means are also proposed. Detection, scanning, and signal processing have been investigated as a function of beam radius, pixel size, and stage operation [73]. By using a birefringent lens that provides different focal lengths for two orthogonal polarization eigenstates (p and s), the surface profile is reported as a function of phase variations of p and s waves [74]. The surface of an aspherical lens is also investigated with two laser diodes (810 and 830 nm wavelength) [75].

10.5.5 REFRACTOMETRY

An optical path length depends on the refractive index of the path. In particular, the refractive index of air negatively affects high-precision measurement as described in Section 10.2.1.

Inversely, by measuring the variation of the length, a refractive index of the sample or several factors hidden by the index, for example, density, pressure, temperature, absorption coefficient, and other physical quantities, are obtained. Therefore, strict conditions must be controlled for whole of measurement system.

For a liquid sample, a refractive index and its thermal coefficient are measured with a phase variation of a probe light passed through a sample. With this principle, the refractive index of the liquid is measured [76].

For a gas sample, more attention is needed. The reason is that the refractive index of gas is 1.0001 (at 1 atm, 15°C) at most, and the refractive index of liquid is at least 1.2. The sample is filled with a cavity, and the cavity is deformed or distorted by a pressure difference between inside and outside. The deformation yields some errors. In particular, the thickness, flatness, and homogeneity of the cavity are important. Therefore, the quality of the cavity should be selected cautiously using the Young modulus and the thermal expansion coefficient of the cavity. Gas density is measured in a high dynamic range of 0.5–700 Torr [77]. As a cavity, Zerodur® glass is used and investigated [41]. Fiber sensor-typed interferometer is used to measure gas density in an engine [78].

10.5.6 Photothermal Interferometry

Thermodynamics shows that heat diffuses until a thermal equilibrium is obtained as a function of the sample's characteristics and thermal conditions. So, when heat is added to the sample locally, heat propagates in the form of a heat wave. The propagation wave brings about a distortion of the sample. The distortion depends on the diffusion coefficient, viscosity coefficient, compressibility, expansion coefficient, density, etc. Therefore, these physical values can be investigated by measuring the distortion. For example, a high-power "pump" laser (CO_2, Ar, etc.) light and a low-power "probe" laser (HeNe) irradiate to a target sample. This sample heating modulates its characteristics and varies the optical path length of the probe light. By measuring the variation as a function of time, thermal characteristics (absorption, dissipation, energy level, structure, etc.) of the sample can be investigated. In [79], pump laser is Ar laser (514 nm wavelength and 1.5 W max), and a gas sample is confined with a cell of 5 cm in length.

When the target is solid, a thermal acoustic wave (ultrasonic wave) is induced. A longitudinal wave or a surface wave propagates in or on the sample diffusively and distorts the surface. The distortion's distribution, amplitude, and their time dependence are obtained as a function of the crystal structure, conductivity, diffusivity, density, shear modulus, and Young modulus of the sample [80].

For a semiconductor target, by using a tunable and pulse laser as a pump laser, gain dynamics in a single-quantum-well laser are investigated [81].

10.5.7 Thermal Expansion Coefficient

The expansion coefficient of the sample is calculated by adding heat to the sample homogeneously and measuring the amount of the stretch or shrink of the sample. In particular, for an extremely low coefficient, the utilization of the cancelable optical circuit and a strict control of environmental conditions are indispensable conditions.

To exclude influences of sample pitching and rolling, ω_1 beam reflects on a sample twice and ω_2 beam reflects on a basement twice as shown in Figure 10.7. The OPD variation of the two beams is basically zero as long as the sample and the basement are a rigid body and the layout is symmetrical. The thermal expansion coefficient of a lithographic glass is measured with the resolution of ppb scale under 0.1°C temperature control [82].

For a FZ silicon single crystal oriented in the <111> direction, the coefficient is measured in a range of 300–1300 K [83].

FIGURE 10.7 Thermal expansion detection system. Fiber collimator's output is divided into two paths. Two beams are modulated by AOMs whose wavelengths are λ_A and λ_B. Beams pass through complex path and finally focus at SIG. detector for P wave and at REF. detector for S wave. On the P wave path, the beam reflects twice at the sample and the basement. Influence of pitching and rolling of them is canceled. And influence of the basement can also be canceled.

10.5.8 YOUNG'S MODULUS OF THIN FILM

By adding pressure to a thin film sample, the film bends. The relation between the pressure and the distortion gives a Young modulus of the film with the relation [84] of

$$p = \frac{t\sigma}{r^2} h + \frac{8tE}{3r^4(1-\nu)} h^3, \tag{10.15}$$

where
    h is the distortion at the center of the film
    p is the difference of the pressures between the front and the back of the film
    E, r, t, σ, and ν are the Young modulus, the diameter, the thickness, the internal stress, and the Poisson ratio of the film, respectively

Using a displacement meter and a microscope, the measurement system is constructed as shown in Figure 10.8 [85]. A vacuum chamber connects to a vacuum pump, and a sample is placed on the upper surface of the chamber. The pressure in the chamber decreases slowly, and the sample distorts. The displacement meter measures the distortion at the center of the sample. A curve fitting analysis using Equation 10.15 gives σ and $E/(1-\nu)$.

10.5.9 BIREFRINGENCE INDEX AND POLARIMETER

Some of crystal, liquid, organic compound, sol, gel, and plasma show birefringence. For two orthogonal polarization eigenstates's lights (p and s waves), their refractive indexes are different. Therefore, the optical path length of p wave is slightly different from that of the s waves. By utilizing the difference, the birefringence of the target is measured as shown in Figure 10.9. Two linear

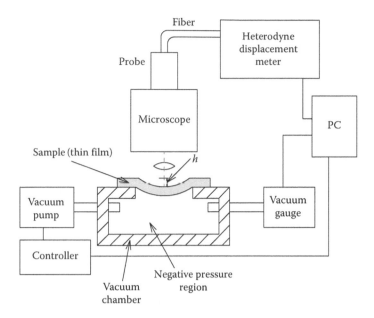

FIGURE 10.8 Measuring system for the Young modulus of thin film. A sample is set on a vacuum chamber that connects to a vacuum pump and a vacuum gauge. Slowly, the operation of the pump makes a negative pressure region in the chamber, and the pressure variation adds a stress to the sample. As a result, the sample distorts. A displacement meter on the microscope measures the distortion at the center h. By analyzing the relation of h and the pressure with Equation 10.15, the Young modulus is derived.

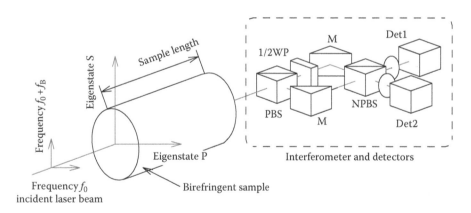

FIGURE 10.9 Measurement of birefringence.

polarization lights (wavelengths λ_1 and λ_2) exactly adjust their polarization directions to the eigenstates's directions of the sample. The adjustment and the extinction ratio of the lights are the main factors that determine the accuracy of the result [86–88]. The analysis of the beat signal's phase gives the birefringence. This basic principle is applied to several fields, and some commercial instruments have been proposed. Also, the polarization dependence of mode dispersion and power loss has been investigated [89].

Two orthogonal circular polarization lights are used to investigate plasma [90]. A very high extinction ratio of about 10^{-9} results in a very high angle resolution of about 10^{-7} rad [91]. Birefringence appears on a mirror of high-finesse (38,000) optical cavity; a distortion based on the photorefractive effect is reported [92].

10.5.10 Fiber Sensor

Fiber sensors are classified into two types: (1) a fiber that is used as a flexible optical guide to a target and (2) a fiber that is used as a long and flexible detector. In the former sensor type, two means are investigated: (1) a sensitive material is set just in front of the fiber and (2) a sensitive material is coated or deposited onto the fiber tip. The selection and construction of the material are key points for obtaining better sensors. In a hydrophone sensor, PVDF, PET [93], and Parylene film [94] are used as the material. By detecting a change in the polarization or a variation in the optical path length, a pressure range of 10 kPa is measured with 25 MHz bandwidth [93]. By detecting a rotation of the polarization of the light passed through the Faraday element [95,96], the current and magnetic field are sensed with a resolution of about 2 G [97]. A near-field fiber-optic probe [98] has also been proposed in which the distance between a sample and the fiber tip is about 1 μm [55].

10.5.11 Dynamical Surface Measurement

For a liquid surface, a longitudinal wave or a surface wave is generated. Hydromechanics can clarify their motions with the equation of continuity, the wave equation, Helmholtz's equation, and other equations. The group velocity, amplitude, and frequency of the waves depend on sample's characteristics. Therefore, their characteristics are analyzed by a measurement of the wave's motion with these equations. For the signal processing, a Fourier-transform analysis is usually used (it is called Fourier-transform heterodyne spectroscopy) [99]. Taylor instability in a rotating fluid is also reported [100].

10.6 OPTICAL DOPPLER MEASUREMENT

10.6.1 Introduction

The optical Doppler method is a noncontact method based on the Doppler effect. (The effect occurs in all waves, sound, ultrasound, surface wave, electromagnetic wave, and all waves having a finite velocity. But in this chapter, only light is discussed as a subject.) For the light, the effect should be discussed fundamentally with the relativity theories. However, a usual velocity of the sample is extremely less than the light velocity. So, the effect can be analyzed by the classical mechanics for usual system.

There are two patterns in the Doppler shift: (1) a sample itself is a light source and the sample moves relatively against a detector, and (2) a light source is fixed and the light reflects on a sample surface that moves relatively against detector. Many instruments and systems are based on the second pattern.

In this section, the following articles are discussed briefly: the principle and defect of the Doppler method and the comparison with the optical heterodyne method. The Doppler method is utilized for several fields, such as astronomy, flowmetry [101], velocimetry [102], vibrometry, astronomy [103], Doppler lidar [104], medical operation [105], and optical CT [106].

10.6.2 Principle and Defect

When a light source, a sample, and a detector are arranged two dimensionally as shown in Figure 10.10 and the sample moves with a velocity V, the Doppler effect shows that the detected light shifts its frequency (i.e., wavelength) from that of the source light. Under the condition of $V \ll c$, the wavelength shift $\Delta\lambda$ (= Doppler shift) is represented approximately by

$$\frac{\Delta\lambda}{\lambda} = \frac{V}{c}\frac{\sin(\theta + \phi)}{\sin\phi}(1 + \cos\varphi) \qquad (10.16)$$

where
  λ, V, c are the wavelength of the source light, the velocity of the sample, and the velocity of the light, respectively
  θ, φ, ϕ are angles shown in Figure 10.1

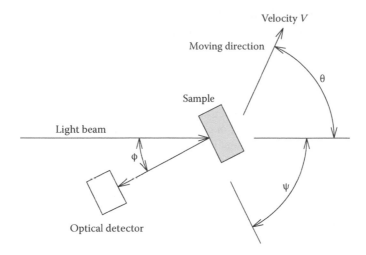

FIGURE 10.10 Doppler effect—the relation of light direction, sample moving direction, and detector position.

The frequency shift Δf is usually observed in the range from 100 [GHz] to 10 [Hz] and gives the velocity V in the range from 100 [km/s] to 10 [μm/s]. By measuring Δf, absolute V is obtained as a function of time.

However, Equation 10.16 shows a defect obviously.

Defect: In the cases of (1) $\theta + \varphi = 0$ or π, (2) $\varphi = 0$ or π, and (3) $1 + \cos \Phi = 0$, Δf cannot be detected.

As V can directly represent with Δf, you may misunderstand that a modulator (AOM or EOM) is not used. However, in a real electric circuit, $1/f$ noise is a critical problem for detection in a low-frequency region. To avoid the problem, a modulator (AOM or EOM) is usually inserted into the optical circuit. As a result of this insertion, a zero velocity gives a nonzero-frequency shift. The fluctuation of the shift mainly determines an influence in the accuracy of the Doppler method. By having the same discussion in the heterodyne method section, the fluctuation in one-AOM system is worse than that in two-AOM system. However, the two-AOM circuit (beat frequency = f_1) limits the measurable region of the velocity for moving that the target comes close to the detector.

10.6.3 COMPARISON BETWEEN DOPPLER METHOD AND OPTICAL HETERODYNE METHOD

10.6.3.1 Accuracy

As the Doppler method directly measures the frequency shift, the accuracy of the measurement is determined by (1) laser frequency fluctuation, (2) modulation fluctuation, (3) frequency resolution in the detection system, (4) arrangement of the sample and the detector (accuracies of θ, φ, and Φ), and (5) flatness of the sample.

Even if the laser stabilization is $\Delta f/f = 10^{-9}$, the frequency fluctuation is about 500 kHz. (See the optical heterodyne method section about the frequency-stabilized HeNe laser.) Even if the modulator (case of center frequency $\cong$ 80 MHz) fluctuation is $\Delta f/f = 10^{-5}$, the frequency fluctuation is about 800 Hz. These fluctuations are too large noises to measure with a high resolution. To solve this problem, attention should be paid to the following points: (1) introduction of a cancelable optical circuit, (2) two detectors, (3) symmetric optical circuit, (4) low-noise electric processing, and (5) averaging of data with a sampling time interval under the assumption that these fluctuations are white noise. Long sampling time makes low noise, but makes low time resolution.

Optical parts and mechanical parts should be selected to obtain high precision by the same reasons mentioned in the optical heterodyne method section.

10.6.3.2 Measurement of Velocity and Displacement

Basically, the physical variable measured in the Doppler method is a velocity. Therefore, the method is applied when the target moves. This point is very attractive in the field of space engineering, automobile industry, high-tech industry, and medical engineering. However, when the target hardly moves, the method is seldom adopted.

In spite of this weak point, by combining with the heterodyne method, an ideal measuring system can be constructed for both a dynamic and a static measurement. For the two methods, many optical parts and electric circuit can be used in common. Therefore, by adding a few parts (including the processing) to the system, an expedient measurement system is built up. By using two methods simultaneously, both velocity and displacement are measured with high resolution.

The detection of the frequency, f, in the Doppler method gives a limitation for the length of the sampling time. The highest time resolution is $1/f$. So, the beat frequency has to be selected by considering the time resolution, the maximum velocity of the target, and the optical circuit condition. However, the selection gives a restriction against the heterodyne method. Therefore, the two methods cannot operate simultaneously with the best condition. That is to say, either the heterodyne method or the Doppler method is the main measurement. When displacement is a target physical variable, the heterodyne method should be operated as the main measuring means, and the Doppler method should be treated as the auxiliary means.

10.7 MEASUREMENT OF TRANSIENT PHENOMENA

10.7.1 APPLICATION OF THE OPTICAL HETERODYNE METHOD

All physical phenomena are divided into two types: static phenomena and dynamic phenomena. In the static type, most of physical quantities are described with static equations or partial differential equations in which the variable of time does not include. Most of physical coefficients or physical constants have been investigated under the static condition theoretically and experimentally. Many suitable measurement methods have proposed as systems to observe the static physical response.

In the dynamic type, most of physical quantities are usually described theoretically with the partial differential equation where the time and positional coordinates are the independent variables. The observation methods for the dynamic response have been investigated. In the observations, the transient recording systems have been a barrier against the development of the measurement method. Especially, it was required to realize a measurement method having many characteristics of high-speed, fine variation, wide dynamic range, etc. Recent scientific technology and manufacturing industry technology extend means of the transient measurement method and the transient recording method. For the optical heterodyne measurement, by adding a transient recording function having high time resolution, many dynamic phenomena will be able to be measured with high time and high space resolution performances.

However, an exact measurement for the dynamical response is difficult. The reason is that any physical value interacts complicatedly with the other physical values, and their interactions among the values and their environmental conditions change with the increase in time. The time dependences of the interaction strengths will change the dominant effects operated among physical values. The dynamical response may not connect directly to the transient characteristics of the object. Moreover, the transient measurement system may distort the dynamic phenomena under the situation that several unexpected effects occur. For these problems, the optical heterodyne method can avoid by introducing the relative measurement method described in the previous section.

In the dynamic response, many partial differential equations for the physical quantity (e.g., expansion, diffusion, concentration, pressure, and elasticity) are described with differential coefficient of first or second order of time, so the equation suggests that the decay and/or the vibration appears in the motion of the object. However, all physical quantities are not always observable quantities, which can be appreciated directly. So, as the second best way measured these nonobservable quantities, the following way will be selected. First, search or check several observable quantities relating to the nonobservable quantity theoretically. Second, create or find a measurement method detecting the observable quantity. For example, Young's modulus is not an observable quantity, but it is expressed with strain and stress. Both the strain and the stress are observable. Therefore, by detecting them transiently, the dynamic characteristic of the Young modulus is obtained.

The transient measurement involves one tough mechanical problem that both the measurement and the analysis must be done continuously at a very short time interval. In addition, the observable quantities must be measured with good signal-to-noise (S/N) ratio even if the spatial variation is very small. Generally, short time interval detection decreases the S/N ratio so that the spatial resolution is poor. However, the heterodyne method is a useful measuring system that keeps high spatial resolution even if the measurement time is very short.

For the transient measurement, the heterodyne method is classified broadly into two detecting ways: (1) detection of the reflected light from the sample and (2) detection of the transmitted light through the sample. Two types have characteristics as shown in Table 10.4. The reflection light detection and transmission light detection are effective to search the property of solid samples and liquid or gaseous samples, respectively.

10.7.2 Attention to the Design of Transient Measuring System

The transient measuring system is classified roughly into two parts: an optical circuit and a processing electric circuit. Since the optical part is discussed in the previous sections, the processing part is studied in this section. First, to get high time resolution in the system, the following six points must be examined or designed carefully as important factors:

1. Measuring target property (frequency, velocity, etc.)
2. Output time interval τ_C: processing operation time on one of output data
3. Data latched up time interval Δt: latching operation time on one of input data
4. Measuring point N_F: output data points
5. Basis beat frequency f_{B0}: the beat frequency formed by two outputs of the frequency shifter
6. Recording specifications (the number of bits, memory type, transfer speed, and interface)

These points correlate with each other complicated manner. The designer of the measuring system should pay attention to the following relations (from (A) to (D)) to construct better circuit.

(A) *Relation between the measuring target property and the limitation of the measurement*

To design better system, the following properties of the target must be estimated correctly: (1) the maximum velocity V_m of the target and (2) the maximum displacement D_m within the output interval τ_C. When the estimation is over the limitation brought with only one essential defect of the heterodyne method, the design must be reconsidered. The defect is written by

<The reflection light detection case>
  "When the displacement within τ_C is greater than the half of wavelength of measuring laser light, the displacement is uncertainty"

<The transmission light detection case>
  "When the displacement within τ_C is greater than the wavelength of measuring laser light, the displacement is uncertainty"

TABLE 10.4

Characteristics of Reflection Light Detection and Transmission Light Detection

	Reflection Light Detection	Transmission Light Detection
Circuit	(Diagram: f_1 light → PBS → 1/4 WP → Sample; reflected f_2 light → NPBS → Det.)	(Diagram: Sample, Mirror; f_1 light through sample; f_2 light → NPBS → Det.)
	The f_1 light reflects at sample. The reflected light interferes with f_2 light at detector. Both 1Ch and 2Ch set up the same circuit. But for 2Ch, a sample surface changes to a reference surface.	The f_1 light passes through sample. The transmission light interferes with f_2 light. Both 1Ch and 2Ch have the same circuit. But for 2Ch, a sample changes to a reference sample or an empty box.
Merit	Direct measurement of the variation on the solid sample surface with high accuracy.	Direct measurement of the variation of the refractive index of a semitransparency sample with high accuracy.
Demerit	Displacement component for the direction perpendicular to the incident light cannot be detected.	Distribution in the path cannot be detected. Large scattering particles become worse S/N ratio.
Example of physical variables	Elastic modulus, pressure, film thickness, vibration, piezoelectric modulus, linear expansion, volume expansion, dielectric constant, sound pressure, etc.	Density, viscosity, compressibility, diffusion, concentration, longitudinal wave strength, dielectric constant, sound velocity, rotatory, solubility, etc.
Example of measurement	Tuning fork and cantilever vibration (Section 10.7.4.1).	Concentration fluctuation (Section 10.7.4.2).

Notes: The f_1 light and the f_2 light have the frequency of $f_1 + f_0$ and $f_2 + f_0$, respectively, where f_0 is the laser frequency, and f_1 and f_2 are shift values produced by AOMs (acousto-optic modulator). PBS, NPBS, and WP are polarizing beam splitter, nonpolarizing beam splitter, and wave plate, respectively. Det is a photo-electric detector whose output is an interference signal having beat frequency. Both f_1 and f_2 lights are linear polarization, but they are perpendicular in the reflection detection case and parallel in the transmission detection case.

These restrictions are expressed by the following relation for the possibility of the measurement:

$$D_m < \alpha\lambda \quad \text{and} \quad V_m < \frac{\alpha\lambda}{\tau_C}, \tag{10.17}$$

where α is 0.5 for the reflection light detection case and is 1.0 for the transmission light detection case, and λ is the wavelength of the used laser. When both relations are not satisfied, the value of τ_C must be reduced.

(B) Relation between the measuring time resolution and the measuring point

The output data interval of τ_C is usually so short that the data cannot directly transport to any external memory through any interface. The memory capacity on the circuit board is not limited so far as the designed circuit can carry out completely. However, even though the electric circuit has very high processing speed, the total necessary time transported from the memory to the external personal computer through the interface is so long that the interval of the experiments increases, and the environmental condition changes unexpectedly. Therefore, the number of the data N_F should be set to a minimum necessary amount. In addition, since the number of bit of the data affects the transfer rate considerably, the measuring points N_F should be considered from the number of bit to suppress error including a selection of an analog–digital (A/D) conversion IC.

(C) Correlation among the latch time interval, the resolution time, and the basis beat frequency

The input data latch time interval Δt is limited mechanically by the following points:

1. The crystal oscillator's frequency on the processing board
2. The operating frequency of the analog to digital converter and some other devices on the board

Moreover, the relation of $\tau_C = N_C \Delta t$ (N_C; integer) is required in the program rule of the digital programming device (e.g., FPGA IC). The value of N_C has a restriction; N_C needs several hundreds or more to compute the frequency and the phase of the signals precisely. (These restrictions will be understood in the subsequent section described about the crossing points.)

Since the heterodyne method checks the variations of the frequencies and the phases, the high-resolution system is required to measure them accurately. To get them with high accuracy, it seems that smaller Δt is a more desirable condition. However, small Δt corresponds to small S/N because of small optical power. Therefore, the designer must search an optimum Δt by discussing the requirements and preconditions: laser selection, wavelength (maximum and minimum) optical power, optical circuit, detector, S/N ratio, fluctuation of the optical path, signal processing circuit, measuring time, output time interval, the number of digital signal bit, etc.

When a period of the noise is comparable to Δt, the digital signal occurs as an unexpected component, and the phase detection may bring unnecessary variations. Therefore, some precautions against noises must be planned, and the selection of the crystal oscillator must be done carefully.

(D) Correlation among the processing method, the latch time interval, and the resolution time

In every stage in a series of operations (from the detection of optical signal to the output of analyzed data), many electric circuits are utilized, and they are usually developed and programmed by user originally. As electric IC devices including optical detectors run with very high-frequency performance, the total processing time is controlled dominantly by an execution time of programming ICs. Recently, as the digital programming ICs are improved rapidly, the resolution time becomes available to operate on about 100 ns.

However, for small resolution time, the condition of $\tau_C = N_C \Delta t$ becomes severe restriction. The value of N_C has to be over 100 as described previously. The requirement means that Δt is less than 1 ns, and the processing circuits have to operate with very high frequency. Since the high-speed processing increases noises and gets worse S/N ratio, small Δt is not always useful to get good resolution time. Therefore, the designs in the circuits are an important criterion for the decision of the latch time interval and the resolution time.

10.7.3 TRANSIENT MEASUREMENT METHOD

10.7.3.1 Signal Processing

In the heterodyne method, two optical interference circuits are constructed in the probe(s). (The circuit is described in the previous section.) Two optical beat signals outputted from the circuits, SIG and REF signals, are processed on the system diagram as shown in Figure 10.11. The SIG

FIGURE 10.11 The signal processing system. Two optical signals, SIG and REF, input to a photoelectric conversion circuits independently. Electric signals are changed to digital form with an A/D converter, and they are calculated in a digital circuit to output the displacement and/or the velocity. The outputted data are stored in a memory and transferred to an external computer through an interface. A crystal oscillator is used as a basic clock in these electric circuits.

signal usually contains the information of the sample target. The REF signal contains the information of a reference, for example, a fixed mirror.

Signals flow in the following steps. First, the optical signals change to electric signals in the photoelectric conversion circuits. Second, the electric analog signals are converted to digital forms in A/D conversions. Third, the digital signals are processed in a digital programming IC, in which data of displacements and/or velocities are calculated at a constant interval continuously, and they are stored in some external (or internal) memories. Fourth, the data are outputted to an external personal computer through an interface board.

Before discussing the careful points in the four steps, the system has to realize the following points. (1) A synchronous parallel processing mode for SIG and REF signals. (2) A symmetrical and an identical design of electric circuits and electric parts for SIG and REF signals. These points should be introduced to reduce all of errors or noises based on the differences between SIG and REF signals.

In the fourth step, the transfer time is determined mainly by a processing speed and a capacity of the interface board. When the number of measuring points is large, this unexpected time extends the repetition time of the measurement beyond tester's expectations. This unanticipated long interval may make the reproducibility loss by change in the environmental condition. The other steps are discussed in the following subsections.

10.7.3.2 Photoelectric Conversion Circuit

Optical signals include several kinds of noise as described in the previous sections. The main factor of the noises is based on the direct fluctuations of the optical path length and on some virtual fluctuations of that. As examples of the direct fluctuations, optical polarization rotations, air flows, instruments' sounds, temperature fluctuations, shaking of the laboratory's building, testers' actions, and testers' conversations are problems awaiting solutions. Examples of virtual fluctuation are weak optical power, damages from sources of electricity, unsymmetrical circuit pattern (time lag needs less than 0.5 ns) for SIG and REF signals, and unsymmetrical electric parts for them. The two types of the fluctuations have to be suppressed by some warrantable ways. For the suppression in the photoelectric conversion circuit, the following points should be considered: (1) use of an APD having a sufficiently wide frequency range, (2) use of a capacitor having a large capacitance for all power sources (0.1 Hz noise needs 0.1 F) to get good smoothing, (3) fixation of all of boards and parts to cut vibrations, (4) use of a shield cover not to detect electromagnetic pulses, (5) use of a symmetrical circuit pattern and layout of symmetrical electric parts, and (6) no action and silence of testers during measurement.

10.7.3.3 Period Detection

The digital signal is expressed by using variables of τ_C and Δt. The number of the digital signal per one output data N_C is determined as described previously. To analyze the data (total output data $= N_F$), all digital signals are classified with a datum number k ($k = 0, 1, 2, ..., N_F - 1$) and a series number in a datum n ($n = 1, 2, ..., N_C$) in this study. (Total output time $= N_F\tau_C$). Since all digital signals are defined at Δt intervals, the signal $W(t(k, n))$ is expressed as a function of the time $t(k,n) = k\tau_C + n\Delta t$. The following study is discussed with these parameters.

In the heterodyne method, a frequency and a phase of the signal are important factors to calculate the displacement and the velocity. They are computed with crossing times, which are defined as times satisfied the relation $W(t) = 0$. The time is hardly expressed with integer values of k and n. So, to show the crossing time $CR(k)$ accurately, the following relation is defined as an approximation:

$$CR(k) = t(k,n) + \frac{|W(t(k,n))|}{|W(t(k,n+1)) - W(t(k,n))|} \Delta t,$$

where the crossing time is assumed to be between $t(k, n)$ and $t(k, n + 1)$, and it is approximated under the assumption that $W(t(k, n))$ varies linearly in the neighborhood of the crossing time. To realize the approximation for a sine wave, the following relation between Δt and the period is adequate as discussed in Section 10.7.2(C):

$$t(k,n+1) - t(k,n) = \Delta t < \text{period}/(\text{several\_hundreds}).$$

There are two types for the crossing, a minus-crossing type: signal decreases from plus domain to minus domain, and a plus-crossing type: signal increases from minus domain to plus domain. Several crossing times are appeared within τ_C, that is, the first plus-crossing time (FPC), the second plus-crossing time (SPC), the third plus-crossing time (TPC), the first minus-crossing time (FMC), the second minus-crossing time (SMC), and the third minus-crossing time (TMC). The signals, the period, and the crossing times are understood as shown in Figure 10.12.

For the period, some expressions, $SPC(k) - FPC(k)$, $SMC(k) - FMC(k)$, etc., can be defined within τ_C. However, they do not always coincide because the signal changes subtly by drift or high-frequency noise. Therefore, to obtain more accurate value, the geometric mean of $P_f(k)$ can be defined as the period of k'th data (the first definition). In addition, the difference between the crossing time of k'th data and that of $(k + 1)$'th data, $FPC(k + 1) - FPC(k)$, etc., must be expressed with an integer multiple of the period. However, the signal variation does not always satisfy this demand because the signals pick up noises and the crossing times shift. Therefore, as the second definition, the period $P_S(FPC, k)$ is proposed by introducing an integer $N(FPC, k)$ for FPC as follows:

$$P_S(FPC, k) = \frac{FPC(k+1) + \tau_C - FPC(k)}{N(FPC,k)}.$$

The periods for other crossing types are defined with the same rule, but these periods $P_S(FPC, k)$, $P_S(FMC, k)$, and the other values) do not always coincide. Since the period gives the frequency,

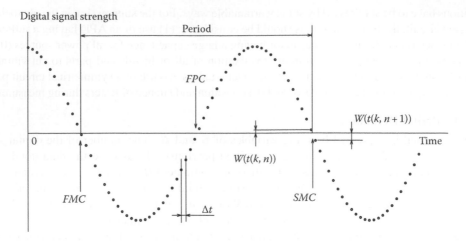

FIGURE 10.12 Digital signal in the processing. All data are memorized at a constant time interval Δt. In the data curve, two types of the crossing appeared. The data time $t(k, n)$ is expressed with a datum section number k and a serial number n in the section. The crossing time is calculated in a function of $W(t(k, n))$. The period is defined by using several crossing points (*FPC*, *FMC*, *SPC*, etc.).

the period's definition has an influence upon the precision of the output data. So, as an available judge, the following average value $P(k)$ will be an applicable period:

$$P(k) = \frac{1}{6}\left(2P_P(k) + P_S(FPC, k) + P_S(SPC, k) + P_S(FMC, k) + P_S(SMC, k)\right).$$

This relation is too complex that the calculation program may not accept in user's circuit. However, user must select the most useful and effective definition within a practicable plan.

In the calculation of the period, either *SPC* or *SMC* must exist necessarily within τ_C. Therefore, the value of τ_C has the minimum and must satisfy the condition $\tau_C \geq 2 \times$ period.

While though long τ_C yields many crossing points and gives more appropriate period, the time resolution decreases. So, the maximum τ_C will be given by the relation of $\tau_C < 3 \times$ period because TPC and TMC are not necessary on purpose.

10.7.3.4 Phase Detection and Displacement

In the datum section k, two typed phases are defined as follows:

$$\phi_P(k) = \frac{FPC(k)}{P(k)} \quad \text{and} \quad \phi_M(k) = \frac{FMC(k)}{P(k)}. \tag{10.18}$$

The phase difference between the phase in the section k and the phase in the section $k + 1$ has also two types, and they are defined as

$$\Delta\Phi_P(k) = \phi_P(k+1) - \phi_P(k) - N(FPC, k) + \frac{\tau_C}{P(k)} \quad \text{(for FPC crossing)}$$

and $\Delta\Phi_M(k)$ is represented with the same rule. The phase difference gives the displacement variation $\Delta R(k)$ in the section k, which has two types classified by the detecting ways:

$$\Delta R(k) = \frac{\lambda}{2} \frac{\Delta\Phi_P(k) + \Delta\Phi_M(k)}{2} \quad \text{(for the reflection measuring case)},$$

$$\Delta R(k) = \lambda \frac{\Delta \Phi_P(k) + \Delta \Phi_M(k)}{2} \quad \text{(for the transmission measuring case)},$$

where λ is the wavelength of the measuring laser. The total displacement $R(N_F)$ is represented with the sum of the displacement variation terms assigned by the section number

$$R(N_F) = \sum_{k=0}^{N_F-1} \Delta R(k), \tag{10.19}$$

where N_F is the number of the output data (the measuring time $= N_F \tau_C$).

10.7.3.5 Analysis of Error Factor

Earlier description for the digital processing clearly indicates an existence of error factors. That is, the following factors are closely connected with an occurrence of the error:

1. Absolute value of the laser wavelength
2. Basis beat frequency (period τ_{B0})
3. Latch time interval (Δt)
4. Output data interval ($\tau_C = N_R \Delta t$; N_R: integer)

The effect of these factors for the displacement is investigated by introducing the fluctuations of these factors: $\lambda = \lambda_0 + \Delta \lambda$, $\tau_B = \tau_{B0} + \Delta \tau_B$, and $\tau_C = \tau_{C0} + \Delta \tau_C$.

In the processing program, the utilized clock time $t(k,n) \equiv t_n^k$ is represented with the genuine time $T(k,n) \equiv T_n^k$ as follows:

$$t(k,n) = T(k,n) + \left(k + \frac{n}{N_R} \right) \times \Delta \tau_C.$$

This time difference increases (or decreases) monotonously with an increase in k and n. The difference induces the shift of the signal strength $W(t(k,n))$. Therefore, the functions of $CR(k)$, $P(k)$, $\phi_P(k)$, $\Delta R(k)$, etc., are also changed; finally, the displacement is expressed with these fluctuations. The changes in the functions are expressed approximately as follows:

$$W\left(t_{n+1}^k\right) - W\left(t_n^k\right) = W\left(T_{n+1}^k\right) - W\left(T_n^k\right) + \frac{\Delta \tau_C}{N_R} \frac{\partial W\left(T_{n+1}^k\right)}{\partial t} + \left(k + \frac{n}{N_R}\right) \frac{\tau_C}{N_R} \Delta \tau_C \frac{\partial^2 W\left(T_n^k\right)}{\partial t^2} + \cdots$$

$$\cong W\left(T_{n+1}^k\right) - W\left(T_n^k\right) + \frac{\alpha}{N_R} \Delta \tau_C,$$

$$FPC(k) = T_n^k + \left(k + \frac{n}{N_R}\right) \times \Delta \tau_C + \frac{-W\left(T_n^k\right) + \left(k + \dfrac{n}{N_R}\right) \Delta \tau_C \dfrac{\partial W\left(T_n^k\right)}{\partial t}}{W\left(T_{n+1}^k\right) - W\left(T_n^k\right) + \dfrac{\alpha}{N_R} \Delta \tau_C}$$

$$\cong T_n^k + \frac{-W\left(T_n^k\right)}{W\left(T_{n+1}^k\right) - W\left(T_n^k\right)} + \gamma_P(k) \Delta \tau_C \equiv FPC_0(k) + \gamma_P(k) \Delta \tau_C$$

$$P_S(FPC,k) = \frac{1}{N(FPC,k)} \{FPC_0(k+1) + \tau_C - FPC_0(k) + (\gamma_P(k+1) - \gamma_P(k))\Delta\tau_C\} + \cdots$$

$$\equiv P_{S0}(FPC,k) + \delta_P(k)\Delta\tau_C + \cdots,$$

$$\phi_P(k) = \frac{FPC_0(k)}{P_0(k)} + \left(\frac{\gamma_P(k)}{P_0(k)} - \frac{FPC_0(k)}{P_0^2(k)} \delta_P(k) \right) \Delta\tau_C + \cdots \equiv \phi_{P0}(k) + \varepsilon_P(k)\Delta\tau_C + \cdots,$$

where $P_0(k)$ is given by $(P_{S0}(SPC, k)$, which are represented with the same rule as $P_{S0}(FPC, k))$:

$$P_0(k) = \frac{1}{6}\left(2P_{F0}(k) + P_{S0}(FPC,\ k) + P_{S0}(SPC,\ k) + P_{S0}(FMC,\ k) + P_{S0}(SMC,\ k)\right).$$

The phase difference and the displacement variation in the section k are also arranged as follows:

$$\Delta\Phi_P(k) = \phi_{P0}(k+1) - \phi_{P0}(k) - N(k) + \frac{\tau_{C0}}{\tau_{B0}} + (\varepsilon_P(k+1) - \varepsilon_P(k))\Delta\tau_C + \frac{\tau_{C0}}{\tau_{B0}}\left(\frac{\Delta\tau_C}{\tau_C} - \frac{\Delta\tau_B}{\tau_{B0}} \right) + \cdots$$

$$\equiv \Delta\Phi_{P0}(k) + \xi_P(k)\Delta\tau_C + \eta\Delta\tau_B + \cdots,$$

where $N(FPC, k) = N(SPC, k) = N(FMC, k) = N(SMC, k) = N(k)$ is used undoubtedly.

$$\Delta R(k) = \frac{\lambda_0 + \Delta\lambda}{2}\left\{ \frac{1}{2}\left(\Delta\Phi_{P0}(k) + \Delta\Phi_{M0}(k)\right) + \frac{1}{2}\left(\xi_P(k) + \xi_M(k)\right)\Delta\tau_C + \eta\Delta\tau_B + \cdots \right\}$$

$$= \Delta R_0(k) + \frac{\Delta R_0(k)}{\lambda_0}\Delta\lambda + \frac{\lambda}{4}\left(\xi_P(k) + \xi_M(k)\right)\Delta\tau_C + \frac{\lambda\eta}{2}\Delta\tau_B + \cdots$$

where the zero mark in the subscript of $\Delta\Phi_{P0}(k)$ and $\Delta R_0(k)$ corresponds to the genuine condition. Therefore, the displacement expressed in Equation 10.19 is arranged with some additional terms as follows:

$$R(N_F) = R_0(N_F) + \sum_{k=0}^{N_F}\Delta R_0(k)\frac{\Delta\lambda}{\lambda_0} + \frac{\lambda_0}{4}\sum_{k=0}^{N_F}\left(\xi_P(k) + \xi_M(k)\right)\Delta\tau_C + \frac{N_F\lambda_0\eta}{2}\Delta\tau_B, \qquad (10.20)$$

where $R_0(N_F)$ is the genuine displacement result. In these relations, parameters of α, $\gamma_P(k)$, $\delta_P(k)$, $\varepsilon_P(k)$, $\xi_P(k)$, and η are introduced to shorten the lengthy expressions. Of course, the quantities having subscript of M, $\gamma_M(k)$, $\delta_M(k)$, $\varepsilon_M(k)$, and $\xi_M(k)$, are described with the same rule as those having the subscript P.

It is shown clearly that three additional terms in Equation 10.20 generate errors in the displacement results. However, the value of $R(N_F)$ is expressed only as an output of SIG (or REF) channel. When the relative measuring optical system is introduced, the displacement is given with a difference expressed by $R_{SIG}(N_F) - R_{REF}(N_F)$. (The subscripts of SIG and REF correspond to SIG and REF channels, respectively.) The relative measuring system brings great noise reduction for three terms as follows:

(First additional term) The factor $\Delta R_0(k)$ includes two error sources mainly: the target movement and the fluctuation of the optical path. When the SIG's optical path is nearly the same to the REF's optical path ("nearly the same" means that two optical axes are satisfied in the following conditions: parallel, small spacing, the same height from the basement, the same

length, the same reflection way for mirrors, the same transmission way for PBS and NPBS, etc.), the fluctuations of two paths are balanced because several causes of the fluctuations including environmental variation induce nearly the same effect on both paths. The balance and the fixation of the REF target make the noises decrease dramatically. Therefore, the displacement of the SIG target can be measured with very small error.

Since the frequency-stabilized HeNe laser is usually kept at the relation $\Delta\lambda/\lambda = 3 \times 10^{-9}$ [107], even if measurement is made under the worst experimental condition and the measuring point is 10^6, the measurement result is held down within 0.3% error. (An unstabilized HeNe laser has usually about $\Delta\lambda/\lambda = 3 \times 10^{-6}$ [108]. The error in 10^6 measuring points is much more than the genuine value in the worst case. This is the reason the free-running (= unstabilized) HeNe laser cannot be used in the heterodyne method).

(Second additional term) The relation of $(\xi_P(k) + \xi_M(k))_{SIG} \approx (\xi_P(k) + \xi_M(k))_{REF}$ is required to compensate the errors between SIG and REF signals. When the measurement system sets up under "nearly the same" condition, the relation is almost satisfied though some lengthy expansions of the terms are need. Therefore, "nearly the same" condition is one of the necessary conditions to obtain the high-precision results.

(Third additional term) The compensation requires the relation of $(\eta)_{SIG} \approx (\eta)_{REF}$. As the η value does not depend on targets' movements, this relation is satisfied under "nearly the same" condition.

Three terms suggest that the relative measurement having "nearly the same" condition for SIG and REF paths is essential for the measurement and that the displacement can suppress its errors by working compensations in all of additional terms. In other words, all designers and programmers should pay attention to the following points:

1. Absolute laser frequency: the frequency-stabilized HeNe laser is indispensable for the high-resolution measurement, because the error is so small even if the worst fluctuation occurs.
2. Environmental temperature and atmospheric pressure: the refractive index of the air varies largely by the temperature and the atmospheric pressure [109]. Especially, when the laser light transfers through the air space, the optical path length is changed largely by fluctuations of the environmental conditions. Therefore, introductions of the vacuum system and/or the temperature control for the whole optical circuits are acceptable condition to obtain higher precision.
3. Stability of crystal oscillator: in the digital circuit, the clock time of the crystal oscillator is usually used as a standard time base. Therefore, the fluctuation of the crystal oscillator directly affects the fluctuation of Δt and τ_C. High-quality crystal oscillator should be used as much as possible.
4. Stability of the frequency shifter: the basis beat frequency is made by the frequency shifter. Therefore, the clock generator of the frequency shifter needs good quality to decrease the fluctuation of the beat frequency. Two absolute frequencies of the shifter do not necessarily keep high quality, but the difference of two output frequencies, the beat frequency f_B, must have high stabilization of about $\Delta f_B/f_B < 1 \times 10^{-6}$.

10.7.4 Examples of Transient Measurement

10.7.4.1 Tuning Fork and Cantilever Vibration

The tuning fork or the cantilever is known as an item vibrating with a constant frequency. Their characteristics have been investigated [110–122]. An example of the measuring system for them

FIGURE 10.13 An example of optical circuit in the measurement system for the tuning fork. Two lights (frequency f_1 and f_2) are outputted from the optical fiber and are reflected repeatedly, and make two optical signals (SIG and REF). SIG signals include the displacement information of the tuning fork, while REF signals do that of the reference surface. By measuring under "nearly the same" condition, the difference of two signals, SIG–REF, indicates only the displacement. The "nearly the same" condition means that two optical axes have the following characteristics: parallel, small spacing, the same height from the basement, the same length, the same reflection, and transmission way for optical parts.

is shown in Figure 10.13. Two lights, $f1$ light and $f2$ light, have the frequencies of $f_0 + f_1$ and $f_0 + f_2$, respectively. The frequency f_0 is the HeNe laser frequency of about 474 THz. The frequencies f_1 and f_2 are the shift values of about 80 MHz made by the frequency shifter. The $f1$ light is divided, and two lights are reflected from a sample surface and a reference surface. The $f2$ light is also divided and comes to optical detectors after reflecting repeatedly. On the optical detectors, f_1 and f_2 lights interfere and yield two electric signals, SIG and REF. Both signals contain all information of the optical path including the sample and the reference surfaces. Especially, the SIG signal includes the displacement information of the sample surface, and the REF signal does that of the reference surface. Therefore, the subtraction between two signals SIG–REF can cancel practically all effects on the optical path excluding the displacement of the sample and the reference surfaces because the optical path length and position of the SIG signal are almost the same as those of the REF signal. (It satisfies "nearly the same" condition.) Therefore, the subtraction value can pick up only the displacement of the sample. (In optical circuit, to suppress the influences of optical parts, mechanical parts, and holders, the same goods are used for both paths.)

An example of results is shown in Figure 10.14a and b [123]. The measurement time is 16 s, the time resolution (τ_C) is 16 μs, and the measuring points are 10^6. In Figure 10.14a, no sine wave can be observed because the measuring point is too large to draw on this scale. However, in the envelope form, a damping is observed. By multiplying the figure as shown in Figure 10.14b, a sine wave is confirmed apparently. Two figures indicate that the tuning fork vibrates in the form of the damping oscillation. The transient characteristics of the tuning fork can be investigated with three strong points of the heterodyne measuring system: (1) the high displacement resolution (having good S/N ratio for small vibration amplitude of about 300 nm), (2) the high time resolution (having good S/N ratio for small phase detection of about 0.008 rad), and (3) the large measuring points (keeping good S/N ratio for wide amplitude).

By analyzing these data, several dynamic characteristics are obtained for the damping oscillation expressed by $B\exp(-\gamma t)\sin(2\pi f t + \varphi)$. (1) The frequency f increases linearly with a decrease in

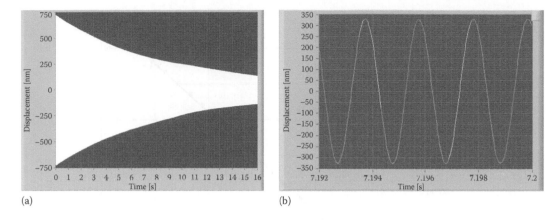

FIGURE 10.14 (a) An example of the transient variation of the tuning fork. Vertical, displacement; horizontal, time. Measurement time, 16 [s]; the time resolution, 16 [μs]; measuring points, 10^6 points. Any sine wave is not recognized because the number of the data is too large to draw all points in this graph. But the envelope of the curve shows damping. (b) Magnification of (a) within the time region from 7.192 to 7.2 [s]. It is confirmed that vibration form is a sine wave, and its frequency is 489.913 [Hz]. High resolution gives good S/N ratio under very small noise of about 1.4 [nm]. Both figures (a) and (b) show clearly that the tuning fork vibrates undoubtedly in the form of a damping oscillation.

FIGURE 10.15 An example of the relation between f_0 and $\sqrt{B}$, where f_0 is the frequency at zero amplitude and B is the vibration amplitude. Experimental error bars' lengths for all data are nearly equal to the dot size in this expression. High correlation coefficient confirms a linear relation of $f_0 = m\sqrt{B} + b$.

the square root of the vibration amplitude $\sqrt{B}$ as shown in Figure 10.15. (2) The logarithm of the damping factor $\log(\gamma)$ is proportional to the logarithm of the vibration amplitude $\log(B)$ as shown in Figure 10.16. (3) Both f and γ depend on the temperature and the measuring position located on the tuning fork's tin. (4) The frequency decreases with an increase in temperature as shown in Figure 10.17.

The measurement for the cantilever gives nearly the same results because it has the basic form of the tuning fork. The Bernoulli–Euler equation [124] is the standard equation for investigations

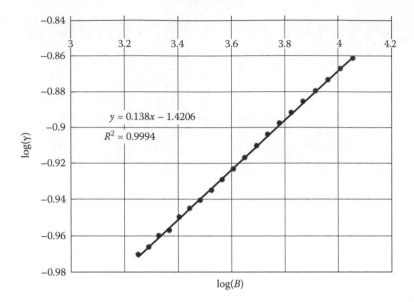

FIGURE 10.16 An example of the relation between the $\log(B)$ and $\log(\gamma)$ is shown under the condition, $X = 15$ mm, $T = 14.7°C$, $P = 991.3$ hPa, and humidity = 39%. Very high linearity is shown (correlation coefficient is over 0.999). The vibration amplitude B is represented with nm unit. The X, T, and P are the distance from the tuning fork's tip, the temperature, and the atmospheric pressure, respectively.

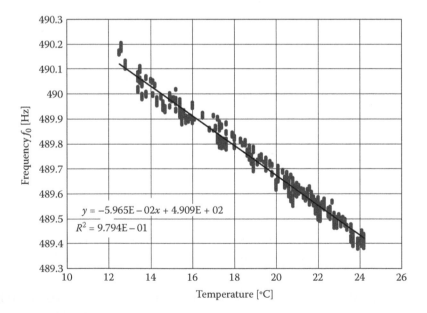

FIGURE 10.17 The temperature dependence of the frequency of the tuning fork. The sizes of data points in this figure are comparable to the spread of the error bar. The data distribution seems to form a belt. The width or extension of the belt depends on the vibration amplitude and/or the measuring position on the tuning fork's tin. The distance between the measurement position and the tip distributes from 5 to 75 mm, the atmospheric pressure does from 981.7 to 1005.1 hPa, and the humidity does from 37% to 68%.

of the cantilever vibration. However, the equation can't explain any damping oscillation transiently. Several modified equations have been proposed and investigated [125–133] to understand the transient oscillation. Details are described in these references. One of them is the following form:

$$EI\left(\frac{\partial^4 y}{\partial x^4} + \mu \frac{\partial^5 y}{\partial x^4 \partial t}\right) + \rho A \frac{\partial^2 y}{\partial t^2} + \lambda \frac{\partial y}{\partial t} = 0, \tag{10.21}$$

where
  E (Young's modulus)
  I (second moment of area),
  ρ (density),
  A (cross section) are constants given by the material and the size of the tuning fork or the cantilever
  y, x, and t are displacement, vibration position, and time, respectively

The damping factors, μ and λ, represent the transient resistance strengths against the shearing stress and the movement, respectively
By analyzing Equation 10.21 and experimental results, it is shown that these damping factors depend on the vibration amplitude, measuring position, and other factors and that they are NOT constant in the damping oscillation of the cantilever.

The modified theoretical equation matches to some of the experimental results [132]. However, some of the dynamic characteristics are hardly understood, especially the relations among f, γ, and vibration amplitude are left as a problem of the transient analyses. In addition, the damping factors have to be investigated transiently including a question of whether they are constants or not.

10.7.4.2 Concentration Fluctuation

Several optical detections for the concentration fluctuation of the fluids have been investigated [134–138]. Generally, when gases or liquids are mixed, a concentration distribution is born and brings about diffusion till it is homogeneous. By measuring the concentration variation transiently, the diffusion as one of dynamic phenomena will be able to be investigated. As one of the optical measurement method for fluid, the concentration dependence of the refractive index of the fluid sample is utilized. Since the concentration varies transiently and spatially in a chamber of packed or filled samples, the variation of the optical path length of the chamber $L(t)$ is measured totally and transiently, which is expressed by the relation

$$L(t) = \int_0^K dy\, n(y,t) = K\bar{n},$$

where
  K is the mechanical length of the chamber
  $n(y, t)$ is the refractive index at space variable y and time t
  $\bar{n}$ is the average refractive index on the optical path

As the optical heterodyne method measures the value of $L(t) - K$, any concentration distribution in the chamber is neglected, and only the average value is used in the analysis. Certainly, this method may not be effective to investigate the distribution variation of the concentration. However, this method has a strong point that very low concentration variation can be detected, for example,

FIGURE 10.18   An example of the optical circuit of the concentration measurement. Both Ch1 and Ch2 chambers are filled with pure water. The refined sugar water dropped into the Cha camber only. The difference between Ch1 and Ch2 detectors' signals corresponds to the variation of the optical path length of Ch1. The variation is generated by a diffusion of the sugar water. Therefore, the diffusion of liquid sample can be observed. To realize a high precision, "nearly the same" condition needs. Especially, the relation of $\left| L_0 - L_1 \right| < 0.5$ mm should be kept.

the refractive index change of 10^{-6} in 10 mm chamber length is detectable. Therefore, the high precision of the heterodyne method can be applied to the concentration variation measurement.

An example of the optical circuits for the concentration variation measurement is shown in Figure 10.18 [139]. A laser light is divided, and both lights are deflected with AOMs. After doing optical interference, two signals are generated. Pure water of 12.8 g is set for both chambers. A refined sugar water ($C_{12}H_{22}O_{11}$, 0.077 mol/L) of 0.15 cm^3 is dropped from 2 cm height above the surface of the pure water for Ch1 chamber only.

An example of the optical path variation (corresponding to the difference between Ch1 chamber and Ch2 chamber) is shown in Figure 10.19. The time resolution is 32 μs, and the data points are 10^6. The sample is dropped at 5.1 s after the starting time of recording. Though there are noises of about 10 nm width generated by several causes over the whole of data, the optical path length starts to vary slowly, and it repeats the rise and fall and settles into a certain quantity finally. In short, the concentration distribution converges at an equilibrium value after dynamic variation. The variation including overshoot will be understood by the following process. The dropped sample goes straightly into the pure water as a lump of sugar water keeping the potential energy and starts to diffuse slowly by the dissipation of the energy. The lump will start to expand and become hazy gradually. The optical path length depends on the lump's position and size. Therefore, the movement of the lump (i.e., diffusion) creates the variation of the optical path length including overshoot and/or undershoot.

And the optical path settles down finally into a constant corresponding to a homogeneous distribution in the chamber. As the optical path length is based on the refractive index, the variation of the optical path length ΔL between the starting time ($t = 0$) and at $t = \infty$ is expressed with the relation

FIGURE 10.19 An example of the transient phenomenon of the concentration diffusion. Measurement time is 32 s. The refined sugar water is dropped at 5.1 s in this graph. Two types of vibration are observed. First type: it decreases monotonously till 11.2 s and repeats to go up and down and finally converges to nearly constant. Second type: it vibrates with an amplitude of 10 nm or less and a period of 0.24 ± 0.07 s after the dropped time. The first type shows the concentration diffusion. The second type may show an impulse wave or a compressed wave.

of $\Delta L = K \times (\bar{n}(t = \infty) - \bar{n}(t = 0))$. In the example, as $\Delta L = 120$ nm, the concentration dependence of the refractive index of the sugar water is obtained [139] as follows:

$$n(w) = n(0) + 0.0065 \times w,$$

where

w is the concentration in mol/L unit
$n(0)$ is the refractive index of the pure water given by 1.3318

The optical heterodyne measurement system shown in Figure 10.18 can apply to all transparent samples and many semitransparent samples, and it will analyze the characteristics of liquid, vapor, and gas with the transient variations. In addition, this absolute detection of the characteristic variation within very short times will be useful in the observations of various physical quantities, for example, temperature, pressure, force, viscosity, magnetic field, radiation quantity, and current.

REFERENCES

1. J.P. King, D.F. Smith, K. Richards, P. Timson, R.E. Epworth, and S. Wright, Development of a coherent OTDR instrument, *J. Lightwave Technol.* 5, 616 (1987).
2. A.V. Jellalian and R.M. Huffacker, Laser Doppler techniques for remote wind velocity measurements, *Digital Specialist Conference on Molecular Radiation* (Huntsville, AL), October 1967.

3. M.S. Fee, K. Danzmann, and S. Chen, Optical heterodyne measurement of pulsed lasers: Toward high precision pulsed spectroscopy, *Phys. Rev. A* 45, 4911 (1992).

4. H. Hockel, M. Lauters, M. Jackson, J.C.S. Moraes, M.D. Allen, and K.M. Evenson, Laser frequency measurement and spectroscopic assignment of optically pumped $^{13}CH_3OH$, *Appl. Phys. B* 73, 257 (2001).

5. J.L. Gottfried, B.J. McCall, and T. Oka, Near-infrared spectroscopy of H_3^+ above the barrier to linearity, *J. Chem. Phys.* 118(24), 10890 (2003).

6. M. Toida, M. Kondo, T. Ichimura, and H. Inaba, Two-dimensional coherent detection imaging in multiple scattering media bared on the directional resolution capability of the optical heterodyne method, *Appl. Phys. B* 52, 391 (1991).

7. M. Sato, Phase-shift-suppression using harmonics in heterodyne detection and its application to optical coherence tomography, *Opt. Commun.* 184, 95 (2000).

8. L.V. Wang, Ultrasound-mediated biophotonic imaging: A review of acousto-optical tomography and photo-acoustic tomography, *Dis. Markers* 19, 123 (2004).

9. C.F. Yen, M.C. Chu, H.S. Seung, R.R. Dasari, and M.S. Feld, Noncontact measurement of nerve displacement during action potential with a dual-beam low-coherence interferometer, *Opt. Lett.* 29(17), 2028 (2004).

10. S.E. Pollack and R.T. Stebbins, Demonstration of the zero-crossing phasemeter with a LISA test-bed interferometer, *Class. Quantum Grav.* 23, 4189 (2006); A demonstration of LISA laser communication, *Class. Quantum Grav.* 23, 4201 (2006).

11. A.M. Jorgensen, D. Mozukewich, J. Muirphy, M. Sapantaie, J.T. Armstrong, G.C. Gilbreath, R. Hindsley, T.A. Pauls, H. Schmitt, and D.J. Hutter, Characterization of the NPOI fringe scanning stroke, *Proc. SPIE Astronomical Telescopes and Instrumentation*, Orland, FL, SPIE 6268 Advances in Stellar Interferometry, 2006.

12. J.D. Monnier, Optical interferometer in astronomy, *Rep. Prog. Phys.* 66, 789 (2003).

13. M.V. Aguanno, F. Lakestani, M.P. Whelan, and M.J. Connelly, Heterodyne speckle interferometer for full-field velocity profile measurements of a vibrating membrane by scanning, *Opt. Laser Eng.* 45, 677 (2007).

14. M.D. Alaimo, M.A.C. Potenza, D. Magatti, and F. Ferri, Heterodyne speckle velocimetry of Poiseuille flow, *J. Appl. Phys.* 102, 073133 (2007).

15. M.I. Dykman, G.P. Golubev, I.K. Kaufman, D.G. Luchinsky, P.V.E. McClintok, and E.A. Zhukov, Noise-enhanced optical heterodyning in all-optical bistable system, *Appl. Phys. Lett.* 67(3), 308 (1995).

16. P.T. Beyersdorf, M.M. Fejer, and R.L. Byer, Polarization Sagnac interferometer with a common path local oscillator for heterodyne detection, *J. Opt. Soc. Am. B* 16(9), 1354 (1999).

17. M.S. Shahriar, G.S. Pati, R. Tipathi, V. Gopal, M. Messall, and K. Salit, Ultrahigh enhancement in absolute and relative rotation sensing using fast and slow light, *Phys. Rev. A* 75, 053807 (2007).

18. C.F. Bharucha, J.C. Robinson, F.L. Moore, B. Sundaram, Q. Niu, and M.G. Raizen, Dynamical localization of ultracold sodium atoms, *Phys. Rev.* 60(4), 3881 (1999).

19. W.X. Ding, D.L. Brower, B.H. Deng, and T. Yates, Electron density measurement by differential interferometry, *Rev. Sci. Instr.* 77, 10F105 (2006).

20. M. Bass (Chief Ed.), Acousto-optic devices and application, *Handbook of Optics*, vol. 2, Chapter 12, McGraw-Hill, New York, 1994.

21. M. Hirano, AOM used in heterodyne method, *Oplus E* 151, 86 (1992).

22. AOM catalogue: Crystal Technology, Inc. (USA), Gooch & Housego, Inc. (Germany), Isomet, Inc. (USA), etc.

23. R. Waynant and M. Ediger, *Electro-Optic Handbook*, McGraw-Hill, New York, 2000.

24. G. Heinzel, The LTP interferometer and phasemeter, *TAMA Symposium*, Osaka, Japan, 2005.

25. C. Leinert et al., MIDI-the 10 μM instrument on the VLTI, *Astrophy. Space Sci.* 286, 73 (2003).

26. S.J.A.G. Coijins, H. Haitjema, and P.H.J. Schollekens, Modeling and verifying non-linearities in heterodyne displacement interferometry, *J. Int. Soc.* 26, 338 (2002).

27. B.A.W.H. Knarren, Application of optical fibers in precision heterodyne laser interferometry, *Einhoven: Technishe Universiteit Eindhoven*, Eindhoven, the Netherlands, 2003.

28. C. Wu, J. Lawall, and R.D. Deslattes, Heterodyne interferometer with subatomic periodic nonlinearity, *Appl. Opt.* 38(19), p. 4089 (1999).

29. J. Guo, Y. Zhang, and S. Shen, Compensation of nonlinearity in a new optical heterodyne interferometer with doubled measurement resolution, *Opt. Commun.* 184(1–4), 49 (2000).

30. T.C. TaeBong et al., A simple method for the compensation of the nonlinearity in the heterodyne interferometer, *Meas. Sci. Technol.* 13, 222 (2002).

31. L.M. Krieg, R.G. Klaver, J.J.M. Braat, Absolute optical path difference measurement with angstrom accuracy over ranges of millimeters, *Proc. SPIE Laser Metrol.* 4398, 116, (2001).

32. D.A. Swyt, Length and dimensional measurement at NIST, *J. Res. Natl. Inst. Stand. Technol.* 106, 1 (2001).

33. S. Hurlebaus and L.J. Jacobs, Dual-probe laser interferometer for structural health monitoring (L), *J. Acoust. Soc. Am.* 119(4), 1923 (2006).

34. F.C. Demarest, High-resolution, high-speed, low data age uncertainty, heterodyne displacement measuring interferometer electronics, *Meas. Sci. Technol.* 9, 1024 (1998).

35. A.K. Alfred et al., Modified optical heterodyne direct conversion system for measuring frequency response of wide-band wavelength-sensitive electro optic devices, *IEEE Photon. Technol. Lett.* 18(20), 2183 (2006).

36. P.C.D. Hobbs, Ultrasensitive laser measurement without tears, *Appl. Opt.* 36(4), 903 (1997).

37. B. Edlén, The refractive index of air, *Metrologia* 2, 71 (1966).

38. K.P. Birch and M.J. Downs, The precise determination of the refractive index of air, NPL report MOM90, July (National Physical Laboratory, Teddington, U.K., 1988).

39. R. Muijlwijk, Update of the Edlén formula for the refractive index of air, *Metrologia* 25, 189 (1988).

40. M.L. Eickhoff and J.L. Hall, Real-time precision refractometry: New approaches, *Appl. Opt.* 36(6), 1223 (1997).

41. S.J.A.G. Coijin, H. Haitjema, and P.H.J. Schellekens, Modeling and verifying non-linearities in heterodyne displacement interferometry, *J. Int. Soc. Presion Eng. Nano.* 26, 448 (2002).

42. M. Kranjcec et al., Acousto-optic modulator with a $(Ga_{0.4}In_{0.6})_2Se_3$ monocrystal as the active element, *Appl. Opt.* 36(2), 490 (1997).

43. W. Lee and S. Chen, Acousto-optical effect induced by ultrasound pulses in a nematic liquid-crystal film, *Appl. Opt.* 40(10), 1682 (2001).

44. T. Kurosawa and M. Hirano, Beat frequency stability of the optical frequency shifter utilized in heterodyne method, *Techn. Rep. IEICE OPE* 94(41), 31 (1994).

45. K. Shimoda and A. Javan, Stabilization of the HeNe maser on the atomic line center, *J. Appl. Phys.* 36(3), 718 (1965).

46. J. Bartl, J. Guttenová, V. Jacko, and R. Ševčík, Circuits for optical frequency stabilization of metrological lasers, *Meas. Sci. Rev.* 7(5), 59 (2007).

47. Frequency stabilization HeNe laser catalogue: Armstrong, Inc. (U.K.); PLASMA Lab. (Russia); SIOS Meßtechnik GmbH (Germany); Melles Griot, Inc. (USA); Coherent Inc. (USA); Photon Probe, Inc. (Japan), etc.

48. Isolator catalogue: Laser 2000 Ltd. (U.K.); Electro-Optics Technology, Inc. (USA), etc.

49. Fiber collimator catalogue: Schäfter+Kirchhoff GmbH (Germany); Silicon Lightwave Technology, Inc. (USA); Division of Thorlabs, Inc. (USA); Photon Probe, Inc. (Japan), etc.

50. Optical parts catalogue: Edmund Optics, Inc. (USA); Photop Technologies, Inc. (China), etc.

51. M.S. Fee, K. Danzmann, and S. Chu, Optical heterodyne measurement of pulsed lasers: Toward high-precision pulsed spectroscopy, *Phys. Rev.* 45(7), 4911 (1992).

52. A. Ahn, C. Yang, A. Wax, G. Popescu, C.F. Yen, K. Badizadegan, R.D. Dasari, and M.S. Feld, Harmonic phase-dispersion microscope with a Mach-Zender interferometer, *Appl. Opt.* 44(7), 1188 (2005).

53. D.V. Baranov, A.A. Egorov, E.M. Zolotov, and K.K. Sviedzinsky, Influence of phase distortion on the response of an optical heterodyne microscope, *Nonlinear Quantum Opt.* 6(4), 753 (1996).

54. J.E. Hall, G.P. Wiederrecht, and S.K. Gray, Heterodyne apertureless near-field scanning optical microsopy on periodic gold nanowells, *Opt. Exp.* 15(7), 4098 (2007).

55. J.M. Flores, M. Cywiak, M. Servin, and L. Juarez, Heterodyne two beam Gaussian microscope interferometer, *Opt. Exp.* 15(13), 8346 (2007).

56. H. Toda, M. Haruna, and H. Nishihara, Integrated optic heterodyne interferometer for displacement measurement, *Light Tech.* 9(5), 683 (1991).

57. F. Tian, R. Ricken, S. Schmid, and W. Sohler, Integrated acousto-optical heterodyne interferometers in LiNbO3, Laser in der Technik, *Proceedings of Congree Lasr'93*, München, Germany, p. 725 (1993).

58. K. Miyagi, M. Nanami, I. Kobayashi, and A. Taniguchi, A compact optical heterodyne interferometer by optical integration and its application, *Opt. Rev.* 4(1A), 133 (1997).

59. A. Rubiyanto, R. Ricken, H. Herrmann, and W. Sohler, Integral acousto-optical heterodyne interferometer operated with a TiEr:LiNbO3 DBR-waveguide laser, *Proceedings od the Ninth European Conf. on Integrated Optics (ECIO'99)*, Torino, Italy, p. 275, April 1999.

60. F. Xia, D. Datta, and S.R. Forrest, Amonolithically suppression using harmonics in heterodyne detection, *Photonics Technol. Lett.* 17(8), 1716 (2005).

61. J. Lawall and E. Kessler, Michelson interferometer with 10 pm accuracy, *Rev. Sci. Instrum.* 71, 2269 (2000).
62. S.J.A.G. Cosijns, *Displacement Laser Interferometer with Sub-nanometer Uncertainty*, Grafisch bedrijf Ponsen & Looijen, Wageningen, the Netherlands, 2004.
63. R.K. Heilmann, P.T. Konkola, C.G. Chen, and M. Schattenburg, Relativistic corrections in displacement measuring interferometry, *44th EIPBN 2000*, Palm Springer, CA.
64. W.C.J. Gardiner, Y. Hidaka, and T. Tanzawa, Refractivity of combustion gases, *Combust Flame* 40, 213 (1980).
65. N. Kawahara, E. Tomita, and H. Kmakura, Unburned gas temperature measurement in a spark-iginition engine using fiber-optic heterodyne interferimetry, *Meas. Sci. Technol.* 13(1), 125 (2002).
66. T. Higashiki, T. Tojo, M. Tabata, T. Nishizaka, M. Matsumoto, and Y. Sameda, A chromic aberration-free heterodyne alignment for optical lithography, *J. Appl. Phys.* 29, 2568 (1990).
67. Stage catalogue; Polytec PI, Inc. (USA), etc.
68. S. Topcu, L. Chassagne, D. Haddad, and Y. Alayli, Heterodyne interferometric technique for displacement control at the nanometric scale, *Rev. Sci. Instrum.* 74, 4876 (2003).
69. V. Protopopov, S. Cho, K. Kim, S. Lee, and H. Kim, Differential heterodyne interferometer for measuring thickness of glass panels, *Rev. Sci. Instrum.* 78, 076101 (2007).
70. L. McMackin and D.G. Voelz, Multiple wavelength heterodyne array interferometry, *Opt. Exp.* 1(11), 332 (1997).
71. A. Wax and J.E. Thomas, Optical heterodyne imaging and Wigner phase space distributions, *Opt. Lett.* 21(18), 1427 (1996).
72. M.A. Aguanno, F. Lakestani, and M.P. Whelan, Full-field heterodyne interferometry using a complimentary metal-oxide semiconductor digital signal processor camera for high-resolution profilometry, *Opt. Eng.* 46(9), 095601 (2007).
73. C. Chou, J. Shyu, Y. Huang, and C. Yuan, Common-path optical heterodyne profilometer: A configuration, *Appl. Opt.* 37(19), 4137 (1998).
74. H.J. Tiziani, A. Rothe, and N. Maier, Dual-wavelength heterodyne differential interferometer for high-precision measurement of reflective aspherical surfaces and step heights, *Appl. Opt.* 35(19), 3525 (1996).
75. D. Su, J. Lee, and M. Chiu, New type of liquid refractometer, *Opt. Eng.* 37(10), 2795 (1998).
76. M.V. Ötügen and B. Ganguly, Laser heterodyne method for high-resolution gas-density measurements, *Appl. Opt.* 40(21), 3502 (2001).
77. N. Kawahara, E. Tomita, and H. Kamakura, Transient temperature measurement of unburned gas using optic heterodyne interferometry, *10th International Symposium on Application of Laser Measurement on Fluid Mechanics*, Lisbon, Portugal, 2000.
78. A.J. Sedlacek, Real time detection of ambient aerosols using photothermal interferometry: Folded Jamin interferometer, *Rev. Sci. Instr.* 77, 064903 (2006).
79. E.B. Cummings, I.A. Leyva, and H.G. Hornung, *Appl. Opt.* 34, 3290 (1995); S. Schlump, E.B. Cummings, and T.H. Sobata, Laser-induced thermal acoustic velocimetry with heterodyne detection, *Opt. Lett.* 25(4), 224 (2000).
80. C.K. Sun, B. Golubovic, J.G. Fujimoto, H.K. Choi, and C.A. Wang, Heterodyne nondegenerate pump-probe measurement technique for guided-wave devices, *Opt. Lett.* 20(2), 210 (1995).
81. Y. Takeuchi, I. Nishiyama, and N. Yamada, High-precision (<1 ppB/C) optical heterodyne interferometric dilatometer for determining absolute CTE of EUVL materials, *Proc. SPIE*, San Jose, CA, p. 61511Z, 2006.
82. M. Okaji, Absolute thermal expansion measurement of single-crystal silicon in the range 300–1300 K with an interferometric dilatometer, *Int. J. Thermophys.* 9, *Symposium on Thermophysical Properties*, Gaithersburg, MD (1988).
83. E.I. Bromlev, J.N. Randall, D.C. Flanders, and R.W. Mountain, A technique for the determination of stress in thin films, *J. Vac. Sci. Technol.* B1, 1364 (1983).
84. M. Hirano, H. Inamura, H. Sekiguchi, and T. Kurosawa, Bulge measuring system,'95 SOMIT, Yokohama, Japan, p. 44, 1995.
85. H.K. Teng, C. Chou, C.N. Chang, C.W. Lyu, and Y.C. Huang, Linear birefringence measurement with differential-phase optical heterodyne polarimeter, *J. Appl. Phys.* 41, 3140 (2002).
86. R.A. Chipman, Polarimetry in Handbook of Optics II (Michael Bass Editor in Chief), Chapter 22, pp. 22.1–22.37, McGraw-Hill, New York, 1995.
87. J.U. Lin and D.C. Su, A new type of optical polarimeter, *Meas. Sci. Technol.* 14, 55 (2003).
88. T. Yamaguchi, K. Oka, and Y. Ohtsuka, Dynamic photoelastic analysis by optical heterodyne polarimetry, *Opt. Rev.* 1(2), 276 (1994).
89. Y. Li and A. Yariv, Solutions to the dynamical equation of polarization-mode dispersion and polarization-dependent losses, *J. Opt. Soc. Am. B* 17(11), 1821 (2000).

90. J.H. Rommers and J. Howard, A new scheme for heterodyne polarimetry with high temporal resolution, *Plasma Phys. Control. Fusion* 38, 1805 (1996).

91. H.H. Mei, S.J. Chen, and W.T. Ni, Suspension of the fiber mode cleaner launcher and measurement of the high extinction-ratio (10-9) ellipsometer for the Q and A experiment, *J. Phys.* 32, 236 (2006).

92. J.L. Hall, J. Ye, and L.S. Ma, Measurement of mirror birefringence at the sub-ppm level: Proposed application to a test of QED, *Phys. Rev.* 62, 013815 (2000).

93. A.J. Coleman, E. Draguioti, R. Tiptaf, N. Shotri, and J.E. Saunders, Acoustic performance and clinical use of fiberoptic hydrophone, *Ultrason. Med. Biol.* 24(1), 143 (1998).

94. P.C. Beard, A.M. Hurrell, and T.N. Mills, Characterization of a polymer film optical fiber hydrophone for use in the range 1 to 20 MHz: A comparison with PVDF needle and membrane hydrophone, *IEEE Trans. Ultrason.* 47(1), 256 (2000).

95. S. Tsuji, T. Akiyama, E. Sato, T. Nozawa, H. Tsutsui, R. Shimada, M. Takahashi, and K. Terai, Fiberoptic heterodyne magnetic field sensor for long-pulsed fusion devices, *Rev. Sci. Instrum.* 72, 413 (2002).

96. A. Krzysztof and M.T. Hubert, Photonic electromagnetic field sensor developments, *Opt. Appl.* 32(1–2), 7 (2002).

97. S. Iio, A. Akiyama, E. Sato, T. Nozawa, H. Tsutsui, R. Shimada, M. Takahashi, and K. Terai, Fiberoptic heterodyne magnetic field sensor for long-pulsed fusion devises, *Rev. Sci. Instrum.* 72, 413 (2001).

98. S.K. Han, K.Y. Kang, M.E. Ali, and H.R. Fetterman, Demonstration of the high-frequency optical heterodyne technology using near-field fiber-optic probes, *Appl. Phys. Lett.* l75(4), 454 (1999).

99. D.S. Chung, K.Y. Lee, and E. Mazur, Fourier-transform heterodyne spectroscopy of liquid and solid surfaces, *Appl. Phys. B* 64, 1 (1997).

100. J.P. Gollub and M.H. Freilich, Optical heterodyne study of the Taylor istability in a rotating fluid, *Phys. Rev. Lett.* 33(25), 1465 (1974).

101. B.S. Rinkevichyus and A.V. Tolkachev, Optical Doppler flowmeter for gases, *J. Quantum Electron.* 4, 1062 (1975).

102. F.G. Omenetto and J.R. Torgerson, Optical velocimetry, *Phys. Opt.* 15, 73 (2004).

103. R.A. Kolt, T.J. Scholl, and S.D. Rosner, Laser-rf saturation spectroscopy: A novel fast-ion-beam sub Doppler method, *Can. J. Phys.* 75(10), 721 (1997).

104. W.L. Eberhand and R.M. Schotland, Dual-frequency Doppler-lidar method of wind measurement, *Appl. Opt.* 19, 2967 (1980).

105. A.K. Murry, A.L. Herrick, and T.A. King, Laser Doppler imaging: A developing technique for application in the rheumatic diseases, *Rheumatology* 43(10), 1210 (2004).

106. R. Maniewski, A. Liebert, M. Kacprzak, and A. Biec, Selected applications of near infrared optical methods in medical gnosis, *Opt. Electronic Rev.* 12(3), 255 (2004).

107. Frequency stabilized HeNe laser catalogue: Melles Griot, Inc., Japan Laser, Corp., Photon Probe, Inc.

108. HeNe laser catalogues: JDS Uniphase, Inc, NEO ARK, Corp., JSC Plasma, Inc.

109. B. Edlen, The refractive index of air, *Metrologia* 2, 71–80 (1966).

110. H. Hölscher, P. Milde, U. Zerweck, L.M. Eng, and R. Hoffmann, The effective quality factor at low temperatures in dynamic force microscopes with Fabry-Perot interferometer detection, *Appl. Phys. Lett.* 94, 223514 (2009).

111. G. Zeleter, J.C. Randel, A.K. Gupta, R. Bashir, S.H. Song, and H.C. Manoharan, Scanning optical homodyne detection of high-frequency picoscale resonances in cantilever and tuning fork sensors, *Appl. Phys. Lett.* 91, 173124 (2007).

112. R.S.E. Kristine, A. Langdon, W.A. Richardson, and J. Lapato, High-speed laser interferometry applied to transient vibration measurements, *Appl. Opt.* 23, 2956–2960 (1984).

113. G. Rajshekhar, S.S. Gorthi, and P. Rastogi, Estimation of dynamically varying displacement derivatives using fringe projection. *Appl. Opt.* 50, 282–286 (2011).

114. D.W. Schindel, D.A. Hutchins, and S.T. Smith, A study of materials at high temperature using miniaturized resonant tuning forks and noncontact capacitance transducers, *J. Acoust. Soc. Am.* 102, 1296–1309 (1997).

115. L. Bates, T. Beach, and M. Arnott, Determination of the temperature dependence of Young's modulus for stainless steel using a tuning fork, *J. Undergraduate Res. Phys.* 18, 9–13 (1999).

116. S. Sizgoric and A.A. Gundjian, An optical homodyne technique for measurement of amplitude and phase subangstrom ultrasonic vibrations, *Proc. IEEE* 57, 1313–1314 (1969).

117. S. Yamada, S. Kurosaki, and Y. Jimbo, Fixed end conditions of a cantilever vibrator and tuning fork vibrator, *IIIJ* 75, 1–15 (1975).

118. K. Kokubun, M. Hirata, H. Murakami, Y. Toda, and M. Ono, Bending and stretching mode crystal oscillation as a vacuum gage, *Vacuum* 34, 731–735 (1984).

119. S. Silja, S. Jan, H. Xia, and F. Holger, Displacement characteristics of piezoelectric langasite transducers at high temperature, *AIP Conf. Proc.* 1253, 164 (2010).
120. T. Pozar, P. Gregorcic, and J. Mozina, Optical measurements of the laser-induced ultrasonic waves on moving objects, *Opt. Exp.* 17, 22906 (2007).
121. J.C. Acosta, G. Hwang, J.P. Polesel-Maris, and S. Regnier, A tuning fork based wide range mechanical characterization tool with nanorobotic manupulators inside a scanning electron microscope, *Rev. Sci. Instrum.* 82, 035116 (2011).
122. J.F. Valley, J.W. Wu, and C.L. Valencia, Heterodyne measurement of poling transient effects in electro-optic polymer thin films, *Appl. Phys. Lett.* 57, 1084–1086, (1990).
123. M. Hirano and Y. Hirano, Transient characteristics of tuning fork investigated with heterodyne displacement meter, *J. Solid Mech. Mater. Eng.* (to be published 2013).
124. C.R. Wylie, *Advanced Engineering Mathematics*, McGraw-Hill, New York, 1966.
125. D.M. Silva, E.A. Barbosa, and N.U. Wetter. Simple and versatile heterodyne whole-field interferometer for phase optics characterization. *Rev. Sci. Instrum.* 83, 103103 (2012).
126. S. Calle, J. Remenieras, O.B. Matar, M.E. Hachmi, and F. Patat, Temporal analysis of tissue displacement induced by a transient ultrasound radiation force, *J. Acoust. Soc. Am.* 118, 2829–2840, (2005).
127. J.H. Lee, S.T. Lee, C.M. Yao, and W. Fang, Comments on the size effect on the microcantilever quality factor in free air space, *J. Micromech. Microeng.* 17, 139–146, (2007).
128. T.D. Rossing, D.A. Russell, and D.E. Brown, On the acoustics of tuning forks, *Am. J. Phys.* 60, 620–626 (1992).
129. K. Hosaka, K. Iitao, and S. Kuroda, Damping characteristics of beam-shaped micro oscillators, *Sensors Actuat.* 49, 87–95 (1995).
130. W.T. Thomson, *Theory of Vibration and Applications*, 4th edn., Prentice Hall, Englewood Cliffs, NJ, 1993.
131. L.E. Goodman, Material damping and slip damping, in *Shock and Vibration Handbook*, Chapter 36, C.M. Harris (ed.), McGraw-Hill, New York, 1991.
132. M. Hirano, Transient characteristics of frequency and damping factor for cantilever investigated with optical heterodyne method having high performances, *J. Solid Mech. Mater. Eng.* (to be published 2013).
133. M. Renier, C. Barriere, and D. Royer. Optical measurements of the self-demodulated displacement and its interpretation in terms of radiation pressure, *J. Acoust. Soc. Am.* 121, 3341–3348, (2007).
134. J.K. Wahlstrand, Y.H. Cheng, and H.M. Milchberg, Absolute measurement of the transient optical nonlinearity in N_2, O_2, N_2O, and Ar, *Phys. Rev. A* 85, 043820 (2011).
135. C.R. Wylie, *Advanced Engineering Mathematics*, McGraw-Hill, New York, 1966.
136. Y. Jimbo and S. Yamada, Vibration of an unsymmetrical tuning fork, *The Horologic. Inst. Jpn.* 64, 3–29 (1972).
137. J. Zhang, C. Dai, X. Su, and S.J. O'Shea, Determination of liquid density with a low frequency mechanical sensor based on quartz tuning fork, *Sensor Actuator* B84, 123–128 (2002).
138. N. Tamada, S. Fuchino, N. Natori, I. Ishii, and M. Okano, Optical measurement of transient phenomenon of liquid helium boiling under very high centrifugal operation, *Cryogenics* 33, 1023–1027, (1993).
139. M. Hirano, Report of concentration detection, *Photon Probe Magazine* 14, 1–6 (2012).

11 Diffraction

Toru Yoshizawa

CONTENTS

11.1 Introduction 343
11.2 Fundamentals of Diffraction 343
11.3 Application of Diffractographic Method 349
11.4 Application of Grating Diffraction Method 351
11.5 Conclusion 353
References 353

11.1 INTRODUCTION

The diffraction phenomenon in which light sneaks around to the back of the shading screen is one of the typical properties inherent to light, and its physical explanation is described well in famous reference books on optics [1–3]. Diffraction is well known and we can easily observe this phenomenon by using a coherent beam from a laser. However, it is not easy to understand the theoretical description of diffraction completely. Here, we focus on diffraction in the Fraunhofer region, because it is much easier to understand fundamental properties of diffraction in this region rather than in the Fresnel region. In addition, most of the practical applications have been derived on the basis of the Fraunhofer diffraction. By using diffraction phenomena, a number of applications have been developed in optical metrology. However, topics such as structural analysis of a crystal using x-ray or neutron diffraction and another recently developed field like diffractive optics are not described here. The readers are referred to Refs. [4,5] to obtain useful knowledge in these fields.

11.2 FUNDAMENTALS OF DIFFRACTION

We start by using the famous and most fundamental expression called the Fresnel–Kirchhoff diffraction formula.

Let us consider the case of Figure 11.1 where an aperture is projected by a light (wavelength λ) from the source and the wave amplitude U_p is observed at the point P. To discuss the influence of this light through the aperture, we assume a domain S consisting of three regions S_1, S_2, and S_3 in Figure 11.1. We need to integrate the contribution from a small area dS over the domains S_1, S_2, and S_3. This domain S is enclosed by a screen S_2 (no contribution from this area, because the light amplitude does not exist), a semi-sphere S_3 with the observation point P at the center (also no contribution under the Sommerfeld radiation condition [2] although intuitive understanding is difficult and theoretical discussion is necessary in detail [3]), and the aperture S_1 which dominates the wave U_p at the point P.

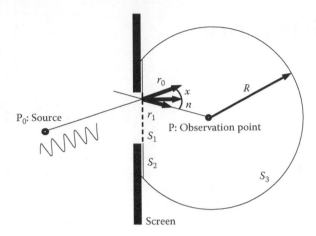

FIGURE 11.1   Derivation of Kirchhoff diffraction formula.

However, if we skip sophisticated discussion, the next equation called the Fresnel–Kirchhoff diffraction formula is obtained using wave number $k = 2\pi/\lambda$ (λ, wavelength) and amplitude A (a constant depending on the strength of the source)

$$U_p = \frac{A}{j2\lambda}\int_{S_1}\frac{\exp[jk(r_0 + r_1)]}{r_0 r_1}[\cos(r_1, \mathbf{n}) - \cos(r_0, \mathbf{n})]dS \qquad (11.1)$$

where the angle between unit vectors r_1 and $\mathbf{n}$ is $(r_1, \mathbf{n})$ and the angle between r_0 and $\mathbf{n}$ is $(r_0, \mathbf{n})$. This is called the Fresnel–Kirchhoff diffraction formula. Here, the factor $[\cos(r_1, \mathbf{n}) - \cos(r_0, \mathbf{n})]$ is introduced by Fresnel to overcome the difficulty (by Huygens' construction) in explaining the direction of propagating light due to diffraction and is called the inclination factor.

If the light source is located far from the screen, $\cos(r_1, \mathbf{n})$ becomes unity, that is, $\cos(r_1, \mathbf{n}) = 1$, and let $\cos(r_0, \mathbf{n})$ be replaced by $-\cos\chi$ (as shown in Figure 11.1). This assumption leads to the next expression:

$$U_p = \frac{A}{j2\lambda}\int_{S_1}\frac{\exp[jk(r_0 + r_1)]}{r_0 r_1}(1 + \cos\chi)dS \qquad (11.2)$$

Now the inclination factor is given by $(1 + \cos\chi)$. In addition, if the observation point is far from the screen, this angle χ becomes nearly equal to zero ($\chi \approx 0$), and the inclination factor is approximated to be 2. This means that the amplitude is given by the following expression:

$$U_p = \frac{A}{j\lambda}\int_{S_1}\frac{\exp[jk(r_0 + r_1)]}{r_0 r_1}dS \qquad (11.3)$$

However, except a limited number of cases, it is still difficult or impossible to solve the formula explicitly because of the integral term of the formula. For the purpose of understanding the diffraction to be applied to metrology, a further approximation is introduced to the formula.

In Figure 11.2, the aperture with the origin O is illuminated by the parallel uniformed beam from the source; then, $g(x_0, y_0)$ plays the role of a light source on the aperture and it produces the diffracted pattern on the observing screen. One point $P_0(x_0, y_0, 0)$ is located on the aperture plane,

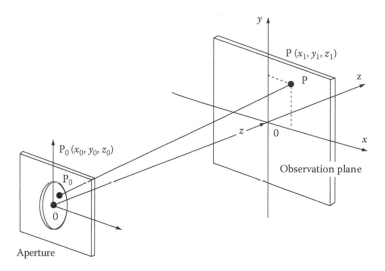

FIGURE 11.2 Diffraction due to aperture.

and the observing point $P(x_1, y_1, z_1)$ is located on the screen. The distance P_0P is expressed by r and the distribution by $U(x_1, y_1)$; then, the following equations hold

$$U(x_1, y_1) = \frac{1}{j\lambda} \int\limits_{-\infty}^{\infty}\!\!\int g(x_0, y_0) \frac{e^{jkr}}{r}\, dx_0 dy_0 \tag{11.4}$$

where

$$r = \sqrt{z^2 + (x_0 - x_1)^2 + (y_0 - y_1)^2} \tag{11.5}$$

However, this r makes the integration difficult.

If $(x_0 - x_1)^2 + (y_0 - y_1)^2$ is much smaller than z^2, that is, the screen is far from the aperture and the size of the aperture is not large, r is approximated as follows:

$$r = z + \frac{x_1^2 + y_1^2}{2z} - \frac{x_1 x_0 + y_1 y_0}{z} + \frac{x_0^2 + y_0^2}{2z} - \frac{\left[(x_1 - x_0)^2 + (y_1 - y_0)^2\right]^2}{8z^3} \tag{11.6}$$

In the case where the first three terms are used (the pattern is observed on the screen located far from the aperture), diffraction can be discussed in the Fraunhofer region, and if the first four terms are necessary (the pattern is observed on the screen near the aperture), we have to discuss the diffraction in the Fresnel region. In fact, mathematical considerations are extremely difficult regarding the Fresnel diffraction. Therefore, the Fraunhofer approximation is preferable for practical applications to metrology, that is, r is approximated by

$$r = z + \frac{x_1^2 + y_1^2}{2z} - \frac{x_1 x_0 + y_1 y_0}{z} \tag{11.7}$$

For example, when such a small aperture as 0.1 mm is illuminated by the beam with the wavelength $\lambda = 0.63\ \mu m$, a distance longer than 60 mm is required; this is quite easy to realize. However, when a larger aperture with a size of 10 mm is illuminated, the distance has to be longer than 600 mm, which becomes difficult to realize. In this case, a convex lens is useful to observe a diffraction pattern in the Fraunhofer region [2]. The thickness of the convex lens (focal length f) decreases from the center to the rim in proportion to $(x_0^2 + y_0^2)/2f$, and this means the fourth term of Equation 11.6 can be canceled. Because, when the lens with the focal length f is inserted, z in Equation 11.6 is replaced by f and the fourth term is canceled by adding the phase $k(x_0^2 + y_0^2)/2f$.

Thus, the next equation is obtained

$$U(x_1, y_1) = \frac{1}{j\lambda z} \exp\left[jk\left(z + \frac{x_1^2 + y_1^2}{2z} \right) \right] \int\int_{-\infty}^{\infty} g(x_0, y_0) \exp\left[-j2\pi\left(\frac{x_1 x_0 + y_1 y_0}{\lambda z} \right) \right] dx_0 dy_0 \quad (11.8)$$

where r is also approximated by z in the denominator.

One application to metrology is suggested in Figure 11.3, where an aperture formed by two edge-like objects produces a diffraction pattern. If diffraction due to this aperture (with narrow width a) is observed on the screen located at the distance D ($= z$, far from the aperture), the pattern is formed in the Fraunhofer region. Thus, the Fresnel–Kirchhoff approximation, Equation 11.2 or 11.6, is expressed as follows. Here the aperture is given by

$$\begin{aligned}
U(x_1, y_1) &= \frac{1}{j\lambda z} \exp\left[jk\left(z + \frac{x_1^2 + y_1^2}{2z} \right) \right] \int_{-\infty}^{\infty} g(x_0) \exp\left[-j2\pi\left(\frac{x_1 x_0}{\lambda z} \right) \right] dx_0 \\
&= \frac{1}{j\lambda z} \exp\left[jk\left(z + \frac{x_1^2 + y_1^2}{2z} \right) \right] \int_{-a/2}^{a/2} 1 \cdot \exp\left[-j2\pi\left(\frac{x_1 x_0}{\lambda z} \right) \right] dx \\
&= \frac{a}{\lambda z} \exp\left[jk\left(z + \frac{x_1^2 + y_1^2}{2z} \right) \right] \cdot \sin c\left[\pi\left(\frac{ax_1}{\lambda z} \right) \right]
\end{aligned} \quad (11.9)$$

where

$$g(x_0) = 1(-a/2 \sim a/2), \quad 0 \text{ (elsewhere)}$$

In this case, calculation for integration is easily possible, but in most of the cases, this kind of mathematical calculation is nearly impossible. Therefore, the Fourier transformation method [6] should be used to know the diffraction pattern in the Fraunhofer region.

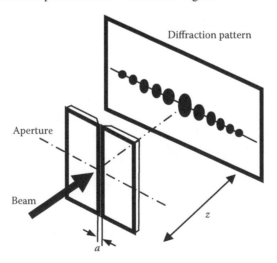

FIGURE 11.3 Diffraction pattern formed by two edges.

In the case of a rectangular aperture with the size of $a \times b$, the next expression is easily guessed right:

$$U(x_1, y_1) = \frac{ab}{j\lambda z} \exp\left[jk\left(z + \frac{x_1^2 + y_1^2}{2z} \right) \right] \sin c\left(\pi \frac{ax_1}{\lambda z} \right) \sin c\left(\pi \frac{by_1}{\lambda z} \right) \quad (11.10)$$

Beautiful photographs of diffraction patterns brought by various shapes of apertures are represented in books with other pictures of interference and polarization [7,8].

Coming back to Figure 11.3, let us remove the suffix from x_1 to make it a simple expression and let the amplitude be expressed as $I(x)$, which is given by $|U(x)|^2$.

Equation 11.7 indicates that the intensity distribution $I(x)$ of the diffraction pattern is dominated by the function $\sin c\left(\pi \dfrac{ax}{\lambda Z} \right)$. The profiles of the amplitude and intensity distributions are represented in Figure 11.4a and b. We should note that the dark points appear at the points $x = \dfrac{\lambda z}{a}, 2\dfrac{\lambda z}{a}, 3\dfrac{\lambda z}{a}, 4\dfrac{\lambda z}{a}$, etc. with the equal interval $\Delta = \dfrac{\lambda z}{a}$, but the interval between the adjacent two bright points is not exactly equal, as shown in Table 11.1, where z is replaced by f because a lens is usually placed near the aperture to realize the Fraunhofer diffraction.

Therefore, if this interval Δ could be known, the gap (aperture size) a between two edges can be calculated by

$$a = \frac{\lambda z}{\Delta} \quad (11.11)$$

Instead of detecting the darkest points, Δ defined by the bright points are also used. Naturally, this Δ is not constant near the central part of the diffraction pattern (in low diffraction order area); then, spots with higher orders of diffraction should be used.

A kind of tensile strength testing is proposed in Ref. [9] where L-shaped metal edges (at low temperature) or ceramic edges (at high temperature) are attached to a specimen for measuring the

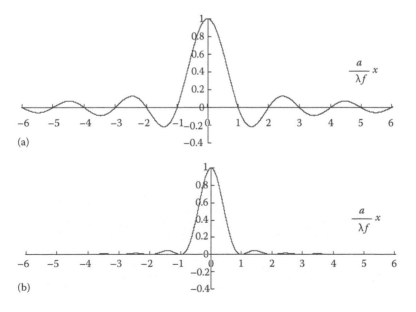

FIGURE 11.4 Distribution profile of a diffraction pattern of a slit aperture: (a) amplitude and (b) intensity.

TABLE 11.1

Maximum Values of Intensity Curve

$\dfrac{a}{\lambda f}x$	1.4306	2.4583	3.4722	4.4778	5.4806
I	0.0472	0.0165	0.0083	0.0050	0.0034
$\dfrac{a}{\lambda f}x$	6.4833	7.4861	8.4889	9.4889	10.4917
I	0.0024	0.0018	0.0014	0.0011	0.0009

strain caused by loading. An aperture and its change formed between opposing edges brings diffraction, and strain is derived from the resultant pattern.

As another application, let us consider one example in which the diameter of a thin wire can be measured using diffraction phenomenon. Nowadays, an optical micrometer is introduced in a textbook on precision measurement of length, and the fundamental principle dates back to the 1920s. At that time, a mercury lamp was used for these experiments, but now with a laser source we can easily confirm this principle.

As shown in Figure 11.5, when such a thin line is projected by a laser beam, we can observe a diffraction pattern, shown in Figure 11.6, on the screen off the distance f from the lens (with the focal length f), which is placed close to the slit. Here, we should note the fact that a diffraction pattern caused by an aperture is exactly the same (except the neighborhood of the optical axis in the Fraunhofer region) as that formed by a mask of the same size and shape. This is explained on the basis of Babinet's principle (see Ref. [4, p. 424]). Here, the equi-linear interval of the pattern on the screen is proportional to the slit-screen distance, that is, the focal length f of the lens is placed

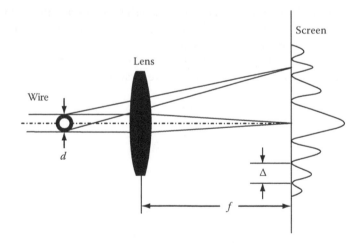

FIGURE 11.5 Diameter measurement of a thin wire.

FIGURE 11.6 Diffraction pattern by an enamel wire.

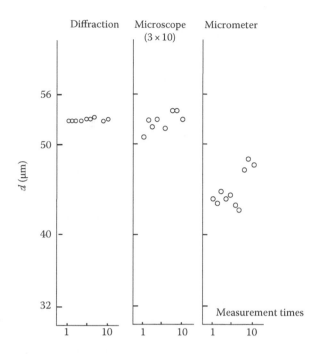

FIGURE 11.7   Measurement of an enamel wire.

close to the slit. The linear distance d between successive minima $\Delta = f\lambda/d$ is exactly the same, but the distance between successive maxima is not constant. We should note that the detection of the minimum point is practically impossible, that is, we are required to find the brightest points. Fortunately, we can use the same equation $\Delta = f\lambda/d$ (now Δ denotes the distance between two successive bright spots), because the brightest spots are equally spaced in the higher-order region of diffraction. Thus, the separation Δ of the higher orders of bright spots is available to determine the diameter d of the wire. Moreover, we should note that the accuracy of measurement becomes higher as the diameter of the wire becomes smaller, because the separation is larger in inverse proportion to the diameter. The result of the measurement by this method is shown in Figure 11.7 together with results using two other methods: a toolmaker's microscope and a mechanical micrometer.

The results of 10 times the measurements of an enamel wire by a beginner are shown in Figure 11.7, where repeatability using the diffraction method brought better results than the two other methods. In the case of a micrometer, the diameter measured is smaller due to contact force, and in the case of the microscopic measurement, the averaged value is nearly the same as that with the diffraction method, but the respective result varies widely. On the contrary, the diffraction method brings high repeatability and correct result. In an industrial application of this method, a thin enamel wire with a diameter of 0.1 to ~0.01 mm is inspected during running for winding onto a reel. A thin wire with an outer diameter of 30 ± 4 μm can be measured at the repeatability of 0.05 μm without any influence of vibration while running. One important fact is that the intensity distribution of the diffraction pattern is stationary, even if the objective vibrates, as long as it remains inside the beam spot. As stated earlier, the line object produces the same diffraction pattern with the slit aperture.

11.3   APPLICATION OF DIFFRACTOGRAPHIC METHOD

Displacement and profile are also easily measured using changes in diffraction pattern, which are caused by a clearance between a sample and the reference. Especially, if the resulting pattern is captured in the Fraunhofer region, a simple calculation is applicable. This kind of technique is named "diffractography," and many practical applications of this technique have been reported [9,10].

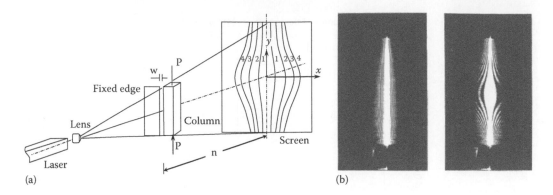

FIGURE 11.8 Diffraction pattern formed by deformation of a column. (a) Optical arrangement and (b) photographed pattern. (After Pryor, T.R. et al., *Appl. Optics*, 11(2), 314, 1972.)

By this technique, the measurement of displacement, especially displacements of a sample object along a line, is easily attainable. Line measurements of displacement are possible at all points along the reference line, with high accuracies over a large range and at one point.

Typically, let us consider Figure 11.8a, which shows the measurement of the slit aperture or change in the aperture formed between the reference edge and the test object. This arrangement is applicable to measure the deflection of a loaded column. Two diffracted patterns are shown in the figure (noises are caused by surface irregularities of the column), and the two results are represented in Figure 11.8b, where a small curvature of the column is detected (in comparison with the straight edge) when not loaded (left) and when loaded (right). This result is brought by the deflection due to a loading of 18.2 kg (178.4 N).

In Figure 11.9, the dynamic accuracy of a tool in ultrasonic welding is checked by this method [12]. In this case, the tool attached to an ultrasonic horn vibrates back and forth and around (Figure 11.9a), and when an aperture between a knife-edge and the tool gives a diffraction pattern

FIGURE 11.9 Vibration analysis of ultrasonic welding tool: (a) optical arrangement and (b and c) diffraction patterns. (From Ono, A. and Komatsu, T., *Bull. Jpn. Soc. Precision Eng.*, 11(2), 101, 1977. With permission.)

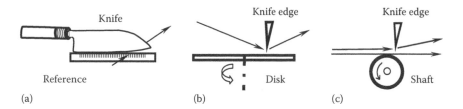

FIGURE 11.10 Some applications of diffractographic method. (a) Blade check of a knife. (b) Bounce of a rotary table. (c) Runout of a rotary shaft.

(Figure 11.9b and c). We can observe the vibrating characteristics of the tool, and at the same time we can find signs of deterioration of the tool.

When the tool vibrates, the pattern and visibility of the fringe change, and transitional vibration as well as amplitude are analyzed. The tool vibration amplitude <1 µm is measured by diffractography and the value is compared with the result by holography to verify the availability of this method.

Similar applications are shown in Figure 11.10, where a change in separation between the reference and a specimen is used. In Figure 11.10a, a traditional knife forged by manual hammering (in a local area of Japan) is checked with the naked eye by observing clearance between the knife edge or surface profile and the reference edge. But for further precise checking, the diffraction pattern should be used. To our surprise, a high sensitivity of 5 µm is necessary to satisfy requirements for professional purposes. Figure 11.10b shows that bouncing of a disk during rotation is captured precisely, and dynamic runout of a rotating shaft of a manufacturing machine is also detected, which is shown in Figure 11.10c.

11.4 APPLICATION OF GRATING DIFFRACTION METHOD

Strain measurement is so important in mechanics and materials science that many papers have been published, especially in the filed of mechanical engineering. There exist numerous principles such as the strain gauge method (contact and point-wise measurement) and moire, speckle, and holographic methods (noncontact and full-field measurement). This situation means that strain measurement is extremely important in industrial applications.

As suggested in Section 11.2, if a laser beam is projected onto a specimen with a grating ruled on the surface, the beam is diffracted to directions depending on the period of the grating. Figure 11.11 shows the angle change during the passage of a shock wave (the thickness of the specimen is exaggerated), which is caused by the impact given at the end of the specimen [13]. When shock is given at the left facet of the sample object, grating lines change the spacing as the shock wave propagates to the right-end facet, and the original directions of the diffraction beams deviate by β_A and β_B, respectively. These two parameters, β_A and β_B, represent the angular displacement of the first-order diffracted patterns, and for small deformations, the longitudinal strain ε and the tangential angle of the surface α arc

$$\varepsilon = \cot\theta_0\,(\beta_A + \beta_B)/2$$
$$\alpha = (\beta_B - \beta_A)/2(1 + \sec\theta_0)$$

(11.12)

where θ is the initial angle of the first-order diffraction. The cylindrical specimens used in this experiment are brass rods with a diameter of 21.5 mm, diffraction gratings with a gauge length 0.127 mm, with a density of 1210 lines/mm, and the incident light wavelength of 0.55 µm.

One result calculated from these equations is shown in Figure 11.11b. Therefore, this diffraction-grating strain gauge method is available to evaluate the dynamic longitudinal strain and local

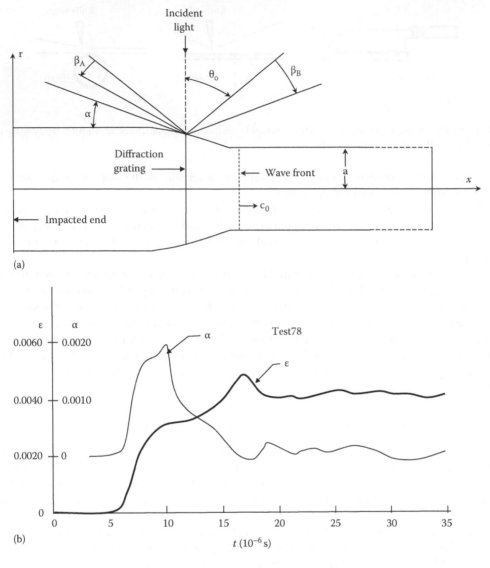

FIGURE 11.11 Diffraction-grating strain gauge method: (a) specimen and (b) testing result. (Courtesy of Dr. Hartman.)

surface inclination of elastically colliding rods, if we rule small gratings over the specimen surface to be checked.

One of the recent applications of this technique is found in a compact microscopic system, which was developed to capture the local strain of a plastic film [14]. A preparatory testing result suggests a high resolution of 442 μm/mm (using a grating of 40–200 lines/mm). Such image processing is also adopted as a sub-pixel technique to define the centroid of the diffracted point with an estimated sensitivity of 0.08 pixel. An example using 80 lines/mm cross grating showed an average strain of 12,000 mm/mm after extension of 0.6 mm by loading.

This technique is not limited to a regularly ruled pattern like a grating, but applicable to any irregular pattern on the surface. One typical system [15] is shown in Figure 11.11, which has a function that automatically checks and evaluates surface quality of a sheet object such as a steel strip and a roll.

11.5 CONCLUSION

In applicative uses of diffraction to the topics described here, one of the problems was detecting a spot or pattern precisely. Although many pioneering studies were reported in the 1970s and 1980s, we had to use detectors such as a photomultiplier, which was a point-wise detector and expensive at that time. Recently, inexpensive CCD cameras have become popular, and, in addition, image and signal processing techniques have made rapid progress. More diffraction-related methods are expected to be explored for metrological applications in industry.

REFERENCES

1. F. Jenkins and H. White, *Fundamentals of Optics*, 4th edn., McGraw-Hill, New York, 1976.
2. K. Iizuka, *Engineering Optics*, 2nd edn., Springer-Verlag, Berlin, Germany, 1987.
3. M. Born and E. Wolf, *Principles of Optics*, 7th edn., Cambridge University Press, Cambridge, U.K., 1999.
4. E. J. Mittemeijer and P. Scardi (eds.), *Diffraction Analysis of the Microstructure of Materials*, Springer-Verlag, Berlin, Germany, 2004.
5. B. Kress and P. Meyrueis, *Digital Diffractive Optics: An Introduction to Planar Diffractive Optics and Related Technology*, John Willy & Sons, Chichester, U.K., 2000.
6. J. Goodman, *Introduction to Fourier Optics*, McGraw-Hill, New York, 1996.
7. M. Cagnet, M. Francon, and J. C. Thrierr, *Atlas of Optical Phenomena*, Springer-Verlag, Berlin, Germany, 1962.
8. M. Cagnet, M. Francon, and J. C. Thrierr, *Atlas of Optical Phenomena (Supplement)*, Springer-Verlag, Berlin, Germany, 1971.
9. H. Pih and K. C. Liu, Laser diffraction methods for high-temperature strain measurements, *Experimental Mechanics* 31(3): 60–64, 1991.
10. T. R. Pryor, O. L. Hageniers, and P. T. North, Diffractographic dimensional measurement, Part 1: Displacement measurement, *Applied Optics* 11(2): 308–313, 1972.
11. T. R. Pryor, O. L. Hageniers, and P. T. North, Diffractographic dimensional measurement, Part 2: Profile measurement, *Applied Optics* 11(2): 314–318, 1972.
12. A. Ono and T. Komatsu, Ultrasonic vibration analysis by diffractometry, *Bulletin of the Japan Society of Precision Engineering* 11(2): 101, 102, and 105, 1977.
13. W. Hartman, Application of the diffraction-grating strain-gage technique for measuring strains and rotations during elastic impact of rods, *Experimental Mechanics* 14(12): 509–512, 1974.
14. B. Zhao and A. Asuni, Strain microscope with grating diffraction method, *Optical Engineering* 38(1): 170–174, 1999.

12 Light Scattering

Lev T. Perelman

CONTENTS

12.1 Introduction 355
12.2 Basic Principles of Light Scattering...357
12.3 Light Scattering Spectroscopy..359
12.4 Early Cancer Detection with Light Scattering Spectroscopy...360
12.5 Confocal Light Absorption and Scattering Spectroscopic Microscopy............................364
12.6 Light Scattering Spectroscopy of Single Nanoparticles...368
12.7 Conclusions...369
Acknowledgment 370
References..370

12.1 INTRODUCTION

Light scattering in biological tissues originates from the tissue inhomogeneities such as cellular organelles, extracellular matrix, blood vessels, etc. This often translates into unique angular, polarization, and spectroscopic features of scattered light emerging from tissue, and therefore information about tissue macroscopic and microscopic structure can be obtained from the characteristics of scattered light. Recognition of this fact has led to a long history of the studies of light scattering by biological structures such as cells and connective tissues.

In 1974, Brunsting et al. performed a series of experiments relating to the internal structure of living cells with the scattering pattern by measuring forward and near-forward scattering in cell suspensions using a rigorous quantitative approach [1]. Later, the researchers used cells of several types such as Chinese hamster's oocytes (CHO), HeLa cells, and nucleated blood cells. They compared the resulting angular distribution of the scattered light with the one predicted by Mie theory, and a very good agreement between the theory and experiment was achieved by approximating a cell as a denser sphere embedded into a larger less-dense sphere. The sizes of these spheres corresponded to the average sizes of the cell nuclei and cells, respectively. The results agree well with scattering theory. Particles that are large compared to a wavelength produce a scattered field that peaks in the forward and near-backward directions in contrast to smaller particles, which scatter light more uniformly. Despite nonhomogeneity and the lack of a perfectly spherical shape of cells and their nuclei, experimental results were well explained using Mie theory, which deals with uniform spheres. These results were supported in the experiments with white blood cells (leukocytes) by Sloot et al. [2], who found that light scattering by the leukocytes in the near-forward direction can be explained if each cell was approximated as being composed of two concentric spheres, one being the cell itself and the other being the nucleus; and Hammer et al. [3], who showed that near-forward scattering of light by red blood cells can be accurately described using the van de Hulst approximation, which is derived for large particles of spherical shapes rather than the actual con-cave–convex disks that are red blood cells.

However, most cell organelles and inclusions are themselves complex objects with spatially vary-ing refractive indices [4,5]. Many organelles such as mitochondria, lysosomes, and nuclei possess an average refractive index substantially different from that of their surrounding. Therefore, an accurate model acknowledges subcellular compartments of various sizes with a refractive index different from that of the surrounding.

Mourant et al. found that the cell structures responsible for light scattering can be correlated with the angle of scattering [6]. These studies showed that when a cell is suspended in a buffer solution of lower refractive index, the cell itself is responsible for small angle scattering. This result has been used in flow cytometry to estimate cell sizes [7]. At slightly larger angles, the nucleus is primarily responsible for scattering. It is the major scatterer in forward directions when the cell is a part of a contiguous layer. Smaller organelles, cell inclusions, suborganelles, and subnuclear inhomogeneities are responsible for scattering at larger angles. Scattering may originate from organelles themselves or their internal components. Angular dependence may elucidate whether the scattering originates from the objects of regular or irregular shape, spherical or elongated, and inhomogeneous or uni-form. In some cases, large angle scattering can be attributed to a specific predominant organelle. Research conducted by Beauvoit et al. [8] provided strong evidence that mitochondria are primarily responsible for light scattering from hepatocytes.

Components of organelles can also scatter light. Finite-difference time-domain (FDTD) simula-tions provide means to study spectral and angular features of light scattering by arbitrary particles of complex shape and density. Using FDTD simulations, Drezek et al. [9] investigated the influence of cell morphology on the scattering pattern and demonstrated that as the spatial frequency of refractive index variation increases, the scattering intensity increases at large angles.

Not only does light scattered by cell nuclei have a characteristic angular distribution peaked in the near-backward directions, but it also exhibits spectral variations typical for large particles. This information has been used to study the size and shape of small particles such as colloids, water droplets, and cells [10]. The scattering matrix, a fundamental property describing the scattering event, depends not only on the scatterer's size, shape, and relative refractive index but also on the wavelength of the incident light. This method is called light scattering spectroscopy or LSS and can be very useful in biology and medicine.

Bigio et al. [11] and Mourant et al. [12] demonstrated that spectroscopic features of elastically scattered light can detect transitional carcinoma of the urinary bladder, adenoma, and adenocarci-noma of the colon and rectum with good accuracy. In 1998, Perelman et al. observed characteristic LSS spectral behavior in the light backscattered from the nuclei of human intestinal cells [13]. Comparison of the experimentally measured wavelength varying component of light backscattered by the cells with the values calculated using Mie theory, and the size distribution of cell nuclei determined by microscopy demonstrated that both spectra exhibit similar oscillatory behavior. The oscillatory behavior of light scattered by a cell nucleus exhibits frequency dependence with size. This was used to obtain the size distribution of the nuclei from the spectral variation of light back-scattered from biological tissues. This method was successfully applied to diagnose precancerous epithelia in several human organs in vivo [14–17].

An important aspect of LSS is its ability to detect and characterize particles smaller than the diffraction limit. Particles much larger than the wavelength of light show a prominent backscatter-ing peak, and the larger the particle, the sharper the peak [15]. Measurement of 260 nm particles was demonstrated by Backman et al. [18,19] and 100 nm particles by Fang et al. [20]. Scattering from particles with sizes smaller than a wavelength dominates at large angles and does not require an assumption that the particles are spherical or homogenous. Not only is submicron resolution achievable, but it can also be done with larger numerical aperture (NA) confocal optics. By combin-ing LSS with confocal scanning microscopy, Itzkan et al. recently identified submicron structures within the cell [21].

12.2 BASIC PRINCIPLES OF LIGHT SCATTERING

Let us consider a particle illuminated by a plane electromagnetic wave

$$\begin{pmatrix} E_{i1} \\ E_{i2} \end{pmatrix} = \begin{pmatrix} E_{01} \\ E_{02} \end{pmatrix} e^{-i(\mathbf{kr}-\omega t)}, \tag{12.1}$$

where
E_{i1} and E_{01} are the components of the wave amplitude perpendicular to the scattering plane
E_{i2} and E_{02} are the components parallel to the scattering plane
$\mathbf{k}$ is the wavevector
ω is the frequency

Then the scattering amplitude matrix relates components of the scattered wave (E_{s1}, E_{s2}) and those of the incident one [22]:

$$\begin{pmatrix} E_{s2} \\ E_{s1} \end{pmatrix} = \frac{e^{-i(\mathbf{kr}-\omega t)}}{ikr} \begin{pmatrix} S_2 & S_3 \\ S_4 & S_1 \end{pmatrix} \begin{pmatrix} E_{i2} \\ E_{i1} \end{pmatrix}, \tag{12.2}$$

where $r = r(\theta, \phi)$ is a direction of propagation of the scattered light given by the polar angles θ and ϕ in the spherical system of reference associated with the particle. The scattering amplitude matrix is the fundamental property that gives complete description of the scattering process. For example, the scattering cross section, σ_s, is given by

$$\sigma_s = k^{-2} \int_0^{2\pi} \int_\pi^0 \left(|S_1 + S_4|^2 + |S_1 + S_4|^2 \right) d\cos\theta\, d\phi. \tag{12.3}$$

To find the matrix elements of the scattering matrix, one needs to solve Maxwell's wave equations with proper boundary conditions for electric and magnetic fields. Such solution is rather difficult to find; as a matter of fact, there are just a few cases when the analytical solution to the wave equation has been found. In 1907, Gustav Mie obtained the solution for the scattering of a plane wave by a uniform sphere. The functions S_1 and S_2 are expressed as infinite series of Bessel functions of two parameters, kd and kmd, with k the wave number, d the diameter of the sphere, and m the relative refractive index of the sphere [23]. Other examples of particles for which the scattering problem has been solved analytically are cylinders, coated spheres, uniform and coated spheroids, strips, and planes [24]. Even for these "simple" cases, the amplitudes can be expressed only as infinite series which are often ill converging.

Difficulties with finding exact solutions of the wave equations have led to the development of the approximate methods of solving the scattering problem. One class of those approximations was originally found by Rayleigh in 1871 [22] and is known as Rayleigh scattering. Rayleigh scattering describes light scattering by particles that are small as compared to the wavelength and is a very important approximation for biomedical optics since a great variety of structures cells organelles are built of, such as the tubules of endoplasmic reticulum, cisternae of Golgi apparatus, etc., fall in this category.

In Rayleigh limit, the electric field is considered to be homogenous over the volume of the particle. Therefore, the particle behaves like a dipole and radiates in all directions. In a most relevant case of isotropic polarizability α of the particle, the scattering amplitude matrix becomes

$$\begin{pmatrix} S_2 & S_3 \\ S_4 & S_1 \end{pmatrix} = ik^3\alpha \begin{pmatrix} \cos\theta & 0 \\ 0 & 1 \end{pmatrix}. \tag{12.4}$$

The scattering cross section in this case becomes simply

$$\sigma_s = \frac{8}{3}\pi k^4 \alpha^2. \tag{12.5}$$

Since α is proportional to the particle volume, the scattering cross section scales with particle's linear dimension a as a^6 and varies inversely with λ^4.

For larger particles with sizes comparable to the wavelength, Rayleigh approximation does not work anymore and one can use another solution called Rayleigh–Gans approximation [23]. It is applicable, if the relative refractive index of the particle is close to unity, and at the same time the phase shift across the particle $2k\alpha |\Delta m - 1|$ is small, where a the linear dimension of the particle and $\Delta m = \dfrac{\max_{r \in V'}[n(r)]}{\min_{r \in V'}[n(r)]}$. Since the refractive index of most cellular organelles ranges from 1.38 to 1.42 [2,5,8] and the refractive index of the cytoplasm varies from 1.34 to 1.36, both conditions of the Rayleigh–Gans approximation are satisfied for the majority of small organelles.

Rayleigh–Gans approximation is derived by applying the Rayleigh's formulas (Equation 12.4) to any volume element dV within the particle. It can be easily shown that

$$\begin{pmatrix} S_2 & S_3 \\ S_4 & S_1 \end{pmatrix} = \frac{ik^3 V}{2\pi} \Re(\theta,\phi) \begin{pmatrix} \cos\theta & 0 \\ 0 & 1 \end{pmatrix} \tag{12.6}$$

with

$$\Re(\theta,\phi) = \frac{1}{V} \int_V (m(\mathbf{r})-1) e^{i\delta(\mathbf{r},\theta,\phi)} d\mathbf{r}, \tag{12.7}$$

where
   $m(\mathbf{r})$ is the relative refractive index at a point $\mathbf{r}$
   δ is the phase of the wave scattered in direction (θ, φ) by the dipole positioned at a point $\mathbf{r}$

If a particle is relatively homogenous, then

$$\begin{pmatrix} S_2 & S_3 \\ S_4 & S_1 \end{pmatrix} = \frac{ik^3(m-1)V}{2\pi} R(\theta,\phi) \begin{pmatrix} \cos\theta & 0 \\ 0 & 1 \end{pmatrix}, \tag{12.8}$$

where m is the relative refractive index averaged over the volume of the particle function $R(\theta, \phi) = 1/V\int e^{i\delta} dV$ is the so-called form factor.

From Equation 12.8 it is easy to see that total intensity of light scattered by a small organelle increases with the increase of its refractive index as $(m - 1)^2$ and with its size as a^6. The angular distribution of the scattered light differs from that of Rayleigh scattering. For $\theta = 0$ the form factor equals unity. In other directions $|R| < 1$, so the scattering has a maximum in forward direction.

Unfortunately, none of the above-mentioned approximations could be applied to the cell nucleus whose size is significantly larger than that of the wavelength. The approximate theory of light scattering by large particles was first proposed by van de Hulst in 1957 [23] and originally formulated for spherical particles only. However, it can be extended to large particles of an arbitrary shape. Although van de Hulst theory does not provide universal rules to find the scattering matrix for all scattering angles even in case of a homogenous sphere, it enables obtaining scattering amplitudes in near-forward direction as well as the scattering cross section.

Let us consider a particle that satisfies the following two conditions: similar to Rayleigh–Gans case the relative refractive index of the particle is close to unity but at the same time opposite to Rayleigh–Gans case, the phase shift across the particle $2ka|\Delta m - 1|$ is large. In this case the phase shift will create constructive or destructive interference, and we can apply the Huygens' principle and find for the components of the scattering matrix [22]

$$S_{1,2}(\theta) = \frac{k^2}{2\pi} \int\int_A \left(1 - e^{-i\xi(\mathbf{r})}\right) e^{-i\delta(r,\theta)} d^2\mathbf{r}, \qquad (12.9)$$

where
   $\mathbf{r}$ is a vector in the plane orthogonal to the direction of propagation of the incident light
   ξ is the phase shift gained by a light ray that enters the particle at the position given by $\mathbf{r}$ and passes through the particle along a straight trajectory relative to the phase shift gained by ray propagating outside the particle
   δ is the phase difference between the rays scattered by different parts of the particle

The integration is performed over the geometrical cross section of the particle, A. The phase shifts depend on the particle shape and refractive index. For example, for a spherical particle of radius a and relative refractive index m, $\xi = 2ka(m - 1)\cos\gamma$ and $\delta = -ka \sin\theta \sin\gamma \cos\phi$, where γ is an angle between the radial direction and the direction of the initial ray, and ϕ is an azimuthal angle of a vector oriented toward an element of the surface of the particle. This expression enables one to obtain the scattering amplitude for a large particle of an arbitrary shape.

A well-known expression for the scattering cross section derived by van de Hulst can then be obtained from Equation 12.9 using the optical theorem [22]

$$\sigma_s \approx 2\pi a^2 \left\{ 1 - \frac{\sin 2x(m-1)}{x(m-1)} + \left[\frac{\sin x(m-1)}{x(m-1)}\right]^2 \right\}. \qquad (12.10)$$

where $x = ka$ is called the size parameter.

It shows that large spheres give rise to a very different type of scattering than small particles considered above do. Both the intensity of the forward scattering and the scattering cross section are not monotonous functions of wavelength. Rather, they exhibit oscillations with the wavelength; frequency of these oscillations is proportional to $x(m - 1)$, so that it increases with the sphere size and refractive index.

12.3   LIGHT SCATTERING SPECTROSCOPY

Strong dependence of the scattering cross section (Equation 12.10) on size and refractive index of the scatterer, such as the cell nucleus, as well as on the wavelength seems to suggest that it should be possible to design a spectroscopic technique that could differentiate cellular tissues by the sizes of the nuclei. Indeed, the hollow organs of the body are lined with a thin, highly cellular surface layer of epithelial tissue, which is supported by underlying, relatively acellular connective tissue. There are four main types of epithelial tissues—squamous, cuboidal, columnar, and transitional—which can be found in different organs of the human body. Depending on the type of the epithelium it consists either of a single layer of cells or might have several cellular layers. Here, to make the treatment of the problem more apparent, we will consider epithelial layers consisted of a single well-organized layer of cells such as simple columnar epithelium or simple squamous epithelium. For example, in healthy columnar epithelial tissues, the epithelial cells often have en-face diameter of 10–20 μm and height of 25 μm. In dysplastic epithelium, the cells proliferate and their nuclei enlarge and appear darker (hyperchromatic) when stained [25].

LSS can be used to measure these changes. The details of the method have been published by Perelman et al. [13] and are only briefly summarized here. Consider a beam of light incident on an epithelial layer of tissue. A portion of this light is backscattered from the epithelial nuclei, while the remainder is transmitted to deeper tissue layers, where it undergoes multiple scattering and becomes randomized before returning to the surface.

Epithelial nuclei can be treated as spheroidal Mie scatters with refractive index, which is higher than that of the surrounding cytoplasm [2,5]. Normal nuclei have a characteristic size of 4–7 μm. In contrast, the size of dysplastic nuclei varies widely and can be as large as 20 μm, occupying almost the entire cell volume. In the visible range, where the wavelength is much smaller than the size of the nuclei, the Van de Hulst approximation (Equation 12.10) can be used to describe the elastic scattering cross section of the nuclei. Equation 12.10 reveals a component of the scattering cross section that varies periodically with inverse wavelength. This, in turn, gives rise to a periodic component in the tissue reflectance. Since the frequency of this variation (in the inverse wavelength space) is proportional to the particle size, the nuclear size distribution can be obtained from that periodic component.

However, single scattering events cannot be measured directly in biological tissue. Because of multiple scattering, information about tissue scatterers is randomized as light propagates into the tissue, typically over one effective scattering length (0.5–1 mm, depending on the wavelength). Nevertheless, the light in the thin layer at the tissue surface is not completely randomized. In this thin region, the details of the elastic scattering process can be preserved. Therefore, the total signal reflected from tissue can be divided into two parts: single backscattering from the uppermost tissue structures, such as cell nuclei, and the background of diffusely scattered light. To analyze the single scattering component of the reflected light, the diffusive background must be removed. This can be achieved either by modeling using diffuse reflectance spectroscopy (DRS) [1,26,27] or by other techniques such as polarization background subtraction [28] or coherence gating method [29,30].

There are several techniques that can be employed to obtain the nuclear size distribution from the remaining single scattering component of the back-reflected light which can be called the LSS spectrum. A good approximation for the nuclear size distribution can be obtained from the Fourier transform of the periodic component as described in Ref. [1]. A more advanced technique based on linear least squares with a nonnegativity constraints algorithm [31] was introduced by Fang et al. [20].

12.4 EARLY CANCER DETECTION WITH LIGHT SCATTERING SPECTROSCOPY

The incidence of adenocarcinoma of the esophagus is increasing more rapidly than any other type of carcinoma in the United States [32]. Almost 100% of cases occur in patients with Barrett's esophagus (BE) [33], a condition in which metaplastic columnar epithelium replaces the normal squamous epithelium of the esophagus. Although the prognosis of patients diagnosed with adenocarcinoma is poor, the chances of successful treatment increase significantly if the disease is detected at the dysplastic stage. The surveillance of patients with BE for dysplasia is challenging in three respects. First, dysplasia is not visible during routine endoscopy [34]. Thus, numerous random biopsy specimens are required. Second, the histopathologic diagnosis of dysplasia is problematic, because there is a poor interobserver agreement on the classification of a particular specimen, even among expert gastrointestinal pathologists [35,36]. Third, reliance on histology imposes a time delay between endoscopy and diagnosis, severely limiting the diagnostic accuracy of the endoscopic procedure.

Once BE has been identified, most gastroenterologists will enroll the patient in an endoscopy/biopsy surveillance program, presuming that the patient is a candidate for surgery should high-grade dysplasia (HGD) be detected. Although the cost effectiveness of this type of surveillance program has not been validated in prospective studies, lack of such studies does not preclude its potential usefulness. Patients with BE who have esophageal carcinoma detected as part of such a surveillance program are more likely to have resectable disease and have an improved 5 year survival as compared with those whose cancer was detected outside of a surveillance program.

Dysplasia in the gastrointestinal tract is defined as neoplastic epithelium confined within an intact basement membrane. Dysplasia in BE can be classified as low grade or high grade, based on the criteria originally defined for dysplasia in inflammatory bowel disease [37]. Low-grade dysplasia (LGD) is defined primarily by cytological abnormalities including nuclear enlargement, crowding, stratification, hyperchromasia, mucin depletion, and mitoses in the upper portions of the crypts. These abnormalities extend to the mucosal surface. HGD is characterized by even more pronounced cytological abnormalities, as well as glandular architectural abnormalities including villiform configuration of the surface, branching and lateral budding of the crypts, and formation of the so-called back-to-back glands. When there is any doubt as to the significance of histological abnormalities because of inflammation, ulceration, or histological processing artifacts, the findings may be classified as indefinite for dysplasia (IND) in order to prevent unnecessary clinical consequences.

Not all patients with BE progress to adenocarcinoma. Some live their entire lives without undergoing malignant or neoplastic transformation. Others demonstrate a rapid progression to carcinoma and will die of esophageal cancer if it is not diagnosed and treated in a timely manner. Several recent attempts at identifying molecular markers that can predict which patients with BE will progress to esophageal cancer have not been proven effective in clinical trials. For example, anti-p53 antibodies have been shown to develop in patients with BE and adenocarcinoma, and may predate the clinical diagnosis of malignancy [38].

At the present time, the standard of care for surveillance of patients with BE remains debated. Although periodic endoscopic surveillance of patients with BE has been shown to detect carcinoma in its earlier stages, surveillance has significant limitations. Dysplastic and early carcinomatous lesions arising in BE are not visible macroscopically; therefore, surveillance requires extensive random biopsies of the esophagus and histologic examination of the excised tissue for dysplasia. Random biopsy is prone to sampling error (missed dysplastic lesions) and significantly increases the cost and risk of surveillance. There is also a significant interobserver disagreement between pathologists in diagnosing dysplasia. An extensive, 10 year observational study in 409 patients with BE published in the *British Medical Journal* [39] concluded that the current random biopsy endoscopic surveillance strategy has very limited value. The optical diagnostic technology described in this chapter should be able to address these issues by providing a safe, fast, reliable way to survey the entire length of BE for endoscopically invisible dysplasia.

Similarly, there is no agreement on the most appropriate management of HGD when it is found [40]. Because of the marked variability (range 0%–73%; most often quoted as 33%) in finding unsuspected carcinoma in patients with HGD, esophagectomy is recommended by many clinicians to eliminate the risk of carcinoma or to detect and treat it at an early and treatable stage [41–43]. However, this approach has been criticized because of the high morbidity and mortality associated with esophagectomy, the lack of a systematic biopsy protocol prior to surgery, and the variable natural history of the disease. An important objective of the optical diagnostic technology research is to develop a reliable and sensitive spectroscopic technique to detect and diagnose dysplasia in patients with BE at endoscopy.

The LSS clinical studies described in Ref. [16] were performed on 16 patients with known BE undergoing standard surveillance protocols. The measurements were performed using the LSS system (Figure 12.1). Immediately before biopsy, the reflectance spectrum from that site was collected using an optical fiber probe. The probe was inserted into the accessory channel of the endoscope and brought into gentle contact with the mucosal surface of the esophagus. It delivered a weak pulse of white light to the tissue and collected the reflected light. The probe tip sampled tissue over a

FIGURE 12.1 Schematic diagram of the system used to perform LSS in human subjects undergoing gastro-enterological endoscopy procedures. (After Wallace, M. et al., *Gastroenterology*, 119, 677, 2000.)

circular spot approximately 1 mm² in area. The pulse duration was 50 ms, and the wavelength range was 350–650 nm. The optical probe caused a slight indentation at the tissue surface that remained for 30–60 s. Using this indentation as a target, the site was then carefully biopsied, and the sample was submitted for histologic examination. This ensured that the site studied spectroscopically matched the site evaluated histologically.

The reflected light was spectrally analyzed, and the spectra were stored in a computer. Example of nuclear size distributions extracted from those spectra for non-dysplastic and dysplastic BE sites are shown in Figure 12.2 [13]. As can be seen, the difference between non-dysplastic and dysplastic sites is pronounced. The distribution of nuclei from the dysplastic site is much broader than that from the non-dysplastic site, and the peak diameter is shifted from ~7 μm to about ~10 μm. In addition, both the relative number of large cell nuclei (>10 μm) and the total number of nuclei are significantly increased. The authors further noted that the method provides a quantitative measure of the density of nuclei close to the mucosal surface.

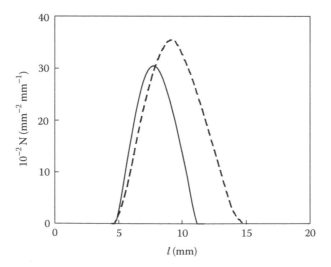

FIGURE 12.2 Nuclear size distributions extracted using LSS technique. Solid line, non-dysplastic site; dashed line, dysplastic site. (After Perelman, L.T. et al., *Phys. Rev. Lett.*, 80, 627, 1998.)

Each site was biopsied immediately after the spectrum was taken. Because of the known large interobserver variation [44], the histology slides were examined independently by four expert gastro-intestinal pathologists. Sites were classified as non-dysplastic Barrett's (NDB), IND, LGD, or HGD. On the basis of the average diagnosis [45,46] of the four pathologists, 4 sites were diagnosed as HGD, 8 as LGD, 12 as IND, and 52 as NDB.

To establish diagnostic criteria, eight samples were selected as a "modeling set," and the extracted nuclear size distributions were compared to the corresponding histology findings. From this, the authors decided to classify a site as dysplasia if more than 30% of the nuclei were enlarged, with "enlarged" defined as exceeding a 10 µm threshold diameter, and classified as non-dysplasia otherwise. The remaining 68 samples were analyzed using this criterion. Averaging the diagnoses of the four pathologists [45], the sensitivity and specificity of detecting dysplasia were both 90%, with dysplasia defined as LGD or HGD, and non-dysplasia defined as NDB or IND, an excellent result, given the limitations of interobserver agreement among pathologists. To further study the diagnostic potential of LSS, the entire data set was then evaluated adding a second criterion, the population density of surface nuclei (number per unit area), as a measure of crowding. The resulting binary plot (Figure 12.3) reveals a progressively increasing population of enlarged and crowded nuclei with increasing histological grade of dysplasia, with the NDB samples grouped near the lower left corner and the HGD samples at the upper right. Using logistic regression [47], the samples were then classified by histologic grade as a function of the two diagnostic criteria.

The percentage agreements between LSS and the average and consensus diagnoses (at least three pathologists in agreement) were 80% and 90%, respectively. This is much higher than that between the individual pathologists and the average diagnoses of their three colleagues, which ranged from 62% to 66%, and this was also reflected in the kappa statistic values [48].

In addition to esophageal epithelium described above, Backman et al. [14] performed a preliminary LSS study for three types of epithelium in three organs: colon (simple columnar epithelium), bladder (transitional epithelium), and oral cavity (squamous stratified epithelium). All studies were performed during routine surveillance procedures. As for the case of BE, all reflectance data were collected in vivo with clinical reflectance instruments. Immediately after the measurement, a biopsy sample was taken from the same tissue site and subsequently examined histologically. The results (Figure 12.4) show a clear distinction between normal and diseased epithelium. The dysplastic sites are seen to have a higher percentage of enlarged nuclei and, on average, a higher population density.

These results show the promise of LSS as a real-time, minimally invasive clinical tool for accurately and reliably classifying invisible dysplasia.

FIGURE 12.3 LSS diagnostic plots of Barrett's esophagus data. NDB—circles; IND—squares; LGD—triangles; HGD—diamonds. The decision threshold for dysplasia is indicated. (After Wallace, M. et al., *Gastroentorolgy*, 119, 677, 2000.)

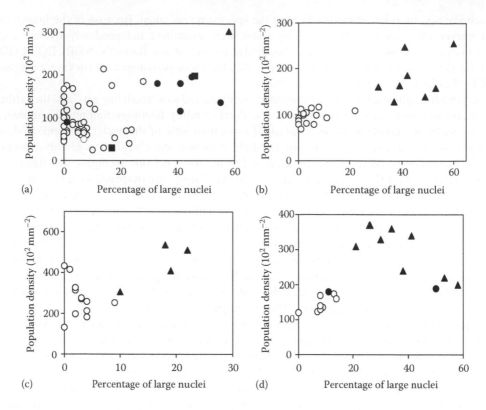

FIGURE 12.4 Dysplasia/CIS classifications for four types of tissues obtained clinically with LSS, compared with histologic diagnosis. In each case, the ordinate indicates the percentage of enlarged nuclei and the abscissa indicates the population density of the nuclei, which parameterizes nuclear crowding. (a) BE: nondysplastic Barrett's mucosa (circles), indefinite for dysplasia (filled circles), LGD (squares), HGD (triangles); (b) colon: normal colonic mucosa (circles), adenomatous polyp (triangles); (c) urinary bladder: benign bladder mucosa (circles), transitional cell carcinoma in situ (triangles); and (d) oral cavity: normal (circles), LGD (filled circles), squamous cell carcinoma in situ (triangles). (After Backman, V. et al., *Nature*, 406, 35, 2000.)

12.5 CONFOCAL LIGHT ABSORPTION AND SCATTERING SPECTROSCOPIC MICROSCOPY

Because of its unique optical sectioning properties and due to high sensitivity and specificity of fluorescence molecular probes, confocal fluorescence microscopy established itself as one of the best techniques for studying living cells [49,50]. However, photobleaching and phototoxication of exogenous fluorescence probes make it often difficult to monitor cells for long periods of time. Exogenous fluorescence probes may also modify the cell normal functioning [51]. To address these problems, significant efforts have been recently made to either develop new imaging methods such as two photon microscopy [52], which are less prone to the above problems, or develop new fluorescent protein probes [53]. At the same time, optical techniques that rely entirely on intrinsic optical properties of tissue for in vivo tissue diagnosis such as confocal reflectance microscopy [54], optical coherent tomography [55], LSS [13], and elastic scattering spectroscopy [11] also play a more and more important role.

Recently, a new type of microscopy that also employs intrinsic optical properties of tissue as a source of contrast has been developed [21]. This technique, called confocal light absorption and scattering spectroscopic (CLASS) microscopy, combines LSS recently developed for early cancer detection [4,13–17] with confocal microscopy. In CLASS microscopy, light scattering spectra are the source of the contrast. Another important aspect of LSS is its ability to detect and characterize

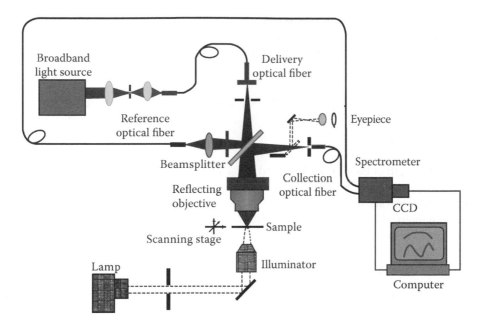

FIGURE 12.5 Schematic of the CLASS microscope. (After Fang, H. et al., *Appl. Optics*, 46, 1760, 2007.)

particles well beyond the diffraction limit. As explained in Perelman and Backman [15], particles much larger than the wavelength of light give rise to a prominent backscattering peak, and the larger the particle, the narrower the angular width of the peak. However, particles with sizes smaller than the wavelength give rise to broad angle backscattering. Thus, the requirement for high NA optics, common in confocal microscopy, is not in conflict with the submicron resolution of LSS. On the contrary, the larger the NA, the larger the contribution of signal from smaller particles.

The CLASS microscope is capable of collecting both spatial and spectroscopic information based on light scattering by submicroscopic biological structures. A schematic of the CLASS microscope is shown in Figure 12.5. Light from the broadband source is delivered through an optical fiber onto a pinhole. The delivery fiber is mounted on a fiber positioner, which allows precise alignment of the fiber relative to the pinhole with the aid of an alignment laser. An iris diaphragm positioned beyond the pinhole is used to limit the beam to match the acceptance angle of the reflective objective. The light beam from the delivery pinhole is partially transmitted through the beam splitter to the sample and partially reflected to the reference fiber. The reflected light is coupled into the reference fiber by the reference collector lens and delivered to the spectrometer. The transmitted light is delivered through an achromatic reflective objective to the sample. Light backscattered from the sample is collected by the same objective and is reflected by the beamsplitter toward the collection pinhole. The collection pinhole blocks most of the light coming from regions above and below the focal plane, allowing only the light scattered from a small focal volume to pass through. The light that passes through the pinhole is collected by a second optical fiber for delivery to an imaging spectrograph with a thermo-electrically cooled CCD detector, which is coupled to a computer.

Since the experimentally measured CLASS spectrum of a cell is a linear combination of the CLASS spectra of various subcellular organelles with different sizes and refractive indices within the cell, in order to extract these parameters, one can express the experimental spectrum as a sum over organelles' diameters and refractive indices. It is convenient to write this in a matrix form $\hat{S} = \hat{I} \cdot \hat{F} + \hat{E}$, where $\hat{S}$ is the experimental spectrum measured at discrete wavelength points, $\hat{F}$ is a discreet size distribution, $\hat{I}$ is the CLASS spectrum of a single scatterer with diameter d and relative refractive index n, and $\hat{E}$ is the experimental noise [56]. Using the scalar wave model similar to the one developed by Weise et al. [57] and Aguilar et al. [58], it is possible to calculate the CLASS spectrum of a single scatterer $\hat{I}$. In this model, the incident and scattering waves are expanded into the set of plane waves with directions limited by the NA of the objective. The amplitude of the signal detected at the center of the focus through the confocal pinhole is expressed as

$$A(\lambda, \delta, n, \mathrm{NA}) = \int\int_{-\infty}^{+\infty} \int\int_{-\infty}^{+\infty} P(-k_X, -k_Y) P(k_X', k_Y') f\left(\frac{\vec{k}}{k}, \frac{\vec{k'}}{k'}\right) dk_X dk_Y dk_X' dk_Y', \qquad (12.11)$$

where

    λ is the wavelength of both the incident and the scattered light
    δ is the diameter of the scatterer
    n is the relative refractive index
    NA is the numerical aperture of the objective
    k' is the wavevector of the incident light
    k is the wavevector of the scattered light
    P is the objective pupil function
    $f\left(\dfrac{\vec{k}}{k}, \dfrac{\vec{k'}}{k'}\right)$ is the far-field Mie scattering amplitude of the wave scattered in direction $\vec{k}$ created by

    the incident wave coming from the $\vec{k'}$ direction

To calculate the CLASS spectrum of a single scatterer, one should calculate the scattering intensity, which is just the square of the amplitude, and relate it to the intensity of the incident light, at each wavelength. This gives the following spectral dependence of the CLASS signal:

$$I(\lambda, \delta, n, \mathrm{NA}) = \frac{[A(\lambda, \delta, n, \mathrm{NA})]^2}{I_0}$$

$$= \frac{\left[\int\int_{-\infty}^{+\infty} \int\int_{-\infty}^{+\infty} P(-k_X, -k_Y) P(k_X', k_Y') f\left(\dfrac{\vec{k}}{k}, \dfrac{\vec{k'}}{k'}\right) dk_X dk_Y dk_X' dk_Y'\right]^2}{\left[\int\int_{-\infty}^{+\infty} \int\int_{-\infty}^{+\infty} P(-k_X, -k_Y) P(k_X', k_Y') dk_X dk_Y dk_X' dk_Y'\right]^2} \qquad (12.12)$$

Since the CLASS spectrum $\hat{I}$ is a highly singular matrix and a certain amount of noise is present in the experimental spectrum $\hat{S}$, it is not feasible to calculate the size distribution $\hat{F}$ by directly inverting the matrix $\hat{I}$. Instead one can multiply both sides of the equation $\hat{S} = \hat{I} \cdot \hat{F} + \hat{E}$ by the transpose matrix $\hat{I}^{\mathrm{T}}$ and introduce the matrix $C = \hat{I}^{\mathrm{T}} \cdot \hat{I}$ [56]. Now it should be possible to compute matrix C eigenvalues $\alpha_1, \alpha_2, \ldots$ and sequence them from large to small. This can be done because $\hat{C}$ is a square symmetric matrix. Then, the authors used the linear least squares with nonnegativity constraints algorithm [31] to solve the set of equations

$$\hat{I}^{\mathrm{T}}\hat{S} - (\hat{C} + \alpha_k \hat{H})\hat{F} \to \min$$
$$\hat{F} \geq 0 \qquad (12.13)$$

where
    $\alpha_k \hat{H}\hat{F}$ is the regularization term
    matrix $\hat{H}$ represents the second derivative of the spectrum

The use of the nonnegativity constraint and the regularization procedure is critical to find the correct distribution $\hat{F}$. Thus, by using this algorithm the authors [56] reconstructed the size and refractive index distributions of scattering particles present in the focal volume of the CLASS microscope.

To confirm the ability of CLASS microscope to monitor unstained living cells on submicrometer scale the researchers in Ref. [56] studied human bronchial epithelial cells undergoing apoptosis. Live 16HBE14o-human bronchial epithelial cells were cultured in minimal essential medium (Gibco, Grand Island, New York) with 10% fetal bovine serum, penicillin 100 unit/mL, and streptomycin 100 µg/mL. Cells (50% confluent) were incubated with 100 µM docosahexaenoic acid (DHA) for 24 h to induce apoptosis. Then the cells were detached with trypsin/EDTA, washed in DMEM solution without phenol red and resuspended in the DMEM/OptiPrep solution.

Figure 12.6 presents CLASS reconstructed images of two 16HBE14o-cells. The left cell is a normal untreated cell and the right cell was treated with DHA for 24 h and is undergoing apoptosis. The diameters of the spheres in the image represent the reconstructed sizes of the individual organelles, and the gray scale represents their refractive index. Individual organelles can easily be seen inside the cell. In the apoptotic cells, the organelles form shell-like structures with an empty space in the middle. The treated and untreated cells show clear differences in organelle spatial distribution.

The results presented here show that CLASS microscopy is capable of reconstructing images of living cells with submicrometer resolution without use of exogenous markers. Fluorescence microscopy of living cells require application of molecular markers that can affect normal cell functioning. In some situations, such as studying embryo development, phototoxication caused by fluorescent tagged molecular markers is not only undesirable but also unacceptable. Another potential problem with fluorescence labeling is related to the fact that multiple fluorescent labels might have overlapping line shapes, and this limits number of species that can be imaged simultaneously in a single cell.

CLASS microscopy is not affected by those problems. It requires no exogenous labels and is capable of imaging and continuously monitoring individual viable cells, enabling the observation of cell and organelle functioning at scales on the order of 100 nm. CLASS microscopy can provide not only size information but also information about the biochemical and physical properties of the cell since light scattering spectra are very sensitive to absorption coefficients and the refractive indices, which in turn are directly related to the organelle's biochemical and physical composition (such as the chromatin concentration in nuclei or the hemoglobin concentration and oxygen saturation in red blood cells).

(a)　　　　　(b)

FIGURE 12.6 Cross-sectional images of two 16HBE14o-human bronchial epithelial cells reconstructed from the CLASS microscope spectra. The left cell (a) is a normal untreated cell and the right cell (b) was treated with DHA for 24 h and is undergoing apoptosis. The diameters of the spheres in the image represent the reconstructed sizes of the individual organelles, and the grayscale represents their refractive index. (After Fang, H. et al., *Appl. Optics*, 46, 1760, 2007.)

12.6 LIGHT SCATTERING SPECTROSCOPY OF SINGLE NANOPARTICLES

Recently, significant attention has been directed toward the applications of metal nanoparticles to medical problems. Both diagnostic and therapeutic applications have been explored. Metal nanoparticles were also suggested as labels for cancer imaging [59]. Gold nanorods have the potential to be employed as extremely bright molecular marker labels for fluorescence, absorption, or scattering imaging of living tissue. However, samples containing a large number of gold nanorods usually exhibit relatively wide spectral lines. This linewidth would limit the use of the nanorods as effective molecular labels, since it would be rather difficult to image several types of nanorod markers simultaneously. In addition, the observed linewidth does not agree well with theoretical calculations, which predict significantly narrower absorption and scattering lines.

As shown in Ref. [60], the discrepancy is explained by the apparent line broadening because of the contribution of nanorods with various sizes and aspect ratios. That suggests that nanorod-based molecular markers with a narrow aspect ratio and, to a lesser degree size distributions, should provide spectral lines sufficiently narrow for effective biomedical imaging.

Nanoparticles with sizes small compared to the wavelength of light made from metals with a specific complex index of refraction, such as gold and silver, have absorption and scattering resonance lines in the visible part of the spectrum. These lines are due to in-phase oscillation of free electrons and are called surface plasmon resonances.

In Ref. [60], the researchers performed optical transmission measurements of gold nanorod spectra in aqueous solutions using a standard transmission arrangement for extinction measurements described in Ref. [61]. Concentrations of the solutions were chosen to be close to 10^{10} nanoparticles per milliliter of the solvent to eliminate optical interference. The measured longitudinal plasmon mode of the nanorods is presented as a dotted curve on Figure 12.7. It shows that multiple nanorods in aqueous solution having width at half maximum of approximately 90 nm. This line is significantly wider than the line one would get from either T-matrix calculations or dipole approximation. Solid line on Figure 12.7 shows plasmon spectral line calculated with the T-matrix calculations for nanorods with the length and width of 48.9 and 16.4 nm, respectively. The theoretical line is also centered at 700 nm but has a width of approximately 30 nm.

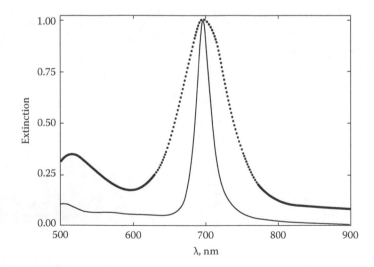

FIGURE 12.7 Optical properties of an ensemble of gold nanorods. Normalized extinction of the sample of gold nanorods in aqueous solution. Dots—experiment; dashed line—T-matrix calculation for a single-size nanorod with length and width of 48.9 and 16.4 nm, respectively. (After Qiu, L. et al., *IEEE J. Selected Top. Quantum Electron.*, 13, 1730, 2007.)

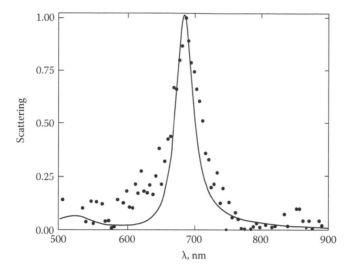

FIGURE 12.8 Normalized scattering spectrum for single gold nanorod. Dots—CLASS measurements, solid lines—T-matrix calculation for nanorods with 3.3 aspect ratio and 16.2 nm width. (After Qiu, L. et al., *IEEE J. Selected Top. Quantum Electron.*, 13, 1730, 2007.)

The CLASS system with the supercontinuum broadband laser source described in Ref. [21] is at present uniquely capable of performing single nanoparticle measurements. It is capable of collecting both spatial (imaging) and spectroscopic information based on light scattering by submicroscopic structures. To illustrate that individual gold nanorods indeed exhibit narrow spectral lines, the authors in Ref. [60] detected single gold nanorods with the CLASS system and measured their scattering spectra.

To measure scattering from individual gold nanorods, diluted aqueous samples of gold nanorods were scanned with the CLASS system and images were acquired, which show locations of the individual gold nanorods. The system measured scattering spectra of several single gold nanorods. One can see (Figure 12.8) that typical single gold nanorod exhibit spectrum that is significantly narrower than the spectrum collected from the nanorods distribution. By comparing these measurements with the numerical calculations based on the T-matrix approach that utilized the complex refractive index of gold from Ref. [62], the researchers found an excellent agreement with the theory [60].

Using the CLASS instrument, the researchers have detected plasmon scattering spectra of single gold nanorods. From these measurements, one can draw the conclusion that single gold nanorods exhibit a scattering line significantly narrower than the lines routinely observed in experiments that involve multiple nanorods. Narrow, easily tunable spectra would allow several biochemical species to be imaged simultaneously with molecular markers which employ gold nanorods of several different aspect ratios as a label. These markers could be used for cellular microscopic imaging where even a single nanorod could be detected. Minimizing the number of nanoparticles should reduce possible damage to a living cell. For optical imaging of tumors, multiple gold nanorods with a narrow aspect ratio distribution might be used. A possible technique for obtaining a narrow aspect ratio distribution might employ devices already developed for cell sorting. These would use the position of the narrow plasmon spectral line for particle discrimination.

12.7 CONCLUSIONS

Emerging light-scattering-based techniques such as LSS and CLASS discussed in this chapter are capable of providing quantitative morphological diagnostic information about tissues and cells *in vivo*. This is a unique capability since methods for providing such information without tissue removal are not currently available. They are nondestructive and do not require the contrast agents common to many other optical techniques since they employ light scattering spectra as a source of the native contrast. Since they rely on multiwavelength spectral information to extract sizes of the scatteres, they are not diffraction limited. Applications for light-scattering-based techniques in such diverse areas as clinical medicine, cell biology, and drug discovery are all linked by the potential of these techniques to observe functional intracellular processes nondestructively.

ACKNOWLEDGMENT

This work was supported by NIH Grants No. R01EB003472 and No. RR017361.

REFERENCES

1. Brunsting A and Mullaney F. Differential light scattering from spherical mammalian cells, *Biophys. J.*, 14, 439–453, 1974.
2. Sloot PMA, Hoekstra AG, and Figdor CG. Osmotic response of lymphocytes measured by means of forward light-scattering—Theoretical considerations, *Cytometry*, 9, 636–641, 1988.
3. Hammer M, Schweitzer D, and Michel B. Single scattering by red blood cells, *Appl. Opt.*, 37, 7410–7418, 1998.
4. Backman V, Gurjar R, Badizadegan K, Dasari RR, Itzkan I, Perelman LT, and Feld MS. Polarized light scattering spectroscopy for quantitative measurement of epithelial cellular structures *in situ*, *IEEE J. Selected Top. Quantum Electron.*, 5, 1019–1027, 1999.
5. Beuthan J, Milnet O, and Helfmann J. The spatial variation of the refractive index in biological cells, *Phys. Med. Biol.*, 41, 369–382, 1996.
6. Mourant JR, Freyer JP, Hielscher AH, Eick AA, Shen D, and Johnson TM. Mechanisms of light scattering from biological cells relevant to noninvasive optical-tissue diagnosis, *Appl. Opt.*, 37, 3586–3593, 1998.
7. Watson JV. *Introduction to Flow Cytometry*, Cambridge University Press, Cambridge, U.K., 1991.
8. Beauvoit B, Kitai T, and Chance B. Contribution of the mitochondrial compartment to the optical properties of rat liver: A theoretical and practical approach, *Biophys. J.*, 67, 2501–2510, 1994.
9. Drezek R, Dunn A, and Richards-Kortum R. Light scattering from cells: Finite-difference time-domain simulations and goniometric measurements, *Appl. Optics*, 38, 3651–3661, 1999.
10. Newton RG. *Scattering Theory of Waves and Particles*, McGraw-Hill Book Company, New York, 1969.
11. Bigio IJ and Mourant JR. Ultraviolet and visible spectroscopies for tissue diagnostics: Fluorescence spectroscopy and elastic-scattering spectroscopy, *Phys. Med. Biol.*, 42, 803–814, 1997.
12. Mourant JR, Boyer J, Johnson T et al. Detection of gastrointestinal cancer by elastic scattering and absorption spectroscopies with the Los Alamos Optical Biopsy System, *Proc. SPIE*, 2387, 210–217, 1995.
13. Perelman LT, Backman V, Wallace M et al. Observation of periodic fine structure in reflectance from biological tissue: A new technique for measuring nuclear size distribution, *Phys. Rev. Lett.*, 80, 627–630, 1998.
14. Backman V, Wallace M, Perelman LT et al. Detection of preinvasive cancer cells. Early-warning changes in precancerous epithelial cells can now be spotted *in situ*, *Nature*, 406(6791), 35–36, 2000.
15. Perelman LT and Backman V. Chapter XII. Light scattering spectroscopy of epithelial tissues: Principles and applications, in: Tuchin V, ed., *Handbook on Optical Biomedical Diagnostics*, SPIE Press, Bellingham, WA, pp. 675–724, 2002.
16. Wallace M, Perelman LT, Backman V et al. Endoscopic detection of dysplasia in patients with Barrrett's esophagus using light scattering spectroscopy, *Gastroenterology*, 119, 677–682, 2000.
17. Gurjar R, Backman V, Perelman LT et al. Imaging human epithelial properties with polarized light scattering spectroscopy, *Nat. Med.*, 7, 1245–1248, 2001.
18. Backman V, Gopal V, Kalashnikov M et al. Measuring cellular structure at submicrometer scale with light scattering spectroscopy, *IEEE J. Selected Top. Quantum Electron.*, 7, 887–893, 2001.
19. Backman V, Gurjar R, Perelman LT et al. Imaging and measurement of cell organization with submicron accuracy using light scattering spectroscopy, in: Alfano RR, ed., *Optical Biopsy IV*, Proceedings of SPIE, 4613, San Jose, CA, pp. 101–110, 2002.

20. Fang H, Ollero M, Vitkin E et al. Noninvasive sizing of subcellular organelles with light scattering spectroscopy, *IEEE J. Selected Top. Quantum Electron.*, 9, 267–276, 2003.

21. Itzkan I, Qui L, Fang H et al. Confocal light absorption and scattering spectroscopic (CLASS) microscopy monitors organelles in live cells with no exogenous labels, *Proc. Natl Acad. Sci. USA*, 104, 17255–17260, 2007.

22. Newton RG. *Scattering Theory of Waves and Particles*, McGraw-Hill Book Company, New York, 1969.

23. van de Hulst HC. *Light Scattering by Small Particles*, Dover Publications, New York, 1957.

24. Kerker M. *The Scattering of Light*, Academic Press, New York, 1969.

25. Cotran RS, Robbins SL, and Kumar V. *Pathological Basis of Disease*, W. B. Saunders Company, Philadelphia, PA, 1994.

26. Zonios G, Perelman LT, Backman V et al. Diffuse reflectance spectroscopy of human adenomatous colon polyps in vivo, *Appl. Optics*, 38, 6628–6637, 1999.

27. Georgakoudi I, Jacobson BC, Van Dam J et al. Fluorescence, reflectance and light scattering spectroscopies for evaluating dysplasia in patients with Barrett's esophagus, *Gastroentorolgy*, 120, 1620–1629, 2001.

28. Backman V, Gurjar R, Badizadegan K et al. Polarized light scattering spectroscopy for quantitative measurement of epithelial cellular structures in situ, *IEEE J. Selected Top. Quantum Electron.*, 5, 1019–1027, 1999.

29. Pyhtila JW, Graf RN, and Wax A. Determining nuclear morphology using an improved angle-resolved low coherence interferometry system, *Opt. Express*, 11, 3473–3484, 2003.

30. Graf R and Wax A. Nuclear morphology measurements using Fourier domain low coherence interferometry, *Opt. Express*, 13, 4693–4698, 2005.

31. Craig IJD and Brown JC. *Inverse Problems in Astronomy: A Guide to Inversion Strategies for Remotely Sensed Data*, Adam Hilger, Boston, MA, 1986.

32. Blot W, Devesa SS, Kneller R, and Fraumeni J. Rising incidence of adenocarcinoma of the esophagus and gastric cardia, *J Am. Med. Assoc.*, 265, 1287–1289, 1991.

33. Antonioli D. The esophagus, in: Henson D and Alobores-Saavdera J, eds., *The Pathology of Incipient Neoplasia*, Saunders, Philadelphia, PA, pp. 64–83, 1993.

34. Cameron AJ. Management of Barrett's esophagus, *Mayo Clin. Proc.*, 73, 457–461, 1998.

35. Reid BJ, Haggitt RC, Rubin CE et al. Observer variation in the diagnosis of dysplasia in Barrett's esophagus, *Hum. Pathol.*, 19, 166–178, 1988.

36. Petras RE, Sivak MV, and Rice TW. Barrett's esophagus. A review of the pathologist's role in diagnosis and management, *Pathol. Annual.*, 26, 1–32, 1991.

37. Riddell RH, Goldman H, Ransohoff DF et al. Dysplasia in inflammatory bowel disease: Standardized classification with provisional clinical applications, *Hum. Pathol.*, 14, 931–968, 1983.

38. Cawley HM, Meltzer SJ, De Benedetti VM et al. Anti-p53 antibodies in patients with Barrett's esophagus or esophageal carcinoma can predate cancer diagnosis, *Gastroenterology*, 115, 19–27, 1998.

39. Macdonald CE, Wicks AC, and Playford RJ. Final results from 10 year cohort of patients undergoing surveillance for Barrett's oesophagus: Observational study, *Br. Med. J.*, 321, 1252–1255, 2000.

40. Falk GW, Rice TW, Goldblum JR et al. Jumbo biopsy forceps protocol still misses unsuspected cancer in Barrett's esophagus with high-grade dysplasia, *Gastrointest. Endosc.*, 49, 170–176, 1999.

41. Altorki NK, Sunagawa M, Little AG et al. High-grade dysplasia in the columnar-lined esophagus, *Am. J. Surg.*, 161, 97–100, 1991.

42. Pera M, Trastek VF, Carpenter HA et al. Barrett's esophagus with high-grade dysplasia: An indication for esophagectomy, *Ann. Thorac. Surg.*, 54, 199–204, 1992.

43. Rice TW, Falk GW, Achkar E et al. Surgical management of high-grade dysplasia in Barrett's esophagus, *Am. J. Gastroenterol.*, 88, 1832–1836, 1993.

44. Reid BJ, Haggitt RC, Rubin CE et al. Observer variation in the diagnosis of dysplasia in Barrett's esophagus, *Hum. Pathol.*, 19, 166–178, 1988.

45. Riddell RH, Goldman H, Ransohoff DF et al. Dysplasia in inflammatory bowel disease: Standardized classification with provisional clinical applications, *Hum. Pathol.*, 14, 931–986, 1983.

46. Haggitt RC. Barrett's esophagus, dysplasia, and adenocarcinoma, *Hum. Pathol.*, 25, 982–993, 1994.

47. Pagano M and Gauvreau K. *Principles of Biostatistics*, Duxbury Press, Belmon, CA, 1993.

48. Landis J and Koch G. The measurement of observer agreement for categorical data, *Biometrics*, 33, 159–174, 1977.

49. Conchello J-A and Lichtman JW. Optical sectioning microscopy, *Nat. Methods*, 2, 920–931, 2005.

50. Lippincott-Schwartz J and Patterson GH. Development and use of fluorescent protein markers in living cells, *Science*, 300, 87–91, 2003.

51. Lichtman JW and Confchello J-A. Fluorescence microscopy, *Nat. Methods*, 2, 910–919, 2005.

52. Zipfel WR, Williams RM, and Webb WW. Nonlinear magic: Multiphoton microscopy in the biosciences, *Nat. Biotechnol.*, 21, 1369–1377, 2003.

53. Shaner NC, Steinbach PA, and Tsien RY. A guide to choosing fluorescent proteins, *Nat. Methods*, 2, 905–909, 2005.

54. Dwyer PJ, Dimarzio CA, Zavislan JM et al. Confocal reflectance theta line scanning microscope for imaging human skin in vivo, *Opt. Lett.*, 31, 942–944, 2006.

55. Fujimoto JG. Optical coherence tomography for ultrahigh resolution in vivo imaging, *Nat. Biotechnol.*, 21, 1361–1367, 2003.

56. Fang H, Qiu L, Zaman MN et al. Confocal light absorption and scattering spectroscopic (CLASS) microscopy, *Appl. Optics*, 46, 1760–1769, 2007.

57. Weise W, Zinin P, Wilson T et al. Imaging of spheres with the confocal scanning optical microscope, *Opt. Lett.*, 21, 1800–1802, 1996.

58. Aguilar JF, Lera M, and Sheppard CJR. Imaging of spheres and surface profiling by confocal microscopy, *Appl. Optics*, 39, 4621–4628, 2000.

59. Durr NJ, Larson T, Smith DK et al. Two-photon luminescence imaging of cancer cells using molecularly targeted gold nanorods, *Nano Lett.*, 7, 941–945, 2007.

60. Qiu L, Larson TA, Smith DK et al. Single gold nanorod detection using confocal light absorption and scattering spectroscopy, *IEEE J. Selected Top. Quantum Electron.*, 13, 1730–1738, 2007.

61. Bohren CF and Huffman DR. *Absorption and Scattering of Light by Small Particles*, John Wiley & Sons, New York, 1983.

62. Johnson PB and Christy RW. Optical constants of noble metals, *Phys. Rev. B*, 6, 4370–4379, 1972.

13 Polarization

Michael Shribak

CONTENTS

13.1 Introduction 373
13.2 Degenerate Polarization States 374
13.3 Polarization Ellipse..374
13.4 Polarization Ellipse Transformations 378
13.5 Linear Retarder between Crossed Linear Polarizer and Analyzer.................................379
13.6 Basic Optical Schematics for Generating Polarized Beam 381
    13.6.1 Complete Polarization State Generator Assembled with Independently Rotating Linear Polarizer and Quarter-Wave Plate...381
    13.6.2 Complete Polarization State Generator Assembled with Jointly Rotating Linear Polarizer and Variable Retarder 382
    13.6.3 Complete Polarization State Generator Assembled with Rotating Linear Polarizer and Fixed Variable Retarder 383
    13.6.4 Complete Polarization State Generator Assembled with Fixed Linear Polarizer and Two Variable Retarders (Type 1) 385
    13.6.5 Complete Polarization State Generator Assembled with Fixed Linear Polarizer and Two Variable Retarders (Type 2) 386
13.7 Conclusion 387
Acknowledgment 387
References..387

13.1 INTRODUCTION

For all electromagnetic radiations, the oscillating components of the electric and magnetic fields are directed at right angle to each other and to the propagation direction. Because a magnetic field vector of the radiation is unambiguously determined by its electric field vector, the polarization analysis considers the last one only. Here, we shall use a right-handed Cartesian coordinate system XYZ, with the Z-axis pointing along the direction of propagation.

Let us assume that we have a plane harmonic wave of angular frequency ω, which is traveling with velocity c in the direction Z. The angular speed is $\omega = 2\pi \, (c/\lambda)$, where λ is the wavelength.

The electric field vector has two components $E_x(z,t)$ and $E_y(z,t)$, which can be written as

$$E_x\left(z,t\right) = E_{0x} \cos\left[\omega\left(t - \frac{z}{c}\right) + \delta_x\right]$$

and

$$E_y\left(z,t\right) = E_{0y} \cos\left[\omega\left(t - \frac{z}{c}\right) + \delta_y\right]$$

(13.1)

where

E_{0x} and E_{0y} are the wave amplitudes
δ_x and δ_y are arbitrary phases
t is the time

The phase difference between two components can be denoted as $\delta = \delta_y - \delta_x$, where $0 \le \delta < 2\pi$.

13.2 DEGENERATE POLARIZATION STATES

Particular cases of propagation of the two components given earlier are shown in Figure 13.1. The central part illustrates a distribution of electric field strength along the Z-axis at some moment of time $t = 0$. Phase of the y-component δ_y is zero and $\delta_x = -\delta$ correspondingly. The axes E_x and E_y have units V/m. The unit length along Z-axis is one wavelength. The wave is moving forward, and the tip of the electric strength vector generates a Lissajous figure in the E_xE_y-plane. The shape traced out by the electric vector when such a plane wave passes over it is a description of the polarization state.

The right part in Figure 13.1 represents the corresponding loci of the tip of the oscillating electric field vector. The shown cases are called degenerate polarization states: (1) linearly horizontal polarized (**LHP**) light, $E_{0y} = 0$; (2) linearly vertical polarized (**LVP**) light, $E_{0x} = 0$; (3) linearly +45° polarized (**L + 45P**) light, $E_{0x} = E_{0y} = E_0$, $\delta = 0$; (4) right circularly polarized (**RCP**) light, $E_{0x} = E_{0y} = E_0$, $\delta = \pi/2$; (5) linearly −45° polarized (**L − 45P**) light, $E_{0x} = E_{0y} = E_0$, $\delta = \pi$; and (6) left circularly polarized (**LCP**) light, $E_{0x} = E_{0y} = E_0$, $\delta = 3\pi/2$. The RCP light rotates clockwise, and the LCP light rotates counterclockwise when propagating toward the observer.

The clockwise sense of the circle describing right circular polarization is consistent with the definition involving a right-handed helix: if a right-handed helix is moved bodily toward the observer (without rotation) through a fixed transverse reference plane, the point of intersection of helix and plane executes a clockwise circle. This classical convention of right and left polarizations is accepted by most authors [1–4].

However, some authors do not agree, and they use the opposite definition of polarization handedness [5–7]. The latter refers to the quantum-mechanical properties of light. In this description, a photon has an intrinsic (or spin) angular momentum of either negative $-h/2\pi$ or positive $h/2\pi$, where $h = 6.626 \times 10^{-34}$ J s is Planck's constant. The angular momentum is parallel to the direction of the photon's propagation. In the classical convention, the positive and negative momentums correspond to LCP and RCP photons, respectively. The alternative approach suggests to designate the positive momentum to the right polarized light. Here, we will follow the classical convention.

13.3 POLARIZATION ELLIPSE

In a general case, the Lissajous figure of two perpendicular sinusoidal oscillations with same frequencies and stable phase difference is an ellipse. Thus, both linear and circular polarized light may be considered as special cases of elliptically polarized light. In order to find an equation of the corresponding locus, we have to eliminate the time–space term $\omega\left(t - (z/c)\right) + \delta_x$ in the system (13.1). Expand the expression for $E_y(z,t)$ and combine it with $E_x(z,t)/E_{0x}$ to yield

$$\sin\left[\omega\left(t - \frac{z}{c}\right) + \delta_x\right]\sin\delta = \frac{E_x}{E_{0x}}\cos\delta - \frac{E_y}{E_{0y}}. \tag{13.2}$$

Then, we multiply both parts of expression (13.1) by $\sin\delta$

$$\cos\left[\omega\left(t - \frac{z}{c}\right) + \delta_x\right]\sin\delta = \frac{E_x}{E_{0x}}\sin\delta. \tag{13.3}$$

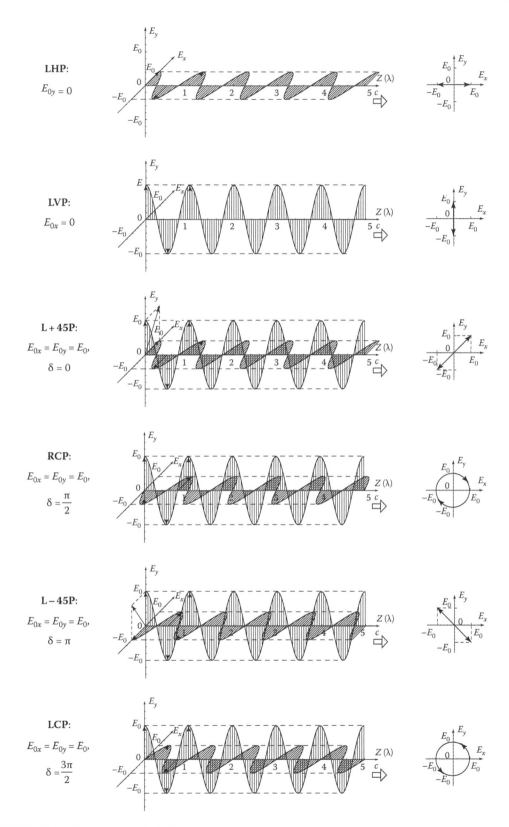

FIGURE 13.1 Degenerate polarization states.

Finally, by squaring and adding Equations 13.2 and 13.3, we have

$$\left(\frac{E_x}{E_{0x}}\right)^2 + \left(\frac{E_y}{E_{0y}}\right)^2 - 2\frac{E_x}{E_{0x}}\frac{E_y}{E_{0y}}\cos\delta = \sin^2\delta. \tag{13.4}$$

This formula describes an ellipse, which is called the polarization ellipse.

The standard form equation of an ellipse with semi-major axis a and semi-minor axis b is the following:

$$\left(\frac{E'_x}{a}\right)^2 + \left(\frac{E'_y}{b}\right)^2 = 1. \tag{13.5}$$

Expression (13.5) can be transformed to the expression (13.4) by a coordinate rotation on angle ψ:

$$E'_x = E_x \cos\psi + E_y \sin\psi,$$
$$E'_y = -E_x \sin\psi + E_y \cos\psi. \tag{13.6}$$

Thus, the equation of ellipse with the major axis at angle ψ becomes

$$E_x^2\left(a^2 \sin^2\psi + b^2 \cos^2\psi\right) + E_y^2\left(a^2 \cos^2\psi + b^2 \sin^2\psi\right) + E_x E_y\left(b^2 - a^2\right)\sin 2\psi = a^2 b^2. \tag{13.7}$$

The formula (13.5) can be written in a similar way:

$$E_x^2 E_{0y}^2 + E_y^2 E_{0x}^2 - 2E_x E_y E_{0x} E_{0y} \cos\delta = E_{0x}^2 E_{0y}^2 \sin^2\delta. \tag{13.8}$$

Comparing coefficients in the quadratic equations (13.7) and (13.8) yields

$$E_{0y}^2 = a^2 \sin^2\psi + b^2 \cos^2\psi,$$
$$E_{0x}^2 = a^2 \cos^2\psi + b^2 \sin^2\psi,$$
$$2E_{0x}E_{0y}\cos\delta = (a^2 - b^2)\sin 2\psi,$$
$$\pm E_{0x}E_{0y}\sin\delta = ab. \tag{13.9}$$

Subtracting the first equation of system (13.9) from second, we obtain

$$E_{0x}^2 - E_{0y}^2 = \left(a^2 - b^2\right)\cos 2\psi. \tag{13.10}$$

The major axis orientation ψ of the polarization ellipse can be found after dividing the third equation in (13.9) by Equation 13.10:

$$\tan 2\psi = \frac{2E_{0x}E_{0y}}{E_{0x}^2 - E_{0y}^2}\cos\delta. \tag{13.11}$$

The orientation is the angle between the major axis of the ellipse and the positive direction of the X-axis. All physically distinguishable orientations can be obtained by limiting ψ to the following range:

$$-\frac{\pi}{2} \leq \psi < \frac{\pi}{2}. \qquad (13.12)$$

The shape of the polarization ellipse can be expressed by a number called the ellipticity of the ellipse, which is the ratio of the length of the semi-minor axis b to the length of its semi-major axis a:

$$e = \frac{b}{a}. \qquad (13.13)$$

Dividing the fourth equation of system (13.9) by the sum of first and second equations, we get

$$\frac{e}{1+e^2} = \pm \frac{E_{0x}E_{0y}}{E_{0x}^2 + E_{0y}^2} \sin \delta. \qquad (13.14)$$

The definition of ellipticity in polarization optics incorporates also the handedness by allowing the ellipticity to assume positive and negative values, to correspond to right-handed and left-handed polarizations, respectively. Using this notation, it is easily seen that all physically distinguishable ellipticity values can be obtained by limiting e to the range

$$-1 \leq e \leq 1. \qquad (13.15)$$

However, according to the classical convention of the polarization handedness, the sign of ellipticity is opposite to the sign of angular momentum of light.

It is convenient to use an ellipticity angle ε such that

$$e = \tan \varepsilon. \qquad (13.16)$$

From Equation 13.15, it follows that ε is limited to the range

$$-\frac{\pi}{4} \leq \varepsilon \leq \frac{\pi}{4}. \qquad (13.17)$$

Now from (13.14), we have

$$\sin 2\varepsilon = \frac{2E_{0x}E_{0y}}{E_{0x}^2 + E_{0y}^2} \sin \delta. \qquad (13.18)$$

The two Equations 13.11 and 13.18 can be rewritten completely in trigonometric terms by introducing an auxiliary angle σ ($0 \leq \sigma \leq \pi/2$) such that

$$\tan \sigma = \frac{E_{0y}}{E_{0x}}. \qquad (13.19)$$

This leads to purely trigonometric equations:

$$\begin{cases} \tan 2\psi = \tan 2\sigma \, \cos\delta \\ \sin 2\varepsilon = \sin 2\sigma \, \sin\delta \end{cases}. \tag{13.20}$$

13.4 POLARIZATION ELLIPSE TRANSFORMATIONS

Formulas (13.20) allow to calculate the orientation and ellipticity angles ψ and ε if we know the ratio and the phase difference of two components $\tan\sigma$ and δ in particular coordinate system.

For example, both the components are either in-phase or antiphase $\delta = 0$ or $\delta = \pi$. Then, formulas (13.20) yield

$$\begin{cases} \psi = \sigma \\ \varepsilon = 0 \end{cases} \text{if } \delta = 0 \quad \text{and} \quad \begin{cases} \psi = -\sigma \\ \varepsilon = 0 \end{cases} \text{if } \delta = \pi.$$

Thus, the light is linearly polarized at the orientation angle ψ determined by the component ratio, and the angle ψ is positive for the in-phase components and negative for the antiphase components.

Let us consider the second example when the phase difference equals to quarter or three quarter wave, $\delta = \pi/2$ or $\delta = 3\pi/2$. According to formulas (13.20), we obtain the following:

$$\text{for } \delta = \frac{\pi}{2}: \begin{cases} \psi = 0 \\ \varepsilon = \sigma \end{cases} \text{if } 0 \le \sigma < \frac{\pi}{4}, \begin{cases} \psi = \frac{\pi}{2} \\ \varepsilon = \frac{\pi}{2} - \sigma \end{cases} \text{if } \frac{\pi}{4} < \sigma \le \frac{\pi}{2}, \text{ and } \begin{cases} \psi \text{ is not defined} \\ \varepsilon = \frac{\pi}{4} \end{cases} \text{if } \sigma = \frac{\pi}{4}.$$

$$\text{for } \delta = \frac{3\pi}{2}: \begin{cases} \psi = 0 \\ \varepsilon = -\sigma \end{cases} \text{if } 0 \le \sigma < \frac{\pi}{4}, \begin{cases} \psi = \frac{\pi}{2} \\ \varepsilon = \sigma - \frac{\pi}{2} \end{cases} \text{if } \frac{\pi}{4} < \sigma \le \frac{\pi}{2}, \text{ and } \begin{cases} \psi \text{ is not defined} \\ \varepsilon = -\frac{\pi}{4} \end{cases} \text{if } \sigma = \frac{\pi}{4}.$$

The light is elliptically polarized with the major axis oriented along to X- or Y-axis. The polarization is right handed if $\delta = \pi/2$ and left handed if $\delta = 3\pi/2$. If amplitudes of the components are equal ($\sigma = \pi/4$), the light is circularly polarized and does not have the major axis. In this case, the first Equation 13.20 contains product of infinity and zero that cannot be defined.

Systems (13.20) allow to solve the direct problem: to compute angles of polarization ellipse ψ and ε if parameters of polarization components σ and δ are known. It is important for practical application to have a solution of the inverse problem: to compute parameters of polarization components σ and δ in particular coordinate system if angles of polarization ellipse ψ and ε are known.

The inverse solution can be found by transformation of Equations 13.20:

$$\begin{cases} \cos 2\sigma = \cos 2\varepsilon \, \cos 2\psi \\ \tan\delta = \tan 2\varepsilon / \sin 2\psi \end{cases}. \tag{13.21}$$

An example of polarization ellipse with corresponding angles is illustrated in Figure 13.2. Here, orientation angle $\psi = 30°$, ellipticity angle $\varepsilon = 22.5°$ ($b/a = 0.41$), and the handedness is clockwise

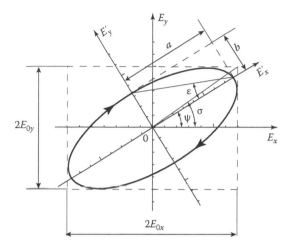

FIGURE 13.2 Polarization ellipse.

(right-handed polarization). Using formulas (13.21), we can compute ratio angle σ and phase differ-ence δ in the *XY*-coordinate system: σ = 34.6° (E_{0y}/E_{0x} = 0.7) and δ = 49°.

In the *X'Y'*-coordinate system, which is oriented along the major and minor axes, orientation angle is zero. The ellipticity angle ε does not change, and it equals 22.5°. The corresponding ratio angle σ' and phase difference δ' will be the following: σ' = 22.5° (E'_{0y} / E'_{0x} = 0.41) and δ' = 90°.

In the majority cases, we will not be interested in the absolute intensity of an optical wave. This justifies the use of a definition for the intensity of the wave that overlooks a constant multiplicative factor. Thus, we may express the intensity *I* simply as the squared amplitudes of the component oscillating along two mutually perpendicular directions:

$$I \equiv E_{0x}^{2} + E_{0y}^{2} = a^{2} + b^{2}. \tag{13.22}$$

The right part of this equation can be easily verified by the addition of the first two equations in system (13.9).

Thus, the direct and inverse formulas (13.20) and (13.21) can be used for a numerical computa-tion of the transformation of the polarization ellipse during passing through an optical system and to obtain the corresponding analytical solution for polarization state of the output beam. Equation 13.22 allows to find the beam intensity.

13.5 LINEAR RETARDER BETWEEN CROSSED LINEAR POLARIZER AND ANALYZER

A linear polarizer (analyzer) is a device that creates a linear polarization state from an arbitrary input. It does this by removing the component orthogonal to the selected state. The transmission axis of a linear polarizer is defined with respect to a beam of linearly polarized light that is incident normally on the prime face of the polarizer. The direction that the electric vibration of the beam must have in order that the actual transmittance be maximized is the transmission axis.

A linear retarder (wave plate) is a polarization element designed to produce a specified phase difference between two perpendicular linear polarization components of the incident beam with-out altering the intensity of the beam appreciably. It conserves the polarization of incident beam having either of two particular linear polarizations but alters the polarization of other types of beams. Associated with a linear retarder are two directions or axes, a fast axis and a slow axis.

FIGURE 13.3   Linear retarder between crossed linear polarizer and analyzer.

The component of incident polarized light whose electric vector is parallel to the slow axis is retarded in phase relative to the component vibration parallel to the fast axis as the beam passes through the retarder. The most commonly used retarders have a phase difference (retardance) of 90° and 180°; these are often called quarter-wave plates and half-wave plates.

Let us consider a simple system consisting of crossed linear polarizer and analyzer and retarder in between. The setup is shown in Figure 13.3, where $\mathbf{P}(0°)$ and $\mathbf{A}(90°)$ are linear polarizer and analyzer with transmission axes at 0° and 90° correspondingly, and $\mathbf{R}(\Delta, \varphi)$ is a retarder with retardance Δ and slow axis orientation φ. The polarization states at the front and rear faces of the retarder and the analyzer are marked as $\vec{E}_1$, $\vec{E}_1'$, $\vec{E}_2$, and $\vec{E}_2'$.

Polarization of the beam after the polarizer is linear and horizontal. In eigen coordinate system of the retarder with X'-axis parallel to the slow axis, the polarization is also linear but oriented at the angle $-\varphi$. Thus, the polarization of the incident beam $\vec{E}_1$ in the retarder eigen coordinates has the ellipticity angle $\varepsilon_1 = 0$ and the orientation angle $\psi = -\varphi$. Using formulas (13.21), we can compute the ratio angle $\sigma_1 = -\varphi$. The phase difference $\delta_1 = 0$ because the polarization is linear.

The retarder introduces an additional phase difference Δ. Therefore, the output beam $\vec{E}_1'$ in the retarder eigen coordinates will be described by the ratio angle $\sigma_1' = -\varphi$ and the phase difference $\delta_1' = \Delta$. The corresponding ellipticity and orientation angles of the output polarization ellipse can be obtained by employing Equation 13.20:

$$\begin{cases} \tan 2\psi_1' = -\tan 2\varphi \, \cos\Delta \\ \sin 2\varepsilon_1' = \sin 2\varphi \, \sin\Delta \end{cases}. \tag{13.23}$$

In order to find how much light will be transmitted by the analyzer, we have to rotate the polarization ellipse back to the initial coordinate system by adding angle φ to the orientation angle ψ_1'. The polarization state $\vec{E}_2$ at the front of the analyzer will be described by the orientation angle $\psi_2 = \psi_1' + \varphi$ and ellipticity $\varepsilon_2 = \varepsilon_1'$. Using expression (13.23), we obtain

$$\begin{cases} \tan 2\psi_2 = \tan 2\left(\psi_1' + \varphi\right) = \dfrac{\tan 2\psi_1' + \tan 2\varphi}{1 - \tan 2\psi_1' \tan 2\varphi} = \dfrac{\left(1 - \cos\Delta\right)\tan 2\varphi}{1 + \tan^2 2\varphi \cos\Delta} \\ \sin 2\varepsilon_2 = \sin 2\varphi \, \sin\Delta \end{cases}. \tag{13.24}$$

Intensity of the beam I passed through the linear analyzer with transmission axis at 90° can be found from the ratio

$$I = \frac{E_{0y}^2}{E_{0x}^2 + E_{0y}^2} I_0 = \frac{\tan^2\sigma}{1 + \tan^2\sigma} I_0 = \frac{1}{2} I_0 \left(1 - \cos 2\sigma\right), \tag{13.25}$$

where I_0 is the intensity of the beam after the polarizer

The ratio angle of the polarization ellipse at the front of the polarizer σ_2 is computed from the first equation of formula (13.21):

$$\cos 2\sigma_2 = \cos 2\varepsilon_2 \, \cos 2\psi_2.$$

In the considered case, after some straightforward trigonometric transformations, we obtain a simple analytical expression for the ratio angle:

$$\cos^2 2\sigma_2 = \cos^2 2\varepsilon_2 \cos^2 2\psi_2 = \frac{1 - \sin^2 2\varepsilon_2}{1 + \tan^2 2\psi_2} = \frac{\left(1 - \sin^2 2\varphi \sin^2 \Delta\right)\left(1 + \tan^2 2\varphi \cos \Delta\right)^2}{\left(\left(1 - \cos \Lambda\right)\tan 2\varphi\right)^2 + \left(1 + \tan^2 2\varphi \cos \Delta\right)^2}$$

$$= \frac{\left(1 - \sin^2 2\varphi \sin^2 \Delta\right)\left(1 + \tan^2 2\varphi \cos \Delta\right)^2}{\left(1 + \tan^2 2\varphi\right)\left(1 + \tan^2 2\varphi \cos^2 \Delta\right)} = \left(\cos^2 2\varphi + \sin^2 2\varphi \cos \Delta\right)^2. \tag{13.26}$$

Finally, the intensity of the transmitted beam I is reduced to the following formula:

$$I = I_0 \sin^2 2\varphi \sin^2 \frac{\Delta}{2}. \tag{13.27}$$

Formula (13.27) can also be derived using Jones or Muller matrix computation [8]. Advantage of the computation with the polarization ellipse is that the technique does not require employing matrix algebra and complex numbers, and it explicitly shows a transformation of the polarization by each optical element. Of course, not every optical system can be described by a simple analytical expression like formula (13.27). In this case, the ellipse transformation formulas allow to create a numerical mathematical model.

13.6 BASIC OPTICAL SCHEMATICS FOR GENERATING POLARIZED BEAM

13.6.1 Complete Polarization State Generator Assembled with Independently Rotating Linear Polarizer and Quarter-Wave Plate

A polarization generator consists of a light source and polarization optical elements to produce a beam of known polarization state. An example of the polarization generator, which is also called the Sénarmont compensator [9], is shown in Figure 13.4. It includes a rotatable linear polarizer **P** with transmission axis at angle θ and rotatable quarter-wave plate **QWP** with the slow axis oriented at angle φ. Let us choose the reference axis X perpendicular to the direction of beam propagation. The polarization states at the front and rear faces of the quarter-wave plate are marked as $\vec{E}$ and $\vec{E}'$.

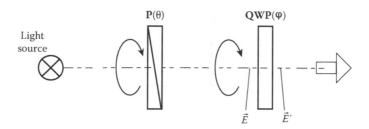

FIGURE 13.4 Complete polarization state generator with independently rotating linear polarizer and quarter-wave plate.

To obtain parameters of the output polarization ellipse $\vec{E}'$, we substitute angle difference $\varphi-\theta$ for angle φ, and $\pi/2$ for Δ in Equation 13.23, and then add angle φ to the computed ellipse orientation:

$$\begin{cases} \psi' = \varphi \\ \varepsilon' = \varphi - \theta \end{cases}. \tag{13.28}$$

As it follows from this result, the principal axis of the ellipse is oriented along the slow axis of the quarter-wave plate, and the ellipticity angle is a difference between azimuths of the quarter-wave plate and polarizer. Thus, we can obtain any desired orientation of the polarization ellipse by rotating the quarter-wave plate and any desired ellipticity by turning the polarizer relatively to the quarter-wave plate. For example, the difference $\varphi-\theta$ has to be $\pi/4$ for the right circular polarization ($\varepsilon' = \pi/4$). Since the generator allows to produce a beam with any polarization state, it is called the complete polarization state generator [10].

Let us consider an interesting case when the quarter-wave plate is oriented at angle $\varphi = \pi/4$. Then, the output polarization ellipse is also oriented at angle $\psi' = \pi/4$. Using inverse Equation 13.21, we can show that the ratio angle $\sigma' = \pi/4$, which means two orthogonal components E_{0x}' and E_{0y}' have equal amplitudes (see Equation 19). According to Equations 13.21, the phase difference between two components δ' is determined by difference in the orientation of the polarizer and the quarter-wave plate:

$$\delta' = 2\varepsilon' = \frac{\pi}{2} - 2\theta. \tag{13.29}$$

Thus, we can change the phase difference linearly by rotating the polarizer. For example, if $\theta = \pi/4$, the output components are in-phase. If $\theta = 0$, then the components are shifted by $\pi/2$ (quarter-wave). If the polarizer is oriented at $-\pi/4$, then the output electric vectors are antiphase (half-wave shift). Polarizer at $-\pi/2$ produces phase shift $3\pi/2$ (three quarter-wave). This feature of the Sénarmont compensator is often employed in phase shifting interferometry and birefringence mapping techniques.

13.6.2 COMPLETE POLARIZATION STATE GENERATOR ASSEMBLED WITH JOINTLY ROTATING LINEAR POLARIZER AND VARIABLE RETARDER

Another optical configuration of polarization state generator is presented in Figure 13.5. Linear polarizer **P** and variable retarder **VR** are mounted in the rotatable body. The angle between the

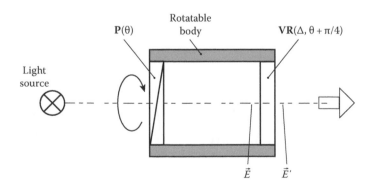

FIGURE 13.5 Complete polarization state generator with jointly rotating linear polarizer and variable retarder.

transmission axis of the polarizer and the slow axis of the retarder is $\pi/4$. As a variable retarder, one can use Babinet–Soleil compensator [2,3,9], Berek compensator [3,9,11] or its analog made of quartz [12], the Ehringhaus compensator [9,13], liquid crystal cell [7,14], electro-optical or piezo-optical modulator [2,5], for instance. A phase difference Δ of the retarder can be adjusted. The variable retarder is turned together with the polarizer. Orientation of the polarizer transmission axis is θ. The retarder slow axis orientation is $\theta + \pi/4$ accordingly. The polarization states at the front and rear faces of the variable retarder are shown as $\vec{E}$ and $\vec{E}'$.

For the analysis of the generator, we use the coordinate system in which X'-axis is parallel to the transmission axis of the polarizer. Parameters of the output polarization ellipse $\vec{E}'$ are determined from formula (13.24) at $\phi = \pi/4$ and then angle θ is added to the computed ellipse orientation:

$$
\begin{cases}
\psi' = \theta \\
\varepsilon' = \Delta/2
\end{cases} .
\tag{13.30}
$$

As one can see, the ellipticity angle of the output beam equals a half of retardance, and the principal axis of the ellipse is oriented along the transmission axis of the polarizer. For example, if the retardance is quarter-wave ($\Delta = \pi/2$), then the output beam is RCP. Rotating the entire setup can produce any desired orientation of the polarization ellipse, and any ellipticity can be obtained by changing the retardance of the variable retarder. Using a fast variable retarder, such as a liquid crystal cell or electro-optic crystal, allows quick control of the ellipticity of the output beam.

13.6.3 COMPLETE POLARIZATION STATE GENERATOR ASSEMBLED WITH ROTATING LINEAR POLARIZER AND FIXED VARIABLE RETARDER

In order to control the introduced retardance, a variable retarder requires either mechanical movement of its components or application of electrical voltage. Therefore, the retarder's rotating mechanism could be bulky and complicated to build. Sometimes, it is necessary to quickly change the polarization ellipse orientation, not its ellipticity. Recently, we proposed a new complete polarization state generator with one variable retarder, which is fixed [15]. New generator allows rapid change in the polarization ellipse orientation.

A schematic of the proposed polarization state generator is shown in Figure 13.6. It contains in-series rotatable polarizer **P** and fixed variable retarder **VR**, which introduces a phase shift Δ and quarter-wave plate **QWP**. Let us choose the slow axis of the quarter-wave plate as X-axis of the Cartesian coordinates. The transmission axis of the polarizer is inclined at an azimuth θ from the X-axis. For simplicity of the description, we consider angle range $-\pi/4 \leq \theta \leq \pi/4$. The polarization states at the front and rear faces of the variable retarder and quarter-wave plate are noted as $\vec{E}_1$, $\vec{E}_1'$, $\vec{E}_2$, and $\vec{E}_2'$, respectively.

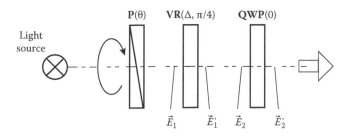

FIGURE 13.6 Complete polarization state generator with rotating linear polarizer and fixed variable retarder.

Electric vector $\vec{E}_1$ represents a linear vibration at angle $\theta - \pi/4$ to the slow axis of the variable retarder, and it is described by the polarization ellipse with the following parameters in the eigen coordinates of the retarder:

$$\begin{cases} \psi_1 = \theta - \pi/4 \\ \quad \varepsilon_1 = 0 \end{cases}. \tag{13.31}$$

The corresponding polarization components σ_1 and δ_1 are obtained with formula (13.21):

$$\begin{cases} \sigma_1 = \theta - \pi/4 \\ \quad \delta_1 = 0 \end{cases}. \tag{13.32}$$

The variable retarder creates phase delay Δ between the slow and fast components. Then, the output electric vector $\vec{E}_1'$ has the next ratio angle and retardation:

$$\begin{cases} \sigma_1' = \theta - \pi/4 \\ \quad \delta_1' = \Delta \end{cases}. \tag{13.33}$$

Shape and orientation of the interim polarization ellipse $\vec{E}_2$ in eigen coordinate of the quarter-wave plate can be obtained by replacing δ with Δ and σ with $\theta - \pi/4$ in Equation 13.20 and then adding $\pi/4$ to the orientation angle:

$$\begin{cases} \tan 2\psi_2 = \tan 2\theta / \cos \Delta \\ \sin 2\varepsilon_2 = -\cos 2\theta \sin \Delta \end{cases}. \tag{13.34}$$

The corresponding ratio σ_2 and phase shift δ_2 at the entrance of the quarter-wave plate can be found with formula (13.21). Then, the quarter-wave plate adds retardation $\pi/2$ to the phase shift δ ($\delta' = \delta + \pi/2$) and keeps the ratio constant ($\sigma' = \sigma$):

$$\begin{cases} \cos 2\sigma_2' = \cos 2\varepsilon_2 \cos 2\psi_2 \\ \tan \delta_2' = -\sin 2\psi_2 / \tan 2\varepsilon_2 \end{cases}. \tag{13.35}$$

The output polarization ellipse $\vec{E}_2'$ can be computed in the following way (see Equation 13.20):

$$\begin{cases} \tan 2\psi_2' = \tan 2\sigma_2' \cos \delta_2' \\ \sin 2\varepsilon_2' = \sin 2\sigma_2' \sin \delta_2' \end{cases}. \tag{13.36}$$

After squaring Equations 13.34 through 13.36 and sequential substitution of the variables, we receive simple straightforward formulas that connect ellipticity and orientation of the output polarization, ε_2' and ψ_2', with the polarizer's orientation and variable retarder's phase shift, θ and Δ:

$$\begin{cases} \psi_2' = \Delta/2 \\ \quad \varepsilon_2' = \theta \end{cases}. \tag{13.37}$$

Thus, the proposed polarization generator can produce any polarization state of the output beam:

$$\begin{cases} \quad 0 \leq \psi_2' \leq \pi, & \text{if } 0 \leq \Delta \leq 2\pi \\ -\pi/4 \leq \varepsilon_2' \leq \pi/4, & \text{if } -\pi/4 \leq \theta_2 \leq \pi/4 \end{cases}.$$

The desired ellipticity angle ε_2' can be achieved by rotating the polarizer, and the major axis orientation ψ_2' is determined by phase shift of the variable retarder. So, using a fast variable retarder, such as a liquid crystal cell or electro-optic crystal, allows quick control of the orientation of the output polarization ellipse.

We have also derived formula (13.37) by multiplication of Jones matrices followed by a complex vector conjugation [15]. The mentioned publication describes techniques for fast and sensitive differential measurements of 2D birefringence distribution, which utilize the complete polarization state generator with single variable retarders.

13.6.4 Complete Polarization State Generator Assembled with Fixed Linear Polarizer and Two Variable Retarders (Type 1)

The earlier described polarization generators have two or one rotatable component. Arrangement of the complete polarization state generator without any mechanical rotation was proposed by Yamaguchi and Hasunuma [2,16]. A schematic of the Yamaguchi–Hasunuma compensator, which consists of two variable retarders, is shown in Figure 13.7. Two liquid crystal cells **LCA** and **LCB** are used as variable wave plate whose retardations α and β are controlled by voltages. The slow axes of **LCA** and **LCB** are chosen at $\pi/4$ and 0 (parallel) orientations from the transmission axis of the polarizer **P**. The polarization states at the front and rear faces of the liquid crystal cells are noted as $\vec{E}_1$, $\vec{E}_1'$, $\vec{E}_2$, and $\vec{E}_2'$. Instead of liquid crystal cells, one can use other kinds of variable retarders as mentioned earlier.

Shape and orientation of the polarization ellipse $\vec{E}_2$ can be obtained by substitution $\varphi = \pi/4$ and replacing Δ with α in Equation 13.24:

$$\begin{cases} \psi_2 = 0 \\ \varepsilon_2 = \alpha/2 \end{cases}. \tag{13.38}$$

The corresponding polarization components σ_2 and δ_2 are obtained with formula (13.21):

$$\begin{cases} \sigma_2 = \alpha/2 \\ \delta_2 = \pi/2 \end{cases}. \tag{13.39}$$

The second variable retarder introduces phase difference β. Then, the output beam $\vec{E}_2'$ will have the following ratio angle and phase difference:

$$\begin{cases} \sigma_2' = \alpha/2 \\ \delta_2' = \beta + \pi/2 \end{cases}. \tag{13.40}$$

FIGURE 13.7 Complete polarization state generator with fixed linear polarizer and two variable retarders (type 1).

The relative amplitude of the X- and Y-polarization components of the output light is controlled by α, whereas the phase difference of the same components is controlled by β. Therefore, by controlling the voltages applied to **LCA** and **LCB**, any state of polarization for the output beam can be generated.

The corresponding ellipticity and orientation angles of the polarization ellipse can be computed from Equation 13.20:

$$\begin{cases} \psi_2' = -\dfrac{1}{2}\arctan\left(\tan\alpha\sin\beta\right) \\ \varepsilon_2' = \dfrac{1}{2}\arcsin\left(\sin\alpha\cos\beta\right) \end{cases}$$

(13.41)

The inverse transformation, which shows what phase shifts α and β should be applied in order to obtain the required ellipticity and orientation angles of the output beam ε_2' and ψ_2', is the following:

$$\begin{cases} \alpha = \arccos\left(\cos 2\psi_2' \cos 2\varepsilon_2'\right) \\ \beta = -\arctan\left(\sin 2\psi_2'/\tan 2\varepsilon_2'\right) \end{cases}.$$

(13.42)

We have also derived formula (13.41) by multiplication of Jones matrices followed by a complex vector conjugation [17]. This publication describes techniques for fast and sensitive differential measurements of 2D birefringence distribution, which utilize the complete polarization state generator with two variable retarders. The corresponding systems for quantitative birefringence imaging are currently manufactured by Perkin Elmer (http://www.perkinelmer.com). The polarization generator is mentioned as a precision liquid crystal compensator in the company's website.

13.6.5 COMPLETE POLARIZATION STATE GENERATOR ASSEMBLED WITH FIXED LINEAR POLARIZER AND TWO VARIABLE RETARDERS (TYPE 2)

Disadvantage of the Yamaguchi–Hasunuma setup is that parameters of the output polarization ellipse are connected with retarder's phase shifts by quite complicated trigonometric Equation 13.41. Also, in order to change the ellipse orientation angle or ellipticity only, we have to change phase shifts of both the retarders (see Equation 13.42).

In order to simplify control of the output polarization ellipse and provide more functionality, we are proposing another type of the complete polarization state generator assembled with fixed linear polarizer and two variable retarders, which is shown in Figure 13.8. It contains, in series, a fixed polarizer **P**, two variable retarders **VR1** and **VR2**, and polymer quarter-wave retardation film **QWF**. The two variable retarders introduce retardations α and β, respectively. The slow axes of **VR1** and **VR2** are chosen at $\pi/4$ and 0 (parallel) orientations from the transmission axis of the polarizer **P**. The slow axis of quarter-wave film **QWF** has direction $\pi/4$. Of course, instead of the quarter-wave film, we could employ a quarter-wave plate, as in the polarization state generators described earlier. Using a quarter-wave retardation film allows to simplify the generator.

The proposed setup could be considered as a sequential combination of two polarization state generators with single variable retarder each (see Section 13.6.3), where retardation of the second retarder consists of two parts, $\pi/2$ and $\beta-\pi/2$. The slow axes of both the parts are oriented at 0 rad. The polarizer **P**, first variable retarder **VR1**, and the first quarter-wave part of the second variable retarder **VR2** form the first polarization state generator, which produces linearly polarized beam $\vec{E}_1$ with orientation angle $\psi_1 = \alpha/2$ (see Equation 13.37). The second part of retarder **VR2** with retardation $\beta-\pi/2$ and quarter-wave film **QWF** creates the second polarization state generator. Polarization plane of beam $\vec{E}_1$ is turned by angle $\theta = \psi_1 - \pi/4 = \alpha/2 - \pi/4$ relatively to the slow axis

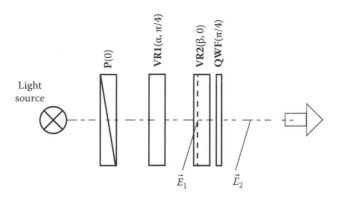

FIGURE 13.8 Complete polarization state generator with fixed linear polarizer and two variable retarders (type 2).

of the quarter-wave film. Then, according to Equation 13.37, the output beam $\vec{E}_2$ has the following ellipticity and orientation angles of the polarization ellipse, ε_2 and ψ_2:

$$\begin{cases} \psi_2 = \beta/2 \\ \varepsilon_2 = \alpha/2 - \pi/4 \end{cases}. \tag{13.43}$$

The inverse transformation, which shows what phase shifts α and β should be applied in order to obtain the required ellipticity and orientation angles of the output beam ε_2 and ψ_2, is the following:

$$\begin{cases} \alpha = 2\varepsilon_2 + \pi/4 \\ \beta = 2\psi_2 \end{cases}. \tag{13.44}$$

The ellipticity and orientation angles of the output beam are directly determined by retardations of the first and second variable retarders, α and β, respectively.

13.7 CONCLUSION

In this chapter, we gave the classical description of the polarization of monochromatic light. The direct and inverse equations for transformations of the polarization ellipse were derived. An example of the application of the obtained formulas for the analysis of beam propagation through crossed polarizers and a wave plate was given. Basic optical schematics for generating and analyzing any polarization states were presented.

ACKNOWLEDGMENT

This publication was made possible by Grant Number R01-GM101701 from the National Institute of General Medical Sciences, the National Institutes of Health. Its contents are solely the responsibility of the author and do not necessarily represent the official views of the National Institute of General Medical Sciences or the National Institutes of Health.

REFERENCES

1. Shurcliff, W.A., *Polarized Light: Production and Use*, Harvard University Press, Cambridge, MA, 1962.
2. Azzam, R.M.A. and Bashara, N.M., *Ellipsometry and Polarized Light*, Elsevier Science, Amsterdam, the Netherlands, 1987.

3. Born, M. and Wolf, E., *Principles of Optics*, 7th edn., Cambridge University Press, New York, 1999.
4. Ramachandran, G.N. and Ramaseshan, S., Crystal optics, in *Handbuch der Physik*, vol. 25/1, S. Flugge, ed., Springer, Berlin, Germany, 1961.
5. Yariv, A. and Yeh, P., *Optical Waves in Crystals: Propagation and Control of Laser Radiation*, John Wiley & Sons, New York, 1984.
6. Yeh, P., *Optical Waves in Layered Media*, John Wiley & Sons, New York, 1988.
7. Robinson, M., Chen, J., and Sharp, G.D., *Polarization Engineering for LCD Projection*, John Wiley & Sons, New York, 2005.
8. Gerrard, A. and Burch, J.M., *Introduction to Matrix Methods in Optics*, John Wiley & Sons, New York, 1975.
9. Hartshorne, N.H. and Stuart A., *Crystals and the Polarizing Microscope*, 4th edn., Edward Arnold, London, U.K., 1970.
10. Hauge, P.S., Recent development in instrumentation in ellipsometry, *Surf. Sci.*, 96, 108–140, 1980.
11. Rinne, F. and Berek, M., *Anleitung zu Optischen Untersuclhungen mit dem Polarizationsmikroskop*, Schweizerbart'sche Verlagsbuchhandlung, Stuttgart, Germany, 1953.
12. Shribak, M., Use of gyrotropic birefringent plate as quarter-wave plate, *Sov. J. Opt. Technol.*, 53, 443–446, 1986.
13. Holmes, D.A., Wave optics theory of rotary compensators, *J. Opt. Soc. Am.*, 54, 1340–1347, 1964.
14. Scharf, T., *Polarized Light in Liquid Crystals and Polymers*, John Wiley & Sons, Hoboken, NJ, 2007.
15. Shribak, M., Complete polarization state generator with one variable retarder and its application for fast and sensitive measuring of two-dimensional birefringence distribution, *J. Opt. Soc. Am. A*, 28, 410–419, 2011.
16. Yamaguchi, T. and Hasunuma, H., A quick response recording ellipsometer, *Sci. Light*, 16(1), 64–71, 1967.
17. Shribak, M. and Oldenbourg, R., Techniques for fast and sensitive measurements of two-dimensional birefringence distributions, *Appl. Opt.*, 42, 3009–3017, 2003.

14 Near-Field Optics

Wenhao Huang, Xi Li, and Guoyong Zhang

CONTENTS

14.1 Introduction 389
14.2 Evanescent Field 390
14.3 Instrumentation.. 390
    14.3.1 Optical Probe... 391
    14.3.2 Probe Height Control System 393
    14.3.3 Calibration of SNOM... 394
        14.3.3.1 Calibration of Resolution 394
        14.3.3.2 Elimination of Optical False Images 395
14.4 Applications.. 396
    14.4.1 Simulation Based on FDTD Method.. 396
    14.4.2 Measurement of the Profile of Micro-Focused Spot 397
    14.4.3 Super-Resolution Imaging 397
    14.4.4 Nanolithography 398
    14.4.5 Near-Field Storage 400
References.. 401

14.1 INTRODUCTION

Looking back, over the past 500 years, the resolution of the optical microscope was not improved in an order of magnitude after its invention because of the diffraction limit.

In order to break through such limit, in 1928, E.H. Synge published a paper in *Philosophical Magazine* titled "A suggested method for extending microscopic resolution into the ultra-microscopic region." He proposed a new idea of microscope that could possibly achieve a resolution within 0.01 μm. However, the technology then available of mechanical precision and the intensity of light source was not advanced enough to satisfy the requirements for the new-type microscope. As a result, his paper was easily forgotten until the invention of the scanning near-field optical microscope (SNOM).

In the beginning of 1980s, Dr. G. Binnig and H. Rohrer in IBM Research Center of Zurich, Switzerland, solved the problem of controlling probe scanning on the sample surface within the nanometer scale, and the first scanning tunneling microscope (STM) was invented [1], winning them a Nobel Prize in 1986. In 1984, Dr. D.H. Pohl in the same research center applied this technology to control the optical probe, and the first SNOM was born [2]. Since then great progress, both theoretical and technological, has been made. The advantages of SNOMs are high resolution, non-invasive nature for various materials, and unique optical spectral information. These make SNOMs a strong and versatile member in the scanning probe microscope (SPM) family, and they are widely used in micro/nano-science and technology fields.

In this chapter, key aspects of near-field optics including fundamental principles, novel methods, instrumental technologies, and applications in nanofabrication and nanomeasurement fields are introduced.

14.2 EVANESCENT FIELD

It is well known that when light irradiates from an optically denser medium (medium 1) into one that is optically less dense (medium 2) with an angle θ_1 larger than the critical angle θ_c, a total reflection occurs. Nevertheless, the electromagnetic field does not disappear in medium 2 and its electric vector can be written as [3]

$$E(x,z) = T_0 \exp\left(-i\omega t + ik_2 \frac{x}{n}\sin\theta_1\right)\exp\left(-k_2 z\sqrt{\frac{1}{n_2}\sin^2\theta_2 - 1}\right), \tag{14.1}$$

where
    n_1 and n_2 are the index of refraction of medium 1 and 2, respectively
    $n = n_1/n_2$ is the relative index of refraction
    ω is the angular frequency
    $k_2 = 2\pi/\lambda_2$ is the wave number
    λ_2 is the wavelength in medium 2

Equation 14.1 implies that there exists a kind of surface wave on the boundary between mediums 1 and 2, namely evanescent wave. It is a kind of inhomogeneous wave along the surface x and its amplitude along z attenuates exponentially. Actually, according to the boundary conditions of electromagnetic field, evanescent wave exists with necessity because electric and magnetic fields cannot discontinue on the surface between two mediums.

Evanescent wave does not only exist in the boundary where the total reflection occurs. When studying the reflection and transmission of a finite object, a formula similar to Equation 14.1 can be obtained using the Fourier analysis method, that is, an evanescent wave attenuates sharply along Z direction. The finite objects generate propagating and evanescent waves at the same time. The propagating waves are the radiation components, while the evanescent waves are non-radiation components. Because radiation components only contain the low-frequency part of spatial frequency, to improve the signal bandwidth of the optical detection system, that is, to improve the optical resolution, non-radiation field components have to be collected. In other words, only by collecting the high spatial-frequency information carried by the evanescent field can the high-resolution that breaks the diffraction limit be obtained [4].

The properties of evanescent field are as follows:

1. Evanescent field only exits within a scale of wavelength. This implies we could only use evanescent field to detect the ultraprecise structures.
2. Evanescent field is a spatial exponential-decay field in a scale of wavelength, which indicates that its application should be limited to a very small spatial range.
3. None of the evanescent fields transfer energy outside, so they are non-radiation fields. However instant energy flux do exist.

14.3 INSTRUMENTATION

Usually, the SNOM consists of several parts such as laser, optical probe, three-dimensional scanner, probe height control system, optoelectric accepter, and the computer (Figure 14.1). There are several work modes as shown in Figure 14.2. The most common one is the transmission mode, which can be

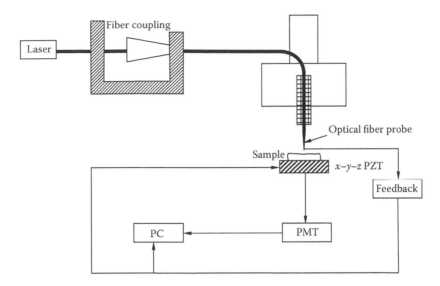

FIGURE 14.1 Schematic diagram of the structure of the near-field optical microscope system.

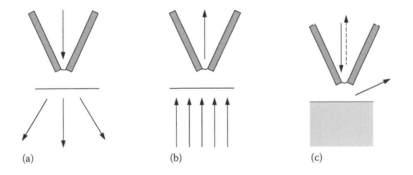

FIGURE 14.2 Work modes of the near-field optical microscope: (a) illumination mode, (b) collection mode, and (c) reflection mode.

divided into the illumination mode (Figure 14.2a) and the collection mode (Figure 14.2b). However, when the object is opaque, the reflection mode (Figure 14.2c) is necessary.

14.3.1 OPTICAL PROBE

In SNOM, the core element is the optical probe with a pinhole, which is used to illuminate the sample at the nanometer scale with its aperture smaller than the wavelength. This aperture determines the upper limit of resolution of the near-field optical microscope. The optical probes are divided into two kinds: optical fiber probe (with pinhole) and apertureless probe.

There are two methods to make the optical fiber probe: heat stretching and chemical etching methods. In the heat stretching method, a carbon dioxide laser is used to heat the fiber; meanwhile, force is applied to stretch the fiber till its abruption, and then two conically shaped probes can easily be obtained (Figure 14.3). While in the chemical etching method, fiber is put into hydrofluoric acid to get etched (Figure 14.4). The diameter of the pinhole on the tip of the fiber is usually tens of nanometers. Sometimes, a metallic film can be coated on the surface of the probe to get better optical signals.

The structure of the typical optical fiber probe is shown in Figure 14.5. The apex of the probe is the detection part, which can be made in nano-order, with the diameter $d = 2a$ conducting

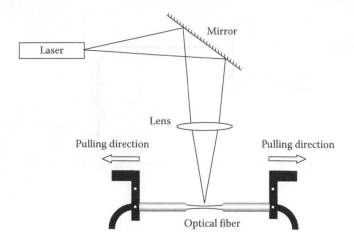

FIGURE 14.3   Schematic diagram of the heating stretching method.

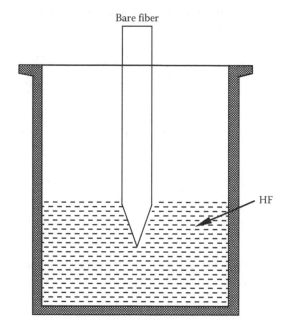

FIGURE 14.4   Schematic diagram of the chemical etching method.

electromagnetic exchanges with sample. The tapered part is simply a cone, whose apex angle is θ. The outside of the probe is covered by the opaque metallic file to avoid the disturbance of external light [5].

However, the commercial SNOM probes of high resolution (<50 nm) are usually very expensive. The probe fabrication, particularly with high-resolution aperture, high reproduction, and low cost, is of great interest to users of SNOM.

Figure 14.6 shows the side view (left) and back view (right) of the probe on a cantilever structure, respectively [6]. This kind of aperture is manufactured by focused ion beam. The bottom part is removed to allow light to transmit through the pinhole. The size of the pinhole can be precisely controlled (120–40 nm) if well designed, to get ideal resolution capability. The optical efficiency passing through the aperture is higher than through the conic fiber.

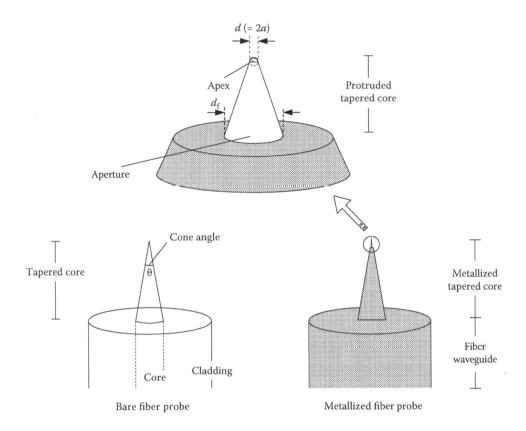

FIGURE 14.5 Schematic diagram of the typical shape of the probe. (From Ohtsu, M. (Ed.), *Near-Field Nano/Atom Optics and Technology*, Springer, Berlin, Germany, 1998. With permission.)

FIGURE 14.6 Structure image of a SNOM aperture.

Another kind of probe is called the apertureless probe. Usually, the probe of atomic force microscope (AFM) with a metal-coated tip is used. The aperture on the AFM tip is much sharper than that on optical fiber. When a laser irradiates onto the metal probe of AFM, the local light field will be enhanced under the probe. The enhancement can reach more than 100 times. So this novel method is especially used in nanofabrication. As shown in Figure 14.7a, when polarized light irradiates onto the surface of the probe, the distribution of electron charges will vary with the polarized directions. Figure 14.7b shows the energy distribution when incident light is P-polarized [7].

14.3.2 PROBE HEIGHT CONTROL SYSTEM

In order to bring the probe to the near-field area to control the distance between probe and sample stably with nanometer sensitivity, probe height detection and control methods must be adopted. Among these, the most widely used one is the so-called shear force method.

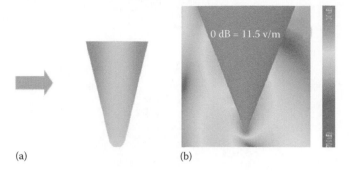

(a)   (b)

FIGURE 14.7   Schematic diagram of an apertureless probe and distribution of transient electric field. (a) Schematic diagram of an apertureless probe and (b) distribution of transient electric field.

When the probe moves close to the sample surface in several tens of nanometers, vibrating with the frequency of mechanical resonance parallel to the surface, lateral shear force will generate from the interaction between the probe and the sample. Therefore, the vibration will be attenuated due to the damping of the shear force. Thus, the amplitude of vibration can reflect the distance value between the probe and the sample. By introducing the feedback system to control the amplitude, the probe height control can be achieved. The methods for detecting the vibration of the probe can mainly be divided into two types: optical methods and non-optical methods.

In optical methods, another light source is adopted to detect the amplitude and phase of the probe vibration to control the distance between the probe and the sample. However, since the intensity of the detecting light is stronger than in the optical probe, it will cause great disturbance to the near-field signal. Now, non-optical methods are more widely adopted in SNOM. For instance, fixing an optical fiber probe on one leg of the quartz tuning fork, and the other leg glued on a piezoelectric ceramic tube. Applying voltage can make the tuning fork vibrate with its resonance frequency parallel to the sample surface. The distance variation between the probe and the sample will cause a change in the amplitude of the tuning fork. As a result, the distance between the probe and the sample can be controlled.

To avoid the short comings of the tuning fork's fragility and difficulty in assembling, in 1996, J. Barenz proposed that optical fiber can be inserted into minute piezoelectric ceramics divided into four split electrodes [8]. When the driving signal is applied on one quadrant, the vibration signal obtained from the other three can be used to detect shear force. However, it is hard to detect the small harmonic peak under the background of huge vibration peaks. In order to detect such harmonic peaks, a novel structure of exciting film-foil detector-probe can be used (Figure 14.8), where two pieces of piezoelectric ceramics are introduced, the thicker one for excitation, while the thinner one with optical fiber probe attached for detecting the vibration of the probe. When resonance happens, higher voltage output can be obtained [9].

14.3.3   CALIBRATION OF SNOM

14.3.3.1   Calibration of Resolution

SNOM is widely used because of its super spatial resolution. It is a great task to evaluate the resolution qualitatively. Now people usually use some methods adopted for calibrating high-resolution optical microscopes. For instance, high-resolution one- or two-dimensional diffraction grating, for which micro-fluorescent balls can be used. Recently, a work group was set up for "Standards on the definition and calibration of spatial resolution of NSOM/SNOM" in ISO/TC201/SC9. This proposal is now under discussion, and it will be standardized in about 2 years.

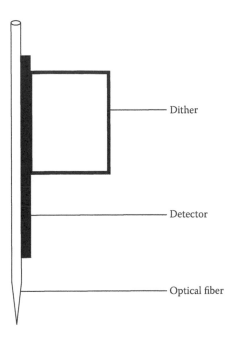

FIGURE 14.8   A kind of experiment scheme for shear force control. (From Wang, K. et al., *J. Chin. Elect. Microsc. Soc.*, 18, 124, 1999. With permission.)

14.3.3.2   Elimination of Optical False Images

Generally speaking, the image obtained from a SNOM carries a lot of information. In other words, there are many parameters that will affect the final SNOM image, such as the optical properties of the sample, the geometric structure of the sample, the scanning parameters, the surrounding environmental conditions, etc. This means we must abstract useful information from the raw data. An interesting example can be found from Ref. [10]. In the simulation, the image shows a valley position, but it corresponds to the assumed peak position. After reconstruction, the image will be more close to the original structure. A brief introduction to this reconstruction is as follows.

Supposing that the surface equation of the contact surface is $z = f(x, y)$, which can be written in Fourier form as follows:

$$f(x,y) = \iint dp dq g(p,q) \exp\left[i(px + qy)\right]. \tag{14.2}$$

By introducing the boundary conditions, we get a coupled algebraic equation that can be solved by iteration. It is known that

$$U(x,y) = H(x,y) * I(x,y),$$

where
  $U(x, y)$ stands for the detected signal where the probe tip is located
  $H(x, y)$ is the convolution function of the detection apparatus
  $I(x, y)$ is the near-field intensity, which is proportional to $|E|^2$
  E is the amplitude of electric vector

FIGURE 14.9 Schematic diagram of the instrumentation of AF/PSTM. (From Wu, S. et al., *Proc. SPIE*, 4923, 21, 2002. With permission.)

and

$$I(x, y) = F^{-1} \left\{ \frac{F[U(p,q)]}{F[H(p,q)]} \right\}, \tag{14.3}$$

where F and F^{-1} stand for Fourier transformation and its inverse transform, respectively. After calculation, the final form of the equation is

$$g(p,q) = \frac{\left| A^0 \right|^2}{A^0 \cdot A^1(p,q) + A^{0*} \cdot A^{1*}(-p,-q)} \times F \left\{ F^{-1} \left[\frac{\tilde{U}(p,q)}{\tilde{H}(p,q)} \right] \middle/ F^{-1} \left[\frac{\tilde{U}_0(p,q)}{\tilde{H}(p,q)} \right] - 1 \right\}. \tag{14.4}$$

Furthermore, hardware can also be used to eliminate optical false images. Professor S.F. Wu proposed a method of illuminating by π-symmetry dual-beam [11]. And the invention of atomic force/photon scanning tunneling microscope (AF/PSTM) is used to obtain two images of AFM (topography and phase image) and another two optical ones (transmitting efficiency and refractive index image), which can eliminate the optical false image effectively. Figure 14.9 shows the scheme of the instrumentation of AF/PSTM; when incident light irradiates from 0-direction and π-direction simultaneously, optical false image can be eliminated.

14.4   APPLICATIONS

14.4.1   SIMULATION BASED ON FDTD METHOD

The algorithm of finite-difference time-domain (FDTD) method proposed by K.S. Yee in 1996, is simply a mathematical method solving Maxwell equations related to the time domain deduced from the finite element method. The FDTD method makes discretization of the Maxwell equation both in the time domain and the spatial domain to substitute the original differential equation with a finite difference scheme.

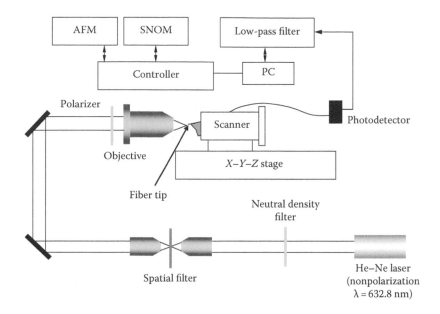

FIGURE 14.10 Schematic diagram of the experimental setup for measurement of the intensity distribution at the focus. (From Fu, Y.H. et al., *J. Microsc.*, 210, 225, 2003. With permission.)

The FDTD method can directly deal with the Maxwell equations avoiding the extra use of mathematical tools. Therefore, the FDTD method is widely used in the simulation of near-field optics. For example, using the FDTD method to simulate the light propagation in an optical fiber probe with aperture, the relationship between the spatial resolution and the collecting efficiency of the aperture diameter can be obtained [12].

14.4.2 MEASUREMENT OF THE PROFILE OF MICRO-FOCUSED SPOT

Near-field optics can also be used to measure the profile of focused laser spots, which is difficult to do by traditional optical method. Y.H. Fu et al. introduced their experimental setup for measuring the intensity distribution of a micro-spot [13]. An object lens with a high numerical aperture (NA) was used to focus the incident light, and a tapered probe of the SNOM system (Figure 14.10) was used to detect the intensity of the focal field. The measured spatial resolution was determined by the size of the near-field optical fiber probe (usually, <100 nm).

The results show an asymmetric distribution of the focused intensity with the linear-polarized laser beam due to the aberration effects of the focusing lens, which means that polarization and aberration play important roles in the focusing of a high-NA lens.

14.4.3 SUPER-RESOLUTION IMAGING

SNOM is famous for its high-resolution power to break through the diffraction limit. However, its weakness is also obvious, due to its fragility and low transmitting efficiency. To avoid these problems, SNOM can be combined with the conventional confocal laser scanning microscope (CLSM) to achieve super resolution [14]. It can be switched between SNOM mode and CLSM mode easily and rapidly, which makes the detection efficiency high enough to investigate a single fluorochrome for it can be applied to biomedical field. The resolution of 300 nm can be reached with this kind of scanning confocal microscope. Figure 14.11 shows the scheme of a SNOM/CLSM.

In order to solve the problems caused by conventional probes, the virtual probe concept, which is based on the principle of evanescent wave interference, was introduced into near field [15–17].

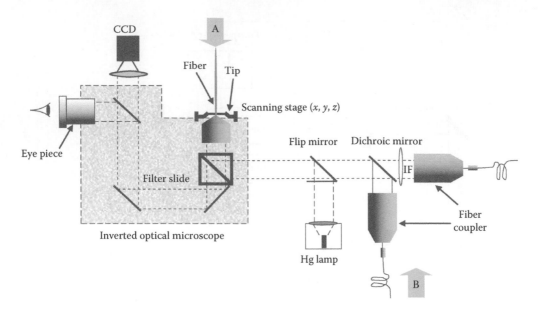

FIGURE 14.11 Schematic diagram of the SNOM/CLSM system. (From Freyland, J.M. et al., *Microelectron. Eng.*, 53, 653, 2000. With permission.)

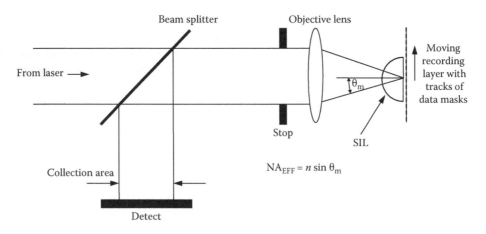

FIGURE 14.12 Schematic diagram of a kind of solid immersion lens (SIL) system. (From Milster, T.D. et al., *Jpn. J. Appl. Phys.*, 40, 1778, 2001. With permission.)

It has been demonstrated by simulation that the transmitting efficiency of virtual probes is about 10 times higher than that of conventional probes; its effective working distance is about 10 times longer than that of conventional probes; and the full width at half maximum (FWHM) of the central peak of the virtual probe remains constant over a certain range. The virtual probe can be used in optical storage, nanolithography, near-field imaging, near-field measurement, etc. Figure 14.12 shows an application example of the virtual probe in SIL near-field recording [18].

14.4.4 NANOLITHOGRAPHY

In the information industry, "smaller, faster, and cheaper" is pursued. Miniaturization is the most important driving force. With the progress of micro-nanotechnology, lithography at nanometer scale resolution is required. Nanolithography is an essential basis for microelectronic-integrated

FIGURE 14.13 Nanofabricated lines on PMMA by femtosecond laser processing.

manufacturing and micro-electro-mechanical system (MEMS), containing all kinds of nontraditional processing and high-energy beam processing.

At the end of the last century, a nanolithography method of near-field-enhanced laser based on the theory of SPM was developed. A laser beam irradiating onto the SPM tip surface, locally enhanced optical field will be generated beneath the probe, to do the processing on sample [7,19,20]. Heat expansion shows little influence on the femtosecond laser processing based on AFM, while the effects on the surface of the sample are mainly caused by the interaction between the locally enhanced optical field and the sample, which will not scratch the surface mechanically. Figure 14.13 shows a series of lines obtained when pulse energy is 6 nJ, and the scanning speed is 0.2 μm/s, with FWHM less than 100 nm and height of lines about 4 nm. All the processing will be done within 3 min. An increase in energy density of manufacturing will increase the height and width of poly (methyl methacrylate) (PMMA) lines.

Recently, a new approach has been developed for nanolithography using nanoaperture. The FTDF method can be introduced for analyzing the excitation of surface plasmon, and designing nanoaperture with the shape of H (Figure 14.14) to gain the resolution at nano-order for nanolithography.

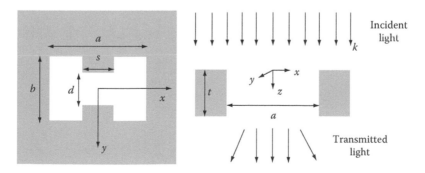

FIGURE 14.14 Schematic diagram of nanoaperture with the H-shape aperture.

14.4.5 NEAR-FIELD STORAGE

Nowadays, the optical method and magneto-optical (MO) method cannot achieve higher storage density on account of the diffraction limit. Since 1992, when E. Betzig first realized the storage of MO method in near field [21], a lot of contributions have been made to increase the density of storage.

In 1996, B.D. Terris et al. applied SIL to the technology of near-field storage [22]. They obtained a spot with the size of 317 nm, by using a SIL in glass whose refractive index is 1.83 illuminated by the light (λ = 780 nm). Figure 14.15 shows the device used in the experiment. It was proved to be possible to achieve 125 nm spots and the storage densities approaching 40 Gb/in.^{2} with shorter wavelength light.

Then, phase-change recording layers based on conductive AFM were developed [23]. The method has provided an opportunity for studying the recorded marks on phase-change rewriteable optical disks in depth, especially for marks on super-resolution near-field structure disks. Moreover, using this method, recording marks within 100 nm has been achieved.

The problem of the storage of high spatial resolution has been settled; however, difficulties still exist in the control of distance between the probe and the medium less than a wavelength, and the rapid speed of scanning. As we know, the speed of near-field recording is in the order of micrometers per second, while that of digital video disk (DVD) is higher than 3.5 m/s.

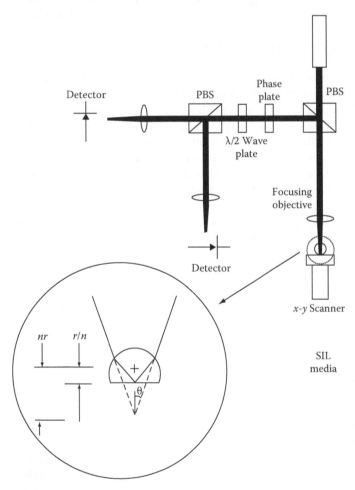

FIGURE 14.15 Schematic diagram of the optical storage test setup with the SIL device. (From Terris, B.D. et al., *Appl. Phys. Lett.*, 65, 388, 1994. With permission.)

Besides, SNOM is also widely used to study the spectroscopy of a single molecule in biomedical field [24], or used to obtain both the morphology and optical properties of the photonic crystals (PCs) simultaneously in the field of metrology [25].

In summary, near-field optics plays an active and important role in nanofabrication and nanometrology fields. It will have a brighter future with the development of novel methods and technologies, especially those for high spatial resolution, high scanning speed, and for parallel and mass production.

REFERENCES

1. G. Binning, H. Rohrer, C. Gerber et al. 1982. Surface studies by scanning tunneling microscopy. *Physical Review Letters* 49: 57–61.
2. D. W. Pohl, W. Donk, and M. Lanz. 1984. Resolution of internal total reflection scanning near-field optical microscopy. *Applied Physics Letters* 44: 651–653.
3. M. Ohtsu and K. Kobayashi. 2003. *Optical Near Field: Introduction to Classical and Quantum Theories of Electromagnetic Phenomena at the Nanoscale*. Berlin, Germany: Springer.
4. D. Courjon, J.-M. Vigoureux, M. Spajer, K. Sarayeddine, and S. Leblanc. 1990. External and internal reflection near field microscopy: Experiments and results. *Applied Optics* 29: 3734–3740.
5. M. Ohtsu (Ed.). 1998. *Near-Field Nano/Atom Optics and Technology*. Berlin, Germany: Springer.
6. E. Jin and X. Xu. 2008. Focussed ion beam machined cantilever aperture probes for near-field optical imaging. *Journal of Microscopy* 229: 503–511.
7. L. Zhu, F. Yan, Y. Chen et al. 2007. Femto-second laser induced nano structuring on PMMA. In *Sixth Asia-Pacific Conference on Near-Field Optics*, Yellow Mountain, China.
8. J. Barenz, O. Hollricher, and O. Marti. 1996. An easy-to-use non-optical shear-force distance control for near-field optical microscopes. *Review of Scientific Instruments* 67: 1912.
9. K. Wang, X. Wang, N. Jin et al. 1999. The study of the height regulation technique of SNOM optical probe tip (in Chinese). *Journal of Chinese Electron Microscopy Society* 18: 124–128.
10. Z. Jin, W. Yi, and W. Huang. 1998. Reconstruction of the surface topography by detected SNOM/PSTM signals. *Applied Physics A* 66: 915–917.
11. S. Wu, J. Zhang, Y. Li et al. 2002. The development of AF/PSTM. *Proceedings of SPIE* 4923: 21–25.
12. H. Nakamura, T. Saiki, H. Kambe et al. 2001. FDTD simulation of tapered structure of near-field fiber probe. *Computer Physics Communication* 142: 464–467.
13. Y. H. Fu, F. H. Ho, W. C. Lin et al. 2003. Study of the focused laser spots generated by various polarized laser beam conditions. *Journal of Microscopy* 210: 225–228.
14. J. M. Freyland, R. Eckert, and H. Heinzelmann. 2000. High resolution and high sensitivity near-field optical microscope. *Microelectronic Engineering* 53: 653–656.
15. T. Hong, J. Wang, L. Sun et al. 2002. Numerical simulation analysis of a near-field optical virtual probe. *Applied Physics Letters* 81: 3452–3454.
16. T. Hong, J. Wang, L. Sun et al. 2004. Numerical and experimental research on the near-field optical virtual probe. *Scanning* 26: 1–6.
17. Q. Sun, J. Wang, T. Hong et al. 2002. A virtual optical probe based on evanescent wave interference. *Chinese Physics* 11: 1022–1027.
18. T. D. Milster, F. Akhavan, M. Bailey et al. 2001. Super-resolution by combination of a solid immersion lens and an aperture. *Japanese Journal of Applied Physics* 40: 1778–1782.
19. E. Jin and X. Xu. 2005. Obtaining super resolution light spot using surface plasmon assisted sharp ridge nanoaperture. *Applied Physics Letters* 86: 111106.
20. E. Jin and X. Xu. 2006. Enhanced optical near field from a bowtie aperture. *Applied Physics Letters* 88: 153110.
21. E. Betzig, J. K. Trautman, R. Wolfe et al. 1992. Near-field magneto-optics and high density data storage. *Applied Physics Letters* 61: 142.
22. B. D. Terris, H. J. Mamin, and D. Rugar. 1994. Near-field optical data storage using a solid immersion lens. *Applied Physics Letters* 65: 388–390.
23. S. K. Lin, I. C. Lin, S. Y. Chen et al. 2007. Study of nanoscale recorded marks on phase-change recording layers and the interactions with surroundings. *IEEE Transactions on Magnetics* 43: 861–863.
24. M. Nirmal, B. O. Dabbousi, M. G. Bawendi et al. 1996. Fluorescence intermittency in single cadmium selenide nanocrystals. *Nature* 383: 802–804.
25. B. Jia, J. Li, and M. Gu. 2006. Near-field characterization of three-dimensional woodpile photonic crystals fabricated with two-photon polymerization. In *90th USA Annual Meeting*, Rochester, NY.

Section III

Practical Applications

Length and Size

René Schödel

CONTENTS

15.1 Introduction 405
15.2 Basic Principle of Length Measurements by Interferometry 406
15.3 Basic Types of Length Measuring Interferometers 407
    15.3.1 Twyman–Green Interferometers 407
    15.3.2 Fizeau Interferometers 409
    15.3.3 Fringe-Counting Interferometers 410
15.4 Requirements for Accurate Length Measurements by Interferometry..................410
    15.4.1 Light Sources 410
    15.4.2 Generation and Alignment of Plane Waves..412
    15.4.3 Determination of the Refractive Index of Air 414
    15.4.4 Importance of Temperature Measurements...417
15.5 Interferometry on Prismatic Bodies 417
    15.5.1 Imaging of the Body onto a Pixel Array 418
    15.5.2 Interferogram Analysis..419
    15.5.3 Length Evaluation...421
    15.5.4 Examples of Precise Length Measurement Applications........................422
15.6 Additional Corrections and Their Uncertainties 427
    15.6.1 Optics Error Correction 427
    15.6.2 Surface Roughness and Phase Change on Reflection.............................429
    15.6.3 Wringing Contact between a Body and a Platen 429
    15.6.4 Platen Flexing 430
15.7 Conclusions..430
References..431

15.1 INTRODUCTION

The provision of length standards and the ability to measure length to a required accuracy are of fundamental importance to any technologically developed society. Throughout history, there have been many standards for length beginning with simple definitions based on the human body, for example, cubit and feet. The continuing refinement of standards led to more specific definitions and more accurate methods of realizing them. As a milestone, in 1887, Michelson proposed the use of optical interferometers for the measurement of length. However, it needed many years until the meter was defined in terms of the vacuum wavelength of light from a krypton lamp. In 1960, this definition replaced the International Prototypes deposited in 1889 at the Bureau International des Poids et Mesures (BIPM), where they remain today. Since 1983, the meter, one of the seven base units of the SI, is defined as the length of the path traveled by light in vacuum during a time interval of 1/299,792,458 of a second [1]. This definition is based on the availability of primary frequency standards (atomic clocks) defining a second accurately. It opens two alternative ways for the realization of length measurements: (1) propagation delay: the length l is the path traveled in vacuum by a plane electromagnetic wave in time t, which is obtained using the relation $l = c_0 t$, and the value of the speed of light in vacuum $c_0 = 299,792,458$ m s^{-1}; (2) interferometry: by means of the vacuum wavelength λ_0 of a plane electromagnetic wave of frequency f, which is obtained from the relation $\lambda_0 = c_0/f$.

The frequency of atomic clocks serving as primary standards can be transferred to frequencies of laser radiation used in length measurements by optical interferometry. This transfer is realized by frequency chains or more modern techniques utilizing femtosecond laser frequency combs. There are two ways for providing laser sources whose frequency is traceable to the SI. The common way is the use of recommended radiations generated by lasers whose frequencies are stabilized to selected hyperfine absorption lines. Currently, 12 reference frequencies covering the visible and infrared regions of the electromagnetic spectrum are recommended by the Comité International des Poids et Mesures (CIPM) [1]. As an alternative to the usage of recommended radiations, the laser light source may be directly synchronized with the primary frequency standard by femtosecond laser frequency combs.

The practical realization of length measurements depends upon the application. The propagation delay method is basically useful for long distances, for example, in space, while for the calibration of secondary length standards, for example, gauge blocks, interferometry by means of known wavelengths is preferable. The length of these primary calibrated material artifacts basically transfers the SI unit of the length to the industry and society to be used subsequently in mechanical calibrations based on comparison length measurements. Almost any relevant usage of length standards is applied under air, that is, not under vacuum conditions. Accordingly, the primary calibration of the material artifacts has to be performed under air; otherwise, the length is affected by the air pressure due to the material's compressibility. This means that the actual relations $l = ct$ and $\lambda = c/f$ have to be considered in the primary calibration measurements. Here, the speed of light is reduced by the refractive index n of air according to $c = c_0/n$. Accordingly, the accurate evaluation of the air refractive index is one of the key points for the transition from the SI definition of the meter to the actually provided length standards. Other key points regard the use of adjustment methods to make sure that the SI definition is realized properly, for example, that plane waves are used. Besides these points, the laser frequency itself is no longer the limiting factor regarding length measurement uncertainty, provided that stabilized lasers are used as mentioned earlier [2]. This conclusion even holds for measurements under vacuum conditions. Therefore, the often found designation of stabilized lasers serving as secondary frequency standards as "primary length standards" is misleading.

15.2 BASIC PRINCIPLE OF LENGTH MEASUREMENTS BY INTERFEROMETRY

The basic principle of length measurement by interferometry is the comparison of a mechanical length (or a distance in space) against a known wavelength of light. Commonly, the optics is arranged such that the light beam double-passes the required length. Therefore, the measurement units are half-wavelengths, and the length being measured is expressed as

$$l = (i+q)\frac{\lambda}{2}, \tag{15.1}$$

where
  λ is the wavelength
  i is the integer order
  q is the fractional order of interference

The idea of expressing the length as multiples of the half-wavelength bases on the observation of interference intensities of two or multiple light waves is written as

$$I = I_0 \left[1 + \gamma \cos\left(\frac{2\pi}{\lambda/2} (z_1 - z_2) \right) \right] \tag{15.2}$$

in the case of two-beam interference. Here, γ is the interference contrast, and $z_1 - z_2$ represents the geometrical distance difference between the pathways of the two beams. Thus, a change of one of the path lengths changes the interference intensity periodically in units of $\lambda/2$. The phase difference $2\pi/(\lambda/2)(z_1 - z_2)$ in Equation 15.2 may be expressed in terms of the vacuum wavelength, that is, by $2\pi/(\lambda_0/2)n(z_1 - z_2)$. Here, n is the (air) refractive index within the path, and $n(z_1 - z_2)$ represents the difference between the two optical pathways instead of difference between the geometrical distances.

While the fractional order of interference, q, is mostly extracted from a measured interferogram, the integer order of interference, i, is usually derived from other measurements or using multiple wavelengths as described in Section 15.5.3.

The wavelength of visible light is typically 400–700 nm, and hence, the basic unit of measurement is 200–350 nm in size. By careful measurement and analysis of the interference orders (fringes), measurement uncertainties of less than 1/1000 of a fringe can be achieved with sophisticated interferometers. Extension of the wavelength range toward smaller wavelengths has the potential of obtaining picometer uncertainties under vacuum conditions. However, limitations in the practical realization, for example, due to interferometer alignment, remain the same, and new limitations may appear such as the nonavailability of appropriate monochromatic light sources or problems with optical components.

15.3 BASIC TYPES OF LENGTH MEASURING INTERFEROMETERS

15.3.1 TWYMAN–GREEN INTERFEROMETERS

The Twyman–Green interferometer is a Michelson interferometer using a collimated beam. The beam is split into two beams: one reaching a reference mirror and the other an object under test. Each beam is reflected and split again at the beam splitter, resulting in interference. The interference pattern (interferogram) is usually observed at the interferometer's output. When the object under test is represented by a flat mirror, and the tilt angles with respect to each incident beam are the same, the entire field of view appears homogeneously illuminated and appears bright in cases of a path difference of $m\lambda/2$, and dark for differences of $(m+(1/2))\lambda/2$, where m is an integer. This situation changes when one of the mirrors is tilted. In this case, straight fringes appear along the field of view, the number and orientation of which are strongly dependent on the tilt. When the object under test is not ideally flat, the fringes are generally bent and no longer equally spaced. Such fringe pattern may be used for qualitative testing of surfaces as well as for quantitative analysis of surface topographies.

For length measurements, a prismatic body can be set in the measurement pathway as indicated in Figure 15.1. The back face of the body is attached to a platen acting as plane mirror as the front face does. The lateral offset of the fringes as shown in Figure 15.1 directly indicates the fractional order of interference, while the integer order is unsearchable because of the interferogram discontinuity induced by the length of the body.

Unwanted secondary reflection from the beam splitter's second side can be suppressed by an antireflection coating. However, even small residual reflections lead to disturbing interferences. An effective way to eliminate the secondary reflection is to use a wedged beam splitter. This makes the direction of the secondary reflection different from the main direction, and the resulting secondary

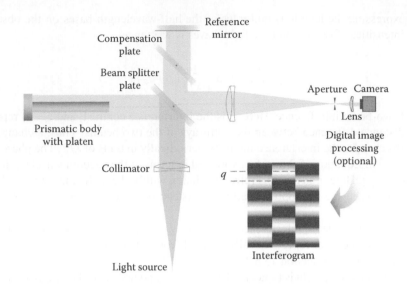

FIGURE 15.1   Diagram of a Twyman–Green interferometer. A prismatic body is placed in the measurement pathway. Unwanted reflections, for example, from the back sides of the beam splitter, can be eliminated by using wedged plates in combination with an aperture in the focal point of the first output lens. The inset shows a typical interferogram covering the body and the platen that illustrates how the fractional order of interference, q, can be extracted visually.

spot in the focal plane of the focusing lens can be blocked by an aperture of appropriate size. The wedge angle ε of the beam splitter causes a small deflection between the interfering beams, that is, $\varepsilon\left(n_{BS}-1\right)$, where n_{BS} is the refractive index of the beam splitter. When using a single wavelength, this deflection is compensated by appropriate adjustment of the reflecting surfaces. However, when multiple wavelengths are used, the dispersion of the beam splitter causes a variation of the number of observed fringes. This dispersion effect can be compensated by inserting a wedged compensation plate into the reference pathway, the wedge angle, its orientation, and the substrate material of which have to be identical with those of the beam splitter. It should be mentioned that the arrangement shown in Figure 15.1 is the only possibility. Here, the reflecting surface of the beam splitter is directed to the reference mirror. Even though in the opposite case the dispersion effect can be compensated as well, both plates together constitute a parallel plate, that is, the separation of the unwanted secondary reflection at the beam splitter is negated causing unwanted secondary interferences.

When the reference pathway is shifted, any fringes within the interferogram will move according to Equation 15.2. This is applied in phase-stepping interferometry [3], which is useful in large field Twyman–Green interferometers equipped with camera systems [4]. In such an approach, a number of interferograms are recorded at different but equally spaced positions along the reference pathway. From such a set of interferograms, the interference phase topography can be calculated by appropriate algorithms. In this way, the fractional order of interference, and therefore the length, can be obtained more accurately as with fringe evaluation techniques using single interferograms [5]. An example of applying phase-stepping interferometry in length measurements is given in Section 15.5.2.

A more compact interferometer design, called "Kösters–Zeiss interference comparator," is obtained when a Kösters prism is used instead of a "normal" beam splitter [6] (see Figure 15.2). Such design is particularly useful for length measurements of relatively long prismatic bodies, for example, gauge blocks of 1 m in length. The vacuum cell shown in Figure 15.2 enables direct comparison of the path length in vacuum with the length of the optical path of air surrounding the cell. Since the air pathway is situated next to the length bar, the refractive index of air resulting from these path length differences almost coincides with that of the light pathway along the bar.

FIGURE 15.2   Diagram of the Kösters–Zeiss interference comparator.

15.3.2   FIZEAU INTERFEROMETERS

Fizeau interferometers are common large field interferometers requiring a minimum of optical components as shown in Figure 15.3. A single beam entering the interferometer is semi-reflected at the first surface, while the other is mostly totally reflecting. Multiple beams are generated by multiple reflections and can be separated from the incident beam by a beam splitter plate for analysis of the multibeam interference. Alternatively, when using two semi-reflecting surfaces, the transmitted interference may be observed, although its contrast is clearly smaller than that of the reflected interference. As in the Twyman–Green interferometer, a number of fringes are observed when there is a tilt between the reflecting faces. However, the intensity distribution of the fringes, described by the well-known airy formula, is sensitive to the reflectivity R of the semi-reflecting surface as indicated in Figure 15.3, that is, the shape is similar to a cosine (see Equation 15.2) for small values of R, while at large values, sharp peaks can be observed. As mentioned earlier, unwanted reflections leading to disturbing interferences can be eliminated by using wedged flats as optical components (including the beam splitter) in combination with an aperture.

Length measurements with a Fizeau interferometer can be performed similarly as with the Twyman–Green interferometer by inserting a prismatic body attached to a platen (see Figure 15.1) and determining the fractional order of interference from a displacement of fringes. When applying phase-stepping interferometry by recording and processing a number of interferograms at

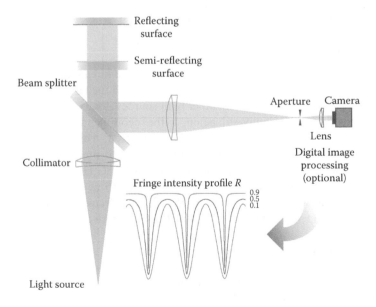

FIGURE 15.3   Diagram of a Fizeau interferometer.

equally spaced faces positions, it is recommended to use a special algorithm [7] for Fizeau interference instead of using an algorithm designed for the two-beam interference. Errors are generated otherwise.

The Fizeau interferometer can be extended to a double-ended version applying two reversely directed light sources [8]. In such a design, a prismatic body with reflecting surfaces (without platen) is set between two semi-reflecting faces. The length of the body can then be determined from the difference of the distance between the semi-reflecting faces and the sum of distances with respect to the body's faces at each side, where each distance is obtained from the relevant interferogram. One application of such length measurement is the determination of the volume of cuboids in the so-called cube interferometer.

Another double-ended Fizeau interferometer is the so-called spheres interferometer [9]. Here, spherical waves are used together with spherical semi-reflecting reference faces to determine a diameter topography from which the volume of spheres can be obtained highly accurately.

15.3.3 FRINGE-COUNTING INTERFEROMETERS

These dynamic types of interferometers provide relative measurements of distances, which is given attention to in Chapter 16. Here, just a few remarks relating to absolute length measurements are made: when such systems are used to define a length, accurate setting at two defined points is required. In the measurement of absolute lengths, the optical path difference of the fringe-counting interferometer must be increased by a distance equal to the length of the object to be measured. This can be done by traversing a mirror on a linear stage between two points coincident with the two ends of the object. Fringe-counting interferometers can be found as heterodyne systems where the beat between two separated laser frequencies is observed [10]. This technique is highly appropriate for the primary calibration of line scales as interpolation errors are reduced. Another variation of the fringe-counting technique is the combination with white light interferometry. In this case, interference can be observed only when the optical path length difference between the interfering beams is around zero. This criterion can be used to define certain positions along a displacement pathway.

15.4   REQUIREMENTS FOR ACCURATE LENGTH MEASUREMENTS BY INTERFEROMETRY

15.4.1   LIGHT SOURCES

The availability of a monochromatic light source is one of the key points when interferometry is applied for length measurements. Before laser radiation entered the laboratories, it was a big challenge to generate monochromatic light from other light sources, for example, from special spectral lamps. Although specific spectral lamps still belong to the recommended light sources, they almost disappeared, these light sources meanwhile almost disappeared, while the list of stabilized laser radiations recommended by the CIPM is growing from time to time. One example of a recommended radiation is the radiation of a frequency-doubled Nd:YAG laser, stabilized with an iodine cell external to the laser, having a cold-finger temperature of $-15°C$. Here, the absorbing molecule is $^{127}I_2$, and the transition is the a_{10} component, R(56) 32-0 at a frequency of 563,260,223,513 kHz corresponding to a vacuum wavelength of 532.245036104 nm with a relative standard uncertainty of 8.9×10^{-12}.

There is a close link between the spectral linewidth of a light source and temporal coherence. Accordingly, it is not surprising that stabilized laser light sources as mentioned earlier have superior long coherence times. Temporal coherence is an important criterion for the observation of interference in length measurements as the interfering beams travel different lengths.

The size of the light source is also important in length measurements by interferometry. This regards different aspects as (1) the divergence of the beam affects the length measurements as discussed in Section 15.4.2 and (2) the emittance of the finite light source is desired to be large. When an aperture or a pinhole is used to reduce the size of the light source, as necessary with spectral lamps, the signal-to-noise ratio of the interference intensity signals is limited by the pinhole size, (3) the spatial coherence of the light beam. This in principle means the ability of the beam to interfere with a shifted copy of the beam. Spatial coherence is necessary when interference is formed between different parts of beams as in a shearing interferometer. A nearly perfect point source, for example, monochromatic light coming from a single-mode fiber, will generate maximum spatial coherence. On the other hand, spatial coherence is not necessary when using a non-sheared Twyman–Green arrangement together with rectangular incidence of the beams to the plane reflecting faces. In fact, in the case of the Twyman–Green interferometer, it is even advantageous that spatial coherence can be reduced. This can be done in order to lower the risk of parasitic interferences generated by any additional reflections within the interferometer's beam path. An advisable light source generating reduced spatial coherence is a multimode fiber end. The mode field intensity distribution in such case reveals a characteristic grain as shown in the photograph in Figure 15.4a. Figure 15.4b shows the intensity distribution observed when the fiber is vibrated at 500 Hz by a speaker's membrane.

In order to illustrate the effect of parasitic interferences, Figure 15.5 shows phase topographies resulting from measurements in a Twyman–Green interferometer equipped with a 16-bit CCD-camera system applying phase-stepping interferometry. The phase topographies each resulting from eight interferograms, recorded at 0.5 s exposure, show the total field of view of approximately 60 mm when an almost ideal optical flat is inserted and aligned so that zero fringes are observed (fluffed out mode). Figure 15.5a shows the interference phase topography when a single-mode fiber is used as light source. In this case, obscure vertical variations appear, the amplitude of which is about ±0.03 interference orders corresponding to about ±8 nm ($\lambda \cong 532$ nm). Figure 15.5b shows the same field of view when the single-mode fiber is replaced by a vibrated multimode fiber as light source. The parasitic interference pattern disappears in the latter case, and the remaining optics error of the interferometer becomes visible.

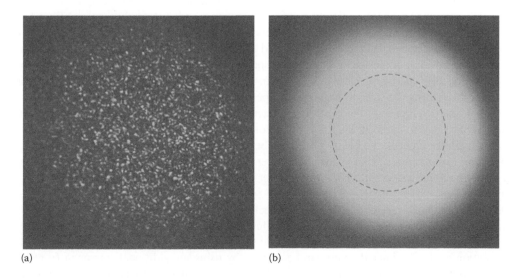

(a)   (b)

FIGURE 15.4 Photographs of light that traveled through a multimode fiber (exposure time: 50 ms); (a) original grain pattern; (b) mixed grain pattern generated by fiber vibration. The dashed circle indicates the selected beam part entering the collimator of a Twyman–Green interferometer.

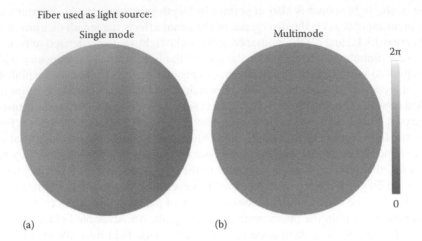

FIGURE 15.5 Phase topographies of an optical flat mirror measured in a Twyman–Green interferometer using phase-stepping interferometry with different fiber types as indicated in the figure. In the case of the single-mode fiber, vertical variations with amplitudes of about ±0.03 interference orders appear.

Parasitic interferences as shown in the earlier example may not appear in general when using single-mode fibers. However, it should be mentioned that their origin could not be identified in the example of Figure 15.5. This means that even when the expected sources of parasitic interferences are switched off (wedged optics, etc.), there still may exist unwanted reflections affecting the measured interference when a spatial coherent light source is used.

15.4.2 GENERATION AND ALIGNMENT OF PLANE WAVES

The realization of the meter in terms of plane electromagnetic waves requires accurate adjustment procedures. A first requirement regards the preparation of the waves itself, that is, the generation of a collimated beam. The collimation of a beam is limited by the divergence of the light source. When laser beams are used directly as a light source, it is recommended to apply a spatial filter for the following reasons: (1) unwanted residual transversal modes of the laser are suppressed effectively, (2) beams can be expanded at the same time, and (3) collimation alignment can easily be adjusted by tuning the position of the second lens of the spatial filter. A wedged shearing plate can be used to generate equally spaced interference fringes in a reflected beam. The beam is collimated properly when the fringes are straight and oriented perpendicular to the wedge orientation. After this adjustment, the divergence of the beam is predetermined by the spatial filter, that is, by the small pinhole size related to the focal length of the second lens acting as a collimator. Instead of observing shearing interferences, the beam profile can be observed and adjusted to be invariant along the propagation direction over a large distance. Alternatively to the usage of spatial filters, the laser may be coupled into an optical fiber, and the fiber output can be set in the focal point position of a collimating lens.

Assuming plane waves; their propagation direction with respect to the direction of the length to be measured must be coincident. Otherwise, as illustrated in Figure 15.6, an angle α causes a cosine error, that is, the length, $\tilde{l} = l \cos \alpha$, is measured to short. In length measurements by interferometry the direction of the collimated beam can be aligned by using an additional mirror in front of the interferometer. Alternatively, when an optical fiber is used as mentioned earlier, the fiber position within the focal plane of the collimator can be adjusted.

In order to adjust the beam direction to be perpendicular to the reflecting faces of a prismatic body, the retroreflected beam is observed in the focal plane of the collimator at the input or at the output of the interferometer. The beam is adjusted in a way that the image of the light source

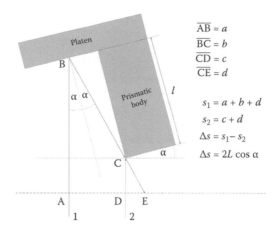

$$\overline{AB} = a$$
$$\overline{BC} = b$$
$$\overline{CD} = c$$
$$\overline{CE} = d$$

$$s_1 = a + b + d$$
$$s_2 = c + d$$
$$\Delta s = s_1 - s_2$$
$$\Delta s = 2L \cos \alpha$$

FIGURE 15.6 Derivation of the cosine error. The path difference of the two plane waves reflected at a platen and at the body's front face is compared with path length $2l$.

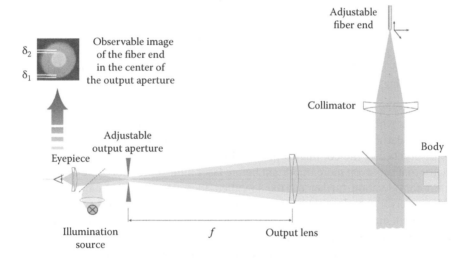

FIGURE 15.7 Visual autocollimation adjustment scheme.

appears exactly in the focal point of the collimator. Such autocollimation adjustment can be realized in different ways. Figure 15.7 illustrates a visual adjustment procedure where the exit pinhole of the interferometer is illuminated via a beam splitter by an additional light source. The light traveling through the pinhole and the collimator lens is reflected by the interferometer mirror and refocused in the focal plane of the collimator forming an image of the pinhole.

The position of the output pinhole is adjusted so that its image completely coincides with the pinhole. In a second step, the beam direction of the interferometer is adjusted so that the image of the light source, for example, the fiber end, is centered within the exit pinhole. A magnifying lens or microscope is used to observe the pinhole. Although an adequate magnification reduces alignment errors, both procedures are tainted with remaining misalignments indicated by δ_1 and δ_2 in Figure 15.7.

A more direct autocollimation adjustment method is useful when a fiber is used as light source of the interferometer. In this case, the quantity of retroreflected light passing the fiber in reverse direction is observed as shown in Figure 15.8. When this method is combined with scanning in x–y-direction, the resulting signal topography can be used to calculate a center position onto which the fiber end can be positioned more accurately as by the visual method. A further advantage of the retroreflection method is that the collimation can be inspected at the same time, that is, the signal

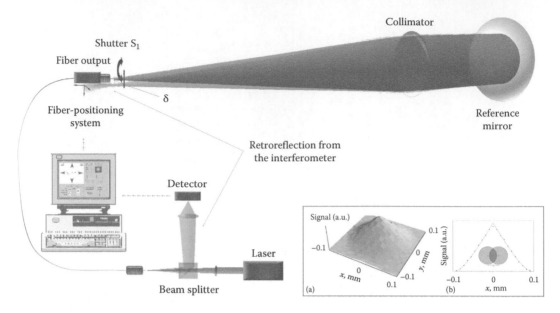

FIGURE 15.8 Scheme of the retroreflection method. The retroreflection signal is detected as a function of the x–y-position of the fiber. The inset shows a measurement example when retroreflection scanning is applied. A section in the x-direction is shown together with a theoretical curve (solid line) obtained from the overlap of two circles representing the fiber and its image.

distribution is sensitive to the z-position of the fiber. Misalignment in z-direction leads to a smooth signal distribution, while otherwise a sharp peak appears as indicated in the inset of Figure 15.8.

As mentioned earlier, the size of the light source related to the focal length of the collimator defines the divergence of the collimated beam. This leads to an effect in length measurements by interferometry that was even more important before laser light sources became available but should be taken into consideration further on: the so-called aperture correction can be derived from an integrated cosine error over the light source area. As an example, the aperture correction for a circular light source with diameter d (e.g., a fiber) positioned in the focal point of the collimator amounts to $+l\left(d/4f\right)^{2}$, where f is the focal length of the collimator.

A further effect is produced by the finite diameter of the collimated beam. Ideally, plane waves are supposed to propagate between the reflecting surfaces in the interferometer. In reality, the wave fronts are distorted by beam diffraction. For Gaussian beam profiles, the length l to be measured is affected by $-l\left(\lambda/2\pi w_0\right)^{2}$, where w_0 is the beam diameter (waist). As an example, a waist of 3 mm causes an effect of about -1 nm m^{-1}. The size of this effect is even increased by a factor of about 10 when the beam profile is homogeneous. When small beams are used in combination with single detectors, the influence of the beam diffraction entering into the integral interference pattern has to be taken into consideration. Thus, this effect is especially important in compact interferometer designs using small beams, while when expanded beams are used, as in large field imaging interferometers, beam diffraction rings can be ruled out from the interferogram analysis.

15.4.3 DETERMINATION OF THE REFRACTIVE INDEX OF AIR

As already mentioned in Section 15.1, the wavelength in air determines the basic length scale in majority of length measurements performed with interferometry. The wavelength in air results from the vacuum wavelength, λ_0, scaled down with the amount of the refractive index of air:

$$\lambda = \frac{\lambda_0}{n}. \tag{15.3}$$

Under atmospheric conditions, n is larger by about 2.7×10^{-4} than unity (vacuum). This deviation, called refractivity, in first-order scales with the atmospheric pressure. Thus, when the atmospheric pressure is determined with a relative uncertainty of 10^{-4}, which is small, the contribution to the relative uncertainty of the length measurement is still 2.7×10^{-8} (27 nm m^{-1}). This is larger than three orders of magnitude than the contribution from the uncertainty of the vacuum wavelength of iodine-stabilized laser ($\approx 10^{-11}$). Consequently, n has to be determined highly precisely in interferometric length measurements in air. However, achievement of uncertainty of the air refractive index smaller than 10^{-8} is possible only under well-defined laboratory conditions using sophisticated instruments for either measurement of environmental parameters together with appropriate equations or optical refractometers.

The physical background of the air refractive index is based, without going into detail, on the Lorenz–Lorentz equation for the refractive index of a mixture of nonpolar gases:

$$\frac{n^2-1}{n^2+2} = \sum_i R_i \rho_i, \tag{15.4}$$

where
  ρ_i is the partial density of the ith component of the mixture
  R_i is the specific refraction, which is invariant to changes of the density to a high degree and
    which is related to

$$R_i = \frac{4}{3}\pi\left(\frac{N_A}{M_i}\right)\alpha_i, \tag{15.5}$$

where
  N_A is the Avogadro constant
  M_i is the molecular weight
  α_i is the polarizability

For atmospheric air, the dispersion curves for N_2, O_2, and Ar are sufficiently similar. Furthermore, it is assumed in Ciddor's equation [11] that each CO_2 molecule replaces a molecule of O_2 and that both have the same molecular refractivity so that all these constituents may be combined in a single term on the right-hand side of Equation 15.4. Thus, only water vapor must be treated separately. Finally, the density equation for moist air, for example, the BIPM-1981/91 equation [12], can be used in order to calculate the partial densities in Equation 15.4. Certainly, the use of the density equation requires an accurate measured set of the environmental parameters such as air pressure, temperature, humidity, and CO_2 content.

Alternatively to the earlier-mentioned formalism, there exist a number of empirical equations for the refractive index of air that are called Edlén equations related to one of the scientists suggesting such equation [13]. The basis of these specialized equations are precise measurements of the air refractive index under well-defined laboratory's standard conditions, that is, these equations are valid exactly under the same standard conditions, only, and may not be used in a wide range of environmental parameters (e.g., in geodetically length measurement) where Ciddor's equation is mandatory. A valid example for an empirical modified Edlén equation can be found in [14]. Accordingly, the input parameters—wavelength, 633 nm; pressure, 10^5 Pa; temperature, 20°C; CO_2 content, 400 ppm; relative humidity, 50% (corresponding to a water vapor pressure of 1170 Pa and a dew point of 9.28°C)—lead to a refractive index of $n = 1.000267794$, which is sensitive to the input parameters as shown in Table 15.1.

The air refractive index may also be obtained directly from using refractometric methods. An example scheme of appropriate refractometer design is shown in Figure 15.9. Here, the optical pathway

TABLE 15.1

Sensitivity of the Air Refractive Index at Wavelength 633 nm to Changes of Environmental Parameters at Standard Conditions

Influence	Sensitivity	Influence	Sensitivity
Temperature	$-9.2E-07$/K	Rel. humidity	$-8.7E-09$/%
Pressure	$+2.7E-09$/Pa	Dew point	$-1.5E-08$/K
CO_2 content	$+1.4E-10$/ppm	Water vapor pressure	$-1.8E-10$/Pa

FIGURE 15.9 Scheme of refractometer design using a vacuum cell along the beam pathway of the interferometer. On the right-hand side, a measurement example of phase topography is shown.

of light traveled in vacuum is compared with the pathway in air. The vacuum cell can be inserted along the same pathway as used in the length measurement, and for highest accuracy, the same collimated light beam is used as the length measurement path as already indicated in Figure 15.2.

In such an approach, the length of the vacuum cell can be written in two alternative ways using Equation 15.1:

$$l_{cell} = \left(i_{vac} + q_{vac}\right)\frac{\lambda_0}{2} = \left(i_{air} + q_{air}\right)\frac{\lambda_0}{2n_{air}}, \qquad (15.6)$$

in which the subscripts "vac" and "air" indicate the corresponding integer and fractional orders of interference. Using Equation 15.6, the difference between the optical pathways in vacuum and in air can then be written as

$$l_{cell}\left(n_{air} - 1\right) = \left(\Delta i - \Delta q\right)\frac{\lambda_0}{2}, \qquad (15.7)$$

where
$\Delta i = i_{air} - i_{vac}$
$\Delta q = q_{air} - q_{vac}$ resulting in

$$n_{air} = \frac{\left(\Delta i - \Delta q\right)\lambda_0/2}{l_{cell}} + 1. \qquad (15.8)$$

The value of Δi may be reversely obtained from Equation 15.7 using an estimate refractive index (e.g., based on environmental parameters using an empirical equation) and setting Δq to zero.

Δq results from the interferometric measurement, for example, as indicated on the right-hand side of Figure 15.9, from the interference phase topography:

$$\Delta q = \frac{1}{2\pi}\left[\frac{1}{2}\left(\phi_{air}^1 + \phi_{air}^2\right) - \phi_{vac}\right], \tag{15.9}$$

where

ϕ_{vac} indicates the mean phase measured within a region of the vacuum pathway

$\frac{1}{2}\left(\phi_{air}^1 + \phi_{air}^2\right)$ is the average of the mean phases in the air pathways in two regions arranged symmetrically around the vacuum pathway

As shown in Figure 15.9, each pathway travels through the transparent plates at both ends of the vacuum cell. These plates should be wedged to avoid parasitic interferences. The wedge angles at both ends should be oriented reversely so that the same pathway through the plates is traveled at each position within the beam's cross section. However, flatness of the plates and the homogeneity of their refractive index are limited. A way to overcome this limitation can be realized by a correction for the influence of optical pathway differences through the plates. Such so-called zero correction can be extracted from measurements using reconciled vacuum and air pathways, for example, when the interferometer is evacuated together with the vacuum cell, or when the same (dry) gas is injected in the cell as found in the air pathway. In any case, the quality/uncertainty of this correction will limit the attainable measurement uncertainty of the air refractive index using such refractometer. It is noticed that applications of helium gas is a suggested alternative to air environment [15].

15.4.4 Importance of Temperature Measurements

Length measurements are always related to the temperature of the body. For this reason, accurate temperature measurement is very important when length measurements are performed. For example, in primary calibrations of gauge blocks by interferometry, the quantity to be determined is the length at a reference temperature t, mostly 20°C. Typically, during the length measurement by interferometry, the temperature is slightly different from the reference temperature. For this reason, a correction is necessary taking into account the (small) temperature deviation resulting in the length at the reference temperature:

$$l_{20°C} = l\left[1 - \alpha_0\left(t - 20°C\right)\right], \tag{15.10}$$

in which α_0 denotes a constant for the coefficient of thermal expansion. For example, a 1 m gauge block of steel with a typical thermal expansion coefficient of $1.2 \times 10^{-5}\,K^{-1}$ will be measured about 0.2 μm shorter at $t = 19.985°C$ than at $t = 20.000°C$. This is even about the half of an interference order and illustrates the importance of accurate temperature measurement. Depending on the body's material and the desired uncertainty of the length measurement, it may be necessary to perform the temperature measurements with sub-millikelvin uncertainty, traceable to the international temperature scale in its current version [16] as outlined at the BIPM website [17].

In the practical realization of the temperature measurements, it is important to provide a good contact of the calibrated sensors with the body, to account for possible self-heating of the sensors and for temperature gradients along the body.

15.5 INTERFEROMETRY ON PRISMATIC BODIES

While the earlier sections are more of general relevance in length measurements by interferometry, the following sections discuss in detail the special case of length measurements on prismatic bodies, for example, gauge blocks. However, the description is restricted to interferometric techniques in which large field phase-stepping interferometry is combined with imaging onto a camera array. The main advantage of this combination is in well-defined interference phase topographies resulting from the measurements. This provides not only highly accurate length/height topographies but also information about the exact lateral position of a sample body related to the camera array and, therewith, enables accurate length measurements even when the body does not exhibit ideal properties with respect to surface flatness and parallelism. The phase topography at the platen, attached to the body by wringing, may be even used to extract a correction for the flexing of the platen.

It should be mentioned that there exist other length measurement techniques based on interference signals from single detectors. Although such methods may provide a high resolution, the absence of topography and position information increases the attainable uncertainty in these techniques.

15.5.1 IMAGING OF THE BODY ONTO A PIXEL ARRAY

When a prismatic body is set in the measurement pathway of a Twyman–Green interferometer, as shown in Figure 15.1, its length causes a fringe discontinuity at the transition from the body's front face to the wrung platen. In addition, diffraction occurs at the side edges of the front face. Therefore, in the interfering beams observed at the output of the interferometer, diffraction rings may be observed around the front face. However, when the front face is imaged, which can be achieved by the addition of an appropriate lens in front of the camera, the edges will appear sharp, and the diffraction rings will disappear. This is even more obvious in the phase topography, which can be obtained by applying phase-stepping interferometry. Figure 15.10 illustrates the effect of the aperture size onto the sharpness of the phase topography for a gauge block shaped body in which reflecting front face is particularly textured. In the case of the large aperture of 5 mm (Figure 15.10b), the sharpness is limited by the pixel resolution of the camera, while at 0.5 mm (Figure 15.10c), the sharpness clearly decreases (aperture in the focal point of a lens with $f = 700$ mm).

Concerning sharpness in the framework of length measurements by interferometry, there are two points to be considered: (1) an aperture is necessary in the focal point of a first lens in the output pathway in order to block unwanted reflections, for example, from the rear side of a wedged beam splitter plate (see Figure 15.1 and the text around), and (2) the image sharpness is reduced when this aperture is undersized. In this case, higher diffraction orders of light deflected at the body's edges, containing the sharpness information, are blocked as well. Combining (1) and (2), it becomes clear

FIGURE 15.10 (a) Photograph of a prismatic body and illustration of image sharpness of the related interference phase topography when the body's front face is imaged using an appropriate set of lenses and using aperture sizes of (b) 5 mm and (c) 0.5 mm.

that the distance between secondary spots with respect to the main spot, scaling the maximum possible aperture size, is the limiting parameter for the attainable image sharpness in the interferogram. At a given wedge angle ε, this distance is approximately given by $2\varepsilon n_{BS} f$, in which n_{BS} is the refractive index of the beam splitter and f is the focal length of the lens. It is mentioned that in the example of Figure 15.10, the main spot size is 0.2 mm (size of a multimode fiber's image), while the distance to the secondary spots is 6 mm.

The lateral position of a body within the phase topography is important information for the further extraction of the length and, for highest accuracy, length correction related to an average sub-pixel position of the body [18].

15.5.2 Interferogram Analysis

Interferogram analysis will give the primary information for further length evaluation, that is, from the measured interferograms, the fractional order of interference is to be extracted as shown in Equation 15.1. This may be done by phase-stepping interferometry as illustrated in Figure 15.11a in which a set of four interferograms each obtained at a different reference pathway is used to calculate a phase topography (Figure 15.11c). Figure 15.11b illustrates the calculation of the phase at a certain pixel position.

Besides interference phase evaluation at each pixel position, it is important to identify the body's position within the interferogram. When using phase-stepping interferometry, this assignment of the body's position related to the camera pixel coordinates can be performed most accurately. Different strategies exist for extracting an average sub-pixel-related position of the body. In each case, a tilt between the body's edges related to the pixel directions of the camera is necessary (as visible in Figure 15.11). One way is to process the discontinuity within the phase topography, which is related to the side edges of the body's front face. This strategy requires a high quality of the body's edges surrounding the front face and a nearly perfect right-angled body. Another effective way for extracting an average sub-pixel-related position bases on the use of physical mask pieces that are connected to the front face edges. For gauge block-shaped bodies, the mask may consist of four right-angled metal pieces, which are held in contact against the body's side edges by a small rubber band as shown in Figure 15.10a. At the pixel positions where these mask pieces are located, no interference is observed. This information can be used to define a mask array in which coordinates of the transitions interference → no interference can be processed around the front face.

FIGURE 15.11 (a) Set of four interferograms each obtained at a different position of the reference pathway; (b) calculation of the phase value at a certain pixel position using an appropriate phase-stepping algorithm; (c) phase topography obtained from the interferograms: the black pixels around the front face indicate the transitions interference → no interference, the black squares show ROIs in which phase values will be averaged (see text for details).

In Figure 15.11c, these transitions are indicated by black pixels forming lines in a certain range around the front face. In the next step, the coordinates left/right and top/down are averaged resulting in the white pixels forming lines. The coordinates of these pixels are averaged for the two pixel directions separately resulting in a highly accurate sub-pixel coordinate for the central position of the front face. This coordinate is used to define a region of interest (ROI) at the front face and two symmetrically arranged ROIs at the platen, indicated as black squares in Figure 15.11c.

In principle, the fractional order of interference is based on average phase values within these well-defined ROIs. However, before averaging the phase values, it is necessary to eliminate 2π discontinuities. This procedure is called unwrapping and can easily be performed by processing the phase values within each ROI with respect to phase differences at neighboring pixel positions. The unwrapped phase values may, for example, be obtained from differences by

$$\phi^{uw}(x, y) = \phi(x, y) - \text{Round}\big[\phi(x, y) - \phi(x-1, y)\big]. \tag{15.11}$$

Then, the averaged phase values are simply obtained from

$$\phi_{p/f} = \frac{1}{(x_f - x_i + 1)(y_f - y_i + 1)} \cdot \sum_{\substack{x=x_i \dots x_f \\ y=y_i \dots y_f}} \phi^{uw}(x, y), \tag{15.12}$$

in which (x_i, y_i) and (x_f, y_f) indicate the initial and final coordinates within each ROI based on the integer of the average position of the body's front face. For highest accuracy, the exact sub-pixel position of the front face, (x_c, y_c), can be incorporated into the calculation by using

$$\phi_{p/f} = \frac{1}{(x_f - x_i + 1)(y_f - y_i + 1)} \sum_{\substack{x=x_i \dots x_f \\ y=y_i \dots y_f}} \phi^{uw}(x, y) + \phi^x \Delta x_c + \phi^y \Delta y_c \tag{15.13}$$

instead of Equation 15.12. Here, ϕ^x and ϕ^y indicate the slopes within an ROI along both pixel directions, and the residuals $\Delta x_c = x_c - \text{Round}[x_c]$ and $\Delta y_c = y_c - \text{Round}[y_c]$ take into account for the sub-pixel residuals of the center position of the front face.

It is necessary that the averages within the ROIs on the platen's left-hand side, $\phi_{p,1}$, and on the right-hand side, $\phi_{p,r}$, are consistent, that is, approximately within the same plane without 2π discontinuities. Such consistency can be obtained by the so-called left–right unwrapping. In this approach, the average phase on the platen's left hand is extrapolated to the right-hand side and compared with the actual value on the right-hand side. Then, the integer difference can be used as "unwrapping correction":

$$\phi_{p,r}^{uw} = \phi_{p,r} - \text{Round}\left[\frac{1}{2\pi}\big(\phi_{p,1} + \phi^x(x_r - x_1) - \phi_{p,r}\big)\right], \tag{15.14}$$

where
$x_r - x_1$ is the pixel distance between the ROIs on the left- and right-hand sides
ϕ^x is the average slope along x-direction (left to right) detected within $\phi_{p,1}$ and $\phi_{p,r}$

From these phase averages, the fractional order of interference can be calculated for each wavelength used in the measurements:

$$q = \frac{1}{2\pi}\left[\frac{1}{2}\big(\phi_{p,1} + \phi_{p,r}^{uw}\big) - \phi_f\right]. \tag{15.15}$$

15.5.3 LENGTH EVALUATION

When the fractional order of interference is provided from the measurements as shown earlier, at a particular wavelength, λ, the body's length can be expressed as a multiple of the used wavelength, that is, by $l = (i+q)\lambda/2$ (Equation 15.1). Consequently, the number of integer interference order i, which is typically large, has to be determined separately to the earlier-described determination of the fractional order q.

When there is only a single wavelength available, the number i must be guessed from Equation 15.1 by setting $q = 0$:

$$i = \text{Round}\left[\frac{l_{\text{est}}}{(\lambda/2)}\right], \tag{15.16}$$

where l_{est} is an estimate of the length that can be obtained from the nominal length and the previous knowledge about the deviation δl of the length from the nominal length l_{nom}. When using only a single wavelength, an error in the determination of i is most likely as, for example, small temperature deviations from the nominal temperature to which l_{nom} refers (typically 20°C) lead to length changes (see Equation 15.10) that, in most cases, cannot be determined accurately enough. One typical reason for the latter is that the thermal expansion coefficient of a sample body is not known exactly.

When more than one wavelength is available as light source, that is, N different wavelengths, $\lambda_k, k = 1...N$, a variation technique referred to as "method of exact fractions" can be used for a more unambiguous extraction of the integer interference orders, $i_k, k = 1...N$. Moreover, using multiple wavelengths provides the opportunity to check the entire length measurement approach, for example, for correct wavelengths λ_k of the light sources and the correct determination of the fractional interference orders q_k. In this approach, Equation 15.1 represents a set of equations resulting in lengths, which can be compared with each other while the integer orders are varied by whole numbers δ_k, that is, for each wavelength, the length is expressed as

$$l_k = \left(i_k + \delta_k + q_k\right)\frac{\lambda_k}{2}, \tag{15.17}$$

in which i_k is calculated from Equation 15.16 for each λ_k. For a set of $\{\delta_1, \delta_2, ..., \delta_N\}$, the mean length $\bar{l}$ and the amount of average deviations Δ are calculated according to

$$\bar{l} = \frac{1}{N}\sum_{k=1}^{N} l_k, \quad \Delta = \frac{1}{N}\sum_{k=1}^{N}\left|\bar{l} - l_k\right|. \tag{15.18}$$

The resulting set $\{\bar{l}, \Delta\}$ can be displayed as scatter plot, which represents a coincidence pattern. Figure 15.12 shows a typical example of such pattern in case of two wavelengths ($N = 2$) in which the data points for $\Delta < 20$ nm are displayed.

In the example shown in Figure 15.12, the two wavelengths are fictively set to $\lambda_1 = 532.3$ nm and $\lambda_2 = 548.6$ nm. The nominal length is set to $l_{\text{nom}} = 10$ mm, and the estimate deviation of the length l from the nominal length $\delta l = l - l_{\text{nom}}$ is set to 2.5 µm. The fractional interference orders q_1 and q_2 are virtually set to $q_k = l\cdot2/\lambda_k - \text{Round}\left(l\cdot2/\lambda_k\right)$.

The mentioned input quantities reflect themselves in the coincidence pattern $\{\bar{l}, \Delta\}$ in the following way (see Figure 15.12):

1. At $\bar{l} - l_{\text{nom}} = \delta l$, there exists a minimum for which Δ is zero (marked by the dotted circle).
2. Further minima exist that are spaced by the half of the so-called synthetic wavelength, $\lambda_{\text{synt}} = \lambda_1\lambda_2/\left(\lambda_2 - \lambda_1\right)$, that is, at positions $\bar{l} - l_{\text{nom}} = \delta l + m\cdot\lambda_{\text{synt}}/2$ in which m is an integer.

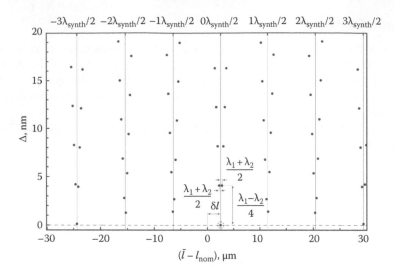

FIGURE 15.12   Typical coincidence pattern when using two wavelengths.

3. The next neighbored minima for which Δ is zero can be found at specific multiples of $\lambda_{\text{synt}}/2$. In this example, the multiples are localized at $m = \pm 3$. It is noticed that this number is strongly dependent on the amounts of wavelengths λ_1 and λ_2, that is, there is no rule for a position of these minima.

4. The lengths that are closest to the minimum $\bar{l} - l_{\text{nom}} = \delta l$ ($\Delta = 0$) are separated by a single interference order, that is, at $\bar{l} - l_{\text{nom}} = \delta l \pm (\lambda_1 + \lambda_2)/2$. The corresponding values of Δ amount to $|\lambda_1 - \lambda_2|/4$.

Conclusion (4) reveals a very important requirement for the applicability of the described variation technique: the two wavelengths must be clearly separated. Otherwise, the distance of neighboring lengths mentioned in (4) becomes too small. When, for example, λ_1 and λ_2 would be just 2 nm apart from each other, for $\{\delta_1 \pm 1, \delta_2 \pm 1\}$, the values of Δ result in 0.5 nm, which is very close to zero (zero is obtained for the variation numbers $\{\delta_1, \delta_2\}$). It is noticed for this case that, in spite of the small values of Δ, the corresponding lengths are apart by $\pm(\lambda_1 + \lambda_2)/2$!

Conclusion (4) illustrates also that the variation method actually requires "exact fractions." The "wrong lengths" correlated with $\{\delta_1 \pm 1, \delta_2 \pm 1\}$ can be discriminated from the "true lengths" only when l_k can be determined much more accurately than $|\lambda_1 - \lambda_2|/4$, that is, when the fractional interference orders (fractions) can be determined "exactly."

While the earlier example is fictive in order to illustrate the inherent features of the "method of exact fractions," real measurement examples, even for more than two wavelengths, can be found in the literature (e.g., [19]). Nevertheless, the general conclusion holds for an arbitrary number of wavelengths: the strength of the "method of exact fractions," that is, the achievable range of unambiguous determination of the integer order of interference, is significantly influenced by the achievable accuracy with which the fractional order of interference can be determined. Moreover, consideration of possible wavelength-dependent effects, for example, phase correction for the surface of the body and chromatic aberration of optical components, might be necessary to be considered.

15.5.4 EXAMPLES OF PRECISE LENGTH MEASUREMENT APPLICATIONS

The first example of length measurements by interferometry focuses on the length measurements separately, that is, influences due to variations of the body's temperature are minimized by using

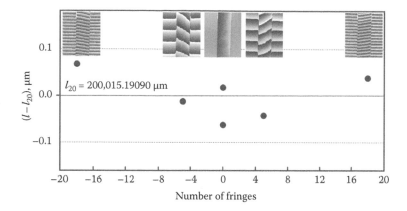

FIGURE 15.13 Length measurements performed at different tilt angles of the reference mirror. The related phase topographies ($\lambda_1 \cong 532$ nm) are shown as insets.

an ultralow thermal expansion glass as body material. For smallest uncertainty, the measurements were performed under vacuum conditions. The scope of the measurements shown in Figure 15.13 was to investigate the linearity of the interference phase. For this purpose, several measurements were performed at different tilt angles of the reference mirror. The resulting phase topographies are shown as insets in Figure 15.13. The resulting lengths obtained from different measurements are well within 0.1 nm around the mean length of 200,015.19090 μm. Accordingly, nonlinearities of the interference phase could be concluded to be negligible in this example. It is mentioned that autocollimation adjustment was performed as described earlier, that is, in each case, the light beam incidents perpendicular to the body's faces. The situation is different when the fringes in Figure 15.13 are generated by tilting the body instead of the reference mirror. In such case, the cosine error leads to systematic deviations of several nanometers, dependent on the number of fringes generated (see Section 15.4.2).

The next example is related to primary length calibrations of long gauge blocks. Here, in addition to the length at a reference temperature, the thermal expansion of a 1 m steel gauge block is investigated by applying a temperature range from 15°C to 25°C. The measurements are performed under atmospheric pressure in a Kösters–Zeiss interference comparator, as shown in Figure 15.2, equipped with phase stepping and a 1 m long vacuum cell dealing as refractometer as outlined in Section 15.4.3. The data points in the lower part of Figure 15.14 show the measured lengths as a function of the temperature. Accordingly, the length varies in a relatively large range around the length at a reference temperature of 20°C (indicated by the zero line). The solid line shows a linear regression fit to the data. Deviations from this linear fit become visible in the middle part of Figure 15.14. The range of these deviations is typical not only for steel material. Applying a second-order fit polynomial to the data results in very small deviations from the fit as shown in the upper part of Figure 15.14.

It is pointed out in this connection that a temperature deviation of only 1 mK leads to a length change of about 12 nm in the 1 m steel gauge block. Accordingly, a very small uncertainty of the temperature measurement is necessary in the example of Figure 15.14. This was achieved by using a sophisticated temperature measurement system with which the surface temperature of the gauge block is transmitted by thermocouples near the front face and near the platen. The advantage of using thermocouples is that self-heating can be neglected. As reference point for the thermocouples, a copper block is used whose temperature is measured by an AC bridge together with a PT-25 standard thermometer. The system is primary calibrated according to ITS90 using the fix points of water and gallium. The temperature measurement uncertainty using the entire calibrated equipment is stated to be 0.5 mK around room temperature.

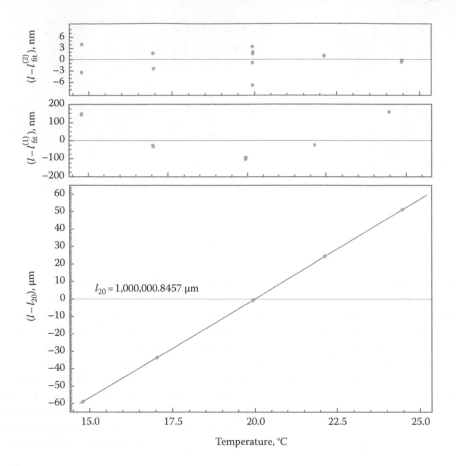

FIGURE 15.14 Length measurements at different temperatures performed on a steel gauge block (data points in the lower part). The solid line shows a linear regression to the data. Deviations from the linear fit and from a second-order polynomial fit become visible in the middle part and in the upper part, respectively.

The second-order polynomial fit, $l^{(2)}$, can be used to calculate the coefficient of thermal expansion according to

$$\alpha^{(2)}(t) = \frac{1}{l^{(2)}} \cdot \frac{dl^{(2)}}{dt} = \frac{a_1 + 2a_2(t - 20°C)}{a_0 + a_1(t - 20°C) + a_2(t - 20°C)^2} \approx \frac{1}{a_0}\left[a_1 + 2a_2(t - 20°C)\right]. \quad (15.19)$$

The approximate expression on the right-hand side of Equation 15.18 gives consideration to the fact that the length change is much smaller than the length $a_0 = l_{20°C}$ itself.

Another example of length measurements at a body made of single-crystal silicon is shown in Figure 15.15. Here, different amounts of air pressure within a Twyman–Green interferometer were set starting from vacuum pressure (10^{-3} hPa). The length of the body is reduced from a value indicated in the figure, with increasing pressure due to the material's compressibility. The measurements were performed at about 20°C, and the small temperature deviations from 20°C are considered in a correction (Equation 15.10). The overall effect between vacuum and atmospheric pressure is about 65 nm. The straight line shows a linear fit to the data. Except for the vacuum pressure, the pressure was measured by a pressure balance that is difficult to be adjusted at lower pressures (resulting in the larger uncertainties at 250 and 500 hPa as indicated in Figure 15.15). As the silicon crystal is isotropic, the relative length change provides the relative volume change by

FIGURE 15.15 Length decrease due to the air pressure measured at a body made from single-crystal silicon (data points). Bars indicate the range of measurement uncertainty. The solid line shows a linear regression fit to the data.

$\Delta V/V = 3 \cdot \Delta l/l$. Therefore, the isotropic volume compressibility, K, can be calculated from the slope of the straight line of Figure 15.15 according to

$$\frac{\Delta l}{l} = -\frac{K}{3} \cdot p, \tag{15.19}$$

in which p is the air pressure within the interferometer.

The next example addresses the so-called length relaxation, which may exist for certain materials but not for single crystals. Length relaxation is observed after the temperature of the material was changed. In contrast to thermal expansion, the temperature of the body may be constant during the length relaxation. In this case, length relaxation can be visualized best. Figure 15.16 shows a measurement cycle performed at a body made of a low-expansion glass ceramics. The initial length (as indicated in the figure) is reduced due to a first stepwise decrease in the temperature toward 10°C. The inset labeled "10°C" shows the length relaxation at 10°C as a function of the time. Residual length relaxation is also visible in the fact that the length at 20°C within the next measurement series toward 30°C (day 10) is smaller than the initial length. The inset labeled "20°C" shows the length relaxation following a single temperature step from 30°C toward 20°C. Accordingly, the length at 20°C is still apart from the initial length even after a period of 10 days. It is pointed out that the functional relationship of the length relaxation is different from a simple exponential course. The solid curves in the insets of Figure 15.16 are of the type

$$\Delta l(t) = c \cdot \exp\left[-\left(\frac{t}{\tau}\right)^b\right], \tag{15.20}$$

in which c, τ, and b are fit parameters, whereas the latter was set to 0.5. Equation 15.20 is not sufficient when the length relaxation is to be fit over a longer period of time (data not shown), where a more complex function is necessary.

The long-term stability of materials, shown in the last example, can be regarded as length relaxation over a much longer period of time, initiated by the manufacturing process itself or by any further treatment. However, the long-term behavior of the length is typically described in a time

FIGURE 15.16 Length relaxation observed in length measurements by interferometry. Data points indicate the measured length at different positions of the temperature cycle. The insets show the length relaxation at the constant temperatures of 10°C and 20°C, respectively.

range outside a first relaxation period only. Figure 15.17 shows length measurements performed on two almost identical bodies cut from the same piece of glass ceramics material. The difference between the bodies is that one was heated to 220°C for about 1 h at a time of −0.4 years related to the timescale in Figure 15.17. Accordingly, the heated body exhibits a shrinking that is about five-fold compared to the unheated body, that is, the heat treatment drastically influences the long-term stability, and this effect is visible even after several years.

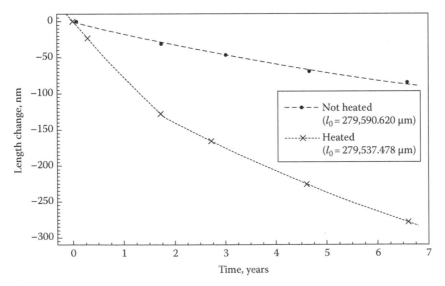

FIGURE 15.17 Length measurements performed on two almost identical bodies cut from the same piece of glass ceramics material: one of the bodies was heated to 220°C for about 1 h at a time of −0.4 years, whereas the other was not heated. In each case, l_0 indicates the length of the body at the first data point ($t = 0$), and the dashed/dotted lines represent second-order fit polynomials.

15.6 ADDITIONAL CORRECTIONS AND THEIR UNCERTAINTIES

In length measurements by interferometry, it is important to take into account the physical relationship between the interference phase, which is the primary result of the measurements, and the geometrical length to be determined. Some of these points are discussed earlier, for example, the basic idea to relate a body's length to differences of optical pathways traveled within an interferometer, the direction of the beam, and the size of the light source (aperture correction), as well as the importance of taking into account for the topography of the interference phase by defining appropriate regions of interest within the phase topography.

Besides these points, there are other important aspects that have to be taken into consideration by additional corrections in length measurements by interferometry. Each of these corrections has its uncertainty, which is related to the uncertainty of its determination. On the other hand, each correction's uncertainty contribution limits the attainable overall combined uncertainty in length measurements by interferometry, which is therefore clearly larger than promised by the interference phase and by the resolution and repeatability of the measurements.

15.6.1 OPTICS ERROR CORRECTION

When an optical flat is inserted in the pathway of an interferometer, normally, a phase topography will be measured that deviates from an ideally plane, even when parasitic interferences are eliminated (see Figure 15.5). In a Twyman–Green interferometer, this deviation still exists when the length of both optical pathways is identical. In this case, deviation of the measured interference topography from an ideal plane represents the sum of optics errors of the interferometer, that is, the imperfectness of the optical components with respect to (1) flatness of optical faces and (2) homogeneity of the optical pathway traveled through the material (including refractive index homogeneity) of the beam splitter (1, 2), compensation plate (2), and the reference mirror (1). An example topography for the optics error of a Twyman–Green interference comparator that is mainly used for primary length calibration of gauge blocks is shown in Figure 15.18. Here, a primary-calibrated optical flat with a maximum flatness deviation of 5 nm is applied. The phase topography was obtained using phase-stepping interferometry from four interferograms measured at a (fourfold) reduced camera resolution of 128 × 128 pixel (x- and y-coordinates) covering the full field of view of about 60 mm in diameter. The data shown in Figure 15.18 are related to a wavelength of 612 nm. Accordingly, the size of deviation from a plane is in the order of 20 nm (according to a phase difference of 0.4 from the side to the center). Averaged values, as displayed by the solid lines along x- and y-directions in the 2D plots of Figure 15.18, are considered as optics error correction for the particular wavelength of 612 nm.

The correction for the optics error of the interferometer can be measured at each of the wavelength used in the measurements. In the interference comparator mentioned earlier, three different wavelengths (612, 633, and 780 nm) are used alternatively. In Figure 15.19, the topographies resulting at different wavelengths are compared by phase differences in which the phases are scaled to

FIGURE 15.18 Phase error due to the interferometer optics measured using an optical flat at a wavelength of 612 nm.

FIGURE 15.19 Consistence check between the optics error corrections obtained at three different wavelengths.

each other by the corresponding wavelengths as indicated in the figure. The phase error topographies are consistent within about 0.02 radiant (corresponding to about 1 nm). Accordingly, the measurement uncertainty of the optics error correction at each particular wavelength is limited by the quality of the optical flat used.

As mentioned earlier, the optics error corrections are measured under the condition that both optical pathways within the interferometer are the same. In this way, wave front errors are excluded. However, in the length measurements, different pathways exist, that is, one at the position of the body's front face and another at positions of the platen. For this reason, wave front errors are relevant, and there is an effect to the measured phase and, thus, the length. Errors of wave front are not only related to the beam entering the interferometer. In the same way as the wave front of the collimated beam may be distorted, there is an effect due to the interferometer optics. The imperfectness of optical components, which is visible in the optics error correction, causes a wave front distortion as well. To cope with this effect, it is useful to investigate the optics error as described earlier but at different distance levels of the optical flat used. Figure 15.20 shows the difference of phase topographies measured at two positions separated by −100 mm (labeled "top") and +100 mm (labeled "bottom") at 612 nm wavelength. Consequently, the difference is up to 0.06 (about 3 nm) and clearly exceeds the differences of the consistence check between the different wavelengths as shown in Figure 15.19.

It is pointed out that the difference shown in Figure 15.20 is clearly influenced by the collimation adjustment of the beam. However, the data shown represent the optimum situation. Consequently, the optics error cannot be corrected completely, but data as shown in Figure 15.20 can be used to estimate an uncertainty of the optics error that is relevant in the length measurements. Certainly, the specific amount of this uncertainty contribution is dependent upon the geometry of the arrangement, that is, definition of the regions of interest within the phase topography. In the example mentioned earlier, it was estimated that the optics error correction is related to an uncertainty contribution of 3 nm, independent of the length of gauge blocks to be calibrated.

FIGURE 15.20 Difference of phase errors due to the interferometer optics measured at different positions of the optical flat separated by 200 mm (wavelength: 612 nm).

15.6.2 SURFACE ROUGHNESS AND PHASE CHANGE ON REFLECTION

The measured length is affected by the roughness of the surface at which the light is reflected. The effect cancels out in length measurements in which a platen is attached to the body, and both have the same surface roughness. However, this is generally not the case, as, for example, the effect due to surface roughness of steel gauge blocks has been found to vary by several nanometers within a set of gauges. There are a number of techniques for the measurement of surface roughness. One common technique uses an integrating sphere with which light is reflected, and the ratio of the diffuse to directly reflected light is measured. It is assumed that the roughness of a test surface is proportional to the square root of this ratio [20]. This technique requires a calibrated surface, for reference. The roughness difference between a body and a platen is used as correction of the uncertainty, which is typically 3 nm.

There exists another effect to the length measurement that has also to do with the reflection. When light is reflected at normal incidence off the surface of a dielectric material, there is a phase change on reflection, which is π. When the reflecting medium is non-dielectric, that is, there exists absorption loss, the phase change is between 0 and π radians, depending on the properties of the media. In this case, the complex refractive index involves the extinction coefficient κ in addition to the refractive index n ($\tilde{n} = n + i\kappa$). For light traveling in air/vacuum, the phase change on reflection is then given by

$$\delta = \arctan\left[\frac{2\kappa}{1 - n^2 - \kappa^2}\right]. \tag{15.21}$$

Length-equivalent values for the phase change on reflection are obtained by multiplying half of the wavelength. Typical values for steel are listed in Table 15.2.

Thus, the light beam appears to be reflected approximately 20 nm inside the surface, even if the surface is perfectly smooth. As for roughness, the effect cancels out in length measurements in which a platen is attached to the body, which has the same optical parameters. However, a variation of 4 nm for the length-equivalent difference of the phase change on reflection between the front face and the platen is assumed as uncertainty contribution in primary length calibrations of gauge blocks.

The corrections for the surface roughness and the phase change on reflection are often combined into a single correction termed as "phase correction."

15.6.3 WRINGING CONTACT BETWEEN A BODY AND A PLATEN

Some length measurement techniques involve a platen to be attached to the body. In this process, often called "wringing," the surfaces of the body and of the platen are carefully cleaned using a solvent, for example, acetone. When both the surfaces are polished/lapped to the highest degree, they may represent almost perfect optical flats. Putting together enables molecular attraction between

TABLE 15.2

Typical Parameters for n, κ, and for the Corresponding Length-Equivalent Phase Change on Reflection for Steel

Wavelength (nm)	n	κ	$\frac{\lambda}{2}\delta$ (nm)
543	2.09	3.00	−19.5
633	2.33	3.21	−20.7
780	2.60	3.54	−22.9

the two surfaces, which is much stronger than that needed to hold the weight of the platen. This is also called "optical contacting." In cases of increased surface roughness when the wringing of completely dry-cleaned surfaces is difficult, a small quantity of wringing fluid may be used, which fills the microscopic gaps. Although, in such a case, the wringing force is somewhat smaller compared to the optical contacting, a so-called wringing film does not contribute to the length as long as the amount of wringing fluid is small.

On the other hand, the wringing influences the length measurements by the fact that the surface topography of the body is influenced. This effect is dependent upon any deviation from flatness, mainly of the body's surface. The difference of length measurements obtained when the platen is wrung to either the one or the other face of the body gives a good indication for this effect. Dependent on the size of the resulting length difference, the wringing influence is considered as uncertainty contribution ranging from 2 to 10 nm for the length measurements.

15.6.4 PLATEN FLEXING

When the length measurement involves a wrung platen that is attached to one of the faces of a prismatic body, the extraction of the length from the interferogram as outlined earlier assumes that the platen represents an ideal flat. Accordingly, the phase on the platen is represented by the mean of averaged unwrapped phase values detected at the platen, for example, by $\frac{1}{2}\left(\phi_{p,1} + \phi_{p,r}^{uw}\right)$ in Equation 15.15. Platen flexing refers to the deformation of the wrung platen that then cannot be considered as ideal reference plane in the length measurements. There are two ways leading to platen flexing, regardless of the flatness of the isolated platen, both involving the thermal expansion of the material: (1) when, during the wringing, the temperatures of the platen and the body are different, a flexing will be observed after thermal equilibrium; (2) a temperature-dependent platen flexing will be observed when the thermal expansion coefficients of the body and the platen are different. The latter mainly regards thermal expansion measurements extracted from the temperature dependence of the length. The platen's flexing can be examined from the phase topography. However, there is no direct access to the amount of platen flexing at the center position of the body as the platen's surface is covered by the body. For this reason, it is useful to extract a flexing correction from the differences in the slopes of unwrapped phases within ROIs on the platen. One suggestion is to assume a parabolic flexing, which leads to a flexing correction for the measured length of

$$\delta l_{\text{flexing}} = \frac{\lambda}{2}\frac{1}{2\pi}\left(\frac{\partial\phi_{Pr}}{\partial x} - \frac{\partial\phi_{Pl}}{\partial x}\right)\frac{x_r - x_l}{4},$$

(15.22)

in which $x_r - x_l$ represents the distance between the ROIs located at the platen. The uncertainty of the flexing correction is dependent on the accuracy of the phase topography and the size and geometry of the ROIs. Uncertainties of up to 0.1 nm could be achieved for the standard design of gauge block-shaped bodies, that is, for $l \times 35$ mm $\times$ 9 mm.

15.7 CONCLUSIONS

This chapter describes the principles of highest-level length measurements by optical interferometry. The main focus is on measurements in which the length of a body is represented by a step height, which allows measurement by probing in one direction. This is advantageous because the penetration of the probe into the surface of the body and platen due to phase effects is then accounted for when the platen and body are of the same material.

The main conclusion of this chapter is that length measurements by interferometry, which provide the link between the lengths of macroscopic bodies to the SI-definition of the meter, can be performed highly accurately when the light waves and their alignment are provided on a very high level. Under certain conditions where additional phase corrections can be regarded as constant, for example, in thermal expansion or stability measurements of materials, the length measurements can be performed with even sub-nanometer uncertainty. However, in the general case of length measurements by interferometry, the attainable uncertainty is clearly larger than $10^{-8}l$. This can be due to limited uncertainty of the temperature measurement but mainly due to necessary corrections, which must be considered carefully and which comprise their uncertainty contributions. The situation is the same even if picometer resolution or repeatability is achieved in the measurements.

REFERENCES

1. Unit of length (metre), SI brochure, Section 2.1.1.1. URL: http://www.bipm.org/en/si/si_brochure/chapter2/2-1/metre.html (accessed August 24, 2013).
2. H. Haitjema. Achieving traceability and sub-nanometer uncertainty using interferometric techniques. *Meas. Sci. Technol.* 19 (2008) 084002.
3. J. H. Bruning, D. R. Herriott, J. E. Gallagher, D. P. Rosenfeld, A. D. White, and D. J. Brangaccio. Digital wave-front measuring interferometer for testing optical surfaces and lenses. *Appl. Opt.* 13 (1974) 2693–2703.
4. K. Creath. Step height measurement using two-wavelength phase-shifting interferometry. *Appl. Opt.* 26 (1987) 2810–2816.
5. G. Bönsch. Gauge blocks as length standards measured by interferometry or comparison length definition, traceability chain and limitations. *Proc. SPIE* 3477 (1998) 199–210.
6. J. E. Decker, R. Schödel, and G. Bönsch. Next generation Kösters interferometer. *Proc. SPIE* 5190 (2003) 14–23.
7. G. Bönsch and H. Böhme. Phase-determination of Fizeau interferences by phaseshifting interferometry. *Optik* 82 (1989) 161–164.
8. J. B. Saunders, Sr. Ball and cylinder interferometer. *J. Res. Nat. Bur. Stand.* 76C (1972) 11–20.
9. R. A. Nicolaus and G. Bönsch. A novel interferometer for dimensional measurement of a silicon sphere. *IEEE Trans. Instrum. Meas.* 46 (1997) 563–565.
10. C. Weichert, P. Köchert, R. Köning, J. Flügge, B. Andreas, U. Kuetgens, and A. Yacoot. A heterodyne interferometer with periodic nonlinearities smaller than $\pm$ 10 pm. *Meas. Sci. Technol.* 23 094005 (2012).
11. P. E. Ciddor. Refractive index of air: New equations for the visible and near infrared. *Appl. Opt.* 35 (1996) 1566–1573.
12. R. S. Davis. Equation for the determination of the density of moist air. (1981/91) *Metrologia* 29 (1992) 67–70.
13. B. Edlén. The refractive index of air. *Metrologia* 2 (1966) 71–80.
14. G. Bönsch and E. Potulski. Measurement of the refractive index of air and comparison with modified Edlen's formulae. *Metrologia* 35 (1998) 133–139.
15. J. A. Stone and A. Stejskal. Using helium as a standard of refractive index: Correcting errors in a gas refractometer. *Metrologia* 41 (2004) 189–197.
16. H. Preston-Thomas. The International Temperature Scale of 1990 (ITS-90). *Metrologia* 77 (1990) 3–10.
17. Unit of thermodynamic temperature (kelvin), SI brochure, Section 2.1.1.5 URL: http://www.bipm.org/en/si/si_brochure/chapter2/2-1/kelvin.html (accessed August 24, 2013).
18. R. Schödel. Compensation of wavelength dependent image shifts in imaging optical interferometry. *Appl. Opt.* 46 (2007) 7464–7468.
19. R. Schödel. Ultra high accuracy thermal expansion measurements with PTB's precision interferometer. *Meas. Sci. Technol.* 19 (2008) 084003.
20. G. Bönsch. Interferometric calibration of an integrating sphere for determination of the roughness correction of gauge blocks. *Proc. SPIE* 3477 (1998) 152–160.

16 Displacement

Akiko Hirai, Mariko Kajima, and Souichi Telada

CONTENTS

16.1 Introduction 433
16.2 Laser Interferometer 434
    16.2.1 Homodyne Interferometer and Heterodyne Interferometer 434
    16.2.2 Optical System of an Interferometer 434
    16.2.3 Factors That Effect the Measurement Result and Uncertainty 435
        16.2.3.1 Refractive Index of Air 435
        16.2.3.2 Stability of the Dead Path 436
        16.2.3.3 Unbalanced Optical Path Lengths and Thermal Change 436
        16.2.3.4 Abbe Error 437
        16.2.3.5 Cosine Error 437
        16.2.3.6 Nonlinear Errors 437
    16.2.4 Other Interferometers 437
16.3 Linear Encoders 438
    16.3.1 Measurement Principle 439
        16.3.1.1 Incremental Type and Absolute Type 440
        16.3.1.2 Scale Mark Detection 440
    16.3.2 Methods of Scale Mark Detection 440
        16.3.2.1 Moiré Method 440
        16.3.2.2 Image Transmission Scale Encoders 441
        16.3.2.3 Diffraction Scale Encoders 441
        16.3.2.4 Hologram Scale Encoders 442
    16.3.3 Direction of Linear Encoders 442
    16.3.4 Calibration of Linear Encoders 443
16.4 Distance Measurements 444
    16.4.1 Introduction 444
    16.4.2 Pulse Method 445
    16.4.3 Modulation Method 446
    16.4.4 Optical Frequency Sweep Method 446
    16.4.5 Air Refractivity for Distance Measurement 447
    16.4.6 Two Color Measurement 448
References 449

16.1 INTRODUCTION

Measurement of displacement is important for precise positioning, processing, and testing. Typical optical techniques for the measurement of displacement use equipment such as laser interferometers, linear encoders, and distance meters. This chapter introduces these measurement techniques.

16.2   LASER INTERFEROMETER

A laser interferometer measures displacement using interferometry. It is widely used in areas such as lithography, precision machine processing, and coordinate measuring machines due to its non-contact, high-speed, and high precision-measurement features.

16.2.1   Homodyne Interferometer and Heterodyne Interferometer

Most laser interferometers have the same configuration as the Michelson interferometer. One of the reflecting mirrors of the Michelson interferometer is attached to the object whose displacement is to be measured. The two main types of laser interferometers are the homodyne type and the heterodyne type.

Consider two linearly polarized light waves on photodetectors E_1 and E_2, whose polarization directions are the same and optical frequencies are f_1 and f_2, respectively. These waves are expressed by

$$E_1(t) = V_1 \exp[i(2\pi f_1 t + \phi_1)], \tag{16.1}$$

$$E_2(t) = V_2 \exp[i(2\pi f_2 t + \phi_2)], \tag{16.2}$$

where
  V_1 and V_2 are the amplitudes
  ϕ_1 and ϕ_2 are the phases of the waves

The interference signal between the two light waves is expressed by

$$I(t) = \left\langle \left| E_1(t) + E_2(t) \right|^2 \right\rangle = V_1^2 + V_2^2 + 2V_1V_2 \cos\left[2\pi(f_1 - f_2)t + (\phi_1 - \phi_2) \right]. \tag{16.3}$$

The case of $f_1 = f_2$ is referred to as a homodyne interferometer. The displacement is measured by the change in the phase of the interference signal. The fluctuation of the DC level of the interference signal that is caused by the change of laser power, for example, deteriorates the accuracy of the phase measurement. The interferometer developed by National Physical Laboratory takes three interference signals whose phases are 0°, 90°, and 180° respectively and eliminates DC component by combining these signals [1]. This interferometer is therefore robust to DC level fluctuation.

When $f_1 \neq f_2$, a beat signal with a beat frequency of $|f_1 - f_2|$ is observed. The case is referred to as a heterodyne interferometer. If the reflecting mirror moves, the light reflected by the mirror undergoes a Doppler shift. When the light with the optical frequency of f_2 is reflected by the moving mirror with a velocity of v, the Doppler shift Δf_2 is $2vf_2/c$, where c is the speed of light. The number of waves counted during time T is $2vT(f_2/c) = 2L/\lambda_2$, where L is the displacement and λ_2 is the wavelength of light with frequency f_2.

For precise interferometry, the stability of the laser frequency is important. The two frequencies are usually generated by a two-mode stabilized laser, a Zeeman laser, an acousto-optic modulator (AOM), or a pair of AOMs. The beat frequency is ~600 MHz in the case of a He–Ne two-mode stabilized laser, in the range of a few hundred kilohertz to a few megahertz in the case of a Zeeman laser and several tens of megahertz in the case of an AOM. A higher beat frequency means a faster measurable movement speed.

16.2.2   Optical System of an Interferometer

Figure 16.1 shows the heterodyne-type laser interferometer used to measure the displacement of a corner reflector [2–4]. Light beams with different frequencies, f_1 and f_2, are linearly polarized, and the directions of polarization are orthogonal to each other. In order to make the reference beat

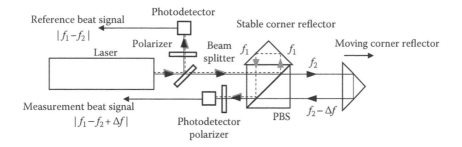

FIGURE 16.1 Heterodyne-type laser interferometer for the displacement measurement of a corner reflector. PD: photodetector.

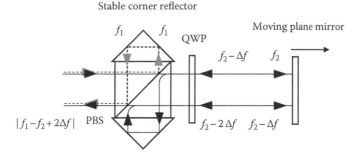

FIGURE 16.2 Heterodyne-type laser interferometer for the displacement measurement of a plane mirror. QWP: quarter-wave plate.

signal with a frequency of $|f_1 - f_2|$, portions of the light beams are reflected by a beam splitter (BS). These light beams pass through a polarizer with a direction of 45° and generate the beat signal on a photodetector. The remaining light beams are incident on a polarization beam splitter (PBS). The reflected light by PBS, whose optical frequency is f_1, traverses the fixed path with a stable corner reflector and is reflected by the PBS again. The transmitted light with an optical frequency of f_2 traverses the changing path with a moving corner reflector and is transmitted to the PBS. These two beams interfere and generate a beat signal on another photodetector after passing through a polarizer. The phase difference between the measurement beat signal and reference beat signal indicates the displacement of the moving corner reflector.

A corner reflector is useful because it reflects the light beam back in the same direction without feedback to the laser light source. However, some applications require the use of a plane mirror instead of a corner reflector. The configuration shown in Figure 16.2 is often used in the case of a plane mirror [5]. By changing the polarization direction of the measurement beam with a quarter-wave plate (QWP), the measurement beam traverses a measurement path twice. As a result, the resolution is doubled compared to that of a corner reflector.

16.2.3 Factors That Effect the Measurement Result and Uncertainty

16.2.3.1 Refractive Index of Air

The refractive index of air should be taken into account when optical interferometry is used for the measurement of length. There are two methods for the correction of the refractive index of air. One is to measure the environmental conditions such as air temperature, air pressure, humidity, and density of carbon dioxide. An empirical equation is then used to calculate and correct the refractive index of air [6]. The other method is to use a length-stabilized cavity, the so-called wavelength tracker [7,8]. The wavelength tracker consists of a stable cavity and a differential interferometer, as illustrated in Figure 16.3. One polarization component traverses the optical path that is reflected

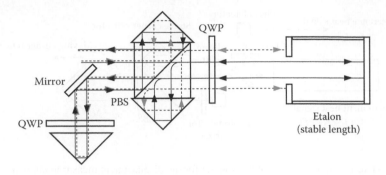

FIGURE 16.3 Schematic representation of a wavelength tracker.

from the front surface of the etalon, and the other is reflected on the back surface. This differential interferometer measures the optical length of the cavity. The cavity is made of a low thermal-expansion material and its geometric length is very stable; therefore, the change in the optical length of the cavity is considered to be the result of the change in the refractive index of air inside the cavity. The change in the refractive index of air is measured by monitoring the optical length of the cavity, and can then be used for the correction of displacement measurement conducted in the same environment.

16.2.3.2 Stability of the Dead Path

The temperature change along the reference and measurement paths induces error. Most commercial laser interferometers set the reference mirror or corner reflector near the beam splitter so that alignment is easier and the measurement is more robust to environmental changes. The distance from the beam splitter to the closest limit of the displacement range is called the dead path, as shown in Figure 16.4. The dead path has no role in the displacement measurement, however, it does have an effect on the measurement results, that is, the fluctuation in the optical length of the dead path caused by temperature change or vibration induces an error in the measurement result.

16.2.3.3 Unbalanced Optical Path Lengths and Thermal Change

The optical length is the product of the refractive index and the geometric thickness. The temperature dependence of the refractive index of general optical glass is larger than that of air. For example, when the total path lengths in the glass of the reference and measurement paths are not balanced due to the use of wavelength plates, the temperature change may have an effect on the result. To avoid an imbalance of the path lengths, optical systems such as that shown in Figure 16.5 are used [9]. In Figure 16.2, the optical path length of the measurement light beam in the PBS is twice that of the reference light beam, while the optical path lengths of the measurement and reference light beams in the PBS are the same in the optical system of Figure 16.5.

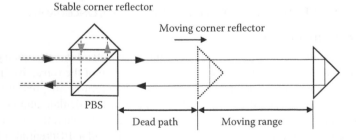

FIGURE 16.4 Schematic representation of the dead path. PBS: polarizing beam splitter.

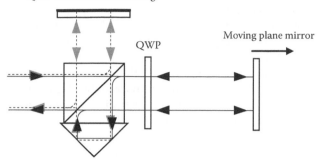

FIGURE 16.5 Interferometer used to balance the optical path lengths.

16.2.3.4 Abbe Error

The measurement system should be coaxial to the length to be measured, according to the Abbe principle. If the measurement laser beam is shifted by d from the moving axis whose displacement is to be measured, then an error, which is called the Abbe error, is induced by the angular error of movement, as shown in Figure 16.6. If the translation stage has a rotation error of θ, then the Abbe error, δ, is $d \times \sin \theta$.

16.2.3.5 Cosine Error

When the measurement axis is not parallel to the moving axis, a cosine error is induced. Alignment of the measurement and moving axes should be performed by moving the corner reflector or reflecting mirror as much as possible in order to minimize the cosine error.

16.2.3.6 Nonlinear Errors

Imperfections of antireflection coatings, polarization optics, and polarization alignment may induce contamination with unwanted reflections or the leakage of polarization in the measurement signal. The phase of the detected signal is not linear to the actual displacement in this case. There are several reports regarding the avoidance of nonlinear errors by spatially separating the light beams [10,11].

16.2.4 OTHER INTERFEROMETERS

The progress of nanotechnology has increased the demand for sub-nanometer displacement measurements and the calibration of such measurement systems. Two-beam interferometers suffer from nonlinear errors and have difficulty achieving sub-nanometer range uncertainties. Techniques that

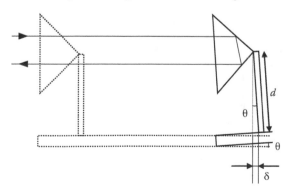

FIGURE 16.6 Schematic representation of Abbe error.

use the Fabry–Perot cavity are attractive for measurements with picometer-order uncertainty over a small range, such as in the order of micrometers. In this technique, the optical frequency of a tunable laser is locked to one of the resonance modes of the Fabry–Perot cavity. The displacement of one of the mirrors, which is the quantity to be measured, causes a change in the resonant frequency of the mode. By measuring the resonant frequency change, the small displacement of the mirror can be precisely measured. The tuning range of the laser limits the measurable range of the displacement. There have been several attempts to expand the measurable displacement range, for example, using successive resonant modes of the cavity [12], optical frequency comb generators [13], and femto-second-comb wavelength synthesizers [14].

16.3 LINEAR ENCODERS

A linear encoder usually consists of a scale unit, signal converting and interpolating unit, and an indicator (Figure 16.7). The scale unit includes a scale (or tape) that has marks to indicate the distance or position, and a reading head that detects the marks. A reading head usually includes optical source and optical system. The mark detection signal is converted to an electrical signal, interpolated by a counter in the interface unit, and then the displacement information is output to the indicator. A schematic diagram of the linear encoder is given in Figure 16.8.

In practical use, such as in the manufacturing industries, linear encoders other than laser interferometers are commonly used for positioning. Linear encoders are more convenient than laser interferometers in manufacturing, because they do not require complicated alignment of optics, and the resolution of linear encoders has improved; the resolution of recent linear encoders is of the same order or better than that of laser interferometers. Linear encoders also have a feature in that they are insusceptible to environmental fluctuations. Moreover, the encoder "encodes" the displacement of the scale and outputs a digital signal, as its name suggests. The output digital signal can be used to measure the position and operation of the working system. This feature has resulted in the widespread use of encoders in manufacturing and machine tools.

Rotary encoders detect the angle of a ball screw rotation, and the position of the stage can be calculated from the number of rotations. However, linear encoders directly detect the displacement by detecting the number of graduations on a scale using the reading head. This reason also improves the usefulness of linear encoders.

There are several kinds of linear encoders. The magnetic-scale type and the magnet–electric type are the oldest types of linear encoders. They have scales with tiny magnet graduations on it. The resolutions of the magnetic type and the magnet–electric type are limited by the spacing of magnets on the scale tape. The optical scale has overcome the limit by making fine marks on glass using micro-fabrication technologies. We concentrate on the optical type linear encoders according to the concept of this book.

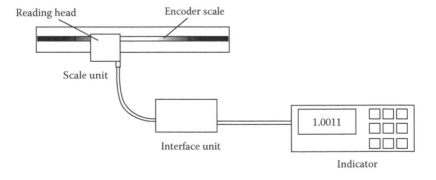

FIGURE 16.7 Example of linear encoder configuration.

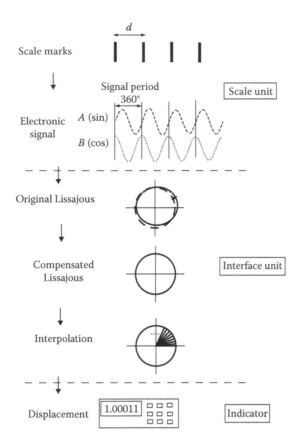

FIGURE 16.8 Process of displacement measurement by a linear encoder.

There are various kinds of optical linear encoders, which are categorized by their appearance: enclosed type (shield type) and exposed type. Enclosed-type linear encoders are mainly used for positioning in machining tools, because they are resistant to dirt or oil contaminants found in the factories due to their enclosing shield. However, exposed-type linear encoders are mainly used in the clean rooms of high-technology industries. Therefore, the resolutions of the exposed-type linear encoders are usually high, about 1 nm. The scale of optical linear encoders is usually made of glass with graduations made of evaporated chromium or gratings on it. There are also different concept exposed-type optical encoders; the long-scale measuring type. The long-scale type has exposed tape scale, which is made of metal and reading head. The resolutions of this type of encoders are not so high compared to the high-resolution-type exposed encoders. However, these types of encoders are used for large dimension manufacturing because the tape can be made longer, and attachment to the machining tool is easier than that for the enclosed type.

The measurement principles of linear encoders with glass scale, details of scale mark detection, view for the near future, and calibration of linear encoders are explained in the following subsections.

16.3.1 MEASUREMENT PRINCIPLE

The scale marks of glass-scale optical linear encoders are made of etched metals, gratings, holographic gratings, etc. The optical source for the optical linear encoders is light-emitting diode (LED) or laser diode (LD). An optical source such as LED or LD and a photodetector are usually built into one reading head. Light that is transmitted, reflected, or diffracted by the scale is detected by the photodetector in the reading head. The intensity of detected light generates a sinusoidal signal (in-phase signal) and the different sinusoidal signal, which has 90° offset signal (quadrature signal). From the two 90° phase-shifted signals, cyclic signals, such as sinusoidal signals or transistor–transistor logic (TTL) signals, are synthesized [15]. The cyclic signal usually has distortion and noise caused by contamination or manufacturing errors of the scale. The cyclic signal with distortion and noise is compensated to a true cyclic signal electrically or by processor, after which, the cyclic signal is interpolated. The signal period of interpolated signal is the resolution of the encoder. Interpolation technology is just as important as the optical system. The interpolation number reaches 4000, therefore, increasing the S/N ratio is a significant issue.

16.3.1.1 Incremental Type and Absolute Type

The incremental-type linear encoder has equal interval marks (or periodic marks) on the scale. The detector in the reading head counts the number of marks the reading head passed through. The distance that the reading head moved, x, is calculated from the interval of marks, d, and the number of marks the reading head passed through, m, as $x = dm$.

Because the absolute position cannot be determined using a primitive incremental-type linear encoder, the origin points are also usually placed on the scale. The absolute position can then be determined by calculating the distance from the origin point. However, this requires an additional process, in that the reading head must go back to the origin point, at the beginning of each measurement.

The absolute-type linear encoder provides the position of the reading head. The absolute-type encoder has equal interval (or periodic) marks on the scale, in addition to serial-absolute-code marks that indicate the absolute position on the scale. Using the serial absolute code marks, the detector can identify the absolute position without passing through the origin point.

16.3.1.2 Scale Mark Detection

Optical linear encoders are categorized by their scale-marking method. The basic type is the image transmission (shutter) type, which uses two scales with slits, a main scale and an index scale. The interval of the slits is about 10 µm. The intensity of the light that passes through the two slits is detected and converted to displacement information.

Diffraction type scale marking is also commonly used, and generally, the resolution is better than the image transmission type. In this type, a diffraction grating is ruled on the scale. The incident light is diffracted by the scale and interferes at the detector. The grating pitch can reach 5 µm using lithographic techniques. In high-resolution-type diffraction encoders, the light beam passes through the slits twice before reaching the detector, and is thus diffracted twice, which allows the resolution to be increased by four times compared to the single diffraction.

The highest resolution is realized by the hologram type, which uses a hologram scale to diffract the light. A hologram grating is memorized using a laser source with almost the same wavelength as the encoder light source, ~0.55 µm. In this type, double diffraction is also used, and the resolution of the detected signal is increased by four times compared with the grating pitch. The period of the detected signal is 0.138 µm.

16.3.2 Methods of Scale Mark Detection

16.3.2.1 Moiré Method

The Moiré method uses two scales that have different pitch marks. Marks spaced with a pitch of p_1 are on the main scale, and on the index scale the marks are spaced with a slightly different pitch of p_2. Parallel light is formed using a condenser lens, passes through the index scale and the main scale, and is then detected by a photodetector. At the detector, the Moiré fringe is detected, which corresponds to the pitch of the scales. When the main scale displaces by $p_1 - p_2$, the Moiré fringe displaces by $p_1 p_2/(p_1 - p_2)$, and the displacement of the scale can be measured by counting the Moiré fringe.

In Moiré scale encoders, the diffracted light and interference of the diffracted light becomes larger as the mark pitch becomes finer. To avoid contamination of the signal by diffraction, the index scale must be set closer to the main scale in proportion to the square of the mark pitch. The limit of the mark pitch is approximately 10 μm.

16.3.2.2   Image Transmission Scale Encoders

An image transmission scale encoder (Figure 16.9) has an index scale and a main scale with marks of the same pitch [16]. When the light is aligned, the light passes through both, the index scale and main scale, and the intensity of light detected by the photodetector is high. When the marks on the index scale overlap the gap on the main scale, no light passes through the scales, and the intensity of light detected by the photodetector is low. In this type of scale, the displacement of the scale is determined by detecting the intensity of light that passes through the scales. In this case, an LED is used as the light source. In a commercial image transmission scale, other marks on the index scale and another photodetector generate quadrature signal, and therefore, the direction of the movement of the scale can be determined [16,17]. The image transmission scale has an advantage in that it is robust to an incline of the reading head, including an incline of the index scale to the main scale. For the same reason as with the Moiré scale, the limit of the scale pitch is determined by diffraction limit and the distance of the index scale and main scale; 20–40 μm. There are several types of commercially available image transmission scales; the limit of resolution of these types of scales is 0.05 μm.

16.3.2.3   Diffraction Scale Encoders

Diffraction scale encoders have a diffraction grating on the main scale, the index scale, or both of these scales. The light from an LD is diffracted by the grating on the main scale, passes through the index scale, and is then detected by the photodetector. The light diffracted by the diffraction grating on the main scale makes a Fresnel diffraction pattern on the index scale. When the bright pattern is matched to the interval of marks on index scale, the intensity of the light detected by the photodetector is high. Movement of one of the scales causes the intensity of the light to become sinusoidal. This method actively utilizes diffracted light to detect the displacement of the scale. The main scale pitch of commercial encoders is 8 or 5 μm.

FIGURE 16.9   Optical system of an image transmission scale encoder.

16.3.2.4   Hologram Scale Encoders

Hologram scale encoders use a hologram diffraction grating on the scale. The light beam from an LD is divided into two optical paths using a PBS, and is incident on the hologram scale. The hologram grating is held between two glass plates in order to avoid blurs or scratch.

Magnescale Co., Ltd. developed a hologram scale encoder using only one scale with a hologram diffraction grating having a pitch of 0.55 μm; almost the same length as the wavelength of the light source (Figure 16.10) [18]. One of the beams, which is going in the right direction, is diffracted to the left direction by the grating. The diffracted light is then reflected by a mirror and goes back to the grating and the PBS. Two times diffraction by the grating results in the phase of the light being shifted by $\Delta\phi_r = 4\pi x/p$, where x is the displacement of the scale and p is the grating pitch. The other beam, which is going in the left direction, is diffracted two times and the phase is shifted by $\Delta\phi_l = -4\pi x/p$. The two beams interfere at the beam splitters in front of the photodetectors. The intensity of the interference signal changes sinusoidally by the displacement of the scale. The period of intensity change is $8\pi x/p$, four times larger than one period of phase change. The period of the detected signal is $p/4$, a four-cycle phase change corresponding to the displacement of the scale by one grating pitch, p. Since the grating pitch is 0.55 μm, the signal pitch is 0.14 μm.

Four photodetectors are used to detect the interference signal. A quarter-wave plate (QWP) and half mirror cause the interfered beam to split into two signals with 90° different phases. By subtracting the reverse phase signals from the four photodetectors, a sine signal and a cosine signal with a 0.14 μm period are electrically induced. Using the resistive divider and an A/D converter, the sine and cosine signals are interpolated 4000 times. The resolution of this encoder is 0.034 nm. This hologram grating scale encoder has not only high resolution but also robustness for vibration, fluctuations of air, and air pressure changes because it is a bilaterally symmetric optical system.

16.3.3   DIRECTION OF LINEAR ENCODERS

According to the current trend for miniaturization of semiconductor manufacturing, sizes of the processing scale have already entered the nanometer scale; therefore, higher resolution is demanded for linear encoders. A linear encoder that has a 34 pm resolution is already commercially available.

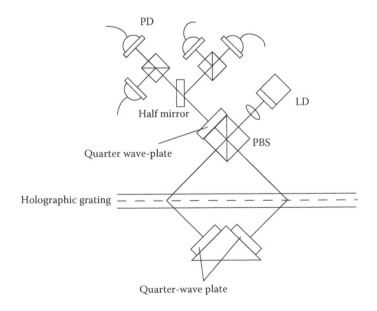

FIGURE 16.10   Optical system of a hologram scale encoder.

However, other trends of developing linear encoders are influenced by features such as convenience of use, environmental resistance, and minimization. Enclosed-type encoders, in which the scale is enclosed in a case and assembled with the reading head, are suitable for use in factory and general industrial environments, where oil or dirt contaminants are common. Absolute measurement type rather than the incremental type encoders are increasing according to the usefulness of absolute type; the absolute encoders do not require the process of passing through the original point.

To make machine tools smaller or to speed up the movement of the stage or machine by reducing weight, minimization of linear encoders is also demanded. Smaller optical components and smaller optical systems are being developed to make the reading head smaller. A reading head in which the optical system and interpolating circuit are mounted together has already been developed.

For real-time measurement in high-speed manufacturing, optical linear encoders that are applicable for reading speeds faster than 3 m/s are used.

16.3.4 CALIBRATION OF LINEAR ENCODERS

As the resolution of linear encoders is increasing, the demand for calibration of linear encoders is also increasing. The resolution of general linear encoders is in the order of 1 nm; the highest resolution of linear encoders has reached 34 pm (34×10^{-12} m). These high-resolution linear encoders are widely used in cutting-edge industries, such as semiconductor manufacturing, optical disk exposure, and high-precision processing. In these industries, the traceability of measurements is a strict requirement with the advance of globalization; therefore, precise calibration of these encoders is very important.

There are various types of linear encoders with different shapes of scales, shapes of reading head, types of optical system, and different types of signal processing for the various scales. This variety makes the development of a calibration system rather difficult. At present, the calibration systems used for linear encoders are individual systems from each manufacturer.

Comparative calibration methods are usually used by the encoder manufacturers in the encoder-manufacturing process or for the product inspection. The encoder that is to be calibrated is set on the measurement bed adjacent to the master standard linear encoder. The reading heads of the calibrated and reference linear encoders are fixed to a movable stage on the bed. As the movable stage displaces, both the encoders measure the displacement. The calibrated encoder can be calibrated compared to the master standard encoder. The reference standard encoder is calibrated by laser interferometer once a year or every several months. In this comparative method, Abbe's principle cannot be satisfied, because the two encoders cannot be set on the same space. Therefore, the Abbe error is compensated by calculating the space difference between the two encoders, and the straightness of the movable stage. The measurement uncertainty of these comparators is generally on the order of 100 nm.

Laser interferometers are used for absolute measurement of linear encoders for the calibration of master linear encoders used for standard manufacturing in factories or calibration laboratories, or in national metrology institutes. Figure 16.11 shows a schematic diagram of the absolute calibrator for linear encoders at the National Metrology Institute of Japan. This calibrator uses a double-pass laser interferometer with a resolution of 1.2 nm. The calibrated linear encoder is set on the center of two corner cubes to satisfy Abbe's condition. The calibration system is covered by an insulation box, and the air temperature fluctuation inside the box is passively stabilized to 50 mK. Air temperature, air pressure, humidity, and the density of carbon dioxide are measured for compensation of the refractive index of air. The scale temperature is also measured and used to compensate for the thermal expansion coefficient. The measurement range is up to 1000 mm. Using this calibrator, the measurement uncertainty is 50 nm ($k = 2$) for encoders with 350 mm length and 1.2 nm resolution.

Vacuum interferometers are the outstanding calibration systems for linear encoders. By encapsulating the optical path of the interferometers of calibrator into a vacuum, fluctuations in the refractive index of air become significantly smaller, so that the uncertainty caused by the calibration system

FIGURE 16.11   Scheme of the absolute calibrator of linear encoder in the National Metrology Institute of Japan.

becomes small. The uncertainty of the linear encoder calibration system used at Physikalisch-Technische Bundesanstalt (PTB) has already reached 10 nm for a 600 mm scale [19,20].

For higher resolution linear encoders, better than approximately 1 nm resolution, several methods of calibration are under development. The accuracy of laser interferometers is limited to approximately 1 nm due to the nonlinearity error. New concepts and ideas for the prevention of nonlinearity in laser interferometers or methods other than laser interferometer are necessary.

At the National Physical Laboratory, United Kingdom, a sub-nanometer calibrator has been developed by combining an optical laser interferometer and an x-ray interferometer [21]. This system uses a silicon lattice as a sub-nanometer graduation. Although in these x-ray interferometer systems sub-nanometer resolution is realized, there are the problems of safety of x-ray and the large size of the x-ray system. The feature that the resolution is determined by the average lattice spacing also makes a problem for the direct calibration of linear encoders in real time.

With regard to direct traceability and real-time calibration, sub-nanometer calibration systems based on laser interferometers are in high demand, and several systems are under development [13,22].

16.4   DISTANCE MEASUREMENTS

16.4.1   INTRODUCTION

Distance is defined as a quantity between two separate points. Although the definition is essentially independent of value, the word distance is also frequently used to mean "long length." In this subsection, we describe how to optically measure lengths over approximately a few meters.

Distance measurements are classified into two types. One is the measurement of a value between two points, and the other is a measurement of a change in long length. The former is an absolute length and the latter a displacement. Optical measurements have an advantage in their fine scale, that is, wavelength. The wavelengths of light range from several hundred nanometers to several micrometers. Moreover, smaller lengths can be obtained by measuring the phase of an interfering fringe. However, this "fine scale" also has a disadvantage in distance measurements, especially in absolute distance measurements, because the wavelengths are too small to measure distances.

For this reason, wavelength is not generally used as a scale for absolute distance measurements. However, wavelength is usually used as a scale for displacement measurements of long lengths. In this case, long baseline interferometers and/or long baseline optical resonators are used. For example, laser strain meters that have observed ground motion and laser interferometer gravitational wave detectors are in existence. With regard to laser strain meters, a 100 m baseline Michelson interferometer has been operated in Kamioka, Japan [23]. An example of a gravitational wave detector, Fabry–Perot cavity interferometers with a 4 km baseline have been operated in the United States [24]. The largest interferometer has been planned as a space gravitational wave detector with a baseline of 5,000,000 km [25]. These large interferometers and long resonators are placed in a vacuum, because variation of the air refractivity limits the sensitivity of optical distance measurements. The influence of air refractivity on distance measurements is described in the following subsection. Although special techniques and knowledge are required for the construction of large interferometers, the principle is the same as that for small or conventional size interferometers. Therefore, we will not give a detailed description of long length displacement measurements, but will describe in detail the measurement of absolute distance.

16.4.2 PULSE METHOD

The most easily understood concept of distance measurements is the method of time of flight of a light pulse. A pulse of light is sent and a receiver receives the pulse. The time difference between sending and receiving provides information on the distance that the light pulse has traveled. Distance can be calculated by dividing the speed of light by the time taken. An illustration of this scheme is shown in Figure 16.12. The resolution of the measured distance is limited by the time resolution. If the time can be measured with a resolution of 1 ns, then the distance can be determined with a resolution of 30 cm. This method is useful in roughly long distance measurements.

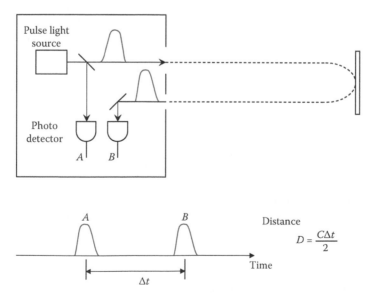

FIGURE 16.12 Diagram of distance meter using pulse method.

$$D = \frac{C}{2f_{\mathrm{m}}} \left(\frac{\Delta\phi}{2\pi} + n \right)$$

Distance

f_{m} Modulation frequency

n Integer

FIGURE 16.13 Diagram of distance meter using modulation method.

16.4.3 MODULATION METHOD

A method that uses amplitude modulated light is usually used in distance measurements. This method is based on phase-sensitive detection. The optical configuration is shown in Figure 16.13. Light to which amplitude modulation (AM) is added or which originally had amplitude modulation is divided into two paths. One is the reference path, which has a short path to a photodetector, and the other travels the measurement path and is detected by another photodetector. The phase difference between the two modulated light detected by each photodetector provides information on the measurement path length. Almost all electric distance meters (EDMs) used for surveying employ this method. For example, when the modulation frequency is 50 MHz, the amplitude modulation wavelength is approximately 6 m. As the light travels both ways between the EDM and the target, the phase is turned 360° with a distance of 3 m. By accurately measuring the phase, some EDMs have a resolution of 0.1 mm. In order to obtain information of an integer multiple of the half modulation wavelength, commercially available EDMs use methods to switch to multi-modulation frequency or to sweep modulation frequency.

Higher modulation frequency, that is, smaller modulation wavelength, may provide better distance resolution. A special EDM with a modulation frequency of 28 GHz and distance resolution of a few micrometers has been developed [26]. Another special EDM has been developed, which has a femto-second laser as the light source [27]. One of the features of a femto-second laser is its ultra-stable repetition rate, which means that a femto-second laser has an amplitude modulation that is a combination of fundamental and higher orders of repetition frequency. If the repetition frequency is 50 MHz, then the femto-second laser has amplitude modulations of 50 MHz, 100 MHz, 150 MHz, …, 10 GHz, …. By extracting only one frequency component of the electric signal, for example 10 GHz, this is the same system as the EDM with a sinusoidal wave of the frequency.

16.4.4 OPTICAL FREQUENCY SWEEP METHOD

In this method, the optical frequency of the light source is continually changed. From the optical frequency of the returned light that was reflected by a target, we can know when the light went out.

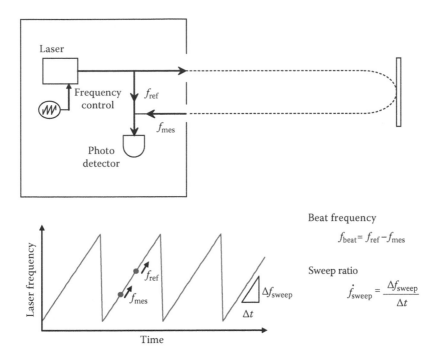

FIGURE 16.14   Diagram of distance meter using serrodyne sweep method.

Generally, the optical frequency is changed using a serrodyne waveform. The light is separated into a short reference path and the measurement path. The light through the reference path and the light through measured path are combined by the beam splitter. The combined light has a beat signal whose frequency is the difference of optical frequencies between the reference light and the measured light. An illustration of the method is given in Figure 16.14 and the calculation is made using the following equation:

$$D = \frac{c}{2} \frac{f_{beat}}{\dot{f}_{sweep}},$$ (16.4)

where
    D is the measured distance
    c is the speed of light
    f_{beat} is the frequency of the beat signal
    $\dot{f}_{sweep}$ is the frequency sweep rate

The beat frequency should be measured during the time range in which the optical frequency monotonically changes. If the measured distance is fixed, a constant beat frequency is obtained. The frequency is dependent on the sweep rate of the optical frequency at the slope on the serrodyne waveform. Features of this system are quick measurements and high resolution. This method is used in coordinate measuring equipment known as Laser Tracker® (Leica Geosystems).

16.4.5   AIR REFRACTIVITY FOR DISTANCE MEASUREMENT

Almost all distance measurements in air are limited in their accuracy by air refractivity. The refractive index of air is close to 1.00027 under standard conditions of air. The value of air refractivity is changed by the wavelength of light, air temperature, air pressure, humidity, and the density of carbon dioxide. Air temperature and air pressure have a significant effect on the change of air refractivity, and their factors are approximately 1 ppm/K and 0.3 ppm/hPa, respectively. Air pressure can be regarded as unity in horizontal distance measurements. However, air temperature is different at any spatial point within the measurement area. Therefore, it is necessary to have a large number of thermometers in order to determine the correct average temperature of the air along the measured optical path. Therefore, it is difficult to measure a long distance in air with good accuracy, especially outdoors.

Regarding the speed of the measurement light, it should be noted that in the modulation method, the modulated light travels with the speed of a group velocity in air. It is several parts per million (ppm) or more than a dozen ppm slower than the phase velocity. The refractive index is equal to the value obtained by dividing the speed of light by the phase velocity in air. Similarly, the value obtained by dividing the speed of light by a group velocity is sometimes referred to as a "group refractive index." This word is expediently used with regard to length measurements in air, notwithstanding that it is an incorrect word. It is because the law of group refraction is nonexistent!

16.4.6 Two Color Measurement

As previously remarked, it is difficult to exactly determine the environmental parameters in long distance measurements. Two color measurements can measure a distance without measuring environmental parameters such as temperature, pressure, humidity, and the density of carbon dioxide. Since differential coefficients of the refractive index for environmental parameter are dependent on the wavelength of light, the values of the refractive index of air can be determined for each wavelength from the difference of two wavelength measurements. The refractive indexes of air as a function of air temperature at wavelengths of 532 and 1064 nm, and the respective differences, are shown in Figure 16.15.

Let us define two words, "geometrical distance" and "optical distance." Geometrical distance is defined as the true distance between two points. Optical distance is defined as the distance that

FIGURE 16.15   Air refractivity and difference of air refractivity.

includes refractivity, that is, it is the product of geometrical distance multiplied by the refractive index of air. The equation of the two color measurement is given as

$$D = L_1 - A(L_1 - L_2),$$ (16.5)

where

D is the geometrical distance

L_1 and L_2 are the optical distances for wavelengths 1 and 2, respectively

A is an essential parameter in two color measurements and is expressed as

$$A = \frac{n(\lambda_1) - 1}{n(\lambda_1) - n(\lambda_2)},$$ (16.6)

where $n(\lambda_1)$ and $n(\lambda_2)$ are the refractive indexes of air for wavelength 1 and 2, respectively. If the humidity is 0%, then the value of A is constant and independent of the other environmental parameters. Thus, the geometrical distance can be determined without measuring the environmental parameters. If humidity is not 0%, the value of A is not exactly constant, but is regarded as being almost constant. Although it is advantageous that the environmental parameters need not be measured, there is a disadvantage in the two color distance measurements. The value of A is several tens to one hundred, and depends on the two wavelengths. From Equation 16.5, we know that the accuracy of the geometrical distance is A times worse than that of measured optical distances. For this reason, the two color measurement is useless for short distance or short length measurements.

REFERENCES

1. Downs, M.J. and Raine, K.W., An unmodulated bi-directional fringe-counting interferometer system for measuring displacement, *Prec. Eng.*, 1, 85, 1979.
2. *Laser and Optics User's Manual*, Chap. 7A, Linear interferometers and retroreflectors, Agilent Technologies, 2002, http://literature.cdn.keysight.com/litweb/pdf/05517-90108.pdf (accessed January 20, 2015).
3. Wuerz, L.J. and Quenelle, R.C., Laser interferometer system for metrology and machine tool applications, *Prec. Eng.*, 5, 111, 1983.
4. Demarest, F.C., High-resolution, high-speed, low data age uncertainty, heterodyne displacement measuring interferometer electronics, *Meas. Sci. Technol.*, 9, 1024, 1998.
5. *Laser and Optics User's Manual*, Chap. 7C, Plane mirror interferometer, Agilent Technologies, 2002, http://literature.cdn.keysight.com/litweb/pdf/05517-90110.pdf (accessed January 20, 2015).
6. Ciddor, P.E., Refractive index of air: New equations for the visible and near infrared, *Appl. Opt.*, 35, 1566, 1996.
7. Siddall, G.J. and Baldwin, R.R., Developments in laser interferometry for position sensing, *Prec. Eng.*, 6, 175, 1984.
8. *Laser and Optics User's Manual*, Chap. 7I, Wavelength tracker, Agilent Technologies, 2002, http://literature.cdn.keysight.com/litweb/pdf/05517-90116.pdf (accessed January 20, 2015).
9. Achieving Maximum Accuracy and Repeatability, Product Note, Agilent Technologies, 2001, http://literature.cdn.keysight.com/litweb/pdf/5952-7973.pdf (accessed January 20, 2015).
10. Wu, C.-M., Lawall, J., and Deslattes, R.D., Heterodyne interferometer with subatomic periodic nonlinearity, *Appl. Opt.*, 38, 4089, 1999.
11. Lawall, J. and Kessler, E., Michelson interferometry with 10 pm accuracy, *Rev. Sci. Instrum.*, 71, 2669, 2000.
12. Haitjema, H., Schellekens, P.H.J., and Wetzels, S.F.C.L., Calibration of displacement sensors up to 300 mm with nanometer accuracy and direct traceability to a primary standard of length, *Metrologia*, 37, 25, 2000.
13. Bitou, Y., Schibli, T.R., and Minoshima, K., Accurate wide-range displacement measurement using tunable diode laser and optical frequency comb generator, *Opt. Exp.*, 14, 644, 2006.

14. Schibli, T.R. et al., Displacement metrology with sub-pm resolution in air based on a fs-comb wavelength synthesizer, *Opt. Exp.*, 14, 5984, 2006.
15. Nagai, R., Saito, H., and Ueda, M., Present and future technology of ultraprecision positioning, Chapter 5, p. 2, Fuji Technosystems Co., Ltd. 2000 (in Japanese).
16. Magnescale Co., Ltd., http://www.mgscale.com.
17. Heidenhain Corp., http://www.heidenhain.com.
18. Mitutoyo Corp., http://www.mitutoyo.co.jp/eng/.
19. Tiemann, I. et al., An international length comparison using vacuum comparators and a photoelectric incremental encoders as transfer standard, *Prec. Eng.*, 32, 1, 2008.
20. Flugge, J., Koning, R., and Bosse, H., Recent activities at PTB nanometer comparator, Recent developments in traceable dimensional measurements II, *Proc. SPIE*, 5190, 391, 2003.
21. Leach, R., Haycocks, J., Jackson, K., Lewis, A., Oldfield, S., and Yacoot, A., Advances in traceable nano-metrology at the National Physical Laboratory, *Nanometrology*, 12, R1, 2001.
22. Kajima, M. and Matsumoto, H., Calibration of linear encoders with sub-nanometer uncertainty using an optical-zooming laser interferometer, *Prec. Eng.*, 38, 769, 2014.
23. Takemoto, S. et al., A 100 m laser strainmeter system installed in a 1 km deep tunnel at Kamioka, Gifu, Japan, *J. Geodynamics*, 38, 477, 2004.
24. Waldman, S.J. et al., Status of LIGO at the start of the fifth science run, *Class. Quantum Grav.*, 23, S653, 2006.
25. Danzmann, K. et al., LISA technology—Concept, status, prospects, *Class. Quantum Grav.*, 20, S1, 2003.
26. Fujima, I., Iwasaki, S., and Seta, K., High-resolution distance meter using optical intensity modulation at 28 GHz, *Meas. Sci. Technol.*, 9, 1049, 1998.
27. Minoshima, K. and Matsumoto, H., High-accuracy measurement of 240-m distance in an optical tunnel by use of a compact femtosecond laser, *Appl. Opt.*, 39, 5512, 2000.

17 Straightness and Alignment

Ruedi Thalmann

CONTENTS

17.1 Definitions and Normative Reference.. 451
    17.1.1 Traceability 452
17.2 Principal Measurement Methods for Straightness... 453
    17.2.1 Comparison to a Straightness Datum 453
    17.2.2 Reversal Technique.. 453
    17.2.3 Integration of Slope 453
17.3 Straightness Measurement Based on an Optical Beam... 453
    17.3.1 Optical Alignment Systems... 453
    17.3.2 Laser Beam Alignment.. 454
    17.3.3 Straightness Interferometer 455
17.4 Straightness Measurement Based on a Mechanical Datum 456
    17.4.1 Error Separation by Reversal... 456
    17.4.2 Experimental Result 457
17.5 Straightness Measurement Using Angle Instruments 457
    17.5.1 Electronic Levels 458
    17.5.2 Autocollimators 458
    17.5.3 Angle Interferometer 459
17.6 Machine Geometry... 459
    17.6.1 Straightness Measurement of a Machine Axis.. 460
    17.6.2 Rotational Error Measurement of a Machine Axis 460
Reference 460

17.1 DEFINITIONS AND NORMATIVE REFERENCE

There is a series of basic geometrical elements used to describe three-dimensional objects. With each of these elements of ruled geometry, a feature is associated describing the form deviation that is subject to geometrical measurements:

- Line—straightness
- Plane—flatness
- Circle—roundness
- Sphere—sphericity
- Cylinder—cylindricity

The straightness is considered to be a two-dimensional feature, that is, the deviation from a straight line confined to the two dimensions of a plane. According to ISO 12780-1 [1], the straightness profile is defined as the intersection of a nominally plane surface with the straightness plane perpendicular to the surface (Figure 17.1). The local straightness deviations are the distances of the straightness profile points to the reference line in the reference plane. The reference line may be

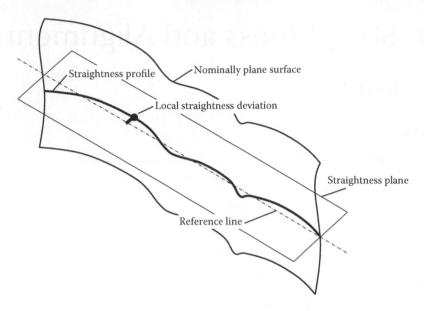

FIGURE 17.1 Definition of the straightness of a profile as the intersection of the surface and a straightness plane perpendicular to the surface.

calculated based on least squares linear regression or minimum zone criteria. The total straightness deviation STRt is the sum of the maximum local deviations from the reference line to both sides, that is, a peak to valley value:

$$STRt = Max[S(x)] - Min[S(x)],$$

where $S(x)$ is the straightness deviation along the x-coordinate.

The ISO definition of the straightness refers to the feature of the surface of an artifact. The term "straightness" may also be applied to describe the deviation of the movement of a point along a straight line, typically a linear guide way. This leads us to the distinction of two fundamentally different characteristics of straightness:

- Straightness of an artifact, such as a straight edge, a work piece; symbol STRt
- Straightness of a movement, such as the carriage of a linear guide way; symbol T_{xy}, T_{xz}

Note that the latter describes a physical movement in three-dimensional space, not confined to a plane. The deviations from a straight line are then expressed as projections into two Cartesian directions resulting in two components, such as T_{xy} and T_{xz}, which mean the deviation of a straight movement along an x-axis in y-direction and z-direction, respectively.

In this chapter too, alignment is discussed, since this is essentially based on a straightness measurement and the same techniques and instruments are involved.

17.1.1 Traceability

The question on how straightness measurements may be made traceable to national standards and thus to the international system of units arises. Since straightness refers to a perfect geometry, there is finally no physical standard. Every primary method of straightness measurement is therefore based on a physical phenomenon known to be straight (such as an optical beam) or on an error separation method. This leads to the classification of the principal measurement methods presented hereafter.

17.2 PRINCIPAL MEASUREMENT METHODS FOR STRAIGHTNESS

17.2.1 COMPARISON TO A STRAIGHTNESS DATUM

The straightness is measured with reference to a straight datum, realized by a mechanical artifact (straight edge, wire, etc.) or an optical datum (line of sight, light beam, etc.). Examples are straightness tester, straightness interferometer, alignment telescope, and laser beam alignment.

17.2.2 REVERSAL TECHNIQUE

The straightness is measured with reference to a nominally straight, but not ideal mechanical datum. The straightness errors of both, the device under test and the reference datum, can be independently separated in a subsequent measurement by reversal of the datum.

17.2.3 INTEGRATION OF SLOPE

The third principal method of straightness determination is based on the measurement of local slopes and reconstruction of the surface profile by summing up the height differences. The measurands are thus angles and distances. Again, for the angle measurements, an independent reference is required. While optical angle measurements refer to a line of sight, as in Section 17.2.1, mechanical slope measurements usually refer to the gravitational field. The inclination angles are measured by an autocollimator, an angle interferometer, or electronic levels.

17.3 STRAIGHTNESS MEASUREMENT BASED ON AN OPTICAL BEAM

17.3.1 OPTICAL ALIGNMENT SYSTEMS

In optical alignment systems, the datum or reference line is defined by the optical axis of a precision optical instrument. In different configurations of a telescope, a collimator, and targets, it may be made sensitive to position or angle:

1. *Alignment telescope*: An alignment telescope establishes an accurate line of sight. The optical system has the essential feature that the direction of the optical axis is precisely conserved during focusing. With the wide setting range of objective distances from the tube ending to infinity, these instruments serve to determine the deviation of targets with respect to the reference line (Figure 17.2) and are used for the alignment of bore holes, guides, axes, planes, etc.
2. *Alignment collimator*: An alignment collimator serves to project the image of a reticle along a reference line over a range of distances. In combination with an alignment telescope (Figure 17.3), this has, in fact, the effect that the target is moved to infinity. Due to the parallel beams between the telescope and the collimator, the system becomes insensitive to a lateral displacement of the optical axes and thus insensitive to the targets' lateral position, but sensitive to the angle and thus to the tilt of the collimator. This configuration therefore serves the alignment of the direction and is used for aligning bore holes, shafts, guides, etc. with respect to a reference line.

FIGURE 17.2 Alignment telescope with a reticle in the front focal plane, which allows the observation of the lateral deviation of a moving target from the reference line.

FIGURE 17.3   Alignment telescope with a reticle in the front focal plane, which allows the observation of the direction (tilt) of a collimator projecting the target to infinity.

FIGURE 17.4   Autocollimator that allows the observation of the direction (tilt) of the mirror reflector.

FIGURE 17.5   Alignment autocollimator as a combination of an alignment telescope and an autocollimator, which allows the observation of the lateral displacement and the direction of the moving target.

3. *Autocollimator*: In an autocollimator setup, the collimator is also used as the observation telescope. This is achieved by introducing a beam splitter close to the front focal plane of the collimator optics (Figure 17.4). In a photoelectric instrument where visual observation is superseded by an electronic readout, the illuminated reticle is replaced by a back-lighted slit or crossed slits for a two-axis instrument or an LED or laser diode. The setup from Figure 17.4 is, in fact, obtained by folding Figure 17.3 and replacing the collimator by a mirror. Consequently, the configuration serves in the alignment of the direction of the mirror mount with respect to an optical reference line.

4. *Alignment autocollimator*: Alignment autocollimators are a combination of the foregoing variants. They offer the possibility to measure both, the lateral displacement (straightness deviation) and the direction (tilting) of the target (Figure 17.5). The target is thus a combination of a mirror reflector and a reticle.

17.3.2   LASER BEAM ALIGNMENT

In modern alignment systems, the line of sight produced by an optical telescope is more and more replaced by a collimated laser beam (Figure 17.6). With the help of an electronic target (position-sensitive detector or four-quadrant detector), the deviation from the line of sight can be precisely measured and indicated on a display or recorded. This can be applied for straightness measurement of a moving carriage or for alignment of shafts or bores along a line of sight. The sensitivity of the target is dependent on its measurement range and the beam diameter and is typically in the order of 1–10 µm.

FIGURE 17.6   Straightness measurement or alignment using a collimated laser beam.

Attention has to be paid to the divergence of the laser beam due to diffraction. The diameter $w(z)$ of a laser beam varies with the distance z by

$$w(z) = w_0 \sqrt{1 + \left(\frac{\lambda z}{\pi w_0^2}\right)^2}, \tag{17.1}$$

where
$w_0 = w(0)$ is the beam diameter at its waist
λ is the optical wavelength

For example, a beam with a waist $w_0 = 1$ mm will double its diameter after approximately 9 m, whereas a 2 mm waist will double after roughly 36 m.

17.3.3 Straightness Interferometer

Another method for straightness measurement based on an optical beam is the straightness interferometer (Figure 17.7). The polarizing interferometer consists of a birefringent prism (Wollaston prism), which splits the incoming beam into two bent beams of orthogonal polarization. A fixed angled mirror reflects the beams for recombination and interference in the prism. The interference signal is usually detected after a beam splitter within the housing of the laser head. A lateral displacement of the prism will change the optical path length between the two polarizations of the beam and induce a linear change in the interference phase. It is thus the prism that plays the role of the moving target.

Note that the straightness interferometer is only sensitive to one transversal direction. The measurement of the straightness deviation in the orthogonal direction necessitates the prism and the reflector to be rotated by 90°. The range of transversal displacement measurement is essentially limited by the size of the prism and is typically in the order of ±1 mm. The sensitivity is dependent on the angular separation of the beams. Given the practical limitations for the reflector size, a higher refracting prism and thus a larger beam angle may be used for shorter longitudinal measurement ranges. Sensitivities of 0.01 and 0.1 µm are typical for short and long range measurements, respectively.

The lateral displacement measurement of a straightness interferometer is intrinsically linear provided the polarizing prism is of sufficient optical quality. The straightness deviation along the longitudinal direction is not only given by the perfectly straight beam line, but also by flatness deviations of the angled reflector: depending on the distance of the interferometer from the reflector, the beam separation on the reflector will change accordingly and the beams will be reflected by another part of the mirrors. A flatness deviation of the mirrors will then be interpreted by the instrument as a straightness deviation. Attention must therefore be paid to the manufacturer's specifications for distance-dependent accuracy.

FIGURE 17.7 Straightness interferometer.

17.4 STRAIGHTNESS MEASUREMENT BASED ON A MECHANICAL DATUM

The straightness of an artifact can be measured by comparison to a mechanical datum, usually realized by a precision guide way with a low noise bearing such as a hydrostatic or an air bearing. The straightness difference of the artifact with respect to the datum is usually recorded by a mechanical displacement transducer such as an inductive probe (Figure 17.8a). The recorded profile $P_1(x)$ is then given by the difference of the straightness $S(x)$ of the unknown artifact and the straightness $S_D(x)$ of the datum:

$$P_1(x) = S(x) - S_D(x) + R_1(x),\qquad(17.2)$$

where $R_1(x)$ is a random noise term due to the non-repeatable behavior of the bearing and the electronic noise of the transducer. The straightness $S(x)$ of the straight edge under test is then given by

$$S(x) = P_1(x) + S_D(x) - R_1(x).\qquad(17.3)$$

The straightness deviation $S_D(x)$ of the datum is ideally sufficiently small or corrected by error mapping, eventually known from another source. The uncertainty $u(S)$ is given by

$$u^2(S) = u^2(S_D) + \sigma^2[R_1(x)],\qquad(17.4)$$

where σ^2 is the variance of the noise term (or σ the root mean square, rms).

17.4.1 ERROR SEPARATION BY REVERSAL

An additional reversal measurement allows to determine the straightness error of the artifact independent of the straightness error of the reference datum. For this, the straight edge under test is reversed in its position such that the measurement surface is oriented toward the opposite direction and has to be probed from the other side (Figure 17.8b). The measured profile is then given by

$$P_2(x) = S(x) + S_D(x) + R_2(x).\qquad(17.5)$$

(a)

(b)

FIGURE 17.8 (a) Straightness measurement using a linear guide way as mechanical datum and (b) reversal measurement for error separation.

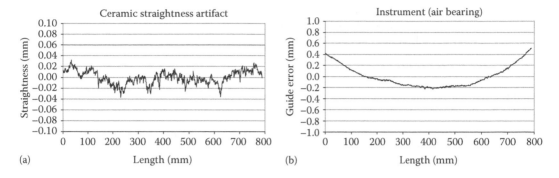

FIGURE 17.9 (a) Measured straightness deviation on a ceramic artifact and (b) guide error of the instrument used, resulting from error separation by reversal.

With this, the straightness error of both, the artifact and the datum, can be determined independently:

$$S(x) = \tfrac{1}{2}[P_1(x) + P_2(x)] + R(x) \quad \text{and}$$
$$S_D(x) = \tfrac{1}{2}[P_2(x) - P_1(x)] + R(x),$$

(17.6)

where $R(x) = \sqrt{R_1^2(x) + R_2^2(x)}$ is the combined noise term assuming the noise to be uncorrelated.

The uncertainty of the measured straightness profile is now only limited by the repeatability and noise $R(x)$ of the bearing and the transducer, described by its rms value, and the stability of the straight edge due to mechanical constraints and thermal gradients, which are both difficult to estimate other than from experimental data.

17.4.2 Experimental Result

Reversal straightness measurements were made using a 1000 mm granite beam with a vacuum-preloaded air-bearing slide of 800 mm travel. The carriage is moved by a low noise DC motor by means of a steel chord. The position of the carriage is measured by an incremental scale. The straightness deviation is measured in horizontal direction by an inductive probe. The measured artifact is a ceramic straight edge of 54 mm × 93 mm × 1016 mm.

The local reproducibility of independent straightness measurements turned out to be in the order of 3 nm standard deviation, whereas the length-dependent reproducibility was smaller than $5 \times 10^{-8} \cdot L$, with L being the length of the straightness profile. This latter term is mainly due to the mechanical stability of the straight edge under test during the measurement process and thermal gradients inducing an irreproducible bending.

Figure 17.9a and b shows the resulting straightness profiles $S(x)$ and $S_D(x)$ of the straightness artifact and the instruments, guide error, respectively. The straightness deviation of the ceramic straight edge was measured to be STRt = 0.07 µm over a length of 800 mm. The short wavelength structure as shown in Figure 17.9 for the ceramic artifact was perfectly reproducible and therefore not caused by noise but due to the surface structure.

17.5 STRAIGHTNESS MEASUREMENT USING ANGLE INSTRUMENTS

The straightness of artifacts may also be determined by integration of local slopes on discrete intervals. These slopes are measured with the help of angle instruments, either as inclinations with respect to the horizontal direction given by the gravitational field or as angles of a target with respect

FIGURE 17.10   Straightness deviation represented by a height profile measured from local slopes on discrete intervals.

to an optical beam. The heights h_i in the discrete points along the measured profile (Figure 17.10) are given by a recursive formula:

$$h_{i+1} - h_i = L \cdot \sin(\alpha_i), \tag{17.7}$$

where L is the length of the base.

For the optimum choice of the base length L, there is a trade-off between the resolution (density of the measurement points) and the accuracy: a short base length L means that only small slope changes are observed. When these small values become comparable with random errors, the integration of these when applying the recursive formula for calculating the heights leads to an undue accumulation of uncertainty. Furthermore, a large number of measurement points increase the measurement time and thus potential drift.

The length Δx of the contact surfaces is usually chosen to be much smaller than the base length but sufficiently large to achieve some averaging over the microstructure of the surface and also to have reduced requirements for the positioning accuracy of the base at the intended positions.

17.5.1   ELECTRONIC LEVELS

Electronic levels are devices that measure the inclination from horizontal direction. The sensor is based commonly on a friction-free pendulum, whose amplitude is measured by a linear differential inductive or capacitive transducer. Electronic levels are available with resolutions down to 0.1″, corresponding to a height difference of 0.05 μm/100 mm base length. Figure 17.11 shows the application of electronic levels for the measurement of straightness along a given direction on a surface plate. The use of a differential mode with two instruments, one of them in a fixed position, removes the influence of a rigid body motion of the whole plate during the measurements. The limiting factors for such measurements are vibrations and thermal drift of the levels during the measurement time due to the continuous handling of the instruments.

17.5.2   AUTOCOLLIMATORS

Autocollimators as described in Section 17.3.1 measure the tilt angle of the reflector mirror and thus the inclination of the base plate supporting the mirror. Their application for straightness measurement is straight forward as shown in Figure 17.12. The angle resolution of autocollimators is potentially much smaller than that for most of the other angle instruments and are in the order of 0.01″–0.1″ for standard commercial instruments. In metrological high-end applications using

FIGURE 17.11   Straightness measurement of a surface plate using electronic levels in the differential mode.

FIGURE 17.12 Straightness measurement of a surface plate using an autocollimator.

FIGURE 17.13 Straightness measurement of a surface plate using an angle interferometer.

highly precise instruments and a fully automatic positioning of the reflector unit, straightness measurement uncertainties in the nanometer region were reported. The quality of the collimator optics (apparent angle deviation with increasing distance) and air turbulence are setting the ultimate limitations for straightness measurement uncertainty.

17.5.3 ANGLE INTERFEROMETER

An angle interferometer is an interferometer where the optical path is folded in such a way that the measurement beams and the reference beams are parallel side by side (Figure 17.13). By its sensitivity to the optical path length difference $\Delta l = \lambda \frac{\varphi}{4\pi}$, where φ is the measured interference phase, the tilt angle α of the reflector is obtained by

$$\sin \alpha = \frac{\Delta l}{S} = \frac{\varphi}{4\pi} \frac{\lambda}{S}, \tag{17.8}$$

where S is the distance between the two corner cube reflectors.

The resolution of commercial angle interferometers is typically in the order of $0.1''$. When taking into account the arcsin correction, angle interferometers are perfectly linear within a large range up to about $\pm 10°$.

The application of angle interferometers to the measurement of artifact straightness is analogous to the use of autocollimators. It has to be noted that the interferometer optics serves for the reference and must be situated on the same surface plate in order to avoid any influence of a rigid body motion of the plate, whereas the position of the laser is not critical and may be on a tripod beside the plate.

17.6 MACHINE GEOMETRY

The carriage of a linear machine axis is ideally described by a rectilinear, perfectly straight movement. Modeling the real geometry of a machine axis implies all 6 degrees of freedom (DOF) of a rigid body in space, that is, three translational and three rotational movements. The longitudinal DOF describes the principal movement of the carriage, whereas the remaining two transversal DOFs and the three rotational DOFs are fixed by the guide way. Their deviation functions describe the guide errors.

Figure 17.14 shows the model of a linear x-axis. The deviation function $T_{xy}(x)$ denotes the transversal deviation of the x-axis in y-direction, that is, the horizontal straightness, and the function

FIGURE 17.14 Representation of a carriage on a linear x-axis showing the geometrical errors of a guide way.

$T_{xz}(x)$ consequently denotes the transversal deviation of the x-axis in z-direction, that is, the vertical straightness. The deviation function $R_{xy}(x)$ is the rotational movement of the x-axis around the y-axis, denoted as pitch error. The rotational movements $R_{xz}(x)$ and $R_{xx}(x)$ around the z-axis and the x-axis are called yaw and roll errors, respectively.

Obviously, there is some dependence between rotational guide errors and straightness deviation, but it is by far not trivial and often impossible to predict or model this. For example, a horizontal straightness error T_{xy} may be caused by a yaw error R_{xz}, but dependent on the offset and its direction of the measurement axis from the center or origin of the axis, the other rotational errors R_{xy} and R_{xx} may contribute as well. In modern machine tools or measuring machines, the relevant guide errors are measured and subsequently mapped for correction.

17.6.1 STRAIGHTNESS MEASUREMENT OF A MACHINE AXIS

The straightness of a machine axis may be measured with any of the following instruments that were described above: alignment telescope (Section 17.3.1), laser beam alignment (Section 17.3.2), straightness interferometer (Section 17.3.3), or in comparison with a known straightness artifact. Care has to be taken when comparing any of the optical straightness measurement methods, where the straightness of a moving point in space is measured with a mechanical measurement using an artifact where in some configurations the mechanical probe may be fixed in space (typically in the tool holder of a machine tool), and the straight edge is moving with the carriage. In the latter case, the measurand is not the straightness of the moving axis, however, it is the function used for error mapping of the axis.

17.6.2 ROTATIONAL ERROR MEASUREMENT OF A MACHINE AXIS

- Pitch and yaw of a horizontal axis can be measured with the help of an autocollimator or an angle interferometer; yaw also with an electronic level. A dual axis autocollimator has the advantage that both rotational errors may be measured at the same time.
- Pitch and yaw of a vertical axis can be measured with electronic levels or with the help of an autocollimator or an angle interferometer if the optical beam is bent into vertical direction.
- Roll of a horizontal axis can be measured with the help of electronic levels.
- Roll of a vertical axis cannot be measured by either of the above mentioned means. Usually, these measurements are carried out using a vertical straightness standard (straight edge or cylinder) and an off-axis probe.

REFERENCE

1. ISO/TS 12780-1, 2003, Geometrical product specifications (GPS)—Straightness. Part 1: Terms, definitions and parameters of straightness.

18 Flatness

Toshiyuki Takatsuji and Youichi Bitou

CONTENTS

18.1 Introduction 461
    18.1.1 Definition of Flatness...461
18.2 Interferometric Flatness Measurements 462
    18.2.1 Fizeau Interferometer 462
        18.2.1.1 Principle of Operation of a Phase-Shifting Fizeau Interferometer...........462
        18.2.1.2 Methods to Determine the Surface Profile of a Reference Flat................464
        18.2.1.3 Error Sources in a Phase-Shifting Fizeau Interferometer 466
        18.2.1.4 Fizeau Interferometer in Practical Applications......................................467
    18.2.2 Oblique Incidence Interferometer...469
18.3 Flatness Measurement Using Angle Sensors..470
    18.3.1 Contact-Type Flatness Measurement System Using Angle Sensors....................470
    18.3.2 Noncontact-Type Flatness Measurement System Using Angle Sensors..................471
        18.3.2.1 Extended Shear Angle Difference Deflectometry 471
        18.3.2.2 Long Trace Profiler 471
References...472

18.1 INTRODUCTION

Because flatness is one of the most important parameters that characterize optical properties of optical devices such as plane mirrors and prisms, a wide variety of measuring techniques has been invented so far. The simplest method is observing the curvature of the interference fringes formed between the sample and an optical flat. This method, however, is capable of measuring only $\lambda/20$ of the curvature, and it is not always sufficient for current applications.

Phase shifting method drastically enables more accurate measurement of the phase of the interference fringes. Additionally, the development of image processing devices and computers improves measurement accuracy as well as instrument facilitation. Under this condition, flatness measurement is becoming more important in industries. Quality inspection of not only optical devices but also silicon wafers, hard disk substrates, and flat panel displays requires accurate flatness measurement. There are other ultra-accurate applications, such as lithography mask for semiconductor manufacturing and optical devices for synchrotron facilities and gravitational wave observatories, and for these applications 1/100 wavelength or smaller measurement accuracy is needed.

In this section, various measurement principles of flatness measurement are introduced. Then, error factors and elimination methods are explained followed by the current situation of measurement uncertainty and traceability.

18.1.1 DEFINITION OF FLATNESS

According to the ISO standards, flatness is defined by the minimum zone method. This definition is practically suitable for mechanical engineering where surfaces of mechanical parts make physical contact with each other. It is not always easy to calculate the flatness from measurement data acquired using optical measuring instruments which are explained in this section.

The most common definition that is used in almost all instruments is based on the least square method. The least square plane is derived from the measurement data and the distance from this plane to each measurement data is calculated. Additionally, there are two definitions based on the least square method: P – V value and rms. The P – V value is the sum of the distances of two points that are the farthest from the least square plane on both sides. The rms is the root mean square of the distances of all measurement data from the least square plane.

The minimum zone flatness and the P – V flatness are not always the same value, and they have the following relation:

$$\text{Minimum zone flatness} \leq \text{P} - \text{V value}$$

The P – V flatness and the rms flatness have the following relationship:

$$\text{P} - \text{V value} \geq \text{rms value}$$

The relation between the P – V and the rms varies for the form of the plane, and the P – V is often more than 10 times larger than the rms. When we discuss the value of the flatness, special caution should be taken for the definition of flatness.

SI unit is recommended to express flatness; however, the wavelength λ of the light source is often used such as λ/20. In this case, caution should be exercised for the value of λ. Recently, 632.8 nm (the wavelength of He–Ne laser) has been commonly used.

In addition to the expression of the value of flatness, the same expression such as λ/20 is used for that of measurement uncertainty (or error). In most cases, it is not clearly stated whether λ/20 means the range or ±λ/20 (=1/10 in range).

When the uncertainty is larger than the measurement value, the probable area where true measurement value exists extends to the negative region. No clear solution for this contradiction has been presented even in the field of metrology.

18.2 INTERFEROMETRIC FLATNESS MEASUREMENTS

Interferometry has been widely employed for optical testing including flatness. Using interferometry, we can obtain the two-dimensional surface profiles of test objects without scanning. Conventionally, a Newton interferometer using a lamp source has been used for practical flatness testing [1]. However, this method requires contact between a Newton standard and the test surface. In at least one case, damage occurred to the test surface as a result. Recently, laser interferometers have become popular for flatness testing because they can perform contact-free measurements.

18.2.1 FIZEAU INTERFEROMETER

18.2.1.1 Principle of Operation of a Phase-Shifting Fizeau Interferometer

A Fizeau interferometer has a simple optical configuration and is commonly used in practical flatness measurements. In a Fizeau interferometer, the interference fringe pattern between a reference flat and the measurement (test) flat is observed, and the phase distribution of the fringe pattern is calculated. A surface profile of the measurement flat can be determined from the calculated phase distribution.

Figure 18.1 shows the basic optical configuration of a Fizeau interferometer using a laser light source. A laser beam is collimated using an objective lens, pinhole, and collimating lens. A reference optical flat is oriented perpendicular to the collimated beam. The reference optical flat is made with a wedge or an antireflection (AR) coating to prevent reflection from its back surface. The interference fringe pattern between the reference and the measurement flats is detected using a television

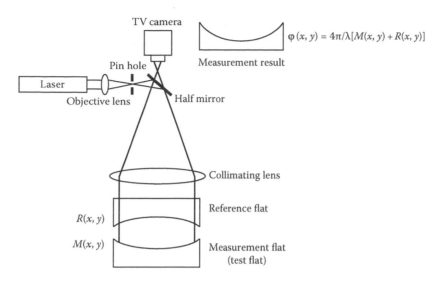

FIGURE 18.1   Configuration of a Fizeau interferometer.

(TV) camera. A beamsplitter or a λ/4 plate and polarized beamsplitter are used to steer the beam to the TV camera. Alternatively, the reference flat can be set at a slight angle and a beamsplitter is not used [2]. In that type of Fizeau interferometer, a collimating lens with a long focal length is required to achieve high accuracy.

Theoretically, the interference fringe pattern $I(x, y)$ is

$$I(x,y) = I_0\{1 + \gamma(x,y)\cos[\varphi(x,y) + \varphi_0]\} \tag{18.1}$$

where
   I_0 is the background intensity
   $\gamma(x, y)$ is the fringe modulation
   $\varphi(x, y)$ is the phase distribution to be measured
   φ_0 is the initial phase due to the optical path difference between the reference and the measurement surfaces

To obtain the phase distribution $\varphi(x, y)$ from the interference fringe pattern $I(x, y)$, the phase-shifting method is well used. In the phase-shifting method, the phase of a fringe is shifted by an optical path length change or a wavelength change. When an additional phase shift φ_i is given, the interference fringe pattern $I_i(x, y)$ becomes

$$I_i(x,y) = I_0\{1 + \gamma(x,y)\cos[\varphi(x,y) + \varphi_0 + \varphi_i]\} \tag{18.2}$$

where i is an integer. Theoretically, i is required to be larger than 3 in order to calculate the phase distribution $\varphi(x, y)$. In the standard phase-shifting method, phase steps of $2\pi/j$ are used, where $j \geq 3$ is an integer such that $\varphi_i - \varphi_{i-1} = 2\pi/j$. In the most popular phase-shifting algorithm (called the four-step algorithm with $i = j = 4$), four interference fringe patterns with a $\pi/2$ phase step are used. In the four-step algorithm, the phase distribution $\varphi(x, y)$ can be calculated as

$$\varphi(x,y) = \arctan[(I_2 - I_4)/(I_1 - I_3)] \tag{18.3}$$

For achieving precise phase distribution, a high degree of phase step accuracy is required. To reduce measurement errors, phase-shifting algorithms with a large number of phase steps have been developed [3–8]. In actual applications, five- ($i = 5, j = 4$) or seven- ($i = 7, j = 4$) step algorithms are often used [3,7]. For example, in a typical five-step algorithm, the phase distribution $\varphi(x, y)$ can be calculated as [3]

$$\varphi(x, y) = \arctan\left[2(I_2 - I_4)/(2I_3 - I_5 - I_1) \right]$$ (18.4)

The arctangent returns a phase value between 0 and 2π. There are discontinuities at 2π that must be corrected to obtain a usable result (phase unwrapping). The TV camera resolution must be at least two pixels per interference fringe period (according to the Nyquist theorem). Many types of unwrapping algorithms have been developed [9] and used in commercially available Fizeau interferometers.

The phase step is realized by an optical path length or wavelength change. A piezoelectric transducer (PZT) can vary the optical path length and is widely employed as a phase shifter in Fizeau interferometers. Although a PZT phase shifter is simple and easy to operate, it is difficult to obtain spatially uniform phase steps over a large aperture. Laser diodes (LDs) are useful light sources in a phase-shifting interferometer (PSI) because of their frequency tunability. The phase of the interference fringe in an unbalanced (Fizeau) interferometer is shifted by changing the LD frequency. Unlike in a PSI that uses a PZT, a PSI without moving parts and having a spatially uniform phase shift can be realized by tuning the frequency of an LD [10]. In both PZT and LD systems, special efforts must be made to calibrate the phase step values for high accuracy.

The surface profile $M(x, y)$ of the measurement flat can be calculated from the phase distribution $\varphi(x, y)$ as

$$M(x, y) = \lambda/4\pi \times \varphi(x, y) - R(x, y)$$ (18.5)

where
λ is the wavelength of the laser source
$R(x, y)$ is the surface profile of the reference flat

To obtain the surface profile $M(x, y)$ of the measurement flat, we have to know not only the phase distribution $\varphi(x, y)$ but also the surface profile $R(x, y)$ of the reference flat. Generally, the flatness of the reference flat is better than $\lambda/10$ and we can neglect its surface profile $R(x, y)$. However, for high-accuracy measurements (better than $\lambda/20$), the surface profile $R(x, y)$ of the reference flat has to be known. Methods to determine the surface profile $R(x, y)$ of the reference are described in the next section.

18.2.1.2 Methods to Determine the Surface Profile of a Reference Flat

For absolute flatness testing, the surface profile of the reference flat is needed. It can be determined using a liquid-flat reference [11] or a three-flat test [12].

A liquid surface can be used to determine the surface profile of another reference flat or it can be directly used as the reference flat in a Fizeau interferometer. Theoretically, the radius of curvature of a liquid surface is equal to that of Earth. For a diameter of 0.5 m, a phase value of the liquid surface can be estimated to better than $\lambda/100$ [1]. Mercury or oil is normally used as the liquid-flat reference [11]. But one of the main problems in a practical system is to isolate vibrations.

On the other hand, the three-flat test is frequently adopted in interferometric flatness measurement machines because the liquid-flat reference is difficult to realize practically. In the three-flat test, one prepares three similar flats (A, B, and C, including the target reference flat) and measures the phase

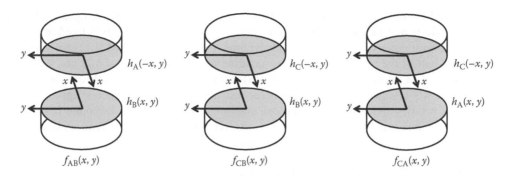

FIGURE 18.2   Three measurement combinations used in the three-flat test.

distributions of every pair of flats (AB, AC, and BC). Figure 18.2 illustrates the concept behind the three-flat test. From the three measurements, one obtains the functions $f_{AB}(x,y), f_{AC}(x,y)$, and $f_{BC}(x,y)$ as

$$h_A(-x,y) + h_B(x,y) = f_{AB}(x,y)$$

$$h_C(-x,y) + h_B(x,y) = f_{CB}(x,y) \qquad (18.6)$$

$$h_C(-x,y) + h_A(x,y) = f_{CA}(x,y)$$

where $h_A(x,y)$, $h_B(x,y)$, and $h_C(x,y)$ are the surface profiles of A, B, and C, respectively. In the measurement procedure, one flat (A in Equation 18.6) has to be inverted across a flip axis (the y-axis in Equation 18.6). From Equation 18.6, we obtain the solution for profiles on the y-axis (at $x = 0$) as

$$h_A(0,y) = \left[f_{AB}(0,y) + f_{CA}(0,y) - f_{CB}(0,y) \right]/2$$

$$h_B(0,y) = \left[f_{AB}(0,y) + f_{CA}(0,y) - f_{CB}(0,y) \right]/2 \qquad (18.7)$$

$$h_C(0,y) = \left[-f_{AB}(0,y) + f_{CA}(0,y) - f_{CB}(0,y) \right]/2$$

To obtain the whole two-dimensional surface profiles $h_A(x,y)$, $h_B(x,y)$, and $h_C(x,y)$, additional measurements with rotations of one surface is necessary [12] (as illustrated in Figure 18.3). One rotation

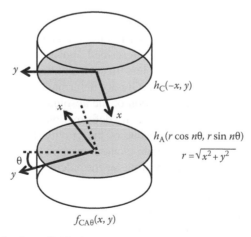

FIGURE 18.3   Rotation in the three-flat test.

angle θ is set to be $360°/n$ and the phase distributions with n rotational steps are measured, where n is an integer. In actual Fizeau interferometer systems, one rotation angle θ is fixed at between $10°$ and $30°$. Several types of reconstruction algorithms have been developed [1,12–16]. Typically, the profiles for half of the rotation angle (corresponding to radial line data of $h_C(r \cos m\theta/2, r \sin m\theta/2)$, where m is an integer) can be determined [16] and Zernike polynomials are then used to approximate the whole surface profiles. In some reconstruction algorithms, special approximations for the surface profiles and for the iteration calculations are adopted. For example, if the flat A satisfies the symmetry condition $h_A(x, y) = h_A(-x, y)$, one can calculate the whole surface profiles $h_A(x, y)$, $h_B(x, y)$, and $h_C(x, y)$ using Equation 18.6 alone.

18.2.1.3 Error Sources in a Phase-Shifting Fizeau Interferometer

Interferometric measurements are sensitive to environmental fluctuations. There are many kinds of systematic errors in a PSI. In this section, some of the principal sources of error are described.

18.2.1.3.1 Environmental Fluctuations

Mechanical vibrations and air turbulence are fundamental error sources in interferometric measurements. Using appropriate prevention systems such as a vibration absorbing table, environmental influences can be reduced to the nanometer or sub-nanometer level. To reduce air turbulence, the measurement and reference flats should be brought as close together as possible.

18.2.1.3.2 Phase-Shifting Errors

Systematic phase errors produce a non-sinusoidal waveform of the interference signal and a periodic error in the calculated phase profile. For example, in a practical Fizeau interferometer system using a PZT phase shifter, residual uncertainties in the phase shift remain because of the nonlinear response of a PZT to the applied voltage and thermal drift in the PZT. In addition, nonlinear response of the TV camera to the light intensity becomes an error source in a phase-shifting Fizeau interferometer [17].

For practical applications, the phase-shifting algorithm given by Equation 18.3 (four-step algorithm) is undesirable because of these errors. Several other kinds of phase-shifting algorithms have been developed instead. Even keeping the same number of phase steps, there are different kinds of algorithms [7,8]. The sensitivity to errors is different for each algorithm. But on the whole, phase-shifting algorithms that use a large number of phase steps are more effective at reducing the errors. Using an appropriate phase shifter and algorithm, one can attain a phase-shifting error of less than $\lambda/100$.

18.2.1.3.3 Determination of the Absolute Profile of a Reference Flat

In the three-flat test process, many interferometric measurements are required to determine the phase profile. In addition to the error sources described above, a gravity deformation effect is a serious problem in vertical Fizeau interferometers. In such an interferometer, both flats are equally deformed by gravity, as shown in Figure 18.4. Gravity deformations therefore cannot be observed and become an error source in the determination of the absolute profile of the reference flat. One solution to reduce this error is to analyze the gravity deformation numerically. Generally speaking, a reference flat has a simple figure and the value of the gravity deformation can be numerically estimated by computer simulation using a finite-element method [18]. The error due to the gravity deformation can then be reduced by compensating for the calculated value.

18.2.1.3.4 Mounting the Measurement Flat

For high-accuracy flatness measurements (better than $\lambda/20$), the mounting method becomes important. The surface profile depends on the figure of the base (for a vertical setting) or on the mounting method (for a horizontal setting). To reduce their influences, a thick plate (more than several centimeters) or an appropriate mounting method is desirable.

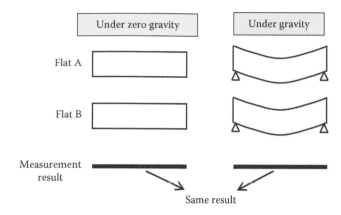

FIGURE 18.4   Influence of gravity deformations in a Fizeau interferometer.

18.2.1.3.5   Coherent Noise

In a laser interferometer system, unwanted interference fringes caused by reflections from the optical elements, such as the lenses, beamsplitter, and TV camera, become error sources in the measured phase profiles. Theoretically, fixed interference fringe noise does not influence the measurement result. However, in a PSI with an LD wavelength change, unwanted interference fringes are shifted and give rise to error in the measured phase profiles. To prevent this coherent noise problem, AR coatings on the optical elements or reduction of the spatial coherence using a rotating diffusion plate are effective. In addition, avoiding unwanted interference fringes from the TV camera is necessary [19].

18.2.1.4   Fizeau Interferometer in Practical Applications

18.2.1.4.1   National Standard Flatness Measurement Machine

Fizeau interferometers are often used as standard flatness measurement machines because of their high accuracy. Figure 18.5a through c shows photographs of the national standard Fizeau interferometer at the National Metrology Institute of Japan (NMIJ), the National Institute of Standards and Technology (NIST), and the Physikalische Technische Bundestadt (PTB), respectively. The absolute profiles of the reference flats in the NMIJ and NIST interferometers have been determined using the three-flat test and that in the PTB interferometer has been determined using a mercury liquid surface. The NIST Fizeau interferometer does not have a gravity deformation problem because it has a horizontal configuration. These three interferometers have a similar accuracy of around 10 nm for 250–300 mm diameter flats. The results of these interferometers guarantee the absolute accuracy of a practical flatness measurement machine and the tractability of flatness measurements [20].

18.2.1.4.2   Measurement of Optical Parallels

In a Fizeau interferometer, it is difficult to measure the surface profile of parallel plates because of unwanted interference fringes due to the back surface reflection, as shown in Figure 18.6a. One method to suppress the back reflection is to use index-matching oil on the back surface. Another method is wavelength-tuned phase-shifting interferometry. A special algorithm can selectively extract the correct interference signal from the noise-overlapped interference signal because the phase shifts of each fringe depend on the optical path difference [21,22]. A low-coherence interferometer provides a third method to remove the unwanted interference fringe due to the back surface reflection [23].

18.2.1.4.3   Measurement of an Optical Flat Having High Reflectivity

In a Fizeau interferometer, it is difficult to measure the surface profile of an optical flat having high reflectivity, such as an aluminum-coated mirror. When the reference plane is the glass surface and

(a)   (b)

(c)

FIGURE 18.5 Photograph of the national standard Fizeau interferometer located at (a) NMIJ, (b) NIST, and (c) PTB.

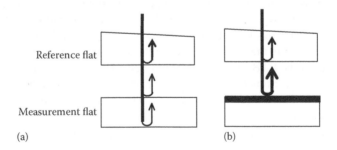

FIGURE 18.6 Measurement conditions for (a) optical parallels and (b) an optical flat having high reflectivity.

the reflectivity of the measurement flat is higher than that of the glass surface, the two reflected beam intensities become unbalanced, as shown in Figure 18.6b. In this case, the interference fringe contrast decreases and the measurement accuracy deteriorates. In addition, sinusoidal interference fringe signals are not observed and consequently, the interference signal is not given by Equation 18.1. The phase distribution calculated by the phase-shifting algorithm includes a systematic error.

A simple method to increase the fringe contrast is to insert an absorption plate, although the phase distortion due to the absorption plate has to be compensated for in high-accuracy measurements. Another method to overcome this problem is to utilize a special coating that absorbs on the reference surface [24,25]. By utilizing such special coatings, high fringe contrast can be obtained even for reflectivities as high as 100%.

18.2.1.4.4 Fourier Transform Method for a Fizeau Interferometer

In a phase-shifting Fizeau interferometer, five or seven interference fringe images are required and so it is not suitable for rapid testing. For faster measurements such as of vibrations, a Fourier transform method is applicable because only one interference fringe image is used to calculate the phase distribution [26]. Typically, in a phase-shifting Fizeau interferometer, the number of interference fringes is set to as few as possible by adjusting the tilt angle between the reference and the measurement flats to reduce the phase-unwrapping error. In contrast, in the Fourier transform method, a spatial carrier fringe is added and the phase distribution is calculated for the carrier fringe. Note that a higher resolution is then required for the TV camera.

18.2.2 OBLIQUE INCIDENCE INTERFEROMETER

In Fizeau interferometry, the measurement sensitivity depends on the wavelength of the laser light source (usually sub-micrometer). For some practical applications, this sensitivity is too high to observe interference fringes. For example, in the flatness measurement with several tens of micrometer phase value, such as in the surface profile measurement of a silicon wafer, the spatial frequency of the interference fringes becomes higher than the resolution of the TV camera and it is difficult to measure the fringe images. In this case, an oblique incidence interferometer is useful [27,28]. Figure 18.7 shows the optical configuration of a typical oblique incidence interferometer. A collimated laser beam incident on grating 1 splits into two beams (due to zeroth- and first-order diffractions). The zeroth-order beam is used as the reference and the first-order beam is incident on the measurement flat at an angle Θ. The reflected beam from the measurement flat and the zeroth-order diffracted beam are combined by grating 2 and the interference fringes are observed on a screen. If the surface profile of the reference beam is negligibly small, the measured phase distribution $\varphi'(x, y)$ is

$$\varphi'(x, y) = 4\pi \cos\Theta / \lambda \times M(x, y) \qquad (18.8)$$

In an oblique incidence interferometer, the measurement sensitivity is $1/2 \cos\Theta$ times that of a Fizeau interferometer. To calculate the phase distribution, the shifting of the interference fringes is needed; it is obtained by moving grating 1 or 2.

Another configuration of an oblique incidence interferometer uses a prism as the beamsplitter. The configuration of such an interferometer is shown in Figure 18.8. The reflection surface of

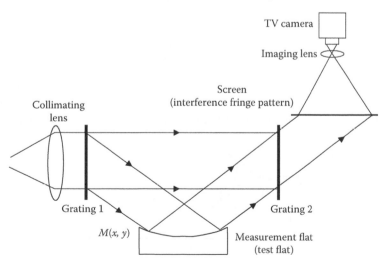

FIGURE 18.7 Configuration of an oblique incidence interferometer.

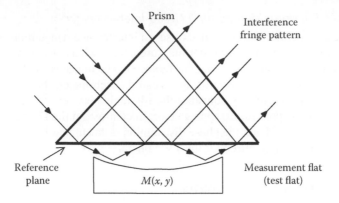

FIGURE 18.8 Configuration of an oblique incidence interferometer using a prism.

the prism serves as the reference plane. The incident beam angle to the measurement flat can be changed and the measurement surface and the reference flat can be brought close together. However, it is difficult to manufacture a large prism.

18.3 FLATNESS MEASUREMENT USING ANGLE SENSORS

Using interferometric methods, it is difficult to measure flatness over a diameter larger than 300 mm. One solution is to use angle sensors. Although a scanning operation is then required to obtain the two-dimensional surface profile, it is possible to expand the measurement range to over 1 m. Similar accuracy to that of a Fizeau interferometer is expected using precise angle sensors.

There are two kinds of angle sensor systems: a contact type and a noncontact type. Contact systems are simpler, while noncontact systems have been developed for ultraprecise surface profile measurements.

18.3.1 CONTACT-TYPE FLATNESS MEASUREMENT SYSTEM USING ANGLE SENSORS

Figure 18.9 shows a flatness measurement system using an autocollimator. In this system, a reflector is scanned across the measurement surface and the variation $\Delta h(x)$ (i.e., local slope) of the surface profile $h(x)$ is deduced from the reflection angle φ using the relation

$$\tan \frac{\varphi(x)}{2} = \frac{\Delta h(x)}{D} \tag{18.9}$$

where D is the length of the reflector. The surface profile $h(x)$ can be obtained by integrating Equation 18.9. To obtain a two-dimensional surface profile, it is necessary to scan the reflector in two directions. This method is simple for *in situ* testing of flatness. However, the spatial resolution depends on the length D and so one cannot obtain high spatial resolution. Furthermore, the positioning

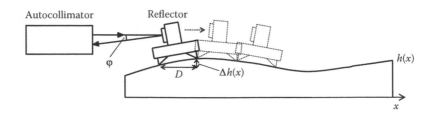

FIGURE 18.9 Configuration of a contact-type flatness measurement system using an autocollimator.

uncertainty of the reflector causes a large measurement error and this method is not suitable for soft or thin materials, such as a silicon wafer. More simple method is to utilize a mechanical indicator.

18.3.2 Noncontact-Type Flatness Measurement System Using Angle Sensors

In this section, two types of noncontact flatness measurement systems are introduced.

18.3.2.1 Extended Shear Angle Difference Deflectometry

Extended shear angle difference (ESAD) was developed for ultraprecise measurement of slope and topography of near-plane surfaces [29]. The configuration of the ESAD system is shown in Figure 18.10. A pentaprism deflects the autocollimator beam by 90° toward the surface of the measurement flat. The change in reflection angle is measured at points on the surface separated by lateral shears. The surface profile is reconstructed from two sets of angles measured by the autocollimator. To realize accurate results, a careful scanning method and integration algorithm have been developed [30,31]. The special resolution of this method depends on the value of the lateral shear and high resolution leads to long measurement times.

Unlike interferometric methods, the ESAD results do not depend on a reference surface. Instead, they depend on the accuracy of the angle and shear value measurements. A pentaprism is one of the key device to realize the high accuracy and long measurement range [32]. ESAD systems have achieved measurement ranges of over 1 m with sub-nanometer repeatability and can be expected to produce nanometer accuracy.

18.3.2.2 Long Trace Profiler

A long trace profiler (LTP) also relies on small-angle measurements. The configuration of such a LTP system is shown in Figure 18.11. The LTP measures the surface slope profile using a pencil beam interferometer [33]. In the LTP system, a pencil beam pair is created by an interferometer using two prisms, as shown in Figure 18.11. Two sets of pencil beam pairs are then made by a polarizing beam splitter and incident to the measurement flat and reference flat, which are rigidly mounted on the optical table, respectively. The beams from the measurement flat and reference flat do not interfere and the reference flat is used to compensate the mechanical error. Both sets of reflection beams are focused on a linear detector array, typically a CCD. On the linear detector array, interference spot patterns are observed. The variation in the surface slope due to the laser beam scanning is measured via the change in the interference spot position, detected using a linear array sensor. Laser beam scanning with small error is realized using a pentaprism, just as in the ESAD system [34]. The surface profile is reconstructed from the measured surface slope profile. The LTP can measure small angles with sub-microradian sensitivity and accuracy, and the surface profile with nanometer accuracy. In addition, both ESAD and LTP systems can measure aspheric profiles.

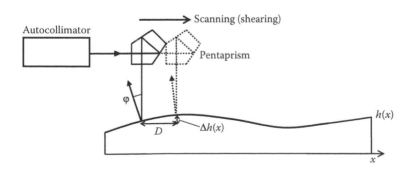

FIGURE 18.10 Geometry for ESAD deflectometry.

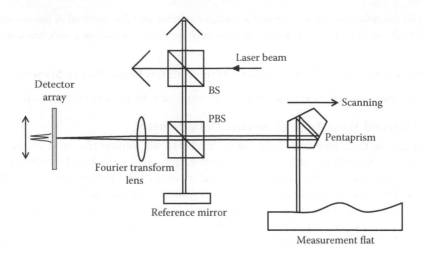

FIGURE 18.11   Configuration of an LTP.

REFERENCES

1. Mantravadi, M. V., Newton, Fizeau, and Haidinger interferometers, in *Optical Shop Testing*, 2nd edn., Malacara, D. (Ed.), John Wiley & Sons Inc., New York, 1992, Chapter 1.
2. Oreb, B. F., Farrant, D. I., Walsh, C. J., Forbes, G., and Fairman, P. S., Calibration of a 300-mm-aperture phase-shifter, *Appl. Opt.* 39, 5161, 2000.
3. Hariharan, P., Oreb, B. F., and Eiju, T., Digital phase-shifting interferometry: A simple error-compensating phase calculation algorithm, *Appl. Opt.* 26, 2504, 1987.
4. Larkin, K. G. and Oreb, B. F., Design and assessment of symmetrical phase-shifting algorithms, *J. Opt. Soc. Am. A* 9, 1740, 1992.
5. Hibino, K., Oreb, B. F., Farrant, D. I., and Larkin, K. G., Phase-shifting for nonsinusoidal waveforms with phase-shift errors, *J. Opt. Soc. Am. A* 12, 761, 1995.
6. Schmit, J. and Creath, K., Extended averaging technique for derivation of error-compensating algorithms in phase-shifting interferometry, *Appl. Opt.* 34, 3610, 1995.
7. Groot, P. D. and Deck, L. L., Numerical simulations of vibration in phase-shifting interferometry, *Appl. Opt.* 35, 2172, 1996.
8. Hibino, K., Oreb, B. F., Farrant, D. I., and Larkin, K. G., Phase-shifting algorithms for nonlinear and spatially nonuniform phase shifts, *J. Opt. Soc. Am. A* 14, 918, 1997.
9. Greivenkamp, J. E. and Bruning, J. H., Phase-shifting interferometers, in *Optical Shop Testing*, 2nd edn., Malacara, D. (Ed.), John Wiley & Sons Inc., New York, 1992, Chapter 14.
10. Ishii, Y., Chen, J., and Murata, K., Digital phase-measuring interferometry with a tunable laser diode, *Opt. Lett.* 12, 233, 1987.
11. Powell, I. and Goulet, E., Absolute figure measurements with a liquid-flat reference, *Appl. Opt.* 37, 2579, 1998.
12. Shultz, G. and Schweider, J., Precise measurement of planeness, *Appl. Opt.* 6, 1077, 1967.
13. Shults, G., Absolute flatness testing by an extended rotation method using two angles of rotation, *Appl. Opt.* 32, 1055, 1993.
14. Griesmann, U., Three-flat test solutions based on simple mirror symmetry, *Appl. Opt.* 45, 5856, 2006.
15. Vannoni, M. and Molesini, G., Absolute planarity with three-flat test: An iterative approach with Zernike polynomials, *Opt. Express* 16, 340, 2008.
16. Takatsuji, T., Bitou, Y., Osawa, S., and Furutani, R., A simple algorithm for absolute calibration of flatness using a three flat test (in Japanese), *J. Jpn. Soc. Prec. Eng.* 72, 1368, 2006.
17. Schödel, R., Nicolaus, A., and Bönsch, G., Phase-stepping interferometry: Methods for reducing errors caused by camera nonlinearities, *Appl. Opt.* 41, 55, 2002.
18. Takatsuji, T., Osawa, S., Kuriyama, Y., and Kurosawa, T., Stability of the reference flat used in a Fizeau interferometer and its contribution to measurement uncertainty, *Proc. SPIE* 5190, 431, 2003.
19. Takatsuji, T., Ueki, N., Hibino, K., Osawa, S., and Kurosawa, T., Japanese ultimate flatness interferometer (FUJI) and its preliminary experiment, *Proc. SPIE* 4401, 83, 2001.

20. Bitou, Y., Takatsuji, T., and Ehara, K., Simple uncertainty evaluation method for an interferometric flatness measurement machine using a calibrated test flat, *Metrologia* 45, 21, 2008.
21. Groot, P. D., Measurement of transparent plates with wavelength-tuned phase-shifting interferometry, *Appl. Opt.* 39, 2658, 2000.
22. Hibino, K., Oreb, B. F., Fairman, P. S., and Burke, J., Simultaneous measurement of surface shape and variation in optical thickness of a transparent parallel plate in a wavelength-scanning Fizeau interferometer, *Appl. Opt.* 43, 1241, 2004.
23. Freischlad, K., Large flat panel profiler, *Proc. SPIE* 2862, 163, 1996.
24. Netterfield, R. P., Drage, D. J., Freund, C. H., Walsh, C. J., Leistner, A. J., Seckold, J. A., and Oreb, B. F., Coating requirement for the reference flat of a Fizeau interferometer used for measuring from uncoated highly reflecting surfaces, *Proc. SPIE* 3738, 128, 1999.
25. Dynaflect coating, ZYGO Corporation, United States Patent 4820049 coating and method for testing plano and spherical wavefront producing optical surfaces and systems having a broad range of reflectivities, 1989.
26. Takeda, M., Ina, H., and Kobayashi, S., Fourier-transform method of Fringe pattern analysis for computer-based topography and interferometry, *J. Opt. Soc. Am.* 72, 156, 1982.
27. Birch, K. G., Oblique incidence interferometry applied to non-optical surfaces, *J. Phys. E* 6, 1045, 1973.
28. Jones, R. A., Fabrication of small nonsymmetrical aspheric surfaces, *Appl. Opt.* 18, 1244, 1979.
29. Geckeler, R. D. and Weingartner, I., Sub-nm topography measurement by deflectometry: Flatness standard and wafer nanotopography, *Proc. SPIE* 4779, 1, 2002.
30. Geckeler, R. D., Error minimization in high-accuracy scanning deflectometry, *Proc. SPIE* 6293, 62930O, 2006.
31. Elster, C. and Weingartner, I., Solution to the shearing problem, *Appl. Opt.* 38, 5024, 1999.
32. Yellowhair, J. and Burge, J. H., Analysis of a scanning pentaprism system for measurements of large mirrors, *Appl. Opt.* 46, 8466, 2007.
33. Bieren, K. V., Pencil beam interferometer for aspheric optical surfaces, *Proc. SPIE* 343, 101, 1982.
34. Qian, S. and Takacs, P., Design of a multiple-function long trace profiler, *Opt. Eng.* 46, 043602, 2007.

19 Surface Profilometry

Toru Yoshizawa and Toshitaka Wakayama

CONTENTS

19.1 Introduction 475
19.2 Historical Background 477
19.3 Point-Wise Techniques 480
    19.3.1 Intensity Detection Method 480
    19.3.2 Focus-Error Detection Method 481
    19.3.3 Confocal Method 486
    19.3.4 Contrast Detection Method 486
    19.3.5 Conoscopic Holography Method 489
    19.3.6 Chromatic Confocal Imaging Method 490
19.4 Full-Field Measurement 491
    19.4.1 Shape from Focus/Defocus Method 491
    19.4.2 Pattern Projection Method 493
        19.4.2.1 Traditional Techniques 495
        19.4.2.2 Digital Techniques in Pattern Projection 496
        19.4.2.3 Trends in Profilometry 498
19.5 Conclusion 513
References 514

19.1 INTRODUCTION

A great number of research papers on profilometry have been presented, but few systems have been commercialized up to now. Especially three-dimensional profilometry (3D profilometry) [1] is one of the most important topics in the present academic study and in industrial research. Conventionally, a coordinate measuring machine (CMM) with a contact type of stylus (so-called touch probe or touch sensor) has been popular in various fields. However, recent rapid and remarkable progress in optical techniques and computer technology has been bringing novel noncontact type of 3D measuring machines. In addition to the traditional contact probe, a lot of non-noncontact probes based on laser triangulation or optical coaxial principle have become popular.

What fields in academic study and/or industry require noncontact 3D measuring systems? Electrical and electronic engineering (semiconductors, various elements and devices, etc.), mechanical engineering (aircrafts, car, machine tool, engines, gears, dies, etc.), optical engineering (lenses, prisms, etc.), civil engineering, medicine (medical and biomedical imaging, live body, internal organs, endoscopic application, etc.), dentistry (teeth, oral cavity, etc.), apparel industry (human body, clothing, etc.), cosmetics industry (human face, body, skin, etc.), security (face, biometrics, etc.), industrial design, and so on.

However, it is extremely difficult to systematically classify various noncontact measurement methods, hence let us consider several methods from the viewpoint of "stand-off-distance (SOD)" (Table 19.1).

A lot of academic principles and practical systems have been reported which are represented in Figure 19.1. There exist numerous methods in addition to these. But this fact is not desirable, because it is meant that the powerful and versatile method that is applicable to any purpose does not exist. All requests cannot be covered by only one principle.

TABLE 19.1
3D Profilometry

Contact Methods	Mechanical Stylus
Noncontact methods	
Optical principles	
Point-wise measurement	Optical stylus and laser probe
Line-wise measurement	Optical section
Full-field measurement	
Close range	Confocal microscopy and interferometry
Short range	Moire method and space encoding method
Middle range	Pattern projection method
Long range	Photogrammetry and laser tracker
Far range	Range finder
Nonoptical methods	Capacitance, magnetism, and ultrasonics

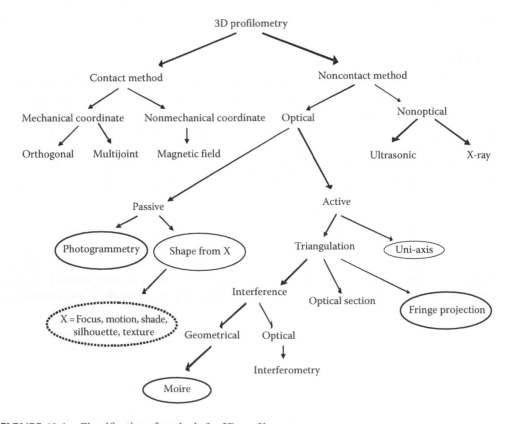

FIGURE 19.1 Classification of methods for 3D profilometry.

Most of these methods are reviewed exhaustively in Chapter 20. Topics such as moire method and photogrammetry as well as interferometry are independently described in other chapters. In the following sections, a limited number of principles will be discussed.

19.2 HISTORICAL BACKGROUND

One of traditional 3D profilometry is stereoscopy which has a long history in practical use. However, until the 1970s, we had to rely on photographic recording, and optical measuring instruments precisely calibrated were indispensable. Consequently, a very long time was required in mathematical calculation for data analysis.

One example is shown in Figure 19.2 where nine astronauts were examined by stereo-photography to check change of body shapes in advance to and after the Skylab flight in space. According to a report [2], they lost their total weight, and region change is found especially in their thighs which became thin. We should pay attention to devices like a tape measure and the reference targets and optical arrangement which are still informative in elementary experimental work.

At present, this kind of stereoscopic photography is revised up to digital photographic method described in detail in Chapter 22. This digital photogrammetry is potentially one of the powerful tools due to rapid progress in a megapixel charge-coupled device (CCD) camera and PC technologies. In my personal and a little dogmatic opinion, digital photogrammetry in addition to range finding method based on time-of-flight principle may grow up to a standard tool in 3D profilometry.

Looking back at the history of 3D profilometry, it was difficult in the 1960s to capture a 3D shape of an objective by a single shot of a photographic picture. "Moire topography," invented by Prof. Hiroshi Takasaki in 1970, brought an innovative progress in 3D profilometry, because he succeeded in expressing the 3D shape by moire contours which give intuitive understanding of 3D shapes [3]. The picture shown in Figure 19.3 was given to the author with a message that "Both of two pictures should be used at the same time because these are stereoscopic pair of moire photos. These pair is

FIGURE 19.2 Experimental setup for stereographic method. (Courtesy of Prof. M.W. Whittle.)

FIGURE 19.3 Stereo-pair of moire pictures. (Courtesy of Prof. Hiroshi Takasaki.)

available to know a hill or a valley on the surface shape." picture is not useful to identify concavity or convexity of the shape. As is easily understood, moire topography can be classified into a stereoscopic principle, because the optical configuration consists of two optical axes: one for projection and the other for observation.

The details of moire method are explained in Chapter 9. The fundamental principle of this moire method was adapted by various pattern projection methods. Here, the term "pattern projection method" implies a concept that totally covers moire topography, fringe projection, structured light method, space encoding, and others.

Today, an optical probe based on optical triangulation is attached to the conventional CMM (usually equipped with a stylus). Figure 19.4 shows an optical scheme for triangulation method, where a beam spot projected onto an objective surface is focused through a lens on the focal plane of a detector (CCD or position-sensitive device [PSD] is popular). The position of the beam spot (A', B', C', etc.) varies in accordance with the position of a beam spot on the sample surface (A, B, C, etc.). By a simple geometrical relations between A, B, C, etc. and A, B, C, etc. we can easily find the distance or displacement from this probe to the target point A, B, C, etc. This optical arrangement looks, at a glance, simple, but the "Scheimpflug rule" (another similar principle called "Hinge rule" is found in discussion on view camera focus [4]) has to be satisfied to make the spot sharply focused. The lens plane LP and the focal plane FP should intersect on the point P along the beam projection axis. This principle is well known in designing a view camera. In order for the beam spot to be in focus, the three fundamental planes, the film plane, the lens plane, and the plane of sharp focus, need to be converged along one common line (Scheimpflug line).

A measuring system of a human body is represented in Figure 19.5 which was developed in 1986 by a project team under financial support of the Japanese government. Using this "measuring capsule unit," the mesh data over the surface points from the shoulder to the thighs can be captured with every 4 mm separation in 3 min. The body was horizontally scanned by a laser beam from six directions, and the light spot was detected by six CCDs. At the same time, these six sensors are moved down from the shoulders to the thighs in 120 s. They intended to use these data to manufacture clothes, shoes, and chairs. However, mainly because of long calculation time by "minicomputer" at

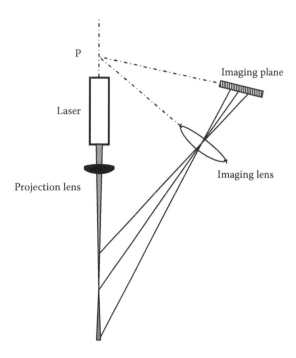

FIGURE 19.4 Principle of triangulation method.

FIGURE 19.5 Body measuring capsule unit. (From the document of human body measurement project in Japan.)

that time, this interesting project did not fully succeed in commercial applications. However, today, owing to 30 years of technological progress, similar systems are commercially successful in the apparel and shoes markets.

19.3 POINT-WISE TECHNIQUES

In this section, fundamental techniques used for two-dimensional profile measurement are described. This kind of optical sensor based on triangulation is in widespread use and sometimes called a "laser displacement meter." One of shortcomings inherent to this principle is in that it has two optical axes: for projection of a beam and for detection of the beam spot. This optical configuration causes "occlusion" that some area over the object is not measured due to blocking of the light. To solve this problem, a uniaxis arrangement (coaxis or line-of-sight arrangement) is required.

19.3.1 INTENSITY DETECTION METHOD

A lot of papers have been reported to meet this requirement. One of fundamental trials is shown in Figure 19.6 where the intensity change due to displacement of a light spot is detected [5]. One result shows such accuracy as 0.25 μm over the range of 12.7 μin. was attainable 30 years ago. This schematic of the system is shown in Figure 19.6. An image of a pinhole source (S) is projected on a surface of a sample (O) through the beam splitter BS_1 and the microscopic objective (L). The light reflected from the sample surface forms the images at P_1 and P_2. A large area photodetector measures the total intensity I_1 at P_1. Another detector with a slit in front of it measures the intensity I_2 of a part of the light through the slit. The intensity I_1 is a measure of the reflectivity of the sample surface, and the intensity I_2 is a measure of the displacement along the optical axis. The output signal of the system should be given by normalizing I_2 by I_1 to be I_2/I_1. If the sample surface is displaced a small amount z longitudinally, the position of the image P_2 shifts by the amount of M^2z

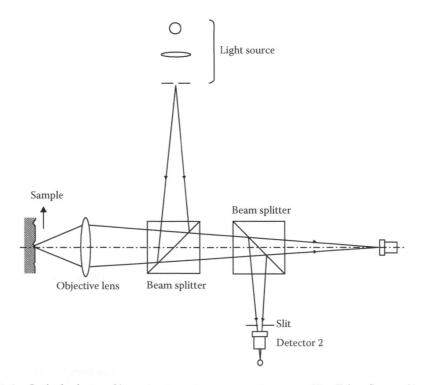

FIGURE 19.6 Optical scheme of intensity detection method. (Courtesy of Dr. Takeo Sawatari.)

TABLE 19.2
Characteristics of the Prototype (1979)

Stand-off distance	1.38 μm
Linear range	12.7 μm
Linearity	±1.0%
Error	Max. 0.25 μm
Minimum spot size	10 μm
Scanning speed	2.3 μm/s

(M, lateral magnification). Using a simple geometrical approximation for $z \ll d/M^2$, the intensity is expressed as a function of a displacement z as follows:

$$I_2(z) = \frac{\Delta s I_1 (b-d)}{\pi d^2 r_0} M^2 z + I_2(0).$$ (19.1)

where

I$_2$(z) and I$_2$(0) are the intensities of the detector Ph$_2$ at the longitudinal surface positions z and 0 (reference position), respectively

Δs is the slit width

r_0 is the radius of the pupil of the objective lens

b is the distance from the pupil plane of the lens to the initial image point P$_2$ (for z = 0)

As the sample is moved in x direction, z is calculated by the equation, that is, the profile of the surface is obtained by this principle.

The profile of a machined surface obtained by this optical transducer correlates well with the result by a mechanical profilometer with a diamond stylus. This paper was published in 1979, and the specification of the prototype instrument shown in Table 19.2 may not be surprisingly high at this point of time. However, this suggestive paper is still useful to know the fundamental principle of intensity detection method in profilometry.

19.3.2 FOCUS-ERROR DETECTION METHOD

A number of methods have been proposed for a noncontact displacement sensor using the focus-error detection system. The most fundamental principles were presented in 1980s including the astigmatism [6], knife-edge [7], and critical-angle methods [8]. These methods can attain high resolution by using simple optical configurations.

In Figure 19.7, focus-error detection using astigmatism method is shown where a cylindrical lens is inserted intentionally to produce anastigmatic aberration.

A quadrant detector (with four elements) is available to judge where the object surface is positioned: whether at the focus position or this side or that side of the original focus point.

The critical-angle method using a right angle prism is presented by Kohno and other three persons in 1988 [8], and this excellent principle was adopted to a commercial instrument called HIPOSS (high precision surface sensor). Inspection of machined surface by single point diamond turning and semiconductors require noncontact precise surface sensors instead of conventional mechanical stylus methods. The main reason was in that the conventional stylus may scratch the surface of soft metal, thin films, and silicon wafer. One result showed such high resolution as 1 nm in the vertical direction and proved to be applicable to a surface with inclination of ±5° by this compact and light weight system.

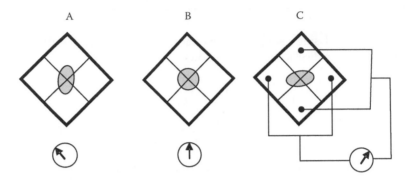

FIGURE 19.7 Principle of astigmatism method.

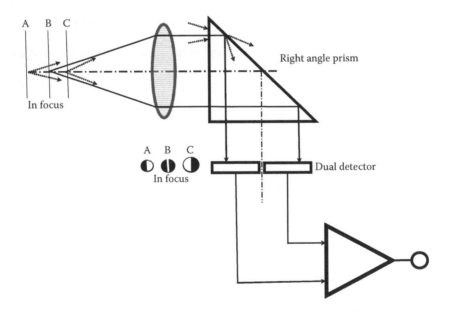

FIGURE 19.8 Principle of critical-angle method. (Courtesy of Prof. Tsuguo Kohno.)

Figure 19.8 shows the principle of the method. After passing through the microscope objective, the reflected light converted into parallel flux if the surface under test is in focus at the position B. A total reflection prism is located to reflect the light at the critical angle, and thus, the same levels of light intensity are incident on the photodetector with dual element. Consequently, the out-of-focus signal becomes zero. When the object surface is at position A, close to the lens, the light flux diverges slightly after passing through the lens. As a result, the light on the upper side of the optical axis shown in the figure strikes the prism at an angle smaller than the critical angle. This causes the

TABLE 19.3

Specifications of Rodenstock Model RM 600 3-D/C

Vertical measuring range	300 μm
Vertical resolution	0.01 μm
Horizontal measuring range	100 × 100 mm
Horizontal resolution	1 μm
Sampling rate	2000 Hz, maximum
Maximum measuring speed	260 mm/min
Maximum positioning speed	800 mm/min
Maximum load	20 kg
Specimen table size	185 × 185 mm

light to be reflected, and it passes out of the prism. The light on the lower side of the optical axis is totally reflected at a large incidence angle, creating a difference in the output of the photodiodes and thereby producing an out-of-focus signal. At position C, far from the lens, the opposite phenomenon occurs and a signal with the reverse sign is obtained (Table 19.3).

The real optical system of a microfigure measuring head is illustrated in Figure 19.9. A laser diode with a 780 nm wavelength is incorporated in this head, and a polarized beam splitter (PBS) and a quarter-wave plate is provided to serve as an optical isolator. This design is employed to prevent noise caused by light radiation process. A half-mirror is used to split the optical paths to avoid any effects of object surface inclination on the measured results. By calculating the next signal

$$E = \frac{(A-B)+(C-D)}{A+B+C+D},$$ (19.2)

displacement in the vertical direction can be exclusively detected. The arrangement that the light is reflected toward inside the prism improves detection sensitivity. Figure 19.10a shows one measurement result together with the result using a stylus method (Figure 19.10b). Another principle is based on Foucault method. One example is shown in Figure 19.7 where the focal point is determined by

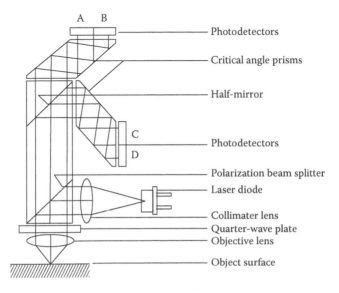

FIGURE 19.9 Optical layout of HIPOSS system. (Courtesy of Prof. Tsuguo Kohno.)

FIGURE 19.10 Comparison of measurement result of diamond turned Al surface: (a) by HIPOSS and (b) by mechanical stylus. (Courtesy of Prof. Tsuguo Kohno.)

using a pair of dual-element detectors. Here the objective lens is controlled to find the focal point by detecting the balance of light volume given to this pair and the up-and-down moving distance of the lens gives the distance from the lens to the measuring point, that is, surface profile [9] (Figure 19.11).

One more similar commercial system is the Rodenstock Model RM 600 3-D/C topography measuring station which is used for mapping the topography of the surface [10]. It is applicable not only to conventional surface but to delicate surfaces which may be damaged by a contact stylus. Also a noncontact optical sensor which uses a dynamically focused infrared laser is incorporated. Focal point diameter is approximately 2 μm. The displacement of the objective lens is measured inductively, and this change is proportional to the vertical displacement of the surface point to be measured.

However, it is frequently indicated that unreasonable displacement signals, which do not appear in the case of a stylus instrument, are superposed on measurement results, depending on the material

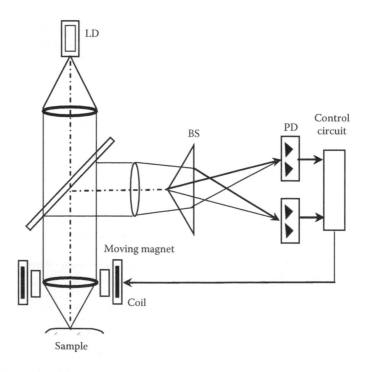

FIGURE 19.11 Principle of focus detection method (UBM, UB16).

and processing method of the surface. One reason of this problem is the speckle noise included in the reflected light due to scattering at the measured surface. To solve this problem, a novel optical stylus for surface profiling system is reported which consists of a displacement measurement optical system based on the knife-edge method [11].

Optical stylus measurement system using any kind of focus-error detection makes use of segmented photodiodes to detect reflected light. The displacement signal is obtained from the difference in an output of each photodiode. The intensity distribution of reflected light is supposed to be uniform or to follow a Gaussian distribution. But the speckle pattern causes unequal intensity distribution, when microirregularity exists on a measuring surface. It can be seen that speckle pattern changes according to location even for the same sample. Consequently, if reflected light with the unequal speckle pattern is received by a segmented photodiode, the displacement signal might be affected by the speckle and measurement error might be arisen. Figure 19.12 shows the configuration of a displacement measurement system for removing the influence of speckle noise (intensity irregularities). In the figure, section A enclosed by the broken line represents the displacement measurement optical system using the knife-edge method, and section B the system for detecting speckle noise. The light source is LD (λ: 670 nm) with a beam diameter of 3.5 field lens L_2, f = 50 mm. Light irradiated from an LD is focused onto the measuring surface by the objective lens L_1. The light reflected back from the surface passes through the beam path in the reverse direction, and is deflected by a PBS. One beam of light is passed through a knife-edge and detected by two-segmented photodiode (2PD-1), and displacement signal Esn is operated. The other reflected beam of light is bent by the beam splitter (BS) and directly detected by 2PD-2, and signal Er is operated. Here, Esn is a displacement signal in which the speckle noise is included, while Er is a signal having the speckle noise only. Operating difference between Esn and Er, that is, $Es = Esn - Er$, can therefore generate a displacement signal with the speckle noise removed.

These principles mentioned earlier are similar or essentially same with the principle that have been used for optical pickups used for digital memory disks such as a compact disk (CD) or a DVD (digital versatile/video disk). This means commercially manufactured CD pickups are potentially available for distance sensor to measure the surface profile.

Bartoli et al. proposed to revise an optical pickup (SONY KSS-210A) to utilize as a displacement sensor [12]. The use of principles for CD pickups as distance sensors in optical profilometry

FIGURE 19.12 Configuration of optical stylus sensor for removing speckle noise. *Notes*: A, focus error detection system by the knife-edge method; B, measurement system for the intensity irregularity. (Courtesy of Prof. Hiroya Fukatsu.)

is a well-established practice. A sophisticated application of this device is: by making use of built-in objective lens-tracking actuator of the pickup to perform one-dimensional beam scanning of the sample. This prototype profilometer is made up of a standalone sensor-actuator optical head that can be easily positioned on samples of complex shapes with every orientation and can be operated by a remote control system. When two-dimensional scanning is unnecessary or even impossible, as for surface quality inspection of bulky mechanical pieces, a device that uses a CD pickup as both the sensor and the actuator is cheaper, faster, lighter, more compact, and more versatile than standard object-scanning setups that use x–y translation stages. A lateral scan range of 800 mm and a vertical scan range of 160 mm are achieved. This example suggests a good application of devices manufactured in large numbers. These products are reliable because of quality control in manufacturing.

19.3.3 CONFOCAL METHOD

Another principle is found in Figure 19.13 where a kind of coaxial confocal microscopic method is found [13]. The waist point of a laser beam projected on the specimen surface is scanned along the longitudinal axis by a moving mirror and timing that the spot size becomes smallest is measured. Generally, in coaxial confocal microscopic method, a beam irradiated from LD is focused on a target point on the surface through the lens.

Then the reflected beam moves back through the lens to the detector with a pinhole in front. In most of the cases, this focused point is controlled by the lens so as to be in focus, that is, to be in conjugation with the pinhole. Therefore, if the lens is moved up and down, this system works as a displacement meter to give a distance to the measuring point. However, moving the lens speedily is not always easy, and here a small and light weight mirror is moved to control the spot position. Schematic construction of a prototype sensor is seen in Figure 19.14, and some specifications are also shown in Table 19.4.

19.3.4 CONTRAST DETECTION METHOD

Several types of optical sensors based on triangulation are widely used for many applications due to their simple technique. Triangulation may be applied to both displacement detection and profile

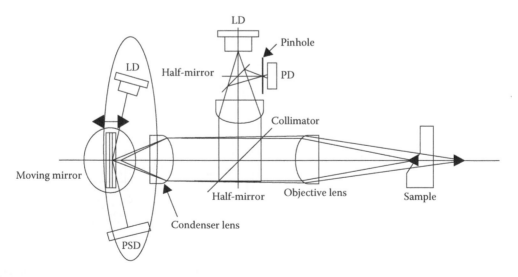

FIGURE 19.13 Principle of confocal coaxial sensor head. (Reproduced on the basis of the Japan Patent No. 3809803.)

High-precision type

Normal type

FIGURE 19.14 Configuration of prototype model. (Reproduced on the basis of the Japan Patent No. 3809803.)

TABLE 19.4
Main Features of Two Prototype Models

	Type 1	Type 2
Stand-off-distance (mm)	13.8	12.0
Measuring range (μm)	±50	±500
Linearity (μm)	±0.5	±2.0
Resolution (μm)	0.01	0.1

TABLE 19.5
Specifications of a Prototype Probe

Stand-off distance	2.3 mm
Linear range	150 mm
Vertical resolution	1 μm
Linearity	±1.3%
Projected pattern size on the object	900 μm

measurement. Many optical displacement sensors based on triangulation use semiconductor PSD or CCD (Table 19.5).

These sensors produce erroneous measurements if the surface of the object is not uniform in color or if it is inclined at an angle. In these cases, the intensity of light received by the detector varies with color or angle and the output signal shows erroneous indication of displacement. Such defects are attributed to the fact that light projecting and receiving units are not on the same optical axis but lie along two different axes. Various kinds of displacement sensors that adjust a focus along a single axis have been developed to overcome this type of error. To solve the problems relating to the influence of color or inclination of an object surface, an apparatus designed to sense the variation of contrast of a projected pattern is developed [14]. This sensor uses the variation of contrast of the edges of a quadrant pattern of light and shade region that is projected on an object along a single optical axis.

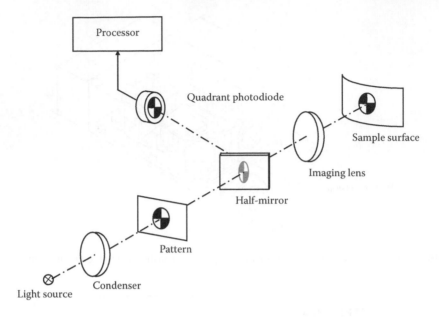

FIGURE 19.15   Optical arrangement for contrast detection method.

The optical system for displacement detection is shown in Figure 19.15. A quadrant pattern consisting of highly distinct light and dark regions is projected by an objective on the object's surface. The quadrant pattern image thus projected on the surface is reimaged by the lens and reflected by a half-silvered mirror onto a quadrant photodiode (hereafter referred to as 4PD) of the same size as the original quadrant pattern. The pattern projected on the surface is such that it can be observed by the naked eye. The size of the pattern useful for detection can be adjusted on the 4PD by selecting the appropriate objective lens. Contrast of the quadrant pattern image becomes higher when the image and the surface of the object coincide. The image on the surface becomes low in contrast as the object is displaced from this ideal image position. Thus, the axial displacement of the surface along the optical axis is determined by sensing the contrast of the image as is shown in Figure 19.15, the contrast of the pattern is obtained as signal S is computed so that it is largely uninfluenced by differences of reflectivity of the surface or variations of the intensity of the light source. Signal S is given by

$$S = \frac{\left(I_1 + I_3\right) - \left(I_2 + I_4\right)}{I_1 + I_2 + I_3 + I_4} \tag{19.3}$$

where
   I_1 and I_3 are the signals due to the bright regions of the pattern
   I_2 and I_4 are the signals due to the dark regions of the pattern

In the case of an objective lens of 20× magnification, the range of measurement of this displacement sensing apparatus is linear from 100 to 250 μm behind the image position.

When the surface is inclined by 30°, errors remain within 3%. Error due to color variations is the same. The measurement technique may be used at inclinations up to 45°. The results obtained with the objective lens of 20× magnification are summarized in Table 19.5, which indicates that vertical resolving power obtained over a range of 150 pm was 1 μm. Although the visual diameter of the pattern projected on the measurement object surface is 900 μm, the actual diameter used is 200 μm, since the region where light is detected on the 4PD corresponds to a central region with the diameter determined within a range of 200 μm.

19.3.5 CONOSCOPIC HOLOGRAPHY METHOD

As is mentioned previously, there are a lot of reports and proposals on a coaxial type of optical probes. Among these sensors, one of the most successful commercial products is known as "ConoProbe" (Optical Metrology Ltd.) [15,16]. Figure 19.16 shows the optical arrangement to represent the fundamental principle. This shows the arrangement used for conoscopic hologram recording. Let a light wave with wavelength λ propagate in a birefringent crystal at an angle φ relative to the optical axis. If the ordinary and extraordinary indices of refraction are n_o and n_E, respectively, and their difference is $\Delta n = n_E - n_o$, then two orthogonally polarized waves will propagate, one with an index of refraction n_o (the ordinary ray), and the second, the extraordinary ray, with an index of refraction n_E given approximately by

$$n_E(\theta) \approx n_o + \Delta n \sin^2 \theta \qquad (19.4)$$

The phase retardation between the extraordinary and ordinary waves is given by

$$\Delta\phi = \left(\frac{2\pi}{\lambda}\right)\left(\frac{L}{\cos\theta}\right)\Delta n \sin^2\theta \approx \left(\frac{2\pi L}{\lambda}\right)\Delta n \theta^2 \qquad (19.5)$$

where L indicates the length of the crystal and $\theta \ll 1$.

This expression means that the incident angle (which is dependent on the distance from the target point) determines this retardation $\Delta\phi$, that is to say, distance from the target point to the crystal is given by $\Delta\phi$. And, due to this phase retardation $\Delta\phi$, interference similar to polarization interference is brought. The interference fringe shown in the figure varies in accordance with the angle θ, that is, the distance to the object. Displacement can be calculated by processing the fringe pattern.

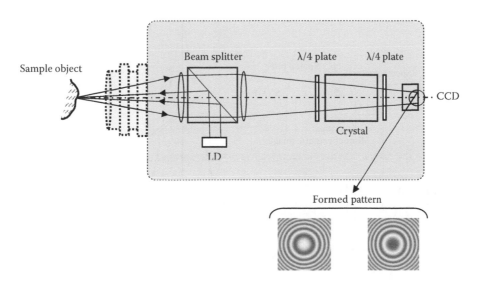

FIGURE 19.16 Principle of ConoProbe sensor. (Courtesy of Dr. Gabriel Y. Sirat.)

19.3.6 Chromatic Confocal Imaging Method

A commercialized optical sensor named "CHR" is constructed on the basis of a quasiconfocal microscopic configuration with extended range in z-direction [17]. This extension, a large range of measurement, is attained by spectral (chromatic) coding in the z-axis. In traditional confocal microscopy, when x–y scanning is added, we obtain the image of a plane located at the same distance from the objective lens, eliminating influence from points outside this thin plane. This property is called "optical sectioning" in z-direction, and it constitutes the principal feature of confocal microscopy. As is easily understood in Figure 19.17, this optically sectioned plane is determined by the wavelength to be used (now the wavelength: λ_0). Then, if we change the wavelength λ continuously (instead of moving some mechanical element like a lens), we can achieve "dynamic focusing" easily.

Details of this sensor are described on the webpage of STL SA, and the characteristics of the CHR sensor are given as follows:

- The principle of confocal imaging yields an excellent spatial resolution regardless of ambient illumination.
- The chromatic coding ensures that measurement is insensitive to reflectivity variations in the sample and allows to work with all types of materials, transparent and opaque, specular or diffusing, polished or not, with no need to treat the sample prior to measurement.
- The use of a white light source and not of a coherent source (laser) eliminates completely all the difficulties associated with speckle.

This principle is so simple that it has been known for a long time; however, it was not easy to produce commercial products. Because the indispensable key devices such as spectrometers were not small and because small optical elements like lenses were not so popular. Due to miniaturization of elements and devices (such as a palm-top spectrometer), pen types of CHR sensors are now available in the market. Some specifications are shown in Table 19.6.

The innovative setup of chromatic confocal imaging opens new ways for simultaneous extended field confocal scanning optical microscopy and 3D profilometry with an absolute reference surface in situ.

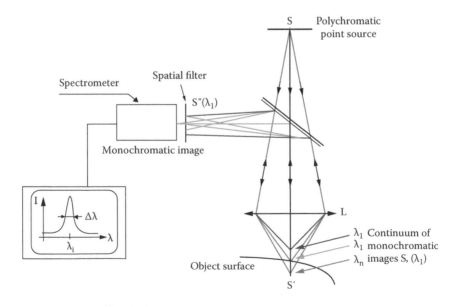

FIGURE 19.17 Principle of CHR sensor. (Courtesy of STILL, SA, France.)

TABLE 19.6

Examples of CHR Sensors for Profilometry

Specifications			Probe for Profilometry			
Measurement range (μm)	300	600	30,000	60,000	10,000	25,000
Resolution (nm)	10	20	100	200	300	800
Accuracy (μm)	0.1	0.2	1	2	3	8
Working distance (mm)	4.5	6.5	22.5	36	70	80
Spot size (μm)	5	4	12	16	24	25

19.4 FULL-FIELD MEASUREMENT

In addition to noncontact measurement, optical methods using an imaging technique have such an excellent feature that one scene can be captured in one moment. Therefore, a full-field measurement (sometimes called "whole-field measurement") is an attractive matter in two- and three-dimensional profile measurements, and this "full-field measurement" cannot be attained by the point-wise measurement techniques described in Section 19.3.

19.4.1 SHAPE FROM FOCUS/DEFOCUS METHOD

One of techniques for three-dimensional profilometry is classified into "shape from X" methods in which only one camera (not one pair of cameras) is incorporated. Many SfX methods have been proposed. For example, X represents "shade," "silhouette," "motion," "texture," and so forth. One principle mentioned already [14] suggests its potential applicability to full-field measurement, because the fact that contrast variation of a projected pattern depends on distance might have been extended to the full-field measurement. In fact, one excellent method called "shape from focus/ defocus" was developed to find applications mainly in semiconductor industry [18].

The methods based on triangulation including pattern projection methods suffer from serious drawbacks. One is occlusion, which brings shadings due to two optical axes. Another is multiple reflection caused by specular parts on the object surface. Furthermore, the fringe analysis method such as phase shift method and Fourier-transform method also have a drawback that measurement results are wrapped in a 2π area. This means these methods have difficulty in measuring the surface with discontinuities such as large steps and/or isolated areas. One method without any of these drawbacks is the confocal method. The confocal method performs surface measurement by using optical sectioning capability. This method allows for measurements with an accuracy of less than 1 μm and does not have any drawback inherent to triangulation, because the axes of lighting optics and imaging optics are coaxial. However, the confocal method is a time-consuming method. Since it takes several tens of seconds per scene for measuring, the method is not appropriate for in-line inspections. In order to accelerate the confocal method, there are three problems to overcome. First, conventional confocal microscopes cannot perform high-speed imaging, because the laser-scanning system takes a lot of time for 2D scanning and the Nipkow-disk scanning system cannot obtain sufficient light intensity by short time exposure. Second, the number of confocal images to be processed is enormous.

Finally, the z-axis stage, which conventional confocal systems use for axial scanning, cannot scan fast in discrete steps.

Figure 19.18 shows the construction of the nonscanning multiple-beam confocal microscope. In several respects, this system is based on the Nipkow-disk scan type confocal microscope. However, this system has no scanning mechanism. Instead of a scanning mechanism, this system has a stationary pinhole-array and a micro-lens-array that consists of the micro-lenses being coaxial to the pinholes in the pinhole-array. The pinhole-array and the micro-lens-array are fabricated on both sides of a glass substrate, and the number of pinholes in the pinhole-array, which is equal to that of

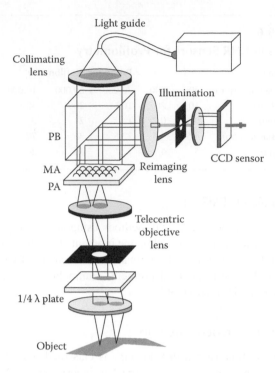

FIGURE 19.18 Nonscanning multiple-beam confocal microscope. *Notes*: PB, polarizing beam splitter; MA, micro-lens-array; PA, pinhole-array. (Courtesy of Dr. Mitsuhiro Ishihara.)

micro-lenses, is almost the same as that of CCD pixels. The role of the reimaging lens is to form an image of the micro-lens-array on the CCD sensor, not to form the image of the pinhole-array onto the CCD sensor such as the Nipkow-disk system. The effects by using the micro-lens-array are as follows: (1) high illumination efficiency, (2) wide divergence angle, (3) no necessity of alignment between the pinholes and CCD sensor pixels. This system uses a halogen white lamp as its source instead of a coherent light source to eliminate noises by simple optics.

Let us describe this procedure referring to Figure 19.19. The confocal images $I_1, I_2, \ldots, I_n$ indicate the images that were obtained at positions $z_1, z_2, \ldots, z_n$ where the image of the pinhole-array of the confocal system is formed. When pixel values at one certain identical location (x_m, y_l) in all images are arranged in the order of the z-axis position, we can see that the pixel values are obtained as sampled values from the axial response curve. The peak position of the response curve (z_{focus}) is estimated by an interpolation technique. When Gaussian function is adopted with the constant sampling interval Δz, the peak position z_{focus} is given by the next expression:

$$z_{focus} = z_k + \frac{\left(\ln I_{k+1} - \ln I_{k-1}\right)\Delta z}{2\left(2\ln I_k - \ln I_{k+1} - \ln I_{k-1}\right)} \tag{19.6}$$

Figure 19.20 is a schematic diagram of the system which incorporates a rotating disk with different thickness of parallel glass plates embedded for adjusting the focus. The image processor with specific pipelined hardware permits to obtain a depth image with 14-bit resolution per pixel. In this system, since the image sequence contains 18 confocal images, it takes about 20 ms to obtain 1 image, and the total measurement time including data processing time is less than 0.4 s. Figure 19.21 shows two measurement results: (a) a Japanese coin, too specular for triangulation measurement and (b) a ceramics workpiece with a lot of poles 0.1–0.5 mm high. The measurement of the calibration standard verified that the system had a measurement speed of 0.4 s for the measurement range of 9.6 × 9.6 × 0.64 mm with an accuracy of 1 μm.

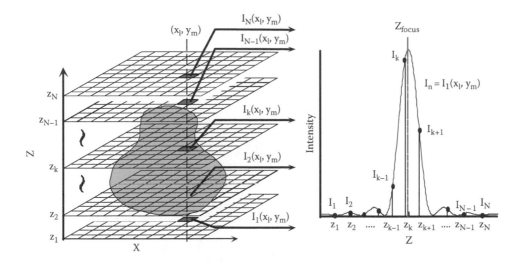

FIGURE 19.19 Principle of fast procedure of confocal images. (Courtesy of Dr. Mitsuhiro Ishihara.)

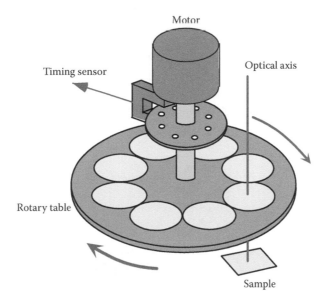

FIGURE 19.20 Fast focus shifting using rotary disk with glass plates. (Courtesy of Dr. Mitsuhiro Ishihara.)

19.4.2 Pattern Projection Method

Here we bundle several principles used for three-dimensional profilometry such as space encoding method, fringe projection method, structured light method, and even moire topographic method in one group called "pattern projection method." When some kind of pattern (line pattern, grating pattern, checker board pattern, circle pattern, colored pattern, etc.) is projected onto an object to be measured, the pattern deformed in accordance to the surface profile of the specimen is captured from a different angle direction, and then this deformed image is analyzed by using some algorithm. We can easily understand the fundamental principle for space coding method in Figure 19.22.

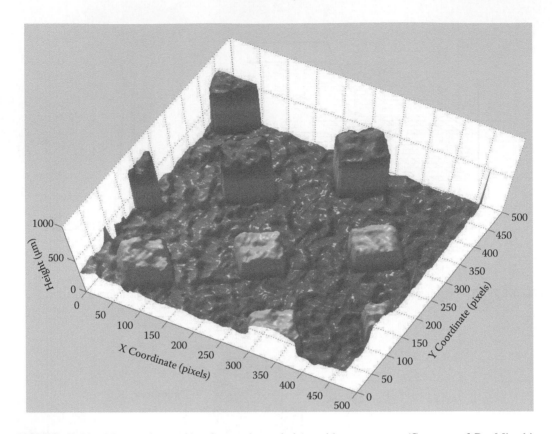

FIGURE 19.21 Measured example of ceramic workpiece with steep steps. (Courtesy of Dr. Mitsuhiro Ishihara.)

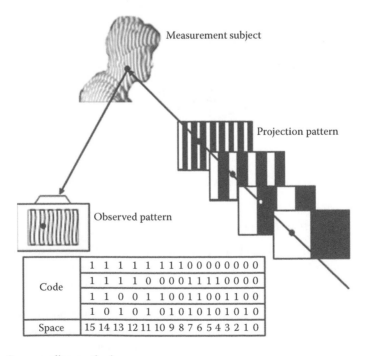

FIGURE 19.22 Space-coding method.

19.4.2.1 Traditional Techniques

Figure 19.23 also represents a fundamental optical arrangement for pattern projection methods. Any kind of reference pattern is projected onto an objective surface through a projection lens. Then, when the pattern is observed on another optical axis, a deformed or modulated pattern is seen. This deformed image involves every information on surface shape of the object. Hitherto a lot of patterns applicable to this technique has been proposed by many researchers. However, we should note two points: (1) the pattern should be formed easily, and (2) the deformed pattern needs be appropriate for analysis. Some examples are shown in Figure 19.24. This pattern should be matched with the analyzing algorithm. For example, a pattern with a sinusoidal intensity distribution is required to bring highly accurate result as far as phase-shifting techniques based on most popular three- or four-step shifting are applied. And sometimes switching a pattern to another one (changing the period of a line pattern) is requested. Figure 19.25 shows a fundamental necessity for changing the period of the reference pattern. To realize this idea, one interesting trial is found in using projection of optical interference fringe produced by acousto-optic technique [19].

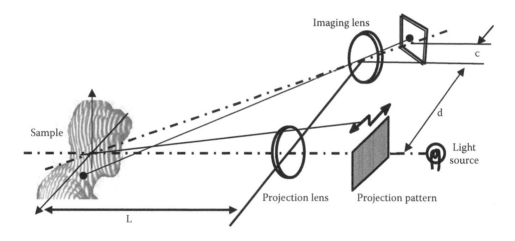

FIGURE 19.23 Fundamental arrangement for pattern projection method.

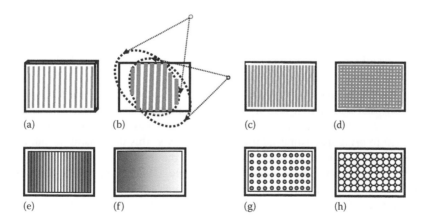

FIGURE 19.24 Various patterns for projection method: (a) Ruled grating, (b) interference fringe, (c) stripe LC grating, (d) DMD-based pattern, (e) colored pattern, (f) gradation pattern, (g) fiber grating, and (h) colored-dot pattern.

(a)

(b)

FIGURE 19.25 Necessity of period control in pattern projection method. (a) Period 15 mm and (b) period 10 mm.

For easy production of changeable pattern and giving phase-shifting, liquid crystal (LC) grating and digital mirror devices (DMD) are available. The architecture and functionality of the two digital technologies, namely, LC grating and digital light processing (DLP) using DMD, were extensively reviewed [20]. They have demonstrated experimentally that LCD projection supersedes DLP projection technology in terms of accurately producing a sinusoidal fringe pattern; however, the image preprocessing of each of the tested projectors was assumed similar. Future work should consider the validity of this assumption. DLP projection sources surpassed LCD in terms of the geometric precision of LCD in regard to the "screen door effect: tiny discontinuities in the digitally projected intensity distribution." In addition, they demonstrated that for dynamic multichannel phase measuring profilometers, LCD projections brings lower color crosstalk relative to DLP projection technology. They occluded that such implications should be considered when designing triangulation-based optical profilometry systems.

19.4.2.2 Digital Techniques in Pattern Projection

Up to now, many methods and applications for three-dimensional shape measurement have been proposed, and recently, the measurement accuracy is becoming higher and higher. And, in most of the cases, phase-shifting technique is applied to analyze the projected fringe pattern. To attain highly accurate result in such measurement systems, phase-shifting technique should be applied under the fundamental assumption that the projected pattern is sinusoidal, not binary, in intensity distribution. However, sinusoidal pattern is not always easy to make and sometimes, defocusing in projection is used to bring quasisinusoidal pattern. However, this defocusing technique is not sufficient for the precise and accurate result. And, even if the pattern is exactly sinusoidal, we have to move the pattern precisely for phase-shifting technique. Conventionally the projected pattern is shifted by mechanically moving some element such as a grating. Even if an electric actuator may be used for moving the grating on a mechanical stage, this way of movement is not satisfactory for practical inspection on shop-floor, because it has problems such as a long measurement time due to mechanical movement and errors caused by unstable repeatability in positioning. In addition, we have necessity that the grating period should be adjusted appropriately in accordance with the measurement range and/or inclination of the sample surface. To overcome these problems, availability

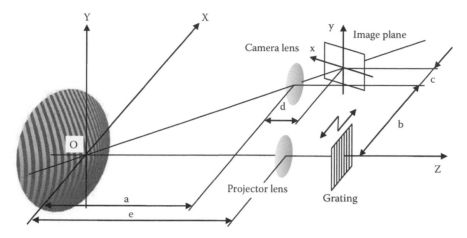

FIGURE 19.26 Principle of pattern projection method.

of a stripe-type LC grating instead of a conventional grating ruled on a glass plate or optical interference fringe has been proposed [21,22].

Figure 19.26 shows a profile measurement system using grating projection method based on phase-shifting technique. Here, the three-dimensional coordinate (X, Y, Z) of the point P on the sample is defined by the following equation using the period p of the original grating, phase $\phi(x, y)$ of the grating pattern on the image plane (x, y) of the CCD camera, the magnification m of the camera lens, and geometrical parameters shown in the figure:

$$X = \frac{e - a + \alpha b}{\alpha + \dfrac{e}{w}},$$

$$Y = -\frac{\sqrt{(b-X)^2 + (a-Z)^2}}{\sqrt{(b-X')^2 + (a-Z')^2}}\, y, \tag{19.7}$$

$$Z = -\frac{X}{w}\, e + e$$

where

$$X' = b + \frac{1}{\sqrt{a^2+b^2}}(1+x), \quad Y' = y, \quad Z' = a + x\,\frac{1}{\sqrt{a^2+b^2}}(1-x), \quad w = \frac{p\phi}{2\pi}, \quad \alpha = \frac{Z'-a}{X'-b} \tag{19.8}$$

Here, when the grating with the period of p is projected onto the object, the point T(X,Y,Z) on the surface of the object is captured at the point t(x,y) on the imaging plane. And the point T corresponds to the point C(w, 0, 0) on the X–Z plane.

The intensity distribution I_i (i = 1 ~ 4) of the projected pattern is given as follows:

$$I_i = I_0 \cdot \left[1 + \gamma(x,y) \cdot \cos\{\phi(x,y) + \delta_i\}\right], \tag{19.9}$$

where
  I_0 is an illuminated intensity
  $\gamma(x, y)$ is a contrast
  $\phi(x, y)$ is the initial phase which expresses the profile
  δ_i (i = 1–4) is given by shifting of 0, $\pi/2$, π, $3\pi/2$, respectively

(a)   (b)

FIGURE 19.27 Example measurement of a lily; expression by (a) contours and (b) shading. (Courtesy of Dr. Masayuki Yamamoto.)

Here, "four-stepping phase-shifting algorithm" is adopted by which the phase $\phi(x, y)$ is determined as follows:

$$\phi(x, y) = \tan \frac{I_4 - I_2}{I_1 - I_3}. \tag{19.10}$$

The phase distribution given by this equation is wrapped in a 2π area; hence, the final result should be produced after unwrapping procedure for smooth connection of phase.

The stripe-type LC grating is suited for applications such as phase-shifting technique, because the projected pattern can be electrically controlled without any mechanical movement of the optics. One example measured by a system using LC grating projection is shown in Figure 19.27.

Here, a newly structured LC grating is illustrated in Figure 19.28. This stripe-type LC grating of 40 mm × 60 mm in size with 960 stripes is developed for monochromatic use (without any colored filters). Owing to the stripe-type of periodical configuration, the intensity shows uniform distribution with smaller noise shown along the line. To realize high contrast of the LC grating, nematic (TN) type of liquid crystal is chosen. The characteristics of TN type are that they require low electric power, and smallness in size, and have a high dynamic range.

One application using this LC grating is shown in Figure 19.29 which indicates an optical configuration with dual-projection optics [23]. This system contains one CCD camera and two projection units with telecentric nondistortion lenses for simplification of calculation. This system was originally developed to explore applicative usage in semiconductor industry. Conventionally such problems were caused as occlusion and phase error due to halation from the surface. The first problems is inevitable as long as triangulation is adopted. The second one is caused mainly by unequal distribution of surface reflectivity of the BGA, CSP, solder bumps, etc. These problems can be solved by using dual-projection system. Two images from both directions are useful for redeeming mutual week points. One measurement result of a BGA is shown in Figure 19.30. Here the field of view is 14 mm × 14 mm and depth of field is approximately 1.0 mm. The three-dimensional expression of the BGA sample (a) and the cross section (b) are represented.

Experimental results verify this system is so flexible that it is appropriate for industrial applications.

19.4.2.3 Trends in Profilometry

In recent years, surface profile measurement has become more and more important in various industrial applications. Two-dimensional and three-dimensional optical profilometry has been

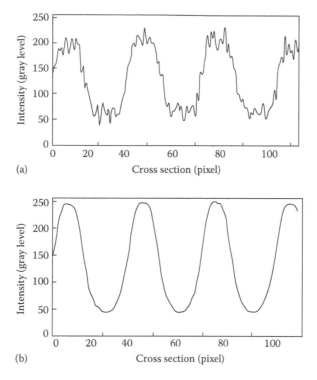

(a)

(b)

FIGURE 19.28 Grating using (a) matrix-type LC and (b) stripe-type LC.

FIGURE 19.29 Dual-projection system with LC grating.

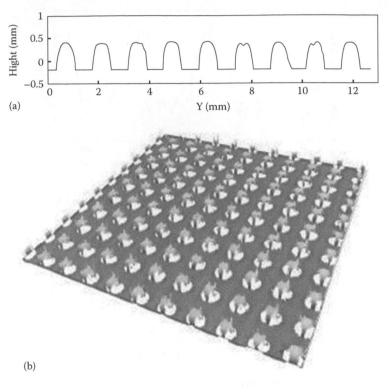

FIGURE 19.30 Measured result of BGA without occlusion. (a) Cross section of BGA and (b) reconstructed image of BGA without occlusion.

successfully applied to various industrial fields such as shape measurement of a soft object including a human body as well as hard objects like a car body. For more prevalent use of commercial equipment, a few more tasks are proposed from the viewpoint of practical use. Let us introduce recent research works to cope with these requirements.

19.4.2.3.1   High-Speed Measurement

Nowadays high-speed measurement systems are too heavy in practical use of optical equipment for profilometry. The pattern projection method, one of the successful principles based on fringe analysis, has been widely applied to various academic and industrial fields. To expand this method to wider purposes such as quality check of manufactured components, three-dimensional sensing for robot vision, and dynamic analysis of a deforming object, high-speed measurement has to be attained immediately soon. There are two points to satisfy this request. First, data need to be acquired at one moment, and, second, the data have to be processed with high speed and accuracy to meet the purpose.

Phase-measuring systems have become increasingly popular in commercial equipment because of an accuracy superior to intensity-based measurement systems depending on optical sectioning or space coding technique. However, in fringe projection using phase-shifting technique, the resultant wrapped phase is modulo 2π. And this unwrapping problem has been a difficult and serious task for a long time. Traditional phase unwrapping algorithms involve the calculation of an arctangent function, which is a bottleneck in the case that real-time processing is required.

Coggrave and Huntley realized a high-speed profile measurement using a commercial projector and a pipeline image processor [24]. In the case of sinusoidal patterns projection using a commercial projector with DMD inside, the resulting fringe patterns are acquired by a CCD camera at a rate of 30 frames/s. The images are analyzed in real time using a pipeline image processor. The combination of four-step phase shifting and temporal phase unwrapping enables discontinuous objects to be profiled as easily as continuous ones. An optimized sequence of fringe pitches is used in which the spatial frequency is reduced exponentially from the maximum value. A total time of 0.87 s is required from the start of the measurement process to the final display of a surface profile consisting of 250,000 coordinates. An accuracy of 1 part in 2000 was achieved with a maximum fringe density of 16 fringes across the field of view. This pioneering research result verified possibility of real-time measurement for robotics and in-line measurement in manufacturing.

As another solution, one trial was proposed which also makes use of digital technique. In conventional grating projection based on analog technique, a pattern with sinusoidal intensity distribution was difficult, and moreover, rectangular pattern, triangle pattern, saw-teeth pattern, etc. were nearly impossible to produce. By taking advantage of a DMD projector, Huang, Zhang, and Chiang proposed a novel (or a little bit tricky) phase-shifting method, namely, the trapezoidal phase-shifting method, that used trapezoidal fringe patterns instead of the traditional sinusoidal ones [25]. By calculating an intensity ratio using a simple function, instead of phase using the arctangent function, they succeeded in increasing the calculation speed by at least 4.6 times with similar resolution, which made the real-time reconstruction of 3D shapes possible (even 40 frames/s at the resolution of 532 × 500 pixels). However, the use of trapezoidal fringe patterns led to some error due to image defocus (as is anticipated, sharp edges of trapezoidal patterns are easily blurred), even though the error is small, especially when compared to the traditional intensity ratio–based methods.

According to their paper [26], in the course of trying to deal with this error, they discovered that if they apply the algorithm developed for the trapezoidal method to sinusoidal patterns, considering the sinusoidal patterns as the defocused trapezoidal patterns, and then compensate for the small error due to defocus, they could preserve the calculation speed of the trapezoidal method while achieving the same accuracy of the traditional sinusoidal phase-shifting algorithm. This new three-step phase-shifting algorithm, which is much faster than the conventional three-step algorithm, can achieve the speedy processing by using a simple intensity ratio function to replace the arctangent function in the traditional algorithm. The phase error caused by this algorithm is compensated by using a lookup table. The experimental results showed that the new algorithm had the same accuracy, but was 3.4 times faster. By implementing this new algorithm in a high-resolution, real-time three-dimensional shape measurement system, a measurement speed of 40 frames/s at a resolution of 532 × 500 pixels has been achieved with an ordinary personal computer. In fact, this technique brought a surprising result, as shown in Figure 19.31, in real-time measurement.

FIGURE 19.31   Real-time measurement of human facial expression. The left figure is a real-time reconstruction of the right person's face. (Courtesy of Prof. Song Zhang.)

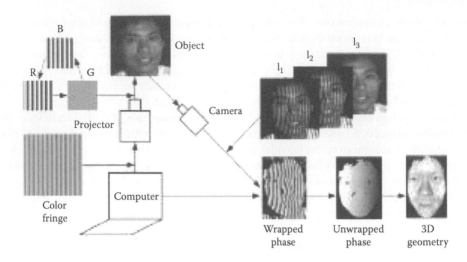

FIGURE 19.32 Schematic diagram for modified two-plus-one algorithm. (Courtesy of Prof. Song Zhang.)

One more result was presented by Zhang, who modified "two plus one phase-shifting algorithm" to alleviate the error due to motion [27]. The schematic diagram of this shape measurement system is shown in Figure 19.32. Their previous work encountered the problem of fast motion such as the facial geometric changes during speaking. In this trial, two fringe images with 90° of phase-shift and a third flat image (not the average of two fringe images with a phase shift of 180°, but a computer-generated uniform flat image) are collected. Their experimental result demonstrated that this system can satisfactorily measure the dynamic geometrical changes. The data acquisition speed attained up to 60 frames/s with an image resolution of 640 × 480 pixels per frame. This system is expected to be applied to online inspection in manufacturing, medical imaging, and computer graphics.

19.4.2.3.2 Reduction in Size and Weight

For the purpose of practical usage, a compact and light weight instrument is preferable. Occasionally, it is difficult to carry the measurement subjects into a measuring room, and more-over, on-line measurement (not off-line measurement) is always necessary in manufacturing process. Recently downsizing has been attainable owing to the rapid progress in device technology. In some cases, compact, portable, and light weight measurement systems are requested, even though the measurement accuracy is not so high. At the same time, high-speed image capturing is indispensable for easy use of such portable systems. Now a handy compact system like, say, a "3D camera" is also required for easy capturing of the objective shape from the fields like robot vision and security checking that do not need such high accuracy as tens of micrometers: One example is a commercial product in range finding (e.g., SwissRanger [28]) on the basis of the time-of-flight method. However, the resolution still remains at some tens of millimeters and beam scanning is required to cover the objective. And the low spatial resolution of 176 × 144 pixels is not available for 3D digitizing.

Hence, to meet the requirement described earlier, a trial that is now under development was reported [29,30]. In this case, the key point is the scanning mechanism. Various methods have been proposed and tried for optical sectioning, and, in most of the cases, devices such as a galvano-mirror, a rotating polygon mirror, or a prism were utilized. Moving elements such as a mirror or a prism may not have the following features: (1) small, compact in size and light in weight; (2) high-speed and stable scanning; and (3) wide scanning angle enough to cover the measuring area.

In considerations of these requirements, one comes to a conclusion that an MEMS scanning mirror is appropriate for the purpose. In fact, a range-finder type apparatus (based on time-of-flight principle) was already reported which made use of an MEMS (Micro-Electro-Mechanical Systems) scanner (manufactured by Nippon Signal Co.), and other prototype range finders (roughly 50×50 mm^2 in size) have been announced to have appeared in the market.

In Figure 19.33, a beam from LD, when reflected by a single MEMS mirror, scans the sample objective to be measured. When the beam is scanned vertically or horizontally by the mirror, the beam makes a sheet of light. This sheet of light is projected onto the object and forms an optical cross section. In the figure, a cylindrical lens is used for making one sheet of light and the sheet is used for scanning. In this case, a MEMS mirror with a size of 4 mm × 3 mm (or 6 mm × 7 mm) is incorporated. Here a MEMS scanner "EcoScan" manufactured by Nippon Signal is used. In recent years, "linear scanning module" fabricated by Venture Forum Mie Ltd. has also been employed to illuminate the fringe pattern. These miniature mirrors are moved magnetoelectrically using resonance phenomenon in torsion bar structure which is effective for making a large scanning angle.

FIGURE 19.33 Optical layout for 3D measurement using one MEMS mirror. (a) Optical configuration of compact 3D camera and (b) prototype compact 3D camera. (From Wakayama, T. and Yoshizawa, T., *Opt. Eng.*, 51, 013601-1, 2012.)

Therefore, even small driving force is enough for attaining vibration with large amplitude angle. Details of this principle are described in the website of the manufacturer [31,32].

If two-dimensional scanning is applied by the MEMS (in truth, Lissajous figures are drawn in 2D scanning, not raster scanning), multiple sections are obtained. This means the various pattern projection methods are also applicable including FFT (Fast Fourier Transform) analysis and phase-shifting techniques.

The experimental results including a pattern formed on a plastic model in Figure 19.34a through d shows that a sharp pattern is possible and that the depth of the measurement field is large enough

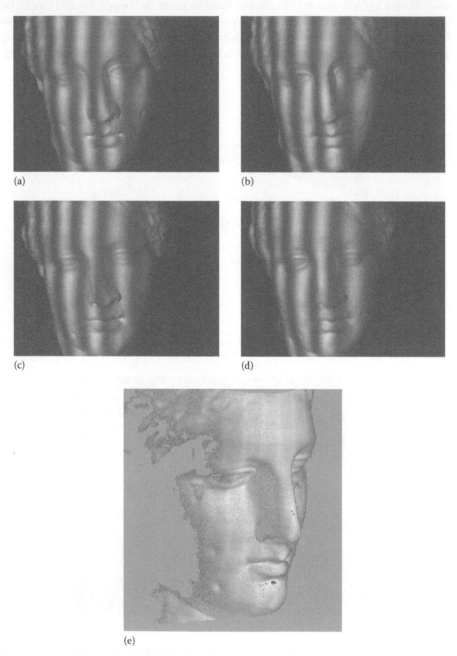

FIGURE 19.34 Measured example: (a) 0°, (b) 90°, (c) 180°, (d) 270°, and (e) 3D representation of a plaster figure. (Courtesy of Mr. Yoshihisa Kamakura.)

without blurring or distortion due to the lens. These figures show precisely shifted phases of fringe pattern at every 90°. Using phase-shifting algorithm, the 3D data are given in the same figure (Figure 19.34e). This technique has the potential not only to produce compact 3D cameras but also to realize high-speed 3D measurements.

Conventional portable 3D cameras (P3D) are presently released from Nikon Corp., as shown in Figure 19.35a. The P3D can easily operate a built-in flat panel display to obtain the 3D data of samples. It captures the 3D shape in a short time with high accuracy. The details of the system is reported [33]. Figure 19.35b shows a result after connecting the 3D data captured from 11 different directions. This 3D camera can also be available in various industrial fields including medical applications [33].

A similar trial was represented in the paper "Cordless hand-held optical 3D sensor," where reflection type with liquid crystal on silicon (LCOS) microprojection technology is incorporated in a digital projection unit together with an LED as the light source [34].

This sensor consists of this projection unit and two cameras in a stereo arrangement which is necessary to use the measurement principle "phasogrammetry." In this technique, shown in Figure 19.36, two sequences of fringes rotated by 90°, generated by a digital projector, are projected onto an objects surface. The projected pattern is captured by two cameras at the same time to

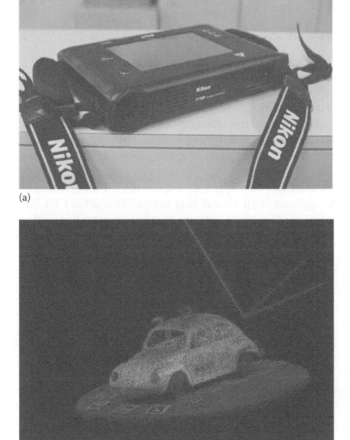

(a)

(b)

FIGURE 19.35 Portable compact camera system for practical use. (a) Portable 3D camera (P3D) and (b) a result of P3D. (Courtesy of Nikon Corp., Tokyo, Japan.)

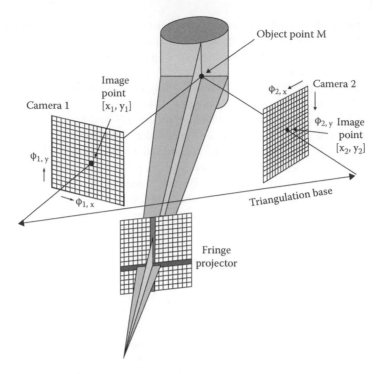

FIGURE 19.36 Fundamental principle of phasogrammetry. (Courtesy of Dr. Christoph Munkelt.)

calculate subpixel accurate pixel corresponding to every pixel from this image sequence, using the principle of phase correlation [35].

The sensor head consisting of two cameras and a miniaturized projector weighs approximately 950 g and measures approximately $130 \times 160 \times 140$ mm³ in volume. One result is shown in Figure 19.37 where a 30×25 mm² patch of a thin paper was glued to demonstrate the sensing resolution in height.

Furthermore, the system can be utilized to acquire the all around shape of objects by using the phasogrammetric approach with virtual land marks. This allows for hassle-free and convenient walk-around scanning. Combined with the traditional advantages of "photogrammetry," it enables high sensitive multiview 3D surface extraction, independent of the objects color, texture,

(a)   (b)

FIGURE 19.37 Hand-held sensor head using LCOSZ (a) and measurement result (b).

and erosion. New application areas can be developed, where conventional 3D sensors did not offer the required flexibility, portability, speed, and resolution.

19.4.2.3.3   Hybrid Sensing in 3D Profilometry

As the commercial equipment based on optical profilometry becomes popular, strong requests for versatile systems that are applicable to any case are increasing. However, it is hard to meet the customer's demands by using only one principle. Recently, demands from industry are accelerating diversification of methods for shape measurement. Hence, hybrid sensing or multisensor systems made up of one main principle and auxiliary methods have become widely used.

Here, one of such systems is introduced with focus on engine blocks measurement [36]. The main optical 3D sensor based on fringe projection method is able to digitize a 120 mm × 150 mm area of an engine block casting into up to one million points in just a few seconds. By mounting this optical 3D sensor onto a special horizontal arm CMM with three-translation and two-rotation axes along with a conventional touch probe working side by side, an engine block casting can be digitized into a huge point cloud data in about 40 min. The measured point cloud data can be compared with its CAD model to generate a color-coded deviation map in about 3 min.

As the heart of a high quality engine, engine blocks need efficient cooling and tight tolerances for crankshaft bearings, piston rings, and camshafts. Automakers have been casting engine blocks with iron for a century and have significantly refined their processes. However, as the heaviest individual component in a car, the engine block has been a major target for weight reduction. As one solution, automakers have been increasing their efforts in the past few years to replace traditional cast iron engine blocks by aluminum engine blocks in order to cut vehicle weight and improve fuel efficiency. As a result, the majority of current passenger cars from sedans to SUVs have aluminum engines.

High-pressure die-casting (HPDC) is the most commonly used process to produce aluminum engine blocks, in which molten aluminum is injected at high pressure into a metal mold by a hydraulically powered piston. HPDC machines conventionally include an aluminum injector assembly, a die assembly, a pressure accumulator assembly, a cast part take-out assembly, and a die spraying assembly.

The die assembly includes a fixed die and a movable die, which has a plurality of movable cores. The pressure accumulator supplies pressurized fluid for moving of the die between an open and closed position, and for clamping the dies in the closed position. After an engine block is formed and removed from the die assembly, it is needed to spray the dies to keep them clean. The die spraying process sometimes also includes application of a heated fluid to maintain portions of the dies at an appropriate elevated temperature to facilitate the flow and distribution of molten aluminum in the next casting operation.

Dimensional repeatability in engine block castings is one of the important quality issues, which every die-caster has to face on a daily basis. Metal shrinkage, casting machine conditions, die conditions, and consistency between different die sets can all contribute to dimensional repeatability in engine block castings. So far, conventional coordinate measurement machines (CMM) have been the standard tools to check the dimensional repeatability of castings. While a touch probe on a CMM measures one point at a time, it will take more than 30 min to check few hundred points on the casting. This has been the only tool so far to check the dimensional defects in the castings. However, a CMM does not automatically guarantee that a large number of bad castings will not happen. With a few hundred points on a casting measured, a CMM leaves a lot of surface areas on a casting unchecked assuming they are problem free. If defects hit on these usually-not-very-important and unchecked areas, it will usually cause a big problem and cost a lot of money. This is the current situation in die casting industry. That is why die-casters are always looking for a better way to improve their inspection abilities. The goal is to have an automatic machine to check all necessary areas on a casting by using the same amount of time as what a CMM does. An optical noncontact 3D measurement machine, which is based on structured lighting method, is selected to bring the casting inspection to the next level. A picture of this machine is shown in Figure 19.38.

FIGURE 19.38   3D measuring machine with hybrid sensors. (Courtesy of Opton Corp., Aichi, Japan.)

This machine has basically an optical noncontact sensor (based on fringe projection method) head mounted on a special horizontal arm CMM with two additional rotary axes. Since this optical sensor head is heavier than a conventional touch probe, the structure of a conventional horizontal arm CMM cannot support such a heavy load. A special machine with much stronger arm has been used. Two additional rotary axes are used for rotating both the sensor head and the casting during the measurements to cover all the necessary casting surfaces. However, this machine is much more powerful than a conventional CMM due to the white light structured lighting sensor, which can digitize a 120 mm × 150 mm area into up to one million XYZ points in just few seconds. Therefore, it can measure one 120 mm × 150 mm patch at a time instead of one point at a time of a conventional CMM. Another important feature is that there is only one global coordinate system built in with the machine as default, which means there is no need to merge different views taken at different sensor head positions and orientations. The overall measurement accuracy of this machine is 0.05 mm, which is adequate for casting inspection. Figure 19.39 gives a close-up look at its sensor head. The white light sensor includes a grating projector and a digital camera. It can also have a conventional touch probe working side by side with the optical sensor. The touch probe equipped with this machine is mainly for setting up a user-defined coordinate system even before carrying out unattended measurement. The user-defined coordinate system is usually the CAD coordinate system. If the original point cloud data is already in the CAD coordinate system, there is no need to align them together when calculating the deviations between the point cloud and its CAD model.

As shown just below the touch probe, an optional laser sensor (ConoProbe sensor; single point measurement) is also included in this hybrid sensor head. This laser sensor is for the measurement of deep holes, which are not that easy to measure by using either the touch probe or the white light sensor. These holes are formed by pins in the die, which could be bent by molten aluminum. So this hybrid sensor head has three different sensors—the touch probe, the white light sensor, and the laser sensor (Figure 19.39 and Table 19.7).

An L4 engine block as shown in Figure 19.40 (left) has been measured. After the engine block is loaded onto the turntable, an operator selects a preset inspection routine to run, which typically

FIGURE 19.39 Auxiliary sensors attached to machine. (Courtesy of Opton Corp., Aichi, Japan.)

TABLE 19.7
Specifications of This Machine in Figure 19.38

Overall accuracy (mm)	0.05	Measurable depth (mm)	±50
Overall repeatability (mm)	0.025	Max. measurable angle (°)	±60
Sensor field of view (mm)	120 × 150	Coordinate calc. time	Up to 5 s/patch
Stand-off distance (mm)	300	Max. number of points/patch	One million

includes the touch probe portion and the white light sensor portion. The touch probe is first used to measure several features on the engine block, such as pads (with a flat surface) and holes to set up a coordinate system to use. After that the white light sensor will be used to measure all the patches defined in the inspection routine. In Figure 19.40, each white dot means one measured XYZ points. After a casting has been measured into a huge set of point cloud, the next step is to compare the measured point cloud data on the casting with its CAD model. Since the measured point cloud data has already been in CAD model's coordinate system due to the use of the touch probe, there is no need to align two datasets together. This simplifies the data processing a lot. As shown in Figure 19.41, all the points inside the tolerance (±0.2 mm in this case) will be shown in colors. Points that are on the positive side of its CAD surface (this means there are extra material) are shown in the red color. On the other hand, the blue color is assigned to those points that are inside its CAD surface (this means there are less material). The operator only needs to focus on these colors when reviewing all surfaces on the casting. If necessary, annotations can be created to see how serious the problem is and appropriate measures can be taken to address the problems found. In most cases, more likely the die service department will be informed to make necessary corrections on the die surfaces.

As shown in Figure 19.41, there are extra material (thickness from 0.885 to 1.674 mm) in area A and E showing in the figure. This is likely caused by die wearing at these edges. To correct them, some die material is needed to weld on and hand grinded back to proper radius. There are approximately 0.6 mm thick material missing in area C and D. Extra materials on the die surface at these locations may cause this. The problem at area C needs special attention, since the top surface of this

FIGURE 19.40   Measured engine and partial result showing points of cloud. (Courtesy of Opton Corp., Aichi, Japan.)

FIGURE 19.41   Measured result in comparison with CAD data. (Courtesy of Opton Corp., Aichi, Japan.)

casting will be machined into a flat surface. It is essential to make sure there is still enough material left in area C for the machining. Area D is on nonmachining surface. In case it does not need any immediate action, however, it may need to be watched continuously to make sure it never gets worse. Actually die repairing or reconditioning is a constant effort in die-casting and die conditions play a key role in the quality of final castings. Same casting usually has more than one set of dies (usually three sets). At the beginning all different sets of dies are identical. After dies are in service and go through several repairing, they will be different from each other. Maintaining dimensional consistency of a casting among different die sets is another reason for more thorough measurement on castings. We can see this color-code deviation map is very useful for monitoring the status of a casting process to assure dimensional repeatability of castings. It is even more important to have such a deviation map quickly after casting. This inspection process has provided die-casters a new powerful tool to monitor dimensional repeatability of castings.

19.4.2.3.4   Reconstruction of Shape

In addition to combination with CAD mentioned earlier, data acquired by 3D profiling technique are utilized in a lot of applications including clothes production in apparel industry based on human body measurement and preproduction samples for industrial designing for various merchandise using rapid prototyping. One sample is shown in Figure 19.42 (top) where a head (approximately

FIGURE 19.42   Measurement sample (Bodhisattva) and reproduced sample in reduced size by micromachining. (Courtesy of Prof. Yoshimi Takeuchi.)

180 mm × 350 mm in size) of Bodhisattva (Buddhist saint) is measured by fringe projection method. This result is first used to fabricate a miniaturized sample model (with 40 mm in diameter) by milling operation. Ultimately, based on the same 3D data of the head part, this system (five-axis control micromilling system using a diamond pseudo-ball [0.03 mm in radius]) end mill succeeded in creating a 1 mm microreplica in height (right), taking account of collision between the tool and the workpiece [37]. Although this image is captured by SEM, it is verified that the system has possibility of producing three-dimensionally complicated shapes, which are required to make micromachines and microrobots in the future.

19.4.2.3.5 Inner Profile Measurement

There are various requirements for measuring the inner diameter and/or inner profile of cylindrical objects like a pipe. Especially in mechanical engineering field, serious problems come from automobile industry, because the inner surface of engine blocks or other cylindrical parts are strongly requested to be inspected and measured by optical methods (not by the naked eyes using a borescope). Another necessity is known that the rehabilitation and maintenance of sewers is of importance to water and waste utility in urban life [38]. For these problems, hitherto, optical inspection using borescopes has been performed. However, observation and inspection and manual annotation of defects are not sufficient, because the method suffers from a lack of quantitative information and is not appropriate for computer analysis of the recorded data. In addition to the applications in mechanical industry, the medical and dental fields request methods for measuring the inner profile of the stomach, trachea, and oral cavity. If the inner diameter is large enough like water pipes or drain pipes, a complicated and large equipment may be applicable. However, small pipes with a diameter ranging from 10 to 200 mm are difficult to be inspected by such a large instrument as is used for sewers inspection.

The first revision is focused on forming a ring beam without such a mechanically moving device as a rotary mirror and on realizing simple and stable "optical sectioning." In Figure 19.43, which shows the basic arrangement for the above-mentioned purpose, a miniature CCD camera checks inner defects like cracks at the same time with the profile of the inner wall [39]. In this figure, when the pencil beam from an LD is reflected at the sharp top of the conical mirror, the beam expands in a circular direction. This circular beam makes the inner profile by optical section. Figure 19.44 shows a result on a pipe used for mixing of two fluids. This part is produced by hydroforming process, the inner profile of which has to be checked to know if it is formed as designed. A measured result is shown in Figure 19.44 where some of optical sections measured by every 5 mm are arranged. On the basis of optically captured sections, inner profiles of the object are analyzed as shown in the figure. Figure 19.44g and h shows 3D inner shapes of the pipe at different view angles. The inner profile measurements are thus available to many practical applications in the fields of industry such as automobiles, aeroplanes, energy plants, and so on. Some other applications for fuel assembly of nuclear reactor can be found in Refs. [40,41].

FIGURE 19.43 Miniaturized sensor for inner diameter and/or profile measurement.

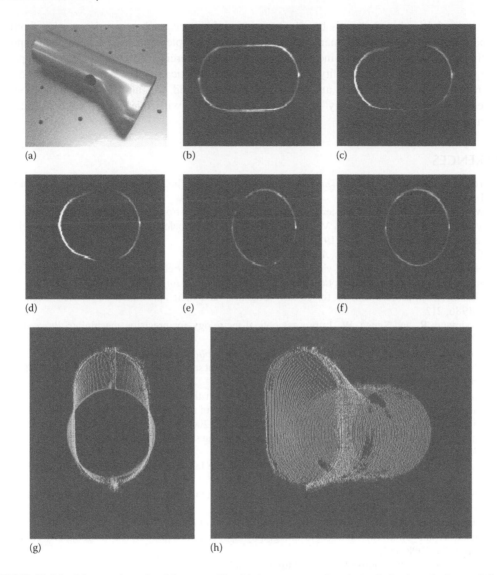

FIGURE 19.44 Measured result of inner profile: (a) Appearance of an irregularly shaped pipe and (b–d) profiles by every 5 mm distance from the end (b, 5 mm; c, 35 mm; d, 65 mm), (e) 95 mm, (f) 125 mm, (g) view angle 0°, and (h) view angle 45°.

In recent years, the inner profile measurement has been tried to such small pipes with the inner diameter smaller than 10 mm. To attain these trials, the ring beam device and a rigid scope camera are combined as shown in Ref. [42]. As a result, in addition to measurement of 3D inner profile, inspection of scratches and flaws on the inner wall of the small holes is successful. The future prospect for the inner profile measurement is becoming increasingly possible in various fields.

19.5 CONCLUSION

Principles for two- and three-dimensional profilometry are described here including a few applications for industrial usage. We are apt to emphasize the accuracy in metrological research works, but reduction of dimensional size and weight of the systems should also be considered for practical use.

The material in this chapter is based on papers and reports such as are impressive from my personal preferences. Naturally many other interesting principles and results have been reported up to this time, but some of the principles and even the commercialized products have disappeared due to severe filtering of availability and convenience in practical usage. While looking back at a lot of research papers, new ideas and rediscoveries were found. Further progress in profilometry is expected to push "3D metrology" to "3D engineering" that will cover industrial measurements, inspection, and manufacturing.

REFERENCES

1. Yoshizawa, T. (Ed.), *Optical Three-Dimensional Metrology* (in Japanese), Asakura Publishing Co., Tokyo, Japan, 2006.
2. Whittle, M.W., The use of biostereoscopic methods to study astronaut body composition, *Proceedings of SPIE*, Vol. 166, Coblentz, A.M. and Herron R.E. (Eds.), Society of Photo-Optical Instrumentation Engineering, Bellingham, WA, 1978, p. 286.
3. Takasaki, H., Moire topography, *Applied Optics*, 9, 1970, 1467.
4. Merklinger, H.M., Principles of view camera focus. http://www.trenholm.org/hmmerk/HMbook18.html.
5. Sawatari, T. and Zipin, R.B., Optical profile transducer, *Optical Engineering*, 18, 1979, 222.
6. Mitsui, K., In-process sensors for surface roughness and their applications, *Precision Engineering*, 8, 1986, 212.
7. Brodmann, R. and Smiga, W., In-process optical metrology for precision machining, *Proceedings of SPIE*, 802, 1987, 165.
8. Kohno, T. et al., High precision optical surface sensor, *Applied Optics*, 27, 1988, 103.
9. Ohta, S., Surface measurement by non-contact displacement sensor based on focus detection (in Japanese), *Optical Engineering Contact*, 26, 1988, 773.
10. Rodenstock GmbH, http://www.html.ornl.gov/mituc/rodenstk.htm.
11. Fukatsu, H., Yanagi, K., and Shimamoto, A., Development of a novel optical stylus sensor for surface profiling instrument, *Proceedings of IIP/ISPS Joint MIPE'03*, 2003, p. 337, held in Yokohama, Japan, 2003.
12. Bartoli, A., Optical profilometer with a standalone scanning sensor head, *Optical Engineering*, 40, 2001, 2852.
13. Figures 19.13 and 19.14 were reconstructed on basis of the Japan Patent No. 3809803 with permission of the holder.
14. Yoshizawa, T. and Tochigi, A., Displacement measurement utilizing contrast variation of a projected pattern, *Optical Engineering*, 31, 1992, 1726.
15. Sirat, G. and Psaltis, D., Conoscopic holography, *Optics Letters*, 10, 1985, 4.
16. Malet, Y. and Sirat, G., Conoscopic holography application: Multipurpose rangefinders, *Journal of Optics*, 29, 1998, 183.
17. STIL SA Chromatic confocal imaging. http://www.stilsa.com/EN/prin.htm.
18. Ishihara, M. and Sasaki, H., High-speed surface measurement using a nonscanning multi-beam confocal microscope, *Optical Engineering*, 38, 1999, 1035.
19. Mermelstein, M.S., Feldkhun, D.L., and Shirley, L.G., Video-rate surface profiling with acousto-optic accordion fringe interferometry, *Optical Engineering*, 39, 2000, 106.
20. Baker, M., Xi, J., Chicharo, J., and Li, E., A contrast between DLP and LCD digital projection technology for triangulation based phase measuring optical profilometers, *Proceedings of SPIE*, 6000, 2005, 60000G-1.
21. Fujita, H., Suguro, A., and Yoshizawa, T., Three-dimensional surface profile measurement by using liquid crystal grating with triangular intensity distribution patterns, *Proceedings of SPIE*, 4902, 2002, 22.
22. Yoshizawa, T., Fujita, H., Otani, Y., and Yamamoto, M., Three-dimensional imaging using liquid crystal grating projection, *Proceedings of SPIE*, 5202, 2003, 30.
23. Yamamoto, M. and Yoshizawa, T., Surface profile measurement by grating projection method with dual-projection optics, *Proceedings of SPIE*, 6000, 2005, 60000I-1.
24. Coggrave, C.R. and Huntley, J.M., High-speed surface profilometer based on a spatial light modulator and pipeline image processor, *Optical Engineering*, 38, 1999, 1573.
25. Huang, P.S., Zhang, S., and Chiang, F.P., Trapezoidal phase-shifting method for 3D shape measurement, *Proceedings of SPIE*, 5606, 2004, 142.

26. Huang, P.S. and Zhang, S., Fast three-step phase-shifting algorithm, *Optical Engineering*, 45, 2006, 5086.
27. Zhang, S. and Yau, S.Z., High-speed three-dimensional shape measurement system using a modified two-plus-one phase-shifting algorithm, *Optical Engineering*, 46, 2007, 113603-1.
28. MESA Imaging AG. http://www.mesa-imaging.ch/index.php.
29. Yoshizawa, T., Wakayama, T., and Takano, H., Applications of a MEMS scanner to profile measurement, *Proceedings of SPIE*, 6762, 2007, 67620B.
30. Wakayama, T. and Yoshizawa, T., Compact camera for three-dimensional profilometry incorporating a single MEMS mirror, *Optical Engineering*, 51, 2012, 013601-1–013601-7.
31. Aoki, H., Development of portable 3D optical measuring system using structured light projection method, *Proc. SPIE* 9110, 91100Q (2014).
32. Nippon Signal Co. Ltd. http://www.signal.co.jp/vbc/mems/index_e.html.
33. Venture Forum Mie Ltd. http://www.vfm.co.jp/.
34. Munkelt, C., Bräuer-Burchardt, C., Kühmstedt, P., Schmidt, I., and Notni, G., Cordless hand-held optical 3D sensor, Optics for Arts, Architecture, and Archaeology, *Proceedings of SPIE*, 6618, 2007, 66180D.
35. Kühmstedt, P., Munkelt, C., Heinze, M., Himmelreich, M., Bräuer-Burchardt, C., Notni, G., 3D shape measurement with phase correlation based on fringe projection, *Proceedings of SPIE*, 6616, 2007, 66160B.
36. Li, Y., Tanaka, H., and Yogo, T., Structured lighting non-contact 3D measurement machines and applications in casting parts inspection, in *Optical Three-Dimensional Metrology* (in Japanese), Yoshizawa, T. (Ed.), Asakura Publishing Co., Tokyo, Japan, 2006. (This content was presented at SPIE Optics East Conferences in 2005, Boston, MA.)
37. Sasaki, T., Sono, M., Ishida, T., Kawai, T., and Takeuchi, Y., Creation of 3-dimensional complicated tiny statue by means of 5-axis control ultraprecision milling, *Journal of the Japan Society for Precision Engineering*, 73, 2007, 1256.
38. Clarke, T.A., The development of an optical triangulation pipe profiling instrument. http://www.optical-metrology-centre.com/Downloads/Papers/Optical%203-D%20Measurement%20Techniques%201995%20Profiler.pdf.
39. Yoshizawa, T., Yamamoto, M., and Wakayama, T., Inner profile measurement of pipes and holes using a ring beam device, *Proceedings of SPIE*, 6382, 2006, 63820D.
40. Inari, T., Takashima, K., Watanabe, M., and Fujimoto, J., Optical inspection system for the inner surface of a pipe using detection of circular images projected by a laser source, *Measurement*, 13 1994, 99.
41. Chugui, Y.V., Optical measurement systems and technologies for scientific and industrial applications, *Pattern Recognition and Image Analysis*, 22, 2012, 10.
42. Wakayama, T. and Yoshizawa, T., Small size probe for inner profile measurement of pipes using optical fiber ring beam device, *Proceedings of SPIE*, 8563, 2012, 85630L.

20 Three-Dimensional Shape Measurement

Frank Chen, Gordon M. Brown, and Mumin Song

CONTENTS

20.1   Introduction 518
20.2   Optical 3D Measurement Techniques... 518
    20.2.1 Time/Light in Flight 518
    20.2.2 Laser Scanning 519
    20.2.3 Moiré.. 519
    20.2.4 Laser Speckle Pattern Sectioning.. 519
    20.2.5 Interferometry.. 520
    20.2.6 Photogrammetry 520
    20.2.7 Laser Tracking System 520
    20.2.8 Structured Light... 521
20.3   General Approach to Measure 360° Shape of an Object................................ 521
20.4   Global and Local Coordinates Translation.. 521
20.5   Structured Light Sources, Image Sensors, Camera Model, and Calibration..................... 522
20.6   Absolute Phase Value Measurement and Discontinuity Ambiguity.............................. 523
20.7   Image Data Patching and CAD Data Comparing... 524
20.8   Sensor Planning 524
    20.8.1 Sensor Planning Examples 525
20.9   Demonstration Examples... 525
    20.9.1 Component Shape Measurement 525
    20.9.2 Vehicle Shape Measurement.. 527
    20.9.3 Vibration 3D 528
    20.9.4 Paint Defects.. 528
20.10 Conclusion and Future Research Trend 530
    20.10.1 Real-Time Computing 530
    20.10.2 Direct Shape Measurement from a Specular Surface without Paint 530
    20.10.3 Shading Problem 531
    20.10.4 Standard Methodology to Evaluate Optical Shape Measurement System............ 531
    20.10.5 Large Measurement Range with High Accuracy............................. 532
    20.10.6 Measurement System Calibration and Optimization and Sensor Planning.......... 532
Acknowledgments.. 532
References.. 532

20.1   INTRODUCTION

In industry, there is need for accurately measuring the three-dimensional (3D) shape of objects to expedite and ensure product development and manufacturing quality. There are a variety of applications of 3D shape measurement, such as control for intelligent robots, obstacle detection for vehicle guidance, dimension measurement for die development, stamping panel geometry checking, and accurate stress/strain and vibration measurement. Moreover, automatic online inspection and recognition issues can be converted to the 3D shape measurement of an object under inspection, for example, body panel paint defect and dent inspection. Recently, with the evolution in computer technologies, coupled with the development of digital imaging devices, electro-optical components, laser, and other light sources, 3D shape measurement is now at the point where some techniques have been successfully commercialized and accepted in the industry. For a small-scale depth or shape, micrometer or even nanometer measurements can be reached if a confocal microscope or other 3D microscope is used. However, the key is the relative accuracy or one part out of the measurement depth that poses a real challenge for a large-scale shape measurement. For example, how accurate can a 0.5 m depth measurement be? Moreover, for a large-scale depth and shape measurement, frequently, more cameras and camera positions are required to obtain several shapes from which the final large shape can be patched. It raises the issue of how to patch these shapes together in a highly accurate manner and perform local and global coordinate transforms. This subsequently generates another problem to be solved, namely to overcome lens distortion and aberration. After the 3D shape is obtained, there is a need to compare this data with a computer-generated computer aided engineering (CAE) model.

This chapter is based on the review paper [1] that provides an overview of 3D shape measurement using various optical methods supplemented with some recent developments and updates. Then it focuses on structured light and photogrammetry systems for measuring relatively large-scale and 360° shapes. Various detail aspects such as absolute phase measurement, structured light sources, image acquisition sensors, camera model, and calibration are discussed. Global and local coordinate translation methods, point cloud patching, and computer aided design (CAD) data comparing are also discussed. Several applications are described. Finally, future research trends such as real-time computing, automating, and optimizing sensor placement, and the need for a common standard for evaluation of optical coordinate measurement systems are presented.

20.2   OPTICAL 3D MEASUREMENT TECHNIQUES

Various optical techniques have recently been developed for measuring 3D shape from one position. A comprehensive overview for some of the techniques can be found in the literature [2].

20.2.1   Time/Light in Flight

The time-of-flight method for measuring shape is based on the direct measurement of the time of flight of a laser or other light source pulse [3]. During measurement, an object pulse is reflected back to the receiving sensor, and a reference pulse is passed through an optical fiber and received by the sensor. The time difference between the two pulses is converted to distance. A typical resolution for the time-of-flight method is around a millimeter. With sub-picosecond pulses from a diode laser and high-resolution electronics, sub-millimeter resolution is achievable. The recently reported time-correlated single-photon counting method has depth repeatability better than 30 μm at a 1 m standoff distance [4]. Another similar technique is called light-in-flight holography where either short temporal coherence light or very short light pulse is used to generate a motion image of a propagating optical wavefront [5,6]. Combined with digital reconstruction and a Littrow setup, the depth resolution may reach 6.5 μm [7].

20.2.2 LASER SCANNING

Point laser triangulation employs the well-known triangulation relationship in optics. The typical measurement range is ±5 to ±250 mm, accuracy is about one part in 10,000, and frequency is up to 40 kHz or higher [8,9]. A charge-coupled device (CCD) or a position sensitive detector (PSD) is widely used to digitize the point laser image. For PSD, the measurement accuracy is mainly dependent on the accuracy of the image on the PSD. The beam spot reflection and stray light will also affect the measurement accuracy. Masanori Idesawa [10] has developed some methods to improve the accuracy of the PSD by using a high-accuracy kaleidoscopic mirror tunnel position-sensing technique and a hybrid type of PSD. CCD-based sensors avoid the beam spot reflection and stray light effects and provide more accuracy because of the single pixel resolution. Another factor that affects the measurement accuracy is the difference in the surface characteristic of a measured object from the calibration surface. Usually, calibration should be performed on similar surfaces to ensure the measurement accuracy. The recently developed confocal technique can tolerate surface color change, transparency difference, and irregularity without calibration [11].

20.2.3 MOIRÉ

The Moiré method can be divided into shadow and the more practical projection techniques [12,13]. The key to the Moiré technique is the use of a master grating and a reference grating, from which contour fringes can be generated and resolved by a CCD camera. Increased resolution is realized since the gratings themselves do not need to be resolved by the CCD camera. However, if the reference grating is computer generated as in the logic-Moiré method [14,15], the camera must resolve the master grating. The penalties for the high resolution are the implementation complexity and the need for a high-power light source as compared with a structured light technique. In order to (1) overcome environmental perturbations, (2) increase image acquisition speed, and (3) utilize phase-shift methods to analyze the fringe pattern, snap shot or multiple image Moiré systems have been developed. Two or more Moiré fringe patterns with different phase shifts are simultaneously acquired using multi-camera or image-splitting methods [16–20]. Reference [21] provides a comparison of some high-speed Moiré contouring methods with particular stress on sources of noise and system error functions. The typical measurement range of the phase-shifting Moiré method is from 1 mm to 0.5 m with the resolution at 1/10–1/100 of a fringe. Some novel applications and related references can be found in Refs. [22–31].

20.2.4 LASER SPECKLE PATTERN SECTIONING

The 3D Fourier transform relationship between optical wavelength (frequency) space and the distance (range) space is used to measure the shape of an object [32–34]. Laser speckle pattern sectioning, also known as speckle pattern sampling or laser radar [35–38], is achieved by utilizing the principle that the optical field in the detection plane corresponds to a two-dimensional (2D) slice of the object's 3D Fourier transform. The other 2D slices of the object's 3D transform are acquired by changing the wavelength of the laser. A speckle pattern is measured using a CCD array at each different laser wavelength, and the individual frames are added up to generate a 3D data array. A 3D Fourier transform is applied on this data array to obtain the 3D shape of an object. When a reference plane method [39] is used, this technique is similar to two-wavelength or multiwavelength speckle interferometry. The measurement range can be from a micrometer to a few meters. The accuracy is dependent on the measurement range. With the current laser technology, 1–10 μm resolutions are attained in the measurement range of 10 mm and 0.5 μm measurement uncertainty is achievable (see HoloMapper in the commercial system list). The advantages of this technique are (1) the high flexibility of the measurement range and (2) phase shifting as in conventional interferometry may not be required. The limitation of this technique is that for relatively large-scale shape measurement, it takes more time to acquire the images with the different wavelengths.

20.2.5 INTERFEROMETRY

The idea behind interferometric shape measurement is that the fringes are formed by variation of the sensitivity matrix that relates the geometric shape of an object to the measured optical phases. The matrix contains three variables; wavelength, refractive index, and illumination and observation directions, from which three methods; two or multiple wavelength [40–44], refractive index change [45–47], and illumination direction variation/two sources [48–52], are derived. The resolution of the two wavelength method depends on the equivalent wavelength (Λ) and the phase resolution of ~Λ/200. For example, two lines of an argon laser (0.5145 and 0.4880 µm) will generate an equivalent wavelength 9.4746 µm and a resolution of 0.09 µm.

Another range measurement technique with high accuracy is double heterodyne interferometry using a frequency shift. Recent research shows it achieves a remarkable 0.1 mm resolution with 100 m range [53]. Interferometric methods have the advantage of being mono-state without the shading problem of triangulation techniques. Combined with phase-shifting analysis, interferometric methods and heterodyne techniques can have an accuracy of 1/100 and 1/1000 of a fringe, respectively. With dedicated optical configuration design, accuracy can reach 1/10,000 of a fringe [54]. The other methods such as shearography [55–57], diffraction grating [58,59], digital wave-front reconstruction and wavelength scanning [60,61], and conoscopic holography [62] are also under development. Both shearography and conoscopic holography are collinear systems, which are relatively immune to mechanical disturbances.

20.2.6 PHOTOGRAMMETRY

Typical photogrammetry employs the stereo technique to measure 3D shapes, although other methods such as defocus, shading, and scaling can also be used. Photogrammetry is mainly used for feature type 3D measurement. It usually needs to have some bright markers such as retroreflective painted dots on the surface of a measured object. In general, photogrammetric 3D reconstruction is established on the bundle adjustment principle in which the geometric model of the central perspective and the orientation of the bundles of light rays in a photogrammetric relationship are developed analytically and implemented by a least square procedure [63]. Extensive research has been done to improve the accuracy of photogrammetry [64–68]. Recent advances make it achieve high accuracy as one part in hundred thousand or even one part in a million [69]. For some fundamental description, one may refer to the website www.geodetic.com/Whatis.htm. There are quite a few of commercially available systems and they may be found by searching the web.

20.2.7 LASER TRACKING SYSTEM

A laser tracker uses an interferometer to measure distances, and two high-accuracy angle encoders to determine vertical and horizontal angles. The laser tracker SMART 310, developed at the National Bureau of Standards, was improved at API (Automated Precision Inc.) to deliver 1 µm range resolution and 0.7^2 angular resolution [70]. The laser tracker is a scanning system and usually is used to track the positions of optical sensors or robots. The Leica LTD 500 system can provide an absolute distance measurement with accuracy of about ±50 µm and angle encoders that permit accuracy of five parts in a million within a 35 m radius measurement volume [71].

With recent progress in laser tracker technology, the measurement range is increasing. For example, FARO laser tracker Xi has a 230 ft range and accuracy on ADM (absolute distance measurement) can be around 0.001″. The LTD840's measurement range reaches 262 ft and ADM can be around 0.001″ accuracy. API TRACKER 3 renders a 400 ft range and ADM can be around 0.001″ accuracy.

20.2.8 STRUCTURED LIGHT

The structured light method, also categorized as active triangulation, includes both projected coded light and sinusoidal fringe techniques. Depth information of the object is encoded into a deformed fringe pattern recorded by an image acquisition sensor [72–75]. Although related to projection Moiré techniques, shape is directly decoded from the deformed fringes recorded from the surface of a diffuse object instead of using a reference grating to create Moiré fringes. One other related technique uses projected random patterns and a trilinear tensor [76–79]. When a LCD/DMD (liquid crystal projector/digital mirror display) based and optimized shape measurement system is used, the measurement accuracy may be achieved at one part in 20,000 [80]. The structured light method has the following merits: (1) easy implementation, (2) phase shifting, fringe density, and direction change can be realized with no moving parts if a computer-controlled LCD/DMD is used, and (3) fast full-field measurement. Because of these advantages, the coordinate measurement and machine vision industries have started to commercialize the structured light method (see the list) and some encouraging applications can be found in Refs. [81–83]. However, to make this method even more accepted in industry some issues have to be addressed, including the shading problem, which is inherent to all triangulation techniques. The 360° multiple-view data registration and defocus with projected gratings or dots promise a solution [84–86]. These are discussed briefly in the following sections. For small objects using a microscope, lateral and depth resolutions of 1 and 0.1 μm, respectively, can be achieved [87,88]. Using confocal microscope for shape measurement can be found in Ref. [89].

20.3 GENERAL APPROACH TO MEASURE 360° SHAPE OF AN OBJECT

The global coordinate system is set up and local coordinates systems are registered during measurement. A structured light imaging system is placed at an appropriate position to measure 3D shape from one view, and the absolute phase value at each object point is calculated. These phase values and a geometric-optic model of the measurement system determine the local 3D coordinates of the object points. Three ways are usually used to measure 360° shape, the object rotation method [90–94], the camera/imaging system transport technique, and the fixed imaging system with multiple cameras approach. For camera transport, which is usually for measuring large objects, the measurement is repeated at different views to cover the measured object. All the local 3D coordinates are transformed into the global coordinate system and patched together using a least squares fit method. The measured final 3D coordinates of the object may be compared with CAD master data in a computer using various methods in which the differentiation comparison technique and least square fit are often used.

20.4 GLOBAL AND LOCAL COORDINATES TRANSLATION

For a 360° 3D measurement of an object, an optical sensor has to be positioned at different locations around the object [95,96]. The point clouds obtained at each position need to be input or transformed into global coordinates from each local coordinate so that these point clouds can be patched together to generate the final data set. This can be achieved by multiple cameras at different positions at the same time or a single camera at different position at different time. In order to accomplish this, each sensor coordinate system location and orientation has to be known or measured. Any error in measuring and calculating the sensor location and orientation will cause a propagation error in the global coordinates which will prevent a high over-all accuracy of the final measurement. There are several approaches to determine the relationship between the global and local coordinate systems: (1) accurate mechanical location and orientation of the sensor (local coordinate system), (2) optical tracking of the location and orientation of the sensor using active or passive targets attached to the sensor, and (3) photogrammetry of markers accurately fixed in the object field and hybrid methods. Figure 20.1 shows these approaches. With the recent development in photogrammetry, projected marks can be used [115].

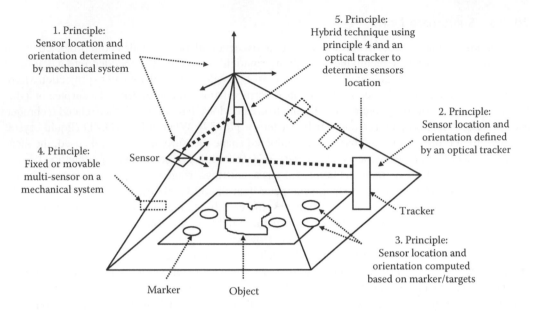

1. Principle:
Sensor location and
orientation determined
by mechanical system

5. Principle:
Hybrid technique using
principle 4 and an
optical tracker to
determine sensors
location

2. Principle:
Sensor location and
orientation defined
by an optical tracker

4. Principle:
Fixed or movable
multi-sensor on a
mechanical system

Sensor

Tracker

3. Principle:
Sensor location and
orientation computed
based on marker/targets

Marker   Object

FIGURE 20.1 Sensor planning diagram showing several approaches to determine the relationship between the global and local coordinate systems.

For the mechanical approach 1, the sensor is attached on a mechanical positioning system of high accuracy. The location and orientation of the sensor are derived from the system coordinate and angle information. The advantage of the mechanical system is that it is robust and of high accuracy. However, the cost for accuracy mechanical devices, overcoming environmental perturbation, and maintenance of the equipment is very high.

For the optical approach 2, the local coordinate system is calculated from the measured global position of reference targets (active or passive, with the known local coordinates) on the frame of the optical sensor using an optical tracker system. The advantage is portability and compactness. However, the sensor targets have to be visible and this limits the flexibility. Moreover, floor vibration effect has to be considered. If a high-accuracy tracking system is used, such as laser tracking system, the cost is also relatively high. Both mechanical and optical methods are prone to angular error.

The photogrammetry approach 3 can provide high-accuracy local coordinate system location and orientation from measurement of the global coordinates of makers accurately fixed in the object field. The accuracy can be as high as one part in a million [69]. Conservatively, accuracy can be as one part in 100,000. However, the key limitation for this method is that registration markers have to be placed on or around the object. This increases the measurement time and automation complexity.

20.5 STRUCTURED LIGHT SOURCES, IMAGE SENSORS, CAMERA MODEL, AND CALIBRATION

The light source is important for the overall accuracy of a 3D shape measurement system. Important parameters include uniformity, weight, intensity profile, and speckle/dot size. The projection of a Ronchi grating slide provides high resolution with bright images and is currently used in some commercial systems. However, to calculate absolute distance, multiple grating slides are needed to apply the phase-shift method and to vary the grating frequency. This in turn results in slow speed and relatively large space for storing different gratings. Around 1991–1993, LCDs using incoherent light [97–104] have been used in which each pixel can be addressed by a computer image-generating system. The advantage of this type of projection is the high speed for phase shifting and variable grating frequency. The disadvantage is that LCDs need powerful light sources resulting in cooling concerns and increased weight. Moreover, the resolution is low compared with the film-slide-based light projection. To overcome the brightness concern of the LCD, the reflective LCD, the gas plasma display (GPD), and DMD [105–107] have been developed. In addition, the gaps between DMD mirrors are smaller than those pixels between the LCD so that the DMD images are relatively sharper. A detailed error analysis and optimization of a shape measurement system using a LCD/DMD-type fringe projector can be found in Ref. [107]. The LCD, GPD, and DMD have RGB color by which simultaneously acquisition of three images or three phase-shifted images can be used, and this makes the phase-shift technique immune to environmental perturbation [108]. The color advantage may also be used for absolute phase determination [109–113]. Other light sources are the two point source laser interferometer using a Mach–Zhender configuration [114], fiber optics [115], birefringence crystal [116], acoustic optical modulator [117], and Lasiris' non-Gaussian-structured light projector using a specially designed prism that can generate 99 lines with inter-beam angle at 0.149° [118].

In optical 3D shape measurement, image acquisition is a key factor for accuracy. Currently, images are acquired using a CCD or a charge injection device (CID) sensor. There are full frame, frame transfer, and interline transfer sensors. The major concerns regarding these sensors are the speed, resolution, dynamic range, and accuracy. Up to 5k × 5k pixel CCD sensors are commercially available such as DALSA IA-D9-5120, Ford Aerospace 4k × 4k, Kodak Model 16.8I (4k × 4k), and Loral CCD481, to name a few. Usually, the high-resolution sensor is a full-frame CCD that does not have storage and needs to have a shutter to allow image transfer that results in relatively slow speed. Combined with micro- and macro-scanning techniques, the image resolution can be 20k × 20k, which is equivalent to the resolution of a 20 cm × 20 cm area photographic film with 100 lines/mm. The CID sensor differs from a CCD sensor in that it does not bloom though overexposure and can be read out selectively since each pixel is individually addressed.

A high-accuracy CCD sensor or a video camera requires high radiometric and geometric accuracy, including both intrinsic parameters, such as lens distortion, and extrinsic parameters, such as the coordinate location and orientation of the camera. A detailed discussion regarding characterization and calibration of radiometric and geometric feature of a CCD sensor can be found in Ref. [119]. The relative accuracy of one part in 1000 can be achieved using on-site automatic calibration during measurement. More accurate calibration, such as 10^{-4} to 10^{-5} accuracy, may be achieved using a formal off-line calibration procedure with a more complex and nonlinear camera model. A high-accuracy camera model is necessary to characterize the lens aberration and to correct the captured image for the distortions caused by the aberration [120–128]. The calibration of an optical measurement system can be further divided into a geometric parameter technique as described above and geometric transformation approach. The geometric parameter technique requires the known parameters of the optical setup including the projector and the image sensor. On the other hand, it is not essential to know the parameters of the image system in the geometric transformation approach [129–134], in which the recently developed projection or image ray tracing technique is discussed [132] and the known position of the object or camera variation approach is provided [133]. Once the imaging system is moved or the measured object size/depth is changed, the calibration procedure may be performed again. This, however, may pose some limitation for this method. Reference [135] developed a self-calibration approach that may reduce the complexity of calibration procedure and increase accuracy.

20.6 ABSOLUTE PHASE VALUE MEASUREMENT AND DISCONTINUITY AMBIGUITY

In general, using phase-shifted structured light to measure the 3D shape of the object only renders relative phase values. Phase shifting determines the fractional order of fringes at any pixel. These fractional orders are connected together using their adjacent integer orders, which is called unwrapping process. However, when the phase difference between the adjacent pixels is larger than 2π, such as occurs at a discontinuity or steep change of shape, the integer fringe order becomes ambiguous. Recently, several methods have been developed to overcome discontinuities [75,136–159]. The basic idea is that changing the measurement system's sensitivity results in fringe or projected structured strip density changes. This means that the integer order of the fringes sweep through the discontinuity. It can be viewed both in spatial and temporal domains, and results in various methods. As mentioned in Ref. [144], the key to overcoming discontinuity is to determine the fringe order n during unwrapping process. These methods, such as two wavelength or parameter change, are used in interferometry to determine absolute fringe fractional and integer orders [160–164]. In plane, rotation of the grating and varying the grating frequency (e.g., fringe projection with two point variable spacing) are useful techniques. Triangulation and stereography can also be employed to determine absolute phase values and overcome discontinuity, although there are limitations since all pixel points may not be covered without changing setup. Some direct phase calculation, such as phase derivative methods without phase unwrapping, may still need continuous condition [165]. However, the same problem can also be solved by phase-locked loop technique [166].

20.7   IMAGE DATA PATCHING AND CAD DATA COMPARING

After processing the 360° local images, the local point cloud patches need to be merged together to obtain a final global point cloud of the object. The accuracy of a measurement system is also determined by the matching accuracy. There is extensive research on the matching methods and algorithms in photogrammetry, which can generally be categorized as area-based matching, feature-based matching, and other methods such as centroid method [167–172]. The area-based matching takes advantage of correlation coefficient maximization and least squares minimization, while feature-based matching exploits all algorithms extracting features of points, lines, and areas. Area-based matching usually employs pixel intensity as constraint, while feature-based matching uses geometric constraint. All the above methods require sub-pixel accuracy to achieve overall accuracy. Under optimized condition, 0.02 pixel accuracy can be achieved [173–175] and in general 0.05 pixel accuracy should be obtained [175]. There is a discussion between sub-pixel accuracy and geometric accuracy where geometric accuracy is more promising [176].

For CAD data comparison, the differentiation and least mean square methods are mainly used. The measured point cloud data are subtracted from CAD data to obtain differences as an error indicator. The comparison to the appropriate CAD model can be used to obtain the best fit of registration between the two. Model matching can start with a selection of a subset of point cloud data. The measured point data in this subset are matched to the CAD data by making the normal vector collinear to the normal vector of the local CAD surface. The distances in the normal direction between the CAD surface and measured point cloud are fed into a least square error function. The best fit is achieved by optimizing the least square error function [95,96,177]. Before the measured data can be compared to CAD master data, it needs to be converted into standard CAD representations [178–180]. This is usually done by first splitting the measured data into major geometric entities and modeling these entities in the nonuniform rational B-spline surface form; this has the advantage of describing the quadric in addition to free-form surface using a common mathematical solution.

20.8   SENSOR PLANNING

Large-scale surface inspection often requires either multiple stationary sensors or relocation of a single sensor for obtaining the 3D data. A multi-sensor system has an advantage on high-volume inspection of similar products, but usually lacks flexibility. In order to improve the flexibility, various portable sensor systems and automated eye-on-hand systems are produced. However, no matter what kinds of sensor systems be used, the first and most critical problem that should be solved is how the sensors can be placed to successfully view the 3D object without missing required information. Given the information about the environment (e.g., the observed objects and the available sensors) and information about the mission (i.e., detection of certain object features, object recognition, scene reconstruction, object manipulation, and accurate, dense enough point clouds), strategies should be developed to determine sensor parameters that achieve the mission with a certain degree of satisfaction. Generally, solving such a problem is categorized as a sensor planning problem. Considerable effort on general techniques has been made in sensor planning. We can collect them into the following four categories.

1. Generate-and-test approach [181–183]: By using this method, sensor configurations are generated first, then evaluated using performance functions and mission constraints. In order to avoid an exhausting search process, the domain of sensor configurations is discretized by tessellating a viewing sphere surrounding the object under observation. This is a time-consuming technique without guaranteeing an optimal result.
2. Synthesis approach [184–190]: This approach is built upon an analytical relation between mission constraints and sensor parameters. It has a beautiful and promising theoretical framework that can determine the sensor configurations for certain cases. The drawback of this approach is that the analytical relations sometimes are missing, especially when constraints are complex.
3. Sensor simulation system [191–193]: This system brings objects, sensors, and light sources into a unified virtual environment. It then uses the generate-and-test approach to find desired sensor configurations. The simulation systems are useful in the sense that operators can actively evolve the process and ensure the results.
4. Expert system approach [194–197]: Rule-based expert systems are utilized to bridge reality and expert knowledge of viewing and illumination. The recommended sensor configurations are the output of the expert system from reality checking. In general, the more complete knowledge we have, the "wiser" the advice we can get.

20.8.1 SENSOR PLANNING EXAMPLES

In fact, these sensor-planning techniques have strong application background. Their goal is aggressively set to improve machine intelligence and reduce human-intensive operations that cause the long development cycle time, high cost, and complexity in modern industry. Several examples are (1) an intelligent sensor-planning system was conceptually defined to be applied in the application of automated dimensional measurements by using CAD models of measured parts [198], (2) an inspection system is able to determine appropriate and flexible action in new situations since online sensor-planning techniques were adopted [199], and (3) the techniques were applied to a robot vision system so that the orientation and position of vision sensors and a light source can be automatically determined [200].

20.9 DEMONSTRATION EXAMPLES

20.9.1 COMPONENT SHAPE MEASUREMENT

During component development cycle, there is a tryout phase that requires measurement of the shape of a component or the die to make the component. The following is a demonstration of using a 3D optical measurement setup to measure a component and compare it to the master CAD data to evaluate the spring back effect. Figure 20.2 shows the point cloud data of a component. Figure 20.3 shows the corresponding CAD data. Figure 20.4 shows the comparison between the measured data and CAD data, in which the two sets of data are compared using the least square fit.

FIGURE 20.2 Measured point cloud data of a component.

FIGURE 20.3 Corresponding CAD data of a component.

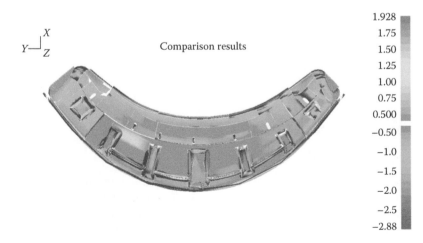

FIGURE 20.4 Comparison between the measured data and CAD data of a component.

20.9.2 VEHICLE SHAPE MEASUREMENT

For rapid prototyping or benchmarking, often the vehicle body shape needs to be measured. The following example uses the structured light method combined with photogrammetry to measure car body shape. Some coded targets were placed on the vehicle body to permit local to global coordinate transformation. Then structured light was projected on the vehicle surface combined with phase shifting and absolute phase measurement technique using fringe frequency change to determine the local coordinate pixel by pixel at one view direction. The 240 view directions were used to cover the whole vehicle surface (it is a real vehicle). Then the 240 point clouds were patched together using a least mean square method. The point cloud data were extracted using 1 out of 8 pixels. The shaded measured data is shown in Figure 20.5 and the point cloud data is shown in Figure 20.6.

FIGURE 20.5 Shaded measured data of a vehicle.

FIGURE 20.6 Measured point cloud data of a vehicle.

20.9.3 Vibration 3D

For accurate analysis of vibration or strain, the geometric information of the tested structure needs to be known. Using the two-wavelength shape measurement method, the vibration amplitude, phase, and geometric information of a tested structure can be measured using a single compact Electronic Speckle Pattern Inter ferometry (ESPI) setup [201]. Figure 20.7a and b shows the four vibration states of a corrugated plate clamped along its boundaries and subjected to harmonic excitation at 550 Hz. The state 1 in Figure 20.7a depicts the original corrugated plate geometric shape and the rest are vibrating states at distinct times. From Figure 20.7 one can clearly see the shape effect on vibration.

20.9.4 Paint Defects

The geometry measurement technique can also be applied to measure paint defects of a body panel of a vehicle, although it is a challenge to detect small flaws in large areas. A methodology has been developed as shown in the flow chart in Figure 20.8, in which structured light generated from a monitor is reflected from the tested panel and digitized into a computer image processing system. Then the digital Fourier transform method is used to extract the global shape of the panel by selecting the structured light frequency. The defect geometry coupled with the global shape of the panel is calculated by selecting half-spatial frequencies. The defect geometry is finally obtained by subtracting the above two results as shown in Figure 20.9, where Figure 20.9a shows the panel with projected structured light and Figure 20.9b shows the final measurement result. One can see that without the calculation, one can only observe the large defects by enhanced fringe modulation. Figure 20.10 shows the optical setup. The measurement area is about 0.25 m × 0.25 m and the minimum defect size is about 250 μm [202]. Some other application examples can be found in Refs. [203,204], to name a few.

(a) Vibration state1   (b) Vibration state 2

(c) Vibration state 3   (d) Vibration state 4

FIGURE 20.7 (a) Vibration state 1 depicts the original corrugated plate geometric shape and (b–d) vibration states 2–4 show the effect of the underlying shape.

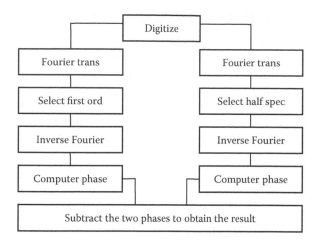

FIGURE 20.8 Structured light generated from a monitor and reflected from the test panel and digitized into a computer image processing system.

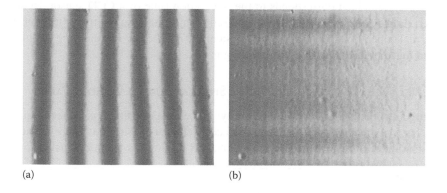

(a)   (b)

FIGURE 20.9 (a) Panel with projected structured light and (b) defects in the final measurement result.

FIGURE 20.10 Optical setup showing structured light generated from a monitor reflected from the panel and digitized into a computer image processing system.

20.10 CONCLUSION AND FUTURE RESEARCH TREND

The principles of triangulation, structured light, and interferometry have been in existence for decades. However, it is only with the recent availability of advanced and low-cost computers, electro-optical elements, and lasers that such techniques have reached the breakthrough point to be commercialized, and are increasingly being applied in industry. To make it even more acceptable in industry and to strive to achieve 10^{-4} to 10^{-5} accuracy [80,135], there are still some challenges that need to be addressed. The following may provide some future trends. Table 20.1 shows some full-field shape measurement commercial systems based on leading edge technologies.

20.10.1 REAL-TIME COMPUTING

Real-time 3D shape measurement is an ongoing request in industry to drive down product cost and increase productivity and quality. The major impact will be in digital design, digital and physical manufacturing, and fast prototyping that streamline and integrate product design and manufacture. Real-time 3D shape measurement is the key for successfully implementing 3D coordinate display and measurement, manufacturing control, and online quality inspection. An encouraging example of this is jigless assembly based on the real-time, simultaneous measurement of different but related components. Described by Hobrough [205], real time is to assign a Z-value or depth for every pixel within a 17 ms cycle, which corresponds to the integration time of a CCD sensor. Recently, over 100 measured points for every 40 ms was achieved using photogrammetry [206], and there is a report on the real-time 3D shape measurement system [207]. The key for real time is a high computational speed that can meet online manufacturing needs.

20.10.2 DIRECT SHAPE MEASUREMENT FROM A SPECULAR SURFACE WITHOUT PAINT

There is an urgent need, but little research activity, in the area of using optical techniques to measure the 3D shape of an object with specular surface, such as a die surface. There are some efforts to develop techniques in this area. References [208,209] proposed a technique using four lights to

---

TABLE 20.1

Some Full Field Shape Measurement Commercial Systems Based on Leading Edge Technologies

System	Principle	Accuracy on Volume
Atos System: Capture 3D, Costa Mesa, CA 92626, Tel.: 1-714-546-7278	Structured light + photogrammetry, 360° view/patching	About 50 μm (2σ) on relatively large volume
Comet/OptoTrak System: 4000 Grand River Ave., mikeb@steinbichler.com	Structured light + optical tracking, 360° view/patching	About 50 μm (2σ) on relatively large volume
Optigo/CogniTens System: US, Tel.: 1-815-637-1926	Random dot pattern + photogrammetry + view/patching	About 20–100 μm (2σ) on medium volume
4DI System: IA, 149 Sidney St., Cambridge, MA 02139, Tel.: 1-617-354-3830	Structured light + real time computing, one view/no patching	About 10^{-3} on medium volume
HoloMapper System: ERIM International, Inc. 1975 Green Rd., Ann Arbor, MI 48105, Tel.: 1-313-994-0287	Laser radar/multiple wavelength, one view/no patching	Uncertainty 0.5 μm on medium volume

measure specular features. Reference [210] employed a simplified version of the Torrance–Sparrow reflectance model to retrieve object curvature; this method relied on prior knowledge of surface reflectance. References [211,212] suggested using multiple (127) point sources to detect specular shape. Reference [213] developed a photometry sampling technique employing a matrix of extended sources to determine the specular shape. References [214–218] used diffusive TV screen as structured light source; however, since the diffusive screen has to be placed close to the measurement surface, the illuminated area is limited. References [216–218] proposed a retroreflective screen coupled with projection of structured light by which a large area may be visualized. However, the retroreflective screen has to be rigid, controlled, or calibrated without deformation or movement during the measurement. References [219,220] developed a coaxial linear distance sensor for measuring specular shape of a die surface; however, it is based on point scanning which is not fast enough for the industrial application of measuring large die surface. The current technique for measuring die surfaces requires painting the surface with powder, which slows measurement speed and reduces measurement accuracy.

20.10.3 SHADING PROBLEM

There is a lack of research activity for overcoming shading problem inherent in triangulation methods, although some other methods [84–86] besides interferometric and laser speckle sectioning show some progress. These methods use a defocus technique similar to the confocal microscope principle and the newly developed diffraction grating technique [58,59].

20.10.4 STANDARD METHODOLOGY TO EVALUATE OPTICAL SHAPE MEASUREMENT SYSTEM

An international standard needs to be established to evaluate optical shape measurement systems. Important parts of this standard should include the following: (1) standard sample parts with known dimensions, surface finishes, and materials, (2) math assumptions and error characterization, (3) measurement speed and volume capability, (4) repeatability and reproducibility procedures, (5) calibration procedures, and (6) reliability evaluation.

Standard specification/terminology (Table 20.2) is needed to define precision/repeatability and accuracy. An example is, what should be used: accuracy, error, or uncertainty [221]. Many coordinate measurement manufacturers specify a range accuracy of $\pm x$ μm. Some manufacturers try to specify measurement accuracy as 1σ, 2σ, or 3σ. The ISO/R 1938-1971 standard employs a 2σ band [222].

TABLE 20.2
Terminology

Accuracy	The closeness of the agreement between the result of a measurement and the value of the quantity subject to measurement, that is, the measurand [221].
Uncertainty	The estimated possible deviation of the result of measurement from its actual value [224].
Error	The result of a measurement minus the measured value [221].
Precision	The closeness of agreement between independent test results obtained under stipulated conditions [221].
Repeatability	The closeness of the agreement between the results of successive measurements of the same measurand carried out under the same conditions of measurement [221].
Reproducibility	The closeness of the agreement between the results of measurements of the same measurand carried out under changed conditions of measurement [221].
Resolution	A measure of the smallest portion of the signal that can be observed [224].
Sensitivity	The smallest detectable change in a measurement. The ultimate sensitivity of a measuring instrument depends both on its resolution and the lowest measurement range [224].

20.10.5 Large Measurement Range with High Accuracy

Most shape measuring systems trade off measurement accuracy for measurement range. However, there is an industrial need for systems that have a large measurement range and high accuracy. Further research needs to be done in this area, though there is encouraging report in which the shape of a 4 m wide area of a brick wall was measured using fringe projection [223].

20.10.6 Measurement System Calibration and Optimization and Sensor Planning

System calibration and optimization are key factors to stretch the measurement accuracy [80] and to achieve 10^{-4} to 10^{-5} accuracy [135]. Reference [80] shows how to use the same system but with optimization to achieve one order higher accuracy and Ref. [135] demonstrates how to use the novel self-calibration to accomplish mathematically estimated approximately 10^{-5} accuracy. Sensor planning will help to fulfill the above goals. Reference [135] also provides a way to eliminate markers using photogrammetry that is one step further to make 3D optical methods be more practical in industry.

ACKNOWLEDGMENTS

The authors thank T. Cook, B. Bowman, C. Xi, E. Liasi, P. Harwood, and J. Rankin for providing test setup, data registration and processing, and test results, and T. Allen for his valuable suggestions.

REFERENCES

1. F. Chen, G. M. Brown, and M. Song, Overview of three-dimensional shape measurement using optical methods, *Opt. Eng.*, **39**(1), 10–22, 2000.
2. H. J. Tiziani, Optical metrology of engineering surfaces—Scope and trends, in *Optical Measurement Techniques and Applications*, P. K. Rastogi (Ed.), Artech House Inc., London, U.K., 1997.
3. I. Moring, H. Ailisto, V. Koivunen, and R. Myllyla, Active 3-D vision system for automatic model-based shape inspection, *Opt. Lasers Eng.*, **10**(3 and 4), 1989.
4. J. S. Massa, G. S. Buller, A. C. Walker, S. Cova, M. Umasuthan, and A. Wallace, Time of flight optical ranging system based on time correlated single photon counting, *Appl. Opt.*, **37**(31), 7298–7304, 1998.
5. N. Abramson, Time reconstruction in light-in-flight recording by holography, *Appl. Opt.*, **30**, 1242–1252, 1991.
6. T. E. Carlsson, Measurement of three dimensional shapes using light-in-flight recording by holography, *Opt., Eng.*, **32**, 2587–2592, 1993.
7. B. Nilsson and T. E. Carlsson, Direct three dimensional shape measurement by digital light-in-flight holography, *Appl. Opt.*, **37**(34), 7954–7959, 1998.
8. Z. Ji and M. C. Leu, Design of optical triangulation devices, *Opt. Laser Technol.*, **21**(5), 335–338, 1989.
9. C. P. Keferstein and M. Marxer, Testing bench for laser triangulation sensors, *Sensor Rev.*, **18**(3), 183–187, 1998.
10. M. Idesawa, High-precision image position sensing methods suitable for 3-D measurement, *Opt. Lasers Eng.*, **10**(3 and 4), 1989.
11. Keyence technical report on sensors & measuring instruments, 1997.
12. H. Takasaki, Moiré topography, *Appl. Opt.*, **9**, 1467–1472, 1970.
13. K. Harding and R. Tait, Moiré techniques applied to automated inspection of machined parts, *SME Vision'86 Conference*, Detroit, MI, 1986.
14. A. Asundi, Computer aided Moiré methods, *Opt. Lasers Eng.*, **17**, 107–116, 1993.
15. C. M. Wong, Image processing in experimental mechanics, MPhil thesis, the University of Hong Kong, Pokfulam, Hong Kong, 1993.
16. Y. Arai, H. Hirai, and S. Yokozeki, High-resolution dynamic measurement using electronic speckle pattern interferometry based on multi-camera technology, *Opt. Lasers Eng.*, **46**(10), 733–738, 2008.
17. A. J. P. van Haasteren and H. J. Frankena, Real time displacement measurement using a multicamera phase stepping speckle interferometer, *Appl. Opt.*, **33**(19), 4137–4142, 1994.

18. M. Kujawinska, L. Salbut, and K. Patorski, Three channels phase stepped system for Moiré interferometry, *Appl. Opt.*, **30**(13), 1991.

19. F. Chen, Y. Y. Hung, and J. Gu, Dual fringe pattern phase shifting interferometry for time dependent phase evaluation, *SEM Proceedings on VII International Congress on Experimental Mechanics*, 1992.

20. L. Bieman and K. Harding, 3D imaging using a unique refractive optic design to combine Moiré and stereo, *Proc. SPIE*, **3206**, 1997.

21. K. Harding, High speed Moiré contouring methods analysis, *Proc. SPIE*, **3520**, 27, 1998.

22. J. J. Dirckx and W. F. Decraemer, Video Moiré topography for in-vitro studies of the eardrum, *Proc. SPIE*, **3196**, 2–11, 1998.

23. K. Yuen, I. Inokuchi, M. Maeta, S. Manabu, and Y. Masuda, Dynamic evaluation of facial palsy by Moiré topography video: Second report, *Proc. SPIE*, **2927**, 138–141, 1996.

24. Y.-B. Choi and S.-W. Kim, Phase shifting grating projection Moiré topography, *Opt. Eng.*, **37**(3), 1005–1010, 1998.

25. T. Matsumoto, Y. Kitagawa, and T. Minemoto, Sensitivity variable Moiré topography with a phase shift method, *Opt. Eng.*, **35**(6), 1754–1760, 1996.

26. T. Yoshizawa and T. Tomisawa, Shadow Moiré topography by means of the phase shift method, *Opt. Eng.*, **32**(7), 1668–1674, 1993.

27. T. Yoshizawa and T. Tomisawa, Moiré topography with the aid of phase shift method, *Proc. SPIE*, **1554B**, 441–450, 1991.

28. J. F. Cardenas-Garcia, S. Zheng, and F. Z. Shen, Projection Moiré as a tool for the automated determination of surface topography, *Proc. SPIE*, **1554B**, 210–224, 1991.

29. T. Matsumoto, Y. Kitagawa, M. Adachi, and A. Hayashi, Laser Moiré topography for 3D contour measurement, *Proc. SPIE*, **1332**, 530–536, 1991.

30. J. E. A. Liao and A. S. Voloshin, Surface topography through the digital enhancement of the shadow Moiré, *SEM Proceedings*, pp. 506–510, 1990.

31. J. S. Lim and M. S. Chung, Moiré topography with color gratings, *Appl. Opt.*, **27**, 2649, 1988.

32. J. Y. Wang, Imaging laser radar—An over view, *Proceedings of the Ninth International Conference Laser'86*, pp. 19–29, 1986.

33. J. C. Marron and K. S. Schroeder, Three dimensional lensless imaging using laser frequency diversity, *Appl. Opt.*, **31**(2), 255–262, 1992.

34. J. C. Marron and T. J. Schulz, Three dimensional, fine resolution imaging using laser frequency diversity, *Opt. Lett.*, **17**, 285–287, 1992.

35. L. G. Shirley, Speckle decorrelation techniques for remote sensing of rough object, *OSA Annu. Meet. Tech. Dig.*, **18**, 208, 1989.

36. L. G. Shirley, Remote sensing of object shape using a wavelength scanning laser radar, *OSA Annu. Meet. Tech. Dig.*, **17**, 154, 1991.

37. G. R. Hallerman and L. G. Shirley, A comparison of surface contour measurements based on speckle pattern sampling and coordinate measurement machines, *Proc. SPIE*, **2909**, 8997, 1996.

38. T. Dressel, G. Häusler, and H. Venzhe, Three dimensional sensing of rough surfaces by coherence radar, *Appl. Opt.*, **31**, 919–925, 1992.

39. L. G. Shirley and G. R. Hallerman, Application of tunable lasers to laser radar and 3D imaging, Technical Report No. 1025, MIT Lincoln Laboratory, Lexington, MA, 1996.

40. K. A. Haines and B. P. Hildebrand, Contour generation by wavefront construction, *Phys. Lett.*, **19**, 10–11, 1965.

41. K. Creath, Y. Y. Cheng, and J. Wyant, Contouring aspheric surface using two-wavelength phase shifting interferometry, *Opt. Acta*, **32**(12), 1455–1464, 1985.

42. R. P. Tatam, J. C. Davies, C. H. Buckberry, and J. D. C. Jones, Holographic surface contouring using wavelength modulation of laser diodes, *Opt. Laser Technol.*, **22**, 317–321, 1990.

43. T. Maack, G. Notni, and W. Schreiber, Three coordinate measurement of an object surface with a combined two-wavelength and two-source phase shifting speckle interferometer, *Opt. Commun.*, **115**, 576–584, 1995.

44. Y. Yu, T. Kondo, T. Ohyama, T. Honda, and J. Tsujiuchi, Measuring gear tooth surface error by fringe scanning interferometry, *Acta Metrologica Sin.*, **9**(2), 120–123, 1986.

45. J. S. Zelenka and J. R. Varner, Multiple-index holographic contouring, *Appl. Opt.*, **8**, 1431–1434, 1969.

46. Y. Y. Hung, J. L. Turner, M. Tafralian, J. D. Hovanesian, and C. E. Taylor, Optical method for measuring contour slopes of an object, *Appl. Opt.*, **17**(1), 128–131, 1978.

47. H. Ei-Ghandoor, Tomographic investigation of the refractive index profiling using speckle photography technique, *Opt. Commun.*, **133**, 33–38, 1997.
48. N. Abramson, Holographic contouring by translation, *Appl. Opt.*, **15**, 1018–1022, 1976.
49. C. Joenathan, B. Franze, P. Haible, and H. J. Tiziani, Contouring by electronic speckle pattern interferometry using dual beam illumination, *Appl. Opt.*, **29**, 1905–1911, 1990.
50. P. K. Rastogi and L. Pflug, A holographic technique featuring broad range sensitivity to contour diffuse objects, *J. Mod. Opt.*, **38**, 1673–1683, 1991.
51. R. Rodrfguez-Vera, D. Kerr, and F. Mendoza-Santoyo, Electronic speckle contouring, *J. Opt. Soc. Am. A*, **9**(1), 2000–2008, 1992.
52. L. S. Wang and S. Krishnaswamy, Shape measurement using additive-subtractive phase shifting speckle interferometry, *Meas. Sci. Technol.*, **7**, 1748–1754, 1996.
53. E. Dalhoff, E. Fischer, S. Kreuz, and H. J. Tiziani, Double heterodyne interferometry for high precision distance measurements, *Proc. SPIE*, **2252**, 379–385, 1993.
54. J. D. Trolinger, Ultrahigh resolution interferometry, *Proc. SPIE*, **2861**, 114–123, 1996.
55. J. R. Huang and R. P. Tatam, Optoelectronic shearography: Two wavelength slope measurement, *Proc. SPIE*, **2544**, 300–308, 1995.
56. C. T. Griffen, Y. Y. Hung, and F. Chen, Three dimensional shape measurement using digital shearography, *Proc. SPIE*, **2545**, 214–220, 1995.
57. C. J. Tay, H. M. Shang, A. N. Poo, and M. Luo, On the determination of slope by shearography, *Opt. Lasers Eng.*, **20**, 207–217, 1994.
58. T. D. DeWitt and D. A. Lyon, Range-finding method using diffraction gratings, *Appl. Opt.*, **23**(21), 2510–2521, 1995.
59. T. D. Dewitt and D. A. Lyon, Moly: A prototype hand-held 3D digitizer with diffraction optics, *Opt. Eng.*, **39**, 2000.
60. S. Seebacher, W. Osten, and W. Jüptner, Measuring shape and deformation of small objects using digital holography, *Proc. SPIE*, **3479**, 104–115, 1998.
61. C. Wagner, W. Osten, and S. Seebacher, Direct shape measurement by digital wave-front reconstruction and wavelength scanning, *Opt. Eng.*, **39**(1), 2000.
62. G. Sirat and F. Paz, Conoscopic probes are set to transform industrial metrology, *Sensor Rev.*, **18**(2), 108–110, 1998.
63. W. Wester-Ebbinghaus, Analytics in non-topographic photogrammetry, *ISPRS Congress, Commission V*, Kyoto, Japan, Vol. 27, Part B11, pp. 380–390, 1988.
64. H. M. Karara, *Non-Topographic Photogrammetry*, 2nd edn., American Society of Photogrammetry and Remote Sensing, Falls Church, VA, 1989.
65. A. Gruen and H. Kahmen (Eds.), *Optical 3-D Measurement Techniques*, Wichmann, Heidelberg, Germany, 1989.
66. A. Gruen and H. Kahmen (Eds.), *Optical 3-D Measurement Techniques II*, SPIE Vol. 2252, 1993.
67. A. Gruen and H. Kahmen (Eds.), *Optical 3-D Measurement Techniques III*, Wichmann, Heidelberg, Germany, 1995.
68. C. C. Slama, *Manual of Photogrammetry*, 4th edn., American Society of Photogrammetry, Falls Church, VA, 1980.
69. C. S. Fraser, Photogrammetric measurement to one part in a million, *Photogrammetric Eng. Remote Sens.*, **58**(3), 305–310, 1992.
70. W. Schertenleib, Measurement of structures (surfaces) utilizing the SMART 310 laser tracking system, *Optical 3-D Measurement Techniques III*, Wichmann, Heidelberg, Germany, 1995.
71. S. Kyle, R. Loser, and D. Warren, *Automated Part Positioning with the Laser Tracker*, Eureka Transfers Technology, 1997.
72. V. Sirnivasan, H. C. Liu, and M. Halioua, Automated phase measuring profilometry of 3D diffuse objects, *Appl. Opt.*, **23**, 3105–3108, 1984.
73. J. A. Jalkio, R. C. Kim, and S. K. Case, Three dimensional inspection using multi-stripe structured light, *Opt. Eng.*, **24**(6), 966–974, 1985.
74. F. Wahl, A coded light approach for depth map acquisition, in Mustererkennung 86, Informatik Fachberichte 125, Springer-Verlag, Berlin, Germany, 1986.
75. S. Toyooka and Y. Iwasa, Automatic profilometry of 3D diffuse objects by spatial phase detection, *Appl. Opt.*, **25**, 1630–1633, 1986.
76. M. Sjodahl and P. Synnergren, Measurement of shape by using projected random patterns and temporal digital speckle photography, *Appl. Opt.*, **38**(10), 1990–1997, 1999.

77. A. Shashua, Trilinear tensor: The fundamental construct of multiple-view geometry and its applications, *International Workshop on Algebraic Frames for the Perception Action Cycle*, Kiel, Germany, September 1997.

78. S. Avidan and A. Shashua, Novel view synthesis by cascading trilinear tensors, *IEEE Trans. Vis. Comput. Graph.*, **4**(4), 1998.

79. A. Shashua and M. Werman, On the trilinear tensor of three perspective views and its underlying geometry, *Proceedings of the International Conference on Computer Vision*, Boston, MA, June 1995.

80. C. R. Coggrave and J. M. Huntley, Optimization of a shape measurement system based on spatial light modulators, *Opt. Eng.*, **39**(1), 2000.

81. H. Gartner, P. Lehle, and H. J. Tiziani, New, high efficient, binary codes for structured light methods, *Proc. SPIE*, **2599**, 4–13, 1995.

82. E. Muller, Fast three dimensional form measurement system, *Opt. Eng.*, **34**(9), 2754–2756, 1995.

83. G. Sansoni, S. Corini, S. Lazzari, R. Rodella, and F. Docchio, Three dimensional imaging based on gray-code light projection: Characterization of the measuring algorithm and development of a measuring system for industrial application, *Appl. Opt.*, **36**, 4463–4472, 1997.

84. K. Engelhardt and G. Häusler, Acquisition of 3D data by focus sensing, *Appl. Opt.*, **27**, 4684, 1988.

85. A. Serrano-Heredia, C. M. Hinojosa, J. G. Ibarra, and V. Arrizon, Recovery of three dimensional shapes by using a defocus structured light system, *Proc. SPIE*, **3520**, 80–83, 1998.

86. M. Takata, T. Aoki, Y. Miyamoto, H. Tanaka, R. Gu, and Z. Zhang, Absolute three dimensional shape measurements using a co-axial optical system with a co-image plan for projection and observation, *Opt. Eng.*, **39**(1), 2000.

87. H. J. Tiziani, Optical techniques for shape measurements, *Proceedings of the Second International Workshop on Automatic Processing of Fringe Patterns, Fringe '93*, W. Jüptner and W. Osten (Eds.), Akademie Verlag, Berlin, Germany, 1993.

88. K. Leonhardt, U. Droste, and H. J. Tiziani, Mircoshape and rough surface analysis by fringe projection, *Appl. Opt.*, **33**, 7477–7488, 1994.

89. H. J. Tiziani and H. M. Uhde, Three dimensional image sensing with chromatic confocal microscopy, *Appl. Opt.*, **33**, 1838–1843, 1994.

90. M. Halioua, R. S. Krishnamurthy, H. C. Liu, and F. P. Chiang, Automated 360° profilometry of 3-D diffuse objects, *Appl. Opt.*, **24**, 2193–2196, 1985.

91. X. X. Cheng, X. Y. Su, and L. R. Guo, Automated measurement method for 360° profilometry of diffuse objects, *Appl. Opt.*, **30**, 1274–1278, 1991.

92. H. Ohara, H. Konno, M. Sasaki, M. Suzuki, and K. Murata, Automated 360° profilometry of a three-dimensional diffuse object and its reconstruction by use of the shading model, *Appl. Opt.*, **35**, 4476–4480, 1996.

93. A. Asundi, C. S. Chan, and M. R. Sajan, 360° profilometry: New techniques for display and acquisition, *Opt. Eng.*, **33**, 2760–2769, 1994.

94. A. Asundi and W. Zhou, Mapping algorithm for 360-deg profilometry with time delayed and integration imaging, *Opt. Eng.*, **38**, 339–344, 1999.

95. C. Reich, Photogrammetric matching of point clouds for 3D measurement of complex objects, *Proc. SPIE*, **3520**, 100–110, 1998.

96. R. W. Malz, High dynamic codes, self calibration and autonomous 3D sensor orientation: Three steps towards fast optical reverse engineering without mechanical CMMs, *Optical 3-D Measurement Techniques III*, Wichmann Verlag, Heidelberg, pp. 194–202, 1995.

97. H. J. Tiziani, High precision surface topography measurement, in *Optical 3-D Measurement Techniques III*, A. Gruen and H. Kahmen (Eds.), Wichmann Verlag, Heidelberg, Germany, 1995.

98. S. Kakunai, T. Sakamoto, and K. Iwata, Profile measurement by an active light source, *Proceedings of the Far East Conference on Nondestructive Testing and the Republic of China's Society for Nondestructive Testing Ninth Annual Conference*, pp. 237–242, 1994.

99. Y. Arai, S. Yekozeki, and T. Yamada, 3-D automatic precision measurement system by liquid crystal plate on moire-topography, *Proc. SPIE*, **1554B**, 266–274, 1991.

100. G. Sansoni, F. Docchio, U. Minoni, and C. Bussolati, Development and characterization of a liquid crystal projection unit for adaptive structured Illumination, *Proc. SPIE*, **1614**, 78–86, 1991.

101. A. Asundi, Novel grating methods for optical inspection, *Proc. SPIE*, **1554B**, 708–715, 1991.

102. Y. Y. Hung and F. Chen, Shape measurement using shadow Moiré with a teaching LCD, Technical Report, Oakland University, Rochester, MI, 1991.

103. Y. Y. Hung and F. Chen, Shape measurement by phase shift structured light method with a LCD, Technical Report, Oakland University, Rochester, MI, 1992.

104. Y. Y. Hung, 3D machine vision technique for rapid 3D shape measurement and surface quality inspection, SAE 1999-01-0418, 1999.
105. H. O. Saldner and J. M. Huntley, Profilometry by temporal phase unwrapping and spatial light modulator based fringe projector, *Opt. Eng.*, **36**(2), 610–615, 1997.
106. G. Frankowski, The ODS 800—A new projection unit for optical metrology, *Proceedings of the Fringe '97, Bremen 1997*, Akademie Verlag, Berlin, Germany, pp. 532–539, 1997.
107. J. M. Huntley and H. O. Saldner, Error reduction methods for shape measurement by temporal phase unwrapping, *J. Opt. Soc. Am. A*, **14**(12), 3188–3196, 1997.
108. P. S. Huang, Q. Hu, F. Jin, and F. P. Chiang, Color-enhanced digital fringe projection technique for high-speed 3D surface contouring, *Opt. Eng.*, **38**(6), 1065–1071, 1999.
109. K. G. Harding, M. P. Coletta, and C. H. Vandommelen, Color encoded Moiré contouring, *Proc. SPIE*, **1005**, 169, 1988.
110. R. A. Andrade, B. S. Gilbert, S. C. Cahall, S. Kozaitis, and J. Blatt, Real time optically processed target recognition system based on arbitrary Moiré contours, *Proc. SPIE*, **2348**, 170–180, 1994.
111. C. H. Hu and Y. W. Qin, Digital color encoding and its application to the Moiré technique, *Appl. Opt.*, **36**, 3682–3685, 1997.
112. J. M. Desse, Three color differential interferometry, *Appl. Opt.*, **36**, 7150–7156, 1997.
113. S. Kakunai, T. Sakamoto, and K. Iwata, Profile measurement taken with liquid crystal gratings, *Appl. Opt.*, **38**(13), 2824–2828, 1999.
114. G. M. Brown and T. E. Allen, Measurement of structural 3D shape using computer aided holometry, Ford Research Report, 1992.
115. C. H. Buckberry, D. P. Towers, B. C. Stockley, B. Tavender, M. P. Jones, J. D. C. Jones, and J. D. R. Valera, Whole field optical diagnostics for structural analysis in the automotive industry, *Opt. Lasers Eng.*, **25**, 433–453, 1996.
116. Y. Y Hung, Three dimensional computer vision techniques for full surface shape measurement and surface flaw inspection, SAE 920246, 1992.
117. M. S. Mermelstein, D. L. Feldkhun, and L. G. Shirley, Video-rate surface profiling with acoustic optic accordion fringe interferometry, *Opt. Eng.*, **39**(1), 2000.
118. Structured light laser, Lasiris, Inc.
119. R. Lenz and U. Lenz, New developments in high resolution image acquisition with CCD area sensors, *Proc. SPIE*, **2252**, 53–62, 1993.
120. M. Ulm and G. Paar, Relative camera calibration from stereo disparities, *Optical 3-D Measurement Techniques III*, Wichmann, Heidelberg, Germany, pp. 526–533, 1995.
121. A. M. G. Tommaselli et al., Photogrammetric range system: Mathematical model and calibration, *Optical 3-D Measurement Techniques III*, Wichmann, Heidelberg, Germany, pp. 397–403, 1995.
122. F. Wallner, P. Weckesser, and R. Dillmann, Calibration of the active stereo vision system KASTOR with standardized perspective matrices, *Proc. SPIE*, **2252**, 98–105, 1993.
123. W. Hoflinger and H. A. Beyer, Characterization and calibration of a S-VHS camcorder for digital photogrammetry, *Proc. SPIE*, **2252**, 133–140, 1993.
124. R. G. Willson and S. A. Shafer, A perspective projection camera model for zoom lenses, *Proc. SPIE*, **2252**, 149–158, 1993.
125. R. Y. Tsai, A versatile camera calibration technique for high accuracy 3D machine vision metrology using off the shelf TV camera and lenses, *IEEE J. Robot. Automat.*, **RA-3**(4), 1987.
126. H. A. Beyer, Geometric and radiometric analysis of a CCD camera based photogrammetric system, PhD dissertation ETH No. 9701, Zurich, Switzerland, May 1992.
127. H. A. Beyer, Advances in characterisation and calibration of digital imaging systems, *IntArchPhRS*, *17th ISPRS Congress*, Washington, DC, Vol. 29, Part B5, pp. 545–555, 1992.
128. S. X. Zhou, R. Yin, W. Wang, and W. G. Wee, Calibration, parameter estimation, and accuracy enhancement of a 4DI camera turntable system, *Opt. Eng.*, **39**(1), 2000.
129. P. Saint-Marc, J. L. Jezouin, and G. Medioni, A versatile PC-based range-finding system, *IEEE Trans. Robot. Automat.*, **7**(2), 250–256, 1991.
130. W. Nadeborn, P. Andrä, W. Jüptner, and W. Osten, Evaluation of optical shape measurement methods with respect to the accuracy of data, *Proc. SPIE*, **1983**, 928–930, 1993.
131. R. J. Valkenburg and A. M. Mclor, Accurate 3D measurement using a structured light system, *Image Vis. Comput.*, **16**, 99–110, 1998.
132. A. Asundi and W. Zhou, Unified calibration technique and its applications in optical triangular profilometry, *Appl. Opt.*, **38**(16), 3556–3561, 1999.

133. Y. Y. Hung, L. Lin, H. M. Shang, and B. G. Park, Practical 3D computer vision techniques for full-field surface measurement, *Opt. Eng.*, **39**(1), 2000.

134. R. Kowarschik, P. Kühmstedt, J. Gerber, W. Schreiber, and G. Notni, Adaptive optical 3D-measurement with structured light, *Opt. Eng.*, **39**(1), 2000.

135. W. Schreiber and G. Notni, Theory and arrangements of self-calibrating whole-body 3D-measurement system using fringe projection technique, *Opt. Eng.*, **39**(1), 2000.

136. R. Zumbrunn, Automatic fast shape determination of diffuse reflecting objects at close range, by means of structured light and digital phase measurement, *ISPRS Intercommission Conference on Fast Proceedings of Photogrammetric Data*, Interlaken, Switzerland, 1987.

137. S. Kakunai, K. Iwata, S. Saitoh, and T. Sakamoto, Profile measurement by two-pitch grating projection, *J. Jpn. Soc. Prec. Eng.*, **58**, 877–882, 1992.

138. R. Malz, Adaptive light encoding for 3D sensing with maximum measurement efficiency, *11th DAGM-Symposium*, Informatik-Fachberichte 219, Springer, Hamburg, Germany, 1989.

139. L. G. Shirley, Three dimensional imaging using accordion fringe interferometry, *Appl. Opt.*, to be submitted.

140. H. Steinbichler, Method and apparatus for ascertaining the absolute coordinates of an object, US Patent 5,289,264, 1994.

141. G. Indebetouw, Profile measurement using projection of running fringes, *Appl. Opt.*, **17**(18), 2930, 1978.

142. X. Xie, M. J. Lalor, D. R. Burton, and M. M. Shaw, Four map absolute distance contouring, *Opt. Eng.*, **36**(9), 2517–2520, 1997.

143. W. Nadeborn, P. Andra, and W. Osten, A robust procedure for absolute phase measurement, *Opt. Lasers Eng.*, **24**, 245–260, 1996.

144. H. Zhao, W. Chen, and Y. Tan, Phase unwrapping algorithm for the measurement of three dimensional object shapes, *Appl. Opt.*, **33**(20), 4497–4500, 1994.

145. M. K. Kalms, W. Juptner, and W. Osten, Automatic adaption of projected fringe patterns using a programmable LCD projector, *Proc. SPIE*, **3100**, 156–165, 1997.

146. J. M. Huntley and H. O. Saldner, Shape measurement by temporal phase unwrapping: Comparison of unwrapping algorithm, *Meas. Sci. Technol.*, **8**(9), 986–992, 1997.

147. H. O. Saldner and J. M. Huntley, Temporal phase unwrapping: Application to surface profiling of discontinuous objects, *Appl. Opt.*, **36**(13), 2770–2775, 1997.

148. J. M. Huntley and H. O. Saldner, Temporal phase unwrapping algorithm for automated interferogram analysis, *Appl. Opt.*, **32**(17), 3047–3052, 1993.

149. M. Takeda, Q. Gu, M. Kinoshita, H. Takai, and Y. Takahashi, Frequency multiplex Fourier transform profilometry: A single short three dimensional shape measurement of objects with large height discontinuities and/or surface isolation, *Appl. Opt.*, **36**(22), 5347–5354, 1997.

150. M. Takada and H. Yamamoto, Fourier transform speckle profilometry: Three dimensional shape measurement of diffuse objects with large height steps and/or spatially isolated surfaces, *Appl. Opt.*, **33**, 7829–7837, 1994.

151. S. Kuwamura and I. Yamaguchi, Wavelength scanning profilometry for real time surface shape measurement, *Appl. Opt.*, **36**, 4473–4482, 1997.

152. H. J. Tiziani, B. Franze, and P. Haible, Wavelength shift speckle interferometry for absolute profilometry using a mode hop free external cavity diode laser, *J. Mod. Opt.*, **44**, 1485–1496, 1997.

153. T. E. Allen, F. Chen, and C. T. Griffen, Multiple wavelength technique for shape measurement, Ford Technical Report, 1995.

154. C. Joenathan, B. Franze, P. Haible, and H. J. Tiziani, Shape measurement by use of temporal Fourier transformation in dual beam illumination speckle interferometry, *Appl. Opt.*, **37**(16), 3385–3390, 1998.

155. S. W. Kim, J. T. Oh, M. S. Jung, and Y. B. Choi, Two-frequency phase shifting projection Moiré topography, *Proc. SPIE*, **3520**, 36–52, 1998.

156. F. Bien, M. Camac, H. J. Caulfield, and S. Ezekiel, Absolute distance measurements by variable wavelength interferometry, *Appl. Opt.*, **20**(3), 400–403, 1981.

157. K. C. Yuk, J. H. Jo, and S. Chang, Determination of the absolute order of shadow Moiré fringes by using two differently colored light sources, *Appl. Opt.*, **33**(1), 130–132, 1994.

158. I. Yamaguchi, A. Yamamoto, and S. Kuwamura, Shape measurement of diffuse surface by wavelength scanning, *Fringe'97 Automatic Processing of Fringe Patterns*, Akademie Verlag, Berlin, Germany, pp. 171–178, 1997.

159. W. Osten, W. Nadeborn, and P. Andra, General hierarchical approach in absolute phase measurement, *Proc. SPIE*, **2860**, 2–13, 1996.

160. Z. Q. Tao and D. Z. Yun, Fringe order identification of shadow Moiré with rotating grating, Technical Report, Dalian Technology University, Dalian, China, 1982.

161. F. Chen and D. Z. Yun, Identification and division of fringe orders in speckle interferometry, Technical Report, Dalian Technology University, Dalian, China, 1984.

162. F. Chen and D. Z. Yun, Identification and division of fringe orders in speckle method, *Acta Opt. Sin.*, **7**(5), 405–409, 1987.

163. N. Plouzennec and A. Lagarde, Two-wavelength method for full-filed automated photoelasticity, *Exp. Mech.*, **39**(4), 274, 1999.

164. K. A. Stetson, Use of sensitivity vector variation to determine absolute displacements in double exposure hologram interferometry, *Appl. Opt.*, **29**(4), 502–504, 1990.

165. H. Singh and J. S. Sirkis, Direct extraction of phase gradients from Fourier transform and phase step fringe patterns, *Appl. Opt.*, **33**(22), 5016–5020, 1994.

166. J. Kozlowski and G. Serra, Complex phase tracking method for fringe pattern analysis, *Appl. Opt.*, **38**(11), 2256–2262, 1999.

167. E. Gulch, Results of test on image matching of ISPRS WG III/4, *IntArchPhRS, Congress Commission III*, Kyoto, Japan, Vol. 27, Part B3, pp. 254–271, 1988.

168. M. Leaders, A survey on stereo matching techniques, *IntArchPhRS, Congress Commission III*, Kyoto, Japan, Vol. 27, Part B3, pp. 11–23, 1988.

169. A. W. Gruen, Geometrically constrained multiphoto-matching, *Photogrammetric Eng. Remote Sens.*, **54**(5), 633–641, 1988.

170. A. Blake, D. McCowen, H. R. Lo, and D. Konash, Epipolar geometry for trinocular active range-sensors, *Br. Mach. Vision Conf.*, **BMVC90**, 19–24, 1990.

171. N. Ayache, *Artificial Vision for Mobile Robots: Stereo Vision and Multisensory Perception*, Massachusetts Institute of Technology, The MIT Press, Cambridge, MA, p. 342, 1991.

172. C. Heipke, A global approach for least squares image matching and surface recognition in object space, *Photogrammetric Eng. Remote Sens.*, **58**(3), 317–323, 1992.

173. I. Maalen-Johansen, On the precision of sub-pixel measurements in videometry, *Proc. SPIE*, **2252**, 169–178, 1993.

174. T. A. Clarke, M. A. R. Cooper, and J. G. Fryer, An estimator for the random error in sub-pixel target location and its use in the bundle adjustment, *Proc. SPIE*, **2252**, 161–168, 1993.

175. C. Heipke, An integral approach to digital image matching and object surface reconstruction, *Optical 3D Measurement Techniques*, Wichmann Verlag, Heidelberg, Germany, pp. 347–359, 1989.

176. C. Graves, Key to success for vision system users, *Sensor Rev.*, **18**(3), 178–182, 1998.

177. R. Gooch, Optical metrology in manufacturing automation, *Sensor Rev.*, **18**(2), 81–87, 1998.

178. C. Bradley and G. W. Vickers, Automated rapid prototyping utilizing laser scanning and free form machining, *Ann. CIRP*, **41**(2), 437–440, 1992.

179. M. Milroy, D. J. Weir, C. Bradley, and G. W. Vickers, Reverse engineering employing 3d laser scanner: A case study, *Int. J. Adv. Manuf. Technol.*, **12**, 111–121, 1996.

180. D. J. Weir, M. Milroy, C. Bradley, and G. W. Vickers, Reverse engineering physical models employing wrap-around b-spline surfaces and quadrics, *Proc. Inst. Mech. Eng. B*, **210**, 147–157, 1996.

181. R. Niepold, S. Sakane, T. Sato, and Y. Shirai, Vision sensor set-up planning for a hand-eye system using environmental model, *Proceedings of the Society of Instrument and Control Engineering Japan*, Hiroshima, Japan, pp. 1037–1040, July 1987.

182. S. Sakane, M. Ishii, and M. Kakikura, Occlusion avoidance of visual sensors based on a hand eye action simulator system: HEAVEN, *Adv. Robot.*, **2**(2), 149–165, 1987.

183. S. Sakane, T. Sato, and M. Kakikura, Planning focus of attentions for visual feedback control, *Trans. Soc. Instrum. Control Eng.*, **24**(6), 608–615, June 1988.

184. D. P. Anderson, An orientation method for central projection program, *Comput. Graph.*, **9**(1), 35–37, 1982.

185. D. P. Anderson, Efficient algorithms for automatic viewer orientation, *Comput. Graph.*, **9**(4), 407–413, 1985.

186. C. K. Cowan, Model based synthesis of sensor location, *Proceedings of the 1988 IEEE International Conference on Robotics and Automation*, pp. 900–905, 1988.

187. C. K. Cowan and A. Bergman, Determining the camera and light source location for a visual task, *Proceedings 1989 IEEE International Conference on Robotics and Automation*, pp. 508–514, 1989.

188. C. K. Cowan and P. D. Kovesi, Automatic sensor placement from vision task requirements, *IEEE Trans. Pattern Anal. Mach. Intell.*, **10**(3), 407–416, May 1988.

189. K. Tarabanis, R. Y. Tsai, and S. Abrams, Planning viewpoints that simultaneously satisfy several feature detectability constraints for robotic vision, *Fifth International Conference on Advanced Robotics ICAR*, 1991.

190. K. Tarabanis, R. Y. Tsai, and P. K. Allen, Automated sensor planning for robotic vision tasks, *Proceedings 1991 IEEE International Conference on Robotics and Automation*, April 1991.

191. M. Huck, J. Raczkowsky, and K. Weller, Sensor simulation in robot applications, *Advanced Robotics Program, Workshop on Manipulators, Sensors and Steps Towards Mobility*, 1997–209, Nuclear Research Center, Karlsruhe, Germany, 1987.

192. K. Ikeuchi and J. C. Robert, Modeling sensors detectability with the VANTAGE geometric/sensor modeler, *IEEE Trans. Robot. Automat.*, **7**, 771–784, December 1991.

193. J. Raczkowsky and K. H. Mittenbuehler, Simulation of cameras in robot applications, *Comput. Graph. Appl.*, **9**(1), 16–25, January 1989.

194. B. G. Batchelor, Integrating vision and AI for industrial application, *Proceedings of the Intelligent Robots and Computer Vision VIII: Systems and Applications, Proc. SPIE*, **1193**, 295–302, 1989.

195. B. G. Batchelor, D. A. Hill, and D. C. Hodgson, *Automated Visual Inspection*, IFS Ltd., Bedford, U.K., 1985.

196. Y. Kitamura, H. Sato, and H. Tamura, An expert system for industrial machine vision, *Proceedings of the 10th International Conference on Pattern Recognition*, Atlantic City, NJ, pp. 771–773, June 1990.

197. A. Novini, Lighting and optics expert system for machine vision, *Proceedings of the Optics, Illumination, and Image Sensing*, pp. 1005–1019, 1998.

198. A. J. Spyridi and A. G. Requicha, Accessibility analysis for the automatic inspection of mechanical parts by coordinate measuring machines, *Proceedings of the 1990 ICRA*, pp. 1284–1289, 1990.

199. J. L. Mundy, Industrial machine vision—Is it practical? in *Advances in Digital Image Processing*, P. Stucki (Ed.), Plenum Press, New York, pp. 235–248, 1979.

200. S. Yi, R. M. Haralick, and L. G. Shapiro, Automatic sensor and light source positioning for machine vision, *Proceedings of the 10th International Conference on Pattern Recognition*, Atlantic City, NJ, pp. 55–59, 1990.

201. F. Chen, C. T. Griffen, T. E. Allen, and G. M. Brown, Measurement of shape and vibration using a single electronic speckle interferometry, *Proc. SPIE*, **2860**, 150–161, 1996.

202. F. Chen, C. T. Griffen, and N. Arnon, Fast paint defect detection using structured light and Fourier transform method, Ford Technical Report, 1995.

203. B. H. Zhuang and W. W. Zhang, Nondestructive profiler for pipe inner wall using triangulation scanning method, *Proc. SPIE*, **3520**, 76–79, 1998.

204. G. Lu, S. Wu, N. Palmer, and H. Liu, Application of phase shift optical triangulation to precision gear gauging, *Proc. SPIE*, **3520**, 52–63, 1998.

205. G. L. Hobrough, A future for real time photogrammetry, *Vermessung, Photogrammetrie, Kulturtechnik Heft*, **9**, 312–315, 1985.

206. T. Clark and R. Gooch, Real time 3D metrology for aerospace manufacture, *Sensor Rev.*, **19**(2), 113–115, 1999.

207. S. J. Gordon and F. Benayad-Cherif, 4DI—A real time three dimensional imager, *Proc. SPIE*, **2348**, 221–226, 1995.

208. E. N. Colenman and R. Jain, Obtaining 3-dimensional shape of texture and specular surfaces using four-source photometry, *Comput. Graph. Image Process.*, **18**(4), 309–328, 1982.

209. F. Solomon and K. Ikeuchi, Inspection specular lobe object using four light sources, *Proceedings of the IEEE Conference on Robotics and Automation*, Nice, France, pp. 1707–1712, 1992.

210. G. Healey and T. O. Binford, Local shape from specularity, *Proceedings of Image Understanding Workshop*, Vol. 2, pp. 874–887, 1987.

211. S. K. Nayar, A. C. Sanderson, L. E. Weiss, and D. D. Simon, Specular surface inspection using structured highlight and Gaussian images, *IEEE Trans. Robot. Automat.*, **6**(2), 208–218, 1990.

212. A. C. Sanderson, L. E. Weiss, and S. K. Nayar, Structured highlight inspection of specular surfaces, *IEEE Trans. Pattern Anal. Mach. Intell.*, **10**(1), 44–55, 1988.

213. S. K. Nayar, Shape recovery using physical models of reflection and interreflection, CMU, 1991.

214. Y. Y. Hung, F. Chen, S. H. Tang, and J. D. Hovanesian, Reflective computer vision technique for measuring surface slope and plate deformation, *SEM Proceeding*, 1993.

215. Y. Y. Hung, F. Chen, and J. D. Hovanesian, DFT-based reflective 3-D computer vision technique for measuring plate deformation, *SEM Proceeding*, 1994.

216. R. Höfling, P. Aswendt, and R. Neugebauer, Phase reflection—A new solution for the detection of shape defects on car body sheets, *Opt. Eng.*, **39**(1), 2000.

217. X. Zhang and W. P. T. North, Retroreflective grating generation and analysis for surface measurement, *Appl. Opt.*, **37**(5), 2624–2627, 1998.
218. X. Zhang and W. P. T. North, Analysis of 3D surface waviness on standard artifacts by retroreflective metrology, *Opt. Eng.*, **39**(1), 2000.
219. Osaka University researchers develop non-contact profile sensor, *Photonics*, **46**, February 1997.
220. Y. K. Ryu and H. S. Cho, New optical sensing system for obtaining the three-dimensional shape of specular objects, *Opt. Eng.*, **35**(5), 1483–1495, 1996.
221. B. N. Taylor and C. E. Kuyatt, Guidelines for evaluating and expressing the uncertainty of NIST measurement results, NIST Technical Note 1297, US Government Printing Office, Washington, DC, 1994.
222. G. Dalton, Reverse engineering using laser metrology, *Sensor Rev.*, **18**(2), 92–96, 1999.
223. M. Lehmann, P. Jacquot, and M. Facchini, Shape measurements on large surfaces by fringe projection, *Exp. Tech.*, **23**(2), 31–35, March/April 1999.
224. Keithley Instruments, Inc., *Low Level Measurements Handbook*, 4th edn., Keithley Instruments, Inc., Cleveland, OH, 1993.

21 Fringe Analysis

Jun-ichi Kato

CONTENTS

21.1 Introduction 541
21.2 Basics of Fringe Analysis 541
    21.2.1 Method Using Multiple-Input Images 543
    21.2.2 Method Using Single-Input Image 545
21.3 Phase Unwrapping 548
21.4 Practical Example of Fringe Analysis in Profilometry.......................................550
21.5 Summary 551
Symbols..551
References...552

21.1 INTRODUCTION

In this chapter, basics of the "fringe analysis," which is one of the major methods for digitizing various physical phenomena, are briefly introduced.

Merits of optical or image measurements are that they can quickly acquire spatial information of objects without contacting them. For example, in interferometry, which uses wave nature of light, a highly accurate distribution of physical information such as surface height, deformation, or indices of objects can be instantaneously obtained as a contour-like fringe pattern. Also, in the fringe projection techniques that typically apply the Moiré or the deformation grating principles, the height map of an object is visualized as a deformed fringe pattern and also a contour map. Although the fringe pattern obtained in these manners visualizes the physical conditions of the object, it is usually necessary to extract the quantitative values of the conditions from these patterns for metrological purposes. This process is called "fringe analysis," which has rapidly developed together with image processing capabilities with computers and has become a large field of modern image analysis.

In many cases, the major purpose of the fringe analysis is an accurate conversion of the above fringe pattern to a distribution of its phase and contrast which can be directly connected with the physical parameters generating the fringes. Therefore, this technique is also called the "phase analysis" of the fringe patterns. Here, the principles and basic procedures and some of their applications in the fringe analysis with a computer are surveyed. Additionally, a distinctive process of converting a wrapped phase map to the final physical distribution, namely, "phase unwrapping," is briefly described.

21.2 BASICS OF FRINGE ANALYSIS

Typical examples of the fringe images treated in the fringe analysis are shown in Figure 21.1a and b. There are mainly two types of fringe images: a general case shown in Figure 21.1a is the fringes that can be interpreted as contour lines and the other case in Figure 21.1b is the fringe pattern that consists of high-frequency fringes (conventionally, we call them "carrier fringes" here) along a certain direction carrying their spatial deformation. For each type, suitable analysis techniques exist, as described in the following section.

FIGURE 21.1 Examples of fringe images with their formation method: (a) a contour-like interference fringe generated by the Michelson type interferometer, (b) a deformed grating image by a surface shape of an object obtained with the fringe projection method. HM, half mirror; L, imaging lens.

Although the intensity profile of the fringe patterns grabbed in a computer from a CCD camera, for example, is not exactly sinusoidal, the sinusoidal representation is conventional in many cases for mathematical treatments. Thus, in general, an intensity distribution of fringe patterns using sine function is expressed as

$$I(x, y) = I_0(x, y)\{1 + \gamma(x, y)\cos[\phi(x, y)]\}$$
$$= a(x, y) + b(x, y)\cos\phi(x, y), \tag{21.1}$$

where $I_0(x, y)$ and $\gamma(x, y)$ are the bias intensity and the modulation depth of the fringes in the field of view, which is namely the background intensity distribution and the contrast heterogeneity, and are henceforth substituted as $a(x, y)$ and $b(x, y)$ for simplicity. In many cases, the required value is the phase distribution $\phi(x, y)$. In the early period of the computer's introduction, an image processing technique for contour-like fringes, including extractions of the fringe centers and their order number evaluations, was mainstream in the fringe analysis where the sub-fringe order was estimated through the interpolation between fringe centers. The use of this analysis principle is rare nowadays.

Alternatively, more sophisticated analysis principles were evolved after the 1970s; they are, namely, the sub-fringe phase analysis represented by the fringe scanning method [1], the phase-shifting method [2], and the Fourier transformation (FT) method [3]. These methods are classified by the number of input fringe images needed for the phase extraction.

1. Method using multiple-input images, in which more than three fringe images, whose initial phases are shifted by a given amount (phase-shifted fringes), are acquired and used for the analysis. The phase-shifting method and the fringe-scanning method belong to this class.
2. Method using single-input image, in which a fringe image introducing the high-frequency carrier fringes is acquired and directly analyzed. The FT method is a typical example.

Next, the principles in each class of the fringe analysis are briefly explained.

21.2.1 METHOD USING MULTIPLE-INPUT IMAGES

The phase-shifting method is the most popular fringe analysis method using multiple-input images. In this method, multiple (N) fringe patterns with the initial phase of a target object shifted by a given value of $\Delta\psi$ are first acquired by some means. These phase-shifted fringe images are obtained at different timings or sampling spaces. In the interferometer as shown in Figure 21.1a, the optical path length of the reference arm is gradually changed by a small displacement of $\lambda \times (\Delta\psi)/(2\pi)$ (λ is the wavelength of a light source) with a mirror translated by, for example, a PZT transducer. In a grating projection method, the phase shifting is realized by giving in-plane shifts to the grating perpendicularly by a distance of $\Lambda \times (\Delta\psi)/(2\pi)$ (Λ is the pitch of gratings) as shown in Figure 21.1b.

For the sampling number $k = 1, 2, \ldots, N$, the phase shift value is usually settled by

$$\Delta\psi = \frac{2\pi}{N}. \tag{21.2}$$

For each shifted phase, the N images of fringe intensity distributions,

$$I_k(x,y) = a(x,y) + b(x,y)\cos[\phi(x,y) + (k-1)\Delta\psi], \tag{21.3}$$

are acquired or sampled. By using all the images, the relation between the phase map $\phi(x, y)$ and the intensity values at each pixel is deduced as their general least square solution:

$$\tan\phi(x,y) = \frac{\displaystyle\sum_{k=1}^{N} I_k(x,y)\sin[(k-1)\Delta\psi]}{\displaystyle\sum_{k=1}^{N} I_k(x,y)\cos[(k-1)\Delta\psi]}. \tag{21.4}$$

In principle, the larger sampling number N decreases the random intensity noise by the factor of $1/\sqrt{N}$ and gives rise to the high accuracy in measurements. In practice, however, from three to seven phase-shifted fringes are used for the analysis. There were various sampling algorithms depending on the number N reported, which are called "N steps" or "N buckets" algorithms [2].

The most basic algorithm corresponds to the case that the sampling number of the phase-shifted fringes is equal to the unknown parameters to be measured, I_0, γ, and ϕ, that is, $N = 3$, with the phase-shifting of $\Delta\psi = \pi/2$. Then, the phase map is retrieved by the following equation:

$$\phi(x,y) = \tan^{-1}\left[\frac{I_3(x,y) - I_2(x,y)}{I_1(x,y) - I_2(x,y)}\right] + \frac{\pi}{4}, \tag{21.5}$$

where $\tan^{-1}$ is the *arctangent*. Furthermore, the contrast map of fringes is also calculated with

$$\gamma(x,y) = \frac{\sqrt{[I_1(x,y) - I_2(x,y)]^2 + [I_2(x,y) - I_3(x,y)]^2}}{2I_0}, \tag{21.6}$$

and the bias intensity of fringes is estimated by

$$I_0(x,y) = \frac{I_1(x,y) + I_3(x,y)}{2}. \tag{21.7}$$

TABLE 21.1

Frequently Used or Improved Phase-Shifting Algorithms

Four-steps method [2]

Phase shift $\Delta\psi = \pi/2$

Phase map
$$\phi(x,y) = \tan^{-1}\frac{I_4(x,y) - I_2(x,y)}{I_1(x,y) - I_3(x,y)}$$

Contrast map
$$\gamma(x,y) = \frac{2\sqrt{[I_1(x,y) - I_3(x,y)]^2 + [I_2(x,y) - I_4(x,y)]^2}}{I_1(x,y) + I_2(x,y) + I_3(x,y) + I_4(x,y)}$$

Carre's method (four steps) [4]

Phase shift $\Delta\psi = \alpha$ (arbitrary constant)

Phase shift α is obtained by (coordinates (x, y) are omitted)

$$\alpha = 2\tan^{-1}\sqrt{\frac{[3(I_2 - I_3) - (I_1 - I_4)]}{[(I_2 - I_3) + (I_1 - I_4)]}}$$

Phase map
$$\phi = \tan^{-1}\frac{\sqrt{[3(I_2 - I_3) - (I_1 - I_4)][(I_2 - I_3) + (I_1 - I_4)]}}{(I_1 + I_3) - (I_1 + I_4)}$$

Contrast map
$$\gamma = \frac{1}{2I_0}\sqrt{\frac{[(I_2 - I_3) + (I_1 - I_4)]^2 + [(I_2 + I_3) - (I_1 + I_4)]^2}{2}}$$

Hariharan's five-steps method [5]

Phase shift $\Delta\psi = \pi/2$

Phase map
$$\phi(x,y) = \tan^{-1}\frac{2[I_2(x,y) - I_4(x,y)]}{2I_3(x,y) - I_5(x,y) - I_1(x,y)}$$

$$\cos\Delta\psi = \frac{I_5 - I_1}{2(I_4 - I_2)} \Rightarrow 0 \text{ for calibrations of the phase shift value}$$

For the most popular algorithm in the case of using the four phase-shifting fringes (four-steps or four-buckets algorithms with $N = 4$ and $\Delta\psi = \pi/2$), the equations shown in Table 21.1 (four-steps method) are normally used.

Notice that in the use of the above algorithms, since exact and known phase-shifting values are presumed, highly accurate phase shifters should be used and that phase-shifting errors are also introduced by various disturbances such as vibration among image acquisitions and could give rise to serious measurement errors. In particular, linear and nonlinear deviations in phase shifts are mainly caused by calibration error and the drift of the phase shifters such as a PZT, which are major sources for the phase-measurement errors as well as the deviation of fringe profiles from the exact sinusoidal shapes. For these problems, several error-correction algorithms have been proposed.

For example, in the case where the phase-shift value $\Delta\psi$ is deviated from $\pi/2$ but is constant, an error-correction method using four phase steps has been proposed by Carré [4], which is shown in Table 21.1 (Carré's method [four steps]). For the fringe profiles containing spatial nonuniformity such as second order frequency components, a five-steps algorithm shown in Table 21.1 (Hariharan's five-steps method) is frequently adopted as an error compensation [5]. Various improved algorithms have been proposed against nonlinear phase-shift errors, vibrations at specific frequency, and others [6–8].

Figure 21.2 shows a practical procedure of the fringe analysis based on three-buckets phase-shifting method. The optical setup is the Michelson type interferometer, the same as that in Figure 21.1a. The reference mirror was displaced stepwise by $\pi/2$ in phase with a PZT transducer and the phase-shifted fringe patterns were stored in a frame grabber as shown in Figure 21.2a. Next, the phase map in Figure 21.2b was derived from the three images through the image operations for

FIGURE 21.2 Fringe analysis procedure in three-buckets phase-shifting method: (a) input fringe patterns with phase shifted by $\pi/2$ rad for each, (b) an obtained phase map and its unwrapped results (the detail described after), and (c) a finally obtained height (optical path difference) map.

each pixel based on Equation 21.5. In the phase map, black (gray level 0) corresponds to $-\pi$ in phase and white (gray level 255) to $+\pi$. To convert the phase map to the actual physical information, for example, maps of the optical path differences or object surface height, an additional operation to the phase map, called "phase unwrapping," is necessary as described later. Finally, the surface height map of the object is deduced using the physical relation between the phase ϕ and the height Δz, $\lambda \times (\phi/2\pi) = 2\Delta z$, as shown in Figure 21.2c.

A merit of the phase-shifting method is the pixel-by-pixel operation for obtaining the phase value to be measured, by which no spatial resolution is sacrificed and the accuracy of phase calculation can be easily improved by increasing the image sampling numbers. On the contrary, because of the necessity for acquiring multiple images at different timings, it is easily influenced by surrounding disturbances like vibrations and is not suitable for an instantaneous measurement of moving objects. To meet such applications, several techniques obtaining the phase-shifted fringes spatially parallel have been proposed, where the multiple phase-shifted fringes are generated, for example, by a parallel polarization interferometer or a grating interferometer, and are simultaneously acquired with single or multiple cameras [9–11].

21.2.2 Method Using Single-Input Image

As described before, the fringe analysis using multiple fringe patterns needs special equipment introducing the phase shift and usually does not suit dynamic measurements. As a complementary

FIGURE 21.3 Introducing way of the carrier fringes in an interferometer and an example of the fringe pattern.

FIGURE 21.4 Signal processing flow in FT method: (a) phase-modulated carrier fringes, (b) filtering and shift in spectrum space, and (c) demodulated phase.

principle, there are proposed fringe analysis methods using a single fringe pattern in which the pattern superposed with equispaced fringes that have higher spatial frequency, namely, carrier fringes, enough for that of the phase gradient to be measured is acquired. The word "carrier fringes" is derived from the concept of a carrier wave used in communication technology and corresponds, for instance, to the dense fringe condition generated by tilting a reference mirror in an interferometer, as shown in Figure 21.3. Here, simply suppose the case introducing the carrier fringes with a spatial frequency of f_0 (1/pixel) along the one-dimensional axis x as in Figure 21.4a. The intensity distribution of this fringe pattern is expressed by

$$I(x, y) = a(x, y) + b(x, y)\cos[2\pi f_0 x + \phi(x, y)], \qquad (21.8)$$

where the phase term $\phi(x, y)$ deforms the equispaced carrier fringes. This situation can be regarded as a spatial phase modulation of carrier fringes with a fundamental frequency of f_0. Considering this aspect, various phase-demodulation algorithms can be applied. The principal way is the FT method which extracts only the modulated phase term through spectrum filtering, using one- or two-dimensional FT and its inversion [3].

In the case of introducing the carrier fringes in the interferometer as shown in Figure 21.3, the reference mirror is tilted by a small angle of θ about the y-axis to generate carrier fringes with a spatial frequency of $f_0 = (2 \tan \theta)/\lambda$ along the x-axis. The fringe intensity distribution expressed by Equation 21.8 can be substituted to the complex form by using a relation $c(x, y) = (1/2)b(x, y)$ $\exp[i\phi(x, y)]$ as

$$I(x,y) = a(x,y) + c(x,y)\exp(2\pi i f_0 x) + c^*(x,y)\exp(-2\pi i f_0 x), \tag{21.9}$$

where c^* represents a complex conjugate of c. By performing one-dimensional FT in terms of the coordinate x to the above equation, the spatial frequency spectrum along the axis is derived as

$$FT[I(x,y)] = \int_{\infty}^{\infty} I(x,y)\exp(-2\pi i f x)\mathrm{d}x$$

$$= A(f,y) + C(f - f_0, y) + C^*(f + f_0, y), \tag{21.10}$$

where $A(f, y)$ and $C(f, y)$ correspond to the Fourier transforms of $a(x, y)$ and $c(x, y)$, respectively. As known from Figure 21.4b, if we select the larger enough carrier frequency f_0 than that of spatial variations of $a(x, y)$ and $b(x, y)$, the above major spectrums A, C, and C^* can be easily separated. Consequently, by applying a sequential processing, which consists of the extraction of the $C(f - f_0, y)$ component with a window filtering, its spectrum shift to the zero frequency position, and the inverse FT of the modified signal, the carrier frequency components are removed and only the desired component $c(x, y)$ is demodulated. Finally, in Figure 21.4c, the phase distribution $\phi(x, y)$ is calculated by taking log $c(x, y)$ or using the arctangential relation for the real and imaginary parts of $c(x, y)$:

$$\phi(x, y) = \tan^{-1}\frac{\mathrm{Im}[c(x,y)]}{\mathrm{Re}[c(x,y)]}. \tag{21.11}$$

The procedure shown in Figure 21.4 can easily be extended to the two-dimensional (2D) FT and filtering in a 2D frequency space, which can be used for the 2D analysis of the fringe pattern superposed with obliquely introduced carrier fringes. An example of phase-extraction procedures for this case is shown in Figure 21.5a through c.

Figure 21.5a is an interferogram obtained by a Fizeau interferometer with slight tilts of the reference mirror around both x- and y-axes in which oblique dense carrier fringes are introduced.

(a)   (b)   (c)

FIGURE 21.5 Phase map extraction in 2D FT method: (a) input fringe patterns with oblique carrier fringes introduced, (b) an FT result of the input pattern is filtered around carrier frequency and shifted to zero frequency position in 2D spectrum space, and (c) a finally extracted phase map obtained after an inverse 2D FT.

After the 2D FT for the input fringes, the 2D spectrum distribution shown in the upper frame in Figure 21.5b is obtained, where the two symmetrical spectrum peaks for the modulated carrier fringes terms, $C^*(f \pm f_0)$, are clearly separated from the peak for the bias terms located at the central zero frequency position. By cutting out one of the carried spectrum peaks, for example, in a circular window (dotted circle in the figure) and shifting to the center, the lower figure in Figure 21.5b is prepared and is again processed with the inverse 2D FT to generate the phase distribution of the object wavefront according to Equation 21.11. Since the fringe intensity profile in the Fizeau interferometer is usually largely deviated from a sinusoidal shape because of multiple reflections, the phase-shifting method described in the last section. On the contrary, this FT method provides accurate results for this case. In the actual processing with computers in the above analysis, the fast FT, whose processing performances have recently become better thanks to the speeding up of the computers and the developments of the speedy and efficient calculation algorithms [12], is usually applied.

The fringe analysis using a single fringe pattern can be performed by the simple or basic optical setup of interferometers or grating projectors and does not require special instruments such as phase shifters. Its one shot measurement capability is suitable for dynamic objects and phenomena. It should be noted that the spatial resolution in this method is variable dependent on the size of the spectrum filtering windows and that aberrations of the used imaging optics and the border of the sampling area cause considerably more phase errors than the former methods using multiple fringe images. As the other phase demodulation methods use a single-input fringe pattern, a variety of techniques, such as the synchronous detection or spatial phase synchronization methods [13–15], the phase-locked loop method [16,17], and the spatial phase-shifting or the electronic Moiré methods, have been proposed mainly for the suppression of the processing time or their implements on hardware [18–20].

The readers are referred to the other literature for a detailed comparison of the principles and accuracies of the two classes of fringe analysis mentioned here [21].

21.3 PHASE UNWRAPPING

In the usual fringe analysis since the fringe order cannot usually be decided a priori, only the fractional phase divided by 2π is "wrapped" in the region between $-\pi$ and $+\pi$. This leads to discontinuous phase jumps between adjacent pixels from $+\pi$ to $-\pi$ or its opposite direction. Therefore, the procedure to connect the discontinuous phase jumps, assuming the spatial continuity of the phase distribution, becomes necessary for obtaining the final physical distribution, as shown in Figure 21.6. This procedure is called as "phase unwrapping."

In most simple cases of phase unwrapping, the phase differences between adjacent pixels along the one-dimensional unwrapping path are estimated and are connected by using the following operations:

$$
\begin{aligned}
\text{If} \quad & \phi(\mathbf{r}_{i+1}) - \phi(\mathbf{r}_i) > \pi \\
& \phi(\mathbf{r}_{i+1}) = \phi(\mathbf{r}_{i+1}) - 2\pi \\
\text{else, if} \quad & \phi(\mathbf{r}_{i+1}) - \phi(\mathbf{r}_i) \leq -\pi \\
& \phi(\mathbf{r}_{i+1}) = \phi(\mathbf{r}_{i+1}) + 2\pi,
\end{aligned}
\tag{21.12}
$$

where $\mathbf{r}_i$ is the vectorial position at pixel number i. If the effective phase area is a rectangular $x - y$ coordinate, a combination of unwrappings on all horizontal paths along x-axis and one along the vertical y-axis gives an appropriate result [22]. However, general fringe patterns to be measured

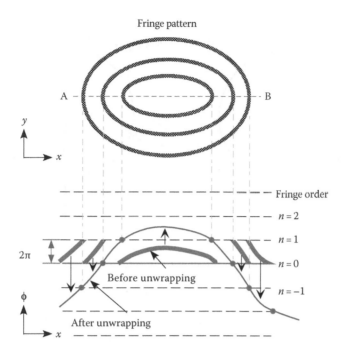

FIGURE 21.6   Concept of the phase unwrapping in the fringe analysis.

might be surrounded by a nonrectangle boundary and might contain pixels considered as singular points, where the fringe modulation is very low or the actual phase differences are larger than 2π against some of the adjacent pixels. In these situations, the above phase-unwrapping path is accidentally interrupted and the unwrapped phase distribution is deviated from the true one. However, several effective and robust phase-unwrapping algorithms have been proposed to avoid these problems as follows:

1. *Algorithms based on the minimum-spanning tree tracking*: Methods to obtain a continuous phase-unwrapping path by spanning the effective phase area with tree-like trajectories that have no localized closed loop. To select the effective path, cost functions such as the spatial phase gradient and the fringe contrast are used to decide the reliability of the pixels. Unreliable pixels are removed from the spanning tree. High-speed algorithms [23,24].
2. *Algorithms based on energy minimization*: Methods to realize the optimum unwrapped state by minimizing the energy function, based on the summation of the phase differences among the neighboring pixels in which the initial phase distribution corresponds to the maximum energy state, are converged to the minimum energy state by varying the phases at all pixels by $\pm 2\pi$ based on a certain rule. This method requires many steps of iteration for convergence and does not depend on the unwrapping path. There are several algorithms proposed such as the simulated annealing method [25], a method using the Euler–Poisson equation [26], and the cellular automata method [27,28].

In the above procedure, the singular points mentioned earlier should be checked in advance and removed from the unwrapping operations. To check the singularity of the pixels, the rotating phase-unwrapping operation for the local four neighboring pixels (Figure 21.7) is frequently applied.

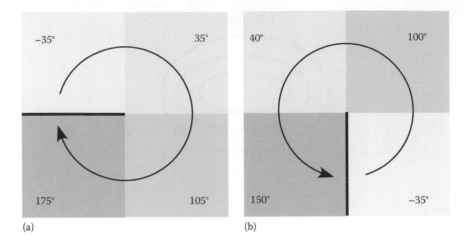

FIGURE 21.7 Singular points generated for four neighboring pixels in phase-unwrapping procedure. (a) Right-handed singular point with abnormal phase jump along horizontal left direction. (b) Left-handed singular point with abnormal phase jump along vertical down direction.

In some cases, a line joining two singular points having different signs (right-handed or left-handed) can be used as a cutline for the phase-unwrapping path finding. A method to directly remove singular points from a phase distribution has also been proposed [29].

21.4 PRACTICAL EXAMPLE OF FRINGE ANALYSIS IN PROFILOMETRY

Finally, we show an actual process of the fringe analysis for profilometry using the deformed grating method with an interference fringe projection. The obliquely projected equispaced fringes on the object are deformed and phase modulated depending on the surface profile by being observed from a different direction. Modulated phases have to be extracted from a single fringe pattern. The overall processing flow of the fringe analysis is shown in Figure 21.8a through f. The image in Figure 21.8a is the input fringe pattern where the carrier-introduced interference fringes are projected on a face of a "Hina" doll and the deformed fringes on the face profile are captured by a CCD camera to be stored as a digitized image of 8 bits grayscale in a frame grabber of 512 × 512 pixels. Here, care should be taken in the matching of the fringe's modulation depth to the digitizing range of the frame grabber through the tuning or exposure condition.

The electronic Moiré technique [21] was used as the fringe analysis method, in which three phase-shifted Moiré fringes are digitally generated in a computer by multiplying the input fringes to three phase-shifted reference fringes prepared in the memory and spatially low-pass filtering them. The distributions of the phase and the modulation depth shown in Figure 21.8b and c are calculated using Equations 21.5 and 21.6 with three Moiré fringes. As the pretreatment before the phase unwrapping, the map of the singular points in Figure 21.8d is extracted from the wrapped phase map as described in the previous section. White- and black-colored pixels in Figure 21.8d correspond to the extracted singular points with different signs. The modulation depth information is also used to remove the pixels with low modulation depth from the phase-unwrapping operation. By using the information on singular points and the modulation depth as the cost function, the phase map in Figure 21.8b is phase unwrapped with the minimum spanning tree method to generate a smooth unwrapped phase, as shown in Figure 21.8e. Finally, the obtained unwrapped phase map is converted to the physical height map of the object surface by taking into account various parameters in the deformation grating method such as the projected fringe pitch, projection angle, and the imaging conditions including the calibration factor [30]. Figure 21.8f shows the 3D representation of the finally measured surface profile of the object.

FIGURE 21.8 Practical processing flow of fringe analysis in the case of the profile measurement by the deformed grating method with an interference fringe projection. (a) Input image obtained from a face of a "Hina" doll under the projection of carrier fringe pattern, (b) phase map extracted with the electronic Moiré technique, (c) singular points map extracted from the phase map, (d) modulation depth map of the fringe pattern, (e) processing result after the phase unwrapping based on the minimum spanning tree method with the removal of unreliable pixels in terms of the singularity and the modulation depth, and (f) finally obtained profile of the target surface.

21.5 SUMMARY

In this chapter, the basic principles and concepts of the generally used fringe analysis are briefly explained. Nowadays, fringe analysis is widely used in many fields where a variety of novel techniques including spatiotemporal modulation of the various parameters like fringe pitch, frequency, and fringe orientation are being developed. Recent rapid evolution of computers, imaging devices, and light sources must further expand the application fields of fringe analysis.

SYMBOLS

I	Fringe intensity
x, y	Coordinates
I_0, γ	Bias intensity and modulation depth of fringes
a, b	Background intensity and contrast of fringes
ϕ	Phase term of fringe pattern
λ	Wavelength of light
Λ	Pitch of projection grating

$\Delta\psi$	Phase-shifting value
N	Number of frames used in phase-shifting method
k	Sampling number of phase-shifted fringes
I_k	Fringe intensity kth sample
α	Constant but unknown phase-shifting value
Δz	Object height
f, f_0	Frequencies of original fringe and carrier fringes
θ	Tilting angle
c, C	Complex amplitude and its Fourier transform
$\mathbf{r}$	Vectorial position

REFERENCES

1. Bruning, J. H., Herriott, D. R., Callagher, J. E., Rosenfeld, D. P., White, A. D., and Brangaccio, D. J. 1974. Digital wavefront measuring interferometer for testing optical surfaces and lenses. *Appl. Opt.* **13**:2693–2703.
2. Creath, K. 1988. Phase-measurement interferometry techniques. In *Progress in Optics* XXVI, E. Wolf (Ed.), pp. 349–393, Amsterdam, the Netherlands: Elsevier.
3. Takeda, M., Ina, H., and Kobayashi, S. 1982. Fourier-transform method of fringe-pattern analysis for computer-based topography and interferometry. *J. Opt. Soc. Am.* **72**:156–160.
4. Carré, P. 1966. Installation et utilisation du comparateur photoelectrique et interferentiel du Bureau International des Poids et Mesures. *Metrologia* **2**:13–23.
5. Hariharan, P., Oreb, B. F., and Eiju, T. 1987. Digital phase-shifting interferometry: A simple error-compensating phase calculation algorithm. *Appl. Opt.* **26**:2504–2506.
6. Hibino, K., Oreb, B. F., Farrant, D. I., and Larkin, K. G. 1997. Phase-shifting algorithms for nonlinear and spatially nonuniform phase shifts. *J. Opt. Soc. Am. A* **14**:918–930.
7. de Groot, P. J. 1995. Vibration in phase-shifting interferometry. *J. Opt. Soc. Am. A* **12**:354–365.
8. Zhu, Y.-C. and Gemma, T. 2001. Method for designing error-compensating phase-calculation algorithms for phase-shifting interferometry. *Appl. Opt.* **40**:4540–4546.
9. Onuma, K., Tsukamoto, K., and Nakadate, S. 1993. Application of real-time phase-shift interferometer to the measurement of concentration field. *J. Crystal Growth* **129**:706–718.
10. Kwon, O. Y. 1984. Multichannel phase-shifted interferometer. *Opt. Lett.* **9**:59–61.
11. Qian, K., Wu, X.-P., and Asundi, A. 2002. Grating-based real-time polarization phase-shifting interferometry: Error analysis. *Appl. Opt.* **41**:2448–2453.
12. http://www.fftw.org/.
13. Ichioka, Y. and Inuiya, M. 1972. Direct phase detecting system. *Appl. Opt.* **11**:1507–1514.
14. Womack, K. H. 1984. Interferometric phase measurement using spatial synchronous detection. *Opt. Eng.* **23**:391–395.
15. Toyooka, S. and Tominaga, M. 1984. Spatial fringe scanning for optical phase measurement. *Opt. Commun.* **51**:68–70.
16. Kato, J., Tanaka, T., Ozono, S., Fujita, K., Shizawa, M., and Takamasu, K. 1988. Real-time phase detection for fringe-pattern analysis using digital signal processing. *Proc. SPIE* **1032**:791–796; Kato, J. et al. 1993. A real-time profile restoration method from fringe patterns using digital phase-locked loop—Improvement of the accuracy and its application to profile measurement of optical surface. *Jpn. J. Prec. Eng.* **59**:1549–1554 (in Japanese).
17. Servin, M. and Rodriguez-Vera, R. 1993. Two-dimensional phase locked loop demodulation of interferograms. *J. Mod. Opt.* **40**:2087–2094.
18. Williams, D. C., Nassar, N. S., Banyard, J. E., and Virdee, M. S. 1991. Digital phase-step interferometry: A simplified approach. *Opt. Laser Technol.* **23**:147–150.
19. Küchel, M. 1990. The new Zeiss interferometer. *Proc. SPIE* **1332**:655–663.
20. Kato, J., Yamaguchi, I., Nakamura, T., and Kuwashima, S. 1997. Video-rate fringe analyzer based on phase-shifting electronic Moiré patterns. *Appl. Opt.* **36**:8403–8412.
21. Takeda, M. 1984. Subfringe interferometry fundamentals, *KOGAKU* **13**:55 (in Japanese).
22. Itoh, K. 1982. Analysis of the phase unwrapping algorithm. *Appl. Opt.* **21**:2470.
23. Judge, T. R., Quan, C.-G., and Bryanstoncross, P. J. 1992. Holographic deformation measurements by Fourier-transform technique with automatic phase unwrapping. *Opt. Eng.* **31**:533–543.

24. Takeda, M. and Abe, T. 1996. Phase unwrapping by a maximum cross-amplitude spanning tree algorithm: A comparative study. *Opt. Eng.* **35**:2345–2351.
25. Huntley, J. M. 1989. Noise-immune phase unwrapping algorithm. *Appl. Opt.* **28**:3268–3270.
26. Kerr, D., Kaufmann, G. H., and Galizzi, G. E. 1996. Unwrapping of interferometric phase-fringe maps by the discrete cosine transform. *Appl. Opt.* **35**:810–816.
27. Ghiglia, D. C., Mastin, G. A., and Romero, L. A. 1987. Cellular-automata method for phase unwrapping. *J. Opt. Soc. Am. A* **4**:267–180.
28. Ghiglia, D. C. and Pritt, M. D. 1998. *Two-Dimensional Phase Unwrapping: Theory, Algorithms, and Software*. New York: Wiley-Interscience.
29. Aoki, T., Sotomaru, T., Ozawa, T., Komiyama, T., Miyamoto, Y., and Takeda, M. 1998. *Opt. Rev.* **5**:374–379.
30. Sansoni, G., Carocci, M., and Rodella, R. 1999. Three-dimensional vision based on a combination of gray-code and phase-shift light projection: Analysis and compensation of the systematic errors. *Appl. Opt.* **38**:6565–6573 (For example).

22 Photogrammetry

Nobuo Kochi

CONTENTS

22.1 Introduction 556
22.2 Principle..556
    22.2.1 Basic Principle 556
        22.2.1.1 Single Photogrammetry 556
        22.2.1.2 Stereo Photogrammetry 557
        22.2.1.3 Photogrammetry with Many Images (Bundle Adjustment)......558
    22.2.2 Flow of Measuring Process 558
        22.2.2.1 Camera Calibration..559
        22.2.2.2 Photographing..560
        22.2.2.3 Exterior Orientation 561
        22.2.2.4 3D Measurement..562
        22.2.2.5 3D Image Reconstruction and Output 562
    22.2.3 Photogrammetric SLAM (Visual SLAM, SFM).....................................563
22.3 Application Examples of 3D Measurement and Modeling 564
    22.3.1 Application of Cultural Heritage, Architecture, and Topography 564
        22.3.1.1 3D Modeling of Wall Relief...564
        22.3.1.2 Byzantine Ruins on Gemiler Island...564
        22.3.1.3 All-Around Modeling of a Sculpture in the Archaeological Museum
                of Messene in Greece...565
        22.3.1.4 All-Around Modeling of the Church of Agia Samarina...........565
        22.3.1.5 Topography Model 566
        22.3.1.6 3D Model Production with Aerial and Ground Photographs ...566
    22.3.2 Application to Human Body Measurement 569
        22.3.2.1 Face Modeling 569
        22.3.2.2 Body Measurement 570
    22.3.3 Application to Automobile Measurement...571
        22.3.3.1 Measuring Car Bottom 571
        22.3.3.2 Measuring Surface of the Entire Body 572
        22.3.3.3 Measuring Surface of the Tire 573
    22.3.4 Indoor Measurement..575
    22.3.5 Photogrammetric SLAM 576
        22.3.5.1 Measurement by Vehicle (Measurement by Image Sequences)...............576
        22.3.5.2 Measurement by UAV (Measurement by Image Sequences and Still
                Images)..577
22.4 Conclusions...579
References...580

22.1 INTRODUCTION

Recent developments in the digital camera with its extremely dense pixel capacity and that of the PC with its high speed and enormous memory capacity has enabled us to process the super-voluminous data involved in point clouds and photo data for three-dimensional (3D) works, computer graphics, and virtual reality (VR) with very high speed and efficiency. Its applied technology is rapidly expanding not only in mapmaking but also in such new areas as archiving and preserving valuable cultural heritages, forensics, civil engineering, architecture construction, industrial measurement, human body measurement, even to such areas as entertainment animation. And it is daily opening new vistas, as it is integrated with devices like global positioning system (GPS), total station, gyro-sensor, accelerometer, laser scanner, and video image [1].

Here, we explain its technological principles and practical applications with our newly developed software for digital photogrammetry (Topcon Image Master, PI-3000) [6,13,23,61].

22.2 PRINCIPLE

22.2.1 BASIC PRINCIPLE

Photogrammetry technology has developed with aerial photogrammetry for mapmaking and terrestrial photogrammetry, which measures and surveys the ground with metric camera. It has also developed with close-range photogrammetry, which measures large constructions with precision in the industrial measuring [2,3]. Digital photogrammetry, therefore, is the digital photo technology, which replaced film-based data with digital data to integrate with the image-processing technology [4].

The basic principle of photogrammetry [2,3] is based on photo image, center of projection, and the geometrical condition called colinearity condition, where the objects should be on the same straight line (Figure 22.1a). In digital photogrammetry, we have three different types: (1) single photogrammetry with one single photograph, (2) stereo photogrammetry with two photographs as one unit, and (3) photogrammetry with many photographs, using (1) or (2) or their combination. These methods are performed through mathematical calculation and image processing; the basic system structure is composed of nothing other than a PC, a specific software and digital camera.

22.2.1.1 Single Photogrammetry

Figure 22.1a shows the geometry of a single photogrammetry. The light coming from any given point in real space always passes through the projection center (O) and creates image on the photograph. To make 3D measurement of an object with this method, we must satisfy the following conditions other than that of colinearity condition:

1. We must have at least one coordinate of 3D coordinates of the object. For example, in Figure 22.1a, when Z1, which is the coordinate of P1 on Z-axis, is given.
2. The point to be identified should be on the plain or curved surface geometrically. For example, in Figure 22.1a, when P2 is on the specific plane $aX + bY + cZ = 0$.

Therefore, for measuring or surveying flat plain or topography or flat wall of a building, we use single photogrammetry under these conditions.

The earlier conditions (1) and (2) diminish the number of unknown quantity, because the dimension is diminished from three to two. As a result, the data between two dimensions of the picture and two dimensions of the real space become interchangeable (2D projective transformation).

FIGURE 22.1 (a) Single photogrammetry and colinearity condition, (b) stereo method and relative orientation, and (c) rectification.

If we can get more than three control points by which the ground coordinates are measured, we can obtain the exterior orientation parameters (the 3D position of the camera and the inclination of the axes). This process to obtain the position and inclination of the camera from the control points is called "orientation of single photograph" or "space resection." There is another method called "DLT" method (direct linear transformation), which is also the 3D projective transformation. Though this DLT method requires at least 11 unknown parameters, it is still convenient and simple, because the calculation is done by the linear transformation.

22.2.1.2 Stereo Photogrammetry

This system is used to obtain 3D coordinates that identifies the corresponding points of more than two images of the same object. In order to make this stereo method workable geometrically, we transform the photographing image in ideal setup as in Figure 22.1b (see dashed line) and calculate or make analytical calculation.

We call a "model" the cubic image obtained by more than two images. We can make our model, if the relation between the image position and inclination is relative. We call it "relative orientation," because it creates a relative model relatively. This means, we obtain the position and inclination of projection center (O_1, O_2) of each right and left camera in such a way that the two light beams would make the forward intersection so that we can obtain the exterior orientation parameters. There are several methods to obtain the same result: the method that uses coplanarity condition, the method that uses y-parallax, the method that uses colinearity condition, etc.

With the stereo photogrammetry, we can rectify the image so that the cubic image may be displayed without y-parallax (rectified image, Figure 22.1c).

The measurement accuracy by stereo photogrammetry can be calculated from the following equation, which deals with the resolution capability (Δxy, Δz) on horizontal (x, y) and in depth (z):

$$\Delta xy = H \times \frac{\Delta p}{f}$$

$$\Delta z = H \times H \times \frac{\Delta p}{(B \times f)} \tag{22.1}$$

where
H is the photo distance
Δp is the pixel size
f is the principal distance (focal length)
B is the distance between two cameras

22.2.1.3 Photogrammetry with Many Images (Bundle Adjustment)

When we measure with more than one picture, we combine bundle adjustment to the process. As the measuring method that uses the more images gets the more discrepancy, we minimize the discrepancy by bundle adjustment. Bundle adjustment process is the process to determine simultaneously, by least-squares method, both the exterior orientation parameters (the 3D position of the camera and the inclination of three axes) and the interior orientation parameters (lens distortion, principal distance, principal point, and others) of each picture, identifying control point, tie point, and pass point of the same object photographed on each different picture (or in the light bundle of each picture).

The basic equation for calculating colinearity condition, essential to digital photogrammetry, is as follows:

$$x = -f \frac{a_1(X - X_0) + a_2(Y - Y_0) + a_3(Z - Z_0)}{a_7(X - X_0) + a_8(Y - Y_0) + a_9(Z - Z_0)} + \Delta x$$

$$y = -f \frac{a_4(X - X_0) + a_5(Y - Y_0) + a_6(Z - Z_0)}{a_7(X - X_0) + a_8(Y - Y_0) + a_9(Z - Z_0)} + \Delta y \tag{22.2}$$

where
a_i is the elements of rotation matrix
X, Y, and Z are the ground coordinates of the object P
X_0, Y_0, and Z_0 are the ground coordinates of projection center
x and y are the image coordinates
Δx and Δy are the correction terms of camera

By applying bundle adjustment to more than one picture, we can analytically obtain accurately not only the external orientation and 3D coordinates, but also the interior orientation of the camera.

22.2.2 Flow of Measuring Process

Figure 22.2 shows the flow of the measuring process. To attain high accuracy in measuring, we have to acquire with precision the interior orientation parameters (camera calibration). And after photographing the object, we have to make exterior orientation to determine its parameters. And then we make rectification image, which enables us to observe by 3D display; we cannot only observe the actual site of photographing in 3D way, but also measure the object in 3D and confirm

FIGURE 22.2 Process flow.

the result of its measurement [9]. The 3D measurement can be performed manually, semiautomatically, or all automatically. The advantage of 3D display is that we can make 3D measurement and plotting (allocating lines and points on 3D space) of an object of complex form and features, which cannot be worked automatically, but manually or semiautomatically. For 3D display, we can use micro-pole system [13].

In the automatic measurement of a surface, it is possible to obtain in short time the 3D data as many as several thousands to several hundred thousands by simply determining the area [4]. And out of these data, we can make polygon and the texture-mapping image pasted with the real picture, which enables us to view the reconstructed image from all angles. All these 3D data can be stored in the files of DXF, CSV, STL, OBJ, and VRML, and used for other purposes, such as creating plan drawing, rapid prototyping, and 3D printing. And if we feed 3D data to point-clouds-based alignment software by the iterative closet point algorithm [15], we can compare the original design data (3D CAD data) with the data obtained by the last measurement and also compare the previous state and the state after the change.

In each following paragraph, each processing is explained according to the process flow of Figure 22.2.

22.2.2.1 Camera Calibration

We can make 3D measurement accurate by obtaining interior orientation parameters through camera calibration [2,3]. In this process, with the camera to be calibrated from different angles and directions, we take plural number of pictures on the control points allocated with precision in 3D (Figure 22.3a). And from each of the plural images thus obtained, we extract, by image processing, the control points, and from their coordinates, we calculate their parameters. For simpler way, however, we can use merely a sheet on which control points are printed (Figure 22.3b). For calculation, we use self-calibrated bundle adjustment method in which calibration program is included [5].

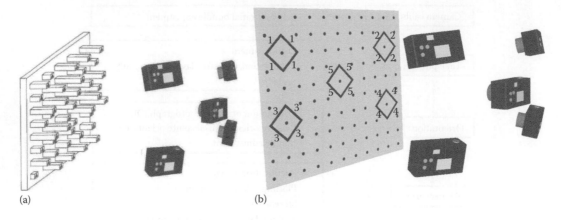

FIGURE 22.3 Camera calibration: (a) 3D field and (b) sheet.

The self-calibrated bundle adjustment is the adjustment that is done by interior orientation parameters that are obtained by integrating correction terms of camera (Equation 22.2: Δx, Δy) simultaneously.

There are different kinds of rectifying model for lens distortion, but common type works in the direction of radiation and tangential line. These equations are as follows [32]:

$$\Delta x = \left(K_1 r^2 + K_2 r^4 + K_3 r^6 \right) X' + P_1 \left(r^2 + 2X'^2 \right) + 2P_2 X'Y'$$

$$\Delta y = \left(K_1 r^2 + K_2 r^4 + K_3 r^6 \right) Y' + 2P_1 X'Y' + P_2 \left(r^2 + 2Y'^2 \right) \tag{22.3}$$

$$X' = X - X_p, Y' = Y - Y_p, r = \sqrt{X'^2 + Y'^2}$$

where
    X, Y are image coordinates
    X_p, Y_p are coordinates of principal point
    K_1, K_2, K_3 are radial lens distortion parameters
    P_1, P_2 are tangential lens distortion parameters
    r is the distance from principal point to image coordinates

By camera calibration, we can come to the precision as close as 1/1,000–1/20,000 (with a photo distance of 10 m and the difference of less than 0.5 mm) in stereo photogrammetry [6]. There is a report that in the area of industrial measurement (not stereo photogrammetry), where they pasted retro-targets on an object, the accuracy as close as to 1/200,000 was attained [7].

22.2.2.2 Photographing

If every camera has already been calibrated, we can make any combination of cameras depending on a project. And if we make shuttering all at the same moment, we can even measure a moving body. And if a scale bar (determined distance between two points) is photographed with an object or if the 3D coordinates of more than three control points are photographed, we can obtain the actual size of the object. Even if there is no control point, we can still make a relative model.

In case of indoor measurement, we take pictures of an object with scale bars (Figure 22.4a). In case of outdoor measurement, we also use surveying instruments such as total station or GPS. If we can place reflective targets (Figure 22.4b) or refractive coded targets (Figure 22.4c) on or around the object, we can introduce automatization, increase accuracy and reliability, and save a lot of time. And the targets would also work as tie points to connect images when we measure a large object or extensive area. It also would serve measuring points of 3D coordinates.

(a)

(b)

(c)

55

FIGURE 22.4 (a) Scale bar, (b) reflective targets, and (c) reflective coded targets (left: centripetal type, middle: dispersing type, and right: color type).

We can also make a 3D model, detecting the feature points of the image without using targets. But if we have to make the measurement with actual scale with the utmost accuracy, it is necessary to input the distance between feature points or to place the targets as control (measured) points.

With stereo photogrammetry, we can make full automatic stereo-matching of the entire object surface. In this case, however, we need some features on the surface in order to identify the corresponding positions on the right and left images. Ordinarily, if we work with an outdoor object, we have enough features, but if it is a man-made object, oftentimes we do not. In such case, we put features or patterns on the object by a projector or put some paint on its surface. An ordinary projector is good enough.

22.2.2.3 Exterior Orientation

Exterior orientation is a process to determine the 3D position of the camera and the inclination of three axes (exterior orientation parameters) at the time of measuring. The stereo photogrammetry is possible if we have at least six corresponding points from both pictures for identification. We can make automatic measurement by detecting the feature points through image processing without placing the targets. However, we cannot only simplify but also automatize the process of identification and make the 3D measurement much more accurate by placing specific targets on or around the object. If we can fix the position of cameras and if we determine the exterior orientation parameters beforehand, the identification of the features is not necessary. Furthermore, once we have integrated the bundle adjustment in orientation, we can make image connections on extensive areas, or even all-around modeling, if the data of a scale or 3D coordinates or model (relative) coordinates space are fed to PC.

There are three types of orientation: manual, semiautomatic, and full automatic. For full automatic, however, we need reflective targets or special targets called coded target (Figure 22.4b and c). On these target images, we make target identification automatically. We can also automatically detect the feature points of the image and make identification process without targets. Based on the data of these identified corresponding points (targets), we make exterior orientation and acquire its parameters and 3D coordinates of targets. Even interior orientation parameters can be obtained by this process.

There are three kinds of coded targets (Figure 22.4c): centripetal type (left) [8], dot-dispersing type (middle) [9,10], and color type (right) [11]. We have developed color-coded target, which enabled us to obtain even texture picture and to measure on the surface [11]. As shown in Figure 22.4c (right), the color-coded target consists of a reflective retro-target in three corners and the color section where the position and combination of colors constitute the code. We have 720 combinations as code.

22.2.2.4 3D Measurement

In the stereo photogrammetry, we can relocate the images in such a way that the corresponding points can line up on the epi-polar line (rectified image: Figure 22.1c). Therefore, the searching work of corresponding points can be done only on one line. And if we integrate stereo-matching function here so that the corresponding points on the right and left images can be automatically obtained, we can easily make 3D measurement by semiautomatic operation. The automatic measurement of a surface is the process, which automatically identifies the corresponding points on the right and left images, calculates the data of height, and visualizes or creates the 3D undulation on the surface. We call this digital surface model. For this, we use "area-based" stereo-matching. For this operation, we use not only coarse-to-fine cross-correlation method, but also least squares matching (LSM [4,14]) method to minimize mismatching on the project distortion or tilted parts. Presently, standard deviation of LSM is reported as 0.26 mm and that of normalized-cross-correlation method is 0.35 mm. There is also report that LSM is more accurate in inclined object than normalized-cross-correlation method [6,35].

In the research of industrial measurement, there is a report of 0.077–0.205 mm [23,33,34]. In addition, combining feature-based matching and area-based matching, we have also developed the stereo-matching method, which is robust to brightness change and occlusion [35,36].

22.2.2.5 3D Image Reconstruction and Output

To make 3D model, we work with the triangle irregular network (TIN) model by using Delaunay triangulation [37], considering the measuring points not as the group of points but as a surface. TIN model (Figure 22.5) can reconstruct the smooth surface, if mismatching is eliminated.

In our method, as the correlation between the photographed image and the 3D position of the points measured from each stereo model is already established, it is not necessary to match texture

FIGURE 22.5 Triangle irregular network (TIN).

with TIN model. As to the texture image, we can make the texture mapping with the same degree of resolution as at the time of measuring, if the image is the image used for measuring, because the texture image makes simultaneously the adjustment of camera lens distortion and project distortion at the time of photographing. And the 3D model with texture can be turned around or enlarged or downsized as we wish and can be studied from all sides on the PC display. If we output the measured result as ortho-photo with texture, it would reflect the actual size. And if we output as DXF, VRML, OBJ, or STL file and feed the molding system, we can make replica made by 3D printer as well as compare with CAD.

22.2.3 Photogrammetric SLAM (Visual SLAM, SFM)

Computer vision is the method to process the enormous amount of still images and video image. They are the methods called structure from motion (SFM [38–40]) or visual simultaneous localization and mapping (SLAM). They are the methods to reconstruct 3D images, simultaneously estimating camera's position and inclination out of the scenes taken by photographs.

The analytic calculation differs in the photogrammetry and the computer vision. But the basic principle is the same as the ones explained in Section 22.2.1. And the method by photogrammetry is called photogrammetric SLAM.

Basic principle (see Figure 22.6)

1. We obtain the feature points detected from the two images. And we calculate the 3D of the corresponding points from the forward intersection. For the matching of the feature points, we use feature-based matching, such as SIFT [41] and SURF [42].
2. We remove mismatchings.
3. We calculate the exterior orientation from 3D coordinates (space resection).
4. We repeat this process in the orderly way.

We make the bundle adjustment during and at the end of the process to minimize the errors. Various studies are made to improve the integration of the bundle adjustment [27,43–47].

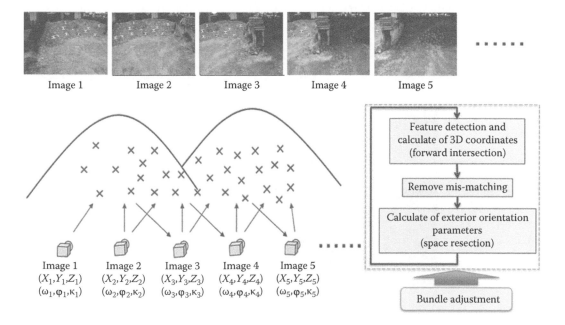

FIGURE 22.6 Photogrammetric SLAM.

22.3 APPLICATION EXAMPLES OF 3D MEASUREMENT AND MODELING

We are striving to develop the technology of measuring various objects in 3D using our software of digital photogrammetry. The explanations of the examples are given in the following sections.

22.3.1 APPLICATION OF CULTURAL HERITAGE, ARCHITECTURE, AND TOPOGRAPHY

22.3.1.1 3D Modeling of Wall Relief

We made 3D modeling of a wall relief at the Acropolis Theatre of Dionysus in Greece [13]. We used an ordinary digital camera (5 million pixels). We took two pictures of the relief. On the relief pictures of left and right, we simply marked 10 corresponding points on each pictures for orientation as in Figure 22.7a and made automatic measuring. As the automatic measurement takes only a few minutes, the modeling of the whole picture was finished in less than 10 min. Figure 22.7b is the image obtained by rendering from the measured result and by adding contour lines. Figure 22.7c is the bird's view. We can see from the result how the relief is successfully 3D modeled in great details.

22.3.1.2 Byzantine Ruins on Gemiler Island

The Byzantine ruins on Gemiler Island, which is located off the Lycian coast of southern west of Turkey, has been studied by the research group of Byzantine Lycia, a Japanese joint research project [16]. The Church II we photographed was about 10 m × 10 m. The camera was Kodak DCS Pro-Back (16 million pixels; body, Hasselblad 555ELD; and lens, Distagon 50 mm). We took two stereo images with the shooting distance 16 m, the base length 6 m, and the focal length 52 mm. To obtain the actual measurement, we used the survey instrument total station to determine three points as the reference points. Under these conditions, the pixel resolution was 3 mm for horizontal direction and 8 mm for

(a)   (b)

(c)

FIGURE 22.7 3D modeling of wall relief: (a) orientation, (b) after rendering and contour lining, and (c) bird's view.

(a) (b) (c)

FIGURE 22.8 Byzantine ruins in Gemiler Island: (a) photographed image of Church II, (b) rendering image (front), and (c) different view.

depth direction. A triangulated irregular network (TIN) model has been automatically created, and the number of produced points was about 260,000. Figure 22.8a is the photographed image of Church II. Figure 22.8b is the rendering image reconstructed on the PC after measuring, and Figure 22.8c illustrates different view of the same model. Furthermore, to evaluate the accuracy, we compared the ortho-photo image with the elevation plan. We were able to superimpose even the fissures within the range of less than 1 cm [6,35]. As for other examples in this site, we made 3D models and digital ortho-photos of the floor mosaic, which we analyzed and compared with the drawing [17,35].

22.3.1.3 All-Around Modeling of a Sculpture in the Archaeological Museum of Messene in Greece

The lion statue was found from Grave Monument K1 in Gymnasium complex buildings and is now kept at the Archaeological Museum of Messene in Greece. The lion is hunting a deer from behind (Figure 22.9a). This lion statue is considered to be on top of the Grave Monument roof. As this sculpture (its front part as in Figure 22.9a) was placed only 1 m away from the back wall, the photographing distance from the back was short. So we had to take many pictures. And as we did not use the light fixture, the brightness of each picture is different. The digital camera we used had 6 million pixels with single lens reflex (SLR) type (Nikon D70), with lens focal length of 18 mm. We took 100 pictures from all around, of which 20 were used for modeling.

The result of the all-around modeling is shown in Figure 22.9b and c. The reason why the color of texture mapping is different is due to the difference in brightness at the places of shooting.

As for other examples of the pictures taken for measuring in this archeological site of ancient city Messene, we have that of an architectural member of a relief, which we analyzed and compared with the drawing [18].

22.3.1.4 All-Around Modeling of the Church of Agia Samarina

The Byzantine church of Agia Samarina is located 5 km south of Ancient Messene (Figure 22.10a). This church was renovated from ancient building (probably a kind of temple) to Byzantium church, so big limestone blocks were used in the lower part of the church. We used the same digital SLR camera with 18 mm wide angle lens. The environment of taking photo was as follows: object distance was about 11 m, distance between viewpoints was about 4 m, and 34 photos were used to make the 3D model. In this situation, 1 pixel was about 4.5 mm in horizontal direction and about 15 mm in vertical direction. Figure 22.10b shows the 3D wire flame model with texture mapping, and Figure 22.10c shows the digital ortho-photo of the 3D model. While we could not get the pictures of higher parts like roof, as taken from the ground, as you see in these pictures, we succeed in making digital ortho-photos of the elevation of this church.

(a)

(b)

(c)

FIGURE 22.9 (a) Lion devouring a deer, (b) texture mapping, and (c) rendering.

22.3.1.5 Topography Model

We took two pictures of topography from a helicopter with a digital camera [19]. Then, we made 3D measurement from the pictures and produced a diorama out of the data. We used a digital camera of SLR type with 6 million pixels and with 20 mm wide angle lens.

As the control points, we used six points measured by GPS. The helicopter altitude was 230 m, and the distance between cameras was 53.4 cm with the resolution capacity 9 cm horizontally and 38 cm in depth. Figure 22.11a shows the picture taken. Figure 22.11b and c shows the picture of the diorama produced by rapid prototyping from the modeling data, with thermoplastics (Figure 22.11b) and paper (Figure 22.11c) as base material.

22.3.1.6 3D Model Production with Aerial and Ground Photographs

We produced 3D model by putting together the data from aerial pictures, the pictures taken from a paraglider, and the pictures taken on the ground by digital cameras. The object is the Topcon main building in Itabashi ward of Tokyo, Japan [20].

The pictures we used were the aerial photo by the film camera of airplane, the photo by digital camera of powered paraglider, and the lateral photo obtained by digital camera on the ground. The area of the each photo we used for modeling was about 300 m × 300 m, 100 m × 50 m, and 20 m × 20 m.

We first produced 3D model for each of them by our software (PI-3000, Image Master) and then fused them together. As for the control points, we first measured the side of the building by reflectorless total station [21,61] and converted the data to GPS coordinate using the points measured by GPS in the building lot.

(a)

(b)

(c)

FIGURE 22.10   (a) Byzantium church of Agia Samarina, (b) 3D model, and (c) digital ortho-photo image.

(a)

(b)

(c)

FIGURE 22.11   Topography model: (a) picture taken, (b) diorama (thermoplastics), and (c) diorama (paper).

TABLE 22.1

Cameras Used for High-, Low-, and Ground-Level Photographs and Conditions

Camera	Airplane WILD RC-20	Paraglider CANON EOS Digital	Ground Minolta Dimage7
Sensor	Film	C-MOS (22.7 mm × 15.1 mm)	CCD (2/3 in.)
Number of pixels	6062 × 5669 (by Scanner)	3072 × 2048	2568 × 1928
Resolution	42.3 μm	7.4 μm	3.4 μm
Focal length	152.4 mm	18 mm	7.4 mm
Photographing Area	2000 m × 2000 m	200 m × 100 m	30 m × 25 m
Modeling area	300 m × 300 m	100 m × 50 m	20 m × 20 m
Altitude	1609 m	163 m	26(side)m
Base length	886 m	18 m	5 m
Resolution: ΔXY	0.89 m	0.06.6 m	0.01 m
Resolution: ΔZ	1.61 m	0.6 m	0.06 m

Table 22.1 shows the capacity of the cameras used for high-, low-, and ground-level photographs and the conditions for analysis by each camera. And Figure 22.12a is the display of PI-3000 (Image Master) showing all of the analysis of each camera. Figure 22.12b shows the digital photograph by powered paraglider, and Figure 22.12c and d shows the results of the 3D model with texture mapping.

(a) (b)

(c) (d)

FIGURE 22.12 (a) All of the analysis of each camera, (b) digital photograph by powered paraglider, and (c, d) different views of the 3D model with texture mapping.

22.3.2 Application to Human Body Measurement

The digital photogrammetry has an advantage to obtain the picture of the exact state of a moving object at a certain given moment [22]. To give an example, we worked on measuring human face and body.

22.3.2.1 Face Modeling

We made 3D modeling of a human face and used it for 3D simulation of ophthalmology equipment designing. Since there are people whose faces have less features than others, we took pictures with random-dot patterns projected by an ordinary projector. We also took pictures without dots for texture mapping. We used two digital SLR cameras of 6 million pixels with 50 mm lens in stereosetting. Photographing conditions were as follows: distance was 1 m, distance between cameras was 30 cm, and resolution capacity was 0.2 mm horizontally and 0.5 mm in depth.

Figure 22.13 shows the result of photo measuring. Figure 22.13a is the texture-mapped picture without patterns. Figure 22.13b is the result of measuring the cross section. The 3D data obtained by

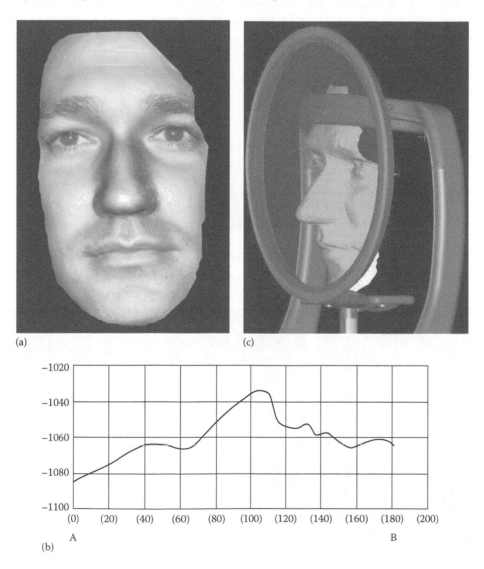

(a)

(c)

(b)

FIGURE 22.13 3D modeling of human face: (a) texture mapping, (b) cross section, and (c) simulation with 3D CAD.

modeling was fed to 3D CAD, and the image of 3D CAD was created, simulating a person putting chin on the ophthalmology equipment (Figure 22.13c). This enables us to determine the optional relation between the equipment and the face in 3D, the data necessary to make the design of the equipment.

22.3.2.2 Body Measurement

We set up four stereo-cameras facing each other at an angle of 90° around the body. This enables us to make all-around measurement without being disturbed by the body movement. Here, we made all-around images on the human body of a female. As shown in Figure 22.14a, we set up a projector behind each stereo-camera. For this photographing, we made two kinds of images: those with projected random patterns and those without them. All four stereo-cameras were identical in structure and in applied conditions. They were all SLR-type digital camera of 6 million pixels with the lens of 28 mm focal length. The photographing conditions were as follows: photographing distance was 1.3 m, distance between cameras was 0.3 m, resolution was horizontally 0.28 mm and in depth 1.3 mm. Figure 22.14b and c shows the measurement results of surface model and texture-mapping model. Thus equipped with easily obtainable cameras and projectors, we were able to measure human body and with considerable precision [36].

FIGURE 22.14 Human body measurement: (a) photographing setup, (b) result of surface model, and (c) result of texture-mapping model.

22.3.3 APPLICATION TO AUTOMOBILE MEASUREMENT

There are many different things to be measured in automobile, for example, small parts to middle-sized parts like seat, as well as large parts like bottom or entire body. So, we have measured them with digital cameras and our software and proved its feasibility [23], as explained in the following sections.

22.3.3.1 Measuring Car Bottom

In auto accident, it is often extremely difficult and complicated to precisely measure and record the deformation and debris and to visualize it in image. Therefore, we photographed and measured the bottom of a car and visualized it. Figure 22.15a shows the bottom width 1.5 m × length 4.5 m. We used one digital SLR camera of 6 million pixels. We used lens of focal length 20 mm.

We lifted a car by crane, placed camera in 18 positions (in 3 lines with 6 positions in each) (Figure 22.15b), and shot upward (Figure 22.15c). The bottom part has lots of dirt, which served as a texture. So, pattern projection was not necessary. Photographing distance was about 1.7 m, and the distance between cameras was 0.5 m. As to the precision, the standard deviation was 0.6 mm in plane and 2 mm in depth. It took 1 h to photograph, and the analysis of 16 models (32 pictures) took almost a day. The results are shown in Figure 22.16.

(a)　　　　　　　　　　(b)　　　　　　　　　　(c)

FIGURE 22.15 Car measurement: (a) bottom, (b) photo position, and (c) photographing.

(a)　　　　　　　　　　　　　　　(b)

FIGURE 22.16 Car measurement result: (a) texture mapping and (b) wire frame.

22.3.3.2 Measuring Surface of the Entire Body

Measuring an entire body is necessary, for example, in auto accident site or industrial production of a clay model. We made all-around measurement of a car by two different ways: with solely point measurement and with surface measurement.

1. Automatic point measurement of a vehicle by coded targets: it is reported that in order to simplify the measuring of a damaged car, coded targets had been pasted on its surface for automatic measurement [24]. Therefore, we placed color-coded targets on a car and made all-round automatic measurement. The number of images or photos was 36, of which stereo-pairs were 32. Figure 22.17a is the result of the target identification detected automatically. You can see three retro-targets labeled with a number.

   As to the accuracy, the error of transformation in the color-code recognition or the error of wrong recognition of target in the place where it does not exist was none. The rate of false detection was zero. Figure 22.17b shows the 3D measuring result of detected points.

2. All-around measurement of a car body by stereo-matching: as shown in Figure 22.18a, we used a stereo-camera and a projector. We photographed from eight different positions around the car, each from both upper angle and lower angle (16 models altogether). We also photographed the roof from above and this in four positions on the left and four positions on the right. All these totaled to 32 models with 64 images altogether. Furthermore, for each of these 64, we took pictures with patterns and without patterns. This means we photographed 128 images altogether for analysis (Figure 22.18b). We used digital cameras Nikon D70 of 6 million pixels and attached a lens of focal length 28 mm. For each model, the photo distance was about 2 m.

(a)   (b)

FIGURE 22.17   (a) Result of target identification and (b) 3D measuring result of detected points.

(a)   (b)

FIGURE 22.18   All-around measurement of a car body by stereo-matching: (a) photographing and (b) photo position.

(a) (b) (c)

FIGURE 22.19 The measured result of Figure 22.18: (a) texture mapping, (b) wire frame, and (c) mesh form.

The distance between cameras was 0.95 m. Given this photographing condition, the measuring area of one model is 1 m × 0.7 m, and the measuring accuracy is about 0.5 mm in plane and 1.2 mm in depth. Including the time to attach the targets, it took about 2 h to photograph and 1.5 days to analyze. Figure 22.19 shows the resulted texture mapping, wire frame, and mesh form. Here, the window is shown simply as a plane, since we could not measure the window glasses. The transparent parts like lights and black parts like door knobs cannot be measured. If we want to measure them, we must paint or put powder on them.

22.3.3.3 Measuring Surface of the Tire

This is to show the new measuring method we have developed.

Area-based matching methods set an image area as a template and search the corresponding match. As a direct consequence of this approach, it becomes not possible to correctly reconstruct the shape around steep edges. Moreover, in the same regions, discontinuities and discrepancies of the shape between the left and right stereo-images increase the difficulties for the matching process. In order to overcome these problems, we have developed the approach of reconstructing the shape of objects by embedding reliable edge line segments into the area-based matching process. We have developed a robust stereo-matching method that integrates edges and which is able to cope with differences in right and left image shapes, brightness changes, and occlusions [49,50].

The method consists of the following three steps: (1) parallax estimation, (2) edge-matching, (3) edge-surface matching.

Here, as an example, we measured a tire, using its algorithm.

We tried to find whether it is possible to show clearly the structural feature of a tire with deep groove. We obtained 144,991 points separated from each other by 1 mm. Figure 22.20a is the original picture. Figure 22.20b shows the result of overlapping the stereo-matching result onto the detected edge image. On the surface of tire, we can see the color-coded targets for orientation and the groove. Figure 22.20c shows the matching result without edge-matching. Figure 22.20d shows the result of the edge-surface matching by detecting 3D edge in our new method. Figure 22.20e shows the line profile of each method. Here we can see clearly the feature of the groove, which cannot be obtained without edge-matching.

We made an accuracy assessment, using the tire (Figure 22.21a), which we measured with the contact 3D measuring device (Zeiss UMC550S: accuracy 10 μm). We compared the cross section of our new method and the contact 3D measuring device. We mounted 35 mm lens to the Nikon SLR camera and took stereo picture with the photographing distance 618 mm and base length 240 mm. The resolution for this work was Δxy: 0.1 mm for plane and Δz: 0.26 mm for depth. Table 22.2 shows the result. Figure 22.21b shows the cross section of the comparison between contact 3D measuring device and edge-surface matching. For line 1, the average was 0.09 mm, standard deviation was 0.1 mm, and maximum difference was 0.4 mm. For line 2, the average was 0.14 mm, standard deviation was 0.14 mm, and maximum was 0.5 mm.

FIGURE 22.20 (a) Original image, (b) edge detection + stereo-matching, (c) result of surface matching (without edge), (d) result of edge-surface matching, and (e) result of measurement compared with surface matching.

FIGURE 22.21 (a) Evaluation line and (b) comparison between contact 3D measuring device and edge-surface matching.

TABLE 22.2
Accuracy Assessment

	Line 1	Line 2
Number of points	51	49
Average	0.09 mm	0.14 mm
Standard deviation	0.10 mm	0.14 mm
Maximum	0.4 mm	0.5 mm

While in the usual method, the groove measuring gauge of tire is ±0.1 mm for point measurement, in the new method, we can measure even the surface as the evaluation of tire wear.

22.3.4 INDOOR MEASUREMENT

This paragraph shows the application of indoor measurement.

Indoor measurement is difficult, as the features are poor and 3D model creation is not easy. Therefore, the effort to develop the model is made, using the edge information and plane information.

For example, the method is recommended that uses the Manhattan world hypothesis [51]. This hypothesis considers basically that the building is composed of three vectors perpendicular to each other. This method is to roughly reconstruct the indoor structure out of many images and often used for walk-through [52].

We propose an automated technique for creating a simplified 3D structure shape model of indoor environments from the sparse 3D point information provided by the photogrammetric method [53].

Figure 22.22 shows the overview of the proposed system.

1. Multiple coded targets (see Figure 22.4c, right) are used for the automatic detection, recognition, and measurement of 3D coordinates of points on the room walls from the acquired stereo images.
2. The sparse 3D information is then processed.

FIGURE 22.22 Overview of the proposed system. (1) Color-coded targets are placed on the walls of a room and stereo images are processed photogrammetrically, (2) sparse 3D points coordinates are obtained. (3) A simplified 3D structure shape model of the room is constructed from the sparse 3D information. (4) The development figure is generated by orthographic projection of the 3D model.

3. A simplified 3D structure shape model of the room is estimated.
4. Additionally, the 3D model is also translated into a development figure that represents multiple orthographic projections of the room.

This method is implemented into a flawless procedure that allows to obtain, by a few operations, a simplified 3D structure shape model of a room and its orthographic representation in drawing sheets. The developed application software can be used for the determination of the shape of rooms easily also by persons without experience in 3D measurement techniques.

22.3.5 PHOTOGRAMMETRIC SLAM

The basic principle of measuring that uses SFM and SLAM in the computer vision was explained in Section 22.2.3. The well-known system that uses the image sequence is PTAM [54], and the one that uses the still image is photo tourism [55]. PTAM makes augmented reality (AR) of small space from a video-camera. Photo tourism is a web system to enjoy the virtual tourism, using the 3D reconstruction out of tourism pictures.

The followings are what we are trying to do as photogrammetric SLAM.

22.3.5.1 Measurement by Vehicle (Measurement by Image Sequences)

Here is an example of measuring big pile of earth by putting a video-camera on top of the car [56]. Figure 22.23a shows the camera and GPS on the car. Figure 22.23b shows the actual operation of measurement.

FIGURE 22.23 Measurement by vehicle: (a) camera and GPS on the car, (b) measuring, (c) flow of image tracking, (d) tracking feature points, and (e) result.

We obtain the feature points by tracking the image sequences from the video-camera, and from the feature points, we obtain the camera's position and the 3D coordinates of the object. Figure 22.23c and d shows the flowchart of image tracking and feature points tracking. The process of image tracking is as follows:

1. Input of video image
2. Detecting the feature points by image processing
3. Matching of adjacent pictures of the feature points detected
4. Removing the errors

We repeat (1)–(4) processes orderly.

We simultaneously seek camera's position and the 3D coordinates of the object from the feature points detected, calculating the exterior orientation parameters and the 3D coordinates in order. Figure 22.23e shows the result. You can see the trajectory of the car and the big pile of the earth (in our example, we did not use GPS).

In the ordinary mobile mapping system, we obtain the 3D model by laser scanner and obtain the car position by using GPS, IMU, and odometer. And we combine these images with measuring instruments.

22.3.5.2 Measurement by UAV (Measurement by Image Sequences and Still Images)

The low-cost unmanned aerial vehicle (UAV) has become useful tool for low-altitude photogrammetry from the rapidly increasing automatic control technique in the past several years. Figure 22.24a shows an example of UAV.

In this paragraph, we show the automatic exterior orientation procedure and reconstruct the 3D model for the low-cost UAV photogrammetry. We have developed the automatic corresponding point detection technique using both video image and still images [57]. The video image is obtained with still images simultaneously. Therefore, the tracking result of video image gives robust correspondence information between each of the still images. Moreover, the exterior orientation

(a)

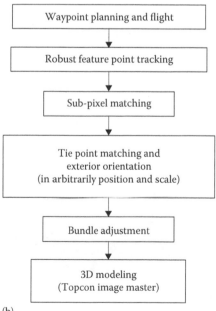

Waypoint planning and flight

↓

Robust feature point tracking

↓

Sub-pixel matching

↓

Tie point matching and
exterior orientation
(in arbitrarily position and scale)

↓

Bundle adjustment

↓

3D modeling
(Topcon image master)

(b)

FIGURE 22.24 (a) UAV and (b) flow of automatic measurement.

procedure using GPS information from UAV and minimum ground control points has been inves-tigated. For model making, we used the method of area-based matching, which can show the edges clearly in details.

The following are the flow of the measurement (Figure 22.24b):

1. The waypoint planning and flight control are performed. High-resolution still images at waypoints, video image during flight, and also positioning information of each waypoint are obtained as a result of a flight.
2. At the beginning of this process, the detection of common feature points from high-res-olution still image and tracking of detected feature points on video image sequences are performed using robust feature point tracking method [57]. The tracking result of feature points are obtained in the image coordinate on high-resolution still image.
3. Then, sub-pixel matching of each feature point is performed. The good results of this sub-pixel matching are used as candidate of correct pass points.
4. In order to perform tie-point matching, the exterior orientation using these candidate pass points is performed. The result of exterior orientation in this step gives the geometry of candidate pass points and still images in arbitrarily position and scale.
5. Next, registration in global coordinate using the positioning information of waypoints and minimum ground control points is performed by using bundle adjustment with robust regression.
6. Finally, we make 3D model reconstruction by edge-surface matching (Section 22.3.3.3), which we have developed [49,50].

We took pictures of the area (50 m × 50 m) from 30 m high using UAV.

Figure 22.25a shows the result of the stereo-matching, using 157 images with the additional texture mapping (without using edge-surface matching). Figure 22.25b through d shows the result of the edge-surface matching. Figure 22.25c is the result of texture mapping, Figure 22.25d is the result of 3D edge detection. As you can see, without using edge-surface matching method (Figure 22.25a), the features are not so clear, while, as shown in Figure 22.25b and c, the features are clear.

(a) (b) (c) (d)

FIGURE 22.25 Measurement result (a) without edge-surface matching, (b) edge-surface matching, (c) tex-ture mapping with edge-surface matching, and (d) 3D edge.

(a)   (b)   (c)

FIGURE 22.26 Measurement result of staircase (a) without edge-surface matching, (b) edge-surface matching, and (c) 3D edge.

Figure 22.26 shows the result of measuring the staircase with 25 pictures from 15 m high. We can clearly see the features (Figure 22.26b and c), which could not be shown clearly by the method without using edge-surface matching (Figure 22.26a).

22.4 CONCLUSIONS

As digital photogrammetry can make 3D measuring and modeling out of plural number of pictures obtained by digital camera, it has an array of advantages as follows:

1. As the equipment requirement is as simple as a digital camera and a PC, we can easily take it out to any place. We can also use as many kinds and numbers of cameras as we wish.
2. Basically, there is no limit to the kind and size of an object from minimum to maximum. It can even measure an object from all around.
3. Utilization for moving object: 3D measuring and modeling of movable object is possible in its state of given instance, including such object as human body.
4. We can use it for a wide variety of purposes, such as plan drawing and model making that require 3D data. It can be also used for reverse engineering and such works as comparing and inspecting with CAD data.
5. Utilization of texture: as we obtain 3D data from pictures, we cannot only grasp the object visually but also make 3D measurement through image processing and produce ortho-image as well.

And from its wide range of possible applications and developments, there are other important spill-outs such as the ones shown as follows:

1. Utilization of moving image: 3D measurement and modeling using a video-camera [25–27,46,47,54].
2. Utilization for the image of electro-microscope: 3D measurement from the scanning electron microscope image [28].
3. Application for image recognition: utilization of 3D data for facial image recognition [29].
4. Utilization for multi-oculus 3D display system: we can use this for the data display of multi-oculus 3D display system (e.g., 128 eyes), which makes it possible for many people to see together the 3D reality without stereo-glasses [30].
5. Utilization for sensor fusion technology: it is a 3D measurement technology to integrate and use the data obtained through cameras and various sensors (GPS, total station, position sensor, laser scanner, etc.) put on a robot or car [1,31].

6. Model synthesizing and automatization of whole process: synthesizing the point cloud obtained by laser scanner and the 3D point cloud obtained from the images. The researches first used the laser scanner to take the dense point cloud of small areas and the photogrammetry of UAV to take the large area [58,59]. And then contrariwise, we took the picture of the large areas by the laser scanner, and the small areas by photogrammetry of handy digital camera [49]. And we synthesize them together. Thus, we achieved the total automatization and speeding up of model production and mapping.

7. Integration of 2D image and 3D point cloud, and processing and recognition: the technology of the registration and the object recognition, using the image features and shape features, is an important goal we are presently challenging [48,49,60].

Though we could not present other examples in these limited pages, we believe that the future of digital photogrammetry is almost unlimited as the presented examples indicate.

REFERENCES

1. Urmson, C., Wining the DARPA urban challenge—The team, the technology, the future, *Fifth Annual Conference on 3D Laser Scanning, Mobile Survey, Lidar, Dimensional Control, Asset Management, BIM/CAD/GIS Integration, SPAR2008*, Spar Point Research, Houston, TX, 2008.
2. Kochi, N. et al., 3 Dimensional measurement modeling system with digital camera on PC and its application examples, *International Conference on Advanced Optical Diagnostics in Fluids, Solids and Combustion*, Visual Society of Japan, SPIE, Tokyo, Japan, 2004, p. V0038-1-10.
3. Kochi, N. et al., PC-based 3D image measuring station with digital camera an example of its actual application on a historical ruin, *ISPRS*, Ancona, Italy, 34, 195–199, 5/W12, 2003.
4. Kochi, N. et al., 3D-measuring-modeling-system based on digital camera and pc to be applied to the wide area of industrial measurement, *The International Society for Optical Engineering (SPIE) Conference on Optical Diagnostics*, San Diego, CA, 2005.
5. TOPCON CORPORATION, Products and applications, 2009–2014, http://www.topconpositioning.com/products/ (accessed November 27, 2014).
6. Karara, H. M., Introduction to metrology concepts, Analytic data-reduction schemes in non-topographic photogrammetry, Camera calibration in non-topographic photogrammetry, *Non-Topographic Photogrammetry*, 2nd edn., McGlone, J. C. (ed.), American Society for Photogrammetry and Remote Sensing, Falls Church, VA, Chapters 2, 4, 5, 1989.
7. Slama, C. C., Basic mathematics of photogrammetry, Non-Topographic photogrammetry, *Manual of Photogrammetry*, 4th edn., Karara, H. M. (ed.), American Society for Photogrammetry and Remote Sensing, Falls Church, VA, Chapters 2, 16, 1980.
8. Schenk, T., *Digital Photogrammetry*, Vol. 1, Terra Science, Laurelville, OH, 1999.
9. Heipke, C., State-of-the-art of digital photogrammetric workstations for topographic applications, *Photogrammetric Engineering and Remote Sensing*, 61(1), 49–56, 1995.
10. Noma, T. et al., New system of digital camera calibration, *ISPRS Commission V Symposium*, Corfu, Greece, 54–59, 2002.
11. Brown, D. C., Close-range camera calibration, *Photogrammetric Engineering*, 37(8), 855–866, 1971.
12. Fraser, C. S., Shotis, M. R., and Ganci, G., Multi-sensor system self-calibration, *Videometrics IV*, 2598, 2–18, SPIE, 1995.
13. Heuvel, F. A., Kroon, R. J. G., and Poole, R. S., Digital close-range photogrammetry using artificial targets, *ISPRS*, Washington, DC, 29(B5), 222–229, 1992.
14. Hattori, S. et al., Design of coded targets and automated measurement procedures in industrial visionmetrology, *ISPRS*, Amsterdam, the Netherlands, 33, WG V/1, 2000.
15. Ganci, G. and Handley, H., Automation in videogrammetry, *ISPRS*, Hakodate, Japan, 32(5), 47–52, 1998.
16. Moriyama, T. et al., Automatic target-identification with color-coded-targets, *ISPRS*, Beijing, China, 21, WG V/1, 2008.
17. Kochi, N., Ito, T., Kitamura, K., and Kaneko, S., Development of 3D image measurement system and stereo-matching method, and its archaeological measurement, *Electronics and Communications in Japan*, 96(6), 9–21, 2013.
18. Grun, A., Adaptive least-squares correlation: A powerful image matching technique, *South Africa Journal of Photogrammetry, Remote Sensing and Cartography*, 14(3), 175–187, 1985.

19. Besl, P. J. and McKay, N. D., A method for registration of 3D shapes, *IEEE Transaction on Pattern Analysis and Machine Intelligence*, 14(2), 239–256, 1992.
20. Bethmann, F. and Luhmann, T., Least-squares matching with advanced geometric transformation models, *ISPRS*, Vol. XXXVIII, Part 5 Commission V Symposium, Newcastle upon Tyne, U.K., 2010, pp. 86–91.
21. Luhmann, T., Bethmann, F., Herd, B., and Ohm, J., Experiences with 3D reference bodies for quality assessment of freeform surface measurements, *ISPRS*, Vol. XXXVIII, Part 5 Commission V Symposium, Newcastle upon Tyne, U.K., 2010, pp. 405–410.
22. Kitamura, K., Kochi, N., Watanabe, H., Yamada, M., and Kaneko, S., Human body measurement by robust stereo matching, *Ninth Conference on Optical 3-D Measurement Techniques*, Vol. 2, 2009, pp. 254–263.
23. Aurenhammer, F., A survey of a fundamental geometric data structure, *ACM Computing Surveys*, 23(3), 345–405, 1991.
24. Beardsley, P., Zisserman, A., and Murray, D., Sequential updating of projective and affine structure from motion, *International Journal of Computer Vision*, 23(3), 235–259, 1997.
25. Tomasi, C. and Kanade, T., Shape and motion from image streams under orthography: A factorization method, *International Journal of Computer Vision*, 9(2), 137–154, 1992.
26. Pollefeys, M., Koch, R., Vergauwen, M., Deknuydt, A., and Gool, L.J.V., Three-dimensional scene reconstruction from images, *Proceedings of the SPIE*, 3958, 215–226, 2000.
27. Lowe, D., Distinctive image features from scale invariant keypoints, *International Journal of Computer Vision*, 60(2), 91–110, 2004.
28. Bay, H., Ess, A., Tuytelaars, T., and Gool, L., Speeded-up robust features (SURF), *Computer Vision and Image Understanding*, 110, 346–359, 2008.
29. Anai, T., Kochi, N., and Otani, H., Exterior orientation method for video image sequences using robust bundle adjustment, *Eighth Conference on Optical 3-D Measurement Techniques*, Vol. 1, Zurich, Switzerland, 2007, pp. 141–148.
30. Mouragnon, E., Lhuillier, M., Dhome, M., Dekeyser, F., and Sayd, P., Real time localization and 3D reconstruction, *Proceedings of the 2006 IEEE Computer Society Conference on Computer Vision and Pattern Recognition*, Vol. 1, New York, NY, 363–370, 2006, pp. 363–370.
31. Eudes, A. and Lhuillier, M., Error propagations for local bundle adjustment, *CVPR'09*, Miami, Florida, 2411–2418, 2009.
32. Mouragnon, E., Lhuillier, M., Dhome, M., Dekeyser, F., and Sayd, P., Generic and real time structure from motion using local bundle adjustment, *Image and Vision Computing*, 27(8), 1178–1193, 2009.
33. Fukaya, N., Anai, T., Sato, H., Kochi, N., Yamada, M., and Otani, H., Application of robust regression for exterior orientation of video images, *ISPRS*, Beijing, China, XXIth Congress, WG III/V, 2008, pp. 633–638.
34. Anai, T., Fukaya, N., Sato, T., Yokoya, N., and Kochi, N., Exterior orientation method for video image sequences with considering RTK-GPS accuracy, *Ninth Conference on Optical 3-D Measurement Techniques*, Vol. 1, Viena, Austria, 2009, pp. 231–240.
35. Tsuji, S. et al., *The Survey of Early Byzantine Sites in Oludeniz Area (Lycia, Turkey)*, Osaka University, Osaka, Japan, 1995.
36. Kadobayashi, R. et al., Comparison and evaluation of laser scanning and photogrammetry and their combined use for digital recording of cultural heritage, *ISPRS*, Istanbul, Turkey, 20th Congress, WG V/4, 2004.
37. Yoshitake, R. and Ito, J., The use of 3D reconstruction for architectural study: The Askleption of ancient Messene, *21st International CIPA Symposium*, Athens, Greece, 2007, Paper No. 149.
38. Smith, C. J., The mother of invention—Assembling a low-cost aerial survey system in the Alaska wilderness, *GEOconnexion International Magazine*, 26–29, December 2007, 20–22, February 2008.
39. Otani, H. et al., 3D model measuring system, *ISPRS*, Istanbul, Turkey, 20th Congress, WG V/2, 2004, pp. 165–170.
40. Ohishi, M. et al., High resolution rangefinder with a pulsed laser developed by an under-sampling method, *SPIE Optical Engineering*, 49(6), 064302-1-8, 2010.
41. D'Apuzzo, N., Modeling human face with multi-image photogrammetry, *Three-Dimensional Image Capture and Applications V*, Corner, B. D., Pargas, R., Nurre, J. H. (eds.), *Proceedings of SPIE*, Vol. 4661, San Jose, CA, 2002, pp. 191–197.
42. Fraser, C. S. and Clonk, S., Automated close-range photogrammetry: Accommodation of non-controlled measurement environments, *Eighth Conference on Optical 3-D Measurement Techniques*, Vol. 1, Zurich, Switzerland, 2007, pp. 49–55.
43. Kochi, N., Kitamura, K., Sasaki, T., and Kaneko, S., 3D modeling of architecture by edge-matching and integrating the point clouds of laser scanner and those of digital camera, *XXII Congress of the International Society for Photogrammetry, Remote Sensing*, Vol. XXXIX-B5, Melbourne, Victoria, Australia, 2012, V/4, pp. 279–284.

44. Kochi, N., Sasaki, T., Kitamura, K., and Kaneko, S., Robust surface matching by integrating edge segments, *ISPRS Annals of the Photogrammetry, Remote Sensing and Spatial Information Sciences*, Volume II-5, 203–210, 2014.

45. Vanegas, C. A., Aliaga, D. G., and Benes, B., Building reconstruction using Manhattan-World grammars, *2010 IEEE Conference on Computer Vision and Pattern Recognition (CVPR)*, San Francisco, CA, June 13–18, 2010, pp. 358–365.

46. Furukawa, Y., Curless, B., and Sieitz, S., Reconstructing building interiors from images, *ICCV*, Kyoto, Japan, 80–87, 2009.

47. Hirose, S., Simple room shape modeling with sparse 3D point information using photogrammetry and application software, *ISPRS—International Archives of the Photogrammetry, Remote Sensing and Spatial Information Sciences*, 2012, XXXIX-B5, pp. 267–272.

48. Klein, G. and Murray, D., Parallel tracking and mapping for small AR workspaces, *ISMAR07*, Nara, Japan, November 2007.

49. Snavely, N., Seitz, S., and Szeliski, R., Photo tourism: Exploring photo collections in 3D, *ACM Siggraph*, 25(3), 835–846, 2006.

50. Anai, T., Kochi, N., Fukaya, N., and D'Apuzzo, N., Application of orientation code matching for structure from motion, *The International Archives of the Photogrammetry, Remote Sensing and Spatial Information Sciences*, Vol. 34, Part XXX Commission V Symposium, Newcastle upon Tyne, U.K., 2010, pp. 33–38.

51. Anai, T., Sasaki, T., Osaragi, K., Yamada, M., Otomo, F., and Otani, H., Automatic exterior orientation procedure for low-cost UAV photogrammetry using video image tracking technique and GPS information, *The International Archives of the Photogrammetry, Remote Sensing and Spatial Information Sciences*, Vol. XXIX-B7, V/I, Melbourne, Victoria, Australia, 2012, pp. 469–474.

52. Anai, T. and Chikatsu, H., Dynamic analysis of human motion using hybrid video theodolite, *ISPRS*, 33(B5), Amsterdam, Netherlands, 25–29, 2000.

53. D'Apuzzo, N., Human body motion capture from multi-image video sequences, *Videometrics VIII*, Elhakim, S. F., Gruen, A., Walton, J. S. (eds.), *Proceedings of SPIE*, Vol. 5013, Santa Clara, CA, 2003, pp. 54–61.

54. Abe, K. et al., Three-dimensional measurement by tilting and moving objective lens in CD-SEM(II), *Micro Lithography XVIII, Proceedings of SPIE*, Vol. 5375, San Jose, CA, 2004, pp. 1112–1117.

55. D'Apuzo, N. and Kochi, N., Three-dimensional human face feature extraction from multi images, *Optical 3D Measurement Techniques VI*, Gruen, A., Kahmen, H. (eds.), Vol. 1, Zurich, Switzerland, 2003, pp. 140–147.

56. Takaki, Y., High-density directional display for generating natural three-dimensional images, *Proceedings of the IEEE*, 93, 654–663, 2006.

57. Asai, T., Kanbara, M., and Yokoya, N., 3D modeling of outdoor scenes by integrating stop-and-go and continuous scanning of rangefinder, CD-ROM *Proceedings of the ISPRS Working Group V/4 Workshop 3D-ARCH 2005: Virtual Reconstruction and Visualization of Complex Architectures*, 36, Mestre-Venice, Italy, 2005.

58. Remondino, F. et al., Multi-sensor 3D documentation of the Maya site of Copan, *22nd CIPA, Symposium*, Kyoto, Japan, October 11–15, 2009.

59. El-Hakim, S.F. et al., Detailed 3D reconstruction of large-scale heritage sites with integrated techniques, *IEEE Computer Graphics and Applications*, 24(3), 21–29, May/June 2004.

60. Rusu, R. B., Marton, Z. C., Blodow, N., Dolha, M., and Beetz, M., Towards 3D Point cloud based object maps for household environments, *Robotics and Autonomous Systems*, 56, 927–941, 2008.

61. Tomono, M., Dense object modeling for 3D map building using segment-based surface interpolation, *Proceedings 2006 IEEE International Conference on Robotics and Automation, 2006*, Orlando, FL, May 15–19, 2006, pp. 2609–2614.

23 Optical Methods in Solid Mechanics

Anand Asundi

CONTENTS

23.1 Introduction 583
23.2 Basic Solid Mechanics...584
23.3 Digital Image Correlation and Its Applications...585
    23.3.1 Principles of 2D Digital Image Correlation...585
    23.3.2 Principles of 3D Digital Image Correlation...586
    23.3.3 Application to Residual Strain Measurement...587
23.4 Low Birefringence Polariscope 589
    23.4.1 Principle of Low Birefringence Polariscope 589
    23.4.2 Four-Step Phase-Shifting Method 590
    23.4.3 Low Birefringence Polariscope Based on Liquid Crystal Polarization Rotator.......591
    23.4.4 Experimental Results...593
23.5 Optical Diffraction Strain Sensor...594
    23.5.1 Principle of the Optical Diffraction Strain Sensor 594
    23.5.2 Multipoint Diffraction Strain Sensor...595
23.6 Digital Holography 597
    23.6.1 Introduction 597
    23.6.2 Reflection Digital Holography 597
    23.6.3 Digital Holographic Interferometry...601
23.7 Conclusions...602
References...602

23.1 INTRODUCTION

Optical methods are finding increasing acceptance in the field of experimental solid mechanics due to the easy availability of novel, compact, and sensitive light sources; detectors; and optical components. This, coupled with the faster data and image acquisition, processing, and display has resulted in optical methods providing information directly relevant to engineers in the form they require. The various methods discussed in the earlier chapters provide a background for the methods described in this chapter. Optical methods principally rely on theories of geometric optics or wave optics to gather information on the deformation, strain, and stress distribution in specimen subject to external loads. Methods based on geometric optics such as moiré and speckle correlation usually have simpler optical systems making them more robust at the expense of resolution. Wave optics based systems are generally very sensitive and hence have a more complex optical system.

In this chapter, after a brief introduction to basic solid mechanics, four techniques are discussed—the digital image correlation method, the optical diffraction strain sensor (ODSS), the low birefringence polarimeter, and the digital holographic system. Each has its own niche in application, which is highlighted in this chapter.

23.2 BASIC SOLID MECHANICS

Solid mechanics [1] deals with the action of external forces on deformable bodies. Consider a generic element inside a body subject to external forces as shown in Figure 23.1a. Without loss of generalization, a two-dimensional element is considered and the resulting stresses on this element in the x and y directions are shown in Figure 23.1b. The stresses, two normal stresses (σ_x and σ_y) and the shear stress (τ_{xy}) can be transformed to a different coordinate system through the stress transformation system or the Mohr's circle shown in Figure 23.1c. The maximum and minimum normal stresses are the principal stresses (σ_1, σ_2) while the maximum shear stress (τ_{max}) is at an angle of $45°$ with the maximum principal stresses. These stresses give rise to strains following the stress–strain equations which are linear for an elastic material. The strains are related to displacement using strain–displacement equations as shown in Figure 23.1d. The normal strain along the x-direction, defined as the change in length divided by the original length can be written as $\varepsilon_x = du/dx$, where dx is the length of the original line segment along the x-direction. Similarly, the normal strain in the y-direction is $\varepsilon_y = du/dy$, where dv is the change in the length of the segment whose initial length in the y-direction is dy. The shear strain is the difference between the right angle before deformation and the angle between the line segments after deformation. As with stresses, a Mohr circle of strain can be drawn and the principal strains and maximum shear strains can be determined.

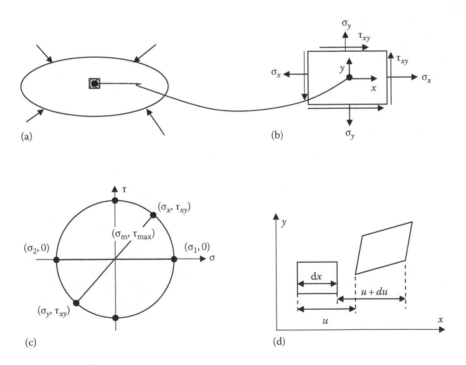

FIGURE 23.1 (a) External forces acting on an object, (b) stress components at a generic point, (c) Mohr's circle for stress, and (d) strain–displacement relationship.

23.3   DIGITAL IMAGE CORRELATION AND ITS APPLICATIONS

23.3.1   PRINCIPLES OF 2D DIGITAL IMAGE CORRELATION

Two-dimensional digital image correlation (2D DIC) directly provides the full-field in-plane displacement of the test planar specimen surface by comparing the digital images of the planar specimen surface acquired before and after deformation. The strain field can be deduced from the displacement field by numerical differentiation. The basic principle of DIC is schematically illustrated in Figure 23.2. A square reference subset of $(2M + 1) \times (2M + 1)$ pixels centered on the point of interest $P(x_0, y_0)$ from the reference image is chosen and its corresponding location in the deformed image determined through a cross-correlation or sum-squared difference correlation criterion. The peak position of the distribution of correlation coefficients determines the location of $P(x_0, y_0)$ in the deformed image, which yields the in-plane displacement components, u and v at point (x_0, y_0). The displacements can similarly be computed at each point on the user-defined grid to obtain the full-field deformation fields.

Following the compatibility of deformation, $Q(x, y)$ around the subset center $P(x_0, y_0)$ in the reference, subset can be mapped to point $Q'(x', y')$ in the target subset according to the following first-order shape function:

$$x' = x_0 + \Delta x + u + u_x \Delta x + u_y \Delta y$$
$$y' = y_0 + \Delta y + v + v_x \Delta x + v_y \Delta y$$

(23.1)

where
  u and v are the displacement components for the subset center P in the x and y directions, respectively
  Δx and Δy are the distance from the subset center P to point Q
  u_x, u_y, v_x, and v_y are the displacement gradient components for the subset as shown in Figure 23.1

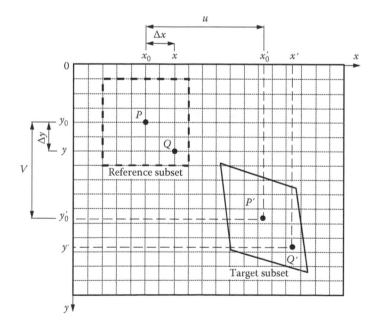

FIGURE 23.2   Schematic figure of digital image correlation method.

To obtain accurate estimation for the displacement components of the same point in the reference and deformed images, the following zero-normalized sum of squared differences correlation criteria [2], which is insensitive to the linear offset and illumination intensity fluctuations, is utilized to evaluate the similarity of reference and target subsets:

$$C_{f,g}(\mathbf{p}) = \sum_{x=-M}^{M} \sum_{y=-M}^{M} \left[\frac{f(x,y) - f_m}{\sqrt{\sum_{x=-M}^{M} \sum_{y=-M}^{M} [f(x,y) - f_m]^2}} - \frac{g(x',y') - g_m}{\sqrt{\sum_{x=-M}^{M} \sum_{y=-M}^{M} [g(x',y') - g_m]^2}} \right]^2 \quad (23.2)$$

where
   $f(x, y)$ is the gray level intensity at coordinates (x, y) in the reference subset of the reference image
   $g(x', y')$ is the gray level intensity at coordinates (x', y') in the target subsets of the deformed image
   $f_m = \frac{1}{(2M+1)^2} \Sigma_{x=-M}^{M} \Sigma_{x=-M}^{M} [f(x,y)]$ and $g_m = \frac{1}{(2M+1)^2} \Sigma_{x=-M}^{M} \Sigma_{y=-M}^{M} [g(x',y')]$ are the mean intensity values of reference and target subsets, respectively
   $\mathbf{p} = (u, u_x, u_y, v, v_x, v_y)^T$ denotes the desired vector with respect to six mapping parameters as given in Equation 23.1.

Equation 23.2 can be optimized to get the desired displacement components in the x and y directions using the Newton–Raphson iteration method. A least square smoothing of the computed displacement field provides the required information for strain calculation.

23.3.2 PRINCIPLES OF 3D DIGITAL IMAGE CORRELATION

The 3D DIC technique is based on a combination of the binocular stereo vision technique and the conventional 2D DIC technique. Figure 23.3 is a schematic illustration of the basic principle of the binocular stereo vision technique where O_{c1} and O_{c2} are optical centers of the left and right cameras, respectively. It can be seen from the figure that a physical point P is imaged as point P_1 in the image plane of the left camera and point P_2 in the image plane of the right camera. The 3D DIC technique aims to accurately recover the 3D coordinates of point P with respect to a world coordinate system from points P_1 and P_2.

To determine the 3D coordinates of point P, there is a need to establish a world coordinate system, which can be accomplished by a camera calibration technique. The world coordinate system

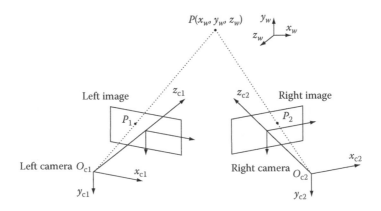

FIGURE 23.3 Schematic diagram of the binocular stereovision measurement.

is the one on which the final 3D shape reconstruction will be based. The second step is to calculate the location disparities of the same physical point on the object surface from the two images. Based on the calibrated parameters of the two cameras and the measured disparities of the point, the 3D coordinates of point P can then be determined.

Camera calibration is the procedure to determine the intrinsic parameters (e.g., effective focal length, principle point, and lens distortion coefficient) and extrinsic parameters (including the 3D position and orientation of the camera relative to a world coordinate system) of a camera. For the 3D DIC technique, the calibration also involves determining the relative 3D position and orientation between the two cameras. As the accuracy of the final measurement heavily depends on the obtained calibration parameters, camera calibration plays an important role in the 3D DIC measurement.

The goal of the stereo matching is to precisely match the same physical point in the two images captured by the left and right cameras. This task, commonly considered the most difficult part in stereovision measurement, can be accomplished by using the well-established subset-based matching algorithm adopted in the 2D DIC technique. The basic concept of the 2D DIC technique is to match the same physical point from two images captured before and after deformation (only one camera is employed, and is fixed during measurements); to ensure a successful and accurate matching, the specimen surface is usually coated with random speckle patterns. Similar to but slightly different from the 2D DIC scheme, the stereo matching used in the 3D DIC technique aims to match two random speckle patterns on specimen surface recorded by the left and right cameras [3].

Based on the obtained calibration parameters of each camera and the calculated disparities of points in the images, the 3D world coordinates of the points in the regions of interest on the specimen surface can be easily determined using the classical triangulation method. By tracking the coordinates of the same physical points before and after deformation, the 3D motions of the each point can be determined. Similarly, the same technique proposed in previous section can be utilized for the strain estimation based on obtained displacement fields.

23.3.3 APPLICATION TO RESIDUAL STRAIN MEASUREMENT

One specific application of the DIC method for the residual plastic deformation of a notched specimen under fatigue test is exemplified to show the capability of the method. A tensile specimen with semicircular notch in the middle as shown in Figure 23.4 is used as the test specimen.

Figure 23.5 shows the surface displacement fields computed using the Newton–Rapshon method after 26,000 fatigue cycles. It is seen that the displacement in x-direction (normal to the loading direction) is much smaller than that of y-direction. Figure 23.6 shows the residual plastic strain distributions computed from the displacement field using a point-wise least squares fitting technique.

FIGURE 23.4 Geometry of the specimen. The region of interest and the invalid area is highlighted.

FIGURE 23.5 Residual displacement fields of the specimen surface after 26,000 fatigue cycles: (a) u displacement field and (b) v displacement field.

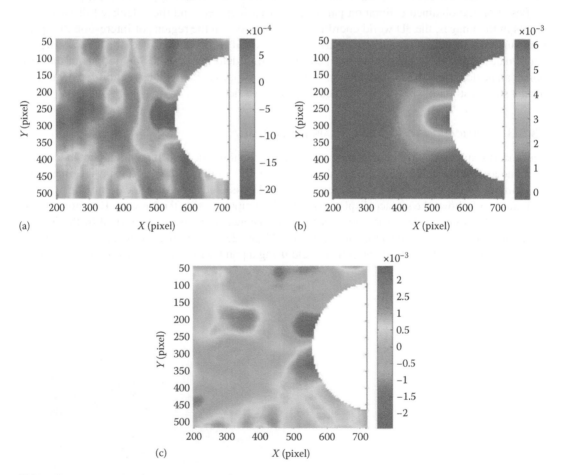

FIGURE 23.6 The residual plastic strain fields derived from the displacement field: (a) ε_x, (b) ε_y, and (c) γ_{xy}.

23.4 LOW BIREFRINGENCE POLARISCOPE

A modified Senarmont polariscope coupled to a phase-shift image-processing system provides a system for measurement of low levels of birefringence. The birefringence is related to the change in phase between the two propagating polarized waves through the specimen and can be related to the principle stress difference. In addition, the difference in the normal stress and the shear stress distribution can be quantitatively determined.

23.4.1 PRINCIPLE OF LOW BIREFRINGENCE POLARISCOPE

In the current system, schematically shown in Figure 23.7, the specimen is illuminated with circularly polarized light. The birefringence in the specimen transforms the circular polarized light to one with elliptic polarization, which is then interrogated by a rotating linear analyzer.

The change in polarization by the specimen can be attributed to a change in the orientation of the fast axis of the light as well as a phase lag between the two polarization components of the input beam. For materials, where the birefringence is attributed to the stress, the change in orientation of the fast axis (β) is related to the direction of the principal stresses while the phase lag (Δ) is related to the principal stress difference as

$$\Delta = \frac{2\pi Cd}{\lambda}(\sigma_1 - \sigma_2) \tag{23.3}$$

where
    d is the material thickness
    λ is the wavelength of light
    C is the stress optic coefficient of the material
    σ_1 and σ_2 are the principal stresses

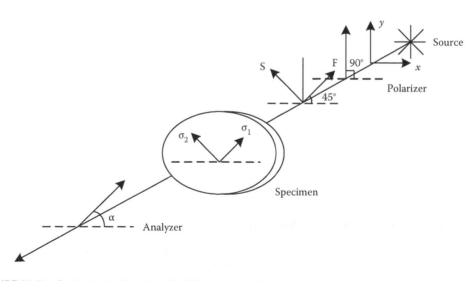

FIGURE 23.7 Optical train for a low birefringence polariscope.

From Jones' calculus, the components of the transmitted light vector perpendicular and parallel to the analyzer axis (U, V) are

$$\begin{bmatrix} U \\ V \end{bmatrix} = J_A J_M J_Q J_P a e^{j\omega t} \qquad (23.4)$$

where J_P, J_q, J_M, and J_A are the Jones' vector for the polarizer, the quarter-wave plate (QWP), the model, and the analyzer respectively. Expanding using matrix multiplication gives

$$\begin{Bmatrix} U \\ V \end{Bmatrix} = \frac{\sqrt{2}}{2} \begin{bmatrix} 0 & 0 \\ \cos\alpha & \sin\alpha \end{bmatrix} \begin{bmatrix} \cos\Delta/2 - i\cos 2\beta \sin\Delta/2 & -i\sin 2\beta \sin\Delta/2 \\ -i\sin 2\beta \sin\Delta/2 & \cos\Delta/2 + i\cos 2\beta \sin\Delta/2 \end{bmatrix} \times \begin{bmatrix} 1 & i \\ i & 1 \end{bmatrix} \begin{bmatrix} 0 \\ 1 \end{bmatrix} a e^{j\omega t}$$

$$= \frac{\sqrt{2}}{2} \begin{bmatrix} 0 \\ \left(\sin\dfrac{\Delta}{2}\cos(\alpha-2\beta) + \sin\alpha\cos\dfrac{\Delta}{2} \right) + i\left(\sin\dfrac{\Delta}{2}\sin(\alpha-2\beta) + \cos\alpha\cos\dfrac{\Delta}{2} \right) \end{bmatrix} a e^{j\omega t} \qquad (23.5)$$

where
 α is the angular position of the rotating polarizer
 β is the orientation of the fast axis (maximum principal stress direction)
 a is the amplitude of the circularly polarized light
 ω is the temporal angular frequency of the wave

The intensity of light emerging from a point on the specimen is then given as

$$I = |V|^2 = \frac{a^2}{2}[1 + \sin\Delta \sin 2(\alpha - \beta)] \qquad (23.6)$$

As the analyzer transmission axis aligns with the maximum intensity axis twice per full rotation, the intensity will be modulated at 2α intervals with a phase of 2β. Equation 23.6, the intensity collected by the detector, can be rewritten as

$$I = I_a + I_{c\alpha}\cos 2\alpha + I_{s\alpha}\sin 2\alpha \qquad (23.7)$$

where

$$I_{c\alpha} = -\left(\frac{a^2}{2}\sin\Delta \right)\sin 2\beta$$

$$I_{s\alpha} = \left(\frac{a^2}{2}\sin\Delta \right)\cos 2\beta$$

$I_a = a^2/2$, which is the average light intensity collected by the detector unit.

23.4.2 FOUR-STEP PHASE-SHIFTING METHOD

In order to determine the phase lag (Δ) and the direction of the fast axis (β) from Equation 23.6, we need at least three more equations. These are obtained by recording phase-shift patterns by changing the angle (α) of the analyzer. Assuming the average light intensity I_a to be constant at a given point during experiment, a four-step phase-shifting method can be used. Table 23.1 summarizes the four optical arrangements and the corresponding intensity equations.

TABLE 23.1
Intensity Equations for Rotating Analyzer Method

Number	Analyzer Angle α	Intensity Equation
1	0	$I_1 = I_a (1 - \sin \Delta \sin 2\beta)$
2	$\pi/4$	$I_2 = I_a (1 + \sin \Delta \cos 2\beta)$
3	$\pi/2$	$I_3 = I_a (1 + \sin \Delta \sin 2\beta)$
4	$3\pi/4$	$I_4 = I_a (1 - \sin \Delta \cos 2\beta)$

The retardation and direction of the fast axis can be deduced as

$$\Delta = \frac{1}{2I_a}\left[(I_3 - I_1)^2 + (I_4 - I_2)^2\right]^{1/2} \tag{23.8a}$$

$$\beta = \frac{1}{2}\tan^{-1}\left(\frac{I_1 - I_3}{I_4 - I_2}\right) \tag{23.8b}$$

From these values, $I_c\alpha$ and $I_s\alpha$ can be calculated through Equation 23.6. From the Mohr's circle, it can be seen that the sine and cosine components of the light intensity are directly related to the stresses along the 0°/90° and ±45° planes. Thus,

$$I_{s\alpha} \propto (\sigma_{xx} - \sigma_{yy})$$
$$I_{c\alpha} \propto \tau_{xy} \tag{23.9}$$

23.4.3 Low Birefringence Polariscope Based on Liquid Crystal Polarization Rotator

In the above polariscope, there is a need to manually rotate the analyzer, which makes it difficult to achieve high stability and repeatability. A new method based on a liquid crystal polarization rotator to get full-field sub-fringe stress distribution is proposed. By changing the applied voltage to the liquid crystal phase plate, the phase-shift images are recorded and processed to get the stress distribution. Figure 23.8 is the schematic of the proposed polariscope using the liquid crystal (LC)

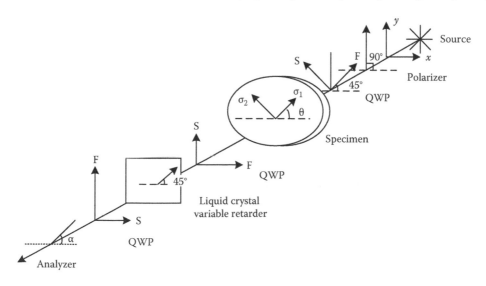

FIGURE 23.8 Schematic of the LC polarization rotator-based polariscope.

polarization rotator as the phase-shifting element. The LC polarization rotator, a key component of the proposed polariscope system, consists of an LC phase plate inserted between two crossed QWPs. The extraordinary (fast) axis of the first QWP is parallel to the x-axis. The extraordinary axis of the LC phase plate is oriented at 45°, and the extraordinary axis of the second QWP is oriented parallel to the y-axis. An LED light source illuminates the object with circularly polarized light and the transmitted light goes through LC polarization rotator before impinging on the analyzer whose orientation is set parallel to the x-axis ($\alpha = 0°$).

Using Jones' calculus, the components of the transmitted light vector perpendicular to and parallel to the analyzer axis are

$$\begin{bmatrix} U \\ V \end{bmatrix} = J_A J_{LC} J_M J_Q J_P a e^{j\omega t} \tag{23.10}$$

where J_P, J_Q, J_M, J_{LC}, and J_A are the Jones' vector for the polarizer, the QWP, the model, the LC polarization rotator, and the analyzer, respectively. Expanding and multiplying the equation

$$\begin{Bmatrix} U \\ V \end{Bmatrix} = \frac{\sqrt{2}}{2} \begin{bmatrix} 0 & 0 \\ 1 & 0 \end{bmatrix} \begin{bmatrix} \cos(\phi/2) & \sin(\phi/2) \\ -\sin(\phi/2) & \cos(\phi/2) \end{bmatrix}$$
$$\times \begin{bmatrix} \cos\Delta/2 + i\cos 2\beta \sin\Delta/2 & i\sin 2\beta \sin\Delta/2 \\ i\sin 2\beta \sin\Delta/2 & \cos\Delta/2 - i\cos 2\beta \sin\Delta/2 \end{bmatrix} \begin{bmatrix} 1 & i \\ i & 1 \end{bmatrix} \begin{bmatrix} 0 \\ 1 \end{bmatrix} a e^{i\omega t} \tag{23.11}$$

The intensity of light emerging from a point on the specimen is thus

$$I = |V|^2 = \frac{a^2}{2}[1 + \sin\Delta \sin(\delta - 2\beta)] \tag{23.12}$$

Table 23.2 summarizes the four optical arrangements and the corresponding intensity equations.

It can be seen that the intensity equations are the same as those of Table 23.1. In this case, the phase retardation is due to changes in voltage applied to the liquid crystal rather than rotation of the analyzer. This can be electronically controlled resulting in much better precision and accuracy for the phase-shift routine.

TABLE 23.2
Intensity Equations Used in the New Method

Number	Phase Retardation δ	Intensity Equation
1	0	$I_1 = I_a(1 - \sin\Delta \sin 2\beta)$
2	$\pi/2$	$I_2 = I_a(1 + \sin\Delta \cos 2\beta)$
3	π	$I_3 = I_a(1 + \sin\Delta \sin 2\beta)$
4	$3\pi/2$	$I_4 = I_a(1 - \sin\Delta \cos 2\beta)$

23.4.4 EXPERIMENTAL RESULTS

An epoxy disc (22 mm diameter and 3 mm thick) under diameteral compression of a load of 0.98 kg is tested because of its well-known principal stress magnitude and direction. Both polariscopes were tested and the images processed using a MATLAB® algorithm [4] to determine the desired phase lag and orientation of the fast axis. The experimental results using the low birefringence polariscope and the polariscope based on liquid crystal polarization rotator are shown in the left images of Figure 23.9 whereas the images on the right were obtained using the rotating analyzer method. The normal stress difference map is symmetric as shown in Figure 23.9c while the shear stress is antisymmetric (Figure 23.9d) about both the x and y axes as expected. These stress distributions also match with theory and show the high sensitivity of the system.

(a)

(b)

FIGURE 23.9 Experimental results using LC polarization rotator (left) and rotating the analyzer (right). (a) Direction map and (b) phase map.   (*Continued*)

FIGURE 23.9 (*Continued*) Experimental results using LC polarization rotator (left) and rotating the analyzer (right). (c) $\sigma_{+45}-\sigma_{-45}$ and (d) $\sigma_x-\sigma_y$.

23.5 OPTICAL DIFFRACTION STRAIN SENSOR

23.5.1 PRINCIPLE OF THE OPTICAL DIFFRACTION STRAIN SENSOR

The basic principle of measurement is illustrated in Figure 23.10. A diffraction grating bonded to the surface of the specimen follows the deformation of the underlying specimen. Consider a thin monochromatic collimated beam that is normal to the grating plane and illuminating a point on a specimen grating. The diffraction is governed by the equation:

$$p\sin\theta = n\lambda \quad n = 0, \pm 1, \pm 2, \pm 3, \dots \tag{23.13}$$

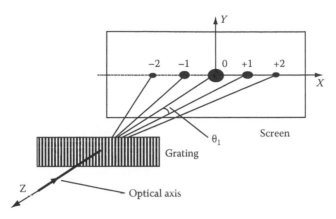

FIGURE 23.10 Sketch of one-dimensional grating diffraction. Grating is illuminated by a collimated beam; a line of diffraction spots is formed on a screen.

where

    p is the grating pitch
    n is the diffraction order
    λ is the light wavelength
    θ is the diffraction angle

When the specimen and hence the grating deforms, the pitch changes to p^*, with a corresponding change in θ according to Equation 23.1. Thus, the strain along the x-direction at the illuminated point on the grating can be determined as

$$\varepsilon_x = \frac{p^*-p}{p} = \frac{X_n^* - X_n}{X_n}\cos\theta_n \tag{23.14}$$

where X_n and X_n^* are the centroids of the undeformed and deformed nth-order diffraction spot. These centroids can be determined precisely with subpixel accuracy. If a cross grating is used instead of a line grating, the strain along y-direction, ε_y can be similarly obtained. The shear strain γ_{xy} can also be evaluated.

By using a high-frequency grating and a sensitive position-sensing detector, a system with capabilities to rival the electrical resistance strain gauge has been developed [5]. The schematic of the setup and typical result of strain distribution along the interface of the die and substrate in an electronic package is shown in Figure 23.11. The display interface directly determines the strain at a particular point and since the diffraction grating can cover a large area, strains at different points can be monitored by translating the specimen. Furthermore, the gauge length that is determined by the size of the laser beam can be readily adjusted and the system has been demonstrated for dynamic strain measurement. Overall, it is a worthy challenger for the electrical strain gauge.

23.5.2 Multipoint Diffraction Strain Sensor

The patented multipoint diffraction strain sensor (MISS) [6] uses a high-frequency diffraction grating along with a micro-lens array-based CCD detector. A reflective diffraction grating is bonded to the surface of the specimen and follows the deformation of the underlying specimen. The grating is illuminated by two symmetric monochromatic laser beams at a prescribed angle such that the first-order diffracted beams emerge normal to the specimen surface. The micro-lens array samples each of the incident beams and focuses them as spots onto the CCD as shown in Figure 23.12a.

(a)   (b)

FIGURE 23.11 (a) Schematic of the ODSS and (b) strain distribution along the interface of die and substrate of an electronic package.

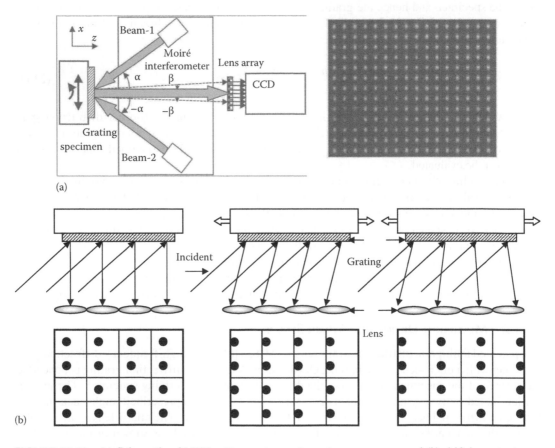

FIGURE 23.12 (a) Schematic of MISS with spot image from the two sources and (b) shift in spots due to uniform and nonuniform strain.

When the specimen is deformed, tilted, or rotated, the diffracted wave fronts emerging from the specimen are distorted and hence the spots shift accordingly. Figure 23.12b shows the simulated spot pattern for a 4×4 array of micro-lenses for one of the beams when the specimen is unstrained, undergoes uniform strain and nonuniform strain. The symmetric beam incident from other direction gives a similar array of spot patterns. Strains at each spot location which corresponds to a small area of the specimen, can then be readily deduced from the shift of the spots as described below. Without loss of generality, spots shifts in a single direction are used in this derivation.

Starting from the well-known diffraction equation, the change in diffraction angle when the specimen is subject to strain, ε_x, and out-of-plane tilt, $\Delta\phi$, can be deduced. Hence the corresponding shift, Δx, of a typical spot on the CCD plane can be related to the strain and tilt as

$$\frac{K_f \Delta x_1}{f} = -\varepsilon_x f \sin\alpha + \Delta\phi(1 + \cos\alpha)$$

$$\frac{K_f \Delta x_2}{f} = +\varepsilon_x f \sin\alpha + \Delta\phi(1 + \cos\alpha)$$

(23.15)

where

f is the focal length of each micro-lens
α is the angle of incidence
K_f is a multiplication factor which is 1 if the two beams are collimated and the subscripts 1 and 2 refer to the two incident beams

Solving Equation 23.15

$$\varepsilon_x = K_f \frac{\Delta x_2 - \Delta x_1}{2f \sin\alpha}$$

(23.16a)

$$\Delta\phi = K_f \frac{\Delta x_2 + \Delta x_1}{2f(1 + \cos\alpha)}$$

(23.16b)

The two incident beams thus enable us to compute the strains and the rigid body tilts of the specimen separately.

Figure 23.13a shows the uniform strain on specimen; note that the shear or rotation component that causes the spots to shift in the perpendicular direction is nearly zero. Similarly, for a specimen subject to pure rotation, the normal component of strain is zero as shown in Figure 23.13b.

23.6 DIGITAL HOLOGRAPHY

23.6.1 INTRODUCTION

In digital holography, the optical hologram is digitally recorded by CCD or CMOS arrays. It is then reconstructed numerically to determine the amplitude and phase of the object beam. The amplitude provides the intensity image while the phase provides the optical path difference between the object and reference beam. This quantitative measurement of phase is one of the key aspects of the digital holographic system leading to its widespread adoption in a variety of applications.

23.6.2 REFLECTION DIGITAL HOLOGRAPHY

Reflection digital holography [7] is widely used for deformation and shape measurement. The schematic of the reflection holographic setup is shown in Figure 23.14. Collimated light from a laser is split into two beams—the object beam which scatters of the object, and a reference beam.

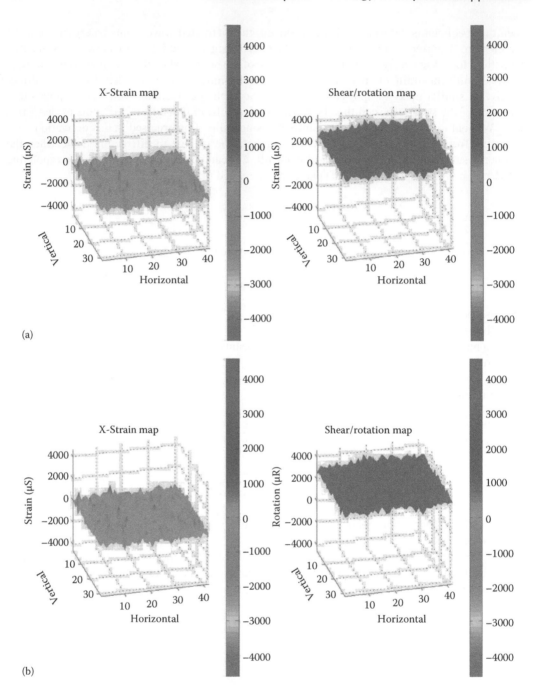

FIGURE 23.13 (a) 2D strain map of specimen subject to uniform strain and (b) 2D strain map of specimen subject to pure rotation.

The reference mirror can be tilted to create an off-axis holographic recording. However, due to the low resolution of the digital camera, the angle needs to satisfy the sampling theorem in order to record the hologram. Alternatively, an in-line setup can also be used. However, in this case, the effect of the zero order and twin image needs to be accounted for. This can readily be done through software [5].

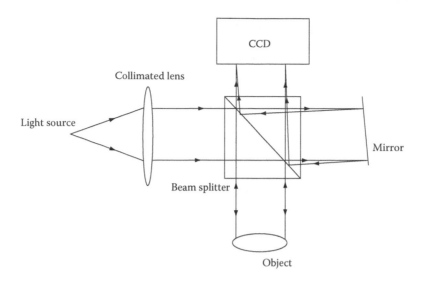

FIGURE 23.14 Schematic of the reflection digital holographic.

The numerical reconstruction of the digital hologram is actually a numerical simulation of the scalar diffraction. Usually, two methods, the Fresnel transform method and the convolution method, are used. Numerical reconstruction by the Fresnel transform method requires simulation of the following equation:

$$U\left(n\Delta x_i, m\Delta y_i\right) = e^{j\pi d\lambda\left(\frac{n^2}{N^2\Delta x_h^2}+\frac{m^2}{M^2\Delta y_h^2}\right)}\sum_{k=0}^{N-1}\sum_{l=0}^{M-1}h\left(k\Delta x_h, l\Delta y_h\right)R*\left(k\Delta x_h, l\Delta y_h\right)e^{\frac{j\pi}{d\lambda}\left(k^2\Delta x_h^2+l^2\Delta y_h^2\right)}e^{-2j\pi\left(\frac{kn}{N}+\frac{lm}{M}\right)}$$

(23.17)

where

$\Delta x_i = \dfrac{d\lambda}{N\Delta x_h}$ and $\Delta y_i = \dfrac{d\lambda}{M\Delta y_h}$ are the pixel sizes in the reconstructed image plane

$h(k\Delta x_h, l\Delta y_h)$ is the hologram
$R*(k\Delta x_h, l\Delta y_h)$ is the reference beam

Numerical reconstruction by the convolution method gives a reconstruction formula of digital holography as follows:

$$\begin{cases} U\left(n\Delta x_i, m\Delta y_i\right) = F^{-1}\left\{F\left[h\left(k\Delta x_h, l\Delta y_h\right)R*\left(k\Delta x_h, l\Delta y_h\right)\right]\cdot G\left(n\Delta\xi, m\Delta\eta\right)\right\} \\ G\left(n\Delta\xi, m\Delta\eta\right) = e^{\frac{2j\pi d}{\lambda}\sqrt{1-\left[\frac{\lambda(n-1)}{N\Delta\xi}\right]^2-\left[\frac{\lambda(m-1)}{M\Delta\eta}\right]^2}} \end{cases}$$

(23.18)

where

$h(k\Delta x_h, l\Delta y_h)$ is the hologram
$R*(k\Delta x_h, l\Delta y_h)$ is the reference beam
$G(n\Delta\xi, m\Delta\eta)$ is the optical transfer function in the frequency domain
$\Delta x_h, \Delta y_h$ are the pixel sizes in the hologram plane
$\Delta\xi, \Delta\eta$ are the pixel sizes in the frequency domain
$\Delta x_i, \Delta y_i$ are the pixel sizes in the image plane

Using Fresnel transform method, the resolution of the reconstructed object wave depends not only on the wavelength of the illuminating light but also on the reconstruction distance and it is always lower than the resolution of the CCD camera. While using the convolution method, one can obtain the reconstructed object wave with the maximum resolution with the same pixel size as that of the CCD camera.

The reconstruction by convolution method is shown in Figure 23.15 where Figure 23.15a is the recorded hologram, and Figure 23.15b is the Fourier transform of the hologram showing the zero order and the two other orders corresponding to the object wave. One of these orders is isolated and moved to the origin to remove the tilt due to the reference beam. It is then multiplied by the transfer function and propagated to the image plane. An inverse Fourier transform of the product gives the reconstructed object wave from which the quantitative amplitude (Figure 23.15c) and phase (Figure 23.15d) can be extracted.

The phase, $\varphi(x, y)$ which is related to the optical path difference between the object and reference waves can thus be used to obtain the height, $d(x, y, z)$ of the object as

$$\varphi(x, y) = \frac{2\pi}{\lambda} d(x, y, z) \tag{23.19}$$

If the phase is $-\pi < \varphi(x, y) < \pi$, the height can be directly determined without the need to unwrap the data. This corresponds to a height between $\pm\lambda/2$. With an 8 bit resolution CCD camera and

(a)   (b)

(c)   (d)

FIGURE 23.15   (a) Digital hologram of USAF positive target, (b) spectra, (c) intensity, and (d) quantitative 3D phase map.

a 532 nm frequency-doubled diode-pumped laser, this results in a sensitivity of about 2 nm. If, however, the phase exceeds this value, then, unwrapping becomes necessary to extract the true height distribution.

23.6.3 DIGITAL HOLOGRAPHIC INTERFEROMETRY

Digital holographic interferometry proceeds in much the same way as traditional holographic interferometry as far as recording is concerned. In the reconstruction step, the phases of the holograms before and after deformation are extracted and subtracted to reveal the modulo 2π fringes. These can be readily unwrapped to reveal the true deformation [8]. In previous cases, magnifying optics was necessary to enlarge the object and increase its spatial resolution. In this approach, a patented [9] system based on the same Michelson interferometer geometry, as shown in Figure 23.14, was adopted. However, instead of a collimated beam, a diverging beam was used as shown in Figure 23.16a. The setup in this configuration can be quite compact and still provide the desired magnification. The reconstruction process is slightly different since the reference beam is not collimated and hence the appropriate divergence of the beam has to be incorporated into the reconstruction equations. A typical reconstruction of 300 μm long and 20 μm wide micro-cantilevers using this setup is shown in Figure 23.16b, which can be readily compared to a microscope image shown in Figure 23.16c.

In holographic interferometry, there are two possibilities—for static deformation measurement, the double exposure method is used. In this, two digital holograms of the object before and after deformation are recorded. They are separately reconstructed and their phase extracted. The phase difference provides the deformation of the object between the two exposures. A MEMS micro-heater was chosen as the sample. Two exposures before and after application of a current, which caused the micro-heaters to deform, were recorded. The resulting out of displacements of the micro-heater is shown in Figure 23.17a. Alternately, if the cantilevers were subject to dynamic loading, a time average hologram, where the exposure time is greater than the period of vibration, is recorded. Reconstruction of this hologram reveals the familiar Bessel-type fringes modulating the amplitude image as shown in Figure 23.17b. The excitation frequency was 417 kHz.

FIGURE 23.16   (a) Schematic of lensless microscopic digital holographic interferometer.

(a)

(b)

FIGURE 23.17 (a) Double exposure lensless microscopic digital holographic interferometery of MEMS micro-heater; and (b) time average digital holographic interferometry of cantilever vibrating at 417 kHz.

23.7 CONCLUSIONS

Optical methods in solid mechanics have shown tremendous growth in the past decade. With the availability of novel light sources and detectors and faster data acquisition and processing speeds, real-time deformation and strain measurement is now possible. In this chapter, four different approaches have been demonstrated for deformation and strain measurement. The digital correlation method appears to be a simple and straightforward method for deformation and strain measurement. However, the strain sensitivity of ~100 με leaves a bit to be desired. Despite this, there are various applications for which this method would be particularly suited. The polariscope system requires transparent birefringent material. This limits its application but with increasing use of polymers, plastics, and glasses, measuring low levels of birefringence becomes particularly relevant and the low birefringence polariscope has significant applications. With the use of an infrared light source, it would also be possible to extend this to measure residual stress in silicon wafers, which would be of great interest to the semiconductor and MEMS industry. The ODSS has tremendous potential as an alternative to the ubiquitous strain gauge. Its specifications match or surpass that of the traditional strain gauge. With its ability to monitor multiple points at the same time, this has particular interest in whole-field strain analysis without the need for numerical differentiation. Finally, digital holography has shown great promise as a novel tool of nanoscale deformation measurement of micron-sized objects. It has specific applications in the fields of static and dynamic MEMS characterization. However, unlike the other methods, digital holography provides out-of-plane displacements and not directly related to in-plane deformation and stress measurement.

REFERENCES

1. Asundi A. Introduction to engineering mechanics, in P. Rastogi (ed.), *Photomechanics*, *Topics in Applied Physics*, Chapter 2, Vol. 77, Springer Verlag, New York, 2000.
2. Pan B, Xie HM, Guo ZQ, and Hua T. Full-field strain measurement using a two-dimensional Savitzky-Golay digital differentiator in digital image correlation, *Optical Engineering*, 46(3), 033601-1–033601-10, 2007.
3. Luo PF, Chao YJ, and Sutton MA, Accurate measurement of three-dimensional displacement in deformable bodies using computer vision, *Experimental Mechanics*, 33(2), 123–132, 1993.
4. Asundi A. *MATLAB® for Photomechanics—A Primer*, Elsevier Science Ltd., Oxford, U.K., 2002.
5. Zhao B, Xie H, and Asundi A. Optical strain sensor using median density grating foil: Rivaling in electric strain gauge, *Review of Scientific Instrument*, 72(2), 1554–1558, 2001.

6. Asundi A. Moire interferometric strain sensor, Patent No. TD/006/05, 2005.
7. Asundi A and Singh VR. Circle of holography—Digital in-line holography for imaging, microscopy and measurement, *Journal of Holography and Speckle*, 3, 106–111, 2006.
8. Xu L, Peng XY, Miao J, and Asundi A. Hybrid holographic microscope for interferometric measurement of microstructures, *Optical Engineering*, 40(11), 2533–2539, 2001.
9. Asundi A and Singh VR, Lensless digital holographic microscope, Nanyang Technological University, Singapore, Technology Disclosure Number, TD/016/07, 2007.

24 Optical Methods in Flow Measurement

Sang Joon Lee

CONTENTS

24.1 Introduction 605
24.2 Laser Doppler Velocimetry 606
    24.2.1 Basic Principle..606
    24.2.2 Optical Arrangement and SNR 608
    24.2.3 LDV System..609
24.3 Particle Image Velocimetry 611
    24.3.1 Principle of PIV Measurement 611
    24.3.2 Classification of PIV Techniques.. 613
    24.3.3 PIV System 614
    24.3.4 Typical Application: Measurement of a Turbulent Boundary Layer 616
24.4 Stereoscopic PIV 616
    24.4.1 Principle.. 616
    24.4.2 Typical Application: Flow behind an Axial-Fan 618
24.5 Holographic PIV 619
    24.5.1 Film-Based Holographic PIV 619
    24.5.2 Digital Holographic PTV 621
24.6 Micro-PIV... 623
References.. 625

24.1 INTRODUCTION

Fluid is the common name given to gas and liquid. Flow cannot be at rest when there is shearing stress. In addition, most flows occurring in nature and engineering applications are turbulent, in which flow velocity varies rapidly with randomness and irregularity in space and time, difficult to predict accurately. To understand the turbulent flow motion in detail, we need to measure the quantitative velocity profile or distribution of the flow.

There are various kinds of flow-velocity measurement techniques. The simplest means is to use a measuring probe such as a pitot tube or a hot-wire probe. The velocity probe is positioned at one point or at multiple points in a flow field to be measured. It usually disturbs the flow and changes the flow downstream. In addition, it requires elaborated effort to obtain velocity components separately and to measure flow reversal. This point-wise measurement technique is difficult to apply to unsteady or transient flows [1].

On the other hand, optical methods do not disturb the flow and measure velocity components precisely. Light or sound waves undergo a Doppler shift when they are scattered off from particles in a moving fluid. The Doppler frequency shift is proportional to the speed of the scattering particles. By measuring the frequency difference between the scattered and unscattered waves, the flow velocity can be measured. Laser doppler velocimetry (LDV) has been used as a reliable and representative velocity-measuring device. Flow velocity is measured precisely by detecting the variation of Doppler frequency of light scattered by tracer particles seeded in the flow at a small measurement volume formed by laser-beam crossover. LDV does not disturb the flow and no velocity calibration is required. In addition, the velocity components of a flow, even with flow reversal, can be measured separately. However, the expensive LDV system requires precise optical arrangement and seeding of tracer particles [2].

Recently, quantitative optical flow visualization has become an indispensable tool in the investigation of complex flow structures. Recent advances in laser, computer, and digital image-processing techniques have made it possible to extract quantitative flow information from visualized flow images of tracer particles [3]. Particle image velocimetry (PIV) is one of the most important achievements of flow diagnostic technologies in the modern history of fluid mechanics. The PIV technique measures the whole velocity field in a plane by dividing the displacements ΔX and ΔY of tracer particles with the time interval Δt during which the particles have been displaced. Since the flow velocity is inferred from the particle velocity, it is important to select tracing particles that will follow the flow motion within an acceptable uncertainty without affecting the fluid properties that are to be measured.

The PIV velocity-field measurement method is a very powerful tool to measure whole flow of field information for various flows and to understand the flow structure, compared with point-wise velocity measurement techniques such as hot wire or LDV. Furthermore, other physical properties such as vorticity, deformation, and forces can also be derived easily from the PIV data. Recently, PIV and particle tracking velocimetry (PTV) have been used as reliable and powerful two-dimensional velocity-field measurement techniques.

In this chapter, the basic principle and typical applications of LDV and PIV methods are explained. As a three-dimensional (3D) velocity-field measurement technique, the stereoscopic PIV (SPIV) and holographic PIV (HPIV) techniques are briefly introduced in this chapter. In addition, a micro-PIV method and its simple application are also discussed.

24.2 LASER DOPPLER VELOCIMETRY

LDV is an optical method used to measure a local velocity of a moving fluid. The LDV utilizes the Doppler effect by detecting the Doppler frequency shift of laser light that has been scattered by tracer particles moving with the fluid. As one of the major advantages of the LDV system, the flow being measured is not affected by a velocity probe. Since the velocity information is derived based on a simple geometric correlation, calibration process is unnecessary. Furthermore, the direction of the flow velocity being measured is determined directly by the optical arrangement. Another advantage is its high-spatial and temporal-resolution qualities. However, the LDV system has drawbacks such as highly complicated optical setups and bothersome signal processing. The LDV signal obtained from a photodetector contains a considerable noise.

24.2.1 BASIC PRINCIPLE

In its basic form, the LDV system consists of a laser, focusing and receiving optics, a photodetector, a signal processor, and data-processing computer as shown in Figure 24.1. At first, a laser provides a collimated and coherent light. A beam splitter divides the laser light into two coherent beams. Two laser beams are crossed at a point where the flow velocity is to be measured. The fluid is seeded with fine particles of typically 0.1 to ~10 μm in diameter. Tracer particles are seeded in the fluid scatter light in all directions while the laser beam goes through the beam crossover. This scattered light can be collected from any direction by a photodetector (or photomultiplier). The frequency of the scattered light collected by a photodetector is Doppler-shifted and the shifted frequency is referred to as the Doppler frequency of the flow. The electrical signal converted by a photodetector goes through

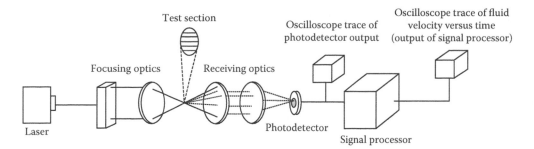

FIGURE 24.1 Basic LDV system configuration.

a signal processor, then the signal processor converts the Doppler frequency into a voltage signal. At last, data processor converts the output of the signal processor into readable flow information [4].

The basic principle of LDV measurement is the Doppler frequency shift of the scattered light from tracer particles seeded in the flow. As a tracer particle moves with the fluid flow, its relative motion to the laser beam determines the degree of the frequency shift, compared with the frequency of the incident laser light. In practice, however, the real value of the frequency shift is very small, compared to the frequency of the incident laser beam. Therefore, a process known as optical heterodyne is commonly used to determine the amount of frequency shift accurately.

The scattered light from a tracer particle that undergoes the optical heterodyne has a form of beat called Doppler burst as shown in the upper left side of Figure 24.2. The measuring volume has an ellipsoidal shape. From the Doppler burst and the fringe pattern, the velocity of a tracer particle, that is, the local velocity of the flow, is determined by the following equation:

$$u_x = \frac{\lambda}{2 \sin k} f_{\mathrm{D}} \tag{24.1}$$

The fringe spacing (d_f) depends on the wavelength of incident laser light (λ) and half angle (κ). The wavelength of lasers is well known accurately (0.01% or better). Therefore, the Doppler frequency (f_{D}) is proportional to the particle velocity.

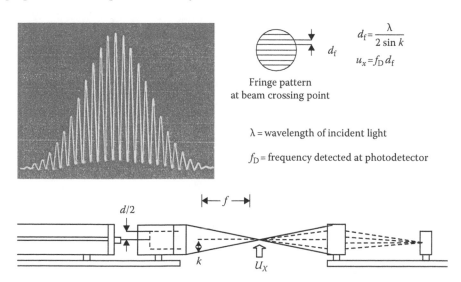

FIGURE 24.2 Doppler burst signal and fringe pattern at the laser beam crossing point.

24.2.2 OPTICAL ARRANGEMENT AND SNR

Depending on the location of receiving optics and photodetector, collecting the scattering light, the LDV system is classified into the forward-scatter mode and the backscatter mode as shown in Figure 24.3. The collection optics with the dual-beam LDV system can be placed at any angle and gives the same frequency. However, the signal quality and intensity of the scattered light is affected greatly by an angle θ between the receiving optics and the incident laser beam. While on-axis backscatter and forward-scatter modes are employed most often, the off-axis arrangement can reduce the effects of flare and reflections and decrease the length of the measuring volume.

As the tracer particles pass through the fringe pattern, they scatter light that appears as a blinking signal. The frequency of this on-and-off signal is proportional to the velocity component perpendicular to the planes of the fringes. As shown in Figure 24.4, the scattering light depends on the location of receiving optics for collecting scattering light with respect to the incident light. The light scattered in the forward direction ($\theta = 0°$) has the largest intensity value, about two orders of magnitude larger than the backscatter mode ($\theta = 180°$). When small tracer particles (about 1 μm) are used, the forward-scatter mode is useful, because the light intensity is much stronger than the backscatter mode. However, the backscatter mode has to be employed in the experimental facilities in which back side has space restrictions to locate the receiving optics or a transparent window. The LDV system of backscatter mode is easy to carry out experiments and its optical alignment is not so difficult. However, since the scattering light intensity is weak, the selection of tracer particles and the signal-processing procedure should be done carefully.

The signal-to-noise ratio (SNR) of the Doppler signal obtained from a photodetector should be increased to acquire accurate information of the flow, since the overall performance of the LDV system is largely affected by the SNR of the LDV signal. The SNR of Doppler signal is related with optical factors such as laser power, optical arrangement, performance of photodetector, and seeding particles as a source of scattering light. Therefore, we need to accurately select proper values of these governing factors accurately. Using the Mie-scattering theory, the SNR of LDV signal can be expressed as follow:

$$SNR = \frac{\pi^2}{256} \frac{\eta_q P_0}{h v_0} \frac{1}{\Delta F} \left(\frac{D_a}{r_a} \frac{ED_{in}}{L_f} \frac{D_P}{\lambda} \right)^2 \overline{GV}^2 \tag{24.2}$$

$$= \underbrace{\left(4 \times 10^{11} \eta_q P_0 \right)}_{[A]} \underbrace{\left(\frac{D_a}{r_a} \frac{ED_{in}}{L_f} \right)}_{[B]} \underbrace{\left(D_P{}^2 \overline{GV}^2 \right)}_{[C]} \underbrace{\left(\frac{1}{\Delta F} \right)}_{[D]} \tag{24.3}$$

(a)   (b)   On-axis
                                                                                        Off-axis

FIGURE 24.3 Optical setups for (a) back scatter mode and (b) forward scatter mode.

FIGURE 24.4 Intensity of the scattered light according to the angle between an incident light and a photodetector.

where

D_a is the collection aperture diameter of receiving lens (mm)

D_p is the particle diameter (μm)

$\overline{G}$ is the average gain of scattering light

h is the Plank's constant (6.6×10^{-34} J s)

P_0 is the power of each laser beam in dual beam LDV (W)

r_a is the distance between the measuring volume and receiving lens (mm)

$\overline{V}$ is the Doppler signal visibility

ΔF is the bandwidth of photomultiplier (MHz)

η_q is the quantum efficiency of photomultiplier

υ_0 is the frequency of the laser light

In the above equation, the terms [A], [B], and [C] represent the effects of laser source and photomultiplier, the effects of the optical arrangement of the LDV system, and the physical properties of seeding particles, respectively. In [A], the SNR is proportional to the power of laser beam, because commercial photodetectors have a nearly fixed value of efficiency. In [B], SNR is increased with the increase of the diameter of focusing lens and with the decrease of the measuring volume.

Since the LDV measures only the velocity of the tracer particles instead of the flow itself, the properties of the tracer particle should be selected carefully to secure the measurement accuracy. The most important property is the particle size. As the size of the tracer particle increases, the intensity of the scattered light increases, that is, increasing SNR of the Doppler signal. As the size of the tracer particle increases, on the other hand, the fidelity of the tracer particles to follow the flow accurately is diminished. In addition, pedestal noise is increased and frequency response of particles to the flow is decreased.

When a particle whose diameter is larger than the fringe spacing passes, the measuring volume or incomplete heterodyne mixing on the photodetector occurs due to phase difference of scattering light. That is, the smaller particle is required for good traceability to the highly turbulent flow; however, the bigger particle is needed to produce strong light scattering. It is difficult to satisfy the above two conditions together.

The particle concentration will influence the data rate, that is, the ability to follow the time history of the flow. The signal characteristics depend on the average number of particles in the measuring volume. Clean Doppler signal can be obtained, when the number of seeding tracer particles is maintained as small as possible, but, enough to operate the signal processor at least. Dropout phenomenon occurs when the number of particles in the measuring volume is too little. On the other hand, if excessive particles, more than two particles in the measuring volume, are supplied, the mutual interaction between particles influences the measured velocity data. Therefore, the size and number of particles should be properly selected according to the given experimental condition. The ideal number of particles is one particle in the measuring volume. For this ideal particle concentration, the output signal of photodetector shows a normal Gaussian distribution.

In addition, the other requisites of scattering particles include spherical shape, good light scattering surface, cheap and easy to generate, noncohesive, nontoxic, nonerosive, nonadhesive on the measuring window, with monodispersity, and are chemically inactive.

The tracer particle can be generated by atomization, fluidization, chemical reaction, combustion, and sublimation. For air seeding techniques, the condensation of liquid such as water, and atomization of low vapor-pressure materials (e.g., olive oil) by supplying compressed air are most popular.

24.2.3 LDV System

The signal processor of the LDV system is required to have a high frequency range and high temporal resolution. The signal processor also needs to have an ability to extract effective Doppler signal from LDV signal embedded in noises.

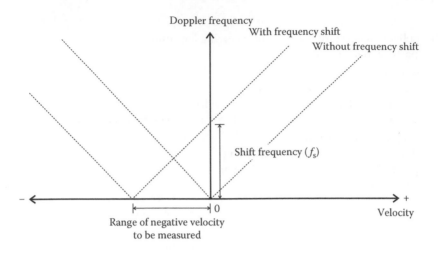

FIGURE 24.5 Frequency shift for measuring negative velocity of flow.

The signal processors are classified into three types, depending on how to extract the Doppler frequency from the LDV signal. The tracker-type processor tracks the instantaneous frequency signal to calculate the Doppler frequency. On the other hand, the counter-type processor measures the frequency of LDV signals by counting accurately the time duration of an integer number of cycles. Lastly, the spectrum analyzer calculates the power spectrum of LDV signal and determines the peak frequency as a Doppler frequency.

LDV system has the ability of measuring negative velocity of flow reversal. This is possible by shifting the frequency of one of the two incident laser beams and translating fringes in the measuring volume as shown in Figure 24.5. Frequency shifting is a very useful tool in LDV measurements for improving the dynamic range and allowing the measurement of flow reversal. In principle, it can be utilized with any scattering mode. In the dual-beam LDV system, a tracer particle in the measuring volume gives the frequency shift (f_s). If the velocity increases in the direction of fringe movement, the Doppler frequency will decrease. If it is opposite to the fringe movement, the frequency will increase.

In practical applications of LDV systems, a fiber-optic LDV system as shown in Figure 24.6 has several advantages. For conventional LDV systems, the laser source is so heavy and bulky that it

FIGURE 24.6 Configuration of a fiber-optic LDV system. (Courtesy of TSI Incorporated, Shoreview, MN.)

is difficult to traverse the whole LDV system to the measurement point. The use of optical fibers connecting the laser source with the optical head of the LDV system makes it easy to locate the optical head to harsh environments such as hot or corrosive fluids. In addition, the LDV probe head is submersible and the electronic components of the LDV system can be separated from the measuring fluid-flows for electrical-noise immunity [5].

24.3   PARTICLE IMAGE VELOCIMETRY

Most flow phenomena encountered in nature or industrial applications are unsteady and 3D turbulent flows. In order to analyze such turbulent flow phenomena, whole velocity-field information is required as a function of time. However, none of the currently available point-wise experimental techniques can provide such flow information. Recent advances in numerical simulations make us understand the sophisticated aspects of flow dynamics [6]. However, it is usually limited to simple geometries and relatively low Reynolds numbers yet. The recent advances in the digital image-processing method, computer science, and optics have opened a new paradigm for measuring whole velocity field of a flow by using a digital flow-assessment technique, such as PIV [3,7]. Nowadays, 2D PIV and PTV methods are widely used as a reliable velocity-field measurement technique. In addition, by statistical analysis of instantaneous 2D velocity fields, it is possible to extract the spatial distributions of turbulent quantities of turbulent flows having complicated flow structures.

24.3.1   PRINCIPLE OF PIV MEASUREMENT

Figure 24.7 shows the basic principle of a PIV velocity-field measurement technique that extracts whole velocity-field information by processing the flow images of tracer particles seeded in the flow. The PIV system consists of a laser, optics, tracer particles, and an image-recording device. To measure the whole velocity field in a measurement plane accurately, at first, tiny tracer particles that have good traceability should be seeded in the flow. Thereafter, the test section is illuminated by a thin laser light sheet and the illuminated light is scattered by tracer particles in the flow. The laser light sheet is formed by passing the laser beam through optical components such as cylindrical or spherical lens.

A particle image illuminated by the thin laser light sheet is captured by an image-recording device such as a digital charge-coupled device (CCD) camera at a time $t = t_0$. Second particle image is obtained at time $t = t_0 + \Delta t$ after a small time interval Δt. The time interval should be adjusted according to the velocity of the measuring flow. From the two consecutive flow images

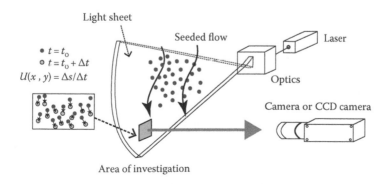

FIGURE 24.7   Basic principle of PIV velocity-field measurement technique.

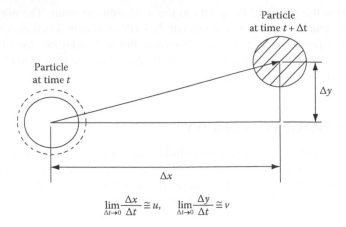

$$\lim_{\Delta t \to 0} \frac{\Delta x}{\Delta t} \cong u, \quad \lim_{\Delta t \to 0} \frac{\Delta y}{\Delta t} \cong v$$

FIGURE 24.8   Calculation of velocity vector of a moving particle.

stored in the computer memory as 2D image data, the displacement vector (ΔS) of each tracer particle can be extracted by using a digital image-processing technique. Here, the displacements of individual particles are presented as a function of space and time, $\Delta S(x, y; t)$. The velocity vector $U(x, y)$ can be obtained by dividing the displacement vector ΔS with the time interval Δt [8].

Figure 24.8 shows how to obtain the velocity vector of tracer particles moved during the time interval Δt of the PIV velocity-field measurement. The stream-wise (x) velocity component u and vertical (y) velocity component v are obtained by dividing the displacement components (Δx, Δy) of the particle, which moved with the flow during the time interval Δt, with the time interval Δt as follows:

$$\lim_{\Delta t \to 0} \frac{\Delta x}{\Delta t} \cong u$$

$$\lim_{\Delta t \to 0} \frac{\Delta y}{\Delta t} \cong v$$

(24.4)

Here, the displacement Δx should be small enough so that $\Delta x/\Delta t$ is a good approximation of the stream-wise velocity component u. This means that the trajectory of the tracer particle should be almost a straight line during the time interval Δt, and the velocity of the tracer particle should be nearly constant along the trajectory. It has been known that these conditions can be satisfied if the time interval Δt is small, compared to the Taylor microscale in the Lagrangian velocity field. However, if the time interval Δt is too small, the particle displacement ΔS also becomes small and the measurement uncertainty of ΔS increases [9].

The velocity field of a flow can be measured by three steps: capturing the particle images, extraction of velocity vectors, and representation of the extracted velocity field. For the particle image-capturing procedure, the proper tracer particles must be selected and the measurement plane is illuminated with a thin laser light sheet. Thereafter, the images of scattered particles are captured by a CCD or complementary metal-oxide semiconductor (CMOS) camera that is installed perpendicular to the laser light sheet. The size and concentration of tracer particles, the exposure time of camera, and the time interval Δt of two consecutive flow images should be set accurately according to the experimental condition and flow information to be measured.

There are several PIV/PTV techniques for measuring velocity vectors from the captured particle images according to the method of determining particle displacements. In order to apply a PIV technique to measure accurately the velocity fields of turbulent flows that change suddenly in space and time, an exact PIV/PTV algorithm should be developed to extract velocity vectors accurately and the effort to reduce the total calculation time is likewise necessary. In addition, error vectors contained in the velocity field measured should be removed in the procedure of extracting velocity vectors.

The post-processing procedure consists of the calculation of vorticity or rate of strain field from the instantaneous velocity field and ensemble averaging of many instantaneous velocity filed samples to obtain various flow parameters such as the turbulence intensity, Reynolds shear stress, and turbulent kinetic energy. In addition, the dissipation rate or spatial distribution of pressure can also be calculated by utilizing the governing equations of the fluid flows.

24.3.2 CLASSIFICATION OF PIV TECHNIQUES

The PIV techniques are classified into particle streak velocimetry (PSV), PIV, PTV, and LSV (laser speckle velocimetry), depending on how to obtain the displacement vector of tracer particles during the time interval Δt.

The PSV method extracts the velocity vectors by dividing the streak lines of tracer particles captured with prolonged exposure of camera. This technique has a straightforward simple mechanism and was developed in the early stage of the velocity-field measurement technique development. In general, there exists a problem of directional ambiguity. It can be solved by utilizing the coding, coloring, or image-shifting method [10]. If the number of seeded particles is too much or the flow has complicated flow structure, the PSV method cannot be applied.

If the particle density of flow image is too high to identify the individual particles, the LSV method is used to extract the velocity vector by image processing the speckle interference pattern generated from the double exposure of laser light sheet. This LSV technique was also developed in the early stage and regarded as an optical solution of the correlation function. The direction of fringes is perpendicular to the flow direction and the spacing in the fringe pattern is inversely proportional to the displacement of tracer particles. However, due to the directional ambiguity problem and complicated procedures for obtaining the velocity fields, this method is not used widely nowadays.

The particle image obtained by a CCD camera can be divided into two groups as shown in Figure 24.9. For the low-particle-density flow image (Figure 24.9a), individual particles are discernable and the displacement of each particle during the time interval Δt is obtained. For this kind of flow images, the PTV method that extracts the displacement by tracing individual particles in the

(a)   (b)

FIGURE 24.9 Two modes of particle image density: (a) low PTV and (b) high PIV.

FIGURE 24.10   Schematic diagram of conventional multi-frame PTV tracking algorithm.

flow image is usually used. The merits of PTV method are the feasibility of extension of the PTV algorithm to 3D velocity-field measurements, relatively inexpensive experimental facility, and high spatial resolution, compared to conventional digital PIV methods. However, it requires clean and low-density particle image to distinguish individual tracer particles [11,12]. The schematic of conventional multi-frame PTV tracking algorithm is shown in Figure 24.10.

When the flow image is of high particle density (Figure 24.9b), the individual particles can be distinguished in each image; however, the displacement pairs of particles during the time interval Δt cannot be identified from the consecutive flow image pairs. In this kind of flow images, each flow image is subdivided into many small interrogation windows and the correlation function of the particle image pair for each interrogation window is numerically calculated. The PIV method extracts the representative velocity vector of each interrogation window from the correlation peak. The cross-correlation PIV algorithm has been widely used for the search of correlation peak due to the merits of nondirectional ambiguity and effectiveness for noise contained flow images. The schematic diagram of the cross-correlation PIV algorithm is represented in Figure 24.11. Nowadays, the PTV and PIV methods are commonly used as a reliable 2D velocity-field measurement technique.

24.3.3 PIV System

A PIV velocity-field measurement system generally consists of a laser light sheet, an image recording camera, tracer particles, and a computer. Figure 24.12 shows the experimental setup of a PIV system applied on a wind tunnel experiment. The time interval Δt should be chosen a priori in the consideration of free-stream velocity and the magnification of the flow image.

The tracer particles seeded in the flow are assumed to move with the fluid without any velocity lag. This implies that the particles must be small enough to minimize velocity lag due to large particle size and its specific weight should be as close as possible to that of the fluid being measured.

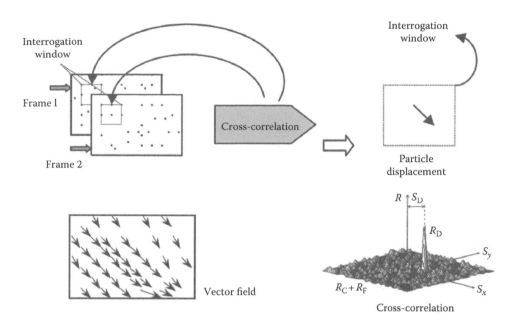

FIGURE 24.11   Schematic of cross-correlation PIV method.

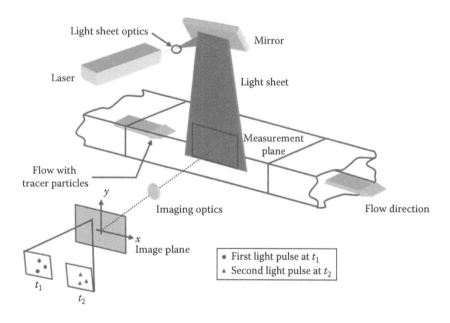

FIGURE 24.12   Schematic for PIV measurement in a wind tunnel.

For liquid flow, fluorescent, polystyrene, silver-coated particles, or other reflective particles are usually used to seed the flow, while olive oil or alcohol droplets are generally used for wind tunnels tests. Thereafter, tracer particles in the measurement plane of the flow are illuminated by a thin laser light sheet twice with a short time interval Δt. A dual-head Nd:YAG pulse laser is widely used as a laser source. The images of scattered particles in the thin laser light sheet are then captured with a digital CCD or CMOS camera. When the flow images are recorded on a negative film, the chemical development and tedious digitizing procedure with a scanner are required. Finally, the velocity fields of the flow are obtained by digital image processing of the captured flow images.

24.3.4 Typical Application: Measurement of a Turbulent Boundary Layer

The PIV velocity-field measurement technique was applied to a turbulent boundary layer formed in a closed return-type wind tunnel. Using the PIV system, 2D in-plane velocity fields (u, v) were measured in the stream-wise-wall-normal (x, y) plane. Olive-oil droplets of 1 µm in mean diameter, generated by a Laskin nozzle, were seeded into the wind tunnel test section as tracer particles. The field of view was illuminated with a thin laser light sheet of approximately 600 µm thickness. The scattered light from tracer particles was captured by a CCD camera of resolution 2048 × 2048 pixels. A delay generator was used to synchronize the laser and camera. The cross-correlation PIV algorithm was applied to the consecutive flow image pairs to calculate instantaneous velocity fields. The size of the interrogation window was 32 × 32 pixels with 50% overlapping. Figure 24.13 shows a typical instantaneous velocity field, subtracted a constant advection velocity from the measured velocity field by using the Galilean decomposition method. The hairpin vortex structures in the turbulent boundary layer are clearly observed as locally swirling vortices in Figure 24.13.

Since most flows of interest are complex 3D turbulent flows, the 2D PIV/PTV techniques were extended to measure the spatial distributions of three velocity components in a plane or a 3D volume. Several SPIV and HPIV methods have been developed as 3D PIV techniques. In addition, micro-PIV technique has been employed to investigate microscale flows such as a flow in a microchannel. They are explained separately in the later parts of this chapter. New PIV/PTV algorithms are still developed worldwide to improve the measurement accuracy and dynamic range. The PIV experiments for 3D turbulent flows should be carried out with care and concern so as to prevent possible sources of error and misinterpretation.

The PIV velocity-field measurement techniques will contribute dominantly to thermo-fluid flow research fields. The PIV data can be compared directly with computational fluid dynamics (CFD) results or theoretical modeling of thermo-fluid flow phenomena.

24.4 STEREOSCOPIC PIV

24.4.1 Principle

Most fluid flows encountered in our lives are 3D turbulent flows. Therefore, simultaneous measurement of three orthogonal velocity components is essential for the analysis of 3D flows such as the flow around a fan or propeller. Conventional 2D PIV velocity-field measurement techniques using a single camera provide only in-plane velocity components. The information on the out-of-plane flow motion is embedded in the in-plane flow field data and the out-of-plane velocity component becomes a source of error in the 2D PIV measurements. This implies that most previous studies carried

FIGURE 24.13 Instantaneous velocity field of a turbulent boundary layer subtracted by a constant advection velocity.

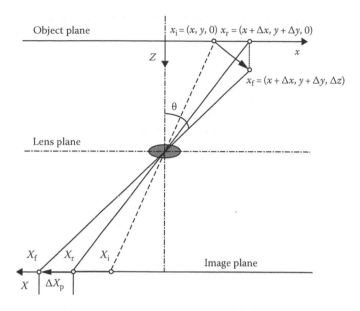

FIGURE 24.14 Perspective error caused by out-of-plane particle motion.

out with a 2D PIV system have perspective errors caused by the out-of-plane velocity component. Figure 24.14 illustrates the perspective error ($\Delta X_p = -\Delta z \times M \times \tan \theta$) caused by the out-of-plane displacement (Δz). In the case of constant magnification M, the perspective error is directly proportional to the viewing angle subtended by the particle position at the optical axis of the recording device. The SPIV method can eliminate the perspective error, yielding accurate 3D velocity information in the measurement plane [13].

The SPIV techniques usually employ two CCD cameras. Each camera simultaneously captures the same particle displacements at a different angle. The measuring volume is confined by the thickness of the laser light sheet and the imaging area. Two cameras at different angles capture two particle images synchronized with a pulsed laser light sheet. The particle displacements inside the flow images captured by each camera are transposed to 3D velocity-field data by intermediate procedures.

Two stereoscopic configurations have been used for SPIV measurements: the translation configuration and the angular displacement configuration. Different camera arrangements are shown in Figure 24.15. In the translation configuration method, the stereoscopic effects are directly related to

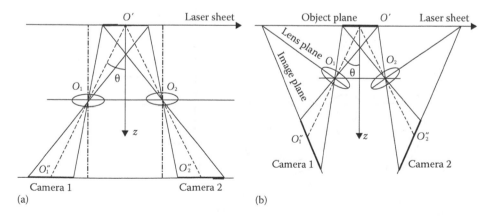

FIGURE 24.15 Different configurations of SPIV system: (a) translation method and (b) angular displacement method.

the distance between the optical axes of the cameras. The optical axis of the first camera is parallel to the optical axis of the second camera. These optical axes are aligned to be perpendicular to the illuminated plane.

However, the two optical axes for the angular configuration are neither parallel nor perpendicular to the measurement plane, because the image planes are tilted to focus the whole measuring volume. The tilting of image planes and camera lenses causes image distortion and varying magnification, requiring an elaborate calibration process to obtain accurate flow data. The translation configuration has advantages of convenient mapping and well-focused image over the whole observation area. Since the off-axis aberration restricts the viewing angle, the optical lens was carefully chosen for reducing the off-axis aberration [14].

24.4.2 TYPICAL APPLICATION: FLOW BEHIND AN AXIAL-FAN

The SPIV velocity-field measurement system consists of two CCD cameras with stereoscopic lens, an Nd:YAG pulse laser, and a delay generator. The stereoscopic lenses that were employed have special features for tilting and shifting the lenses without any additional adaptors or stages. They were specially designed to minimize optical aberrations and showed high optical performance throughout the experiments.

SPIV measurements were carried out in a transparent water tank (340 mm × 280 mm × 280 mm). Figure 24.16 shows schematic diagrams of the axial-fan and the field of view for axial plane measurements. The axial-fan tested in this application has five forward-swept blades. The tip-to-tip diameter and hub diameter of the fan are 50 and 14.3 mm, respectively, leading to a hub-to-tip ratio of 0.286. The fan is a 1/10 scale-down model of an outdoor cooling fan used for commercial air conditioners. The axial-fan was installed inside a rectangular basin.

The SPIV system was calibrated using a rectangular calibration target on which white dots were evenly distributed. A translation stage equipped with a micrometer was used to traverse the calibration target aligned with the laser light sheet along the z-direction. The calibration images were captured by two CCD cameras at different z-planes by translating the calibration target along the z-axis. The mapping function $F(\mathbf{x})$ was calculated using a centroid detection algorithm and least square method.

Figure 24.17 represents a typical 3D reconstructed instantaneous velocity field in the measurement plane. During experiments, the fan was rotating at 180 rpm (revolution per minute). The 500 instantaneous velocity fields were measured and they were ensemble averaged to obtain the phase-averaged mean flow structure. By combining the vorticity contours measured at four consecutive

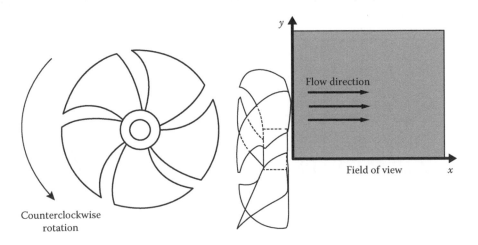

FIGURE 24.16 Schematic diagram of axial-fan and measurement planes.

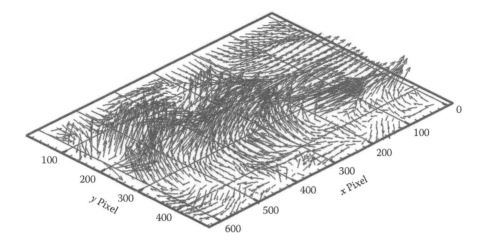

FIGURE 24.17 Typical instantaneous three-component velocity vectors measured by SPIV.

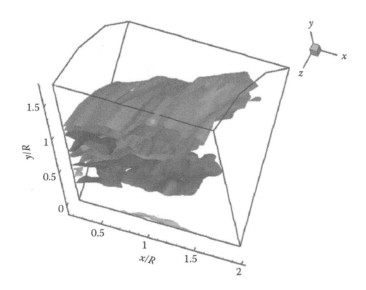

FIGURE 24.18 3D iso-vorticity structure of spanwise vorticity.

phases, isovorticity contours having the same spanwise vorticity (ω_z) were derived. The contour plot of the equi-spanwise vorticity surface is depicted in Figure 24.18. It shows the quasi 3D distribution of positive tip vortices shed from the blade tip, trailing vortices, and negative vortices induced from the pressure side of fan blade. The positive tip vortices and negative trailing vortices interact strongly in the region $0.6 < y/R < 0.8$.

24.5 HOLOGRAPHIC PIV

24.5.1 FILM-BASED HOLOGRAPHIC PIV

The HPIV method is a 3D velocity-field measurement technique for a whole measurement volume, different from the SPIV method for a measurement plane. However, its practical application to real flows is not easy to apply to real flows because of the complexity of the optical arrangement and experimental setup [15].

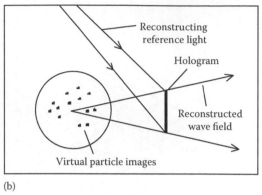

(a)   (b)

FIGURE 24.19 Off-axis HPIV measurement of tracer particles: (a) recording and (b) reconstruction of the virtual image.

The HPIV measurement has two steps: recording procedure and reconstruction procedure, as shown in Figure 24.19. Like a conventional PIV method, a pulse laser is required for HPIV system to illuminate a test volume. But, the HPIV system differs from conventional method in that it needs holographic film or holographic plate, instead of a CCD camera to record particle images seeded in the test volume. First of all, all particles in the measurement volume are illuminated by an expanded object wave made by passing through a spatial filter. The interference pattern is recorded in a holographic film or plate by the superposition of object wave and reference wave. When the reference and object waves have an oblique angle, it is called an off-axis holography. The coherent length of the laser source should be long enough to form clear interference patterns [16].

The interference pattern, hologram, is used for reconstructing the corresponding virtual particle image. The virtual image is reconstructed by illuminating the hologram with the original reference wave (Figure 24.19b). Since the reconstructed hologram includes the 3D original particle distribution at the initial recoding time, various velocity-field measurement techniques can be applied for a 2D slice selected in the test volume by adjusting the depth of the focusing point of the hologram. The resolution angle in the depth direction depends on the magnitude of the hologram.

Another advantage of the HPIV method is that the real image can be acquired without any lens system. For this real-image acquisition, the hologram should be illuminated by the complex conjugation of reference wave.

In this case, the hologram produces a backward traveling wave, which makes real image of planar laser light. The particle information at any planar section of the test volume can be taken by a CCD camera. The object wave may have optical aberrations due to low-quality windows or optical lens. This distortion would be compensated in the backward traveling path of the conjugated wave. Finally, the original 3D particle image is reconstructed by illuminating the original wave front.

For obtaining high-quality holograms of tracer particles, the wavelength of the laser used for reconstruction should be the same as that of the recording laser, and the geometrical arrangement for the reconstruction procedure should be organized like that of the recording procedure. For example, the wavelength of ruby laser differs from that of He–Ne laser about 10%. This difference can cause disturbance to the complete reconstruction. Therefore, the Nd:YAG pulse laser is much convenient to use due to the presence of a continuous wave (CW) Nd:YAG laser of the same wavelength [17].

Synchronized
Nd:YAG laser

Seeded flow

Objectwave
beam

Hologram
plate

Reference wave
beam

FIGURE 24.20 Schematic of typical holographic installation for PIV recording.

Figure 24.20 shows the schematic diagram of experimental setup for a HPIV experiment. In this figure, an off-axis type reference wave is used for distinguishing the virtual image of particle field from the real image transmitted indirectly. The hologram is brought to completion by double exposure to the same holographic film.

The information of 3D particle location can be obtained by scanning a CCD camera through the reconstructed hologram. Finally, 3D velocity-field information is given by applying the correlation PIV algorithm to the 3D particle information.

In addition to the off-axis HPIV method, many other HPIV methods such as in-line HPIV method, multiple light sheet holography, and hybrid HPIV have been introduced. In the in-line holography, the object and reference waves are overlapped. Compared to a SPIV system, the HPIV technique provides much more flow information, because it measures the 3D velocity components of tracer particles in the whole 3D fluid volume. The film-based HPIV technique has the directional ambiguity problem, when it adopts the single-frame double-exposure method. It also requires cumbersome 3D mechanical scanning and elaborate optical arrangement. Due to the disadvantages and long processing time of the film-based HPIV, the digital HPIV methods have been introduced recently.

24.5.2 DIGITAL HOLOGRAPHIC PTV

The digital holographic PTV (HPTV) technique is basically similar to digital HPIV method, except the procedure to acquire 3D velocity vectors. The basic configuration of digital HPTV system is very simple and similar to conventional 2D PIV systems. A holographic film is replaced by a digital image sensor, eliminating the chemical-processing routine. In other words, the digital holography consists of digital recording and numerical reconstruction procedures. In a film-based HPIV, the off-axis configuration is usually used. However, the digital HPIV does not need to adopt the off-axis configuration, because the directional ambiguity and twin-image problems can be resolved easily using the special intrinsic feature of the digital CCD camera.

Figure 24.21 shows a schematic diagram of the experimental setup for measuring a vertical turbulent jet flow using a digital HPTV technique. The in-line digital HPTV system consists of a CCD camera, a laser, a spatial filter, and a beam expander. He–Ne laser ($\lambda = 632.8$ nm) was used as a light source. An AOM (acoustic optical modulator) was used to chop the laser beam into a kind of laser

FIGURE 24.21   Experimental setup for HPTV measurement of a jet flow.

pulse. The hologram images were acquired with a digital CCD camera (PCO SensiCam), which has a spatial resolution of 1280 × 1024 pixels and 12-bit dynamic range. The physical size of each pixel is 6.7 μm. A delay generator was used to synchronize the AOM chopper and CCD camera. For each experimental condition, 100 instantaneous velocity-vector fields were obtained. This single-beam configuration with simple optical arrangement uses the forward scattering method, giving strong light scattering.

Figure 24.22 shows the spatial distributions of tracer particles reconstructed with different particle number density. The number of reconstructed particles (N_r) for the two comparative cases is 671 and 3471, respectively. As shown in Figure 24.22, most of the particles are uniformly distributed. Open circles indicate the particle position in the first frame captured at the first exposure and closed circles denote particle position in the second frame at the second exposure.

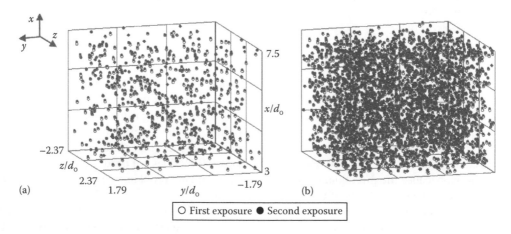

FIGURE 24.22   Distribution of tracer particles reconstructed at different particle concentration C_o: (a) $C_o = 2$ ($N_r \sim 671$) and (b) $C_o = 13$ ($N_r \sim 3471$).

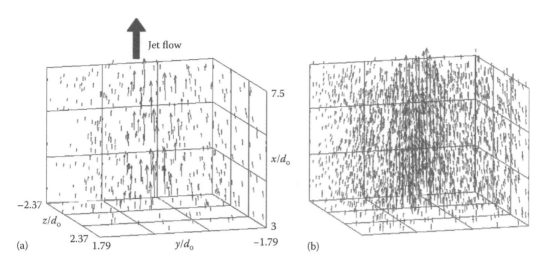

FIGURE 24.23 Instantaneous velocity vectors measured for two different particle concentrations: (a) $C_0 = 2$ ($N_c \sim 379$) and (b) $C_0 = 13$ ($N_c \sim 2152$).

Figure 24.23 shows instantaneous velocity-vector fields measured at the two different particle number densities of $C_0 = 2$ and 13 particles/mm³. In case of $C_0 = 13$, however, the number of recovered vectors (N_c) is 2152 and the corresponding spatial resolution is about 115 μm. The instantaneous velocity vectors do not show drastic variation in the low-speed flow.

24.6 MICRO-PIV

Recently, progress in the bio- and nanotechnology fields has led to heightened interest in the miniaturization and integration of chemical and biological analysis techniques. Work in this area could potentially revolutionize our life style and medical environment through the development of micro-fluidic devices. In fact, many micro-fluidic devices such as biochips, micro-pump, micro-mixer, and micro-heat-exchanger are commercially available already. New researches have been created widely in the field of in micro-fluidics and are receiving large attention nowadays. Such micro-fluidic devices, especially biochips, have various kinds of micro-channels inside. A detailed understanding on the flow inside the micro-channels is essential for the design of the micro-fluidic devices. A micro-PIV system has been employed usefully to measure microscale flow in micro-fluidic devices.

The micro-PIV system consists of an Nd:YAG laser, a CCD camera, a synchronizer, optics, and a computer. Since the field of view is very small, an objective lens of high magnification is usually attached in front of the CCD camera and tracer particles of submicron size are seeded in the flow. In micro-PIV measurements, fluorescent particles of submicron size are commonly used instead of conventional tracer particles because the flow images captured by the volume illumination method suffered from severe light scattering from the channel walls or adjacent peripheral structures. Scattered light can be removed by installing an optical filter in front of the image recorder.

Figure 24.24 shows a schematic diagram of the micro-PIV system used for a typical micro-channel experiment. A dual head Nd:YAG laser of wavelength 532 nm illuminates particles seeded inside the micro-channel. The seed particles are fluorescent polystyrene particles of diameter 530 nm that absorb light at $\lambda = 542$ nm and emit light at $\lambda = 612$ nm. The fluorescent particle images were passed through an optical filter that transmits wavelength longer than 570 nm and then were captured by a cooled CCD camera. As an image recording device, a cooled CCD camera is generally recommended because the light intensity emitted from fluorescent particles is relatively low.

FIGURE 24.24   Experimental setup of micro-PIV system.

This kind of diachronic mirror method has been commonly used for taking fluorescent images from a microscale field of view. In this micro-PIV measurement, a diachronic mirror or emission filter is attached in front of the microscope so that scattered reflection of laser light can be removed effectively and clear fluorescent particle images are usually obtained. The laser beam enters perpendicularly to the micro-channel to excite tracer particles.

As an alternative method, as shown in Figure 24.24, the laser beam directly illuminates the micro-channel from top of the channel by reflecting it with a mirror. Using this method, the image intensity is intensified and light scattering from adjacent structures are removed without any obstacles against the laser beam. Therefore, much clear images can be obtained, compared to the diachronic mirror method.

Figure 24.25 shows the mean stream-wise velocity fields at the entrance region (332 μm × 267 μm) of the six micro-channels having different inlet corner shape. These velocity-field distributions were measured by attaching a 40× objective lens with a numerical aperture of 0.75 in front of the camera. The interrogation window was 48 × 48 pixels in size and 50% overlapped, yielding a spatial resolution of 7.78 μm × 7.78 μm. The mean velocity-field data was obtained by ensemble averaging 80 instantaneous velocity fields. During the experiments, the flow rate supplied by the syringe pump was preset to 1085 μL/h; the corresponding Reynolds number based on the channel hydraulic diameter and mean velocity is about 1.

All PIV measurements were carried out in the mid-depth plane of the channel and the flow direction is from left to right. Figure 24.25 shows that, irrespective of the radius of curvature of the channel inlet corner, the stream-wise mean velocity distribution is almost symmetric with respect to the channel centerline. As the flow enters the channel inlet region, the stream-wise mean velocity is gradually increased. As the radius of curvature of the inlet corner decreases, the stream-wise velocity is noticeably increased in the entrance region and rapidly approaches to a fully developed state. With increasing the radius of curvature of the inlet corner, on the other hand, the flow entering the inlet region becomes smoother and the fully developed stream-wise velocity is increased due to reduction of flow resistance. For the case of $r > 0.8w$ (Figure 24.25e and f), however, the increase of the radius of curvature does not cause noticeable change in the magnitude of the fully developed stream-wise velocity.

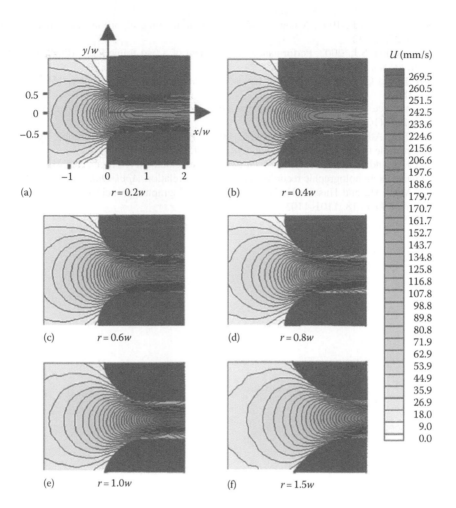

FIGURE 24.25 Stream-wise mean velocity fields of flow in the entrance region of micro-channels having different radius of curvature.

REFERENCES

1. Goldstein R.J., 1996, *Fluid Mechanics Measurements*, 2nd edn., Taylor & Francis, Washington, DC.
2. Guenther R., 1990, *Modern Optics*, John Wiley & Sons, New York.
3. Adrian R.J., 1991, Particle-imaging techniques for experimental fluid mechanics, *Ann. Rev. Fluid Mech.*, 23: 261–304.
4. Franz M. and Oliver F., 2001, *Optical Measurements: Techniques and Applications*, Springer-Verlag, New York.
5. Richard S.F. and Donald E.B., 2000, *Theory and Design for Mechanical Measurements*, Wiley, New York.
6. Choi H.C., Moin P., and Kim J., 1993, Direct numerical simulation of turbulent flow over riblets, *J. Fluid Mech.*, 255: 503–539.
7. Adrian R.J., 1986, Multi-point optical measurements of simultaneous vectors in unsteady flow—A review, *Int. J. Heat Fluid Flow*, 7: 127–145.
8. Becker R.J., 1999, Particle image velocimetry: A practical guide, M. Raffel et al. (eds.), *Appl. Mech. Rev.*, 52: B42.
9. Gharib M. and Willert C.E., 1990, Particle tracing revisited, in *Lecture Notes in Engineering: Advances in Fluid Mechanics Measurements 45*, M. Gadel-Hak (ed.), Springer-Verlag, New York, pp. 109–126.
10. Adrian R.J., 1986, Image shifting technique to resolve directional ambiguity in double-pulsed velocimetry, *Appl. Opt.*, 25: 3855–3858.

11. Baek S.J. and Lee S.J., 1996, A new two-frame particle tracking algorithm using match probability, *Exp. Fluids*, 22: 23–32.
12. Kim H.B. and Lee S.J., 2002, Performance improvement of a two-frame PTV using a hybrid adaptive scheme, *Meas. Sci. Technol.*, 13: 573–584.
13. Arroyo M.P. and Greated C.A., 1991, Stereoscopic particle image velocimetry, *Meas. Sci. Technol.*, 2: 1181–1186.
14. Yoon J.H. and Lee S.J., 2002, Direct comparison of 2-D and 3-D stereoscopic PIV measurement, *Meas. Sci. Technol.*, 13(10): 1631–1642.
15. Barnhart D.H., Adrian R.J., and Papen G.C., 1994, Phase-conjugate holographic system for high-resolution particle image velocimetry, *Appl. Opt.*, 33: 7159–7170.
16. Coupland J.M. and Halliwell N.A., 1992, Particle image velocimetry: Three-dimensional fluid velocity measurements using holographic recording and optical correlation, *Appl. Opt.*, 31: 1005–1007.
17. Zimin V., Meng H., and Hussain F., 1993, Innovative holographic particle velocimeter: A multibeam technique, *Opt. Lett.*, 18: 1101–1103.

25 Polarimetry

Baoliang (Bob) Wang

CONTENTS

25.1 Introduction to Basic Concepts 629
    25.1.1 Stokes Parameters, Stokes Vector, and Light Polarization 629
    25.1.2 Mueller Matrices and Optical Devices 629
    25.1.3 Retardation and Diattenuation 630
    25.1.4 Photoelastic Modulator 631
25.2 Stokes Polarimeter 632
    25.2.1 Early PEM-Based Stokes Polarimeters in Astronomy 632
    25.2.2 Imaging Stokes Polarimeters Using the PEM 633
    25.2.3 Laboratory Dual PEM Stokes Polarimeters 634
    25.2.4 Dual PEM Stokes Polarimeter Used in the Tokamak 636
    25.2.5 Commercial Dual PEM Stokes Polarimeters 638
        25.2.5.1 Visible Stokes Polarimeters 638
        25.2.5.2 Near-Infrared Fiber Stokes Polarimeters 640
        25.2.5.3 Spectroscopic Stokes Polarimeter 642
25.3 Mueller Matrix Polarimeter 645
    25.3.1 Dual PEM Mueller Matrix Polarimeter 645
    25.3.2 Four Modulator Mueller Polarimeter 647
25.4 Special Polarimeters 648
    25.4.1 Linear Birefringence Polarimeters and Their Applications in the Optical
        Lithography Industry 648
        25.4.1.1 Linear Birefringence Polarimeter 648
        25.4.1.2 Linear Birefringence in Photomasks 652
        25.4.1.3 Effect on Image Quality Due to Linear Birefringence in a Photomask 655
        25.4.1.4 Deep Ultraviolet Linear Birefringence Polarimeter 656
        25.4.1.5 Measuring Linear Birefringence at Different DUV Wavelengths 658
        25.4.1.6 DUV Polarimeter for Measuring Lenses 660
    25.4.2 Near-Normal Reflection Dual PEM Polarimeter and Applications in the
        Nuclear Fuel Industry 662
    25.4.3 Special Polarimeters for the FPD Industry 662
        25.4.3.1 Measuring Glass Substrate 662
        25.4.3.2 Measuring LCD-FPD Components at RGB Wavelengths 663
        25.4.3.3 Measuring Retardation Compensation Films at Normal
            and Oblique Angles 666
        25.4.3.4 Measuring Retardation at a Fast Speed 666

25.4.4 Special Polarimeters for Chemical, Biochemical, and Pharmaceutical
         Applications 668
         25.4.4.1 Circular Dichroism Spectrometer 668
         25.4.4.2 Optical Rotation Polarimeter 668
         25.4.4.3 Polarization Analyzer for Fluorescence and Scattering 669
25.4.5 Stress Birefringence Polarimeter for Si Used in the Solar
         Photovoltaic Industry 669
25.4.6 PEM–Camera Imaging Polarimeter 671
References 673

A dictionary definition of polarimetry is "the art or process of measuring the polarization of light" [1]. A more scientific definition is that polarimetry is the science of measuring the polarization state of a light beam and the diattenuating, retarding, and depolarizing properties of materials [2]. Similarly, a polarimeter is an optical instrument for determining the polarization state of a light beam, or the polarization-altering properties of a sample [2].

There are several ways to categorize polarimeters. By definition [2], polarimeters can be broadly categorized as either light-measuring polarimeters or sample-measuring polarimeters. A light-measuring polarimeter is also known as a Stokes polarimeter, which measures the polarization state of a light beam in terms of the Stokes parameters. A Stokes polarimeter is typically used to measure the polarization properties of a light source. A sample-measuring polarimeter is also known as a Mueller polarimeter, which measures the complete set or a subset of polarization-altering properties of a sample. A sample-measuring polarimeter is used to measure the properties of a sample in terms of diattenuation, retardation, and depolarization. There are functional overlaps between light-measuring and sample-measuring polarimeters. For example, one can direct a light beam with known polarization properties onto a sample and then use a Stokes polarimeter to study the polarization-altering properties of the sample.

Polarimeters can also be categorized by whether they measure the complete set of polarization properties. If a Stokes polarimeter measures all four Stokes parameters, it is called a complete Stokes polarimeter, or an incomplete or a special Stokes polarimeter. Similarly, there are complete and incomplete Mueller polarimeters. Nearly all sample-measuring polarimeters are incomplete or special polarimeters, particularly for industrial applications. These special polarimeters bear different names. For example, a circular dichroism (CD) spectrometer, which measures the differential absorption between left and right circularly polarized light ($\Delta A = A_{\mathrm{L}} - A_{\mathrm{R}}$), is a special polarimeter for measuring the circular diattenuation of a sample; a linear birefringence measurement system is a special polarimeter for measuring the linear retardation of a sample.

Another way to classify polarimeters is by how the primary light beam is divided. The primary light beam can be divided into several secondary beams by beam splitters and then each secondary beam can be sent to an analyzer–detector assembly (amplitude division) for the measurement of its polarization properties. The primary light beam can be partitioned by apertures in front of an analyzer–detector assembly (aperture division) so that different polarization properties can be measured from different parts of the primary beam. The primary light beam can also be modulated by polarization modulators, and different polarization properties are then determined from different frequencies or harmonics of the modulators.

Polarimeters have a great range of applications in both academic research and industrial metrology. Polarimeters are applied to chemistry, biology, physics, astronomy, material science, and many other scientific areas. Polarimeters are used as metrology tools in the semiconductor, fiber telecommunication, flat panel display (FPD), pharmaceutical, and many other industries. Different branches of polarimetry have established their own scientific communities, within which regular conferences are held [3–7]. Tens of thousands of articles have been published on polarimeters and their applications, including books and many review articles [2,8–16]. It is beyond the scope of this chapter to review the full range of polarimetry. Instead, I will introduce necessary general concepts for polarimetry and focus on polarization modulation polarimeters using the photoelastic modulator (PEM) [17,18].

25.1 INTRODUCTION TO BASIC CONCEPTS

25.1.1 STOKES PARAMETERS, STOKES VECTOR, AND LIGHT POLARIZATION

The polarization state of a light beam can be represented by four parameters that are called the Stokes parameters [19–22]. There are two commonly used notations for the Stokes parameters, $(I, Q, U,$ and $V)$ [20] and $(S_0, S_1, S_2,$ and $S_3)$ [8,9]. However, Shurcliff, in his polarization book [22], used the $(I, M, C,$ and $S)$ notation. I will use the $(I, Q, U,$ and $V)$ notation in this chapter.

The four Stokes parameters are defined as follows:

- $I \equiv$ total intensity
- $Q \equiv I_0 - I_{90} =$ difference in intensities between horizontal and vertical linearly polarized components
- $U \equiv I_{+45} - I_{-45} =$ difference in intensities between linearly polarized components oriented at $+45°$ and $-45°$
- $V \equiv I_{rcp} - I_{lcp} =$ difference in intensities between right and left circularly polarized components

The four Stokes parameters can be grouped in a column to form a Stokes vector. The Stokes vector can represent both completely polarized light and partially polarized light. If the light beam is completely polarized,

$$I = \sqrt{Q^2 + U^2 + V^2} \tag{25.1}$$

If the light beam is partially polarized,

$$I > \sqrt{Q^2 + U^2 + V^2} \tag{25.2}$$

The degree of polarization (DOP) of a light beam is

$$\text{DOP} = \frac{\sqrt{Q^2 + U^2 + V^2}}{I} \tag{25.3}$$

In addition to the Stokes vector, the Jones vector [23] and Poincare sphere [20,21] are other commonly used methods to represent light polarization. I will only use the Stokes vector in this chapter.

25.1.2 MUELLER MATRICES AND OPTICAL DEVICES

When the Stokes vector is used to represent the polarization state of a light beam, the Mueller matrix [24] is typically used to represent an optical device (such as a linear polarizer, a quarter-wave plate, an elliptical retarder, etc.) and to describe how it changes the polarization state of a light beam. A Mueller matrix is a 4×4 matrix that is defined as follows:

$$\begin{pmatrix} I' \\ Q' \\ U' \\ V' \end{pmatrix} = \begin{pmatrix} m_{11} & m_{12} & m_{13} & m_{14} \\ m_{21} & m_{22} & m_{23} & m_{24} \\ m_{31} & m_{32} & m_{33} & m_{34} \\ m_{41} & m_{42} & m_{43} & m_{44} \end{pmatrix} \begin{pmatrix} I \\ Q \\ U \\ V \end{pmatrix}$$

(25.4)

where

I, Q, U, and V are the Stokes parameters of the light beam entering the optical device
I', Q', U', and V' are the Stokes parameters of the light beam exiting the optical device

The derivations and tabulated Mueller matrices for optical devices commonly used in polarization optics are available [2,8–10,20–22,25]. For example, the Mueller matrix of a general linear retarder with the linear retardation of δ and the angle of fast axis of ρ is

$$\begin{bmatrix} 1 & 0 & 0 & 0 \\ 0 & \cos(4\rho)\sin^2\left(\frac{\delta}{2}\right)+\cos^2\left(\frac{\delta}{2}\right) & \sin(4\rho)\sin^2\left(\frac{\delta}{2}\right) & -\sin(2\rho)\sin\delta \\ 0 & \sin(4\rho)\sin^2\left(\frac{\delta}{2}\right) & -\left(\cos(4\rho)\sin^2\left(\frac{\delta}{2}\right)\right)+\cos^2\left(\frac{\delta}{2}\right) & \cos(2\rho)\sin\delta \\ 0 & \sin(2\rho)\sin\delta & -\cos(2\rho)\sin\delta & \cos\delta \end{bmatrix}$$

The Mueller matrix of a general linear retarder can be simplified to more specific cases. For instance, when $\delta = \pi/2$ and $\rho = 0$, the Mueller matrix becomes

$$\begin{bmatrix} 1 & 0 & 0 & 0 \\ 0 & 1 & 0 & 0 \\ 0 & 0 & 0 & 1 \\ 0 & 0 & -1 & 0 \end{bmatrix}$$

that represents a quarter-wave plate with its fast axis oriented at $0°$.

Notice that most available Mueller matrices are derived for ideal optical elements. For example, the aforementioned Mueller matrix represents an ideal linear retarder. This Mueller matrix contains no diattenuation, which is evident from the null values of the matrix elements in the first row and the first column except m_{11}.

25.1.3 RETARDATION AND DIATTENUATION

Polarization-altering properties of a sample include diattenuation, retardation, and depolarization. Depolarization, referring to changing polarized light to unpolarized light, is perhaps the least understood. My favorite definition of unpolarized light is that "light is unpolarized only if we are unable to find out whether the light is polarized or not [26]." One may say the definition of unpolarized light is "less scientific." Furthermore, all the applications described in this chapter have negligible depolarization. I will only introduce retardation and diattenuation.

Linear birefringence is the difference in the refractive indices for two orthogonal components of linearly polarized light. The linear birefringence of an optical element produces a relative phase shift between the two linear polarization components of a passing light beam. The net phase shift integrated along the path of the light beam through the optical element is called linear retardation or linear retardance. Linear retardation requires the specification of both magnitude and angle of fast axis. The fast axis corresponds to the linear polarization direction for which the speed of light is a maximum (or for which the refractive index is a minimum). Linear birefringence, linear retardation, and linear retardance are sometimes loosely used interchangeably.

Circular birefringence is the difference in the refractive indices for right and left circularly polarized light. The circular birefringence of a sample produces a relative phase shift between the right and left circularly polarized components of a passing light beam. The net phase shift integrated along the path of the light beam through the sample is called circular retardation or circular retardance. Since a linearly polarized light can be represented by a linear combination of right and left circularly polarized light, circular retardation will produce a rotation of the plane of linear polarization (optical rotation) as the light beam goes through the sample. Circular birefringence, circular retardation, circular retardance, and optical rotation are sometimes loosely used interchangeably.

Linear diattenuation is defined as $Ld = (T_{max} - T_{min})/(T_{max} + T_{min})$ where T_{max} and T_{min} are the maximum and minimum intensities of transmission, respectively, for linearly polarized light. The angle of the axis with maximum transmission (the bright axis) for linearly polarized light is represented by θ.

Circular diattenuation is defined as $Cd = (T_{RCP} - T_{LCP})/(T_{RCP} + T_{LCP})$, where T_{RCP} and T_{LCP} are the intensities of transmission for right and left circularly polarized light, respectively.

In summary, there are six polarization parameters that represent the retardation and diattenuation of a nondepolarizing sample. These six parameters are the magnitude of linear retardation, the fast axis of linear retardation, circular retardation, the magnitude of linear diattenuation, the angle of linear diattenuation, and circular diattenuation. These 6 polarization parameters comprise 6 of the 16 freedoms in a Mueller matrix. (The other 10 freedoms in a Mueller matrix include 1 for intensity and 9 for depolarization properties.)

25.1.4 Photoelastic Modulator

The PEM is a resonant polarization modulator operating on the basis of the photoelastic effect. The photoelastic effect refers to the linear birefringence in a transparent dielectric solid that is induced by the application of mechanical stress. This is a long-known effect that dates back to Sir David Brewster's work in 1816 [27].

The PEM was invented in the 1960s [17,28–30]. The most successful PEM design [17,18] employs a bar-shaped fused silica optical element and a piezoelectric transducer made of single crystal quartz. This type of PEM operates in the so-called sympathetic resonance mode where the resonant frequency of the transducer is precisely tuned to that of the optical element. In a bar-shaped PEM, the lengths of the optical element and the transducer are each half of the length of the standing ultrasound wave in resonance. Figure 25.1a illustrates, with great exaggeration, the mechanical vibrational motions of the optical element and the transducer of a bar-shaped PEM. Figure 25.1b shows the retardation and light polarization states during a full PEM modulation cycle for two special cases where the peak retardation is a half-wave ($\lambda/2$ or π) and a quarter-wave ($\lambda/4$ or $\pi/2$), respectively.

In 1975, Kemp invented the symmetric PEM, a design that provides a higher range of retardation modulation, a larger optical aperture, and more symmetric retardation distribution [31]. Figure 25.2 depicts, again with great exaggeration, the mechanical vibrational motions of an "octagonal" (more accurately, square with the four corners cut off) PEM optical element.

The PEM is made of isotropic optical materials, such as fused silica and polycrystal ZnSe, in contrast to the birefringent materials used in electro-optic modulators. It is this property of the PEM, along with its operation at resonance, that affords the PEM high modulation purity and efficiency, broad spectral range, high power handling capability, large acceptance angle, large useful aperture,

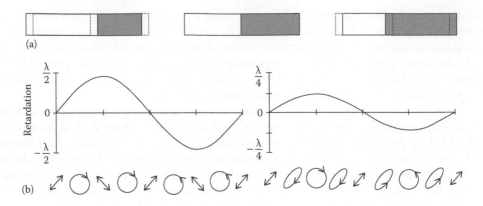

FIGURE 25.1 (a) Optical element and piezoelectric transducer with greatly exaggerated motions in modulation and (b) polarization modulation at peak retardation of half-wave and quarter-wave, respectively.

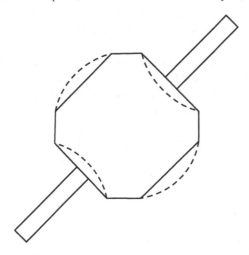

FIGURE 25.2 An "octagonal" PEM with greatly exaggerated motions of the optical element in modulation.

and excellent retardation stability [17,18,32]. The quality of the PEM has been further improved in a new model, PEM–ATC (advanced thermal control) by dynamically stabilizing its working temperature [18,90]. These superior optical properties of the PEM are particularly suited to high-sensitivity polarimetric measurements, which we will discuss in detail in the following sections.

25.2 STOKES POLARIMETER

25.2.1 EARLY PEM-BASED STOKES POLARIMETERS IN ASTRONOMY

Kemp, one of the inventors of the PEM, was an astronomical physicist. He employed the PEM in astronomical polarimeters immediately after its invention in order to improve the measurement sensitivity of polarization [33–36], which he certainly did. Using a PEM solar polarimeter, Kemp and coworkers made it possible to measure accurately the small magnetic circular polarization of light rays emanating from different astronomical objects in the early 1970s [33,34]. In subsequent work, they developed a modified PEM solar polarimeter that achieved an absolute polarization sensitivity of parts per 10 million [35], which may still be the record today for the highest sensitivity obtained by a polarimeter.

In their solar polarimeter, Kemp and coworkers used the simplest PEM arrangement to achieve their extremely high sensitivity in polarization measurements. The Stokes parameters,

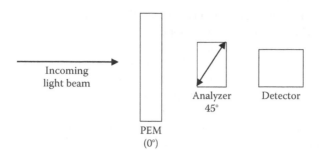

FIGURE 25.3 A single PEM setup for Stokes polarimeter.

I, Q, U, and V, were measured using a single PEM followed by a polaroid analyzer as shown in Figure 25.3. Using this setup, Q and U cannot be measured simultaneously. One of the two sets of Stokes parameters, (I, Q, and V) or (I, U, and V), is measured first, then the other is measured after rotating the PEM–analyzer assembly by 45°.

Kemp's success in achieving high sensitivity was the result of both using the PEM and meticulously carrying out the experiments. For example, Kemp and coworkers analyzed and eliminated the surface-dichroism effect of the PEM, etched edge faces of the PEM to remove strain, used average of measurements at several cardinal positions, and carefully considered the alignment of the PEM and the analyzer [33,35]. They claimed that no optical components of any kind should intervene between the observed source and the principal polarimeter element, not even the telescope. Researchers today still find value in their advice for obtaining high-sensitivity polarization measurements.

25.2.2 Imaging Stokes Polarimeters Using the PEM

Since Kemp's applications of the PEM in point-measurement astronomical Stokes polarimeters [33–36], researchers have investigated applying the PEM to imaging polarimeters. The main challenge to using the PEM with the charge-coupled device (CCD) is that the readout of the CCD is too slow for the PEM modulation (tens of kHz). Several methods have been proposed and tested for PEM–CCD applications [37–42]. I will briefly describe two examples for solving this problem.

The group at the Institute for Astronomy, ETH Zurich, Switzerland, has long been involved in developing imaging Stokes polarimeters using PEMs and other modulators [37–40]. The ZIMPOL (Zurich imaging polarimeter) I and II are two generations of state-of-the-art imaging polarimeters. This group, led by Prof. Stenflo, has used the ZIMPOL with different telescopes. Combined with large aperture telescopes, the ZIMPOL provides reliable polarization images with a sensitivity of better than 10^{-5} in DOP, and thus obtains a rich body of polarization information for non–solar systems, the solar system, and second solar spectrum [43,44].

The ETH Zurich group overcame the incompatibility between the fast PEM modulation and slow CCD readout in an innovative way. Their approach was to use a commercially available CCD that allows the charges to be shifted laterally at a very fast rate. They created a mask to expose only desired rows of pixels on the CCD in synchrony with the modulation frequency (or harmonics) of the PEM. The unexposed pixels thus acted as hidden buffer storage areas within the CCD sensor. The CCD readout was done after thousands of cycles of photocharge between different rows of pixels on the CCD. In ZIMPOL I, they had every second pixel row masked to record two Stokes parameters simultaneously. In ZIMPOL II, they had three rows masked and one row open in each group of four pixel rows, which allowed all four Stokes parameters to be recorded simultaneously.

It is worth noting that ZIMPOL can be used with different types of modulators, such as PEMs, Pockels cells, and ferroelectric liquid crystal modulators. To measure all four Stokes parameters simultaneously, two modulators are required to work in synchronization [37]. The resonant feature of the PEM, a key in providing superior optical properties, makes it difficult to synchronize two

PEMs in phase and with long-term stability. In fact, the long-term stability of two synchronized PEMs has not been achieved successfully despite significant effort spent in electronic design [45]. However, practically all published scientific results from ZIMPOL are based on the use of PEMs. Prof. Stenflo in a recent review paper [44] explained why his group chose the PEM over other modulators for their scientific observations. He attributed the success of the PEM to its superior optical properties in polarimetric applications as compared to other types of modulators.

In ZIMPOL II, the synchronized PEM approach was dropped and a single PEM was used. The polarization modulation module in ZIMPOL II is similar to the configuration shown in Figure 25.3. The images of I, Q, and V are recorded first; then the images of I, U, and V are recorded after rotating the polarization modulation module by 45°. The ETH Zurich group has reported a large collection of images of Stokes parameters that were recorded by the ZIMPOL using different telescopes for the sun and other astronomical objects.

Another method of using PEMs in imaging polarimeters has been developed by Diner (Jet Propulsion Laboratory, JPL) and coworkers. The goal of this group was to develop a multiangle spectropolarimetric imager (MSPI) for aerosol satellite remote sensing. High sensitivity in measuring degree of linear polarization (DOLP) is again one of the major challenges. After evaluating different types of polarization modulators, including rotating wave plate and ferroelectric liquid crystal modulators, this group selected the PEM for its application [41,42].

The approach selected by Diner and coworkers involves using two PEMs with a small difference in frequencies. When such a PEM pair is placed in the optical path, the detector signal contains both a high-frequency (PEM modulated) component and a low-frequency component (~10 Hz, this being the beat frequency of the two PEMs) in the waveform. The line readout integration time (1.25 ms) is long enough to average out the high-frequency signal (tens of kHz) and is fast enough to acquire plenty of samples in a beat frequency cycle [41,42]. The signal at the beat frequency is related to Stokes parameters Q and U. The Stokes parameter V is not measured in this application due to its insignificance (<0.1%) in aerosol remote sensing.

The polarization modulation module contains two quarter-wave plates, two PEMs at slightly different modulating frequencies, and an analyzer that can be set at either 0° or 45° [41,42]. In this optical configuration, the Stokes parameters Q and U have the following relationship to I_0 and I_{45}, which are the intensities measured when the analyzer is set at 0° and 45°, respectively:

$$I_0 \approx \frac{1}{2}\left[I + J_0\left[2\delta_0 \cos\left(\omega_b t - \eta\right)\right]Q\right]$$

$$I_{45} \approx \frac{1}{2}\left[I + J_0\left[2\delta_0 \cos\left(\omega_b t - \eta\right)\right]U\right]$$

(25.5)

where ω_b is the beat frequency of the two PEMs. Therefore, normalized Stokes parameters Q/I and U/I can be measured at the beat frequency. The first results of an imaging Stokes polarimeter using this approach are promising [42]. The program integrating this imaging polarimeter into the satellite remote sensor is still ongoing.

25.2.3 Laboratory Dual PEM Stokes Polarimeters

For a laboratory point-measurement Stokes polarimeter, while different instrumental configurations have been used [46], the most common dual PEM configuration is shown in Figure 25.4. The key polarization modulation components of the Stokes polarimeter are two PEMs oriented at 45° apart and an analyzer that bisects the axes of the two PEMs [47]. Using this dual PEM configuration, Stokes parameters Q, U, and V can each be measured at a specific harmonic of a PEM. Namely, Q and U can be measured at the second harmonics (2F) of the two PEMs and V can be measured at the first harmonic (1F) of the first PEM. An advantage of this approach is that no cross-terms are used in measuring the Stokes parameters.

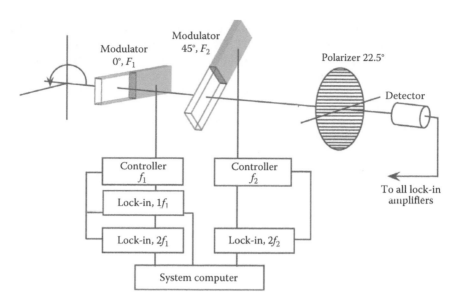

FIGURE 25.4 A typical dual PEM Stokes polarimeter.

The optical train of the dual PEM Stokes polarimeter shown in Figure 25.4 can be analyzed using Mueller matrix calculus. The time varying signal generated on the detector is

$$i(t) = I + \frac{\sqrt{2}}{2} \left\{ Q\cos(\delta 2) + U\left[\cos(\delta 1) + \sin(\delta 1)\cdot\sin(\delta 2)\right] + V\left[\sin(\delta 1) - \sin(\delta 2)\cdot\cos(\delta 1)\right] \right\} \quad (25.6)$$

where
   ω_1 and ω_2 are the modulation frequencies of the two PEMs
   $\delta 1_0$ and $\delta 2_0$ are the peak retardation amplitudes of the two PEMs
   $\delta 1$ and $\delta 2$ are the time-varying phase retardations of the two PEMs ($\delta 1 = \delta 1_0 \sin \omega_1 t; \delta 2 = \delta 2_0 \sin \omega_2 t$)

When sin $\delta 1$, cos $\delta 1$, sin $\delta 2$, and cos $\delta 2$ in Equation 25.6 are expanded with the Bessel functions of the first kind, we have

$$i(t) = I + \frac{\sqrt{2}}{2} \left\{ \begin{array}{l} Q\left[J_0\left(\delta 2_0\right) + 2J_2\left(\delta 2_0\right)\cdot\cos\left(2\omega_2 t\right)\right] \\ + U\left[J_0\left(\delta 1_0\right) + 2J_2\left(\delta 1_0\right)\cdot\cos\left(2\omega_1 t\right) + 2J_1\left(\delta 1_0\right)\cdot\sin\left(\omega_1 t\right)\cdot 2J_1\left(\delta 2_0\right)\cdot\sin\left(\omega_2 t\right)\right] \\ + V\left[2J_1\left(\delta 1_0\right)\cdot\sin\left(\omega_1 t\right) - 2J_1\left(\delta 2_0\right)\cdot\sin\left(\omega_2 t\right)\cdot\left(J_0\left(\delta 1_0\right) + 2J_2\left(\delta 1_0\right)\cos\left(2\omega_1 t\right)\right)\right] \end{array} \right\}$$

$$(25.7)$$

where the subscripts of the Bessel functions indicate their orders. As seen in Equation 25.7, Q, U, and V can be determined by detecting the 2F signals of both PEMs (the $\cos(2\omega_1 t)$ and $\cos(2\omega_2 t)$ terms) and the 1F signal of the first PEM (the $\sin(\omega_1 t)$ term).

The DC signal can be derived from Equation 25.7 to be

$$V_{DC} = I + \frac{\sqrt{2}}{2}\left[Q\cdot J_0(\delta 2_0) + U\cdot J_0(\delta 1_0)\right] \quad (25.8)$$

where all AC terms that vary as functions of the PEMs' modulation frequencies are omitted, because they have no net contribution to the averaged DC signal. In practice, a low-pass electronic filter can be used to eliminate such oscillations. As seen from Equation 25.8, it is convenient to choose the peak retardation amplitudes of both PEMs to be $\delta 1_0 = \delta 2_0 = 2.405$ rad (0.3828 waves) so that $J_0(\delta 1_0) = J_0(\delta 2_0) = 0$. Then the DC signal will be independent of light polarization (Q, U, or V) and thus will directly represent the light intensity from the light source.

When the polarization state of a light beam is the primary concern, normalized Stokes parameters are commonly used. In the dual PEM Stokes polarimeter, the ratios of PEM modulated signals to the average DC signal are used to calculate normalized Stokes parameters, as shown in Equation 25.9:

$$Q_N = \frac{Q}{I} = \frac{1}{J_2(\delta 2_0)} \cdot \frac{V_{2,2F}}{V_{DC}}$$

$$U_N = \frac{U}{I} = \frac{1}{J_2(\delta 1_0)} \cdot \frac{V_{1,2F}}{V_{DC}} \qquad (25.9)$$

$$V_N = \frac{V}{I} = \frac{1}{J_1(\delta 1_0)} \cdot \frac{V_{1,1F}}{V_{DC}}$$

where $V_{2,2F}$, $V_{1,2F}$, and $V_{1,1F}$ represent the strengths of the 2F signals of both PEMs and the 1F signal of the first PEM, respectively.

The electronic signals generated at the detector have both AC and DC components. When lock-in amplifiers are used to demodulate the signals, three lock-in amplifiers are needed to detect Q, U, and V simultaneously. When it is not necessary to measure Q, U, and V simultaneously, one or two lock-in amplifiers can be used for sequential measurements. Alternatively, one can use a fast digitizer to capture the waveform generated at the detector. The waveform is the combined result of all the modulation frequencies generated by both PEMs. Fourier analysis of the digitized waveform (hereafter the digital signal process, or DSP method) gives the signals at the first and second harmonics of both PEMs, thus the Stokes parameters.

25.2.4 Dual PEM Stokes Polarimeter Used in the Tokamak

The dual PEM Stokes polarimeter has been successfully applied to Tokamak plasma (ionized gas) diagnostics. A Tokamak is a machine for confining the plasma using a magnetic field. The Tokamak was originally conceived in 1950 by then Soviet scientists I. Tamm and A. Sakharov. The term "Tokamak" is the translation of a Russian word which itself is an acronym of Russian words [48]. There are several Tokamaks in the world including the DIII-D in the United States, JET in the United Kingdom, and the upcoming ITER in France [49]. The Tokamak is the most researched nuclear fusion reactor and is considered to be a promising fusion power generator.

In a Tokamak, a relatively constant electric current in the toroidal coils creates a relatively constant toroidal magnetic field. The moving ions and electrons in the plasma (toroidal plasma current) create a poloidal magnetic field. The total magnetic field at any location inside the plasma is the combination of the toroidal and poloidal fields as shown in Figure 25.5. The so-called magnetic field pitch angle is defined as

$$\gamma_p = \tan^{-1}\left(\frac{B_p}{B_T}\right) \qquad (25.10)$$

The toroidal magnetic field is known from the electric current supplied to the toroidal coils. If the magnetic field pitch angle is measured, the poloidal magnetic field and thus the plasma current can be calculated. The plasma current profile can be determined from the magnetic field pitch angles measured at different locations in the plasma. To achieve high stability and confinement in a Tokamak, it is critical to monitor and control the distribution of plasma current.

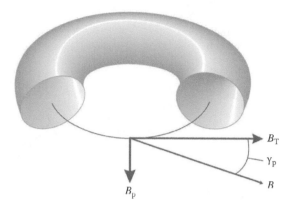

FIGURE 25.5 A diagram showing the relationship between the toroidal and poloidal magnetic fields in a Tokamak.

The motional Stark effect (MSE) polarimeter was first developed by Levinton and coworkers to determine the magnetic field pitch angle in the plasma of a Tokamak [50,51]. The MSE polarimeter was extended by Wroblewski and Lao to a multichannel polarimeter to determine the plasma current profile in the DIII-D Tokamak [52]. This method has now been applied to most of the Tokamaks in the world as a standard diagnostic instrument.

In a Tokamak, if a high-energy neutral deuterium beam is injected into the plasma, as the injected deuterium beam propagates inside the plasma, the deuterium atoms are subjected to an electric field that is generated by their motion across the magnetic field. Namely, $\mathbf{E} = \mathbf{V}_{beam} \times \mathbf{B}$ where $\mathbf{V}_{beam}$ is the velocity of injected beam, $\mathbf{B}$ is the local magnetic field, and $\mathbf{E}$ is the local electric field generated by the motion across the magnetic field. The Stark effect [53] refers to spectral splittings due to an external electric field. The electric field generated by the motion of deuterium atoms across a magnetic field produces splittings in the spectrum of the Balmer-α emission of the injected deuterium beam. This effect is called the MSE.

The MSE polarimeter takes advantage of polarized emission due to the Stark effect to measure the magnetic field pitch angle. In addition to the spectral splittings, the MSE produces polarized emissions according to the selection rules for atomic electronic transitions. Emission spectral lines from the $\Delta m = \pm 1$ transitions, or the σ components, are particularly useful to the MSE polarimeter. The σ component emissions are linearly polarized, and the polarization direction is perpendicular to the electric field and parallel to the local magnetic field. By measuring the two linear polarization parameters, Q and U, of the Balmer-α emissions from the injected deuterium atoms, the MSE polarimeter determines the magnetic field pitch angle at the intersection of the injected beam and the observation direction. In a multichannel MSE polarimeter, the magnetic field pitch angles at multiple locations inside the plasma are measured simultaneously. Figure 25.6 shows a block diagram of using an MSE polarimeter in the Tokamak.

The PEMs used in the MSE polarimeter have some special features unique to this application. They have symmetrical optical elements (square with the four corner cut off for mounting). A symmetrical optical element in the PEM provides symmetrical modulation distribution over its aperture. The PEMs for this application have large optical apertures. Since the PEM is a resonant modulator, its size is inversely proportional to its resonant frequency. Modern MSE polarimeters use 20 and 23 kHz PEMs specially designed into a single enclosure as shown in Figure 25.7. Since MSE polarimeters use optical fibers to link to multiple measurement locations in the plasma, the symmetrical retardation distribution and large optical aperture of the 20/23 kHz PEM pair allow simultaneous, multichannel measurements using the same MSE polarimeter. Finally, the PEM electronics are magnetically shielded and separated 10–15 m from the strong magnetic field in the plasma.

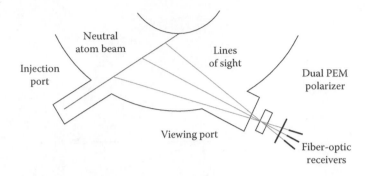

FIGURE 25.6 A diagram showing an MSE polarimeter used in a Tokamak.

FIGURE 25.7 A photo picture of the 20/23 kHz PEM pair used in a modern MSE polarimeter.

25.2.5 COMMERCIAL DUAL PEM STOKES POLARIMETERS

25.2.5.1 Visible Stokes Polarimeters

Several models of dual PEM Stokes polarimeters have become commercially available from Hinds Instruments, including laser polarization analyzers and spectroscopic Stokes polarimeters in the UV–vis, near IR, and mid-IR spectral regions [18]. In one model, a DSP method is used to calculate Stokes parameters using Fourier analysis. Using this DSP method, the dual PEM Stokes polarimeter measures all normalized Stokes parameters simultaneously at a rate of >350 sets of data per second [54].

To test this polarimeter, a mechanical "chopper wheel" was modified to simulate a light source with controlled, yet fast changing polarization states. This "chopper wheel" consists of a polymer film retarder in one half and an open space in the other half. The retarder film has an average linear retardation of ~280 nm with significant variations over its aperture. The spin rate of the wheel is regulated by the controller of a mechanical chopper. In an experiment, the polarization of a linearly polarized He–Ne laser (632.8 nm) was set at approximately −45°. The light polarization was further purified with a calcite prism polarizer that was precisely oriented at −45.0°. The modified "chopper wheel" was placed between the laser–polarizer combination and the dual PEM Stokes polarimeter and was spun at 2 turns per second during the experiment. A typical set of data obtained in this experiment is depicted in Figure 25.8a through c.

Figure 25.8a shows five repeated patterns of the normalized Stokes parameters (five cycles) measured in 2.5 s. The rate of data collection for this experiment was 371 sets of normalized Stokes parameters per second. As the "chopper wheel" rotated one full turn (one cycle of repeated pattern in Figure 25.8a), the laser beam passed through the opening (i.e., no retardation) for half of the time. During this half-cycle, the polarization of the laser was unchanged. As shown in Figure 25.8a, we observed for this half-cycle that $U_N = -1$, $Q_N = 0$, $V_N = 0$, which represents linearly polarized

FIGURE 25.8   (a) Normalized Stokes parameters measured as the "chopper wheel" was spun at 2 turns per second; (b) Figure 25.8a expanded to show a half-cycle when the laser beam passes the retarder; (c) DOP during the half-cycle illustrated in Figure 25.8b.

light at $-45°$. For the other half-cycle, Q_N, U_N, and V_N all changed significantly as the laser beam passed through different locations of the rotating retarder.

Figure 25.8b, which is an expansion of Figure 25.8a, illustrates in detail how each normalized Stokes parameter varied during the half-cycle when the laser beam passed through the retarder. The polarization state of the light beam after passing the polymer film was dominated by the linear components, Q_N and U_N. This is because the polymer film used in the "chopper wheel" was fairly close to a half-wave retarder. If an exact half-wave retarder was used, it would rotate just the plane of the linearly polarized light beam. There would have been no circular polarization component. The weak circular component (V_N) seen in Figure 25.8b is due to the deviation from half-wave retardation of the polymer retarder.

During the other half-cycle, when the light beam goes through the opening, we typically observed an U_N of -1.00 with a standard deviation of ~0.0004. For example, the data in the next half-cycle following the data shown in Figure 25.8b have an average U_N of -0.9922 and a standard deviation of 0.00036.

Figure 25.8c displays the DOP during the half-cycle illustrated in Figure 25.8b. Although Q_N, U_N, and V_N all varied significantly, the DOP changed little during this half-cycle, since the depolarization of the retarder is negligible. The data plotted in Figure 25.8c have an averaged DOP of 0.994 and a standard deviation of 0.0054.

Notice the significant difference in the standard deviations of DOP for the two halves (0.0054 vs. 0.00036) in a full cycle as the "chopper wheel" spun a full turn. The lower standard deviation, which was measured during the half-cycle when the light beam passed the opening, represents the repeatability of the instrument. The higher standard deviation of DOP is primarily due to "mechanical noise." The polymer film retarder used in the "chopper wheel" had some floppiness. The floppiness, coupled with the mechanical spin of the chopper, introduced higher noise to the data. This was confirmed by observing increased standard deviation of DOP, 0.0089 and 0.0097, when the spin rate of the "chopper wheel" was increased to 4 and 7 turns per second, respectively.

Finally, the 370 sets of data per second is a fast speed for collecting data, particularly with the high repeatability and accuracy obtained at this rate. However, this rate is still far below the limit imposed by the PEMs.

As an alternative to the DSP method, lock-in amplifiers can be used for signal processing in the same dual PEM Stokes polarimeter. When lock-in amplifiers are used for signal processing, each one can be used to detect a signal at a specific modulation frequency. The full dynamic range of the lock-in amplifier can thus be used in demodulating that particular signal. High-quality lock-in amplifiers are still the best demodulators currently available for many PEM-based instruments, particularly when the measurement speed is not critical. The extremely high sensitivity in measuring low DOP that was achieved by Kemp attests to the powerful measurement capability of the PEM, lock-in amplifier, and careful experimentation.

One application of the polarimeter described in this section is to monitor the polarization quality of a laser. Figure 25.9a through e shows the measured result of a randomly polarized laser (sometimes mistakenly called an unpolarized laser). As seen in Figure 25.9a, the DC signals exhibit sharp and frequent intensity variations even though the total percentage of the variations is small.

Figure 25.9b through d illustrates fairly significant variations in all three normalized Stokes parameters. In particular, Q_N and U_N have a much higher magnitude than V_N, which indicates a much higher degree of varying linear polarization in the laser. Figure 25.9b and c shows that Q_N and U_N reach "0" at the same time where the laser beam is nearly unpolarized. Q_N and U_N also reach maxima or minima at the same time, though with opposite signs. At these data points the laser beam has a fairly high DOLP. The variation in the DOP of this randomly polarized laser, shown in Figure 25.9e, reveals that this laser does not output randomly polarized light even when the integration time is on the order of 1 s, and it does not produce a light beam with a constant DOP. For this reason, a randomly polarized laser is avoided in building polarization sensitive instruments.

25.2.5.2 Near-Infrared Fiber Stokes Polarimeters

Based on the same principle, a near-infrared (NIR) Stokes polarimeter is made commercially available for applications using optical fibers [91]. In the light source module shown in Figure 25.10a, there is a 1550 nm diode laser followed by a linear polarizer. A single mode fiber connects the light source module to the Stokes polarimeter. The polarimeter is used to measure the normalized Stokes parameters for the light beam entering it from the fiber. Three manual fiber stretchers are attached to the fiber to generate different polarization states for testing the polarimeter. During a test of this polarimeter, the three fiber stretchers were manually operated in a random fashion to generate arbitrary polarization states for the light beam entering the polarimeter. A typical set of data is shown in Figure 25.10b. In the middle part of the dataset, all three normalized Stokes parameters changed as the fiber was stretched by the stretchers. Although the polarization states of the light beam changed, the DOP remained constant at 1.00, which indicates

FIGURE 25.9 Measured Stokes parameters for a randomly polarized He–Ne laser (632.8 nm). (a) DC; (b) Q_N; (c) U_N; (d) V_N; and (e) degree of polarization.

negligible depolarization in the stretched fiber. At both ends of the dataset shown in Figure 25.10b, all normalized Stokes parameters remained constant when the fiber stretchers were not moved.

In addition to testing optical fibers, this Stokes polarimeter can be used in other applications. In fact, this polarimeter was originally built for monitoring the level of a weak magnetic field. In that application, a special fiber, made from a material with a high Verdet constant, was placed in a weak magnetic field. The strength of the magnetic field was determined from the Faraday rotation measured by the polarimeter.

(a)   (b)

FIGURE 25.10   (a) A photo picture of an NIR Stokes polarimeter for fiber applications and (b) a typical set of data obtained on the NIR Stokes polarimeter.

25.2.5.3   Spectroscopic Stokes Polarimeter

Spectroscopic polarimetry, combining the rich information from both spectroscopy and polarimetry, is one of the most exhaustive methods for analyzing light properties and light–matter interaction. It is applied to a wide range of scientific disciplines. In searching for life in the universe, we need a sensitive remote sensor capable of identifying a universal biosignature. Since all known living materials contain chiral molecules (a molecule that does not superimpose with its mirror image) possessing almost exclusively a single handedness in their biological processes, homochirality is likely to be a universal biosignature for all biochemical life whether similar to terrestrial or not. A PEM-based spectroscopic Stokes polarimeter is well suited to both remote sensing and detecting weak circular polarization signals that may arise from homochirality of astrobiological samples.

The PEM-based spectroscopic Stokes polarimeter [92] is designed to work at low light levels that are typical in astronomical applications. It is optimized to detect a circular polarization signal that is orders of magnitude weaker than a linear polarization signal. This polarimeter can work in different laboratory configurations. In the application described in this section, the polarimeter is mounted to an integrating sphere to allow operation in either transmissive or reflective mode. The integrating sphere randomly diffuses the illumination so that it is averaged over incident directions to minimize possible artifacts from linearly polarized light. A block diagram of this polarimeter and the integrating sphere is shown in Figure 25.11. For future applications, this polarimeter is designed to be mounted to telescopes.

Using this polarimeter, we studied whether photosynthetic microbes produce a macroscopic circular polarization signature in scattered light [93,94]. If photosynthetic microbes do produce such a signature, it could be used in remote sensing as a powerful indicator of the presence of a universal biosignature, namely, homochirality. In a series of experiments, we used this polarimeter to detect the small circular polarization present in light scattered by photosynthetic microorganisms and macroscopic vegetation. We obtained cultures of photosynthetic marine cyanobacteria that possess chlorophyll *a* and antenna pigments phycocyanin and phycoerythrin. For context, we measured examples of macroscopic vegetation chosen largely by availability near the laboratory, and control minerals iron oxide, sulfur (which has a strong spectral "edge"), and a Mars regolith analog. The detailed results may be found in Refs. [93,94].

A comparison of the transmission and absorption polarization spectra for specimen *Synechococcus* WH8101 is presented in Figure 25.12 [93].

FIGURE 25.11 Block diagram of the spectroscopic Stokes polarimeter and the integrating sphere to allow operation in either transmissive or reflective mode.

FIGURE 25.12 Circular polarization spectra of cyanobacteria WH8101. (a) Transmission polarization spectrum of *Synechococcus* WH8101. The line above shows the degree of circular polarization, with $\pm 1\sigma$ uncertainty; the solid line below shows a scaled version of the absorbance spectrum. (b) Reflection polarization spectrum as for Figure 25.12a except that the solid line is scaled $-\log_{10}$(Reflectance) and the dashed line is a scaled plot of linear polarization degree.

The strongest circular polarization features in both transmission and absorption are related to the electronic absorption bands of the photosynthetic pigments. This result characterizes all the results we obtained on cyanobacteria, as presented in Refs. [93,94]. The circular polarization peaks across the well-defined photosynthesis absorption bands, and in the case of chlorophyll *a*, there is a very distinctive sign change at the location of the absorption maximum. This has important consequences for the likelihood of a large-scale macroscopic polarization being present. Oceanic concentrations of microbes are low compared to our laboratory specimens, but with modest volume, the optical depth can be unity and dominated by chlorophyll and cyanobacteria. Hence, there is the possibility that the oceans will yield a measurable circular polarization signal.

We also acquired polarization spectra of a variety of leaves. Leaf circular polarization reflection spectra show substantial diversity, but the average brings out common features. Figure 25.13a and b

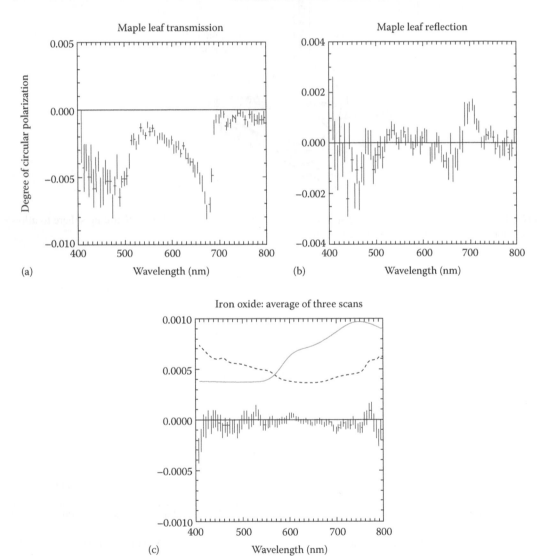

FIGURE 25.13 Circular polarization spectra for a leaf and a mineral. (a) A maple leaf transmission polarization spectrum. (b) Corresponding maple leaf reflection polarization spectrum. (c) A control iron oxide polarization spectrum. The data points with error bars are the degree of circular polarization in each panel. The solid line of Figure 25.13c shows the reflection spectrum of iron oxide and the dashed line, the degree of linear polarization, both arbitrarily scaled.

exhibits a strong polarization signal associated with chlorophyll absorption between 650 and 700 nm and a broader spectral dependence of polarization to the blue. Polarization is low at wavelengths longer than the chlorophyll red edge at 700 nm. In a planetary context, it will be of interest to establish the polarization characteristics of large-scale, sometimes heterogeneous, vegetation-dominated scenery with empirical measurements.

As controls we made measurements of two minerals exhibiting significant strength "edges" in their spectra: iron oxide and sulfur [93,94]. This also served to guard against potential instrumental artifacts in the vicinity of rapidly changing light levels. Figure 25.13c shows the polarization spectrum of the iron oxide. The mineral spectra show no relationship between polarization and intensity, unlike the vegetation, and polarization is low overall. Additional polarization spectra of a variety of other minerals were presented in Ref. [93], and this was true in all cases.

Using this instrument, we have quantified the circular polarization signal produced by astrobiologically relevant microorganisms and compared the results to macroscopic vegetation (such as leaves) and abiotic minerals. We see unambiguous circular polarization from photosynthetic microbes. Therefore, circular polarization spectroscopy offers the prospect of remotely sensing life's unique chiral signature. Beyond the laboratory, we hope to test such a polarimeter in an airborne instrument looking at the Earth.

25.3 MUELLER MATRIX POLARIMETER

25.3.1 Dual PEM Mueller Matrix Polarimeter

The Mueller matrix polarimeter can be used to determine all 16 elements of the Mueller matrix, thus the complete polarization-altering properties of a sample. The most studied Mueller matrix polarimeter is perhaps based on the dual rotating wave plate design, for which there are many published articles [2,55–57]. When PEMs are employed to obtain a higher measurement sensitivity [58–61], the dual PEM configuration, shown in Figure 25.14, is the setup most

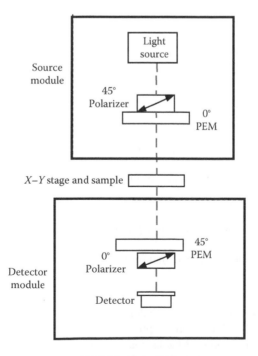

FIGURE 25.14 A block diagram of a dual PEM Mueller polarimeter.

commonly used in a Mueller matrix polarimeter. The key elements for polarization modulation in the setup are the polarizer–PEM and PEM–analyzer assemblies.

If a normalized Mueller matrix is used for a sample, the signal reaching the detector, I_{det}, in the setup shown in Figure 25.14, can be derived by using the Mueller matrix calculus:

$$I_{det} = \frac{KI_0}{2}(1 + \cos(\delta 2) \cdot m21 + \sin(\delta 2) \cdot m41$$

$$+ (m13 + \cos)(\delta 2) \cdot m23 + \sin(\delta 2) \cdot m43) \cdot \cos\delta 1 - (m14 + \cos(\delta 2) \cdot m24 + \sin(\delta 2) \cdot m44) \cdot \sin \delta 1)$$

(25.11)

where
  I_0 represents the light intensity after the first polarizer
  K is a constant representing transmission efficiency of the optical system after the first polarizer

The "DC" signal from Equation 25.11 is

$$V_{DC} = \frac{KI_0}{2}\left[1 + J_0(\delta 2_0)m21 + J_0(\delta 1_0)m13 + J_0(\delta 1_0)J_0(\delta 2_0)m23\right]$$ (25.12)

where any AC term that varies as a function of the PEMs' modulation frequencies is omitted, because they have no net contribution to the averaged DC signal. At $J_0(2.405) = 0$ PEM setting, the "DC" term is simplified to

$$V_{DC} = \frac{KI_0}{2}$$ (25.13)

The useful "AC" terms from Equation 25.11, which are related to eight normalized Mueller matrix elements, are summarized in Equation 25.14.

Equations	AC Terms
$P_{21} = m21 \cdot 2J_2(\delta 2_0)\cos(2\omega_2 t)$	$(2\omega_2)$
$P_{41} = m41 \cdot 2J_1(\delta 2_0)\sin(\omega_2 t)$	(ω_2)
$P_{13} = m13 \cdot 2J_2(\delta 1_0)\cos(2\omega_1 t)$	$(2\omega_1)$
$P_{23} = m23 \cdot 2J_2(\delta 1_0)\cos(2\omega_1 t) \cdot 2J_2(\delta 2_0)\cos(2\omega_2 t)$	$(2\omega_2 \pm 2\omega_1)$
$P_{43} = m43 \cdot 2J_2(\delta 1_0)\cos(2\omega_1 t) \cdot 2J_1(\delta 2_0)\sin(\omega_2 t)$	$(2\omega_1 \pm \omega_2)$
$P_{14} = -m14 \cdot 2J_1(\delta 1_0)\sin(\omega_1 t)$	(ω_1)
$P_{24} = -m24 \cdot 2J_1(\delta 1_0)\sin(\omega_1 t) \cdot 2J_2(\delta 2_0)\cos(2\omega_2 t)$	$(2\omega_2 \pm \omega_1)$
$P_{44} = -m44 \cdot 2J_1(\delta 1_0)\sin(\omega_1 t) \cdot 2J_1(\delta 2_0)\sin(\omega_2 t)$	$(\omega_2 \pm \omega_1)$

(25.14)

The eight Mueller matrix elements that can be determined from this optical configuration [**P1(45°)-M1(0°)-S-M2(45°)-P2(0°)**] and three other optical configurations are provided in Table 25.1.

In Table 25.1, each "*x*" indicates a Mueller matrix element that cannot be determined using the corresponding optical configurations. A total of four optical configurations is required to determine

TABLE 25.1

Dual PEM Optical Configurations and Measured Mueller Matrix Elements

Optical Configurations	Measured Mueller Matrix Elements
P1(45°)-M1(0°)-S-M2(45°)-P2(0°)	$\begin{pmatrix} 1 & x & m_{13} & m_{14} \\ m_{21} & x & m_{23} & m_{24} \\ x & x & x & x \\ m_{41} & x & m_{43} & m_{44} \end{pmatrix}$
P1(0°)-M1(45°)-S-M2(0°)-P2(45°)	$\begin{pmatrix} 1 & m_{12} & x & m_{14} \\ x & x & x & x \\ m_{31} & m_{32} & x & m_{34} \\ m_{41} & m_{42} & x & m_{44} \end{pmatrix}$
P1(45°)-M1(0°)-S-M2(0°)-P2(45°)	$\begin{pmatrix} 1 & x & m_{13} & m_{14} \\ x & x & x & x \\ m_{31} & x & m_{33} & m_{34} \\ m_{41} & x & m_{43} & m_{44} \end{pmatrix}$
P1(0°)-M1(45°)-S-M2(45°)-P2(0°)	$\begin{pmatrix} 1 & m_{12} & x & m_{14} \\ m_{21} & m_{22} & x & m_{24} \\ x & x & x & x \\ m_{41} & m_{42} & x & m_{44} \end{pmatrix}$

all Mueller matrix elements. Of course, one may use −45° and 90° instead of 45° and 0° in the aforementioned configurations. The results will be the same after calibrating the signs of the "AC" signals.

Jellison championed this dual PEM design and its applications [58–61]. He developed a dual PEM system for ellipsometric studies and called it two-modulator generalized ellipsometry (2-MGE). He has applied the 2-MGE to measuring retardation and diattenuation properties of both transmissive and reflective samples [60–63].

25.3.2 Four Modulator Mueller Polarimeter

If one wishes to measure all 16 Mueller matrix elements simultaneously, 4 modulators are required [64,95]. There are many optical configurations to arrange the four PEMs and two polarizers in a complete Mueller polarimeter. Figure 25.15 depicts one of such arrangements of 4 PEMs in order to measure all 16 Mueller matrix elements.

The optical setup in Figure 25.15 generates many modulation frequencies from different combinations of the harmonics of the four PEMs. All 16 Mueller matrix elements can be determined from

FIGURE 25.15 A block diagram of a four PEM Mueller polarimeter.

TABLE 25.2

Possible Modulation Frequencies to Determine

Each Mueller Matrix Element

Mueller Matrix Elements	PEM Frequencies
m_{12}	$\omega_0 + \omega_1$
m_{13}	$2\omega_0$
m_{14}	$-\omega_0 - 2\omega_1$
m_{21}	ω_0
m_{22}	$\omega_2 + \omega_3$
m_{23}	$\omega_2 + \omega_3$
m_{24}	$\omega_0 + \omega_1 - \omega_2 + \omega_3$
m_{31}	$2\omega_3$
m_{32}	$-\omega_0 + \omega_1 - 2\omega_3$
m_{33}	$2\omega_0 - 2\omega_3$
m_{34}	$\omega_0 - 2\omega_1 - 2\omega_3$
m_{41}	$-\omega_2 + \omega_3$
m_{42}	$\omega_0 + \omega_1 - 2\omega_2 + \omega_3$
m_{43}	$2\omega_0 + 2\omega_2 - \omega_3$
m_{44}	$\omega_0 + 2\omega_1 - 2\omega_2 + \omega_3$

Note: The four frequencies of the modulators are ω_0, ω_1, ω_2, and ω_3.

one or more of the combined modulation frequencies. Table 25.2 lists one modulation frequency corresponding to each Mueller matrix element. Detailed analysis of the four PEM Mueller polarimeter and measured data in both spectroscopic and spatial domains is available in Ref. [95].

25.4 SPECIAL POLARIMETERS

While it is true that a complete Mueller polarimeter can be used to measure all 16 elements in a Mueller matrix, it is complicated to detect all polarization-altering properties, in terms of retardation, diattenuation, and depolarization, from a measured Mueller matrix [2]. The difficulty is also due to the less well-understood depolarization [65]. Currently, applications of the complete Mueller polarimeter are still rather limited. On the other hand, specific industrial applications often require the determination of just one or two polarization parameters. Hence, special polarimeters have gained a wide range of applications in industrial metrology. In this section, I provide several examples of PEM-based special polarimeters and their applications.

25.4.1 LINEAR BIREFRINGENCE POLARIMETERS AND THEIR APPLICATIONS
IN THE OPTICAL LITHOGRAPHY INDUSTRY

25.4.1.1 Linear Birefringence Polarimeter

Figure 25.16 depicts the block diagram of a linear birefringence polarimeter (Exicor®) that the author coinvented [66–68]. In this special polarimeter, the optical bench contains a polarization modulation module (a light source, a polarizer, and a PEM), a sample holder mounted on a computer-controlled X–Y stage, and a dual channel detecting assembly.

Each detecting channel contains an analyzer and a detector. Channel 1 (crossed-polarizer) measures the linear retardation component that is parallel to the PEM's optical axis (0°), and channel 2 measures the linear retardation component that is oriented 45° from the PEM's optical axis.

FIGURE 25.16 A block diagram of a linear birefringence polarimeter (Exicor).

Using the Mueller matrix calculus, the light intensities reaching the detectors of both channels are derived to be

$$I_{ch1} = \frac{I_0}{2}\left\{ 1 + \cos(4\rho)\sin^2\left[\frac{\delta}{2}\right]\cos\Delta - \cos^2\left[\frac{\delta}{2}\right]\cos\Delta + \cos(2\rho)\sin\delta\sin\Delta \right\}$$

$$I_{ch2} = \frac{I_0}{2}\left\{ 1 + \sin(4\rho)\sin^2\left[\frac{\delta}{2}\right]\cos\Delta + \sin(2\rho)\sin\delta\sin\Delta \right\} \tag{25.15}$$

Substituting the Bessel expansions into Equation 25.15 and taking only up to the second order of the Bessel functions, we have

$$I_{ch1} = \frac{I_0}{2}\left\{ \begin{array}{l} 1 + J_0(\Delta_0)\left\{\cos(4\rho)\sin^2\left[\frac{\delta}{2}\right] - \cos^2\left[\frac{\delta}{2}\right]\right\} + 2J_1(\Delta_0)\cos(2\rho)\sin\delta\sin(\omega t) \\ \\ + 2J_2(\Delta_0)\left\{\cos(4\rho)\sin^2\left[\frac{\delta}{2}\right] - \cos^2\left[\frac{\delta}{2}\right]\right\}\cos(2\omega t) + \cdots \end{array} \right\}$$

$$I_{ch2} = \frac{I_0}{2}\left\{ \begin{array}{l} 1 + J_0(\Delta_0)\sin(4\rho)\sin^2\left[\frac{\delta}{2}\right] + 2J_1(\Delta_0)\sin(2\rho)\sin\delta\sin(\omega t) \\ \\ + 2J_2(\Delta_0)\sin(4\rho)\sin^2\left[\frac{\delta}{2}\right]\cos(2\omega t) + \cdots \end{array} \right\} \tag{25.16}$$

The "DC" signals from both detectors are

$$DC_{ch1} = \frac{K_{ch1}I_0}{2}(1 - J_0(\Delta_0))$$

$$DC_{ch2} = \frac{K_{ch2}I_0}{2}$$

(25.17)

The ratios of 1F "AC" signals to the corresponding DC signals from both detectors are

$$\frac{V_{ch1}(1F)}{DC_{ch1}} = \sin\delta\cos(2\rho) \times \frac{\sqrt{2}J_1(\Delta_0)}{1 - J_0(\Delta_0)}$$

$$\frac{V_{ch2}(1F)}{DC_{ch2}} = \sin\delta\sin(2\rho) \times \sqrt{2}J_1(\Delta_0)$$

(25.18)

where $V_{ch1}(1F)$ and $V_{ch2}(1F)$ are the 1F components of the detector electronic signals from both detectors.

Defining R_{ch1} and R_{ch2} as corrected ratios for both channels, Equation 25.18 becomes

$$R_{ch1} = \frac{V_{ch1}(1F)}{DC_{ch1}} \times \frac{1 - J_0(\Delta_0)}{2J_1(\Delta_0)} \times \sqrt{2} = \sin\delta\cos(2\rho)$$

$$R_{ch2} = \frac{V_{ch2}(1F)}{DC_{ch2}} \times \frac{1}{2J_1(\Delta_0)} \times \sqrt{2} = \sin\delta\sin(2\rho)$$

(25.19)

When the PEM retardation amplitude is selected to be $\Delta_0 = 2.405$ rad (0.3828 waves) so that $J_0(\Delta_0) = 0$, the magnitude and angular orientation of the linear retardation of the sample are expressed as

$$\delta = \sin^{-1}\left(\sqrt{(R_{ch1})^2 + (R_{ch2})^2} \right)$$

$$\rho = \frac{1}{2}\tan^{-1}\left[\frac{R_{ch2}}{R_{ch1}}\right] \quad \text{or} \quad \rho = \frac{1}{2}ctg^{-1}\left[\frac{R_{ch1}}{R_{ch2}}\right]$$

(25.20)

where δ, represented in radians, is a scalar. When measured at a specific wavelength (i.e., 632.8 nm), it can be converted to retardation in "nm."

The accuracy of this instrument was tested using a Soleil-Babinet compensator [70]. Figure 25.17 shows the agreement between the measured data and the values of a calibrated Soleil-Babinet compensator. The experimental data shown in Figure 25.17 lead to a near-perfect linear fit. In the ideal case, the instrumental readings and the corresponding compensator values will be identical ($Y = X$).

This instrument provides a high repeatability of measurements and a low noise level. Figure 25.18a displays the data of 3000 repeated measurements at a fixed spot of a compound quartz wave plate [69]. These 3000 data points have an average retardation of 11.34 nm with a standard deviation of 0.0078. When the instrument is properly calibrated, the linear retardation readings of the instrument without any sample should be a representation of the overall instrumental noise level. Figure 25.18b displays a collection of ~18,000 data points that was recorded in about 8 h (automatic offset correction was used at 20 min intervals) [69]. This dataset has an average of 0.0016 nm and standard deviation of 0.0009.

When the ratios of the AC and DC signals are used, linear retardation measurements will not be affected by light intensity changes caused by light source fluctuation, sample absorption, partial

$$y = 1.005x + 0.160$$
$$R^2 = 0.9998$$

FIGURE 25.17   Measured values on Exicor vs. calibrated values of a Soleil-Babinet compensator.

FIGURE 25.18   (a) A typical dataset for instrumental stability and (b) a typical dataset for instrumental noise.

reflection, and others. In a test, we purposely varied the light intensity to demonstrate this feature of the instrument. The raw data of AC and DC signals for both detecting channels at different light intensities, along with the retardation values determined from AC/DC, are presented in Figure 25.19a and b. As seen, the retardation magnitude and fast axis angle of the sample remain constant as the light intensity is reduced [69]. As expected, the measurement errors for retardation increase when the light intensity becomes very small.

FIGURE 25.19   (a) Raw data of AC and DC signals from both detecting channels and (b) retardation magnitude and angle calculated from AC/DC with decreasing light intensities.

25.4.1.2   Linear Birefringence in Photomasks

In the semiconductor industry, fused silica is the standard material for the lenses and photomask substrates used in modern optical lithography step and scan systems. To follow Moore's law (doubling the number of transistors per integrated circuit every 18 months) [71], the semiconductor industry continuously moves toward finer resolution by adopting shorter wavelengths and other resolution enhancement means. Linear birefringence in optical components can degrade the imaging quality of a lithographic step and scan system through several effects including bifurcation, phase front distortion, and alternating light polarization. Consequently, the requirement for low-level residual birefringence in high-quality optical components, including photomasks, has become stringent.

The linear birefringence polarimeter described earlier, with its high accuracy and sensitivity (<0.005 nm), is particularly suited to the quality control of photomask substrates, or photomask blanks. Photomask blanks are made of high-quality fused silica that has low levels of linear retardation, low color centers, and well-polished surfaces. They have negligible circular birefringence, depolarization, and diattenuation (at normal incidence). Low-level linear birefringence due to residual stress in photomask blanks is the primary concern of this industry.

Figure 25.20 shows the birefringence images of four fused silica photomask blanks (6″ × 6″; thickness: 0.25″) [72,73]. In Figure 25.20a through d, the magnitude of retardation is color-coded. Three of the four samples (Figure 25.20a, b, and d) display a retardation level generally below 3 nm. The angle of fast axis at each data cell is described with a short bar. Figure 25.20a through d shows different patterns of residual linear retardation in these photomask blanks. The different patterns and levels of residual linear retardation in the mask blanks reveal the residual strain formed during annealing and other manufacturing processes.

For over 20 photomask blanks measured, the birefringence pattern of Figure 25.20a was observed most often. Photomask blanks exhibiting such a birefringence pattern have a wide range of residual retardation levels. Three other examples are shown in Figure 25.21a through c with maximum

FIGURE 25.20 Overlaid images of retardation magnitude and fast axis angle for photomask blanks (a–d; fused silica; 6″ × 6″ × 0.25″; spatial resolution: 3 mm).

retardation levels at ~20, ~10, and ~4 nm, respectively, as measured at 632.8 nm [72,73]. All three photomask blanks have a similar birefringence angle pattern as shown in Figure 25.21d.

The images of linear birefringence shown in Figure 25.21a through d have some common features. In each photomask blank, four areas along the edges of the blank exhibit the highest levels of retardation. Fortunately, these areas are at the periphery of a photomask blank. It is unlikely that die patterns will overlap these regions when a photomask is made. Therefore, the residual birefringence in those areas, although high, may not affect the image projected onto the wafer. There are four inner regions, shown in Figure 25.21a through c, with high levels of residual linear retardation. These four peak areas, as shown in the birefringence surface plots, are fairly close to the center of the photomask. They would overlap with the die patterns on a photomask, especially if a photomask contains more than one die. Furthermore, Figure 25.21d shows that the fast axis angles at those four inner high retardation areas are close to ±45°.

Figure 25.22a and b illustrates two typical linear birefringence images of photomasks with low density of features [73]. For both samples, the die patterns are located in the central part of the photomasks. The retardation map shown in Figure 25.22a exhibits a maximum retardation that is below 5 nm and an angular pattern of fast axis that closely resembles that shown in Figure 25.21d. In Figure 25.22a, there are two gray bars that carry no birefringence information. The two gray bars represent the approximate location of two "chrome" rails on this photomask that blocked the

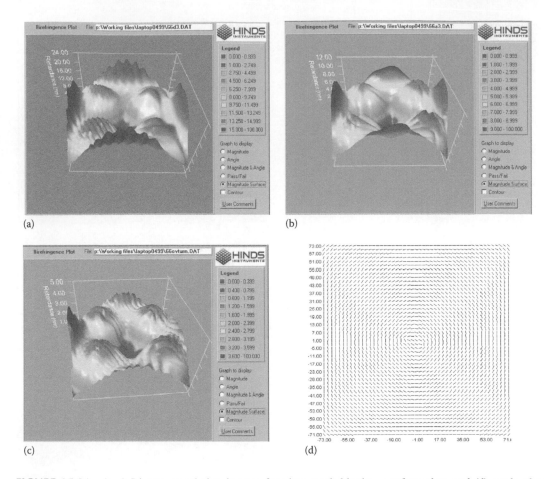

(a)　　　　　(b)

(c)　　　　　(d)

FIGURE 25.21　(a–c) Linear retardation images for photomask blanks—surface plots and (d) angle plot (fused silica; $6'' \times 6'' \times 0.25''$; spatial resolution: 3 mm).

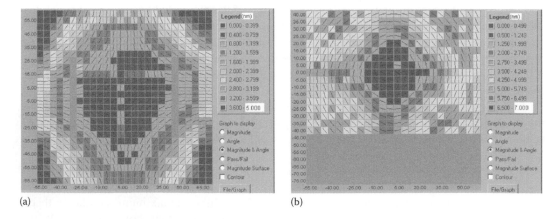

(a)　　　　　(b)

FIGURE 25.22　(a and b) Linear retardation images of two fused silica photomasks (fused silica; $6'' \times 6'' \times 0.25''$; spatial resolution: 5 mm).

light beam completely during measurement. The smooth flow of the birefringence angle pattern inside and outside of the die area demonstrates that the "chrome" features on the photomask have no significant effect on linear birefringence, at least at the retardation level shown in Figure 25.22a.

Figure 25.22b is the linear retardation image of a photomask with four die patterns located in the central area of this photomask. Figure 25.22b shows a maximum retardation below 7 nm and a birefringence angular pattern that closely resembles the patterns depicted in Figure 25.21d. It further confirms that the "chrome" features on a photomask have little effect on the residual birefringence at the retardation level shown here.

The experimental results for the two photomasks measured here indicate that the residual linear birefringence in a photomask is primarily due to the contribution from the substrate, rather than from the "chrome" features formed on the substrate. When the residual linear birefringence in a photomask blank is reduced to a much lower level, one would expect that the "chrome" features should have some effect on the mechanical stress pattern of the photomask substrate.

25.4.1.3 Effect on Image Quality Due to Linear Birefringence in a Photomask

In general, a high level of residual birefringence in an optical component can lead to aberrations in polarization ray tracing. Low linear birefringence in a photomask is particularly important for producing high-quality wafers. In an optical lithographic step and scan system, there may be over 20 optical components adding to a total thickness of optical materials exceeding 1 m. The residual linear birefringence integrated over the entire path-length of the optical components should be much larger than the birefringence in a photomask. However, a photomask is the only component in the optical path that will be changed from time to time in a given step and scan system. Assuming that a calibration can effectively eliminate the effect of residual linear birefringence in all fixed optical components in a step and scan system, the residual birefringence in a photomask would become a significant factor affecting the imaging quality formed on the wafer.

There are a variety of different designs in optical lithographic step and scan systems. For example, a design using all refractive lenses is called the refractive design; a design using mixed reflective and refractive optical components is called the catadioptric design [74]. The same residual linear birefringence in a photomask could have a different impact for step and scan systems with different designs. Of the various types of step-scan systems, linear birefringence in a photomask makes the largest impact on imaging quality in a particular catadioptric system [75].

The key component of this catadioptric design [75] is a polarization beam splitter. Its basic function is illustrated in Figure 25.23. Imagine that a linearly polarized light beam enters the beam

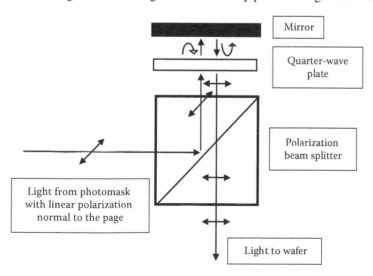

FIGURE 25.23 A block diagram for the polarization beam splitter used in a catadioptric design.

splitter with the polarization direction of the beam normal to the plane defining the page you are reading. The light beam is reflected up at the interface of the polarization beam splitter; then the light beam exits the beam splitter with no change in polarization. Passing the quarter-wave plate, the light beam becomes right circularly polarized. Reflected by the mirror, the light beam changes to left circularly polarized. Passing the quarter-wave plate again, the light beam becomes linearly polarized along the horizontal direction in the plane defining the page you are reading. The horizontally polarized light beam enters the beam splitter again, passes the interface with no further change in its polarization, and finally exits the beam splitter. The die pattern on the photomask is imaged to the photoresist layer on the wafer. Note that the elaborately designed lens groups for an optical lithographic system [75] are omitted in Figure 25.23.

As illustrated in Figure 25.23, the light intensity pattern created by the photomask is perfectly imaged onto the wafer if there is no linear birefringence in the photomask substrate. What happens if the photomask substrate has the linear retardation pattern as shown in Figure 25.21a through c. Let us assume that the four inner peak areas have linear retardation values ~10 nm with the fast axes at either 45° or −45°. The linearly polarized light, passing through the four inner high birefringence areas, will become elliptically polarized. The elliptically polarized light can be decomposed into a linear polarization component and a circular polarization component. The linear component will pass through the beam splitter as illustrated in Figure 25.23. A retardation of 10 nm with its fast axis angle 45° from the incident polarization will result in a circular polarization component of ~10% at 193 nm. Half the intensity of the circular component will be lost when the light beam goes through the polarizing beam splitter. Therefore, a specific light intensity pattern correlated to the linear birefringence pattern in the photomask substrate is imaged onto the wafer, which significantly distorts the image of the photomask.

However, if the linear birefringence in a photomask substrate is controlled to below 2 nm, the maximum circular polarization component produced at 193 nm will be <0.5%. The distortion of the image due to the linear birefringence in the photomask substrate will not exceed 0.25%. Furthermore, the retardation pattern shown in Figure 25.21 generates the maximum image distortion in this catadioptric step and scan system. Such a retardation pattern is likely caused by an imperfect process for annealing the photomask substrate. When the manufacturing process is refined and the maximum value of residual retardation in a photomask is controlled to below 2 nm, the angular pattern of residual birefringence may be significantly different from the pattern shown in Figure 25.21d, which will have an even smaller impact on degrading the imaging quality.

Finally, fused silica photomask blanks containing levels of residual retardation as high as that shown in Figure 25.21 are unusual in the industry today. Once the problem in photomask blanks was identified, the industry moved quickly to correct the problem. Consequently, using the special polarimeter, the material suppliers have the linear birefringence in photomask substrates well controlled.

25.4.1.4 Deep Ultraviolet Linear Birefringence Polarimeter

The 633 nm linear birefringence polarimeter does not measure birefringence at the working wavelengths of the optical lithography industry. So, the industry needed a deep ultraviolet (DUV) linear birefringence polarimeter to provide at-wavelength measurements.

The need for a DUV linear birefringence polarimeter was further accelerated by the development of a 157 nm lithographic tool several years ago. In a 157 nm step and scan system, calcium fluoride would become the primary optical material. CaF_2 is a single crystal that belongs to the cubic group with high degrees of symmetry (i.e., fourfold and threefold rotation axes). It is generally thought that single crystals in the cubic group have isotropic optical properties including index of refraction [76]. However, Burnett and coworkers reported [77,78] the measurement of intrinsic birefringence (spatial-dispersion-induced birefringence) in CaF_2 at UV wavelengths. They found that the intrinsic birefringence of CaF_2, $(n_{[-110]} - n_{[001]})$, is $-11.2 \pm 0.4 \times 10^{-7}$ at 157.6 nm. This birefringence corresponds to a retardation value of 11.2 ± 0.4 nm/cm (normalized to the thickness of a sample) when the light beam propagates along the [110] crystalline axis. The discovery of intrinsic bire-fringence in CaF_2 piqued the industry's interest in measuring birefringence at 157 nm and other lithographic wavelengths.

In the DUV linear birefringence polarimeter, a dual PEM design, similar to what is shown in Figure 25.11, is selected [79,80]. The light source is a deuterium lamp for optimizing the DUV wavelengths. The wavelength is selected by a monochromator. The choice of a deuterium lamp is due to both economics (high cost of suitable lasers) and flexibility (a deuterium lamp provides a range of DUV wavelengths). The light beam exiting the monochromator is collimated by a calcium fluoride lens. The two PEMs have modulation frequencies of 50 and 60 kHz, respectively. Both the polarizer and analyzer are MgF_2 Rochon polarizers. The sample is mounted on an XY scanning stage (450 mm × 450 mm) that is controlled by a PC. The detector is a CsI photomultiplier tube (PMT). The electronic signals generated at the PMT are processed using either lock-in amplifiers or a waveform analysis method developed by Jellison and coworkers [58–60].

The theoretical analysis of this optical configuration using the Mueller matrix calculus yields

$$V_{DC} = \frac{KI_0}{2} \qquad (25.21)$$

when the zero-order Bessel function of either PEM is set to 0.

The useful AC terms for determining linear retardation are:

the $(2\omega_1 + \omega_2)$ and $(\omega_1 + 2\omega_2)$ terms:

$$V_{2\omega_1 + \omega_2} = \frac{KI_0}{2} 2J_2(\delta 1_0) \cdot 2J_1(\delta 2_0) \cos(2\rho) \sin\delta \qquad (25.22a)$$

$$V_{2\omega_2 + \omega_1} = \frac{KI_0}{2} 2J_2(\delta 2_0) \cdot 2J_1(\delta 1_0) \sin(2\rho) \sin\delta \qquad (25.22b)$$

and the $(\omega_1 + \omega_2)$ term:

$$V_{\omega_1 + \omega_2} = \frac{KI_0}{2} 2J_1(\delta 1_0) \cdot 2J_1(\delta 2_0) \cdot \cos\delta \qquad (25.22c)$$

Defining R_1, R_2, and R_3 as corrected ratios of the AC signals to the DC signal, we have

$$\frac{V_{2\omega_1 + \omega_2}}{V_{DC}} \frac{1}{2J_2(\delta 1_0) \cdot 2J_1(\delta 2_0)} = R_1 = \cos(2\rho) \sin\delta \qquad (25.23a)$$

$$\frac{V_{2\omega_2 + \omega_1}}{V_{DC}} \frac{1}{2J_2(\delta 2_0) \cdot 2J_1(\delta 1_0)} = R_2 = \sin(2\rho) \sin\delta \qquad (25.23b)$$

$$\frac{V_{\omega_1 + \omega_2}}{V_{DC}} \frac{1}{2J_1(\delta 1_0) \cdot 2J_1(\delta 2_0)} = R_3 = \cos\delta \qquad (25.23c)$$

The linear retardation magnitude and angle of fast axis are expressed as

$$\delta = \tan^{-1}\left(\sqrt{\left(\frac{R_1}{R_3}\right)^2 + \left(\frac{R_2}{R_3}\right)^2}\right) \quad \text{or} \quad \delta = \cos^{-1} R_3 \qquad (25.24a)$$

FIGURE 25.24 Accuracy test data at 157 nm using a Soleil-Babinet compensator: (a) retardation and (b) angle of fast axis.

$$\rho = \frac{1}{2} \tan^{-1} \left[\frac{R_2}{R_1} \right] \quad \text{or} \quad \rho = \frac{1}{2} ctg^{-1} \left[\frac{R_1}{R_2} \right] \tag{25.24b}$$

Equations 25.24a and b give the correct values of magnitude and angle of fast axis of linear retardation in the range of $0 - \pi$. When the actual retardation is between π and 2π, the polarimeter will still report a retardation value between 0 and π but an angle of fast axis shifted by 90°. This is because the Mueller matrices are identical for both linear retarders (δ, ρ) and $(2\pi - \delta, 90° + \rho)$. Consequently, this seemingly large error has no impact for optical systems that can be modeled by Mueller matrices.

The accuracy of this linear birefringence polarimeter was tested using a magnesium fluoride Soleil-Babinet compensator. This process involves simply measuring a dozen different retardation values in the range of $0-2\pi$ (λ, the measuring wavelength) and linearly fitting the data from $0-\pi$ ($\lambda/2$) to $\pi-2\pi$. Figure 25.24a and b displays the measured linear retardation and fast axis values, respectively, when the Soleil-Babinet compensator was dialed to successive settings at a regular interval. The retardation values shown in Figure 25.24a form a linear relationship in each half ($Y = 21.217X - 49.599$, $R^2 = 1.0000$, and $Y = -21.419X + 209.08$, $R^2 = 1.0000$). The linear-fit equations for the two halves give nearly identical slopes with opposite signs, which indicate correct calibration and proper alignment of the optical components in the polarimeter. Ideally, the linear fit lines of the measured retardation values in the first and second halves should intersect at exactly one half of the measuring wavelength ($\lambda/2$), or 78.8 nm, as compared to the measured values of 79.2 nm.

This is a simpler procedure than what was used to obtain Figure 25.17. The Soleil-Babinet compensator used here does not need to be calibrated. Simpler compensators such as a Babinet compensator can also be used with this procedure. This simple procedure can be repeated at other lithographic wavelengths and when the compensator is set at different angles. Table 25.3 summarizes the test results of a few experiments. The instrument is estimated to have an accuracy error of <1%. Furthermore, the error found by this simple process can be used to correct it if necessary.

25.4.1.5 Measuring Linear Birefringence at Different DUV Wavelengths

The DUV linear birefringence polarimeter provides at-wavelength (157, 193, and 248 nm) measurement for CaF_2, fused silica, and other UV materials used in the optical lithography industry. A variety of CaF_2 samples were measured at those three DUV wavelengths with the measuring light beam propagating along the [111] crystal axis. For comparison, the same samples were also measured on a 633 nm linear birefringence polarimeter. The birefringence maps for most of the CaF_2 samples measured at 157 nm showed a similar pattern of fast axis angle and a higher linear retardation value when compared with the results measured at 633 nm. This is as expected from the dispersion of stress-birefringence with wavelengths.

However, a number of CaF_2 samples exhibited little correlation in the maps of fast axis angle measured at 157 and 633 nm [80]. Those samples also gave very different maps of linear retardation

TABLE 25.3

Accuracy Error of the DUV Polarimeter

SBC[a] Oriented at	Linear Fit of Measured Retardation Data (Y: Retardation; X: SBC Micrometer Reading)			Determined $\lambda/2$ Retardation	Relative Error (%)
157.6 nm					
0°	1st half	$Y = 21.392X - 52.409$	$R^2 = 0.9999$	79.19	+0.47
	2nd half	$Y = -21.119X + 209.1$	$R^2 = 1.0000$		
45°	1st half	$Y = 21.217X - 49.599$	$R^2 = 1.0000$	79.13	+0.40
	2nd half	$Y = -21.419X + 209.08$	$R^2 = 1.0000$		
193.4 nm					
0°	1st half	$Y = 20.339X - 50.383$	$R^2 = 1.0000$	96.89	+0.53
	2nd half	$Y = -20.419X + 244.31$	$R^2 = 1.0000$		
45°	1st half	$Y = 20.334X - 48.213$	$R^2 = 1.0000$	97.21	+0.20
	2nd half	$Y = -20.656X + 244.94$	$R^2 = 1.0000$		
248.3 nm					
0°	1st half	$Y = 19.53X - 48.697$	$R^2 = 1.0000$	125.2	+0.81
	2nd half	$Y = -19.628X + 300.47$	$R^2 = 0.9999$		
45°	1st half	$Y = 19.31X - 46.317$	$R^2 = 0.9999$	125.4	+0.97
	2nd half	$Y = -19.721X + 300.46$	$R^2 = 1.0000$		

[a] SBC, Soleil-Babinet compensator.

value measured at 157 and 633 nm. One such example is shown in Figure 25.25a and d. The two linear retardation maps shown in Figure 25.25a and d for the same CaF_2 sample are clearly different in both magnitude and angular patterns. The linear retardation maps measured at 193 and 248 nm are shown in Figure 25.25b and c for comparison. There is a gradual progression in the birefringence angular and magnitude patterns from 157 to 193, 248, and 633 nm.

Since the photoelastic coefficient of a material depends on the wavelength, a given stress in a material will exhibit different values of linear retardation at different wavelengths. From this stress-birefringence dispersion, one would expect to observe different values of retardation when measuring the same sample at different wavelengths. However, stress-birefringence dispersion would not explain a change in the angle of fast axis measured at different wavelengths. After all, the fast axis of stress-birefringence is determined by the direction of the stress-force; this does not change with measuring wavelengths. There must be a different factor to explain the angular pattern changes shown in Figure 25.25a through d.

In addition to stress-birefringence, intrinsic birefringence is also a factor in CaF_2, especially at short wavelengths. Intrinsic birefringence has an acute angular dependence in CaF_2 [77,78]. While it is zero when a light beam propagates along the [111] crystal axis, the intrinsic birefringence reaches a maximum when a light beam propagates along the [110] crystal axis. A perfect CaF_2 crystal has no intrinsic birefringence along the [111] axis. However, this would not be the case if a CaF_2 sample has serious crystal defects where the crystal axis may vary significantly in different domains. When a low-quality [111] CaF_2 sample is measured at 157 nm, variations in crystal axis can cause the intrinsic birefringence to contribute to the measured retardation.

When measured at 633 nm, intrinsic birefringence in CaF_2 is negligible due to its sharp decrease at longer wavelengths. Therefore, the 633 nm linear retardation result shown in Figure 25.25d is entirely due to stress birefringence. However, when this sample is measured at 157 nm, there is an additional contribution to total linear retardation from intrinsic birefringence. This is caused by crystal axis "wandering" in poor quality CaF_2 samples. X-ray imaging data of CaF_2 samples with poor crystal quality supports this argument [81].

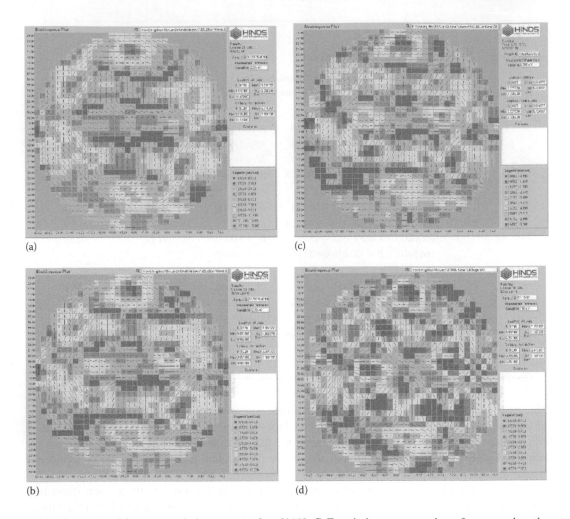

(a)   (c)

(b)   (d)

FIGURE 25.25 Linear retardation maps of a [111] CaF$_2$ window measured at four wavelengths. (a) 157 nm (display scale for retardation: 0–15 nm; averaged retardation: 5.77 nm; number of data points: 1020); (b) 193 nm (display scale for retardation: 0–10 nm; averaged retardation: 3.89 nm; number of data points: 1020); (c) 248 nm (display scale for retardation: 0–8 nm; averaged retardation: 3.01 nm; number of data points: 1020); and (d) 633 nm (display scale for retardation: 0–7 nm; averaged retardation: 2.47 nm; number of data points: 1000).

25.4.1.6 DUV Polarimeter for Measuring Lenses

The DUV special polarimeter described earlier has fixed source and detector modules. It was designed to measure samples with parallel surfaces. To measure a lens, the polarimeter must fulfill several additional requirements.

A lens will bend a light beam by refraction. The detector module must first be moved to where the refracted beam is located after the lens and then be tilted normal to the refracted beam for polarization measurement. In addition, to model how a lens is used in an optical system, the incident angle of the measuring beam to the lens being tested will be controlled. Therefore, both the detector and source modules require linear and rotational controls.

Figure 25.26 shows an engineering diagram of the so-called Exicor OIA (oblique incident angle) birefringence measurement system, which is specifically designed for measuring lenses. In this system, the sample stage provides three degrees of freedom of motion. A lens under test can be moved in the horizontal plane (*XY* plane) with two stages allowing translation up to 500 mm in both *X* and *Y* axes. A lens sample can also be rotated by 360° in the *XY* plane during test. Additionally, each of

FIGURE 25.26   An engineering diagram of the Exicor OIA birefringence measurement system.

the source module and detector module has two degrees of freedom of motion, translational in the X-axis up to 1250 mm and tilt from −50° to 50°. All motions are accurately controlled by a computer. The motion control allows us to measure residual birefringence in a lens at different incident angles. Note that all components inside each module remain fixed relative to one another.

The light source used in this system provides a well-collimated beam from 190 to 800 nm. This instrument can provide birefringence mapping at fixed wavelengths, such as 193, 248, 365, 633 nm and any user defined wavelength from DUV to NIR. With an added option, the Exicor OIA system can also provide spectroscopic measurement from DUV to NIR at a fixed position on the sample.

We have measured several lenses at 193 nm using this system. One of them is an optical lithographic lens. The photo of this lens is shown in Figure 25.27a. There is a significant curvature on the top surface of this lens. The incident angles to the top surface of this lens were so chosen that the measuring beam after the refraction of the top surface is parallel to the vertical direction for all measurements. The measured residual birefringence map is shown in Figure 25.27b. Excluding the peripheral region, nearly all measured values of residual linear retardation for this lens are below 0.4 nm. This is a very high-quality lens with extremely low residual birefringence.

(a)   (b)

FIGURE 25.27   (a) A photo picture of a lithographic lens and (b) map of measured linear retardation values of the same lens with normal incidence to the top surface (measuring wavelength: 193 nm).

25.4.2 Near-Normal Reflection Dual PEM Polarimeter and Applications in the Nuclear Fuel Industry

Previously, I mentioned that Jellison championed the dual PEM design and applied it to measuring retardation and diattenuation properties of both transmissive and reflective samples. In this section, I describe one particular example of Jellison's recent work.

Jellison and coworkers recently extended the 2-MGE to a two-modulator generalized ellipsometry microscope (2-MGEM) [61–63]. The 2-MGEM instrument simultaneously measures eight elements of the Mueller matrix in near-normal reflection mode with 4-μm spatial resolution. One of the applications of the 2-MGEM is to measure the linear diattenuation of the pyrocarbon layers in the cross-section of TRistructural-ISOtropic (TRISO) nuclear fuel particles [63]. Traditionally, the nuclear industry has measured the OPTical Anisotropy Factor (OPTAF), which is related to the linear diattenuation (defined as N in Refs. [58–63]), where OPTAF = $R_{max}/R_{min} = (1 + Ld)/(1 - Ld)$.

TRISO nuclear fuel is used in many of the fourth-generation nuclear power reactor designs, such as the advanced gas reactor (AGR). Each TRISO particle is ~1 mm in diameter and contains a kernel of radioactive material and four layers designed to contain the radioactive material during irradiation. The four layers are (1) a buffer layer, consisting of porous carbon, (2) an inner pyrocarbon layer (IPyC), (3) a silicon carbide (SiC) layer for structural integrity, and (4) an outer pyrocarbon layer. The optical anisotropy (expressed either as the OPTAF or the linear diattenuation) of the two pyrocarbon layers is a critical quality control factor, which can be accurately measured by the 2-MGEM. Furthermore, the 2-MGEM provides a "map" of the diattenuation over the entire TRISO particle cross-section, making enhanced characterization possible.

This 2-MGEM OPTAF system was commercialized in 2007 through a collaboration between Oak Ridge National Laboratory and Hinds Instruments, Inc. Two commercial systems have been built so far. Figure 25.28a shows the reflected light intensity of a TRISO nuclear fuel particle obtained using this system, while Figure 25.28b shows the magnitude of the diattenuation N for the same particle. The kernel (containing the nuclear material) has been removed and replaced with potting epoxy. Clearly, the IPyC and OPyC layers show significant diattenuation, while the other layers do not. Other results are shown in Refs. [61–63].

Comparisons between measurements taken on the system at Oak Ridge and the commercial version obtain nearly identical results, as do measurements taken on the same samples at different times. Therefore, the 2-MGEM system provides reproducible data that is useful for the quality assurance requirements of the nuclear fuel industry. It is my hope that scientists and engineers within the nuclear fuel industry can read the diattenuation data and decipher the quality control information to improve the performance of TRISO fuel particles.

25.4.3 Special Polarimeters for the FPD Industry

The liquid crystal display (LCD) has become the dominant FPD technology in consumer electronics, including digital cameras, cell phones, PC monitors, and TVs. Since an LCD depends on the birefringence properties of an LC material to achieve its display functions, the birefringence properties are critical to performance for all key optical components used in an LCD. Consequently, special polarimeters have been developed for a variety of applications in the LCD industry.

25.4.3.1 Measuring Glass Substrate

The glass substrate of an LCD is usually thinner than 1 mm. The linear retardation is typically below 1 nm in high-quality LCD glass substrates. The PEM linear birefringence polarimeter, which serves the optical lithography industry well with its high sensitivity, is also suited to measuring the stress birefringence (and stress) in LCD glass substrates.

(a)

(b)

FIGURE 25.28 (a) Grayscale image of the reflected light intensity; (b) a map of the linear diattenuation from a cross-section of a TRISO nuclear fuel particle. Notice that the outmost layer shown was an epoxy layer introduced when making the cross-section of the TRISO particle.

25.4.3.2 Measuring LCD-FPD Components at RGB Wavelengths

The LCD-FPD industry also requires at-wavelength measurement. For this industry, the wavelengths are red, green, and blue (RGB). One way to build an RGB linear birefringence polarimeter is to use a Hg–Xe lamp as the light source. This source emits strong peaks at both blue and green wavelengths in addition to the broad spectral irradiation in the visible spectrum [82]. The strong blue and green mercury emissions improve the accuracy and sensitivity of measurements at those wavelengths. Other light sources include LEDs, fluorescent lamps, lasers, and black-body thermal light sources.

Figure 25.29 shows the linear retardation maps of a stretched polymer film measured at the blue, green, and red wavelengths. The same retardation pattern was observed at all three wavelengths.

FIGURE 25.29 Retardation maps of a stretched polymer film sample measured at top middle, and bottom wavelengths.

The average retardation values measured at RGB increase from 8.38 to 8.66 nm and 9.33 nm, respectively, as expected due to the birefringence dispersion of this polymer film sample.

An advantage of the RGB polarimeter described earlier is that it eliminates possible interference effects resulting from the use of a coherent laser to measure thin films. It is true that many of the film samples I have measured exhibit little laser interference. However, strong laser interference has been observed in some thin film samples. Two examples, one with and the other without laser interference effect, are shown in Figures 25.30 and 25.31.

Figure 25.30a and b displays the linear retardation maps of the same thin film sample measured on the RGB polarimeter and a polarimeter using a He–Ne gas laser, respectively. The averaged retardation values for those two measurements are in good agreement (6.29 and 6.21 nm). The retardation patterns displayed in Figure 25.30a and b are very similar. One visual difference between the two maps is their spatial resolutions. Namely, Figure 25.30b exhibits a higher spatial resolution than Figure 25.30a does, but that is caused by the smaller beam size of the laser used. This thin film sample exhibits no observable laser interference.

Figure 25.31a and b illustrates an example with significant laser interference. Figure 25.31a and b displays the linear retardation maps of a thin film sample measured on the RGB polarimeter and a He–Ne gas laser polarimeter, respectively. Although the averaged retardation values for those two measurements are in good agreement (10.89 and 10.81 nm), the retardation pattern depicted in Figure 25.30b clearly shows laser interference. It is, therefore, advantageous to use a noncoherent light source in a polarimeter for thin film applications.

(a)   (b)

FIGURE 25.30 Linear retardation maps of a thin film sample measured on (a) the RGB polarimeter and (b) a polarimeter using a He–Ne gas laser.

(a)

(b)

FIGURE 25.31 Linear retardation maps of a thin film sample measured on (a) the RGB polarimeter and (b) a polarimeter using a He–Ne gas laser, showing laser interference effect.

25.4.3.3 Measuring Retardation Compensation Films at Normal and Oblique Angles

Retardation compensation films are key optical components used in the LCD to improve the viewing angle. Different types of sophisticated compensation films have been developed recently, including biaxially stretched polymers, discotic compensation films, and photo-aligned anisotropic films [83–86]. Ideally, a compensation film should have uniform in-plane retardation and a well-defined pattern for out-of-plane (vertical) retardation. The retardation value and uniformity in a compensation film directly affect the performance of an LCD at normal and oblique viewing angles. A polarimeter is required to provide both in-plane and out-of-plane measurements for these films.

A polarimeter using a tilt stage can provide retardation measurements at normal and different oblique incident angles. A typical set of data is shown in Figure 25.32a for a compensation film measured at different incident angles [87]. This compensation film has an in-plane retardation value of ~14 nm. To avoid complicated additions of in-plane and out-of-plane linear retardation, the sample was oriented so that the fast axes of both the normal and oblique retardations were approximately aligned to 0°. Figure 25.32a and b shows the measured retardation and angles of fast axis in the tilt range from −50° to 50° in 5° increments. The pattern of oblique retardation with tilt angle is used to calculate vertical birefringence, traditionally the $\left(\left(R_{\mathrm{th}}=\left|n_Z-\left((n_X+n_Y)/2\right)\right|d\right)\right)$ parameter, and ultimately to improve the viewing angle in LCD designs.

25.4.3.4 Measuring Retardation at a Fast Speed

For LCD-FPD industrial applications, birefringence polarimeters should be able to operate in two modes. In a so-called Lab mode, PC-controlled XY linear stages provide precise motion control to accurately map the birefringence property of a sample. This is an "off-line" application, and a data

(a)

(b)

FIGURE 25.32 Birefringence data measured at normal and different oblique angles of a stretched polymer film: (a) oblique retardation values and (b) angles of fast axis.

collection rate of several measurements per second is usually sufficient. In the "Lab" mode, precise motion control may consume most of the time in a measurement cycle. In a so-called Fab mode, sample movement is normally handled in a production process. The speed of a fast moving sample demands a fast data collection rate. An "in-line" application such as this is typically involved in monitoring the quality of polymer films or thin glass sheets in production. In the "Fab" (semiconductor fabrication plant) mode, a PEM-based polarimeter, with fast modulation of the PEM (tens of kHz), easily affords a rate of over 100 data points per second.

To simulate an in-line film monitoring process in our laboratory, we made a motor-driven rolling loop using a polymer film [88]. In this experiment, the polarimetric function was separated from the mechanical function of moving the polymer film. Figure 25.33a and b shows the repeatable patterns of retardation magnitude and angle of fast axis measured in two rolling cycles. The sampling rate for the experiment was ~140 data points per second.

The PEM-based special polarimeter gives a distinctive advantage in the "Fab" mode operation over other polarization modulation methods. The superior optical properties of the PEM deliver polarimetric measurements characterized by high sensitivity and accuracy. At the same time, the fast modulation frequency of the PEM (tens of kHz) furnishes a sampling rate of hundreds of data points per second. For an "in-line" application, polarimeters using rotating polarizers or wave plates are too slow. Electro-optic modulators can modulate very fast, but they do not provide sufficient sensitivity for many polarimetric measurements. A PEM-based special polarimeter affords both high sensitivity and fast data collection, and thus, it is best suited to "in-line" monitoring of glass sheet and thin film productions.

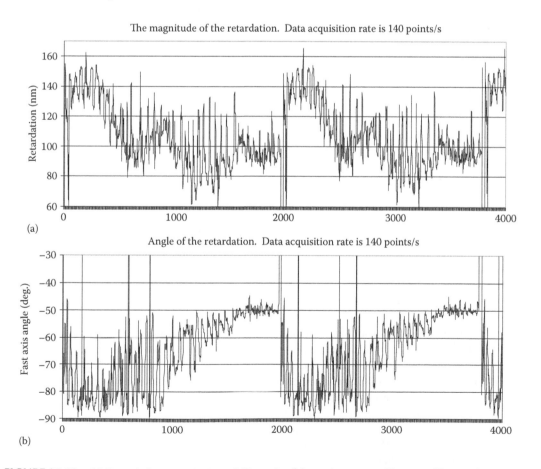

FIGURE 25.33 (a) Retardation magnitude and (b) angle of fast axis measured in two rolling cycles.

25.4.4 SPECIAL POLARIMETERS FOR CHEMICAL, BIOCHEMICAL, AND PHARMACEUTICAL APPLICATIONS

In chemical, biochemical, and pharmaceutical applications, a polarimeter is primarily used to characterize chiral molecules. A molecule is chiral when it cannot be superimposed with its mirror image. Since many molecules in the bodies of human beings and animals, including amino acids and proteins, are chiral, most drugs used to treat diseases are also chiral. The chiral branches of the chemistry, biochemistry, and pharmaceutical industries are critical to the health of living beings.

For a chiral molecule, the two nonsuperimposable mirror images are called enantiomers. Under normal conditions, enantiomers have identical physical and chemical properties, except for how they interact with circularly polarized light. Polarimetry is the key optical means for distinguishing between enantiomers and studying chiral molecules. It is beyond the scope of this chapter to discuss chiral molecule applications in detail. However, I will briefly introduce three major types of polarimeters used in chemical, biochemical, and pharmaceutical applications. Interested readers are referred to a collection of articles in natural optical activity [89].

25.4.4.1 Circular Dichroism Spectrometer

A CD spectrometer measures the differential absorption of a chiral sample between left and right circularly polarized light ($\Delta A = A_L - A_R$). The CD signals of a chiral sample are usually very small. It is impossible to determine a typical CD signal by measuring A_L and A_R separately and then calculating the difference. High-quality polarization modulation must be used in a CD spectrometer. Although there are different types of polarization modulators, all modern CD spectrometers use the PEM due to its superior optical properties.

In a CD spectrometer, the polarization modulation module contains a polarizer followed by a PEM where the optical axes of the two components are 45° apart. The sample is placed between the PEM and the detector. Note that there is no analyzer in front of the detector in a CD spectrometer. The PEM is set to modulate between left and right circularly polarized light. When a sample exhibits no differential absorption between left and right circularly polarized light, the detector receives a constant light intensity with no modulated signal. When a chiral sample absorbs left and right circularly polarized light differently, the detector will receive a modulated CD signal.

CD spectrometers are common in UV–vis, near IR, and mid-IR spectral regions. In academic research, CD spectrometers are primarily used to extract different levels of molecular structural information about the molecule and its interactions with other molecules. In industrial applications, CD spectrometers are often used to detect the concentration of a desired enantiomer or the enantiomeric excess of a chiral compound. CD spectrometers in both UV–vis and IR spectral regions have been commercially available for decades from a number of manufacturers.

25.4.4.2 Optical Rotation Polarimeter

Optical rotation is also called circular birefringence. An optical rotation polarimeter measures the rotation of the polarization plane of a linearly polarized light beam when it passes through a chiral sample. The simplest optical rotation polarimeter is constructed by using a polarizer and a crossed analyzer. The optical rotation is the angular difference in the null positions on the analyzer with and without a chiral sample. Simple optical rotation polarimeters have been used in the sugar industry for nearly two centuries. An optical rotation polarimeter is perhaps the most commonly used special polarimeter in industrial and analytical laboratories. In fact, when people use the term "polarimeter," most refer to an optical rotation polarimeter.

When a PEM is used in an optical rotation polarimeter, the polarization modulation module may contain a polarizer at 0°, a PEM at 0° and an analyzer at 45°. The sample can be placed between the polarizer and the PEM. When there is no optical rotation, the detector receives no modulated signal, because the polarizer and the optical axis of the PEM are parallel to each other. When the sample rotates the polarization plane away from 0°, the detector will receive a modulated signal at the second harmonic of the PEM. The PEM-based optical rotation polarimeter provides high sensitivity.

25.4.4.3 Polarization Analyzer for Fluorescence and Scattering

Fluorescence and light scattering from chemical and biochemical molecules contains polarization information that is useful in studying the properties and chemical dynamics of the molecules. While the chemical and biochemical processes can be complicated, the polarimetric applications are similar to what has been described in previous sections. A special Stokes polarimeter using a single PEM, as shown in Figure 25.3, is often sufficient to measure the polarization state of the fluoresced or scattered light beam, although complete Stokes and Mueller polarimeters are used for some applications.

25.4.5 STRESS BIREFRINGENCE POLARIMETER FOR Si USED
IN THE SOLAR PHOTOVOLTAIC INDUSTRY

Silicon (Si)-based solar cell technology dominates the commercial photovoltaic (PV) market. Controlling the cost of Si material is one of the most significant factors in the production of Si solar panels. Material losses occur throughout the manufacturing process, mostly due to stress and defects that arise from the crystal growth process. Stress in Si crystals can remain undetected long into the fabrication process, until ultimately the end product (finished cell) fails, or its performance is severely compromised. Measuring stress in Si components can improve the yield of high-quality solar panels.

Several models of birefringence polarimeter using the Exicor technology are available to measure stress in Si components used in producing solar panels. A special polarimeter as described in Section 25.4.1 using NIR light source and detector can be used to map the stress in solar wafers and other samples (<160 mm × 160 mm). A modular polarimeter as described in Section 25.4.3.4 can be adopted to NIR for in-line measurement of stress in Si wafers at a high speed. A birefringence polarimeter with large stages can be used to measure stress in large Si ingots. It is advantageous to measure stress in Si ingots early in the production process so that low-quality segments in an Si ingot can be identified before sawing them into wafers, thus avoiding the subsequent processing costs incurred in the solar cell fabrication process. Figure 25.34 illustrates an example of stress measurement in a squared monocrystal Si ingot [96].

The test segment of an Si ingot was cut from p-type Czochralski (CZ)-grown Si crystals. The as-grown ingot was shaped into 156 mm × 156 mm pseudosquares. The surfaces of the squared ingot were finished using a grinding process. The Si ingot was measured with the measuring beam parallel to the [100] crystal direction. To calculate stress from measured linear retardation, we use the stress-optic law

$$\sigma = \frac{\delta}{t \cdot C_{\mathrm{B}}} \tag{25.25}$$

where
  σ is the differential stress (in MPa; 10^6 pascal)
  δ is the linear retardation (in nm)
  t is the thickness of the sample (in mm)
  C_{B} is the stress-optical constant (in Brewsters or 10^{-12} Pa^{-1}; $C_{\mathrm{B}} = 24$ for Si$_{(100)}$ as an average of four reported values) [97]

Figure 25.34a depicts the 2D retardation map for a unique segment of an Si ingot (about 310 mm in length). The false color retardation image indicates multiple orders of retardation along the crystal growth direction of the ingot. Figure 25.34b and c illustrates, respectively, the measured raw retardation and calculated stress values for a line of data in the middle of the 2D map.

FIGURE 25.34 (a) 2D linear retardation map of stressed Si ingot; (b) line data for measured linear retardation (The right end of the Si ingot is labeled "0" on the horizontal axis. Each data point represents a 1 mm spacing.); (c) calculated stress; and (d) bowing of Si wafers produced from the same ingot.

Moving in the crystal growth direction from right to left, the 2D map starts with retardation values that are low and uniform, increases significantly and continuously as concave-curved striation patterns form, and finally develops into more complicated patterns of multiple orders of retardation. The appearance of the pronounced concave-shaped pattern and its evolution to a rather complex stress distribution indicate enhanced defect densities due to radial stress relaxation toward the outside.

Shown in Figure 25.34b and c, from right to left, the measured raw retardation values start low, which correspond to a very low level of stress (≤0.1 MPa) for the first 50 mm in length from the right end of the Si ingot. The measured raw retardation values then go through several peak-to-valley cycles until 210 mm from the end (right side) of the Si ingot. Using the measured angles of fast axis, we can convert the measured raw retardation values to corrected retardation values, then to stress in MPa. As seen in Figure 25.34c, the stress increases almost continuously from 0.1 MPa at 50 mm (from the right end of the Si ingot) to about 1.4 MPa at 210 mm. The stress remains fairly constant around 1.4 MPa beyond 210 mm for the left-end segment of the Si ingot.

We have sliced this ingot into 200 μm (thickness) wafers. Figure 25.34d shows the average bow of these wafers with respect to their axial position in the crystal. We can clearly see a correlation between the levels of stress in the Si ingot and the magnitudes of bowing in the wafers sliced from that ingot. The results shown in Figure 25.34 suggest that stress above 0.5 MPa in monocrystal Si ingots could result in strong wafer bow, thus severely lowering the quality of a wafer.

25.4.6 PEM–CAMERA IMAGING POLARIMETER

As previously mentioned, the main challenge to employing PEMs in a camera-based imaging instrument is that CCD or CMOS (complementary metal–oxide–semiconductor) cameras have a much slower readout rate than the modulation frequency of a PEM (tens of kHz), such that the specific polarization information carried on the modulation frequency of the PEM will be averaged out. Generally speaking, there are three basic approaches to overcome this challenge: slowing down the modulation, pulsing the light source, and gating the camera according to the modulation of the PEM.

The first approach is to slow down polarization modulation using the beat frequency of two PEMs. This approach, developed by Diner et al. [41,42], is briefly described in Section 25.2.2.

The second approach is to periodically expose, or gate, the camera in synchronization with the modulation of the PEM. Masking selected rows of the CCD according to the modulation of the PEM, which was developed by the ETH Zurich group [37–40], is one such example. This second approach is also briefly described in Section 25.2.2.

With today's CCD and CMOS technologies, gating the camera becomes easier. For example, an intensified CCD (ICCD) camera has an image intensifier that can act as both a fast (nanoseconds) optical shutter and a signal amplifier [98]. Such an ICCD may be directly used in a PEM–camera polarimeter, with the reference signal of the PEM triggering the intensifier. Lack of industrial applications and the high cost of ICCD cameras may be the primary reasons for not having a commercially available PEM–ICCD polarimeter. In addition to ICCD, Vitkin and coworkers used field programmable gate array (FPGA) to trigger the camera according to the relative phase of two PEMs [99]. Subsequently, the same group has extended the FPGA triggering method to a complete Mueller polarimeter using four PEMs [100].

The third approach is to generate short pulses of a light source in synchronization with the modulation frequency of a PEM (so-called stroboscopic illumination) so that the camera only receives light pulses carrying specific polarization information. Gleyzes, Boccara, and Saint-Jalmes used this approach in 1997 to build an imaging interferometer by integrating a PEM into a commercial microscope [101]. Dubois, Boccara, and coworkers subsequently developed a full-field microscope for optical coherence tomography (OCT) applications using stroboscopic illumination where an infrared LED was pulsed at four specific phases during a modulation cycle of the PEM [102]. Chao and coworkers further applied stroboscopic illumination to PEM imaging ellipsometry [103–105].

In our laboratory, we have built several PEM–camera imaging polarimeters using stroboscopic illumination. In one of the PEM–camera imaging polarimeters, we use a green (530 nm) LED array that is pulsed according to the modulation phases of two PEMs. The field of view of this polarimeter is about 60 mm × 60 mm. Figure 25.35 displays the measured linear retardation images of a plastic ruler and a quartz wedge coupled to a matching fused silica wedge.

The quartz wedge provides linearly varying retardation values in the horizontal direction. Since we use a 530 nm light source, this PEM–camera imaging polarimeter has an upper measurement limit of 265 nm. Consequently, the reported retardation values show a saw-tooth pattern with peaks (265 nm) and valleys (0 nm) along the horizontal direction of the quartz wedge. In Figure 25.35c, a single row of data from the image shown in Figure 25.35a is plotted to exhibit the saw-tooth pattern more clearly. Note that the good linearity of measured retardation values between 0 and 265 nm is an indication of the good accuracy of this PEM imaging polarimeter. Furthermore, the measured angles of fast axis switch 90° in every peak or valley cycle as shown in Figure 25.35b. As is well known, using two or more wavelengths in the measurements can resolve multiple orders of retardation.

The plastic ruler was a poor choice of sample in terms of industrial application. We have measured many cell phone panels (one example of an outdated cell phone panel is shown in Figure 25.35d),

FIGURE 25.35 Measured retardation images of a mid-field PEM imaging polarimeter. (a) and (b) retardation magnitude and angle images of a plastic ruler and a quartz wedge; (c) a single row of data from image (Figure 25.35a); (d) retardation image of a cell phone panel.

FIGURE 25.36 Preliminary retardation image of a slice of mouse brain. (The tissue section is a courtesy of S. Segal, Saint Louis University School of Medicine, St. Louis, MO.)

waveplates, and special wafers that are more relevant to industrial applications. However, I particularly like this plastic ruler, because it has a large variation of retardation, it directly displays the field size, and its image indicates how effectively the algorithm eliminates certain measurement artifacts.

Another PEM–camera system that we built in our laboratory is a microscopic polarimeter. This polarimeter contains a simple, home-made microscope with a low magnification (2× or 5×) objective, two PEMs, a modest resolution (4 megapixels) camera, and an FPGA-triggered LED light source. This particular device has proven effective in resolving birefringence found in biological specimens. One such birefringent substance is myelin, which is a lipoprotein found in the nervous system, particularly in the white matter. Myelin wraps around the elongated part of neuronal cells known as axons, forming cable-like structures coated with an insulating (myelinated) cuff. Axons (also termed fibers) tend to form thicker bundles with millions of myelinated axonal fibers known as white matter tracts. White matter tracts containing myelin are characteristically birefringent. Since other parts of the brain tissue are not as rich in myelin as the white matter tracts, there is significant contrast between myelinated and nonmyelinated brain tissue in birefringence imaging [106]. The demonstration of the white matter tracts birefringence at cross section (thickness = 100 μm) in a sample from the mouse brain is shown in Figure 25.36.

Our goal is to provide a series of polarization imaging tools to research communities and industrial applications.

REFERENCES

1. http://dictionary.webster.us/polarimetry; *Webster's Revised Unabridged Dictionary*, 1913.
2. R. A. Chipman, Polarimetry, in *Handbook of Optics I*, 3rd edn., M. Bass, editor in chief, McGraw-Hill, New York, Chapter 15, 2010.
3. D. H. Goldstein, D. B. Chenault et al., Polarization: Measurement, analysis, and remote sensing I–XI, *Proc. SPIE* from 1997 to 2014.
4. R. M. A. Azzam, Optical polarimetry: Instrumentation and applications, *Proc. SPIE* **112** (1977).
5. H. Mott and W. -M. Boerner, Radar polarimetry, *Proc. SPIE* **1748** (1994).
6. A. Adamson, C. Aspin, C. J. Davis, and T. Fujiyoshi, Astronomical polarimetry: Current status and future directions, *ASP Conference Series* **343** (2005).
7. S. Fineschi, Polarimetry in astronomy, *Proc. SPIE* **4843** (2003).
8. R. M. A. Azzam and N. M. Bashara, *Ellipsometry and Polarized Light*, North-Holland, Amsterdam, the Netherlands, reprinted 1989.
9. D. H. Goldstein *Polarized Light*, 3rd edn., Taylor & Francis, New York, 2010.
10. J. Tinbergen, *Astronomical Polarimetry*, Cambridge University Press, Cambridge, MA, 2005.
11. J. Schellman and H. P. Jensen, *Chem. Rev.* **87**, 1359 (1987).
12. S.-M. F. Nee, Polarization measurement, in *The Measurement, Instrumentation and Sensors Handbook*, J. G. Webster, ed., CRC Press, Boca Raton, FL, 1999.
13. P. S. Hauge, *Surf. Sci.* **96**, 108 (1980).
14. J. M. Bueno, *J. Opt. A: Pure Appl. Opt.* **2**, 216 (2000).
15. J. S. Tyo, D. H. Goldstein, D. B. Chenault, and J. A. Shaw, *Appl. Opt.* **45**, 5453 (2006).
16. C. U. Keller, Instrumentation for astrophysical spectropolarimetry, in *Astrophysical Spectropolarimetry*, J. Trujillo-Bueno, F. Moreno-Insertis, and F. Sanchez, eds., Cambridge University Press, Cambridge, MA, 2002.
17. J. C. Kemp, *J. Opt. Sci. Am.* **59**, 950 (1969).
18. www.hindsinstruments.com.
19. G. G. Stokes, *Trans. Camb. Philos. Soc.* **9**, 399 (1852).
20. D. S. Kliger, J. W. Lewis, and C. E. Randall, *Polarized Light in Optics and Spectroscopy*, Academic Press, San Diego, CA, 1990.
21. C. Brosseau, *Fundamentals of Polarized Light—A Statistical Optics Approach*, John Wiley & Sons, New York, 1998.
22. W. A. Shurcliff, *Polarized Light: Production and Use*, Harvard University Press, Cambridge, MA, 1962.
23. R. Clark Jones, *J. Opt. Soc. Am.* **31**, 488; 493; 500 (1941); **32**, 486 (1942).
24. H. Mueller, *J. Opt. Soc. Am.* **38**, 661 (1948).

25. P. S. Theocaris and E. E. Gdoutos, *Matrix Theory of Photoelasticity*, Springer-Verlag, Berlin, Germany, 1979.
26. R. P. Feynman, R. B. Leighton, and M. Sands, *The Feynman's Lectures on Physics*, Vol. I, p. 3202, Addison-Wesley Publishing, Menlo Park, CA, 1977.
27. D. Brewster, *Philos. Trans.* **106**, 156 (1816).
28. M. Billardon and J. Badoz, *C. R. Acad. Bc. Paris Ser. B* **262**, 1672 (1966).
29. L. F. Mollenauer, D. Downie, H. Engstrom, and W. B. Grant, *Appl. Opt.* **8**, 661 (1969).
30. S. N. Jasperson and S. E. Schnatterly, *Rev. Sci. Instrum.* **40**, 761 (1969); Errata **41**, 152 (1970).
31. J. C. Kemp, US Patent No. 3,867,014 (1975).
32. B. Wang and J. List, *Proc. SPIE* **5888**, 436 (2005).
33. J. C. Kemp, J. B. Swedlund, and B. D. Evans, *Phys. Rev. Lett.* **24**, 1211 (1970).
34. J. C. Kemp, R. D. Wolstencroft, and J. B. Swedlund, *Astrophys. J.* **177**, 177 (1972).
35. J. C. Kemp, G. D. Henson, C. T. Steiner, and E. R. Powell, *Nature* **326**, 270 (1987).
36. J. C. Kemp, *Proc. SPIE* **307**, 83 (1981).
37. J. O. Stenflo and H. Povel, *Appl. Opt.* **24**, 3893 (1985).
38. H. P. Povel, C. U. Keller, and A. Yadigaroglu, *Appl. Opt.* **33**, 4254 (1994).
39. A. M. Gandorfer and H. P. Povel, *Astron. Astrophys.* **328**, 381 (1997).
40. A. M. Gandorfer, H. P. Povel, P. Steiner, F. Aebersold, U. Egger, A. Feller, D. Gisler, S. Hagenbach, and J. O. Stenflo, *Astron. Astrophys.* **422**, 703 (2004).
41. D. J. Diner, A. Davis, B. Hancock, G. Gutt, R. A. Chipman, and B. Cairns, *Appl. Opt.* **46**, 8428 (2007).
42. D. J. Diner, A. Davis, B. Hancock, S. Geier, B. Rheingans, V. Jovanovic, M. Bull, D. M. Rider, R. A. Chipman, A. -B. Mahler, and S. C. McClain, *Appl. Opt.* **49**, 2929–2946 (2010).
43. J. O. Stenflo, *Mem. S. A. It.* **78**, 181 (2007).
44. J. O. Stenflo, *Rev. Mod. Astron.* **17**, 269 (2004).
45. S. S. Varnum, US Patent No. 5,744,721 (1998).
46. G. R. Boyer, B. F. Lamouroux, and B. S. Prade, *Appl. Opt.* **18**, 1217 (1979).
47. B. Wang, J. List, and R. R. Rockwell, *Proc. SPIE* **4819**, 1 (2002).
48. http://en.wikipedia.org/wiki/Tokamak.
49. www.tokamak.info.
50. F. M. Levinton, R. J. Fonck, G. M. Gammel, R. Kaita, H. W. Kugel, E. T. Powell, and D. W. Roberts, *Phys. Rev. Lett.* **63**, 2060 (1989).
51. F. M. Levinton, G. M. Gammel, R. Kaita, H. W. Kugel, and D. W. Roberts, *Rev. Sci. Instrum.* **61**, 2914 (1990).
52. D. Wroblewski and L. L. Lao, *Rev. Sci. Instrum.* **63**, 5140 (1992).
53. H. Friedrich, *Theoretical Atomic Physics*, Springer-Verlag, Berlin, Germany, 1990.
54. B. Wang, R. R. Rockwell, and A. Leadbetter, *Proc. SPIE* **5888**, 33 (2005).
55. R. M. A. Azzam, *Opt. Lett.* **2**, 148 (1977).
56. D. H. Goldstein and R. A. Chipman, *J. Opt. Soc. Am. A* **7**, 693 (1990).
57. M. H. Smith, *Appl. Opt.* **41**, 2488 (2002).
58. G. E. Jellison Jr. and F. A. Modine, *Appl. Opt.* **36**, 8184; 8190 (1997).
59. G. E. Jellison Jr. and F. A. Modine, US Patent No. 5,956,147 (1999).
60. G. E. Jellison Jr., C. O. Griffiths, D. E. Holcomb, and C. M. Rouleau, *Appl. Opt.* **41**, 6555 (2002).
61. G. E. Jellison Jr., D. E. Holcomb, J. D. Hunn, C. M. Rouleau, and G. W. Wright, *Appl. Surf. Sci.* **253**, 47 (2006).
62. G. E. Jellison Jr., J. D. Hunn, and C. M. Rouleau, *Appl. Opt.* **45**, 5479–5488 (2006).
63. G. E. Jellison Jr., J. D. Hunn, and R. A. Lowden, *J. Nucl. Mater.* **352**, 6 (2006).
64. R. C. Thompson, J. R. Bottiger, and E. S. Fry, *Appl. Opt.* **19**, 1323 (1980).
65. R. A. Chipman, *Proc. SPIE* **6682**, 66820I-1 (2007).
66. B. Wang and T. C. Oakberg, *Rev. Sci. Instrum.* **70**, 3847 (1999).
67. B. Wang, T. C. Oakberg, and P. Kadlec, US Patent No. 6,473,179 (2002).
68. B. Wang, T. C. Oakberg, and P. Kadlec, US Patent No. 6,697,157 (2004).
69. B. Wang, *Proc. SPIE* **4103**, 12 (2000).
70. B. Wang and W. Hellman, *Rev. Sci. Instrum.* **72**, 4066 (2001).
71. G. E. Moore, *Electronics*, **38**(8), 114–117 (April 19, 1965).
72. B. Wang and P. M. Troccolo, *Proc. SPIE* **3873**, 544 (1999).
73. B. Wang, *J. Microlith. Microfab. Microsyst.* **1**, 43 (2002).
74. H. J. Levinson and W. H. Arnold, in *Handbook of Microlithography, Micromachining, & Microfabrication*, Vol. 1, P. Rai-Choudhury, ed., IET, Chapter 1, pp. 11–138, 1997.

75. D. M. Williamson, US Patent No. 5,537,260 (1996).
76. E. Hecht and A. Zajac, *Optics*, Addison-Wesley, London, U.K., p. 251, 1974.
77. J. H. Burnett, Z. H. Levine, E. L. Shirley, and J. H. Bruning, *J. Microlith. Microfab. Microsys.* **1**, 213 (2002).
78. J. H. Burnett, Z. H. Levine, and E. L. Shirley, *Phys. Rev. B* **64,** 241102(R) (2001).
79. B. Wang, C. O. Griffiths, R. Rockwell, J. List, and D. Mark, *Proc. SPIE* **5192**, 7 (2003).
80. B. Wang, R. Rockwell, and J. List, *J. Microlith. Microfab. Microsyst.* **3**, 115 (2003).
81. B. Wang and W. Rosch, *Proc. SPIE* **6682**, 36 (2007).
82. B. Wang, A. Leadbetter, R. R. Rockwell, and D. Mark, *Proc. IDRC* **26**, P4 (2006).
83. H. Mori and P. J. Bos, *Jpn. J. Appl. Phys.* **38**, 2837 (1999).
84. H. Mori, Y. Itoh, Y. Nishiura, T. Nakamura, and Y. Shinagawa, *Jpn. J. Appl. Phys.* **36**, 143 (1997).
85. T. Ishikawa, J. F. Elman, and D. M. Teegarden, US Patent No. 7,236,221 (2007).
86. K.-H. Kim, Y. Jang, and J.-K. Song, US Patent No. 7,321,411 (2008).
87. B. Wang and J. List, *SID Digest*, P55 (2004).
88. B. Wang, R. R. Rockwell, and A. Leadbetter, *Proc. SPIE* **5531**, 367 (2004).
89. A. Lakhtakia, ed., Selected papers on natural optical activity, *SPIE Milestone Series*, Vol. MS 15, B. J. Thompson, General ed., SPIE, 1990.
90. L. Hirschy, B. Wang, J. Wolf, R. Lakanen, and B. Hartmann, *Proc. SPIE* **8486**, 848619 (2012).
91. B. Wang and A. Leadbetter, *Proc. SPIE* **7853**, 78530C (2010).
92. B. Wang, W. B. Sparks, T. A. Germer, and A. Leadbetter, *Proc. SPIE* **7441**, 744108 (2009).
93. W. B. Sparks, J. Hough, T. A. Germer, F. Chen, S. DasSarma, P. DasSarma, F. T. Robb et al., *Proc. Natl. Acad. Sci. USA* **106**, 7816–7821 (2009).
94. W. B. Sparks, J. Hough, L. Kolokolova, T. A. Germer, F. Chen, S. DasSarma, P. DasSarma et al., *JQSRT* **110**, 1771–1779 (2009).
95. O. Arteaga, J. H. Freudenthal, B. Wang, and B. Kahr, *Appl. Opt.* **51**, 6805 (2012).
96. B. Wang, A. Leadbetter, and B. Seipel, *Energy Procedia* **38**, 959–967 (2013).
97. V. Ganapati et al., *J. Appl. Phys.* **108**, 063528 (2010).
98. For example, iStar gated ICCD from Andor Technology, www.andor.com.
99. S. Alali, T. Yang, and I. Alex Vitkin, *Opt. Lett.* **38** (16), 2997–3000 (2013).
100. S. Alali and I. A. Vitkin, *Opt. Eng.* **52**(10), 103114 (2013).
101. P. Gleyzes, A. C. Boccara, and H. Saint-Jalmes, *Opt. Lett.* **22**(20), 1529 (1997).
102. A. Dubois, L. Vabre, A.-C. Boccara, and E. Beaurepaire, *Appl. Opt.* **41**(4), 805–812 (2002).
103. C. Y. Han and Y. F. Chao, *Rev. Sci. Instrum.* **77**, 023107 (2006).
104. H. M. Tsai and Y. F. Chao, *Opt. Lett.* **34** (15), 2279–2281 (2009).
105. H. M. Tsai, C. W. Chen, T. H. Tsai, and Y. F. Chao, *Rev. Sci. Instrum.* **82**, 035117 1-5 (2011).
106. Segal, S. (2013). Personal communication.

26 Birefringence Measurement

Yukitoshi Otani

CONTENTS

26.1 Introduction 677
26.2 Applications of Retardance and Birefringence Measurement... 678
26.3 About Polarized Light 679
26.4 Polarization State... 679
26.5 What Is Birefringence?.. 682
26.6 Notation of Polarization States 684
    26.6.1 Jones Vector... 684
    26.6.2 Stokes Parameter 685
    26.6.3 Poincaré Sphere 686
26.7 Jones Method and Mueller Method 687
26.8 Birefringence Measurement 688
    26.8.1 Linear Polariscope.. 688
    26.8.2 Circular Polariscope 691
    26.8.3 Senarmont Method 692
    26.8.4 Photoelastic Modulator Method 693
    26.8.5 Optical Heterodyne Method... 696
    26.8.6 Two-Dimensional Birefringence Measurement by Phase-Shifting Method 698
References... 703

26.1 INTRODUCTION

A number of polarization and birefringence users, also known as a polarization society, have shown remarkable development in the last few decades. Most of the recent advanced technologies are based on the polarization technology including birefringence [1–4]. Many new interesting topics such as liquid crystal (LC) displays, LC projectors, optical disks, and optical communications are proposed using the polarization effect. A birefringence measurement is one of the visualizing methods for internal conditions of materials. We can understand information on stress distribution and macromolecular orientation distribution inside materials from birefringence mapping.

This chapter describes the polarization situation mathematically and explains how birefringence distribution is measured. As we have to handle the two parameters, birefringence perplexes us, because birefringence has a magnitude and an azimuthal direction and is not a scalar.

Birefringence measurement has a long history that originated from photoelasticity [5]. A polariscope for the photoelastic phenomenon was first discovered by Sir David Brewster in 1816. We can know the stress distribution from the fringe intensity. Photoelasticity has been expanded widely to the field of experimental mechanics. However, its activities have been broken down after computer simulation as the famous finite element method. Moreover, fringe counting is difficult because precise measurements are required. We have to measure birefringence that is less than one fringe. Polarimetric measurements, especially Mueller matrix polarimetry, are powerful tools for mapping birefringence [6]. It is also possible to analyze birefringence parameters from elements of a partial matrix. Moreover, if we can also neglect the diattenuation, which means no absorption dependences, we could propose many unique methods for mapping birefringence.

26.2 APPLICATIONS OF RETARDANCE AND BIREFRINGENCE MEASUREMENT

Recently, many optical instruments have used polarization effects. Table 26.1 shows remarkable applications of polarization and birefringence effects. We can divide features of optical applications into two groups as information and energy. Most of the applications are used for information such as manufacture, optical communication, biotechnology, nanotechnology, astronomy, and remote sensing, but some reports for energy application become increasingly unique applications of polarization. Polarization has contributed to the technological advances of recent years.

Representative examples are an LC display, an optical pickup for data storage, an optical isolator, and an optical multiplexer for optical communications. Birefringence mapping is an important method to visualize and analyze crystal orientation, stress distribution, polymer molecule orientation, and thin films for any kind of material. In optically isotropic materials, we have a problem with birefringence when the materials are applied to a stress field. This can possibly be explained by the fact that the speed of light changes depending on the stress fields caused by the force inside

TABLE 26.1
Remarkable Applications of Polarization and Birefringence

Feature	Area	Applications	Characteristics
Information	Manufacturing technology (LC)	Phase film, polarization film	Polarization
		Development, orientation	Birefringence
		Substrate	Birefringence
	Manufacturing technology (pickup)	Optical elements	Birefringence
	Manufacturing technology (injection process)	Orientation of polymer molecule	Birefringence
	Manufacturing technology (semiconductor photolithography)	Optical elements	Birefringence
	Manufacturing technology (metrology)	White-light interferometry	Geometric phase
	Nanotechnology	Polarization elements	Structural birefringence
	Optical communications	Polarization analysis	Polarization
		Optical isolator	Optical rotation
		Polarization control	Polarization
	Biotechnology	Glucose sensor	Optical rotation
		Cancer diagnosis	Polarization
		Cell observation	Birefringence
		Visualization	Polarization
		Orientation of polymer molecule	Fluorescent polarization
	Astronomy	Coronagraph	Geometric phase
		Observation of radiation mechanism	Zeeman effect
	Remote sensing	Visualization	Polarization
		Earth observation (surface, aerosol, cloud, sea)	Polarization
Energy	Manufacturing process	Laser processing	Polarization
	Manipulation	Laser trapping	Polarization

the material because of mass density. External forces make the atomic distance narrower along the force and broaden the distance perpendicular to it. In recent years, many optical instruments have become small and precise. Deterioration in quality causes birefringence, even if it is small. For example, birefringence in pickup lens is caused by astigmatism. Some experimental results are shown later.

Major optical devices using polarized light are LC displays in optical displays, pickups in optical memories, and multiplexing in optical communications or isolators. In addition, the acquisition of internal information is possible in the field of measurement. Nowadays, the acquisition of an object's internal information is highly required in the field of optoelectronics performing in various fields such as material development including glasses and transparent plastics, LC displays, optical parts such as polymers and crystals, or optical disks and thin film products and also in the development of new products. In addition, in practical use, because the shape of optoelectronics is ever more complex, specialized, and large-sized, the material's original double refraction and the double refraction due to residual stresses created inside while processing cause various problems. In addition, orientation information, important in the field of polymers, can be taken as double refraction.

In recent years, polarized light has been used in a wide variety of fields. Especially, its importance is increasing due to the recent discovery of high-resolution optical parts. The term polarized light has many meanings such as double refraction, optical rotating power, dichroism, and depolarization. Here, we set our goal to the active use of polarizers. Thus, we begin with the questions "What is polarized light? and What is double refraction?" and then discuss the basis to actively use polarized light. It would be a pleasure if this would help in the understanding of research papers and catalogs on polarized light or the introduction of polarimeters.

26.3   ABOUT POLARIZED LIGHT

Although light is said to have wave nature and particle nature, only particle nature is considered when dealing with polarized light. That is, light can be treated as a lateral electromagnetic wave. Consequently, light consists of electric vibration and magnetic vibration, but generally, only electric vibration is considered. In this case, we deal with electric vibration by decomposing it in two perpendicular vibrating directions. Suppose the z-axis to be in the direction of light in the right-hand coordinate system, x-axis in the horizontal direction, and y-axis in the vertical direction. Just for reference, the wavelength of visible light is 400–700 nm.

Since light is a lateral wave, an electric vibrating plane is formed on the xz surface as shown in Figure 26.1. This plane is called plane of polarization. The H-axis represents the magnetic vibrating plane. If a polarizer made with small enough gratings compared to light wave or with absorptiveness in a particular plane of polarization is installed, light wave penetrates or is intercepted depending on the vibrating direction. That is, light that has penetrated a polarizer will have only one vibrating direction. Here, since the vibrating direction is linear, it is called linear polarization, and the plane in the vibrating direction is called plane of polarization. Generally, p-polarized light (parallel in German) and s-polarized light (senkrecht) are used to express the polarization state. When light incidents upon a medium, the angle between the incident light's direction and normal to the surface is called the angle of incidence. The vibrating components inside the surface are called p-polarized light, and those normal to the surface, s-polarized light.

26.4   POLARIZATION STATE

Polarization states are linear polarization, circular polarization, and elliptical polarization. Let us consider these polarization states. As mentioned earlier, a light wave can be decomposed into xy-axis electric vibration vector components. Figure 26.2 shows the relationship between the electric vibration vectors E_x and E_y before and after passing through a sample with the retardance

(a)

(b)

FIGURE 26.1   Electric vector of light. (a) Polarized light and (b) electric vector of light.

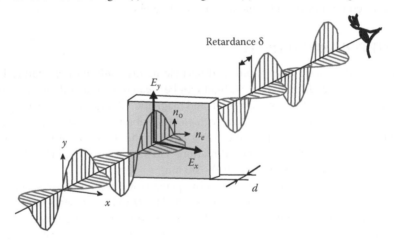

FIGURE 26.2   Electric field before and after passing through a sample.

$\delta(-2\pi(n_o - n_e)d/\lambda)$. Suppose A to be the amplitude, t the time, ω the angular frequency, and λ the wavelength, the electric vibration components E_x and E_y can be expressed as

$$E_y = A_y \exp i\left(\omega t - \frac{2\pi}{\lambda}z + \varphi_y\right)$$

$$E_x = A_x \exp i\left(\omega t - \frac{2\pi}{\lambda}z + \varphi_x\right)$$

(26.1)

provided that the phase difference δ is

$$\delta = \varphi_y - \varphi_x$$

(26.2)

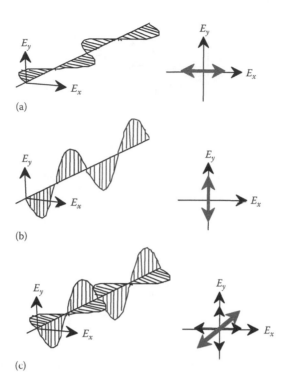

FIGURE 26.3 Composition of the electric vibration vectors E_x and E_y. (a) Linear polarization in case of $A_y = 0$, (b) linear polarization in case of $A_x = 0$, and (c) linear polarization $A_x = A_y$, $\delta = 0$.

It is possible to determine the statuses of linear polarization, circular polarization, and elliptical polarization from the amplitude A and the phase difference δ.

As shown in Figure 26.3, linear polarization takes place when the phase difference is 0 or π. As shown in Figure 26.3a, horizontal linear polarizations vibrating in the x-axis direction are obtained when the amplitude is $A_x = 0$, vertical linear polarization when $A_y = 0$ as shown is Figure 26.3b, and linear polarization with azimuth 45° when $A_x = A_y = 1$, as shown in Figure 26.3, because it is the composition of the amplitude vectors of x and y directions.

Next, circular polarization takes place when δ is 90° as shown in Figure 26.4. In this case, the relationship between the phase differences of E_x and E_y is given as a composition of vectors as shown in Figure 26.4a and traces the locus shown in Figure 26.4c. When observed from the direction of light, the vector's rotation is counterclockwise as shown in Figure 26.4b, and so the locus will be as shown in Figure 26.4c. Thus, this kind of polarization state is called left-hand circular polarization. In addition, this is when E_y has a phase delay of 90°(−90°) when compared with E_x. On the other hand, if E_y has a phase lead of 90°(+90°), the locus is clockwise as opposed to Figure 26.4. This kind of polarization state is called right-hand circular polarization.

With these facts considered, Equation 26.1 can be expressed as

$$\left(\frac{E_y}{a_y}\right)^2 + \left(\frac{E_x}{a_x}\right)^2 - 2\left(\frac{E_y}{a_y}\right)\left(\frac{E_x}{a_x}\right)\cos \delta = \sin^2 \delta \tag{26.3}$$

Linear polarization is obtained when the phase difference is 0° or 180°. When δ is 90° and $A_x = A_y$, Equation 26.3 is expressed in the equation of a circle. When the phase difference is anything other than these, the polarization state will have its vectors directed elliptically, and its locus will be as shown in Figure 26.4c. This kind of polarization state is called elliptical polarization.

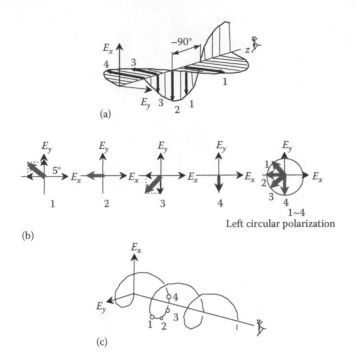

FIGURE 26.4 Left circular polarization light. (a) Polarization state, (b) direction of polarization vector, and (c) trajectory.

26.5 WHAT IS BIREFRINGENCE?

Birefringence has, as it is written, two indexes of refraction. That is, the propagation velocity differs depending on the polarization direction. Figure 26.5 shows how an optical wave transmits double refractive objects with indexes of refraction $n_{//}$ and $n_\perp$, respectively. Figure 26.5b separates and aligns vertically each plane of polarization of Figure 26.5a. Phase difference Δ depending on the double refraction occurs between orthogonal polarizations. In Figure 26.5, since light E_x vibrating in the x direction is transmitted through the sample slowly, birefringent phase difference (retardation) depending on the transmitting velocity arises in the light emitted from the sample. In other words, the transmitting velocity in the x direction differs from that in the y direction. That is, their indexes of refraction are different. Since the refractive index is the ratio between the velocity of light c' transmitted through a substance and the velocity of light c in vacuum,

$$n = \frac{c}{c'} \tag{26.4}$$

the double refraction Δn is given as

$$\Delta n = n_\perp - n_{//} \tag{26.5}$$

Generally, birefringent phase difference Δ,

$$\Delta = 2\pi l / \lambda \cdot \Delta n \tag{26.6}$$

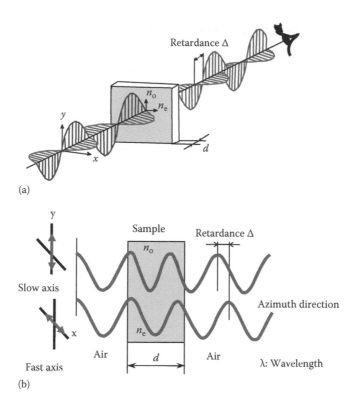

(a)

(b)

FIGURE 26.5 An example that the light passes though a birefringence sample. (a) Birefringence sample and (b) difference of light speed along the polarization difference refractive index.

TABLE 26.2

Determination of the Retardation Unit

Waves	1 Wave (λ)
Degrees	360°
Radians	2π rad
Nanometers	λ nm
	λ: wavelength

since phase display ($2\pi = 360°$) can be expressed as follows:

$$\Delta = 360\,\Delta n \; l/\lambda \tag{26.7}$$

and can be standardized by wavelength (nm) in Table 26.2.

The direction in which light is fastly transmitted (phase lead) is called the fast axis. On the other hand, the direction in which light is slowly transmitted (phase delay) is called the slow axis. The general term for the fast axis and the slow axis is the principal axis. In addition, samples with double refraction are called anisotropic and those without double refraction, such as glasses, are called isotropic in optics.

Thus, determination of the size (birefringent phase difference) and the principal axis direction (the fast axis and the slow axis) of a double refraction is needed to measure inner information of a sample.

The s-polarization is called (transverse electric) TE-mode, and p-polarization is called (transverse magnetic) TE-mode.

26.6 NOTATION OF POLARIZATION STATES

The notation of polarization states can be classified into those using visual techniques and those using mathematical techniques. The Poincaré sphere is a visual technique, and the Mueller method and the Jones method are mathematical methods.

26.6.1 JONES VECTOR

To express the polarization state mathematically, let us refer to Equation 26.1, which expresses optical waves decomposed in x and y components. When this equation is expressed in a matrix, it becomes

$$\mathbf{E} = \begin{bmatrix} \dfrac{A_x}{\sqrt{A_x^{\,2} + A_x^{\,2}}}\exp\left(-i\dfrac{\delta}{2}\right) \\ \dfrac{A_x}{\sqrt{A_x^{\,2} + A_y^{\,2}}}\exp\left(i\dfrac{\delta}{2}\right) \end{bmatrix} \tag{26.8}$$

For, only relative phases should be considered when dealing with general polarization states:

$$\mathbf{E} = \begin{bmatrix} E_x \\ E_y \end{bmatrix} = \exp i\left(\omega t - \dfrac{2\pi}{\lambda} + \varphi_x\right)\begin{bmatrix} A_x \\ A_y \exp i\delta \end{bmatrix} \tag{26.9}$$

In addition, if the polarization status is the only one to be considered, the amplitude must be standardized. When the phase is also transformed,

$$\mathbf{E} = \begin{bmatrix} A_x' \\ A_y' \end{bmatrix} = \begin{bmatrix} A_x \\ A_y \cdot \exp i(\delta) \end{bmatrix} \tag{26.10}$$

For example, since in horizontal linear polarizations $A_x = 1$, $A_y = 0$, and $\delta = 0$, Jones vector $\mathbf{J}$ is obtained as follows by using the above equation.

$$\mathbf{J} = \begin{bmatrix} 1 \\ 0 \end{bmatrix} \tag{26.11}$$

Also, for instance, in clockwise circular polarizations, $A_x = 1$, $A_y = 1$, and $\delta = 90°$. This is when E_y has a lead of 90°(+90°), and Jones vector $\mathbf{J}$ is expressed as follows by using the complex number i, the same as the notation of an electric current.

$$\mathbf{J} = \dfrac{1}{\sqrt{2}}\begin{bmatrix} 1 \\ i \end{bmatrix} \tag{26.12}$$

All polarization states can similarly be expressed with Jones vectors. The Jones vectors of representative polarization states are shown in Table 26.3.

TABLE 26.3

Various Forms of Standard Normalized Jones Vectors

0° Linear	90° Linear	45° Linear	−45° Linear	General Linear	Right Circular	Left Circular	General Elliptical
$\begin{bmatrix} 1 \\ 0 \end{bmatrix}$	$\begin{bmatrix} 0 \\ 1 \end{bmatrix}$	$\dfrac{1}{\sqrt{2}}\begin{bmatrix} 1 \\ 1 \end{bmatrix}$	$\dfrac{1}{\sqrt{2}}\begin{bmatrix} 1 \\ -1 \end{bmatrix}$	$\begin{bmatrix} \cos\theta \\ \pm\sin\theta \end{bmatrix}$	$\dfrac{1}{\sqrt{2}}\begin{bmatrix} -i \\ 1 \end{bmatrix}$	$\dfrac{1}{\sqrt{2}}\begin{bmatrix} i \\ 1 \end{bmatrix}$	$\begin{bmatrix} e^{-i\beta/2}\cos\theta \\ e^{i\beta/2}\sin\theta \end{bmatrix}$

26.6.2 STOKES PARAMETER

Here, unlike the Jones vector, which uses two elements to express a polarization state, we use the Stokes parameter made up of four elements, s_0 to s_3.

Let us refer to Equation 26.1, which expresses optical waves decomposed in x and y components, just like the Jones vector. When this equation is expressed in matrices, it becomes

$$s_0 = \langle E_x E_x^* + E_y E_y^* \rangle = A_x^2 + A_y^2$$

$$s_1 = \langle E_x E_x^* - E_y E_y^* \rangle = A_x^2 - A_y^2$$

$$s_2 = \langle E_x^* E_y + E_x E_y^* \rangle = 2A_x A_y \cos\Delta \tag{26.13}$$

$$s_3 = i\langle E_x^* E_y - E_x E_y^* \rangle = 2A_x A_y \sin\Delta \quad \text{where } i^2 = -1$$

The four elements could also be expressed with I, M, C, S or I, Q, U, V. When these are put together in the matrix $\mathbf{S}$,

$$\mathbf{S} = \begin{bmatrix} s_0 \\ s_1 \\ s_2 \\ s_3 \end{bmatrix} = \begin{bmatrix} I \\ M \\ C \\ S \end{bmatrix} = \begin{bmatrix} I \\ Q \\ U \\ V \end{bmatrix} \tag{26.14}$$

In the Stokes parameter, s_0 stands for the light intensity, s_1 stands for the horizontal linear polarization component, s_2 stands for the 45° straight line, and s_3 stands for the right-hand circular polarization. Generally, light intensity is treated as the unit intensity.

Based on these, we deal with representative polarization states. For example, since $A_x = 1$, $A_y = 0$, and $\delta = 0$ in horizontal linear polarization,

$$S = \begin{bmatrix} 1 \\ 1 \\ 0 \\ 0 \end{bmatrix} \tag{26.15}$$

Likewise, in right-handed circular polarization, $A_x = 1$, $A_y = 1$, and $\delta = 90°$.

$$S = \begin{bmatrix} 1 \\ 0 \\ 0 \\ 1 \end{bmatrix} \tag{26.16}$$

All these forms are shown in Table 26.4.

TABLE 26.4

Various Forms of Standard Normalized Stokes Vectors

0° Linear	90° Linear	45° Linear	−45° Linear	General Linear	Right Circular	Left Circular	General Elliptical	General Elliptical, Partially Polarized	Unpolarized
$\begin{bmatrix}1\\1\\0\\0\end{bmatrix}$	$\begin{bmatrix}1\\-1\\0\\0\end{bmatrix}$	$\begin{bmatrix}1\\0\\1\\0\end{bmatrix}$	$\begin{bmatrix}1\\0\\-1\\0\end{bmatrix}$	$\begin{bmatrix}1\\\cos 2\theta\\\sin 2\theta\\0\end{bmatrix}$	$\begin{bmatrix}1\\0\\0\\1\end{bmatrix}$	$\begin{bmatrix}1\\0\\0\\-1\end{bmatrix}$	$\begin{bmatrix}1\\\cos 2\alpha\cos\gamma\\\cos 2\alpha\sin 2\gamma\\\sin 2\alpha\end{bmatrix}$	$\begin{bmatrix}\langle a_x^2+a_y^2\rangle\\\langle a_x^2-a_y^2\rangle\\2a_x a_y\cos\beta\\2a_x a_y\sin\beta\end{bmatrix}$	$\begin{bmatrix}1\\0\\0\\0\end{bmatrix}$

The relationship between the Stokes parameter's four elements s_0 to s_3 depends on the polarization state.

In totally polarized lights,

$$s_0^2 = s_1^2 + s_2^2 + s_3^2 \tag{26.17}$$

In partially polarized lights,

$$s_0^2 > s_1^2 + s_2^2 + s_3^2 \tag{26.18}$$

Thus, the degree of polarization V is defined as an index of how much light is polarized.

$$V = \frac{\sqrt{s_1^2 + s_2^2 + s_3^2}}{s_0} \tag{26.19}$$

26.6.3 POINCARÉ SPHERE

The possibility of expressing the polarization states mathematically was characteristic to the Jones vector and the Stokes parameter. However, the polarization states, hard enough to perceive, are even more difficult to understand when expressed in matrices. The Poincaré sphere is a method to visually understand the polarization states.

As mentioned in the section dealing with the Stokes parameter, since Equation 26.17 is satisfied in totally polarized lights, when each axis of spatial Cartesian coordinates is expressed as s_1, s_2, and s_3, a sphere with a radius of s_0 is obtained. In addition, if the coordinates of point S, located on the sphere, are expressed with latitude 2β and longitude 2θ, the following equations are obtained.

$$s_1 = s_0 \cos 2\beta \cos 2\theta$$

$$s_2 = s_0 \cos 2\beta \sin 2\theta \tag{26.20}$$

$$s_3 = s_0 \sin 2\beta$$

We will not go into details, but the correspondence to the azimuth θ of the long axis and to the ellipticity angle β in elliptical polarizations can be considered. This situation is illustrated in

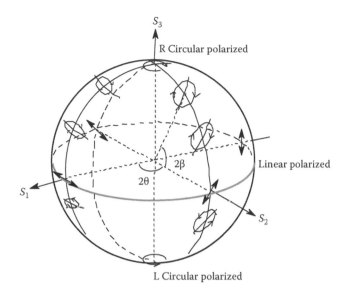

FIGURE 26.6 Poincaré sphere representation of polarization state.

Figure 26.6. It is easier to understand when the Poincaré sphere is treated as the Earth. A right-hand polarization and a left-hand polarization are created in the upper and lower hemispheres respectively. On the equator, a linear polarization is obtained by the ellipticity of 0°. A right-hand and a left-hand circular polarization are acquired in the upper and the lower poles, respectively, and elliptical polarizations are distributed between each pole and the equator. Polarization states symmetrical about the center of the sphere have a difference of 90° in the azimuth angle of the major axis.

26.7 JONES METHOD AND MUELLER METHOD

All polarization states could be described using the Jones vector and the Stokes parameter. As shown in Figure 26.7, once the incident light and the output light's polarization states are identified, the polarization elements of an optical device, which is a black box, can be determined subsequently.

The Jones calculus is written as

$$\mathbf{J} \cdot \mathbf{E} = \mathbf{E}'$$

$$\begin{bmatrix} J_{xx} & J_{xy} \\ J_{yx} & J_{yy} \end{bmatrix} \begin{bmatrix} E_x \\ E_y \end{bmatrix} = \begin{bmatrix} E'_x \\ E'_y \end{bmatrix} \tag{26.21}$$

FIGURE 26.7 Jones matrices and Mueller calculus.

The Mueller calculus is indicated as

$$\mathbf{M} \cdot \mathbf{S} = \mathbf{S}'$$

$$\begin{bmatrix} M_{00} & M_{01} & M_{02} & M_{03} \\ M_{10} & M_{11} & M_{12} & M_{13} \\ M_{20} & M_{21} & M_{22} & M_{23} \\ M_{30} & M_{31} & M_{32} & M_{33} \end{bmatrix} \begin{bmatrix} s_0 \\ s_1 \\ s_2 \\ s_3 \end{bmatrix} = \begin{bmatrix} s_0' \\ s_1' \\ s_2' \\ s_3' \end{bmatrix} \tag{26.22}$$

The polarization elements that can be obtained from these methods are put together in Table 26.5.

Only multiplication of the matrices is needed if the polarization elements are lined together. Additionally, the light intensity obtained at the end in the case of Jones matrix is

$$I = \langle E_x + E_y \rangle^2 \tag{26.23}$$

In the case of Mueller matrix, the light intensity is directly given by s_0.

To conclude, the polarization degrees of freedom of the Jones and the Mueller matrices are as shown in Figure 26.7. Both the Jones method and the Mueller method give double refraction, double absorption (dichroism), circular double refraction (optical rotating power), circular double absorption (circular double dichroism), and others.

Their major difference is that as opposed to the Jones vector treating only totally polarized lights, the Mueller matrix can deal with all polarization states including depolarization.

26.8 BIREFRINGENCE MEASUREMENT

26.8.1 LINEAR POLARISCOPE

In this section, we progress to observe and measure stress distribution. Birefringence distribution can be observed from an intensity distribution inserted in a sample between two polarizers as in Figure 26.8. If you analyze this fringe, you can determine the quantity of stress.

In this chapter, three nonmodulated methods and four modulated methods are introduced in the following pages.

Figure 26.9 shows two polarizers and a light source. The optical axes of the polarizer are set perpendicular to each other. A sample is inserted between the polarizers.

The relation among the polarization elements is given using a Mueller matrix and Stokes vector by

$$S' = P(0°) \cdot R(\Delta, \theta) \cdot P(90°) \cdot S \tag{26.24}$$

where the Stokes vector S' and S are the output and input of the light, $P(0°)$, $R(\Delta, \theta)$, and $P(90°)$ are Mueller matrix of polarizer with $0°$ of azimuthal direction, sample with Δ retardance and θ azimuthal direction and polarizer with $90°$ of azimuthal direction, respectively.

The retardance Δ indicates stress-induced birefringence by σ_1 and σ_2.

$$\begin{bmatrix} s_0 \\ s_1 \\ s_2 \\ s_3 \end{bmatrix} = \frac{1}{2}\begin{bmatrix} 1 & 1 & 0 & 0 \\ 1 & 1 & 0 & 0 \\ 0 & 0 & 0 & 0 \\ 0 & 0 & 0 & 0 \end{bmatrix}\begin{bmatrix} 1 & 0 & 0 & 0 \\ 0 & \cos^2 2\theta + \sin^2 2\theta \cos\Delta & (1-\cos\Delta)\sin 2\theta \cos 2\theta & -\sin 2\theta \sin\Delta \\ 0 & (1-\cos\Delta)\sin 2\theta \cos 2\theta & \sin^2 2\theta + \cos^2 2\theta \cos\Delta & \cos 2\theta \sin\Delta \\ 0 & \sin 2\theta \sin\Delta & -\cos 2\theta \sin\Delta & \cos\Delta \end{bmatrix}$$

$$\times \frac{1}{2}\begin{bmatrix} 1 & -1 & 0 & 0 \\ -1 & 1 & 0 & 0 \\ 0 & 0 & 0 & 0 \\ 0 & 0 & 0 & 0 \end{bmatrix}\begin{bmatrix} 1 \\ 0 \\ 0 \\ 0 \end{bmatrix} = \frac{1}{4}\begin{bmatrix} \sin^2 2\theta(1-\cos\Delta) \\ \sin^2 2\theta(1-\cos\Delta) \\ 0 \\ 0 \end{bmatrix} = \frac{1}{2}\begin{bmatrix} \sin^2 2\theta \sin^2(\Delta/2) \\ \sin^2 2\theta \sin^2(\Delta/2) \\ 0 \\ 0 \end{bmatrix} \tag{26.25}$$

TABLE 26.5
Polarization Elements of Jones Matrix and Mueller Matrix

Polarization Elements	Jones Matrix	Mueller Matrix
Linear polarizer azimuthal direction θ	$\begin{bmatrix} \cos^2\theta & \sin\theta\cos\theta \\ \sin\theta\cos\theta & \sin^2\theta \end{bmatrix}$	$\dfrac{1}{2}\begin{bmatrix} 1 & \cos2\theta & \sin2\theta & 0 \\ \cos2\theta & \cos^2 2\theta & \sin2\theta\cos2\theta & 0 \\ \sin2\theta & \sin2\theta\cos2\theta & \sin^2 2\theta & 0 \\ 0 & 0 & 0 & 1 \end{bmatrix}$
Optical rotator azimuthal direction θ	$\begin{bmatrix} \cos\theta & \sin\theta \\ -\sin\theta & \cos\theta \end{bmatrix}$	$\begin{bmatrix} 1 & 0 & 0 & 0 \\ 0 & \cos2\theta & \sin2\theta & 0 \\ 0 & -\sin2\theta & \sin2\theta & 0 \\ 0 & 0 & 0 & 1 \end{bmatrix}$
Half-wave plate azimuthal direction θ	$\begin{bmatrix} \cos2\theta & \sin2\theta \\ \sin2\theta & -\cos2\theta \end{bmatrix}$	$\begin{bmatrix} 1 & 0 & 0 & 0 \\ 0 & \cos4\theta & \sin4\theta & 0 \\ 0 & \sin4\theta & -\sin4\theta & 0 \\ 0 & 0 & 0 & 1 \end{bmatrix}$
Quarter-wave plate azimuthal direction θ	$\dfrac{1}{\sqrt{2}}\begin{bmatrix} 1+\cos2\theta & i\sin2\theta \\ i\sin2\theta & 1+\cos2\theta \end{bmatrix}$	$\begin{bmatrix} 1 & 0 & 0 & 0 \\ 0 & \cos^2 2\theta & \sin2\theta\sin2\theta & -\sin2\theta \\ 0 & \sin2\theta\cos2\theta & \sin^2 2\theta & \cos2\theta \\ 0 & \sin2\theta & -\cos2\theta & 0 \end{bmatrix}$
Linear retarder azimuthal direction θ	$\begin{bmatrix} e^{i\Delta}\cos^2\theta + \sin^2\theta & (e^{i\Delta}-1)\sin\theta\cos\theta \\ (e^{i\Delta}-1)\sin\theta\cos\theta & \exp(i\Delta)(\sin^2\theta+\cos^2\theta) \end{bmatrix}$	$\begin{bmatrix} 1 & 0 & 0 & 0 \\ 0 & \cos^2 2\theta + \sin^2 2\theta\cos\Delta & (1-\cos\Delta)\sin2\theta\cos2\theta & -\sin2\theta\sin\Delta \\ 0 & (1-\cos\Delta)\sin2\theta\cos2\theta & \sin^2 2\theta + \cos^2 2\theta\cos\Delta & \cos2\theta \\ 0 & \sin2\theta\sin\Delta & -\cos2\theta\sin\Delta & \cos\Delta \end{bmatrix}$
Birefringence Δ		

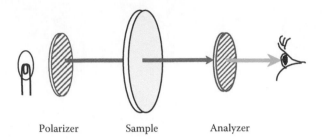

FIGURE 26.8 Visualization of birefringence.

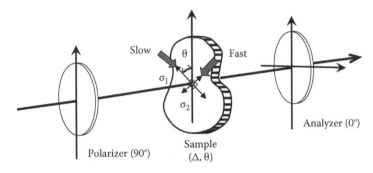

FIGURE 26.9 Plane polariscope.

The intensity I observed at the eye is given from the first term of s_0

$$I = s_0 = \frac{1}{2}\sin^2 2\theta \sin^2(\Delta/2) = \frac{1}{4}\sin^2 2\theta \cdot (1 - \cos\Delta) \tag{26.26}$$

This means the intensity changes sinusoidally along the retardance Δ and the azimuth angle makes the intensity of the periodic circle 90° independently.

Figure 26.10 shows stress mapping by the plane polariscope [16] (http://www.luceo.co.jp). A plastic disk substrate and square plate are used as samples.

In Figure 26.10a, the intensity distribution of a disk substrate changes bright to dark radially because of the birefringence distribution. The intensity distribution is modulated sinusoidally along the circumferential direction and this period is 90°. This means the azimuth direction of the birefringence indicates concentricity.

The three pictures in Figure 26.10b through d mean merely rotating the sample at the same stress condition. The stress around the surrounding part is larger than that at the center section. This intensity distribution is also modulated sinusoidally in respect of the sample rotation.

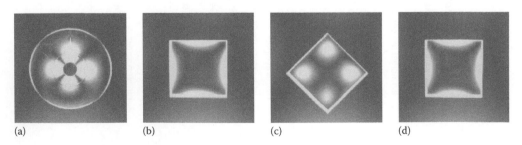

FIGURE 26.10 Stress mapping by the plane polariscope. (a) Plastic disk substrate, (b) stressed plastic plate stress direction is 0°, (c) rotation angle of 45°, and (d) rotation angle of 90°.

It is difficult to determine both the retardance and azimuth direction simultaneously due to the image of the plane polariscope.

26.8.2 Circular Polariscope

Figure 26.11 shows a circular polariscope. It consists of a circular polarizer and a circular analyzer. A sample is set between quarter wave plates. The relation among the polarization elements is given using a Mueller matrix and Stokes vector by

$$S' = P(0°) \cdot Q(-45°) \cdot R(\Delta, \theta) \cdot Q(45°) \cdot P(90°) \cdot S \tag{26.27}$$

where
   the Stokes vector S' and S are the output and input of the light
   P and Q are Mueller matrix of polarizer and quarter wave plate with azimuthal direction angle between parentheses, respectively
   $R(\Delta, \theta)$ is a Mueller matrix of the sample with Δ retardance and θ azimuthal direction

The retardance Δ indicates stress-induced birefringence by σ_1 and σ_2. All Stokes parameters and Mueller matrices are shown as

$$\begin{bmatrix} s_o \\ s_1 \\ s_2 \\ s_3 \end{bmatrix} = \frac{1}{2} \begin{bmatrix} 1 & 1 & 0 & 0 \\ 1 & 1 & 0 & 0 \\ 0 & 0 & 0 & 0 \\ 0 & 0 & 0 & 0 \end{bmatrix} \begin{bmatrix} 1 & 0 & 0 & 0 \\ 0 & 0 & 0 & 1 \\ 0 & 0 & 1 & 0 \\ 0 & -1 & 0 & 0 \end{bmatrix}$$

$$\times \begin{bmatrix} 1 & 0 & 0 & 0 \\ 0 & \cos^2 2\theta + \sin^2 2\theta \cos \Delta & (1 - \cos \Delta) \sin 2\theta \cos 2\theta & -\sin 2\theta \sin \Delta \\ 0 & (1 - \cos \Delta) \sin 2\theta \cos 2\theta & \sin^2 2\theta + \cos^2 2\theta \cos \Delta & \cos 2\theta \sin \Delta \\ 0 & \sin 2\theta \sin \Delta & -\cos 2\theta \sin \Delta & \cos \Delta \end{bmatrix}$$

$$\times \begin{bmatrix} 1 & 0 & 0 & 0 \\ 0 & 0 & 0 & -1 \\ 0 & 0 & 1 & 0 \\ 0 & 1 & 0 & 0 \end{bmatrix} \frac{1}{2} \begin{bmatrix} 1 & -1 & 0 & 0 \\ -1 & 1 & 0 & 0 \\ 0 & 0 & 0 & 0 \\ 0 & 0 & 0 & 0 \end{bmatrix} \begin{bmatrix} 1 \\ 0 \\ 0 \\ 0 \end{bmatrix} = \frac{1}{4} \begin{bmatrix} 1 - \cos \Delta \\ 1 - \cos \Delta \\ 0 \\ 0 \end{bmatrix} \tag{26.28}$$

FIGURE 26.11   Circular polariscope.

(a)   (b)   (c)   (d)

FIGURE 26.12   Stress mapping by the circular polariscope. (a) Plastic disk substrate, (b) stressed plastic plate stress direction is 0°, (c) rotation angle of 45°, and (d) rotation angle of 90° (http://www.luceo.co.jp).

Finally, the intensity I observed at the eye is given from the first term of s_0

$$I = s_0 = \frac{1}{4}(1 - \cos \Delta) = \frac{1}{2}\sin^2(\Delta/2) \tag{26.29}$$

The intensity distribution means the retardation is changing sinusoidally along the retardance Δ but the azimuth angle is independent of the intensity change.

Figure 26.12 shows stress mapping by the circular polariscope (http://www.luceo.co.jp), the same as that in Figure 26.10.

26.8.3 Senarmont Method

The requirement of birefringence measurement includes stress mapping to measure the state of polarization following the sample, to measure the state of the polarization of incident beam to the sample, or to determine the state of polarization in advance. In addition, it measures the alignment of direction of the sample to incident light or measures both retardation and azimuthal direction. Below are some measurement methods including the Senarmont method, the optical heterodyne method, and the two-dimensional birefringence measurement, using the phase-shifting technique.

The Senarmont method is a classical method for stress mapping. Because it is economical, it is often used for industrial inspection. It has features such as easy alignment, an equal method of ellipsometry, and high accuracy even when using a low-quality wave plate. An optical arrangement of the Senarmont method is shown in Figure 26.13. It consists of a polarizer, a quarter wave plate with 0°, and a rotating analyzer. A sample is set between quarter wave plates with the fixed value of

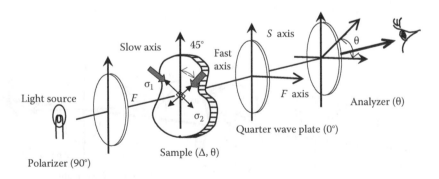

FIGURE 26.13   Optical arrangement of Senarmont method.

azimuthal direction as 45°. The relation among the polarization elements is given using a Mueller matrix and Stokes vector by

$$S' = P(\theta) \cdot Q(0°) \cdot R(\Delta, 45°) \cdot P(90°) \cdot S \qquad (26.30)$$

where
   the Stokes vector S' and S are the output and input of the light
   P and Q are Mueller matrices of polarizer and quarter wave plate with azimuthal direction angle between parentheses, respectively
   θ is the azimuthal direction of the analyzer
   $R(\Delta, 45°)$ is a Mueller matrix of the sample with Δ retardance and 45° azimuthal direction

The retardance Δ indicates stress-induced birefringence by σ_1 and σ_2 as shown in Equation 26.11. All Stokes parameters and Mueller matrices are shown as

$$
\begin{bmatrix} s_o \\ s_1 \\ s_2 \\ s_3 \end{bmatrix} = \frac{1}{2} \begin{bmatrix} 1 & \cos 2\theta & \sin 2\theta & 0 \\ \cos 2\theta & \cos^2 2\theta & \sin 2\theta \cos 2\theta & 0 \\ \sin 2\theta & \sin 2\theta \cos 2\theta & \sin^2 2\theta & 0 \\ 0 & 0 & 0 & 0 \end{bmatrix} \begin{bmatrix} 1 & 0 & 0 & 0 \\ 0 & 1 & 0 & 0 \\ 0 & 0 & 0 & 1 \\ 0 & 0 & -1 & 0 \end{bmatrix} \begin{bmatrix} 1 & 0 & 0 & 0 \\ 0 & \cos \Delta & 0 & -\sin \Delta \\ 0 & 0 & 1 & 0 \\ 0 & \sin \Delta & 0 & \cos \Delta \end{bmatrix}
$$

$$
\times \frac{1}{2} \begin{bmatrix} 1 & -1 & 0 & 0 \\ -1 & 1 & 0 & 0 \\ 0 & 0 & 0 & 0 \\ 0 & 0 & 0 & 0 \end{bmatrix} \begin{bmatrix} 1 \\ 0 \\ 0 \\ 0 \end{bmatrix} = \frac{1}{4} \begin{bmatrix} 1 - \cos \Delta \cos 2\theta - \sin \Delta \cos 2\theta \\ \cos 2\theta - \cos^2 2\theta \cos \Delta - \sin 2\theta \cos 2\theta \sin \Delta \\ \sin 2\theta - \sin 2\theta \cos 2\theta \cos \Delta - \sin^2 2\theta \sin \Delta \\ 0 \end{bmatrix} \qquad (26.31)
$$

Finally, the intensity I observed at the eye is given from the first term of s_0

$$I = s_0 = 1 - \cos(\Delta - 2\theta) \qquad (26.32)$$

The intensity modulates sinusoidally along the retardance Δ and the azimuthal direction θ of the analyzer. This means we can determine the retardation Δ from the azimuthal angle θ of the analyzer if we find the dark fringe by the rotating analyzer. The retardation is

$$\Delta = 2\theta \qquad (26.33)$$

26.8.4 PHOTOELASTIC MODULATOR METHOD

One of the popular methods used for many kinds of industries by Hinds is a birefringence measurement system using a photoelastic modulator (PEM) [7–10]. The optical arrangement consists of a polarizer, a PEM with 0° azimuthal direction, and an analyzer with 0° and 45° azimuthal angle (Figure 26.14). The PEM modulates the retardance δ with temporal modulation. A sample is set

FIGURE 26.14 Optical arrangement of birefringence measurement.

between PEM and the analyzer. The relation among the polarization elements is given using Mueller matrices and Stokes vectors by

$$S' = P(-45°) \cdot X(\Delta, \theta) \cdot PEM(\delta, 0°) \cdot P(45°) \cdot S \tag{26.34}$$

where
  the azimuthal angle of the analyzer is 45°
  the Stokes vector S' and S are the output and input of the light, respectively
  P is the Mueller matrix of the polarizer
  $PEM(\delta, 0)$ is the Mueller matrix of the polarizer with the temporal phase modulation δ
  $X(\Delta, \theta)$ is the Mueller matrix of the sample with Δ retardance and θ azimuthal direction

The retardance Δ also indicates stress-induced birefringence by σ_1 and σ_2 as shown in Equation 26.11. All Stokes parameters and Mueller matrices are shown as

$$
\begin{bmatrix} s_0 \\ s_1 \\ s_2 \\ s_3 \end{bmatrix} = \frac{1}{2} \cdot
\begin{bmatrix}
1 & 0 & -1 & 0 \\
0 & 0 & 0 & 0 \\
-1 & 0 & 1 & 0 \\
0 & 0 & 0 & 0
\end{bmatrix}
\begin{bmatrix}
1 & 0 & 0 & 0 \\
0 & \cos^2 2\theta + \sin^2 2\theta \cos\Delta & (1-\cos\Delta)\sin 2\theta \cos 2\theta & -\sin 2\theta \sin\Delta \\
0 & (1-\cos\Delta)\sin 2\theta \cos 2\theta & \sin^2 2\theta + \cos^2 2\theta \cos\Delta & \cos 2\theta \sin\Delta \\
0 & \sin 2\theta \sin\Delta & -\cos 2\theta \sin\Delta & \cos\Delta
\end{bmatrix}
$$

$$
\times
\begin{bmatrix}
1 & 0 & 0 & 0 \\
0 & 1 & 0 & 0 \\
0 & 0 & \cos\delta & \sin\delta \\
0 & 0 & -\sin\delta & \cos\delta
\end{bmatrix}
\begin{bmatrix} 1 \\ 0 \\ 1 \\ 0 \end{bmatrix}
\tag{26.35}
$$

Finally, the detected intensity I is given from the first term of s_0

$$
\begin{aligned}
I = s_0 &= I_0 \left[1 + \sin\delta \cdot (\cos 2\theta \cdot \sin\Delta) - \cos\delta \cdot (\sin^2 2\theta + \cos^2 2\theta \cos\Delta) \right] \\
&= I_0 \left[1 + \sin(\delta_0 \sin\omega t) \cdot \cos 2\theta \cdot \sin\Delta - \cos(\delta_0 \sin\omega t) \cdot (\sin^2 2\theta + \cos^2 2\theta \cos\Delta) \right] \\
&= I_0 \left\{ 1 + \cos 2\theta \cdot \sin\Delta \cdot [2J_1(\delta_0) \sin\omega t] - (\sin^2 2\theta + \cos^2 2\theta \cos\Delta) \cdot [J_0(\delta_0) + 2J_2(\delta_0) \cos 2\omega t] \right\}
\end{aligned}
\tag{26.36}
$$

where J_n is the nth order's Bessel function.

The phase modulation δ is

$$\delta = \delta_0 \sin\omega t \tag{26.37}$$

where

t is the time
δ_0 is the maximum amplitude of retardance
ω is the angular frequency

Using appropriate modulation δ_0 we can eliminate the J_0 and J_1 terms. If the azimuthal angle θ is $0°$, the retardation is obtained using a lock-in amplifier:

$$\Delta = \sin^{-1}\left[\frac{kI(\omega)}{J_1(\delta_0)}\right] \tag{26.38}$$

In this case, the calibration process is to determine the parameter k.

One more equation is required to determine both retardation and azimuthal direction, simultaneously. The different angles of the analyzer such as $0°$ are written as

$$S' = P(0°) \cdot X(\Delta,\theta) \cdot \text{PEM}(\delta,0°) \cdot P(45°) \cdot S \tag{26.39}$$

The detected intensity I is given from the first term of s_0,

$$\begin{aligned}
I_{45°} &= I_0\left\{1 + \sin(\delta_0\sin\omega t)\cdot\sin 2\theta\cdot\sin\Delta + \cos(\delta_0\sin\omega t)\cdot((1-\cos\Delta)\sin 2\theta\cos 2\theta - \sin 2\theta\sin\Delta)\right\} \\
&= I_0\left\{1 + \sin 2\theta\cdot\sin\Delta\cdot[2J_1(\delta_0)\sin\omega t] \right. \\
&\quad \left. -((1-\cos\Delta)\sin 2\theta\cos 2\theta - \sin 2\theta\sin\Delta)\cdot[J_0(\delta_0)+2J_2(\delta_0)\cos 2\omega t]\right\}
\end{aligned} \tag{26.40}$$

Using appropriate modulation δ_0 we can eliminate the J_0 and J_1 terms. The amplitude signal of a lock-in amplifier is $\Delta_{0°}$ at the $0°$ of the analyzer and $\Delta_{45°}$ at the $45°$ of the analyzer. Therefore, retardation is observed by

$$\theta = \frac{1}{2}\tan^{-1}\frac{\Delta_{45°}}{\Delta_{0°}} \tag{26.41}$$

and the azimuthal direction is

$$\Delta = \sqrt{\Delta_{0°}^2 + \Delta_{45°}^2} \tag{26.42}$$

Figure 26.15 is a measured result of birefringence distribution of CaF_2 [111]. CaF_2 is only a lens material for UV lithography. A He–Ne laser with 632.8 nm wavelength is used for a light source. The size of material is 40 mm diameter and 7 mm thickness. Retardation distribution is 0.5–7 nm. This distribution is caused by residual stress. If you change a light source to the vacuum ultraviolet such as 157 nm, intrinsic birefringence could be observed.

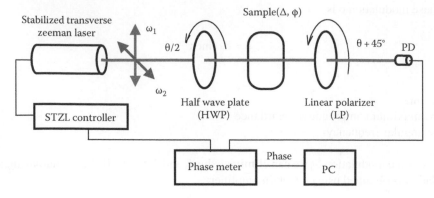

FIGURE 26.15 Optical arrangement of birefringence measurement by optical heterodyne method.

26.8.5 OPTICAL HETERODYNE METHOD

The third method is a birefringence measurement using optical heterodyne detection [10]. Figure 26.15 is an optical arrangement. It consists of a stabilized transverse Zeeman laser (STZL), a half wave plate, and the linear polarizer. The laser STZL has two orthogonal linearly polarized modes and their angular frequencies are ω_1 and ω_2, and $\omega_b = |\omega_1 - \omega_2|$. The frequency difference is $\omega_b \approx 10^5$ rad. The half wave plate and linear polarizer are mounted on a rotating stage. The rotating angle of HWP is 2θ and the polarizer is always $45°$ between two beams so that they interfere with the sample; when two linear polarizations are aligned with the sample fast axis, the phase shifts one way. About $90°$ later, the phase shifts in the opposite direction.

The relation among the polarization elements is given using Mueller matrices and Stokes vectors by

$$\mathbf{S}' = \mathbf{LP}_{2\theta+45} \cdot \mathbf{X}_{\Delta,\varphi} \cdot \mathbf{HW}_\theta \cdot \mathbf{S} \tag{26.43}$$

The Stokes vector is expressed by

$$\mathbf{S} = \begin{bmatrix} a_x^2 + a_y^2 \\ a_x^2 - a_y^2 \\ 2a_x a_y \cos \omega_b t \\ 2a_x a_y \sin \omega_b t \end{bmatrix} \tag{26.44}$$

All Stokes parameters and Mueller matrices are shown as

$$\begin{bmatrix} s_0 \\ s_1 \\ s_2 \\ s_3 \end{bmatrix} = \frac{1}{2} \begin{bmatrix} 1 & \cos(4\theta - 90°) & \sin(4\theta + 90°) & 0 \\ \cos(4\theta + 90°) & \cos^2(4\theta + 90°) & \sin(4\theta + 90°)\cos(4\theta + 90°) & 0 \\ \sin(4\theta + 90°) & \sin(4\theta + 90°)\cos(4\theta + 90°) & \sin^2(4\theta + 90°) & 0 \\ 0 & 0 & 0 & 0 \end{bmatrix}$$

$$\cdot \begin{bmatrix} 1 & 0 & 0 & 0 \\ 0 & \cos^2 2\phi + \sin^2 2\phi \cdot \cos\Delta & (1-\cos\Delta)\sin 2\phi \cdot \cos 2\phi & -\sin 2\phi \sin\Delta \\ 0 & (1-\cos\Delta)\sin 2\phi \cdot \cos 2\phi & \sin^2 2\phi + \cos^2 2\phi \cdot \cos\Delta & \cos 2\phi \sin\Delta \\ 0 & \sin 2\phi \sin\Delta & -\cos 2\phi \sin\Delta & \cos\Delta \end{bmatrix}$$

$$\times \begin{bmatrix} 1 & 0 & 0 & 0 \\ 0 & \cos 4\theta & \sin 4\theta & 0 \\ 0 & \sin 4\theta & -\cos 4\theta & 0 \\ 0 & 0 & 0 & -1 \end{bmatrix} \begin{bmatrix} a_x^2 + a_y^2 \\ a_x^2 - a_y^2 \\ 2a_x a_y \cos \omega_b t \\ 2a_x a_y \sin \omega_b t \end{bmatrix} \tag{26.45}$$

Finally, the detected intensity I is given from the first term of s_0

$$I = S_0' = a_x^2 + a_y^2 - 2a_x a_y \cos\left[\omega_b t - \Delta \cdot \cos(4\theta - 2\phi)\right] \quad (26.46)$$

where ω_b is called the beat signal of the optical heterodyne method and $\Delta \ll 1$.

The optical heterodyne technique is one of the famous interferometers, as shown in Figure 26.16. The electric fields with ω_1 and ω_2 are shown in Figure 26.16a. The interference signal at the reference

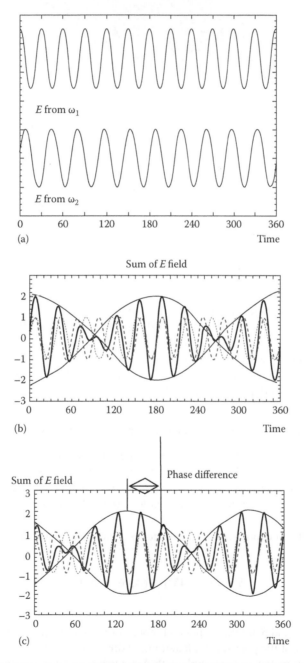

FIGURE 26.16 Optical heterodyne detection. (a) Reference detector measures signal coupled from back of He–Ne laser through LP (45°), (b) from reference detector, and (c) from measurement detector.

(a)

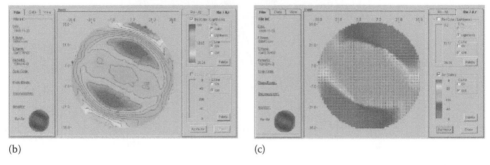

(b)   (c)

FIGURE 26.17   Stress mapping of CaF_2. (a) CaF_2, (b) retardation, and (c) azimuth direction.

detector and measurement detector are shown in Figure 26.16b and c, respectively. The phase difference in Figure 26.16 means the phase $\Delta \cdot \cos(4\theta - 2\varphi)$ of Equation 26.46. This retardance Δ and azimuth angle φ are easy to determine by using a discrete Fourier transform. The resolution of the birefringence measurement is 0.01 nm of the retardation at 0.2° of the azimuthal direction.

Figure 26.17 is birefringence distribution of CaF_2. The size of this sample is 70 mm in diameter. The distribution of the retardance is less than 25°.

26.8.6   Two-Dimensional Birefringence Measurement by Phase-Shifting Method

Figure 26.18 represents a birefringence distribution measurement by the phase-shifting technique [11,12]. Figure 26.18 is an optical arrangement. It consists of a quarter wave plate, a retarder that uses a Babinet-Soleil compensator, and a linear polarizer. A sample with Δ retardation and φ azimuthal direction is set after the retarder. The retarder can work to modulate both retardation δ and azimuthal direction θ.

The relation among the polarization elements is given using Mueller matrices and Stokes vectors by

$$S' = P(\theta + 45°) \cdot X(\Delta, \varphi) \cdot R(\delta, \theta) \cdot S \tag{26.47}$$

where
the Stokes vector S' and S are the output and input of the light, respectively
P is Mueller matrix of the polarizer with azimuthal direction angle θ
$X(\Delta, \varphi)$ is a Mueller matrix of the sample with Δ, retardance and φ azimuthal direction
$R(\delta, \theta)$ is a Muller matrix of the retarder

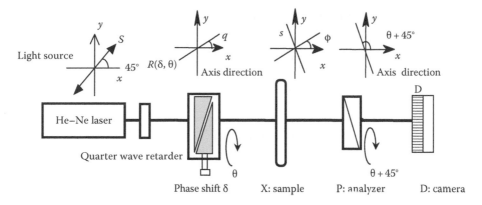

FIGURE 26.18 Birefringence measurement by phase-shifting technique.

All Stokes parameters and Mueller matrices are calculated as

$$
\begin{bmatrix} s_0 \\ s_1 \\ s_2 \\ s_3 \end{bmatrix} = \frac{1}{2} \cdot
\begin{bmatrix}
1 & -\sin 2\theta & \cos 2\theta & 0 \\
-\sin 2\theta & \sin^2 2\theta & -\sin 2\theta \cos 2\theta & 0 \\
\cos 2\theta & -\sin 2\theta \cos 2\theta & \cos^2 2\theta & 0 \\
0 & 0 & 0 & 0
\end{bmatrix}
\begin{bmatrix}
1 & 0 & 0 & 0 \\
0 & 1 & 0 & -\Delta \sin 2\phi \\
0 & 0 & 1 & \Delta \cos 2\phi \\
0 & \Delta \sin 2\phi & -\Delta \cos 2\phi & 1
\end{bmatrix}
$$

$$
\times
\begin{bmatrix}
1 & 0 & 0 & 0 \\
0 & \cos^2 2\theta + \sin^2 2\theta \cdot \cos\delta & (1-\cos\delta)\sin 2\theta \cdot \cos 2\theta & -\sin 2\theta \sin\delta \\
0 & (1-\cos\delta)\sin 2\theta \cdot \cos 2\theta & \sin^2 2\theta + \cos^2 2\theta \cdot \cos\delta & \cos 2\theta \sin\delta \\
0 & \sin 2\theta \sin\delta & -\cos 2\theta \sin\delta & \cos\delta
\end{bmatrix}
\begin{bmatrix} 1 \\ 0 \\ 0 \\ 1 \end{bmatrix}
\tag{26.48}
$$

Finally, the detected intensity I is determined by the first term of s_0

$$
I = s_0 = \frac{1}{2} \cdot I_0 [1 - \cos(\tan^{-1}[\Delta\cos(2\theta - 2\phi)] + \delta)]
\tag{26.49}
$$

where retardation $\Delta \ll 1$.

The intensity modulates sinusoidally by changing the phase δ of the retarder in Equation 26.49. If we apply the phase-shifting technique with $\pi/2$ of each phase shift shown in Figure 26.19 [13,14], we can determine the initial phase by

$$
\Delta\cos(2\theta_M - 2\phi) = \frac{I_3 - I_1}{I_0 - I_2} = \Phi_M
\tag{26.50}
$$

The retardation Δ and azimuthal direction ϕ analytically substitute the phase Φ_M corresponding to the azimuthal angle θ.

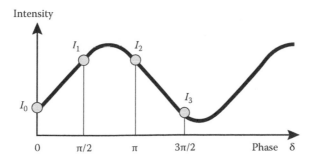

FIGURE 26.19 Four steps of phase-shifting technique.

The birefringence phase difference is given by

$$\Delta = \frac{1}{2}\sqrt{(\Phi_2 - \Phi_0)^2 + (\Phi_3 - \Phi_1)^2}$$

(26.51)

The azimuthal direction of

$$\phi = \frac{1}{2}\tan^{-1}\frac{\Phi_3 - \Phi_1}{\Phi_0 - \Phi_2}$$

(26.52)

This means 16 images are enough to determine birefringence mapping. The resolution of retardance is ±1°.

Figure 26.20 shows a birefringence distribution measurement using two LC retarders for high-speed measurement. Three LC retarders are required to determine all polarization states, generally. However, two retarders are enough to determine the limited retardation in this experiment. Two LC cells whose azimuthal directions are set at 0° and 45°, respectively, are used. If the incident light to the LC retarder is circular polarization, phase φ (ellipticity) and azimuthal direction θ by LC retarders are expressed as

$$\delta_1 = \sin^{-1}(\sin\phi \cdot \sin 2\theta)$$
$$\delta_2 = \tan^{-1}(\tan\phi \cdot \cos 2\theta)$$

(26.53)

where δ_1 and δ_2 are retardation of the LC modulator. Figure 26.21 represents a two-dimensional accuracy check using a 15° retarder at three different orientations with 30°, 0°, and −30°. The dots indicate the measurement points, and the directions of the line indicate the azimuthal direction.

Now, a process to determine birefringence distribution of a plastic disk is shown in Figure 26.22. The original image in Figure 26.22a is the same as that of a polariscope. After analyzing the birefringence distribution using 16 images, the birefringence distribution such as retardation and azimuthal angle are shown in Figure 26.22b and c. A vector map shown in Figure 26.22d indicates both magnitude of retardance and azimuthal direction as a whole. A cross section of AA' in Figure 26.22b is displayed in Figure 26.22e.

FIGURE 26.20 Birefringence distribution measurement using two LC retarders.

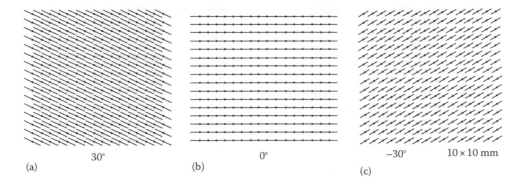

FIGURE 26.21 Two-dimensional accuracy check; same 15° retarder at three orientations. (a) 30° of azimuthal direction, (b) 0° of azimuthal direction, and (c) –30° of azimuthal direction.

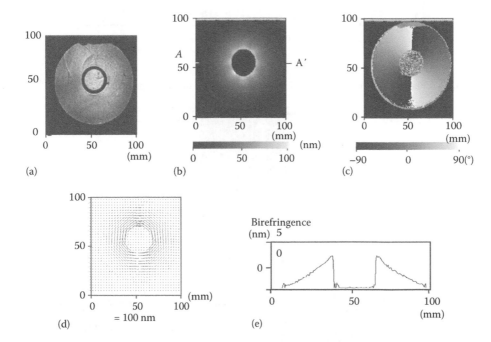

FIGURE 26.22 Process to determine birefringence distribution of plastic disk (3.5 in.). (a) Original image, (b) birefringence distribution, (c) azimuthal angle distribution, (d) vector map, and (e) cross section *AA*.

This birefringence distribution is mainly caused by polymer orientation. The residual stress mapping of the plastic disk is shown in Figure 26.23. We can compare the affective mapping in Figure 26.23a to the defective one in Figure 26.23b. Both cross sections are shown at the bottom. The birefringence distribution of the defective sample indicates the birefringence change in concentric circles and two parts of the radial direction. The part corresponds to the water channel for cooling water, indicating the residual stress caused by thermal stress.

Figure 26.24 demonstrates stress-induced birefringence by uniaxial stretching. A sample is a gelatin seat measuring 10 × 10 mm. An experimental setup for uniaxial stretching is shown in Figure 26.24a. The stress is indicated by a moving micrometer. Birefringence and azimuthal direction distributions along the length of extension are shown in Figure 26.24b and c, respectively. This relation is shown in Figure 26.25. There are not many changes until 50 $\propto$m because of the bio-sample [15].

FIGURE 26.23   Plastic disk inspection. (a) Good product and (b) defective product.

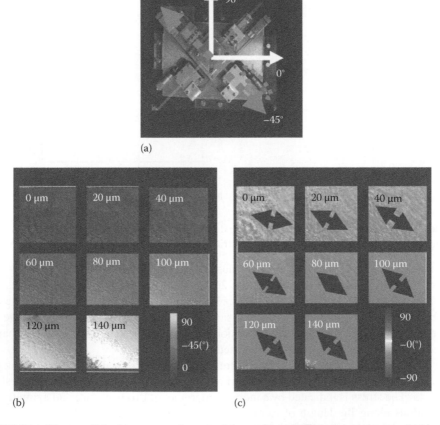

FIGURE 26.24   Change of birefringence under uniaxial stretching. (a) Experimental setup, (b) birefringence distribution, and (c) azimuth distribution.

FIGURE 26.25 Change of birefringence under uniaxial stretching.

REFERENCES

1. P.S. Theocaris and E.E. Gdoutos: *Matrix Theory of Photoelasticity*, Springer, Berlin, Germany, pp. 105–131, 1979.
2. R.M.A. Azzam and N.M. Bashara: *Ellipsometry and Polarized Light*, North-Holland, New York, pp. 153–268, 1987.
3. D. Goldstein: *Polarized Light,* Marcel Dekker, New York, pp. 553–623, 2003.
4. W.A. Shurclif: *Polarization Light,* Harvard University Press, Cambridge, U.K., pp. 124–148, 1962.
5. H. Aben and C. Guillement: *Photoelasticity of Glass*, Springer, Berlin, Germany, pp. 51–77, 1993.
6. R.A. Chipman: Polarimetry, *Handbook of Optics II*, Michael Bass (Editor in Chief) McGraw-Hill, New York, Chapter 22, pp. 22.1–22.37, 1995.
7. Y. Shindo and H. Hanabusa: An improved highly sensitive instrument for measuring optical birefringence, *Polym. Commun.*, 25, 378–382, 1984.
8. B. Wang: Linear birefringence measurement at 157 nm, *Rev. Sci. Instrum.*, 74, 1386, 2003.
9. B. Wang and T.C. Oakberg: A new instrument for measuring both the magnitude and angle of low level linear birefringence, *Rev. Sci. Instrum.*, 70, 3847, 1999.
10. N. Umeda, S. Wakayama, S. Arakawa, A. Takayanagi, and H. Kohwa: Fast birefringence measurement using right and left hand circularly polarized laser, *Proc. SPIE*, 2873, 119–122, 1996.
11. Y. Otani, T. Shimada, T. Yoshizawa, and N. Umeda: Two-dimensional birefringence measurement using the phase shifting technique, *Opt. Eng.*, 33(5), 1604–1609, 1994.
12. Y. Otani, T. Shimada, and T. Yoshizawa: The local-sampling phase shifting technique for precise two-dimensional birefringence measurement, *Opt. Rev.*, 1(1), 103–106, 1994.
13. J.E. Greivenkamp and J.H. Bruning: Phase shifting interferometry, *Optical Shop Testing*, Daniel Malacara, (Ed.), John Wiley, New York, Chapter 14, pp. 501–598, 1992.
14. K. Creath: Temporal phase measurement methods, *Interferogram Analysis*, David W. Robinson and T. Reid (Eds.), Institute of Physics Publishing, London, U.K., Chapter 4, pp. 94–140, 1993.
15. M. Ebisawa, Y. Otani, and N. Umeda: Microscopic birefringence imaging by the local-sampling phase shifting technique, *Proc. SPIE,* 5462, pp. 45–50, (Biophotonics Micro and Nano Imaging, Strasbourg, France), 2004.
16. Luceo Co., Ltd, Circular polarimetry, http://www.luceo.co.jp/technical/pdf/enhenko.pdf (accessed on January 5, 2015).

27 Ellipsometry

Hiroyuki Fujiwara

CONTENTS

27.1 Principles of Ellipsometry 705
27.2 Ellipsometry Measurement..709
27.3 Optical Model 712
27.4 Dielectric Function 714
27.5 Data Analysis Examples...717
References...722

During the 1990s, ellipsometry technology developed rapidly due to rapid advances in computer technology that allowed the automation of ellipsometry instruments as well as ellipsometry data analyses. As a result, the application area of the ellipsometry technique has expanded drastically, and now ellipsometry is applied to wide research areas from semiconductors to organic materials [1]. Nevertheless, principles of ellipsometry are often said to be difficult because the meaning of (ψ, Δ) obtained from ellipsometry measurements is not straightforward. For the understanding of overall ellipsometry technique, this chapter provides general descriptions for principles, measurement, and data analysis method employed widely in the ellipsometry field. In particular, physical backgrounds and examples of ellipsometry analysis are described in detail in this chapter.

27.1 PRINCIPLES OF ELLIPSOMETRY

Basically, ellipsometry is an optical measurement technique that characterizes light reflection (or transmission) from samples [1–6]. Specifically, ellipsometry measures the change in polarized light upon light reflection on a sample (or light transmission by a sample). Table 27.1 summarizes the features of ellipsometry. In principle, ellipsometry measures the two values (ψ, Δ) that represent the amplitude ratio ψ and phase difference Δ between light waves known as p- and s-polarized light waves (see Figure 27.2 later in the chapter). In spectroscopic ellipsometry, (ψ, Δ) spectra are measured by changing the wavelength of light in the ultraviolet (UV)/visible region or infrared region [1,4–6]. The ellipsometry technique has also been applied as a tool for real-time monitoring of film processing because light is employed as the probe. Unlike reflectance/transmittance measurement, ellipsometry allows the direct measurement of the refractive index n and extinction coefficient k, which are also referred to as optical constants. From the two values (n, k), the complex refractive index defined by $N \equiv n - ik$ ($i = \sqrt{-1}$) is determined. The complex dielectric constant $\varepsilon \equiv \varepsilon_1 - i\varepsilon_2$ and absorption coefficient α can also be obtained from N using the simple relations expressed by

$$\varepsilon_1 = n^2 - k^2, \quad \varepsilon_2 = 2nk \left(\varepsilon = N^2 \right), \tag{27.1}$$

$$\alpha = \frac{4\pi k}{\lambda}, \tag{27.2}$$

TABLE 27.1

Features of Ellipsometry

Measurement Probe	Light
Measurement value	(ψ, Δ)
	Amplitude ratio ψ and phase difference Δ between p- and s-polarized light waves
Measurement region	Mainly in the visible/ultraviolet region
Characterized value	Complex refractive index $N \equiv n - ik$ (refractive index n, extinction coefficient k), absorption coefficient $\alpha = 4\pi k/\lambda$, complex dielectric constant $\varepsilon \equiv \varepsilon_1 - i\varepsilon_2$, and film thicknesses
General restriction	1. Surface roughness of samples has to be small
	2. Measurement has to be performed at oblique incidence
Advantages	1. High precision (thickness sensitivity: ~0.01 nm)
	2. Nondestructive and fast measurement
	3. Real-time monitoring (feedback control) is possible
Disadvantages	1. Necessity of an optical model in data analysis (indirect characterization)
	2. Difficulty in the characterization of low absorption coefficients ($\alpha < 100$ cm^{-1})

where λ is the wavelength of light. When samples have thin film structures, the film thicknesses of the samples can be estimated. Moreover, from optical constants and film thicknesses obtained, the reflectance R and transmittance T are calculated.

However, there are two general restrictions on the ellipsometry measurement: (1) surface roughness of samples has to be rather small and (2) the measurement can be performed at oblique incidence. When light scattering by surface roughness reduces a reflected light intensity severely, the ellipsometry measurement becomes difficult as ellipsometry determines a polarization state from its light intensity. In ellipsometry, an incidence angle is chosen at the Brewster angle so that the sensitivity for the measurement is maximized. For semiconductor characterization, the incidence angle is typically from 70° to 80°. It should be noted that at normal incidence the ellipsometry measurement becomes impossible since p- and s-polarization cannot be distinguished anymore at this angle.

As shown in Table 27.1, one of the remarkable features of ellipsometry is high precision for the measurement, and very high thickness sensitivity (~0.01 nm) can be obtained even for conventional instruments. Moreover, as the ellipsometry measurement is nondestructive and takes only a few seconds, real-time observation and feedback control of processing can be performed relatively easily by this technique [1,4]. The one inherent drawback of the ellipsometry technique is the indirect nature of this characterization method. Specifically, ellipsometry data analysis requires an optical model defined by optical constants and layer thicknesses of a sample (see Figure 27.5 later in the chapter). In addition, this ellipsometry analysis using an optical model tends to become complicated, which can be considered as another disadvantage of the technique. Nevertheless, once data analysis method is established, very fast and high-precision characterization becomes possible by this technique. As shown in Table 27.1, characterization of small absorption coefficients ($\alpha < 100$ cm^{-1}) is difficult in ellipsometry since light reflection is rather insensitive to small light absorption.

As mentioned earlier, ellipsometry basically characterizes the polarization state of light. When electric fields of light waves (electromagnetic waves) are oriented in specific directions, such light is referred to as polarized light. For light polarization, the most important fact is that "all the polarization states can be expressed by superimposing two light waves propagating in orthogonal directions." For example, if a light wave is traveling along the z-axis, the polarization state is described by superimposing two electric fields whose directions are parallel to the x- and y-axes. Figure 27.1 shows the polarization states referred to as linear polarization and right-circular

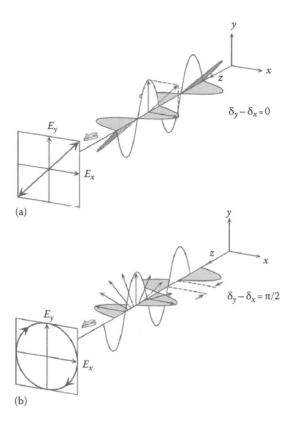

FIGURE 27.1 Representation of (a) linear polarization and (b) right-circular polarization. Phase differences between the electric fields parallel to the x- and y-axis ($\delta_y - \delta_x$) are (a) 0 and (b) $\pi/2$.

polarization. The light waves oscillating in the x and y directions (E_x and E_y) can be treated as one-dimensional waves traveling along the z-axis at a time t:

$$E_{x,y} = E_{0x,0y} \ \exp\left[i\left(\omega t - Kz + \delta_{x,y}\right)\right] \tag{27.3}$$

where

ω and K are the angular frequency and propagation number

$E_{0x}(E_{0y})$ and $\delta_x(\delta_y)$ denote the amplitude and initial phase for the light wave oscillating in the $x(y)$ direction

For the representation of the polarization state, relative amplitude and phase difference between E_x and E_y are important. As shown in Figure 27.1a, when $\delta_y - \delta_x = 0$, there is no phase difference between E_x and E_y, and the oscillating direction is always 45° in the x–y plane. In other words, the light wave oriented at 45° can be resolved into the two waves vibrating parallel to the x- and y-axis. When the phase difference is 90° ($\delta_y - \delta_x = \pi/2$), the synthesized electric filed rotates in the x–y plane as the light propagates. For $\delta_y - \delta_x = 0$ or π, the polarization state becomes linear polarization, while the polarization state is circular polarization for $\delta_y - \delta_x = \pi/2$ or $3\pi/2$. These are special cases, and all the other polarization states are called elliptical polarization.

Figure 27.2 illustrates the measurement principle of ellipsometry. When light is reflected or transmitted by samples at oblique incidence, the light is classified into the p- and s-polarized light waves depending on the oscillatory direction of its electric field. In Figure 27.2, $E_{ip}(E_{is})$ and $E_{rp}(E_{rs})$ show the incident and reflected light waves for the p-polarization (s-polarization).

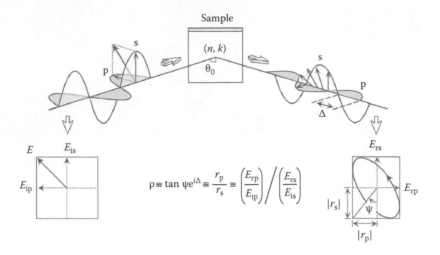

FIGURE 27.2   Measurement principle of ellipsometry.

In p-polarization, the electric fields of E_{ip} and E_{rp} oscillate within the same plane, and this particular plane is called the plane of incidence. In ellipsometry measurement, the polarization states of incident and reflected light waves are defined by the coordinates of p- and s-polarization. Notice that the vectors on the incident and reflection sides overlap completely when the incident angle is $\theta_0 = 90°$ (straight-through configuration). In Figure 27.2, the incident light is the linear polarization oriented at +45° relative to the E_{ip}-axis. In particular, $E_{ip} = E_{is}$ holds for this polarization since the amplitudes of p- and s-polarization are the same and the phase difference between the polarizations is zero.

In general, upon light reflection on a sample, p- and s-polarization shows quite different changes in amplitude and phase [1,2,7]. Ellipsometry measures the two values (ψ, Δ) that represent the relative amplitude ratio and phase difference between p- and s-polarization, respectively. In ellipsometry, therefore, the variation of light reflection with p- and s-polarization is measured as the change in a polarization state. In particular, when a sample structure is simple, the amplitude ratio ψ is characterized by the refractive index n while Δ represents light absorption described by the extinction coefficient k. In this case, the two values (n, k) can be determined directly from the two ellipsometry parameters (ψ, Δ). This is the basic principle of ellipsometry measurement.

In Figure 27.2, each wave $(E_{ip}, E_{is}, E_{rp}, \text{and } E_{rs})$ is expressed by Equation 27.3. The ratio of reflected light to incident light is called the amplitude reflection coefficient r, and the ratios for the p-polarization (r_p) and s-polarization (r_s) are given by

$$r_p \equiv \frac{E_{rp}}{E_{ip}} = |r_p| \ \exp(i\delta_p), \qquad r_s \equiv \frac{E_{rs}}{E_{is}} = |r_s| \ \exp(i\delta_s) \tag{27.4}$$

Equation 27.4 represents the relative change induced by light reflection. Accordingly, upon light reflection the amplitude changes from 1 to $|r_p|$ with a relative phase change of δ_p for p-polarization. The (ψ, Δ) measured from ellipsometry are defined from the ratio of the amplitude reflection coefficients for p- and s-polarization:

$$\rho \equiv \tan \psi \ e^{i\Delta} \equiv \frac{r_p}{r_s} \tag{27.5}$$

It follows from Equation 27.4 that

$$\rho \equiv \tan \psi \; e^{i\Delta} \equiv \frac{r_{\mathrm{p}}}{r_{\mathrm{s}}} \equiv \left(\frac{E_{\mathrm{rp}}}{E_{\mathrm{ip}}} \right) \bigg/ \left(\frac{E_{\mathrm{rs}}}{E_{\mathrm{is}}} \right) \tag{27.6}$$

As confirmed from Equation 27.4, r_{p} and r_{s} are originally defined by the ratios of reflected electric fields to incident electric fields and $\tan \psi \; e^{i\Delta}$ is defined further by the ratio of r_{p} to r_{s}. In the case of Figure 27.2, Equation 27.6 can be simplified to $\tan \psi \; e^{i\Delta} = E_{\mathrm{rp}}/E_{\mathrm{rs}}$ since $E_{\mathrm{ip}} = E_{\mathrm{is}}$. By using Equations 27.4 and 27.6, we get

$$\tan \psi = \frac{|r_{p}|}{|r_{s}|} \quad (0° \le \psi \le 90°), \quad \Delta = \delta_{p} - \delta_{s} \quad (-180° \le \Delta \le 180°) \tag{27.7}$$

Thus, ψ represents the angle determined from the amplitude ratio between reflected p- and s-polarization while Δ expresses the phase difference between reflected p- and s-polarization, as shown in Figure 27.2. On the other hand, the reflectance R obtained in conventional measurements is given by

$$R_{\mathrm{p}} = \frac{|E_{\mathrm{rp}}|^{2}}{|E_{\mathrm{ip}}|^{2}} = |r_{\mathrm{p}}|^{2}, \quad R_{\mathrm{s}} = \frac{|E_{\mathrm{rs}}|^{2}}{|E_{\mathrm{is}}|^{2}} = |r_{\mathrm{s}}|^{2} \tag{27.8}$$

From Equation 27.8, ψ is redefined as

$$\psi = \tan^{-1}\left(|\rho| \right) = \tan^{-1}\left(\frac{|r_{\mathrm{p}}|}{|r_{\mathrm{s}}|} \right) = \tan^{-1}\left[\left(\frac{R_{\mathrm{p}}}{R_{\mathrm{s}}} \right)^{1/2} \right] \tag{27.9}$$

Accordingly, ψ represents the angle basically determined from $R_{\mathrm{p}}/R_{\mathrm{s}}$.

As confirmed from Equation 27.6, ellipsometry measures the ratio of the amplitude reflection coefficients $(r_{\mathrm{p}}/r_{\mathrm{s}})$. Since the difference between r_{p} and r_{s} is maximized at the Brewster angle [1,2,7], sensitivity for the measurement also increases at this angle. Thus, ellipsometry measurement is generally performed at the Brewster angle. The Brewster angle θ_{B} can be obtained from $\theta_{\mathrm{B}} = \tan^{-1}(n_{\mathrm{s}})$, where n_{s} is the refractive index of samples. If $n_{\mathrm{s}} = 3$, for example, we obtain $\theta_{\mathrm{B}} = 71.6°$. Accordingly, the choice of the incidence angle varies according to the optical constants of samples.

27.2 ELLIPSOMETRY MEASUREMENT

Until the early 1970s, only an ellipsometry instrument called the null ellipsometry had been used for measurements [1,2]. Ellipsometry instruments that are used now can be classified into two major categories: instruments that use rotating optical elements [1,6,8–19] and instruments that use a photoelastic modulator [1,18–24]. The rotating-element ellipsometers can further be classified into the rotating-analyzer ellipsometry (RAE) [1,6,8–13] and rotating-compensator ellipsometry [1,14–18]. In this section, as the simplest ellipsometry instrument, the principle of RAE is explained in detail.

Figure 27.3 shows the schematic diagram of RAE that has been used widely up to now. This instrument consists of (light source)-polarizer-sample-(rotating analyzer)-detector. Although a polarizer and an analyzer are the same optical element, these are called separately due to the difference in their roles. Basically, the polarizer (analyzer) transmits only linearly polarized light whose direction is parallel to the transmission axis [1,4,7]. A polarizer is placed in front of a light source and is utilized to extract linearly polarized light from the unpolarized light source. On the other hand, an analyzer is placed in front of a light detector and the state of polarization is determined

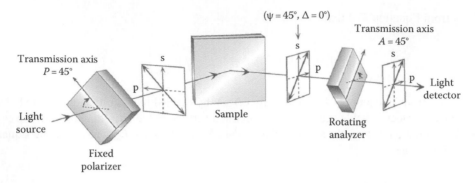

FIGURE 27.3 Schematic diagram of the measurement in RAE. In this figure, the rotation angles of the polarizer and analyzer are $P = 45°$ and $A = 45°$, respectively, and the incident wave and reflected wave from a sample are linear polarizations of $+45°$ ($\psi = 45°$, $\Delta = 0°$).

from the intensity of light transmitted through the analyzer. In RAE instrument, the rotation angle of the polarizer (P) is fixed ($P = 45°$ in Figure 27.3). On the other hand, the analyzer rotates toward the direction indicated by the arrow and the rotation angle of the analyzer (A) changes continuously ($A = 45°$ in Figure 27.3). When looking against the direction of the beam, counterclockwise rotation is the positive direction for the rotation.

Now consider that the analyzer rotates continuously with time at a speed of $A = \omega t$, where ω is the angular frequency of the analyzer. In this case, variation of light intensity $I(t)$ in RAE is expressed as follows [1,6,18]:

$$I(t) = I_0 \left(1 + \alpha \cos 2\omega t + \beta \sin 2\omega t\right) \tag{27.10}$$

where
  I_0 represents the proportional constant of the reflected light whose intensity is proportional to incident light intensity
  α and β are the normalized Fourier coefficients of $\cos 2A$ and $\sin 2A$ ($A = \omega t$), respectively

It can be seen from Equation 27.10 that the light intensity varies as a function of the analyzer angle $2A$. This implies that there is no distinction between the upper and lower sides of the transmission axis and the 180° rotation of the analyzer corresponds to one optical rotation.

In RAE instrument, the normalized Fourier coefficients (α, β) in Equation 27.10 are given by

$$\alpha = \frac{\tan^2 \psi - \tan^2 P}{\tan^2 \psi + \tan^2 P}, \quad \beta = \frac{2 \tan \psi \cos \Delta \tan P}{\tan^2 \psi + \tan^2 P} \tag{27.11}$$

Solving Equation 27.11 for (ψ, Δ) gives the following equations [1,6,19]:

$$\tan \psi = \sqrt{\frac{1 + \alpha}{1 - \alpha}} \left|\tan P\right|, \quad \cos \Delta = \frac{\beta}{\sqrt{1 - \alpha^2}} \tag{27.12}$$

Figure 27.4 shows the normalized light intensity ($I_0 = 1$) calculated from Equation 27.10 using $P = 45°$, plotted as a function of the angle of the rotating analyzer $A = \omega t$ ($0° \leq A \leq 180°$). In this figure, Δ is varied from 0° to 90° with a constant value of $\psi = 45°$. As shown in "a" in Figure 27.4, when reflected light is linear polarization ($\psi = 45°$, $\Delta = 0°$), a light intensity is maximized at $A = 45°$ while the light intensity becomes zero at $A = 135°$. This result can be understood easily from an optical configuration shown in Figure 27.3. In this figure, incident light is the linear polarization of $+45°$ ($P = 45°$),

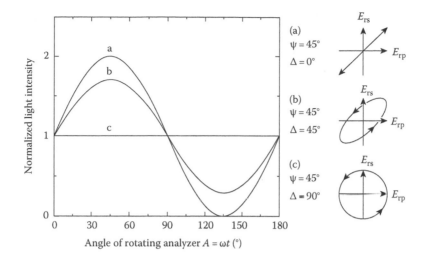

FIGURE 27.4 Normalized light intensity in the RAE, plotted as a function of the angle of rotating analyzer $A = \omega t$. This figure summarizes the calculation results when the polarization states of reflected light are (a) $\psi = 45°$, $\Delta = 0°$; (b) $\psi = 45°$, $\Delta = 45°$; and (c) $\psi = 45°$, $\Delta = 90°$.

and the polarization state does not change upon light reflection on a sample ($\psi = 45°$, $\Delta = 0°$). When $A = 0°$, only the p-polarized component transmits the analyzer. At $A = 45°$, however, the oscillatory direction of the reflected light is parallel to the transmission axis of the analyzer (Figure 27.3), and consequently the light intensity measured by the detector is maximized, as shown in "a" in Figure 27.4. When the analyzer rotates further to $A = 135°$, the transmission axis of the analyzer becomes perpendicular to the polarization direction of the reflected light and the light intensity becomes zero. On the other hand, when reflected light is circular polarization, the light intensity is independent of the analyzer angle, as shown in "c" in Figure 27.4. This result can be understood from the propagation of circular polarization, as shown in "b" in Figure 27.1. When reflected light is elliptical polarization, normalized light intensities are intermediate between linear and circular polarizations.

From the above results, it can be seen that the shape of $I(t)$ slides in the horizontal direction (analyzer angle) depending on a value of ψ and the amplitude of $I(t)$ reduces as the state of polarization changes from linear to elliptical polarizations. In RAE, therefore, the polarization state of reflected light is determined from a variation of light intensity with the analyzer angle. In this method, however, right-circular polarization ($\delta_y - \delta_x = \pi/2$), shown in Figure 27.1b, cannot be distinguished from left-circular polarization ($\delta_y - \delta_x = 3\pi/2$) since these polarizations show the same light intensity variation versus the analyzer angle. Accordingly, in RAE, the measurement range for A becomes half ($0° \leq \Delta \leq 180°$) [1]. If we introduce a compensator (retarder) between the sample and the analyzer in Figure 27.3, circular polarizations can be distinguished and more accurate measurement in the range of $-180° \leq \Delta \leq 180°$ becomes possible [1,14–18].

In RAE measurement, we first determine (α, β) from the Fourier analysis of measured light intensities and then extract (ψ, Δ) values by substituting the measured (α, β) into Equation 27.12.

It should be emphasized that ellipsometry allows the high-precision measurements for (ψ, Δ) because ellipsometry measures relative light intensities modulated by optical elements instead of the absolute light intensities of reflected p- and s-polarization, as shown in Figure 27.4. Accordingly, measurement errors induced by various imperfections of instruments become very small in ellipsometry measurement if we compare with absolute reflectance measurements. This is the reason why optical constants and film thickness can be estimated with high precision in ellipsometry technique.

In conventional single-wavelength ellipsometry, a He–Ne laser and photomultiplier tube are employed as a light source and detector, respectively [10]. In spectroscopic ellipsometry, the wavelength of incident light is changed using a monochromator and the monochromatic light is detected by a photomultiplier tube [6,11]. In spectroscopic ellipsometry instruments that allow real-time monitoring, white light is illuminated to a sample and all the light waves at different wavelengths are detected simultaneously using a photodiode array [13,16] or charge coupled device (CCD) detector. Up to now, the capability of spectroscopic ellipsometry measurement has been extended from the vacuum ultraviolet (VUV) region to the infrared region. For VUV ellipsometry [4,25], a deuterium lamp is used as a light source in addition to a xenon lamp used in conventional measurements in the visible/UV region. For infrared ellipsometry [1,4,5,25–27], Fourier-transform infrared spectrometer has been employed as a light source.

27.3   OPTICAL MODEL

In order to evaluate the optical constants and thickness of samples from ellipsometry, data analysis using an optical model is necessary. The optical model is represented by the complex refractive index and layer thickness of each layer. Figure 27.5 shows an example of the optical model consisting of an air/thin layer/substrate structure. In this figure, N_0, N_1, and N_2 denote the complex refractive indices of air ($N_0 = 1$), thin layer (thickness d), and substrate, respectively. The transmission angles (θ_1 and θ_2) can be calculated from the angle of incidence θ_0 by applying Snell's law expressed as $N_0 \sin \theta_0 = N_1 \sin \theta_1 = N_2 \sin \theta_2$ [1,7]. As shown in Figure 27.5, when light absorption in a thin layer is small, optical interference occurs by multiple light reflections within the thin layer. In this case, the total amplitude reflection coefficient of the structure is given by

$$r_{012} = \frac{r_{01} + r_{12}\,e^{-i2\beta}}{1 + r_{01}r_{12}\,e^{-i2\beta}} \tag{27.13}$$

Here, r_{01} and r_{12} show the amplitude reflection coefficients at the air/thin layer and thin layer/substrate interfaces in Figure 27.5, respectively, and the phase thickness β is described by $\beta = (2\pi d N_1 \cos \theta_1)/\lambda$ [1,7]. Note that Equation 27.13 can be applied for the calculation of both p-polarization ($r_{012,p}$) and s-polarization ($r_{012,s}$). From Fresnel equations, the amplitude reflection coefficients for the p-polarization ($r_{jk,p}$) and s-polarization ($r_{jk,s}$) at an interface jk are obtained [1,7]:

$$r_{jk,p} = \frac{N_k \cos \theta_j - N_j \cos \theta_k}{N_k \cos \theta_j + N_j \cos \theta_k}, \quad r_{jk,s} = \frac{N_j \cos \theta_j - N_k \cos \theta_k}{N_j \cos \theta_j + N_k \cos \theta_k} \tag{27.14}$$

Finally, (ψ, Δ) of the structure in Figure 27.5 are calculated using Equations 27.5 and 27.13:

$$\rho = \frac{r_{012,p}}{r_{012,s}} \left(\frac{r_{01,p} + r_{12,p}\,e^{-i2\beta}}{1 + r_{01,p}r_{12,p}\,e^{-i2\beta}} \right) \bigg/ \left(\frac{r_{01,s} + r_{12,s}\,e^{-i2\beta}}{1 + r_{01,s}r_{12,s}\,e^{-i2\beta}} \right) \tag{27.15}$$

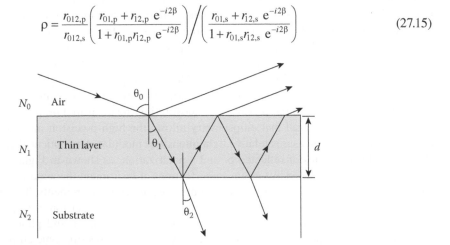

FIGURE 27.5   Optical model consisting of an air/thin layer/substrate structure.

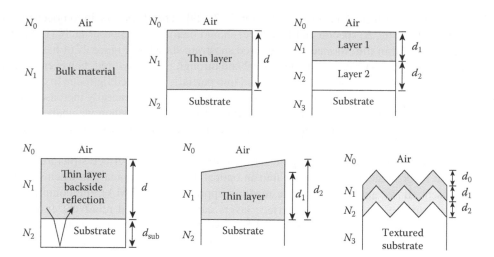

FIGURE 27.6 Optical model used in data analysis of ellipsometry.

Figure 27.6 summarizes optical models used in ellipsometry data analysis. In the optical model shown in Figure 27.6a, only the light reflection at an air/bulk material interface is taken into account, and $\rho = \tan \psi \, e^{i\Delta}$ of this optical model is given by $\rho = r_{jk,p}/r_{jk,s}$ using Equation 27.14. When we apply this model, the thickness of the bulk material must be thick enough so that there is no light reflection from the substrate backside. In the case of this model, ε of the bulk material can be estimated directly from measured (ψ, Δ) using the following equation [1,2]:

$$\varepsilon \equiv \varepsilon_1 - i\varepsilon_2 = \sin^2 \theta_0 \left[1 + \tan^2 \theta_0 \left(\frac{1-\rho}{1+\rho} \right)^2 \right] \tag{27.16}$$

The optical model in Figure 27.6b shows the same structure as Figure 27.5. In this optical model, the infinite thickness of a substrate is also assumed. If two thin layers are formed on a substrate (Figure 27.6c), we first calculate the amplitude reflection coefficients for layer 2 and substrate by applying Equation 27.13:

$$r_{123} = \frac{r_{12} + r_{23} \, e^{-i2\beta_2}}{1 + r_{12}r_{23} \, e^{-i2\beta_2}} \tag{27.17}$$

The phase variation β_2 is given by $\beta_2 = (2\pi d_2 N_2 \cos \theta_2)/\lambda$, where d_2 is the thickness of layer 2. From r_{123}, we obtain the amplitude reflection coefficients for the multilayer as follows:

$$r_{0123} = \frac{r_{01} + r_{123} \, e^{-i2\beta_1}}{1 + r_{01}r_{123} \, e^{-i2\beta_1}} \tag{27.18}$$

In Equation 27.18, $\beta_1 = (2\pi d_1 N_1 \cos \theta_1)/\lambda$, where d_1 is a thickness of layer 1. The (ψ, Δ) of this optical model can then be calculated as $\rho = r_{0123,p}/r_{0123,s}$. In this manner, the calculation can be performed upward from the substrate even if there are many layers in a multilayer structure.

As shown in Figure 27.6d, when the light absorption in a substrate is small ($\varepsilon_2 \sim k \sim 0$) and the backside reflection of the transparent substrate is present, we need to employ an analytical model that incorporates the effect of backside reflection [1,28,29]. The thickness inhomogeneity of a thin layer also affects (ψ, Δ) spectra and this effect should be modeled appropriately [1,28,30]. So far, the analysis of a thin film formed on a textured substrate shown in Figure 27.6f has also been reported [31]. When the samples shown in Figure 27.6d through f are measured, however, polarized light used as a probe in ellipsometry is transformed into partially polarized light due to depolarization effect of the samples [1,28–31]. In such cases, measurement errors generally increase, although this effect depends completely on the type of instrument [1,16]. Thus, when we analyze samples having depolarization effects, extra care is required for the data analysis as well as the measurement itself.

Basically, ellipsometry technique is quite sensitive to surface and interface structures. Thus, it is necessary to incorporate these structures into an optical model in data analysis when such structures are present. If we apply the effective medium approximation (EMA), the complex refractive indices of surface roughness and interface layers can be calculated relatively easily [1,32]. Furthermore, from ellipsometry analysis using EMA, we can characterize each volume fraction in a composite material [33]. If the film is a two-phase composite, EMA is expressed by the following equation [1,32]:

$$f_a \frac{\varepsilon_a - \varepsilon}{\varepsilon_a + 2\varepsilon} + \left(1 - f_a\right) \frac{\varepsilon_b - \varepsilon}{\varepsilon_b + 2\varepsilon} = 0 \tag{27.19}$$

In Equation 27.19, f_a and $(1 - f_a)$ represent the volume fractions of the phases a and b whose complex dielectric constants are ε_a and ε_b, respectively. This model can be extended easily to describe a material consisting of many phases:

$$\sum_{i=1}^{n} f_i \frac{\varepsilon_i - \varepsilon}{\varepsilon_i + 2\varepsilon} = 0 \tag{27.20}$$

When a thin layer on substrates has surface roughness, the optical model in Figure 27.6c can be employed. In this case, layers 1 and 2 express surface roughness layer and thin layer, respectively. The surface roughness layer is originally composed of the layer material (N_2) and air ($N_0 = 1$). Thus, if we apply EMA for these two phases, N of surface roughness layer (N_1) can be estimated relatively easily. For the calculation of N_1, the volume fraction of voids (f_{void}) within the surface roughness layer is required, although the analysis can also be performed assuming $f_{void} = 0.5$ (void volume fraction is 50 vol.%). By substituting $\varepsilon_a = N_0^2$, $\varepsilon_b = N_2^2$, and $f_a = f_{void} = 0.5$ into Equation 27.19, ε of the surface roughness layer can be obtained. From ε, N_1 is determined using $N_1 = \sqrt{\varepsilon}$. When $N_0 = 1$, $N_2 = n_2 = 5$ ($N_2 = n_2 - ik_2$, $k_2 = 0$), and $f_{void} = 0.5$, for example, we get $N_1 = n_1 = 2.83$. Accordingly, n_1 roughly becomes half of n_2 due to the presence of voids (50 vol.%) within the surface roughness layer. Notice that only N_2 is required for the calculation of N_1 if $N_0 = 1$ and $f_{void} = 0.5$ are assumed. Thus, data analysis is simplified considerably if we apply EMA. Furthermore, the complex refractive index of an interface layer can also be determined from similar calculation.

27.4 DIELECTRIC FUNCTION

The optical constants of materials generally change depending on photon energy or wavelength of light. This optical (dielectric) response is generally referred to as the dielectric function or dielectric dispersion [1,7]. For spectroscopic ellipsometry, the dielectric function of a sample is required in the data analysis. When the dielectric function of a sample is not known, modeling of the dielectric

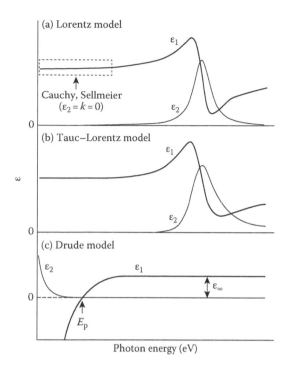

(a) Lorentz model

ε_1

Cauchy, Sellmeier
$(\varepsilon_2 = k = 0)$

ε_2

0

(b) Tauc–Lorentz model

ε_1

ε_2

0

(c) Drude model

ε_2 ε_1

ε_∞

0

E_p

Photon energy (eV)

FIGURE 27.7 Dielectric functions calculated from (a) Lorentz model, (b) Tauc–Lorentz model, and (c) Drude model.

function is necessary. Figure 27.7 illustrates the dielectric functions calculated from Lorentz model, Tauc–Lorentz model, and Drude model. It can be seen from the figure that the dielectric function shows complicated variations. When there is no light absorption in materials ($k = 0$), it is obvious from Equation 27.1 that $\varepsilon_1 = n^2$ and $\varepsilon_2 = 0$. Thus, in the region of $\varepsilon_2 = k = 0$, ε_1 basically shows the contribution of n. Moreover, since $\varepsilon_2 = 2nk$ (Equation 27.1), ε_2 is proportional to k. Accordingly, the ε_2 peaks in Figure 27.7a and b show that light is absorbed in specific regions for photon energy or wavelength.

The Lorentz model in Figure 27.7a is a classical model and can be derived from a physical model assuming forced oscillation by light [1]. The expression for the Lorentz model is given by

$$\varepsilon = 1 + \sum_j \frac{A_j}{En_{0j}^2 - En^2 + i\Gamma_j\, En} \tag{27.21}$$

where
  En is the photon energy
  A is the oscillator strength that determines the amplitude of the ε_2 peak

The peak position and half width of the ε_2 peak are described by En_0 and Γ, respectively. In Equation 27.21, the dielectric function is expressed as the sum of different oscillators and the subscript j denotes the jth oscillator.

As shown in Figure 27.7a, in a transparent region ($\varepsilon_2 = k = 0$), the Sellmeier or Cauchy model can be used. The Sellmeier model corresponds to a region where $\varepsilon_2 = 0$ in the Lorentz model and this model can be derived by assuming $\Gamma \to 0$ at $En \ll En_0$ in Equation 27.21. By using the relation $\omega/c = 2\pi/\lambda$ where c is the speed of light, the Sellmeier model is expressed by

$$\varepsilon_1 = n^2 = A + \sum_j \frac{B_j \lambda^2}{\lambda^2 - \lambda_{0j}^2}, \quad \varepsilon_2 = 0 \tag{27.22}$$

where A and B_j represent analytical parameters used in data analysis and λ_0 corresponds to En_0. On the other hand, the Cauchy model is given by

$$n = A + \frac{B}{\lambda^2} + \frac{C}{\lambda^4} + \cdots, \quad k = 0 \tag{27.23}$$

The above equation can be obtained from the series expansion of Equation 27.22. Although the Cauchy model is an equation relative to the refractive index n, this is the approximate function of the Sellmeier model.

The Tauc–Lorentz model in Figure 27.7b has been employed widely to model the dielectric function of amorphous materials including a-Si [34], a-SiN:H [35], and HfO_2 [36]. Recently, this model has also been applied to dielectric function modeling for transparent conductive oxides such as SnO_2 [31], In_2O_3:Sn (ITO) [37], and ZnO [37]. As shown in Figure 27.7a, the shape of a ε_2 peak calculated from the Lorentz model is completely symmetric. However, the ε_2 peaks of amorphous materials generally show asymmetric shapes. In the Tauc–Lorentz model [34], therefore, ε_2 is modeled from the product of a unique bandgap of amorphous materials (Tauc gap [38]) and the Lorentz model:

$$\varepsilon_2 = \frac{AEn_0 C (En - E_g)^2}{(En^2 - En_0^2)^2 + C^2 En^2} \frac{1}{En} (En > E_g)$$
$$= 0 (En \le E_g) \tag{27.24}$$

where
   A is the amplitude
   En_0 is the peak position
   C is the broadening parameter
   E_g is the Tauc gap

Although not shown here, the equation for ε_1 has also been derived using the Kramers–Kronig relations [34]. If we introduce an additional parameter $\varepsilon1(\infty)$ to express ε_1 (generally $\varepsilon1(\infty) = 1$), the Tauc–Lorentz model is described by total five parameters ($\varepsilon1(\infty)$, A, C, En_0, and E_g) [34]. So far, the dielectric function of amorphous materials has also been expressed using other models including the Cody–Lorentz model [39], Forouhi–Bloomer model [40], model dielectric function (MDF) [41], and band model [42]. Furthermore, the tetrahedral model [43] and a model that extends the tetrahedral model using the EMA [44,45] have also been proposed.

On the other hand, free electrons in metals and free carriers in semiconductors absorb light and alter dielectric functions. To express the contribution of such free-carrier absorption, the Drude model has been applied widely. The Drude model is described by the following equation [1,37]:

$$\varepsilon = \varepsilon_\infty - \frac{A}{En^2 - i\Gamma En} \tag{27.25}$$

where
   ε_∞ is the high-frequency dielectric constant
   A is the amplitude
   Γ is the broadening parameter

FIGURE 27.8 Dielectric function of Si crystal. (Data from Adachi, S., *Optical Constants of Crystalline and Amorphous Semiconductors: Numerical Data and Graphical Information,* Kluwer Academic Publishers, Norwell, MA, 1999.)

In the case of semiconductors, A can be related to carrier concentration in materials while Γ is inversely proportional to carrier mobility, and we can evaluate these values from detailed ellipsometry analysis [1,37]. When free-carrier concentration in a sample is high (typically $>10^{18}$ cm^{-3}), we can simply add the second term on the right in Equation 27.25 to other dielectric function models. In the analysis of transparent conductive oxides, dielectric function modeling has been performed $\varepsilon = \varepsilon_{\text{Tauc–Lorentz}} + \varepsilon_{\text{Drude}}$ [37].

As shown in Figure 27.7c, when the free-carrier absorption is present, the value of ε_1 reduces and the free-carrier absorption described by ε_2 increases rapidly at lower energies. At the energy referred to as plasma energy E_p, ε_1 becomes zero and ε_1 shows negative values at $En < E_p$. When $\varepsilon_1 < 0$, the electric field of the incident light is screened by the free carriers present at a material surface. Consequently, the electric field of the light cannot penetrate into the material and the light reflection becomes quite strong.

Figure 27.8 shows the dielectric function of Si crystal (c-Si) [46]. As confirmed from this figure, actual dielectric functions show rather complicated structures. Since the bandgap E_g of c-Si is 1.1 eV, $\varepsilon_2 = k = 0$ at $En < 1.1$ eV. In this region, the optical constant shows constant values of $\varepsilon_1 = n^2 = 11.6$ ($n = 3.41$). In Figure 27.8, there are two sharp ε_2 peaks at 3.4 and 4.25 eV that represent optical transitions at Λ and X points in the c-Si band structure [47]. If we perform the analysis using a transparent region of $En < 2.5$ eV ($\varepsilon_2 \sim k \sim 0$), we can model the dielectric function of c-Si rather easily [48]. So far, in order to express the dielectric functions of semiconductors, various models including the Lorentz model, harmonic oscillator approximation (HOA) [49], band model [42], MDF [41,50], and parametric semiconductor model [51] have been developed.

27.5 DATA ANALYSIS EXAMPLES

For simple ellipsometry characterization, ellipsometry measurement at a single wavelength using a He–Ne layer ($\lambda = 632.8$ nm) has been performed widely [10]. When a thin SiO$_2$ layer formed on c-Si is evaluated, the single wavelength ellipsometry has commonly been used [10,52]. For the structure, the optical model shown in Figure 27.6b can be employed assuming no surface roughness and interface layers. In this case, (ψ, Δ) are calculated from Equation 27.15 and the parameters used in the calculation are described as

$$\tan \psi \, e^{i\Delta} = \rho \left(N_0, N_1, N_2, d, \theta_0 \right) \tag{27.26}$$

FIGURE 27.9 $\psi - \Delta$ trajectory in an air/thin layer/c-Si structure. In this calculation, $\theta_0 = 70°$, and the refractive index of the thin layer ($N_1 = n_1$) was changed from 1.46 to 3.0. Open circles show the values at every 10 nm.

Since values of N_2 and θ_0 are usually known and $N_0 = 1$, there are total three unknown parameters in Equation 27.26: (n_1, k_1) of $N_1 = n_1 - ik_1$ and d. If a thin layer does not show any light absorption at the measured wavelength ($k_1 = 0$), the unknown parameters become only n_1 and d. In this case, (n_1, d) of the layer can be estimated directly from two measured values (ψ, Δ) obtained by a single wavelength measurement.

Figure 27.9 shows the $\psi - \Delta$ trajectory when the refractive index of a thin transparent layer ($N_1 = n_1$) on a c-Si substrate is varied from 1.46 (SiO_2) to 3.0 in the calculation of Equation 27.15. For the calculation, $\theta_0 = 70°$, $N_0 = 1$, and $N_2 = 3.87 - i0.0146$ [53] (c-Si substrate at $\lambda = 632.8$ nm) were used. The open circles in Figure 27.9 represent the values at every 10 nm, and the (ψ, Δ) values move toward the direction of arrows with increasing layer thickness. It can be seen from Figure 27.9 that the (ψ, Δ) values change largely according to the (n_1, d) values and, from measured (ψ, Δ) values, (n_1, d) of a thin layer can be evaluated analytically. In Figure 27.9, however, ψ is almost constant in the region of $d < 10$ nm, although Δ shows large changes. This arises from the fact that R_p and R_s show almost no changes at $d < 10$ nm, even when n_1 is varied in the wide range. Recall from Equation 27.9 that $\psi = \tan^{-1}[(R_p/R_s)^{1/2}]$. Consequently, the evaluation of n_1 becomes difficult when the thickness of a transparent layer is quite thin ($d < 10$ nm), although d can still be estimated if n_1 is known. Thus, when thin layers are characterized, we determine d using n_1 obtained from the analysis of thick layers.

In contrast to single-wavelength ellipsometry, (ψ, Δ) spectra are measured in spectroscopic ellipsometry. From the analysis of these spectra, various characterizations including bandgap and phase structures become possible [1]. Furthermore, even when a thin layer on substrates shows light absorption ($k > 0$) or samples have complicated layer structures, ellipsometry analyses can be performed [1]. Figure 27.10 shows a simple example of spectroscopic ellipsometry analysis. The sample is a SiO_2 thermal oxide formed on a c-Si substrate. The optical model of this sample also corresponds to the one shown in Figure 27.6b. Here, we use $\theta_0 = 75°$, $N_0 = 1$, and the reported value for N_2 [53]. Thus, the analysis parameters are only (n_1, d) as described earlier. Since SiO_2 is a transparent material, the Sellmeier or Cauchy model can be applied to the analysis. As shown in Equation 27.23, the Cauchy model is expressed by the three parameters (A, B, C). Accordingly, when the Cauchy model is applied to calculate $N_1(n_1)$, the final analysis parameters in the air/SiO_2/c-Si structure become (A, B, C, d).

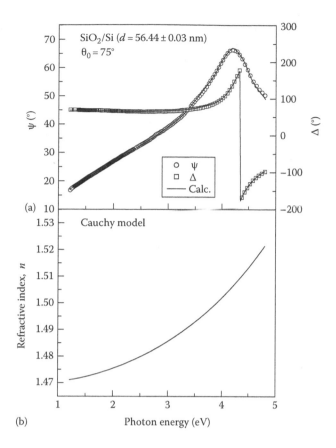

FIGURE 27.10   (a) (ψ, Δ) spectra obtained from a SiO_2/c-Si substrate structure and (b) refractive index spectrum of the SiO_2 layer deduced from the Cauchy model. In (a), the incidence angle of the measurement is $\theta_0 = 75°$.

Figure 27.10a shows the (ψ, Δ) spectra of a sample having the SiO_2/c-Si structure. The solid lines in the figure show the calculated spectra obtained from the fitting analysis. From the analysis, $A = 1.469$, $B = 2.404 \times 10^{-3}$ μm^2, $C = 7.357 \times 10^{-5}$ μm^4, and $d = 56.44 \pm 0.03$ nm are estimated. Figure 27.10b shows the refractive index spectrum of the SiO_2 layer calculated from these (A, B, C) values. In spectroscopic ellipsometry, the construction of an optical model and modeling of dielectric function are performed in this manner, and from the fitting analysis, the optical constants and thickness of a sample are evaluated. In particular, since the measurement of spectroscopic ellipsometry is performed in a wide wavelength region, even if there are several analysis parameters, these values can be estimated from the analysis. Accordingly, more reliable analysis can be performed from (ψ, Δ) spectra measured in a wider wavelength region with a larger number of data points. This example shows clearly that we can determine film thickness very precisely from spectroscopic ellipsometry. It has been reported that SiO_2 thicknesses estimated from ellipsometry show excellent agreement with those characterized by transmission electron microscope [54], x-ray photoemission spectroscopy [55], and capacitance-voltage measurements [55].

When we characterize the refractive index spectrum and thickness of SiO_2 layers, the above analysis is generally sufficient. If we employ the reported dielectric function for SiO_2 [53], d is determined more easily. The fitting in the analysis can be improved by increasing the number of parameters in the Cauchy or Sellmeier model. Nevertheless, for the accurate characterization of SiO_2 layers, it is necessary to take the influence of an interface layer into account. In particular, it has been confirmed that the overall refractive index of SiO_2 layers decreases with increasing d [53,54].

This implies that there exists the interface layer with high refractive index at the SiO_2/c-Si interface [53,54,56]. In other words, the effect of the interface layer becomes insignificant as d increases since the high refractive index of the interface layer is averaged out by thicker bulk layers. The presence of the interface layer has been attributed to the formation of suboxide (SiO_x [$x < 2$]) at the SiO_2/c-Si interface [56]. Thus, for accurate characterization of the SiO_2 layer, the optical model consisting of air/SiO_2 layer/interface layer/substrate has been employed [53,56].

As we have seen in the above example, when the dielectric function of a sample is not known, dielectric function modeling is required. Nevertheless, complete modeling of the dielectric function in a wide wavelength range is often difficult. In this case, if we use a method known as mathematical inversion, the dielectric function in a whole measured range can be extracted rather easily. F   s example, the optical model of a sample is given by an air/thin layer/substrate structure. If known parameters are assumed to be measured (ψ, Δ), N_0, N_2, and θ_0, unknown parameters become N_1 and d. Now suppose that the dielectric function of the thin layer changes smoothly in a region f   t model using the parameters (A, B, C). In this case, the fitting at $En = a \sim b$ can be performed eas- i   d sis, now the unknown parameter becomes only $N_1 = n_1 - ik_1$. Thus, if we solve Equation 27.26,

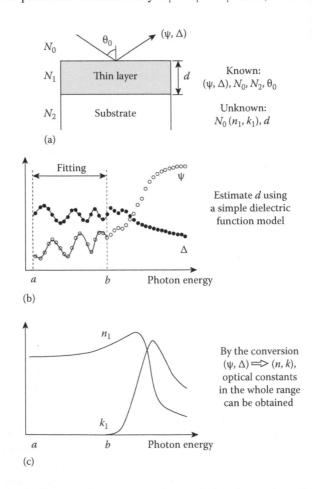

(a)

(b)

(c)

FIGURE 27.11   Schematic diagram of ellipsometry data analysis using mathematical inversion: (a) optical model, (b) fitting in selected region, and (c) optical constants of thin film.

the measured (ψ, Δ) can be converted directly into (n_1, k_1) (Figure 27.11c). This procedure is referred to as the mathematical inversion. Actual mathematical inversion, however, is performed for each wavelength using linear regression analysis. Thus, the mathematical inversion is also called the optical constant fit or point-by-point fit. As shown in Figure 27.11c, from mathematical inversion, the optical constants in the whole measured range can be determined. This data analysis procedure is quite effective in determining the dielectric function of a sample, particularly when dielectric function modeling is difficult in some specific regions.

Figure 27.12 shows an example of the ellipsometry analysis using mathematical inversion. Here, the sample is a ZnO:Ga thin layer formed on a substrate. The substrate is c-Si covered with SiO_2 (50 nm), and the analysis of this substrate can be performed by the same procedure shown in Figure 27.10. If we assume that the ZnO:Ga layer has surface roughness, the optical model of this sample is expressed by the two layer model shown in Figure 27.6c, that is, air/layer 1 (surface roughness layer)/layer 2 (ZnO:Ga layer)/substrate (SiO_2/c-Si) structure. If we calculate N_1 (surface roughness layer) from N_2 (ZnO:Ga layer) by applying EMA ($f_{void} = 0.5$) using Equation 27.19 and determine the SiO_2 layer thickness ($d_{SiO} \sim 50$ nm) prior to the ZnO:Ga formation, the unknown parameters of this optical model become (N_2, d_1, d_2). Although the ZnO layer is doped with Ga, the effect of free-carrier absorption in this layer is negligible, and we employ the Tauc–Lorentz model to represent the dielectric function of the ZnO:Ga layer. The free parameters in the Tauc–Lorentz model are (A, C, En_0, E_g) if we assume $\varepsilon1(\infty) = 1$. Accordingly, the final analysis parameters of this sample are ($A, C, En_0, E_g, d_1, d_2$).

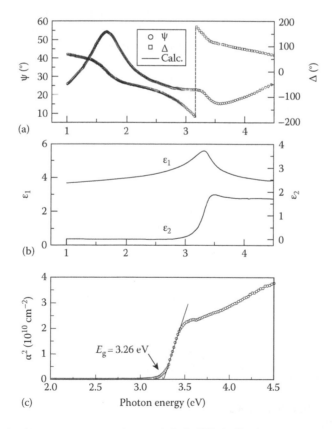

FIGURE 27.12 (a) (ψ, Δ) spectra obtained from a ZnO:Ga/SiO_2/c-Si substrate structure, (b) dielectric function of the ZnO:Ga layer, and (c) α^2 obtained from the ZnO:Ga dielectric function in (b). The bandgap is estimated from the intercept at $\alpha^2 = 0$.

Figure 27.12a shows the (ψ, Δ) spectra obtained from the sample. The ellipsometry measurement of this sample was carried out at $\theta_0 = 70°$. In this figure, the peak position of ψ represents the ZnO:Ga layer thickness and shifts toward lower energies with increasing layer thickness. The solid lines show the fitting result obtained from the calculation. In this analysis, however, the analyzed energy region is limited at $En < 3.0$ eV in order to avoid the complicated structures observed in the dielectric function of ZnO:Ga. From this analysis, the thickness parameters can be estimated ($d_1 = 3.5 \pm 0.1$ nm and $d_2 = 57.8 \pm 0.1$ nm). In this case, we can perform the mathematical inversion to extract the dielectric function of ZnO:Ga in the whole measured region. Figure 27.12b shows the dielectric function of the ZnO:Ga layer extracted from the mathematical inversion. At lower energies ($En < 3.0$ eV), however, the dielectric function calculated from the Tauc–Lorentz model is shown to eliminate spectral noise. It can be confirmed from Figure 27.12a and b that only the transparent region ($\varepsilon_2 \sim 0$) in ZnO:Ga has been used in the analysis to estimate d_1 and d_2. As shown by this example, the mathematical inversion is quite effective in determining dielectric functions of unknown layers.

Since ZnO is a direct bandgap material [57], the absorption coefficient can be approximated by $\alpha = A(En - E_g)^{1/2}$ [1]. Thus, the bandgap E_g of ZnO is estimated by plotting α^2 versus En. Figure 27.12c shows α^2 versus En for the ZnO:Ga. The absorption coefficient can be obtained easily from the dielectric function using Equations 27.1 and 27.2. The solid line shows the linear fit to the experimental data, and E_g is estimated from the intercept at $\alpha^2 = 0$ ($E_g = 3.26$ eV).

From spectroscopic ellipsometry, various information including optical constants (properties) and thickness parameters can be obtained. From the analysis of free-carrier absorption, we can even evaluate electrical properties accurately [37]. As mentioned earlier, however, ellipsometry is an indirect characterization technique, and the analysis is performed using an optical model. It should be emphasized that an optical model used in ellipsometry analysis merely represents an approximated sample structure, and obtained results are not necessarily correct even when the fitting is sufficiently good. Accordingly, when the optical constants or layer structures of a sample are not well known, we must justify ellipsometry results using other measurement methods. This is the greatest disadvantage of ellipsometry technique. However, once an analytical method is established, it becomes possible to perform various high-precision characterizations in a short time using ellipsometry.

REFERENCES

1. H. Fujiwara, *Spectroscopic Ellipsometry: Principles and Applications*, Wiley, West Sussex, U.K., 2007.
2. R. M. A. Azzam and N. M. Bashara, *Ellipsometry and Polarized Light*, North-Holland, Amsterdam, the Netherlands, 1977.
3. H. G. Tompkins and W. A. McGahan, *Spectroscopic Ellipsometry and Reflectometry: A User's Guide*, John Wiley & Sons, New York, 1999.
4. H. G. Tompkins and E. A. Irene (eds.), *Handbook of Ellipsometry*, William Andrew, New York, 2005.
5. M. Schubert, *Infrared Ellipsometry on Semiconductor Layer Structures: Phonons, Plasmons, and Polaritons*, Springer, Heidelberg, Germany, 2004.
6. R. W. Collins, Automatic rotating element ellipsometers: Calibration, operation, and real-time applications, *Rev. Sci. Instrum.* **61**: 2029–2062, 1990.
7. E. Hecht, *Optics*, 4th edn., Addison Wesley, San Francisco, CA, 2002.
8. W. Budde, Photoelectric analysis of polarized light, *Appl. Opt.* **1**: 201–205, 1962.
9. B. D. Cahan and R. F. Spanier, A high speed precision automatic ellipsometer, *Surf. Sci.* **16**: 166–176, 1969.
10. P. S. Hauge and F. H. Dill, Design and operation of ETA, an automated ellipsometer, *IBM J. Res. Dev.* **17**: 472–489, 1973.
11. D. E. Aspnes and A. A. Studna, High precision scanning ellipsometer, *Appl. Opt.* **14**: 220–228, 1975.
12. R. H. Muller, Present status of automatic ellipsometers, *Surf. Sci.* **56**: 19–36, 1976.
13. I. An, Y. M. Li, H. V. Nguyen, and R. W. Collins, Spectroscopic ellipsometry on the millisecond time scale for real-time investigations of thin-film and surface phenomena, *Rev. Sci. Instrum.* **63**: 3842–3848, 1992.

14. P. S. Hauge and F. H. Dill, A rotating-compensator Fourier ellipsometer, *Opt. Commun.* **14**: 431–437, 1975.

15. P. S. Hauge, Generalized rotating-compensator ellipsometry, *Surf. Sci.* **56**: 148–160, 1976.

16. J. Lee, P. I. Rovira, I. An, and R. W. Collins, Rotating-compensator multichannel ellipsometry: Applications for real time Stokes vector spectroscopy of thin film growth, *Rev. Sci. Instrum.* **69**: 1800–1810, 1998.

17. R. W. Collins, J. Koh, H. Fujiwara, P. I. Rovira, A. S. Ferlauto, J. A. Zapien, C. R. Wronski, and R. Messier, Recent progress in thin film growth analysis by multichannel spectroscopic ellipsometry, *Appl. Surf. Sci.* **154–155**: 217–228, 2000.

18. P. S. Hauge, Recent developments in instrumentation in ellipsometry, *Surf. Sci.* **96**: 108–140, 1980.

19. D. E. Aspnes, Spectroscopic ellipsometry of solids, in *Optical Properties of Solids: New Developments*, B. O. Seraphin (ed.), Chapter 15, pp. 801–846, North-Holland, Amsterdam, the Netherlands, 1976.

20. S. N. Jasperson and S. E. Schnatterly, An improved method for high reflectivity ellipsometry based on a new polarization modulation technique, *Rev. Sci. Instrum.* **40**: 761–767, 1969.

21. S. N. Jasperson, D. K. Burge, and R. C. O'Handley, A modulated ellipsometer for studying thin film optical properties and surface dynamics, *Surf. Sci.* **37**: 548–558, 1973.

22. B. Drévillon, J. Perrin, R. Marbot, A. Violet, and J. L. Dalby, Fast polarization modulated ellipsometer using a microprocessor system for digital Fourier analysis, *Rev. Sci. Instrum.* **53**: 969–977, 1982.

23. O. Acher, E. Bigan, and B. Drévillon, Improvements of phase-modulated ellipsometry, *Rev. Sci. Instrum.* **60**: 65–77, 1989.

24. W. M. Duncan and S. A. Henck, In situ spectral ellipsometry for real-time measurement and control, *Appl. Surf. Sci.* **63**: 9–16, 1993.

25. J. N. Hilfiker, C. L. Bungay, R. A. Synowicki, T. E. Tiwald, C. M. Herzinger, B. Johs, G. K. Pribil, and J. A. Woollam, Progress in spectroscopic ellipsometry: Applications from vacuum ultraviolet to infrared, *J. Vac. Sci. Technol. A* **21**: 1103–1108, 2003.

26. A. Röseler, IR spectroscopic ellipsometry: Instrumentation and results, *Thin Solid Films* **234**: 307–313, 1993.

27. A. Canellas, E. Pascual, and B. Drévillon, Phase-modulated ellipsometer using a Fourier transform infrared spectrometer for real time applications, *Rev. Sci. Instrum.* **64**: 2153–2159, 1993.

28. J.-Th. Zettler, Th. Trepk, L. Spanos, Y.-Z. Hu, and W. Richter, High precision UV-visible-near-IR Stokes vector spectroscopy, *Thin Solid Films* **234**: 402–407, 1993.

29. R. Joerger, K. Forcht, A. Gombert, M. Köhl, and W. Graf, Influence of incoherent superposition of light on ellipsometric coefficients, *Appl. Opt.* **36**: 319–327, 1997.

30. G. E. Jellison, Jr. and J. W. McCamy, Sample depolarization effects from thin films of ZnS on GaAs as measured by spectroscopic ellipsometry, *Appl. Phys. Lett.* **61**: 512–514, 1992.

31. P. I. Rovira and R. W. Collins, Analysis of specular and textured SnO_2:F films by high speed four-parameter Stokes vector spectroscopy, *J. Appl. Phys.* **85**: 2015–2025, 1999.

32. D. E. Aspnes, Optical properties of thin films, *Thin Solid Films* **89**: 249–262, 1982.

33. M. Wakagi, H. Fujiwara, and R. W. Collins, Real time spectroscopic ellipsometry for characterization of the crystallization of amorphous silicon by thermal annealing, *Thin Solid Films* **313–314**: 464–468, 1998.

34. G. E. Jellison, Jr. and F. A. Modine, Parameterization of the optical functions of amorphous materials in the interband region, *Appl. Phys. Lett.* **69**: 371–373, 1996; Erratum, *Appl. Phys. Lett.* **69**: 2137, 1996.

35. G. E. Jellison, Jr., F. A. Modine, P. Doshi, and A. Rohatgi, Spectroscopic ellipsometry characterization of thin-film silicon nitride, *Thin Solid Films* **313–314**: 193–197, 1998.

36. Y. J. Cho, N. V. Nguyen, C. A. Richter, J. R. Ehrstein, B. H. Lee, and J. C. Lee, Spectroscopic ellipsometry characterization of high-k dielectric HfO_2 thin films and the high-temperature annealing effects on their optical properties, *Appl. Phys. Lett.* **80**: 1249–1251, 2002.

37. H. Fujiwara and M. Kondo, Effects of carrier concentration on the dielectric function of ZnO:Ga and In_2O_3:Sn studied by spectroscopic ellipsometry: Analysis of free-carrier and band-edge absorption, *Phys. Rev. B* **71**: 075109, 2005.

38. J. Tauc, R. Grigorovici, and A. Vancu, Optical properties and electronic structure of amorphous germanium, *Phys. Stat. Sol.* **15**: 627–637, 1966.

39. A. S. Ferlauto, G. M. Ferreira, J. M. Pearce, C. R. Wronski, R. W. Collins, X. Deng, and G. Ganguly, Analytical model for the optical functions of amorphous semiconductors from the near-infrared to ultraviolet: Applications in thin film photovoltaics, *J. Appl. Phys.* **92**: 2424–2436, 2002.

40. A. R. Forouhi and I. Bloomer, Optical dispersion relations for amorphous semiconductors and amorphous dielectrics, *Phys. Rev. B* **34**: 7018–7026, 1986.

41. S. Adachi, *Optical Properties of Crystalline and Amorphous Semiconductors: Materials and Fundamental Principles*, Kluwer Academic Publishers, Norwell, MA, 1999.

42. J. Leng, J. Opsal, H. Chu, M. Senko, and D. E. Aspnes, Analytic representations of the dielectric functions of crystalline and amorphous Si and crystalline Ge for very large scale integrated device and structural modeling, *J. Vac. Sci. Technol. A* **16**: 1654–1657, 1998.

43. H. R. Philipp, Optical properties of non-crystalline Si, SiO, SiO_x and SiO_2, *J. Phys. Chem. Solids* **32**: 1935–1945, 1971.

44. D. E. Aspnes and J. B. Theeten, Dielectric function of $Si\text{-}SiO_2$ and $Si\text{-}Si_3N_4$ mixtures, *J. Appl. Phys.* **50**: 4928–4935, 1979.

45. K. Mui and F. W. Smith, Optical dielectric function of hydrogenated amorphous silicon: Tetrahedron model and experimental results, *Phys. Rev. B* **38**: 10623–10632, 1988.

46. S. Adachi, *Optical Constants of Crystalline and Amorphous Semiconductors: Numerical Data and Graphical Information*, Kluwer Academic Publishers, Norwell, MA, 1999.

47. U. Schmid, N. E. Christensen, and M. Cardona, Relativistic band structure of Si, Ge, and GeSi: Inversion-asymmetry effects, *Phys. Rev. B* **41**: 5919–5930, 1990.

48. G. E. Jellison, Jr. and F. A. Modine, Optical functions of silicon at elevated temperatures, *J. Appl. Phys.* **76**: 3758–3761, 1994.

49. M. Erman, J. B. Theeten, P. Chambon, S. M. Kelso, and D. E. Aspnes, Optical properties and damage analysis of GaAs single crystals partly amorphized by ion implantation, *J. Appl. Phys.* **56**: 2664–2671, 1984.

50. T. Suzuki and S. Adachi, Optical properties of amorphous Si partially crystallized by thermal annealing, *Jpn. J. Appl. Phys.* **32**: 4900–4906, 1993.

51. B. Johs, C. M. Herzinger, J. H. Dinan, A. Cornfeld, and J. D. Benson, Development of a parametric optical constant model for $Hg_{1-x}Cd_xTe$ for control of composition by spectroscopic ellipsometry during MBE growth, *Thin Solid Films* **313–314**: 137–142, 1998.

52. E. A. Irene, Application of spectroscopic ellipsometry to microelectronics, *Thin Solid Films* **233**: 96–111, 1993.

53. C. M. Herzinger, B. Johs, W. A. McGahan, J. A. Woollam, and W. Paulson, Ellipsometric determination of optical constants for silicon and thermally grown silicon dioxide via a multi-sample, multi-wavelength, multi-angle investigation, *J. Appl. Phys.* **83**: 3323–3336, 1998.

54. A. Kalnitsky, S. P. Tay, J. P. Ellul, S. Chongsawangvirod, J. W. Andrews, and E. A. Irene, Measurements and modeling of thin silicon dioxide films on silicon, *J. Electrochem. Soc.* **137**: 234–238, 1990.

55. M. L. Green, E. P. Gusev, R. Degraeve, and E. L. Garfunkel, Ultrathin (<4 nm) SiO_2 and Si–O–N gate dielectric layers for silicon microelectronics: Understanding the processing, structure, and physical and electrical limits, *J. Appl. Phys.* **90**: 2057–2121, 2001.

56. D. E. Aspnes and J. B. Theeten, Optical properties of the interface between Si and its thermally grown oxide, *Phys. Rev. Lett.* **43**: 1046–1050, 1979.

57. Y. Mi, H. Odaka, and S. Iwata, Electronic structures and optical properties of ZnO, SnO_2 and In_2O_3, *Jpn. J. Appl. Phys.* **38**: 3453–3458, 1999.

28 Optical Thin Film and Coatings

Cheng-Chung Lee and Shigetaro Ogura

CONTENTS

28.1 Introduction 725
28.2 Basic Theory 726
28.3 Some Typical Filters and Their Design Principles 729
    28.3.1 Antireflection Coating 729
    28.3.2 High Reflection Coating 730
    28.3.3 Long-Wave Pass and Short-Wave Pass Filters 731
    28.3.4 Band-Pass Filter 732
    28.3.5 Band-Stop Filter (Rugate Filter) 732
    28.3.6 Polarizer 733
28.4 Optical Monitoring 734
28.5 Characterization of a Coated Film from Optical Measurements 737
Symbols 739
References 740

28.1 INTRODUCTION

Optical thin films are widely used in optics and photonics. Not only do they improve the optical performance of high-quality optical devices but they are also, of a key part their operations. Indeed, without optical thin films it might be impossible for such devices to perform properly or even function at all.

The optical thin films discussed in this chapter have a thickness on the order of wavelengths so that optical interference takes place. The film is generally deposited on a glass, crystal, semiconductor, plastic, or flexible foil substrate such that the characteristics of the incident light are changed. The films can be composed of metallic or dielectric materials or a combination of the two. The characteristics changed include the transmittance, reflectance, polarization, absorption, scattering, and phase.

It is customary to call this type of optical thin film an "interference filter," a "thin film optical filter," or simply a "filter." Nowadays, most such filters are designed with the aid of computer-utilizing sophisticated software. However, the coating process needed to realize the design is still a challenge. This is because there are very few nonabsorbing materials suitable for the coating process. Usually the coated film is also not as isotropic, homogeneous, or nonstructural as assumed in the filter design. Therefore, research into coating material, coating techniques, layer monitoring, and film measurement continues.

Our intent here is not to discuss the field of optical thin films and coating technologies in detail, but to give a general review of some selected topics of interest such as monitoring of film deposition and measurement of the optical properties. The basic theory of how a thin film works is explained in Section 28.1. In Section 28.2, some typical filters and their design principles are discussed. The important thickness monitoring during the coating procedure needed to realize the filter design is described in Section 28.3. In the last section, some optical measurement techniques for characterization of the coated films are introduced.

28.2 BASIC THEORY

Light propagating through these films can be treated as an electromagnetic wave. The light interacts with the media for the most part via the electric field E, rather than the magnetic field, expressed as [1,2]

$$E = E_0 \exp[i(\omega t - \varphi)] \propto E_0 \exp(-i\varphi), \tag{28.1}$$

where

E_0 is the amplitude of the electric field
ω is the angular frequency of the light
φ is the phase shift $\varphi = (2\pi/\lambda)nz$

It is assumed that the light propagates in the z-direction. As the light propagates from medium n_0 to n_s, as shown in Figure 28.1, it is divided into two parts, reflection and transmission. The amplitude reflection coefficient and transmission coefficient, ρ and τ respectively, are expressed as [2]

$$\rho = \frac{n_0 - n_s}{n_0 + n_s} = \left| \frac{n_0 - n_s}{n_0 + n_s} \right| \exp(i\psi), \tag{28.2}$$

$$\tau = \frac{2n_0}{n_0 + n_s}, \tag{28.3}$$

where ψ is the phase change during reflection.

Figure 28.2 depicts a single layer with a refractive index n and a thickness d, coated on a substrate with a refractive index n_s. This means the film has an optical thickness nd and a phase thickness $\delta = (2\pi/\lambda)nd$, where λ is the wavelength of the incident light in a vacuum. There are two interfaces

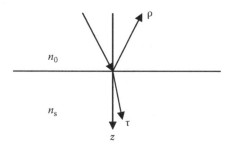

FIGURE 28.1 Light propagates from medium n_0 to n_s.

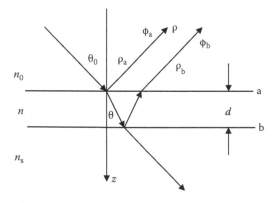

FIGURE 28.2 Single-layer film with a refractive index n and a thickness d coated on a substrate with a refractive index n_s.

a and b. The incident medium can be air ($n_0 = 1$), water ($n_0 = 1.33$), or cement ($n_0 = 1.4$–1.5). When the incident light reaches the first interface a, it is divided into reflected and transmitted parts with the amplitude reflection and transmission coefficients being ρ_a and τ_a, respectively. The transmitted light is reflected back to the incident medium when it reaches the second interface b. Its amplitude is denoted by ρ_b. Let $\psi_a = \varphi_a$ and ψ_b be the phase changes as the light is reflected by interfaces a and b. Since ρ_b travels twice the thickness of the film, which introduces a phase delay -2δ, there is a phase change $\varphi_b = -2\delta + \psi_b$ in the reflected beam ρ_b.

Thus ρ_a and ρ_b can now be expressed as follows [2,3]:

$$\rho_a = \frac{n_0 - n}{n_0 + n} = \left| \frac{n_0 - n}{n_0 + n} \right| \exp(i\varphi_a), \tag{28.4}$$

$$\rho_b = \frac{n - n_s}{n + n_s} \times \exp(-i2\delta) = \left| \frac{n - n_s}{n + n_s} \right| \exp(i\varphi_b). \tag{28.5}$$

The resulting amplitude of the reflectance ρ is approximately the sum of the two beams ρ_a and ρ_b so that

$$\rho = \rho_a + \rho_b \tag{28.6}$$

The reflectance R represents the radiance of the detectable light [2,3]

$$R = \rho \times p^*. \tag{28.7}$$

Let the incident medium be air. In this case, n_0 is always less than n. According to Equation 28.4, there is a phase shift $\phi_a = \pi$ for the reflected light ρ_a since $-1 = \exp(i\pi)$. Depending on the magnitudes of the refractive indices of the film and the substrate, that is $n > n_s$ or $n < n_s$, the two beams ρ_a and ρ_b will have constructive interference or destructive interference. The maximum interference happens at $\delta = (2m + 1)\pi/2$ when the optical thickness of the film nd is an odd multiple of the quarter wave $(2m + 1)\lambda/4$, where m is an integer. Figure 28.3a and b depicts the two extreme cases: (1) the destructive maximum, where $n < n_s$ and ρ_a and ρ_b are out of phase π and (2) the constructive maximum, where $n > n_s$ and ρ_a and ρ_b are in phase.

When δ is not exactly a multiple of $\pi/2$, it means nd is not exactly a multiple of the quarter wave; the reflectance is still different from a bare substrate even though it is not as extreme. Figure 28.4 shows changes in the reflectance R as the film thickness increases. From Equation 28.7, we can derive the extreme of R which is at $nd = (2m + 1)\lambda/4$. Its value is

$$R = \left(\frac{1 - n^2/n_s}{1 + n^2/n_s} \right) \left(\frac{1 - n^2/n_s}{1 + n^2/n_s} \right)^*. \tag{28.8}$$

If we let the reflectance of a bare substrate be R_s, then $R_s = [(1 - n_s)/(1 + n_s)][(1 - n_s)/(1 + n_s)]^*$. Therefore, it is clear that $R < R_s$ when $n < n_s$ and $R > R_s$ when $n > n_s$. Please note that $R = R_s$ at a multiple of the half-wave thickness $nd = m\lambda/2$. For this reason we call a layer with a half-wave thickness an absentee layer.

For the case shown in Figure 28.3a, if we set $|\rho_a| = |\rho_b|$ or using Equation 28.8, we get $R = 0$ when $n = \sqrt{n_s n_0}$, at the wavelength $\lambda = 4nd$. Let us denote this wavelength as the reference wave λ_0. This

FIGURE 28.3 Two-beam interference from Figure 28.1: (a) destructive interference at $\lambda = 4nd$; and (b) constructive interference at $\lambda = 4nd$.

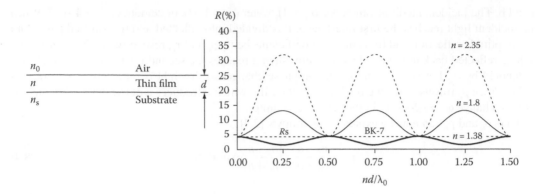

FIGURE 28.4 Changes of reflectance R as the film thickness increases.

is the so-called single-layer antireflection coating. For a BK-7 glass, $n_s = 1.52$ and $R_s = 4.26\%$, therefore, to have $R = 0$, n must be 1.23.

On the other hand, for the case in Figure 28.3b, when the film has a refractive index $n > n_s$, the larger the n, the higher the R. The maximum reflectance is again at the wavelength $\lambda = 4nd$, expressed by Equation 28.8. For ZnS, $n = 2.35$, deposited on BK-7, $R = 32\%$ and is a potential beam splitter.

For an oblique incident light at the angle θ, the optical thickness nd behaves as $nd \cos \theta$, so the phase thickness is changed to $\delta = (2\pi/\lambda) nd \cos \theta$ [2,3]. This explains why the colors of soap bubbles and oil films vary with the thickness of the layer and the incident angle.

The above derivation is approximate, but it is generally correct for a single-layer coating. It has the advantage of making it easy to see how a thin film affects a substrate optically. However, for multilayer analysis, it is more convenient and precise to use a matrix formulation where each layer is represented by a characteristic matrix M [2–4],

$$M = \begin{bmatrix} \cos\delta & \dfrac{i}{n}\sin\delta \\ in\sin\delta & \cos\delta \end{bmatrix}. \tag{28.9}$$

The matrix method has no approximation. When a layer with refractive index n and phase thickness $\delta = (2\pi/\lambda)nd$ is deposited on a substrate, the refractive index n_s and interfaces "a" and "b" can be replaced by a single interface, as shown in Figure 28.5, with an equivalent refractive index n_E expressed as follows:

$$\begin{bmatrix} B \\ C \end{bmatrix} = \begin{bmatrix} \cos\delta & \dfrac{i}{n}\sin\delta \\ in\sin\delta & \cos\delta \end{bmatrix} \begin{bmatrix} 1 \\ n_s \end{bmatrix}, \tag{28.10}$$

$$n_E = \frac{C}{B}. \tag{28.11}$$

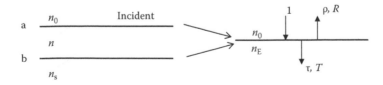

FIGURE 28.5 Two interfaces of a single-layer coating are replaced by a single interface with an equivalent refractive index n_E.

We can now treat Figure 28.2 as in Figure 28.1 with the amplitude reflection and transmission coefficients ρ and τ expressed similar to Equations 28.2 and 28.3, that is,

$$\rho = \frac{n_0 - n_E}{n_0 + n_E}, \tag{28.12}$$

$$\tau = \frac{2n_0}{n_0 + n_E}. \tag{28.13}$$

The reflectance and transmittance become

$$R = \frac{(n_0 B - C)(n_0 B - C)^*}{(n_0 B + C)(n_0 B + C)^*},$$
$$T = \frac{4n_0 n_s}{(n_0 B + C)(n_0 B + C)^*}. \tag{28.14}$$

It is easy to see that if the film thickness is an odd multiple of the quarter wave, that is, $\delta = (2m + 1)\pi/2$, then, according to Equation 28.11, R is the maximum or minimum with the equivalent index $n_E = n^2/n_s$. If the film thickness is a multiple of the half wave, that is, $\delta = m\pi$, M becomes a unity matrix. The reason has already been mentioned that a layer with an integral number of half-wave thicknesses is called as an absentee layer. This absentee layer is very useful in filter design, as we will see in the next section. For such special characterization in thin-film design we use the notations H and L to indicate a quarter-wave layer with the refractive indices higher and lower than the substrate, respectively, and the one with a refractive index in between is denoted as M, I, or J.

The matrix formulation can be extended to multilayer coatings, that is, m layers with the mth layer next to the substrate, as follows:

$$\begin{bmatrix} B \\ C \end{bmatrix} = \prod_{j=1}^{m} \begin{bmatrix} \cos\delta_j & \dfrac{i}{n_j}\sin\delta_j \\ in_j \sin\delta_j & \cos\delta_j \end{bmatrix} \begin{bmatrix} 1 \\ n_s \end{bmatrix}. \tag{28.15}$$

A multilayer coating with multi-interfaces is then replaced by a single interface with an equivalent refractive index $n_E = C/B$, the same expression as in Equation 28.11; the reflectance and transmittance are expressed as in Equation 28.14.

28.3 SOME TYPICAL FILTERS AND THEIR DESIGN PRINCIPLES

Some typical filters and their design principles are described briefly in this section. Interested readers can refer to several sources for more details [2–4].

28.3.1 ANTIREFLECTION COATING

We have already looked at a single-layer antireflection coating (AR coating) in the last section. For BK-7, $n_s = 1.52$, we need a film with a refractive index $n = \sqrt{n_s} = 1.23$ to make $R = 0$. However, there is no robust material where $n = 1.23$. The closest one is MgF_2 (with $n = 1.38$) that, according

FIGURE 28.6 Single-layer, two-layer, and three-layer AR coatings.

to Equation 28.8, has a residual reflectance of $R = 1.26\%$. To reduce R to zero, we should precoat a quarter-wave layer, named as M layer, with a material that has a refractive index of 1.7 such that n_s is increased to the equivalent index $n_E = n^2/n_s = 1.9$ and then $\sqrt{n_E} = 1.38$, therefore $R = 0$. This describes a two-layer AR coating, customarily called a quarter-quarter coating, symbolized here by S/ML/A, where S and A stand for the substrate and the incident medium, in this case air. If we add a high refractive index half-wave layer 2H after the M layer, the R at the reference wavelength λ_0 does not change, but the other wavelength is reduced. For this reason, a half-wave layer is called a broadening layer or a flattening layer. Figure 28.6 shows a comparison of several AR layering systems: S/M2HL/A (three-layer quarter-half-quarter), S/ML/A (two-layer, quarter-quarter), and S/L/A (single quarter-wave layer), where the high index material is assumed to be Ta_2O_5 with $n = 2.15$.

Usually an operator likes to use only two high-quality materials with refractive indices that are as different as possible. Let us say, for example, that these two materials are Ta_2O_5 and SiO_2 with refractive indices of 2.15 and 1.46, respectively. In this case, the M layer will be replaced by two non-quarter-wave layers, denoted as H'L', with optical thicknesses of $0.084\lambda_0$ and $0.073\lambda_0$, respectively, where SiO_2 is the last layer. The layering system thus becomes S/0.336H0.292L2HL/A.

28.3.2 High Reflection Coating

A simple metallic film such as Ag, Al, Au, or Rh can also act as a high-reflection coating. However, to have low absorption it is necessary to have an all-dielectric multilayer. The basic design is to use a quarter-wave stack: S/H(LH)^p/A. The equivalent refractive index from Equation 28.15 can now be expressed as in Equation 28.16. We thus find that the higher the p, the higher the reflectance

$$n_E = \left(\frac{n_H}{n_L}\right)^{2p} \frac{n_H^2}{n_s}. \tag{28.16}$$

The width of the high-reflection zone $2\Delta g$ shown in Figure 28.7 can now be expressed as [2,3]

$$2\Delta g = \frac{4}{\pi} \sin^{-1}\left(\frac{n_H - n_L}{n_H + n_L}\right), \tag{28.17}$$

where $g = \lambda_0/\lambda$ is the relative wavelength. It can thus be seen that the higher the index ratio, the wider the $2\Delta g$.

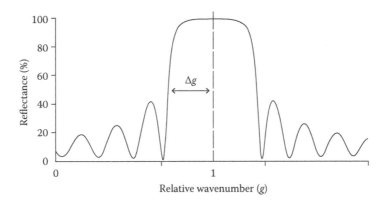

FIGURE 28.7 High-reflection coating composed of quarter-wave stacks with width $2\Delta g$.

28.3.3 LONG-WAVE PASS AND SHORT-WAVE PASS FILTERS

A long-wave pass filter allows the transmission of long wavelengths but reflects short wavelengths; a short-wave pass filter allows the opposite. Color filter, cold mirror, heat mirror, UV-cut filter, and IR-cut filter are some examples. The basic design depends on symmetrical layers [4,5]. For example, S/(0.5HL0.5H)^x/A and S/(0.5LH0.5L)^x/A represent a long-wave pass filter and a short-wave pass wave filter, respectively. Usually we have to add some matching layers to increase the transmittance in the pass band; see Figure 28.8 [3]. The matching layer acts as an antireflection layer. The optical performance of long-wave pass filters and short-wave pass filters, with and without matching layers coated on germanium, where H and L indicate the Ge and ZnS quarter-wave layers, can be seen in Figures 28.9 and 28.10 [3]. One can use a computer simulation to increase the transmission in the pass-band by adjusting the thickness of some layers.

FIGURE 28.8 Symmetrical layers with an equivalent index E and matching layers n_{mS} and n_{mA}.

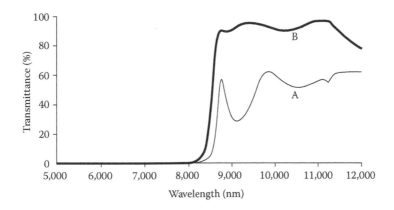

FIGURE 28.9 A: Ge/(0.5LH0.5L)^7/A, $\lambda_0 = 6200$ nm; B: Ge/1.47H(0.5LH0.5L)^{7}1.47L/A. (Adapted from Lee, C.C., *Thin Film Optics and Coating Technology*, 5th edn., Yi Hsien, Taipei, Taiwan, 2006.)

FIGURE 28.10 A: Ge/(0.5LH0.5L)^7/A, λ_0 = 5000 nm; B: Ge/0.79L(0.5LH0.5L)^7/A. (Adapted from Lee, C.C., *Thin Film Optics and Coating Technology*, 5th edn., Yi Hsien, Taipei, Taiwan, 2006.)

28.3.4 BAND-PASS FILTER

The combination of a long-wave pass and a short-wave pass filter gives a fairly wide band-pass filter. To narrow the bandwidth (BW), we need a Fabry–Perot type filter (HR-Sp-HR) where HR is a high reflector that is made of an Al or Ag film or quarter stacks and Sp is a multiple half-wave layer called a spacer. Taking the quarter-wave stacks as HRs and an mth order spacer, we obtain a filter, S/H(LH)^{p}m2L(LH)^{p}HL/A, where the last layer L is an antireflection coating layer. The half peak BW $\Delta\lambda_h$ of the filter is determined by p and m [2,3] and can be expressed as

$$\frac{\Delta\lambda_h}{\lambda_0} = \frac{4n_L^{2x-1}n_s}{m\pi n_H^{2x}}\left[\frac{n_H - n_L}{n_H - n_L + \left(\dfrac{n_L}{m}\right)}\right] \quad \text{for a low-index spacer } m2L; \qquad (28.18)$$

$$\frac{\Delta\lambda_h}{\lambda_0} = \frac{4n_L^{2x}n_s}{m\pi n_H^{2x+1}}\left[\frac{n_H - n_L}{n_H - n_L + \left(\dfrac{n_L}{m}\right)}\right] \quad \text{for a high-index spacer } m2H. \qquad (28.19)$$

This design is called the single-cavity filter and is approximately triangular in shape. To have a square top we need multiple cavities, as is the case, for example, of a three-cavity filter: | HR | Sp | HR | L | HR | Sp | HR | L | HR | Sp | HR |, where L is the coupling layer. Figure 28.11 depicts the spectra of the single-, two-, and three-cavity narrow band-pass filters.

28.3.5 BAND-STOP FILTER (RUGATE FILTER)

A band-stop filter, also called as a rugate filter, is opposite to the band-pass filter in that it reflects one or more specified bands. It can be realized by using an inhomogeneous film with index variation, as shown in Figure 28.12 [6–8]. The matching layers act as an antireflector to reduce ripples and suppress the sidelobes in the transmission regions. The half peak BW $\Delta\lambda_h$ and the optical density of the stop band D can be expressed by Equations 28.20 and 28.21 where λ_c, n_a, n_p, and S are

FIGURE 28.11 Spectra of the single- (S), two- (D), and three-cavity (T) narrowband-pass filters.

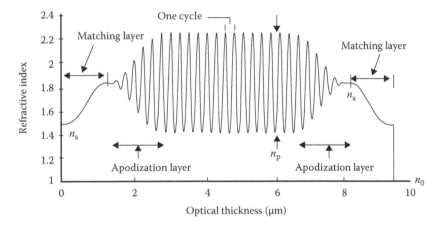

FIGURE 28.12 Index profile of a band-stop filter with matching layers for antireflection.

the center wavelength of the stop band, the average refractive index, the difference between the high and low indices, and the cycle of the inhomogeneous layer, respectively.

$$\frac{\Delta\lambda_h}{\lambda_0} = \frac{n_p}{2n_a}, \tag{28.20}$$

$$D = \log_{10}\left(\frac{1}{1-R}\right) = 0.6822\frac{n_p}{n_a}S - \log_{10}\left(\frac{4n_0}{n_s}\right). \tag{28.21}$$

It is clear that the smaller the n_p, the narrower the $\Delta\lambda_h$, and the larger the S, the higher the reflection of the stop band. The spectrum of a typical rugate filter is shown in Figure 28.13.

28.3.6 POLARIZER

A polarizer allows the passage of P-polarized light but reflects S-polarized light. For example, look at a tilted quarter-wave stack (S/(HL)^{p}H/A) which acts as a plate polarizer for a small region

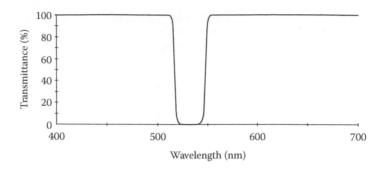

FIGURE 28.13 Spectrum of a rugate filter with matching and apodization layers center at 532 nm.

of wavelengths, particularly in high-power lasers. To have a wide spectral region, the film stack has to be embedded between prisms (Prism/(HL)^{p}H/Prism) designed at the Brewster angle so that there is no reflection of P-polarized light. A polarizer for light normally incident to the film can be fabricated utilizing sculptured film [9].

28.4 OPTICAL MONITORING

Various monitoring methods such as time counting, color recognition, quartz monitoring, and optical monitoring have been developed to control thickness during thin-film deposition. Optical monitoring is generally used in the manufacturing of optical thin-film filters [10,11]. Optical monitoring is based on changes of reflectance/transmittance as the film thickness increases; see Figure 28.4. The turning point method (TPM) is the most commonly used, and has the advantage of error compensation for quarter-wave stacks [11,12]. For nonquarterwave stacks, the wavelength of the monitoring system must be changed to an appropriate wavelength for monitoring each individual layer. Another method based on a single wavelength is called the level monitoring method [13]. This method is suitable for monitoring nonquarterwave multilayers. Error analysis shows that the monitoring plate must be precoated for error compensation. Several numerical methods and the optimum trigger point method have been developed to improve the performance, sensitivity, and compensation associated with the TPM and the level methods [14–18]. However, all of those monitoring methods use runsheet diagrams to control the thickness of the films based on their transmittance or reflectance. The refractive index n, the extinction coefficient k, and the thickness d are not independently determined in real time which makes phase extraction difficult. In a growing thin-film stack, the refractive index of the materials may change with time due to variations in the coating parameters or environmental fluctuations in the chamber, meaning we cannot terminate the current layer at the point specified in the original design. Therefore, real-time determination of n, k, and d is necessary for precision coatings. Ellipsometric and phase extraction methods have been proposed to solve this problem. Ellipsometric monitoring [19,20] requires algebraic computation to obtain the optical constants that fit the measurements, but require so many parameters that they are difficult to solve analytically. In the case of multilayer coatings, this could lead to the accumulation of errors, which would cause the final optical performance to differ from that of the original design.

To avoid this type of error accumulation, a new method, in situ sensitivity monitoring with proper error compensation, has been proposed to find the correct refractive index and physical thickness of each deposited layer [21]. An operator can choose the sensitivity monitoring wavelength that offers better error compensation for the reference wavelength. The result is better performance. Figure 28.14 shows the experimental results of the monitoring of a narrow band-pass filter with a center wavelength at 800 nm by this new method, the TPM, and the level method. The better performance of the new method is clear. The half peak BW and the position of the maximum transmittance are closer to the designed values than are those obtained using the other two methods. This is because TPM offers error compensation only for previous layers, but lacks sensitivity at termination

FIGURE 28.14 Spectra of narrow band-pass filters monitored by the three methods. (Adapted from Lee, C.C. and Wu, K., *Opt. Lett.*, 32, 2118, 2007.)

points which results in a broadening of the BW. The level monitoring method has higher sensitivity, but no error compensation which results in a shift of the output spectrum. The new method on the other hand, combines the advantages of both methods, the sensitivity at termination points and the error compensation for previous layers.

Recently, a new polarization Fizeau interferometer-based monitoring system has been developed [22] which can get rid of the influence of the variations in the coating parameters caused by environmental fluctuations in the chamber. This is a vibration-insensitive in situ system, whereby one can instantly obtain the real-time phase and magnitude of the reflection coefficient and/or the transmission coefficient. The layout of a reflection monitoring system with a short coherence light source is shown in Figure 28.15. The test surface is located on one side of thin film growth,

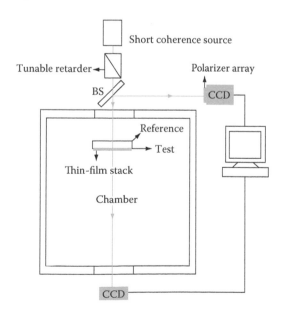

FIGURE 28.15 Schematic representation of the reflection optical monitoring system. (Adapted from Lee, C.C. et al., *Opt. Express*, 15, 17536, 2007.)

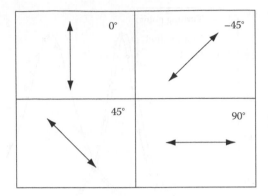

FIGURE 28.16   Layout of the polarizer detector array.

while on the other side of the substrate is the reference surface. The tunable retarder is composed of two wedge x-cut uniaxial birefringent crystals. The crystal-induced phase difference between two orthogonally polarized beams has to match the difference in the optical phase induced by the reference surface and the test surface. The interference will therefore occur only between either the S-polarized reference beam and the P-polarized test beam or the P-polarized reference beam and the S-polarized test beam. Other reflections from the interfaces will be suppressed because of the short coherence length of the light source [23]. Interference will occur when the transmission axes of the polarizers are oriented 45° and −45° to the fast axis of the birefringent crystal. These two polarizers as well as the other two (whose transmission axes are oriented 0° and 90° to the fast axis of the birefringent crystal) are combined and detected by the compact polarizer detector array, as shown in Figure 28.16. The magnitude of the reflection coefficient and the phase can be acquired from the phase-shifted interferograms of two orthogonal polarization beams on the detector array (after mathematical analysis using Equations 28.22 and 28.23). We now have complete information about the growing film stack and are able to accurately monitor the thickness of the whole stack.

$$I_{0°} = I_s \left|\rho_r\right|^2 + \frac{I_s\left(1-\left|\rho_r\right|\right)^4 \left|\rho_{st}\right|^2}{1-\left|\rho_{st}\right|^2\left|\rho_r\right|^2}, \quad I_{90°} = I_p \left|\rho_r\right|^2 + \frac{I_p\left(1-\left|\rho_r\right|\right)^4 \left|\rho_{st}\right|^2}{1-\left|\rho_{st}\right|^2\left|\rho_r\right|^2},$$

$$I_{+45°} = \frac{1}{2}(I_{0°} + I_{90°}) + \sqrt{I_{0°}I_{90°}} \left|\rho_r\right|\left|\rho_{st}\right|\left(1 - \frac{\left|\rho_r\right|^2 + \left(1-\left|\rho_r\right|\right)^2}{\sqrt{I_{0°}I_{90°}}}\right)\cos\theta, \qquad (28.22)$$

$$I_{-45°} = \frac{1}{2}(I_{0°} + I_{90°}) - \sqrt{I_{0°}I_{90°}} \left|\rho_r\right|\left|\rho_{st}\right|\left(1 - \frac{\left|\rho_r\right|^2 + \left(1-\left|\rho_r\right|\right)^2}{\sqrt{I_{0°}I_{90°}}}\right)\cos\theta,$$

where

I_j is the reflectance of $j = 0°, -45°, +45°,$ and 90° on the detector array

ρ_r and ρ_{st} are the reflection coefficients from the substrate to air and the deposited film stack, respectively

$$\theta = \phi_{st} + \phi_0,$$

FIGURE 28.17 Optical performance of an antireflection coating obtained by admittance diagram and the conventional runsheet method. (Adapted from Chen, Y.R. and Lee, C.C., Monitoring of multilayer by admittance diagram, *Conference on Optical Interference Coatings*, Technical Digest, PWC7, Optical Society of America, Washington, DC, 2007.)

ϕ_0 is the phase shift caused by the optical path difference between two orthogonal polarizations as they pass through the substrate and the retarder. ϕ_0 can be acquired before the deposition of the thin film. Therefore, the phase of the thin-film stack's reflection coefficient can be obtained. We now have complete information about the whole stack and each layer can be deposited with the correct thickness.

$$\phi_{st} = \left\{ \arccos\left[\frac{(I_{+45°} - I_{-45°})}{2\sqrt{I_{0°}I_{90°}}\,|\rho_r||\rho_{st}|\left(1 - \frac{|\rho_r|^2 + \left(1 - |\rho_r|\right)^2}{\sqrt{I_{0°}I_{90°}}}\right)} \right] \right\} - \phi_0. \tag{28.23}$$

Better-monitoring performance can be achieved from an admittance diagram than a runsheet diagram because of the higher sensitivity and larger amount of physical and visual information. It has therefore been proposed that they be used in a new optical monitoring method [24–26]. The diagram is obtained by plotting the locus of the effective admittance during the deposition process. Changes in the refractive index during the coating process can be easily observed and immediately compensated for from layer to layer. Figure 28.17 shows plots of the optical performance of an antireflection coating monitored using a real-time admittance diagram and the conventional runsheet method.

28.5 CHARACTERIZATION OF A COATED FILM FROM OPTICAL MEASUREMENTS

Characterizing the optical properties of a coated film is very important for improving the quality and the functionality of the film and the resultant filter. There are several methods that have been used to do this and they can be classified into two categories: ellipsometric and photometric.

Ellipsometry is based on the measurement of the complex amplitude reflections in the P and S directions, r_p and r_s, respectively. We obtain the ratio ρ and the phase difference Δ for r_p and r_s,

$$\rho = \frac{r_p}{r_s} = \frac{|r_p|e^{i\delta_p}}{|r_s|e^{i\delta_s}} = \tan\Psi e^{i\Delta}, \tag{28.24}$$

$$\Delta = \delta_p - \delta_s, \tag{28.25}$$

$$\Psi = \tan^{-1}\left|\frac{r_p}{r_s}\right|. \tag{28.26}$$

ψ and Δ are called the ellipsometric angles and are used for calculating the refractive index n, the extinction coefficient k, and the thickness d of the coated film through a numerical fitting algorithm [27–29].

Photometry is based on the measurement of the reflectance $R(\lambda)$ and the transmittance $T(\lambda)$. The $R(\lambda)$ and $T(\lambda)$ of a coated film can be obtained from spectrophotometric measurements. The techniques for measuring $R(\lambda)$ and $T(\lambda)$ have been described in several articles, interested readers can refer to Ref. [30] for more details. The refractive index $n(\lambda)$, the extinction coefficient $k(\lambda)$, and the thickness d of a coated film can be determined from $R(\lambda)$ and $T(\lambda)$. For example, look at the envelope method first proposed by Manifacier et al. [31] and then by Swanepoel [32]. First, the refractive index of the substrate n_s is calculated using Equation 28.27; n, k, and d can now be derived from T_M and T_m using Equations 28.28 through 28.30, as shown in Figure 28.18, where T_M and T_m represent the maximum and minimum of the spectrum fringe, respectively. Each T_M and T_m pair shall be at the same wavelength during the calculation.

$$n_s = \frac{1}{T_s} + \left(\frac{1}{T_s^2} - 1\right)^{1/2}. \tag{28.27}$$

$$n = \left[Q + \left(Q^2 - n_s^2\right)^{1/2}\right]^{1/2}, \tag{28.28}$$

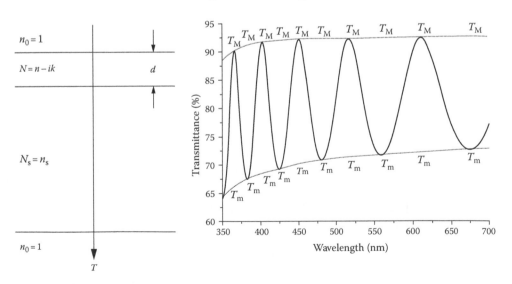

FIGURE 28.18   Envelope method for calculating the n, k, and d of a coated film.

$$Q = 2n_s \left(\frac{T_M - T_m}{T_M T_m} \right) + \frac{n_s^2 + 1}{2}.$$

$$k = -\frac{\lambda}{4\pi d} \ln x, \tag{28.29}$$

$$x = \frac{F - \left[F^2 - \left(n^2 - 1 \right)^3 \left(n^2 - n_s^4 \right) \right]^{1/2}}{(n-1)^3 (n - n_s^2)},$$

$$F = \frac{8n^2 n_s}{T_i},$$

$$T_i = \frac{2T_M T_m}{T_M + T_m},$$

$$d = \frac{\lambda_1 \lambda_2}{2(n_2 \lambda_1 - n_1 \lambda_2)}. \tag{28.30}$$

Note that the $n(\lambda)$, $k(\lambda)$, and d obtained from above equations still need to be refined to fit an optimum solution, especially for an inhomogeneous film [33,34].

There are several other methods for measuring $n(\lambda)$, $k(\lambda)$, and d developed based on optical metrology, for example, the wave guide technique [35] and attenuated total reflection [36], which are also valuable references for those interested in thin film research. Accurate measurements of the optical constants of thin films by these techniques are indispensable for thin-film-related applications.

SYMBOLS

Δ	phase thickness
θ	angle of incidence in a medium
λ	wavelength
λ_0	reference wavelength
λ_c	center wavelength
$\Delta\lambda_h$	half peak bandwidth
ρ	reflection coefficient; ratio of r_p and r_s
τ	transmission coefficient
φ, ϕ	phase shift
ω	angular frequency
Δ	ellipsometry angle
ψ	ellipsometry angle; phase change
d	film thickness
n	refractive index
k	extinction coefficient
n_E	equivalent refractive index
g	relative wavelength
r_p	complex amplitude reflection in the P direction
r_s	complex amplitude reflection in the S direction
Δ	phase difference of r_p and r_s
BW	half peak bandwidth
D	optical density

E	amplitude of electric field
H	quarter-wave layer of high index
HR	high reflector
I	intensity
L	quarter-wave layer of low index
M	characteristic matrix; quarter-wave layer of intermediate index
R	reflectance
S	substrate
Sp	spacer
T	transmittance
TPM	turning point method

REFERENCES

1. M. Born and E. Wolf, *Principles of Optics*, 7th edn., Cambridge University Press, New York, 1999.
2. H. A. Macleod, *Thin-Film Optical Filters*, 3rd edn., Institute of Physics Publishing, Bristol, U.K., 2001.
3. C. C. Lee, *Thin Film Optics and Coating Technology*, 5th edn., Yi Hsien, Taipei, Taiwan, 2006.
4. A. Thelen, *Design of Optical Interference Coatings*, McGraw-Hill Book Company, New York, 1989.
5. L. I. Epstein, The design of optical filter, *J. Opt. Soc. Am.* 42, 806–810, 1952.
6. W. H. Southwell, Spectral response calculations of rugate filters using coupled-wave theory, *J. Opt. Soc. Am.* A5, 1558–1564, 1988.
7. W. H. Southwell and R. L. Hall, Rugate filter sidelobe suppression using quintic and rugated quintic matching layers, *Appl. Opt.* 28, 2949–2951, 1989.
8. W. H. Southwell, Using apodization functions to reduce sidelobes in rugate filters, *Appl. Opt.* 28, 5091–5094, 1989.
9. I. J. Hodgkinson and Q. H. Wu, *Birefringent Thin Films and Polarizing Element*, World Scientific, Singapore, 1997.
10. B. Vidal, A. Fornier, and E. Pelletier, Wideband optical monitoring of nonquarter wave multilayer filter, *Appl. Opt.* 18, 3851–3856, 1979.
11. H. A. Macleod, Monitor of optical coatings, *Appl. Opt.* 20, 82–89, 1981.
12. H. A. Macleod, Turning value monitoring of narrow-band all-dielectric thin-film optical filters, *Opt. Acta* 19, 1–28, 1972.
13. F. Zhao, Monitoring of periodic multilayer by the level method, *Appl. Opt.* 24, 3339–3343, 1985.
14. C. J. van der Laan, Optical monitoring of nonquarterwave stacks, *Appl. Opt.* 25, 753–760, 1986.
15. B. Bobbs and J. E. Rudisill, Optical monitoring of nonquarterwave film thickness using a turning point method, *Appl. Opt.* 26, 3136–3139, 1987.
16. C. Zang, Y. Wang, and W. Lu, A single-wavelength monitoring method for optical thin-film coatings, *Opt. Eng.* 43, 1439–1443, 2004.
17. C. C. Lee, K. Wu, C. C. Kuo, and S. H. Chen, Improvement of the optical coating process by cutting layers with sensitive monitor wavelength, *Opt. Express* 13, 4854–4861, 2004.
18. A. V. Tikhonravov and M. K. Trubetskov, Eliminating of cumulative effect of thickness errors in monochromatic monitoring of optical coating production: Theory, *Appl. Opt.* 46, 2084–2090, 2007.
19. J. Lee and R. W. Collins, Real-time characterization of film growth on transparent substrates by rotating-compensator multichannel ellipsometry, *Appl. Opt.* 37, 4230–4238, 1998.
20. S. Dligatch, R. Netterfield, and B. Martin, Application of in-situ ellipsometry to the fabrication of multi-layered coatings with sub-nanometre accuracy, *Thin Solid Films* 455–456, 376–379, 2004.
21. C. C. Lee and K. Wu, In situ sensitive optical monitoring with error compensation, *Opt. Lett.* 32, 2118–2120, 2007.
22. C. C. Lee, K. Wu, S. H. Chen, and S. J. Ma, Optical monitoring and real time admittance loci calculation through polarization interferometer, *Opt. Express* 15, 17536–17541, 2007.
23. B. Kimbrough, J. Millerd, J. Wyant, and J. Hayes, Low coherence vibration insensitive Fizeau interferometer, *Proc. SPIE* 6292, 62920F, 2006.
24. Y. R. Chen, Monitoring of film growth by admittance diagram, Master thesis, National Central University, Jhongli, Taiwan, 2004.
25. B. J. Chun, C. K. Hwangbo, and J. S. Kim, Optical monitoring of nonquarterwave layers of dielectric multilayer filters using optical admittance, *Opt. Express* 14, 2473–2480, 2006.

26. Y. R. Chen and C. C. Lee, Monitoring of multilayer by admittance diagram, *Conference on Optical Interference Coatings*, Technical Digest, Optical Society of America, PWC7, Washington, DC, 2007.
27. R. M. A. Azzam and N. M. Bashara, *Ellipsometry and Polarized Light*, North-Holland Publishing Company, Oxford, U.K., 1977.
28. C. W. Chu, The research on the calculation of optical constant of optical thin film, PhD thesis, Institute of Optical Sciences, National Central University, Jhongli, Taiwan, 1994.
29. J. Rivory, Ellipsometric measurements, in *Thin Films for Optical System*, Chapter 11, edited by Flory, F. R., Marcel Dekker, Inc., New York, 1995.
30. J. P. Borgogno, Spectrophotometric methods for refractive index determination, in *Thin Films for Optical System*, Chapter 10, edited by Flory, F. R., Marcel Dekker, Inc., New York, 1995.
31. J. C. Manifacier, J. Gasiot, and J. P. Filland, Simple method for determination of the optical constant n, k and the thickness of weekly absorbing thin films, *J. Phys. E: Sci. Instrum.* 9, 1002–1004, 1976.
32. R. Swanepoel, Determination of the thickness and optical constant of amorphous silicon, *J. Phys. E: Sci. Instrum.* 16, 1214–1222, 1983.
33. J. P. Borgogno, F. Flory, P. Roche, B. Schmitt, G. Albert, E. Pelletier, and H. A. Macleod, Refractive index and inhomogeneity of thin films, *Appl. Opt.* 23, 3567–3570, 1984.
34. Y. Y. Liou, C. C. Lee, C. C. Jaing, and C. W. Chu, Determination of the optical constant profile of thin weekly absorbing inhomogeneous films, *Jpn. J. Appl. Phys.* 34, 1952–1957, 1995.
35. F. R. Flory, Guided wave techniques for the characterization of optical coatings, in *Thin Films for Optical System*, Chapter 15, edited by Flory, F. R., Marcel Dekker, Inc., New York, 1995.
36. T. Lopez-Rios and G. Vuye, Use of surface plasmon excitation for determination of the thickness and optical constants of very thin surface layers, *Surf. Sci.* 81, 529–538, 1979.

29 Film Surface and Thickness Profilometry

Katsuichi Kitagawa

CONTENTS

29.1 Introduction 744
29.2 Profiling of a Thick Transparent Film by White-Light Interferometry 744
    29.2.1 White-Light Interferometry.. 744
    29.2.2 KF Algorithm 745
        29.2.2.1 Interferogram 745
        29.2.2.2 Calculation Procedure.. 746
    29.2.3 Film Profiler with the KF Algorithm 747
    29.2.4 Range of Measurable Film Thickness 748
    29.2.5 Examples of Measurements.. 748
        29.2.5.1 Steps in Film Thickness... 748
        29.2.5.2 Resist Film with an Unknown Refractive Index.................................... 748
        29.2.5.3 Repeatability of Measurement 748
29.3 Profiling of a Thin Transparent Film... 751
    29.3.1 Measurement of an Oxide Thin Film Step 751
    29.3.2 Measurement of CMP Samples 751
29.4 Film Thickness Profiling by Pseudo-Transmission Interferometry.................................. 753
    29.4.1 Principle of the TF Method 753
    29.4.2 Experiments... 754
29.5 Simultaneous Measurement of Thickness and Refractive Index..................................... 755
29.6 Film Thickness Profiling by Three Wavelength Interference Color Analysis 757
    29.6.1 Introduction to the Interference Color Analysis Technique 757
    29.6.2 Principle of the GMFT Method... 757
        29.6.2.1 GMFT Method.. 758
        29.6.2.2 ACOS Algorithm 761
    29.6.3 Experiments... 762
        29.6.3.1 Test Method.. 762
        29.6.3.2 Test Results 764
    29.6.4 Examples of Measurements.. 765
        29.6.4.1 Thin Film on a Wafer... 765
        29.6.4.2 Soap Film 765
    29.6.5 Features.. 766
29.7 Conclusion 767
References.. 767

29.1 INTRODUCTION

The technique of surface profile measurement using optical interference is widely used in industry as a noncontact and nondestructive method of quickly and accurately measuring the profile of microscopic surfaces. However, its application to transparent thin films has been limited because it is difficult to obtain an exact surface profile measurement if the surface to be measured is covered with a transparent film. This is because the interference beam coming from the back surface of the film creates disturbance.

Thickness profile measurement is also of great importance in industry; for example, the measurement of various kinds of thin film layers on substrates of semiconductors or flat panel display devices. There are several different techniques used to measure these films, of which the most widely used are reflection spectrophotometry and ellipsometry.

In the case of freestanding films such as plastic films, there are also several nondestructive measurement techniques, such as methods using β-rays, infrared (IR) or nuclear gauges, and spectrophotometry.

However, most of these techniques are limited to single-point measurement. To obtain an aerial thickness profile, it is necessary to mechanically scan the sample or the sensing device. Another limitation inherent in these techniques is their large spot sizes. For example, spectrophotometers are limited to spot sizes of a few micrometers or greater because of available light budgets.

Many techniques have been proposed to resolve these problems in the surface and thickness measurement of substrate-supported films and freestanding films. This chapter introduces some of the techniques to realize surface and thickness profile measurement of a transparent film by means of optical interferometry or interference color analysis.

29.2 PROFILING OF A THICK TRANSPARENT FILM BY WHITE-LIGHT INTERFEROMETRY

29.2.1 WHITE-LIGHT INTERFEROMETRY

Figure 29.1 shows the optical section of a microscopic surface profiler using white-light interferometry.

FIGURE 29.1 Schematic illustration of an interference microscope.

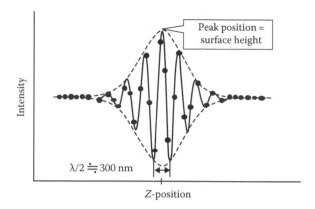

FIGURE 29.2 An example of a white-light interferogram. The black dots indicate sampled values, and the dotted line denotes the envelope.

The equipment uses white light as the light source, and when it performs a vertical scan along the Z-axis of the interference microscope, the intensity signal of one charge-coupled device (CCD) pixel changes, as shown in Figure 29.2, maximizing the modulation of the interference fringe at the point where the difference in the path length between the measurement light path and the reference light path becomes zero. By determining the Z-axis height that corresponds to the largest interference fringe modulation at every point in the TV camera, this technique measures en bloc the three-dimensional profile in the field of view (FOV) of the camera.

Various algorithms have been proposed to detect the peak position of modulation [1–5], most of which assume that the peak is single. Therefore, if a film-covered surface is measured, superposition occurs between the reflected beams from the front surface and the reflected beams from the back surface, hampering correct measurement of the surface profile.

It has long been known that if the thickness of a transparent film is greater than the coherent length, two peaks of reflection corresponding to the front and back surfaces appear on the interferogram, allowing simultaneous measurement of film thickness and surface height variations from those peak positions [6–10].

However, the methods reported to date either require visual determination of the peak positions [1–7] or use a complex calculation algorithm [9,10], and no automatic measuring device was available on the market for practical use until our development of such a device in 2002 [11].

29.2.2 KF Algorithm

We developed an algorithm, the thicK Film (KF) algorithm [12], that is able to automatically detect the positions of two contrast peaks in an interferogram generated by white-light vertical scanning interferometry (VSI). This algorithm calculates the interference amplitudes from an interferogram. The results are interpreted as a histogram, and the automatic threshold selection method [13] then finds the optimal threshold for peak separation. Finally, the peak position is determined for each of the distinct peak regions. We will explain this algorithm using a calculation example in the sections that follow.

29.2.2.1 Interferogram

An oxide film with a thickness of approximately 1 µm formed on an Si wafer was measured. A halogen lamp was used as the light source, and the Z-axis scanning speed was set at 2.4 µm/s with a sampling interval of 0.08 µm. Figure 29.3 shows the obtained interferogram.

FIGURE 29.3 Interferogram used for the KF algorithm test. An oxide film with a thickness of approximately 1 μm on an Si wafer was measured. The sampling interval was 80 nm.

29.2.2.2 Calculation Procedure

1. Step 1: Calculation of contrast

   First, determine the average value of the intensity data and then calculate the AC components by subtracting the average value from each piece of data. By squaring the AC components for the purpose of rectification, we obtain the contrast data of interference shown in Figure 29.4.

2. Step 2: Separation of peaks

   Taking the contrast data shown in Figure 29.4 as a histogram, determine the optimal threshold for peak separation using Otsu's automatic threshold selection method in the field of processing to binary value imagery [13].

   More specifically, given a set of frequency values $n(i)$, $(i = 1, L)$, where the number of observed value levels is L, and the observed value level is i, we take the intraclass variance, expressed by the following equation, as an objective function and choose the threshold that minimizes the function:

$$f = \omega_1 \sigma_1^2 + \omega_2 \sigma_2^2, \tag{29.1}$$

where

   ω_1 and ω_2 are the frequencies of the classes
   σ_1^2 and σ_2^2 are the variances of the classes

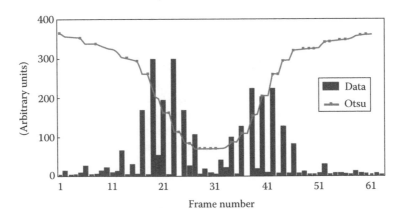

FIGURE 29.4 Contrast data and Otsu's objective function for automatic threshold selection.

The values of the objective function are shown by the solid-line curve in Figure 29.4, with frame number 30 being the threshold.

3. Step 3: Detection of peak position
   Frame number 30 separates the distribution into left and right peak regions. For each region, the peak positions were detected by conventional methods.
4. Step 4: Calculation of two surface heights and one film thickness
   Let the heights converted from the peak positions be z_{p1} and z_{p2}, and let the refractive index of the film be n. Then,
   a. The surface height of the transparent film = z_{p1}
   b. The film thickness $t = (z_{p1} - z_{p2})/n$
   c. The back-surface height of the film = $z_{p1} - t$

29.2.3 FILM PROFILER WITH THE KF ALGORITHM

Using the KF algorithm, Toray Engineering Co. developed the noncontact Film Profiler SP-500F shown in Figure 29.5, which allows simultaneous measurement of the front and back surface topography and the thickness variation of a transparent film.

This development allows the following measurements, which are difficult to perform under conventional methods:

1. Profiling of a surface covered with a transparent film
2. Profiling of a bottom surface through a transparent film
3. 2-D measurement of film thickness distribution

This system has already been introduced and used effectively in the semiconductor, LCD, and film manufacturing processes [11].

FIGURE 29.5 SP-500F film profiler.

FIGURE 29.6   Interferogram of an oxide film on an Si wafer. The measured thickness was 0.77 μm.

29.2.4   RANGE OF MEASURABLE FILM THICKNESS

The KF method of measurement has a lower limit of measurable film thickness because it is based on the assumption that interferogram peaks can be separated. Oxide films with a thickness of 0.5–2 μm on an Si wafer were measured, and the minimum measurable thickness was found to be approximately 0.8 μm. Figure 29.6 shows the interferogram of a 0.77 μm oxide film, which is the lowest measurable range to date.

29.2.5   EXAMPLES OF MEASUREMENTS

29.2.5.1   Steps in Film Thickness

A step in an oxide film formed on an Si wafer was measured, and the results are shown in 3-D representation in Figure 29.7. The measured surface step was 2.9 μm, and the film thickness was 4.0 μm in the higher area and 1.1 μm in the lower area. These thickness values agreed fully with the measurements obtained by ellipsometry. It was also found that the back surface was correctly measured as a flat surface with a supposed refractive index of 1.46.

29.2.5.2   Resist Film with an Unknown Refractive Index

The thickness distribution of a resist film on a glass plate was measured. Since the refractive index of the resist film was unknown, part of the film was peeled off, and the film profile was then measured. The surface step, which is equal to the physical thickness (t), was approximately 1.6 μm at the peeled portion, and the optical thickness ($n{\cdot}t$) was about 3.0 μm. These measurements allowed us to estimate the refractive index (n) at approximately 1.9.

The accuracy of the refractive index can be justified by the fact that the plate surface of the peeled portion and the back surface of the film are flush on the same plane. The back surface profiles calculated with three different refractive indexes are shown in Figure 29.8. A visual estimation gives a result of $n = 1.90 \pm 0.02$.

The film thickness and back surface height were then recalculated with the obtained refractive index; the results are shown in Figure 29.9.

29.2.5.3   Repeatability of Measurement

An oxide film on an Si wafer was measured repeatedly, giving the results shown in Figure 29.10. The measurement of film thickness produced the following data: mean value = 4.609 μm; standard deviation (σ) = 17 nm; and CV value = 0.36%.

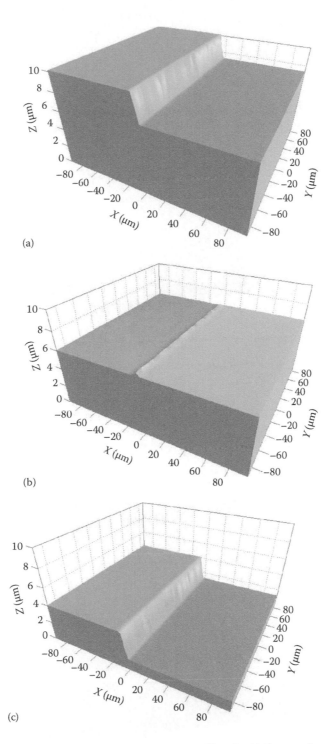

FIGURE 29.7 Measurement results of an SiO$_2$ film step on an Si wafer. (a) Front surface, (b) back surface, and (c) thickness.

FIGURE 29.8 Back surface profiles calculated with different refractive indexes (n). The peeled area is located at center left. (a) $n = 1.88$, (b) $n = 1.90$, and (c) $n = 1.92$.

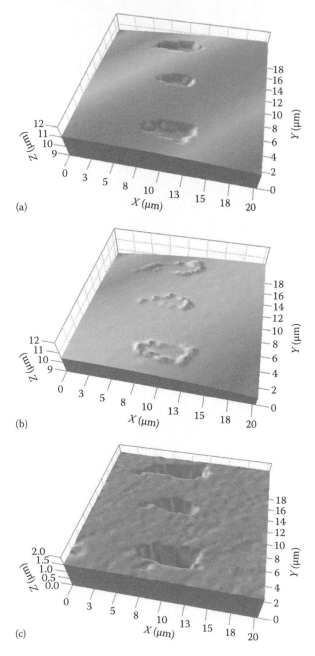

FIGURE 29.9 Measurements of a resist film. (a) Front surface, (b) back surface, and (c) thickness.

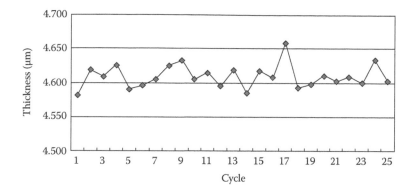

FIGURE 29.10   Repeatability of film thickness measurement.

29.3   PROFILING OF A THIN TRANSPARENT FILM

When a film becomes very thin, the KF algorithm can no longer be applied due to the overlap of two interference peaks. To solve this problem, several methods have been proposed [14–16], but as yet, there is no appropriate commercial product on the market. We developed an algorithm, the thiN Film (NF) algorithm, which can measure the 3-D surface profiles of a surface covered by a thin film and is robust enough for practical use in industry. Using this algorithm, Toray Engineering Co. developed the noncontact Surface Profiler SP-700 for thin film–covered surfaces. Oxide films with a thickness of 0–2000 nm on an Si wafer were measured, and the minimum measurable thickness was found to be 10–50 nm.

29.3.1   MEASUREMENT OF AN OXIDE THIN FILM STEP

An oxide thin film step with thicknesses of 100 and 300 nm was prepared on an Si wafer. Its cross section is shown in Figure 29.11. The test results given in Figure 29.12 show good agreement between the obtained and predicted values.

The repeatability of the film thickness measurement is shown in Figure 29.13. The thickness was an average of 10×10 pixels. The average and sigma of 25 repeated measurements were 301.5 ± 0.21 nm for the thicker film and 99.3 ± 0.30 nm for the thinner film.

29.3.2   MEASUREMENT OF CMP SAMPLES

Figure 29.14 shows the measurement results of oxide chemical mechanical polishing (CMP) samples with three different polish times (0, 20, and 60 s). The front surface profiles show rapid change

FIGURE 29.11   Cross section of the thin film sample.

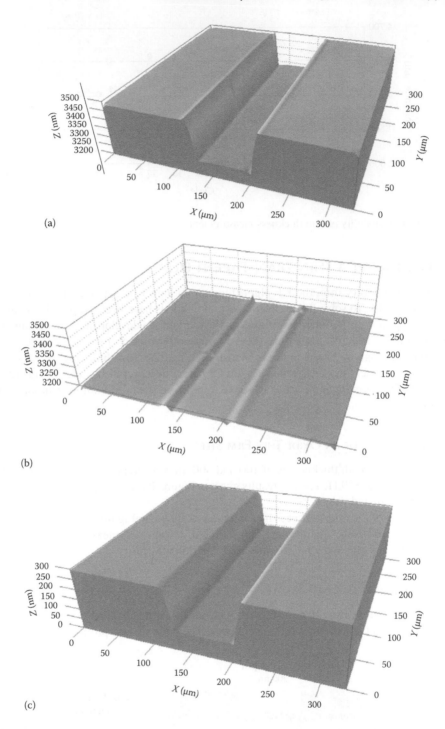

FIGURE 29.12 Thin film measurement results. (a) Front surface, (b) back surface, and (c) thickness.

due to the polish. The back surface profiles, on the other hand, show no change because polishing occurs only within the oxide film layer. Figure 29.15 shows the detailed top surface profile of the 60 s sample. The film thicknesses underneath are 466 nm at the center area and 740 nm in the surrounding area. From this simultaneous measurement of surface topography and film thickness distribution, we can obtain fast, vital, and unique information on the CMP process.

FIGURE 29.13 Repeatability of thin film thickness measurement.

FIGURE 29.14 Measurement results of three CMP samples.

29.4 FILM THICKNESS PROFILING BY PSEUDO-TRANSMISSION INTERFEROMETRY

29.4.1 PRINCIPLE OF THE TF METHOD

When we measure a freestanding film like a plastic film, there is another approach. This method, the transmission film (TF) method [17], measures a reasonably flat mirror surface with a sample film placed on it halfway into the FOV. The surface through the film is then measured at a point lower than the surface without the film by $(n - 1)t$, as shown in Figure 29.16, because of the optical

FIGURE 29.15 Detailed top surface profile of the 60 s CMP sample measured by the SP-700. The height of the pattern is on the order of 10 nm, and the film thickness underneath is 460–740 nm.

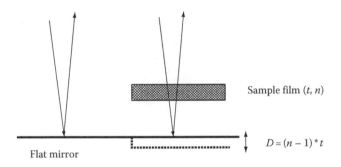

FIGURE 29.16 Principle of the TF method.

path difference (OPD) (D) between air and the film. If the refractive index (n) is known, the thickness (t) is calculated from the OPD.

If the working distance is not long enough to insert the sample halfway in the FOV, we can obtain the OPD by two successive measurements, the first obtained without the film and the second obtained with the sample.

Since any conventional technique for measuring a flat surface, such as phase-shift interferometry, can be used in this technique, measurement of sub-nanometer thickness is theoretically possible. Another advantage of this technique is its robustness. Since the optical path length does not depend on the film position, its vibration or bending does not affect the measurement.

29.4.2 Experiments

Polyester (PET) films with thicknesses of 0.9–6.0 μm were measured using this technique. As the reference mirror, we used a metal substrate for hard disk drive. The surface was measured by white-light VSI with a scanning speed of 2.4 μm/s. The sample was inserted only in the right half of the FOV. Figure 29.17 shows the measured mirror surface profile, the right half through a PET film with a nominal 0.9 μm thickness.

The average height difference between the left and right areas was 0.58 μm. From this value and the nominal refractive index of 1.65, we obtained the film thickness of 0.89 μm.

Figure 29.18 shows the correlation between the measured thicknesses and the nominal values. The obtained film thickness values agree well with the nominal values.

FIGURE 29.17 Measured surface profile of a mirror surface, with a PET film of a nominal 0.9 μm thickness inserted in the right half.

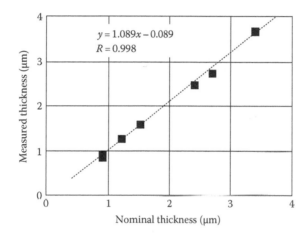

FIGURE 29.18 Correlation between the measured values and the nominal values.

29.5 SIMULTANEOUS MEASUREMENT OF THICKNESS AND REFRACTIVE INDEX

A freestanding polyester film nominally 1.5 μm in thickness was measured using the KF and TF methods. Figures 29.19 and 29.20 show the thickness profiles obtained by the KF and TF methods, respectively. These data were obtained using the nominal refractive index value of 1.65. The upper-right corner is marker ink. The results are largely consistent with each other with the exception of the error along the film edge in the TF method, which is due to the defocus effect of the microscope.

Figure 29.21 shows a comparison of two thickness profiles along a certain horizontal line. Note that there is a bias between them, probably due to an error in the nominal refractive index. From this, we observe that it is possible to obtain the thickness and the refractive index simultaneously by combining the results of two measurements: the optical thickness T by the KF method, and the OPD D by the TF method [18].

Since $T = nt$ and $D = (n - 1)t$, the thickness (t) and the refractive index (n) can be obtained by the following equations:

$$t = T - D, \tag{29.2}$$

FIGURE 29.19   Film thickness profile by the KF method.

FIGURE 29.20   Film thickness profile by the TF method.

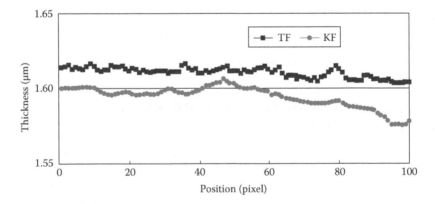

FIGURE 29.21   Film thickness profiles obtained by the TF and KF methods with a nominal refractive index.

and

$$n = \frac{T}{(T-D)} \tag{29.3}$$

Figure 29.22 shows the obtained thickness and refractive index profiles along a horizontal line. This technique was mentioned briefly in the description of an old patent [19] but seems to have remained almost unknown until now. Compared with other proposed techniques [20,21], this technique is simple and easily accomplished by a commercial surface profiler.

FIGURE 29.22 Film thickness and refractive index profiles calculated from the results of the KF and TF methods.

29.6 FILM THICKNESS PROFILING BY THREE WAVELENGTH INTERFERENCE COLOR ANALYSIS

29.6.1 INTRODUCTION TO THE INTERFERENCE COLOR ANALYSIS TECHNIQUE

The interference color phenomenon of thin films is seen in soap bubbles, and the relation between the color and the thickness of a film has been investigated for many years and is represented in the Newton color chart. The knowledge of this relation has been used for measuring semiconductor film thickness [22,23], the flying height of magnetic heads [24], and lubricant film thickness [25,26]. A typical example is the lubricant film thickness measurement system described by Eguchi and Yamamoto [26], in which a color camera captures the interference image, and the film thickness at each pixel is estimated from its hue by using the calibration data obtained in advance.

However, techniques based on the use of the interference color have rarely been used in the industry. The most serious problem with these techniques is that they require frequent calibration because the color thickness relation depends on a variety of environmental conditions such as the illumination and the target film structure. Another problem is their narrow unambiguous measurement range of a few hundred nanometers, resulting from the cyclic repetition of colors depending on the thickness.

To overcome these problems, we have developed a novel 2-D film thickness measurement method, named global model fitting for thickness (GMFT) [27], which is an extension of the global model fitting algorithm developed for three-wavelength interferometric surface profiling [28]. We validated the GMFT method by using both simulations and experiments.

29.6.2 PRINCIPLE OF THE GMFT METHOD

The algorithm consists of two steps, and its flowchart is shown in Figure 29.23.

Step 1 consists of the GMFT method, which is used to estimate the thickness of each pixel from its color information. Because of the high computational cost, this method is applied to a limited number of pixels during actual use. Step 2 consists of another method, named ACOS, which calculates the thickness of the other pixels within a shorter period of time by using the information obtained from step 1. This information is also applicable to other images captured under the same optical conditions. For this reason, it can be referred to as a "recipe."

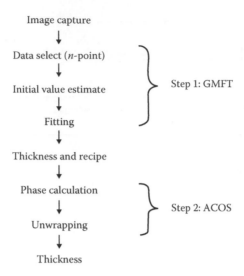

FIGURE 29.23   Flowchart of the GMFT algorithm.

29.6.2.1 GMFT Method

29.6.2.1.1 Relationship between Color and Thickness

Consider light incident on a thin film and reflected by both the upper and lower boundaries. After ignoring multiple reflections, we can obtain the following intensity of the sum of the two reflected waves:

$$I(\lambda) = I_1(\lambda) + I_2(\lambda) + 2\sqrt{I_1(\lambda)I_2(\lambda)} \cos\delta(\lambda), \tag{29.4}$$

where
    I_1 and I_2 are the intensities of the waves
    λ is the wavelength
    δ is the phase difference between them

Assuming a uniform refractive index, $n(\lambda)$, and normal incidence of light, the OPD between the two reflections from a thin film is OPD = $2n(\lambda)t$, where t is the thickness of the layer. Therefore, the phase difference for light reflected from the two surfaces is $\delta(\lambda) = 2\pi \, \text{OPD}/\lambda = 4\pi n(\lambda)t/\lambda$. For a nondispersive medium (i.e., $n(\lambda) = n$), we obtain

$$I(\lambda) = I_1(\lambda) + I_2(\lambda) + 2\sqrt{I_1(\lambda)I_2(\lambda)} \cos\left(\frac{4\pi nt}{\lambda}\right). \tag{29.5}$$

Let us consider a case of three-wavelength (B, G, and R) illumination. By assuming that $I_1 = I_2 = 1/2$, $\lambda_B = 470$ nm, $\lambda_G = 560$ nm, and $\lambda_R = 600$ nm, we calculated the theoretical intensities of each wavelength for the optical thickness range 0–1000 nm and then obtained the synthesized color chart. The results are shown in Figure 29.24.

It should be noted that the color change is due to the cyclic variation in BGR signals according to the film thickness. This means that we can estimate the film thickness from the BGR signals instead of from the color information.

Three-wavelength BGR signals can be obtained by an interference color imaging system, which is shown in Figure 29.25.

This is almost the same configuration as that of the three-wavelength single-shot interferometry reported by Kitagawa [29]. The details of the actual apparatus used in the experiment are described in Section 29.6.3.

FIGURE 29.24 BGR intensities and synthesized color chart as a function of film thickness.

FIGURE 29.25 Microscopic optical setup for three-wavelength interference color imaging.

29.6.2.1.2 GMFT Algorithm

Although the case of three wavelengths was considered in Figure 29.24, in the following equations, we assume a more generalized expression for m wavelengths.

When we capture interference images of m wavelengths, the observed intensity $g(i, j)$ at the point i ($i = 1, 2, \ldots, n$) and for wavelength j ($j = 1, 2, \ldots, m$) is given by the following model:

$$g(i, j) = a(j)[1 + b(j)\cos \varphi(i, j)], \tag{29.6}$$

where $a(j)$ and $b(j)$ are the DC bias and the modulation of the waveform, respectively, and $\varphi(i, j)$ is the phase given by

$$\varphi(i, j) = \frac{4\pi n t(i)}{\lambda_j}, \tag{29.7}$$

where
   n is the known refractive index
   $t(i)$ is the thickness
   λ_j is the wavelength of the jth wavelength

Inserting Equation 29.7 into Equation 29.6 gives the following model:

$$g(i, j) = a(j)\left\{1 + b(j)\cos\left[\frac{4\pi nt(i)}{\lambda(j)}\right]\right\}. \tag{29.8}$$

This model is derived under the assumption that the waveform parameters $a(j)$ and $b(j)$ are constant in the FOV and dependent only on the wavelength. This assumption is almost always valid when the target surface is homogeneous.

The unknown parameters $a(j)$, $b(j)$, and $t(i)$ can be estimated using the following least-squares fitting equation:

$$J[a(j), b(j), t(i)] = \sum_{i=1}^{n}\sum_{j=1}^{m}\left[g(i, j) - g_{ij}\right]^2, \tag{29.9}$$

where
  $g(i, j)$ is the model intensity defined by Equation 29.8
  g_{ij} is the observed intensity

This nonlinear least-squares problem can be solved by many numerical methods. In our experiments, we used the Davidon–Fletcher–Powell algorithm [30].

29.6.2.1.3 Necessary Conditions

Let us consider the necessary conditions to obtain the unknown parameters $a(j)$, $b(j)$, and $t(i)$. When the number of wavelengths is m and the number of points is n, the total number of unknown parameters is $2m + n$. Since m values are observed at one point, the necessary condition for the solution is $mn \geq 2m + n$. Then, the number of points must be

$$n \geq \frac{2m}{m-1}. \tag{29.10}$$

This means that $n \geq 4$ in the case of $m = 2$, and $n \geq 3$ in the case of $m = 3$. When $n = 2m/(m - 1)$, then the problem becomes a $(2m + n)$-order nonlinear simultaneous equation, and when $n > 2m/(m - 1)$, then the problem becomes a $(2m + n)$-order nonlinear least-squares problem.

Figure 29.26 illustrates the principle of this algorithm in the case of three wavelengths and n-points. From $3n$ observed intensities, we can estimate $(n + 6)$ unknown parameters, that is, n-point thicknesses and six waveform parameters.

It should be noted that to avoid the problem from becoming ill-conditioned and to obtain a good estimation, it is advisable to select the points such that their thickness distribution becomes wide.

29.6.2.1.4 Initial Estimates

To find solutions to the nonlinear least-squares problem described in the previous section, we use an iterative technique, which requires initial estimates that allow us to search for the minimum. Since the model function of Equation 29.8 contains a cosine function, the error function has many local minima. Therefore, it is essential to make good initial estimates.

In our experiments, $a(j)$ is set to be the average of the observed values, and the modulation $b(j)$ is set to be the range of the observed values (i.e., the difference between the maximum and minimum observed values) divided by $2a(j)$. The thickness, $t(i)$, is a rough estimate, which is generally given by *a priori* knowledge of the target sample.

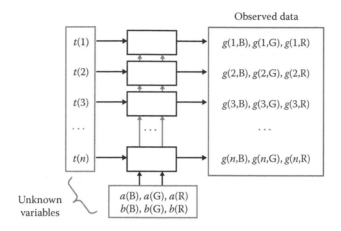

Observed data

FIGURE 29.26 Principle of the GMFT method in the case of three wavelengths and *n*-point fitting.

29.6.2.2 ACOS Algorithm

The computational cost of the nonlinear least-squares problem is very high. Hence, the method becomes impractical when the number of points is large. Therefore, we use the GMFT algorithm with a small number of points (e.g., under 100), and then the thicknesses of the other points are calculated by the following method, named the ACOS method, which uses the estimated waveform parameters from the first step.

29.6.2.2.1 Phase Estimation

When the waveform parameters are given, the phase is obtained from the observed intensity by the following equation derived from Equation 29.6:

$$\varphi(i,j) = \cos^{-1}\left[\frac{g_{ij} - a(j)}{a(j)b(j)}\right], \qquad (29.11)$$

where $\cos^{-1}$ is the arccosine function and its value is in the range $[0, \pi]$. When the argument of the function is not within the domain $[-1, 1]$, the function is undefined. In this case, the argument is approximated as -1 or 1.

29.6.2.2.2 Phase Unwrapping

From the phase data, the thickness, $t(i,j)$, is obtained by

$$t(i,j) = \frac{1}{n}\left[\pm\frac{\varphi(i,j)}{2\pi} + N(i,j)\right]\frac{\lambda_j}{2}, \qquad (29.12)$$

where $N(i,j)$ is the fringe order (integer), which is estimated by the coincidence method. The principle of this method is the same as the so-called exact fractions method [31] used for gauge block length measurement by multiwavelength interferometry.

Figure 29.27 shows an experimental example of the use of this method. For each wavelength, the thicknesses with different orders are plotted. The unknown orders are determined so that the three candidate thicknesses match best. In this case, the thickness is estimated to be 510 nm. It should be noted that the phase is obtained by the arccosine function, not by the arctangent function. Therefore, there are two candidate thicknesses for each fringe order, as shown in Figure 29.27.

FIGURE 29.27   Phase unwrapping by the coincidence method.

29.6.3   Experiments

29.6.3.1   Test Method

The experimental apparatus is shown in Figure 29.28. Its optics consists of a white LED illumination unit and a color camera with a multi-bandpass filter inside, as shown in Figure 29.29.

The spectral transmittance of the filter is shown in Figure 29.30.

The central wavelengths are 470, 560, and 600 nm, and each of their bandwidths is 10 nm. The camera is an ordinary CMOS color camera with 640 × 480 pixels. Its spectral sensitivity is shown in Figure 29.31 for the wavelengths of the illumination.

The captured image was stored in a PC memory with a color depth of 24 bits per pixel. The color crosstalk was compensated by the crosstalk compensation algorithm reported in [28]. We wrote a program in the C language to implement the GMFT algorithm on the Windows PC.

In our experiments, a calibrated step wafer (Mikropack, Germany) was used, which consists of a 100 mm Si-wafer with six silicon dioxide (SiO_2; $n = 1.46$) steps between 0 and 500 nm, as shown in the captured image (Figure 29.32).

FIGURE 29.28   Experimental apparatus.

FIGURE 29.29 Optics of the experimental apparatus.

FIGURE 29.30 Spectral transmittance of the multi-bandpass filter.

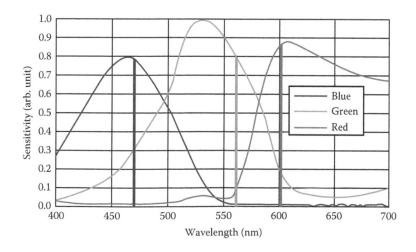

FIGURE 29.31 Spectral sensitivity of the color camera, with the three wavelengths used in the experiment.

FIGURE 29.32   Captured color image of thickness standard, and 12 points for GMFT fitting (shown in white numbers).

The nonlinear least-squares equation in the GMFT algorithm was solved using the Davidon–Fletcher–Powell method [30]. There were 12 data points to be fitted, as shown in Figure 29.32. The initial thickness values were set to be the nominal values. The thickness range of each point was set to be the nominal value, ±50 nm.

In the second step, we calculated the thickness distribution over the whole 640 × 480 pixel area by using the parameters obtained in the first step.

29.6.3.2   Test Results

From the 12-point intensity data, the corresponding thicknesses were estimated by the GMFT method. The results are shown in Table 29.1.

Then, the thickness distribution of the whole area was measured by the ACOS method. As shown in Figure 29.33, good agreement between the measured and predicted thickness distributions was obtained. Figure 29.34 shows a good correlation between the nominal and measured thicknesses.

The total calculation time for 640 × 480 pixels was about 220 ms, including 0.5 ms of the GMFT fitting using a C language program running on a Windows PC (2.5 GHz Intel Core i7 CPU).

TABLE 29.1
Results of GMFT Fitting

Variable	Estimated	Variable	Estimated
$t(1)$	0	$a(B)$	160
$t(2)$	117	$a(G)$	85
$t(3)$	208	$a(R)$	135
$t(4)$	306	$b(B)$	0.49
$t(5)$	403	$b(G)$	0.53
$t(6)$	506	$b(R)$	0.58
$t(7)$	0		
$t(8)$	118		
$t(9)$	208		
$t(10)$	307		
$t(11)$	403		
$t(12)$	506		

FIGURE 29.33   Estimated thickness profile by the ACOS method.

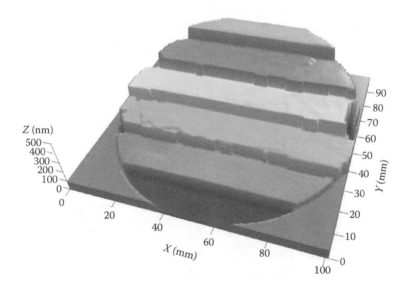

FIGURE 29.34   Correlation between the nominal and average estimated thicknesses of six areas.

29.6.4   Examples of Measurements

29.6.4.1   Thin Film on a Wafer

The thickness profile of an epitaxial film on a wafer was measured. The color image and the measured thickness profile are shown in Figure 29.35.

29.6.4.2   Soap Film

The thickness profiles of a down-flowing soap film were measured continuously. The time-series results are shown in Figure 29.36. The measurement speed was approximately 9 fps with the resolution of 160×120 pixels.

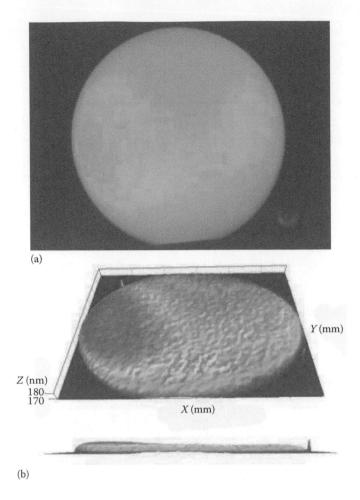

FIGURE 29.35 (a) Captured image and (b) estimated thickness profile of an epitaxial film on a wafer.

FIGURE 29.36 Time-series captured images and estimated thickness profiles of a flowing soap film.

29.6.5 FEATURES

The features of this technique include

- Single-shot thickness distribution measurement
- Low cost and simple optics
- Wide measurement range of 20–8000 nm

- Revolutionarily fast measurement speed of 2,000,000 pixels/s
- Superior repeatability of 1 nm
- No precalibration required
- Vibration resistant
- Applicable to moving objects

29.7 CONCLUSION

In this chapter, we presented five techniques to measure film thickness profiles using optical interferometry: (1) profiling of a thick transparent film, (2) profiling of a thin transparent film, (3) thickness profiling of a freestanding film, (4) simultaneous measurement of the thickness and refractive index of a freestanding film, and (5) thickness profiling by interference color analysis. All techniques are suitable for practical use.

Recent rapid advances in electronics and computer technology now allow the incorporation of newly developed, sophisticated algorithms in interferometric profilers, with dramatic results. These advances indicate the superb future potential of this technology and signal the beginning of a "second generation of interferometry."

REFERENCES

1. Caber, P.J., Interferometric profiler for rough surface, *Applied Optics*, 32, 3438, 1993.
2. Chim, S.S.C. and Kino, G.S., Three-dimensional image realization in interference microscopy, *Applied Optics*, 31, 2550, 1992.
3. Deck, L. and de Groot, P., High-speed non-contact profiler based on scanning white light interferometry, *Applied Optics*, 33, 7334, 1994.
4. Dändliker, R., Zimmermann, E., and Frosio, G., Electronically scanned white-light interferometry: A novel noise-resistant signal processing, *Optics Letters*, 17, 679, 1992.
5. Hirabayashi, A., Ogawa, H., and Kitagawa, K., Fast surface profiler by white-light interferometry by use of a new algorithm based on sampling theory, *Applied Optics*, 41, 4876, 2002.
6. Flournoy, P.A., McClure, R.W., and Wyntjes, G., White-light interferometric thickness gauge, *Applied Optics*, 11, 1907, 1972.
7. Tsuruta, T. and Ichihara, Y., Accurate measurement of lens thickness by using white-light fringes, *Proc. ICO Conf. Opt. Methods in Sci. and Ind. Meas.*, Tokyo, Japan, p. 369372, 1974; *Japanese Journal of Applied Physics, Supplement*, 14-1, 369, 1975.
8. Lee, B.S. and Strand, T.C., Profilometry with a coherence scanning microscope, *Applied Optics*, 29, 3784, 1990.
9. Nakamura, O. and Toyoda, K., Coherence probing of the patterns under films, *Kougaku*, 21, 481, 1992 (in Japanese).
10. Itoh, M. et al., Broad-band light-wave correlation topography using wavelet transform, *Optical Review*, 2, 135, 1995.
11. Toray Inspection Equipment Home Page, Toray Engineering Co., 2013, http://www.scn.tv/user/torayins/.
12. Kitagawa, K., 3-D profiler of a transparent film using white-light interferometry, *Proceedings of SICE Annual Conference 2004*, Sapporo, Japan, p. 585, 2004.
13. Otsu, N., An automatic threshold selection method based on discriminant and least squares criteria, *Transactions of IECE*, 63-D, 349, 1980 (in Japanese).
14. Kim, S.W. and Kim, G.H., Thickness-profile measurement of transparent thin-film layers by white-light scanning interferometry, *Applied Optics*, 38, 5968, 1999.
15. Kim, D. et al., Fast thickness profile measurement using a peak detection method based on an acousto-optic tunable filter, *Measurement Science and Technology*, 13, L1, 2002.
16. Akiyama, H., Sasaki, O., and Suzuki, T., Sinusoidal wavelength-scanning interferometer using an acousto-optic tunable filter for measurement of thickness and surface profile of a thin film, *Optics Express*, 13, 10066, 2005.
17. Kitagawa, K., Film thickness profiling by interferometric optical path difference detection, *Proceedings of DIA (Dynamic Image Processing for Real Application) WS 2007*, Sapporo, Japan, p. 137, 2007 (in Japanese).

18. Kitagawa, K., Simultaneous measurement of refractive index and thickness of transparent films by white-light interferometry, *Proceedings of JSPE Autumn Conference*, 461, 2007 (in Japanese).
19. Kalliomaki, K.J. et al., U.S. Patent 4,647,205, 1987.
20. Fukano, T. and Yamaguchi, I., Simultaneous measurement of thicknesses and refractive indices of multiple layers by a low-coherence confocal interference microscope, *Optics Letters*, 21, 1942, 1996.
21. Maruyama, H. et al., Low-coherence interferometer system for the simultaneous measurement of refractive index and thickness, *Applied Optics*, 41, 1315, 2002.
22. Parthasarathy, S., Wolfe, D., Hu, E., Hackwood, S., and Beni, G., A color vision system for film thickness determination, *Proceedings of IEEE Conference on Robotics and Automation*, Vol. 4, p. 515, 1987.
23. Birnie, D., Optical video interpretation of interference colors from thin transparent films on silicon, *Materials Letters* 58, 2795, 2004.
24. Kubo, M., Ohtsubo, Y., Kawashima, N., and Marumo, H., Head slider flying height measurement using a color image processing technique, *JSME International Journal, Series* 3, 32, 264, 1989.
25. Hartl, M., Krupka, I., Poliscuk, R., and Liska, M., Computer-aided evaluation of chromatic interferograms, *Proceedings of the Fifth International Conference in Central Europe on Computer Graphics and Visualization*, Pizen, Czech Republic, p. 45, 1997.
26. Eguchi, M. and Yamamoto, T., Film thickness measurement using white light spacer interferometry and its calibration by HSV color space, *Proceedings of the Fifth International Conference on Tribology*, Parma, Italy, p. 20, 2006.
27. Kitagawa, K., Thin-film thickness profile measurement by three-wavelength interference color analysis, *Applied Optics* 52, 1998, 2013.
28. Kitagawa, K., Single-shot surface profiling by multi-wavelength interferometry without carrier fringe introduction, *Journal of Electronic Imaging* 21, 021107, 2012.
29. Kitagawa, K., Fast surface profiling by multi-wavelength single-shot interferometry, *International Journal of Optomechatronics* 4, 136, 2010.
30. Fletcher, R. and Powell, M.J.D., A rapidly convergent descent method for minimization, *Computational Journal* 6, 163, 1963.
31. Tilford, C., Analytical procedure for determining lengths from fractional fringes, *Applied Optics* 16, 1857, 1977.

30 Optical Coherence Tomography for Industrial Applications

Tatsuo Shiina

CONTENTS

30.1 Introduction 769
30.2 Current Needs and Solutions of Industrial Applications for Optical Sensors 770
30.3 Principle of OCT 772
30.4 Industrial OCT 777
    30.4.1 Concept 777
    30.4.2 Portable OCT Scanner 778
30.5 Applications 781
    30.5.1 Industrial Use 781
    30.5.2 Basic Health Care 784
30.6 Future Technology 787
    30.6.1 Multifunction 787
    30.6.2 Variation of Industrial OCT Products 788
    30.6.3 Application for Plant Measurements 788
    30.6.4 Fundamental Medicine 788
30.7 Summary 788
References 789

30.1 INTRODUCTION

Many kinds of optical sensors are used in industry. Charge coupled device (CCD) cameras, lighting devices, and mesoscopes are currently popular optical devices. Optical devices have evolved into sensors used to check and measure industrial products, adding values to the products. Optical sensors have the advantage of high precision in comparison with other sensors such as ultrasonic, electromagnetic, and electrostatic sensors. This high precision comes from utilizing the wavelength of the optical light as a scale, which is shorter than 1 μm.

This innovative progress led to the invention of the interferometer, which gained maturity in the nineteenth century. In that period, the sources of light were emission lamps such as the mercury lamps, sodium lamps, and so on. The emission lamp was not convenient from the viewpoints of intensity, stability, and coherence. This stimulated scientists to contrive new ways of using optics. In the twentieth century, after the invention of the laser, interferometers gained popularity and their applications expanded widely. The laser interferometer had many variables that could be adapted to the target situation. With the growth of the industry, optical sensors are now being installed in factories, laboratories, and in various establishments related to the scientific fields.

The white light interferometer uses white light sources such as mercury lamps and halogen lamps. It is used as a measurement tool for sample surfaces. The low coherence of the white light source is utilized for higher accuracy. The coherence is based on the spectrum width of the white light on which is based the coherent length. It is very useful in checking products in terms of quality, damage, accuracy, and so on. It is widely used in industrial applications.

Optical coherence tomography (OCT) is also a low coherence interferometer. The principle is the same as the white light interferometer, but OCT provides a cross-sectional image of the sample. The white light interferometer measures the sample plane, while OCT measures points of the sample in depth direction reconstructs a cross-sectional image. To make its use easier, fiber-optics is installed into the OCT structure. A superluminescent diode (SLD) is utilized to render light to the fiber-optic device with high efficiency. The OCT technique was invented in the late 1980s [1]. It mainly evolved in the medical field, especially in the field of ophthalmology in the 1990s [2–6]. From the 2000s, many commercialized products were released and new technologies installed into the OCT to obtain higher accuracy and high-speed measurement [7–13].

OCT is a noninvase and nondestructive measurement method. It has attracted attention for application in other fields as well, including the medical fields (dentistry, dermatology, and internal medicine) [14–18]. Its application also extended to various plants [19–21] and industries [22–27]. It has been commercialized, however, only for applications in ophthalmology. It is currently suitable for eye measurement, but has not been optimized for applications in other parts of the human body.

From the viewpoint of industrial applications, OCT technology will have a high affinity to industrial samples because the white light interferometer has expanded in this field. Nonetheless, OCT has not seen popular growth in the industrial field because of the problem of flexibility and cost. Industrial samples have a variety of morphologies and sizes. The perspective of OCT measurement in industrial application is not only on structure analysis, but also on the condition and the change it effects on the sample. Industrial OCT should be compact, low-cost, and have the flexibility to adapt to the sample.

Section 30.2 describes the needs and demands from optical sensors in industrial applications.

Section 30.3 explains the principle of OCT technology.

Industrial OCT is explained in Section 30.4.

Its applications are mentioned in Section 30.5.

In Section 30.6, the future directions of industrial OCT are stated.

The summary and references are included in Section 30.7.

30.2 CURRENT NEEDS AND SOLUTIONS OF INDUSTRIAL APPLICATIONS FOR OPTICAL SENSORS

Optical sensors have been utilized to measure and diagnose the condition and appropriate structure of industrial materials for a long time. These devices are interferometer, optical displacement meter, moire inspection, stereoscopy, holography, and so on. Especially after the innovation of laser, these devices and methods have had an advantage of high resolution in comparison with other sensing methods such as ultrasonic or electrical devices.

Interferometers using laser have become an essential tool to examine the roughness of a material surface on the order of a few micrometers. Figure 30.1a shows a fundamental laser interferometer called a Michelson interferometer. The beam of a coherent light source was collimated and divided into two halves by a beam splitter (BS): one going to the reference mirror (M) (reference path) and the other to the target (measurement path). The interference is caused because of the concave–convex shape of the target surface when each reflected beam of light goes back to the BS and combines on it. The interference fringes appear on a screen representing the concave–convex shape of the target with the resolution of the beam's wavelength. This means that the interference fringes show the difference of the optical path on the scale of the beam's wavelength. The coherency of the laser beam is high, and the coherent length becomes long. Even if the difference of optical path between the reference and the measurement paths becomes large, the interference will occur thanks

FIGURE 30.1 Optical interferometers. (a) Interferometer (b) white light interferometer (c) time-domain OCT (d) spectral-domain OCT.

to the long coherent length of the laser. On the other hand, the light reflected from the interior structure of the target also interferes with the reference light, and it is hard to distinguish it. The laser interferometer is used for the surface profile with high precision on the order of optical wavelength.

The white light interferometer is a low-coherence interferometer and is shown in Figure 30.1b. The low coherency is characteristic of a light source with a wide spectrum. The wide-spectrum light source has a short coherent length. The white light interferometer utilizes this short coherent length to measure the concave–convex shape of the target surface. The interference occurs only when the optical length of the reference path is just matched with that of the measurement path. The difference between the paths is allowed to be only a few micrometers. The laser beam has speckle because of light coherency, and the result measured by the laser interferometer is always influenced by the speckle. A wide-spectrum light source has fewer speckles because the beam has a wide spectrum. The obtained waveform data become clear without the speckle effect. This is one of the main features of the white light interferometer. Industrial materials with rough surface, polished surface, and so on are targets.

The interior monitoring of the targets or materials is no less important than their external form. The interior structure is the first target, while clack, bubble, and uneven concentration are the other targets of interior monitoring.

OCT is a form of low-coherence interferometer. The principle of measurement in a laser interferometer and a white light interferometer is image measurement, while the principle of OCT measurement is fixed-point measurement. The former interferometer mainly measures the external form and displacement, while OCT obtains the cross-sectional image. The low-coherence light is directed into an optical fiber so that the influence of background light may be avoided. The output beam of the optical fiber is circular and its collimation is easy. It is hard to make the white light enter into the optical fiber with high efficiency. OCT utilizes an SLD, which has the same structure as a laser diode but lacks one of the mirrors on the edge surface. An SLD has the characteristics of a laser-like collimation property, while its emission spectrum is like that of a light-emitting diode (LED), that is, wide spectrum. The optical paths of the interferometer consist of an optical coupler. The output beams are collimated or focused with a collimator or lens. The scanning in the direction of depth is conducted by a movable reference mirror. This is called time-domain OCT and is shown in Figure 30.1c. Due to the movement of the reference mirror, the interference position shifts in the direction of depth. The interference occurs only when both lengths of the optical paths are matched. To obtain the cross-sectional image, the measurement probe or the target is moved to a vertical direction against the optical axis.

OCT has made progress in the field of ophthalmology. Its development started in the 2000s. Its technology and product value are sufficiently mature. In the initial stage of OCT development, the time-domain OCT (TD-OCT) was evolved. The piezo transducer was used to generate a stable interferogram, which is the interference waveform, and the another stage was installed for depth scanning [2,28,29]. A change in optical path length is obtained by stretching the coiled optical fiber, utilizing, for example, an off-axis Galvano mirror [30–32]. In the 2000s, non-scanning OCTs became the mainstream. These were called spectral-domain OCTs (SD-OCTs), which included Fourier-domain OCT (FD-OCT) and swept-source OCT (SS-OCT) [7,9,11,12]. This method does not use the movable reference mirror, that is, the nonmechanical scanner, while the signal information from the depth direction is obtained as it is piled up and analyzed in the frequency domain. The nonmechanical scanning method progresses the repetition speed of measurement upto the CCD imaging in real time. FD-OCT installs a monochrometer in place of a detector [7]. The SS-OCT utilizes a variable-wavelength laser in place of the low-coherence light source [12]. This is shown in Figure 30.1d. SD-OCT obtains the depth information by analyzing the detected signal in the frequency domain with the signal-processing method. The high speed and high precision will be achieved by these methods, while these high specifications need long memory and high-speed signal processing. The system should be impregnable against vibration.

This ophthalmologic OCT, however, has not been able to dominate industrial applications up to now. Industrial OCT has several requirements [22–27]. Although it is too expensive to install medical OCT in industrial fields, it is not the only reason preventing its prevalence. The ophthalmologic target, that is the human eye, can fix the specification of OCT. In other words, the system specification of OCT is optimized due to the target. By this consideration, high speed and high resolution are the current trends of OCT development. In the industrial fields, there are various kinds of targets. Size, material, structure, and measurement point are different for each. The optical probe, which throws the incident beam onto the target, should be changed because of the target. There are many restrictions in adapting ophthalmologic OCT to industrial fields. To apply the OCT technology in industrial use, the system should have the flexibility to adapt it to various kinds of targets. It is better that the system be simple, compact and low-cost.

The authors finalized such a concept of industrial OCT by 2000, and its portable version was developed in 2009. The system has a unique feature that can meet industrial needs. The short-term targets are structure monitoring of the exterior and interior of industrial materials, such as glass, plastic, and polymeric films. In half a decade, these targets have spread not only to industrial fields, but have also expanded to include the plant/food field and basic health-care.

30.3 PRINCIPLE OF OCT

This chapter explains the OCT technique and describes its principle including the theoretical expansions, devices, and signal processing in detail.

The amplitude of the beam is given by the sine or cosine curve as follows:

$$E_i = A_i \cos(k_0 z - \omega_0 t) \tag{30.1}$$

where A_i, k_0, and ω_0 are optical amplitude, wave number, and angular velocity of the incident beam, respectively. At first, we consider the interference of a single-wavelength beam such as a laser beam. The optical system is a Michelson interferometer as shown in Figure 30.2. The beam transmitted by a light source is separated into two beams with the same amplitude. The two beams are sent in quite

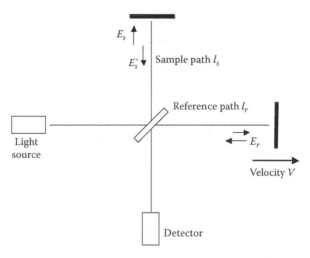

FIGURE 30.2 Michelson interferometer.

different directions against plane mirrors, where they are brought together again to form interference fringes. The intensity of the interfered beams is expressed by the following equation:

$$i_d = \langle |E_d|^2 \rangle$$

$$= \langle |E_l' + E_s'|^2 \rangle$$

$$= i_{dc} + \tilde{i}_d \qquad (30.2)$$

where
E_r, E_s, E_r' and E_r', are optical amplitude on the reference path, reflected amplitude, and amplitude on the sample path, respectively
i_d, i_{dc}, and $\tilde{i}_d$ are the interference intensity, the direct current part, and the alternating current part, which give the following expressions

The time average is expressed as "$\langle \; \rangle$".

$$i_{dc} = \frac{A_r^2}{2} + \frac{A_s^2}{2} \qquad (30.3)$$

$$\tilde{i}_d(\Delta l) = A_r A_s \cos(2k_0 \Delta l)$$

where
A_r and A_s are optical amplitudes on the reference path l_r and sample path l_s, respectively
Δl is the path difference between them

The interference intensity repeats blinks due to the change in the path difference. The ideal single-wavelength beam never attenuates the interference intensity with any path difference.

Next, we suppose the light source of the interferometer has a spectrum of a Gaussian distribution (Figure 30.3), which is written as the following equation where Δk is the spectrum width:

$$S(k) \approx \frac{2\sqrt{\ln 2}}{\Delta k \sqrt{\pi}} \exp\left\{ -\left[\frac{k - k_0}{\left(\frac{\Delta k}{2\sqrt{\ln 2}} \right)} \right]^2 \right\} \qquad (30.4)$$

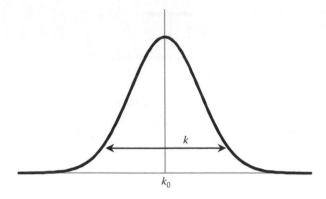

FIGURE 30.3 Gaussian distribution of incident beam.

The alternating current part of the interference intensity is rewritten as follows:

$$G(\Delta l) = \int_0^\infty S(k) \cdot \exp(-jk \cdot 2\Delta l) dk \tag{30.5}$$

Equations 30.1 through 30.5 express the convolution between the two beams. Here, using the following relations,

$$k - k_0 = \omega$$

$$\frac{1}{\left(\dfrac{\Delta k}{2\sqrt{\ln 2}}\right)^2} = \lambda$$

we may then write

$$G(\Delta l) = \frac{\sqrt{\lambda}}{\sqrt{\pi}} \int_0^\infty \exp(-\lambda\omega^2) \cdot \exp\{-j(k_0 + \omega) \cdot 2\Delta l\} d\omega \sum_{i=1}^n X_i Y_i$$

$$= \frac{\sqrt{\lambda}}{\sqrt{\pi}} \exp(-jk_0 \cdot 2\Delta l) \cdot 2\pi \cdot \frac{1}{2\pi} \int_0^\infty \exp(-\lambda\omega^2) \cdot \exp(-j\omega \cdot 2\Delta l) d\omega$$

$$= \frac{\sqrt{\lambda}}{\sqrt{\pi}} \exp(-jk_0 \cdot 2\Delta l) \cdot 2\pi \cdot \frac{1}{\sqrt{4\pi\lambda}} \exp\left(-\frac{(2\Delta l)^2}{4\lambda}\right)$$

$$= \exp\left(-\frac{(\Delta l)^2}{\lambda}\right) \cdot \exp(-jk_0 \cdot 2\Delta l) \tag{30.6}$$

$$G(\Delta l) = \exp\left(-\left[\frac{\Delta l}{\dfrac{2\sqrt{\ln 2}}{\Delta k}}\right]^2\right) \cdot \cos(k_0 \cdot 2\Delta l) \tag{30.7}$$

In this expression, the coherence length is defined as

$$l_c = \frac{4(\ln 2)}{\Delta k} \tag{30.8}$$

As a result, the alternating current part of the interference intensity has the following expression:

$$G(\Delta l) = \exp\left(-\left[\frac{\Delta l}{\frac{l_c}{2\sqrt{\ln 2}}}\right]^2\right) \cdot \cos(k_0 \cdot 2\Delta l) \tag{30.9}$$

The coherence length is decided by the spectrum width λk. Using the wavelength, this is rewritten as follows:

$$l_c = \frac{4(\ln 2)}{\Delta k} = \frac{2(\ln 2)\lambda_0^2}{\pi \Delta \lambda} \tag{30.10}$$

where λ_0 and $\Delta\lambda$ are, respectively, the center and the width of the spectrum of the light source.

Equations 30.1 through 30.9 explain how an optical light source with a Gaussian spectrum generates an interference intensity that is spindle-shaped, as shown in Figure 30.4. The waveform of the interference intensity is called an "interferogram." The Gaussian spectrum is applied not only to low-coherence light but also to laser and LED.

When the mirror on the reference path moves with a constant speed V, the path difference can be rewritten with the required time t, which would cause it:

$$\Delta l = Vt$$

That is, it depends on the velocity. The reflected beam from the moving mirror undergoes a Doppler shift. Due to the speed of the moving mirror, the wavelengths of the two beams from each of the paths are subject to be turned away. As a result, the interference intensity is exponentially small due to the moving speed of the mirror.

The interferogram shows the coherency, and its coherence length depends on the spectrum width of the light source. From another viewpoint, the spectrum width assumed directly depends on the time interval Δt in order to maintain a certain frequency $v_0(= c/\lambda_0)$ of the single-wavelength beam,

FIGURE 30.4 Alternating current part of interference intensity.

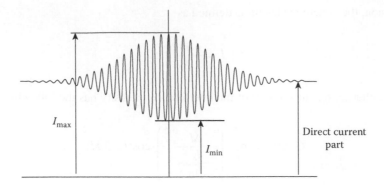

FIGURE 30.5 Interference intensity.

where c is the speed of light. The relationship between the time interval Δt and the spectrum width Δv is as follows:

$$\Delta t \cdot \Delta v \approx 1 \tag{30.11}$$

From this point of view, the coherence length l_c is defined as the path difference Δl ($=|l_s - l_r|$) when the visibility of the interference intensity is equal to $\sqrt{1/2}$. The visibility is defined by the following equation and its meaning is depicted in Figure 30.5:

$$\text{Visibility} \equiv \frac{I_{\max} - I_{\min}}{I_{\max} + I_{\min}} \tag{30.12}$$

That is, the alternating current part of the interference intensity (30.1 through 30.7) is equivalent to $|G(\Delta l)| = 1/2$. Then,

$$\exp\left[-\left(\frac{\pi \Delta v}{2\sqrt{\ln 2}} \cdot \frac{l_c}{c}\right)^2\right] = \frac{1}{2} \tag{30.13}$$

Solving (30.1 through 30.12), the coherence length is represented as

$$l_c = \frac{2c \cdot \ln 2}{\pi \Delta v} \tag{30.14}$$

By the following relationship

$$\frac{\Delta v}{v} = \frac{\Delta \lambda}{\lambda} = \frac{\Delta k}{k}$$

the coherence length is matched with Equations 30.1 through 30.8. This can be described in various ways as follows:

$$l_c = \frac{2\ln 2 c}{\pi \Delta v} = \frac{2\ln 2 \lambda_0^2}{\pi \Delta \lambda} = \frac{4\ln 2}{\Delta k} \approx 0.44 \frac{c}{\Delta v} \tag{30.15}$$

The actual OCT signal includes the noninterference signal and the interference signal, which are given in Equations 30.1 through 30.3 as i_{dc}: direct current part and $\tilde{i}_d(\Delta l)$: alternate current part.

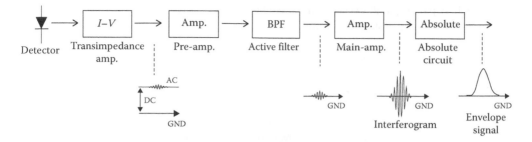

FIGURE 30.6   Circuit diagram of signal processing on OCT.

The noise part will be involved in the signal. To obtain an interference signal with adequate amplitude, signal processing is essential. The circuit diagram of signal processing on simple OCT is illustrated in Figure 30.6. The optical signal from here the interferometer, a Michelson interferometer, is changed to the electrical signal via a detector (PD). The current signal is transformed to voltage through a transimpedance amplifier. The signal intensity is magnified by a pre-amplifier and the direct current part and noise part are eliminated by an active filter used as a band-pass filter. The active filter has the center frequency, sometimes called beat or carrier frequency, which is synchronized with the speed of the movable mirror on the reference path. The relationship between the beat frequency λ_0 and the mirror speed V are as follows:

$$f = \frac{2V}{\lambda_0} \tag{30.16}$$

The selected signal's amplitude is modified by a main amplifier. The interferogram is obtained here. As the cross-sectional image of OCT often treats the interferogram signal as a signal point, the envelope waveform is estimated by an absolute circuit. The interferogram has a spindle shape, which swings its amplitude to a plus or minus value equally, and the opposite value is turned up and smoothed through a low-pass filter. The absolute circuit functions in this procedure. Each interference signal is changed to the envelope waveform owing to the movable mirror's motion. The cross-sectional image is reconstructed by shifting the observation point in vertical direction against the optical axis.

30.4   INDUSTRIAL OCT

30.4.1   Concept

OCT is a cross-section imager of the target. Now its techniques have some variations for interior diagnosis. SD-OCTs including FD-OCT and SS-OCT are technical methods to abandon the scanning in the depth direction, while the depth information is derived from analyzing the obtained signal. With nonmechanical scanning, these techniques become high-speed and high-resolution. These methods, however, fix their measurement range and resolution according to the light source, that is, depending on its spectrum characteristics, and it is hard to change them for each target. For high resolution the optical probe needs to be solid. The flexibility of the measurement will be less. Ophthalmologic OCT is specialized for the human eye, while flexibility of the system's configuration and measurement is essential for industrial application, in which the targets will have different size and specification.

Ophthalmologic OCT has a robust optical probe to refrain from oscillation. Ball sizes of the human eye are almost the same, and its optical probe has a fixed focus on the retina. On the contrary, industrial OCT should be able to cover various kinds of targets. The targets in industrial application are transparent materials such as glass, plastic, and polymeric film, and scattering or absorbing materials such as paper, paint, crystals, and biological tissue. It will be desirable to change the following parameters in the system according to the target:

Measurement range and accuracy
Numerical aperture and working distance of the optical probe
Scanning speed and repetition rate

Measurement range should be lengthened or shortened depending on the target. Biological tissues may not need high accuracy at all because these tissues are soft and have elasticity. On the other hand, industrial materials are mostly hard, and sometimes need high accuracy, especially for high positioning accuracy. It is preferable to have a flexible working distance of the optical probe in order to expand or shorten it as the measuring situation may demand. The scanning speed will be adjusted according to the observation method depending on whether it is fixed-point measurement or cross-sectional imaging. The former gives information of length, while the latter is an imager. The sensitivity of OCT measurement may sometimes be too high, and the strong reflection from the target surface may lower or influence the dynamic range of a detector. The background optical noise, which does not contribute to the interference, is also reflected by the target surface, and should be removed.

From the viewpoints of redundancy, compactness, and low cost, we have selected TD-OCT. The resolution of its measurement is uniquely decided by the spectrum width of the low-coherence light source. The resolution of SD-OCT is linked to the observation range. The measurement range and the working distance of TD-OCT can be fixed without relying on the resolution. As the measurement data directly reflects the interior structure, it makes the ensuring judgment easy.

30.4.2 PORTABLE OCT SCANNER

TD-OCT needs mechanical scanning to shift the interference point in the depth direction. Typical TD-OCT utilizes the piezoelectric transducer to generate the beat frequency of the interferogram. The scan in the depth direction is conducted at another scanning stage. This method is hard to speed up with a stable scanning technique. The stability of scanning depends on the stage accuracy. In consequence, the long-path scanning stage must have a solid platform to maintain accuracy. For industrial use, it may be often difficult to fix the position of the measurement sample precisely. It is also hard to shorten the working distance between the sample and the measurement probe. As a result, industrial OCT should have flexibility in the working distance and measurement range designs. The probe design for the long depth and the scanning speed should be optimized for each target.

The repetitive scanning motion should also be stable. In this study, the long optical path scanning mechanism was developed as shown in Figure 30.7 [33]. This consists of a rotating corner reflector and a fixed mirror. The scanning range depends on the rotation radius of the reflector. The scanning speed (repetition rate) is variable according to the rotation speed of the reflector. The optical path length is derived by the following equations. The following examples of the optical path change and the beat frequency of the interferogram are shown in Figure 30.8a and b, respectively:

$$\text{Path\_Difference} = 2l_1 + l_2(1 - \sin 2\theta)$$

$$l_1 = (r+s)\sin\theta - \frac{(1+r)(1-\cos\theta)}{\tan(\pi/4+\theta)}$$

$$l_2 = \frac{l_3}{\cos(\pi/4+\theta)}$$

$$l_3 = \sqrt{2s} + \frac{(1+r)(1-\cos\theta)}{\sin(\pi/4+\theta)} \tag{30.17}$$

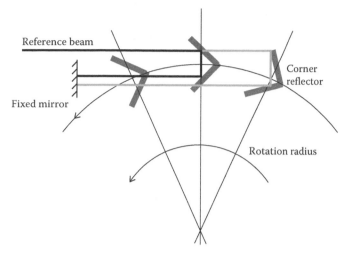

FIGURE 30.7 Long-path scanning algorism.

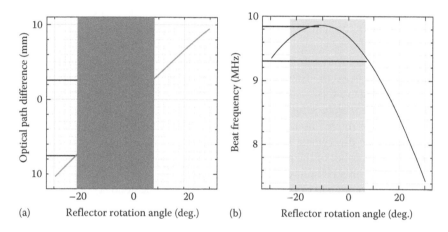

FIGURE 30.8 Optical characteristics of scanning algorism. (a) Optical path difference. (b) Beat frequency shift.

The rotating radius and speed of the reflector are 10 mm and 50 rps (= 3000 rpm), respectively. The gray area in each graph indicates the optical path difference of 10 mm. The rotating corner reflector generates the quasi-linear motion. The divergence from the linear motion is 1%–3% within the rotation angle of ±20°. The interferogram changes its frequency by about 1 MHz within the above rotation range.

The portable OCT scanner has a sophisticated structure. Its current compact structure is composed of an SLD light source, an optical fiber unit, a motor unit for optical path change, and a detector circuit as shown in Figure 30.9. The SLD light source is of a fiber pigtail type, witch is specially fabricated by Anritsu Co. Ltd. Its power is up to 3 mW[max]. Its wavelength is 0.8 or 1.3 μm and its spectrum width is <60 nm. The resolution of TD-OCT is defined by the spectrum width of the light source, which is 6–8 μm for the portable OCT. This SLD light source has an isolator and does not have a cooler. A radiator can make the light source stable when the output of its optical power is refrained. A cooling fan is used for the radiator for long operations where the output is close to the maximum (a few milliwatts). The fiber assembly is specially developed. It consists of a 2 × 2 optical fiber coupler and two collimators in its fundamental structure. Measurement and reference paths have collimators. The optical probe is attached on the measurement path and consists of other lens optics to control the numerical aperture (NA) of the probe and the depth of field. The detector circuit is inserted to transfer the optical signal to the electrical signal with adequate modification. The signals are gathered in a PC

FIGURE 30.9   Portable OCT scanner.

through an A/D converter. To obtain a stable result, the multiple signals are averaged and the cross-sectional image is constructed to visualize the data. The reflector is fixed on the rotating disc and the disc is directly connected to a servo motor, which can control its rotation speed with a dedicated driver.

An example of the concrete setup is summarized in Table 30.1. The scanning range is about 12 mm when a reflector rotates at a radius of 10 mm. The scanning speed is 25 scan/s. Position accuracy is <1 μm. The divergence from the linear motion is corrected by using Equation 30.1. Figure 30.10 shows a snapshot of the portable OCT scanner. Its size is 120 mm × 70 mm × 160 mm. It is handy and works with a DC power supply or a battery.

The wavelength, measurement range, and scanning speed are decided according to the target. The optical probe should be also designed according to individual requirements. The optical probe defines the working distance between the probe edge and the target surface. It receives the backscattering light from the target with an adequate NA. It is designed from 0 (parallel beam) to 0.3 according to the target. A plate-like target is measured with a narrow NA, while a high-scattered target such as a biological tissue is observed with a wide NA. The spot size of the incident beam becomes 0.5 mmϕ (NA = 0) and 5 μmϕ (NA = 0.3). The NA of the optical probe controls the depth of the field. A narrow NA has a long depth of field. On the contrary, a wide NA has a short depth of field. Measurement

TABLE 30.1
Portable OCT Scanner

SLD Light Module	Anritsu Co. LTD
Wavelength	1.3 μm
Spectral width	61.2 nm
Fiber module	Tatsuta Electric Wire & Cable Co., LTD
Scanning mechanism	Maxon DC Servo motor
Rotation	25 scan/s (1500 rpm)
Rotation radius	10 mm
Resolution	7 μm
Position accuracy	<1 μm
Scanning range	12 mm

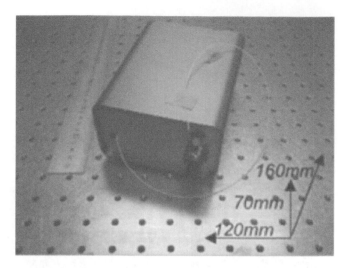

FIGURE 30.10 Portable OCT scanner, smallest one.

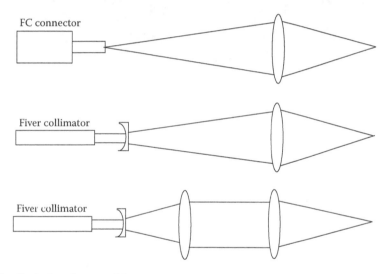

FIGURE 30.11 Optical probe assembly.

data will be obtained by combining the measurement range depending on the reflector motion and the depth of field on the optical probe. The design of the optical probe has several assemblies of lenses as shown in Figure 30.11. The upper case uses only a single lens, that is, it utilizes the NA of the optical fiber. The middle case adds a concave lens to expand the beam size. The control of the NA and the depth of field is easier in the middle than in the upper case. The bottom one adds an extra lens to make a parallel beam in the probe. This structure is useful to adjust the interference point on the focal point. As a result, the differences among them are ease of optical arraignment, simpleness, and probe size.

30.5 APPLICATIONS

30.5.1 INDUSTRIAL USE

The portable OCT can be installed in a factory as embedded equipment on the production line. The targets comprise interior monitoring such as crack and air bubble detection of plastic products, uneven fluctuation of materials and paints, and accuracy evaluation of casting. For industrial use, the cross-sectional image is not always essential. In many cases, manufacturing products are

FIGURE 30.12 Evaluation of laser fusion on plastic plates.

moving in a factory line. The fixed-point measurement and its multichannel measurement can accommodate the demands of industrial use.

Figure 30.12 shows one of the measurement results of an industrial product, which is the evaluation of laser fusion on plastic plates [24]. The laser fusion machine focuses on the boundary between the upper and lower plastic plates and melts it to bring the two plates together. The portable OCT can check and evaluate its condition by following the fusion process. The spot size of the OCT beam is equal to the beam spot of the laser fusion, and there is no need of a cross-sectional image. Here, however, the cross-sectional image is obtained by moving the probe in a perpendicular direction to the optical axis (depth direction) to visualize the quality of the fusion. When laser fusion is conducted perfectly, the boundary surface between the upper and bottom plastic plates disappears. Then the scanner does not receive any signal from the fusion part as shown in the left OCT image of Figure 30.12. The gradation of the reflected interferogram intensity is observed on the edges of the fusion part (circular mark on the left OCT image). It is dependent on the spot profile of the laser fusion; in other words, the intensity distribution of the laser beam. On the other hand, when the fusion is incomplete, the scanner receives the reflected interferogram signal from the boundary surface as shown in the right OCT image. The obtained OCT image clearly shows the difference. In the actual operation, the alert from the quality check of the OCT fixed-point measurement is enough to inform the operator regarding the mis-fusion.

Figure 30.13 shows the measurement of a transparent laminated tube (mayonnaise tube). Thin films of six to seven layers are detected. As the difference between the refractive indices of these files was small, the interference intensities inside the tube were small. The layer structure is clearly distinguished. The thickness of each layer is different at a part of the laminated tube. The individual difference is a target of the portable OCT.

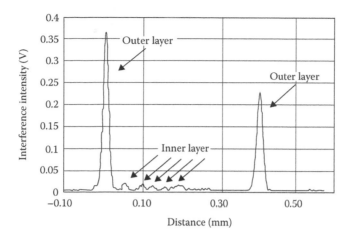

FIGURE 30.13 Transparent laminated tube (mayonnaise tube).

FIGURE 30.14 Tablet coating (thickness measurement).

There is a large demand for methods of thickness measurement for glass, polymeric film, plastic, and so on. Thickness fluctuation, precision, and layer structures are targets of industrial OCT measurement. Tablet coating (wafer) of medicine was observed as shown in Figure 30.14. The merit of portable OCT is not only direct measurement of a tablet, but also indirect measurement. The portable OCT can measure tablets through the packaging because the measurement range is wide enough. There is no need of precise positioning of the targets. The wavelength of portable OCT is near infrared light (0.8–1.6 μm), and there is little influence of color. Industrial materials come in a wide variety of sizes (thickness). The portable OCT can adjust its measurement range to the target size easily. The flexibility of the portable OCT is suitable for industrial use. The portable OCT can ne expanded to accommodate new functions to fulfill industrial demands. In a factory, a certain procedure can be conducted on multiple lines following the same routine. In such a case, the portable OCT can separate the optical probe into the number of lines functioning. To realize this, an optical switch is installed in the portable OCT. The optical switch produced by LEONI Co. Ltd. was so small that it could be installed in the chassis of a portable OCT as shown in Figure 30.15. When all the probes have the same length of optical fiber, the optical switch changes its channel at every reflector rotation.

Large targets such as combination lens, crystal block, and liquid tank need a long measurement range, while traditional optical sensors use a long linear stage to scan a long range. Such a sensor would be large, heavy, and high cost. The portable OCT can expand the measurement range by

FIGURE 30.15   Multichannel probe on a portable OCT.

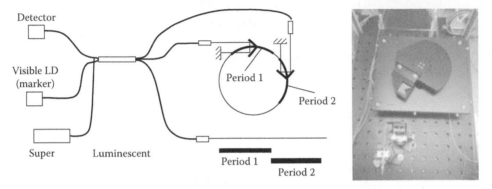

FIGURE 30.16   Long-path scanning on a portable OCT.

enlarging the radius of the rotating reflector. The radius of 60 mm, which is the same as that of a compact disc, has the measurement range of 100 mm. To rotate a large disc, the jitter of its rotation and its wobbling should be controlled. Figure 30.16 shows a high-precision rotating disc considering the mentioned task. The rotary encorder can correct the distortion from the linear motion. It is not easy to get the accuracy of 1 mm in the measurement range on the order of >100 mm, but the portable OCT has the ability to achieve this requirement. Furthermore, the 3×3 fiber coupler can add a reference port. When these two reference ports have different optical path lengths, the measurement range can expand to twice as long as for a single reference port without impairing the accuracy [35]. The arrangement of the reference path lengths can add the function not only of enlarging the measurement range, but also of increasing the repetition rate at one rotation.

30.5.2 Basic Health Care

Medical OCTs, including in ophthalmology, internal medicine, dentistry, and dermatology, are productive applications [5,12,13,25]. The ophthalmologic OCT can be successfully installed in a hospital, whereas other installations are still in the study phase. The main impediment is the cost. The introduction of OCT apparatus requires space, cost, and specialists. The portable OCT could be applied to the medical field because it achieved the same sensitivity as the medical OCT (the dynamic range is lower than the current medical OCT), although not in the same direction as medical OCTs. To fulfill basic health care is the purpose of the portable OCT.

The portable OCT for dentistry was specialized for its optical probe, which was a stick-like body with a diameter of 10 mmφ having a one-directional auto-stage. The NA of the optical probe was controlled to be 0.14. The depth of field was designed as 5 mm. The caries check was demonstrated with this portable OCT and the result is shown in Figure 30.17. Caries occurs just under the surface. The caries of CO, the initial stage of caries, could be easily monitored for depth and area.

FIGURE 30.17 Caries monitoring.

Another OCT measurement of dentistry is the boundary layer detection. The boundary layer between enamel and dentin is called the enamel–dentin junction (EDJ) and its recognition is important for artificial tooth and implant placement. The depth of EDJ is about 2–3 mm. The portable OCT can visualize it with a long measurement range as shown in Figure 30.18. However, intricate positions of teeth cannot be easily reached using the optical probe in an adequate manner.

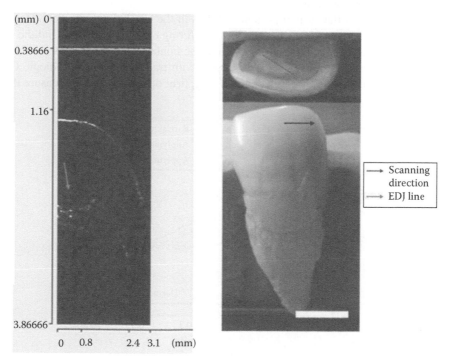

FIGURE 30.18 EDJ monitoring.

To use the portable OCT beneficially, the form of the optical probe and measurement method should be defined in the medical scenario.

The human skin is a scattering material for visible to infrared light to penetrate. The propagation depth to obtain the interference signal on medical OCT measurement is about 0.7 mm. The cross-sectional image is often reported for studies in dermatology [14,15,17]. On another front, the application of the portable OCT for skin measurement is unique. Basic health care and the cosmetic perspective are the main targets of the portable OCT. As the portable OCT can be compact and low cost, it is easy to install in a cosmetics counter and a medical interview scene. To adapt the portable OCT to skin measurement, the optical probe should have a high NA of 0.3 because the measurement depth is short and the scattered light should be gathered with a wide angle. The depth of the field of the high-NA probe should be about 1 mm. In this application, the portable OCT does not construct a cross-sectional image, but a fixed-point measurement is conducted. The optical probe is lightly vibrated to average out the individual speckle-like interior signals. An accumulation of about 1000 times was conducted for a stable evaluation. The measurement depth was thus about 0.5–0.6 mm.

Figure 30.19 shows an example of skin measurement by the portable OCT. the effect of cosmetics for skin care is visualized. The interference intensity was corrected by multiplying the square of the distance. The vertical axis was represented by the log-scale. The normal skin of the arm has a two-layer structure, of which the first is the epidermal layer and the second is the dermal layer. The normal thickness of the epidermal layer is about 0.2 mm. The patient in Figure 30.19 has dry skin. The peak of the second layer is therefore not clear. After cosmetic care by applying a foundation and a lotion to the skin, the second peak from the dermal layer becomes clear. It is clearly seen that the water retentivity of the lotion is more effective than that of the foundation.

Atopic dermatitis causes severe damage to the skin structure. This damage is not exclusive to the epidermal layer; it also seeps into the dermal layer. Figure 30.20 shows such damage of atopic dermatitis. The second peak of the dermal layer was spoiled. It is now clear compared with the normal skin. The epidermal layer was also subject to influence. There may be no need to visualize such skin conditions with a cross-sectional image. Fixed-point observation such as with a stethoscope is useful for a medical interview scene.

The other application of the current portable OCT is in the ophthalmologic field. Here too, the portable OCT is directed toward basic health care. The fitting of the contact lens and the combination of the optical power between the naked eye and the contact lens are targets for the portable OCT. Of course it is possible to evaluate such targets with traditional ophthalmologic OCT. The current sophisticated medical OCT, however, has a long, firm optical probe. To measure the optical

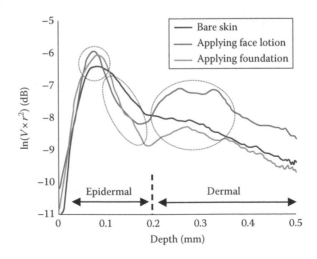

FIGURE 30.19 Skin measurement through cosmetics.

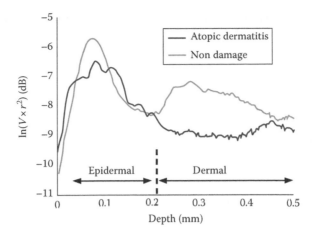

FIGURE 30.20 Atopic dermatitis measurement.

power, the eye should be unblocked, that is, the eye should be focused on a target. The portable OCT has the flexibility to design the optical probe to unblock the eye because the design of the optical probe is separated from the resolution factor and has a long measurement range.

Some of the applications mentioned are now in clinical practice. To install the portable OCT for clinical use as a basic health care tool or a cosmetic counter, further research on its usability will have to be pursued. This is because the users will not be professionals. From this perspective, simple structure of the device and direct manifestation of the observed information are important in medical care, especially basic health care. The portable OCT can be operated with a battery. It is useful in rural clinics and may be installed in developing countries.

30.6 FUTURE TECHNOLOGY

Up to now, a wide variety of targets have been evaluated by the portable OCT for industrial use. Of course, not all of these targets are suitable for the portable OCT. Details of these, however, are summarized in Table 30.2. This shows the direction of the targets sorted by categories. Now several Japanese companies are producing commercial products of the portable OCT. They may come as a built-in system, stand-alone apparatus, or custom-made device.

The future perspective of the industrial OCT is described in the following terms.

30.6.1 MULTIFUNCTION

Some of the industrial targets do not have a precise position in the production process. For example, to monitor the manufacturing process, the thickness of the target must be tracked by a sensor. In such a case, the industrial OCT with tracking can be installed. Its applications can be for injection stroke, chemical vapor deposition process, polishing process, and so on.

TABLE 30.2

OCT Applications of Industrial Use

Industrial use
　　Displacement, thickness, reflective index, solution concentration, crack check, volume change, time response, vibration

Plant and food processing
　　Plant factory, growth monitoring, food processing, freshness check, seal check

Basic health care and cosmetics
　　Ophthalmology, internal medicine, visceral examination, dentistry, dermatology, cosmetics

---

Sample point measurement of large targets is a strong need in industry. The large area of the targets does not always need whole scanning. The optical probe of the industrial OCT on a large positioning system will satisfy such needs. A large glass plate, thick acrylic material, thin film sheet, and so on are examples of large targets for industrial application.

Variation and yield ratio among production lines can be evaluated by multichannel OCT. This is mentioned in Section 30.5.

30.6.2 VARIATION OF INDUSTRIAL OCT PRODUCTS

The targets of industrial OCT are spread over a wide variety of fields. To keep pace with all of them by means of custom-made OCTs is not efficient because it requires great care. Modulation of the optical path change mechanism, optical fiber assembly, electrical circuits for the SLD driver, signal processing, and PC interface are necessory.

Some targets are visualized as a cross-sectional image, but the current portable OCT has the ability to produce two to three images in 1 s. When the operator needs higher speed, the multichannel OCT can be installed to conduct high-speed processing; otherwise, the light source should be changed into an SS light. The cheep and compact SS light will be selectable.

Monitoring not only the interior structure but also the deterioration and deformation with time will become the target of the industrial OCT. Painting material, plastic plate, and ABS resin are such targets. The extinction coefficient has information on the material quality [35–37].

30.6.3 APPLICATION FOR PLANT MEASUREMENTS

Plants vary considerably in size, brand, and so on. To evaluate them, statistical measurement is effective. The calibration curve method will also be installed. Furthermore, it is helpful to tie up specialists in biology with the practical farmer to analyze the results of plant OCT measurement. Environmental response, growth response, freshness assessment, and food processing are good applications for the portable OCT. To install the portable OCT in the field of growth of a plant, the system should be "portable" with battery-driven facility, operability, and resistance to dust, vibration, and so on.

30.6.4 FUNDAMENTAL MEDICINE

The measurement method should be selected depending on whether fixed-point measurement or cross-sectional imaging is required. On the medical interview scene or cosmetic counter, simple system and direct result are important for nonprofessional operators and subjects. Restriction of the application to only necessary information leads to its low cost and simplicity of use.

30.7 SUMMARY

The current portable OCT is a TD-type OCT. It does not involve the latest technology in comparison with the medical OCT. Its scanning speed, especially, is much slower than that of the current SD-OCT. The high scanning speed of the SD-OCT, however, is for real-time observation of three-dimensional or four-dimensional structures of the target. High-speed signal processing is also needed to calculate the huge memorized data. The portable OCT is simple for analyzing the obtained data because the obtained interferogram directly represents the condition of the interior structure of the target. The scanning speed is not so high, but it is enough for most of the industrial applications. Instead, the portable OCT has a feature of wide flexibility for the designs of the optical probe and the measurement range. It is adequate for the variety of industrial targets.

The applications of the portable OCT for medical use are directed toward basic health care and cosmetics. From the viewpoint of simple operation, small apparatus, and low cost, the portable

OCT can be easily installed in such fields. Cross-sectional images of the teeth are useful to check caries and EDJ, while skin condition can be evaluated by fixed-point observations. An easy-to-use approach and direct representation of the desired information are important for the medical interview scene and cosmetics formulation.

In the near future, the medical OCT will continue to develop state-of-the-art technology to pursue the new demands of medicine in the current direction, while it becomes a low-cost and single-function technique in another direction. On the other hand, the industrial OCT will expand across wide fields by improving the formulation of the system and modular technology. The portable OCT in fundamental medicine will be installed in areas separate from the medical OCT.

REFERENCES

1. H. H. Gilgen, R. P. Novak, R. P. Salathe, W. Hodel, and P. Beaud, Submillimeter optical reflectometry, *J. Lightwave Technol.*, **7**, 1255, 1989.
2. D. Huang, E. A. Swanson, C. P. Lin, J. S. Schuman, W. G. Stinson, W. Chang, M. R. Hee et al., Optical coherence tomography, *Science*, **254**, 1178–1181, 1991.
3. M. R. Hee, J. A. Izatt, E. A. Swanson, D. Huang, J. S. Schuman, C. A. Puliafito, and J. G. Fujimoto, Optical coherence tomography of the human retina, *Arch. Ophthalmol.*, **113**, 325–327, 1995.
4. A. Can, J. S. Duker, T. H. Ko, J. S. Schuman, and J.G. Fujimoto, Ultra-high resolution optical coherence tomography of retinal pigment epithelial tear following blunt trauma, *Arch. Ophthalmol.*, **124**, 281–283, 1996.
5. B. W. Colston, M. J. Evertt Jr., L. B. D. Silva, L. L. Otis, P. Stroeve, and H. Nathel, Imaging of hard- and soft-tissue structure in the oral cavity by optical coherence tomography, *Appl. Opt.*, **37**, 3582–3585, 1998.
6. J. M. Schmitt, Optical coherence tomography (OCT): A review, *IEEE J. Quantum Electron.*, **5**, 1205–1215, 1999.
7. M. Wojtkowski, R. Leitgeb, A. Kowalczyk, A. F. Fercher, and T. Bajraszewski, In vivo human retinal imaging by Fourier domain optical coherence tomography, *J. Biomed. Opt.* **7**, 457–463, 2002.
8. W. Drexler, Ultrahigh-resolution optical coherence tomography, *J. Biomed. Opt.*, **9**, 47–74, 2004.
9. S. Jiao, R. Knighton, X. Huang, G. Gregori, and C. Puliafito, Simultaneous acquisition of sectional and fundus ophthalmic images with spectral-domain optical coherence tomography, *Opt. Exp.*, **13**, 444–452, 2005.
10. C. C. Rosa, J. Rogers, J. Pedro, R. Rosen, and A. Podoleanu, Multiscan time-domain optical coherence tomography for retina imaging, *Appl. Opt.*, **46**, 1795–1808, 2007.
11. I. Grulkowski, M. Gora, M. Szkulmowski, I. Gorczynska, D. Szlag, S. Marcos, A. Kowalczyk, and M. Wojtkowski, Anterior segment imaging with spectral OCT system using a high-speed CMOS camera, *Opt. Exp.*, **17**, 4842–4858, 2009.
12. Y. Shimada, A. Sadr, A. Nazari, H. Nakagawa, M. Otsuki, J. Tagami, and Y. Sumi, 3D evaluation of composite resin restoration at practical training using swept-source optical coherence tomography (SS-OCT), *Dent. Mater. J.*, **31**, 409–417, 2012.
13. C. K. S. Leung and R. N. Weinreb, Anterior chamber angle imaging with optical coherence tomography, *Eye*, **25**, 261–267, 2011.
14. V. R. Korde, G. T. Bonnema, W. Xu, C. Krishnamurthy, J. R. Moore, K. Saboda, L. D. Slayton et al., Using optical coherence tomography to evaluate skin sun damage and precancer, *Lasers Surg. Med.*, **39**, 687–695, 2007.
15. M. Mogensen and L. Thrane, OCT imaging of skin cancer and other dermatological disease, *Biophotonics*, **2**, 442–451, 2009.
16. R. Steiner, K. Kunzi-rapp, and K. Scharffetter-Kochanek, Optical coherence tomography: Clinical applications in dermatology, *Med. Laser Appl.*, **18**, 249–259, 2003.
17. K. Koenig, Hybrid multiphoton multimodal tomography of in vivo human skin, *Intra Vital*, **1**, 11–26, 2012.
18. H. Yabushita, B. E. Bouma, S. L. Houser, H. T. Aretz, I. Jang, K. H. Schlendorf, C. R. Kauffman et al., Characterization of human atherosclerosis by optical coherence tomography, *Circulation*, **106**, 1640–1645, 2002.
19. J. W. Hettinger, M. P. Mattozzi, W. R. Myers, M. E. Williams, A. Reeves, R. L. Parsons, R. C. Haskell, D. C. Petersen, R. Wang, and J. I. Medford, Optical coherence microscopy. A technology for rapid, in vivo, non-destructive visualization of plants and plant cells, *Plant Physiol.*, **123**, 3–16, 2000.

20. T. H. Chow, K. M. Tan, S. G. Razul, C. M. Tay, T. F. Chia, and W. T. Poh, Diagnosis of virus infection in orchid plants with high-resolution optical coherence tomography, *J. Biomed. Opt.*, **14**, 014006, 2009.
21. V. V. Sapozhnikova, V. A. Kamenskii, and R. V. Kuranov, Visualization of plant tissues by optical coherence tomography, *Russ. J. Plant Physiol.*, **50**, 282–286, 2003.
22. M. Wurm, K. Wiesauer, K. Nagel, M. Pircher, E. Götzinger, C. K. Hitzenberger, and D. Stifter, Spectral-domain optical coherence tomography: A novel and fast tool for NDT, *Fourth International Workshop NDT in Progress*, Prag, Czech Republic, pp. 225–232, 2007.
23. B. G. Goode, OCT aims for industrial application, *Laser Focus World*, **45**, 41–45, 2009.
24. T. Shiina, M. Satoshi, and T. Honda, Factory built-in type simplified OCT system for industrial application, *IEEE J. Trans. EIS*, **129**, 1276–1281, 2009 (written in Japanese).
25. C. Todea, M. L. Negrutiu, C. Balabuc, C. Sinescu, F. I. Topala, C. Marcauteanu, S. Canjau, G. Semez, and A. G. Podoleanu, Optical coherence tomography applications in dentistry, *TMJ*, **60**, 5–17, 2010.
26. P. Merken, R. Vandersmissen, and G. Yurtsever, Optical coherence tomography: OCT supports industrial nondestructive depth analysis, *Laser Focus World*, **47**, 2–8, 2011.
27. G. Song and K. Harding, OCT for industrial applications, *SPIE Proc.*, 8563, 2012.
28. J. A. Izatt, M. D. Kulkarni, H. Wang, K. Kobayashi, and M. V. Sivak, Jr., Optical coherence tomography and microscopy in gastrointestinal tissues, *IEEE J. Quantum Electron.*, **2**, 1017–1028, 1999.
29. M. Leahy, J. Hogan, C. Wilson, H. Subhash, and R. Dsouza, Multiple reference optical coherence tomography (MR-OCT) System, *Proceeding of SPIE* 8580, 8, 2013.
30. J. A. Izatt, M. R. Hee, G. M. Owen, E. A. Swanson, and J. G. Fujimoto, Optical coherence tomography in scattering media, *Opt. Lett.*, **19**, 590–592, 1994.
31. G. J. Tearney, M. E. Brezinski, B. E. Bouma, S. A. Boppart, C. Pitris, J. F. Southern, and J. G. Fujimoto, In vivo endoscopic optical biopsy with optical coherence tomography, *Science*, **276**, 2037–2039, 1997.
32. G. J. Tearney, B. E. Bouma, and J. G. Fujimoto, High-speed phase- and group-delay scanning with a grating-based phase control delay line, *Opt. Lett.*, **22**, 1811–1813, 1997.
33. T. Shiina, Y. Moritani, M. Ito, and Y. Okamura, Long optical path scanning mechanism for optical coherence tomography, *Appl. Opt.*, **42**, 3795–3799, 2003.
34. PCT/JP2010/070844.
35. A. B. Vakhtin, K. A. Peterson, D. J. Kane, E. H. Jordan, G. Hansen, and M. Teicholz, Combination of Fourier-domain optical coherence tomography and photo-stimulated luminescence piezo-spectroscopy as an NDE tool for thermal barrier coatings, *ASME Proceedings, Ceramics*, pp. 275–279, 2007.
36. G. Popescu and A. Dogariu, Scattering of low coherence radiation and applications, *Eur. Phys. J. Appl. Phys.*, **32**, 73–93, 2005.
37. G. Yao and L. V. Wang, Monte Carlo simulation of an optical coherence tomography signal in homogeneous turbid media, *Phys. Med. Biol.*, **44**, 2307–2320, 1999.

31 Interference Microscopy for Surface Structure Analysis

Peter J. de Groot

CONTENTS

31.1 Overview 791
    31.1.1 Introduction 791
    31.1.2 Historical Perspective 792
    31.1.3 Interference Microscopy 792
31.2 Principles 794
    31.2.1 Interference Signals and Patterns 794
    31.2.2 Phase Shifting Interferometry 795
    31.2.3 Coherence Scanning Interferometry 797
31.3 Instrument Design for Interference Microscopy 802
    31.3.1 Light Source and Detection 802
    31.3.2 Optical System 802
    31.3.3 System Configurations 804
    31.3.4 Postprocessing 807
31.4 Signal Modeling 807
    31.4.1 Modeling of Interference Microscopy Fringe Patterns 807
    31.4.2 The Obliquity Effect 808
    31.4.3 Transparent Films 809
    31.4.4 Model-Based Analysis 810
31.5 Performance 812
    31.5.1 Good Practice and Instrument Evaluation 812
    31.5.2 Surface Topography Repeatability 812
    31.5.3 Vibration and Environmental Effects 814
    31.5.4 Optical Effect of Surface Materials 815
    31.5.5 Lateral Resolution 816
    31.5.6 Field of View and Slope Acceptance 819
    31.5.7 Height Response 819
    31.5.8 Profile Fidelity 822
31.6 Future Perspectives 823
References 824

31.1 OVERVIEW

31.1.1 INTRODUCTION

This chapter considers the principles of interferometric dimensional metrology applied to surface features best viewed in a microscope. Interference microscopy encompasses a wide variety of techniques for measuring surface texture, shape, step heights, lateral positions and dimensions, transparent film structure, and the scattering behavior of optically unresolved patterns. In addition to applications that range from semiconductor wafer analysis to fuel injection valve measurements, interference microscopy encompasses a variety of basic geometries from Mirau to Michelson and light sources from lasers to light-emitting diodes (LEDs). As with companion techniques such as confocal microscopy, interference microscopy enjoys the advantages of rapid, noncontact operation and rapid data acquisition, often at data acquisition rates in the millions of 3D image points per second. Unique among optical areal measurements instruments, interferometry has the benefit of high sensitivity to surface structure irrespective of magnification.

31.1.2 HISTORICAL PERSPECTIVE

The physical concepts and fundamental instrument geometries underlying interference microscopy have a history reaching back to the nineteenth century, but the more recent developments of the technique link closely to rapidly advancing enabling technologies, particularly in digital data processing. While the modern age of interferometry for testing of optical components was born with the helium–neon laser, automated interference microscopy owes its existence to the desktop computer. The differences relate to the relative complexity of the interference signals generated in the high-magnification optical systems required to see small features and to the kinds of surface structures that interest us in microscopy. These factors modify the interference patterns with effects related to focus position, incoherent imaging, discrete surface structure, roughness, and semi-transparent films. Making good use of the resulting interference patterns requires sophisticated data acquisition and analysis commanded by a computer.

The modern age of interference microscopy began therefore in the 1980s, initially with single-point systems [1], then with single-line optical profilers based on phase shifting interferometry (PSI) [2,3]. Full-field areal topography imaging using computerized phase shifting completed the basic functions of automated 3D imaging on the microscopic scale [4], but applications remained limited, primarily by the requirement that the surface be optically smooth, so as to generate clear, continuous interference patterns across the surface with a light source filtered to a narrow spectral bandwidth. This limitation fell away with the discovery that meaningful topographical imaging was possible even when the interference image was a seemingly indecipherable speckle pattern, using the full spectral bandwidth of white light illumination [5] and a technique known as white light or coherence scanning interferometry (CSI) [6]. Coherence scanning is now the dominant method for interferometric 3D surface structure analysis on the microscopic scale, with expanding commercial offerings and tens of thousands of instruments installed worldwide. Modern systems have overcome previous limitations of environmental robustness, slope acceptance, profile fidelity, and correlation to other surface topography methods.

31.1.3 INTERFERENCE MICROSCOPY

The most common form of interference microscope today for areal surface structure requires that the object be in focus and operates by comparing the sample surface to a physical or virtual reference surface by two-beam division of amplitude interferometry. The schematic geometry of Figure 31.1 fits this description. The overall structure is that of a common microscope with a digital camera, but equipped with a specialized interference objective, which in most cases is mechanically and optically interchangeable with a conventional noninterferometric imaging objective. Figure 31.2 illustrates a typical interference fringe pattern for a roughness specimen, while Figure 31.3 shows the resulting topography image generated by automated modulation of the fringe pattern followed by detailed computer analysis. Figure 31.2 illustrates how interference fringes follow the surface contours, as well as the effects of limited coherence that modulate the fringe contrast—an important characteristic of modern coherence scanning methods. This chapter concentrates on reflection-mode surface measuring interferometers of this general type.

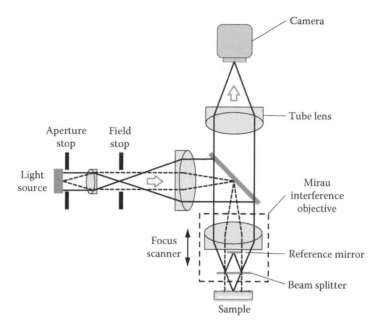

FIGURE 31.1   Interference microscope for areal surface structure analysis.

FIGURE 31.2   Fringe pattern for a reflection interference microscope as in Figure 31.1, showing interference fringes that follow the surface height contours a roughness specimen.

FIGURE 31.3   Surface topography measurement corresponding to Figure 31.2.

31.2 PRINCIPLES

31.2.1 INTERFERENCE SIGNALS AND PATTERNS

The simplified Michelson interferometer illustrated in Figure 31.4 with a laser source provides a foundation for understanding the basics of interferometric surface topography measurement. Following the usual two-beam interference analysis, the interference signal at the square-law detector can be written [7,8] as

$$I(h,\zeta) = I_{DC} + I_{AC}\cos[(h-\zeta)K+\gamma] \qquad (31.1)$$

where I_{DC} and I_{AC} are fixed coefficients. The small inset graph in Figure 31.4 shows the sinusoidally modulated intensity as a function of the difference $(h-\zeta)$. The fringe frequency

$$K = \frac{4\pi}{\lambda} \qquad (31.2)$$

is the rate at which the interference signal oscillates as a function of changes in sample surface height h or in position ζ of the reference mirror. This frequency depends on the wavelength λ but, as we shall see later, geometric factors as well when we consider the more complex and interesting geometries of interference microscopes. The phase offset γ relates among other factors to the reflection and transmission properties of the interferometer components.

In a surface topography measurement, a camera replaces the detector in Figure 31.4 and associated imaging optics bring the sample surface into focus, allowing for intensity values to correspond to specific points on the surface. In surface structure analysis, most often we are interested in the height h, which we extract by inversion of the detected phase

$$\theta = hK \qquad (31.3)$$

that appears inside the argument of the cosine in Equation 31.1. For the special case where all factors in Equation 31.1 are constant with the exception of surface height, then there are straightforward ways to infer the relative phase θ as it varies across a surface by inspection of the intensity value I. In Figure 31.2, the interference fringes, corresponding to lines of equal intensity, follow the surface topography as contours of equal surface height h. In the earliest applications of interference microscopy, skilled observers would estimate surface texture to a small fraction of a micron based on the appearance of these fringes, first with the aid of reticles and later by means of photometric tools.

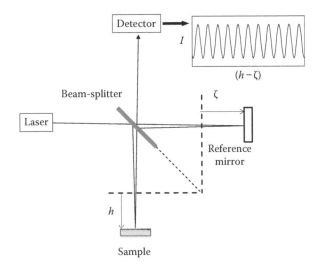

FIGURE 31.4 Conceptual diagram of a Michelson interferometer.

31.2.2 PHASE SHIFTING INTERFEROMETRY

Increased precision and greater surface structure complexity require something more automated than visual inspection of lines of equal interference intensity. Electronic data acquisition together with computerized data analysis allows for quantification of surface texture at the nanometer level, over surfaces that may have highly variable reflectivity and structure. With the exception of carrier fringe analyses similar to holographic methods [9], computerized fringe analysis almost invariably relies on digitization of image data acquired during a controlled phase shift, most often over time, and introduced by controlled mechanical oscillation of an interference objective in the focus direction indicated in Figure 31.1.

The earliest and still one of the most precise and widespread computerized fringe analysis methods is PSI. Here, the extraction of this phase involves heterodyning the signal by means of a controlled phase shift φ, for example, by modulation of the reference length ζ:

$$I = I_{DC} + I_{AC}\cos(\theta - \varphi) \tag{31.4}$$

where $\varphi = \zeta K$ and the constant phase offset γ has been dropped as unimportant here. In practice, these phase shifts correspond to a sequence of discrete sample intensities

$$I_j = I_{DC} + I_{AC}\cos(\theta - \varphi_j). \tag{31.5}$$

Given that the constants I_{DC}, I_{AC} are typically unknown *a priori,* we need at least three values for φ to solve for θ.

In the most common case, these discrete phase shifts φ_j are evenly spaced by a phase increment α

$$\varphi_{j+1} - \varphi_j = \alpha, \tag{31.6}$$

as illustrated in Figure 31.5 for three data samples indexed $j = 1, 2, 3$. These data points represent either instantaneous acquisitions or phase steps, or more commonly, the center points for intensity measurements integrated over the period of time between sample points, in a method commonly referred to as the integrating bucket method [10,11].

Once we have these samples over at least one full fringe cycle, a phase detection or PSI algorithm provides the phase θ. A simple and historically significant example of such an algorithm uses just three samples [12]:

$$\tan(\theta) = \frac{\sqrt{3}(I_2 - I_3)}{I_2 + I_3 - 2I_1} \tag{31.7}$$

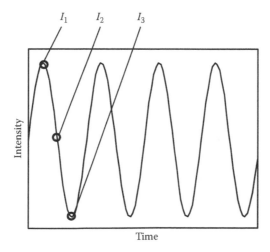

FIGURE 31.5 Intensity sampling for PSI with a linear phase shift (constant phase step between acquisitions).

where the phase shift between data acquisitions is $\alpha = 2\pi/3$. Improving camera and data processing technologies together with increased demands on performance have driven the number of data samples and corresponding phase shifts to larger numbers, and it is now common to encounter PSI algorithms of a dozen or more sample points with floating-point coefficients. The methodology of PSI algorithm design is perhaps the single most thoroughly analyzed feature of modern computerized interferometry for surface measurements [13–15] and has been extensively reviewed, for example, in the excellent book chapter by Schreiber and Bruning [16].

The phase shifts ϕ need not necessarily be evenly spaced to be useful for PSI. An alternative for continuous data acquisition and averaging is sinusoidal phase shifting, a technique that has the advantage of being less demanding on phase shift mechanisms than repeated linear ramps [17]. This can be especially interesting for interference microscopes, where the phase shifting mechanism may be a compliant structure with a collection of turret-selectable objectives [18]. The intensity pattern resulting from a sinusoidal phase shift is a superposition of multiple harmonics of the original phase modulation. Although the signal as shown in Figure 31.6 is very different from a linear phase shift, the appearance of sinusoidal phase shift algorithms is similar, involving the arctangent of a ratio of weighted sums, as in the following 12-sample algorithm [19,20]:

$$\tan(\theta) = \frac{1.2461(I_1' - I_6') - 1.5525(I_2' - I_5') - 2.5746(I_3' - I_4')}{0.2707(I_1' + I_6') - 2.6459(I_2' + I_5') + 2.3753(I_3' + I_4')} \tag{31.8}$$

where

$$I_j' = I_j + I_{13-j} \tag{31.9}$$

$$\varphi_j = 3.384\cos\left(\frac{j\pi}{6}\right) \tag{31.10}$$

for $j = 1, 2, \ldots, 6$.

The evaluation of phase $\theta = hK$ using PSI involves the inverse of an arctangent function, which has a domain of $-\pi < \theta \leq \pi$ or a height range of $2\pi/K$, which for a simple low-NA interferometer such as in Figure 31.4 corresponds to one-half the wavelength of light. Given that the vast majority of sample surfaces require a larger metrology range, if for no other reason than to accommodate part positioning, it is standard practice to augment this range by a process known as phase connecting or unwrapping. The basic premise is that at least locally there is continuity in the surface structure to within the $2\pi/K$ limit, allowing for integration of the phase changes across the surface.

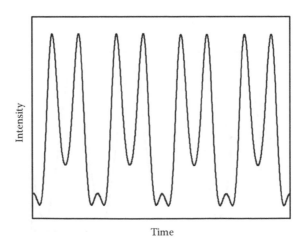

Time

FIGURE 31.6 Example intensity pattern for PSI with a sinusoidal phase shift. The shape of such patterns varies with phase and with the amplitude of the sinusoidal phase shift function.

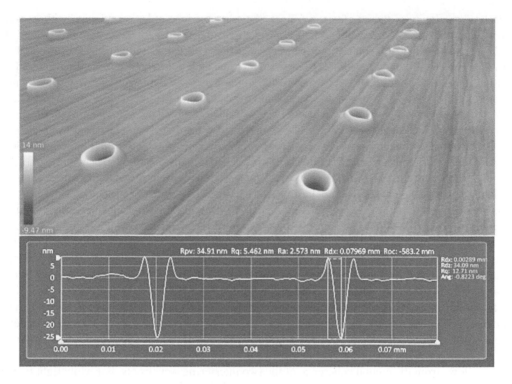

FIGURE 31.7 Example 3D PSI image of a laser textured disk.

This allows for removal of what is commonly referred to as the *fringe-order ambiguity*, where *fringe order* refers to an integer multiple of the 2π *unambiguous range* of the interference phase. One of the most common failure modes for interferometers is a mistaken fringe order, resulting in false steps or sharp features in the topography data. Techniques for unwrapping fringes robustly have been extensively studied in view of the desire to accommodate difficult surface structures and noisy interference patterns [21] (Figure 31.7).

31.2.3 COHERENCE SCANNING INTERFEROMETRY

For surface structures that have discrete features larger than the 2π limit, the issue of fringe order ambiguity becomes a fundamental limitation for many traditional interferometric methods of surface structure analysis. The canonical example is a step height feature illustrated in Figure 31.8a. Because of

(a)　630 nm, 10 nm BW　　(b)　570 nm, 90 nm BW

FIGURE 31.8 Tilt fringes on a 1.80 μm step height viewed with narrow spectral bandwidth illumination (a) and with white light illumination (b).

the lack of continuity in the interference fringes, the relative fringe order is uncertain without some approximate prior knowledge (within a half wavelength) of the step height value. For generations, the solutions have been either to employ multiple wavelengths and variations of the method of excess fractions, as in gage block interferometry [22], or to employ white light or low coherence interferometry [23]. In visual interference microscopes, prior to the era of computerized data analysis, white light fringes appear colored to the eye because of the wavelength dependence of the interference fringe periodicity. Skilled observers were able to infer step heights well beyond the 2π limit by observation of these fringes, referring to interference color charts as a guide. In black and white digital cameras, the effect reduces to one of fringe contrast; but this is still effective over a limited range, as is clear from Figure 31.8b. From the white light interferometry image, it is possible to establish visually that the step height in Figure 31.8 is about midway between 6 and 7 fringes tall, which for a 570 nm wavelength and taking into account the double-pass on reflection is 1.85 µm; not far from the calibrated value of 1.80 µm.

As interference microscopes became automated, it was natural to consider first using multiple wavelengths [24–26] and then the contrast modulation of interferometry in white light [27–29] to increase measurement range. The significance of this evolution was not fully realized until the early 1990s when it was found that computer analysis of white light fringes could be effective on surfaces that are so rough that continuous interference fringes in the traditional sense are no longer discernible anywhere in the image because of surface texture [5,30].

CSI, which encompasses white light interferometry, generally refers to any method for which the localization of interference fringes during a scan of optical path length provides a means to determine a surface topography map [6,31,32]. Figure 31.9 shows a typical CSI signal, illustrating the term *localization* by means of a calculated fringe-contrast modulation envelope. The fringe localization follows from the mutual coherence of the two interfering wavefronts. For white light fringes at low magnifications, the position of peak coherence is near the position of zero group-velocity optical path difference (OPD)—also known as the stationary phase point [33]. More generally, the fringe localization is the result of a combination of spectral and geometric factors that limit both the temporal and spatial coherence [34]. The dominant geometric effects are associated with the numerical aperture (NA) of the objective, defined in Figure 31.10.

Over a full imaging field on a smooth object, the localization of fringe contrast creates a compelling visual concept for the measurement principle. As shown in Figure 31.11, CSI relies on identifying surface heights by the mere presence of interference fringes during a mechanical scan of the interference objective shown in Figure 31.2. For those familiar with confocal microscopy, the peak fringe contrast is analogous to the position of best focus and the concept of 3D sectioning by scanning the objective. At NA values characteristic of high magnification, the position of peak fringe contrast is indeed quite close to the position of best imaging focus. This last point is a critical advantage of CSI over nonscanning methods such as PSI and multiple-wavelength interferometry—in CSI,

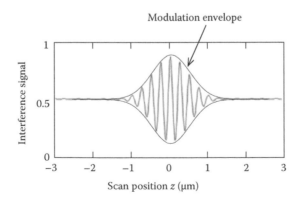

FIGURE 31.9 CSI signal for a single pixel showing the modulation envelope.

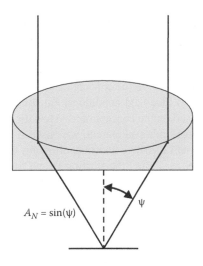

FIGURE 31.10 Definition of numerical aperture (NA), denoted A_N.

FIGURE 31.11 Images of interference fringes on a highly textured surface structure using spectrally broadband illumination. Interference phenomena localize to a common surface height, allowing for a topography measurement simply by the detection of an interference effect.

the data are always acquired at the position of best focus and best image contrast for every image point, even at high magnification and over a large range of surface heights [27].

At low NA, a serviceable conceptual model of a CSI signal modifies Equation 31.1 to read

$$I(h,\zeta) = I_{DC} + I_{AC} \exp\left[\frac{-(h-\zeta)^2}{2\sigma^2}\right]\cos[(h-\zeta)K + A] \qquad (31.11)$$

where now the magnitude of the cosine modulation is a Gaussian envelope. The constant A is the *phase gap* between the modulation peak position and the central bright fringe of the interference pattern, resulting from the combined phase shift and phase dispersion effects of the optical properties of the sample material as well as the interferometer optics themselves. The standard deviation σ of the modulation envelope is one-half the inverse of the standard deviation of the optical spectrum, assuming again for the sake of this conceptual model that the light source has a perfectly Gaussian spectral distribution with wavenumber. Although the concept of a slowly modulated carrier signal is only an approximation corresponding to idealized conditions, Equation 31.11 is a useful picture for many signal processing strategies based on the concept of a systematic search for the peak fringe contrast.

As a starting point for algorithm development, we observe that any PSI algorithm can be rearranged to provide a measurement of fringe contrast, allowing for the calculation of the modulation envelope shown in Figure 31.9 by sliding the starting frame for the algorithm along the scan position for the recorded signal [35,36]. This sliding-window PSI method consists of taking the sum of the squares of the numerator and denominator of the PSI algorithm, which are proportional to the sine and the cosine of the interference phase, respectively, so as to calculate the square of the fringe contrast or modulation strength M. For the 3-frame algorithm of Equation 31.7, we then have

$$M^2 = 3(I_2 - I_3)^2 + (I_2 + I_3 - 2I_1)^2 \tag{31.12}$$

The value of M determined from repeated application of this formula traces out the modulation envelope. The next step is to locate the envelope peak, which we identify as the scan position ζ for which $\zeta = h$ and consequently we have a measure of local surface height. Clearly we can expect improved results with PSI algorithms that employ more data frames, ultimately leading to algorithms that are nearly as long as broad as the CSI signals themselves [37]. Alternative approaches include hardware [38] or software demodulation [39] specifically oriented toward rapid evaluation of signal strength.

In part because the coherence-based method is based on an evaluation of the modulation envelope, which varies slowly with scan position, it has inherently greater height-equivalent noise than conventional PSI, which leverages the rapidly changing interference fringes themselves. To overcome this limitation, we can calculate the interference phase $\theta = hK$ at the envelope peak by interpolation of the evolving phase values over the range of scan positions ζ. This technique provides two measurements of the same height h: one that leverages the high precision of phase measurement, and the other, based on fringe contrast, that is free of any height range limitations related to the 2π problem. Indeed, for most practical configurations, the phase measurement is an order of magnitude lower in noise on smooth surfaces. So it is natural to consider combining the two measurements into a single, final topography map having the benefits of both approaches [36,40].

In practice, combining phase and coherence data for higher precision is not as straightforward as one might think. The first issue encountered, which is actually quite an interesting one, is that the fundamental metric of measurement is different for the two methods. On the one hand, the phase data scales to height according to the inverse of the fringe frequency K, which we usually calculate based on our knowledge of the wavelength and optical geometry. The coherence-based topography, on the other hand, is calculated based on the scan position ζ, which has nothing whatsoever to do with the wavelength. The two calculations are sufficiently different that many early CSI instruments separated the two methods into two distinct measurement sequences, the first using broadband light and the fringe contrast method, and the second using narrowband light and traditional PSI [30,41].

A way out of this dilemma is to link both the coherence and phase calculations together to the same length scale by using a common data reduction method for both. In the frequency domain analysis (FDA) method commercialized in the early 1990s, this is achieved by setting aside the calculation of the fringe contrast envelope and working entirely in the frequency domain by means of a discrete Fourier transform of the CSI signal [33,42]. The result is a series of Fourier coefficients of phases ϕ_v associated with fringe frequencies K_v scaled according to the known or calibrated scan step between camera data acquisitions. These Fourier phases are

$$\phi_v = hK_v \tag{31.13}$$

and the coherence-based estimation h_C of surface height h is now given by the phase slope

$$h_C = \frac{dK}{d\phi} \tag{31.14}$$

estimated from a least-squares linear fit to those discrete values ϕ_ν, K_ν for which the Fourier magnitude is the strongest (see Figure 31.12). This same linear fit provides a mean phase value at a chosen center frequency K_0

$$\theta = hK_0 \tag{31.15}$$

equivalent to a PSI measurement using all of the data frames that have nonzero signal modulation. To calculate the final height value, we apply

$$h = \frac{\theta}{K_0} + \frac{2\pi}{K_0} \text{ round}\left(\frac{A - \langle A \rangle}{2\pi}\right) \tag{31.16}$$

where

$$A = \theta - K_0 h_C \tag{31.17}$$

In Equation 31.16, the *round* function returns the nearest integer to its argument, and the brackets $\langle \, \rangle$ represent a field average [33]. The shape of the source spectrum, the mean wavelength, and the specifics of the optical geometry are no longer needed for scaling surface heights.

As is the case with PSI, CSI has enjoyed a great deal of attention at the algorithm level and there are many data processing alternatives to the example PSI and FDA methods described here.

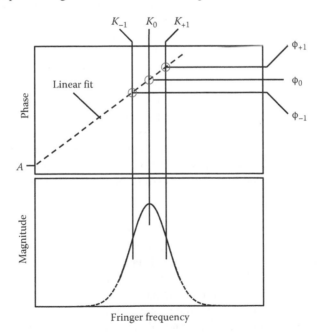

FIGURE 31.12 Illustration of the frequency-domain analysis method. A digital Fourier transform of the CSI signal (Figure 31.9) yields a magnitude and phase for each Fourier coefficient as a function of fringe frequency. A linear fit to the unwrapped phase data weighted by magnitude provides the surface topography information using the scan speed to scale the surface heights.

Somewhat confusingly, these developments have led to a pantheon of pseudonyms for CSI according to the specifics of the data reduction. ISO 25178-604 list 16 alternative names for CSI, ranging from high-definition vertical scanning interferometry (HDVSI) to scanning white light interferometry (SWLI) [32]. The most recent entry in the nomenclature library is full-field time-domain optical coherence tomography (OCT), a method derived from single-point OCT systems but now using a camera in a manner that is in many ways parallels traditional CSI.

31.3 INSTRUMENT DESIGN FOR INTERFERENCE MICROSCOPY

31.3.1 LIGHT SOURCE AND DETECTION

The illumination for interference microscopes may be incandescent, such as a halogen lamp, a discharge lamp, or more commonly in modern instruments, an LED or other solid-state device. Some but not all instruments have light stops for adjusting the size of the illumination field as well as the illumination aperture in a Köhler geometry, which images the light source into the pupil of the objective.

Microscopes for quantitative microscopy employ a two-dimensional array camera for electronic image capture and computer processing. Areal measuring systems may use conventional cameras of the CCD or CMOS type, with a format ranging from 640 × 480 pixels to over 4 million pixels [43]. In addition to field size, the data acquisition speed in camera frames per second, the linearity, and quantum well size are of primary importance. Additional capabilities of interest include pixel binning for lower noise or faster acquisition speeds.

31.3.2 OPTICAL SYSTEM

Modern microscopes are dominantly of the infinite conjugate type, with telecentric imaging, and magnification determined by the combined action of the objective and the tube lens [44]. The net magnification figure relates to the relative size of an image feature on the camera to the actual or object feature, and is the product of the objective magnification and the magnification of the imaging optics, often referred to as either the tube lens or the optical zoom lens. Both the objective and the imaging lens magnifications are given by the ratio of the focal length to a common fixed value: conventional imaging objectives from Meiji, Leica, and Nikon all use 200 mm, while Olympus uses 180 mm, and Zeiss 165 mm. Interference objectives often use these lenses and will carry these or system-corrected magnifications as engraved indications. The physical size of the camera chip determines the FOV corresponding to a given magnification. For example, if the product of the objective and tube lens magnifications is 10×, then the FOV will be 10× smaller than the physical size of the camera-sensing area.

The Mirau interference objective in Figure 31.7 has a partially transmitting beamsplitter and a small reflecting reference surface far outside the focal plane of the objective. This interferometer is specifically designed to work with diffuse illumination that fills the pupil of the microscope objective: the illumination and imaging light paths go around the reference mirror, which is sometimes referred to as a central obscuration.

Modern instruments allow for interchanging objectives of various magnifications and interferometer type. The Mirau objective of Figure 31.1 is common for magnifications of 10× and above, while at lower magnifications, the Michelson type shown in Figure 31.13 is preferred. Some high-magnification systems that demand larger working distance than the Mirau are served by the Linnik objective [45], similar to the Michelson but constructed from two matched microscope objectives, one for the reference and one for the measurement path. Figure 31.14 illustrates a collection of modern interference objectives operating over a range of magnifications.

As shown in Figure 31.1, adjusting the position of the objective vertically brings the sample into focus. In such a configuration, the focus adjustment is synchronous with an adjustment of the OPD between the measurement and reference paths. The objective itself must be engineered or adjusted so that the OPD is zero and the measurement and reference paths are equal at the position of best focus. At high magnification, this may require making use of a fine-adjustment ring on the outside

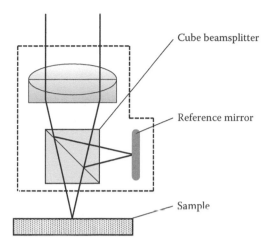

FIGURE 31.13 Michelson-type interference objective.

FIGURE 31.14 Modern commercial interference objectives. (a) Michelson objectives, (b) Mirau objectives, and (c) Turreted objectives [46].

of the objective. High-magnification objectives often automatically compensate for thermal effects to obviate the need for frequent manual adjustment.

Instrument software typically offers fine focus as part automated procedure, although it is still very common to require the user to perform some sort of initial setup involving both focus and tip and tilt adjustment of the sample. There are many alternatives to the focus mechanism of Figure 31.1. Often there is a coarse adjustment mechanism, sometimes referred to as the z stage, which moves the entire microscope assembly up or down; while fine focus is provided by a PZT stage that may also provide the mechanical scanning that is integral to most automated data acquisition and analysis methods. A less common method involves independent actuation of the reference mirror, decoupled from the focus, so that optical path length adjustment becomes independent of focus position [47,48].

An additional consideration in interferometer design for spectrally broadband illumination is the need to balance the amount of dispersive material in the reference and measurement paths. In the Michelson objective of Figure 31.13, this is achieved by using a cube beamsplitter in place of a plate, which would require a compensating element. The cube is fabricated from glass components that are nominally of the same shape and thickness. Similar considerations drive the design of Mirau objectives, which are usually manufactured with the beamsplitter and the reference mirror cut from the same piece of glass to match their thicknesses. Alternatively, the beamsplitter may

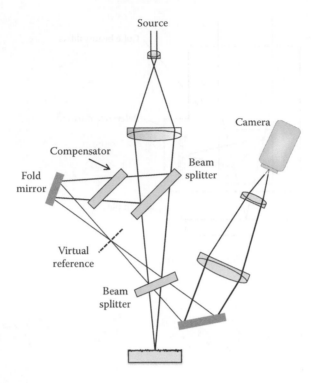

FIGURE 31.15 Interferometer without a physical reference surface in focus.

be made of a sandwich of two plates of equal thickness [49]. If the sample is inside a chamber with a transparent window, specialized objectives are available with adjustable compensation in the reference path [50–52].

For all of the interference objectives described so far, the reference mirror is always kept at the focal plane of the objective, while the sample is scanned along the axis orthogonal to the part surface. Proper setup of the instrument requires that the focus for the objective be coincident with the zero OPD position. High-magnification objectives (50×, 100×) are often designed to compensate for shifts in focus position caused by temperature changes.

The sharp focus on the reference mirror means that the measurement is in reality a comparison of the sample with the reference surface, including the high-spatial frequency roughness of the reference. High-precision measurements usually involve calibration of the residual flatness of the form of reference surface [53,54], but the fine-scale roughness is difficult to compensate and can become a source of uncertainty when measuring super-smooth samples. One solution is to arrange the optics so that there is actually no physical surface in the reference arm that is in focus, an example of which is illustrated in Figure 31.15 [55,56].

31.3.3 SYSTEM CONFIGURATIONS

Complete interferometric microscope systems for surface characterization have a wide variety of configurations depending on the end application. Perhaps the most common is a general purpose or *benchtop* system, as shown in Figures 31.16 and 31.17. These systems accommodate a wide variety of part types and shapes, and are common in academic, R&D, and general purpose applications. Benchtop configurations include a computer, display, joystick control, and other essential peripherals, often integrated into a custom platform as in Figure 31.18.

For targeted applications in precision manufacturing, where the interferometer must operate close to machinery, an industrial enclosure as shown in Figure 31.19 is appropriate. The enclosure

FIGURE 31.16 Bench-top interference microscope with a single objective [57].

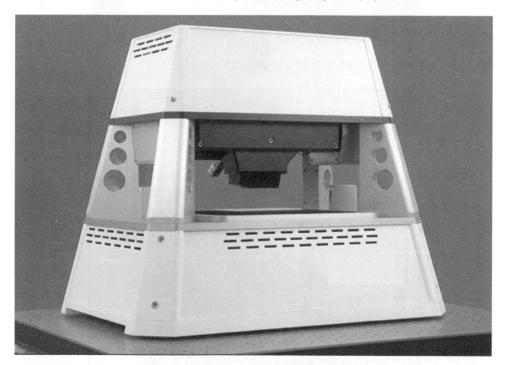

FIGURE 31.17 Bench-top configuration designed for high stability [43].

protects the instrument from oils and debris and provides additional acoustic and environmental isolation to improve the metrology. For many such applications, instrument users and providers customize the configuration for specific routine tasks requiring specialized fixtures and staging [58]. The computer interface reduces to a simple pass–fail reporting system, with results reported to an external database that tracks production quality and yield. Semiconductor wafer metrology requires the most demanding and complete automation, with the enclosure and part handling often comprising the largest part of the system cost (Figure 31.20).

FIGURE 31.18 Integrated platform for general-purpose applications.

FIGURE 31.19 Industrial interferometry for precision manufacturing.

FIGURE 31.20 Fully automated semiconductor wafer metrology using interference microscopy.

- No dominant angle
- Random look to finish
- ISO Str*= 1.0

- Single dominant angle
- Horizontal finish
- ISO Str = 0.0

- Two dominant angles
- Cross hatch finish
- ISO Str = 0.5

* "ISO Str" measures the texture aspect ratio

FIGURE 31.21 Examples of surface finish having texture characteristics that are evident in the surface topography maps generated by interference microscopy.

31.3.4 POSTPROCESSING

One of the strengths of any computerized surface analysis tool is the ability to extract surface details and texture parameters digitally. Common postprocessing steps include the application of filters to highlight particular spatial frequency ranges [59], advanced texture analysis as in Figure 31.21, and extraction of specific form and roughness parameters [60–62]. Systems capable of transparent films analysis also provide 3D maps of thickness and refractive index homogeneity.

31.4 SIGNAL MODELING

31.4.1 MODELING OF INTERFERENCE MICROSCOPY FRINGE PATTERNS

The conceptual model of a Michelson interferometer in Figure 31.4 and the accompanying signal models are sufficient for basic ideas, but in reality, the interference signal in a microscope configuration is much more complex and interesting. The spatial extent of the incoherent illumination in the pupil of Michelson and Mirau objectives, the broadband optical spectrum, and the surface structure contribute to shaping the interference signals. These realities can complicate metrology in some cases,

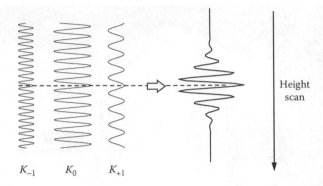

K_{-1} K_0 K_{+1}

FIGURE 31.22 Conceptual diagram of the incoherent superposition method of synthesizing interference signals for interference microscopy.

while at other times they make it possible to extract additional information beyond surface topography, including material composition, transparent film layers, and optically unresolved surface structure.

A common physical model of the interference process assumes incoherent superposition of ray bundles of differing wavelengths [63]. The signal is the incoherent sum of the interference contributions of the ray bundles passing through the pupil plane of the objective and reflecting from the object and reference surfaces [27,64,65]. Figure 31.22 provides a qualitative understanding of the signal generation process using the concept of incoherent superposition. The summation over a range of frequencies results in a peak signal at a specific scan position where all of the individual interference contributions are mutually in phase, sometimes referred to as the *stationary phase point*. This is the peak of the modulation envelope shown in Figure 31.9. This concept maps to the frequency-domain analysis method of Figure 31.12, where now each complex Fourier magnitude and phase in Figure 31.12 corresponds to one of the interference contributions of Figure 31.22. The integrations may be performed numerically as part of a mathematical model, as detailed in Ref. [34]. Computational methods such as these allow for prediction of instrument response with variations in optical geometry and surface structure, while allowing for new measurement modes based on model-based analysis [66,67].

31.4.2 THE OBLIQUITY EFFECT

One of the uses for signal modeling is the determination of the net fringe frequency of the interference signal for the case where there is a broadband spectrum or a range of incident angles in the illumination. The effect of incident angles is of particular importance to PSI, which relies on the fringe frequency K as the fundamental metric for scaling heights. Traditionally, the effects of the source geometry contribute to an *effective wavelength* or *equivalent wavelength* starting from the actual mean wavelength of the light source and using the *obliquity factor* as a multiplicative correction.

The obliquity factor has been extensively analyzed, and there are a range of approximation formulas for its calculation based uniquely on geometric factors and assumptions regarding the pupil plane light distribution [64,68–73]. Signal modeling provides an alternative calculation of the effective wavelength that incorporates the optical spectrum, the system geometry, the shape of the light source [73], and potentially the special characteristics of the object material. The procedure involves direct numerical calculation of the Fourier components followed by centroiding the resulting Fourier spectrum, which provides an accurate determination of the mean wavelength of the final interference signal. Results in Table 31.1 show that the obliquity effect is easily detectable with common low-magnification objectives and is of critical importance at high magnifications, where the obliquity effect can change the equivalent wavelength by 30%.

TABLE 31.1

Obliquity Factor Calculated by Modeling of the Interference Signal, Assuming a Fully Filled Objective Pupil, a Dielectric Sample, and Circular Polarization

Magnification	Type	A_N	Obscuration	Obliquity Factor
2.5×	Michelson	0.075	0	1.001415
5×	Michelson	0.13	0	1.004256
10×	Mirau	0.3	0.09	1.025546
20×	Mirau	0.4	0.07	1.044264
50×	Mirau	0.55	0.04	1.087317
100×	Mirau	0.85	0.27	1.301778

31.4.3 TRANSPARENT FILMS

Many high-value devices comprise thin films, some of which may be partially transparent in the wavelength range of an interference microscope. Consequently, a complete surface characterization may entail mapping both the film thickness and the top-surface height variation. At a minimum, we require surface topography–measuring instruments to be compatible with the presence of thin transparent films.

For a PSI instrument in near-monochromatic illumination at low NA, a transparent film on the sample surface presents something of a challenge: the light reflects partially from both the top surface and the underlying substrate, or in the case of multilayer films, from several places within the semi-transparent surface structure. The measurement principle as described so far does not account for this. The potential errors are not intuitive or easy to interpret—the measured surface topography is not the average of the top and underlying surfaces, but rather is some nonlinear function of the film thicknesses and refractive indices. Corrections are feasible if the film thickness and index are well known, but these *a priori* data are not always available (Figure 31.23).

From the earliest implementations of CSI, it was recognized that a unique benefit of a low-coherence illumination is the ability to distinguish between the multiple surface reflections of a

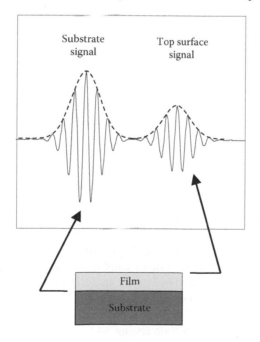

FIGURE 31.23 Multiple CSI signals from a surface film structure.

FIGURE 31.24 Method for locating top-surface reflections using a truncated signal model and a least-squares fit [77].

transparent film, provided that the film thickness is greater than the width of the modulation envelope [74–76]. For these so-called "thick" films, the multiple reflections from the top surface, substrate, and possibly intermediate layers result in readily distinguishable signals. The location of these signals in the scan history identifies the optical thickness of the film, at least for low-NA applications where we can neglect the competing effect of spatial coherence. Knowledge of the film's refractive index allows for simultaneous topography and film thickness mapping.

The same concept of signal separation extends with advanced processing to films that are too thin for full signal separation. Figure 31.24 illustrates a method for isolating the top-surface contribution to a CSI signal even when the substrate and top-surface signals are partially merged. The technique consists of a least-squares fit of a signal model that contains only the leading-edge portion of the signal. This method allows for top-surface profiling in the presence of transparent films that are too thin for complete separation of the CSI signals, but still thick enough to identify portions of the CSI signal that are dominated by specific surface reflections [77]. Figure 31.25 compares a 3D top-surface interferometry image with atomic force microscopy (AFM) for a flat-panel display pixel. This sample has transparent films in several areas, with a minimum thickness of 600 nm. A nice feature of this approach is that it does not require any advance knowledge of the material properties of film or embedded features.

31.4.4 MODEL-BASED ANALYSIS

Provided that we know the optical properties of our instrument well enough, an effective approach to analyzing thin transparent films is to model the expected response of the instrument to a range of possible film parameters, creating a *library* of model signals that we compare in their entirety with the experimental signal to find the best match (Figure 31.26). A model-based approach has the advantage of simultaneously providing a direct 3D measurement of film thickness as well as the upper and lower interface profiles. Most systems employ frequency-domain methods, wherein the experimental CSI signals are transformed by Fourier analysis and the complex coefficients are compared to stored library signals to evaluate surface structure [33,78–81]. Although the library search technique requires detailed system and surface structure modeling, this approach has found applications for specialized high-value production testing, as in the TFT LCD example of Figure 31.27. Film thickness measurement is possible down to 10 nm using this technique.

(a)

(b)

FIGURE 31.25 Comparison of CSI with AFM measurements of the top surface of a thin film transistor (TFT) of a liquid-crystal display (LCD). The images are 40 μm². (a) Coherence scanning interferometry and (b) atomic force microscopy [77].

FIGURE 31.26 Model based films analysis. Experimental CSI signals are matched to precomputed theoretical signals that depend on instrument calibration data and a model of the sample under test, in this example a layer of silicon dioxide on silicon [66].

FIGURE 31.27 Map of photoresist thickness near a transistor gate on a TFT LCD panel using model-based analysis. The horseshoe-shaped trench is nominally 400 nm thick and 5 μm wide [66].

The same basic concepts of a model-based library search enable analysis of surface structures smaller than the lateral resolution limits of conventional microscopy. For these techniques, the mathematical approach is necessarily more sophisticated than the simple incoherent superposition model, which implicitly assumes only a nominally flat structure complicated at most by uniform layers of transparent films. The presence of sub-wavelength features requires modeling methods such as rigorous coupled wave analysis (RCWA). These advanced techniques have found application in the semiconductor industry, in some cases using specialized optical geometries where the objective pupil is imaged onto the camera in place of the object surface itself [67,82,83].

31.5 PERFORMANCE

31.5.1 GOOD PRACTICE AND INSTRUMENT EVALUATION

Interference microscopes measure a wide variety of surface textures, and real-world performance depends on influence factors such as surface texture, surface slope, thermal drifts, and form distortions related to the interaction of the surface structure with the illumination. Additional factors include measurement procedure and relevant calibrations and adjustments. Consequently, meaningful statements of instrument performance are closely tied to the specifics of the measurement task [84]. Quality control experts have understood this for some time, which is the reason for empirical tests such as the classic test for gage repeatability and reproducibility (GR&R) performed on the part of interest close to the actual conditions of measurement. These tests combine with correlation studies that relate the functional properties of the part to the detected surface structure.

Instrument providers are the primary resource regarding how to get the best performance from interference microscopes for specific tasks, supplemented by good practice guides [85,86], international standards documents [32], and a growing body of literature. A general caution is that published instrument specifications can be misleading and often use nonstandard terminology. Statements related to performance should include a clear procedure for validation, either as a footnote or reference to a national or international standards document that specifies these procedures.

31.5.2 SURFACE TOPOGRAPHY REPEATABILITY

Perhaps the most basic performance characteristic of any metrology task is the *measurement noise*. As with any metrological characteristic, the measurement noise has several influence factors, some of which may be controllable, such as environmental vibration, air turbulence, acoustic noise, and thermal instability. The *instrument noise* is usually specified and evaluated according to an example measurement under ideal conditions—a quiet lab, a part chosen to provide good results, such as a polished optical flat with just the right reflectivity for optimal results. This is a close approximation to a test of the intrinsic additive noise that cannot be reduced by control of external influence factors [87]. It is common to see terms such as *precision* as surrogate specifications for noise, consistent with established vocabulary (VIM Section 2.15 [88]).

A quantitative evaluation of noise consists of evaluating the measurement consistency with repeated trials, without changing anything. For 3D interference microscopes, the most fundamental output is an aerial surface topography map, which shows the height over an array of perhaps a million different surface points corresponding to camera pixels. The *surface topography repeatability* tells us how close we can expect the indicated height value for a specific sample point (or camera pixel) will repeat if we measure it over and over again [32,87]. The value usually is expressed as a root-mean-square (RMS) value or standard deviation in units of height, and is readily computed from statistics over a full image of surface points.

The simplest repeatability test is to measure a suitable flatness artifact twice, subtract the difference of the two resulting topography maps, and calculate the standard deviation S_q of the *difference map* [53,89]. The surface topography repeatability, denoted here as the ISO measurement noise N_M, is then

$$N_M = \frac{S_q}{\sqrt{2}}. \tag{31.18}$$

For situations where intermittent disturbances such as vibrations may influence the repeatability, or where a more authoritative value is essential, a more reliable statistical approach is to take a large number of repeated measurements of the artifact. Several methods treat the statistics of these multiple measurements to provide a robust estimate of repeatability [90,91].

For interference microscopes, we can often estimate instrument noise based on first principles, at least for the major noise contributors. For most automated interference microscopes in a calm environment, this is random shot noise from the camera. For PSI, the resulting phase noise depends to some degree on the specific PSI algorithm coefficients, but as a general rule, scales inversely with the square root of the number of camera frames used to calculate the phase, as one would expect from random fluctuations [92,93]. For a single data acquisition using the 12-frame sinusoidal phase shifting algorithm of Equation 31.8, the theoretically predicted noise level for a 600 nm wavelength without any lateral filtering is

$$N_M = 28 \text{ nm } \sigma \tag{31.19}$$

where σ is the signal to noise ratio, defined here as the standard deviation of the random intensity noise divided by the amplitude of the interference signal [20]. A typical intensity noise σ is 1.5%, which, after 25 averages provides a surface topography repeatability of 0.1 nm, independent of magnification. This is a common specification value for PSIs, consistent with the qualitative "sub-nanometer" characterization of the technique.

For noise contributions that are entirely uncorrelated between image pixels, lateral smoothing or S-filter as described in ISO 25178-3 can considerably improve the surface topography repeatability value [59,60]. Surface filtering is a legitimate postprocessing step in areal surface topography and may even be mandated for correlation to other measurement techniques that may have intrinsically different spatial frequency response [94]. For a formal specification, however, it is essential to clearly state any lateral filtering operations as these may have an influence on other specifications such as lateral sampling and resolving power.

Surface topography repeatability improves with averaging or by using a PSI algorithm that requires more camera frames, following the observation that random noise declines as the square root of the number of sample values. Increased averaging comes at the cost of data acquisition time; therefore, a specification for the surface topography repeatability should include the corresponding time to measure. This is consistent with common engineering practice, which expresses random noise as a value divided by $\sqrt{\text{Hz}}$. For example, a stated surface topography repeatability number for a phase-shifting type interference microscope might be 0.1 nm. This number may require 25 averages of individual measurements, each one of which requires 12 camera frames, for a total of 300 frames. With a 75 Hz camera, the fastest possible measurement using continuous data acquisition would be 4 s, or $0.2 \text{ nm}/\sqrt{\text{Hz}}$. The 0.25 Hz frequency associated with the 0.1 nm specification allows the user to predict behavior for longer or shorter data acquisition times, following the square root rule, and to make adjustments to achieve a specific noise target consistent with throughput requirements.

For CSI, specifying instrument noise for a given time interval is complicated somewhat by the availability of multiple modes and selectable height-scanning range. For any single measurement, representative performance values without lateral filtering and with a data sampling rate of four camera frames per interference fringe are 3 nm for methods based on the modulation envelope

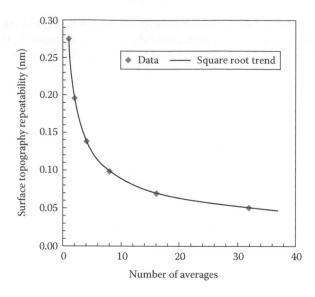

FIGURE 31.28   Noise reduction in a CSI instrument using measurement averaging [91].

and 0.4 nm for measurements based on a phase evaluation to improve precision [33]. The phase performance is consistent with the example calculation following Equation 3.19, and the larger CSI value is consistent with the differential nature of this measurement. The acquisition time for a single measurement scales inversely with the scan rate, which for a 100 Hz camera and a 600 nm wavelength would be 7.5 μm/s. These values together with the desired height range allow for an estimate of the surface topography repeatability as a function of total data acquisition time. As is the case with any measurement dominated by random noise, the CSI result can be further reduced with averaging of final height maps, as shown by the experimental data of Figure 31.28.

31.5.3 Vibration and Environmental Effects

Not usually found on an instrument specification sheet are several important error sources that can significantly influence the measurement performance and that potentially far outweigh the basic surface topography repeatability specification in the evaluation of measurement uncertainty. One of these contributors is vibration. Instruments based on interferometry are almost invariably equipped with systems to dampen or isolate unexpected mechanical motions to overcome the vibration sensitivity of time-sequenced phase shifting [95,96]. The residual errors depend on the vibrational frequency [6,32,95,97], with low-frequency vibrations associated with form errors and high-frequency vibrations at specific frequencies contributing to ripple errors at twice the spatial frequency of the interference fringes. Figure 31.29 illustrates these effects using experimental CSI data at 20× magnification for a flat mirror. Figure 31.29a shows the orientation of the interference fringes. Without vibration, the measured surface roughness shown in the upper-right topography image is 1 nm rms. The errors caused by environmental vibrations shown in the lower-left and lower-right images for the same surface are 4 nm rms for low- and high-frequency vibrations, respectively, although the effect on the measured surface form is clearly different. Even though the direction of vibration is dominantly along the scanning or z direction, the resulting errors are phase dependent and therefore follow the orientation of the fringes in the intensity image. This is particularly evident in the high-frequency vibration ripple errors shown in Figure 31.29d. The magnitude of the error for the same vibrational amplitude varies significantly with vibrational frequency, with peak sensitivity for ripple errors at one-half the camera frame rate when the camera sampling rate is four frames per interference fringe.

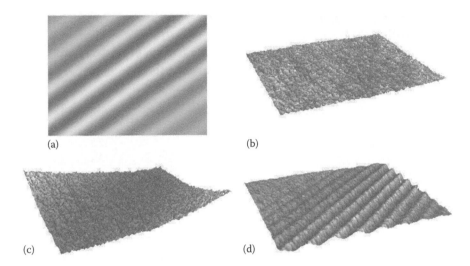

(a)   (b)

(c)   (d)

FIGURE 31.29 Experimental demonstration of potential errors caused by vibration in a traditional CSI instrument. Newer instrument employ technologies to reduce these effects and make it easier to use interference microscopy in production environments. (a) Intensity image, (b) 3D topography, (c) low frequency vibrations, and (d) high frequency vibrations [98,99].

Knowledge of vibrational frequency spectrum for a given environment can lead to strategies for avoiding these frequencies and improving performance. More ambitiously, several nontraditional approaches to PSI data analysis do not require precision in the phase shifts and are more tolerant of vibrations and other unexpected disturbances [100–102], Vibration-tolerant software relies heavily on the ever-increasing data processing power using statistical or maximum likelihood analyses. These techniques, when combined with additional hardware and software mitigation strategies, provide improvements over conventional interferometry of more than a factor of 100. Vibration robustness allows optical instruments to function in environments where conventional interferometers and precision stylus tools would fail, helping to bring interferometers from the quality control laboratory closer to the production floor [98,103].

Further to the goal of reducing vibration sensitivity, it has been proposed to use multiple detectors to acquire phase-shifted images simultaneously in an interference microscope using carrier fringes as in holography [104], or polarization encoding of the reference and measurement beams [105,106]. Such methods duplicate PSI functionality without time-dependent phase shift, providing near-instantaneous 3D surface topography measurements. Such methods allow for video-rate imaging of moving objects such as live cells [107], but have some limitations related to the need for polarized optics.

31.5.4 Optical Effect of Surface Materials

The optical properties of the materials that make up the object surface influence the shape and position of the interference signal within the data acquisition scan. As a general rule of thumb, the phase of the interference pattern depends on the phase change on reflection (PCOR) of the surface material, while the rate of change of PCOR or dispersion influences the position of the modulation envelope [32,42,108]. The shift in the envelope position is the same phenomenon as the offset experienced in multiple-wavelength interferometry when there is refractive index dispersion [109,110]. If the object surface material is uniform, there is generally no obvious way to distinguish between variations in PCOR and actual height differences.

Table 31.2 provides examples of the change in apparent surface height as determined from signal modeling, for a 560 nm light source with a 110 nm bandwidth, at low NA. In some cases, the envelope

TABLE 31.2

Effect of Material Phase Change on Reflection on Measured Surface Heights at Low NA for a Wavelength of 560 nm

Material	Envelope Shift (nm)	Phase Shift (nm)
Bare glass	0.0	0.0
Silicon	−2.4	−0.2
Aluminum	−9.5	−12.5
Chrome	−15.5	−13.4
Platinum	−11.0	−17.9
Copper	−1.1	−30.4
Cobalt	−11.0	−18.0
Silver	−25.9	−25.1

shift can be small while the phase shift is large, as is the case for example with copper. The amount of shift can vary significantly depending on the wavelength and the bandwidth, particularly for materials such as gold that have absorptions and nonlinear dispersion in the visible wavelength range. For other materials such as silver, the envelope shift is nearly identical to the phase shift. The amount of phase shift is only weakly dependent on the spectral bandwidth for most materials, but the envelope shift can be a strong function of both bandwidth and NA. In the limit of very high NA (above 0.6), the envelope position is dominated by focus effects that are less sensitive to material properties. This last observation has been proposed as a means of detecting and quantifying the PCOR [71].

Some software packages allow for calculation of the PCOR using a known refractive index and a simple analytical formula [31], although these calculations are valid only for low NA and are of no value in determining the envelope shift. The only accurate way of calculating dissimilar materials effects is through numerical modeling that includes the full geometry of measurement and accurate data regarding the wavelength-dependent index of refraction. Alternatively, it is often expedient to evaluate the effects of PCOR empirically, using known structures [111]. This experimental determination may involve metrology methods such as stylus profilometry that are less sensitive to PCOR and transparent film effects [112].

31.5.5 LATERAL RESOLUTION

The resolving power or lateral resolution of a 3D metrology instrument quantifies its ability to separate adjacent features on the surface. As an example, Figure 31.30 shows on the right two closely spaced trenches formed by patterning silicon on a quartz substrate [113]. The interference microscopy 3D image on the left-hand side of the figure shows that there are indeed two lines present, although they appear blurred at this high magnification. As the center-to-center separation between the lines decreases, the resolving power of the instrument becomes a limiting factor in determining if the two lines are clearly separated in the 3D image. If the resolving power is insufficient, the two lines will simply appear as one large line.

In conventional noninterferometric imaging systems with incoherent illumination, it is common for a single number to define the resolving power. According to the Sparrow criterion borrowed from optical imaging, as features get closer together, the lateral resolution is the center-to-center separation between two surface features at which the instrument is no longer able to distinguish these features as separate images. For a diffraction-limited microscope with an incoherent light source, the Sparrow limit of a conventional imaging microscope for two source points is

$$R_l = \frac{0.47\lambda}{A_N} \tag{31.20}$$

FIGURE 31.30   (a) 3D CSI microscope image of two parallel trenches of a 200 nm linewidth standard from Supracon AG, using a 100× Mirau objective with an NA of 0.85. (b) Corresponding scanning electron microscope image, showing the linewidth and the sample structure. The center-to-center separation of the lines is 440 nm [114].

where R_l is the center-to-center spacing of the features, and as previously defined previously in Figure 31.10, A_N is the NA. To take advantage of this diffraction-limited resolution, the camera pixel size in object space must be much smaller than R_l, otherwise the resolution is said to be *camera limited*.

Turning now to areal surface topography measurement using optical methods, it is not obvious that the traditional imaging definitions such as the Sparrow criterion should apply [114]. The principle of interferometric surface metrology, for example, relies on a simple relationship between phase and surface height, as in Equation 31.3. When we add together optical wavefront contributions from closely spaced surface features, the phase shifts caused by the various surface heights do not directly add together. Instead, it is the complex wavefront amplitudes that are summed. When the feature separations are close to the traditional resolving limit of the optical imaging system, we cannot assume a linear response.

The concept of imaging resolution as applied to 3D optical instruments is recovered if we restrict ourselves to small feature heights, defined as heights for which the instrument response is linear. The definition of "small" follows from approximations similar to the small-angle limit of the sine function and translates to a few tens of nanometers for common visible-wavelength interference microscopes [116]. This limitation does not restrict the overall useable height measurement range of the instrument, which may be several tens or even hundreds of microns; rather it applies to sharp edges and features that are closely spaced on the scale of the lateral resolution of the instrument. If we respect this limitation of small heights for closely spaced surface features, we can leverage traditional and familiar 2D imaging definitions to describe the 3D topographical measurement properties. Indeed, it can be shown that in the limit of small heights, the well-known modulation transfer function (MTF) has a meaningful analog in the instrument transfer function (ITF) for 3D metrology [115]. Figure 31.31 illustrates this correspondence for a CSI instrument for a 20× Mirau objective, a filled 0.4 NA pupil, 570 nm wavelength. The spatial frequency for which the MTF reaches zero is the Abbe limit, which is close to the Sparrow limit. The close similarity between MTF and ITF breaks down for highly aberrated or defocused optical systems, for which 3D topography may be distorted in a way that has no direct analog in conventional optical imaging.

It is evident that paths to higher resolving power lie with smaller wavelengths λ, which may be difficult to realize in practice, or with increased NA, which is more commonly how instruments

FIGURE 31.31 Comparison of the theoretical optical MTF and the measured 3D topography ITF for a CSI instrument at 20×. The Sparrow criterion limit is 670 nm or 1490 cycles/mm [115].

TABLE 31.3

Specifications of Selected Modern Commercial Interferometer Objectives, Including the Lateral Resolution as Calculated according to the Sparrow Criterion of Equation 31.20 and Pixel Size for a 2× Imaging Lens

Magnification	Type	NA (A_N)	Lateral Resolution (μm)	Lateral Sampling (μm)
1×	Michelson	0.03	8.8	4.1
2.75×	Michelson	0.08	3.3	1.5
5×	Michelson	0.13	2.0	0.8
10×	Michelson	0.28	0.94	0.41
10×	Mirau	0.30	0.88	0.41
20×	Mirau	0.40	0.66	0.21
50×	Mirau	0.55	0.48	0.08
100×	Mirau	0.85	0.31	0.04

Source: Zygo Corporation, NexView Optical Profiler, Specification sheet, 2013.

---

achieve high-resolving powers. High-NA objectives are more common at higher magnifications, and therefore there is a sensible tradeoff between FOV and NA for available interference objectives. Table 31.3 provides a list of common objectives and their resolving powers.

   In addition to these optical effects, real instruments introduce additional influences on the ITF related to the electronic detection, conversion to 3D images, and postprocessing. The magnification and camera format define the lateral sampling, usually expressed in terms of pixel spacing in object space. Table 31.3 includes lateral sampling for a camera and imaging system providing an 8.2 μm pixel size in object space when using a 1× imaging lens. Ideally, the lateral sampling is fine enough to avoid aliasing high frequencies beyond the Nyquist frequency, a requirement that translates to a lateral sampling period that is equal to or less than half the lateral resolution. Particularly at low magnifications, the pixel size itself further attenuates the ITF with frequency according to a sinc function, with the first zero at a spatial period equal to the pixel size.

Postprocessing, including in particular lateral filtering, has an obvious and significant effect on the ITF. Instrument software invariably includes options to extract textures or surface form. ISO 25178-3 defines areal filters according to some basic rules on how to set the size of the primary surface and the sampling distance [60]. In addition to attenuating selected frequencies, given sufficient signal to noise, postprocessing can amplify portions of the ITF to sharpen details that are otherwise only marginally resolved [117]. ITF correction is also feasible for derived parameters such as the measured power spectral density (PSD) by means of multiplicative corrections to the measured results [118].

A number of methods are available for evaluating lateral resolution experimentally, using well-characterized sample structures. One approach involves an artifact similar to that shown in Figure 31.30, with a selection of line pairs to establish the distance at which the line images merge. Some linewidth artifacts [119] can serve this purpose; although it is essential to not confuse the width of individual lines with feature separation. Referring to Figure 31.30, the center Si line is 200 nm wide, much smaller than the Sparrow limit identified in Table 31.3; but the center-to-center spacing of the trenches that form this line is 440 nm, consistent with the calculated 310 nm resolution limit for 0.85 NA at visible wavelengths. Another caution is that the traditional Sparrow number as calculated from Equation 31.20 applies to source points, and line pairs are somewhat easier to resolve [120].

For ITF characterization, it is often more reliable to measure gratings or arrays of multiple lines at a calibrated spatial period rather than to examine isolated line pairs. Variations on this concept include chirped-frequency gratings [121] and the 3D version of the Siemens star [122,123]. Another approach is an edge feature, which allows for direct measurement of the edge-spread function, which by Fourier methods can be expressed as an ITF [114,116,124,125]. This method provided the experimental ITF data shown in Figure 31.31 [115].

31.5.6 Field of View and Slope Acceptance

In some applications, a larger FOV is more important than lateral resolution, leading to the use of low-magnification objectives with correspondingly reduced NA. The FOV follows from Table 31.3 by multiplying the lateral sampling by the number of pixels. For example, if the camera has a square array of 1000 × 1000 pixels, the field of view at 1× would be 8.2 mm. For these low-NA objectives, it may not be possible to capture specularly reflected light from high slopes.

A first option for large areas and steep surface slopes is to take a sequence of relatively high-magnification, small FOV images and reconstruct or *stitch* a larger image from these multiple measurements [126]. The benefit of this method is that it simultaneously provides a large FOV and detailed, high-resolution information, as illustrated in Figure 31.32.

As a second option, many modern systems have sufficient sensitivity to measure surface geometry using only the small amount of scattered light from highly inclined, smooth surfaces, as illustrated in Figure 31.33. The required amount of scattered light in these systems can be surprisingly small—a return of 0.05% of the incident light is sufficient, enabling a slope acceptance of up to 86° even at low magnifications [127].

31.5.7 Height Response

The overall uncertainty of the measured surface topography depends on the height response of the instrument (Figure 31.34). The importance of calibration and verification of the height response depends to a large degree on the principal application for the interferometer: If the metrology task is to characterize roughness or texture, an uncertainty in the overall scaling of surface heights of a few percent may not be a significant factor in the usefulness of the result. However, if the principal task is to measure discrete feature heights of several microns with an uncertainty of a nanometer, then the height scale is of primary importance.

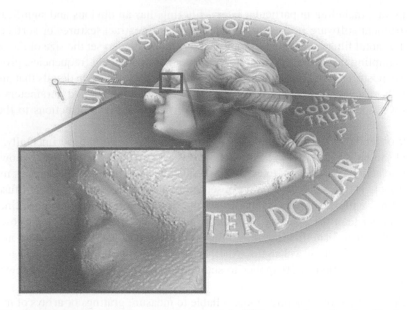

FIGURE 31.32 Large FOV image constructed from multiple CSI images of smaller size. The inset shows the level of detail provided by this stitching method.

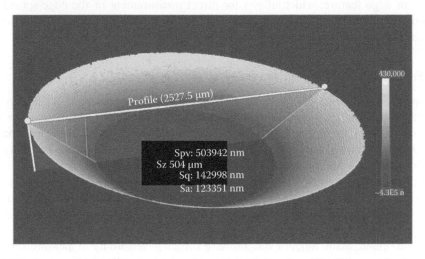

FIGURE 31.33 Image with high slopes using only scattered light at low NA. This machined-metal conical valve seal has a depth is 0.5 mm while the diameter is 2.5 mm.

Common practice is to establish the scaling factor or *amplification coefficient* for surface heights during manufacture. For a PSI instrument, for example, the illumination wavelength combined with the calculated or experimentally determined obliquity factor (Table 31.1) provides the equivalent wavelength for converting interference phase to surface height. For a CSI instrument, a capacitance or similar high-stability gage governs the scan rate. The manufacturer calibrates the capacitance gage, the associated electronics, and the resulting closed-loop mechanical response during assembly using traceable length metrology such as a laser displacement interferometer.

Standard artifacts such as step heights or coarse gratings facilitate verification of the height scale by end users and are common accessories for areal surface topography instruments [128–130]. These artifacts or *material measures* carry a certification from a national metrology institute

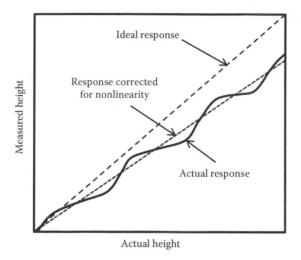

FIGURE 31.34 Height response curves, showing the actual response, the response corrected for nonlinearity, and the ideal linear response with the correct amplification coefficient.

or commercial enterprise equipped with reference metrology [131]. If the measured step height consistently falls outside the certified uncertainty boundaries of the step height standard, this may warrant adjustment of the height scale, or alternatively, the service or recalibration by the manufacturer.

The linearity of the height response of an interferometric microscope is an intrinsic property of the instrument and is more difficult to calibrate and adjust in the field than the height scale. For PSI instruments, the main source of nonlinearity is cyclic error, caused for example by an incorrect phase shift between camera frames, resulting in an error that is periodic with interference phase. For CSI instruments, nonlinearity corresponds to a variation in the OPD scan velocity as a function of scan position. Nonlinearity is usually characterized and corrected during manufacture, and the outcome is a specification such as maximum permissible error. Measurements of several step heights of different size [129], measurements of a small step at various positions in the scan, or external measurements of scan velocity using reference metrology [132] can validate the linearity specification or indicate a problem that requires service.

Another approach to traceable calibration of the height response is to embed the realization of the meter directly into the instrument design. In the evaluation of optical surfaces using laser Fizeau interferometry, for example, there is no need for factory calibration or for transfer artifacts, as the measurement principle is linked to the HeNe or other laser wavelength with a known uncertainty [133]. In some instruments, additional hardware integrates a laser displacement gage into the mechanical scanning system of a CSI microscope [132,134,135], allowing for continuous, independent calibration of the amplification factor and correction for nonlinear response.

With some notable exceptions [119], interference microscopes measure surface topography with respect to a flat datum surface, established and influenced by the combined effects of the reference surface and the optical system. The imaginary topography detected by the instrument when viewing a hypothetical perfect flat part is the *residual flatness*, which essentially is the height bias as a function of image position. Instrument makers do not usually specify the residual flatness, with the expectation that the end user will calibrate for this error source *in situ* using a prescribed procedure on a standard artifact such as a silicon carbide reference flat [54]. The result is a *system error file* that is subtracted from all subsequent measurements [136]. The procedure may require a sequence of multiple measurements with small lateral displacements of the artifact so as to average out the artifact roughness [53].

31.5.8 Profile Fidelity

So far, we have established that interference microscopes under well-controlled conditions have height-equivalent noise levels well below 1 nm for less than a second of data acquisition time and can be calibrated in height response using primary standards based on the wavelength of light, transfer indicators such as capacitance gages, or standard artifacts from NMIs. Keeping these facts in mind, it can be surprising that the lateral resolution and height response for some surface shapes and textures differs from by amounts far greater than the calculated uncertainties after calibration [137]. In some cases, researchers have observed large differences in interference microscope results depending on the data processing mode [138]. In other cases, average amplitude or Ra values measured with CSI have disagreed with stylus measurements on roughness specimens by as much as a factor of two [139].

It has been shown that anomalous results can be overcome in many cases by switching from CSI to PSI measurement modes where feasible [138] or by using CSI data analyses that make use of phase information to improve measurement results on optically smooth parts [42,140]. Figure 31.35 illustrates the improvements afforded by using phase information to reduce the noise level and eliminate measurement errors near the sharp edges of a square profile grating. Modern instruments increasingly incorporate methods to reduce or to compensate for CSI distortions for all surface textures without changing measurement modes, to simplify the instrument setup and operation. The surface profile of Figure 31.36, for example, compares a CSI profile directly with a mechanical stylus profile on a standard roughness specimen, after application of identical filter limits. Figure 31.37 shows the results on Rubert sinusoidal roughness specimen numbers 543, 542, 529, 531, 528, 530, 527, and 525 at 20× and 50× magnifications [99]. Somewhat more ambitious are promising efforts to model the origins of profile distortions and to correct them in postprocessing in a general way [141].

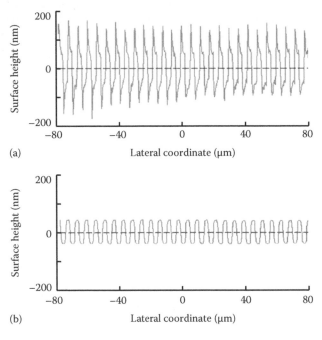

FIGURE 31.35 (a) A particularly alarming example of profile distortions in a traditional CSI microscope for a square-profile 44 nm amplitude grating, viewed at 40× magnification and at 0.4 NA. (b) Improved profile fidelity achieved by incorporating interference phase.

FIGURE 31.36 Comparison of mechanical stylus and CSI optical results at 50× for a Rubert & Co number 502 specimen [99].

FIGURE 31.37 Comparison of interferometric optical measurements of average roughness Ra for a series of standard roughness specimens. The diagonal line represents perfect correlation [99].

31.6 FUTURE PERSPECTIVES

Automated interference microscopy is an established technique for 3D surface topography, transparent films analysis, and texture characterization. Chief attributes of the technique are high sensitivity to surface topography over all magnifications, detailed surface structure analysis including transparent films, and advanced digital postprocessing. The technology continues to evolve, with particular areas of interest in semiconductor yield enhancement [142], industrial quality control on the production floor [58], and biological imaging [107,143]. Further advancements for current research include improving lateral resolution, environmental robustness, measurement speed, profile fidelity, correlation, and ease of use.

REFERENCES

1. Sommargren, G. E., Optical heterodyne profilometry, *Applied Optics* 20, 610, 1981.
2. Wyant, J. C., Koliopoulos, C. L., Bhushan, B. et al., An optical profilometer for surface characterization of magnetic media, *Transactions ASLE* 27, 101–113, 1984.
3. Koliopoulos, C. L., Interferometric optical phase measurement techniques, PhD dissertation, University of Arizona, Tucson, AZ, 1981.
4. Wyant, J. C., Koliopoulos, C. L., Bhushan, B. et al., Development of a three-dimensional noncontact digital optical profiler, *Journal of Tribology* 108, 1, 1986.
5. Dresel, T., Häusler, G., and Venzke, H., Three-dimensional sensing of rough surfaces by coherence radar, *Applied Optics* 31, 919–925, 1992.
6. de Groot, P., Coherence scanning interferometry, in *Optical Measurement of Surface Topography*, R. Leach (ed.), Chapter 9, pp. 187–208 (Springer Verlag, Berlin, Germany, 2011).
7. Hariharan, P., *Optical Interferometry* (Academic Press, San Diego, CA, 2003).
8. Page, D. A., Interferometry, in *Handbook of Optical Metrology*, T. Yoshizawa (ed.), Chapter 6, pp. 191–218 (CRC Press, Boca Raton, FL, 2009).
9. Pedrini, G., Holography, in *Handbook of Optical Metrology*, T. Yoshizawa (ed.), Chapter 7, pp. 219–240 (CRC Press, Boca Raton, FL, 2009).
10. Wyant, J. C., Use of an ac heterodyne lateral shear interferometer with real-time wavefront correction systems, *Applied Optics* 14, 2622, 1975.
11. Greivenkamp, J. E., Generalized data reduction for heterodyne interferometry, *Optical Engineering* 23, 234350, 1984.
12. Bruning, J. H., Herriott, D. R., Gallagher, J. E. et al., Digital wavefront measuring interferometer for testing optical surfaces and lenses, *Applied Optics* 13, 2693, 1974.
13. Freischlad, K. and Koliopoulos, C. L., Fourier description of digital phase-measuring interferometry, *Journal of the Optical Society of America A* 7, 542, 1990.
14. Surrel, Y., Design of algorithms for phase measurements by the use of phase stepping, *Applied Optics* 35, 51–60, 1996.
15. Strojnik, M., Hanayama, R., Hibino, K. et al., Error estimation of phase detection algorithms and comparison of window functions, *Proc. SPIE 8511*, 84930J-1–84930J-8, 2012.
16. Schreiber, H. and Bruning, J. H., Phase shifting interferometry, in *Optical Shop Testing*, D. Malacara (ed.), pp. 547–666 (John Wiley & Sons, Inc., Hoboken, NJ, 2006).
17. Sasaki, O., Okazaki, H., and Sakai, M., Sinusoidal phase modulating interferometer using the integrating-bucket method, *Applied Optics* 26, 1089–1093, 1987.
18. Dubois, A., Phase-map measurements by interferometry with sinusoidal phase modulation and four integrating buckets, *Journal of the Optical Society of America A* 18, 1972, 2001.
19. de Groot, P., Design of error-compensating algorithms for sinusoidal phase shifting interferometry, *Applied Optics* 48, 6788, 2009.
20. de Groot, P., Error compensation in phase shifting interferometry, Patent 7,948,637, 2011.
21. Ghiglia, D. C. and Pritt, M. D., *Two-Dimensional Phase Unwrapping, Theory, Algorithms, and Software* (John Wiley & Sons, New York, 1998).
22. Poole, S. P. and Dowell, J. H., Application of interferometry to the routine measurement of block gauges, in *Optics and Metrology*, P. Mollet (ed.) (Pergamon Press, New York, 1960).
23. Pluta, M., Microinterferometry, in *Advanced Light Microscopy*, Vol. 3, *Measuring Techniques*, Chapter 16, pp. 317–506 (Elsevier, Amsterdam, the Netherlands, 1993).
24. Creath, K., Step height measurement using two-wavelength phase-shifting interferometry, *Applied Optics* 26, 2810, 1987.
25. de Groot, P., Three-color laser-diode interferometer, *Applied Optics* 30, 3612, 1991.
26. Deck, L. and Demarest, F., Two-color light-emitting-diode source for high-precision phase-shifting interferometry, *Optics Letters* 18, 1899, 1993.
27. Davidson, M., Kaufman, K., and Mazor, I., The coherence probe microscope, *Solid State Technology* 30, 57–59, 1987.
28. Balasubramanian, N., Optical system for surface topography measurement, USPTO Patent 4,340,306, 1982.
29. Kino, G. S. and Chim, S. S. C., Mirau correlation microscope, *Applied Optics* 29, 3775, 1990.
30. Caber, P. J., Martinek, S. J., and Niemann, R. J., New interferometric profiler for smooth and rough surfaces, *Proceedings of SPIE* 2088, 195–203, 1993.
31. Schmit, J., White-light interference 3D microscopes, in *Handbook of Optical Dimensional Metrology*, K. Harding (ed.), Chapter 10, pp. 395–418 (Taylor & Francis, 2013).

32. ISO, 25178-604: Geometrical product specification (GPS)—Surface texture: Areal—Nominal characteristics of non-contact (coherence scanning interferometric microscopy) instruments (International Organization for Standardization, Geneva, Switzerland, 2013).

33. de Groot, P. and Deck, L., Surface profiling by analysis of white-light interferograms in the spatial frequency domain, *Journal of Modern Optics* 42, 389–401, 1995.

34. de Groot, P. and Colonna de Lega, X., Signal modeling for low-coherence height-scanning interference microscopy, *Applied Optics* 43, 4821, 2004.

35. Larkin, K. G., Efficient nonlinear algorithm for envelope detection in white light interferometry, *Journal of the Optical Society of America A* 13, 832, 1996.

36. Cohen, D. K., Caber, P. J., and Brophy, C. P., Rough surface profiler and method, US Patent 5,133,601, 1992.

37. de Groot, P., Method and system for analyzing low-coherence interferometry signals for information about thin film structures, Patent 7,321,431, 2008.

38. Caber, P. J., Interferometric profiler for rough surfaces, *Applied Optics* 32, 3438–3441, 1993.

39. Chim, S. S. and Kino, G. S., Three-dimensional image realization in interference microscopy, *Applied Optics* 31, 2550–2553, 1992.

40. Chim, S. S. C. and Kino, G. S., Phase measurements using the Mirau correlation microscope, *Applied Optics* 30, 2197, 1991.

41. Ai, C. and Caber, P. J., Combination of white-light scanning and phase-shifting interferometry for surface profile measurements, US Patent 5,471,303, 1995.

42. de Groot, P., Colonna de Lega, X., Kramer, J. et al., Determination of fringe order in white-light interference microscopy, *Applied Optics* 41, 4571–4578, 2002.

43. Zygo Corporation, ZeMapper HS high-speed low-noise 3D optical profiler, Specification sheet, Middlefield, CT, 2013.

44. Gerhardt, U., Eine Anordnung zur optischen Tiefentastung, *Optik* 15, 132, 1958.

45. Linnik, V. P., Ein apparat fur mikroskopisch-interferometrische untersuchung reflektierender objekte (mikrointerferometer), *Akademiya Nauk S.S.S.R. Doklady* 21, 18–23, 1933.

46. Zygo Corporation, NewView objective chart, Specification sheet Middlefield, CT, SS-0083 06/11, 2011.

47. Lega, X. C. D., Grigg, D., and Groot, P. D., Surface profiling using a reference-scanning Mirau interference microscope, *Proceedings of SPIE* 5532, 106–116, 2004.

48. Kim, M. Y., Veluvolu, K. C., and Lee, S.-G., Lateral scanning linnik interferometry for large field of view and fast scanning: Wafer bump inspection, *International Journal of Optomechatronics* 5, 271–285, 2011.

49. Mirau, A. H., Interferometer, US Patent 2,612,074, 1949.

50. Han, S., Novak, E., Wissinger, J. et al., Surface profiler for fixed through-glass measurement, *Proceedings of SPIE* 5716, 189–197, 2005.

51. Hedley, J., Harris, A. J., Burdess, J. S. et al., Development of a workstation for optical testing and modification of IMEMS on a wafer, *Proc. SPIE 4408*, 402–408, 2001.

52. Zygo Corporation, Using Glass Compensated (GC) objectives, Operating manual OMP-0553A, Middlefield, CT, 2009.

53. Creath, K. and Wyant, J. C., Absolute measurement of surface roughness, *Applied Optics* 29, 3823, 1990.

54. Giusca, C. L., Leach, R. K., Helary, F. et al., Calibration of the scales of areal surface topography-measuring instruments: Part 1. Measurement noise and residual flatness, *Measurement Science and Technology* 23, 035008, 2012.

55. Freischlad, K., Sub-angstrom surface metrology with a virtual reference interferometer, *Proceedings of SPIE* 84930B–84930B-14, 2012.

56. Krug, W., Rienitz, J., and Schulz, G., Reflected-light interference microscopy without a reference plane, in *Contributions to Interference Microscopy: Virtual Reference*, Chapter 8.7, pp. 224–234 (Hilger & Watts, London, U.K., 1964).

57. Zemetrics, ZeGage 3D optical profiler, Specification sheet, Middlefield, CT, 2012.

58. Sachs, R. and Stanzel, F., Interference microscopy for clean air—How optical metrology is improving quality control of fuel injection systems, in *Fringe 2013*, W. Osten (ed.), Chapter 96, pp. 535–538 (Springer, Berlin, Germany, 2014).

59. Seewig, J., Areal filtering methods, in *Characterisation of Areal Surface Texture*, R. Leach (ed.), Chapter 4, pp. 67–106 (Springer, Berlin, Germany, 2013).

60. ISO, 25178-3: Geometrical product specifications (GPS)—Surface texture: Areal—Part 3: Specification operators (International Organization for Standardization, Geneva, Switzerland, 2012).

61. ISO, 25178-2: Geometrical product specifications (GPS)—Surface texture: Areal—Part 2: Terms, definitions and surface texture parameters (International Organization for Standardization, Geneva, Switzerland, 2012).

62. Blateyron, F., The areal field parameters, in *Characterisation of Areal Surface Texture*, R. Leach (ed.), Chapter 2, pp. 15–43 (Springer, Berlin, Germany, 2013).

63. Born, M. and Wolf, E., Elements of the theory of interference and interferometers, in *Principles of Optics*, M. Born and E. Wolf (eds.), Chapter 7, pp. 256–369 (Pergamon Press, New York, 1980).

64. Sheppard, C. J. R. and Larkin, K. G., Effect of numerical aperture on interference fringe spacing, *Applied Optics* 34, 4731, 1995.

65. Abdulhalim, I., Spectroscopic interference microscopy technique for measurement of layer parameters, *Measurement Science and Technology* 12, 1996, 2001.

66. Colonna de Lega, X., Model-based optical metrology, in *Optical Imaging and Metrology*, W. Osten and N. Reingand (eds.), Chapter 13, pp. 283–304 (Wiley-VCH Verlag GmbH & Co. KGaA, Weinheim, Germany, 2012).

67. de Groot, P., Colonna de Lega, X., and Liesener, J., Model-based white light interference microscopy for metrology of transparent film stacks and optically-unresolved structures, in *Fringe 2009*, W. Osten and M. Kujawinska (eds.), pp. 236–243 (Springer Verlag, Berlin, Germany, 2009).

68. Ingelstam, E., Problems related to the accurate interpretation of microinterferograms, *Proceedings of National Physical Laboratory Symposium* 11, 141–163, 1960.

69. Biegen, J. F., Calibration requirements for Mirau and Linnik microscope interferometers, *Applied Optics* 28, 1972, 1989.

70. Creath, K., Calibration of numerical aperture effects in interferometric microscope objectives, *Applied Optics* 28, 3333, 1989.

71. Biegen, J. F., Determination of the phase change on reflection from two-beam interference, *Optics Letters* 19, 1690, 1994.

72. Schulz, G. and Elssner, K.-E., Errors in phase-measurement interferometry with high numerical apertures, *Applied Optics* 30, 4500, 1991.

73. Wan, D.-S., Schmit, J., and Novak, E., Effects of source shape on the numerical aperture factor with a geometrical-optics model, *Applied Optics* 43, 2023, 2004.

74. Lee, B. S. and Strand, T. C., Profilometry with a coherence scanning microscope, *Applied Optics* 29, 3784–3788, 1990.

75. Petitgrand, S., Bosseboeuf, A., Gilles, J. P. et al., Mesures 3D de topographies et de vibrations à l'échelle (sub)micrométrique par microscopie optique interférométrique, *Proceedings of Club CMOI, Méthodes et Techniques Optiques pour l'Industrie*, 2002.

76. de Groot, P., Colonna de Lega, X., and Grigg, D., High-NA interference microscopy of complex surface structures, *Proceedings of ASPE 2003 Annual Meeting*, Portland, OR, 2003.

77. de Groot, P. J. and Colonna de Lega, X., Transparent film profiling and analysis by interference microscopy, *Proceedings of SPIE* 7064, 706401-1–706401-6, 2008.

78. Kim, S.-W. and Kim, G.-H., Thickness-profile measurement of transparent thin-film layers by white-light scanning interferometry, *Applied Optics* 38, 5968, 1999.

79. de Groot, P. and de Lega, X. C., Angle-resolved three-dimensional analysis of surface films by coherence scanning interferometry, *Optics Letters* 32, 1638, 2007.

80. Colonna de Lega, X. and de Groot, P., Optical topography measurement of patterned wafers, in *Characterization and Metrology for ULSI Technology*, D. G. Seiler, A. C. Diegold, C. R. McDonald, C. R. Ayre, R. P. Khosla, S. Zollner, and E. M. Secula (eds.), Vol. CP 788, pp. 432–436 (American Institute of Physics, College Park, MD, 2005).

81. Mansfield, D., Extraction of film interface surfaces from scanning white light interferometry, *Proceedings of SPIE*, 7101, 71010U-1–71010U-8, 2008.

82. Davidson, M. P., Interferometric back focal plane scatterometry with Koehler illumination, US Patent 7,061,623, 2006.

83. Colonna De Lega, X. and de Groot, P., Interferometer for determining characteristics of an object surface, US Patent 7,446,882, 2008.

84. JCGM, Evaluation of measurement data—Guide to the expression of uncertainty in measurement (GUM) (Joint Committee for Guides in Metrology, Geneva, Switzerland, 2008).

85. Leach, R., Brown, L., Jiang, X. et al., *Guide to the Measurement of Smooth Surface Topography using Coherence Scanning Interferometry*, in *Good Practice Guide*, Vol. 108 (National Physical Laboratory, Teddington, U.K., 2008).

86. Petzing, J., Coupland, J., and Leach, R., The measurement of rough surface topography using coherence scanning interferometry, in *Good Practice Guide*, Vol. 116 (National Physical Laboratory, Teddington, U.K., 2010).

87. ISO, 25178-600: Geometrical product specifications (GPS)—Surface texture: Areal—Part 600: Metrological characteristics for areal-topography measuring methods (International Organization for Standardization, Geneva, Switzerland, 2013).
88. JCGM, International vocabulary of metrology—Basic and general concepts and associated terms (VIM) (Joint Committee for Guides in Metrology, Geneva, Switzerland, 2008).
89. VDI/VDE, 2655 Part 1.1: Optical measurement and microtopographies; calibration of interference microscopes and depth setting standards for roughness measurement (Beuth-Verlag, Berlin, 2011).
90. Haitjema, H. and Morel, M. A. A., Noise bias removal in profile measurements, *Measurement* 38, 21–29, 2005.
91. Zygo Corporation, Measuring sub-Angstrom surface texture, application note AN-0002, Middlefield, CT, 2008.
92. Brophy, C. P., Effect of intensity error correlation on the computed phase of phase-shifting interferometry, *Journal of the Optical Society of America A* 7, 537, 1990.
93. Servin, M., Estrada, J. C., Quiroga, J. A. et al., Noise in phase shifting interferometry, *Optics Express* 17, 8789–8794, 2009.
94. Boedecker, S., Bauer, W., Krüger-Sehm, R. et al., Comparability and uncertainty of shape measurements with white-light interferometers, *Proceedings of SPIE* 7718, 77180J–77180J-12, 2010.
95. de Groot, P. J., Vibration in phase-shifting interferometry, *Journal of the Optical Society of America A* 12, 354, 1995.
96. de Groot, P., Phase shifting interferometry, in *Optical Measurement of Surface Topography*, R. Leach (ed.), Chapter 8, pp. 167–186 (Springer Verlag, Berlin, Germany, 2011).
97. Schmit, J., High-speed measurements using optical profiler, *Proceedings of SPIE* 5144, 46–56, 2003.
98. Evans, C. J., Troutman, J., Ganguly, V. et al., Performance of a vibration desensitized scanning white light interferometer, *Proc. Int. Conf. on Metrology and Properties of Engineering Surfaces,* Paper TS1-08, IOP Publishing Ltd, Bristol, UK, 2013.
99. Badami, V. G., Liesener, J., Evans, C. J. et al., Evaluation of the measurement performance of a coherence scanning microscope using roughness specimens, *Proceedings of ASPE*, Denver, CO, pp. 23–26, 2011.
100. Deck, L. L., Phase-shifting interferometry in the presence of vibration, Patent 7,796,273, 2010.
101. Schmit, J. and Munteanu, F., Iterative algorithm for phase shifting interferometry with finite bandwidth illumination, *Proc. SPIE* 7063, 70630L–70630L-12, 2008.
102. Munteanu, F. and Schmit, J., Iterative least square phase-measuring method that tolerates extended finite bandwidth illumination, *Applied Optics* 48, 1158–1167, 2009.
103. de Groot, P., The expanding role of optical metrology in precision engineering, *Proceedings of EUSPEN* 1, 31–34, 2012.
104. Sykora, D. M. and Holmes, M. L., Dynamic measurements using a Fizeau interferometer, *Proceedings of SPIE* 8082, 80821R-1–80821R-10, 2011.
105. Massig, J. H., Interferometric profilometer sensor, US Patent 5,166,751, 1992.
106. Schmit, J. and Hariharan, P., Improved polarization Mirau interference microscope, *Optical Engineering* 46, 077007, 2007.
107. Creath, K. and Goldstein, G., Dynamic phase imaging and processing of moving biological organisms, *Proceedings of SPIE* 8227, 82270M–82270M-10, 2012.
108. Doi, T., Toyoda, K., and Tanimura, Y., Effects of phase changes on reflection and their wavelength dependence in optical profilometry, *Applied Optics* 36, 7157–7161, 1997.
109. de Groot, P., Chromatic dispersion effects in coherent absolute ranging, *Optics Letters* 17, 898, 1992.
110. Harasaki, A., Schmit, J., and Wyant, J. C., Offset of coherent envelope position due to phase change on reflection, *Applied Optics* 40, 2102, 2001.
111. Bennett, J. M., Precise method for measuring the absolute phase change on reflection, *Journal of the Optical Society of America* 54, 612, 1964.
112. Schmit, J., Creath, K., and Wyant, J. C., Surface profilers, multiple wavelength, and white light interferometry, in *Optical Shop Testing*, D. Malacara (ed.), pp. 667–755 (John Wiley & Sons, Inc., Hoboken, New Jersey, 2006).
113. Huebner, U., Morgenroth, W., Boucher, R. et al., Development of a nanoscale linewidth-standard for high-resolution optical microscopy, *Proceedings of SPIE* 5965, 59651W–59651W-9, 2005.
114. de Groot, P., Colonna de Lega, X., Sykora, D. M. et al., The meaning and measure of lateral resolution for surface profiling interferometers, *Optics and Photonics News* 23, 10–13, 2012.
115. Colonna de Lega, X. and de Groot, P., Lateral resolution and instrument transfer function as criteria for selecting surface metrology instruments, *OSA Proceedings of Optical Fabrication and Testing* OTu1D, Monterey, CA, 2012.

116. de Groot, P., Interpreting interferometric height measurements using the instrument transfer function, in *Proceedings of FRINGE 2005*, W. Osten (ed.), pp. 30–37 (Springer Verlag, Berlin, Germany, 2006).

117. Novak, E., Schmit, J., and Munteanu, F., Advanced 3d metrology to enhance manufacturing of precision components, *Proceedings of ASPE Annual Meeting*, Atlanta, GA, 2010.

118. Deck, L. L., High precision interferometer for measuring mid-spatial frequency departure in free form optics, *SPIE Technical Digest TD04*, TD040M, 2007.

119. Yu, G. and Tortonese, M., Metrology standards for semiconductor manufacturing, *Proceedings of Seventh International Conference on Solid-State and Integrated Circuits Technology*, Vol. 1, pp. 588–593, 2004.

120. Barakat, R. and Levin, E., Application of apodization to increase two-point resolution by the sparrow criterion. II. Incoherent illumination, *Journal of the Optical Society of America* 53, 274–282, 1963.

121. Pehnelt, S., Osten, W., and Seewig, J., Vergleichende Untersuchung optischer Oberflächenmessgeräte mit einem Chirp-Kalibriernormal, *Technisches Messen* 78, 457–462, 2011.

122. Giusca, C. L. and Leach, R. K., Calibration of the scales of areal surface topography measuring instruments: Part 3. Resolution, *Measurement Science and Technology* 24, 105010, 2013.

123. Weckenmann, A., Tan, Ö., Hoffmann, J. et al., Practice-oriented evaluation of lateral resolution for micro- and nanometre measurement techniques, *Measurement Science and Technology* 20, 065103, 2009.

124. Takacs, P. Z., Li, M. X., Furenlid, K. et al., Step-height standard for surface-profiler calibration, *Proc. SPIE 1995*, 235–244, 1993.

125. Sykora, D. M. and de Groot, P., Instantaneous measurement Fizeau interferometer with high spatial resolution, *Proceedings of SPIE* 8126, 812610–812610-10, 2011.

126. Roth, J. and de Groot, P., Wide-field scanning white light interferometry of rough surfaces, *Proceedings of ASPE Spring Topical Meeting on Advances in Surface Metrology*, Annapolis, MD, pp. 57–60, 1997.

127. Zygo Corporation, NexView Optical Profiler, Specification sheet, Middlefield, CT, 2013.

128. Zygo Corporation, Step Height MetroPro Application, Operating manual OMP-0368E, Middlefield, CT, 2011.

129. Giusca, C. L., Leach, R. K., and Helery, F., Calibration of the scales of areal surface topography measuring instruments: Part 2. Amplification, linearity and squareness, *Measurement Science and Technology* 23, 065005, 2012.

130. ISO, 25178-70: Geometrical product specification (GPS)—Surface texture: Areal—Part 70: Material measures (International Organization for Standardization, DRAFT, Geneva, Switzerland, 2013).

131. National Physical Laboratory, *Areal Calibration Set* (Pamphlet, Teddington, U.K., 2013).

132. Kiyono, S., Gao, W., Zhang, S. et al., Self-calibration of a scanning white light interference microscope, *Optical Engineering* 39, 2720, 2000.

133. Quinn, T. J., Practical realization of the definition of the metre, including recommended radiations of other optical frequency standards (2001), *Metrologia* 40, 103–133, 2003.

134. de Groot, P. J., Colonna de Lega, X., and Grigg, D. A., Step height measurements using a combination of a laser displacement gage and a broadband interferometric surface profiler, *Proceedings of SPIE* 4778, 127–130, 2002.

135. Olszak, A. and Schmit, J., High-stability white-light interferometry with reference signal for real-time correction of scanning errors, *Optical Engineering* 42, 54–59, 2003.

136. Zygo Corporation, NewView 7200 & 7300, Operating manual OMP-0536F, Middlefield, CT, 2011.

137. Hillmann, W., Brand, U., and Krystek, M., Capabilities and limitations of interference microscopy for two- and three-dimensional surface-measuring technology, *Measurement* 19, 95–102, 1996.

138. Rhee, H.-G., Vorburger, T. V., Lee, J. W. et al., Discrepancies between roughness measurements obtained with phase-shifting and white-light interferometry, *Applied Optics* 44, 5919–5927, 2005.

139. Gao, F., Leach, R. K., Petzing, J. et al., Surface measurement errors using commercial scanning white light interferometers, *Measurement Science and Technology* 19, 015303, 2008.

140. Harasaki, A., Schmit, J., and Wyant, J. C., Improved vertical-scanning interferometry, *Applied Optics* 39, 2107–2115, 2000.

141. Mandal, R., Palodhi, K., Coupland, J. et al., Application of linear systems theory to characterize coherence scanning interferometry, *Proceedings of SPIE* 8430, 84300T–84300T-10, 2012.

142. Colonna de Lega, X., Fay, M., Kruse, R. et al., Interference microscopy for semiconductor back end patterning metrology, *Proceedings of Frontiers of Characterization and Metrology for Nanoelectronics* Albany, NY, CP1173, 359–363, 2009.

143. Toomre, D. and Bewersdorf, J., A new wave of cellular imaging, *Annual Review of Cell and Developmental Biology* 26, 285–314, 2010.

32 Noncontact Dimensional and Profile Metrology by Video Measurement

Hiroo Tsumuraya and Shuichi Sakai

CONTENTS

32.1 Introduction: Background of Expansion of Metrology Applications by Video Measuring Machine 830
32.2 Features of Video Measuring Machines 833
32.3 Basic Structure of Video Measuring Machines 833
    32.3.1 Structure 833
    32.3.2 Coordinate System in Video Measurement 834
32.4 Optics of Video Measuring Machines 836
    32.4.1 Performance of Optical Systems 836
        32.4.1.1 Distortion 836
        32.4.1.2 Telecentric Optics 836
    32.4.2 Variable Magnification Optics 837
    32.4.3 Illumination Units 838
        32.4.3.1 Diascopic (Back Light/Contour) Illuminator 839
        32.4.3.2 Coaxial Episcopic (Top Light/Incident) Illumination 840
        32.4.3.3 Oblique Illumination 840
32.5 Image Processor Unit 841
    32.5.1 Edge Point Detection 841
    32.5.2 Image Rectification 843
    32.5.3 Pattern Matching 843
    32.5.4 Vision Auto-Focus 844
32.6 Profile Measurement Method 846
32.7 Height Measurement and Auto-Focus Technologies 847
    32.7.1 SFF Sensor 847
    32.7.2 Confocal Sensor 848
    32.7.3 Laser Auto-Focus Sensor 849
32.8 Conclusion: Standardization of Optical Measurement Sensors 852
Acknowledgment 852
References 852

32.1 INTRODUCTION: BACKGROUND OF EXPANSION OF METROLOGY APPLICATIONS BY VIDEO MEASURING MACHINE

Instruments that capture optically magnified images and measure objects with image processing technology are called video measuring machines, vision metrology instruments, noncontact optical metrology system, etc. The popularity of these machines along with their applications has been expanded into the industrial metrology market in the past 20 years. These video measuring machines are used to reduce human errors in measurements, metrology automation, and as a replacement for conventional optical measuring machines such as profile projectors and measuring microscopes. With these conventional measuring machines, human eyes are mainly used for recognition of measuring points with manual stage motion by handles. Video measuring machines, equipped with optics, camera, and image processor, provide automated noncontact measurement of parts. Figure 32.1 shows auto-edge detection of object measuring point from an optically magnified video image under proper illumination condition.

In addition to the automatic edge detection technology, the three axes (sometimes equipped with additional tilt and rotation axes) drive structure is one of the most important design factors of the video measuring machines for faster and microstructure object measurement abilities. Video measuring machines have improved measurement repeatability and reproducibility (gauge repeatability and reproducibility, known as GR&R*) and measurement inspection time resulting in improved productivity both for production and quality assurance departments. As to the GR&R validation method, it usually requires 3 measurement trials of each 10 parts representing actual production process by 3 appraisers. Figure 32.2 shows reduction of validation process time and improvement of GR&R by utilizing automated video measuring machines as compared with conventional manual

FIGURE 32.1   Edge detection of measuring point by image processing technology.

---

* GR&R is one of the measurement system validation methods. In accordance with proper analytical procedure, GR&R shows repeatability and reproducibility of gauging dimensions of parts. The former is derived from precision of measurement system and the latter from skill and knowledge level of equipment operators. GR&R is usually validated by an index, which divides variation of repeatability and reproducibility (5.15 × sigma) by either process variation or tolerance range. When the index called %GRR is less than 10%, the analyzed measurement system variation is sufficiently smaller than process variation and is considered valid. For details of GR&R, refer to *Measurement Systems Analysis, Reference Manual*, Third Edition, AIAG: ISO/TS16949, 2002 [1].

	Video measuring system	Manual measuring microscope
Measurement time per part	15 s	50 s
Total work time	45 min	2 h 40 min
%GRR True position	1.32% <10%: Pass Measurement system is valid	28.20% >10%: Fail Reconsider measurement method
%GRR Diameter	1.56% <10%: Pass Measurement system is valid	32.60% >10%: Fail Reconsider measurement method

Measurement data distributions of total 90 dimensions
(10 pcs of parts, 3 times repeat measurement by 3 appraisers)

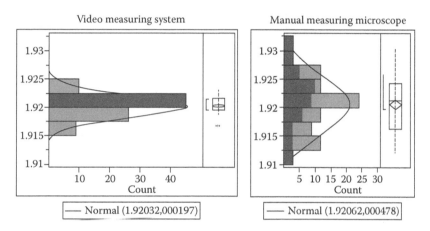

FIGURE 32.2 An example of GRR validation sheet and comparison between manual and automated measurement.

measurement techniques. In an example of hole true position and diameter measurement including datum setting, video measurement machines greatly reduce %GRR and measurement time compared with manual measurement.

Automation of measurements requires part programming and program optimization. Therefore, as long as whole inspection time of a small-sized lot is concerned, there will not be much difference between manual and automated measurements. However, in the case of long-term process monitoring of dimensional measurements by part sampling from production of injection molding, press, electronic devices, and medical devices, automation of metrology provides advantages and merits for process improvement. Figure 32.3 shows that automated measurement of injection molded parts realized reduction of part inspection time by 80% or higher, as compared with manual measurement.

As shown in Figure 32.4, there are many applications for video measuring machines such as electronic devices, medical devices, smartphone parts, injection molding parts, pressed parts, die and molds, flat panel displays, and so on.

In the following sections, we will explain features of video measuring machines (Section 32.2), followed by basic structure of typical video measuring machines (Section 32.3), optics required for video measuring machines (Section 32.4), image processing technology (Section 32.5), profile form measurement (Section 32.6), and noncontact height sensing technologies (Section 32.7).

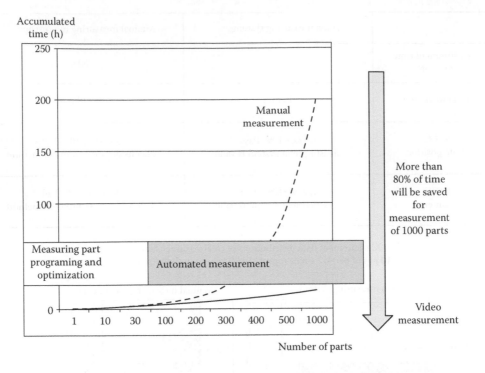

FIGURE 32.3   Inspection time comparison between manual and automated measurement.

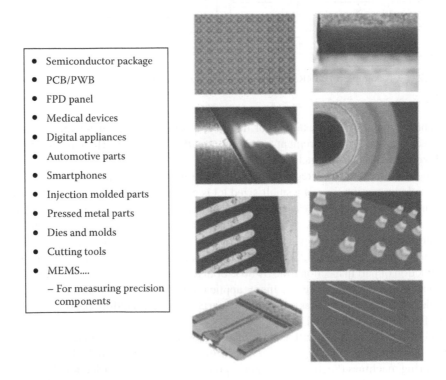

FIGURE 32.4   Application examples of video measuring machines.

TABLE 32.1
Comparison between Tactile Probe and Vision Probe

	Tactile Sensor (Touch Probe)	Noncontact Sensor(Vision Probe)
Data	A series of points (after probing correction)	2D Data from captured image by camera
Probing accuracy	Touch probing mechanism accuracy (mechanical, strain gauge, etc.)	Accuracy derived from optics, magnification and image processing
Operation	Touch probe tip on object surface	Bring measurement object inside field of view
Applications	Large size, 3D features	Microstructure, fragile features
Cautions	Deformation of parts by tactile force, optimum probing sensor length and direction	Critical focusing on objects, removal of burrs and dusts, optimum edge detection conditions

32.2   FEATURES OF VIDEO MEASURING MACHINES

As mentioned earlier, optical measurement machines are capable of noncontact measurement of optically magnified images of objects and high-speed and precise measurements with image processing technology. Main features of video measuring machines are as follows:

1. *Magnification*: High resolution (able to measure minimum 1 μm size line width)
2. *Noncontact*: No deformation or scratches on inspection objects
3. *Image processor*: Human error free, high-precision and high-speed measurements, reduction of human fatigue

Video measuring machines are often compared with tactile measurement machines. In order to characterize the features of vision probe, Table 32.1 summarizes differences between vision and tactile probes.

32.3   BASIC STRUCTURE OF VIDEO MEASURING MACHINES

32.3.1   STRUCTURE

Most video measuring machines consist of two major parts: one is main body with XYZ drive and linear scales, and the other is video head including magnification optics and camera. In order to expand measurement ability, some video heads incorporate built-in height detection sensors and/or tactile probes. Figure 32.5 shows an optical component layout inside a video head.

The illumination system of a video measuring machine consists of diascopic (contour or back light) illumination, coaxial episcopic (incident or top light) illumination as well as oblique ring light illumination. Images of objects being measured are optically magnified by objective lens components and projected on a camera sensor, followed by image processing of captured images. Most basic image processing technology is automated edge detection. As the video edge detection data is based on camera coordinate system, it is necessary to translate it into stage coordinate system by adding it to linear scale reading, so that a measurement result is obtained. Also, with combination of measurement point inputs, measurement application program is capable of feature calculation and construction such as line, circle (hole), distance, angle, etc. Also, a typical measurement application program has the functionality to measure two-dimensional (2D) forms by a series of edge points detected from object profiles. Because of these features of the video measuring machines, they are usually operated by a dedicated measurement application program installed on a host PC with Joystick and mouse controls (Figure 32.6).

As shown in Figure 32.7, the dedicated application software will provide full control of the main measurement hardware, consisting of a video monitor window, illumination and magnification controls, measurement tools, list, and graphic window showing programming history, etc.

FIGURE 32.5   Optical component layout in a video measuring machine.

FIGURE 32.6   Nikon Computer Numerical Control video measuring system VMZ-R4540.

32.3.2   COORDINATE SYSTEM IN VIDEO MEASUREMENT

As explained in Section 32.3.1, because the video measuring machines have XYZ mechanical coordinate system from linear encoder scales and XY video coordinate system based on camera sensor, dimensional measurement data requires proper composition of these two coordinate systems. In order to ensure precise measurement ability from linear encoder scale, the hardware design of main body needs to minimize six degrees of freedom (DOF) errors—pitch, yaw, and roll errors—in XYZ drive as well as non-abbe position errors caused by scale position and measurement position, as shown in Figure 32.8a and b.

FIGURE 32.7   Main window of metrology application program.

FIGURE 32.8   Hardware design requirements (a) 6DOF of motion control and (b) linear scale layout.

Furthermore, in order to combine measurement of main body and video coordinate systems, it is necessary to compensate and correct position of both coordinate systems as well as camera rotation angle (Figure 32.9). The position is given by adding detection points on a video coordinate system to the main body coordinate system. The camera rotation is a camera tilt angle correction against XY mechanical coordinate. Thus, an XY coordinate measurement result is given by the following formula (32.1):

(XY measurement value) = (main body coordinate system at optical axis center)
                + (Rotation correction (θ)) × (video coordinate system)

$$\begin{pmatrix} x \\ y \end{pmatrix} = \begin{pmatrix} xm \\ ym \end{pmatrix} + \begin{pmatrix} \cos\theta & -\sin\theta \\ \sin\theta & \cos\theta \end{pmatrix} \begin{pmatrix} xc \\ yc \end{pmatrix} \tag{32.1}$$

where
    xm, ym are the X and Y main body coordinate value at optical axis center
    xc, yc are the X and Y video coordinate value

FIGURE 32.9 Camera calibration.

In addition to XY coordinate measurement, when video measuring machines have height measurement sensors such as vision auto-focus (AF), laser AF, confocal optical system, and so on, Z measurement values are calculated from main body Z axis scale reading and height displacement value from height sensors (32.2).

$$(\text{Z measurement value}) = (\text{main body Z coordinate}) + (\text{displacement of height sensor}) \quad (32.2)$$

32.4 OPTICS OF VIDEO MEASURING MACHINES

32.4.1 Performance of Optical Systems

One of the most important design factors of video measuring machines is their optical system performance. It is desired to have optical systems with minimized optical aberrations. Any optical system requires optimized design minimizing the five Seidel aberrations concerning monochromatic light aberrations—spherical aberration, coma, astigmatism, curvature of field, and distortion. In case of color imaging systems, chromatic aberrations must be considered as well. In particular, video measuring machines must meet requirements of "distortion-free optics" and "telecentric optics" to provide high-precision optical measurement performance.

32.4.1.1 Distortion

When the optics of a video measuring machine has distortion factors, an ideal square shape might be observed as pincushion or barrel shape at the image plane. The distortion is a monochromatic aberration related to object dimensional size and proportional to the third power of image height (image length from the optical axis), while other four aberrations are related to optical image sharpness. The image distortion causes significant errors of image shape measurement inside a field of view. Therefore, it is desirable that the optical system of measurement machines minimizes distortion to $\pm 0.1\%$ or less (Figure 32.10).

32.4.1.2 Telecentric Optics

Under normal optics, even at slightly defocused position, the object image size may vary and cause dimensional size error. In order to cope with this kind of error, by inserting an aperture at the back

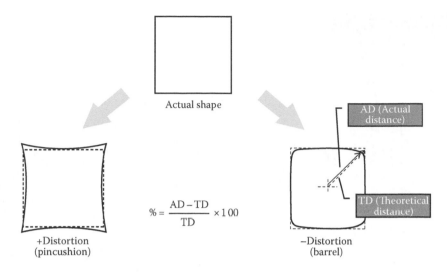

$$\% = \frac{AD - TD}{TD} \times 100$$

FIGURE 32.10 Distortion.

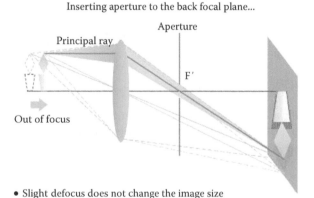

Inserting aperture to the back focal plane...

• Slight defocus does not change the image size

FIGURE 32.11 Telecentric optics.

focal plane, the telecentric optics makes main rays from an object parallel to the optical axis, resulting in the image size consistency at slightly defocused position (Figure 32.11).

32.4.2 VARIABLE MAGNIFICATION OPTICS

Because the design conditions of video measuring machines must meet measurement requirements both for high and low magnification, these instruments typically include variable magnification optics such as zoom optical system and/or objective lens changer. Most video measuring machines have programmable variable magnification controls for automated measurement. In this case, as the object size varies at different magnifications, the video measuring machines perform magnification compensation (pixel size correction) for accurate dimensional measurement. By combination of these technologies, the video measuring machine provides wider field of view at lower magnification and guaranteed measurement accuracy at higher magnification. Figure 32.12 is an example of magnification position images covering 120× zoom range (Nikon Z120X optical head).

1× to 120× (33× to 4000× on screen)

FIGURE 32.12 Nikon Z120X optical head image.

The variable magnification systems in the video measuring machines have mechanisms to minimize measurement errors caused by magnification changes. At least, the following two compensation factors are needed: one is magnification compensation (pixel size calibration), and the other one is optical axis center compensation (parcentricity calibration).

1. *Magnification compensation*: Magnification (or pixel size) is corrected by measuring patterns of which size are already known and traceable to national standards. The calibration patterns are dedicated to each measurement machine.
2. *Optical axis center (parcentricity) compensation*: The eccentricity of optical axis center at each magnification position adds to positional error factor. In order to compensate it, the video measuring machines measure XY coordinate position of a circle pattern at different magnifications and obtain coordinate differences as correction values.

In addition to these, when the video measuring machine provides high-precision height measurement ability, focal position reproducibility at different magnification positions—parfocality compensation—must also be considered.

32.4.3 ILLUMINATION UNITS

Video measuring machines usually have three illumination systems: a diascopic (back light/contour) illuminator, a coaxial-episcopic (top light/incident) illuminator, and oblique light illuminators, which can visualize edge of small height gaps. In recent years, instead of halogen light source, white LED illuminators are incorporated in video measuring machines, due to their high performance, low energy consumption, and light intensity.

32.4.3.1 Diascopic (Back Light/Contour) Illuminator

The dimensions and forms of parts such as pressed parts, gears, semiconductor leadframes, die, and mold parts are measured by back light illumination. In order to measure these parts precisely, it is critical to have even illumination on measurement plane (Figure 32.13).

High-intensity LED light source is now widely used for diascopic illumination providing lower power consumption, longer life, stability, and faster light control response, as compared with halogen light source. Also, new illumination optics has been developed for LED light sources using fly-eye lenses, which provides even illumination and higher lighting efficiency (Figure 32.14).

FIGURE 32.13 An image by diascopic (back light/contour) illumination.

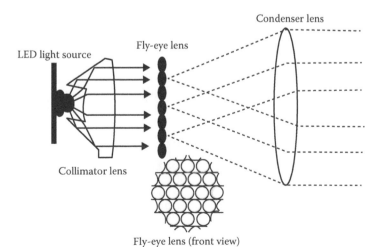

FIGURE 32.14 Light source optics using fly-eye lens.

32.4.3.2 Coaxial Episcopic (Top Light/Incident) Illumination

The episcopic illumination is used for illuminating part surfaces such as metal surface, integrated circuit patterns, etc. While the structure of episcopic illumination system is similar to that of diascopic illumination, the episcopic illumination may have a built-in 1/4λ plate on the top of objective lens to prevent unnecessary light reflection or glare from mirror surfaces (Figure 32.15).

32.4.3.3 Oblique Illumination

When edge recognition is difficult with episcopic lighting, oblique LED light illumination with independent segment lighting control system is used for better edge recognition (Figure 32.16). Some video measuring machines have also lighting angle control to produce optimum illumination angle for visualization of edges (Figure 32.17). Although the oblique illumination will produce high contrast edges, it is necessary to pay attention to the fact that it may sometimes shift the measuring points because of shadows on chamfered edges.

FIGURE 32.15 Episcopic lighting image.

(a)   (b)

FIGURE 32.16 Effects of oblique illumination. (a) Episcopic illuminator and (b) oblique ring LED illuminator.

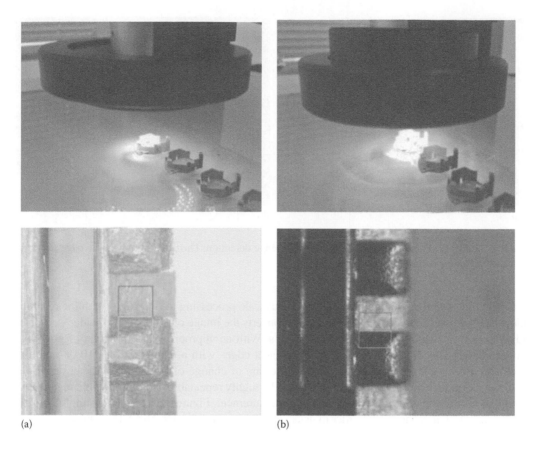

(a)   (b)

FIGURE 32.17 (a) Normal lighting angle and (b) oblique light angle is set to middle position.

32.5 IMAGE PROCESSOR UNIT

Image processing technologies have been widely adopted for industrial inspection and measurement applications with the evolution of semiconductor devices since the late 1980s. In the early stage, the image processor units were external hardware processor units and/or frame grabbers for workstations/PCs. Nowadays, thanks to improvement of CPU performance including digital signal processor, camera data transfer protocols, and so on, it is possible to handle the image processing on the application programs running on personal computers. The image processing technologies of the video measuring machines can be classified into the following categories:

1. *Edge point detection*: input measuring points
2. *Image rectification*: enhancement of edges for easier and stable detection
3. *Pattern matching*: part alignment, measurement target search
4. *Vision AF*: AF by image processing

These technologies are discussed in detail in the following sections.

32.5.1 EDGE POINT DETECTION

Figure 32.18 shows examples of edge point detections. Figure 32.18a shows averaged edge point detection. As seen in Figure 32.18a, the technique of averaged edge detection can be used for dust/burr removals. Figure 32.18b shows multipoint edge detection on a circle feature.

(a)   (b)

FIGURE 32.18 Video edge detection. (a) Average edge detection. Dust/burr removals by edge averaging. (b) Multipoint edge detection (circle feature).

General image processing technologies use grayscale processing to find edges with the resolution below camera pixel size. Grayscale processing converts the image on each pixel of video sensor into gray values from white to black (e.g., 256) levels. With an appropriate interpolation of gray levels between pixels after filtering, it is possible to detect edges with a resolution of 1/10 or smaller of pixel size. Because the edges are detected by the rate of change of brightness, instead of absolute brightness level difference, the measurement will be highly repeatable and reproducible regardless of illumination and/or part deviations in each part measurement. Figure 32.19 shows a simplified model of edge detection mechanism. In order to detect a peak point as an edge, the A/D (Analog to Digital),

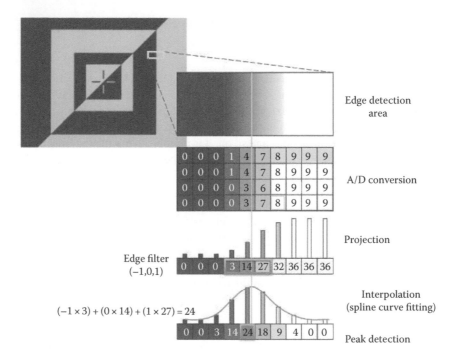

FIGURE 32.19 Principle of edge detection (simplified model into 10 gray levels).

converted data on every pixel is accumulated in one direction (projection), followed by edge filtering (derivative) and interpolation such as spline curve fitting, which is called "subpixel processing." In the case of reasonably high magnification (small pixel size), submicrometer repeatability can be achieved.

32.5.2 IMAGE RECTIFICATION

When edge points are still ambiguous even after magnification adjustment and lighting optimization, image filters can be used to improve edge detection. For measurement purposes, it is not advisable to apply image filters such as erode or dilate, which may change edge positions. Figure 32.20 shows an example of edge enhancement by FFT (Fast Fourier Transform) filter to extract an edge. The signal peak level becomes higher after applying the edge enhancement filter.

32.5.3 PATTERN MATCHING

When video measuring machines are used to find automatically measurement targets and/or to find the part orientation skewed in XY plane, precise object search functions are required. The search functions by pattern matching technologies are indispensable in metrology automation (Figure 32.21).

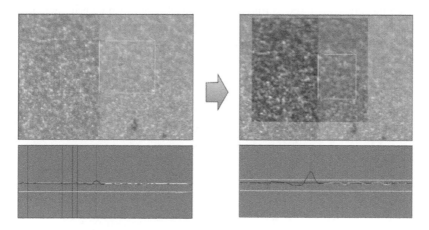

FIGURE 32.20 Image rectification by FFT filter.

FIGURE 32.21 Recorded image (right) and pattern matching (left) in a field of view.

(a)   (b)

FIGURE 32.22 Geometrical feature pattern matching: normalized crossed-correlation method fails in search (a), while geometrical feature pattern matching searches skewed patterns (b).

Pattern matching can be achieved by finding the image position, which is higher correlation coefficient (closer to 1) after normalized cross-correlation to avoid effects of brightness level difference between actual target image and recorded image. The normalized cross-correlation formula of pattern matching can be explained as in the following (32.3):

$$r = \frac{\sum_{n=0}^{n-1}\left(f_n - \overline{f}\right)\left(g_n - \overline{g}\right)}{\sqrt{\sum_{n=0}^{n-1}\left(f_n - \overline{f}\right)^2}\sqrt{\sum_{n=0}^{n-1}\left(g_n - \overline{g}\right)^2}} \tag{32.3}$$

where
  f is the brightness of each pixel of a target image
  g is the brightness of each pixel of a recorded image
  $\overline{f}, \overline{g}$ are the averaged brightness levels of the search area
  n is the number of pixels

After normalized cross-correlation search is performed, its results are X and Y coordinate value of matched location. Search score (correlation coefficient) is also obtained. While this pattern matching technique is widely used for industrial metrology fields, it may fail in appropriate matching with recorded image, if the target image has a large skewed or rotated angle and/or its size is scaled. In order to find objects under such conditions, new pattern matching algorithms using geometrical feature characterization have been developed and implemented for vision metrology (Figure 32.22).

32.5.4 VISION AUTO-FOCUS

In automated optical measurements, it is critical to have precise AF functions to correct for images that are not in proper focus due to variations in part thickness and/or part warpage. Video measuring machines incorporate video AF function to find a focus position from image contrast at different heights. Video AF function can be performed at edge positions and at part surfaces. The principle of vision AF is (1) capture images at a constant sampling step in Z axis direction, (2) find contrast values in an area of interest, and (3) estimate the contrast peak point with an interpolation calculation

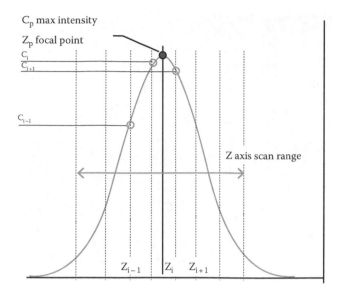

FIGURE 32.23 Height detection by vision AF.

such as Gaussian fitting from maximum contrast point and its neighborhood data points (two points or more). Fitting formula can be expressed as follows (32.4) (Figure 32.23):

$$C_i = C_p \times e^{\frac{Z_i - Z_p}{2\sigma^2}}$$ (32.4)

where

C_p is the estimated value of contrast peak
Z_p is the Z axis value in C_p (focused position)
C_i, C_{i+1}, C_{i-1} are the maximum contrast value and its neighborhood data points
Z_i, Z_{i+1}, Z are the Z axis values in C_i, C_{i+1}, C_{i-1}, respectively

Some video measuring machines incorporate pattern (Ronchi grid) projection mechanism to add contrast on mirror and/or transparent surfaces. The accuracy of video AF is dependent on the depth of focus defined by a value expressing optical performance, called numerical aperture (NA). Therefore, it is advisable to use high NA optics to meet requirements of height measurement. The video AF normally produces repeatability and reproducibility below the depth of focus with interpolation processing. The depth of focus can be expressed by the following formula (32.5) and can be used as a guide value for vision AF accuracy. For instance, in the case of NA = 0.5 and 500 nm wavelength for observation, the depth of focus will be 1 μm.

$$DOF = \frac{\lambda}{2NA^2}$$ (32.5)

where λ is the wavelength used for observation.

More advanced video AF function supports simultaneous height detections in a field of view by placing multiple detection areas for height measurement purposes. The functionality is suitable for continuous height difference measurements in a single Z axis scan. Figure 32.24 shows the example of maximum height detection of a wire bonding loop in a semiconductor device.

FIGURE 32.24   Loop height measurement by multiple-vision AF.

32.6   PROFILE MEASUREMENT METHOD

In order to measure curved shapes such as spur gear teeth shape defined by involute spline curve or cam profiles consisting of many different arcs, the video measuring machines measure these curve features by a series of points with the constant point-spacing, followed by comparison with nominal shape from CAD (Computer Aided Design) data making it possible to calculate the error between actual measured shape and the ideal shape. This type of measurement is typically known as profile (contour) measurement. There are two methods used in the measurement process with constant point-spacing: (1) Auto-edge tracing along profile shapes: this is a method used to find edge points at constant point spacing. The next edge point can be estimated from the points, which have been already measured. (2) Measurement points generated from CAD data: edge detection tools can be automatically generated from feature data in CAD (Figure 32.25).

After the profile measurement, the application program calculates errors between the actual part profile data and the CAD nominal profile data. When the profile of line tolerance [2] in a drawing

FIGURE 32.25   Gear teeth profile measurement.

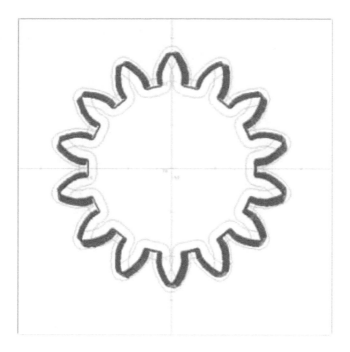

FIGURE 32.26 Gear teeth profile evaluation.

allows for parallel shift and/or rotation to minimize the part-to-CAD errors, the application program will perform the best-fit correction. Figure 32.26 shows an example of gear teeth profile measurement report. The report is helpful in knowing errors graphically with tolerance areas defined by CAD data.*

32.7 HEIGHT MEASUREMENT AND AUTO-FOCUS TECHNOLOGIES

In this chapter, we will explain a couple of examples of noncontact height measurement technologies: shape from focus (SFF) probe, confocal sensor, and laser AF sensor. The SFF sensor can be regarded as an expansion of vision AF technologies. The confocal sensor is used for precise height measurements of features in a wide field of view. The laser AF sensor can be used as an accurate and easy height measurement tool, regardless of optical magnification.

32.7.1 SFF Sensor

As explained in Section 32.5.4, video measuring machines have the video AF function, which focuses on the part surfaces by image contrast. Based on the video AF focusing algorithm, some video measuring machines have an area height measurement sensor to capture the height information in each pixel on the image sensor. This measurement sensor is called SFF sensor. By adding the video image to the area height information from SFF sensor, the video measuring machines can also produce an extended depth of focus (EDF) image (Figure 32.27).

Similar to the video AF, the SFF sensor is a passive sensor, which is dependent on the image contrast derived from the surface texture (roughness) of a part. Therefore, it is not suitable for a surface without texture such as mirrors, highly polished surfaces, etc. In order to measure surfaces without texture by the SFF sensor, some video measuring machines use a built-in pattern projector inside the optics. By projecting the patterns actively on the part surfaces, the SFF sensor measures the height information at the peak contrast of the patterns on each pixel (Figure 32.28).

---

* For details, refer to ASME Y14.5-2009.

FIGURE 32.27   An example of EDF image captured by SFF method.

(a)   (b)

FIGURE 32.28   Pattern projection on surface (a) and resulting 3D height image by SFF sensor (b) of Nikon multisensor metrology system, HN-6060.

As in the case of the video AF, the accuracy of SFF sensor depends on the NA of the optics. In general, with the increase of NA, the field of view becomes smaller, resulting in longer measurement time. Therefore, it is necessary to optimize the measurement accuracy and speed by adjusting magnification and image contrast.

32.7.2   CONFOCAL SENSOR

In order to meet the precise measurement requirements for small height and surface texture, some video measuring machines feature a precise height measurement sensor, which is known

(a)   (b)

FIGURE 32.29   (a) Nikon Confocal Nexiv VMZ-K3040. (b) Confocal optics with Nipkow spinning disk.

as confocal optics, as shown in Figure 32.29a. Confocal optics with many pinholes on a built-in Nipkow spinning disk is designed to guide image rays only at the focal position. Therefore, the resulting image has a very narrow depth of field as compared with a normal optical image. By this optically sliced image from the confocal optics, the video measuring machine is able to measure height information within the FOV (field of view) from continuously captured images with Z axis scan at constant steps and dedicated image processing. Similar to the SFF sensor, the image processor detects Z axis height on each pixel on a camera sensor as shown in Figure 32.29b. Interpolation is then used to calculate its maximum contrast point from sampled confocal images by Z axis scanning at constant steps.

With this measurement method, it is possible to measure height dimensions and form as well as 2D object locations and dimensions in a field of view. In metrology applications, video measuring machines equipped with a confocal head are used for the measurement of high density bump location and height of semiconductor devices (Figure 32.30) and surface texture (roughness) evaluation by surface analysis software (Figure 32.31).*

32.7.3   Laser Auto-Focus Sensor

For a displacement measurement method to measure part thickness, flatness, etc., some video measurement machines use an active AF sensor using a laser light source. In many cases, the laser AF sensor is used with 2D video measurement. Therefore, it is advisable to adopt the through-the-lens (TTL) design, enabling to guide and detect the laser beam through the optics such that the focusing operation can be done in the same field of view as the video measurement. Figure 32.32a explains the structure and principle of the "knife edge" methodology for TTL laser AF. The detector uses

---

* MountainsMap X software based on Digital Surf's Mountains Technology was used for surface analysis in this chapter.

FIGURE 32.30   Simultaneous bump height and location measurement in an FOV by the confocal method.

FIGURE 32.31   Three-dimensional view and roughness evaluation of the confocal image by surface analysis software.

a two-segment sensor,* A and B, and monitors the signal difference level between the sensors (signal S = A − B) during the focus motion. The focus position is adjusted at the zero-cross position of signal S. When the NA of objective lens used for the TTL laser AF is 0.35, the height measurement repeatability of standard sample will be 0.5 μm (2σ) or lower.

   In addition to the single point height measurement, the video measuring machine with the TTL laser AF provides scanning measurement ability to measure part surfaces continuously. The scanning measurement can be achieved by XY stage motion and Z axis continuous focusing motion with real-time feedback and compensation of S signal level. Figure 32.33 is an example of measurement of surface texture of a camera grip. It is possible to observe the structure of surface texture and depth/height after removing form by the surface analysis software.

---

* Today, Nikon proposes a new TTL laser AF structure with additional sensor segments of the detector to avoid effects from reflection of back-side surface of a transparent part.

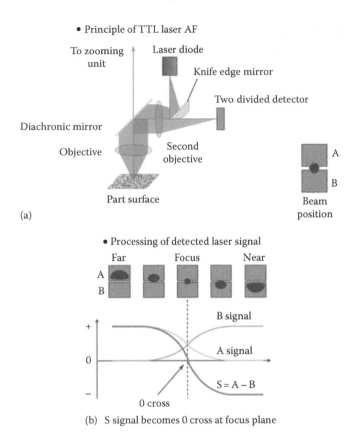

• Principle of TTL laser AF

(a)

FIGURE 32.32 (a) Knife edge style active laser auto-focus and (b) laser signal processing.

Parameters	Unit	Step 1	Step 2
Width	mm	0.179	0.159
Maximum height	μm	45.0	59.6
Mean height	μm	43.2	57.5

FIGURE 32.33 Measuring result of surface texture of a camera grip. After form removal of camera grip, it is possible to evaluate the surface texturing.

32.8   CONCLUSION: STANDARDIZATION OF OPTICAL MEASUREMENT SENSORS

We have introduced the up-to-date technologies of the video measuring machines in this chapter and explained expansion of their measurement applications. The evolution and development of the video measuring machines will continue with optical measurement technologies and solution proposals from different manufactures. Today, the popularity and practical applications of video sensors require their classifications and standardization, resulting in the ISO 10360-7:2011 geometrical product specifications (GPSs)—Acceptance and reverification tests for coordinate measuring machines (CMM)—Part 7: CMMs equipped with imaging probing systems [3]. In this ISO standard, the video measurement probe is positioned as per the CMM standard. In addition to this standard, as explained in Section 32.7, some video measuring machines, having noncontact precise height measurement sensors, might exceed the standard category of the video measuring machines. These noncontact height measurement sensors will be classified and standardized in the ISO25178-6 Surface Texture—Area—Part 6: Classification of methods for measuring surface texture [4].

ACKNOWLEDGMENT

We thank Mr. Kambiz "Tom" Banafshe, business development manager, Nikon Metrology, Inc., for his assistance and suggestions to write this chapter.

REFERENCES

1. AIAG (Automotive Industry Action Group) Copyright 2010 Chrysler Group LLC, Ford Motor (2002).
2. ASME (American Society of Mechanical Engineers) Y14.5–2009 Dimensioning and Tolerancing.
3. ISO 10360-7:2011 Geometrical product specifications (GPS)—Acceptance and reverification tests for coordinate measuring machines (CMM)—Part 7: CMMs equipped with imaging probing systems.
4. ISO 25178-2:2012 Geometrical product specifications (GPS)—Surface texture: Areal—Part 2: Terms, definitions and surface texture parameters.

33 Optical Metrology in Manufacturing Technology

Rainer Tutsch

CONTENTS

33.1 Introduction 853
33.2 Machine-Integrated Encoders 853
33.3 Checking and Adjusting of Production Machines... 857
    33.3.1 Accuracy Assessment of Machine Tools 857
    33.3.2 Vibration Measurement 857
    33.3.3 Adjustment of Repetitively Moving Machine Parts 858
33.4 Quality Control Loops.. 858
33.5 Qualitative Inspection.. 859
    33.5.1 Automated Visual Inspection 860
    33.5.2 Completeness Check.. 860
    33.5.3 Surface Check.. 860
33.6 Quantitative Inspection.. 861
    33.6.1 Measurement Uncertainty 861
    33.6.2 Measurement Capability.. 861
33.7 Optical Techniques Frequently Used in Production Metrology 862
    33.7.1 One-Dimensional Techniques.. 862
    33.7.2 Two-Dimensional Techniques 863
    33.7.3 Three-Dimensional Techniques... 864
33.8 Techniques to Minimize the Effects of Environmental Light............................ 865
    33.8.1 Shading 865
    33.8.2 Optical Filtering.. 865
    33.8.3 Modulation Techniques 866
References.. 866

33.1 INTRODUCTION

Optical metrology plays an important role in modern production. Numerically controlled (NC) machines and robots depend on encoders for length and angle that are integrated in the respective axes. Optical techniques are applied for checking and adjusting production machines. Robots and automated floor-borne vehicles are guided by optical sensors. Optical techniques are used to check the quality of manufactured workpieces and to establish quality control loops. In the following sections, typical applications will be discussed.

33.2 MACHINE-INTEGRATED ENCODERS

Most NC machines depend on incremental encoders integrated in the linear and rotary axes, and in the majority of cases, these are optical encoders. Incremental encoders render a periodic output signal when the respective axis is moving at constant (angular) velocity. The (angular) distance is determined by adding the periods of this signal. In principle, there are two shortcomings in this technique:

- The measurement is only relative, and an additional reference signal is required to get absolute position data.
- The periodic signal does not reveal the direction of movement.

As can be seen from Figure 33.1, a movement starting from position x_1 in the rightward direction gives the same output signal than a movement starting from position x_2 in the leftward direction. This ambiguity is solved by using two phase-shifted output signals. A phase shift of $\pi/2$ is usually chosen, and the two signals are commonly referred to as sin/cos-signal or quadrature signals. The additional signal B in Figure 33.2 in combination with signal A allows an unambiguous determination of the movement.

Furthermore, it is possible to generate additional signals with precisely determined phase shift as a linear combination of signals A and B, which can easily be performed in real-time by electronic hardware. For sinusoidal signals, the following equation holds:

$$\sin(\varphi + \Delta\varphi) = \cos\Delta\varphi \sin\varphi + \sin\Delta\varphi \cos\varphi = a\sin\varphi + b\cos\varphi.$$

A number of such phase-shifted copies of the original signals can be used to determine the shift of the object to a fraction of the period of the original signals A and B. Other interpolation techniques

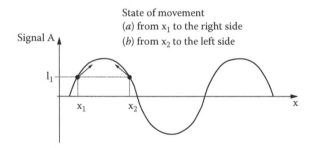

FIGURE 33.1 Ambiguity of direction for a single incremental encoder signal.

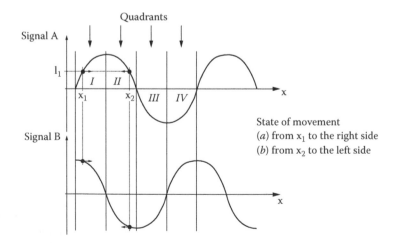

FIGURE 33.2 Unambiguous evaluation of a pair of quadrature signals.

make use of phase locked loop (PLL) oscillators that generate a synchronized measurement signal with a multiple frequency. Depending on the S/N ratio of the original signals, interpolation by a factor of 1000 or even more is possible.

A typical example of a linear optical incremental encoder, the so-called four-field scanning technique, is shown in Figure 33.3. The scale grating is made of glass with a chromium coating that is structured using a precision lithographic process. A common value of the grating period is 20 µm. As this is also the period of the output signal, an interpolation technique is required to get submicron resolution. The scale grating is fixed to the mechanical axis, while a readout head is mounted to the moving slide. This head contains a light source (e.g., an LED), a condenser lens, and several scanning reticles with the same grating period as the scale grating. Moving such a reticle along the scale gives a periodically varying transmitted light intensity. Four reticles are mounted in a way that four phase-shifted signals are generated by the light gathered by the four respective photodetectors. The signals $I_1, \ldots, I_4$ can be designated with sin*, cos*, (−sin*), (−cos*). Subtracting (−sin*) from sin* and (−cos*) from cos* cancels the constant background and results in the desired pair of quadrature signals that can be further processed by an interpolation technique. An additional track on the scale grating contains a reference pattern that is read out with a matched reticle to generate the reference signal. When the power is turned on, the axis should first perform a search travel (usually with slow speed), until the reference signal appears. From then on, the absolute position of the slide can be measured by counting the increments of the measurement signal relative to the reference position.

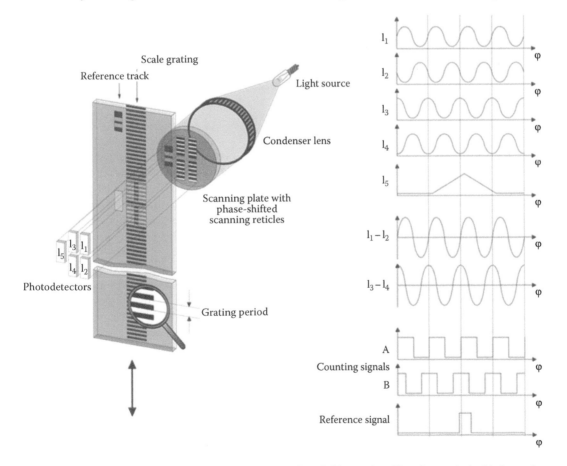

FIGURE 33.3 Linear optical incremental encoder with four-field scanning. (Based on technical information of company Dr. Johannes Heidenhain, GmbH.)

FIGURE 33.4   Moire pattern of the superposition of two gratings with slightly different periods.

In principle, the grating could be directly scanned using photodiodes smaller than the grating period. Using the reticles, however, has several advantages:

- Large photodiodes with a good S/N-ratio can be applied.
- Local defects of the scale or dust have only little effect on the signal quality, because the intensity is integrated over several signal periods.
- While the gratings can only be manufactured with a binary transmission profile, because of the moire effect and diffraction, the signals are nearly sinusoidal, making interpolation easier.

Alternative scanning techniques are available. The single-field scanning technique uses a single scanning reticle with a slightly different grating period. The superposition of two gratings with different period results in a moire pattern with a larger period λ_{beat}, as shown in Figure 33.4.

$$\lambda_{\text{beat}} = \frac{\lambda_1 \lambda_2}{|\lambda_1 - \lambda_2|}$$

A linear array of photodiodes is used to read out the moire pattern. These photodiodes are arranged at a distance of $\lambda_{\text{beat}}/4$ so that their signals can be combined corresponding to the four signals of the four-field scanning technique. Compared to the four-field scanning technique, the larger number of photodiodes and the evaluation of the full width of the grating improve the S/N ratio and make the single-field scanning technique more robust against local defects of the scale grating. Linear encoders based on these or similar evaluation techniques are available with a resolution in the order of 10 nm and an accuracy in the order of 1 μm.

Scale gratings made of glass are limited in length to about 4 m. If longer encoders are required, reflecting scale gratings based on polished steel bands with etched line structures are used. These are manufactured up to lengths of about 30 m. The readout heads are similar to those of the glass scales, but the reflected light is measured instead of the transmitted. Scale gratings made of steel are more robust against vibration and mechanical shocks and their thermal expansion equals that of the machine bed, making thermal compensation easier.

Similar to the linear encoders, angle encoders are based on a radial scale grating with one or more radial scanning reticles. Angle encoders are not only used for the control of rotary axes, but also for linear axes where the rotation of the lead screw is used as a measure of linear movement. While this often is less expensive than a linear encoder, the accuracy is in general worse, because pitch-errors of the lead screw and backlash contribute to the error budget.

For applications with highest demands of accuracy, for example, wafer steppers or ultraprecision machine tools, linear encoders with an interferometric readout scheme are available. In this case, the grating period is in the order of a few micrometers and the scale and the reticle are used as diffraction gratings. Several different diffraction orders are combined and evaluated. A shift of the reticle relative to the scale does not affect the direction of the diffraction orders, but it introduces a phase shift in the diffracted light. This position-dependent phase shift is evaluated interfero-metrically by the superposition of different diffraction orders, and in that way, a pair of quadrature signals is generated [1].

33.3 CHECKING AND ADJUSTING OF PRODUCTION MACHINES

33.3.1 ACCURACY ASSESSMENT OF MACHINE TOOLS

Machine tools have at least three axes that form their coordinate system, but machines with four or five axes are also widely used. For each linear axis, three deviations of position (x, y, z) and three angular deviations (pitch, yaw, and roll) have to be considered. For each rotary axis, the radial (two directions) and axial run-out, the angle error, and two directions of drunkenness error also add up to six degrees of freedom. In the machine, the deviation of orthogonality of different axes has to be added. Apparently, the determination of the errors of a machine tool is a demanding task.

In practice, usually, a laser interferometer is used to measure linear deviations. Manufacturers of these instruments offer optional optics for the measurement of straightness deviations or small tilt angles. Alternatively autocollimators are applied for the measurement of straightness deviation and tilt angle. Calibrated precision pentagonal prisms are used as a reference for the check of the orthogonality of the axes. These are applicable to both laser interferometers and autocollimators. A detailed discussion of straightness measurement and alignment can be found in Chapter 17.

Since a few years, a self-tracking laser-interferometer is commercially available [2]. This so-called laser tracer is a specially designed laser interferometer that can be rotated in two directions and can be controlled to track a retroreflector moving freely in space. A retroreflecting target is fixed to the moving part of the machine under examination and is simultaneously tracked by at least four laser tracers from fixed positions. Using a multilateration algorithm, the 3D-position of the retrore-flector can be calculated. Based on the kinematic model of the machine, the errors of all degrees of freedom of a multiaxes machine can be determined from a set of comparisons of measured versus nominal positions.

33.3.2 VIBRATION MEASUREMENT

A common problem with production machines are vibrations. These cause annoying noise and can deteriorate the precision of the machine. A prerequisite to apply effective countermeasures is the knowledge of the vibration modes and the positions of nodes and antinodes. Vibrations can be mea-sured optically using either laser vibrometers or holographic techniques.

Laser Doppler vibrometers in principle are two-beam laser interferometers. The vibrating sur-face is used as the reference mirror. If the surface is not specularly reflecting, an imaging lens with large aperture is used to gather the backscattered light. Applying a heterodyne technique, a Bragg cell is used to shift the frequency of the measurement light by a carrier frequency f_c. Reflection from a surface moving with velocity $v(t)$ adds a frequency shift of $2v(t)/\lambda$. Usually, the probing beam and the optical axis of the imaging lens are coaxially directed normal to the vibrating surface. $v(t)$ is then the out-of-plane component of the local velocity vector $v(t)$. Superposition of the measurement beam and the reference beam in the vibrometer head gives a beat signal

$$I(t) = I_0 \sin\left(f_c + \frac{2v(t)}{\lambda}\right)t.$$

This can be interpreted as a frequency modulated signal with the carrier frequency f_c. It can be demodulated using techniques well known from radio technology. Adding a 2D-scanning optics to such a single point laser vibrometer allows the sequential measurement of the distribution of ampli-tude and phase of out-of-plane vibration over the surface of an object.

In-plane vibration can be measured by adopting the working principle of laser Doppler velocimetry. Two mutually inclined laser beams are brought to a common focus on the vibrating surface. The focus spot is then structured with a sinusoidal interference pattern with a period depending on the wavelength and the inclination angle of the beams. If there are some inhomogeneities on the surface, the backscattered light will be modulated in intensity, when there is an in-plane-movement orthogonal to the line-pattern; 3D laser vibrometers that combine both techniques to measure the three components of a local velocity vector simultaneously are commercially available.

In contrast to laser vibrometer that scan the surface point by point holographic techniques can be used for integral vibration measurement. In classical holographic interferometry, two main approaches exist: the time-average technique and the double-pulse technique. In both cases, a holographic setup is applied where the coherent light backscattered from the measurement object is superimposed with a reference wave on a photoplate.

For the time-average technique, the exposure time is chosen to be much longer than the period of vibration. If the resulting hologram is reconstructed, a 3D image of the object is visible covered by interference fringes. The contrast of these fringes is highest at the nodes and declines with increasing local amplitude. A double-pulse hologram is taken by using a pulse laser and triggering it at two different states of the vibration, for example, at the two opposite reversal points. In this case, the fringe system covering the 3D image of the object will have constant contrast. As an advantage compared to the time-averaging technique, the double-pulse technique measures not only the amplitude but also the phase of vibration. As a disadvantage the information about the node positions is lost.

In both cases, the distribution of the local amplitude of vibration in the direction of the sensitivity vector, given as the half-angle between the directions of illumination and observation, can be determined with interferometric accuracy. The complete amplitude vector can be measured by combining three setups with noncoplanar sensitivity vector. Holographic interferometry is a powerful tool, but the handling of photographic plates is awkward and time consuming. With the availability of high-resolution electronic imagers and high-power computers, it has been replaced by speckle interferometry and digital holography. Both the time-average technique and the double-pulse technique can be transferred to these new methods. A detailed discussion is beyond the scope of this survey, but can be found in Chapters 7 and 8 and in literature, for example, [3].

33.3.3 Adjustment of Repetitively Moving Machine Parts

There are many applications in production, where precisely synchronized repetitive movements of different parts at high speed are vital, for example, pick and place robots, printing and packaging machines, and automated assembly machines. When it comes to repetition rates of 10 Hz or more and to precision demands of 1/10 mm or less, the analysis of these procedures is beyond the capability of the human eye. In these cases, stroboscopic illumination is a common approach to freeze repetitive movements by illumination with a synchronized flashlight. This can be combined with an electronic camera with short exposure time to get digital images for quantitative evaluation. If a high-speed camera is applied, videos can be taken at a frame rate of 1,000 fps at full frame and more than 10,000 fps at reduced resolution. With this tool, even transient effects can be made visible and measurable. A demanding task, however, is the setting of trigger marks to identify the interesting images in a pile of thousands of images gathered within a few seconds.

33.4 QUALITY CONTROL LOOPS

The main task of production metrology is the supply of methods of measurement and testing to secure the quality of the manufactured products. In general, it is checked whether certain properties that are vital for the function of a product are lying within given tolerance ranges or not. A simple implementation is the end-of-line test with the aim to identify and reject products that do not meet the specifications (Figure 33.5a).

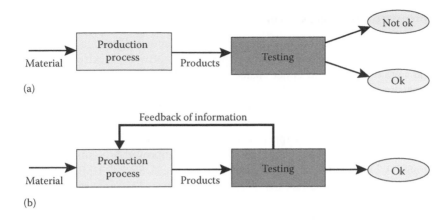

FIGURE 33.5 (a) Product-oriented testing procedure to remove defective products, (b) process-oriented testing procedure; a quality control loop prevents the production of defective products.

FIGURE 33.6 A chain of quality control loops. *Note*: PS, production step; T, test.

According to modern production philosophy, however, quality must not be gained by testing, but has to be immanent to the production processes. On each production process, a variety of disturbing influence factors are acting, for example, changing environmental conditions, variations of raw materials, tool wear, limited precision of machines, or human interaction. Production processes, therefore, have to be designed to be robust against these disturbances. Technical systems are made robust by implementing control loops. In a generalized way, production processes are stabilized by implementing quality control loops (Figure 33.5b). The basic difference to product-oriented testing is that the test is not aimed at sorting the products but to get information about the process and, by using a feedback control, make sure to produce only good products. Typical production processes can be separated in one or more chains of production steps. In many cases, these production steps are quality-controlled individually (Figure 33.6). This has the advantages of a faster response and in general less complex inspection tasks compared to the end-of-line test.

33.5 QUALITATIVE INSPECTION

Quality inspection tasks can be subdivided into qualitative or quantitative inspection. Only for quantitative inspection, measurements are required. Typical examples of qualitative inspection are the check of the completeness of an assembly and the search for irregularities on a surface (scratches, dirt, dents, etc.). Often these inspections are carried out by human workers, because the tasks may be quite complex and the flexibility and brainpower of humans is still superior to any machine. However, human inspectors cannot guarantee for a continuous high level of quality, because humans tend to get distracted or tired when doing repetitive routine jobs for longer time. On the other hand, the constant reliable operation is a strength of machine vision systems and in many applications, these have been implemented successfully.

33.5.1   AUTOMATED VISUAL INSPECTION

The main components of a machine vision system are at least one electronic camera, an illumination system, a computer hardware, and an image-processing software. Electronic cameras are built around a solid state image sensor, either change-coupled device (CCD) or complementary metal-oxide-semiconductor (CMOS) type. Typical machine vision cameras contain a monochromatic image sensor with a resolution of about 1 Mpixel that is sufficient for most applications and—compared to consumer cameras with more than 10 Mpixels and RGB-color-coded output—reduces the load on interface bandwidth, memory space, and computation power. The pixel size typically is considerably larger than in consumer cameras, resulting in an improved S/N ratio. Most cameras feature an integrated A/D converter and a standardized interface (USB 2.0, USB 3.0 or IEEE 1394a).

The images are evaluated based on the detection of contrast. The most simple case is intensity contrast that can be treated by defining thresholds, either fixed or adaptive. Color contrast is usually transferred to intensity contrast by applying appropriate color filters either in the illumination or in the imaging path. The most difficult task is the evaluation of pattern contrast that requires special image-processing operations like spatial frequency filtering or filters that detect preferred orientations of edges (Figure 33.7).

As a general rule, much effort should be invested in the design of the illumination to enhance the details of interest already in the raw images. Thus, the resulting system will be much more robust and reliable than one depending only on image-processing algorithms. The contrast between different materials often can be enhanced by making use of physical effects. The plastic parts of an electronic board under ultraviolet illumination may look bright against metallic parts because of fluorescence. In other cases, polarized light might be appropriate to enhance the contrast.

33.5.2   COMPLETENESS CHECK

A typical example of a high-volume completeness check is the automated optical inspection (AOI) of electronic boards. Because of their flat and regular geometry, the boards can be easily fixed in a well-defined position for inspection. The most simple evaluation is based on a digital reference image of a completely assembled board that is stored in the computer memory. For each serial board, a digital image is taken with an electronic camera under the same illumination conditions and this image is subtracted pixel by pixel from the reference image. A missing component can be easily identified as an object in the difference image.

This simple approach is limited to applications without variations in the components. Some assemblies, for example, a dashboard of a car, may vary in configuration and color from car to car. In such a case, the inspection has to be much more complex. Pattern recognition algorithms are used to identify and localize the different components of the assembly, and the result is compared to the basic schedule of the inspected product.

33.5.3   SURFACE CHECK

Human inspectors searching for irregularities on a surface make use of variation of illumination, for example, by looking at the surface from different directions under controlled directed illumination. This approach is adapted by the designers of machine vision systems. Small irregularities on a surface like scratches or dirt are very sensitively detected by using a directed illumination in dark

Intensity contrast

Pattern contrast

FIGURE 33.7   Intensity and pattern contrast.

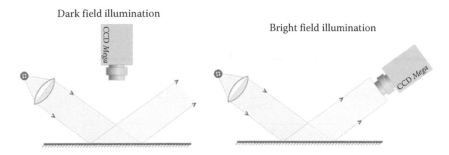

FIGURE 33.8   Different techniques to apply directed illumination for contrast enhancement.

field configuration (Figure 33.8). In case of a perfect surface, the light is reflected past the camera lens, resulting in a dark image. At irregularities, the light is scattered and bright spots or lines are visible in the image. Small dents or waviness on sheet metal on the other hand can very sensitively be detected using directed illumination in bright field configuration and grazing incidence (Figure 33.8). Changes in surface slope are translated to intensity contrast that can be detected automatically in the image.

33.6   QUANTITATIVE INSPECTION

Each workpiece has to fulfill certain requirements, for example, to fit in an assembly. Being aware of the variations unavoidable in every production process, the designer defines not only nominal values for the critical quantities of the workpiece but in addition a tolerance region between a lower and an upper specification limit, L_{SL} and U_{SL}, respectively. In quantitative inspection, all critical quantities of a manufactured part have to be measured and the measured value has to be compared to the related tolerance region. However, not only the production process is disturbed by influence factors, but also the measurement process. Therefore, each measurement value may deviate to some extent from the true value.

The measurement deviations can be divided into systematic and stochastic ones. Systematic deviations are repeatable under constant conditions, and therefore, they can be determined by appropriate calibration procedures. Measurement values can be corrected for known systematic deviations, but not for stochastic ones. For these, a statistical analysis is required.

33.6.1   Measurement Uncertainty

According to the international vocabulary of metrology [4], measurement uncertainty is a nonnegative parameter characterizing the dispersion of the quantity values being attributed to a measurand, based on the information used. The GUM (*Guide to the Expression of Uncertainty in Measurement*) describes a procedure to systematically identify the parameters that have an influence on the measurement, describe this in a mathematical model, estimate the variation range of the values of all relevant influence factors, and—by applying the mathematical model—estimate the confidence interval, that is, the interval within which the measurement value is to be expected with a probability of 95%. The half-width of this interval is taken as the measurement uncertainty U. The documentation of a complete measurement result always has to include the measurement uncertainty.

33.6.2   Measurement Capability

In many cases, the tolerance requirements are quite tough and the measurement uncertainty cannot be neglected compared to the width of the tolerance region. In Figure 33.9, a symmetrical tolerance band around a nominal value N is depicted, and it is assumed that the measurement

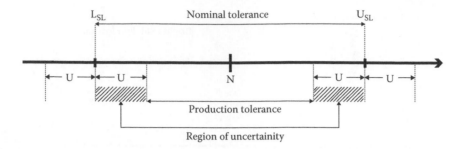

FIGURE 33.9 Reduction of the available tolerance range caused by measurement uncertainty. *Notes*: N, nominal value; LSL, lower specification limit; USL, upper specification limit; U, measurement uncertainity.

uncertainty has the value U. It is important to be aware that only measurement values between the boundaries $(L_{SL} + U)$ and $(U_{SL} - U)$ indicate a workpiece within the specified tolerance limits with a probability >95%. Measurement values within $\pm U$ around the specification limits cannot be interpreted with sufficient reliability. Due to the nonperfect measurement process, the width of the tolerance band in effect is reduced by 2U. The investment cost of sophisticated measurement devices with small uncertainty U might therefore be more than compensated by reduced demands on production equipment because of a larger effective production tolerance range.

Similar to the process capability index C_p, a measurement capability index C_g has been introduced to evaluate the appropriateness of a measurement process for a given application. It is defined as the width of the tolerance region related to the width of the confidence interval:

$$C_g = \frac{U_{SL} - L_{SL}}{2U}.$$

In general, only measurement processes with $C_g > 5$ are considered to be capable.

33.7 OPTICAL TECHNIQUES FREQUENTLY USED IN PRODUCTION METROLOGY

The variety of optical measurement techniques can be arranged according to the dimensionality. Some of the most frequently used techniques are presented in Section 33.7.1.

33.7.1 ONE-DIMENSIONAL TECHNIQUES

A large number of simple devices like light barriers and switching devices are used for counting and checking the presence of workpieces in automated machines.

A very versatile optical distance sensor is the triangulation sensor, shown in Figure 33.10.

The bright spot illuminated by the laser is imaged on a position-sensitive diode (PSD) or a linear CCD image sensor. The distance of the surface is translated to the position of the image of the spot on the sensor. Usually, the sensor is tilted according to the Scheimpflug condition to secure a sharp image within the complete measurement range. Triangulation sensors are available with measurement ranges between 10 mm and 1 m and a relative accuracy of up to 10^{-5} of the measurement range.

An important task in production metrology is the thickness measurement of continuous products like wire or extruded profiles. A special class of optical instruments applied in this field is called

FIGURE 33.10   Triangulation sensor.

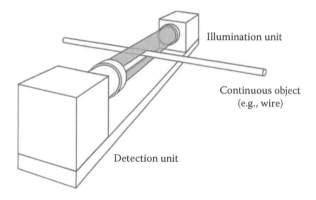

FIGURE 33.11   Optical micrometer.

optical micrometer (Figure 33.11). These consist of an illumination unit and a detection unit oppo-site to one another. The light traveling from the illumination unit is precisely collimated so that the shadow cast by an object keeps its width till it is analyzed by the detection unit. Different kinds of such instruments are available with either continuous illumination or scanned laser beams. If the diameter of the workpiece is large, a so-called optical caliper is built by combining two sensor heads, each detecting one of two opposite edge positions of the workpiece. The diameter is then calculated from both sensor outputs and their distance from one another.

33.7.2   Two-Dimensional Techniques

For two-dimensional geometrical measurements, usually, electronic cameras are applied. Best con-trast of the edges is obtained when using backside illumination. In this case, the workpiece is placed on a glass plate so that no mechanical fixtures are needed. Special instruments incorporating this principle are profile projectors for large- and medium-sized parts and measuring microscopes for small parts. Care should be taken concerning the objective lens, because the accuracy is dependent on its property. Even good lenses show distortion errors of about 10^{-4}; therefore, a calibration of the imaging system is required for precision measurements. Conventional lenses show a distance-dependent magnification factor; therefore, often so-called telecentric lenses are preferred that are designed to have a fixed magnification factor within a certain distance range.

33.7.3 THREE-DIMENSIONAL TECHNIQUES

A variety of optical techniques for three-dimensional measurement are summarized under the term "triangulation." As a common attribute, all these techniques have at least two optical systems with mutually inclined optical axes. In the case of active triangulation, one of the optical systems is a pattern projector and the other is a camera as shown in Figure 33.12a. Most often, a parallel line grid is projected onto the surface. The surface profile causes a deformation of the pattern when viewed by the camera. The information about the surface is coded in these deformed fringes. Usually, either a sinusoidal pattern is used and a phase-shifting evaluation is applied or the measurement is based on a set of binary line patterns using the Gray-code approach. Details of fringe processing can be found in Chapter 21. Both the optical systems of the projector and the camera have to be calibrated. Two sets of parameters are determined: The parameters of external orientation carry the information about the position and orientation of camera and projector in space. The properties of the two imaging systems are described by their parameters of internal orientation.

Figure 33.12b shows the principle of a passive triangulation setup. In this case, two cameras are combined. If an object point can be identified in both camera images, then its position in space can be uniquely calculated. In this case, the external and internal orientations of the two cameras have to be determined by calibration. Passive triangulation with two cameras is also known as stereophotogrammetry. Often a pattern projector is also used in passive triangulation. The application of special patterns can make the evaluation procedure easier, faster, and more precise.

The basic difference between active and passive triangulation is that in active triangulation, the projector is part of the reference frame. Because of the light source inside of the projector, it is more likely to be influenced by thermal effects than a camera. The passive triangulation technique therefore has better preconditions for longtime stability and accuracy.

The performance of fringe projection techniques—active as well as passive ones—is strongly dependent on the surface properties of the workpiece. Diffusely scattering surfaces, for example, etched metal, are ideal for these techniques. Many ceramics and plastics are translucent, that is, the light is not scattered from the surface but from somewhere below the surface. Such volume scattering causes deviations in the measured geometry.

Specularly reflecting surfaces on the other hand will reflect the measurement light according to the law of reflection, and only by chance, some reflected rays may enter the camera lens. Transparent surfaces are also not suited for fringe projection, because most of the projected light is transmitted. In practice, reflecting and transparent surfaces are prepared for the application of fringe projection

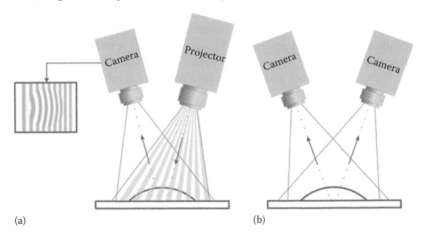

(a)   (b)

FIGURE 33.12 Basic triangulation setups (a) active triangulation and (b) passive triangulation.

techniques by covering the surfaces with diffusely scattering white powder. But this changes the surface profile slightly, and it causes additional expenditure of work for covering and subsequent cleaning of the surface.

It is a common experience that images generated by curved mirrors show characteristic deformations that carry information about the geometry of the reflective surface. This property is used since long time to check the homogeneity of polished surfaces by looking at the mirror images of line grids. In modern test setups, the grid pattern often is displayed by a flat screen monitor, so that the pattern can be adapted quickly and easily to different testing tasks. This so-called deflectometric technique can be modified to measure the absolute geometry of reflecting and transmitting surfaces. In [5], a correspondence of different deflectometric approaches to the basic principles of active and passive triangulation is outlined.

33.8   TECHNIQUES TO MINIMIZE THE EFFECTS OF ENVIRONMENTAL LIGHT

Many optical measurement instruments like triangulation and light-sectioning sensors, 3D-sensors with structured illumination, speckle-pattern-interferometers, or laser vibrometers depend on the detection and evaluation of light that is projected onto the surface of a workpiece. Any additional light components might disturb the measurement.

While in optical research laboratories experiments can be performed in darkness to prevent erroneous effects of environmental light, optical measurement instruments in industrial applications have to work under varying lighting conditions. Some amount of daylight—dependent on daytime and weather conditions—may be mixed with artificial illumination from incandescent lamps, fluorescent lamps, or LED lamps. In addition, there might be some radiation from hot parts of a machine or the workpiece itself. A number of countermeasures are available to minimize the negative effects of environmental light.

33.8.1   SHADING

Environmental light sources have to be shielded, when possible. A lens hood can prevent part of the light from entering the optical system. A further reduction of the erroneous light is often possible by inserting baffles into the optical system. In some cases, it is possible to put the complete optical system into an enclosure.

33.8.2   OPTICAL FILTERING

Environmental light most often is unpolarized and spectrally broadband. A considerable reduction of erroneous effects is often possible by making use of a polarized small-band light source for the measurement instrument. A polarizing filter in front of the imaging lens will block the fraction of the unwanted light that is orthogonally polarized to the measurement light. A color filter matched to the spectrum of the measurement light will reduce the effect of broadband environmental light by an order of magnitude or even more. If the interfering light is limited to a certain spectral region, then measurement light from a different spectral region should be applied. A red glowing hot workpiece, for example, would be difficult to measure with a light-sectioning sensor with red laser light. A blue laser diode would give a much better contrast and its light can be extracted easily with an appropriate filter.

Because of their narrow linewidth, lasers are ideally suited for stray light suppression by spectral filtering. On the other hand, their coherent light causes speckle-structures that cause problems in many metrological applications. For camera-based measurement techniques, LED-illumination or spectrally filtered incandescent light therefore is a better choice.

33.8.3 MODULATION TECHNIQUES

In many optical sensors, the illumination light is electrically modulated, for example, sinusoidally. A matched bandpass filter is applied to the electric output signal to extract the measurement information from the noise caused by the environmental light. The suppression of unwanted signal components is improved with a reduction of the bandwidth of the filter, while on the other hand, the dynamic properties of the sensor are degraded. The S/N ratio of a measurement system with modulated light can be further improved by phase-sensitive filtering (lock-in technique). In camera-based systems, the modulation technique is applied by alternatively taking pictures with illumination switched on and off and by subtracting the adjacent digital images. A combination of shading, optical filtering, and modulation techniques makes optical measurement instruments applicable even under unfavorable lighting conditions.

REFERENCES

1. Guan, J. Interferometric encoders for linear displacement metrology, PhD thesis, Technische Universitaet Braunschweig, Braunschweig, Germany, 2013.
2. Schneider, C.T. Lasertracer—A new type of self tracking laser interferometer. *Proceedings of IWAA 2004*, CERN, Geneva, Switzerland, October 4–7, 2004.
3. Kreis, T. *Handbook of Holographic Interferometry: Optical and Digital Methods*, Wiley-VCH Verlag GmbH & Co. KgaA, Weinheim, Germany, FRG, 2005. ISBN: 9783527405466.
4. International Vocabulary of Metrology (VIM). *ISO/IEC Guide 99:2007*, 3rd edn., Beuth Verlag, Berlin, Germany, 2010.
5. Tutsch, R., Petz, M., and Fischer, M. Optical 3D metrology with structured illumination. *Opt. Eng.* 50 (10), 101507, 2011. doi:10.1117/1.3578448.

34 On-Machine Measurements

Takashi Nomura and Kazuhide Kamiya

CONTENTS

34.1 Introduction 867
34.2 Zone-Plate Interferometer 867
    34.2.1 Zone-Plate Interferometer of A-Type 867
    34.2.2 Zone-Plate Interferometer of B-Type...871
    34.2.3 Zone-Plate Interferometer of C-Type...875
34.3 Lateral-Shearing Interferometer...881
References...886

34.1 INTRODUCTION

Precision mirrors for optical instruments are manufactured by grinding, polishing, or cutting substrates. To manufacture the mirrors with higher precision shape, many cycles of measurements and corrections by the polishing or cutting are repeated in the manufacturing process. The manufacturing time is, therefore, long. And furthermore, the setting error of the mirrors is caused by attaching and detaching the mirrors. To avoid the error and to reduce the time of machining, the shape measurements of the mirrors on the manufacturing machine are essential. As a geometrical optical measurement, a new type of Hartmann test was developed using an optical fiber grating and applied on the manufacturing machine [1]. However, the data obtained by the method are only located at individual spots. In interference methods, on the other hand, the data are obtained at each pixel. Therefore, the spatial resolution of the interference methods is higher than that of the geometrical optical method. This chapter discusses some on-machine shape measurements by using interferometer.

34.2 ZONE-PLATE INTERFEROMETER

A zone-plate interferometer, which is a common-path instrument, was proposed by Murty [2] to measure the shape error of a large concave mirror and zone-plate interferometers of several mode types were developed by Smartt [3]. To test aspherical surfaces, a holographic zone-plate interferometer was presented by Nakajima [4]. A zone-plate interferometer is not much affected by air turbulence and machine vibrations. The zone-plate interferometer needs no real reference surfaces because the reference wavefront produced by a zone-plate is used as a reference surface. Spherical or aspherical surfaces, therefore, can be measured easily by using the zone-plate interferometer. In this section, zone-plate interferometers of three types are explained.

34.2.1 ZONE-PLATE INTERFEROMETER OF A-TYPE

Test surfaces used for the measurement are symmetrical with respect to an optical axis z. A zone-plate is a circular grating such as a Newton ring, as shown in Figure 34.1. The zone-plate must be arranged perpendicularly with respect to the optical axis. Figure 34.2 shows that illumination beams are transmitted and diffracted through the zone-plate [5]. The beams transmitted toward the

FIGURE 34.1 Part of zone-plate.

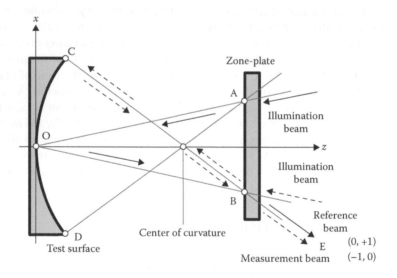

FIGURE 34.2 Diffracted and transmitted beams through a zone-plate.

test surface focus at the center O of the test surface. Part of the beams diffracted to many directions is used for the zone-plate interferometer. The direction of the beams used for the interferometer is perpendicular to the test surface. The measurement of an arbitrary point C on the test surface is performed as follows. Points C and D are symmetrical with respect to the optical axis. Points A and B on the zone-plate are symmetrical with respect to the optical axis. The arrow in the solid line passes through the zone-plate at A, reflects at the center O of the test surface, diffracts at B to the direction E. We denote the diffracted diverging light beam and the diffracted converging light beam of the first order as +1 and −1, respectively. The light beam of (0, +1), of which a typical path is indicated as A–O–B–E in Figure 34.2, generates a wavefront, does not involve the shape error, and acts as a reference beam.

The (−1) beam diffracted at B, as shown by the arrow in the broken line, goes to the test surface and reflects at C. When the shape of the test surface is manufactured precisely as the designed shape, and the test surface and the zone-plate are located in position, the arrow in the broken line reflects perpendicularly at C, passes through B to the direction E. The light beam (−1, 0), of which a typical path is indicated as B–C–B–E in Figure 34.2, generates a wavefront, involves the shape error, and acts as a measurement beam. Reference and measurement light beams make a fringe pattern on an image plane. We called the zone-plate interferometer, zone-plate interferometer of A-type.

A zone-plate has the following transmittivity distribution $T(x)$ by which the reference beam reconstructs a designed surface:

$$T(x) = \frac{1}{2}\left\{1 + \cos\left[\frac{2\pi(AO - AD)}{\lambda}\right]\right\} \tag{34.1}$$

where
   λ is the wavelength of a light source
   AO is the distance between A and O
   AD is the distance between A and D

Because the designed shape of the test surface is given, $T(x)$ is calculated by a computer. The reference and the measurement beams pass through the aperture and interfere on the image plane. The interference fringes indicate the shape difference between the designed surface and the test surface. Schematic diagram of the zone-plate interferometer is shown in Figure 34.3.

If the zone-plate is not located precisely in the right position, precise measurements cannot be performed. In the interferometer, the arrangement of the positioning of the zone-plate is performed by moire technique: A moire fringe pattern appears when the interference fringes formed by beams 2 and 3 interfere with the zone-plate grating. When the zone-plate is located in the right place, the moire fringe pattern of a spherical test surface is null and the pattern of an aspherical surface is concentric circles.

To observe the interference fringes on the image plane, only beam 2 is used whereas beam 3 is cut off with mirror 5 inserted between beam splitter 1 and lens 3. Beam 2 is reflected on the test surface and becomes beam 4 which is shown as reference and measurement beams in Figure 34.2. Beam 4 is guided with mirrors 5 and 6 to the rear surface of the interferometer. Unnecessary diffraction beams shown by broken lines in Figure 34.3 are eliminated by an aperture in order to observe clear interference fringes on the image plane. The fringe pattern on the image plane is taken by a charge coupled device (CCD) camera.

Figure 34.4 is a photograph of the zone-plate interferometer mounted on an ultraprecision cutting machine. The zone-plate used for the experiment is a photo plate and has a pattern calculated to reproduce the shape of the designed surface.

Two test surfaces are spherical concave mirrors with a curvature radius of 140 mm and a measuring diameter of 10 mm and are manufactured by the precision cutting machine. One mirror with very small error is called the mirror without shape error. The other is called the mirror with shape error. The mirrors were measured during machine running at a speed of 900 rpm, the same as the cutting speed of the mirrors.

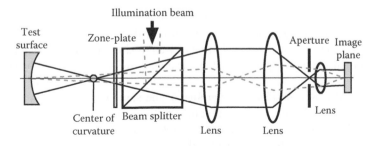

FIGURE 34.3 Schematic diagram of the zone-plate interferometer of A-type.

FIGURE 34.4   Photograph of the zone-plate interferometer of A-type.

FIGURE 34.5   Interference fringes obtained by moving the zone-plate interferometer perpendicularly to the optical axis (spherical mirror without shape error).

Experimental result of the mirror without shape error is shown in Figure 34.5. The fringe pattern is modulated by a carrier component. A fringe pattern of the spherical mirror with shape error is shown in Figure 34.6. The shape error is obtained by analyzing the fringe distortion. Figure 34.7a and b is a three-dimensional (3D) plot and a contour map of the shape error of the spherical mirror, respectively.

To verify the results by zone-plate interferometer, the same test surfaces were measured with the Fizeau interferometer. Fringe patterns obtained with the Fizeau interferometer are shown in Figures 34.8 and 34.9. Figure 34.10a and b is a 3D plot and a contour map of the shape error of the spherical mirror, respectively. They are calculated by a fringe-scanning method [6]. Comparison of Figure 34.7 with Figure 34.10 shows that the results obtained by two different methods agree well.

FIGURE 34.6 Interference fringes obtained by moving the zone-plate interferometer perpendicularly to the optical axis (spherical mirror with shape error).

(a)

(b)

FIGURE 34.7 (a) Three-dimensional plot of the shape error of a spherical mirror analyzing by the zone-plate interferometer and (b) contour map of the shape error of a spherical mirror analyzing by the zone-plate interferometer. The height between adjacent contour lines is 0.05 μm.

34.2.2 Zone-Plate Interferometer of B-Type

In the zone-plate interferometer of A-type, a diffracted light beam of $(-1, 0)$ is used as a test light beam and that of $(0, +1)$ is used as a reference light beam. Numerical aperture (NA) of the lens, which is used as an observation lens, depends on the NA of the mirror being tested because the lens has the focal point at the center of the curvature of the mirror and the light emanating from the entire surface of the mirror to the lens expands over the lens. When a mirror with a large NA is measured, an observation lens with an NA larger than that of the mirror should be used. Furthermore, when the mirror surface has a steep asphericity, the light focused by an imaging lens is astigmatic. The light does not converge to a small point but forms a blurred point. Use of an aperture stop with a large diameter does not result in a clear fringe pattern because extraneous diffraction wavefronts are not effectively blocked out by the aperture stop. In the former interferometer, therefore, it is difficult to measure the shape error of mirrors with a large NA and a steep asphericity.

FIGURE 34.8   Fizeau interference fringes (spherical mirror without shape error).

FIGURE 34.9   Fizeau interference fringes (spherical mirror with shape error).

Honda et al. proposed a zone-plate interferometer to measure the primary mirror of an astronomical telescope with a large aperture; a collimated laser light beam is used as the incident light to a zone-plate [7]. A diffracted light beam of (+1, +1) and that of (0, 0) are used as the test light beam and the reference light beam, respectively. Because the mirror has a hole at its center, another small flat mirror is placed in the hole as a reference mirror. A mirror with a steep asphericity can be measured by the interferometer. The fringes obtained by the interferometer are, however, affected by the machine vibration because the reference mirror and the mirror under test have respective vibrations.

(a) (b)

FIGURE 34.10 (a) Three-dimensional plot of the shape error of a spherical mirror analyzing by the Fizeau interferometer and (b) contour map of the shape error of a spherical mirror analyzing by the Fizeau interferometer. The height between adjacent contour lines is 0.05 μm.

To measure the shape error of mirrors with a large NA and a steep asphericity on the machine, a zone-plate interferometer applying a diffracted light beam of (−1, −1) and of (0, 0) is developed [8].

The zone-plate is illuminated by a laser light beam converging on the center O of the mirror surface as shown in Figure 34.11. An illumination system, not shown in this figure, is inserted between lenses L3 and L6. The zone-plate is made in such a way that one of the first-order diffracted lights illuminates the entire surface of the mirror. The diffracted light beam of −1, of which a typical path is indicated as E–P–B–C in Figure 34.11, generates a wavefront impinging on the mirror surface perpendicularly to the designed one.

The light beam is reflected by the mirror surface being tested and impinges on the zone-plate. It is again diffracted by the zone-plate. The diffracted light beam travels along a path indicated

FIGURE 34.11 Schematic diagram of the zone-plate interferometer of B-type. *Note:* AS, aperture stop; BS, beam splitter; L, lens; M, mirror. (From Nomura, T. et al., *Opt. Rev.*, 3, 34, 1996. With permission.)

as B–P–E. The wavefront of the diffracted light beam involves the shape error of the mirror being tested and, therefore, the diffracted light beam of (–1, –1) is used as the test light beam. The diffracted light beam of (0, 0), actually non-diffracted light beam, of which typical path is indicated as F–Q–A–O–B–P–E, generates the spherical wavefront coming from the center of the mirror surface being tested so that the wavefront does not involve the shape error and is used as the reference light beam. Fringes obtained by the interferometer are scarcely affected by the machine vibrations, because the reference position is located at the center O of the mirror being tested.

Lens L3 used as an illumination lens is also used as an observation lens. The lens has the focal point at the center of the mirror surface and receives only lights reflected on the area between P and Q, as shown in Figure 34.11, that is, the concave mirror under test lens does not depend on the NA of the test mirror. The shape error of a mirror with a large NA is, therefore, measured by the improved interferometer. Because the wavefront from lens L3 to lens L6 is a plane, a small spot appears at aperture stop AS2. Extraneous wavefronts are, therefore, blocked out almost completely by the aperture stop. Clear interference fringes are observed even though the mirror surface has a large steep asphericity. The zone-plate interferometer of B-type is shown in Figure 34.12.

The dimensions of the zone-plate interferometer of B-type are $300 \times 220 \times 315$ mm. Optically parallel glass plates with a diameter of 50.0 mm and a thickness of 3.00 mm were used as substrates of the zone-plates. The zone-plates used for the experiments were of phase type and were manufactured with an instrument consisting of an ultraprecision lathe and an Ar laser with a wavelength of 458 nm [9]. Pitch manufactured with the instrument is limited to a minimum of 2.0 μm. The accuracy of the zone-plate grating is 0.1 μm. The zone-plates were manufactured by the following procedure: (1) the glass plates were coated with photoresist so that they could be used as photographic plates; (2) the photographic plates were mounted in the chuck of the lathe and rotated; (3) grating lines were drawn on the plates by exposing them to a beam from an Ar laser mounted on the lathe; (4) the photoresist exposed to the laser beam was removed by development; (5) the entire surfaces of the plates were coated by SiO_2; and (6) the plates were submerged in a solvent and the unexposed photoresist and its SiO_2 coating were removed, leaving behind the SiO_2 coated on the exposed grating lines. Therefore, the zone-plate gratings have a binary phase structure [7]. The back surfaces of the zone-plates were coated with SiO_2 to prevent

FIGURE 34.12 Photograph of the zone-plate interferometer of B-type. (From Nomura, T. et al., *Opt. Rev.*, 3, 34, 1996. With permission.)

FIGURE 34.13 Fringe pattern of a spherical mirror obtained by the improved interferometer. (From Nomura, T. et al., *Opt. Rev.*, 3, 34, 1996. With permission.)

reflections from the surfaces. The intensity of the test light beam should be equal to that of the reference light beam to obtain interference fringes with a high visibility. The zone-plates were made in such a way that the first-order diffraction efficiency of the zone-plates was equal to the zero-order diffraction efficiency.

Two metallic mirrors were manufactured by the ultraprecision cutting machine and used as mirrors to be tested. One was a spherical mirror with a curvature radius r of 140 mm and a diameter d of 30 mm (NA: $d/r = 0.21$). The other was a parabolic mirror with a focal length f of 70 mm, a diameter of 50 mm (NA: $d/2f = 0.36$), and the shape difference from the same sphere was 4.45 μm maximum at the margin.

The interferometer was moved perpendicularly to the optical axis to obtain straight interference fringes. Figure 34.13 shows a fringe pattern formed on the spherical mirror; this fringe pattern was observed on the entire surface of the mirror in spite of the large NA. The fringe pattern of the parabolic mirror was observed in spite of its steep asphericity (Figure 34.14).

34.2.3 Zone-Plate Interferometer of C-Type

A simple type of zone-plate interferometer to measure precisely the positioning error of a cutting tool is discussed [10]. We called the interferometer, zone-plate interferometer of C-type. The zone-plate is made in such a way that the first-order diffracted light beam that diverges from the center line travels in a direction perpendicular to the mirror surface and the diffracted light beam that converges is directed to the vertex of the mirror. The diverging diffracted light beam is reflected from the entire surface of the mirror and is nominally recollimated at the zone-plate. Therefore the light beam (+1, +1), of which a typical path is indicated as B–D–B–F in Figure 34.15, generates a wavefront, involves the shape error, and acts as a measurement beam. On the other hand, the converging diffracted light beam is reflected at the vertex of the mirror and is recollimated at the zone-plate. Therefore the light beam of (−1, −1), of which a typical path is indicated as A–O–B–F in Figure 34.15, generates a wavefront, does not involve the shape error, and acts as a reference beam. Other unnecessary rays of diffracted light beam are removed by an aperture stop. Reference and measurement light beams make a fringe pattern on an image plane.

FIGURE 34.14 Fringe pattern of a parabolic mirror obtained by the zone-plate interferometer of B-type. (From Nomura, T. et al., *Opt. Rev.*, 3, 34, 1996. With permission.)

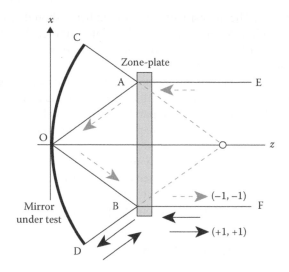

FIGURE 34.15 Diffracted and transmitted beams through a zone-plate.

Schematic diagram of the zone-plate interferometer C-type is shown in Figure 34.16. A zone-plate is set at the midposition between the vertex of the spherical mirror being tested and the center of curvature of the mirror. The error in the shape of the mirror and the positioning error of the tool can be determined by analyzing the interference fringes.

To determine the position of the mirror, reference flat mirror M5 is used. The collimated light reflected on beam splitter BS goes to mirror M5, is reflected by it, and goes to the CCD camera. When mask 1, which has a hole in it, is set in front of the mirror under test, the light reflected by mirror M5 and the light reflected at the vertex of the mirror under test make interference fringes on the image plane of the CCD camera. The mirror under test is adjusted until null interference fringes appear, and the mirror is then in the correct position. When the mirror is being measured, mask 1 in front of the mirror is removed and mask 2 is inserted in front of reference flat mirror M5.

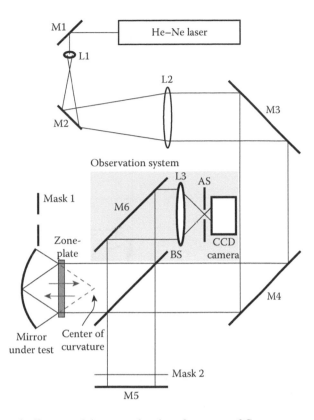

FIGURE 34.16 Schematic diagram of the zone-plate interferometer of C-type.

The location of the tool is indicated by Cartesian coordinates, as shown in Figure 34.17. The z-axis is along the spindle of the lathe and the x- and y-axes are perpendicular to the z-axis. The cutting tool moves in the xz-plane from the minus side of the x-axis to the vertex of the mirror. We assume that the feed error of the tool is negligible. The surface of a mirror is concave and symmetrical with respect to the z-axis. The positioning error in the y-direction causes an error in the

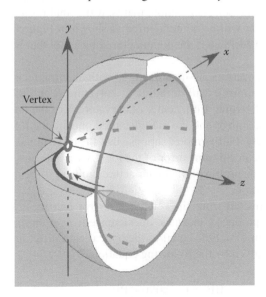

FIGURE 34.17 Location of a tool in Cartesian coordinates.

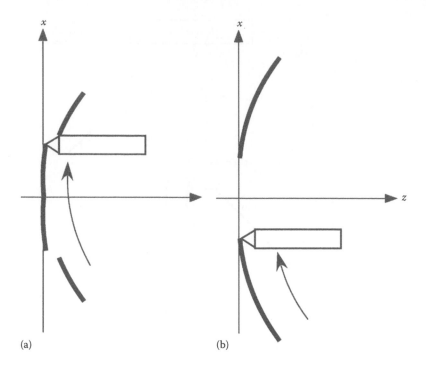

FIGURE 34.18 Profiles of concave mirrors manufactured with an ultraprecision lathe; (a) tool is mounted on the plus side of the correct position and (b) tool is mounted on the minus side of the correct position.

shape near the vertex. This positioning error can be determined by observing the shape near the vertex with a microscope and removed easily. The positioning error in the z-direction causes an error in the thickness of the mirror but does not cause an error in the shape of the mirror surface. The positioning error in the x-direction influences the entire surface of the mirror, and is discussed in this chapter.

Let us suppose that a spherical concave mirror is being manufactured with a tool that moves from the minus side of the x-axis to the vertex of the mirror. If the tool is positioned on the positive side of the correct position side of the correct position, that is, if the final position of the tool is on the plus side of the x-axis, the mirror will be manufactured as shown in Figure 34.18a. On the other hand, the mirror shown in Figure 34.18b will be manufactured by a tool that is positioned on the negative side of the correct position.

Spherical and aspherical mirrors are manufactured with an ultraprecision lathe. Suppose that the designed spherical concave mirror with a radius of curvature R and the manufactured mirror are illuminated by laser light diverging from center of curvature A of the reference mirror, as shown in Figure 34.19. The designed mirror and the manufactured one are the reference mirror and the mirror being tested, respectively. Furthermore, suppose that the error in the shape of the manufactured mirror is caused only by the positioning error of the tool. Arc a indicates the profile of the reference mirror and arc b indicates the profile of the manufactured mirror, if the tool is attached on the negative side of the correct position.

The reference light reflected from the reference mirror and the test light reflected from the manufactured mirror will produce concentric interference fringes on the mirror surface. Twice the distance BB′ shown in Figure 34.19 is equal to the optical path difference between the two rays of light. The distance BB′ is given by the following equation:

$$\mathrm{BB}' = \mathrm{BC}\cos\theta = \mathrm{BC}\frac{(R^2 - r^2)^{1/2}}{R} \tag{34.2}$$

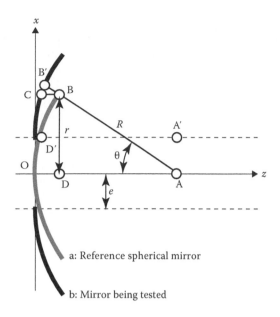

a: Reference spherical mirror

b: Mirror being tested

FIGURE 34.19 Optical path difference between the reference spherical mirror and mirror being tested.

where BC is given by Equation 34.3.

$$BC' = A'D' - AD$$

$$= [R^2 - (r - e)]^{1/2} - (R^2 - r^2)^{1/2} \tag{34.3}$$

If N is the fringe order and λ is the wavelength of the laser light, BB′ can be expressed by Equation 34.4.

$$BB' = \frac{N\lambda}{2} \tag{34.4}$$

The error e is derived from Equations 34.2 through 34.4 and is given by the following equation:

$$e = r - \left\{ R^2 - \left[\frac{NR\lambda}{2(R^2 - r^2)^{1/2}} + (R^2 - r^2)^{1/2} \right]^2 \right\}^{1/2} \tag{34.5}$$

When r is sufficiently small compared to R and λ is sufficiently small compared to r, the above equation becomes approximately

$$e = \frac{NR\lambda}{2r} \tag{34.6}$$

This equation is also valid when the tool is mounted on the positive side of the correct position, as shown in Figure 34.18a.

The absolute value of the error is calculated by Equation 34.6. The sign of the positioning error is determined by the following method. The manufactured mirror is moved to the minus side of the x-axis or the interferometer is moved to the plus side until the interference fringes without the carrier component are observed along the x-axis on half of the mirror. If the fringes without the carrier component are observed on the right side of the mirror, the sign of the error is minus because the upper part of the arc a is superimposed on the upper part of the arc b. On the other hand, if the fringes without the carrier component are observed on the left side of the mirror, the sign of the error is plus. The distance between the original position of the mirror and the position after the movement of the mirror is equal to the absolute value of the positioning error.

The zone-plate interferometer of C type is shown in Figure 34.16. That is $410 \times 300 \times 320$ mm in size and is mounted on a moving stage of the ultraprecision lathe. Two spherical concave mirrors with the interferometer are measured. Test mirror 1 had been manufactured with a tool that had some positioning error on the minus side of the x-direction. Test mirror 2 had been manufactured with the tool set at the correct position. The design dimensions of the test mirrors 1 and 2 were a radius of curvature of 140 mm, a diameter of 30 mm, and a thickness of 50 mm. The zone-plate used for the experiments was a computer-generated hologram manufactured with an instrument that consists of an ultraprecision lathe. Optically parallel glass plate with a diameter of 50.0 mm and a thickness of 3.00 mm was used as the substrate of the zone-plate. The minimum pitch of the zone-plate gratings was 3 μm. The first-order diffraction efficiency of the zone-plate was 38%.

When spherical concave mirrors are measured, measurement errors can be caused by a number of factors: the setting of the zone-plate, the optical elements of the zone-plate interferometer, the wavefront transmitted through the substrate of the zone-plate, and the analysis of the fringes. The total error with respect to the shape measurement was less than 0.18 μm, that is, the fringe order was 0.57. By substituting 0.57 into N in Equation 34.6, the error e caused by the tool positioning can be measured with an accuracy of less than 1.7 μm.

Figure 34.20a shows the fringe pattern of the test mirror 1 which was manufactured with the tool with some positioning error. The maximum order N of the interference fringes was nine. When the order of the interference fringes is substituted in Equation 34.4, the positioning error e is 27 μm.

The mirror was moved to the minus side of the x-axis and interference fringes without the carrier component were obtained on the x-axis for half of the mirror. Figure 34.20b shows the fringe pattern when the mirror was moved −27 μm in the x-direction. The positioning error is, therefore, 27 μm on the minus side of the x-axis. Figure 34.20a and b is the interference fringes obtained by the Fizeau interferometer.

Simulations were performed by the ray tracing method for the same conditions as those of the experiments, and the results are shown in Figure 34.21a and b. The fringe patterns obtained in the

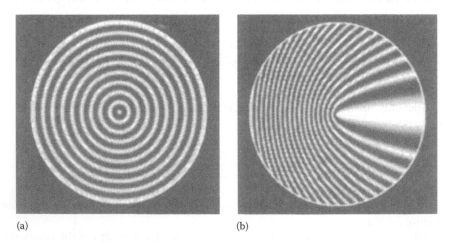

    (a)   (b)

FIGURE 34.20 Fringe patterns obtained by computer simulation for test mirror 1: the positioning error of the cutting tool was −27 μm; (a) fringe pattern without a carrier component and (b) fringe pattern with a carrier component.

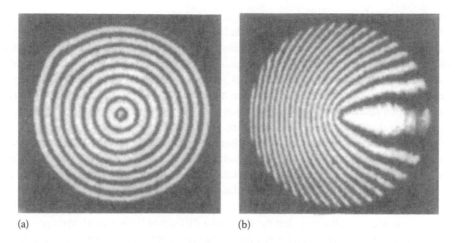

(a)   (b)

FIGURE 34.21   Fringe patterns obtained by the zone-plate interferometer for test mirror 1: (a) fringe pattern without a carrier component and (b) fringe pattern with a carrier component on the half of the mirror along the x-axis.

(a)   (b)

FIGURE 34.22   Experimental results for test mirror 2: (a) fringe pattern obtained by the zone-plate interferometer of C type and (b) fringe pattern obtained by the Fizeau interferometer.

experiments agreed well with the fringe patterns obtained by the computer simulation of the experiments (Figure 34.20). After the experiments, the positioning error of the tool was corrected and test mirror 2 was manufactured with the tool set at the correct position (Figure 34.22).

34.3   LATERAL-SHEARING INTERFEROMETER

Lateral-shearing interferometry has been used extensively in diverse applications such as the testing of optical components and systems and the study of flow and diffusion phenomena in gases and liquids [11]. The method of lateral-shearing interferometry consists of displacing the defective wavefront laterally by a small amount and obtaining the interference pattern between the original and the displaced wavefronts. A lateral-shearing interferometer proposed by Yatagai and Kanou have employed a fringe-scanning method using a piezoelectric-driven mirror to measure the interference fringes obtained by the interferometer precisely [12]. The interferometer is, however, affected by mechanical vibrations and air turbulence because it is not a common-path type. Murty proposed a lateral-shearing interferometer in which a wavefront under test can be sheared by a single parallel flat or a slightly wedge-shaped plate [13]. To produce a compact and lightweight interferometer,

Shakher et al. proposed a shearing interferometer using holo-lenses that are made by the interference between diverging spherical and plane wavefronts [14]. A common path lateral-shearing interferometer with a minimum number of optical components is developed that is not affected by the mechanical vibrations and the air turbulence [14]. In the lateral-shearing interferometer, a plane parallel glass plate is used to shear the wavefront under test. A fringe-scanning method using a slight tilt of the glass plate is used to measure the fringes obtained by the interferometer precisely. In the interferometer discussed in this section, similarly, zone-plates that are computer-generated holograms are used to produce a compact interferometer for measuring spherical and aspherical surfaces.

Figure 34.23 is a schematic diagram of the common-path lateral-shearing interferometer. A laser beam collimated by lenses 1 and 2 becomes a spatially coherent plane wavefront beam and illuminates a zone-plate. The beam is diffracted at the zone-plate and goes to the mirror being tested. In the interferometer, two zone-plates are used. One zone-plate is used to determine the design position of the mirror being tested. The other zone-plate is used to measure the error in the shape of the mirror. The first zone-plate is made in such a way that the first-order diffracted light beam converges at the vertex of the mirror surface. The diffracted light beam reflected from the vertex of the mirror is diffracted at the zone-plate and nominally recollimated. The recollimated light beam is reflected at the beam splitter and goes to a plane parallel glass plate. The glass plate is used to shear the wavefront of the recollimated light, that is, lateral-shearing interference fringes are produced by the optical path difference between the wavefront reflected from the front surface of the parallel glass plate and that from the rear surface.

When the mirror being tested is located at the design position, null interference fringes are obtained. After the mirror has been placed at the design position, the zone-plate is replaced with the other zone-plate to measure the error in the shape of the mirror. The zone-plate is made in such a way that the first-order diffracted light beam converges on the optical axis and travels in a direction perpendicular to the surface of the mirror under test and illuminates the entire surface of the mirror. The light is reflected from the mirror, is diffracted at the zone-plate, and is nominally recollimated. Therefore, the zone-plate functions as a null lens. Extraneous light beams are blocked out almost completely by the aperture stop located at the focal point of the zone-plate.

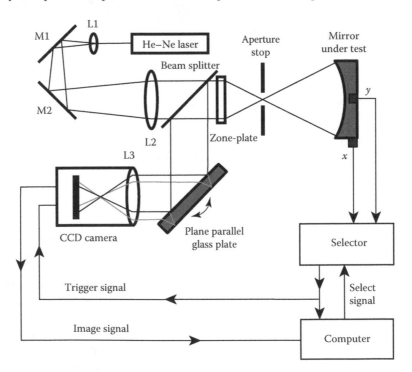

FIGURE 34.23 Schematic diagram of a lateral-shearing interferometer.

The wavefront under test may be expressed as $W(x, y)$, where (x, y) are coordinates of a point of measurement P. When this wavefront is sheared in the x-direction by an amount S, the sheared wavefront at the same point is $W(x - S, y)$. The resulting optical path difference $\Delta W(x, y)$ at P between the original and the sheared wavefronts is $W(x, y) - W(x - S, y)$. When S is small, the optical path difference is written as follows:

$$\Delta W(x, y) = W(x + S, y) - W(x, y)$$

$$\cong \left(\frac{\partial W}{\partial x} \right) S = n\lambda \tag{34.7}$$

where
λ is the wavelength of the light source
n is the order of the interference fringe

The interference fringes obtained by the lateral-shearing interferometer represent the derivative of the aberration in the laterally sheared direction of the wavefront under test. By integrating the phase distribution, the shapes of cross sections of the mirror under test in the laterally sheared direction are obtained. To get the two-dimensional (2-D) wavefront shape, integration is performed in two mutually perpendicular directions, that is, in the x-direction and y-direction. Because the mirror under test is mounted on the chuck of the lathe and revolves on the axis of the lathe spindle, the mirror can be measured at one position and then measured again when the mirror has revolved through 90°. Consequently, mutually perpendicularly sheared interference fringes can be obtained. The shapes of the wavefront cross sections in the two directions are calculated independently. The center line parallel to the y-axis is used for comparing relative heights in the shape of the cross section in the direction. When the shear distance is long, the measurement area becomes small and high-frequency components of the shape error obtained by the fringes are lost. On the other hand, when the shear distance is too short, the wavefront cannot be analyzed because no fringes are obtained. Therefore, the shear ratio is generally from 10% to 20% of the diameter of the wavefront under test. Photograph of the lateral-shearing interferometer is shown in Figure 34.24.

FIGURE 34.24 Photograph of the lateral-shearing interferometer.

The fringe-scanning method [13,14] is applied to analyze the fringes precisely. A slight tilt of the glass plate produces an optical path difference between the two wavefronts, that is, the wavefront reflected from the front surface of the glass plate and that from the rear surface, and gives a change in the phase of the fringes. Four laterally sheared fringe patterns are obtained by a CCD camera. The phase distribution of the fringe pattern is analyzed from these four patterns. Consequently, the error in the shape of the mirror under test is obtained by these calculations.

Two metallic concave mirrors were measured by the interferometer shown in Figure 34.24. One was a concave spherical mirror with a diameter of 100 mm, a radius of curvature of 500 mm, and a thickness of 30 mm. The other was a concave parabolic mirror with a diameter of 50 mm, a focal length of 70 mm, and a thickness of 60 mm. These mirrors had been manufactured on an ultraprecision lathe in advance, were reattached to the chuck of the lathe, and were revolved at 400 rpm. Figure 34.25 shows a photograph of the experiments. The position sensors shown in Figure 34.23 detect the positions of the mirrors, that is, in the x-direction and the y-direction.

When the mirrors reach the directions required to measure, the sensors feed trigger signals to the computer and the CCD camera and fringe patterns are recorded with the computer. Optically plane parallel glass plates with a diameter of 50.0 mm and a thickness of 3.00 mm were used as the substrates of the zone-plates and as the plate used to shear the wavefront under test. The zone-plates used for the experiments were the same zone-plates measured by the zone-plate interferometer of C-type.

Four interference fringes of the concave spherical mirror were obtained with this lateral-shearing interferometer in which the wavefront was sheared in the x-direction. In the measurement, phase shifts were introduced in steps of $\lambda/2$ between each interference fringes; similarly, another four interference fringes of the mirror were obtained with the interferometer in which the wavefront was sheared in the y-direction. The phase distribution analyzed by the fringe-scanning method is shown in Figure 34.26a and b. After the phase unwrap procedure of the phase distribution [9] had been carried out, the phase distributions were integrated along the two sheared directions. The error in the shape of the mirror under test was analyzed from the results. Figure 34.27a and b shows the error in the shape of the concave spherical mirror as measured by the lateral-shearing interferometer.

The spherical mirror was measured by a Fizeau interferometer for comparison with the experimental results as measured by the lateral-shearing interferometer [15]. In the Fizeau interferometer, the fringe-scanning method using a piezo electric driven reference mirror was used. Figure 34.28a and b shows the error in the shape of the mirror as measured by the Fizeau interferometer.

FIGURE 34.25 Photograph of the measurement system.

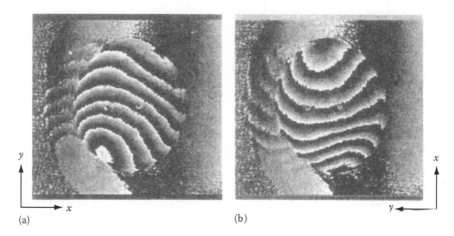

(a)   (b)

FIGURE 34.26   Phase distribution of the concave spherical mirror as obtained by the lateral-shearing inter-ferometer: (a) phase distribution in the x-direction and (b) phase distribution in the y-direction.

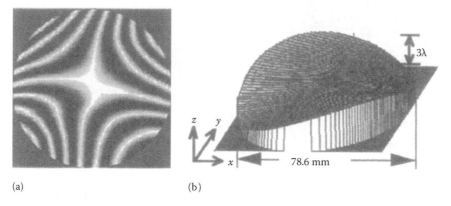

(a)   (b)

FIGURE 34.27   Error in the shape of the spherical concave mirror obtained by the lateral-shearing interfer-ometer: (a) contour map indicating the error in the shape of the concave spherical mirror and (b) 3D plot of the error in the shape sheared directions. The error in the shape of the mirror under test was analyzed from the results.

(a)   (b)

FIGURE 34.28   Error in the shape of the spherical concave mirror obtained by a Fizeau interferometer: (a) interference fringes obtained with a Fizeau interferometer and (b) 3-D plot of the error in the shape.

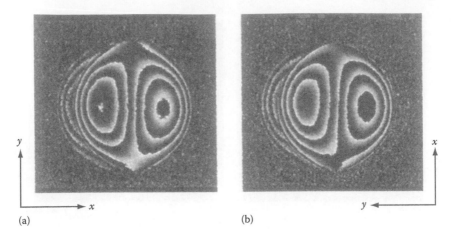

(a)   (b)

FIGURE 34.29 Phase distribution of the parabolic mirror as obtained by the lateral-shearing interferometer: (a) phase distribution in the x-direction and (b) phase distribution in the y-direction.

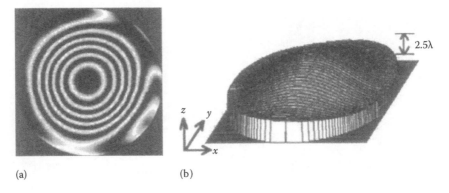

(a)   (b)

FIGURE 34.30 Error in the shape of the parabolic mirror obtained by the lateral-shearing interferometer: (a) contour map indicating the error in the shape of the parabolic mirror and (b) 3-D plot of the error in the shape.

The peak-to-valley value obtained by the Fizeau interferometer was 3.18λ inside the 79 mm diameter circle, indicated as a white circle. These results agreed well with those obtained with the lateral-shearing interferometer. The concave parabolic mirror was also measured by the lateral-shearing interferometer. Figure 34.29a and b shows the phase distribution of the parabolic mirror as measured by the lateral-shearing interferometer. Figure 34.30a and b shows the error in the shape of the parabolic mirror as measured by the lateral-shearing interferometer. A pattern of concentric contour lines with equal interval is observed. The positioning error of a cutting tool can contribute to the formation of the pattern.

REFERENCES

1. Kohno, T. and Tanaka, S., Figure measurement of concave mirror by fiber-grating Hartmann test, *Opt. Rev.*, 1, 118, 1994.
2. Murty, M. V., Common path interferometer using Fresnel zone-plates, *J. Opt. Soc. Am.*, 53, 568, 1963.
3. Smartt, R. N., Zone-plates interferometer, *Appl. Opt.*, 13, 1093, 1974.
4. Nakajima, K., Measurement of quadric surface by using zone-plate, *Jpn. J. Opt.*, 14, 365, 1985 (in Japanese).
5. Nomura, T. et al., Shape measurement of workpiece surface with zone-plate interferometer during machine running, *J. Prec. Eng.*, 15, 86, 1993.

6. Burning, J. H., Digital wavefront measuring interferometer for testing optical surfaces and lenses, *Appl. Opt.*, 13, 2693, 1974.

7. Honda, T. et al., Zone-plate null interferometer for measuring aspherical mirror with large aperture, *Proceedings of SPIE, International Symposium on Optical Fabrication, Testing and Surface Evaluation,* Tokyo, Japan, (Optical Fabrication, Testing and Surface Evaluation), Vols. 1720–1736, p. 305, 1992.

8. Nomura, T. et al., Zone-plate interferometer for measuring the shape error of mirrors with large numerical aperture and steep asphericity, *Opt. Rev.*, 3, 34, 1996.

9. Nomura, T. et al., An instrument for manufacturing zone-plates by using a lathe, *J. Prec. Eng.*, 16, 290, 1994.

10. Nomura, T. et al., Zone-plate interferometer to measure the positioning error of a cutting tool, *J. Prec. Eng.*, 20, 112, 1997.

11. Malacara, D., Lateral shearing interferometers, In *Optical Shop Testing*, New York: Wiley, p. 105, 1992.

12. Yatagai, T. and Kanou, Y., Aspherical surface testing with shearing interferometer using fringe scanning detection method, *Opt. Eng.*, 23, 357, 1984.

13. Murty, M. V., The use of a single plane parallel plate as a lateral-shearing interferometer with a visible gas laser source, *Appl. Opt.*, 3, 531, 1964.

14. Shakher, C., Godbolr, P. B., and Gupta, B. N., Shearing interferometer using holo-lenses, *Appl. Opt.*, 25, 2477, 1986.

15. Nomura, T. et al., Shape measurements of mirror surfaces with a lateral shearing interferometer during machine running, *J. Prec. Eng.*, 22, 185, 1998.

Index

A

Abbe error, 437, 443
Abbe prism, 104
ACOS algorithm
    estimated thickness profile, 764–765
    phase estimation, 761
    phase unwrapping, 761–762
Acousto-optic modulator (AOM), 434
    frequency fluctuation, 306–307
    RF driver, 310–311
AF, *see* Auto-focus (AF)
Air refractive index, 406
    Ciddor's equation, 415
    Edlén equations, 415
    interference phase topography, 417
    Lorenz–Lorentz equation, 415
    refractometric methods, 415
    vacuum cell, 416
    wavelength, 414
Amplitude modulation (AM), 313, 446
Antireflection coating (AR coating), 208, 244, 312, 407,
        437, 729–730
APD, *see* Avalanche photodiode (APD)
Apertureless probe, 391, 393–394
Aspheric surfaces, 98
Astigmatism method, 91–92, 99, 220, 481–482
Atomic force microscope (AFM), 393, 396, 810–811
Atomic force/photon scanning tunneling
        microscope (AF/PSTM), 396
Attenuator, 140, 143, 148, 214
Auto-edge detection technology, 830, 833
Auto-focus (AF), 844–846, 849–851
Avalanche photodiode (APD), 313
    avalanche multiplication, 114–115
    breakdown voltage, 114
    Geiger-mode APD
        equivalent circuit, 115
        MPPC, 115–116
    impact ionization, 114

B

Babinet–Soleil compensator, 383, 698
Babinet's principle, 348
Band-pass filter, 732
Band-stop filter, 732–733
Barrett's esophagus (BE), 360–361
Beam splitter (BS), 435
Benchtop configuration, 804–806
Berek compensator, 383
Birefringence mapping techniques, 382
Birefringence measurement
    applications, 678–679
    circular polariscope, 691–692
    Jones method, 687–689

linear polariscope, 688–691
Mueller method, 687–689
optical heterodyne method, 696–698
PEM, 693–696
phase-shifting method
    detected intensity, 699
    LC retarders, 700
    Mueller matrices, 698–699
    phase difference, 700
    plastic disk inspection, 701–702
    Stokes parameters, 698–699
    uniaxial stretching, 701–703
polarization society, 677
polarization states
    circular polarization, 681–682
    electric field, 679–680
    elliptical polarization, 681–682
    Jones vector, 684–685
    linear polarization, 681
    Poincaré sphere, 686–687
    stokes parameter, 685–686
polarized light, 679–680
refraction index, 682–683
Senarmont method, 692–693
Blackbody radiators, 28, 36–37
Brewster angle, 179–181, 706, 709, 734
Bureau International des Poids et Mesures (BIPM), 405

C

Calibration specimen, 156
Camera shutter, 148
Catadioptric design, 655–656
Cathodoluminescence, 28, 33
Cauchy model, 715–716, 718–720
CCD-digital radiography (CCD-DR), 134
Charge-coupled device (CCD), 231, 519, 523
    back-illuminated CCD, 57, 124, 126–127
    CCD-DR, 134
    vs. CMOS, 126
    dark current of, 126–127
    EM-CCD, 127–128
    front-illuminated CCD, 126–127
    I-CCDs, 130
    interline-transfer and frame-transfer layouts, 124–125
    multi-port, 130
    TDI-CCD, 130–131
    vacuum chamber, 127–128
    ZIMPOL, 633–634
Chemical etching method, 391–392
Chemiluminescence, 28, 33–35
Chinese hamster's oocytes (CHO), 355
Chromatic confocal imaging method, 490–491
Circular birefringence, 631, 652, 668
Circular diattenuation, 628, 631
Circular polariscope, 691–692

Circular retardance, 631
Circular retardation, 631
CMOS image sensor
   vs. CCD, 126
   schematic diagram, 124–126
   sCMOS, 128–130
Coaxial confocal microscopic method, 486–487
Coaxial-episcopic illuminator, 840
Coherence gating method, 360
Coherence scanning interferometry (CSI)
   AFM measurements, 811
   coherence-based method, 800
   Fourier phases, 800
   frequency-domain analysis method, 801, 808
   highly textured surface structure, 799
   modulation envelope, 798, 800, 808
   numerical aperture (NA), 799
   OCT, 802
   serviceable conceptual model, 799
   surface film structure, 809
   tilt fringes, 797
   top-surface location method, 810
Colorimetry, 5
   chromaticity coordinates, 24–25
   CIE color matching functions, 23–24
   CIE test-color method, 25
   colorimetric shift, 25
   CRI, 25
   spectroradiometers, 25
   tristimulus colorimeters, 25
   tristimulus values, 23
   U–B/B–V color index, 26
Color rendering index (CRI), 25
Coma, 91, 94, 100, 220
Comité International des Poids et Mesures (CIPM), 406, 410
Compact fluorescent lamps (CFLs), 39
Complementary metal-oxide-semiconductor sensors (CMOS)
   acquisition speeds, 243
   vs. CCD, 126
   digital holography, 597
   ICCD camera, 671
   MOS transistors, 124–125
   readout electronics, 121
   sCMOS, 128, 130
   spectral sensitivity, 763
Confocal laser scanning microscope (CLSM), 397
Confocal light absorption and scattering
      spectroscopic (CLASS) microscopy, 364–367
Confocal sensor, 847–850
Contrast detection method, 486–489
Corner reflector, 312, 434–437, 778–779
Correlated color temperature (CCT), 28, 36, 58
Cosine error, 412–414, 423, 437
CSI, *see* Coherence scanning interferometry (CSI)
Cube interferometer, 410

D

Deep ultraviolet (DUV) linear birefringence polarimeter
   accuracy error, 658–659
   CsI photomultiplier tube, 657
   deuterium lamp, 643, 657
   intrinsic birefringence, 656–657
   lens measurement, 660–661
   linear retardation maps, 658–660
   Mueller matrix calculus, 657–658
   Soleil–Babinet compensator, 658
   stress-birefringence dispersion, 658–660
Degrees of freedom (DOF), 459
Diascopic illuminator, 833, 838–840
Diattenuation, 628, 630–631, 647–648, 652, 662–663, 678
Diffraction
   diffractographic method, 349–351
   Fraunhofer region, 343
   fundamentals of
      aperture, 345
      Babinet's principle, 348
      Fourier transformation method, 346
      Fresnel–Kirchhoff diffraction formula, 343–344
      mechanical micrometer method, 349
      pattern, 346
      tensile strength testing, 347
      toolmaker's microscope method, 349
   grating diffraction method, 351–352
Diffractography technique, 349
Diffuse reflectance spectroscopy (DRS), 360
Digital holography, 226
   arrangements
      in-line setup, 235–236
      large objects, 234, 236
      lensless off-axis, 234–235
   CCD, 231
   CMOS, 231
   convolution method, 599
   defect recognition, 242–243
   diffraction integral, 231–233
   electro-optical sensors, 231
   ESPI, 241
   FFT, 231, 233–234
   Fresnel transform method, 599–600
   high-resolution 3D representations, 231
   interferometry, 601–602
   metrological application, 241
   microscopy
      bright-field, 237–238
      scanning electron microscope, 237, 239
      spatial resolution, 237
      3D profile, 237
   microsystems investigation, 244–245
   object beam, 597–598
   object deformations measurement, 241–242
   quantitative amplitude and phase, 600
   reference beam, 597–598
   short coherence light source
      light source, 236, 239
      OCT, 237
      phase-shifting method, 239
      plane selection, 239
      reconstructed images, fly, 239
      rough metallic cone, 240–241
   2D discrete FT, 233
   vibrations analysis, 242–243
Digital mirror device (DMD), 145, 496, 500–501, 523
Digital photogrammetry
   automobile measurement
      car bottom, 571
      entire body, 572–573
      tire, 573–575

cultural heritage, architecture, and topography application
  aerial and ground photographs, 566, 568
  Archaeological Museum of Messene (Greece), 565–566
  Byzantine church of Agia Samarina (Ancient Messene), 565, 567
  Byzantine ruins on Gemiler Island, 564–565
  topography model, 566–567
  wall relief 3D modeling, 564
human body measurement
  face modeling, 569–570
  photographing setup, 570
  SLR-type digital camera, 570
  surface model, 570
  texture-mapping model, 570
indoor measurement, 575–576
photogrammetric SLAM
  SFM, 563
  UAV, 577–579
  vehicle, 576–577
  visual SLAM, 563
principle
  bundle adjustment process, 558
  single photogrammetry, 556–557
  stereo photogrammetry, 557–558
process flow measurement, 559
  camera calibration, 559–560
  exterior orientation, 561–562
  photographing, 560–561
  3D image reconstruction and output, 562–563
  3D measurement, 562
Digital speckle pattern interferometry (DSPI)
  fringe analysis
    "difference-of-phases" method, 261–263
    out-of-plane displacement, 257–259
    "phase-of-differences" method, 261
    phase ramp/carrier, 259
    phase unwrapping method, 262
    PZT, 261
    SC fringe patterns, 259–260
    spatial carrier, 261
    SPS, 260–261
    TPS, 259–261
  MEMS, 264–266
  optical systems
    in-plane displacement, 262–263
    out-of-plane displacement, 262–264
    shearing devices, 264
    thermal stress, 263–264
  3D profile measurement, 257
  uses, 257
Digital speckle photography (DSP), 251, 636, 638, 640
Digital speckle shear pattern interferometry (DSSPI), 257
Discrete cosine transform (DCT), 262
Dispersion, light
  anomalous, 188–189
  color of objects, 190
  in dielectrics
    dipole moment, 185–186
    linear complex dielectric susceptibility, 187
    macroscopic polarization, 186–187
    refractive index, 185–187
  group velocity of wave packet, 189–190
  normal, 187–188

Displacement
  distance measurements
    air refractivity, 447–448
    fine scale, 445
    modulation method, 446
    optical frequency sweep method, 446–447
    optical measurements, 444
    pulse method, 445
    two color measurement, 448–449
  laser interferometer
    Abbe error, 437
    cosine error, 437
    dead path, 436
    Fabry–Perot cavity, 438
    homodyne and heterodyne interferometer, 434
    nonlinear errors, 437
    optical path lengths and thermal change, 436–437
    optical system, 434–435
    refractive index of air, 435–436
  linear encoders
    calibration of, 443–444
    diffraction scale encoders, 441
    direction of, 442–443
    enclosed and exposed type, 439
    hologram scale encoders, 442
    image transmission scale encoders, 441
    incremental-type and absolute-type, 440
    magnetic type and the magnet-electric type, 438
    manufacturing and machine tools, 438
    Moiré method, 440–441
    optical scale, 438
    rotary encoders, 438
    scale mark detection, 440
Distortion, primary aberrations, 88, 90–92
Divergence, lasers, 78–79
DMD, see Digital mirror device (DMD)
Doppler method
  defect, 321
  Doppler shift, 320
  vs. heterodyne method
    accuracy, 321
    velocity and displacement measurement, 322
  principle, 320–321
Drude model, 715–717
DSP, see Digital speckle photography (DSP)
DSPI, see Digital speckle pattern interferometry (DSPI)
Dysplasia
  BE, 361
  classifications, 364
  diagnosis of, 360
  gastrointestinal tract, 361
  inflammatory bowel disease, 361
  LGD/HGD, 363

E

Edge-emitting LEDs, 52–53, 80
Edge point detection, 841–843
Effective medium approximation (EMA), 714, 716, 721
Ehringhaus compensator, 383
Eigen coordinate system, 380
Electric distance meters (EDMs), 446
Electroluminescence, 28, 32, 41–42
Electronic speckle interferometry, 241, 257, 528

Electronic speckle pattern interferometry (ESPI), 253, 257
Electron-multiplying CCD (EM-CCD), 127–128
Electro-optic crystal, 383, 385
Electro-optic modulator (EOM), 306–307, 321
Ellipsometry
    definition, 705
    dielectric constant, 705
    dielectric function
        definition, 714
        Drude model, 715–717
        Lorentz model, 715–717
        Si crystal, 717
        Tauc–Lorentz model, 715–716
    drawback, 706
    features, 705–706
    linear polarization, 707
    measurement principle, 707–708
    optical constants, 705
    optical model
        air/thin layer/substrate structure, 712
        data analysis, 713–714
        EMA, 714
    p- and s-polarization, 708–709
    plane of incidence, 708
    polarized light, 706
    RAE
        Fourier coefficients, 710
        light intensity, 710
        normalized light intensity, 710–711
        schematic diagram, 709–710
    right-circular polarization, 707
    single-wavelength ellipsometry
        air/thin layer/c-Si structure, 718
        mathematical inversion, 720–722
        SiO₂/c-Si structure, 719–720
        spectroscopic ellipsometry, 718
    VUV, 712
Elliptical polarization, 168, 679–682, 707, 711
Ellipticity, 377–380, 382–387, 686–687, 700
EMA, see Effective medium approximation (EMA)
Epithelial nuclei, 360
ESAD, see Extended shear angle difference (ESAD)
ESPI, see Electronic speckle interferometry; Electronic
        speckle pattern interferometry (ESPI)
Evanescent field, 162, 390
Evanescent wave, 184–185, 390, 397
Exicor OIA (oblique incident angle), 660–661
Extended shear angle difference (ESAD), 471

F

Fab mode, 667
Fabry–Perot cavity, 72, 438, 445
Fast Fourier transform (FFT), 231, 286, 504, 843
FD-OCT, see Fourier-domain OCT (FD-OCT)
Femtosecond laser processing, 399
Field curvature, 91–92
Film thickness profiling
    pseudo-transmission interferometry, 753–755
    refractive index measurement, 755–757
    thick transparent films
        KF algorithm, 745–747
        measurable range, 748
        refractive index effect, 748, 750

        repeatability, 748, 751
        SP-500F film profiler, 747
        steps/process, 748–749
        white-light interferometry, 744–745
    thin transparent films
        CMP samples, 751–754
        cross section, 751
        measurement results, 751–752
        repeatability, 751, 753
    three wavelength interference color analysis
        ACOS algorithm, 761–762
        Davidon–Fletcher–Powell method, 764
        experimental apparatus, 762–763
        GMFT (see Global model fitting for
            thickness (GMFT))
        spectral sensitivity, 762–763
        spectral transmittance, 762–763
Finite-difference time-domain (FDTD), 356, 396–397
Finite element method (FEM), 265
Fizeau interferometers, 409–410
    Fourier transform method, 469
    interference fringe pattern, 462–463
    laser, 208
    national standard Fizeau interferometer, 467–468
    optical arrangement, 208
    optical configuration, 462–463
    optical flat, 462, 467–468
    optical parallels, 467–468
    phase distribution, 463–464
    PSI
        coherent noise, 467
        environmental fluctuations, 466
        gravity deformations, 466–467
        mounting method, 466
        phase-shifting errors, 466
    PZT, 464
    surface of transmission sphere, 209
    surface profile determination, 464–466
Flash tubes, 40–41
Flatness measurements
    angle sensor systems
        contact systems, 470–471
        ESAD, 471
        LTP, 471–472
    definition, 461–462
    Fizeau interferometer (see Fizeau interferometers)
    oblique incidence interferometer, 469–470
Flat panel display (FPD)
    glass substrate, 662
    in-line application, 667
    off-line application, 666–667
    retardation compensation films, 666
    RGB wavelengths, 663–665
Flexible LEDs, see Polymer LEDs (PLEDs)
Flow measurement
    LDV
        data-processing computer, 606
        Doppler burst, 607
        Doppler frequency, 607
        fiber-optic system, 610
        optical arrangement and SNR, 608–609
        photodetector, 606
        signal processor, 606
    PIV (see Particle image velocimetry (PIV))

Fluorescence, 29–30, 367, 669
Fluorescent lamps, 38–39, 65
Focus-error detection method
    astigmatism method, 481–482
    CD pickups, 485–486
    critical-angle method, 481–482
    focus detection method, 484
    HIPOSS system, 483
    optical stylus method, 483–485
    Rodenstock Model RM 600 3-D/C,
        483–484
Fourier analysis method, 390
Fourier-domain OCT (FD-OCT), 772, 777
Fourier transform (FT)
    convolution theorem, 230
    plane reference wave, 229
    spectrums of, 229
    2D function, 228
Fourier transformation method, 346, 542
Four-stepping phase-shifting algorithm, 498
FPD, *see* Flat panel display (FPD)
Fraunhofer diffraction, 343, 347
Free-space communication, 59
Frequency sweeping method
    initial phase, 286
    measurement result, 286–287
    temporal carrier frequency, 286
    virtual wave number, 284
    wavelength shift method, 284
Fresnel formulae
    electric vector
        parallel to plane of incidence, 176
        perpendicular to plane of incidence, 175
    energy, 177
    incident, reflected, and refracted waves, 174
    reflectivity and transmissivity, 177–179
Fresnel–Kirchhoff diffraction formula, 343–344
Fresnel's biprism, 204
Fringe analysis
    definition, 541
    fringe images, 541–542
    intensity distribution, 542
    multiple-input images, 542
        phase-shifting method, 543–544
        three-buckets phase-shifting method,
            544–545
    phase analysis, 541
    phase unwrapping, 548–550
    profilometry, 550–551
    single-input image, 542
        carrier fringes, 546
        Fizeau interferometer, 547–548
        FT method, 546–547
        2D FT method, 547–548
Fringe-counting interferometers, 410
Fringe evaluation techniques, 408
Full width at half maximum (FWHM), 398

G

Gas discharge lamps
    high-pressure discharge lamps, 39–41
    ionized gas, 37
    low-pressure discharge lamps, 38–39

Geiger-mode APD
    equivalent circuit, 115
    MPPC, 115–116
Geometrical distance, 407, 448–449
Global model fitting for thickness (GMFT)
    algorithm, 759–760
    flowchart, 757–758
    initial estimates, 760
    necessary condition, 760
    principle of, 760–761
    synthesized color chart, 758–759
Global positioning system (GPS), 556, 566
GMFT, *see* Global model fitting for thickness (GMFT)
Goos–Hänchen shift, 184–185, 192–193
Grating monochromator
    dispersion process, 18
    geometrical optical process, 18
    selection criteria, 18–19
    system configurations, 19–20
Group refractive index, 448

H

Half-wave plate (HWP), 380, 696
Harmonic waves, 164
Heat stretching method, 391–392
HeLa cells, 355
Heterodyne-type laser interferometer,
    434–435
High-brightness LEDs, 48, 51
High-grade dysplasia (HGD), 360–361, 363
High-intensity discharge (HID) lamps, 39–40, 58–59;
        see also High-pressure discharge lamps
High-pressure die-casting (HPDC), 507
High-pressure discharge lamps, 39–41
High-pressure mercury-vapor (HMPV) lamps,
    39–40
High-pressure sodium (HPS) discharge lamps, 40
High-reflection coating, 730–731
Holo-graphic method, 351
Holographic optical element (HOE), 257
Holographic particle image velocimetry (HPIV) method
    digital holographic PTV technique, 621–623
    off-axis holography, 619–621
Holography
    applications, 225
    computer-generated holography, 226
    concept, 225
    digital holography (*see* Digital holography)
    hologram
        Fourier transform (*see* Fourier transform (FT))
        reconstruction, 227–228
        recording, 226–227
    off-axis reference beam method, 225
    3D images, 225
    3D television and information storage, 226
HPDC, *see* High-pressure die-casting (HPDC)
HPIV method, *see* Holographic particle image velocimetry
        (HPIV) method
Huygens' principle, 359
Hybrid photodetector (HPD), 118–119
Hybrid sensing system
    CAD model, 508–510
    coordinate measurement machine, 507–508

die assembly, 507, 509–511
engine block, 507–510
HPDC, 507
laser sensor, 508–509

I

Image intensifier
    MCP, 129–130
    structure and operating principle, 128–129
Image processing technique, 353, 542
Image sensors
    and cameras
        back-illuminated CCD, 126–127
        dark current of CCD, 126–127
        EM-CCD, 127–128
        front-illuminated CCD, 126–127
        high-speed cameras, 130–131
        image intensifier, 128–130
        sCMOS, 128–129
        vacuum chamber, 127–128
    types
        CCD, 124–126
        CMOS, 124–126
Indefinite for dysplasia (IND), 361
Infrared (IR) interferometry
    focused and unfocused pupil, 212–213
    fringe pattern, 213
    infrared optical systems, 211
    MTF testing, 213
    phase-shifting systems, 213
    reflection configurations, 212
Inner profile measurement, 512–513
Instrumentation, near-field optics
    optical probe, 391–394
    probe height control system, 393–395
    SNOM, calibration of
        optical false images, elimination of,
            395–396
        resolution, 394
Instrument transfer function (ITF), 817–819
Intensified CCD (I-CCD), 130, 671
Intensity detection method, 480–481
Interference microscopy
    CSI
        coherence-based method, 800
        Fourier phases, 800
        frequency-domain analysis method, 801
        highly textured surface structure, 799
        modulation envelope, 798
        numerical aperture, 798–799
        OCT, 802
        serviceable conceptual model, 799
        tilt fringes, 797
    FOV images, 819–820
    height response, 819–821
    history, 792
    instrument design
        light source and detection, 802
        optical system, 802–804
        postprocessing, 807
        system configurations, 804–807
    Michelson interferometer, 794–795
    PCOR, 815–816

profile fidelity, 822–823
PSI
    algorithm design, 796
    arctangent function, 796
    discrete sample intensities, 795
    intensity sampling, 795
    linear phase shift, 796
    3D image, 797
roughness specimen, 792–793
signal modeling
    conceptual diagram, 808
    model-based analysis, 810–812
    obliquity factor, 809
    transparent film, 809–810
surface structure analysis, 792–793
surface topography
    measurement, 792–793
    repeatability, 812–814
3D metrology, 816–819
vibration and environmental effects, 814–815
Interference orders, 407, 417, 421–422
Interferogram, 205, 235, 241, 407–408, 419–420, 745–746,
        775, 777, 782
Interferometers
    additional corrections
        optics error correction, 427–428
        platen flexing, 430
        surface roughness and phase change on reflection, 429
        wringing contact, body and platen, 429–430
    BIPM, 405
    CIPM, 406
    length measurements
        Fizeau interferometers, 409–410
        Fringe-counting interferometers, 410
        light sources, 410–412
        plane waves, 412–414
        principles, 406–407
        refractive index of air, 414–417
        temperature measurement, 417
        Twyman–Green interferometer, 407–409
    prismatic bodies
        imaging of body, pixel array, 418–419
        interferogram analysis, 419–420
        length evaluation, 421–422
        length measurement applications,
            422–426
Interferometry
    aberration
        astigmatism, 220
        coma, 220
        spherical, 220
    Fizeau interferometer
        laser, 208
        optical arrangement, 207
        surface of transmission sphere, 209
    fringe analysis
        computer-aided, 218
        extrapolation, 219
        MTF, 219
        phase-shifting software, 219–220
        static fringe digitization software,
            218–220
        Strehl ratio, 219
        wavefront error, 218

interference
    Fresnel's biprism, 204
    Lloyd's experiment, 205
    Young's slits, 204
interfering waves
    angular subtense, 201
    complex amplitude summation, 198
    contrast, 199–200
    high- and low-contrast fringes, 199
    non-equal amplitude summation, 198–199
    path difference, 201–202
    sag of wavefront, 202
    two beam interference, 199–200
    wavefront aberration, 203
IR (see Infrared (IR) interferometry)
optical sources, 205–206
simple interferometer, 206–207
test arrangements
    concave surfaces, 214
    convex surfaces, 214
    domes, 217
    flats, 214
    lens, 215
    material blanks, 215
    telescope, 216–218
Twyman–Green interferometer, 209
    laser sources, 209
    Mach–Zender interferometer, 210
    shear plate interferometer, 210–211
Interpolation technology, 440

J

Jones matrices, 386, 687
Jones method, 687–689
Jones/Muller matrix computation, 381
Jones vector, 590, 592, 684–685

K

Kodak DCS Pro-Back, 564
Kösters prism, 408
Kösters–Zeiss interference comparator, 408–409

L

Lab mode, 666–667
Lanthanum-modified lead zirconate titanate (PLZT)
        shutter, 148
Laser auto-focus sensor, 849–851
Laser diode (LD), 439, 464
Laser Doppler velocimetry (LDV), 605
    data-processing computer, 606–607
    Doppler burst, 607
    Doppler frequency, 607
    fiber-optic system, 610
    optical arrangement and SNR, 608–609
    photodetector, 606
    signal processor, 607
Lasers
    cooling system, 72
    gain medium, 69–70
    gas active medium, 79–80
    laser cavity, 71

light amplification, 69
light characteristics
    divergence, 78–79
    irradiance, 79
    laser beam focus, 79
    monochromaticity, 75
    temporal coherence, 76–77
    transverse spatial coherence, 77–78
liquid active medium, 80
operating characteristics
    axial modes, 72
    continuous wave and pulsed operation, 75
    Gaussian laser beams, 73–75
    resonant modes, 72–73
    transverse modes, 73
properties, 66
pump, 70–71
semiconductor active medium, 80–81
solid-state active medium, 80
stimulated emission, 67–69
Laser Tracker® (Leica Geosystems), 447
LDV, see Laser Doppler velocimetry (LDV)
LEDs, see Light-emitting diodes (LEDs)
Left circularly polarized (LCP), 374–375
Length measurement, interferometry
    applications
        glass ceramics material, 426
        length relaxation, 425–426
        single-crystal silicon, 424–425
        steel gauge block, 423–424
        thermocouples, 423
    principles, 406–407
    requirements, 410–417
    types, 407–410
Lenses
    aspheric surfaces, 98
    elements, 94
    optical materials (see Optical materials)
    photogrammetry, 105
    production tolerances, 97–98
    relay, 104
    surfaces, 94–95
    telecentric, 104–105
    testing and single-wavelength systems, 99
Light control device, 145
Light deflection principle, 147
Light emission
    bioluminescence, 34–35
    blackbody radiation
        CCT, 28
        incandescence, 26
        luminosity of spherical star, 28
        luminosity of Sun, 28
        spectral exitance, 27
        Stefan–Boltzmann constant, 28
        total radiant exitance, 27
    cathodoluminescence, 33
    chemiluminescence, 33–34
    electroluminescence, 32
    fluorescence, 29
    multiphoton photoluminescence
        bandgap energy, 31–32
        modified Roosbroeck–Shockley equation, 31
        PL spectra, 31

three-dimensional micro-lithography, 30
two-photon absorption, 30
two-photon PL, 29–30
upconversion fluorescence process, 30
phosphorescence, 29
photoluminescence, 28
scintillation, 35
sonoluminescence, 33
triboluminescence, 35
Light-emitting diodes (LEDs)
applications
blue LED fishing lamp, 66
emission spectral range, 57–58
high-speed pulsed LED light generator, 64
illumination, 58–59
LED/LED two-way communication method, 60, 62
in medicine and dentistry, 64–65
mold's growth suppression, 66
optical communication, 59–60
photo-coupler/photo-relay, 62–63
photovoltaic effect, 60–61
plant growth, 65–66
solar cell, 62
translator, 63
dominant wavelength, 43
edge-emitting, 52–53
electron–hole pairs, 42
emission output, 44
external quantum efficiency, 46
extraction efficiency, 45
FWHM, 42–43
intensity of photon energy, 42
internal quantum efficiency, 44
Lambertian pattern, 46–47
light intensity, 44, 46
linear encoders, 439
material systems, 46–48
metrology
luminous flux measurements, 55–56
luminous intensity measurements, 54–55
mapping spatial radiation pattern, 56
measurement instrumentation, 56–57
OLEDs, 53–54
p–n junction, 41–42
refractive index, 45
surface-emitting, 51–52
white
luminous efficiency and CRI, 49
lunar white, 49
phosphor-based white LED, 49–50
phosphorescence, 48
PRS-LED, 49–51
wavelength conversion materials, 49
Light polarization
in dielectrics, 172
electric and magnetic field, 167–168
in metamaterials (see Metamaterials (MMs))
reflection and refraction
boundary conditions, 172–173
Brewster angle, 179–180
degree of polarization, 179
external reflection, 180–181
Fresnel formulae (see Fresnel formulae)
internal reflection, 181

laws of, 173–174
normal incidence, 179
Light scattering
angle scattering, 356
basic principles, 357–359
CLASS
apoptosis, 367
CLASS microscope, 365
CLASS spectrum, 366
confocal reflectance microscopy, 364
elastic scattering spectroscopy, 364
exogenous fluorescence probes, 364
fluorescence microscopy, 367
optical coherent tomography, 364
scalar wave model, 365
early cancer detection, LSS
adenocarcinoma, 361
anti-p53 antibodies, 361
Barrett's esophagus, 360
dysplasia, 360–361
dysplasia/CIS classifications, 364
esophagectomy, 361
gastro-enterological endoscopy procedures, 362
HGD, 360–361
IND, 363
LGD, 361
NDB, 363
nuclear size distributions, 362
optical diagnostic technology, 361
FDTD simulations, 356
LSS, 356, 359–360
Mie theory, 355
refractive index, 356
scattering pattern, 355
single nanoparticles, LSS of, 368–369
van de Hulst approximation, 355–356
Light scattering spectroscopy (LSS), 356
Light sources
F-number, 16, 18
gas discharge lamps
high-pressure discharge lamps, 39–41
ionized gas, 37
low-pressure discharge lamps, 38–39
grating monochromator
dispersion process, 18
geometrical optical process, 18
selection criteria, 18–19
system configurations, 19–20
incandescent lamps, 36–37
LEDs (see Light-emitting diodes (LEDs))
parasitic interference, 411
size of, 411
source power, 16
source size and shape, 18
spatial coherence, 411
specific spectral lamps, 410
spectral distribution, 16–17
temporal coherence, 410
tungsten halogen lamps, 36–37
Linear birefringence
computer-controlled X–Y stage
AC signals, 650–652
Bessel function, 649
channel detectors, 648–649

DC signals, 650–652
  instrumental noise, 650–651
  instrumental stability, 650–651
  magnitude and angular orientation, 650
  Mueller matrix calculus, 649
  retardation values, 651–652
  Soleil–Babinet compensator,
    650–651
DUV linear birefringence polarimeter
  accuracy error, 658–659
  CsI photomultiplier tube, 657
  deuterium lamp, 643, 657
  intrinsic birefringence,
    656–657
  lens measurement, 660–661
  linear retardation maps, 658–660
  Mueller matrix calculus, 657–658
  Soleil–Babinet compensator, 658
  stress-birefringence dispersion,
    658–659
photomasks
  image quality, 655–656
  Moore's law, 652
  photomask blanks, 652–654
  requirements, 652
  substrates, 652, 654–655
retardation and diattenuation, 630–631
Linearly horizontal polarized (LHP), 374
Linearly vertical polarized (LVP), 374–375
Linear polariscope, 688, 690–691
Linear retardation, *see* Linear birefringence
Liquid crystal cell, 383, 385
Liquid crystal display (LCD), *see* Flat panel display (FPD)
Liquid crystal (LC) polarization rotator
  intensity equations, 592
  Jones' calculus, 592
  MATLAB® algorithm, 593
  polariscope system, 591–592
  stress distributions, 593–594
Lloyd's experiment, 205
Longitudinal spatial coherence, *see* Temporal coherence
Long trace profiler (LTP), 471–472
Long-wave pass filter, 731–732
Lorentz model, 715–716
Low-grade dysplasia (LGD), 361, 363–364
Low-pressure discharge lamps, 38–39
Low-pressure mercury discharge lamps, 39
Low-pressure sodium (LPS) arc lamps, 39
Luminescent radiators, *see* Nonthermal radiators
Luminous efficiency
  fluorescent lamps, 38
  gas discharge lamps, 37
  HPS lamps, 40
  human eye and, 15–16
  incandescent/tungsten halogen lamps, 36
  metal-halide lamps, 40
  white-light LEDs, 49, 51

M

Machine geometry, 459–460
Mach–Zender interferometer, 210
Magnescale Co., Ltd., 442
Magnetic field pitch angle, 636–637

Magneto-optical (MO) method, 400
Maxwell's equations
  Ampere's law for free space, 162–163
  electric permittivity and permeability,
    163–164
  Faraday's induction law for free space, 162–163
  FDTD method, 496–497
  forces, 163
  Gauss's law for electric field, 162
  Gauss's law for magnetic field, 162
  vector divergence and curl, 164
Mechanical micrometer method, 349
Mechanism element, 149
Mercury–xenon arc lamps, 16, 40
Mesopic, 6
Metal-halide lamps, 40
Metamaterials (MMs)
  amplitude reflection coefficients, 192
  applications, 190
  Goos–Hänchen shift, 192–193
  negative refraction, 191–192
  permittivity and permeability,
    191–192
  phase velocity, 191–192
  total internal reflection, 192
Method of exact fractions, 421–422
Michelson interferometer
  interference signal, 794, 807–808
  laser interferometer, 434
  OCT technique, 770–773
  shear interferometer, 263–264
  Twyman–Green interferometer, 407
Michelson-type interference objective, 802
Micro-electro-mechanical system (MEMS)
  fabrication technique, 264–265
  nanolithography, 398–399
  out-of-plane deformation, 244–245
  pressure sensor, 265–266
  scanning mirror, 502–504
Micrometer, 153–154, 349, 862–863
Micromotion element, 153–154
Micromotor, 155
Micro-PIV system, 616, 623–625
Mie theory, 355–356
Mirror mount, 149–150
MMs, *see* Metamaterials (MMs)
Modulation transfer function (MTF),
  213, 219, 817
Moiré method, 440–441, 478, 519, 550–551
Moiré metrology
  applications, 273
  diffraction grating, 288–289
  dot arrays and grid lines, 273–274
  flatness measurement system, 289–290
  in-plane
    linear strain measurement,
      276–277
    pattern formation, 274–275
    shear strain measurement, 277–278
  interference pattern, 288–289
  moiré-based optical tweezers, 291
  out-of-plane
    projection moiré methods, 278, 287–288
    shadow moiré method (*see* Shadow moiré method)

overlapping line grating/random patterns, 290
reflection moiré method, 288
sampling moiré method
    principle, 295–296
    rotating tire, 3D shape measurement, 296
    spatial phase-shifting methods, 295
    specified camera system, 295
    strain distribution measurement, 296–297
shape measurement
    liquid crystal grating and CCD camera, 292, 294–295
    moiré fringes, 292, 294
    optical configuration, 292–293
    phase map, 292, 294
    projection moiré, 292
    reconstructed coin image, 292, 294
    sample coin and measured results, 292, 294
    stylus method, 292, 294
Monochromaticity, 75
Motional Stark effect (MSE) polarimeter, 637–638
Mueller matrix polarimeter
    dual PEM system, 645–647
    four modulator Mueller polarimeter, 647–648
    optical devices, 629–630
Mueller method, 687–689
Multiangle spectropolarimetric imager (MSPI), 634
Multiphoton photoluminescence
    bandgap energy, 31–32
    modified Roosbroeck–Shockley equation, 31
    PL spectra, 31
    three-dimensional micro-lithography, 30
    two-photon absorption, 30
    two-photon PL, 29–30
    upconversion fluorescence process, 30
Multi-pixel photon counter (MPPC), 115–116
Multi-pixel sensors, 121–122, 135
Multipoint diffraction strain sensor (MISS), 595–597

N

National Metrology Institute of Japan (NMIJ), 443–444,
    467–468
Nature of light
    electric and magnetic vectors, 165, 167
    electromagnetic spectrum, 171–172
    energy flow, 168–169
    irradiance, 169–170
    light polarization (see Light polarization)
    light speed in vacuum, 166
    photons, radiation pressure, and momentum, 170–171
    Poynting vector, 169
    three-dimensional plane electromagnetic wave, 166
    transverse waves, 166
Near-field optics
    applications
        FDTD method, simulation based on, 396–397
        micro-focused spot, 397
        nanolithography, 398–399
        near-field storage, 400–401
        super resolution imaging, 397–398
    evanescent field, 390
    instrumentation
        optical probe, 391–393
        probe height control system, 393–394
        SNOM, calibration of, 394–396

        SNOM, 389
Newton–Raphson method, 587
NMIJ, see National Metrology Institute of Japan (NMIJ)
Non destructive testing (NDT), 264
Non-dysplastic Barrett's (NDB), 363
Nonlinear errors, 437
Nonpolarized beam splitter (NPBS), 312, 331
Nonthermal radiators, 28
Nucleated blood cells, 355
Numerical mathematical model, 381

O

Oblique incidence interferometer, 469–470
Oblique light illuminator, 840–841
OCT, see Optical coherence tomography (OCT)
Optical alignment systems
    alignment autocollimator, 454
    alignment collimator, 453–454
    alignment telescope, 453
    autocollimator, 454
Optical bench, 140, 150–151, 648
Optical coherence tomography (OCT)
    operating principle
        coherence length, 775–776
        Gaussian distribution, 773–774
        interference intensity, 775–776
        Michelson interferometer, 770–771
        signal processing, 777
        white light interferometer, 771
    ophthalmologic OCTs, 770, 772, 777, 786
    portable version (see Portable OCT scanner)
    SD-OCT and TD-OCT, 772, 777–778
    short coherence digital holography, 237, 239
Optical contacting, 430
Optical design
    theory
        paraxial approximation, 90
        Snell's law, 89–90
        surface refraction, 90
        thin lens approximation, 90
        thin lens layout, 91
    tools, 89, 97
Optical distance, 448–449, 862
Optical element
    infrared, 140–141
    mount, 149–150
    optical flat, 156–157
    optical parallel, 157–158
    ultraviolet, 141–142
    x-ray, 142–143
Optical fibers
    classification, 146
    communications, 59
    light propagation, 146–147
    mode, 147
    probe, 361, 391–393, 397
    step-index-type optical fiber, 147
Optical flat, 156–157, 411–412, 427–428, 467–468
Optical heterodyne method
    accuracy and noise reduction
        AOM and RF driver, 310–311
        detectors, 313–314
        environment, 309

frequency-stabilized HeNe laser, 311–312
interference circuit and parts, 312–313
sample surface and sample holder, 312–313
application
    birefringence index and polarimeter,
        318–319
    displacement measurement, 315
    dynamical surface measurement, 320
    femto-second pulse laser, 315
    fiber sensors, 320
    instruments and systems production, 315
    photothermal interferometry, 317
    positioning with stage, 315
    profilometry, 315–316
    refractometry, 316–317
    thermal expansion coefficient, 317–318
    thick measurement, 315–316
    vibration amplitude and frequency, 315
    Young's modulus, thin film, 318–319
birefringence measurement, 696–698
cancelable optical circuit
    AOM, 306–307
    circuit layouts, 303–304
    environment fluctuation, 305–306
    EOM, 306–307
    frequency fluctuation, 305
    simultaneous measurement and basement
        displacement, 304
    symmetrical layout, 307–308
    thickness variation measurement, 304–305
    two signal phases, 303
    wavelength and optical path length,
        302–303
Doppler method
    accuracy, 321
    defect, 321
    Doppler shift, 320
    principle, 320–321
    shift value and measuring time, 308
    velocity and displacement measurement, 308, 322
dynamic measurement, 314
high-precision measurement, 300–302
static measurement, 314
transient measurement
    cantilever vibration, 333, 335
    concentration fluctuation, 335–337
    designing, 323–325
    dynamic phenomena, 322–323
    error factors, 329–331
    period detection, 327–328
    phase detection and displacement, 328–329
    photoelectric conversion circuit, 326–327
    reflection light detection, 323–324
    signal processing, 325–326
    static phenomena, 322
    transmission light detection, 323–324
    tuning fork (see Tuning fork)
Optical integrated circuit, 147–148
Optical materials
    Abbe number, 95–96
    dispersion, 95
    glass types, 95–96
    n–V glass chart, 95
    refractive index, 95–96

Optical modulator, 140, 145
Optical monitoring
    antireflection coating, 737
    band-pass filters, 734–735
    polarizer detector array, 736
    reflection monitoring system, 735
    TPM, 734
Optical parallel, 157–158, 467–468
Optical path difference (OPD),
        see Wave-front aberration
Optical polariscope, 147
Optical scanner, 147
Optical thin films
    antireflection coating, 729–730
    band-pass filter, 732
    band-stop filter, 732–733
    changes in reflectance, 727–728
    electric field, 726
    ellipsometry, 738
    equivalent refractive index, 728
    high-reflection coating, 730–731
    interference filter, 725
    light propagation, 726
    long-wave pass filter, 731–732
    optical monitoring
        antireflection coating, 737
        band-pass filters, 734–735
        polarizer detector array, 736
        reflection monitoring system, 735
        TPM, 734
    photometry, 738–739
    polarizer, 733–734
    reflectance and transmittance, 729
    short-wave pass filter, 731–732
    single-layer film, 726
    two-beam interference, 727–728
Optoelectronic integrated circuit (OEIC), 148
Optoelectronic sensors
    image sensors (see Image sensors)
    multi-pixel sensors, 121–122
    photons–electrons conversion,
        108–110
    point sensors (see Point sensors)
    position-sensitive sensors, 122–124
    readout electronics recording methods, 120
    selection
        factors, 108
        guide, 135
    spectral response, 111–113
    x-ray/gamma ray sensors
        physics of detection, 131–132
        practical sensors, 132–134
Organic LEDs (OLEDs), 53–54

P

Particle image velocimetry (PIV)
    classification, 613–615
    experimental setup, 614–615
    flow dynamics, 610
    HPIV method
        digital holographic PTV technique,
            621–623
        off-axis holography, 619–621

micro-PIV system, 623–625
principle, 611–613
SPIV method
axial-fan and measurement planes,
618–619
principle, 616–618
tracer particles, 614–615
turbulent boundary layer, 616
Particle streak velocimetry (PSV), 613–614
Particle tracking velocimetry (PTV), *see* Particle image
velocimetry (PIV)
Pattern matching, 841, 843–844
Pattern projection method
digital techniques
BGA, 498, 500
four-stepping phase-shifting, 498
LC grating, 498–499
profile measurement system, 497
profilometric techniques
downsizing/compact system, 502–507
high-speed measurement, 500–502
hybrid sensing (*see* Hybrid sensing system)
inner profile measurement, 512–513
shape reconstruction, 511–512
traditional techniques
optical arrangement, 495
patterns, 495
period control, 495–496
space coding method, 493–494
P3D cameras, *see* Portable 3D (P3D) cameras
Pechan prism, 103
PEM, *see* Photoelastic modulator (PEM)
Petzval curvature/sum, *see* Field curvature
Phase change on reflection (PCOR), 815–816
Phase-shifting interferometry (PSI)
algorithm design, 796
arctangent function, 796
coherent noise, 467
digital/electronic speckle and speckle shear, 257
discrete sample intensities, 795
environmental fluctuations, 466
gravity deformations, 466–467
intensity sampling, 795
linear phase shift, 796
mounting method, 466
phase-shifting errors, 466
3D image, 797
Phase-shifting method
detected intensity, 699
four-step algorithm, 260–261, 282, 291–292
grating, vertical movement and rotation,
283–284
intensity modulation, 284–285
LC retarders, 700
limitation, 284
Mueller matrices, 698–699
phase difference, 700
plastic disk inspection, 701–702
shape measurement
coin surface arrangement, 291
mutual phase difference, 291–292
score cap checking, 291–293
Stokes parameters, 698–699
uniaxial stretching, 701–703

Phase-shifting speckle interferometry (PSSI), 257
Phase-stepping interferometry, 408–409, 411–412,
418–419, 427
Phasogrammetry principle, 505–506
Phosphorescence, 29, 48–49
Photodiode (PD), 66, 108–109, 313, 483, 485, 488, 712, 856
Photoelastic modulator (PEM); *see also* Mueller matrix
polarimeter; Stokes polarimeter
design, 631
detected intensity, 695
optical arrangement, 693–694, 696
optical element and piezoelectric transducer, 631–632
phase modulation, 695
photoelastic effect, 631
polarization modulation, 631–632
quality, 632
retardance, 694
Photogrammetry lenses, 105
Photographic filter, 143
Photoluminescence (PL), 28
Photometric measurements
astronomical object brightness
absolute photometry, 23
aperture photometry, 23
CCD cameras, 22–23
differential photometry, 23
Johnson–Cousins photometric system, 22
Johnson–Morgan UBV system, 22
magnitudes, 21
red shift, 22
relative photometry, 23
trigonometric parallax, 22
goniophotometers, 21
rotating mirror photometers, 21
spherical photometers, 21
Photometry
definition, 5
frequency range, 5
human eye, 5–6
measurements (*see* Photometric measurements)
quantities and units, 6–7
illuminance, 12–14
light source (*see* Light sources)
lumens to watts conversion, 15–16
luminous flux, 11–14
luminous intensity, 12
watts to lumens conversion, 13–15
Photomultiplier, 22, 35, 110, 115, 350, 353, 606, 609–610,
643, 711–712
Photomultiplier tubes (PMTs), 110
applications, 118
flat panel, 117
gain characteristics, 116–117
miniature, 118
sectional diagram, 116–117
Photonic crystals (PCs), 53, 91, 401
Photons–electrons conversion
photoconductive effect, 109–110
photoelectric effect, 110
photovoltaic effect, 108–109
Physikalisch-Technische Bundesanstalt (PTB), 444, 467
Piezoelectric element, 149, 154–155
Piezoelectric transducer (PZT), 261, 265, 464, 466, 544, 803
PIV, *see* Particle image velocimetry (PIV)

Planckian radiator, 28, 36, 56
Plane polariscope, 690–691
Plane waves, generation and alignment
    collimated beam, 412, 414
    cosine error, 413
    laser beams, 412
    retroreflection method, 413–414
    visual autocollimation adjustment scheme, 413
Plastic optical fiber, 52, 59, 147
PMTs, *see* Photomultiplier tubes (PMTs)
Pneumatic spring, 152
Poincaré sphere, 629, 684, 686–687
Point sensors
    high-light-level applications, 113–114
    low-light-level applications
        APD, 114–116
        HPD, 118–119
        PMT, 116–118
Polarimetry
    camera imaging polarimeter, 671–673
    chemical, biochemical, and pharmaceutical
        applications
        circular dichroism spectrometer, 668
        fluorescence and light scattering, 633, 669
        optical rotation, 668–669
    FPD industry
        glass substrate, 662
        in-line application, 667–668
        off-line application, 666–667
        retardation compensation films, 666
        RGB wavelengths, 663–665
    linear birefringence, optical lithography industry (*see*
        Linear birefringence)
    Mueller matrix polarimeter (*see* Mueller matrix
        polarimeter)
    near-normal reflection, nuclear fuel industry, 662–663
    PEM
        design, 631
        optical element and piezoelectric transducer,
        631–632
        photoelastic effect, 631
        polarization modulation, 631–632
        quality, 632
    retardation and diattenuation, 630–631
    silicon (Si)-based solar cell technology, 669–671
    Stokes polarimeter (*see* Stokes polarimeter)
Polarization
    circular polarization, 681–682
    complete polarization state generator
        fixed linear polarizer and two variable retarders
        (type 2), 386–387
        fixed linear polarizer and two variable retarders
        (type I), 385–386
        independently rotating linear polarizer and quarter-
        wave plate, 381–382
        jointly rotating linear polarizer and variable
        retarder, 382–383
        rotating linear polarizer and fixed variable retarder,
        383–385
    degenerate polarization states, 374
    electric and magnetic fields, 373
    electric field, 679–680
    ellipse, 374–378
    ellipse transformations, 378–379

    elliptical polarization, 681–682
    Jones vector, 684–685
    linear polarization, 681
    linear retarder, 379–381
    Poincaré sphere, 686–687
    society, 677
    stokes parameter, 685–686
Polarization background subtraction technique, 360
Polarization beam splitter (PBS), 312, 435, 483, 655–656
Polymer LEDs (PLEDs), 53
Porro prisms
    chamfer, 102
    paper-strip diagram, 101
    tunnel diagram, 102–103
Portable 3D (P3D) cameras, 505
Portable OCT scanner
    industrial applications
        laser fusion, 782
        multichannel probe, 783–784
        multifunctioning system, 787–788
        plant measurement, 788
        tablet coating, 783
        transparent laminated tube, 782–783
    medical applications
        atopic dermatitis, 786–787
        caries monitoring, 784–785
        EDJ monitoring, 785–786
        future technology, 788
        ophthalmologic field, 784, 786
        skin measurement, 786
    operating principle
        beat frequency shift, 778–779
        long-path scanning, 778–779, 784
        optical path difference, 778–779
        probe design/assembly, 780–781
        SLD light source, 779–780
Positioning elements
    micromotor, 155
    PZT element, 154–155
    stepping motor, 155
Position-sensitive detector (PSD), 122–123, 478, 486–487,
    519, 596, 819
Power LEDs, 51–52, 58–59
Primary aberrations, 88
    astigmatism, 91–92
    coma, 91
    distortion, 92
    field curvature, 92
    spherical aberration, 91
    thin lens Seidel aberration
        Abbe number, 94
        biquadratic equations, 92
        Fraunhofer doublet, 94
        Hammer and Global optimization, 94
        lens shape/bend, 92–93
        Steinheil doublet, 94
Prisms
    Abbe, 104
    image orientation, 100–101
    vs. mirror, 99
    Pechan, 103
    Porro (*see* Porro prisms)
Production tolerances, 97–98, 862
Propagation delay method, 406

PSD, *see* Position-sensitive detector (PSD)
Pseudo-transmission interferometry, 753–755
Pulse height spectrum, 119–120
PZT, *see* Piezoelectric transducer (PZT)

Q

Quantitative birefringence imaging, 386
Quantum efficiency (QE), spectral response, 111–112
Quarter-wave film (QWF), 386
Quarter-wave plate (QWP), 380–381, 435, 442, 590, 592

R

Radiometry
    definition, 4–5
    optical range, 5
    quantities and units, 6–7
        irradiance, 8, 10
        Lambert's law and Lambertian sources, 10–11
        light source (*see* Light sources)
        lumens to watts conversion, 15–16
        radiance, 9
        radiant efficiency, 8
        radiant exitance, 8
        radiant flux, 9–10
        radiant intensity, 8
        solid angle, 7–9
        watts to lumens conversion, 13–15
RAE, *see* Rotating-analyzer ellipsometry (RAE)
Rayleigh–Gans approximation, 358
Rayleigh scattering, 357–358
Red, green, and blue (RGB) wavelengths, 663–665
Refractive design, 655
Region of interest (ROI), 419–420
Relay lenses, 104
Resonant-cavity LEDs (RCLEDs), 52
Right circularly polarized (RCP), 374–375
Rodenstock Model RM 600 3-D/C, 483–484
Rotating-analyzer ellipsometry (RAE)
    Fourier coefficients, 710
    light intensity, 710
    normalized light intensity, 710–711
    schematic diagram, 709–710
Rugate filter, *see* Band-stop filter

S

Sampling moiré method
    principle, 295–296
    rotating tire, 3D shape measurement, 296
    spatial phase-shifting methods, 295
    specified camera system, 295
    strain distribution measurement, 296–297
SBLEDs, *see* Superbright LEDs (SBLEDs)
Scanning near-field optical microscope (SNOM), 389, 394–397
Scanning probe microscope (SPM), 389, 399
Scanning tunneling microscope (STM), 389
Scientific CMOS (sCMOS), 128–129
Scintillation, 35, 132–133
Scintillators, 35, 132–134
SD-OCT, *see* Spectral-domain OCT (SD-OCT)
Sellmeier model, 715–716

Semiconductor sensors, 108, 113–114, 121
Sénarmont compensator, 381–382
Senarmont method, 692–693
Serrodyne sweep method, 447
SFF sensor, *see* Shape from focus (SFF) sensor
Shadow moiré method
    contour, 279–280
    intensity distribution
        height-dependent, 280
        vs. phase, 282
        reference grating, 280–282
        sinusoidal grating, 280
    problem, 286–287
    3D profile measurement, 278
        frequency sweeping method (*see* frequency sweeping method)
        phase-shifting method (*see* Phase-shifting method)
Shape from focus (SFF) sensor, 491–494, 847–849
Shear force method, 393
Shear plate interferometer, 210–211
Short-wave pass filter, 731–732
Shutter array, 148
Signal processing technique, 353
Signal-to-noise ratio (SNR), 111, 254, 608–609
Silicon photomultiplier (Si-PM), *see* Multi-pixel photon counter (MPPC)
Single-mode-type optical fiber, 147
Single photogrammetry, 556–557
Single-wavelength ellipsometry
    air/thin layer/c-Si structure, 718
    mathematical inversion, 720–722
    SiO_2/c-Si structure, 719–720
    spectroscopic ellipsometry, 718
Sinusoidal signals, 440, 854
SLDs, *see* Superluminescent diodes (SLDs)
Solid immersion lens (SIL) system, 398, 400
Solid mechanics
    digital holography
        convolution method, 599
        Fresnel transform method, 599–600
        interferometry, 601–602
        object beam, 597
        quantitative amplitude and phase, 600
        reference beam, 597–598
    digital image correlation
        3D DIC technique, 586–587
        2D DIC technique, 585–586
    external forces, 584
    low birefringence polariscope
        four-step phase-shifting method, 590–591
        Jones' vector, 590
        LC polarization rotator (*see* Liquid crystal (LC) polarization rotator)
        principal stresses, 589
    Mohr's circle, 584
    optical diffraction strain sensor
        MISS, 595–598
        ODSS schematic, 595–596
        one-dimensional grating diffraction, 594–595
    residual strain measurement, 587–588
    stress components, 584
Sonoluminescence, 33
Space coding method, 493–494
Spatial light modulator (SLM), 145

Spatial phase shifting (SPS), 260–261
Spatial/spatial frequency filter, 143–144
Speckle methods
    advantages, 250
    CCD and image processing system, 249
    DSPI (*see* Digital speckle pattern
            interferometry (DSPI))
    speckle interferometry
        in-plane displacement measurement, 253–255, 259
        object deformation, 252
        out-of-plane displacement measurement,
            252–253
        schematic of, 251–252
    speckle pattern, 249
    speckle photography, 250–251
    speckle shear interferometry/shearography, 255
        curvature measurement, 256–257
        slope measurement, 255–256, 259
Spectral-domain OCT (SD-OCT), 771–772, 777–778
Spectral luminous efficiency function, 5
Spectral power distribution, 13, 23, 42, 44, 56
Spheres interferometer, 410
Spherical aberration, 91, 203, 212, 220
SS-OCT, *see* Swept-source OCT (SS-OCT)
Step-index-type optical fibers, 146–147
Stepping motor, 155
Stereo photogrammetry, 557–558, 560–562
Stereoscopic particle image velocimetry (SPIV) method
    axial-fan and measurement planes, 618–619
    principle, 616–618
Stock tools, 97
Stokes parameter
    PEM, 632–634, 638–639
    polarimeters, 628–629
    polarization states, 685–686
Stokes polarimeter
    aerosol remote sensing, 634
    astronomy, 632–633
    laboratory point-measurement, 634–636
    light polarization, 629
    NIR, 640–642
    normalized Stokes parameters, chopper wheel
        data collection, 638–639
        DOP, 639–640
        half-wave retarder, 639–640
        spin rate, 638
    parameters, 629
    spectroscopic polarimetry
        iron oxide, 644–645
        leaf circular polarization, 644–645
        photosynthetic microbes, 642
        Synechococcus WH8101, 642–643
        transmissive/reflective mode, 642–643
    Tokamak, 636–638
    vector, 629
    ZIMPOL, 633–634
Straightness measurement
    angle instruments
        angle interferometer, 459
        autocollimators, 458–459
        electronic levels, 458
    definition, 452
    geometrical measurements, 451
    integration of slope, 453

    machine axis, 460
    mechanical datum
        error separation, reversal measurement, 456–457
        experimental result, 457
    optical beam
        laser beam alignment, 454–455
        optical alignment systems, 453–454
        straightness interferometer, 455
    reversal technique, 453
    rotational error measurement, machine axis, 460
    straight datum, 453
    traceability, 452
Strain gauge method, 351
Strain measurement
    grating diffraction, 351
    linear strain, 276–277
    residual plastic deformation, 587–588
    shear strain, 277–278
Stress mapping
    CaF_2, 698
    circular polariscope, 692
    plane polariscope, 690
    plastic disk inspection, 701–702
    Senarmont method, 692
Structure from motion (SFM), 563, 576
Sub-pixel technique, 352
Superbright LEDs (SBLEDs)
    high-speed pulsed LED light generator, 64
    photo-coupler, 62–63
    photovoltaic effect, 60
    solar cell, 62
    translator, 63
    two-way LED/LED system, 60–61
Superluminescent diodes (SLDs), 52–53, 770–772, 779
Superluminescent LEDs (SLEDs), 58–60
Surface-emitting LEDs, 51–52, 80
Swept-source OCT (SS-OCT), 772, 777

T

Tabletop, optical, 151–152
Tauc–Lorentz model, 715–716, 721–722
TD-OCT, *see* Time-domain OCT (TD-OCT)
Telecentricity, 104–105
Temperature measurements, 417, 423–424
Temporal coherence, 76–77
Temporal phase shifting (TPS), 259–261
Thermocouples, 423
Thick film (KF) algorithm
    calculation procedure, 746–747
    film thickness profile, 755–757
    interferogram, 745–746
    SP-500F film profiler, 747
Thin-film flip chip light-emitting diode (TFFC) LED, 58
Three-dimensional profilometry (3D profilometry)
    classification, 476
    full-field measurement
        pattern projection (*see* Pattern projection method)
        shape from focus method, 491–494
    historical background
        measuring capsule unit, 478–479
        Moire topography, 477–478
        stereographic method, 477
        triangulation method, 478–480

point-wise technique
  chromatic confocal imaging, 490–491
  coaxial confocal microscopic method, 486–487
  conoscopic holography method, 489
  contrast detection method, 486–489
  focus-error detection
    (*see* Focus-error detection method)
  intensity detection method, 480–481
  stand-off-distance, 476
Three-dimensional (3D) shape measurement
  accuracy, 532
  applications, 518
  CAD data comparison, 524
  calibration, 523
  camera model, 523
  CCD sensors, 523
  component shape measurement, 525–526
  direct shape measurement, 530–531
  discontinuity ambiguity, 523–524
  full-field measurement, 530
  global and local coordinate system, 521–522
  image data patching, 524
  interferometry, 520
  laser scanning, 519
  laser tracking system, 520
  LCDs, 522–523
  Moiré method, 519
  paint defects, 528–529
  phase shifting, 523–524
  photogrammetry, 520
  real-time computing, 530
  sensor planning, 524–525
  shading problem, 531
  speckle pattern sampling, 519–520
  structured light method, 521
  system calibration and optimization, 532
  terminology, 531
  360° shape, 521
  time-of-flight method, 518
  vehicle shape measurement, 527
  vibration, 528
Three wavelength interference color analysis
  ACOS algorithm
    estimated thickness profile, 764–765
    phase estimation, 761
    phase unwrapping, 761–762
  experimental analysis
    apparatus, 762–763
    Davidon–Fletcher–Powell method, 764
    spectral sensitivity, 762–763
    spectral transmittance, 762–763
  features, 766–767
  GMFT method
    algorithm, 759–760
    flowchart, 757–758
    initial estimates, 760
    necessary condition, 760
    principle of, 760–761
    synthesized color chart, 758–759
  thin film/soap film, 765–766
Time-delayed integration (TDI), 130–131
Time-domain OCT (TD-OCT), 772, 777–779

Toolmaker's microscope method, 349
Total internal reflection (TIR)
  critical angle, 181
  evanescent wave, 184–185
  prisms, 95
  reflected wave
    amplitude ratios, 181–182
    phase relation, reflected and incident waves, 182–183
    relative phase difference, 183
Transistor–transistor logic (TTL), 440
Transmission film (TF) algorithm
  film thickness profile, 755–757
  measured *vs.* nominal values, 754–755
  polyester films, 754
  principle of, 753–754
Transverse spatial coherence, 77–78
Triangle irregular network (TIN) model, 562–563, 565
Triangle type optical benches, 150
Triboluminescence, 35
Tuning fork
  damping oscillation, 332–333
  optical circuit, 331–332
  temperature dependence and frequency, 333–334
  transient characteristics, 332
  vibration amplitude, 333–334
Turning point method (TPM), 734–735
TV holography (TVH), 257
Two-dimensional digital image correlation (2D DIC), 585–586
Two-modulator generalized ellipsometry (2-MGE), 645–647
Two-photon laser scanning fluorescence microscopy, 30
Two plus one phase-shifting algorithm, 502
Twyman–Green interferometer, 209
  Fizeau interferometer, 409–410
  Kösters–Zeiss interference comparator, 409–410
  laser sources, 209
  light sources, 411
  Mach–Zender interferometer, 210
  measurement pathway, 407–408
  optics error correction, 412, 427
  reference pathway, 408
  shear plate interferometer, 210–211

U

Ultrasonic welding tool, 350
Unmanned aerial vehicle (UAV), 577–579

V

Vacuum interferometers, 443–444
Vacuum ultraviolet (VUV) ellipsometry, 712
van de Hulst approximation, 355–356
van de Hulst theory, 358
Vibration isolation system, 152–153
Video measuring machines
  applications, 831–832
  auto-edge detection, 830, 833
  features of, 833
  GRR validation method, 830–831
  height measurement
    confocal sensor, 848–850
    laser AF sensor, 849–851
    SFF sensor, 847–848

image processing
    edge point detection, 841–843
    image rectification, 843
    pattern matching, 843–844
    vision auto-focus, 844–846
inspection time, 831–832
ISO standard, 852
optical system
    distortion factor, 836–837
    illumination, 838–841
    magnification, 837–838
    telecentric optics, 836–837
profile measurement, 846–847
structural analysis
    application software, 833, 835
    coordinate system, 834–836
    illumination system, 833
    optical component layout, 833–834
Visual simultaneous localization and mapping, 563

W

Wave equation, 165
Wave-front aberration, 88, 203
Wave fronts, 87–88, 165, 198–199

Wave number, 165, 772
Wave vector, 165, 174, 366
White light interferometer, 744–745,
    769–771

X

Xenon arc lamps, 40
X-ray/gamma ray sensors
    physics of detection, 131–132
    practical sensors, 132–134
X-type optical benches, 150–151

Y

Yamaguchi–Hasunuma compensator, 385
Young's slits, 204
Y-type optical benches, 150

Z

Zeeman laser, 434
Zernike/Buchdahl notations, 88
Zurich imaging polarimeter (ZIMPOL), 633–634